The Physics and Astrophysics of Neutron Stars

Astrophysics and Space Science Library

More information about this series at http://www.springer.com/series/5664

Luciano Rezzolla • Pierre Pizzochero •
David Ian Jones • Nanda Rea • Isaac Vidaña
Editors

The Physics
and Astrophysics
of Neutron Stars

 Springer

Editors

Luciano Rezzolla
Institut für Theoretische Physik
Goethe University Frankfurt
Frankfurt
Hessen, Germany

Pierre Pizzochero
Universita' degli Studi di Milano
Dipartimento di Fisica
Milan, Italy

David Ian Jones
University of Southampton
Mathematical Sciences
Southampton, United Kingdom

Nanda Rea
Institute of Space Sciences (ICE)
Consejo Superior de Investigaciones
Científicas (CSIC)
Barcelona, Spain

Isaac Vidaña
Istituto Nazionale di Fisica Nucleare
(INFN)
Dipartimento di Fisica, Università di
Catania
Catania, Italy

ISSN 0067-0057   ISSN 2214-7985   (electronic)
Astrophysics and Space Science Library
ISBN 978-3-319-97615-0   ISBN 978-3-319-97616-7   (eBook)
https://doi.org/10.1007/978-3-319-97616-7

Library of Congress Control Number: 2018960864

Cover illustration: Composite image of the Crab nebula with X-rays from Chandra (blue and white), NASA's Hubble Space Telescope (purple) and NASA's Spitzer Space Telescope (pink). Also shown with a cartoon is the structure of the compact star at the center of the nebula and showing the various parts of a neutron star: the core, the outer core and the crust. Credit: X-ray: NASA/CXC/SAO; Optical: NASA/STScI; Infrared: NASA-JPL-Caltech; Cartoon: L. Rezzolla (GU)

This Springer imprint is published by the registered company Springer Nature Switzerland AG
The registered company address is: Gewerbestrasse 11, 6330 Cham, Switzerland

Foreword

Compact stars, such as neutron stars, strange stars, and hybrid stars, are unique laboratories that allow us to probe the building blocks of matter and their interactions at regimes complementary to terrestrial laboratories. These exceptionally complex astrophysical sources have already led to breakthrough discoveries in nuclear and subnuclear physics, QCD, general relativity, and high-energy astrophysics. The most recent landmark was the first gravitational wave detection from the merging of two neutron stars on August 17, 2017 (GW170817), 2 years after the first ever direct detection of gravitational waves, which has closed a chapter initiated almost one century ago, with the first paper of Albert Einstein on general relativity. With the observation of coalescing neutron stars, the gravitational-wave detectors LIGO and Virgo have demonstrated their potential to directly probe the properties of matter at the extreme conditions found in this scenario. Indeed, constraints have been placed on the tidal effects of the coalescing stars, which in turn constrain the neutron star radii and the equation of state of matter at twice the nuclear saturation density. These observations also mark the beginning of the multi-messenger astronomy era with the simultaneous detection of gravitational waves, gamma-ray burst, and electromagnetic emission from the same source. In the coming years, the predicted large number of new events will greatly improve our understanding of neutron star mergers and of neutron star internal structure. During 2019, the observatory NICER (Neutron Star Interior Composition Explorer) hosted by the International Space Station (ISS) will release its first one-year campaign of X-ray observations of a few neutron stars aiming at determining the radius of these stars within less than 10% accuracy. We shall also mention the GAIA satellite which is measuring the position of billions of objects in our universe with an unprecedented accuracy. We are confident that these measures will considerably help understanding the multiple faces of neutron stars and will provide a better picture of the neutron star jigsaw puzzle.

The nuclear and subnuclear experimental facilities, such as FAIR and GANIL in Europe, offer a complementary approach to explore the phase diagram of dense matter as well as reaction rates and transport properties. They have nurtured innovative and fundamental discoveries thanks to the synthesis of new radioactive elements, on one side, and to the production of new states of matter in heavy ion collisions, on the other side. Theoretical approaches in nuclear physics have also been greatly renewed over the last decade, mainly thanks to the new chiral EFT methods which have allowed to establish constraints on the properties of low-density nuclear matter—up to 1−2 times the nuclear saturation density. The description of the nuclear equation of state at zero and finite temperature, in uniform or non-uniform matter, as well as reaction rates and transport properties has also made a big step during the last decade. The more systematic use of nuclear physics inputs in compact star modeling has contributed to reduce the uncertainties in the model predictions. The various constraints coming from both nuclear physics and astrophysics have been instrumental in improving our understanding of the dense matter equation of state. Detailed questions related to phase transitions and transport properties are now emerging, and future experiments and observations will provide even better data to keep improving the models.

The outstanding variety of observational properties of compact stars challenged observers and requires quite dedicated instruments. NICER is an example, but observations in the whole electromagnetic range, from radio to gamma rays, are crucial to understand the properties of compact stars. Future instruments, e.g., FAST, LOFT, ATHENA, and SKA, will provide an unprecedented amount of data on compact stars. One particular issue that has received significant attention is the astonishing large magnetic fields exhibited by neutron stars. From both the microphysics and the macrophysics sides, the relevance of strong magnetic fields in different processes has been acknowledged. Understanding their impact, for instance, in core-collapse supernovae, the long-term cooling of strongly magnetized neutron stars, the bursting activity of magnetars, the merger of neutron stars binaries, and the crustal cooling of quiescent low-mass X-ray binaries, among other scenarios, has been the object of many theoretical studies, improving our knowledge of these fascinating objects.

The book is meant to address a large spectrum of readers, from first-year PhD students up to senior researchers, who will find a thorough overview of the various facets of the physics and astrophysics of compact stars. The aim is that of summarizing the recent progress in the field and the many challenging questions which still remain to be answered; most importantly, the book aims at identifying effective strategies to explore, both theoretically and observationally, the open problems. To accomplish this goal, each of the 13 chapters of the book written by internationally renowned experts includes a brief overview of the historical context, a detailed review of the main theoretical achievements, experimental and observational results, and finally the present challenges, future prospects, and open questions.

The book is organized into three thematic blocks: the first part (Chaps. 1–4) covers the astrophysical context where neutron stars are formed (core-collapse supernovae), discusses the impact of strong magnetic fields in neutron stars (magnetars and transient phenomena), reviews tests of gravity with compact stars and the detection of gravitational waves, and studies the complexity of binary systems.

The second part (Chaps. 5–9) is dedicated to nuclear physics. Several decades ago, it was suggested that the neutron star equation of state could be directly determined from observations. As a transition from astrophysics to nuclear physics, a review is made of the constraints from electromagnetic observations. Then the nuclear equations of state at both zero and finite temperature are presented, followed by the connection with low-energy QCD and super-dense matter. Interestingly, since the nuclear interaction is attractive at long range, a part of the volume of neutron star core and crust could host superfluid neutrons, or charged particles in superconducting state. Pairing is believed to be responsible for glitches and can affect neutron star oscillations. Finally, the microphysics part of the book is concluded by addressing the questions related to transport phenomena and reaction rates in hadronic and quark matter.

The last part (Chaps. 10–13) is mainly focused on gravitational physics and kilonovae phenomena; the gravitational wave emission from merging neutron star binaries as well as the post-merging dynamics is discussed, and the electromagnetic emission of kilonovae and its engine—the nucleosynthesis—is described. Gravitational waves emitted by a single neutron star due to its quadrupolar "mountain" deformation are also discussed, as these could be the next new signals to look for in advanced terrestrial interferometers. Finally, this section is concluded with a presentation of universal relations arising in general relativity—the so-called I-Love-Q relations—and of alternative theories of gravity and their universal relations.

This book, the *NewCompStar White Book*, is the final deliverable of the MP1304 COST Action, which ran from 2013 to 2017 and was the natural extension of the ESF-funded RPN "CompStar" (2008–2013), which coordinated various initiatives at the national level such as EMMI in Germany and TeonGrav in Italy (still continuing), SN2NS continuing as MODE in France, and many more. After more than 10 years of continued networking at the highest European level, research on compact stars is more active than ever before and has reached maturity. The near future activities are also guaranteed by the new COST Action PHAROS which will operate until 2021. But NewCompStar did not end in Europe. It was also connected to research groups in non-EU countries, such as Armenia, Australia, the Russian Federation, and the USA. Many of the world leading experts in the field have participated in our international conferences and schools. NewCompStar has therefore been an important nexus acting as a global reference for compact star physics.

At this point, the *NewCompStar White Book* provides a timely summary of the enormous progress in our field during the last decade. As illustrated through the chapters of this book, progress has been possible thanks to the multidisciplinary interaction among astrophysicists, nuclear and particle physicists, and experts in gravitational physics. NewCompStar was the framework which allowed leading experts in these fields to work together and jointly address fascinating and challenging problems. The deep and rich interconnectivity between areas was stressed and exploited since the beginning of our European networking efforts, being one of the keystones of our project. In addition to the pure research agenda, Compstar and

NewCompStar also provided a dedicated training program for a new generation of scientists, which grew with a broader view and better skills than previous generations. This book also represents a synthesis of the many ideas transmitted to young researchers during the international schools.

To conclude, it is worth stressing again that, although this white book is a collection of contributions from NewCompStar community members, most of the discussed results have been obtained by large international, worldwide collaborations trying to understand the physics of compact stars. Given its pedagogical purposes and the very broad range of topics covered, we have no doubt that this volume will find its place among the few general books on neutron stars, and we hope it will further stimulate research on these fascinating celestial objects, accompanying a new generation of physicists.

Roma, Italy   Valeria Ferrari
Seattle, WA, USA   Jerome Margueron
Alicante, Spain   Jose A. Pons
June 2018

Preface

Astrophysical objects have always inspired in us a deep sense of wonder, awe and inspiration. Astrophysical *compact* objects—such as neutron stars—obviously belong to this class and are, arguably, among the most exquisite example of what drives human fascination.

This sense of fascination has pervaded "NewCompStar", a network that between 2013 and 2017 has collected scientists across Europe with very different backgrounds—astrophysics and gravitational and nuclear physics—in the common quest for a better understanding of neutron stars and of the fundamental physics behind them.

Numerous are the achievements of these scientists, who have met in a number of meetings—small and large—and who have collectively written hundreds of papers over 4 years only. More importantly, a legacy of collaborations, synergies and friendships has been passed over to "PHAROS", the new and improved incarnation of this spirit.

As a way to cast on paper a small part of what NewCompStar has ultimately achieved, we have organised this collection of chapters that aim at providing the exciting portrait of our present understanding of neutron stars.

The authors of these chapters have been chosen for their extraordinary expertise and to reflect NewCompStar's natural geographic and gender balance. More importantly, these authors are not only very active in their corresponding fields, but they will probably shape them in a permanent manner.

We trust that this book will be useful both as a reference for researchers working in the field and as a first introduction to the subject for the generation of young scientists willing to join the exciting adventure of understanding compact stars.

Frankfurt, Germany   Luciano Rezzolla
Milan, Italy   Pierre Pizzochero
Southampton, UK   David Ian Jones
Barcelona, Spain   Nanda Rea
Catania, Italy   Isaac Vidaña

Contents

Chapter 1
Neutron Stars Formation and Core Collapse Supernovae

Pablo Cerda-Duran and Nancy Elias-Rosa

Abstract In the last decade there has been a remarkable increase in our knowledge about core-collapse supernovae (CC-SNe), and the birthplace of neutron stars, from both the observational and the theoretical point of view. Since the 1930s, with the first systematic supernova search, the techniques for discovering and studying extragalactic SNe have improved. Many SNe have been observed, and some of them, have been followed through efficiently and with detail. Furthermore, there has been a significant progress in the theoretical modelling of the scenario, boosted by the arrival of new generations of supercomputers that have allowed to perform multidimensional numerical simulations with unprecedented detail and realism. The joint work of observational and theoretical studies of individual SNe over the whole range of the electromagnetic spectrum has allowed to derive physical parameters, which constrain the nature of the progenitor, and the composition and structure of the star's envelope at the time of the explosion. The observed properties of a CC-SN are an imprint of the physical parameters of the explosion such as mass of the ejecta, kinetic energy of the explosion, the mass loss rate, or the structure of the star before the explosion. In this chapter, we review the current status of SNe observations and theoretical modelling, the connection with their progenitor stars, and the properties of the neutron stars left behind.

P. Cerda-Duran (✉)
Departamento de Astronomía y Astrofísica, Universitat de València, Valencia, Spain
e-mail: pablo.cerda@uv.es

N. Elias-Rosa
INAF - Osservatorio Astronomico di Padova, Padova, Italy
e-mail: nancy.elias@oapd.inaf.it

© Springer Nature Switzerland AG 2018
L. Rezzolla et al. (eds.), *The Physics and Astrophysics of Neutron Stars*,
Astrophysics and Space Science Library 457,
https://doi.org/10.1007/978-3-319-97616-7_1

1

1.1   Introduction

1.1.1   Core-Collapse Supernovae and Their Importance

Neutron stars (NS) appear at the end point of the evolution of massive stars ($M >$ 8 $M_\odot$). At this stage, most of these stars have grown an iron core that cannot be supported by hydrostatic pressure and collapses. In most of the cases it produces a supernova (SN), releasing a gravitational energy of about $\sim 10^{53}$ ergs (mainly in form of neutrino radiation), and leading to the complete destruction of the progenitor star (e.g. see Janka 2012). SNe leave behind compact remnants, such as neutron stars (NS) or black holes (BH) (e.g. see Heger et al. 2003). The first interpretation of a SN as a transition between massive stars and neutron stars was introduced by W. Baade and F. Zwicky in the 1930s (Baade and Zwicky 1934) in order to explain the extraordinary amount of energy liberated. It still holds nowadays (see Sect. 1.2.3 for a detailed description about how NS are born).

SNe are studied from multiple perspectives. They are crucial for a complete understanding of stellar evolution, being associated with NS and black holes (BH), as well as with other extreme events such as gamma-ray bursts (GRBs; e.g. Galama et al. (1998) or Woosley and Bloom (2006)) and X-ray flashes (XRFs; Pian et al. 2006), mainly connected to stripped-envelope core-collapse SNe. SNe are also among the most influential events in the Universe regarding their energetic and chemical contribution to the interstellar medium in galaxies (Thielemann et al. 1996). In fact, SNe are the major "factories" of heavy elements synthesized along the progenitors life and in the SN explosion itself, as well as sources of gravitational waves, neutrinos, and cosmic rays (e.g. Andersson et al. 2013; Hirata et al. 1987; Koyama et al. 1995). SNe also produce dust (e.g. Todini and Ferrara 2001), and induce star formation (e.g., Krebs and Hillebrandt 1983) since the shock waves generated in their explosion heat and compress interstellar molecular clouds. Finally, they can be used as cosmological distance indicators, and to set constraints on the equation of state of Dark Energy (e.g. Riess et al. 1998; Perlmutter et al. 1999; Hamuy and Pinto 2002).

1.1.2   Brief History of Supernova Observations

Temporary stars, comets and novae, as well as occasional supernovae, were fairly frequently recorded in East Asian history (Stephenson and Green 2005). Possibly, the first supernova for which written reports exist is SN 185, which took place in the year AD 185 in the Milky Way Galaxy (precisely it occurred in the direction of Alpha Centauri).

The birth of modern SN astronomy occurred in 1885, when the first extragalactic SN (S Andromedae or SN 1885A) was detected with a telescope (Hartwig 1885). In the 1920s scientists begun to realise that there was a particular class of very bright novae, and a decade later, Baade and Zwicky (1934) named this kind of event "supernova". The distinction between novae and supernovae was at first based on twelve objects discovered between 1895 and 1930, plus the galactic SN observed by Tycho in 1572. Between 1936 and 1941, the first systematic supernova search was started by W. Baade and F. Zwicky using the Palomar 18-in. Schmidt telescope and led to the discovery of 19 SNe. In the same years, SNe were first classified as Type I or Type II based on the lack or presence of hydrogen spectral lines, respectively (Minkowski 1941). This subdivision is still the basis of the SN classification used today (Sect. 1.2.1 and Turatto et al. 2003; Gal-Yam 2016).

In the following years, the improvement of the instrumentation, the construction of new telescopes and also the numerous progresses in the understanding of the stellar evolution, stimulated the research and cataloguing of the SNe. By the Eighties, with the introduction of the CCD and the construction of larger diameter telescopes, not only did the number of SNe observed grow but it was possible also to obtain spectra with better resolution and to study their luminosity evolution for long times. SNe classification became consequently more complex due mainly to more careful comparison among the SNe, and the previous types of SNe were further sub-categorised attending to the presence/absence of chemical elements other than hydrogen.

Computer controlled search programs of SNe were initiated in the following decade. Thanks to past and ongoing surveys (e.g. the All-Sky Automated Survey for SuperNovae—ASAS-SN; Shappee et al. (2014), the ESA Gaia transient survey—Hodgkin et al. (2013), or the Panoramic Survey Telescope & Rapid Response System—Pan-STARRRS; Kaiser et al. (2002), among numerous others), and to the efforts of amateur astronomers, the rate of SN discoveries has dramatically increased, going from less than 20 SNe at the beginning of the twentieth century, to more than $\sim$200 SNe per year in the first decade of the twenty-first century, and finding today up to 1000s of events per year.

These technological advances also allowed SN searches at $z > 0.2$ (e.g. Dark Energy Survey—DES[1]), and the multi-wavelength observation of SNe. The optical band has played a fundamental role in the knowledge and classification of the SNe, but with the technological progress, it has been possible to observe the SNe in the IR bands, and in the radio, ultraviolet (UV) and X-ray. In particular, with the explosion of the SN 1987A in the LMC, the closest extragalactic SN observed (50 kpc), it has been possible also to directly identify for the first time the SN progenitor in archival images and to detect the neutrino flux produced during the explosion (see e.g. Arnett et al. 1989; McCray 1993 for reviews).

To further complicate this scenario, these new generation of deep, and wide surveys, are discovering new types of transients with unprecedented observational characteristics, as we will describe in Sect. 1.3.1. For example, it has been found extreme SNe types such as superluminous SNe, hundreds of times brighter than those found over the last 50 years ($M_V < -21$ mag), whose energy regime is not

---

[1] http://www.darkenergysurvey.org/.

explained by the standard core-collapse and neutrino-driven mechanisms. Or ultra-faint SNe ($M_V > -14$ mag), which are characterized by a very low explosion energy ($\sim 10^{49}$ ergs) and small amount of ^{56}Ni mass ($< 10^{-3}$ M$_\odot$).

The wide variety of classes of SNe has also shown that core-collapse supernovae (CC-SNe) present observational heterogeneity, consequence of the different properties of the progenitor star at the moment of the explosion, their energetic, angular momentum, and environment. Thus, in the last decades SNe have been studied in order to better establish the link to their progenitor stars and thereby to clarify the evolution of massive stars. This make the SNe a very valuable probe of mass loss, circumstellar structure, and star formation rates. For nearby CC-events it has been possible to directly image the precursor star in pre-explosion images (mainly from the archive of the Hubble Space Telescope, *HST*) and to verify its disappearance years after the SN exploded (Smartt et al. 2009; Smartt 2015 and Sect. 1.2.2.2).

1.1.3 The Theoretical Perspective

In parallel to the observations, our theoretical understanding of the processes leading to the formation of a neutron star as a result of a supernova explosion has grow significantly in the last century. Although the basic scenario was set by Baade and Zwicky (1934), it was not until the 1960s, with the appearance of the first modern computers, that a significant progress was achieved. It was established that the collapsing core should bounce when reaching nuclear matter density (Colgate and Johnson 1960; Colgate et al. 1961) and suggested that the primary energy source in supernova explosions should be in form of neutrino radiation (Colgate and White 1966). The basics behind most advanced models nowadays was set by Bethe and Wilson (1985) in the so-called delayed neutrino-heating mechanism. They realised that, when included the most sophisticated microphysics, a prompt explosion after bounce was not possible, but the energy deposited by neutrino radiation could revive the shock and power the explosion. We know now that multidimensional effects play a primary role in the explosion, and that a comprehensive numerical modelling of the scenario including all relevant physical effects is a computational challenge (see e.g. Janka et al. 2007; Janka 2012; Burrows 2013; Müller 2016). Therefore, our current understanding of the supernova mechanism relies heavily in the results of numerical modelling and the progress in the field has tracked closely the improvements of modern supercomputers and the development of high performance computing.

However, in order to understand the plethora of observations it is not sufficient to be able to model numerically the collapse of massive stars. One also has to be able to link these explosions with their progenitor stars, make predictions of how common or uncommon each type of event is, and what are the consequences for the galactic environment in which they live. The complete picture can only be achieved with the help of the complementary discipline of stellar evolution (see e.g. Woosley et al. 2002). Even if more than 2500 CC-SNe have been discovered so far, it is still missing a complete picture of this progenitor-explosion link. In order to better understand the way in which these stars end their evolution, it is necessary to get stronger constrains of the physical characteristics of the progenitor stars alongside the explosion parameters through the observation of SNe, and thus test and constrain the models. At the same time, studying compact remnants like neutron stars, we can determinate explosion conditions and understand associated phenomena such as mass loss, r-process nucleosynthesis, gravitational waves and neutrino emission.

1.1.4 Aim

The aim of this work is to review the current knowledge about supernovae as the place of birth of neutron stars from both an observational and a theoretical perspective. In Sect. 1.2 we review the status of observations of core-collapse supernovae and our current knowledge about the scenario in which these explosions are produced. In Sect. 1.3 we discuss the present challenges and future perspectives in the field.

1.2 Current Status of Supernova Observations and Modelling

1.2.1 Traditional Supernova Types

As we see in Sect. 1.1.2, the first fundamental classification was given by Minkowski (1941) who distinguished the SNe in two different classes based on the lack (SNe I) or presence (SNe II) of hydrogen lines such as Hα 6563 Å and Hβ 4861 Å in their early spectra. Since SNe can be very different one from another as to spectral features (i.e. chemical composition, physical conditions), photometry, overall SED (spectral energy distribution), time evolution, radio and X-ray properties, in the 1980s sub-classes were introduced such as Ib, Ic, II-P, II-L, IIn, IIb which are related to characteristics of their spectra (small letter) or light curves (capital letter).

The SNe of type I are subdivided in three subclasses, depending basically on the presence or absence of Si II and He I in the spectra (see left panel of Fig. 1.1). Type Ia SNe spectra present the line of Si II (rest wavelength 6347, 6371, and 6150 Å) in absorption, while the spectra of the SNe Ib do not have these features but are characterized by pronounced lines of He I, such as those at 5876, 6678 and 7065 Å. Finally, the SNe Ic do not show Si II nor He I lines (or He I is very weak). Within this last subclass, we can distinguish events with fast-expanding ejecta (v $\sim$ 20,000 km s^{-1}), named broad-line SNe Ic (Ic-BL) or hypernovae. These transients are sometimes related to long gamma-ray burst or X-ray flashes (SN 1998bw is the prototypical SN Ic-BL, discovered at the same time and location as GRB 980425; Galama et al. 1998). Currently, it is known that type

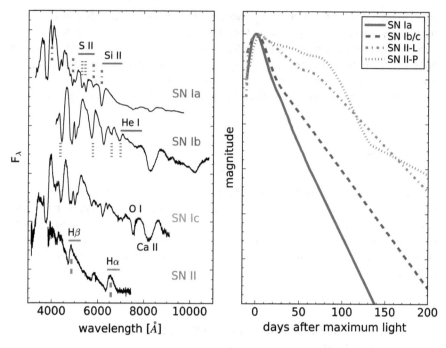

Fig. 1.1 *Left:* Representative spectra near maximum light of the main SN types (SN Ia 2011fe (Pereira et al. 2013), Type Ib iPTF13bvn (Cao et al. 2013), Type Ic SN 2007gr (Valenti et al. 2008), and Type II-P SN 1999em (Leonard et al. 2002)). The most prominent spectral features are indicated. The spectra are available in the public WISeREP repository (Yaron and Gal-Yam 2012). *Right:* Average light curves of the main SN types (Li et al. 2011)

Ia SNe arise from the thermonuclear runaway of an accreting white dwarf (WD), whereas Type Ib and Ic originate from the core collapse of massive stars that had lost their hydrogen, or hydrogen plus helium envelopes before explosion, respectively.

The SNe II are also believed to be CC-SNe whose progenitors have retained (at least part of) their outer envelopes before exploding, reason for which their spectra are dominated by hydrogen lines all the time. As shown in the right panel of Fig. 1.1, SNe II are sub-classified as SNe II-P and SNe II-L, based on their light curve shape. The former exhibit light curves that decrease slowly after maximum for about three months displaying a "plateau". The light curve of SNe II-L shows instead, a linear decline starting shortly past maximum. Recently, it has been argued that these two subclasses are at the extremes of a continuous distribution of SNe with different light-curve slopes (e.g. Anderson et al. 2014).

This classification, based on early phase spectra, is normally used when a new SN candidate is confirmed, but it is not always accurate, as it can also be that the appearance of a SN can change in time due to the characteristics of the progenitor or to those of the circumstellar material (CSM).

During the last part of the 1980s and the 1990s, the family of CC-SNe begun to grow with the identification of various sub-types. The SNe IIn (named after the detection of narrow emission lines in their spectra) present blue continuum spectra, with Balmer emission lines formed by several components that evolve in the time in various ways (Schlegel 1990). The narrow component, with inferred full-width-at-half-maximum (FWHM) velocities from a few tens to a few hundreds $km\,s^{-1}$, are believed to arise from photoionised, slowly expanding gas which recombines and emits photons. This gas, most likely expelled by the progenitor star during the last phases of its evolutions, is located in the outer CSM and is not perturbed by the SN ejecta, at least during the early phases of the SN evolution. At later phases, intermediate-velocity line components (FWHM of a few thousands $km\,s^{-1}$) may be generated when the high velocity SN ejecta (a few 10^{-4}) collides with the dense pre-existing CSM. Although the interaction may mask the innermost ejecta (as well as the explosion mechanism; Chevalier and Fransson 1994), in some cases, particular geometric configurations may also favour the detection of high-velocity components (a few $10^{-4}\,km\,s^{-1}$) arising from the photoionised SN ejecta. SNe IIn light curves are instead quite heterogeneous, showing both slow and fast declining SNe, as well as faint ($M_R \gtrsim -16$ mag) and very bright ($M_R \lesssim -19$ mag) objects (e.g. Kiewe et al. (2012) for a sample of SNe IIn). Note that recently it has been discussed that SNe IIn are not really a SN type, but an external phenomena where any type of SN (due to thermonuclear or core-collapse explosion) or not terminal outburst with fast ejecta and sufficient energy, interact with a slower and denser CSM. It produces a phenomena which appears or mimic what we know as a SN IIn. Therefore this sub-class is more commonly named "interacting SNe" (e.g. see Smith 2016).

Finally, the spectrum of the SNe IIb is similar to that of the SNe II-P and II-L during maximum light, i.e. it has strong lines of H, but in the following week it metamorphoses to that of SNe Ib. This points out a physical link between these two classes (SN 1993J represents the prototypical object of this subclass; Richmond et al. 1994), suggesting that SNe II and SNe Ib/c share a common origin, i.e., the CC of a massive star, but with just different amounts of stripping on the progenitors' outer layers.

Together the Type II-P and II-L represent the majority of CC-SNe (considering a volume-limited rate in the local Universe; Li et al. 2011). Almost 9% is formed by interacting SNe. H-poor SNe (SNe IIb, Ib, Ic) constitute instead the remaining $\sim$37%. Recently, Cappellaro et al. (2015) find similar rates for CC-SNe groups considering a redshift range $0.15 < z < 0.35$. Rare events like SN 1987A-like objects are estimated to form $\sim$3% of the CC-SN population (Pastorello et al. 2012).

1.2.2 Observational Constraints on the Progenitors of Core-Collapse Supernovae

In the last decades SNe have been studied in order to better establish the link to their progenitor stars and thereby to clarify the evolution of massive stars.

1.2.2.1 Observables of Core-Collapse Supernovae

Indirect clues about the SN origin can be derived from the interaction of the material dismissed during the explosion with dense circumstellar medium lost by the progenitor during its turbulent life, or as said before, from their light curve and spectral evolution.

Light Curves

The first observable of electromagnetic radiation from a SN is the shock-breakout. It is a short time scale (minutes for SNe with compact progenitors) in which the shock wave produced during the collapse of the stellar core, reaches the stellar surface. In this instant a flash of soft X-rays and ultra-violet (UV) photons are released. The shock-breakout was directly observed in X-rays for the type Ib SN 2008D (Soderberg et al. 2008), whereas the fast cooling tail (due to the adiabatic expansion of the ejecta) after the shock-breakout has been observed in optical and UV bands for a few other objects (e.g. SNe 1993J and 2013df (Morales-Garoffolo et al. 2014)). This post shock-breakout phase mainly depends on the progenitor radius (e.g. Chevalier 1992).

Successively, the photons gradually leak out of the photosphere in a diffusion time scale ($t_{diff} \propto \rho \kappa R^2$, where ρ is the density, κ the opacity, and R the SN radius) shorter than the expansion time scale ($t_{exp} = R/V$, where V is the expansion velocity). During this period (tens to days) the thermal shock energy decreases as the ejecta expands, and the radioactive decay of ^{56}Ni (produced in the explosion, which has a lifetime of 8.8 days) and ^{56}Co (lifetime of 111.3 days) becomes important, reaching a peak of luminosity. After this maximum the SN ejecta continue to expand and cool, eventually arriving to the hydrogen recombination temperature of the ejecta. While the recombination wave moves inward through the ejecta, the temperature remains practically constant. Hence, depending on the mass of the hydrogen envelope and the radius of the SN progenitor star, the SN luminosity could show a constant or plateau phase (e.g. the case of the SNe Type II-P), or a steep decline after maximum light (e.g. the SNe Type II-L).

Once all the hydrogen has recombined, i.e. at late time (t > 100 days), the light curve declines at the rate of the decay of ^{56}Co $\rightarrow$ ^{56}Fe (0.0098 mag day^{-1}). The late time light curve observations are useful to determine the mass of ^{56}Co (and hence synthesized ^{56}Ni). At times later than 1000 days past explosion, other radioactive elements with longer half-lives such as ^{56}Co and ^{56}Ti power the luminosity evolution of the CC-SNe.

Physical parameters from the SN explosion like the kinetic energy of the explosion, the total ejected mass, the mass of ^{56}Ni synthesized and the radius of the progenitor, can be estimated through the fit of the SN bolometric light curves by semi-analytic functions (e.g. Arnett 1982) or more sophisticated hydrodynamical models (e.g. Blinnikov et al. 2000). More details about the SN modelling can be found in Sect. 1.2.3.2.

Spectral Evolution

SN spectral time evolution are a scan through the SN ejecta. They are important to study the chemical compositions of SNe and their progenitors, and the kinematic of the SN ejecta. A clear example is their utility is to classify SNe.

During the early phases or photospheric phase, the SN ejecta are not completely transparent and only the outer layers of the ejecta are observed. Consequently, spectra are characterized by showing a black-body continuum superposed by absorption and/or emission lines (e.g. see Filippenko (1997) for a detailed review). These features are broad because the SN ejecta is expanding at high velocity. They often show P-Cygni profiles with blue-shifted absorption and red-shifted emission components. Precisely via the absorption component of the P-Cygni profile it is possible to measured the velocity of the region at which the line predominantly forms (through the Doppler effect).

Late-time (nebular) SN spectra, taken several months after the explosion, are a unique way to peer into the very centre of the exploded stars. At those phases the opacity for optical photons has dropped substantially due to the expansion of the ejecta, so that the innermost parts, which were previously shielded by an effective photosphere, are now uncovered. The inner ejecta composition, unveiled by the nebular emission lines, is one of the most powerful tools to constrain the mechanism that gave rise to an explosion, since they all have a characteristic and widely unique nucleosynthesis. Late time spectra are characterized by strong emission lines on top of a faint continuum. Also the profile of a nebular emission line carries important information. The width of an emission line is a measure for the radial extent of the emitting species, and the detailed shape encodes asymmetries in its spatial distribution (e.g. Taubenberger et al. 2009).

When the SN ejecta interact with the CSM, the spectra present a blue continuum with superposed narrow emission lines, arisen from the CSM ionized by the shock interaction emission. If the CSM is thin, it is also possible to observe broader emission lines from the ionized ejecta.

The radiative-transport modelling of the SN late-time spectra can give us information about the kinematics of the ejecta and constrain the SN progenitor masses (e.g. Jerkstrand 2017). For example, considering the sensitive dependency of the oxygen nucleosynthesis with the main-sequence mass of the star, the modelling of the [O I] 6300, 6364 Å features helps us to constrain the progenitor mass (e.g. Morales-Garoffolo et al. 2014).

Supernova Remnants

Studying the observed properties of young SN remnants (SNR) is also an indirect path to connect with the SN progenitors (e.g. see the reviews Chevalier (2005), Patnaude and Badenes (2017)). The remnant phase begins when, after the light from a SN fades away, the ejecta in expansion, cools down, and strongly interacts with the surrounding material, either the interstellar medium (ISM) or a more or less extended CSM modified by the SN progenitor. Thus SNR provide detailed information on the chemical composition of the ejecta, the explosion dynamics, and the progenitor star mass loss distribution. The most notable cases are the Galactic SNR Cassiopeia A, and the youngest know remnant, SN 1987A, in the Large Magellanic Cloud. The advantage of the study of Galactic SNR is that they can be resolved in fine detail. However, their link with CC-SN is complex given the large diversity of the CC-SN explosions and circumstellar environment, and the large mass range of the progenitors.

1.2.2.2 Searching for SN Progenitor Stars

Still, the killer case is made with the *direct identification* of the star prior to explosion. Until the early 90s, only nearby events such SNe 1987A ($\sim$50 kpc; White and Malin 1987) and 1993J ($\sim$3.6 Mpc; Aldering et al. 1994) have allowed the direct progenitor identification in pre-explosion images. In recent years over a dozen CC-SN progenitors have been identified based on the inspection of archival, pre-explosion images. The identification relies on the positional coincidence between the candidate precursor and the SN transient. This requires high spatial resolution and very accurate astrometry because, at the typical distance of the targets (>30 Mpc[2]), source confusion becomes an issue. In practice, only *HST* or 8-m ground-based telescopes mainly provided by adaptive optics images can be used to accurately pin-point (with a typical uncertainty of a few tens mas) the progenitor candidate. Even so, there is always the chance of mis-identification with foreground sources or associated companion stars. Thus, a final approach is visiting the SN field when the SN has weakened: if the candidate star has disappeared, then it was indeed the progenitor, otherwise it was a mis-identification (e.g. see Maund and Smartt 2009, Van Dyk et al. 2013, and Fig. 1.2).

---

[2]This distance limit is based on practical experience. Smartt et al. (2009) and Eldridge et al. (2013) set to 28 Mpc the distance limit for a feasible search of SN progenitors, although there are exceptions such as the massive progenitor of SN 2005gl at 60 Mpc (Gal-Yam et al. 2007).

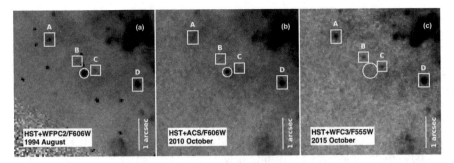

Fig. 1.2 Subsections of the pre-explosion (*panel* **a**), post-explosion (*panel* **b**), and late-time (*panel* **c**) *HST* image of the SN 2010bt site. SN 2010bt is likely a Type IIn SN and its progenitor was a massive star that experienced a powerful outburst. The positions of the SN candidate progenitor and SN are indicated by *circles* each with radius of 3 pixels (between 0.08 and 0.15 arcsec), except for *panel* **c**, for which the radius is 6 pixels (∼0.23 arcsec). The positions of three neighbouring sources of SN 2010bt, "A", "B", "C", and "D" are also indicated

Once a progenitor star candidate is identified, its initial mass and evolutionary state before the CC can be estimated by comparing its brightness and colour (if multi-colour imaging is available) measured with stellar evolution models. These models are chosen taking into account the metallicity in the SN environment, and the distance and the extinction to the star, derived from detailed light curves and spectral evolution of the SN with ground-based data (see Sect. 1.2.4 for a discussion about stellar evolution models).

Taken together, the availability and depth of archive images of nearby galaxies is a determining factor that delimits the rate of SN progenitor stars identified. There is an approximate probability of about 25% to find an image of the host galaxy of a nearby SNe in the HST archive (Smartt et al. 2009).

Following the above or similar steps, direct detections or upper mass limits have been established for progenitors of some types of SNe:

- *Type II-P:* Based on the statistics of around 15 SNe II-P, it appears that all of these progenitors exploded in the red supergiant phase from stars with initial mass range of 8–18 $M_\odot$, as we would theoretically expect (see Sect. 1.2.4.2). However there has been no detection of a higher mass stars in the range 20–40 $M_\odot$, which should be the most luminous and brightest stars in these galaxies. This has led to the intriguing possibility that higher mass stars undergo core-collapse, but form black-holes which prevents much of the stellar mass escaping the explosion (Reynolds et al. 2015). Theoretically, such quenched, low energy explosions have been proposed. Our lack of detection of high mass progenitors could be evidence for this missing population (Kochanek et al. 2008; Smartt et al. 2009). But exceptions exist as the case of SN 1987A, considered a peculiar SN Type II because in spite of exhibiting prevalent hydrogen P-Cygni profiles in the early spectra, it had slow rise to maximum, faint, and broad light curves. Its progenitor was instead found to be a blue supergiant of around 14–20 $M_\odot$, with a hydrogen envelope of around 10 $M_\odot$.

- *Type II-L:* Only in a couple of cases, e.g. SNe 2009hd (Elias-Rosa et al. 2011) and 2009kr (Elias-Rosa et al. 2010; Fraser et al. 2010), the progenitor has been identified and related with more yellow stars than expected by theoretical stellar evolution models, with estimated initial mass of between 18 and 25 $M_\odot$. However, a recent work argues that the progenitor of SN 2009kr is most probably a small compact cluster (Maund et al. 2015). It should be noted that it is no clear the separation between the historically defined SNe II-P and SNe II-L.

- *Type IIb:* These SNe likely originate from binary stripped progenitors with masses in the range 12–18 $M_\odot$. The precursor of the prototype SN 1993J (exploded in M81) was a K-type supergiant with ∼15 $M_\odot$ with a B-type star as companion (e.g. Maund and Smartt 2009). In addition, it has been identified the progenitor of other four SNe IIb. SN 2011dh was found to have a yellow supergiant precursor (Maund et al. 2011; Van Dyk et al. 2011; Bersten et al. 2012), likely part of a binary system with a compact companion (Benvenuto et al. 2013), and with relatively low initial mass (∼13 $M_\odot$). This result was confirmed by the disappearance of the bright yellow star (Van Dyk et al. 2013), and the detection of a UV source at the SN position (Folatelli et al. 2014). The progenitors of SNe 2013df (Van Dyk et al. 2014) and 2016gkg (Tartaglia et al. 2017) have been also associated with moderate mass yellow supergiant stars of 13–17 $M_\odot$. Only in the case of SN 2008ax, the colours of the progenitor candidate might be those of a young Wolf-Rayet (WR) star (single massive stars that have lost their hydrogen envelopes and in some cases also helium layers), which had retained a thin, and low-mass shell of H, or with a system which has contaminated flux from an associated cluster or nearby stars (Crockett et al. 2008).

- *Type Ib/c:* It has been proposed that these SNe arise from either massive stars with initial mass > 25–30 $M_\odot$ (which have lost their outer hydrogen envelopes during a WR phase through radiatively driven winds or intense mass loss) or lower mass stars $8 < M < 30\ M_\odot$ in a binary system (which have been stripped of mass by their companion stars). Deep, high resolution images of pre-explosion sites of about 14 type Ib/c SNe have been analysed so far (Eldridge et al. 2013; Elias-Rosa et al. 2013), but with not detections of progenitors, or progenitors systems in any of these. The probability of not finding a progenitor if these SNe originate from WR stars with initial masses > 25 $M_\odot$ similar to those seen in the Milky Way and Magellanic Clouds is only 12%. Although some authors argue that the parameters derived for the observed WR population do not correspond to those of WR stars at the point of collapse (Yoon et al. 2012; Groh et al. 2013). In short, it is still not possible to set range of masses, or discriminate unambiguously between single WR stars or interacting massive stars in binary systems as the progenitors of Ib/c. Only for the case of iPTF13bvn (Cao et al. 2013) a possible WR was detected at the position of the SN. However, further detailed stellar evolution modelling indicates that the progenitor candidate is a better match to a low mass progenitor in a binary system (Bersten et al. 2014; Eldridge et al.

2015). The recent analysis of post-explosion observations of the iPTF13bvn site finds the disappearance of the progenitor of iPTF13bvn, which is most likely a helium giant of 10–12 $M_\odot$ rather than a WR star (Eldridge and Maund 2016).

- *Interacting SNe:* Some successful detection of progenitors of interacting SNe related these with either high luminosity or high-mass progenitors. A clear progenitor detection is the case of SN IIn 2005gl (Gal-Yam et al. 2007; Gal-Yam and Leonard 2009) which exploded at the same location as a luminous blue variable (LBV) star. For SN 2010jl (Smith et al. 2011) it was also found blue flux coincident with the SN, likely from a very young cluster or a single massive star. Smith et al. (2011) argues that the initial mass of the progenitor should be, in any case, $>30 \, M_\odot$. In addition, the SN Ibn 2006jc precursor was observed as a carbon-oxygen WR star embedded within a helium circumstellar medium with a mass between 60 and 100 $M_\odot$ (Pastorello et al. 2007). It has been found pre-explosion outbursts for some of these objects. In fact, Ofek et al. (2014) allege that this activity is common in SNe IIn. As we will discuss in more detail in Sect. 1.3.1, some interacting SNe show pre-SN photometric variability of the precursor stars which are invaluable to characterise the final stages of the progenitors. The most famous example is SN 2009ip for which were observed regular, but temporally sporadic, series of stellar eruptions in 2009 and 2012, before a final outburst in late 2012 (e.g. Pastorello et al. 2013; Fraser et al. 2013; Mauerhan et al. 2013; Margutti et al. 2014). A luminous progenitor for this transient ($M_V = -10$) was detected in *HST* archival images taken in 1999 (Smith et al. 2010; Foley et al. 2011). The nature of this transient or those of similar objects remains debated: some are undoubtedly genuine core-collapse SNe, while others may be giant non-terminal outbursts from LBV-like stars. The LBV progenitor scenario for SNe IIn is currently the favoured one, but given the large variety of this SN class it is natural to think about the existence of multiple precursor channels. For example, it has been suggested alternative scenarios such as a merger burst event in a close binary system (e.g. Soker and Kashi 2013), or the collision of the SN ejecta of a red supergiant with the surrounded photoionization-confined shell formed through repeated mass loss events (Mackey et al. 2014).

Figure 1.3 shows a Hertzsprung-Russell diagram with a summary of the detected progenitor stars and estimated upper limits of the main SN types (Smartt 2015). As we can see, although it is possible to delimit observationally SNe with their stellar progenitors, there are still many unknowns. Each case we study provides new clues.

1.2.3 Theory of Core Collapse Supernovae

Massive stars, with masses between $\sim$8 and $\sim$100 $M_\odot$, end their lives collapsing under their own gravity. However, the most common kind of progenitors for nearby supernovae and the likely origin of most observed neutron stars consist of isolated

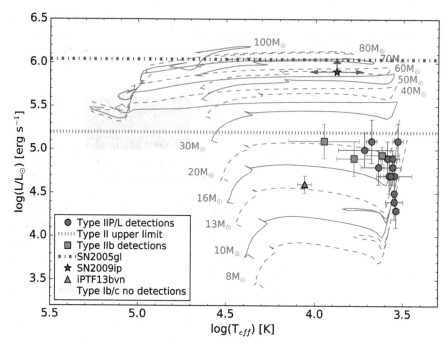

Fig. 1.3 Hertzsprung Russell diagram showing the temperature and luminosity of the detected progenitor stars and upper limits of the main type of SNe presented in Sect. 1.2.2.2 and Smartt (2015). For comparison, model stellar evolutionary tracks from Eldridge and Tout (2004) are also illustrated (figure reproduced from Smartt (2015))

(or non-interacting) massive stars of a mass between 9 and 30 $M_\odot$ and solar metallicity. We focus next on the case of a typical star of this kind while other kind of progenitors are explored in the next sections.

1.2.3.1 The Collapse of a Typical Star

In a typical star, the original material from which it was built, mainly hydrogen, has experienced a series of thermonuclear reactions leading to a stratified onion-like shell structure result of the intermediate ashes of fusion chain reactions. The innermost region, the core, is composed by elements of the iron family, which posses the highest nuclear binding energy per nucleon and hence are unable to fusion further regardless of the density and temperature. The iron core is fed by the surrounding silicon burning shell and it grows until it reaches a mass of 1.2–2 $M_\odot$, which is just below its Chandrasekhar mass (Chandrasekhar 1938). Typical conditions in the iron core, which determine its Chandrasekhar mass, are an electron fraction of $Y_e \sim 0.42$, a specific entropy of $s \sim 1k_B$ per baryon (k_B being the Boltzmann constant), a temperature of $T \sim 10^{10}$ K (about 1 MeV) and a central density of $\rho \sim 10^{10}$ g cm^{-3}. The Fermi energy of this gas of partially degenerate electrons is $\epsilon_F \sim 8$ MeV, therefore, finite temperature corrections are important for the estimation of the "effective" Chandrasekhar mass (Woosley et al. 2002), which reads

$$\frac{M_{Ch}}{M_\odot} = 5.83Y_e^2 \left[1 + \left(\frac{\pi k_B T}{\epsilon_F} \right)^2 \right] \approx 1.03 \left(\frac{Y_e}{0.42} \right)^2 \left[1 + 0.15 \left(\frac{kT}{1\,\text{MeV}} \right)^2 \right]$$

(1.1)

At this point the iron core starts collapsing due to two processes: first, electron captures by nuclei reduce the electron pressure while the neutrinos produced carry away energy from the core. Second, at densities above 10^{10} g cm^{-3}, photo-disintegration of iron nuclei into alpha particles cools down the core. Both processes reduce the value of the Chandrasekhar mass until the pressure is not able to counteract the self gravity of the object and the collapse becomes inevitable (Fig. 1.4a). The detailed structure of the progenitor, in particular the iron core, determines largely the outcome of the collapse and deserves a detailed discussion (see Sect. 1.2.4).

In this case, the collapse progresses in dynamical timescales of several 100 ms, enhanced by deleptonization and cooling processes. Beyond densities of $\rho \sim 10^{12}$ g cm^{-3} neutrinos become trapped preventing further cooling and deleptonization. From this stage on, the collapse proceeds adiabatically inside the neutrinosphere, where matter reaches rapidly beta equilibrium. At $\rho \sim 2 \times 10^{14}$ g cm^{-3} nuclear interactions between nucleons become the fundamental source of pressure. This stops the collapse abruptly as the equation of state stiffens, i.e. as matter becomes less compressible. The supersonically infalling matter forms a shock at about 10 km from the centre that propagates outwards (Fig. 1.4b). At the time of bounce, the shock encloses $\sim 0.5 M_\odot$ of nuclear matter the seed that will form a neutron star, which at this stage is known as proto-neutron star (PNS). The initial mass of the PNS is largely fixed by the deleptonization and cooling processes driving the collapse, which determine the local sound speed of the fluid and this, in turn, the location of the shock formation. As a result, massive stars always end their collapse with a bounce and the formation of a gravitationally supported object, a PNS, and never with a prompt collapse to a black hole.[3]

As the shock propagates against the infalling material of the remaining iron core, nuclei are photo-dissociated into free nucleons. About 10^{51} ergs per $0.1 M_\odot$ of infalling material is used in this process. This weakens the shock that stalls at about 100 km from the centre in a time of a few 10 ms (Fig. 1.4c). The standing shock can survive for timescales of several 100 ms and is effectively a gigantic

---

[3]Note however that pop III stars have very massive ($M > 100 M_\odot$) iron cores with high entropy ($s \sim 8 k_B$ per baryon) and may form black holes promptly (Sekiguchi and Shibata 2011).

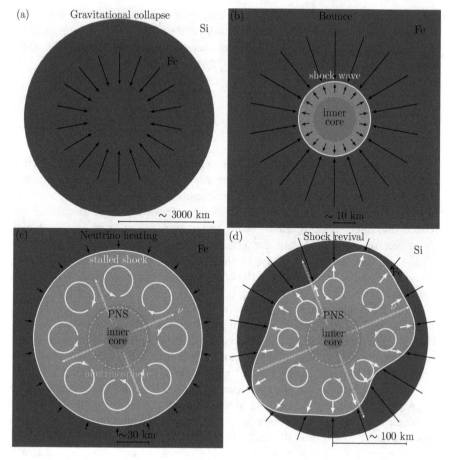

Fig. 1.4 Stages in neutrino-driven core collapse explosion: gravitational collapse (**a**), core bounce (**b**), shock stagnation and neutrino heating (**c**) and shock revival (**d**). Figure approximately at scale

transducer; during the collapse of the iron core, a gravitational binding energy of

$$E_{\text{bind}} \sim \frac{GM_{\text{PNS}}^2}{R_{\text{PNS}}} \sim 10^{53} \left(\frac{M_{\text{PNS}}}{1M_{\odot}}\right)^2 \left(\frac{30\,\text{km}}{R_{\text{PNS}}}\right) \text{erg} \qquad (1.2)$$

is released in form of kinetic energy of the infalling material, which is then transformed into thermal energy at the shock. This is the main source of energy of all the events that will develop next. At this time the PNS is growing rapidly in mass and has a size of about 30 km. It consists of an cold inner core of ∼10 km radius, composed of the unshocked material, and a thick hot envelope. The electron neutrinosphere is located right below its surface, preventing a rapid cooling of the PNS, which is now in thermal and beta equilibrium with neutrinos and stores about

10^{53} erg of thermal energy. Most of this energy is released as thermal neutrinos and antineutrinos streaming out in diffusion timescales of a few $\sim$10 s. However, this neutrino cooling process will not stop at that point but will continue for the next 10^6 years, acting as the main cooling process in neutron stars. Under the appropriate conditions, a small fraction of the neutrino energy, of about 10^{51} erg, is deposited behind the stalled shock, which is revived disrupting the rest of the star and forming a supernova explosion (Fig. 1.4d). This is the so-called delayed neutrino-heating mechanism (Bethe and Wilson 1985). If this mechanism fails, the PNS will grow in size until its maximum mass is reached and will collapse to a black hole (see Sect. 1.2.4.2).

In the region between the PNS surface and the shock, the temperature radial profile is such that neutrino emission rates, responsible for the cooling, have a steeper radial decline than neutrino absorption rates, responsible for the heating. This leads to the formation of a gain radius, above which there is a net absorption of energy by the fluid. This region is called the gain layer. Neutrino heating is a self-regulated process because fluid elements being heated in the gain layer expand away, which in turn decreases the ability of this fluid to absorb energy. This process limits the amount of energy being released in a supernova explosion to $\sim$$10^{51}$ ergs and makes the outcome of the events invoking only the delayed neutrino-heating mechanism rather homogeneous (supernovae with additional sources of energy are discussed in the next sections). As simple as it appears to be, heating proceeds in a rather complicated way due to multidimensional processes. Ultimately, what sets the amount of energy transferred to the shock is determined by the so-called dwell time, which is the average time that takes for a fluid element to cross the gain layer (Janka et al. 2008; Buras et al. 2006; Marek et al. 2009). Two processes increase the residence time of matter in the gain layer and are crucial for the development of a successful explosion: convection and the standing accretion shock instability (SASI). On the one hand, convection overturn produces plumes of low-entropy matter that fall rapidly towards the PNS while hot bubbles of high entropy, which would otherwise fall into the PNS, rise constantly up into the gain layer. On the other hand, the SASI (Blondin et al. 2003) produces a sloshing motion of the shock due to an unstable acoustic-advective cycle that couples the PNS surface with the shock (Foglizzo and Tagger 2000). This instability enhances non-radial motions and helps to expand the shock, which increases its energy deposition efficiency (Marek et al. 2009). Which of both processes is the dominant effect leading to the supernova explosion is still matter of intense debate in the supernova modelling community. The analysis of Yamasaki and Yamada (2007) suggests that SASI would be more favourable in cases with lower neutrino luminosity while convection in cases with higher luminosities. Regardless of this details, the explosion begins whenever the dwell time of matter in the gain layer is longer that the time required to transfer the amount of energy necessary to unbind this matter from the star. At this stage, the shock expands rapidly at the same time as it is continuously being heated by neutrinos. In this runaway situation, the globally expanding shock will gain energy for the next few seconds until reaching the final explosion energy. The shape of the expanding shock is highly asymmetric at this stage, dominated by either the sloshing motions produced by the SASI or by rising hot bubbles produced by convection.

1.2.3.2 Numerical Modelling

While we start having a clear picture of the scenario in which neutron stars are formed, the exact conditions under which successful supernova explosions are produced are still not completely well understood. The modelling of core collapse supernova requires a wide variety of physical ingredients, including among others a nuclear physics motivated finite temperature equation of state, a detailed description of neutrino interactions and general relativity (see e.g. Janka et al. 2007; Janka 2012; Burrows 2013; Müller 2016, for recent reviews). The numerical modelling of this scenario is computationally challenging and even today, with the use of the largest scientific supercomputing facilities available, we cannot afford solving the full set of 6-dimensional Boltzmann equations, necessary for the neutrino transport, or to use sufficiently high numerical resolution to resolve small-scale three-dimensional turbulence that may affect the dynamics of the system. Additionally, the problem at hand, results in a set of non-linear equations associated with the evolution of a fluid interacting with neutrino radiation, which, even if we could manage to solve them accurately, lead to a complex dynamics that may behave in a stochastic and chaotic way, very sensitive to small changes in the initial conditions. Nevertheless, a huge progress have been made in the last four decades. We focus here on the current status of numerical modelling and mostly in the developments of the last decade, which we hope will help the reader understanding the current uncertainties in the theoretical knowledge of the scenario in which neutron stars are born. A more complete and historical perspective of the numerical modelling of supernova can be found e.g. in Janka (2012).

Most of the efforts have been focused in developing an accurate description for the neutrino transport in simulations, since this is a crucial ingredient for the revival of the supernova explosion. So far, state-of-the-art multi-energy group solvers for three-flavour neutrino transport, including energy-bin coupling terms and velocity-dependent corrections, have only been possible in spherically symmetric (1D) simulations (Yamada et al. 1999; Liebendörfer et al. 2004). Under such restrictive symmetry conditions, the energy deposited by neutrinos is not sufficient to revive the shock and simulations fail to produce successful supernova explosions. The exception are stars with O-Ne-Mg cores (see Sect. 1.2.4). In order to tackle the multidimensional case, one has to consider approximations of the Boltzmann transport equations. The two-momentum approximation is a popular choice, in which only equations for the neutrino number, energy and momentum are solved, supplemented by an Eddington factor closure. This approach has been used in 1D simulations (Burrows et al. 2000; Müller et al. 2010) and extended to multidimensional (2D and 3D) simulations in the so-called ray-by-ray-plus (RbR+) scheme (Rampp and Janka 2002; Buras et al. 2006). In this scheme a set of radial 1D problems is solved for each angle in a spherical polar coordinates grid. Angular couplings appear through neutrino pressure gradients and the advection of neutrinos trapped with the fluid in the optically thick regions. Truly multidimensional, multi-energy schemes have been only possible in 2D. The most sophisticated version of those are multi-group flux-limited (MGFD) schemes (Burrows et al. 2006, 2007b,c; Swesty and Myra 2009), specially schemes including multi-angle treatment (Ott et al. 2008; Brandt et al. 2011).

As an alternative to those computationally expensive Boltzmann solvers, a series of additional approximations have appeared in the literature: In the M1 scheme (Obergaulinger et al. 2014), a local algebraic closure was used in the two-momentum transport equations yielding a set of hyperbolic equations. In the isotropic diffusion source approximation (IDSA) (Liebendörfer et al. 2009), neutrino distribution is decomposed into a trapped and a free streaming component. The fast multi-group transport (FMT) method (Müller and Janka 2015; Müller 2015) solves the stationary neutrino transport problem in a ray-by-ray fashion. The so-called neutrino leakage scheme (O'Connor and Ott 2010) is a simple prescription to describe cooling and heating of energy-averaged (grey) neutrinos, which can be easily applied in multidimensional scenarios with a ray-by-ray approach (Ott et al. 2012). The leakage scheme has also been extended to the multi-energy group case (Perego et al. 2016). Although the results of the latter series of approximations cannot be truly compared to RbR+ or MGFLD methods, they are useful to explore a large space of physical parameter or to study in details certain processes taking place during the explosion (e.g. instabilities, 3D effects and magnetic fields).

Also crucial is an accurate and complete treatment of neutrino interactions with matter. These set the emission and absorption rates, which are essential to determine neutrino luminosity and energy deposition rates at the shock driving the supernova explosion. The most relevant interactions included in numerical simulations (see e.g. Buras et al. 2006; Lentz et al. 2012) are: β processes, including neutrino and antineutrino absorption and emission processes by nucleons (Burrows and Sawyer 1998) and nuclei (Langanke et al. 2003); scattering of neutrinos with nucleons (Burrows and Sawyer 1998), nuclei (Horowitz 1997), electrons and positrons (Mezzacappa and Bruenn 1993); nucleon-nucleon bremsstrahlung (Hannestad and Raffelt 1998); and neutrino-antineutrino annihilation to produce either electron-positron pairs (Bruenn 1985; Pons et al. 1998) or different flavour neutrino-antineutrino pairs (Buras et al. 2003). Special care has to be taken in the computation of these interactions to include inelastic terms in the scattering that produce an interchange of neutrino energy with their targets, Pauli blocking factors, high-density nucleon-nucleon correlations and weak magnetism corrections, among others (for a more complete description see Janka (2012) and Chap. 9; for a particular implementation description see Rampp (2000)). The use of incomplete sets of interactions with different corrections implemented has been one of the sources of disagreement in the results of numerical simulations among different supernova groups. This tendency has changed in the last few years as the main groups have adopted a complete and accurate set of interactions, which has facilitated direct comparisons (see e.g. Burrows et al. 2016).

General relativity (GR) also plays an important role in the dynamics of the supernova explosion (Müller et al. 2012a). In comparison to Newtonian gravity, GR deepens the gravitational potential well leading to more compact PNSs. This increases the amount of gravitational binding energy released during the collapse, which adds up to the energy budget of the PNS and has a positive impact in triggering the explosion. Full GR simulations of the core collapse scenario (e.g. Ott et al. 2007b, 2013; Mösta et al. 2014; Abdikamalov et al. 2015) have been performed in the BSSN formulation (Shibata and Nakamura 1995; Baumgarte and Shapiro 1999). To avoid the use of computationally expensive full GR codes, several approximations have been developed. The conformally flat condition (CFC) approximation (Isenberg 2008; Wilson et al. 1996) is a waveless approximation to GR, which is exact in spherical symmetry, and has been used in core collapse simulations codes (see e.g. Dimmelmeier et al. 2002a; Müller et al. 2010; Cerdá-Durán et al. 2013). Direct comparisons of the CFC approach with full GR simulations have shown that differences in the dynamics are minute (Shibata and Sekiguchi 2004; Ott et al. 2007b,a). The CFC approximation has been reformulated (XCFC, Cordero-Carrión et al. 2009) to overcome uniqueness problems; with this improvement it is possible to study the formation of black holes (Cerdá-Durán et al. 2013) and, with the use of excision techniques (Cordero-Carrión et al. 2014), its posterior evolution. Second post-Newtonian corrections to the CFC metric (CFC+, Cerdá-Durán et al. 2005) showed only quantitatively small differences ($<1\%$) in the dynamics. Another popular approximation is the use of an effective pseudo-Newtonian potential to mimic GR effects (Rampp and Janka 2002; Marek et al. 2006). This approximation is widely used by the core-collapse community (see e.g. Buras et al. 2006; Kitaura et al. 2006; Obergaulinger et al. 2006; Scheidegger et al. 2008; Wongwathanarat et al. 2013; Hanke et al. 2013; Bruenn et al. 2013; O'Connor and Couch 2015; Bruenn et al. 2016; Summa et al. 2016; Burrows et al. 2016) and it produces an excellent agreement with GR in spherical symmetry. However, it agrees only qualitatively in the presence of fast rotation, for which either CFC of full GR are better suited. Regarding the computation of gravitational waves, the approximate quadrupole formula is used almost exclusively in all simulations, regardless of the gravity treatment. This approximations has been shown to be accurate for the mildly relativistic gravity of these system even when compared with more sophisticated waveform extraction techniques (Reisswig et al. 2011).

Additionally, simulations need realistic equations of state (EOS) able to handle a wide range of conditions present in the core collapse scenario. These general purpose EOS, also known as Supernova EOS, typically can handle densities from $\sim 10^8$ to 10^{15} g cm^{-3} and finite temperature dependence to cover all steps in the collapse from the iron core to the final neutron star. The composition is parametrised by the value of the electron fraction because nuclear statistical equilibrium (NSE) is considered. This is possible since, for the typical conditions inside the iron core and in the PNS, nuclear reactions occur in timescales much shorter than the dynamical evolution timescales. These EOSs consider the fluid composed by a mixture of heavy nuclei, alpha particles, photons, neutrons and protons. Several families of EOS are available including all these conditions: LS (Lattimer and

Douglas Swesty 1991), STOS (Shen et al. 1998a,b, 2011c), FYSS (Furusawa et al. 2011, 2013), HS (Hempel and Schaffner-Bielich 2010), SFH (Steiner et al. 2013) and SHO/SHT (Shen et al. 2011a,b). Additional details of those EOS and their variants can be found in the recent review by Oertel et al. (2017) and on Chap. 6. The main uncertainties in EOS calculations are the properties of nuclear matter, parametrised in terms of constants such as the incompressibility modulus K and the symmetry energy J, among others. Their values are currently constrained by nuclear physics experiments and neutron stars observations (Oertel et al. 2017). The impact of these EOS uncertainties in the dynamics of the collapse is only moderate. It mainly affects three aspects: first, below nuclear density, most supernova EOS use one average nucleus to describe all nuclei. This has an impact on electron capture rates, which in turn affects the shock formation location and the size of the inner core (Lattimer and Prakash 2000; Langanke et al. 2003; Sumiyoshi et al. 2005; Suwa et al. 2013; Steiner et al. 2013). Secondly, the properties of matter at supranuclear densities influence the structure of the PNS. For a "soft" EOS, i.e. that producing more compressible matter at typical post-bounce densities, more compact PNSs are formed. This enhances neutrino emission and favours supernova explosions (Marek et al. 2009; Hempel et al. 2012; Suwa et al. 2013). Lastly, the EOS has an impact on the maximum mass that a neutron star can support, which is relevant for the possible formation of a black hole (see Sect. 1.2.4.2).

For the treatment of the matter outside the iron core, NSE is no longer a good approximation and nuclear burning happens in longer timescales than dynamics ones. The EOS at densities below $\sim 10^8$ g cm^{-3} depends on the full composition of the fluid in terms of the mass fractions of different isotopes, which have to be advected with the fluid during the evolution. Therefore, at low densities, the supernova EOS described above is matched to a full composition EOS. A popular choice is the Timmes EOS (Timmes and Arnett 1999; Timmes and Swesty 2000). To avoid computationally intensive calculations of nuclear reaction networks (see e.g. Timmes 1999), phenomenological "flashing" prescriptions have been considered (see e.g. Rampp and Janka 2002; Buras et al. 2006) or simply NSE as a crude approach (see e.g. O'Connor and Ott 2010).

The most advanced core collapse simulations include all the physics described above using state-of-the-art numerical methods and the best available approximations in each case. In axial symmetry (2D) recent simulations including full sets of neutrino interactions, three-flavour neutrinos and general relativistic corrections to the gravitational potential, have resulted in successful explosions for a wide range of progenitors (12–25 $M_\odot$) using both the RbR+ approach (Müller and Janka 2014; Summa et al. 2016; Bruenn et al. 2016) and 2D-MGFLD transport (Burrows et al. 2016). Comparison among different groups shows a qualitative agreement in the results but still differences in the time of the onset of the explosion and its energy (Burrows et al. 2016). Analogous simulations with similar physical content have been performed in 3D by Lentz et al. (2015) and Melson et al. (2015a) using a ray-by-ray approach for the neutrino transport. In both cases only low mass progenitors were considered (15 and 9.6 $M_\odot$ respectively) and although both cases produce successful explosions, 3D effects weakened or enhanced the explosion with respect to 2D, depending on the work. Similar work for a 20 $M_\odot$ progenitor did not produced successful explosions unless strange-quark contributions to neutrino-nucleon scattering were considered. 2D and 3D simulations with less sophisticated neutrino transport and incomplete neutrino physics Takiwaki et al. (2016) (3D Newtonian, IDSA transport), Roberts et al. (2016) (3D GR, M1 transport), Dolence et al. (2015) (2D Newtonian, MGFLD transport), Nagakura et al. (2017) (2D Newtonian, Boltzmann transport), O'Connor and Couch (2015) (2D GR, M1 transport) and Suwa and Müller (2016) and Pan et al. (2016) (2D Newtonian, IDSA), have also shown successful explosions in some cases but not in all models.

Overall, results from numerical 2D simulations show that, when an appropriate and complete description of the physics involved is considered, it is possible to produce consistently successful supernova explosions. However, the more realistic 3D case still proves to be a challenge. This is an indication that either the approximations used in the simulations are not sufficiently accurate (specially regarding neutrino transport) or that some important physical ingredient is missing. Unresolved hydrodynamic turbulence could be also be a problem in 3D simulations (Melson et al. 2015b; Abdikamalov et al. 2015; Couch and Ott 2015) and should be investigated in more detail in the future.

1.2.4   Progenitor Dependence

In Sect. 1.2.3 we have described the scenario in which most neutron stars are born, what is thought to be the typical progenitor star of most supernova explosions. Here we discuss how this scenario changes when different progenitor are considered. There are several factors in stellar evolution that lead to differences in the progenitors of core collapse supernovae. The main one is its mass at birth, the so-called zero-age main-sequence (ZAMS) mass. Additionally, there are variations in the metallicity of the environment in which the star was born, its rotation rate, and the presence of a companion star.

1.2.4.1   O-Ne-Mg Cores

Stars with ZAMS mass of 8–10 $M_\odot$, can burn carbon to produce O-Ne-Mg cores. However, for such low mass stars, temperatures are not sufficiently high to ignite Ne. Instead, the core, which is close to electron degeneracy, grows to a mass of about $\sim$1.34 $M_\odot$, close to the Chandrasekhar mass. At this point, electron captures, the dominant process under this conditions, take over and trigger the collapse of the core. The resulting SNe are the so-called electron capture supernovae (ECSNe) (Nomoto 1984, 1987). The main feature of these progenitors is that they have a very steep density gradient outside the core (see Fig. 1.5), with very thin carbon and helium layers and a rapid transition to the hydrogen envelope. Due to this structure, once the core bounces, only a small fraction of the neutrino energy is necessary to

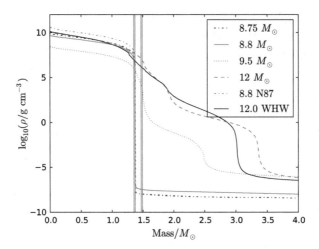

Fig. 1.5 Density profile as a function of mass coordinate for 8.75–12 $M_\odot$ pre-supernova models of Jones et al. (2013), the 8.8 $M_\odot$ model of Nomoto (1987) (N87), and a 12 $M_\odot$ model from Woosley et al. (2002) (WHW). From these models, those with a mass of 9.5 or 12 $M_\odot$ have formed an iron core and the rest an O-Ne-Mag core. Vertical red lines show derived pre-collapse masses for the two peaks in the observed neutron star distribution of Schwab et al. (2010) (figure reproduced from Jones et al. (2013))

power the explosion, shortly after bounce. Numerical simulations of this scenario show that it is possible to obtain successful explosions even in 1D models (Kitaura et al. 2006; Fischer et al. 2010). This has been confirmed by multidimensional simulations (Janka et al. 2008). The resulting supernova explosions are weak ($\sim 10^{50}$ erg) and Ni-poor. They are possibly associated with some subluminous type II-P supernovae e.g. see Smartt (2009), Botticella et al. (2009). The Crab remnant associated with SN 1054 is likely the result of such an explosion (Hillebrandt 1982; Nomoto 1982).

For solar metallicity stars, the range in which O-Ne-Mg cores are formed is limited to a small interval ($\sim 0.2\ M_\odot$) close to 9 $M_\odot$ (Poelarends et al. 2008; Jones et al. 2013). Depending on metallicity and the interaction with a close companion the range of masses may shift and widen (Podsiadlowski et al. 2004; Pumo et al. 2009). As a consequence, the ECSNe represent a contribution of $\sim 4\%$ to all supernovae in the local universe (Poelarends et al. 2008). Comparable results are obtained when comparing abundances of r-process elements in ECSNe simulations with observations of stars in the Galactic halo ($\sim 4\%$ of all core-collapse SNe, Wanajo et al. 2011). This results are consistent with an estimated rate of low-luminosity type II SNe of 4–5% (Pastorello et al. 2004).

Neutron stars formed in this scenario are expected to have a lower mass than typical neutron stars formed from progenitors with iron cores and appear in a very narrow range of masses fixed by the Chandrasekhar mass of the O-Ne-Mg core (Nomoto 1987; Podsiadlowski et al. 2004). Velocity kicks are also expected to be low (Podsiadlowski et al. 2004), since there are neither large asymmetries involved or a large amount of ejected mass (see Sect. 1.2.4.3). This theoretical expectation matches some evidence of a bimodal distribution in the observed masses of neutron stars that have not undergone accretion (van den Heuvel 2004; Schwab et al. 2010). The interpretation is that the lower and higher mass populations would correspond to a ECSNe and iron-core SNe, respectively. However, this bimodal distribution has not been found in a more recent analysis of the observations (Özel et al. 2012). Schwab et al. (2010) also indicated that the observed orbital eccentricity of neutron stars in binaries is lower in cases with lower NS masses, which is consistent with the low velocity kicks resulting of ECSNe.

In the particular case of double neutron stars (DNS), population synthesis calculations favour systems with an ECSNe (Andrews et al. 2015). The reason is that ECSNe produce weaker explosions with lower mass ejecta than iron-core SNe. This prevents the binary to unbind, leading to a low eccentricity binary neutron star. These results are qualitatively compatible with the narrow observed distribution of NS masses in DNS systems, with a mean mass of 1.33 $M_{\odot}$ and a dispersion < 0.1 $M_{\odot}$ (Özel et al. 2012; Özel and Freire 2016). Interestingly, these are precisely the candidates for neutron star mergers. Therefore, neutron stars formed from O-Ne-Mg cores may be crucial for understanding phenomena such as short GRBs and to estimate merger rates relevant for GW detectors.

1.2.4.2 Iron Cores in Solar Metallicity Stars: Mass Dependence and Black Hole Formation

We consider next mass dependence in solar metallicity stars forming an iron core, i.e. those with a mass above $\sim 9 M_{\odot}$. The evolution of massive stars with solar metallicity to their pre-supernovae stage has been studied by a number of authors (Woosley et al. 2002; Limongi and Chieffi 2006; Nomoto et al. 2006; Woosley and Heger 2007; Chieffi and Limongi 2013; Sukhbold and Woosley 2014; Woosley and Heger 2015). These models are the result of 1D simulations in which multidimensional effects (convection, rotation, magnetic fields, mass transfer in binaries) are included in a phenomenological way (see e.g. Woosley et al. 2002). Generally speaking, more massive stars reach higher temperatures at the core and thus have higher specific entropies. The thermal contribution increases the Chandrasekhar mass, and, as a consequence, more massive stars can host larger iron cores (Woosley et al. 2002). This can be seen in the left panel of Fig. 1.6, which shows the dependence of the pre-collapse iron core mass with the ZAMS mass for solar metallicity progenitors (red and black symbols). The relation is however non monotonic due different processes taking place at different mass ranges.

Stars with masses in the range 9–12 $M_{\odot}$ form iron cores. However, their stellar structure is somewhat similar to the O-Ne-Mg cores described in Sect. 1.2.4.1, with a steep decline in density outside the core (see 9.5 $M \odot$ model in Fig. 1.5). Numerical simulations of the collapse of these low mass iron cores in 1D (Müller et al. 2012b), 2D (Müller et al. 2013) and 3D (Melson et al. 2015b), show that the resulting SN explosions are weak, resembling to ECSNe. Although 1D simulations produce

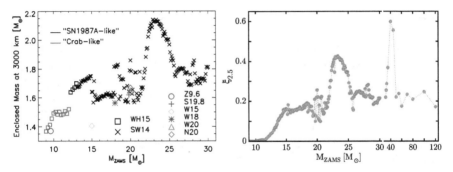

Fig. 1.6 (Left) Mass inside a radius of 3000 km for progenitors up to 30 $M_\odot$, at the time when the central density reaches the same value of 3×10^{10} g cm^{-3}, as a function of ZAMS mass. Black and red symbols denote models forming an iron core ("SN1987-like") or an O-Ne-Mg core ("Crab-like"), respectively. Progenitor models are the results from simulations by Woosley and Heger (2015) (WH15, squares) and Sukhbold and Woosley (2014) (SW14, crosses). The rest of the symbols are calibration models to SN 1987A (see Sukhbold et al. (2016), for details). (Right) Compactness parameter $\xi_{2.5}$ (see Eq. (1.3)) as a function of ZAMS mass for the 200 models between 9.0 and 120 $M_\odot$ of Sukhbold and Woosley (2014), Woosley and Heger (2015) used in Sukhbold et al. (2016). Note the scale break above 32 $M_\odot$ (both figures reproduced from Sukhbold et al. (2016))

successful explosions, multidimensional effects play here a more important role than in ECSNe, increasing the explosion energy in about a factor 5 (Melson et al. 2015b) with respect to spherical symmetry.

In the range 12–20 $M_\odot$ the stars are highly convective during their carbon burning phase. Convection enhances neutrino cooling, which drives the core towards degeneracy. This in turn produces light and compact iron cores. In the range 20–30 $M_\odot$ radiative (non-convective) carbon burning occurs, which hinders neutrino cooling leading to large and hot iron cores. Around 20 $M_\odot$, in the limit between convective and radiative carbon burning, there is a high variability in the resulting iron core masses and compactness (Sukhbold and Woosley 2014). However, not all features can be explained in terms of carbon burning alone. Details in burning of O, Ne and Si, as well as shell burning introduce a non-trivial dependence of the structure of the final iron core as a function of the initial mass (see Woosley et al. 2002, for a review on the topic). At the pre-supernova stage, these stars appear as blue or red supergiants, depending on the details of hydrogen shell burning and mixing.

In the range 30–100 $M_\odot$ density profiles outside the iron core become shallower, due to the presence of thick shells of heavy elements. These stars experience an important mass loss due to winds and as a result their pre-supernova masses decrease for increasing ZAMS mass, reaching pre-supernova values of 6–8 $M_\odot$. Above 35 $M_\odot$ these stars appear as blue variables or WR stars. If winds are able to strip away the hydrogen envelope they may explode as type Ib/Ic SNe. Rotation may also play an important role in the evolution and explosion of these stars, by adding an additional source of energy (see Sect. 1.2.4.5). In that case they may be the origin of long GRBs.

Multidimensional core-collapse simulations with the most advanced microphysics and transport have focused mainly in progenitors in the mass range 12–30 $M_\odot$ (Müller and Janka 2014; Summa et al. 2016; Bruenn et al. 2016; Burrows et al. 2016; Lentz et al. 2012; Melson et al. 2015a). The lower side of this range of masses (12–18 $M_\odot$) is the main contributor to all core collapse supernovae observed in the local universe (see Smartt (2015) and Sect. 1.3.3.1). SN 1987A is a peculiar example of this class, with a blue supergiant progenitor that has been estimated to have a mass of 14–20 $M_\odot$ and solar metallicity (Shigeyama and Nomoto 1990; Woosley et al. 1988, 2002). In this mass range, simulations show that stars with masses below 20 $M_\odot$ result "easier" to explode (shorter time to explosion, higher energies), while above this limit they become "harder" to explode (longer time to explosion, lower energies). By performing a wide set of 1D simulations using a simplified neutrino leakage scheme, O'Connor and Ott (2011) identified that the relevant parameter determining the "explodability" of a progenitor in the compactness of the core, $\xi_{2.5}$ defined as

$$\xi_M = \frac{M/M_\odot}{R(M)/1000\ \mathrm{km}}. \tag{1.3}$$

A small value of $\xi_{2.5}$ indicates that the layers surrounding the core are extended and that there is a steep density profile outside the core (as e.g. in O-N-Mg cores). On the contrary, a large value of $\xi_{2.5}$ indicates that the core is surrounded by a thick envelope, where the density profile is shallow. As a result, cores with high values of $\xi_{2.5}$ are harder to explode than those with lower values.[4] Right panel of Fig. 1.6 shows the compactness of the core as a function of the progenitor ZAMS mass. The compactness of the core is highly correlated with the mass of the iron core (compare both panels of Fig 1.6, below 30 $M_\odot$). Above 40 $M_\odot$, compactness declines to a nearly constant value, due to the significant mass loss of these stars. The compactness of the iron core is thus relevant to determine if a progenitor star will produce a neutron star or a black hole as a result of the collapse.

Traditionally, the onset of black hole formation has been thought to occur for progenitors with masses above certain threshold of $\sim$30 $M_\odot$. But recent numerical simulations suggest that this simple picture is likely wrong and that there are interleaved intervals of masses forming either neutron stars or black holes. A number of works have addressed this issue by exploring systematically the dependence of the supernova properties with the initial progenitor mass (Zhang et al. 2008; Ugliano et al. 2012; Pejcha and Thompson 2015; Ertl et al. 2016; Sukhbold et al. 2016). To be able to cover a wide range of progenitor masses and to evolve

---

[4]The reader should not confuse the parameter $\xi_{2.5}$ with the compactness of a neutron star (M_{NS}/R_{NS}) often appearing in the literature, despite of its similarity in name and definition. While more compact neutron stars have higher values of M_{NS}/R_{NS}, a higher value of $\xi_{2.5}$ indicates that the core of the star is more extended (less compact).

the supernova explosion to very late times, including the fallback into the neutron star, these simulations were performed in spherical symmetry (1D) and with a very simplified parametrised explosion engine. Despite of these simplifications important information can be extracted from these simulations. We review the main results next.

Explodability Below 15 $M_\odot$ and in the range 25–28 $M_\odot$, coinciding with low compactness progenitors (see right panel of Fig. 1.6) the outcome is preferentially a neutron star accompanied by a supernova explosion. Similarly, for masses of about 15–16 $M_\odot$ and 22–25 $M_\odot$ the outcome is preferentially a black hole. In the range 16–22 $M_\odot$ and above $\sim$28$M_\odot$ both neutron stars and black holes can appear. Although the $\xi_{2.5}$ provides a rough guide to explain this behaviour, it is clearly insufficient to explain all features observed (Ertl et al. 2016; Sukhbold et al. 2016); for example a star in the range 25–28 $M_\odot$ has a larger value of $\xi_{2.5}$ than one of about 15 $M_\odot$ and nevertheless, the latter is more likely to collapse to a black hole than the former. The introduction of a second parameter has proven to be sufficient to explain the phenomenology of all the 1D simulations (Ertl et al. 2016).

Black Hole Production Rate When these results are weighted with the initial mass function (IMF), the resulting fraction of massive stars forming black holes is about 20–30% (Ugliano et al. 2012; Sukhbold et al. 2016), increasing significantly previous estimates of $\sim$10% based on a single mass threshold model (Woosley et al. 2002). These results partially solve the so-called "SN rate problem" (Horiuchi et al. 2011), namely that the SN rate predicted from the star formation rate is higher than the SN rate measured by SN surveys. It is currently unclear what would be the observational signature of BH forming events. Even if no supernova explosion is produced there are suggestions that some weak transient could be associated with these events (Nadezhin 1980; Lovegrove and Woosley 2013; Piro 2013) rather than an unnovae (Kochanek et al. 2008) (also know as failed SNe or dark SNe; see Sect. 1.3.3.1). A non-negligible fraction of the dim core-collapse SNe in the very local Universe (10 Mpc), could in fact be related to BH forming events (Horiuchi et al. 2011). Additionally, there are observational indications of a paucity of potential SN progenitors (red supergiants) in the mass range 16.5–25 $M_\odot$ (Kochanek et al. 2008), which could be associated with BH forming progenitors. Direct observations of SNe progenitors are also consistent with most of the stars above $\sim$18 $M_\odot$ forming BHs (see Smartt (2015) and Sect. 1.3.3.1). Fast rotation has a strong influence in the behaviour of these systems and is treated separately in the next sections.

Explosion Energy One could expect that, since $\xi_{2.5}$ is related to the explodability of the star, stars with lower value of this parameter, and thus "easier" to explode, would produce more energetic events. However, all stars producing supernova explosions with initial masses above $\sim$10 $M_\odot$, have very similar explosion energies of $\sim$10^{51} erg (Ugliano et al. 2012; Pejcha and Thompson 2015; Sukhbold et al. 2016), due to the self-regulation of the delayed neutrino-heating mechanism (see Sect. 1.2.3.1). Below 10 $M_\odot$, regardless of whether they have an iron core or an

O-Ne-Mg core, the resulting explosions are weaker, with energies of $\sim 10^{50}$ erg in the lowest end of the progenitor masses. The exception are progenitors very close to the onset of black hole formation, which are very hard to explode and produce explosion a factor of a few weaker than typical explosions. In general, there is a positive weak correlation between Nickel mass production and explosion energy (Zhang et al. 2008; Pejcha and Thompson 2015; Sukhbold et al. 2016).

Remnant Mass All simulations show that progenitors with initial masses larger than $\sim 10 \, M_\odot$ form neutron stars with an average mass of $M_{NS} \sim 1.4 M_\odot$, while, below this threshold, lighter neutron stars are formed with ($M_{NS} \sim 1.2 M_\odot$) (Pejcha and Thompson 2015; Sukhbold et al. 2016). When these results are weighted with the IMF (Sukhbold et al. 2016), the resulting mass distribution (see Fig. 1.7) is weakly bimodal, resembling the observed distribution of NS masses (van den Heuvel 2004; Schwab et al. 2010) (see discussion in Sect. 1.2.4.1). In this distribution, most neutron stars with masses above/below 1.3 $M_\odot$ belong to progenitors with masses above/below 10 $M_\odot$. The higher end of neutron star masses, above 1.6 $M_\odot$, belong to progenitors mostly above $\sim 20 \, M_\odot$ (Pejcha and Thompson 2015; Sukhbold et al. 2016), very close to the onset of black hole formation, and that produce weak explosions. As a general result it is fair to say that, according to current numerical modelling, the most common type of neutron stars was formed with a mass of $\sim 1.4 \, M_\odot$ in supernova explosions of progenitors in the range 10–20 $M_\odot$. However,

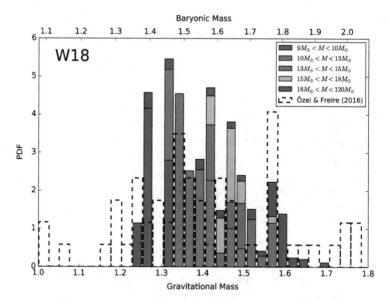

Fig. 1.7 Distributions of neutron star masses for the explosions calculated by Sukhbold et al. (2016) (colour coded), plotted against the observational data from Özel and Freire (2016). Different colour indicate the ZAMS mass of the progenitor star. The calibration W18 was used in these calculations (see Sukhbold et al. (2016) for additional details) (figure reproduced from Sukhbold et al. (2016))

these numerical simulations still lack of a complete modelling of the neutrino-heating mechanism and of important multidimensional effects, so the predicted NS mass distribution functions have to be taken with care. Regarding systems forming black holes, the work of Ugliano et al. (2012), Pejcha and Thompson (2015), Sukhbold et al. (2016) predicts masses above $\sim 5 M_\odot$, producing a gap in the range ~ 2–5 $M_\odot$ in the remnant mass distribution.

Fallback Even if the explosion is sufficiently energetic to disrupt the star, some of the material of the star may fall back onto the neutron star, increasing its mass and potentially forming a black hole (Colgate 1971; Chevalier 1989). This process occurs mainly as the supernova shock crosses composition interfaces, e.g. that between the He shell and the H envelope, with steep density declines, and a reverse shock is formed. The formation of such a reverse shock has been observed in multidimensional simulations of the supernova shock propagation through the star in 2D (Kifonidis et al. 2003; Scheck et al. 2006) and 3D (Hammer et al. 2010; Joggerst et al. 2010; Wongwathanarat et al. 2015). However, estimating the total amount of mass accreted into the NS by this process is difficult, since it implies following the matter falling back for timescales of hours, which is numerically challenging. This limits fallback mass estimates to 1D numerical simulations (Ugliano et al. 2012; Ertl et al. 2016; Sukhbold et al. 2016). In general, the amount of fallback material is expected to be small, mostly in the range 10^{-4}–10^{-2} $M_\odot$, progenitors with lower initial mass experiencing a lower fallback accretion. In simulations by Ertl et al. (2016) and Sukhbold et al. (2016), a few stars of ~ 30 $M_\odot$ formed black holes as a result of the fallback accretion, however this mechanism does not appear to be a significant channel for BH formation. There have been suggestions that, for weak explosions with strong fallback, it would produce an under-luminous SN, that could explain some observed peculiar Type Ia SN (Moriya et al. 2010). Fallback has also been invoked to explain the low magnetic field observed in central compact objects (CCOs), young neutron stars located near the centre of SN remnants. In this "hidden magnetic-field" scenario (Young and Chanmugam 1995; Muslimov and Page 1995; Geppert et al. 1999; Shabaltas and Lai 2012), the fallback material is able to bury the NS magnetic field explaining the observations of young NSs. In a longer timescale of 1–10^7 kyr, the magnetic field is able to re-emerge explaining magnetic fields in older objects (Young and Chanmugam 1995; Muslimov and Page 1995; Geppert et al. 1999). Numerical simulations of the accretion of matter onto magnetised material have shown that this mechanism is indeed feasible (Payne and Melatos 2004, 2007; Bernal et al. 2010; Mukherjee et al. 2013a; Bernal et al. 2013; Mukherjee et al. 2013b). It has been estimated that it is sufficient an accreted mass of 10^{-3}–10^{-2} $M_\odot$ to bury typical NSs magnetic fields (Torres-Forné et al. 2016). As a consequence, it is expected that neutron stars are born with hidden magnetic fields (CCOs) while other appear as regular pulsars (e.g. Crab). This scenario could also explain the lack of NS detected in SN 1987A.

1.2.4.3 Supernova Kicks and Binary Disruption

Considering that the most of stars live in binaries (Sana et al. 2012), if the supernova explosion is able to impart some kick velocity on the newly formed neutron star, then this could be ejected or at least introduce some eccentricity in the resulting system. Kicks are genuinely multidimensional effects that cannot be accounted for in the 1D numerical simulations reviewed in the previous section, so we deal with them separately here. There are two mechanism that produce kicks in neutron stars: asymmetries in the supernova explosion and the ejection of large amounts of mass by the explosion. These kicks have been observed in young neutron stars that show typical velocities of several 100 km/s, with some neutron stars moving at more than 1000 km/s (see e.g. Hobbs et al. 2005; Arzoumanian et al. 2002).

As we show in Sect. 1.2.3.1, supernova explosions are likely to occur in a very asymmetric way, tracing the multidimensional instabilities that helped to revive the stalled shock. Therefore, the expanding shock it is likely to have some net linear momentum in a random direction. This produces a reaction on the neutron star that, being less massive than the envelope, can experience a sudden increase of its velocity, the so called kick, in the opposite direction. Since the kick is the result of an instability breaking the initial symmetry of the star, the resulting kick direction and velocity are highly stochastic. Numerical simulations (Janka and Mueller 1994; Fryer and Young 2007) have shown that this mechanism alone is not able to accelerate NSs to more than a few 100 km/s, which is insufficient to explain observations. However, this initial kick is not the final velocity of the neutron star. In longer time-scales the gravitational interaction between the remnant and the slow moving massive ejecta accelerates further the neutron star in the so called "tug-boat" mechanism (Nordhaus et al. 2010, 2012). A series of 2D (Scheck et al. 2004, 2006; Nordhaus et al. 2010, 2012) and 3D (Wongwathanarat et al. 2010, 2013) simulations have shown that this mechanism can explain natal kick velocities of more than 1000 km/s. The extensive study of Wongwathanarat et al. (2013) showed that, not only the kick velocities are consistent with the observed velocity distribution of NSs, but also the same mechanism would impart a spin in the NS (see also Spruit and Phinney 1998). The resulting NS periods, in the range 100–8000 ms, are similar to those encountered in pulsars.

The second mechanism is specific of binaries. During a supernova explosion, the stars loses most of its mass in very short time, compared to the orbital period of the binary. Even if the explosion is perfectly spherically symmetric with respect to the remnant compact object, there is always a strong asymmetry with respect to the centre of mass of the system. That leads inevitably to a recoil of the compact object, which acquires a kick velocity. This case can be studied analytically (Blaauw 1961; Boersma 1961) and gives very interesting predictions. If more than half of the mass of the binary is ejected during the explosion, the binary is disrupted, and the compact remnant flies away (see e.g. Postnov and Yungelson 2014). For lower ejected masses, the kick is not able to disrupt the system, but can introduce a significant eccentricity to the binary. Note that, unless the previous kick mechanism, this one has low degree of stochasticity, and it can be predicted, which kind of binaries are likely to survive a SN explosion and which are not. This systematic effect has important consequences in stellar evolution that are beyond of the scope of this review (see e.g. Postnov and Yungelson 2014, for more information).

1.2.4.4 Metallicity

Metallicity and rotation play an important role in stellar evolution. They may be responsible for the wide variety of observed properties in supernova explosions and neutron stars. There is often an interplay between both rotation and metallicity, so we start summarising the results for non-rotating stars, and we deal with the effect of metallicity in rotating stars in the next section.

The metallicity of the environment in which a star was born has a significant impact in processes of mass loss. Metallicity increases the opacity of the envelope of the star and allows for the formation of a radiation driven wind. As a consequence, stars with higher metallicity have stronger winds and hence a higher mass loss. In the previous sections we have dealt exclusively with solar metallicity stars, which have a significant mass loss above 20 $M_\odot$, and become WR stars (basically bare He cores) above ~30 $M_\odot$. In low metallicity stars these mass limits are shifted upwards and the typical iron cores formed are more massive, specially for the most massive stars (see e.g. Woosley et al. 2002). A number of authors have performed 1D stellar evolution calculations of low metallicity stars (Woosley and Weaver 1995; Marigo et al. 2001; Heger and Woosley 2002; Eldridge and Tout 2004; Umeda and Nomoto 2005; Hirschi et al. 2006; Tominaga et al. 2007; Limongi and Chieffi 2012; Sukhbold and Woosley 2014). The more extreme cases can be found in population III stars. In very low metallicity environments, stars may form with masses above 100 $M_\odot$. These stars produce a copious amount of $e^- - e^+$ pairs after central carbon burning, which cools down the star and leads to a gravitational pair instability. The outcome is either a black hole or a thermonuclear explosion (pair-instability SN) (see Woosley et al. 2002; Heger et al. 2003 for details), but no neutron stars are formed.

Below ~100 $M_\odot$, simplified 1D core collapse simulations (Zhang et al. 2008; Pejcha and Thompson 2015) show that, since the iron core masses are in average larger for low metallicity, in stars with initial masses above ~25 $M_\odot$ the most common outcome are black holes. In this mass range, those stars not forming a black hole, have larger masses than in the solar metallicity case (Zhang et al. 2008; Pejcha and Thompson 2015). In fact, many of these neutron stars experience a significant fallback, which is favoured under low-metallicity conditions. For initial masses $M < 25\ M_\odot$, the properties of the supernova explosions and the resulting neutron stars are similar, regardless of metallicity, because these stars do not experience a significant mass during their lives. Taking the IMF into account, the fraction of stars forming black holes could be as large as 50% at low metallicities (Pejcha and Thompson 2015). Multidimensional simulations of low metallicity stars have been performed mainly for fast rotating progenitors, which are discussed in the next sections.

1.2.4.5   Rotation and Magnetic Fields

Main sequence stars rotate rapidly, with typical observed surface velocities of the order of 200 km/s (Fukuda 1982). Such a high rotation rate can have important implications on the evolution of the star and in the spin period of the remnant compact object after the supernova explosion. Depending on whether the core is able to retain its angular momentum during its evolution or not, the resulting iron core will produce a neutron star with a period of about 1 ms, very promising as a progenitor of GRBs and magnetars, or a slowly rotating neutron star, compatible with measurements of spin periods in pulsars (see Sect. 1.2.4.3 for the case of a non-rotating core). It is thus clear that there has to be important differences in the rate of angular momentum loss among different stars, depending on initial mass, rotation, and metallicity, to produce the variety of events observed.

Loss of angular momentum in massive stars occurs mainly during the red supergiant phase, for stars with $M < 30 M_\odot$, or WR phase, above this limit (see e.g. Woosley et al. 2002). As the hydrogen envelope expands it spins down, due to angular momentum conservation, and starts rotating differentially with respect to the core. The angular momentum of the envelope can be extracted from the star if there is mass loss due to winds. This process is specially important in WR stars that will lose the whole hydrogen envelope during this phase. If there is some process transporting angular momentum efficiently, coupling the core with the outer layers of the star, the wind will extract angular momentum from the core as well (Langer 1998). Therefore, the final total angular momentum of the core in the pre-supernova stage depends on both, the efficiency of angular momentum transport processes and the amount mass lost by winds. Mass loss depends mainly on metallicity and has been discussed in the previous section. We focus next on angular momentum transport.

Multiple hydrodynamic instabilities contribute to the transport of angular momentum, both in convective and radiative regions (Heger et al. 2000; Maeder and Meynet 2000a,b). However, in the absence of magnetic fields, this transport is rather inefficient and leads to rapidly rotating cores (Heger et al. 2000; Hirschi et al. 2004; Chieffi and Limongi 2013). The main mechanism responsible for the transport of angular momentum are magnetic fields (Spruit and Phinney 1998; Spruit 1999). Regardless of the initial magnetic field, instabilities in combination with rotation can lead to the formation of a dynamo, able to support magnetic fields during the life of the star (Spruit 2002). A detailed understanding of magnetic field dynamos in stars is a long standing problem, which is not even completely solved for the most studied star, the Sun (see e.g. Charbonneau 2013). The incorporation of magnetic fields in stellar evolution codes has only been attempted using simplified 1D models for the magnetic torques so far (Heger et al. 2005). In this case, for stars with $M < 30 M_\odot$, magnetic torques are able to enforce rigid rotation in the star and spin down the core by a factor 30–50 with respect to the case in which magnetic fields are not considered, producing progenitors of pulsar-like objects. This leads to the conclusion that, for the most common type of progenitor of core-collapse supernovae, the iron core is likely to have lost most of its angular momentum during its evolution, and the outcome of its collapse would be very similar to the case of non-rotating progenitors, as described in Sect. 1.2.3. The remaining spin of the iron core, in combination with the spin imparted by the SN explosion (see Sect. 1.2.4.3), would explain the variety of periods observed in most pulsars.

Furthermore, rotation accelerates burning by introducing additional mixing. This has consequences in the mass of the core, the amount of mass loss, the colour of the star in the late evolution (blue instead of red supergiant) and the type of supernova producing (Type Ib/c instead of Type II) (Hirschi et al. 2004; Chieffi and Limongi 2013).

1.2.4.6 Fast Rotation: Hypernovae, Long GRBs and Magnetic Field Amplification

Massive stars with very large initial rotation velocity ($\sim$400 km/s) may undergo a completely different evolutionary path to that described in the previous sections. Rotationally induced mixing produce stars that are chemically homogeneous and that are able to burn efficiently all the hydrogen and skip the red supergiant phase (Maeder 1987). The result is a bare helium core without a hydrogen envelope, very similar to a WR star. Similar situation could be reached if the star is stripped down to a helium core as the result of mass transfer in a binary system (Woosley and Heger 2006). In solar metallicity environments, this helium core can still lose a significant amount of angular momentum due to winds. However, for low metallicities winds cannot spin down the star and the core is able to retain its angular momentum until the pre-supernova stage (Yoon and Langer 2005; Woosley and Heger 2006). These stars can potentially form neutron stars with periods of $\sim$1 ms or fast spinning black holes, and are potential candidates for long GRB progenitors. Additionally, the absence of a hydrogen envelope would explain the observed association of long GRBs with Type Ib/c supernovae.

Although it is not well known the exact evolutionary path that leads to the formation of high spinning cores, it is clear that they are necessary to clarify the phenomenology associated with long GRBs and hypernovae. For those cases, the neutrino heating mechanism is clearly insufficient to explain the presence of a highly collimated jet or the energetics of the explosion in hypernovae. The current understanding is that, in the presence of magnetic fields, the energy stored in the rotation of the star can be extracted and used to drive a powerful magnetorotational explosion. In some cases a relativistic jet may be produced by black hole accretion (collapsar model, MacFadyen and Woosley 1999) or by the presence of a millisecond magnetar (Usov 1994; Wheeler et al. 2000). Fast rotation could also be an explanation for the recently discovered class of superluminous SNe (see Nicholl et al. (2013) and Sect. 1.3.1)

Multidimensional numerical simulations of the collapse of rapidly rotating magnetised cores have been performed both in 2D (Shibata et al. 2006; Obergaulinger et al. 2006; Ott et al. 2006; Burrows et al. 2007a; Cerdá-Durán et al. 2008; Takiwaki and Kotake 2011; Sawai et al. 2013; Sawai and Yamada 2016; Obergaulinger and Ángel Aloy 2017) and 3D (Mikami et al. 2008; Kuroda and Umeda 2010; Scheidegger et al. 2010; Mösta et al. 2014, 2015; Winteler et al. 2012), with different degree of sophistication in the treatment of the neutrinos. For sufficiently high initial magnetic field in the progenitor, numerical simulations show that magneto-rotational effects help in the explosion, producing more energetic events exploding at earlier times than the corresponding non-magnetized models. Explosions are usually highly asymmetric and aligned with the rotation axis. In 2D simulations mildly relativistic outflows have been observed along the axis, but 3D simulations have shown that this effect is likely exaggerated by the axisymmetry imposed to the system (Mösta et al. 2014). Recent 2D simulations have shown that, the asymmetries in the magneto-rotational explosion make it possible to keep the accretion onto the neutron star, even after the shock expands (Obergaulinger and Ángel Aloy 2017). This would allow for the formation of a black hole after the supernova explosion, which could in turn serve as central engine for a long GRB. This would solve the problem that, if a black hole is formed too early, no supernova explosion would be formed.

All the simulations assume a large scale initial magnetic field in the iron core in the range 10^9–10^{12} G, consistent with stellar evolution results by Heger et al. (2005). However, this magnetic field is expected to be amplified once the proto-neutron star has formed by the action of the magneto-rotational instability (MRI) (Akiyama et al. 2003). The MRI (Velikhov 1959; Chandrasekhar 1960) is an instability that appears in differentially rotating magnetised fluids and was proposed as the main mechanism driving accretion in discs (Balbus and Hawley 1991). The MRI is able to amplify the magnetic field and generate turbulence, so it has been invoked in numerous times in the literature as a justification to start numerical simulations with an artificially enhanced magnetic field strength, using as an argument that the MRI would be responsible for this effect. However, the direct simulation of the MRI in core collapse simulations is challenging, because it develops in very small length-scales that require an amount of numerical resolution not feasible with present day supercomputers (see discussion in Rembiasz et al. 2016a). Nevertheless, a few attempts have been carried out using artificially enlarged magnetic fields (Cerdá-Durán et al. 2008; Sawai et al. 2013; Sawai and Yamada 2016; Mösta et al. 2015); by increasing the magnetic field the MRI length-scale grows being easier to resolve numerically. However, it is unclear how close to reality are these simulations. Increasing artificially the global magnetic field may not be equivalent to the turbulent state that would be expected form the development of MRI.

To explore the amplification of the magnetic field due to the MRI, local and semi-local simulations have been performed with the conditions present in proto-neutron stars (Obergaulinger et al. 2009; Guilet and Müller 2015; Rembiasz et al. 2016b,a). The result of the simulations show that the amplification of the magnetic field by MRI is severely limited by the presence of parasitic instabilities of the Kelving-Helmholtz type. These instabilities can limit the magnetic field amplification to a factor of about ∼10 (Rembiasz et al. 2016a). Additionally, the development of the MRI could be severely affected by the presence of neutrinos inside the neutrinosphere (Guilet et al. 2015). Nevertheless, MRI is able to create and sustain turbulence with non-zero kinetic helicity. In presence of differential rotation, this turbulence could create a large scale dynamo (see e.g. Brandenburg and Subramanian 2005) capable of generating a large scale field similar to what is needed to power a hypernova or a GRB. The results by Mösta et al. (2015) may be indicative that this is the case although more detailed simulations will be required in the future to explore this possibility.

1.3 Challenges and Future Prospects

1.3.1 New Supernova Types

As discussed previously, thanks to the current sky surveys, astronomers are collecting unprecedented samples of known classes of objects, as well as discovering novel types of stellar transients that cannot be explained with traditional explosion channels. Some examples are described below:

- *Superluminous SNe (SLSNe):* Their absolute magnitude exceed -20 mag, and may occasionally reach -22 mag, which is one order of magnitude brighter than Type Ia SNe. They have a slow photometric evolution, and are preferentially hosted in very faint, likely metal-poor, dwarf galaxies (Quimby et al. 2011). Most of them are H-poor (Pastorello et al. 2010), though some others show evidence of H spectral lines originated in the stellar CSM (Benetti et al. 2014). In addition, due to their intrinsic luminosity, some SLSNe have been proposed as standardisable candle candidates to redshifts $z \approx 3$–4 (Inserra and Smartt 2014). However, the sample is still small (only eight to ten objects currently have enough data to test this method) and there are still some concerns about the unknown progenitor systems or the explosion physics. SLSNe seems to be incompatible with models powered only by the radioactive decay chain $^{56}Ni \rightarrow {}^{56}Co \rightarrow {}^{56}Fe$ and to favour alternative explosion scenarios, in which the additional energy is provided by the spin-down of a rapidly rotating young magnetar, pair-instability and/or strong interaction of SN ejecta with a very massive opaque CSM (e.g. see Nicholl et al. (2013) and Sect. 1.2.4 for more details). No models yet provide an excellent and unique fit with observations.
- *Type Ibn SNe:* There is wide heterogeneity in SNe showing strong interaction with their CSM. They can be generated in a H-rich environment as we have seen before (Type IIn), or also by stars exploding in a He-rich medium, being labelled as Type Ibn SNe (Pastorello et al. 2008). These Ibn SNe can be luminous objects (peaking at ~ -19 mag), usually followed by quickly declines (~ 0.1 mag day^{-1}), and show a spectroscopically heterogeneity (Pastorello et al. 2016). Since the discovery of its prototypical object, SN 2006jc (Pastorello et al. 2007; Foley et al. 2007), other ~ 30 Type Ibn SNe has been observed at this time.

- *Faint stripped-envelope SNe:* These SNe show faint luminosity peaks (M < −15 mag), fast-evolving light curves and weird spectroscopic properties (narrow features, very weak Si and S lines, sometimes evidence of He, and prominent [Ca II] in the nebular phases; e.g. Valenti et al. 2009). The possible explanation of these weird observables are exotic explosions involving white dwarfs below the Chandrasekhar mass limit (Perets et al. 2010), failed SN explosions, or unusual CC-SNe (e.g. fallback SNe from very massive stars or electron-capture SNe from super-AGB stars; Pumo 2010 and Sect. 1.2.4).
- *SN impostors:* Sometimes there are sneak cases of powerful eruptions of LBVs that mimic the true appearance of an interacting SN. The latter are commonly known as "SN impostors" (e.g. Van Dyk and Matheson 2012). This misclassification used to happen because the spectra of these giant eruptions are characterized by incipient narrow hydrogen lines in emission, like those of type IIn SNe. However, their luminosities are considerably lower (fainter than −14 mag). Although SN impostors may not be SNe, they are included here because it is not clear yet their nature. A grown number of SN impostors have interestingly heralded the terminal SN explosion, from weeks to a few years after the outburst episode (e.g. SN 2009ip (Pastorello et al. 2013), LSQ13zm (Tartaglia et al. 2016), or SN 2015bh (Elias-Rosa et al. 2016)), but some others show repeated intermediate-luminosity without leading (so far) to a SN explosion (e.g. SN 2000ch (Pastorello et al. 2010), or SN 2007sv (Tartaglia et al. 2015)). LBV stars are the most usual channel to explain the bursty activity of the SN impostors, however, these outburst have been also linked to lower mass stars, interacting massive binary system, or pre-SN nuclear burning instabilities. Nevertheless, the mechanisms triggering this class of outbursts are still very poorly understood.
- *Other unusual SNe:*

  - There are a handful of events that have broad light curves with much *longer rise times* from estimated explosion to peak, like SN 2011bm (Valenti et al. 2012), a Type Ic SN with a rise time of 35 days (instead of 10–22 days for normal SNe of this class), or Type II SN 1987A with an exceeding long rise rime of >80 days (Arnett et al. 1989). The long rise times suggest much larger ejected masses.
  - Other objects show *faster decline rates*, such as SN 2005ek (Drout et al. 2013). It is spectroscopically a Type Ic with a decline of 2.5–3 mag in 15 days.
  - The wide range of magnitudes during the plateau phase of type II-P SNe is a well established fact (Hamuy and Pinto 2002). However, *very faint and also luminous Type II-P* exist (e.g. Spiro et al. 2014; Inserra et al. 2013). These rare objects can allow us to better constrain the correlation between ^{56}Ni and progenitor mass, and to test the reliability of other tight relations among physical parameters of Type II-P SNe, on the most and least energetic events. An examples is SN 1997D (Turatto et al. 1998), which showed a small amount of ^{56}Ni and narrow P-Cygni profiles, suggesting a low explosion energy.

1.3.2 Multi-Messenger Astronomy

The core-collapse of massive stars has been associated with SNe (and long GRBs, >2 s). CC-SNe are canonical examples of multi-messenger astrophysical sources since the gravitational energy driven by the CC-SN at the time of explosion is released as ~99% of neutrinos, ~1% is converted to ejecta kinetic energy, ~0.01% becomes photons, and an uncertain, though likely smaller fraction is carried away by gravitational waves (GW).

So far, we have focused this work on electromagnetic observation of core-collapse supernovae. However, there are additional observational channels that will provide in the future complementary information about the scenario in which neutron stars are formed: cosmic rays, neutrinos and GWs. While electromagnetic observations mainly carry information about the matter ejected during the explosion, and on the central regions at late time, both neutrinos and GWs will allow us to infer directly the properties of the compact remnant, and provide information about the thermodynamics and dynamics of the SN engine. In the last decade a big effort has been done to improve current neutrino and GW observatories to reach sensitivities that would allow us to do multi-messenger astronomy of nearby SN explosions, within ~100 kpc.

1.3.2.1 Cosmic Rays

The shockwaves of CC-SNe accelerate charged particles such as protons, some of which end up raining on Earth as *cosmic rays*. The importance in detecting cosmic rays is that they are the only particles detected on Earth which have traversed a considerable distance through the ISM and which were accelerated in events such as SN in a relatively recent past. In fact, data from Fermi Large Area Telescope (Atwood et al. 2009) have been interpreted as a fraction of primary cosmic rays originate from a SN. However, the astrophysical source identification is complicated without any other messenger information.

1.3.2.2 Neutrinos

CC-SNe emit low-energy neutrinos, as confirmed by the observations of the only recent nearby supernova, SN 1987A (a few tens of MeV; e.g. see Hirata et al. 1987). The searches for neutrinos are based on three kind of analysis: searches for significant neutrino excess with respect to the atmospheric neutrino background, searches for anisotropies on the sky to identify point sources, and multi-messenger searches in time and spatial coincidence with other messenger signals. Having information from neutrino observations, we may constrain the exact time of the CC bounce (with just EM information we have an uncertainty of many hours), and establish a tight correlation with GWs. Moreover, with information from neutrinos, we will be able to set limits on neutrino mass, and if this signal is more copious (for example a SN in the Milky Way) we will be able to make distinctions between different theoretical models of CC-SNe explosion. Current neutrino detectors (e.g. SuperKamiokande (Ikeda et al. 2007), IceCube (Karle 2009) and Antares (Ageron et al. 2012)) are capable of measuring neutrinos from a supernova as far as in the Large Magellanic Cloud.

1.3.2.3 Gravitational Waves

The last multi-messenger channel that has been opened are gravitational waves, with the first detection of a binary black hole merger, GW150914 (Abbott et al. 2016a), by the Advanced Laser Interferometer Gravitational-Wave Observatory (aLIGO). In order to be detectable by GW observatories, these waves should be produced by a very dynamic and compact relativistic system, associated to high energy processes. Sources of GWs can be classified according to their frequency (see e.g. Cutler and Thorne 2002, for a review on GW sources). High frequency sources (10–2000 Hz) include binary mergers of compact objects of stellar origin (NS and BH), asymmetric core-collapse of massive stars, and rotating isolated neutron stars. Low frequency sources (10^{-9}–1 Hz) include the inspiral phase of binary NS, BH and white dwarfs, as well as the merger of WD and supermassive BHs, at cosmological distances. Observations in the high frequency range are accessible to ground-based laser interferometers such as aLIGO (LIGO Scientific Collaboration et al. 2015) and Virgo (Acernese et al. 2015), and the upcoming KAGRA (Aso et al. 2013) experiment, while low frequency observations will be performed by space-based laser interferometers such as LISA (Amaro-Seoane et al. 2017) in the range 10^{-4}–0.1 Hz, and by the pulsar timing array (PTA) at the lowest frequencies, 10^{-9}–10^{-8} Hz (Hobbs et al. 2010).

   Differently from the binary black hole merger case, for which reliable and accurate GW templates have been developed during the last decade, the GW signal from the core collapse case still presents many theoretical uncertainties related to the complexity in the numerical modelling of the scenario (see Sect. 1.2.3). The signal can be divided in two parts. *The core bounce* is the part of the waveform which is best understood (Dimmelmeier et al. 2002b). Its frequency (at about 800 Hz) can be directly related to the rotational properties of the core (Dimmelmeier et al. 2008; Abdikamalov et al. 2014; Richers et al. 2017). However, fast-rotating progenitors are uncommon (see Sect. 1.2.4) and this bounce signal is not likely to be observed in non-rotating galactic events. More interesting is the signal related to the *post-bounce evolution* of the newly formed proto-neutron star, which is produced by convection and the excitation of highly damped modes in the PNS (Murphy et al. 2009; Müller et al. 2013; Cerdá-Durán et al. 2013; Kuroda et al. 2016; Andresen et al. 2017). Typical duration of this signal is of ~0.5 s in the case of successful SN explosions (see e.g. Müller et al. 2013) but can last for seconds if the final outcome is a BH (Cerdá-Durán et al. 2013). Typical frequencies raise monotonically with time due to the contraction of the PNS, whose mass is steadily increasing. Characteristic frequencies of the PNS can be as low as $\sim$100 Hz, specially those related to g-modes, which make them a perfect target for ground-based interferometers with the highest sensitivity at those frequencies. In the future, it may be possible to infer the properties of PNS based on the identification of mode frequencies in their waveforms (see e.g. Sotani and Takiwaki 2016). Depending on their rotation rate, supernovae could be detected as far as 1 Mpc for extreme events, but typical distances are more likely 1–100 kpc (Abbott et al. 2016c). The rate of CC of massive stars in the Milky Way is around 2 per century (e.g. see Ott 2009).

One of the main challenges is the difficulty of detecting an electromagnetic counterpart inside the localisation error box of the GW signal. With only two GW detectors (LIGO interferometers in Hanford and Livingston), typical sky localisation is so far poor (hundreds of square degrees) (Abbott et al. 2016b). However, this will change in the near future by the addition of new detectors to the network (advanced Virgo, KAGRA and LIGO India), improving the sky localisation to an error box of a few 10 deg^2. The next challenge is then to detect the electromagnetic counterparts inside the error box of GW signals. Finding these counterparts is a proof of the veracity of the origin of the signal.

1.3.3 Challenges and Future Prospects from an Observational Point of View

1.3.3.1 Progenitor Hunting

In previous sections we have reviewed the progenitor properties derived from the SN observations and the direct observations of the SN site before the explosion. This connection is crucial to test our understanding of stellar evolution. However, it has not been found all what expected.

There is information of about 30 SN progenitors between detections and limits. Still, none of these has an estimated luminosity above of log $L/L_\odot \simeq 5.1$, which corresponds to an initial evolutionary mass of about 18 $M_\odot$ (except for interacting SNe, whose progenitors seems to have high masses, but in some cases the nature of that transients is still debated). This estimate is significantly lower than the stellar evolution textbook value (about 25 $M_\odot$, and Sect. 1.2.4), or than the red supergiants found in the Local Universe, with luminosity up to log $L/L_\odot \simeq 5.5$ (approximately initial mass of 30 $M_\odot$; Levesque et al. 2009). What is more, according a typical Salpeter initial mass function (of slope $\alpha = -2.35$), a $\sim$30% of stars between 8 and 100 $M_\odot$ have masses > 18 $M_\odot$. Thus one would expect to have found some SN progenitor stars with higher masses by now, which is not the case.

A good review exploring possible bias and explanations for this deficit can be found in Smartt (2015). They discussed how the circumstellar dust or the systematic errors in the analysis can affect the luminosity and mass estimates. However, they also hold that these biases, although may affect our estimates, do not seem to explain the missing mass progenitor stars.

Furthermore, it is natural to think that depending on the stellar evolution code used for estimating the progenitor star mass, more uncertainties could be added to the measure. Still, Smartt (2015) compare the end points of three stellar evolution models (STARS models (Eldridge and Tout 2004), rotating Geneva models (Hirschi et al. 2004); and KEPLER models (Woosley and Heger 2007)) and show that while the use of one model or another could affect the lower mass limit to produce a CC-SN (masses between 7 and 10 $M_\odot$), the high mass upper limit appears be secure.

It has been proposed as a possible explanation that massive stars above 18 $M_\odot$ evolve into WR stars and evade the detection because they are too hot and faint at the point of core-collapse. In this case they would produce Ib/c SNe, although this disagrees with the growing evidence that the majority of Ib/c SNe comes from lower mass stars in interacting binaries.

Another plausible explanation could be that these missed massive stars may have collapsed without producing a detectable SN. Such events may happen when the shock wave dies out before reaching the stellar surface, or because the stellar mantle fallback onto the nucleus. These failed explosions generate a black hole and eject negligible amounts of radioactive isotopes (see also Sect. 1.2.4.2). While these events cannot be directly observed, it is possible to detect the disappearance of massive stars through the comparison (subtraction) of reference "template" images (Kochanek et al. 2008; Reynolds et al. 2015; Adams et al. 2016).

Advances in this field could be achieved from the collection of deep, multi-wavelength, wide-field imaging of nearby galaxies for future SN progenitor characterization. This feat is plausible by using high resolution images from *HST*, or from the next generation of ground-based and space telescopes. In the near future, key information will be obtain thanks to the wide field of the 8-m Large Synoptic Survey Telescope (expected to start operations on 2021–2022), the fully adaptive and diffraction-limited optics of the 39-m European Extremely Large Telescope (planned first light for 2024), or the infrared domain of the 6.5 m mirror of the James Webb Space Telescope (scheduled to launch in October 2018).

1.3.3.2   Pre-SN Outbursts in SN Impostors/IIn SNe

In the last years there has been some success in detecting the progenitors of interacting transients, including Type IIn SNe (Sects. 1.2.2 and 1.3.1). The nature of these transients remains debated: some are undoubtedly genuine core-collapse SNe, while others may be giant non-terminal outbursts from LBVs. Observational constraints on their progenitors can help light on their true nature. We note that classical LBVs are not expected to explode directly as CC-SNe, and so these results are challenging models of massive stellar evolution.

A lot of work still needs to be done to improve our knowledge on the mechanisms triggering luminous stellar outbursts and major eruptions. Observational time is needed to trace the photometric history of a these transients and to keep relaxed monitoring of SN impostors to pinpoint any possible re-brightening of the object. In short, only the discovery of a large number of similar transients, extensive follow-up campaigns in a wide wavelength range, and the availability of rich archives with deep images at different domains will allow us to give more robust conclusions on the variety of properties of SN impostors, and how this heterogeneity is connected to stellar parameters (mass, radius, chemical composition, rotation, binarity).

1.3.3.3 Flash Spectroscopy

The SNe require rapid and intense follow-up to characterize their explosions. Thanks to the current high-cadence optical surveys dedicated to the observations of transients, it has been possible to discover new SNe only few hours after explosion. This technique is called flash spectroscopy (Gal-Yam 2014). Some Type II SNe, like 2014G (Terreran et al. 2016), have shown narrow emission lines superimposed to a hot blue continuum, which disappears in few days after the explosion. These emissions arise from the photoionized surrounding CSM distributed around the exploding star, which was previously originated through winds or violent eruption. The analysis of these early spectra can provide unique information about the physical distribution of gas around each event, and therefore from the stellar evolution during the final phases before the explosion.

Some current surveys such as ePESSTO (extension of the Public ESO Spectroscopic Survey of Transient Objects—PESSTO; Smartt et al. 2015) are dedicating part of their observational time to the classification and intensive early monitoring of Type II SNe discovered after 2–3 days (at most) from the explosion.

1.3.4 Future Prospects in Numerical Modelling

1.3.4.1 Advances in High Performance Computing

The main challenge in the numerical modelling of core collapse supernovae is associated with the huge amount of computational resources needed for the completion of realistic simulations. Current three-dimensional state-of-the-art numerical simulations use millions of CPU hours distributed among a few 1000 to 10,000 cores and may take months to finish. This kind of simulations can only be performed in high performance computing (HPC) facilities, and CPU time has to be requested in a similar fashion as observational time in observatories. The number of simulations and the numerical resolution affordable by dedicated groups is limited by the availability of CPU time and the size and number of such HPC facilities. Best supercomputers have a few 100 thousands of cores, a few petaflops (10^{15} operations per second) of computing power and are distributed all around the world in USA (e.g. Titan, Sequoia), Japan (e.g. K Computer), China (e.g. Tiahne2) and Europe (e.g. FZJ, SuperMUC, Marconi).

Next generation of supercomputers will be available during the next decade and will reach the exaflop scale, improving by a factor 1000 previous machines.

In these facilities, it will be possible to perform numerical 3D simulations with unprecedented numerical resolution, which will allow us to study in detail the impact of turbulence in the explosion mechanism and its role in the amplification of the magnetic field due to dynamos. It will also help to increase the realism in the neutrino transport by getting closer to solve the full 6-dimensional Boltzmann equations. Additionally, these detailed simulations will allow to settle down the issue of whether our current understanding of the core-collapse mechanism is sufficient to explain typical SN events without practically any numerical uncertainty, or if there is some unknown physical ingredient missing. Furthermore, it will allow us to explore the connection between progenitor stars and explosion properties by performing parametric studies involving hundreds to thousands of 3D simulations. The challenge here consist in developing numerical algorithms and codes that are capable to scale in parallel to a few 10^5 cores. This will involve an even closer collaboration with applied mathematicians and computer scientists.

1.3.4.2   Subgrid Modelling

For the high Reynolds number conditions present in PNSs (see e.g. Thompson and Duncan 1993), fluid motions can break up into smaller scales in a turbulent cascade that goes down to the dissipative scale. Resolving turbulence numerically in global numerical simulations is a challenge, due to the smallness of the dissipative scale. In the presence of magnetic fields, the problem is even harder, because, under the right conditions, small scale turbulence can develop an inverse cascade giving raise to large scale magnetic field (see e.g. Brandenburg and Subramanian 2005). Next generation of supercomputers, will help us to understand better magneto-hydrodynamic (MHD) turbulence and dynamos by means of local simulations, but it may not allow us to resolve MHD turbulence properly in global simulations of core-collapse supernovae. A possibility would be to do subgrid modelling, i.e. incorporate the effect of the small scales (e.g. turbulence) in a phenomenological way to the equations for the large scales (bulk fluid motions). Subgrid modelling of non-magnetized turbulence is common in other fields of science (e.g. meteorology) and engineering (e.g. aerodynamics). Although the case of MHD turbulence is comparably harder to model, there have been some attempts along this line. The most well know cases are the alpha-viscosity model (Shakura and Sunyaev 1973) describing angular transport in disks by MRI-driven turbulence, and the mean field dynamo formalism (Moffatt 1978), popular in the study of solar dynamos (see e.g. Charbonneau 2013) and accretion disks (e.g. Bugli et al. 2014; Sądowski et al. 2015). There has been some recent progress in this respect in the context of binary neutron stars (Giacomazzo et al. 2015; Shibata et al. 2017). However, applications in the core-collapse scenario are yet to come. The challenge in subgrid modelling is to find the proper closure relations that link small and large scales. In this respect, local simulations may be crucial to fix the parameters and dependencies appearing in such closures.

1.3.4.3 Multidimensional Stellar Evolution

To a great extent, the link between supernova progenitors and explosion properties depends on the accuracy of the stellar evolution models. Nowadays, pre-supernova models are the result of 1D calculations (see Sect. 1.2.4), where multidimensional effects, such as convection and dynamos, are incorporated in a phenomenological way. However, the increase in computing power will allow in the future to perform 3D simulations of the evolution of stars, or at least of the relevant phases and regions, where multidimensional effects are most important (see Viallet et al. 2011 and examples therein). This has lead to an increasing interest in the numerical study of the last stages of the star before collapsing (see e.g. Müller et al. 2016 and references therein). In fact, the asymmetries at the pre-supernova stage, e.g. induced by convection, have been suggested to help the shock revival after bounce (Couch and Ott 2013; Couch et al. 2015; Müller and Janka 2015).

Multidimensional effects also play a role in interacting binaries. They can induce mass transfer, go through common envelope phases, or strip down the envelope of stars. This influences the evolution of stars and can lead to completely different evolutionary paths, e.g. to rapidly rotating cores (see Sect. 1.2.4.6). There has been some work in recent years trying to address the problem using multidimensional simulations of the scenario (see e.g. Motl et al. 2002; Lajoie and Sills 2011; Lombardi et al. 2011) and in the future we foresee a great advances in the field.

Acknowledgements This work is supported by the New Compstar COST action MP1304. N.E.R. acknowledges financial support by the 1994 PRIN-INAF 2014 (project 'Transient Universe: unveiling new types of stellar explosions with PESSTO'). N.E.R. acknowledges the hospitality of the "Institut de Ciències de l'Espai" (CSIC), where part of this work has been done. P.C.D. acknowledges the financial support from the Spanish Ministerio de Economía y Competitividad (grant AYA2015-66899-C2-1-P) and from the Generalitat Valenciana (grant PROMETEO-II-2014-069).

References

Abbott, B.P., Abbott, R., Abbott, T.D., Abernathy, M.R., Acernese, F., Ackley, K., Adams, C., Adams, T., Addesso, P., Adhikari, R.X., et al.: Phys. Rev. Lett. **116**(6), 061102 (2016a). https://doi.org/10.1103/PhysRevLett.116.061102

Abbott, B.P., Abbott, R., Abbott, T.D., Abernathy, M.R., Acernese, F., Ackley, K., Adams, C., Adams, T., Addesso, P., Adhikari, R.X., et al.: Living Rev. Relativ. **19**, 1 (2016b). https://doi.org/10.1007/lrr-2016-1

Abbott, B.P., Abbott, R., Abbott, T.D., Abernathy, M.R., Acernese, F., Ackley, K., Adams, C., Adams, T., Addesso, P., Adhikari, R.X., et al.: Phys. Rev. D **94**(10), 102001 (2016c). https://doi.org/10.1103/PhysRevD.94.102001

Abdikamalov, E., Gossan, S., DeMaio, A.M., Ott, C.D.: Phys. Rev. D **90**(4), 044001 (2014). https://doi.org/10.1103/PhysRevD.90.044001

Abdikamalov, E., Ott, C.D., Radice, D., Roberts, L.F., Haas, R., Reisswig, C., Mösta, P., Klion, H., Schnetter, E.: Astrophys. J. **808**, 70 (2015). https://doi.org/10.1088/0004-637X/808/1/70

Acernese, F., Agathos, M., Agatsuma, K., Aisa, D., Allemandou, N., Allocca, A., Amarni, J., Astone, P., Balestri, G., Ballardin, G., et al.: Class. Quantum Gravity **32**(2), 024001 (2015). https://doi.org/10.1088/0264-9381/32/2/024001

Adams, S.M., Kochanek, C.S., Gerke, J.R., Stanek, K.Z., Dai, X.: ArXiv e-prints (2016)

Ageron, M., et al.: Astropart. Phys. **35**, 530 (2012). https://doi.org/10.1016/j.astropartphys.2011.11.011

Akiyama, S., Wheeler, J.C., Meier, D.L., Lichtenstadt, I.: Astrophys. J. **584**, 954 (2003). https://doi.org/10.1086/344135

Aldering, G., Humphreys, R.M., Richmond, M.: Astrophys. J. **107**, 662 (1994). https://doi.org/10.1086/116886

Amaro-Seoane, P., et al.: ArXiv e-prints (2017)

Anderson, J.P., et al.: Astrophys. J. **786**, 67 (2014). https://doi.org/10.1088/0004-637X/786/1/67

Andersson, N., et al.: Class. Quantum Gravity **30**(19), 193002 (2013). https://doi.org/10.1088/0264-9381/30/19/193002

Andresen, H., Müller, B., Müller, E., Janka, H.T.: Mon. Not. R. Astron. Soc. **468**, 2032 (2017). https://doi.org/10.1093/mnras/stx618

Andrews, J.J., Farr, W.M., Kalogera, V., Willems, B.: Astrophys. J. **801**, 32 (2015). https://doi.org/10.1088/0004-637X/801/1/32

Arnett, W.D.: Astrophys. J. **253**, 785 (1982). https://doi.org/10.1086/159681

Arnett, W.D., Bahcall, J.N., Kirshner, R.P., Woosley, S.E.: Annu. Rev. Astron. Astrophys. **27**, 629 (1989). https://doi.org/10.1146/annurev.aa.27.090189.003213

Arzoumanian, Z., Chernoff, D.F., Cordes, J.M.: Astrophys. J. **568**, 289 (2002). https://doi.org/10.1086/338805

Aso, Y., Michimura, Y., Somiya, K., Ando, M., Miyakawa, O., Sekiguchi, T., Tatsumi, D., Yamamoto, H.: Phys. Rev. D **88**(4), 043007 (2013). https://doi.org/10.1103/PhysRevD.88.043007

Atwood, W.B., Abdo, A.A., Ackermann, M., Althouse, W., Anderson, B., Axelsson, M., Baldini, L., Ballet, J., Band, D.L., Barbiellini, G., et al.: Astrophys. J. **697**, 1071 (2009). https://doi.org/10.1088/0004-637X/697/2/1071

Baade, W., Zwicky, F.: Phys. Rev. **46**, 76 (1934). https://doi.org/10.1103/PhysRev.46.76.2

Balbus, S.A., Hawley, J.F.: Astrophys. J. **376**, 214 (1991). https://doi.org/10.1086/170270

Baumgarte, T.W., Shapiro, S.L.: Phys. Rev. D **59**(2), 024007 (1999). https://doi.org/10.1103/PhysRevD.59.024007

Benetti, S., et al.: Mon. Not. R. Astron. Soc. **441**, 289 (2014). https://doi.org/10.1093/mnras/stu538

Benvenuto, O.G., Bersten, M.C., Nomoto, K.: Astrophys. J. **762**, 74 (2013). https://doi.org/10.1088/0004-637X/762/2/74

Bernal, C.G., Lee, W.H., Page, D.: Rev. Mex. Astron. Astr. **46**, 309 (2010)

Bernal, C.G., Page, D., Lee, W.H.: Astrophys. J. **770**, 106 (2013). https://doi.org/10.1088/0004-637X/770/2/106

Bersten, M.C., Benvenuto, O.G., Nomoto, K., Ergon, M., Folatelli, G., Sollerman, J., Benetti, S., Botticella, M.T., Fraser, M., Kotak, R., Maeda, K., Ochner, P., Tomasella, L.: Astrophys. J. **757**, 31 (2012). https://doi.org/10.1088/0004-637X/757/1/31

Bersten, M.C., Benvenuto, O.G., Folatelli, G., Nomoto, K., Kuncarayakti, H., Srivastav, S., Anupama, G.C., Quimby, R., Sahu, D.K.: Astrophys. J. **148**, 68 (2014). https://doi.org/10.1088/0004-6256/148/4/68

Bethe, H.A., Wilson, J.R.: Astrophys. J. **295**, 14 (1985). https://doi.org/10.1086/163343

Blaauw, A.: Bull. Astron. Inst. Neth. **15**, 265 (1961)

Blinnikov, S., Lundqvist, P., Bartunov, O., Nomoto, K., Iwamoto, K.: Astrophys. J. **532**, 1132 (2000). https://doi.org/10.1086/308588

Blondin, J.M., Mezzacappa, A., DeMarino, C.: Astrophys. J. **584**, 971 (2003). https://doi.org/10.1086/345812

Boersma, J.: Bull. Astron. Inst. Neth. **15**, 291 (1961)

Botticella, M.T., et al.: Mon. Not. R. Astron. Soc. **398**, 1041 (2009). https://doi.org/10.1111/j.1365-2966.2009.15082.x

Brandenburg, A., Subramanian, K.: Phys. Rep. **417**, 1 (2005). https://doi.org/10.1016/j.physrep. 2005.06.005

Brandt, T.D., Burrows, A., Ott, C.D., Livne, E.: Astrophys. J. **728**, 8 (2011). https://doi.org/10. 1088/0004-637X/728/1/8

Bruenn, S.W.: Astrophys. J. Suppl. Ser. **58**, 771 (1985). https://doi.org/10.1086/191056

Bruenn, S.W., Mezzacappa, A., Hix, W.R., Lentz, E.J., Messer, O.E.B., Lingerfelt, E.J., Blondin, J.M., Endeve, E., Marronetti, P., Yakunin, K.N.: Astrophys. J. Lett. **767**, L6 (2013). https://doi. org/10.1088/2041-8205/767/1/L6

Bruenn, S.W., Lentz, E.J., Hix, W.R., Mezzacappa, A., Harris, J.A., Messer, O.E.B., Endeve, E., Blondin, J.M., Chertkow, M.A., Lingerfelt, E.J., Marronetti, P., Yakunin, K.N.: Astrophys. J. **818**, 123 (2016). https://doi.org/10.3847/0004-637X/818/2/123

Bugli, M., Del Zanna, L., Bucciantini, N.: Mon. Not. R. Astron. Soc. **440**, L41 (2014). https://doi. org/10.1093/mnrasl/slu017

Buras, R., Janka, H.T., Keil, M.T., Raffelt, G.G., Rampp, M.: Astrophys. J. **587**, 320 (2003). https:// doi.org/10.1086/368015

Buras, R., Rampp, M., Janka, H.T., Kifonidis, K.: Astron. Astrophys. **447**, 1049 (2006). https:// doi.org/10.1051/0004-6361:20053783

Burrows, A.: Rev. Mod. Phys. **85**, 245 (2013). https://doi.org/10.1103/RevModPhys.85.245

Burrows, A., Sawyer, R.F.: Phys. Rev. C **58**, 554 (1998). https://doi.org/10.1103/PhysRevC.58.554

Burrows, A., Young, T., Pinto, P., Eastman, R., Thompson, T.A.: Astrophys. J. **539**, 865 (2000). https://doi.org/10.1086/309244

Burrows, A., Livne, E., Dessart, L., Ott, C.D. , Murphy, J.: Astrophys. J. **640**, 878 (2006). https:// doi.org/10.1086/500174

Burrows, A., Dessart, L., Livne, E., Ott, C.D., Murphy, J.: Astrophys. J. **664**, 416 (2007a). https:// doi.org/10.1086/519161

Burrows, A., Livne, E., Dessart, L., Ott, C.D., Murphy, J.: Astrophys. J. **655**, 416 (2007b). https:// doi.org/10.1086/509773

Burrows, A., Dessart, L., Livne, E., Ott, C.D., Murphy, J.: Astrophys. J. **664**, 416 (2007c). https:// doi.org/10.1086/519161

Burrows, A., Vartanyan, D., Dolence, J.C., Skinner, M.A., Radice, D.: ArXiv e-prints (2016)

Cao, Y., et al.: Astrophys. J. Lett. **775**, L7 (2013). https://doi.org/10.1088/2041-8205/775/1/L7

Cappellaro, E., et al.: Astron. Astrophys. **584**, A62 (2015). https://doi.org/10.1051/0004-6361/ 201526712

Cerdá-Durán, P., Faye, G., Dimmelmeier, H., Font, J.A., Ibáñez, J.M., Müller, E., Schäfer, G.: Astron. Astrophys. **439**, 1033 (2005). https://doi.org/10.1051/0004-6361:20042602

Cerdá-Durán, P., Font, J.A., Antón, L., Müller, E.: Astron. Astrophys. **492**, 937 (2008). https://doi. org/10.1051/0004-6361:200810086

Cerdá-Durán, P., DeBrye, N., Aloy, M.A., Font, J.A., Obergaulinger, M.: Astrophys. J. Lett. **779**, L18 (2013). https://doi.org/10.1088/2041-8205/779/2/L18

Chandrasekhar, S.: Stellar Structure. Dover, New York (1938)

Chandrasekhar, S.: Proc. Natl. Acad. Sci. **46**, 253 (1960). https://doi.org/10.1073/pnas.46.2.253

Charbonneau, P.: Solar and Stellar Dynamos: Saas-Fee Advanced Course 39 Swiss Society for Astrophysics and Astronomy. Saas-Fee Advanced Courses, vol. 39. Springer, Berlin (2013). ISBN: 978-3-642-32092-7. https://doi.org/10.1007/978-3-642-32093-4

Chevalier, R.A.: Astrophys. J. **346**, 847 (1989). https://doi.org/10.1086/168066

Chevalier, R.A.: Astrophys. J. **394**, 599 (1992). https://doi.org/10.1086/171612

Chevalier, R.A.: Astrophys. J. **619**, 839 (2005). https://doi.org/10.1086/426584

Chevalier, R.A., Fransson, C.: Astrophys. J. **420**, 268 (1994). https://doi.org/10.1086/173557

Chieffi, A., Limongi, M.: Astrophys. J. **764**, 21 (2013). https://doi.org/10.1088/0004-637X/764/1/ 21

Colgate, S.A.: Astrophys. J. **163**, 221 (1971). https://doi.org/10.1086/150760

Colgate, S.A., Johnson, M.H.: Phys. Rev. Lett. **5**, 235 (1960). https://doi.org/10.1103/PhysRevLett. 5.235

Colgate, S.A., White, R.H.: Astrophys. J. **143**, 626 (1966). https://doi.org/10.1086/148549

Colgate, S.A., Grasberger, W.H., White, R.H.: Astrophys. J. **66**, 280 (1961). https://doi.org/10.1086/108573

Cordero-Carrión, I., Cerdá-Durán, P., Dimmelmeier, H., Jaramillo, J.L., Novak, J., Gourgoulhon, E.: Phys. Rev. D **79**(2), 024017 (2009). https://doi.org/10.1103/PhysRevD.79.024017

Cordero-Carrión, I., Vasset, N., Novak, J., Jaramillo, J.L.: Phys. Rev. D **90**(4), 044062 (2014). https://doi.org/10.1103/PhysRevD.90.044062

Couch, S.M., Ott, C.D.: Astrophys. J. Lett. **778**, L7 (2013). https://doi.org/10.1088/2041-8205/778/1/L7

Couch, S.M., Ott, C.D.: Astrophys. J. **799**, 5 (2015). https://doi.org/10.1088/0004-637X/799/1/5

Couch, S.M., Chatzopoulos, E., Arnett, W.D., Timmes, F.X., Astrophys. J. Lett. **808**, L21 (2015). https://doi.org/10.1088/2041-8205/808/1/L21

Crockett, R.M., Eldridge, J.J., Smartt, S.J., Pastorello, A., Gal-Yam, A., Fox, D.B., Leonard, D.C., Kasliwal, M.M., Mattila, S., Maund, J.R., Stephens, A.W., Danziger, I.J.: Mon. Not. R. Astron. Soc. **391**, L5 (2008). https://doi.org/10.1111/j.1745-3933.2008.00540.x

Cutler, C., Thorne, K.S.: ArXiv General Relativity and Quantum Cosmology e-prints (2002)

Dimmelmeier, H., Font, J.A., Müller, E.: Astron. Astrophys. **388**(3), 917 (2002a)

Dimmelmeier, H., Font, J.A., Müller, E.: Astron. Astrophys. **393**, 523 (2002b). https://doi.org/10.1051/0004-6361:20021053

Dimmelmeier, H., Ott, C.D., Marek, A., Janka, H.T.: Phys. Rev. D **78**(6), 064056 (2008). https://doi.org/10.1103/PhysRevD.78.064056

Dolence, J.C., Burrows, A., Zhang, W.: Astrophys. J. **800**, 10 (2015). https://doi.org/10.1088/0004-637X/800/1/10

Drout, M.R., et al.: Astrophys. J. **774**, 58 (2013). https://doi.org/10.1088/0004-637X/774/1/58

Eldridge, J.J., Maund, J.R.: Mon. Not. R. Astron. Soc. **461**, L117 (2016). https://doi.org/10.1093/mnrasl/slw099

Eldridge, J.J., Tout, C.A.: Mon. Not. R. Astron. Soc. **353**, 87 (2004). https://doi.org/10.1111/j.1365-2966.2004.08041.x

Eldridge, J.J., Fraser, M., Smartt, S.J., Maund, J.R., Crockett, R.M.: Mon. Not. R. Astron. Soc. **436**, 774 (2013). https://doi.org/10.1093/mnras/stt1612

Eldridge, J.J., Fraser, M., Maund, J.R., Smartt, S.J.: Mon. Not. R. Astron. Soc. **446**, 2689 (2015). https://doi.org/10.1093/mnras/stu2197

Elias-Rosa, N., et al.: Astrophys. J. Lett. **714**, L254 (2010). https://doi.org/10.1088/2041-8205/714/2/L254

Elias-Rosa, N., Van Dyk, S.D., Li, W., Silverman, J.M., Foley, R.J., Ganeshalingam, M., Mauerhan, J.C., Kankare, E., Jha, S., Filippenko, A.V., Beckman, J.E., Berger, E., Cuillandre, J.C., Smith, N.: Astrophys. J. **742**, 6 (2011). https://doi.org/10.1088/0004-637X/742/1/6

Elias-Rosa, N., Pastorello, A., Maund, J.R., Takáts, K., Fraser, M., Smartt, S.J., Benetti, S., Pignata, G., Sand, D., Valenti, S.: Mon. Not. R. Astron. Soc. **436**, L109 (2013). https://doi.org/10.1093/mnrasl/slt124

Elias-Rosa, N., et al.: Mon. Not. R. Astron. Soc. **463**, 3894 (2016). https://doi.org/10.1093/mnras/stw2253

Ertl, T., Janka, H.T., Woosley, S.E., Sukhbold, T., Ugliano, M.: Astrophys. J. **818**, 124 (2016). https://doi.org/10.3847/0004-637X/818/2/124

Filippenko, A.V.: Ann. Rev. Astron. Astrophys. **35**, 309 (1997). https://doi.org/10.1146/annurev.astro.35.1.309

Fischer, T., Whitehouse, S.C., Mezzacappa, A., Thielemann, F.K., Liebendörfer, M.: Astron. Astrophys. **517**, A80 (2010). https://doi.org/10.1051/0004-6361/200913106

Foglizzo, T., Tagger, M.: Astron. Astrophys. **363**, 174 (2000)

Folatelli, G., Bersten, M.C., Benvenuto, O.G., Van Dyk, S.D., Kuncarayakti, H., Maeda, K., Nozawa, T., Nomoto, K., Hamuy, M., Quimby, R.M.: Astrophys. J. Lett. **793**, L22 (2014). https://doi.org/10.1088/2041-8205/793/2/L22

Foley, R.J., Smith, N., Ganeshalingam, M., Li, W., Chornock, R., Filippenko, A.V.: Astrophys. J. Lett. **657**, L105 (2007). https://doi.org/10.1086/513145

Foley, R.J., Berger, E., Fox, O., Levesque, E.M., Challis, P.J., Ivans, I.I., Rhoads, J.E., Soderberg, A.M.: Astrophys. J. **732**, 32 (2011). https://doi.org/10.1088/0004-637X/732/1/32

Fraser, M., et al.: Astrophys. J. Lett. **714**, L280 (2010). https://doi.org/10.1088/2041-8205/714/2/L280

Fraser, M., et al.: Mon. Not. R. Astron. Soc. **433**, 1312 (2013). https://doi.org/10.1093/mnras/stt813

Fryer, C.L., Young, P.A.: Astrophys. J. **659**, 1438 (2007). https://doi.org/10.1086/513003

Fukuda, I.: Publ. Astron. Soc. Pac. **94**, 271 (1982). https://doi.org/10.1086/130977

Furusawa, S., Yamada, S., Sumiyoshi, K., Suzuki, H.: Astrophys. J. **738**, 178 (2011). https://doi.org/10.1088/0004-637X/738/2/178

Furusawa, S., Sumiyoshi, K., Yamada, S., Suzuki, H.: Astrophys. J. **772**, 95 (2013). https://doi.org/10.1088/0004-637X/772/2/95

Gal-Yam, A.: American Astronomical Society Meeting Abstracts #223, vol. 223, p. 235.02 (2014)

Gal-Yam, A.: ArXiv e-prints (2016)

Gal-Yam, A., Leonard, D.C.: Nature **458**, 865 (2009). https://doi.org/10.1038/nature07934

Gal-Yam, A., Leonard, D.C., Fox, D.B., Cenko, S.B., Soderberg, A.M., Moon, D.S., Sand, D.J., Caltech Core Collapse Program, Li, W., Filippenko, A.V., Aldering, G., Copin, Y.: Astrophys. J. **656**, 372 (2007). https://doi.org/10.1086/510523

Galama, T.J., et al.: Nature **395**, 670 (1998). https://doi.org/10.1038/27150

Geppert, U., Page, D., Zannias, T.: Astron. Astrophys. **345**, 847 (1999)

Giacomazzo, B., Zrake, J., Duffell, P.C., MacFadyen, A.I., Perna, R.: Astrophys. J. **809**, 39 (2015). https://doi.org/10.1088/0004-637X/809/1/39

Groh, J.H., Meynet, G., Georgy, C., Ekström, S.: Astron. Astrophys. **558**, A131 (2013). https://doi.org/10.1051/0004-6361/201321906

Guilet, J., Müller, E.: Mon. Not. R. Astron. Soc. **450**, 2153 (2015). https://doi.org/10.1093/mnras/stv727

Guilet, J., Müller, E., Janka, H.T.: Mon. Not. R. Astron. Soc. **447**, 3992 (2015). https://doi.org/10.1093/mnras/stu2550

Hammer, N.J., Janka, H.T., Müller, E.: Astrophys. J. **714**, 1371 (2010). https://doi.org/10.1088/0004-637X/714/2/1371

Hamuy, M., Pinto, P.A.: Astrophys. J. Lett. **566**, L63 (2002). https://doi.org/10.1086/339676

Hanke, F., Müller, B., Wongwathanarat, A., Marek, A., Janka, H.T.: Astrophys. J. **770**, 66 (2013). https://doi.org/10.1088/0004-637X/770/1/66

Hannestad, S., Raffelt, G.: Astrophys. J. **507**, 339 (1998). https://doi.org/10.1086/306303

Hartwig, E.: Astron. Nachr. **112**, 285 (1885)

Heger, A., Woosley, S.E.: Astrophys. J. **567**, 532 (2002). https://doi.org/10.1086/338487

Heger, A., Langer, N., Woosley, S.E.: Astrophys. J. **528**, 368 (2000). https://doi.org/10.1086/308158

Heger, A., Fryer, C.L., Woosley, S.E., Langer, N., Hartmann, D.H.: Astrophys. J. **591**, 288 (2003). https://doi.org/10.1086/375341

Heger, A., Woosley, S.E., Spruit, H.C.: Astrophys. J. **626**, 350 (2005). https://doi.org/10.1086/429868

Hempel, M., Schaffner-Bielich, J.: Nucl. Phys. A **837**, 210 (2010). https://doi.org/10.1016/j.nuclphysa.2010.02.010

Hempel, M., Fischer, T., Schaffner-Bielich, J., Liebendörfer, M.: Astrophys. J. **748**, 70 (2012). https://doi.org/10.1088/0004-637X/748/1/70

Hillebrandt, W.: Astron. Astrophys. **110**, L3 (1982)

Hirata, K., Kajita, T., Koshiba, M., Nakahata, M., Oyama, Y.: Phys. Rev. Lett. **58**, 1490 (1987). https://doi.org/10.1103/PhysRevLett.58.1490

Hirschi, R., Meynet, G., Maeder, A.: Astron. Astrophys. **425**, 649 (2004). https://doi.org/10.1051/0004-6361:20041095

Hirschi, R. et al.: In: Roeser, S. (ed.) Reviews in Modern Astronomy, vol. 19, p. 101 (2006). https://doi.org/10.1002/9783527619030.ch5

Hobbs, G., Lorimer, D.R., Lyne, A.G., Kramer, M.: Mon. Not. R. Astron. Soc. **360**, 974 (2005). https://doi.org/10.1111/j.1365-2966.2005.09087.x

Hobbs, G., et al.: Class. Quantum Gravity **27**(8), 084013 (2010). https://doi.org/10.1088/0264-9381/27/8/084013

Hodgkin, S.T., Wyrzykowski, L., Blagorodnova, N., Koposov, S.: Philos. Trans. R. Soc. Lond. Ser. A **371**, 20120239 (2013). https://doi.org/10.1098/rsta.2012.0239

Horiuchi, S., Beacom, J.F., Kochanek, C.S., Prieto, J.L., Stanek, K.Z., Thompson, T.A.: Astrophys. J. **738**, 154 (2011). https://doi.org/10.1088/0004-637X/738/2/154

Horowitz, C.J.: Phys. Rev. D **55**, 4577 (1997). https://doi.org/10.1103/PhysRevD.55.4577

Ikeda, M., et al.: Astrophys. J. **669**, 519 (2007). https://doi.org/10.1086/521547

Inserra, C., Smartt, S.J.: Astrophys. J. **796**, 87 (2014). https://doi.org/10.1088/0004-637X/796/2/87

Inserra, C., Pastorello, A., Turatto, M., Pumo, M.L., Benetti, S., Cappellaro, E., Botticella, M.T., Bufano, F., Elias-Rosa, N., Harutyunyan, A., Taubenberger, S., Valenti, S., Zampieri, L.: Astron. Astrophys. **555**, A142 (2013). https://doi.org/10.1051/0004-6361/201220496

Isenberg, J.A.: Int. J. Mod. Phys. D **17**, 265 (2008). https://doi.org/10.1142/S0218271808011997. This article appeared for the first time in 1978 as a University of Maryland preprint

Janka, H.T.: Annu. Rev. Nucl. Part. Sci. **62**, 407 (2012). https://doi.org/10.1146/annurev-nucl-102711-094901

Janka, H.T., Mueller, E.: Astron. Astrophys. **290**, 496 (1994)

Janka, H.T., Langanke, K., Marek, A., Martínez-Pinedo, G., Müller, B.: Phys. Rep. **442**, 38 (2007). https://doi.org/10.1016/j.physrep.2007.02.002

Janka, H.T., Müller, B., Kitaura, F.S., Buras, R.: Astron. Astrophys. **485**, 199 (2008). https://doi.org/10.1051/0004-6361:20079334

Jerkstrand, A.: ArXiv e-prints (2017)

Joggerst, C.C., Almgren, A., Woosley, S.E.: Astrophys. J. **723**, 353 (2010). https://doi.org/10.1088/0004-637X/723/1/353

Jones, S., Hirschi, R., Nomoto, K., Fischer, T., Timmes, F.X., Herwig, F., Paxton, B., Toki, H., Suzuki, T., Martínez-Pinedo, G., Lam, Y.H., Bertolli, M.G.: Astrophys. J. **772**, 150 (2013). https://doi.org/10.1088/0004-637X/772/2/150

Kaiser, N., et al.: In: Tyson, J.A., Wolff, S. (eds.) Survey and Other Telescope Technologies and Discoveries. Proc. SPIE, vol. 4836, pp. 154–164 (2002). https://doi.org/10.1117/12.457365

Karle, A., IceCube Collaboration: Nucl. Instrum. Methods Phys. Res. A **604**, S46 (2009). https://doi.org/10.1016/j.nima.2009.03.180

Kiewe, M., Gal-Yam, A., Arcavi, I., Leonard, D.C., Emilio Enriquez, J., Cenko, S.B., Fox, D.B., Moon, D.S., Sand, D.J., Soderberg, A.M., CCCP, T.: Astrophys. J. **744**, 10 (2012). https://doi.org/10.1088/0004-637X/744/1/10

Kifonidis, K., Plewa, T., Janka, H.T., Müller, E.: Astron. Astrophys. **408**, 621 (2003). https://doi.org/10.1051/0004-6361:20030863

Kitaura, F.S., Janka, H.T., Hillebrandt, W.: Astron. Astrophys. **450**, 345 (2006). https://doi.org/10.1051/0004-6361:20054703

Kochanek, C.S., Beacom, J.F., Kistler, M.D., Prieto, J.L., Stanek, K.Z., Thompson, T.A., Yüksel, H.: Astrophys. J. **684**, 1336–1342 (2008). https://doi.org/10.1086/590053

Koyama, K., Petre, R., Gotthelf, E.V., Hwang, U., Matsuura, M., Ozaki, M., Holt, S.S.: Nature **378**, 255 (1995). https://doi.org/10.1038/378255a0

Krebs, J., Hillebrandt, W.: Astron. Astrophys. **128**, 411 (1983)

Kuroda, T., Umeda, H.: Astrophys. J. Suppl. Ser. **191**, 439 (2010). https://doi.org/10.1088/0067-0049/191/2/439

Kuroda, T., Kotake, K., Takiwaki, T.: Astrophys. J. Lett. **829**, L14 (2016). https://doi.org/10.3847/2041-8205/829/1/L14

Lajoie, C.P., Sills, A.: Astrophys. J. **726**, 66 (2011). https://doi.org/10.1088/0004-637X/726/2/66

Langanke, K., Martínez-Pinedo, G., Sampaio, J.M., Dean, D.J., Hix, W.R., Messer, O.E., Mezzacappa, A., Liebendörfer, M., Janka, H.T., Rampp, M.: Phys. Rev. Lett. **90**(24), 241102 (2003). https://doi.org/10.1103/PhysRevLett.90.241102

Langer, N.: Astron. Astrophys. **329**, 551 (1998)

Lattimer, J.M., Douglas Swesty, F.: Nucl. Phys. A **535**, 331 (1991). https://doi.org/10.1016/0375-9474(91)90452-C

Lattimer, J.M., Prakash, M.: Phys. Rep. **333**, 121 (2000).   https://doi.org/10.1016/S0370-1573(00)00019-3

Lentz, E.J., Mezzacappa, A., Messer, O.E.B., Liebendörfer, M., Hix, W.R., Bruenn, S.W.: Astrophys. J. **747**, 73 (2012). https://doi.org/10.1088/0004-637X/747/1/73

Lentz, E.J., Bruenn, S.W., Hix, W.R., Mezzacappa, A., Messer, O.E.B., Endeve, E., Blondin, J.M., Harris, J.A., Marronetti, P., Yakunin, K.N.: Astrophys. J. Lett. **807**, L31 (2015). https://doi.org/10.1088/2041-8205/807/2/L31

Leonard, D.C., et al.: Publ. Astron. Soc. Pac. **114**, 35 (2002). https://doi.org/10.1086/324785

Levesque, E.M. , Massey, P., Plez, B., Olsen, K.A.G.: Astrophys. J. **137**, 4744 (2009). https://doi.org/10.1088/0004-6256/137/6/4744

Li, W., Leaman, J., Chornock, R., Filippenko, A.V., Poznanski, D., Ganeshalingam, M., Wang, X., Modjaz, M., Jha, S., Foley, R.J., Smith, N.: Mon. Not. R. Astron. Soc. **412**, 1441 (2011). https://doi.org/10.1111/j.1365-2966.2011.18160.x

Liebendörfer, M., Messer, O.E.B., Mezzacappa, A., Bruenn, S.W., Cardall, C.Y., Thielemann, F.K.: Astrophys. J. Suppl. Ser. **150**, 263 (2004). https://doi.org/10.1086/380191

Liebendörfer, M., Whitehouse, S.C., Fischer, T.: Astrophys. J. **698**, 1174 (2009). https://doi.org/10.1088/0004-637X/698/2/1174

LIGO Scientific Collaboration, Aasi, J., Abbott, B.P., Abbott, R., Abbott, T., Abernathy, M.R., Ackley, K., Adams, C., Adams, T., Addesso, P., et al.: Class. Quantum Gravity **32**(7), 074001 (2015). https://doi.org/10.1088/0264-9381/32/7/074001

Limongi, M., Chieffi, A.: Astrophys. J. **647**, 483 (2006). https://doi.org/10.1086/505164

Limongi, M., Chieffi, A.: Astrophys. J. Suppl. Ser. **199**, 38 (2012). https://doi.org/10.1088/0067-0049/199/2/38

Lombardi, J.C. Jr., Holtzman, W., Dooley, K.L., Gearity, K., Kalogera, V., Rasio, F.A.: Astrophys. J. **737**, 49 (2011). https://doi.org/10.1088/0004-637X/737/2/49

Lovegrove, E., Woosley, S.E.: Astrophys. J. **769**, 109 (2013). https://doi.org/10.1088/0004-637X/769/2/109

MacFadyen, A.I., Woosley, S.E.: Astrophys. J. **524**, 262 (1999). https://doi.org/10.1086/307790

Mackey, J., Mohamed, S., Gvaramadze, V.V., Kotak, R., Langer, N., Meyer, D.M.A., Moriya, T.J., Neilson, H.R.: Nature **512**, 282 (2014). https://doi.org/10.1038/nature13522

Maeder, A.: Astron. Astrophys. **178**, 159 (1987)

Maeder, A., Meynet, G.: Annu. Rev. Astron. Astrophys. **38**, 143 (2000a). https://doi.org/10.1146/annurev.astro.38.1.143

Maeder, A., Meynet, G.: Astron. Astrophys. **361**, 159 (2000b)

Marek, A., Dimmelmeier, H., Janka, H.T., Müller, E., Buras, R.: Astron. Astrophys. **445**, 273 (2006). https://doi.org/10.1051/0004-6361:20052840

Marek, A., Janka, H.T., Müller, E.: Astron. Astrophys. **496**, 475 (2009). https://doi.org/10.1051/0004-6361/200810883

Margutti, R., et al.: Astrophys. J. **780**, 21 (2014). https://doi.org/10.1088/0004-637X/780/1/21

Marigo, P., Girardi, L., Chiosi, C., Wood, P.R.: Astron. Astrophys. **371**, 152 (2001). https://doi.org/10.1051/0004-6361:20010309

Mauerhan, J.C., Smith, N., Filippenko, A.V., Blanchard, K.B., Blanchard, P.K., Casper, C.F.E., Cenko, S.B., Clubb, K.I., Cohen, D.P., Fuller, K.L., Li, G.Z., Silverman, J.M.: Mon. Not. R. Astron. Soc. **430**, 1801 (2013). https://doi.org/10.1093/mnras/stt009

Maund, J.R., Smartt, S.J.: Science **324**, 486 (2009). https://doi.org/10.1126/science.1170198

Maund, J.R., Fraser, M., Ergon, M., Pastorello, A., Smartt, S.J., Sollerman, J., Benetti, S., Botticella, M.T., Bufano, F., Danziger, I.J., Kotak, R., Magill, L., Stephens, A.W., Valenti, S.: Astrophys. J. Lett. **739**, L37 (2011). https://doi.org/10.1088/2041-8205/739/2/L37

Maund, J.R., Fraser, M., Reilly, E., Ergon, M., Mattila, S.: Mon. Not. R. Astron. Soc. **447**, 3207 (2015). https://doi.org/10.1093/mnras/stu2658

McCray, R.: Annu. Rev. Astron. Astrophys. **31**, 175 (1993). https://doi.org/10.1146/annurev.aa. 31.090193.001135

Melson, T., Janka, H.T., Bollig, R., Hanke, F., Marek, A., Müller, B.: Astrophys. J. Lett. **808**, L42 (2015a). https://doi.org/10.1088/2041-8205/808/2/L42

Melson, T., Janka, H.T., Marek, A.: Astrophys. J. Lett. **801**, L24 (2015b). https://doi.org/10.1088/2041-8205/801/2/L24

Mezzacappa, A., Bruenn, S.W.: Astrophys. J. **405**, 669 (1993). https://doi.org/10.1086/172395

Mikami, H., Sato, Y., Matsumoto, T., Hanawa, T.: Astrophys. J. **683**, 357–374 (2008). https://doi.org/10.1086/589759

Minkowski, R.: Publ. Astron. Soc. Pac. **53**, 224 (1941). https://doi.org/10.1086/125315

Moffatt, H.K.: Magnetic Field Generation in Electrically Conducting Fluids, 353 pp. Cambridge University Press, Cambridge (1978)

Morales-Garoffolo, A., Elias-Rosa, N., Benetti, S., Taubenberger, S., Cappellaro, E., Pastorello, A., Klauser, M., Valenti, S., Howerton, S., Ochner, P., Schramm, N., Siviero, A., Tartaglia, L., Tomasella, L.: Mon. Not. R. Astron. Soc. **445**, 1647 (2014). https://doi.org/10.1093/mnras/stu1837

Moriya, T., Tominaga, N., Tanaka, M., Nomoto, K., Sauer, D.N., Mazzali, P.A., Maeda, K., Suzuki, T.: Astrophys. J. **719**, 1445 (2010). https://doi.org/10.1088/0004-637X/719/2/1445

Mösta, P., Richers, S., Ott, C.D., Haas, R., Piro, A.L., Boydstun, K., Abdikamalov, E., Reisswig, C., Schnetter, E.: Astrophys. J. Lett. **785**, L29 (2014). https://doi.org/10.1088/2041-8205/785/2/L29

Mösta, P., Ott, C.D., Radice, D., Roberts, L.F., Schnetter, E., Haas, R.: Nature **528**, 376 (2015). https://doi.org/10.1038/nature15755

Motl, P.M., Tohline, J.E., Frank, J.: Astrophys. J. Suppl. Ser. **138**, 121 (2002). https://doi.org/10.1086/324159

Mukherjee, D., Bhattacharya, D., Mignone, A.: Mon. Not. R. Astron. Soc. **430**, 1976 (2013a). https://doi.org/10.1093/mnras/stt020

Mukherjee, D., Bhattacharya, D., Mignone, A.: Mon. Not. R. Astron. Soc. **435**, 718 (2013b). https://doi.org/10.1093/mnras/stt1344

Müller, B.: Mon. Not. R. Astron. Soc. **453**, 287 (2015). https://doi.org/10.1093/mnras/stv1611

Müller, B.: Publ. Astron. Soc. Aust. **33**, e048 (2016). https://doi.org/10.1017/pasa.2016.40

Müller, B., Janka, H.T.: Astrophys. J. **788**, 82 (2014). https://doi.org/10.1088/0004-637X/788/1/82

Müller, B., Janka, H.T.: Mon. Not. R. Astron. Soc. **448**, 2141 (2015). https://doi.org/10.1093/mnras/stv101

Müller, B., Janka, H.T., Dimmelmeier, H.: Astrophys. J. Suppl. Ser. **189**, 104 (2010). https://doi.org/10.1088/0067-0049/189/1/104

Müller, B., Janka, H.T., Marek, A.: Astrophys. J. **756**, 84 (2012a). https://doi.org/10.1088/0004-637X/756/1/84

Müller, B., Janka, H.T., Heger, A.: Astrophys. J. **761**, 72 (2012b). https://doi.org/10.1088/0004-637X/761/1/72

Müller, B., Janka, H.T., Marek, A.: Astrophys. J. **766**, 43 (2013). https://doi.org/10.1088/0004-637X/766/1/43

Müller, B., Viallet, M., Heger, A., Janka, H.T.: Astrophys. J. **833**, 124 (2016). https://doi.org/10.3847/1538-4357/833/1/124

Murphy, J.W., Ott, C.D., Burrows, A.: Astrophys. J. **707**, 1173 (2009). https://doi.org/10.1088/0004-637X/707/2/1173

Muslimov, A., Page, D.: Astrophys. J. Lett. **440**, L77 (1995). https://doi.org/10.1086/187765

Nadezhin, D.K.: Astrophys. Space Sci. **69**, 115 (1980). https://doi.org/10.1007/BF00638971

Nagakura, H., Iwakami, W., Furusawa, S., Okawa, H., Harada, A., Sumiyoshi, K., Yamada, S., Matsufuru, H., Imakura, A.: ArXiv e-prints (2017)

Nicholl, M., et al.: Nature **502**, 346 (2013). https://doi.org/10.1038/nature12569

Nomoto, K.: Astrophys. J. **277**, 791 (1984). https://doi.org/10.1086/161749

Nomoto, K.: Astrophys. J. **322**, 206 (1987). https://doi.org/10.1086/165716

Nomoto, K., Sugimoto, D., Sparks, W.M., Fesen, R.A., Gull, T.R., Miyaji, S.: Nature **299**, 803 (1982). https://doi.org/10.1038/299803a0

Nomoto, K., Tominaga, N., Umeda, H., Kobayashi, C., Maeda, K.: Nucl. Phys. A **777**, 424 (2006). https://doi.org/10.1016/j.nuclphysa.2006.05.008

Nordhaus, J., Brandt, T.D., Burrows, A., Livne, E., Ott, C.D.: Phys. Rev. D **82**(10), 103016 (2010). https://doi.org/10.1103/PhysRevD.82.103016

Nordhaus, J., Brandt, T.D., Burrows, A., Almgren, A.: Mon. Not. R. Astron. Soc. **423**, 1805 (2012). https://doi.org/10.1111/j.1365-2966.2012.21002.x

Obergaulinger, M., Ángel Aloy, M.: ArXiv e-prints (2017)

Obergaulinger, M., Aloy, M.A., Dimmelmeier, H., Müller, E.: Astron. Astrophys. **457**, 209 (2006). https://doi.org/10.1051/0004-6361:20064982

Obergaulinger, M., Cerdá-Durán, P., Müller, E., Aloy, M.A.: Astron. Astrophys. **498**, 241 (2009). https://doi.org/10.1051/0004-6361/200811323

Obergaulinger, M., Just, O., Janka, H.T., Aloy, M.A., Aloy, C.: In: Pogorelov, N.V., Audit, E., Zank, G.P. (eds.) 8th International Conference of Numerical Modeling of Space Plasma Flows (ASTRONUM 2013). Astronomical Society of the Pacific Conference Series, vol. 488, p. 255 (2014)

O'Connor, E., Couch, S.: ArXiv e-prints (2015)

O'Connor, E., Ott, C.D.: Class. Quantum Gravity **27**(11), 114103 (2010). https://doi.org/10.1088/0264-9381/27/11/114103

O'Connor, E., Ott, C.D.: Astrophys. J. **730**, 70 (2011). https://doi.org/10.1088/0004-637X/730/2/70

Oertel, M., Hempel, M., Klähn, T., Typel, S.: Rev. Mod. Phys. **89**(1), 015007 (2017). https://doi.org/10.1103/RevModPhys.89.015007

Ofek, E.O., Sullivan, M., Shaviv, N.J., Steinbok, A., Arcavi, I., Gal-Yam, A., Tal, D., Kulkarni, S.R., Nugent, P.E., Ben-Ami, S., Kasliwal, M.M., Cenko, S.B., Laher, R., Surace, J., Bloom, J.S., Filippenko, A.V., Silverman, J.M., Yaron, O.: Astrophys. J. **789**, 104 (2014). https://doi.org/10.1088/0004-637X/789/2/104

Ott, C.D.: Class. Quantum Gravity **26**(6), 063001 (2009). https://doi.org/10.1088/0264-9381/26/6/063001

Ott, C.D., Burrows, A., Thompson, T.A., Livne, E., Walder, R.: Astrophys. J. Suppl. Ser. **164**, 130 (2006). https://doi.org/10.1086/500832

Ott, C.D., Dimmelmeier, H., Marek, A., Janka, H.T., Zink, B., Hawke, I., Schnetter, E.: Class. Quantum Gravity **24**, S139 (2007a). https://doi.org/10.1088/0264-9381/24/12/S10

Ott, C.D., Dimmelmeier, H., Marek, A., Janka, H.T., Hawke, I., Zink, B., Schnetter, E.: Phys. Rev. Lett. **98**(26), 261101 (2007b). https://doi.org/10.1103/PhysRevLett.98.261101

Ott, C.D., Burrows, A., Dessart, L., Livne, E.: Astrophys. J. **685**, 1069–1088 (2008). https://doi.org/10.1086/591440

Ott, C.D., Abdikamalov, E., O'Connor, E., Reisswig, C., Haas, R., Kalmus, P., Drasco, S., Burrows, A., Schnetter, E.: Phys. Rev. D **86**(2), 024026 (2012). https://doi.org/10.1103/PhysRevD.86.024026

Ott, C.D., Abdikamalov, E., Mösta, P., Haas, R., Drasco, S., O'Connor, E.P., Reisswig, C., Meakin, C.A., Schnetter, E.: Astrophys. J. **768**, 115 (2013). https://doi.org/10.1088/0004-637X/768/2/115

Özel, F., Freire, P.: Annu. Rev. Astron. Astrophys. **54**, 401 (2016). https://doi.org/10.1146/annurev-astro-081915-023322

Özel, F., Psaltis, D., Narayan, R., Santos Villarreal, A.: Astrophys. J. **757**, 55 (2012). https://doi.org/10.1088/0004-637X/757/1/55

Pan, K.C., Liebendörfer, M., Hempel, M., Thielemann, F.K.: Astrophys. J. **817**, 72 (2016). https://doi.org/10.3847/0004-637X/817/1/72

Pastorello, A., Zampieri, L., Turatto, M., Cappellaro, E., Meikle, W.P.S., Benetti, S., Branch, D., Baron, E., Patat, F., Armstrong, M., Altavilla, G., Salvo, M., Riello, M.: Mon. Not. R. Astron. Soc. **347**, 74 (2004). https://doi.org/10.1111/j.1365-2966.2004.07173.x

Pastorello, A., et al.: Nature **447**, 829 (2007). https://doi.org/10.1038/nature05825

Pastorello, A., et al.: Mon. Not. R. Astron. Soc. **389**, 113 (2008). https://doi.org/10.1111/j.1365-2966.2008.13602.x

Pastorello, A., et al.: Mon. Not. R. Astron. Soc. **408**, 181 (2010). https://doi.org/10.1111/j.1365-2966.2010.17142.x

Pastorello, A., et al.: Astron. Astrophys. **537**, A141 (2012). https://doi.org/10.1051/0004-6361/201118112

Pastorello, A., et al.: Astrophys. J. **767**, 1 (2013). https://doi.org/10.1088/0004-637X/767/1/1

Pastorello, A., et al.: Mon. Not. R. Astron. Soc. **456**, 853 (2016). https://doi.org/10.1093/mnras/stv2634

Patnaude, D., Badenes, C.: ArXiv e-prints (2017)

Payne, D.J.B., Melatos, A.: Mon. Not. R. Astron. Soc. **351**, 569 (2004). https://doi.org/10.1111/j.1365-2966.2004.07798.x

Payne, D.J.B., Melatos, A.: Mon. Not. R. Astron. Soc. **376**, 609 (2007). https://doi.org/10.1111/j.1365-2966.2007.11451.x

Pejcha, O., Thompson, T.A.: Astrophys. J. **801**, 90 (2015). https://doi.org/10.1088/0004-637X/801/2/90

Perego, A., Cabezón, R.M., Käppeli, R.: Astrophys. J. Suppl. Ser. **223**, 22 (2016). https://doi.org/10.3847/0067-0049/223/2/22

Pereira, R., et al.: Astron. Astrophys. **554**, A27 (2013). https://doi.org/10.1051/0004-6361/201221008

Perets, H.B., et al.: Nature **465**, 322 (2010). https://doi.org/10.1038/nature09056

Perlmutter, S., et al.: Astrophys. J. **517**, 565 (1999). https://doi.org/10.1086/307221

Pian, E., et al.: Nature **442**, 1011 (2006). https://doi.org/10.1038/nature05082

Piro, A.L.: Astrophys. J. Lett. **768**, L14 (2013). https://doi.org/10.1088/2041-8205/768/1/L14

Podsiadlowski, P., Langer, N., Poelarends, A.J.T., Rappaport, S., Heger, A., Pfahl, E.: Astrophys. J. **612**, 1044 (2004). https://doi.org/10.1086/421713

Poelarends, A.J.T., Herwig, F., Langer, N., Heger, A.: Astrophys. J. **675**, 614–625 (2008). https://doi.org/10.1086/520872

Pons, J.A., Miralles, J.A., Ibanez, J.M.A.: Astron. Astrophys. Suppl. **129**, 343 (1998). https://doi.org/10.1051/aas:1998189

Postnov, K.A., Yungelson, L.R.: Living Rev. Relativ. **17**(1), 3 (2014). http://dx.doi.org/10.12942/lrr-2014-3

Pumo, M.L.: Mem. Soc. Astron. Ital. Suppl. **14**, 115 (2010)

Pumo, M.L., Turatto, M., Botticella, M.T., Pastorello, A., Valenti, S., Zampieri, L., Benetti, S., Cappellaro, E., Patat, F.: Astrophys. J. Lett. **705**, L138 (2009). https://doi.org/10.1088/0004-637X/705/2/L138

Quimby, R.M., et al.: Nature **474**, 487 (2011). https://doi.org/10.1038/nature10095

Rampp, M.: Radiation hydrodynamics with neutrinos: stellar core collapse and the explosion mechanism of type ii supernovae. Dissertation, Technische Universitt Mnchen, Mnchen (2000)

Rampp, M., Janka, H.T.: Astron. Astrophys. **396**, 361 (2002). https://doi.org/10.1051/0004-6361:20021398

Reisswig, C., Ott, C.D., Sperhake, U., Schnetter, E.: Phys. Rev. D **83**(6), 064008 (2011). https://doi.org/10.1103/PhysRevD.83.064008

Rembiasz, T., Guilet, J., Obergaulinger, M., Cerdá-Durán, P., Aloy, M.A., Müller, E.: Mon. Not. R. Astron. Soc. **460**, 3316 (2016a). https://doi.org/10.1093/mnras/stw1201

Rembiasz, T., Obergaulinger, M., Cerdá-Durán, P., Müller, E., Aloy, M.A.: Mon. Not. R. Astron. Soc. **456**, 3782 (2016b). https://doi.org/10.1093/mnras/stv2917

Reynolds, T.M., Fraser, M., Gilmore, G.: Mon. Not. R. Astron. Soc. **453**, 2885 (2015). https://doi.org/10.1093/mnras/stv1809

Richers, S., Ott, C.D., Abdikamalov, E., O'Connor, E., Sullivan, C.: ArXiv e-prints (2017)

Richmond, M.W., Treffers, R.R., Filippenko, A.V., Paik, Y., Leibundgut, B., Schulman, E., Cox, C.V.: Astrophys. J. **107**, 1022 (1994). https://doi.org/10.1086/116915

Riess, A.G., et al.: Astrophys. J. **116**, 1009 (1998). https://doi.org/10.1086/300499

Roberts, L.F., Ott, C.D., Haas, R., O'Connor, E.P., Diener, P., Schnetter, E.: Astrophys. J. **831**, 98 (2016). https://doi.org/10.3847/0004-637X/831/1/98

Sądowski, A., Narayan, R., Tchekhovskoy, A., Abarca, D., Zhu, Y., McKinney, J.C.: Mon. Not. R. Astron. Soc. **447**, 49 (2015). https://doi.org/10.1093/mnras/stu2387

Sana, H., de Mink, S.E., de Koter, A., Langer, N., Evans, C.J., Gieles, M., Gosset, E., Izzard, R.G., Le Bouquin, J.B., Schneider, F.R.N.: Science **337**, 444 (2012). https://doi.org/10.1126/science.1223344

Sawai, H., Yamada, S.: Astrophys. J. **817**, 153 (2016). https://doi.org/10.3847/0004-637X/817/2/153

Sawai, H., Yamada, S., Suzuki, H.: Astrophys. J. Lett. **770**, L19 (2013). https://doi.org/10.1088/2041-8205/770/2/L19

Scheck, L., Plewa, T., Janka, H.T., Kifonidis, K., Müller, E.: Phys. Rev. Lett. **92**(1), 011103 (2004). https://doi.org/10.1103/PhysRevLett.92.011103

Scheck, L., Kifonidis, K., Janka, H.T., Müller, E.: Astron. Astrophys. **457**, 963 (2006). https://doi.org/10.1051/0004-6361:20064855

Scheidegger, S., Fischer, T., Whitehouse, S.C., Liebendörfer, M.: Astron. Astrophys. **490**, 231 (2008). https://doi.org/10.1051/0004-6361:20078577

Scheidegger, S., Käppeli, R., Whitehouse, S.C., Fischer, T., Liebendörfer, M.: Astron. Astrophys. **514**, A51 (2010). https://doi.org/10.1051/0004-6361/200913220

Schlegel, E.M.: Mon. Not. R. Astron. Soc. **244**, 269 (1990)

Schwab, J., Podsiadlowski, P., Rappaport, S.: Astrophys. J. **719**, 722 (2010). https://doi.org/10.1088/0004-637X/719/1/722

Sekiguchi, Y., Shibata, M.: Astrophys. J. **737**, 6 (2011). https://doi.org/10.1088/0004-637X/737/1/6

Shabaltas, N., Lai, D.: Astrophys. J. **748**, 148 (2012). https://doi.org/10.1088/0004-637X/748/2/148

Shakura, N.I., Sunyaev, R.A.: Astron. Astrophys. **24**, 337 (1973)

Shappee, B.J., et al.: Astrophys. J. **788**, 48 (2014). https://doi.org/10.1088/0004-637X/788/1/48

Shen, H., Toki, H., Oyamatsu, K., Sumiyoshi, K.: Nucl. Phys. A **637**, 435 (1998a). https://doi.org/10.1016/S0375-9474(98)00236-X

Shen, H., Toki, H., Oyamatsu, K., Sumiyoshi, K.: Prog. Theor. Phys. **100**, 1013 (1998b). https://doi.org/10.1143/PTP.100.1013

Shen, G., Horowitz, C.J., Teige, S.: Phys. Rev. C **83**(3), 035802 (2011a). https://doi.org/10.1103/PhysRevC.83.035802

Shen, G., Horowitz, C.J., O'Connor, E.: Phys. Rev. C **83**(6), 065808 (2011b). https://doi.org/10.1103/PhysRevC.83.065808

Shen, H., Toki, H., Oyamatsu, K., Sumiyoshi, K.: Astrophys. J. Suppl. Ser. **197**, 20 (2011c). https://doi.org/10.1088/0067-0049/197/2/20

Shibata, M., Nakamura, T.: Phys. Rev. D **52**, 5428 (1995). https://doi.org/10.1103/PhysRevD.52.5428

Shibata, M., Sekiguchi, Y.I.: Phys. Rev. D **69**(8), 084024 (2004). https://doi.org/10.1103/PhysRevD.69.084024

Shibata, M., Liu, Y.T., Shapiro, S.L., Stephens, B.C.: Phys. Rev. D **74**(10), 104026 (2006). https://doi.org/10.1103/PhysRevD.74.104026

Shibata, M., Kiuchi, K., Sekiguchi, Y.I.: Phys. Rev. D **95**(8), 083005 (2017). https://doi.org/10.1103/PhysRevD.95.083005

Shigeyama, T., Nomoto, K.: Astrophys. J. **360**, 242 (1990). https://doi.org/10.1086/169114

Smartt, S.J.: Annu. Rev. Astron. Astrophys. **47**, 63 (2009). https://doi.org/10.1146/annurev-astro-082708-101737

Smartt, S.J.: Publ. Astron. Soc. Aust. **32**, e016 (2015). https://doi.org/10.1017/pasa.2015.17

Smartt, S.J., Eldridge, J.J., Crockett, R.M., Maund, J.R.: Mon. Not. R. Astron. Soc. **395**, 1409 (2009). https://doi.org/10.1111/j.1365-2966.2009.14506.x

Smartt, S.J., et al.: Astron. Astrophys. **579**, A40 (2015). https://doi.org/10.1051/0004-6361/201425237

Smith, N.: ArXiv e-prints (2016)

Smith, N., Miller, A., Li, W., Filippenko, A.V., Silverman, J.M., Howard, A.W., Nugent, P., Marcy, G.W., Bloom, J.S., Ghez, A.M., Lu, J., Yelda, S., Bernstein, R.A., Colucci, J.E.: Astrophys. J. **139**, 1451 (2010). https://doi.org/10.1088/0004-6256/139/4/1451

Smith, N., Li, W., Miller, A.A., Silverman, J.M., Filippenko, A.V., Cuillandre, J.C., Cooper, M.C., Matheson, T., Van Dyk, S.D.: Astrophys. J. **732**, 63 (2011). https://doi.org/10.1088/0004-637X/732/2/63

Soderberg, A.M., et al.: Nature **453**, 469 (2008). https://doi.org/10.1038/nature06997

Soker, N., Kashi, A.: Astrophys. J. Lett. **764**, L6 (2013). https://doi.org/10.1088/2041-8205/764/1/L6

Sotani, H., Takiwaki, T.: Phys. Rev. D **94**(4), 044043 (2016). https://doi.org/10.1103/PhysRevD.94.044043

Spiro, S., et al.: Mon. Not. R. Astron. Soc. **439**, 2873 (2014). https://doi.org/10.1093/mnras/stu156

Spruit, H.C.: Astron. Astrophys. **349**, 189 (1999)

Spruit, H.C.: Astron. Astrophys. **381**, 923 (2002). https://doi.org/10.1051/0004-6361:20011465

Spruit, H., Phinney, E.S.: Nature **393**, 139 (1998). https://doi.org/10.1038/30168

Steiner, A.W., Hempel, M., Fischer, T.: Astrophys. J. **774**, 17 (2013). https://doi.org/10.1088/0004-637X/774/1/17

Stephenson, F.R., Green, D.A.: In: Turatto, M., Benetti, S., Zampieri, L., Shea, W. (eds.) 1604-2004: Supernovae as Cosmological Lighthouses. Astronomical Society of the Pacific Conference Series, vol. 342, pp. 63–70 (2005)

Sukhbold, T., Woosley, S.E.: Astrophys. J. **783**, 10 (2014). https://doi.org/10.1088/0004-637X/783/1/10

Sukhbold, T., Ertl, T., Woosley, S.E., Brown, J.M., Janka, H.T.: Astrophys. J. **821**, 38 (2016). https://doi.org/10.3847/0004-637X/821/1/38

Sumiyoshi, K., Yamada, S., Suzuki, H., Shen, H., Chiba, S., Toki, H.: Astrophys. J. **629**, 922 (2005). https://doi.org/10.1086/431788

Summa, A., Hanke, F., Janka, H.T., Melson, T., Marek, A., Müller, B.: Astrophys. J. **825**, 6 (2016). https://doi.org/10.3847/0004-637X/825/1/6

Suwa, Y., Müller, E.: Mon. Not. R. Astron. Soc. **460**, 2664 (2016). https://doi.org/10.1093/mnras/stw1150

Suwa, Y., Takiwaki, T., Kotake, K., Fischer, T., Liebendörfer, M., Sato, K.: Astrophys. J. **764**, 99 (2013). https://doi.org/10.1088/0004-637X/764/1/99

Swesty, F.D., Myra, E.S.: Astrophys. J. Suppl. Ser. **181**, 1 (2009). https://doi.org/10.1088/0067-0049/181/1/1

Takiwaki, T., Kotake, K.: Astrophys. J. **743**, 30 (2011). https://doi.org/10.1088/0004-637X/743/1/30

Takiwaki, T., Kotake, K., Suwa, Y.: Mon. Not. R. Astron. Soc. **461**, L112 (2016). https://doi.org/10.1093/mnrasl/slw105

Tartaglia, L., et al.: Mon. Not. R. Astron. Soc. **447**, 117 (2015). https://doi.org/10.1093/mnras/stu2384

Tartaglia, L., et al.: Mon. Not. R. Astron. Soc. **459**, 1039 (2016). https://doi.org/10.1093/mnras/stw675

Tartaglia, L., et al.: Astrophys. J. Lett. **836**, L12 (2017). https://doi.org/10.3847/2041-8213/aa5c7f

Taubenberger, S., et al.: Mon. Not. R. Astron. Soc. **397**, 677 (2009). https://doi.org/10.1111/j.1365-2966.2009.15003.x

Terreran, G., et al.: Mon. Not. R. Astron. Soc. **462**, 137 (2016). https://doi.org/10.1093/mnras/stw1591

Thielemann, F.K., Nomoto, K., Hashimoto, M.A.: Astrophys. J. **460**, 408 (1996). https://doi.org/10.1086/176980

Thompson, C., Duncan, R.C.: Astrophys. J. **408**, 194 (1993). https://doi.org/10.1086/172580

Timmes, F.X.: Astrophys. J. Suppl. Ser. **124**, 241 (1999). https://doi.org/10.1086/313257

Timmes, F.X., Arnett, D.: Astrophys. J. Suppl. Ser. **125**, 277 (1999). https://doi.org/10.1086/313271

Timmes, F.X., Swesty, F.D.: Astrophys. J. Suppl. Ser. **126**, 501 (2000). https://doi.org/10.1086/313304

Todini, P., Ferrara, A.: Mon. Not. R. Astron. Soc. **325**, 726 (2001). https://doi.org/10.1046/j.1365-8711.2001.04486.x

Tominaga, N., Umeda, H., Nomoto, K.: Astrophys. J. **660**, 516 (2007). https://doi.org/10.1086/513063

Torres-Forné, A., Cerdá-Durán, P., Pons, J.A., Font, J.A.: Mon. Not. R. Astron. Soc. **456**, 3813 (2016). https://doi.org/10.1093/mnras/stv2926

Turatto, M., et al.: Astrophys. J. Lett. **498**, L129 (1998). https://doi.org/10.1086/311324

Turatto, M., Benetti, S., Cappellaro, E.: In: Hillebrandt, W., Leibundgut, B. (eds.) From Twilight to Highlight: The Physics of Supernovae, p. 200 (2003). https://doi.org/10.1007/10828549_26

Ugliano, M., Janka, H.T., Marek, A., Arcones, A.: Astrophys. J. **757**, 69 (2012). https://doi.org/10.1088/0004-637X/757/1/69

Umeda, H., Nomoto, K.: Astrophys. J. **619**, 427 (2005). https://doi.org/10.1086/426097

Usov, V.V.: Mon. Not. R. Astron. Soc. **267**, 1035 (1994). https://doi.org/10.1093/mnras/267.4.1035

Valenti, S., et al.: Astrophys. J. Lett. **673**, L155 (2008). https://doi.org/10.1086/527672

Valenti, S., et al.: Nature **459**, 674 (2009). https://doi.org/10.1038/nature08023

Valenti, S., et al.: Astrophys. J. Lett. **749**, L28 (2012). https://doi.org/10.1088/2041-8205/749/2/L28

van den Heuvel, E.P.J.: In: Schoenfelder, V., Lichti, G., Winkler, C. (eds.) 5th INTEGRAL Workshop on the INTEGRAL Universe. ESA Special Publication, vol. 552, p. 185 (2004)

Van Dyk, S.D., Matheson, T.: In: Davidson, K., Humphreys, R.M. (eds.) Eta Carinae and the Supernova Impostors. Astrophysics and Space Science Library, vol. 384, p. 249 (2012). https://doi.org/10.1007/978-1-4614-2275-4_11

Van Dyk, S.D., et al.: Astrophys. J. Lett. **741**, L28 (2011). https://doi.org/10.1088/2041-8205/741/2/L28

Van Dyk, S.D., Zheng, W., Clubb, K.I., Filippenko, A.V., Cenko, S.B., Smith, N., Fox, O.D., Kelly, P.L., Shivvers, I., Ganeshalingam, M.: Astrophys. J. Lett. **772**, L32 (2013). https://doi.org/10.1088/2041-8205/772/2/L32

Van Dyk, S.D., Zheng, W., Fox, O.D., Cenko, S.B., Clubb, K.I., Filippenko, A.V., Foley, R.J., Miller, A.A., Smith, N., Kelly, P.L., Lee, W.H., Ben-Ami, S., Gal-Yam, A.: Astrophys. J. **147**, 37 (2014). https://doi.org/10.1088/0004-6256/147/2/37

Velikhov, E.: Sov. Phys. JETP **36** (1959)

Viallet, M., Baraffe, I., Walder, R.: Astron. Astrophys. **531**, A86 (2011). https://doi.org/10.1051/0004-6361/201016374

Wanajo, S., Janka, H.T., Müller, B.: Astrophys. J. Lett. **726**, L15 (2011). https://doi.org/10.1088/2041-8205/726/2/L15

Wheeler, J.C., Yi, I., Höflich, P., Wang, L.: Astrophys. J. **537**, 810 (2000). https://doi.org/10.1086/309055

White, G.L., Malin, D.F.: Nature **327**, 36 (1987). https://doi.org/10.1038/327036a0

Wilson, J.R., Mathews, G.J., Marronetti, P.: Phys. Rev. D **54**, 1317 (1996). https://doi.org/10.1103/PhysRevD.54.1317

Winteler, C., Käppeli, R., Perego, A., Arcones, A., Vasset, N., Nishimura, N., Liebendörfer, M., Thielemann, F.K.: Astrophys. J. Lett. **750**, L22 (2012). https://doi.org/10.1088/2041-8205/750/1/L22

Wongwathanarat, A., Janka, H.T., Müller, E.: Astrophys. J. Lett. **725**, L106 (2010). https://doi.org/10.1088/2041-8205/725/1/L106

Wongwathanarat, A., Janka, H.T., Müller, E.: Astron. Astrophys. **552**, A126 (2013). https://doi.org/10.1051/0004-6361/201220636

Wongwathanarat, A., Müller, E., Janka, H.T.: Astron. Astrophys. **577**, A48 (2015). https://doi.org/10.1051/0004-6361/201425025

Woosley, S.E., Bloom, J.S.: Annu. Rev. Astron. Astrophys. **44**, 507 (2006). https://doi.org/10.1146/annurev.astro.43.072103.150558

Woosley, S.E., Heger, A.: Astrophys. J. **637**, 914 (2006). https://doi.org/10.1086/498500

Woosley, S.E., Heger, A.: Phys. Rep. **442**, 269 (2007). https://doi.org/10.1016/j.physrep.2007.02.009

Woosley, S.E., Heger, A.: Astrophys. J. **810**, 34 (2015). https://doi.org/10.1088/0004-637X/810/1/34

Woosley, S.E., Weaver, T.A.: Astrophys. J. Suppl. Ser. **101**, 181 (1995). https://doi.org/10.1086/192237

Woosley, S.E., Pinto, P.A., Ensman, L.: Astrophys. J. **324**, 466 (1988). https://doi.org/10.1086/165908

Woosley, S.E., Heger, A., Weaver, T.A.: Rev. Mod. Phys. **74**, 1015 (2002). https://doi.org/10.1103/RevModPhys.74.1015

Yamada, S., Janka, H.T., Suzuki, H.: Astron. Astrophys. **344**, 533 (1999)

Yamasaki, T., Yamada, S.: Astrophys. J. **656**, 1019 (2007). https://doi.org/10.1086/510505

Yaron, O., Gal-Yam, A.: Publ. Astron. Soc. Pac. **124**, 668 (2012). https://doi.org/10.1086/666656

Yoon, S.C., Langer, N.: Astron. Astrophys. **443**, 643 (2005). https://doi.org/10.1051/0004-6361:20054030

Yoon, S.C., Gräfener, G., Vink, J.S., Kozyreva, A., Izzard, R.G.: Astron. Astrophys. **544**, L11 (2012). https://doi.org/10.1051/0004-6361/201219790

Young, E.J., Chanmugam, G.: Astrophys. J. Lett. **442**, L53 (1995). https://doi.org/10.1086/187814

Zhang, W., Woosley, S.E., Heger, A.: Astrophys. J. **679**, 639–654 (2008). https://doi.org/10.1086/526404

Chapter 2
Strongly Magnetized Pulsars: Explosive Events and Evolution

Konstantinos N. Gourgouliatos and Paolo Esposito

Abstract Well before the radio discovery of pulsars offered the first observational confirmation for their existence (Hewish et al., Nature 217:709–713, 1968), it had been suggested that neutron stars might be endowed with very strong magnetic fields of 10^{10}–10^{14} G (Hoyle et al., Nature 203:914–916, 1964; Pacini, Nature 216:567–568, 1967). It is because of their magnetic fields that these otherwise small ed inert, cooling dead stars emit radio pulses and shine in various part of the electromagnetic spectrum. But the presence of a strong magnetic field has more subtle and sometimes dramatic consequences: In the last decades of observations indeed, evidence mounted that it is likely the magnetic field that makes of an isolated neutron star what it is among the different observational manifestations in which they come. The contribution of the magnetic field to the energy budget of the neutron star can be comparable or even exceed the available kinetic energy. The most magnetised neutron stars in particular, the magnetars, exhibit an amazing assortment of explosive events, underlining the importance of their magnetic field in their lives. In this chapter we review the recent observational and theoretical achievements, which not only confirmed the importance of the magnetic field in the evolution of neutron stars, but also provide a promising unification scheme for the different observational manifestations in which they appear. We focus on the role of their magnetic field as an energy source behind their persistent emission, but also its critical role in explosive events.

K. N. Gourgouliatos (✉)
Department of Applied Mathematics, University of Leeds, Leeds, UK

Department of Mathematical Sciences, Durham University, Durham, UK
e-mail: Konstantinos.Gourgouliatos@durham.ac.uk

P. Esposito
Anton Pannekoek Institute for Astronomy, University of Amsterdam, Amsterdam, The Netherlands
e-mail: P.Esposito@uva.nl

© Springer Nature Switzerland AG 2018
L. Rezzolla et al. (eds.), *The Physics and Astrophysics of Neutron Stars*,
Astrophysics and Space Science Library 457,
https://doi.org/10.1007/978-3-319-97616-7_2

2.1 Introduction

While the vast majority of the known pulsars exhibits an exceptionally predictable behaviour, with minimal variations in their observable quantities (Hewish et al. 1968), a subset of pulsars have rapid changes in their timing properties and radiative behaviour. These pulsars have stronger magnetic fields than most radio pulsars and have been historically identified as Anomalous X-Ray Pulsars (AXPs) and Soft Gamma-ray Repeaters (SGRs). They are collectively referred to as magnetars, as it is generally believed that their strong magnetic field causes this behaviour. The magnetic field drives changes in the magnetosphere and crust, which are then reflected in their observable quantities as variations in their thermal and high energy emission and timing irregularities. Apart from explosive events, strongly magnetised pulsars have higher temperatures compared to rotation powered pulsars and are generally observable in X-rays.

The first event that has been associated to a magnetar was a burst of gamma-rays observed in 1979 and originated in an X-ray bright pulsar (Mazets et al. 1979). This flare was attributed to changes in the magnetic field structure in the neutron star causing the rapid release of energy in the form of gamma-rays (Duncan and Thompson 1992; Paczynski 1992). Historically, SGRs were discovered as hard X- and gamma-ray transients, while AXPs emerged as a class of persistent X-ray pulsars with X-ray luminosity exceeding that available from spin-down (whence the name 'anomalous'; Mereghetti and Stella 1995). Once the spin-down dipole magnetic fields of SGRs and AXPs were determined through timing measurements and deep observations excluded stellar companions, it was realised that such sources hosted exceptionally strong magnetic fields (Kouveliotou et al. 1998). The magnetic fields were recognised to be responsible for the pulsar's variations in X-ray luminosity and timing behaviour, and the magnetar model started to become more and more popular. Subsequent detections of bursting and flaring events originating from AXPs (Gavriil et al. 2002) unified the SGR and AXP classes, and enriched the available data. Over the course of the last 15 years, and following in particular the launches of Swift (2004) and Fermi (2008), the population of magnetars has more than doubled, with more than 20 confirmed sources (Olausen and Kaspi 2014).

There are a few examples of neutron stars, which, despite being observed as SGRs, have relatively weak magnetic fields inferred from timing measurements overlapping with normal rotational powered pulsars (Rea et al. 2010, 2012, 2014). On the other hand, there are some neutron stars whose magnetic field is comparable with that of magnetars, but their behaviour is that of rotation powered pulsars (Camilo et al. 2000; Gonzalez et al. 2004), with some occasional magnetar-like outbursts (Archibald et al. 2016). Thus, while the magnetic field is the key parameter behind explosive events in neutron stars, it is not the only parameter. The overlap between normal pulsars and magnetars is illustrated in Fig. 2.1, which shows the distribution of magnetic field values inferred in neutron stars from their timing parameters.

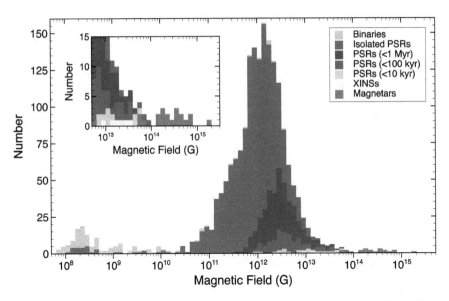

Fig. 2.1 Histogram showing the magnetic field distribution of all neutrons stars inferred from timing measurements. The insert on the top left zooms in the magnetar area. Neutron stars with magnetic fields $B > 10^{14}$ G are all magnetars. However, there is a significant overlap between normal pulsars, X-ray dim isolated neutron stars (XDINSs) and magnetars for $B \gtrsim 10^{13}$ G. Figure from Olausen and Kaspi (2014)

This observational diversity has motivated theoretical endeavours for the interpretation of the data. Several studies have focused on the role of the magnetic field in triggering outbursts and flares, either through internal magnetic field evolution or via a major reconfiguration of the external magnetospheric field (Thompson and Duncan 1995). Furthermore, the long-term evolution of the magnetic field has attracted the attention of theoretical modelling with reference to magnetar quiescent behaviour (Turolla et al. 2015; Kaspi and Beloborodov 2017).

The plan of the chapter is as follows: in Sect. 2.2, we discuss the generation of the magnetic field in strongly magnetised neutron stars; in Sect. 2.3, we focus on its evolution; we discuss the explosive events and their impact on pulsar radiative and timing profile in Sect. 2.4; in Sect. 2.5, we discuss the radiation mechanisms; in Sect. 2.6 we place strongly magnetised neutron stars in the context of the global neutron star population and we summarise and discuss the open questions and future prospects in Sect. 2.7.

2.2 Origin of the Magnetic Field

Even under the assumption that most of the magnetic flux in a strongly magnetised massive star is conserved during the process of the formation of a neutron star, it is unlikely that a newborn neutron star will have a magnetic field higher than 10^{14} G

(Spruit 2008). Magnetars typically host magnetic fields that exceed this value (but it is interesting to notice that it may be enough to give rise to magnetar-like activity in other neutron stars, see Sect. 2.6). Furthermore, the number of massive stars with high enough magnetic field ($\gtrsim 1$ kG) seems too small to account for the magnetar birth rate in the Galaxy (Spruit 2008).

Thus, more complicated processes amplifying the progenitor's seed field probably take place during the formation of a strongly magnetised neutron star. Suggestions to resolve this puzzle have focused on two main directions: either strongly magnetised neutron stars are the offspring of coalescing neutron stars or white dwarfs, or their progenitors are massive stars, which during their collapse underwent strong dynamo that amplified their magnetic field to magnetar levels.

Price and Rosswog (2006) have examined the coalescence scenario in which a double neutron star binary merges into a single neutron star producing a short gamma-ray burst (Rosswog et al. 2003). During the merger process, Kelvin–Helmholtz instability, occurring in the shear layer between the neutron stars, amplifies the field to 10^{17} G. Provided that the remnant does not collapse to a black hole, it has sufficiently strong magnetic field to power a magnetar. It is also feasible that magnetar level fields could be generated through the magneto-rotational instability in such mergers (Giacomazzo and Perna 2013).

Should magnetars form in a core-collapse supernova explosion, an amplification mechanism is required to generate such strong magnetic fields. Duncan and Thompson (1992) proposed that efficient helical dynamo action can operate if the rotation period of the proto-neutron star is ~ 1 ms, while the system may experience an ultra-long gamma-ray burst (Greiner et al. 2015). The effect of a dynamo operating during a core collapse supernova has been explored in detail by Mösta et al. (2015), who found that it can generate a mixed poloidal and toroidal field with strength of 10^{16} G. In general, a combination of toroidal and poloidal fields is favoured by simulations based on stability arguments, Fig. 2.2 (Braithwaite and Spruit 2004, 2006; Lasky et al. 2011; Ciolfi and Rezzolla 2012, 2013).

Observations of young supernova remnants associated with magnetars (Gaensler et al. 2005; Davies et al. 2009; Olausen and Kaspi 2014), support that their progenitors are massive stars. Vink and Kuiper (2006) and Martin et al. (2014) studied three X-ray supernova remnants associated with magnetars but did not find any evidence for a particularly energetic supernova. This would be expected in the case of a rapidly spinning proto-neutron star, which should transfer a large fraction of its rotational energy $E_{rot} \sim 3 \times 10^{52}(P/\text{ms})^{-2}$ erg to the ejecta (Allen and Horvath 2004; Bucciantini et al. 2007). A possible explanation for the lack of any particular signature has been provided by Dall'Osso et al. (2009), who noted that if the internal toroidal field of the proto-neutron star is $\approx 10^{16}$ G, most of its rotational energy is released through gravitational waves and therefore does not supply additional energy to the ejecta.

Both the fossil-field and the dynamo-amplification hypotheses require the progenitors to be *particularly massive* stars, with masses greater than $\sim 20 M_\odot$. For the fossil field, this is due to the positive correlation of magnetic field strength in main sequence stars with mass (Ferrario and Wickramasinghe 2008 and references

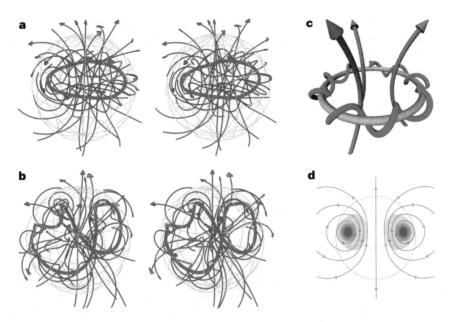

Fig. 2.2 Magnetic field structure of a neutron star in the twisted torus configuration. Stereographic view of the twisted torus configuration, when the torus is confined inside the star, panel **a**; and once it reaches the surface and develops non-axisymmetric modes, panel **b**. Schematic view of the magnetic field structure with the open field lines (red) and the ones in the torus confined inside the star (blue), panel **c**. Azimuthal average of the toroidal field (blue) and the poloidal field lines (red), panel **d**. Figure from Braithwaite and Spruit (2004)

therein). In the dynamo scenario, this is because a star of more than $\sim$20–35 $M_\odot$ star is required to produce neutron stars with rotation period sufficiently short to generate a magnetar field (Heger et al. 2005). The link between magnetars and massive stars is supported by observations of the spatial distribution of magnetars with respect to Galactic plane: their scale height is in fact smaller than that of OB stars, suggesting that magnetars originate from the most massive O stars (Olausen and Kaspi 2014). Furthermore, there are some possible associations between magnetars and clusters of massive stars. The most compelling case is that of CXOU J164710.2–455216 in the young cluster Westerlund 1 (Muno et al. 2006). Since the age of the cluster is (4 ± 1) Myr, the minimum mass for the progenitor is around 40 $M_\odot$.

Such massive progenitors in standard evolutionary models are expected to produce black holes instead of neutron stars. A viable mechanism has been suggested by Clark et al. (2014), who considered a compact binary system consisting of two massive stars. The interaction strips away the hydrogen-rich outer layers of the primary, which is driven to a Wolf–Rayet phase. In this way, the star reduces its mass through powerful stellar winds to the point where the formation of a neutron star is possible and skips the supergiant phase during which the core could lose angular momentum because of the core–envelope coupling. Clark et al. (2014) identified the possible pre-supernova companion of CXOU J164710.2–455216's progenitor in

Wd1–5, a $\sim$9-$M_\odot$ runaway star that is escaping Westerlund 1 at high velocity and has an enhanced carbon abundance that may have resulted from binary evolution. They modelled the precursor binary as a system of 41 $M_\odot$ + 35 $M_\odot$ stars with period shorter than 8 days. Although a lower mass of $\sim$17 $M_\odot$ has been inferred for the progenitor of SGR 1900+14 (Clark et al. 2008; Davies et al. 2009), suggesting that magnetars may form from stars with a wide spectrum of initial masses, binarity may nonetheless be an important ingredient for the birth of a magnetar as it helps the stellar core to maintain the angular momentum necessary for the dynamo mechanisms.

2.3   Magnetic Field Evolution

There are two main types of behaviour associated to magnetic field evolution in strongly magnetised neutron stars: Firstly, there are events caused by changes occurring in short timescales, ranging from a fraction of a second up to a few months, seen as variations in their radiative and timing properties. Secondly, there are observations that hint towards a longer-term evolution of the magnetic field on time-scales comparable to the life of a neutron star.

Short term variations of strongly magnetised neutron stars indicate that the magnetic field evolves in an impulsive manner. Magnetars are notorious for the bursting and flaring activity which is often accompanied by changes in their timing behaviour. Such changes have the characteristics of torque variations indicating alterations in the structure and strength of their magnetic field which most likely is ultimately the main responsible for their spin-down (Archibald et al. 2015). Outbursts, bursts and flares are sudden events where the luminosity of the neutron star rises suddenly in the high energy part of the spectrum (X-rays and soft gamma-rays). These events are likely to be triggered by sudden reordering of the magnetic field structure, which either deforms drastically the crust or leads to a major reconfiguration of the magnetosphere and consequently to the rapid release of energy (Thompson and Duncan 2001).

Regarding the long term evolution, there is a strong correlation between the X-ray luminosity of magnetars and the strength of their magnetic field, with the ones that host stronger magnetic fields having higher thermal L_X (Fig. 2.3). In particular, it is possible to fit their keV part of the electromagnetic spectrum with a black body radiation model implying that part of their X-ray luminosity has a thermal origin. The rapid cooling that should take place during the first few 100 years in a hot newly formed neutron star, according to thermal evolution models (Yakovlev and Pethick 2004), suggests that an extra source of energy must be present to maintain temperatures above than 5×10^6 K for several kyr. The reservoir of this energy can be the magnetic field that provides thermal energy through Ohmic dissipation (Pons et al. 2009). Finally, studies of pulsar populations suggest that there is a trend towards a magnetic field decay (Popov et al. 2010; Gullón et al. 2014, 2015); however, the multitude of parameters that appear in the population problem allows

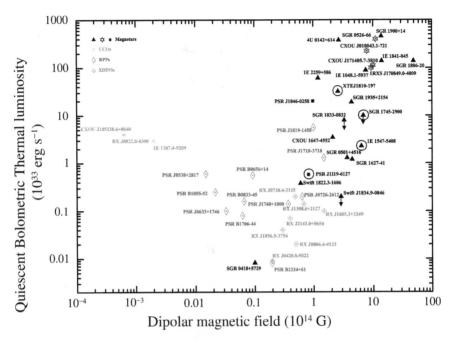

Fig. 2.3 Quiescent bolometric thermal luminosity versus the magnetic dipole strength for all X-ray pulsars with clear thermal emission. Except for central compact objects (see Sect. 2.6), there is a clear trend for X-ray pulsars with stronger dipole magnetic fields to have a higher quiescent bolometric thermal luminosity. Figure from Coti Zelati et al. (2018)

also the reproduction of the observed population without magnetic field evolution (Faucher-Giguère and Kaspi 2006).

2.3.1 Hall and Ohmic Evolution

The question of magnetic field evolution in neutron stars was theorised in the seminal work by Goldreich and Reisenegger (1992). They proposed that three basic processes drive the magnetic field evolution in the interior of a neutron star. In the crust, the magnetic field evolves because of the Hall effect, essentially the field is advected by the moving charges. In the deeper part of the crust and the core, where neutrons are abundant, ambipolar diffusion takes place, where the magnetic field and charged particles drift with respect to neutrons. Finally, Ohmic dissipation, due to finite conductivity, leads to the decay of the field in the crust.

Following the birth of a neutron star, its outer part freezes in what becomes shortly afterwards the crust. The crust consists of a highly conducting ion lattice. Free electrons carry the electric current and the magnetic field is coupled to the electron fluid. That way, the electric current is directly linked to the electron velocity which is also the fluid velocity. As long as the crust is strong enough to absorb any stresses developed by the magnetic field (Horowitz and Kadau 2009), the system remains in dynamical equilibrium. Despite the crust being in dynamical equilibrium, the magnetic field does evolve due to the advection of magnetic flux by the electron fluid. This evolution is also referred to in the literature as the electron-magnetohydrodynamics (E-MHD). Translating these into mathematical language, we obtain:

$$j = -en_e v_e,$$ (2.1)

where j is the electric current density, e the electron charge, n_e the electron number density and v_e the electron velocity. We then write Ohm's law, assuming a finite conductivity σ:

$$E = -\frac{v_e}{c} \times B + \frac{j}{\sigma},$$ (2.2)

where E is the electric field, c the speed of light, and B the magnetic field. As the electron velocities inside the neutron star are non-relativistic, the electric current is

$$j = \frac{c}{4\pi}\nabla \times B.$$ (2.3)

Combining Eqs. (2.1)–(2.3) and substituting into the magnetic induction equation, we obtain the equation that describes the evolution of the magnetic field under the Hall effect and Ohmic dissipation:

$$\frac{\partial B}{\partial t} = -\frac{c}{4\pi}\nabla \times \left[\frac{\nabla \times B}{en_e} \times B + \frac{c}{\sigma}\nabla \times B \right].$$ (2.4)

The first term in the right-hand-side describes the evolution of the magnetic field due to the Hall effect. The second term in the-right-hand-side of this equation is due to Ohmic dissipation. The time-scale of the Hall effect depends on the strength of the magnetic field, the density of the crust and the magnetic field scale height. A reasonable approximation for this length scale is the crust thickness ∼1 km and for electron number density $n_e \sim 10^{35}\,\mathrm{cm}^{-3}$, leading to a Hall time-scale $\tau_H \sim \frac{60}{B_{14}}$ kyr, where B_{14} is the strength of the magnetic field divided by 10^{14}G. Assuming a conductivity $\sigma \sim 10^{24}\,\mathrm{s}^{-1}$ the Ohmic time-scale is $\tau_O \sim 4$ Myrs. Thus, for these choices of magnetic field strength Ohmic evolution is two orders of magnitude slower than Hall.

While Hall evolution can become rather fast and depends directly on the strength of the magnetic field, it conserves magnetic field energy and merely redistributes the magnetic field. On the other hand, the Ohmic term that can lead to magnetic field decay and generation of heat, only depends on the crust physical characteristic and not on the magnetic field strength. Nevertheless, the Hall effect can lead to faster

Ohmic decay. This happens through the generation of smaller scale magnetic fields. There, stronger currents develop and decay faster.

Studies of the magnetic field evolution have focused on three main avenues for the magnetic field decay under the Hall effect and Ohmic dissipation: instability, turbulent cascades and secular evolution.

2.3.1.1 Instabilities

A magnetic field satisfying the relation

$$\nabla \times \left[\frac{\nabla \times \boldsymbol{B}}{n_e} \times \boldsymbol{B} \right] = 0 \tag{2.5}$$

is in equilibrium under the Hall effect (Gourgouliatos et al. 2013). However, such a field may be susceptible to instabilities should a small perturbation be induced. Two main types of instabilities have been identified: ideal and resistive. An example of an ideal instability is the E-MHD density shear instability (Rheinhardt et al. 2004; Wood et al. 2014; Gourgouliatos et al. 2015). This instability requires a covarying electron number density and magnetic field, whose direction is perpendicular to the electron number density gradient. Following the instability, the magnetic field structure gets severely deformed. A neutron star crust can provide appropriate conditions should the field have a strong tangential component, given that the electron number density decreases radially.

Resistive instabilities operate under the combined effect of the Hall and the Ohmic term. In this case, the instability is more likely to appear in regions where strong currents exist, especially where natural discontinuities form (Rheinhardt and Geppert 2002; Pons and Geppert 2010). This could be the inner and outer surface of the crust, where conductivity changes drastically. Moreover, simulations of magnetic field evolution show that the magnetic field may form current sheets (Vainshtein et al. 2000; Reisenegger et al. 2007). The result of this type of instability is the generation of the characteristic islands of magnetic reconnection, in a manner similar to that of the tearing instability (Gourgouliatos and Hollerbach 2016), and subsequent magnetic field decay.

2.3.1.2 Cascades

The presence of the non-linear term in Eq. (2.4) and its resemblance to the vorticity equation from fluid dynamics, has motivated studies of turbulent cascades. In this scenario, the magnetic field develops small-scale structures and the energy dissipates rapidly due to the Ohmic term. The efficiency of this mechanism depends on the shape of the power spectrum of the cascade: if the spectrum is steep, more energy will be concentrated in large scales that dissipate slower, whereas in the opposite case, it could provide a viable path for magnetic field decay. Wareing and Hollerbach (2009) reported a power-law $E_k \propto k^{-2}$ for the energy spectrum as suggested analytically in Goldreich and Reisenegger (1992), with other works reporting values of this index close to 7/3 (Biskamp et al. 1996; Cho and Lazarian 2004). Compared to normal fluid turbulence, the resulting magnetic field forms structures that remain unchanged with time (Wareing and Hollerbach 2010). This may decrease the efficiency of turbulent cascades with respect to magnetic field decay at later times.

2.3.1.3 Secular Evolution

The above mentioned approaches focus on specific characteristics of magnetic field evolution and make little use of the known properties of neutron stars. Over the past 15 years, it has been possible to simulate the magnetic field evolution making realistic assumptions for the density and conductivity of the crust in the appropriate spherical geometry. Moreover, compared to the studies of instabilities, which require the simulations to start from a state of equilibrium, studies of secular evolution are not subject to this constraint. After all, there is no reason to expect that the magnetic field of a newborn neutron star satisfies Eq. (2.5). Given the complicated processes taking place during the formation of a neutron star, a wide range of initial conditions has been explored.

Several works have explored the evolution of the magnetic field using axially symmetric simulations. Hollerbach and Rüdiger (2002) demonstrated that the field undergoes helicoidal oscillations with the energy being transferred between the various modes and also found that the magnetic field can develop current sheets because of the Hall evolution (Hollerbach and Rüdiger 2004). Pons and Geppert (2007) found that the magnetic field undergoes a rapid decay over the first 10^4 year in a strongly magnetized neutron star, and later it adopts a quasi-equilibrium state. This result was confirmed by other various simulations (Kojima and Kisaka 2012; Gourgouliatos and Cumming 2014b). This quasi-equilibrium corresponds to a Hall-attractor, a state where the electrons isorotate with the magnetic field, and is independent of the initial choice of the magnetic field (Gourgouliatos and Cumming 2014a; Marchant et al. 2014; Igoshev and Popov 2015).

Three-dimensional simulations showed that the magnetic field is susceptible to instabilities. A predominantly poloidal magnetic field remains axisymmetric (Wood and Hollerbach 2015), however, if a toroidal field is included, instabilities are induced and non-axisymmetric structure appears, see Fig. 2.4. This leads to an even faster magnetic field decay, especially during the first few 10^4 years of the life of a neutron star. During this stage, the structure of the magnetic field changes drastically. If a very strong toroidal field is included then spots of very strong magnetic field form, as it has been noted in the axisymmetric case (Geppert and Viganò 2014), with the additional feature that that magnetic dipole axis drifting with respect to the surface of the neutron star in the three dimensional study of this setup (Gourgouliatos and Hollerbach 2017).

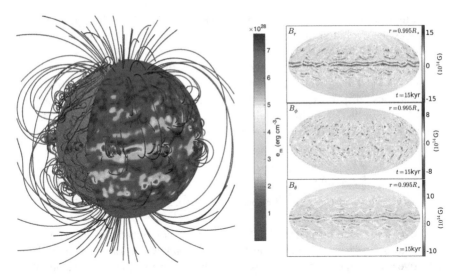

Fig. 2.4 The magnetic field structure from a 3-D simulation of the Hall drift in a neutron star 15 kyr after its birth. The initial conditions contained an axisymmetric poloidal dipole and toroidal quadrupole field. Left: Magnetic field lines in red, and magnetic energy density on the surface in colour. Right: The radial, azimuthal and meridional components of the magnetic field (top to bottom) at $r = 0.995R_*$, where R_* is the neutron star radius. The development of non-axisymmetric structures and patches of strong magnetic field is evident. Figure from Gourgouliatos et al. (2016)

2.3.2 Magnetothermal Evolution

As high temperatures correlate with strong magnetic fields in magnetars, it is crucial to quantify their relation. Temperature and magnetic field evolution are interweaved in the following ways. Magnetic field is supported by electric current whose decay provides Joule heating increasing the temperature of the star. Moreover, it affects the transport of heat. Heat can be mainly transferred either through electrons or phonons. While phonons propagate isotropically in the crust, electrons can move predominantly along the direction of the magnetic field, thus in strongly magnetised neutron stars heat will be mostly transferred along magnetic field lines. Finally, electric and thermal conductivity are functions of temperature, therefore, as temperature changes these parameters evolve as well. Consequently, magnetic field evolution is affected, too. Essentially, the process described above can be summarised in the following equation for temperature (T) evolution:

$$c_V \frac{\partial T}{\partial t} - \nabla \cdot \left[\hat{\kappa} \cdot \nabla T \right] = Q_\nu + Q_h , \qquad (2.6)$$

where c_V is the volumetric heat capacity, $\hat{\kappa}$ is the thermal conductivity tensor, Q_ν is the energy emitted by neutrinos per unit volume and Q_h is the Joule heating, resulting from the magnetic field evolution (Pons et al. 2009). The equation above describes the evolution of the system in a flat space-time, the general relativistic modification of the equations can be found in Viganò et al. (2013).

In general, the thermal conductivity tensor has the following components $\hat{\kappa} = \hat{\kappa}_e + \hat{\kappa}_n + \hat{\kappa}_p + \hat{\kappa}_{ph}$, which are due to conduction by electrons, neutrons, protons and phonons, respectively. In the strong magnetic field regime, the dominant contribution is that of electrons, which flow along magnetic field lines (Yakovlev and Urpin 1981; Aguilera et al. 2008a). The expression for this term is

$$\hat{\kappa}_e = \kappa_e^\perp \left(\hat{I} + (\omega_B \tau)^2 \begin{pmatrix} b_{rr} & b_{r\theta} & b_{r\phi} \\ b_{r\theta} & b_{\theta\theta} & b_{\theta\phi} \\ b_{r\phi} & b_{\theta\phi} & b_{\phi\phi} \end{pmatrix} + \omega_B \tau \begin{pmatrix} 0 & b_\phi & -b_\theta \\ -b_\phi & 0 & b_r \\ b_\theta & -b_r & 0 \end{pmatrix} \right), \quad (2.7)$$

where $\hat{I}$ is the identity matrix, b_r, b_θ, b_ϕ are the components of the unit vector projected on the magnetic field, and $b_{ij} = b_i b_j$, $\omega_B = eB/(m_e^* c)$ with m_e^* is the effective mass of the electron and τ is the electron relaxation time. The values of $\kappa_e^\perp$ depends on the density and temperature of the crust, see figure 5 in Aguilera et al. (2008a).

Given the complexity of this problem, it has been tackled in stages. First, the question of heat transport in a magnetised neutron star was addressed by Geppert et al. (2004). They found that an axially symmetric poloidal magnetic field reaching the core leads to a practically isotropic surface temperature, provided its strength is below 10^{15} G. If the field does not reach the core however, the surface will be significantly hotter at the poles compared to the equator. The inclusion of a toroidal field (Geppert et al. 2006), effectively blankets the heat from escaping towards the equator and channels it even more efficiently towards the poles. When Joule heating is taken into account (Aguilera et al. 2008a,b), the dissipated magnetic field keeps the star warm for a longer time and affects the cooling history of the star both in the early neutrino phase and later on. In the work by Pons et al. (2009), comparisons against individual strongly magnetised neutron stars were made favouring the hypothesis that energy stored in the magnetic field is sufficient to heat the neutron star.

Viganò et al. (2013) addressed the full problem taking into account all three channels of magnetothermal interaction (heat transport, Joule heating and thermal feedback), assuming axial symmetry, see Fig. 2.5. In that work, Eqs. (2.4) and (2.6) are solved simultaneously with electric conductivity being a function of temperature and thermal conductivity depending on the magnetic field and evolving in time. Their work further supported the unification between magnetars and X-ray dim isolated neutron stars. Evolutionary models were provided, which could account for the diversity of magnetars, high magnetic field radio pulsars, and isolated nearby neutron stars varying only their initial magnetic field, mass and envelope composition.

From a different perspective, temperature gradients can lead to the generation or amplification of magnetic fields in neutron stars through thermoelectric effects

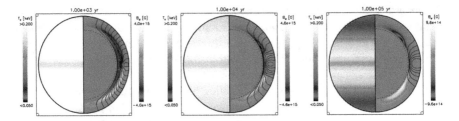

Fig. 2.5 Temperature profile (left side) and magnetic field structure (right side) of a neutron star at 1, 10 and 100 kyr. The initial poloidal field generates a strong toroidal field, leading to non-linear evolution and concentration of the poloidal field towards the equator. Evidently the temperature is higher near the equator compared to the poles. Figure from Viganò et al. (2013)

(Urpin and Yakovlev 1980). This could take place within the first few thousand years of the neutron star life and the magnetic field could reach strengths of 10^{14} G (Blandford et al. 1983; Urpin et al. 1986). However, it has been argued that thermoelectric amplification of the magnetic field saturates at strengths of 10^{11} G, due to non-linear effects (Wiebicke and Geppert 1992). Nevertheless, a strong temperature gradient, which could plausibly appear near the polar cap of a pulsar, may perpetuate a strong toroidal field in this region (Geppert 2017).

2.3.3 Core Magnetic Field

Moving deeper into the star, the abundance of neutrons increases and the material becomes superconductive and superfluid. This difference in composition and physical properties determines the evolutionary mechanism of the magnetic field. A detailed discussion of the processes taking place in the core regarding superconductivity and superfluidity can be found in Chap. 8 of this volume. Here, we briefly discuss the effects in the core that have an impact in the magnetic field global evolution. The magnetic field in this part of the star evolves due to ambipolar diffusion (Goldreich and Reisenegger 1992; Shalybkov and Urpin 1995; Hoyos et al. 2010; Glampedakis et al. 2011; Passamonti et al. 2017; Castillo et al. 2017), which can be visualised as the interaction of the charged particles with the neutral ones. Timescales for ambipolar diffusion compare with magnetar ages ($\sim 10^4$ year) for strong magnetic solenoidal fields, thus they affect the magnetic field evolution in strongly magnetized neutron stars, whereas they become much longer for the irrotational component, reaching or even exceeding the Hubble time even for strongly magnetised neutron star (Goldreich and Reisenegger 1992).

The expulsion of magnetic flux from the core (Elfritz et al. 2016) suggests that strong magnetic fields residing at the core of neutron stars may lead to much longer survival of the magnetic field of a neutron star. Recently, Bransgrove et al. (2017) studied the axially symmetric evolution of the crustal magnetic field allowing it to exchange twist with the core and accounting for the elastic deformation in the crust.

Their results agree with the established picture of the crustal magnetic field from axially symmetric simulations, with much longer timescales though. In addition to that, they found that the superfluid flux drift along with Ohmic decay may deplete the neutron star from its magnetic field within 150 Myr, if it is hot ($T \sim 2 \times 10^8$ K), but at a much slower pace for cold neutron stars. Finally, the outward motion of superfluid vortices in a young rapidly rotating highly magnetised pulsar may expel the flux from the core, provided the magnetic field is $\lesssim 10^{13}$ G. Consequently, the timescales quoted at the magneto-thermal models considering only the evolution of the crustal magnetic field need to be revised.

2.3.4 Magnetospheric Field

The magnetosphere of strongly magnetised neutron stars is the site of much of the observed activity. In normal pulsars the magnetospheric field may range from a rapidly rotating dipole in vacuum (Deutsch 1955), to a plasma-rich corotating magnetosphere (Goldreich and Julian 1969).

Strongly magnetized neutron stars, however, have some critical differences. Firstly, they are typically slow rotators, consequently the effects of corotation are secondary to the overall evolution. Secondly, their magnetic fields are most-likely non dipolar and highly twisted, thus in the inner part of the magnetosphere evolution would be dominated by the multipoles and twisted flux ropes. Finally, strong magnetic fields along with their complex structure give rise to substantially distinct behaviour compared to normal pulsars: Magnetohydrodynamical instabilities of twisted fields can be related to giant flares. Indeed, in the case of the giant-flare in magnetar SGR 1806–20, the initial rising time is in the scale of milliseconds (Palmer et al. 2005), thus the evolution is in the range of relativistic magnetohydrodynamics. Gradual untwisting of the magnetic field lines and formation of j-bundles (bundles of magnetic field lines with $j = (4\pi/c)\nabla \times \boldsymbol{B} \neq 0$) and cavities, where no current forms ($\nabla \times \boldsymbol{B} = 0$), impact the non-thermal emission of strongly magnetised neutron star (Beloborodov 2009).

We will see in more detail the role of the magnetospheric field in Sect. 2.4, where explosive phenomena are discussed.

2.3.5 Overview of Magnetic Field Evolution

Studies of magnetic field evolution have focused into the different paths outlined above. This approach reflects more a mathematical and computation convenience in modelling methods rather than a physical reality. The evolution of the magnetic field in a neutron star can be due to a combination of instabilities and cascades, interaction of magnetic field from different parts of the star and changes in temperature. While it has been possible to provide estimations of magnetic field, temperature and physical parameters of the neutron star and identify which processes will be dominant in each area of the parameter space, a detailed quantification is still missing. It is one of the key challenges of the community to provide a global study of the magnetic field evolution in the neutron star by combining the evolution in the different parts of the star, while accounting for as many as possible different physical processes (Andersson et al. 2017; Gusakov et al. 2017).

2.4 Explosive Events

Magnetars show variations of their radiative emission on almost all intensity and time scales, but their behaviour is generally outlined in three main categories: bursts, outbursts, and giant flares. The signature 'short' or 'SGR-like' bursts of hard-X/soft-γ ray photons have duration from a few milliseconds to a few seconds, and luminosity spanning from $\approx 10^{36}$ to 10^{43} erg s^{-1} (Aptekar et al. 2001; Götz et al. 2006; Israel et al. 2008; van der Horst et al. 2012). Their spectra are generally well described in the 1–100 keV range by a double-blackbody model with temperatures between $kT \sim 2$ and ~ 12 keV, with a bimodal distribution behaviour in the kT–R^2 plane, suggesting two emitting regions: a cold and larger one and a hot and smaller one (Israel et al. 2008); physical interpretations of this behaviour are however unsatisfactory (see e.g. Turolla et al. 2015 and references therein). Short bursts, in particular the brightest ones (sometimes called 'intermediate flares'), can be followed by long-lived tails (seconds to minutes or hours) that may arise from a burst-induced heating of a region on the surface of the neutron star or, at least in some instances, from dust-scattering of the burst emission (Lenters et al. 2003; Esposito et al. 2007; Göğüş et al. 2011; Pintore et al. 2017). Even in the same source, bursts can occur sporadically and sparsely or come in 'storms', with hundreds or thousands of events clustered in days or weeks. Since they are bright enough to be detectable almost anywhere in the Galaxy by large field-of-view, sensitive X-ray monitors such as those onboard *CGRO, Swift, INTEGRAL* and *Fermi*, short bursts have been instrumental in the discovery of most of the known magnetars (see e.g. Olausen and Kaspi 2014).

The short bursts most often announce that a source has entered an active phase commonly referred to as an outburst. During these events, the luminosity suddenly increases up to a factor $\sim 10^3$, generally showing a spectral hardening, and then decays and softens over a time scale from several weeks to months/years (e.g. Rea and Esposito 2011). Outbursts are usually accompanied also by changes in the spin-down rate, in the pulse profile/pulsed fraction and other timing anomalies such as higher-than-usual timing noise level and glitches; outbursts often affect also the radiative properties at wavelengths other than soft X-ray with enhanced hard emission and changes in the infrared/optical counterparts; remarkably, all the magnetars that have shown pulsed radio emission were detected at radio frequencies during an outburst (Camilo et al. 2006, 2007a; Anderson et al. 2012; Eatough et al. 2013; Rea et al. 2013a). Some magnetars have not displayed an outburst over decades, while others have undergone multiple episodes; some sources have low flux when in quiescence and were noticed or detected for the first time only after the onset of an outburst that broke their hibernation: they are often referred to as 'transient' magnetars. This behaviour makes it very difficult to estimate the total Galactic magnetar population.

The pinnacle of the magnetar activity is represented by the giant flares, in which $\approx 10^{44}$–10^{46} erg can be released in a fraction of a second. They are so bright to have detectable effects on the Earth's magnetic field (Mandea and Balasis 2006) and ionosphere (Inan et al. 1999, 2007), and possible extragalactic giant flares could appear at Earth as short gamma-ray bursts (Hurley et al. 2005). The extreme properties of the giant flares were what prompted the idea of neutron stars endowed with an ultra-strong magnetic field and then the magnetar model (Paczynski 1992; Thompson and Duncan 1995) and they still both provide some of the strongest evidences for the existence of such objects and represent the benchmark against which alternative models need to be tested. Only three such events have been observed so far, all from different sources and—curiously—each from one of the 'historical' SGRs: on 1979 March 5 from SGR 0526–66 in the Large Magellanic Cloud (Mazets et al. 1979), on 1998 August 27 from SGR 1900+14 (Hurley et al. 1999), and on 2004 December 27 from SGR 1806–20 (Hurley et al. 2005). All three events started with a $\sim$0.1–0.2-s flash that reached a luminosity of $\gtrsim 10^{45}$–10^{47} erg s^{-1}, followed by a long-lived (few minutes) tail modulated at the spin period of the neutron star. The spectra of the initial spikes were very hard, an optically-thin thermal bremsstrahlung with characteristic temperatures of several hundreds of keV, while the tails had typical temperatures of a few tens of keV, further softening as the flux decayed. This behaviour is very broadly interpreted in terms of a fraction of the realised energy that escapes directly in the spike, while the remaining energy slowly leaks from a pair–plasma fireball trapped by the neutron star magnetic field. In this respect, it is interesting to note that while the peak luminosity of the giant flare of SGR 1806–20 was ≈ 100 time higher than in the first two, the energy emitted in the pulsating afterglows was approximately the same ($\approx 10^{44}$ erg), suggesting a similar magnetic field (of $\approx 10^{14}$ G in the magnetosphere) in the three SGRs. Quasi-periodic oscillations (QPOs) of various main frequencies and durations were also detected in the tails of the SGR 1900+14 and SGR 1806–20's giant flares (and possibly also in the event from SGR 0526–66); although the details remain to be worked out, they were likely associated to 'seismic' oscillations in the stars produced by the flare, potentially offering glimpses into the structure of the neutron star and its magnetic field (see Turolla et al. 2015 and references therein), but also providing another piece of evidence in favour of the presence of a super-strong magnetic field. In fact, Vietri et al. (2007) observed that the fastest QPOs imply a luminosity variation well above the Cavallo–Fabian–Rees luminosity variability limit for a compact source (Fabian 1979), and that only a $\approx 10^{15}$ G magnetic field at the star surface could make this possible.

2.4.1 Bursts and Outbursts

The amounts of energy released in bursts and outbursts imply that a sudden, but not global, change takes place in the neutron star. The prevailing explanation behind such events is that a rapid change in the magnetic field structure allows the release of large amounts of the previously stored energy. The main challenges of theoretical modelling in bursts and outbursts are the process behind the efficient conversion of magnetic energy into heat, the triggering mechanism, and, eventually, the transportation of heat to the surface.

Let us assume that a burst is powered by magnetic energy stored inside the star. The energy of the magnetic field in the volume of star V participating in the event is

$$E_{mag} = \frac{B^2}{8\pi} V .$$
(2.8)

If part of this magnetic energy is channeled to the burst with an energy efficiency factor η, we obtain the following expression for the energy of the burst:

$$E_b = 4 \times 10^{40} \, \eta B_{15}^2 S_4^3 \text{ erg,}$$
(2.9)

where B_{15} is the magnetic field scaled to 10^{15} G and S_4 is the size of the region where the burst takes place scaled to 10^4 cm. In the starquake model, Thompson and Duncan (1995) suggested that the crust fails once the shear stress exceeds a critical value (Thompson et al. 2017). Molecular simulations of the crystal lattice of the crust found that the breaking strain of the crust is 0.1 (Horowitz and Kadau 2009; Chugunov and Horowitz 2010). This implies that a local magnetic field of 2.4×10^{15} G will be sufficient to break the crust (Lander et al. 2015). Closer to the surface of the crust, magnetic fields as low as $\sim 10^{14}$ G can lead to crust breaking due to the smaller density (Gourgouliatos and Cumming 2015). The formation of faults, as in terrestrial materials, is unlikely, given the extreme pressure of the crust (Jones 2003). Moreover, the anisotropy induced by the magnetic field disfavours a deformation normal to the magnetic field lines (Levin and Lyutikov 2012), but allows it to be parallel to them. Given these constraints, a likely mechanism for the development of the burst is that of thermoplastic instability (Beloborodov and Levin 2014). In this model, the stressed material forms a wavefront that flows plastically, similarly to the deflagration of a burning material. The plastic flows may transport magnetic flux accelerating the magnetic field evolution (Lander 2016).

The models discussed above describe the initiation of the bursting event once the magnetic field stresses have led to strains close to the breaking one. In the long run, magnetic field should reach this stage at some point during its evolution. This scenario has been explored by axially symmetric simulations of magnetic field evolution (Perna and Pons 2011; Pons and Perna 2011). Tracing the building up of magnetic stresses while the field evolves, one can identify when a critical value is reached. At this instance a burst will occur, releasing the magnetic energy in the affected area. It was found that bursts are expected to be more frequent and energetic in young magnetars, whereas older neutron stars, even if they sustained a strong magnetic field, were less active. Li et al. (2016) explored the interaction between Hall waves and crust failures through a 1-D model, considering thermal evolution, plastic flow and neutrino cooling. They found that this formation of crust failures leads to avalanches: Hall waves originating in the failure accelerate the dynamics.

Next, the propagation of the burst energy to the surface and its eventual radiation need to be considered. To model that, it is assumed that a sudden deposition of thermal energy occurs inside the crust, which then thermally relaxes. It then appears as a flux enhancement event accompanying the outburst (Brown and Cumming 2009; Scholz et al. 2012; Rea et al. 2012; Scholz et al. 2014). The process followed in these models is similar to the one outlined in Sect. 2.3.2; however, due to the very short timescales involved ($\sim 10^2$ days), the magnetic field does not evolve and only thermal evolution is accounted for. These models can reproduce the observed light curves with good accuracy for few hundred days following the event.

The sudden crustal motions taking place during a starquake also affect the structure of the magnetosphere. The magnetic field lines that are anchored to the displaced region will become twisted with currents flowing along them. These currents will form j-bundles (Beloborodov 2009), surrounded by a cavity region where the magnetic field is potential. In this picture, a hotspot forms at the footprints of the j-bundle, where particles are accelerated and bombard the surface of the star (Beloborodov and Thompson 2007). Eventually, as the electric current of the bundle gets dissipated, the hotspot shrinks and its luminosity gradually fades out.

2.4.2   Giant Flares

While bursts and outbursts are common in all magnetars, there have been three cataclysmic events that clearly stand out from any other form of magnetar activity. The enormous amount of energy emitted in magnetar giant flares implies that a global event takes place, involving a major reconfiguration of the magnetic field. Two main mechanisms for giant flare triggering have been proposed, one internal and one external. In the internal scenario, magnetic energy is stored inside the neutron stars, most likely in the form of stresses in the crust-core boundary (Thompson and Duncan 1995, 2001). Once the crust fails, there is nothing to balance these stresses. The stored energy is then released to the outer parts of the star. This energy mainly powers the original spike of the giant flare. In the external trigger scenario, the magnetospheric field is slowly twisted because of the evolution of the internal magnetic field, while this twist is supported by strong currents. Provided that the dissipation rate of the electric current is slower than the twist rate, energy will be stored in the magnetosphere. Strongly twisted magnetic fields are likely to become unstable, either through ideal or resistive MHD instabilities, accompanied by rapid release of magnetic energy. This scenario is based on the process leading to the eruption of a solar flare (Uzdensky 2002; Lyutikov 2003; Gill and Heyl 2010).

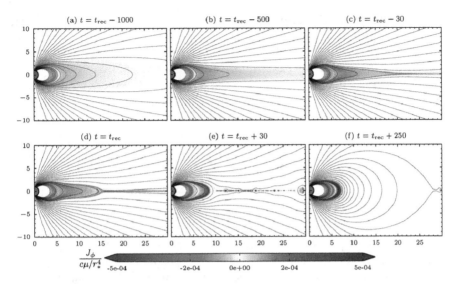

Fig. 2.6 The formation of a current sheet and onset of reconnection. A polar cap of $\theta = 0.15\pi$ is sheared at a rate $2.5 \times c/R_*$, where R_* is the neutron star radius. The toroidal current is plotted in colour and the poloidal field lines are shown in black. As the polar cap is sheared a strong toroidal current builds up, panels (**a**), (**b**) and (**c**). The system then becomes unstable and reconnection occurs at the equator, panel (**d**). Once the current sheet forms plasmoids appear (**e**) and eventually the field relaxes to the untwisted stage (**f**). Figure from Parfrey et al. (2013)

In the external scenario, the magnetospheric field, prior to the flare, is modelled as a series of equilibria corresponding to force-free configurations subject to the slowly changing boundary conditions. A force-free magnetic field satisfies the condition $(\nabla \times \mathbf{B}) \times \mathbf{B} = 0$ implying that the current flows along magnetic field lines. The forces due to the thermal pressure are negligible compared to the Lorentz force, given the strength of the magnetic field. As the field is getting twisted, timing changes can be anticipated: a twisted magnetic field will exert a stronger torque to the magnetar leading to a more efficient spin-down shortly before the flare, see Fig. 2.6 (Thompson et al. 2002; Parfrey et al. 2012, 2013; Akgün et al. 2016). Analytical and semi-analytical studies have demonstrated that the field cannot be twisted indefinitely, due to the strong current that will develop. These currents are prone to ideal or resistive instabilities initiating a flare (Priest et al. 1989; Lynden-Bell and Boily 1994; Gourgouliatos and Vlahakis 2010; Lynden-Bell and Moffatt 2015; Akgün et al. 2017). At this stage, explosive reconnection, driven by the tearing instability (Komissarov et al. 2007), is a possible avenue for the rapid conversion of the magnetic energy into thermal energy powering the flare. In the mean time, hot plasma is concentrated in the closed magnetic field lines, which form a "trapped fireball". The energy from the trapped fireball then gets released within a few tens of seconds which is seen as a tail in the light curve. This fireball corotates with the magnetar, and its luminosity is modulated by the period of the star. As the external magnetic field changes its topology rapidly, it launches Alfvén waves that propagate throughout the magnetosphere. The waves, once reflected onto the star, transfer energy back to the crust. This leads to the excitation of modes, which give rise to the quasi periodic oscillations (Israel et al. 2005). Depending on whether the crust or the core participates in this oscillatory motion, the structure of the external magnetic field different profiles and frequencies are expected to be excited and information on the internal structure of the magnetar can be extracted (Duncan 1998; Watts and Strohmayer 2007; Levin 2007; Steiner and Watts 2009; Cerdá-Durán et al. 2009; Colaiuda et al. 2009; van Hoven and Levin 2011; Gabler et al. 2011, 2013; Passamonti and Lander 2013; Li and Beloborodov 2015).

2.4.3 Timing Behaviour

Similar to normal pulsars, strongly magnetised ones exhibit timing irregularities. These irregularities may appear as sudden changes in frequency, the so-called glitches (Kaspi et al. 2000, 2003; Kaspi and Gavriil 2003; Dib et al. 2008; Dib and Kaspi 2014; Şaşmaz Muş et al. 2014; Antonopoulou et al. 2015), which are also present in normal rotating powered radio pulsars (Anderson and Itoh 1975; Hobbs et al. 2004). Magnetars also exhibit slower torque variations (Gavriil and Kaspi 2004; Camilo et al. 2007b; Archibald et al. 2015). In this case, the torque changes within a factor of a few, usually in conjunction with an outburst, in timescales of 100 days. In general the population of strongly magnetised neutron stars exhibit higher levels of timing noise compared to normal pulsars (Hobbs et al. 2010; Esposito et al. 2011a; Tsang and Gourgouliatos 2013).

Glitches in magnetars can be either radiatively loud or silent (Dib and Kaspi 2014). However, since except for the changes in radiative properties, both silent and loud glitches are otherwise similar, the same mechanism is expected to operate and most likely it is related to the evolution of the superfluid material (Haskell and Melatos 2015). Rare anti-glitch events have been observed in magnetars (Archibald et al. 2013; Şaşmaz Muş et al. 2014). Unlike normal glitches, an anti-glitch is a sudden spin-down which is not compatible with the established theory of pulsar glitches. Most probably, the evolution of the internal magnetic field creates spinning slower than the rest of the star (Mastrano et al. 2015). Sudden exchange of angular momentum can lead to this type of event.

Torque variations are more likely to be related to the external magnetic field twisting. As discussed in the giant-flare external model, twist can slowly build up in the magnetosphere (Beloborodov 2009). The twisted magnetic field lines are capable of exerting stronger torque, as they reach greater altitudes, increasing the number of the ones crossing the light cylinder (Parfrey et al. 2013). In this scenario, prior to some radiative event, a long term increase on the torque occurs.

2.5   Persistent Emission and Spectral Features

The persistent emission of magnetars across the electromagnetic spectrum is discussed in depth in the recent review by Turolla et al. (2015). In general, in X-rays below 10 keV, where the largest amount of data is available, good phenomenological fits are provided by a blackbody with $kT \approx 0.5$ keV and a power law with photon index $\Gamma \approx$ 2–4; sometimes a second blackbody is preferable to the power law (e.g. Tiengo et al. 2008). A non thermal-like component seems to be always present when the sources are detected at hard X-rays (up to ~150–200 kev).

The soft X-ray spectral shape is broadly interpreted in the context of a twisted magnetosphere that supports electric currents. The (first) blackbody arises from the thermal emission from the surface of the neutron star, which also provides seed photons for resonant cyclotron scattering involving the magnetospheric charges, producing in this way the second, harder component (since the charged particles populate vast regions of the magnetosphere, with different magnetic field intensities, the scattering produces a hard tail instead than a narrow line). This basic idea is also successful in explaining the correlation in magnetars between spectral hardness and intensity of the spin down first noticed by Marsden and White (2001): the stronger the twist, the larger the spin down rate and the density of charges in the magnetosphere. Starting from this interpretation of the X-ray emission, many efforts to provide more physically motivated models for the spectra of magnetars have been undertaken; we refer again to Turolla et al. (2015) for a comprehensive overview of the situation.

As we have seen, several pieces of evidence point to a twisted magnetic field in magnetars, including the above mentioned power-law-like tail, when interpreted in terms of resonant cyclotron scattering of thermal photons from the neutron-star surface, and periods of enhanced torque, and even more observations indicate the presence of particularly strong magnetic fields. The most direct probes of magnetic fields in neutron stars, however, are the spectral features that arise from resonant cyclotron scattering by magnetospheric particles. For magnetic fields above 10^{14} G, proton cyclotron lines are in the classic soft X-ray range (0.1–10 keV): the cyclotron energy for a particle of mass m and charge e is

$$E_{\text{cycl}} = \frac{11.6}{1+z} \left(\frac{m_e}{m}\right) B_{12} \text{ keV},$$

where $z \approx 0.8$ is the gravitational redshift, m_e is the mass of the electron, and B_{12} is the magnetic field in units of 10^{12} G. Despite several claims of possible detection in both quiescent and burst spectra, no clear spectral feature was observed in a magnetar until recently, when Tiengo et al. (2013) reported the discovery of a phase-dependent absorption feature in the 'low-magnetic-field magnetar' (see Sect. 2.6) SGR 0418+5729. At the time, the source was particularly bright after a short burst-active episode (Esposito et al. 2010). The feature was more noticeable in

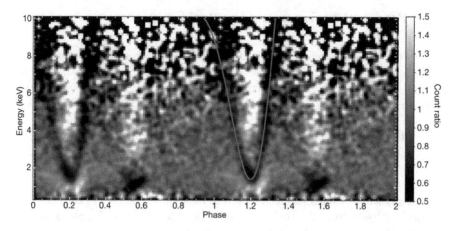

Fig. 2.7 The absorption feature in the XMM–Newton data of SGR 0418+5729 (from Tiengo et al. 2013). The normalised energy versus phase image was obtained by binning the source photons into 100 phase bins and 100-eV width energy channels and normalising the counts first by the phase-averaged energy spectrum and then by the pulse profile (normalised to the average count rate). The V-shaped absorption feature is apparent. For one of the two displayed cycles, the red line shows the results from the proton cyclotron model in Tiengo et al. (2013)

a deep XMM–Newton observation, but was confirmed also by data collected with RossiXTE and Swift in the first months after the outset of the outburst (Tiengo et al. 2013).

The phase-resolved spectroscopy and analysis of the XMM–Newton data showed that the energy of the line changes between ∼1 and 5 keV or more, in approximately one-fifth of the rotation cycle (Fig. 2.7). The dependence of the line energy on the phase rules out atomic transitions as the origin of the source, while an electron cyclotron feature seems unlikely because, considered the dipolar field of the source derived from the rotational parameters ($B \sim 6 \times 10^{12}$ G), the electrons should be somehow trapped in a small volume at a few stellar radii from the neutron star. Tiengo et al. (2013) proposed that the absorption feature is a cyclotron line from thermal photons crossing protons localised close to the surface of the star, in a magnetic loop with field of the order of 10^{14}–10^{15} G. If this interpretation is correct, the feature brings evidence of the presence of strong nondipolar magnetic field components in magnetars, as expected in the magnetar model (Thompson and Duncan 1995, 2001).

A similar, albeit less conspicuous, feature was reported for another magnetar with a relatively low dipole magnetic field [(1–3) $\times$ 10^{13} G], Swift J1822.3–1606 (Rodríguez Castillo et al. 2016). It is interesting to note that SGR 0418+5729 and Swift J1822.3–1606 are the two magnetars with the smallest inferred dipole fields (Olausen and Kaspi 2014): if their spectral features are indeed proton cyclotron lines, it may be the great disparity between the large-scale dipole field and the stronger localised components that makes the lines to stand out in these sources.

2.6 Magnetars and the Other NSs: The Top–Right Part of the $P–\dot{P}$ Diagram

When only a dozen or so magnetars were known, they all could be found tightly packed and aloof in the upper right part of the $P–\dot{P}$ diagram for the population of non-accreting pulsars (Fig. 2.8). Since the magnetic field is held to be the ultimate responsible of the activity of magnetars, it was perhaps natural at that time to identify magnetars as the pulsars with the highest inferred dipole magnetic fields. Traditionally, the magnetic field threshold was set at the electron quantum magnetic field $B_Q = m_e^2 c^2/(\hbar e) \simeq 4.4 \times 10^{13}$ G—even though this value has no direct physical implications for pulsars.

The first clue that things might be more complicated was the discovery of a handful of 'high-magnetic-field' (high-B, $B \gtrsim B_Q$) pulsars overlapping with magnetars in the $P–\dot{P}$ diagram (e.g. Kaspi and McLaughlin 2005; Kaspi 2010). No additional source of power other than rotational energy loss was required to explain their properties, which where similar to those of lower-magnetic-field 'ordinary' pulsars of comparable age, apart from a somewhat higher-than-expected blackbody temperature in some objects (e.g. Olausen et al. 2013), and they did not show

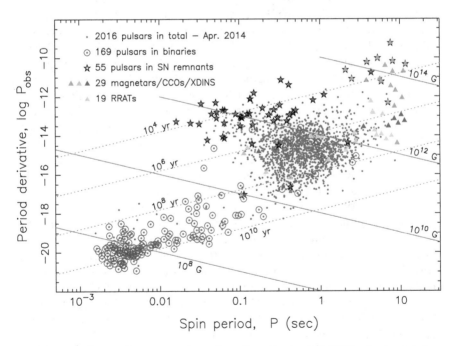

Fig. 2.8 $P–\dot{P}$ diagram for non-accreting pulsars (from Tauris et al. 2015 and using data from Manchester et al. 2005). Lines of constant characteristic age (dotted), $P/(2\dot{P})$, and dipole spin-down luminosity, are drawn in grey, while the main observational manifestations of pulsars are plotted with different symbols

any bursting/outbursting activity over the ~25 year during which they had been observed. The possibility of a connection was strengthened by the discovery of pulsed radio emission from magnetars (Camilo et al. 2006, 2007a; Levin et al. 2010), but the divides between the apparently different classes of isolated neutron stars really started to get blurrer and blurrer when magnetar-like activity began to trickle down in the diagram from the canonical magnetar area.

Firstly, the young, high-B, radio-quiet X-ray pulsar PSR J1846–0258 (period: 0.3 s, $B \simeq 5 \times 10^{13}$ G) in the supernova remnant Kes 75 exhibited a few-week-long X-ray outburst with several short (SGR-like) bursts in 2006 (Gavriil et al. 2008). Then, in 2009, the first of the so-called 'low-B magnetars' (Turolla and Esposito 2013) was discovered: SGR 0418+5729 (Esposito et al. 2010; Rea et al. 2010; Turolla et al. 2011; Rea et al. 2013b). This source, in spite of having an inferred dipole magnetic field as low as ~6×10^{12} G (Rea et al. 2013b, but see also Tiengo et al. 2013 and Sect. 2.5) has showed all the hallmarks of magnetars, demonstrating that an ultra-strong surface dipolar magnetic field is not necessary for magnetar-like activity. In 2016, another young pulsar PSR J1119–6127 (period of 0.4 s and $B \simeq 4 \times 10^{13}$ G) underwent a magnetar-like outburst, and this time it was a *radio* pulsar (Göğüş et al. 2016; Archibald et al. 2016). Interestingly, the radio emission was temporary off during the event (Burgay et al. 2016a,b; Majid et al. 2017). Very recently, Archibald et al. (2017), using simultaneous X-ray (with XMM–Newton and NuSTAR) and radio (at the Parkes radio telescope) of PSR J1119–6127, found that the coherent radio emission was off in coincidence with the short X-ray bursts and recovered on a time scale of ~1 min. They tentatively explain this behaviour suggesting that the positron-electron plasma produced in bursts can engulf the gap region where the radio-emitting particles are accelerated, disrupt the acceleration mechanism and, as a consequence, switch off the coherent radio emission.

The 2016 also saw the detection of an SGR-like burst accompanied by an X-ray outburst from the central neutron star in the 2-kyr-old supernova remnant RCW 103, 1E 161348–5055 (D'Aì et al. 2016; Rea et al. 2016). This source was already known to produce X-ray outbursts, but it defied assignment to magnetars (or any other class of X-ray sources), mainly because of its unusual 6.67-h spin period with a variable pulse profile (De Luca et al. 2006). No period derivative has been measured in 1E 161348–5055 yet, so we have no idea of the strength of its external dipole magnetic field (Esposito et al. 2011b). However, it is interesting to notice that while the slow rotation is unlikely to be due to simple magnetic braking alone (which would require an astonishing field of $B \approx 10^{18}$ G), almost all the explanations put forward for the long spin period involve accretion from fall-back disk material on an ultra-magnetised neutron star born with a normal initial period (e.g. De Luca et al. 2006; Esposito et al. 2011b; Ho and Andersson 2017). Since apart from the abnormal period, 1E 161348–5055 has now shown all the distinguishing features of magnetars (Rea et al. 2016), it is tempting to associate it to these sources.

Interestingly, 1E 161348–5055 was one of the prototypes of the central compact objects in supernova remnants (CCOs), but later the discovery of its unusual properties set it apart from the other objects of the class. CCOs are steady X-ray sources with seemingly thermal spectra which are observed close to centres of young non-plerionic supernova remnants; they have no counterparts in radio and gamma wavebands and their periods—when known—are in the 0.1–0.5 s range. An emerging, but still incomplete in detail, unifying scenario for CCOs is that of 'anti-magnetars': NSs either born with weak magnetic fields ($B < 10^{11}$ G) or with a normal field 'buried' beneath the surface by a post-supernova hypercritical accretion stage of debris matter (Ho 2011; Viganò and Pons 2012; Gotthelf et al. 2013). In the latter case, even sources with magnetar-like magnetic field could in principle be present in CCOs (Viganò and Pons 2012).

Other sources bordering the magnetar cloud in the P–$\dot{P}$ diagram are the so-called X-ray dim isolated neutron stars (XDINSs; e.g. Turolla 2009): They have similar periods and inferred surface magnetic fields of $\approx 10^{13}$ G, at the lower end of the distribution of those of SGRs/AXPs, but have different properties (in particular, a thermal-like emission with blackbody spectra with $kT \approx 50$–100 eV, which is larger than the spin-down power, and relatively low timing noise) and have not shown any bursting activity in the ~ 20 years during which they have been known. Because of their properties, it has been often suggested that XDINSs are aged magnetars that have drained most of their magnetic energy. In particular, the magneto-thermal evolutionary tracks computed by Viganò et al. (2013) show that their observational characteristics are compatible with those expected from neutron stars born with dipolar field of a few 10^{14} G after a few millions of years, a timespan consistent with the characteristic ages of XDINSs. Borghese et al. (2015, 2017) inspected the phase resolved spectra of all XDINSs and found in two of them narrow phase-dependent absorption features that, if interpreted as proton cyclotron lines, akin in SGR 0418+5729, indicate the presence close to the star surface of nondipolar magnetic field components ≈ 10 times stronger than the values derived from the timing parameters, tightening the links between XDINSs and magnetars.

2.7 Summary and Final Remarks

Over the last decade, great progress has been made in our understanding of strongly magnetised neutron stars, thanks to the synergy of observation and theory. The number of confirmed magnetars has more than doubled and different types of behaviour have been studied in detail: from bursts and outbursts to flares and glitches.

While the magnetic field has always been considered a constituent element of neutron stars (Hoyle et al. 1964; Pacini 1967), it has now been established that it is the motive force behind major events in the lives of magnetars. Various aspects of these events have been modelled in detail by the use of numerical simulations through high-performance computing, testing and refining previous analytical and semi-analytical models. At this stage, the structure of the magnetic field in the core, crust and magnetosphere has been explored, improving our physical understanding of the available observational data. Viable scenarios for explosive events have been proposed, involving physical mechanisms active in the interior and the exterior of the neutron stars. Finally, the role of the magnetic field in the generation and transport of heat within the star has been assessed; this proved to be a major step towards the unification of the population of neutron stars and the understanding of their evolutionary links.

Despite these achievements there is still space for progress. A global model for the magnetic field structure and evolution is still missing. Such a model will be taking into account the feedback of the magnetic evolution in different parts of the star by coupling appropriately the magnetospheric, crustal and core field. This task is highly demanding as, even under gross simplifications, the timescales involved in the various parts of the neutron star differ by several orders of magnitude as we explore the core, crust and the magnetosphere, severely challenging the prospect of a single numerical model to simulate global evolution. This is not unique to neutron stars: several systems in nature have a slow built up of energy which is then released in a cataclysmic event and techniques have already been developed to tackle this type of problems.

Unification schemes of neutron stars have explored in detail the role of the initial magnetic field of the star and its evolution. It has been appreciated that it is not only the dipole component that rules the phenomena linked to the magnetic field, but also the presence—or the lack—of strong nondipolar and higher multipole components is crucial. In addition to that, the role of further parameters is worth exploring. First and foremost, the mass of the star, which relates directly to the star's radius, crust thickness, and gravitational potential, leading to different timescales for the evolution of the magnetic field but also affecting the value of the moment of inertia, which is critical for the estimation of the dipole component of the magnetic field. Given that the mass–radius relation is one of the most active areas of research in neutron stars, it is essential to incorporate the latest results in the current studies. Moreover, the chemical composition and the temperature of the crust are critical for the observable properties of the star. While part of this vast parameter space has been explored so far, it is important to quantify this in further detail. From the observational point of view, many of the numerous paradigm-changing discoveries of the recent years were possible thanks to large surveys, systematic analysis of public data archives and, perhaps most importantly, through all-sky monitors coupled to fast-response observing strategies. It is fundamental to continue these efforts, also in view of forthcoming facilities such as eROSITA, Athena and SKA. New hybrid or transitional objects, as well as unexpected events from known sources, are bound to be observed in the future.

The word "magnetar" has been showing up more and more in many branches of astrophysics since a few years, and in particular in researches about ultraluminous X-ray sources, high-mass X-ray binaries, gamma-ray bursts, superluminous supernovae, fast radio bursts, and sources of gravitational waves, but magnetars cannot longer be identified simply with dipolar magnetic fields stronger than a certain value. It is now proved that magnetar-like activity occurs in pulsars with a range of magnetic fields much wider than previously thought. Magnetars can behave like 'normal' pulsars and pulsars can behave like magnetars; moreover, strong nondipolar magnetic field components are probably more diffuse in neutron star than generally acknowledged (some examples are discussed in Sect. 2.6, but see also Mereghetti et al. 2016 and references therein for a possible explanation of the correlated radio and X-ray changes in the mode-switching pulsar PSR B0943+10 involving thermal emission form a small polar cap with a $\sim 10^{14}$ G nondipolar field).

Some authors view this proliferation of bursting activity in isolated neutron stars as evidence that many of them are 'dormant' magnetars. We prefer to see the things from a slightly different angle and observe that the magnetar activity and the frequency with which it manifests are most likely related to the amount of magnetic energy stored in the internal field (Rea et al. 2010; Turolla and Esposito 2013; Tiengo et al. 2013; Turolla et al. 2015); while a huge reservoir is naturally associated with the ultra-magnetised pulsars in the P–$\dot{P}$ diagram, it is not necessary well traced by the external dipolar field—the only one that can be measured directly from the timing properties of the source; therefore, magnetar behaviour may manifest sporadically also in sources that do not boast an exceptionally intense external field. Perhaps, adapting what E. H. Gombrich said about art (*"There really is no such thing as art. There are only artists."*; Gombrich 1995), we may conclude that *magnetars do not exist, only magnetar activity does.*

Acknowledgements KNG acknowledges the support of STFC Grant No. ST/N000676/1. PE acknowledges funding in the framework of the NWO Vidi award A.2320.0076.

References

Aguilera, D.N., Pons, J.A., Miralles, J.A.: 2D Cooling of magnetized neutron stars. Astron. Astrophys. **486**, 255–271 (2008a)

Aguilera, D.N., Pons, J.A., Miralles, J.A.: The impact of magnetic field on the thermal evolution of neutron stars. Astrophys. J. Lett. **673**, L167 (2008b)

Akgün, T., Miralles, J.A., Pons, J.A., Cerdá-Durán, P.: The force-free twisted magnetosphere of a neutron star. Mon. Not. Roy. Astron. Soc. **462**, 1894–1909 (2016)

Akgün, T., Cerdá-Durán, P., Miralles, J.A., Pons, J.A.: Long-term evolution of the force-free twisted magnetosphere of a magnetar. Mon. Not. Roy. Astron. Soc. **472**, 3914–3923 (2017)

Allen, M.P., Horvath, J.E.: Influence of an internal magnetar on supernova remnant expansion. Astrophys. J. **616**, 346–356 (2004)

Anderson, P.W., Itoh, N.: Pulsar glitches and restlessness as a hard superfluidity phenomenon. Nature **256**, 25–27 (1975)

Anderson, G.E., Gaensler, B.M., Slane, P.O., Rea, N., Kaplan, D.L., Posselt, B., Levin, L., Johnston, S., Murray, S.S., Brogan, C.L., Bailes, M., Bates, S., Benjamin, R.A., Bhat, N.D.R., Burgay, M., Burke-Spolaor, S., Chakrabarty, D., D'Amico, N., Drake, J.J., Esposito, P., Grindlay, J.E., Hong, J., Israel, G.L., Keith, M.J., Kramer, M., Lazio, T.J.W., Lee, J.C., Mauerhan, J.C., Milia, S., Possenti, A., Stappers, B., Steeghs, D.T.H.: Multi-wavelength observations of the radio Magnetar PSR J1622–4950 and discovery of its possibly associated supernova remnant. Astrophys. J. **751**, 53 (2012)

Andersson, N., Dionysopoulou, K., Hawke, I., Comer, G.L.: Beyond ideal magnetohydrodynamics: resistive, reactive and relativistic plasmas. Classical Quantum Gravity **34**, 125002 (2017)

Antonopoulou, D., Weltevrede, P., Espinoza, C.M., Watts, A.L., Johnston, S., Shannon, R.M., Kerr, M.: The unusual glitch recoveries of the high-magnetic-field pulsar J1119–6127. Mon. Not. Roy. Astron. Soc. **447**, 3924–3935 (2015)

Aptekar, R.L., Frederiks, D.D., Golenetskii, S.V., Il'inskii, V.N., Mazets, E.P., Pal'shin, V.D., Butterworth, P.S., Cline, T.L.: Konus catalog of soft gamma repeater activity: 1978 to 2000. Astrophys. J. Supp. Ser. **137**, 227–277 (2001)

Archibald, R.F., Kaspi, V., Ng, C.-Y., Gourgouliatos, K.N., Tsang, D., Scholz, P., Beardmore, A.P., Gehrels, N., Kennea, J.A.: An anti-glitch in a magnetar. Nature **497**, 591–593 (2013)

Archibald, R.F., Kaspi, V.M., Ng, C.-Y., Scholz, P., Beardmore, A.P., Gehrels, N., Kennea, J.A.: Repeated, delayed torque variations following X-ray flux enhancements in the Magnetar 1E 1048.1–5937. Astrophys. J. **800**, 33 (2015)

Archibald, R.F., Kaspi, V.M., Tendulkar, S.P., Scholz, P.: A Magnetar-like outburst from a high-B radio pulsar. Astrophys. J. Lett. **829**, L21 (2016)

Archibald, R.F., Burgay, M., Lyutikov, M., Kaspi, V.M., Esposito, P., Israel, G., Kerr, M., Possenti, A., Rea, N., Sarkissian, J., Scholz, P., Tendulkar, S.P.: Magnetar-like X-ray bursts suppress pulsar radio emission. Astrophys. J. Lett. (2017, in press). (eprint: astro-ph.HE/1710.03718)

Beloborodov, A.M.: Untwisting magnetospheres of neutron stars. Astrophys. J. **703**, 1044–1060 (2009)

Beloborodov, A.M., Levin, Y.: Thermoplastic waves in magnetars. Astrophys. J. Lett. **794**, L24 (2014)

Beloborodov, A.M., Thompson, C.: Corona of magnetars. Astrophys. J. **657**, 967–993 (2007)

Biskamp, D., Schwarz, E., Drake, J.F.: Two-dimensional electron magnetohydrodynamic turbulence. Phys. Rev. Lett. **76**, 1264–1267 (1996)

Blandford, R.D., Applegate, J.H., Hernquist, L.: Thermal origin of neutron star magnetic fields. Mon. Not. Roy. Astron. Soc. **204**, 1025–1048 (1983)

Borghese, A., Rea, N., Coti Zelati, F., Tiengo, A., Turolla, R.: Discovery of a strongly phase-variable spectral feature in the isolated neutron star RX J0720.4–3125. Astrophys. J. Lett. **807**, L20 (2015)

Borghese, A., Rea, N., Coti Zelati, F., Tiengo, A., Turolla, R., Zane, S.: Narrow phase-dependent features in X-ray dim isolated neutron stars: a new detection and upper limits. Mon. Not. Roy. Astron. Soc. **468**, 2975–2983 (2017)

Braithwaite, J., Spruit, H.C.: A fossil origin for the magnetic field in A stars and white dwarfs. Nature **431**, 819–821 (2004)

Braithwaite, J., Spruit, H.C.: Evolution of the magnetic field in magnetars. Astron. Astrophys. **450**, 1097–1106 (2006)

Bransgrove, A., Levin, Y., Beloborodov, A.: Magnetic field evolution of neutron stars I: basic formalism, numerical techniques, and first results.Mon. Not. Roy. Astron. Soc. (2017, in press). Eprint: astro-ph.HE/1709.09167

Brown, E.F., Cumming, A.: Mapping crustal heating with the cooling light curves of quasi-persistent transients. Astrophys. J. **698**, 1020–1032 (2009)

Bucciantini, N., Quataert, E., Arons, J., Metzger, B.D., Thompson, T.A.: Magnetar-driven bubbles and the origin of collimated outflows in gamma-ray bursts. Mon. Not. Roy. Astron. Soc. **380**, 1541–1553 (2007)

Burgay, M., Possenti, A., Kerr, M., Esposito, P., Rea, N., Zelati, F.C., Israel, G.L., Johnston, S.: Pulsed radio emission from PSR J1119–6127 disappeared. Astron. Tel. **9286** (2016a)

Burgay, M., Possenti, A., Kerr, M., Esposito, P., Rea, N., Zelati, F.C., Israel, G.L., Johnston, S.: Pulsed radio emission from PSR J1119–6127 re-activated. Astron. Tel. **9366** (2016b)

Camilo, F., Kaspi, V.M., Lyne, A.G., Manchester, R.N., Bell, J.F., D'Amico, N., McKay, N.P.F., Crawford, F.: Discovery of two high magnetic field radio pulsars. Astrophys. J. **541**, 367–373 (2000)

Camilo, F., Ransom, S.M., Halpern, J.P., Reynolds, J., Helfand, D.J., Zimmerman, N., Sarkissian, J.: Transient pulsed radio emission from a magnetar. Nature **442**, 892–895 (2006)

Camilo, F., Ransom, S.M., Halpern, J.P., Reynolds, J.: 1E 1547.0–5408: a radio-emitting magnetar with a rotation period of 2 seconds. Astrophys. J. Lett. **666**, L93–L96 (2007a)

Camilo, F., Cognard, I., Ransom, S.M., Halpern, J.P., Reynolds, J., Zimmerman, N., Gotthelf, E.V., Helfand, D.J., Demorest, P., Theureau, G., Backer, D.C.: The magnetar XTE J1810–197: variations in torque, radio flux density, and pulse profile morphology. Astrophys. J. **663**, 497–504 (2007b)

Castillo, F., Reisenegger, A., Valdivia, J.A.: Magnetic field evolution and equilibrium configurations in neutron star cores: the effect of ambipolar diffusion. Mon. Not. Roy. Astron. Soc. **471**, 507–522 (2017)

Cerdá-Durán, P., Stergioulas, N., Font, J.A.: Alfvén QPOs in magnetars in the anelastic approximation. Mon. Not. Roy. Astron. Soc. **397**, 1607–1620 (2009)

Cho, J., Lazarian, A.: The anisotropy of electron magnetohydrodynamic turbulence. Astrophys. J. Lett. **615**, L41–L44 (2004)

Chugunov, A.I., Horowitz, C.J.: Breaking stress of neutron star crust. Mon. Not. Roy. Astron. Soc. **407**, L54–L58 (2010)

Ciolfi, R., Rezzolla, L.: Poloidal-field instability in magnetized relativistic stars. Astrophys. J. **760**, 1 (2012)

Ciolfi, R., Rezzolla, L.: Twisted-torus configurations with large toroidal magnetic fields in relativistic stars. Mon. Not. Roy. Astron. Soc. **435**, L43–L47 (2013)

Clark, J.S., Muno, M.P., Negueruela, I., Dougherty, S.M., Crowther, P.A., Goodwin, S.P., de Grijs, R.: Unveiling the X-ray point source population of the Young Massive Cluster Westerlund 1. Astron. Astrophys. **477**, 147–163 (2008)

Clark, J.S., Ritchie, B.W., Najarro, F., Langer, N., Negueruela, I.: A VLT/FLAMES survey for massive binaries in Westerlund 1. IV. Wd1–5 - binary product and a pre-supernova companion for the magnetar CXOU J1647–45? Astron. Astrophys. **565**, A90 (2014)

Colaiuda, A., Beyer, H., Kokkotas, K.D.: On the quasi-periodic oscillations in magnetars. Mon. Not. Roy. Astron. Soc. **396**, 1441–1448 (2009)

Coti Zelati, F., Rea, N., Pons, J.A., Campana, S., Esposito, P.: Systematic study of magnetar outbursts. Mon. Not. Roy. Astron. Soc. **474**, 961–1017 (2018)

D'Aì, A., Evans, P.A., Burrows, D.N., Kuin, N.P.M., Kann, D.A., Campana, S., Maselli, A., Romano, P., Cusumano, G., La Parola, V., Barthelmy, S.D., Beardmore, A.P., Cenko, S.B., De Pasquale, M., Gehrels, N., Greiner, J., Kennea, J.A., Klose, S., Melandri, A., Nousek, J.A., Osborne, J.P., Palmer, D.M., Sbarufatti, B., Schady, P., Siegel, M.H., Tagliaferri, G., Yates, R., Zane, S.: Evidence for the magnetar nature of 1E 161348–5055 in RCW 103. Mon. Not. Roy. Astron. Soc. **463**, 2394–2404 (2016)

Dall'Osso, S., Shore, S.N., Stella, L.: Early evolution of newly born magnetars with a strong toroidal field. Mon. Not. Roy. Astron. Soc. **398**, 1869–1885 (2009)

Davies, B., Figer, D.F., Kudritzki, R.-P., Trombley, C., Kouveliotou, C., Wachter, S.: The progenitor mass of the magnetar SGR1900+14. Astrophys. J. **707**, 844–851 (2009)

De Luca, A., Caraveo, P.A., Mereghetti, S., Tiengo, A., Bignami, G.F.: A long-period, violently variable X-ray source in a young supernova remnant. Science **313**, 814–817 (2006)

Deutsch, A.J.: The electromagnetic field of an idealized star in rigid rotation in vacuo. Annales d'Astrophysique **18**, 1 (1955)

Dib, R., Kaspi, V.M.: 16 yr of RXTE monitoring of five anomalous X-ray pulsars. Astrophys. J. **784**, 37 (2014)

Dib, R., Kaspi, V.M., Gavriil, F.P.: Glitches in anomalous X-ray pulsars. Astrophys. J. **673**, 1044–1061 (2008)

Duncan, R.C.: Global seismic oscillations in soft gamma repeaters. Astrophys. J. Lett. **498**, L45–L49 (1998)

Duncan, R.C., Thompson, C.: Formation of very strongly magnetized neutron stars - implications for gamma-ray bursts. Astrophys. J. Lett. **392**, L9–L13 (1992)

Eatough, R.P., Falcke, H., Karuppusamy, R., Lee, K.J., Champion, D.J., Keane, E.F., Desvignes, G., Schnitzeler, D.H.F.M., Spitler, L.G., Kramer, M., Klein, B., Bassa, C., Bower, G.C., Brunthaler, A., Cognard, I., Deller, A.T., Demorest, P.B., Freire, P.C.C., Kraus, A., Lyne, A.G., Noutsos, A., Stappers, B., Wex, N.: A strong magnetic field around the supermassive black hole at the centre of the Galaxy. Nature **501**, 391–394 (2013)

Elfritz, J.G., Pons, J.A., Rea, N., Glampedakis, K., Viganò, D.: Simulated magnetic field expulsion in neutron star cores. Mon. Not. Roy. Astron. Soc. **456**, 4461–4474 (2016)

Esposito, P., Mereghetti, S., Tiengo, A., Sidoli, L., Feroci, M., Woods, P.: Five years of SGR 1900+14 observations with BeppoSAX. Astron. Astrophys. **461**, 605–612 (2007)

Esposito, P., Israel, G.L., Turolla, R., Tiengo, A., Götz, D., De Luca, A., Mignani, R.P., Zane, S., Rea, N., Testa, V., Caraveo, P.A., Chaty, S., Mattana, F., Mereghetti, S., Pellizzoni, A., Romano, P.: Early X-ray and optical observations of the soft gamma-ray repeater SGR 0418+5729. Mon. Not. Roy. Astron. Soc. **405**, 1787–1795 (2010)

Esposito, P., Israel, G.L., Turolla, R., Mattana, F., Tiengo, A., Possenti, A., Zane, S., Rea, N., Burgay, M., Götz, D., Mereghetti, S., Stella, L., Wieringa, M.H., Sarkissian, J.M., Enoto, T., Romano, P., Sakamoto, T., Nakagawa, Y.E., Makishima, K., Nakazawa, K., Nishioka, H., François-Martin, C.: Long-term spectral and timing properties of the soft gamma-ray repeater SGR 1833–0832 and detection of extended X-ray emission around the radio pulsar PSR B1830–08. Mon. Not. Roy. Astron. Soc. **416**, 205–215 (2011a)

Esposito, P., Turolla, R., de Luca, A., Israel, G.L., Possenti, A., Burrows, D.N.: Swift monitoring of the central X-ray source in RCW 103. Mon. Not. Roy. Astron. Soc. **418**, 170–175 (2011b)

Fabian, A.C.: Theories of the nuclei of active galaxies. Proc. Roy. Soc. Lond. Ser. A **366**, 449–459 (1979)

Faucher-Giguère, C.-A., Kaspi, V.M.: Birth and evolution of isolated radio pulsars. Astrophys. J. **643**, 332–355 (2006)

Ferrario, L., Wickramasinghe, D.: Origin and evolution of magnetars. Mon. Not. Roy. Astron. Soc. **389**, L66–L70 (2008)

Gabler, M., Cerdá Durán, P., Font, J.A., Müller, E., Stergioulas, N.: Magneto-elastic oscillations and the damping of crustal shear modes in magnetars. Mon. Not. Roy. Astron. Soc. **410**, L37–L41 (2011)

Gabler, M., Cerdá-Durán, P., Stergioulas, N., Font, J.A., Müller, E.: Imprints of superfluidity on magnetoelastic quasiperiodic oscillations of soft gamma-ray repeaters. Phys. Rev. Lett. **111**(21), 211102 (2013)

Gaensler, B.M., Kouveliotou, C., Gelfand, J.D., Taylor, G.B., Eichler, D., Wijers, R.A.M.J., Granot, J., Ramirez-Ruiz, E., Lyubarsky, Y.E., Hunstead, R.W., Campbell-Wilson, D., van der Horst, A.J., McLaughlin, M.A., Fender, R.P., Garrett, M.A., Newton-McGee, K.J., Palmer, D.M., Gehrels, N., Woods, P.M.: An expanding radio nebula produced by a giant flare from the magnetar SGR 1806–20. Nature **434**, 1104–1106 (2005)

Gavriil, F.P., Kaspi, V.M.: Anomalous X-ray pulsar 1E 1048.1–5937: pulsed flux flares and large torque variations. Astrophys. J. Lett. **609**, L67–L70 (2004)

Gavriil, F.P., Kaspi, V.M., Woods, P.M.: Magnetar-like X-ray bursts from an anomalous X-ray pulsar. Nature **419**, 142–144 (2002)

Gavriil, F.P., Gonzalez, M.E., Gotthelf, E.V., Kaspi, V.M., Livingstone, M.A., Woods, P.M.: Magnetar-like emission from the young pulsar in Kes 75. Science **319**, 1802– (2008)

Geppert, U.: Magneto-thermal evolution of neutron stars with emphasis to radio pulsars. J. Astrophys. Astron. **38**, 46 (2017)

Geppert, U., Viganò, D.: Creation of magnetic spots at the neutron star surface. Mon. Not. Roy. Astron. Soc. **444**, 3198–3208 (2014)

Geppert, U., Küker, M., Page, D.: Temperature distribution in magnetized neutron star crusts. Astron. Astrophys. **426**, 267–277 (2004)

Geppert, U., Küker, M., Page, D.: Temperature distribution in magnetized neutron star crusts. II. The effect of a strong toroidal component. Astron. Astrophys. **457**, 937–947 (2006)

Giacomazzo, B., Perna, R.: Formation of stable magnetars from binary neutron star mergers. Astrophys. J. Lett. **771**, L26 (2013)

Gill, R., Heyl, J.S.: On the trigger mechanisms for soft gamma-ray repeater giant flares. Mon. Not. Roy. Astron. Soc. **407**, 1926–1932 (2010)

Glampedakis, K., Jones, D.I., Samuelsson, L.: Ambipolar diffusion in superfluid neutron stars. Mon. Not. Roy. Astron. Soc. **413**, 2021–2030 (2011)

Goldreich, P., Julian, W.H.: Pulsar electrodynamics. Astrophys. J. **157**, 869 (1969)

Goldreich, P., Reisenegger, A.: Magnetic field decay in isolated neutron stars. Astrophys. J. **395**, 250–258 (1992)

Gombrich, E.H.: The Story of Art, 16th edn. Phaidon Press, London (1995)

Gonzalez, M.E., Kaspi, V.M., Lyne, A.G., Pivovaroff, M.J.: An XMM-Newton observation of the high magnetic field radio pulsar PSR B0154+61. Astrophys. J. Lett. **610**, L37–L40 (2004)

Gotthelf, E.V., Halpern, J.P., Alford, J.: The spin-down of PSR J0821–4300 and PSR J1210–5226: confirmation of central compact objects as anti-magnetars. Astrophys. J. **765**, 58 (2013)

Götz, D., Mereghetti, S., Molkov, S., Hurley, K., Mirabel, I.F., Sunyaev, R., Weidenspointner, G., Brandt, S., del Santo, M., Feroci, M., Göğüş, E., von Kienlin, A., van der Klis, M., Kouveliotou, C., Lund, N., Pizzichini, G., Ubertini, P., Winkler, C., Woods, P.M.: Two years of INTEGRAL monitoring of the soft gamma-ray repeater SGR 1806–20: from quiescence to frenzy. Astron. Astrophys. **445**, 313–321 (2006)

Göğüş, E., Lin, L., Kaneko, Y., Kouveliotou, C., Watts, A.L., Chakraborty, M., Alpar, M.A., Huppenkothen, D., Roberts, O.J., Younes, G., van der Horst, A.J.: Magnetar-like X-ray bursts from a rotation-powered pulsar, PSR J1119–6127. Astrophys. J. Lett. **829**, L25 (2016)

Gourgouliatos, K.N., Cumming, A.: Hall attractor in axially symmetric magnetic fields in neutron star crusts. Phys. Rev. Lett. **112**(17), 171101 (2014a)

Gourgouliatos, K.N., Cumming, A.: Hall effect in neutron star crusts: evolution, endpoint and dependence on initial conditions. Mon. Not. Roy. Astron. Soc. **438**, 1618–1629 (2014b)

Gourgouliatos, K.N., Cumming, A.: Hall drift and the braking indices of young pulsars. Mon. Not. Roy. Astron. Soc. **446**, 1121–1128 (2015)

Gourgouliatos, K.N., Hollerbach, R.: Resistive tearing instability in electron MHD: application to neutron star crusts. Mon. Not. Roy. Astron. Soc. **463**, 3381–3389 (2016)

Gourgouliatos, K.N., Hollerbach, R.: Magnetic axis drift and magnetic spot formation in neutron stars with toroidal fields. Astrophys. J. **852**(1), article id. 21, 12 (2018)

Gourgouliatos, K.N., Vlahakis, N.: Relativistic expansion of a magnetized fluid. Geophys.Astrophys. Fluid Dyn. **104**, 431–450 (2010)

Gourgouliatos, K.N., Cumming, A., Reisenegger, A., Armaza, C., Lyutikov, M., Valdivia, J.A.: Hall equilibria with toroidal and poloidal fields: application to neutron stars. Mon. Not. Roy. Astron. Soc. **434**, 2480–2490 (2013)

Gourgouliatos, K.N., Kondić, T., Lyutikov, M., Hollerbach, R.: Magnetar activity via the density-shear instability in Hall-MHD. Mon. Not. Roy. Astron. Soc. **453**, L93–L97 (2015)

Gourgouliatos, K.N., Wood, T.S., Hollerbach, R.: Magnetic field evolution in magnetar crusts through three-dimensional simulations. Proc. Nat. Acad. Sci. **113**, 3944–3949 (2016)

Göğüş, E., Woods, P.M., Kouveliotou, C., Finger, M.H., Pal'shin, V., Kaneko, Y., Golenetskii, S., Frederiks, D., Airhart, C.: Extended tails from SGR 1806–20 bursts. Astrophys. J. **740**, 55 (2011)

Greiner, J., Mazzali, P.A., Kann, D.A., Krühler, T., Pian, E., Prentice, S., Olivares E., F., Rossi, A., Klose, S., Taubenberger, S., Knust, F., Afonso, P.M.J., Ashall, C., Bolmer, J., Delvaux, C., Diehl, R., Elliott, J., Filgas, R., Fynbo, J.P.U., Graham, J.F., Guelbenzu, A.N., Kobayashi, S., Leloudas, G., Savaglio, S., Schady, P., Schmidl, S., Schweyer, T., Sudilovsky, V., Tanga, M., Updike, A.C., van Eerten, H., Varela, K.: A very luminous magnetar-powered supernova associated with an ultra-long γ-ray burst. Nature **523**, 189–192 (2015)

Gullón, M., Miralles, J.A., Viganò, D., Pons, J.A.: Population synthesis of isolated neutron stars with magneto-rotational evolution. Mon. Not. Roy. Astron. Soc. **443**, 1891–1899 (2014)

Gullón, M., Pons, J.A., Miralles, J.A., Viganò, D., Rea, N., Perna, R.: Population synthesis of isolated neutron stars with magneto-rotational evolution - II. From radio-pulsars to magnetars. Mon. Not. Roy. Astron. Soc. **454**, 615–625 (2015)

Gusakov, M.E., Kantor, E.M., Ofengeim, D.D.: On the evolution of magnetic field in neutron stars. ArXiv e-prints (2017)

Haskell, B., Melatos, A.: Models of pulsar glitches. Int. J. Mod. Phys. D **24**, 1530008 (2015)

Heger, A., Woosley, S.E., Spruit, H.C.: Presupernova evolution of differentially rotating massive stars including magnetic fields. Astrophys. J. **626**, 350–363 (2005)

Hewish, A., Bell, S., Pilkington, J., Scott, P., Collins, R.: Observation of a rapidly pulsating radio source. Nature **217**, 709–713 (1968)

Ho, W.C.G.: Evolution of a buried magnetic field in the central compact object neutron stars. Mon. Not. Roy. Astron. Soc. **414**, 2567–2575 (2011)

Ho, W.C.G., Andersson, N.: Ejector and propeller spin-down: how might a superluminous supernova millisecond magnetar become the 6.67 h pulsar in RCW 103. Mon. Not. Roy. Astron. Soc. **464**, L65–L69 (2017)

Hobbs, G., Lyne, A.G., Kramer, M., Martin, C.E., Jordan, C.: Long-term timing observations of 374 pulsars. Mon. Not. Roy. Astron. Soc. **353**, 1311–1344 (2004)

Hobbs, G., Lyne, A.G., Kramer, M.: An analysis of the timing irregularities for 366 pulsars. Mon. Not. Roy. Astron. Soc. **402**, 1027–1048 (2010)

Hollerbach, R., Rüdiger, G.: The influence of Hall drift on the magnetic fields of neutron stars. Mon. Not. Roy. Astron. Soc. **337**, 216–224 (2002)

Hollerbach, R., Rüdiger, G.: Hall drift in the stratified crusts of neutron stars. Mon. Not. Roy. Astron. Soc. **347**, 1273–1278 (2004)

Horowitz, C.J., Kadau, K.: Breaking strain of neutron star crust and gravitational waves. Phys. Rev. Lett. **102**(19), 191102 (2009)

Hoyle, F., Narlikar, J.V., Wheeler, J.A.: Electromagnetic waves from very dense stars. Nature **203**, 914–916 (1964)

Hoyos, J.H., Reisenegger, A., Valdivia, J.A.: Asymptotic, non-linear solutions for ambipolar diffusion in one dimension. Mon. Not. Roy. Astron. Soc. **408**, 1730–1741 (2010)

Hurley, K., Cline, T., Mazets, E., Barthelmy, S., Butterworth, P., Marshall, F., Palmer, D., Aptekar, R., Golenetskii, S., Il'Inskii, V., Frederiks, D., McTiernan, J., Gold, R., Trombka, J.: A giant periodic flare from the soft γ-ray repeater SGR1900+14. Nature **397**, 41–43 (1999)

Hurley, K., Boggs, S.E., Smith, D.M., Duncan, R.C., Lin, R., Zoglauer, A., Krucker, S., Hurford, G., Hudson, H., Wigger, C., Hajdas, W., Thompson, C., Mitrofanov, I., Sanin, A., Boynton, W., Fellows, C., von Kienlin, A., Lichti, G., Rau, A., Cline, T.: An exceptionally bright flare from SGR 1806–20 and the origins of short-duration γ-ray bursts. Nature **434**, 1098–1103 (2005)

Igoshev, A.P., Popov, S.B.: Magnetic field decay in normal radio pulsars. Astron. Nachr. **336**, 831 (2015)

Inan, U.S., Lehtinen, N.G., Lev-Tov, S.J., Johnson, M.P., Bell, T.F., Hurley, K.: Ionization of the lower ionosphere by γ-rays from a magnetar: detection of a low energy (3–10 keV) component. Geophys. Res. Lett., **26**, 3357–3360 (1999)

Inan, U.S., Lehtinen, N.G., Moore, R.C., Hurley, K., Boggs, S., Smith, D.M., Fishman, G.J.: Massive disturbance of the daytime lower ionosphere by the giant γ-ray flare from magnetar SGR 1806–20. Geophys. Res. Lett., **34**, L08103 (2007)

Israel, G.L., Belloni, T., Stella, L., Rephaeli, Y., Gruber, D.E., Casella, P., Dall'Osso, S., Rea, N., Persic, M., Rothschild, R.E.: The discovery of rapid X-ray oscillations in the tail of the SGR 1806–20 hyperflare. Astrophys. J. Lett. **628**, L53–L56 (2005)

Israel, G.L., Romano, P., Mangano, V., Dall'Osso, S., Chincarini, G., Stella, L., Campana, S., Belloni, T., Tagliaferri, G., Blustin, A.J., Sakamoto, T., Hurley, K., Zane, S., Moretti, A., Palmer, D., Guidorzi, C., Burrows, D.N., Gehrels, N., Krimm, H.A.: A Swift Gaze into the 2006 March 29 Burst Forest of SGR 1900+14. Astrophys. J. **685**, 1114–1128 (2008)

Jones, P.B.: Nature of fault planes in solid neutron star matter. Astrophys. J. **595**, 342–345 (2003)

Kaspi, V.M.: Grand unification of neutron stars. Proc. Nat. Acad. Sci. **107**, 7147–7152 (2010)

Kaspi, V.M., Beloborodov, A.: Magnetars. Ann. Rev. Astron. Astroph. **55**, 261–301 (2017)

Kaspi, V.M., Gavriil, F.P.: A second glitch from the "Anomalous" X-Ray Pulsar 1RXS J170849.0–4000910. Astrophys. J. Lett. **596**, L71–L74 (2003)

Kaspi, V.M., McLaughlin, M.A.: Chandra X-ray detection of the high magnetic field radio pulsar PSR J1718–3718. Astrophys. J. Lett. **618**, L41–L44 (2005)

Kaspi, V.M., Lackey, J.R., Chakrabarty, D.: A glitch in an anomalous X-ray pulsar. Astrophys. J. Lett. **537**, L31–L34 (2000)

Kaspi, V.M., Gavriil, F.P., Woods, P.M., Jensen, J.B., Roberts, M.S.E., Chakrabarty, D.: A major soft gamma repeater-like outburst and rotation glitch in the no-longer-so-anomalous X-ray pulsar 1E 2259+586. Astrophys. J. Lett. **588**, L93–L96 (2003)

Kojima, Y., Kisaka, S.: Magnetic field decay with Hall drift in neutron star crusts. Mon. Not. Roy. Astron. Soc. **421**, 2722–2730 (2012)

Komissarov, S.S., Barkov, M., Lyutikov, M.: Tearing instability in relativistic magnetically dominated plasmas. Mon. Not. Roy. Astron. Soc. **374**, 415–426 (2007)

Kouveliotou, C., Dieters, S., Strohmayer, T., van Paradijs, J., Fishman, G.J., Meegan, C.A., Hurley, K., Kommers, J., Smith, I., Frail, D., Murakami, T.: An X-ray pulsar with a superstrong magnetic field in the soft gamma-ray repeater SGR 1806–20. Nature **393**, 235–237 (1998)

Lander, S.K.: Magnetar field evolution and crustal plasticity. Astrophys. J. Lett. **824**, L21 (2016)

Lander, S.K., Andersson, N., Antonopoulou, D., Watts, A.L.: Magnetically driven crustquakes in neutron stars. Mon. Not. Roy. Astron. Soc. **449**, 2047–2058 (2015)

Lasky, P.D., Zink, B., Kokkotas, K.D., Glampedakis, K.: Hydromagnetic instabilities in relativistic neutron stars. Astrophys. J. Lett. **735**, L20 (2011)

Lenters, G.T., Woods, P.M., Goupell, J.E., Kouveliotou, C., Göğüş, E., Hurley, K., Frederiks, D., Golenetskii, S., Swank, J.: An extended burst tail from SGR 1900+14 with a thermal X-ray spectrum. Astrophys. J. **587**, 761–770 (2003)

Levin, Y.: On the theory of magnetar QPOs. Mon. Not. Roy. Astron. Soc. **377**, 159–167 (2007)

Levin, Y., Lyutikov, M.: On the dynamics of mechanical failures in magnetized neutron star crusts. Mon. Not. Roy. Astron. Soc. **427**, 1574–1579 (2012)

Levin, L., Bailes, M., Bates, S., Bhat, N.D.R., Burgay, M., Burke-Spolaor, S., D'Amico, N., Johnston, S., Keith, M., Kramer, M., Milia, S., Possenti, A., Rea, N., Stappers, B., van Straten, W.: A radio-loud magnetar in X-ray quiescence. Astrophys. J. Lett. **721**, L33–L37 (2010)

Li, X., Beloborodov, A.M.: Plastic damping of Alfvén waves in magnetar flares and delayed afterglow emission. Astrophys. J. **815**, 25 (2015)

Li, X., Levin, Y., Beloborodov, A.M.: Magnetar outbursts from avalanches of Hall Waves and crustal failures. Astrophys. J. **833**, 189 (2016)

Lynden-Bell, D., Boily, C.: Self-similar solutions up to flashpoint in highly wound magnetostatics. Mon. Not. Roy. Astron. Soc. **267**, 146 (1994)

Lynden-Bell, D., Moffatt, H.K.: Flashpoint. Mon. Not. Roy. Astron. Soc. **452**, 902–909 (2015)

Lyutikov, M.: Explosive reconnection in magnetars. Mon. Not. Roy. Astron. Soc. **346**, 540–554 (2003)

Majid, W.A., Pearlman, A.B., Dobreva, T., Horiuchi, S., Kocz, J., Lippuner, J., Prince, T.A.: Post-outburst radio observations of the high magnetic field pulsar PSR J1119–6127. Astrophys. J. Lett. **834**, L2 (2017)

Manchester, R.N., Hobbs, G.B., Teoh, A., Hobbs, M.: The Australia telescope national facility pulsar catalogue. Astron. J. **129**, 1993–2006 (2005)

Mandea, M., Balasis, G.: FAST TRACK PAPER: The SGR 1806–20 magnetar signature on the Earth's magnetic field. Geophys. J. Int. **167**, 586–591 (2006)

Marchant, P., Reisenegger, A., Alejandro Valdivia, J., Hoyos, J.H.: Stability of hall equilibria in neutron star crusts. Astrophys. J. **796**, 94 (2014)

Marsden, D., White, N.E.: Correlations between spectral properties and spin-down rate in soft gamma-ray repeaters and anomalous X-ray pulsars. Astrophys. J. Lett. **551**, L155–L158 (2001)

Martin, J., Rea, N., Torres, D.F., Papitto, A.: Comparing supernova remnants around strongly magnetized and canonical pulsars. Mon. Not. Roy. Astron. Soc. **444**, 2910–2924 (2014)

Mastrano, A., Suvorov, A.G., Melatos, A.: Interpreting the AXP 1E 2259+586 antiglitch as a change in internal magnetization. Mon. Not. Roy. Astron. Soc. **453**, 522–530 (2015)

Mazets, E.P., Golentskii, S.V., Ilinskii, V.N., Aptekar, R.L., Guryan, I.A.: Observations of a flaring X-ray pulsar in Dorado. Nature **282**, 587–589 (1979)

Mereghetti, S., Stella, L.: The very low mass X-ray binary pulsars: a new class of sources? Astrophys. J. Lett. **442**, L17–L20 (1995)

Mereghetti, S., Kuiper, L., Tiengo, A., Hessels, J., Hermsen, W., Stovall, K., Possenti, A., Rankin, J., Esposito, P., Turolla, R., Mitra, D., Wright, G., Stappers, B., Horneffer, A., Oslowski, S., Serylak, M., Grießmeier, J.-M.: A deep campaign to characterize the synchronous radio/X-ray mode switching of PSR B0943+10. Astrophys. J. **831**, 21 (2016)

Mösta, P., Ott, C.D., Radice, D., Roberts, L.F., Schnetter, E., and Haas, R.: A large-scale dynamo and magnetoturbulence in rapidly rotating core-collapse supernovae. Nature **528**, 376–379 (2015)

Muno, M.P., Clark, J.S., Crowther, P.A., Dougherty, S.M., de Grijs, R., Law, C., McMillan, S.L.W., Morris, M.R., Negueruela, I., Pooley, D., Portegies Zwart, S., Yusef-Zadeh, F.: A neutron star with a massive progenitor in Westerlund 1. Astrophys. J. Lett. **636**, L41–L44 (2006)

Olausen, S.A., Kaspi, V.M.: The McGill magnetar catalog. {Astrophys. J. Supp. Ser. **212**, 6 (2014)

Olausen, S.A., Zhu, W.W., Vogel, J.K., Kaspi, V.M., Lyne, A.G., Espinoza, C.M., Stappers, B.W., Manchester, R.N., McLaughlin, M.A.: X-Ray observations of high-B radio pulsars. Astrophys. J. **764**, 1 (2013)

Pacini, F.: Energy emission from a neutron star. Nature **216**, 567–568 (1967)

Paczynski, B.: GB 790305 as a very strongly magnetized neutron star. Acta Astron. **42**, 145–153 (1992)

Palmer, D.M., Barthelmy, S., Gehrels, N., Kippen, R.M., Cayton, T., Kouveliotou, C., Eichler, D., Wijers, R.A.M.J., Woods, P.M., Granot, J., Lyubarsky, Y.E., Ramirez-Ruiz, E., Barbier, L., Chester, M., Cummings, J., Fenimore, E.E., Finger, M.H., Gaensler, B.M., Hullinger, D., Krimm, H., Markwardt, C.B., Nousek, J.A., Parsons, A., Patel, S., Sakamoto, T., Sato, G., Suzuki, M., Tueller, J.: A giant γ-ray flare from the magnetar SGR 1806-20. Nature **434**, 1107–1109 (2005)

Parfrey, K., Beloborodov, A.M., Hui, L.: Twisting, reconnecting magnetospheres and magnetar spindown. Astrophys. J. Lett. **754**, L12 (2012)

Parfrey, K., Beloborodov, A.M., Hui, L.: Dynamics of strongly twisted relativistic magnetospheres. Astrophys. J. **774**, 92 (2013)

Passamonti, A., Lander, S.K.: Stratification, superfluidity and magnetar QPOs. Mon. Not. Roy. Astron. Soc. **429**, 767–774 (2013)

Passamonti, A., Akgün, T., Pons, J.A., Miralles, J.A.: The relevance of ambipolar diffusion for neutron star evolution. Mon. Not. Roy. Astron. Soc. **465**, 3416–3428 (2017)

Perna, R., Pons, J.A.: A unified model of the magnetar and radio pulsar bursting phenomenology. Astrophys. J. Lett. **727**, L51 (2011)

Pintore, F., Mereghetti, S., Tiengo, A., Vianello, G., Costantini, E., Esposito, P.: The effect of X-ray dust scattering on a bright burst from the magnetar 1E 1547.0–5408. Mon. Not. Roy. Astron. Soc. **467**, 3467–3474 (2017)

Pons, J.A., Geppert, U.: Magnetic field dissipation in neutron star crusts: from magnetars to isolated neutron stars. Astron. Astrophys. **470**, 303–315 (2007)

Pons, J.A., Geppert, U.: Confirmation of the occurrence of the Hall instability in the non-linear regime. Astron. Astrophys. **513**, L12 (2010)

Pons, J.A., Perna, R.: Magnetars versus high magnetic field pulsars: a theoretical interpretation of the apparent dichotomy. Astrophys. J. **741**, 123 (2011)

Pons, J.A., Miralles, J.A., Geppert, U.: Magneto-thermal evolution of neutron stars. Astron. Astrophys. **496**, 207–216 (2009)

Popov, S.B., Pons, J.A., Miralles, J.A., Boldin, P.A., Posselt, B.: Population synthesis studies of isolated neutron stars with magnetic field decay. Mon. Not. Roy. Astron. Soc. **401**, 2675–2686 (2010)

Price, D.J., Rosswog, S.: Producing ultrastrong magnetic fields in neutron star mergers. Science **312**, 719–722 (2006)

Priest, E.R., Hood, A.W., Anzer, U.: A twisted flux-tube model for solar prominences. I - General properties. Astrophys. J. **344**, 1010–1025 (1989)

Rea, N., Esposito, P.: Magnetar outbursts: an observational review. In: Torres, D.F., Rea, N. (eds.) High-energy emission from pulsars and their systems. In: Astrophysics and Space Science Proceedings, pp. 247–273. Springer, Heidelberg (2011)

Rea, N., Esposito, P., Turolla, R., Israel, G.L., Zane, S., Stella, L., Mereghetti, S., Tiengo, A., Götz, D., Göğüş, E., Kouveliotou, C.: A low-magnetic-field soft gamma repeater. Science 330, 944 (2010)

Rea, N., Israel, G.L., Esposito, P., Pons, J.A., Camero-Arranz, A., Mignani, R.P., Turolla, R., Zane, S., Burgay, M., Possenti, A., Campana, S., Enoto, T., Gehrels, N., Göğüş, E., Götz, D., Kouveliotou, C., Makishima, K., Mereghetti, S., Oates, S.R., Palmer, D.M., Perna, R., Stella, L., Tiengo, A.: A new low magnetic field magnetar: the 2011 outburst of swift J1822.3–1606. Astrophys. J. 754, 27 (2012)

Rea, N., Esposito, P., Pons, J.A., Turolla, R., Torres, D.F., Israel, G.L., Possenti, A., Burgay, M., Viganò, D., Papitto, A., Perna, R., Stella, L., Ponti, G., Baganoff, F.K., Haggard, D., Camero-Arranz, A., Zane, S., Minter, A., Mereghetti, S., Tiengo, A., Schödel, R., Feroci, M., Mignani, R., Götz, D.: A strongly magnetized pulsar within the grasp of the Milky Way's supermassive black hole. Astrophys. J. Lett. 775, L34 (2013a)

Rea, N., Israel, G.L., Pons, J.A., Turolla, R., Viganò, D., Zane, S., Esposito, P., Perna, R., Papitto, A., Terreran, G., Tiengo, A., Salvetti, D., Girart, J.M., Palau, A., Possenti, A., Burgay, M., Göğüş, E., Caliandro, G.A., Kouveliotou, C., Götz, D., Mignani, R.P., Ratti, E., Stella, L.: The outburst decay of the low magnetic field magnetar SGR 0418+5729. Astrophys. J. 770, 65 (2013b)

Rea, N., Viganò, D., Israel, G.L., Pons, J.A., Torres, D.F.: 3XMM J185246.6+003317: Another low magnetic field magnetar. Astrophys. J. Lett. 781, L17 (2014)

Rea, N., Borghese, A., Esposito, P., Coti Zelati, F., Bachetti, M., Israel, G.L., De Luca, A.: Magnetar-like activity from the central compact object in the SNR RCW103. Astrophys. J. Lett. 828, L13 (2016)

Reisenegger, A., Benguria, R., Prieto, J.P., Araya, P.A., Lai, D.: Hall drift of axisymmetric magnetic fields in solid neutron-star matter. Astron. Astrophys. 472, 233–240 (2007)

Rheinhardt, M., Geppert, U.: Hall-drift induced magnetic field instability in neutron stars. Phys. Rev. Lett. 88(10), 101103 (2002)

Rheinhardt, M., Konenkov, D., Geppert, U.: The occurrence of the Hall instability in crusts of isolated neutron stars. Astron. Astrophys. 420, 631–645 (2004)

Rodríguez Castillo, G.A., Israel, G.L., Tiengo, A., Salvetti, D., Turolla, R., Zane, S., Rea, N., Esposito, P., Mereghetti, S., Perna, R., Stella, L., Pons, J.A., Campana, S., Götz, D., Motta, S.: The outburst decay of the low magnetic field magnetar SWIFT J1822.3–1606: phase-resolved analysis and evidence for a variable cyclotron feature. Mon. Not. Roy. Astron. Soc. 456, 4145–4155 (2016)

Rosswog, S., Ramirez-Ruiz, E., Davies, M.B. High-resolution calculations of merging neutron stars - III. Gamma-ray bursts. Mon. Not. Roy. Astron. Soc. 345, 1077–1090 (2003)

Şaşmaz Muş, S., Aydın, B., Göğüş, E.: A glitch and an anti-glitch in the anomalous X-ray pulsar 1E 1841–045. Mon. Not. Roy. Astron. Soc. 440, 2916–2921 (2014)

Scholz, P., Ng, C.-Y., Livingstone, M.A., Kaspi, V.M., Cumming, A., and Archibald, R.F.: Post-outburst X-ray flux and timing evolution of swift J1822.3–1606. Astrophys. J. 761, 66 (2012)

Scholz, P., Kaspi, V.M., Cumming, A.: The long-term post-outburst spin down and flux relaxation of magnetar swift J1822.3–1606. Astrophys. J. 786, 62 (2014)

Shalybkov, D.A., Urpin, V.A.: Ambipolar diffusion and anisotropy of resistivity in neutron star cores. Mon. Not. Roy. Astron. Soc. 273, 643–648 (1995)

Spruit, H.C.: Origin of neutron star magnetic fields. In: Bassa, C., Wang, Z., Cumming, A., Kaspi, V.M. (eds.) 40 Years of Pulsars: Millisecond Pulsars, Magnetars and More. American Institute of Physics Conference Series, vol. 983, pp. 391–398. American Institute of Physics, New York (2008)

Steiner, A.W., Watts, A.L.: Constraints on neutron star crusts from oscillations in giant flares. Phys. Rev. Lett. 103(18), 181101 (2009)

Tauris, T.M., Kaspi, V.M., Breton, R.P., Deller, A.T., Keane, E.F., Kramer, M., Lorimer, D.R., McLaughlin, M.A., Possenti, A., Ray, P.S., Stappers, B.W., Weltevrede, P.: Understanding the neutron star population with the SKA. In: Advancing Astrophysics with the Square Kilometre Array (AASKA14). Proceedings of Science, vol. 215, p. 39 (Trieste: SISSA) (2015)

Thompson, C., Duncan, R.C.: The soft gamma repeaters as very strongly magnetized neutron stars - I. Radiative mechanism for outbursts. Mon. Not. Roy. Astron. Soc. **275**, 255–300 (1995)

Thompson, C., Duncan, R.C.: The Giant Flare of 1998 August 27 from SGR 1900+14. II. Radiative mechanism and physical constraints on the source. Astrophys. J. **561**, 980–1005 (2001)

Thompson, C., Lyutikov, M., Kulkarni, S.R.: Electrodynamics of magnetars: implications for the persistent X-ray emission and spin-down of the soft gamma repeaters and anomalous X-ray pulsars. Astrophys. J. **574**, 332–355 (2002)

Thompson, C., Yang, H., Ortiz, N.: Global crustal dynamics of magnetars in relation to their bright X-ray outbursts. Astrophys. J. **841**, 54 (2017)

Tiengo, A., Esposito, P., Mereghetti, S.: XMM-Newton observations of CXOU J010043.1–721134: the first deep look at the soft X-ray emission of a magnetar. Astrophys. J. Lett. **680**, L133–L136 (2008)

Tiengo, A., Esposito, P., Mereghetti, S., Turolla, R., Nobili, L., Gastaldello, F., Götz, D., Israel, G.L., Rea, N., Stella, L., Zane, S., Bignami, G.F.: A variable absorption feature in the X-ray spectrum of a magnetar. Nature **500**, 312–314 (2013)

Tsang, D., Gourgouliatos, K.N.: Timing noise in pulsars and magnetars and the magnetospheric moment of inertia. Astrophys. J. Lett. **773**, L17 (2013)

Turolla, R.: Isolated neutron stars: the challenge of simplicity. In: Becker, W. (ed.) Neutron Stars and Pulsars. Astrophysics and Space Science Proceedings, vol. 357, pp. 141–163. Springer, Heidelberg (2009)

Turolla, R., Esposito, P.: Low-magnetic magnetars. Int. J. Mod. Phys. D **22**, 1330024–163 (2013)

Turolla, R., Zane, S., Pons, J.A., Esposito, P., Rea, N.: Is SGR 0418+5729 indeed a waning magnetar? Astrophys. J. **740**, 105 (2011)

Turolla, R., Zane, S., Watts, A.L.: Magnetars: the physics behind observations. A review. Rep. Prog. Phys. **78**(11), 116901 (2015)

Urpin, V.A., Yakovlev, D.G.: Thermogalvanomagnetic effects in white dwarfs and neutron stars. Sov. Astron. **24**, 425 (1980)

Urpin, V.A., Levshakov, S.A., Iakovlev, D.G.: Generation of neutron star magnetic fields by thermomagnetic effects. Mon. Not. Roy. Astron. Soc. **219**, 703–717 (1986)

Uzdensky, D.A.: Shear-driven field-line opening and the loss of a force-free magnetostatic equilibrium. Astrophys. J. **574**, 1011–1020 (2002)

Vainshtein, S.I., Chitre, S.M., Olinto, A.V.: Rapid dissipation of magnetic fields due to the Hall current. Phys. Rev. E. **61**, 4422–4430 (2000)

van der Horst, A.J., Kouveliotou, C., Gorgone, N.M., Kaneko, Y., Baring, M.G., Guiriec, S., Göğüş, E., Granot, J., Watts, A.L., Lin, L., Bhat, P.N., Bissaldi, E., Chaplin, V.L., Finger, M.H., Gehrels, N., Gibby, M.H., Giles, M.M., Goldstein, A., Gruber, D., Harding, A.K., Kaper, L., von Kienlin, A., van der Klis, M., McBreen, S., Mcenery, J., Meegan, C.A., Paciesas, W.S., Pe'er, A., Preece, R.D., Ramirez-Ruiz, E., Rau, A., Wachter, S., Wilson-Hodge, C., Woods, P.M., Wijers, R.A.M.J.: SGR J1550–5418 bursts detected with the fermi gamma-ray burst monitor during its most prolific activity. Astrophys. J. **749**, 122 (2012)

van Hoven, M., Levin, Y.: Magnetar oscillations - I. Strongly coupled dynamics of the crust and the core. Mon. Not. Roy. Astron. Soc. **410**, 1036–1051 (2011)

Vietri, M., Stella, L., Israel, G.L.: SGR 1806–20: evidence for a superstrong magnetic field from quasi-periodic oscillations. Astrophys. J. **661**, 1089–1093 (2007)

Viganò, D., Pons, J.A.: Central compact objects and the hidden magnetic field scenario. Mon. Not. Roy. Astron. Soc. **425**, 2487–2492 (2012)

Viganò, D., Rea, N., Pons, J.A., Perna, R., Aguilera, D.N., Miralles, J.A.: Unifying the observational diversity of isolated neutron stars via magneto-thermal evolution models. Mon. Not. Roy. Astron. Soc. **434**, 123–141 (2013)

Vink, J., Kuiper, L.: Supernova remnant energetics and magnetars: no evidence in favour of millisecond proto-neutron stars. Mon. Not. Roy. Astron. Soc. **370**, L14–L18 (2006)

Wareing, C.J., Hollerbach, R.: Forward and inverse cascades in decaying two-dimensional electron magnetohydrodynamic turbulence. Phys. Plasmas **16**(4), 042307 (2009)

Wareing, C.J., Hollerbach, R.: Cascades in decaying three-dimensional electron magnetohydrody-
namic turbulence. J. Plasma Phys. **76**, 117–128 (2010)
Watts, A.L., Strohmayer, T.E.: High frequency oscillations during magnetar flares. Astrophys.
Space Sci. **308**, 625–629 (2007)
Wiebicke, H.-J., Geppert, U.: Amplification of neutron star magnetic fields by thermoelectric
effects. III - Growth limits in nonlinear calculations. Astron. Astrophys. **262**, 125–130 (1992)
Wood, T.S., Hollerbach, R.: Three dimensional simulation of the magnetic stress in a neutron star
crust. Phys. Rev. Lett. **114**(19), 191101 (2015)
Wood, T.S., Hollerbach, R., Lyutikov, M.: Density-shear instability in electron magneto-
hydrodynamics. Phys. Plasmas **21**(5), 052110 (2014)
Yakovlev, D.G., Pethick, C.J.: Neutron star cooling. Ann. Rev. Astron. Astroph. **42**, 169–210
(2004)
Yakovlev, D.G., Urpin, V.A.: The cooling of neutron stars and the nature of superdense matter.
Sov. Astron. Lett. **7**, 88 (1981)

Chapter 3
Radio Pulsars: Testing Gravity and Detecting Gravitational Waves

Delphine Perrodin and Alberto Sesana

Abstract Pulsars are the most stable macroscopic clocks found in nature. Spinning with periods as short as a few milliseconds, their stability can supersede that of the best atomic clocks on Earth over timescales of a few years. Stable clocks are synonymous with precise measurements, which is why pulsars play a role of paramount importance in testing fundamental physics. As a pulsar rotates, the radio beam emitted along its magnetic axis appears to us as pulses because of the lighthouse effect. Thanks to the extreme regularity of the emitted pulses, minuscule disturbances leave particular fingerprints in the times-of-arrival (TOAs) measured on Earth with the technique of pulsar timing. Tiny deviations from the expected TOAs, predicted according to a theoretical timing model based on known physics, can therefore reveal a plethora of interesting new physical effects. Pulsar timing can be used to measure the dynamics of pulsars in compact binaries, thus probing the post-Newtonian expansion of general relativity beyond the weak field regime, while offering unique possibilities of constraining alternative theories of gravity. Additionally, the correlation of TOAs from an ensemble of millisecond pulsars can be exploited to detect low-frequency gravitational waves of astrophysical and cosmological origins. We present a comprehensive review of the many applications of pulsar timing as a probe of gravity, describing in detail the general principles, current applications and results, as well as future prospects.

D. Perrodin
INAF - Osservatorio Astronomico di Cagliari, Selargius, CA, Italy
e-mail: delphine@oa-cagliari.inaf.it

A. Sesana (✉)
School of Physics and Astronomy and Institute of Gravitational Wave Astronomy, University of Birmingham, Birmingham, UK
e-mail: asesana@star.sr.bham.ac.uk

© Springer Nature Switzerland AG 2018
L. Rezzolla et al. (eds.), *The Physics and Astrophysics of Neutron Stars*,
Astrophysics and Space Science Library 457,
https://doi.org/10.1007/978-3-319-97616-7_3

3.1 Introduction to Pulsar Timing

Pulsars are highly-magnetized and fast-rotating neutron stars. In particular, radio pulsars emit beams of radio waves, which, thanks to the lighthouse effect, appear to distant observers as *pulses* for every rotation of the pulsar. So far we have discovered more than 2000 pulsars in our own Galaxy and the neighbouring Magellanic Clouds (Manchester et al. 2005). Of particular interest, millisecond pulsars (MSPs) are pulsars with very short rotation periods (1–30 ms) and are often found in binaries. It is now understood (Bhattacharya and van den Heuvel 1991) that these pulsars have been spun-up during the *recycling* process in which a companion star transfers angular momentum to the neutron star. Their very regular pulsations make them extremely stable clocks. Indeed, through the process of *pulsar timing*, which consists in monitoring the times-of-arrival (TOAs) of the pulsars' observed pulses over several years of observations, the rotation period of these pulsars can be estimated to 15 significant figures. The monitoring of MSPs therefore allows us to perform *high-precision pulsar timing*, with which we can precisely determine the properties of pulsars and their environment, and study the composition of the interstellar medium between Earth and each pulsar (Lorimer and Kramer 2012).

A newly-discovered pulsar is initially determined by its approximate rotation period P, dispersion measure DM (representing the integrated column density of free electrons along the line of sight between pulsar and Earth) and its position in the sky. Through pulsar timing, additional parameters characterizing the pulsar and its environment can be determined. A typical pulsar timing campaign consists of the regular monitoring of TOAs from a known pulsar over several years and with a weekly to monthly cadence. Each pulsar observation is divided into a number of time intervals (sub-integrations) and frequency channels (sub-bands). Since radio pulsars are faint and single pulses are rarely directly observable, it is necessary to integrate (*fold*) the radio pulses over many rotations of the pulsar to obtain integrated pulse profiles for each sub-integration and each sub-band. In addition, since the dispersion of the radio signal in the interstellar medium means that the higher-frequency signals arrive at the telescope before the lower-frequency signals, it is necessary to perform the process of *de-dispersion* of the radio signals within each sub-band. After *folding* and *de-dedispersing* the radio signals, topocentric TOAs are obtained by comparing the observed pulse profiles with high signal-to-noise standard profiles obtained from observations of the same pulsar over a time span of several years. The precision of our pulsar timing observations is characterized by the precision of the obtained TOAs (TOA error). Meanwhile, we can calculate expected TOAs based on our best-known models for the pulsar parameters. By subtracting the observed TOAs from the expected TOAs, we obtain *timing residuals* that are expected to be scattered around a zero mean, and which are characterized by a root-mean-square (rms) value. An excellent match between timing observations and timing model corresponds to a small rms residual.

Studying the pulsar timing residuals and improving the fitting of pulsar parameters enable us to refine our pulsar models. These models include parameters related to the pulsar's rotation (e.g. the period derivative $\dot{P}$) and orbit (when the pulsar is in a binary), which allow us to test gravity in the strong-field regime. Other parameters describe the dispersion of the radio signal in the interstellar medium as well as its time variations. Finally, and of great interest to Pulsar Timing Arrays (PTAs), we could also find, in the resulting timing residuals, the signature for low-frequency gravitational waves (GWs), such as those emitted by supermassive black hole binaries. In particular, in order to detect a background of low-frequency GWs, PTAs study the correlation of timing residuals for an array of pulsars, which are used as *cosmic clocks*. It is therefore crucial for PTAs to use pulsars with very high precision, or equivalently low rms residuals. In order to extract a low-frequency GW signal from the timing residuals, we also need to properly account for both pulsar timing noise, which is most likely related to instabilities in pulsar magnetospheres (Lyne et al. 2010), and time variations of the dispersion measure.

We note that topocentric TOAs, which are measured with Earth's telescopes, are not in an inertial frame. They need to be converted to *barycentric* TOAs, as if they were observed at the Solar System Barycentre (SSB). To transform topocentric TOAs to SSB TOAs, that is to perform the process of *barycentric correction*, we need to take several time delays related to Earth's orbit within the Solar System into account. There are also delays due to the pulsar's orbit if the pulsar is in a binary. The SSB TOAs t_{SSB} are related to the topocentric TOAs t_{topo} in this way:

$$t_{\mathrm{SSB}} = t_{\mathrm{topo}} + t_{\mathrm{clock}} - k \times \mathrm{DM}/f^2 \qquad (3.1)$$

$$+ (\Delta_{R,\odot} + \Delta_{S,\odot} + \Delta_{E,\odot}) + (\Delta_{R,\mathrm{bin}} + \Delta_{S,\mathrm{bin}} + \Delta_{E,\mathrm{bin}}), \qquad (3.2)$$

where t_{clock} refers to clock correction terms, k is a constant, DM is the dispersion measure and f is the observing frequency. The dominant term in the barycentric correction is the Roemer delay $\Delta_{R,\odot}$, which is the time delay due to light travel across the Earth's orbit. Second, we have the Shapiro delay $\Delta_{S,\odot}$, which is due to the curved gravitational field of the Sun and planets such as Jupiter. Finally, we have the Einstein relativistic time delay $\Delta_{E,\odot}$, which is due to the time dilation from the motion of the Earth, as well as the gravitational redshift from to the Sun and planets in the Solar System. Additionally, if the pulsar is in a binary, there are equivalent time delays due to the orbit of the pulsar and its companion: $\Delta_{R,\mathrm{bin}}$, $\Delta_{S,\mathrm{bin}}$, and $\Delta_{E,\mathrm{bin}}$. In fact, because of their strong-field dynamics, binary pulsars are extremely interesting for performing tests of strong-field gravity.

In Sect. 3.2, we will review the science and main results in the use of radio pulsars (and pulsar timing techniques) in testing gravity in the strong-field regime. In particular, relativistic binaries such as double neutron star (DNS) binaries provide great laboratories for testing General Relativity (GR), while neutron star—white dwarf (NS-WD) binaries are particularly suitable for tests of alternative theories of gravity. In Sect. 3.3, we discuss the science and main results in the use of radio pulsars as 'cosmic clocks' for detecting gravitational waves from distant supermassive black hole binaries and the limits already placed on such a background of gravitational waves. In Sect. 3.4, we discuss future prospects for both tests of strong gravity and gravitational wave detection, especially in light of the Square Kilometre Array (SKA). Finally we summarize our results in Sect. 3.5.

3.2 Tests of Gravity with Radio Pulsars

One hundred years have passed since Einstein presented his theory of gravity known as General Relativity (GR) in 1915. Much progress has been made since then to test the validity of GR. The most stunning confirmations of Einstein's theory include the indirect detection of gravitational waves (GWs) through timing observations of the Hulse-Taylor pulsar (Hulse and Taylor 1975), and the recent, direct detections of GWs from black hole binaries by the advanced Laser Interferometer Gravitational Observatory (LIGO), as predicted by Einstein (Abbott et al. 2016). Most of the earlier astrophysical tests of GR were done in the Solar System, which corresponds to the weak-field limit of gravity (Will 1993), that is a regime where the gravitational potential $\epsilon = GM/(Rc^2)$ around a test body of mass M and radius R (where G is the gravitational constant and c is the speed of light) is small. GR has thus far passed all tests with flying colours in the weak-field limit (Will 2010, 2014). However, the strong-field limit of gravity (where the gravitational potential ϵ is close to unity) has not been extensively tested, and gravity could possibly deviate from GR in this regime, such as in the environments around compact objects like neutron stars and black holes. We note that while the "strength" of gravity is usually characterized by the gravitational potential ϵ, a more thorough approach also includes the spacetime curvature $\xi \equiv GM/(R^3c^2)$ (Psaltis 2008). GR has also passed all tests conducted so far in the strong-field regime, including the recent LIGO observations of black hole binaries (Abbott et al. 2016). A number of *alternative theories of gravity*, which deviate from GR in the strong-field limit, but not in the weak-field limit which has been extensively tested, have been proposed (Damour and Esposito-Farèse 1996). Current tests of gravity seek to better constrain GR and alternative theories of gravity (ruling out some theories in the process), in the absence of any GR violation; or to potentially find deviations from GR in the strong-field limit. Why look for a breakdown of GR if it has thus far passed all tests with flying colours? As we know, GR is not compatible with quantum mechanics and could break down at small scales, such as in the interior of black holes where the concept of a black hole singularity is not physical. In addition, the evolution of the universe cannot be properly described by GR unless one adds the concept of *dark energy*, which could be modelled as a cosmological constant in Einstein's equations. The idea is then that GR is not a complete theory and that by testing gravity in the strong-field limit, we might find deviations from it.

Pulsars are ideal laboratories for testing GR and alternative theories of gravity. Their environments involve strong gravitational fields ($\epsilon \sim 0.2$ at the surface of a neutron star), and they provide us with much information in the form of extremely regular radio pulses. Pulsar binaries, which involve strong gravitational fields in the vicinity of the neutron star as well as high orbital velocities, are especially interesting for testing gravity, since the orbital dynamics depend on the underlying theory of gravity. Through the fitting of post—Keplerian parameters (see Sect. 3.2.1 below) in the pulsar TOAs, the orbital dynamics can be determined and the deformation of spacetime around the pulsar can be constrained (Weisberg and Taylor 1981; Damour and Taylor 1992; Edwards et al. 2006). Pulsars that are in orbit with a compact object provide even more constraining tests of gravity, especially when the two compact objects are in a close orbit. Therefore, by finding systems with companions in closer orbits, we are able to test the limits of GR. In GR, the *self-energy* of the neutron star does not affect the orbital dynamics. This is not the case in most alternative theories of gravity, where additional scalar, vector or tensor fields affect the spacetime curvature (Will 1993, 2010, 2014). We could therefore observe a breakdown of the predictions of GR in these systems.

Recent and comprehensive reviews have been published on the topics of: experimental gravity (Will 2014); astrophysical tests of gravity (Berti et al. 2015); tests of gravity with radio pulsars (Wex 2014). Recent reviews on tests of gravity with radio pulsars also include Possenti and Burgay (2016), Kramer (2016), and (Shao et al. 2015) discusses in particular all of the ways in which the Square Kilometre Array (SKA) will improve current gravity tests with pulsars. In this section, we outline the methods used to constrain GR and alternative theories of gravity with radio pulsars and present the most important results (best constraints) achieved thus far. Future prospects, in particular with the SKA, will be discussed in Sect. 3.4. The main methods with which radio pulsars can probe gravity involve: the Parametrized Post-Keplerian (PPK) formalism in pulsar binaries, including relativistic spin effects, as discussed in Sect. 3.2.1, and the Parametrized Post-Newtonian (PPN) formalism which quantifies deviations from GR (Sect. 3.2.2). We outline the best constraints on GR using the Double Pulsar in Sect. 3.2.1.3 and the best constraints on scalar-tensor theories of gravity (using mostly pulsar—white dwarf binaries) in Sect. 3.2.3.

3.2.1 Testing Gravity with the PPK Formalism

In the context of Newtonian physics, binary systems can be described by five Keplerian parameters: the orbital period P_b, the orbital eccentricity e, the projected semi-major axis $x \equiv a \sin i$, the longitude of periastron ω, and the time of periastron passage T_0. The *mass function* depends on the Keplerian parameters P_b and x:

$$f(M) \equiv \frac{(M_c \sin i)^3}{(M_P + M_c)^2} = \frac{4\pi^2 x^3}{G P_b^2}, \tag{3.3}$$

where M_P is the mass of the pulsar, M_c is the mass of the companion, and G is the gravitational constant. In the context of GR however, we will see below that we also need to include Post-Keplerian (PK) parameters that describe the relativistic effects beyond keplerian orbits, and which constitute excellent tools for testing gravity in binary pulsars.

Since GR is highly non-linear, it does not provide an exact, analytic description of the motion of two bodies. When compact objects move at less than relativistic speeds (the orbital velocity v/c is small), the dynamics of the system can be described by the Post-Newtonian (PN) approximation. In this formalism, the equations of motion are described by a series expansion based on powers of the small parameter $(v/c)^{2n}$, where n is the order of the PN expansion and the 0-th term corresponds to Newtonian dynamics. In fact, the motion of relativistic binaries is adequately described by the PN approximation for most of the binary's inspiral (the orbital velocity is high enough that PN terms are necessary to account for relativistic corrections; however when the velocity is too close to the speed of light right before the merger, the PN expansion breaks down). The 1PN dynamics in binaries—first order in the PN expansion, which corresponds to terms up to $(v/c)^2$—is described by the quasi-Keplerian parametrization of Damour and Deruelle (1985, 1986). Furthermore, Damour and Taylor proposed the Parametrized Post-Keplerian (PPK) formalism, which is a phenomenological parametrization based on the quasi-Keplerian parametrization (Damour 1988; Damour and Taylor 1992): it parametrizes the effects observed in both pulsar timing and pulse structure data. It is theory-independent, which allows us to test both GR and alternative theories of gravity, and consists of a Post-Keplerian (PK) set of parameters that describe the dynamics of relativistic binaries.

PK parameters are a function of known Keplerian parameters (supposedly already known to high precision), leaving only the two masses as unknowns: the pulsar's mass M_p and the companion's mass M_c. Therefore the measurement of two PK parameters leads to the determination of the two masses. By constraining more PK parameters, we can also constrain (or exclude) theories of gravity. N PK parameters will yield $N - 2$ tests for any chosen gravity theory. These PK parameters, which are included in pulsar timing models and therefore determined with years of pulsar data (always gaining higher precision with longer data spans), are best plotted in a M_p–M_c diagram. If PK constraints overlap in a mass-mass plot for a particular gravity theory, the particular theory of gravity is still considered a possible valid theory of gravity. If the PK constraints do not overlap, that theory is excluded (Possenti and Burgay 2016; Wex 2014).

In GR, the most important PK parameters are: the variations of two Keplerian parameters ω and P_b defined in Sect. 3.1, i.e. the relativistic precession of periastron $\dot{\omega}$ and the change in the orbital period due to the back-reaction of gravitational wave emission on the binary motion $\dot{P}_b$. Additionally, we have the time delays such as the Einstein delay related to the changing time dilation of the pulsar clock (due to variations in orbital velocity) and gravitational redshift γ, and the range r and shape s of the Shapiro delay related to a changing gravitational redshift in the gravitational field of the companion (Shapiro 1964). Their expressions as a function of the Keplerian parameters P_b, x, e and the two masses M_p and M_c are shown below (Damour and Deruelle 1986; Taylor and Weisberg 1989; Damour and Taylor 1992):

$$\dot{\omega} = 3\, T_\odot^{2/3} \left(\frac{P_b}{2\pi}\right)^{-5/3} \frac{1}{1-e^2}\, (M_p + M_c)^{2/3}, \tag{3.4}$$

$$\gamma = T_\odot^{2/3} \left(\frac{P_b}{2\pi}\right)^{1/3} e\, \frac{M_c(M_p + 2M_c)}{(M_p + M_c)^{4/3}}, \tag{3.5}$$

$$r = T_\odot\, M_c, \tag{3.6}$$

$$s \equiv \sin i = T_\odot^{-1/3} \left(\frac{P_b}{2\pi}\right)^{-2/3} x\, \frac{(M_p + M_c)^{2/3}}{M_c}, \tag{3.7}$$

$$\dot{P_b} = -\frac{192\pi}{5} T_\odot^{5/3} \left(\frac{P_b}{2\pi}\right)^{-5/3} \frac{\left(1 + \frac{73}{24}e^2 + \frac{37}{96}e^4\right)}{(1-e^2)^{7/2}} \frac{M_p M_c}{(M_p + M_c)^{1/3}}, \tag{3.8}$$

where masses are expressed in solar units, $T_\odot \equiv GM_\odot/c^3 = 4.925490947\,\mu s$, G is Newton's gravitational constant and c is the speed of light. Additional PK parameters of interest include the change in orbital eccentricity $\dot{e}$ and the change in the projected semi-major axis $\dot{x}$. The relativistic precession of periastron $\dot{\omega}$ is easiest to measure in eccentric orbits, while the Shapiro parameters r and s are measurable in nearly edge-on binary systems. In alternative theories of gravity, the expressions for the PK parameters are slightly different and include theory-dependent parameters that can be constrained (Will 1993, 2014).

3.2.1.1 Double Neutron Star Binaries

The first real test of gravity in the strong-field regime was accomplished by Hulse and Taylor in 1974 with the discovery of PSR B1913+16 (dubbed the *Hulse-Taylor pulsar*), which was the first *binary pulsar* ever discovered in the radio band. It consists of a pulsar in a double neutron star (DNS) binary (Hulse and Taylor 1975). The measurement of two PK parameters ($\dot{\omega}$ and γ) enabled the precise determination of the two neutron star masses (assuming GR was correct) (Weisberg et al. 2010). Having fully determined the binary system, any additional test would constitute a test of GR. In fact, the measurement of the decrease in the orbital period $\dot{P}$, associated with a loss of orbital energy, was found to be consistent with GR's predictions (Taylor et al. 1979). Specifically, it is consistent with GR's *quadrupole formula* that describes the backreaction of GW emission on the binary motion (Peters and Mathews 1963). This confirmed GR's predictions and provided the first indirect detection of GWs as predicted by Einstein. The agreement between the measured $\dot{P_b}$ and the predicted GR value is currently at the 0.2% level (Weisberg et al. 2010).

The pulsar PSR J0737–3039, discovered at Parkes in 2003 (Burgay et al. 2003; Lyne et al. 2004) is, like the Hulse-Taylor pulsar, composed of a DNS binary. In addition, the second neutron star has been observed as a pulsar; this system is therefore dubbed *the Double Pulsar* with two pulsars: PSR J0737–3039A (psr A with a period of 22 ms) and PSR J0737–3039B (psr B with a period of 2.7 s). The Double Pulsar is a profoundly unique system for testing gravity, since the radio pulses from both stars provide two clocks that can be monitored with pulsar timing. It is also characterized by large orbital velocities and a closeness of the orbit, which both amplify the importance of relativistic effects, and the high orbital inclination makes its timing easier. In this system, five PK parameters have been determined: $\dot{\omega}$, γ, r and s and $\dot{P}_b$ (Burgay et al. 2003). Additionally, the sizes of both pulsars' orbits were estimated and the mass ratio R, which is independent of the theory of gravity, was measured for the first time in a DNS system (Lyne et al. 2004).

DNS binaries are ideal systems for testing GR. In recent years, an increasing number of DNS systems have been discovered: so far, more than 15 DNS systems are known (Tauris et al. 2017). In the next few years, more pulsar surveys (in particular with the SKA, see Sect. 3.4) will discover new DNS binaries and further constrain GR.

3.2.1.2 Relativistic Spin Effects

Not all relativistic effects can be described at the 1PN level with the PK parameters. For example, tests of relativistic gravity can be done at 2PN (Damour and Esposito-Farèse 1996) or 2.5PN (Mirshekari and Will 2013). In addition, binary pulsars can have spin. The spin terms appear at higher orders in the Post-Newtonian expansion (Barker and O'Connell 1975; Damour 1987; Porto 2006; Brumberg 1991). In particular, the presence of spin-orbit coupling terms (the coupling of the spin of one pulsar with the binary's angular momentum) in the binary's equations of motion leads to the *Lense-Thirring precession of the orbit* or *frame-dragging*, as well as a change in the projected semi-major axis $\dot{x}$ (Barker and O'Connell 1975; Damour and Schafer 1988; Stella and Possenti 2009). Additionally, time-dependent spin terms in the equations of motion lead to changes in the orientations of the pulsar spins (which we refer to as *relativistic spin precession* or *geodetic precession*) (Damour and Ruffini 1974; Barker and O'Connell 1975; Boerner et al. 1975). The precession of the pulsar's rotation axis is essentially being caused by the curvature of spacetime from the companion star. This effect can be seen in changes in the pulsar emission: changes in the spin axis of the pulsar makes different regions of the magnetosphere visible to the observer, thus affecting the observed pulse profile.

In the Double Pulsar, the contribution to the Lense-Thirring precession is dominated by the fast-rotating psr A. However, $\dot{x}$ is difficult to measure because of the near alignment of pulsar spin and orbital angular momentum (Ferdman et al. 2013). Future measurements of the Lense-Thirring precession with the Double Pulsar is discussed in Kehl et al. (2016). The Double Pulsar is however the best system we know so far for testing relativistic spin precession (Stella and Possenti 2009). Indeed, the relativistic precession of psr B's spin axis can be determined thanks to the eclipses of psr A (that is when psr A passes behind psr B). Its precession rate was measured and found to be compatible with GR with an uncertainty of 13%: $\Omega_B = (4.77\pm^{0.66}_{0.65})°$ year^{-1} (Breton et al. 2008; Perera et al. 2010). Relativistic spin precession has also been observed in the following binary pulsars: PSR B1913+16 (Kramer 1998; Weisberg and Taylor 2002; Clifton and Weisberg 2008), PSR B1534+12 (Stairs et al. 2004; Fonseca et al. 2014), J1141–6545 (Manchester et al. 2010) and J1906+0746 (Lorimer et al. 2006). J0737–3039B and PSR B1534+12 are the only two pulsars for which we have a direct measurement of the precession rate (and which matches GR predictions) (Stairs et al. 2004; Fonseca et al. 2014; Breton et al. 2008; Perera et al. 2010).

3.2.1.3 Best Test of GR: The Double Pulsar

In the Double Pulsar, we have a total of seven mass constraints, thanks to the determination of five PK parameters, the mass ratio R (see Sect. 3.2.1.1), and the precession rate Ω_B (Kramer and Stairs 2008) (see Sect. 3.2.1.2). In addition, there are constraints related to the Newtonian mass function, one for each pulsar (see Eq. (3.3)). The masses of both pulsars are determined with high precision, leaving us with an additional five tests of GR, as shown in Fig. 3.1. The Double Pulsar provides

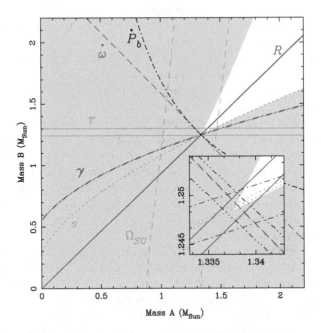

Fig. 3.1 Mass-mass diagram for the Double Pulsar J0737–3039. Shaded regions are excluded by the Newtonian mass functions (one for each pulsar). The five PK parameters ($\dot{P}_b$, $\dot{\omega}$, γ, r and s), the mass ratio R and the precession rate of psr B (Ω_{SO}) constrain the remaining parameter space, providing multiple tests of GR. So far GR is verified with an uncertainty of 0.05% (figure courtesy of Michael Kramer)

thus far the most stringent test of GR, with an uncertainty of 0.05% (Kramer et al. 2006). The longer we continue to monitor this system, the more precise the TOAs, and the better the GR constraints we will obtain. In particular, $\dot{\omega}$ could be determined up to the 2PN order, and the spin of psr A could be determined. With an even better determination of $\dot{P}_b$ (such as that expected thanks to the interferometric determination of the parallax Deller et al. 2009), the Double Pulsar will also provide stringent constraints on alternative theories of gravity that predict the presence of dipolar gravitational radiation. Through a measurement of its moment of inertia (Damour and Schafer 1988), the Double Pulsar could also constrain the equation of state of nuclear matter in neutron star interiors (Watts et al. 2015).

3.2.2 Testing Gravity Using the PPN Formalism

As we have seen in the previous sections, the fitting of PK parameters in the timing data of pulsar binaries allows us to determine the masses of the binary companions (if at least two PK parameters are measured) and to constrain gravity theories (if more than two PK parameters are measured). In addition, the study of the variations in pulse profiles allows us to determine changes in the spin precession of pulsars. These tools can also be applied to test the Strong Equivalence Principle (SEP), Lorentz invariance or conservation of momentum. The SEP is unique to GR: any violation of the SEP is a violation of GR (Will 1993). The SEP has been tested extensively in the Solar System, that is in the weak-field limit, using the PPN formalism (Will 2010; Will and Nordtvedt 1972; Will 1993, 2014). We refer the reader to Will (2014) for a full description of the formalism. The main idea is that in any metric theory of gravity, the dynamics (i.e. the equations of motion) of objects in a gravitational field depends exclusively on the structure of the metric. Therefore, any measurable departure from GR for a given theory has to be characterized by some difference in its metric compared to the GR one. In the weak-field limit, the most general metric can be written as an expansion of the Minkowski spacetime with the addition of ten (small) PPN parameters. Pulsars can test the SEP using the same formalism, providing in this way complementary tests to Solar System tests, since they can test gravity (GR and alternative theories of gravity) in the strong-field limit (Stairs 2003). This however requires a modification of the original ten PPN parameters to account for strong-field effects (Damour and Esposito-Farèse 1996; Damour and Esposito-Farese 1992) (the original PPN expansion is valid in the weak-field limit). The information we collect from pulsars with the determination of PK parameters can be translated into constraints on PPN parameters (PK parameters describe small variations in the motion of compact binaries, which can be mapped into small variations of the underlying metric). The ten (modified) PPN parameters describe the existence of preferred frames, preferred locations, the non-conservation of momentum, the non-linear superposition of gravitational effects, or the space-time curvature produced by a unit mass (for a full definition and physical interpretation of each individual parameter, see Will (2014); Possenti and Burgay (2016)).

3.2.2.1 SEP Violation and Orbital Dynamics

The SEP includes both the Weak Equivalence Principle (WEP) and the Einstein Equivalence Principle (EEP). The WEP tests the universality of free fall, stating that the trajectory of a free-falling body in a gravitational field should be independent of its internal structure. A first test of the SEP can therefore be accomplished by comparing the trajectories of two massive objects in a gravitational field, for example by looking for a *polarization* in the direction of the gravitational potential (this is the Nordtvedt effect or *gravitational Stark effect*, Nordtvedt (1968). Lunar Laser Ranging (LLR) experiments have tested the Nordtvedt effect by comparing the Earth and the Moon's free falls in the Sun's gravitational potential, and have imposed strong constraints on PPN parameters for the Solar System (Hofmann et al. 2010). Similarly, we can look at the two companions of a pulsar binary and how they *fall* in the gravitational potential of the Galaxy. It works best if the two companions are different in mass and composition, therefore double neutron star binaries (DNS) are not ideal laboratories for testing SEP violations. Instead, a sample of pulsars with white dwarf companions (PSR-WD) can impose strong constraints on SEP violations (Stairs et al. 2005; Gonzalez et al. 2011; Freire et al. 2012), in particular on the following parameter:

$$\Delta = \left(\frac{M_{\mathrm{grav}}}{M_{\mathrm{inertial}}} \right)_1 - \left(\frac{M_{\mathrm{grav}}}{M_{\mathrm{inertial}}} \right)_2 , \tag{3.9}$$

where M_{grav} is the gravitational mass and M_{inertial} is the inertial mass of each body. So far the best constraint on Δ is from a study of 27 PSR-WD binaries (Gonzalez et al. 2011): $\Delta < 4.6 \times 10^{-3}$ (see Table 3.1).

The discovery of an MSP (PSR J0337+1715) in a triple system with two white dwarf companions (Ransom et al. 2014) will allow us to greatly improve the constraint on the SEP. The masses of the three bodies have all been determined. The two inner masses (the pulsar and inner WD), of different masses and composition, are moving in the gravitational field of the outer WD, which is larger than that of the Galaxy by at least six orders of magnitude, therefore the SEP violation would be greatly magnified. This system could therefore be the best laboratory we have so far to constrain the SEP, with an estimated constraint on the parameter Δ of four orders of magnitude better than current constraints, most likely with the use of future telescopes (Ransom et al. 2014; Shao et al. 2015; Berti et al. 2015; Shao 2016).

3.2.2.2 SEP Violation: Violation of LLI and LPI

The EEP states that local, non-gravitational experiments are independent of the frame. The EEP consists of the Local Lorentz Invariance (LLI) and Local Position Invariance (LPI). Violations of LLI correspond to the observation of a preferred frame, while violations of LPI correspond to the observation of preferred positions, and may also lead to variations in fundamental constants such as the gravitational

Table 3.1 Best constraints on PPN parameters characterizing deviations from the Strong Equivalence Principle (SEP); on the spatial anisotropy $|\Delta G/G|$ and time variation $|\dot{G}/G|$ of the gravitational constant; on dipolar radiation via the parametrization $|\alpha_P - \alpha_0|$

Parameter	Upper limit	Method
Δ	5.6×10^{-3} (95% CL)	PSR-WD binaries (Stairs et al. 2005)
	4.6×10^{-3} (95% CL)	PSR-WD binaries (Gonzalez et al. 2011) (see Wex (2014) for discussion)
$\hat{\alpha}_1$	$\left(-0.4^{+3.7}_{-3.1}\right) \times 10^{-5}$ (95% CL)	Timing analysis of PSR J1738+0333
		better than solar system (Shao and Wex 2012; Antoniadis et al. 2012; Freire et al. 2012)
$\hat{\alpha}_2$	1.6×10^{-9} (95% CL)	Timing analysis of PSR J1738+0333
		+ pulse profile data of B1937+21/J1744–1134
		better than solar system (Antoniadis et al. 2012; Freire et al. 2012; Shao et al. 2013)
$\hat{\alpha}_3$	4×10^{-20} (95% CL)	PSR-WD binaries (better than solar system) (Stairs et al. 2005)
$\hat{\xi}$	3.9×10^{-9} (95% CL)	Pulse profile data of B1937+21/J1744–1134
		better than solar system (Shao and Wex 2013)
$|\frac{\Delta G}{G}|^{\text{anis.}}$	4×10^{-16}	Derived from $\hat{\xi}$ constraint (Shao and Wex 2013)
$\hat{\zeta}_2$	4×10^{-5}	Non-conservation of momentum from B1913+16 (Will 1992)
$|\frac{\dot{G}}{G}|$	$[(-0.6 \pm 1.1)] \times 10^{-12}$ year^{-1} (95% CL)	J1713+0747 (Zhu et al. 2015)
dipolar	0.002 (95% CL)	J1738+0333 (Freire et al. 2012)
$|\alpha_P - \alpha_0|$	0.005 (95% CL)	J0348+0432 (Antoniadis et al. 2013) interesting because of massive NS

constant G. Violations of LLI and LPI both involve changes in the orbital dynamics of binary pulsars and the spin precession of solitary pulsars, which are characterized by PK parameters such as the changes in orbit eccentricity $\dot{e}$, inclination $\dot{x}$, and the periastron advance rate $\dot{\omega}$. Testing of LPI can in particular be done by looking at the spin precession of pulsars: a violation of LPI could be seen if we observe changes in the expected pulsar spin precession around the acceleration toward the galactic centre. This would be evident by studying the stability of the pulse profiles of solitary pulsars.

The violations of LLI and LPI, which, for binary pulsars, are determined by changes in the aforementioned PK parameters, are characterized by the following PPN parameters: $\hat{\alpha}_1$, $\hat{\alpha}_2$ and $\hat{\alpha}_3$, where the ˆ refers to the strong-field generalization of the associated PPN parameter. The parameters $\hat{\alpha}_1$ and $\hat{\alpha}_2$ involve the existence of a preferred frame (i.e. non-zero values would imply a violation of LLI). $\hat{\alpha}_2$ also includes the spin precession of the pulsar, which can be seen from changes in pulse profiles. A non-zero value of the PPN parameter $\hat{\alpha}_3$ involves both the existence of a preferred frame (a violation of LLI) and a violation of conservation of momentum (Will 1993). The parameter $\hat{\xi}$, which is the strong-field equivalent of the Whitehead PPN parameter ξ, characterizes LPI violation through measurements of the spin precession; a limit on $\hat{\xi}$ can be converted into a constraint on the spatial anisotropy of the gravitational constant G (Shao and Wex 2013). The parameter $\hat{\zeta}_2$ characterizes non-conservation of momentum through the measurements of the polarization of the orbit and the spin precession. Additionally, if gravitational dipole radiation were to be observed, the SEP would be automatically violated, and thus GR would be violated. This would also lead to stringent constraints on alternative theories of gravity.

We find that the timing analysis of the PSR-WD binary PSR J1738+0333 leads to some of the best constraints on PPN parameters. Additionally, it is also the best pulsar so far to constrain scalar-tensor gravity (see Sect. 3.2.3). Other interesting and complementary constraints are obtained from the pulse profile analysis of isolated MSPs PSR B1937+21 and PSR J1744–1134. The best constraints on $\hat{\alpha}_1$, $\hat{\alpha}_2$, $\hat{\alpha}_3$, $\hat{\xi}$ and $\hat{\zeta}_2$ are listed in Table 3.1.

3.2.2.3 Varying Gravitational Constant

Violation of LPI can lead to variations in fundamental constants such as the gravitational constant G. PSR J0437–4715 is one of the best pulsars for high precision timing because of its closeness to Earth and its brightness (Johnston et al. 1993; Deller et al. 2009; Kopeikin 1995, 1996; van Straten et al. 2001). The inclination angle can be determined independently of the theory of gravity and compared to the expected Shapiro delay ($s = \sin i$). They are in good agreement. Until recently, this pulsar provided the best test of $\dot{G}$ (using pulsar binaries): $|\frac{\dot{G}}{G}| < 23 \times 10^{-12}\,\text{year}^{-1}$ (Verbiest et al. 2008). Recent measurements of the pulsar PSR J1713+0747 however show a tighter constraint: $|\frac{\dot{G}}{G}| < (-0.6 \pm 1.1) \times 10^{-12}\,\text{year}^{-1}$ at 95% CL (Zhu et al. 2015). This is the best limit on $\dot{G}/G$ using pulsar binaries, which are now close to competing with Solar System tests. Indeed, Lunar Laser Ranging currently provides the most stringent constraint with $|\frac{\dot{G}}{G}| < (-0.7 \pm 3.8) \times 10^{-13}\,\text{year}^{-1}$ (Hofmann et al. 2010).

3.2.3 Tests of Alternative Theories of Gravity

Pulsars allow us to test both GR and alternative theories of gravity: we may either detect a *breakdown* of GR; or we could confirm GR and place limits on alternative theories of gravity, such as scalar-tensor theories. These theories involve additional degrees of freedom (in the form of scalar fields) in the Einstein-Hilbert action, which describes how gravity is mediated. In the particular case of *tensor-mono-*

scalar theories Damour and Esposito-Farese (1993); Damour and Esposito-Farèse (1996), gravity is mediated by the Einstein metric field $g^*_{\mu\nu}$ as well as a scalar field ϕ, while matter is coupled to the physical metric $g_{\mu\nu}$. The physical metric $g_{\mu\nu}$ and Einstein metric $g^*_{\mu\nu}$ are related through the coupling function $a(\phi)$ between matter and scalar field, such that $g_{\mu\nu} = g^*_{\mu\nu} a^2(\phi)$. In these theories, the coupling function $a(\phi)$ takes the form:

$$a(\phi) = \alpha_0\phi + 1/2\beta_0\phi^2 \qquad (3.10)$$

This formalism includes GR in the case where $\alpha_0 = \beta_0 = 0$. It also includes the Jordan-Fierz-Brans-Dicke theory (Brans and Dicke 1961; Will and Zaglauer 1989) in the case where $\beta_0 = 0$ and $\alpha_0 = \sqrt{1/(2\omega_{BD} + 3)}$, where ω_{BD} is the Brans-Dicke parameter. Variations in the scalar field ϕ could produce observable effects such as a gravitational constant varying with space and time (non-zero $\dot{G}$) (Damour and Esposito-Farese 1992; Will 1993) or the detection of gravitational dipole radiation in a pulsar binary, either of which would constitute a violation of the SEP and a *breakdown* of GR. The existence of a varying gravitational constant or dipole gravitational radiation would affect the PK parameters, most particularly the orbital decay $\dot{P}_b$ (Damour 1988; Wex 2014). In the absence of an obvious breakdown of GR (no detection of $\dot{G}$ or GW dipole radiation), the (α_0, β_0) parameter space can be constrained by binary pulsar observations (Freire et al. 2012). We note that non-perturbative effects such as spontaneous scalarization could also affect the dynamics of the binary system (Damour and Esposito-Farese 1993). While the DNS systems such as B1913+16, B1534+12 and J0737−3039 provide constraints on scalar-tensor theories, they are not the best sources for testing alternative theories of gravity such as scalar-tensor gravity. Indeed, in the case of two identical neutron stars, the dipolar gravitational radiation term essentially vanishes. Pulsars with WD companions, with different masses and compositions, can better constrain these theories. In most PSR-WD binaries, only two PK parameters can be determined (the Shapiro delay parameters r and s), allowing a determination of the two binary masses, however that is not enough for constraining gravity theories (Antoniadis et al. 2012, 2013; Wex 2014). Interestingly, the following PSR-WD binaries allow for the determination of more than 2PK parameters: J1141−6545, J1738+0333, J0437−4715 and J0348+0432. They provide tests that are complementary to the GR tests using DNS J0737−3039 and B1913+16 (Possenti and Burgay 2016).

- In PSR J1141−6545, three PK parameters can be determined: $\dot{\omega}$, γ, and $\dot{P}_b$ (Kaspi et al. 2000). This has led to the determination of both masses and one test of GR at the 10% level (Bhat et al. 2008). The pulsar's relativistic spin precession can also be observed (Manchester et al. 2010), but is not as well measured as for the Double Pulsar or B1913+16. It is however useful for constraining scalar-tensor theories and possibly detecting dipolar gravitational radiation.
- PSR 1738+0333 is so far the most useful pulsar for constraining scalar tensor theories (Jacoby 2005; Freire et al. 2012; Wex 2014), as it provides a precise determination of $\dot{P}_b$ (in good agreement with GR), as well as proper motion and parallax, giving the best upper limit on dipolar GW (see Table 3.1).

- PSR J0348+0432, discovered in 2013 (Boyles et al. 2013; Lynch et al. 2013), has the highest mass of any pulsar observed so far: $2.01 \pm 0.04\, M_\odot$. It provides a stringent constraint on $\dot{P}_b$, which is currently at the 82% agreement with GR, leading to a constraint on dipolar GW radiation, though its upper limit is not as high as J1738+0333 (see Table 3.1). Spontaneous scalarization in such a massive system creates an important amount of gravitational dipolar radiation, which rules out an important part of the parameter space in alternative theories; this pulsar also places constraints on a long-range field (Antoniadis et al. 2013). Finally, thanks to its high mass, J0348+0432 constrains the equation of state of nuclear matter, favouring a stiff equation of state (Watts et al. 2015).

So far, PSR J1738+0333 and PSR J0348+0432 provide the best constraints on scalar-tensor gravity theories (including Jordan-Brans-Dicke theory for which $\beta_0 = 0$); their constraints are comparable to solar system tests such as the Cassini probe (see Fig. 3.2) (Freire et al. 2012). They also provide the best constraints on quadratic scalar-tensor gravity (for for $\beta_0 < -3$ and $\beta_0 > 0$ (Freire et al. 2012; Berti et al. 2015; Kramer 2016; Wex 2014). J1738+0333 also excludes

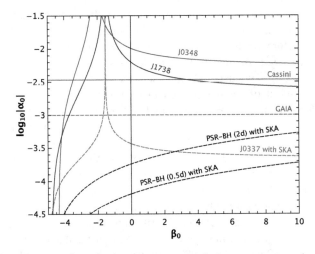

Fig. 3.2 Constraints on the coupling parameters α_0 and β_0 in tensor-mono-scalar theories with coupling $\alpha(\phi) = \alpha_0\phi + 1/2\beta_0\phi^2$ (figure courtesy of Norbert Wex). GR is located at the intersection of $\alpha_0 = 0$ and $\beta_0 = 0$, while Jordan-Brans-Dicke theories, for which $\beta_0 = 0$, are along the y-axis. The allowed parameter space is constrained to the area below all of the solid lines. The PSR-WD binaries J1738+0333 (purple solid line) and PSR J0348+0342 (blue solid line) provide the best constraints so far, and are comparable to Solar System constraints such as with the Cassini spacecraft (grey solid line) and the future GAIA astrometric satellite (grey dashed line). The triple system PSR J0337+1715 (red dashed line) will likely impose stronger constraints in the near future, especially as observed with the SKA. In addition, we show the expected constraints (black dashed lines) from two hypothetical PSR-BH systems we expect to find with the SKA (with orbital periods $P_b = 2$d and $P_b = 0.5$d, respectively). We see that the pulsar timing of PSR-WD systems (such as the triple system PSR J0373+1715) and PSR-BH systems are complementary: the former imposes strong constraints at positive β_0's, while the latter imposes strong constraints at negative β_0's. These estimates are based on a stiff NS equation of state (MPA1), making the constraints rather conservative (Kramer 2016; Shao et al. 2017)

TeVeS-like theories (Freire et al. 2012). Massive Brans-Dicke theories are best constrained by PSR J1141–6545 (Alsing et al. 2012), while Einstein-Aether theories are best constrained by a combination of pulsars: the PSR-WD binaries J1141–6545, J1738+0333, J0348+0432 together with the Double Pulsar J0737–3039 (Yagi et al. 2014). We note that the triple system PSR J0337+1715 will likely impose even stronger constraints in the near future (Ransom et al. 2014; Shao 2016; Berti et al. 2015). The discovery of a pulsar—black hole (PSR-BH) system would also further constrain the parameter space of scalar-tensor theories (Liu et al. 2012, 2014; Wex et al. 2013).

3.3 Gravitational Wave Detection with Radio Pulsars

Because of the exquisite stability of MSPs, pulse TOAs are extremely sensitive to any type of perturbation affecting the photon path from the source to Earth, such as variations in the interstellar medium (ISM), solar wind, etc. (see Sect. 3.3.3.1 below). This makes MSPs formidable tools for detecting GWs. In fact, the passage of a GW between a pulsar and the Earth modifies the null geodesic along which the photons propagate, resulting in small alterations of the pulse TOAs. This was realized even before the discovery of the first MSP (Sazhin 1978; Detweiler 1979), by applying the mathematical formalism developed by Estabrook and Wahlquist (1975) for detecting GWs using Doppler spacecraft tracking to pulsars. Early work based on a handful of regular pulsars made use of the technique to constrain a putative low-frequency GW background (GWB) of cosmic origin to the level of about $\Omega_{gw} \approx 10^{-4}$ times the critical density of the Universe (Hellings and Downs 1983; Bertotti et al. 1983; Romani and Taylor 1983). In particular, Hellings and Downs (1983) proposed that the effect of a GWB is encoded in the peculiar correlation of TOAs collected from pairs of pulsars at different sky locations, and worked out the analytical form of the pattern, which is now known as the Hellings and Downs curve and is at the heart of current GWB searches with PTAs. The idea was elaborated by Foster and Backer (1990), who proposed the concept of a Pulsar Timing Array (PTA), consisting in the regular monitoring of a number of the newly-discovered MSPs (Backer et al. 1982). By just monitoring two MSPs, Kaspi et al. (1994) improved the limit on a stochastic GWB to $\Omega_{gw} = 6 \times 10^{-8}$. In the early 2000 s, three major collaborations formed with the goal of providing systematic timing residuals on a sizable ensemble of MSPs: the European Pulsar Timing Array (EPTA Desvignes et al. 2016), the Parkes Pulsar Timing Array (PPTA Reardon et al. 2016) and the North American Nanohertz Observatory for Gravitational Waves (NANOGrav, Arzoumanian et al. 2015). The three collaborations also share data under the aegis of the International Pulsar Timing Array (IPTA, Verbiest et al. 2016), with the goal of obtaining a combined, more sensitive dataset. Altogether, the three PTAs are timing approximately fifty of the best MSPs with a weekly cadence (Δt) and for a timespan T of several years (more than 20 in some cases), with a timing precision ranging from a few microseconds to a few tens of nanoseconds.

PTAs are therefore sensitive to GWs in the frequency range $1/T < f < 1/(2\Delta t)$, corresponding to a few to a few hundred nanohertz. Putative GW signals in this frequency range include those from cosmological stochastic backgrounds from inflation, phase transitions or cosmic strings (Lasky et al. 2016), but the loudest GW source is expected to be the cosmic population of inspiralling supermassive black hole binaries (SMBHBs), formed following galaxy mergers (Sesana et al. 2008).

3.3.1 Detection Principle

To elucidate the detection principle of PTAs, we follow the derivation in Maggiore (2018). Let us consider a pulsar p pulsating regularly as a perfect clock. A modification in the photon path will result in the pulses arriving slightly earlier or later. The net result is therefore a change in the pulsation frequency $v(t)$ observed on Earth, i.e. a redshift (or Doppler shift):

$$z(t) = \frac{v(t) - v_0}{v_0} = \frac{\delta v(t)}{v_0}, \tag{3.11}$$

where v_0 is the intrinsic pulsar frequency. To establish the potential of PTAs as GW detectors, we need to compute the redshift that is induced by a GW crossing the line of sight to the pulsar. In analogy with spacecraft Doppler tracking studied by Estabrook and Wahlquist (1975), it can be demonstrated that in a conformal flat spacetime, for a wave $h_{ij}(t)$ incident on a pulsar located in direction $\hat{p}$, the observed redshift at time t is

$$z(t) = \frac{1}{2} p^i p^j \int_{t_p}^{t} dt' \frac{\partial}{\partial t'} h_{ij}[t', (t - t_p)\hat{p}]. \tag{3.12}$$

Equation (3.12) is obtained by integrating the time component of the photon null geodesic equation in the weak field limit (see Lee et al. (2010) and referenceses therein). Let us now take, without any loss of generality, the case of a wave incident in the z direction $\hat{\Omega} = (0, 0, -1)$, and a pulsar located in the (x, z) plane in direction $\hat{p} = (\sin\theta_p, 0, \cos\theta_p)$, so that $\pi - \theta_p$ is the angle between the direction to the pulsar and the propagation direction of the incoming wave (thus θ_p is the angle subtended by the pulsar and the GW source, located at $-\hat{\Omega} = (0, 0, 1)$). We restrict our discussion to GR, so that the wave only has tensor components identified by $h_{xx} = -h_{yy}$, $h_{xy} = h_{yx}$. In this case, we have $p^i p^j h_{ij} = \sin^2\theta_p h_{xx}$ and, after some manipulations, Eq. (3.12) gives

$$z(t) = \frac{1}{2}(1 + \cos\theta_p)[h_{xx}(t_p) - h_{xx}(t)]. \tag{3.13}$$

Here $t - t_p \equiv \tau = (L/c)(1 + \hat{\Omega} \cdot \hat{p})$ is the difference in TOAs of the incident wave at the pulsar and at the Earth, where L is the Earth-pulsar distance. We note that the redshift z is given by the difference between the metric perturbations at the pulsar at the time of radio emission t_p and the metric perturbations at the Earth at the time of observation t. Equation (3.13) provides some insight about the response of a pulsar to an incoming wave. If the GW source and the pulsar are located on opposite sides (as seen from Earth), then $\theta_p = \pi$ and the resulting redshift vanishes. If, on the other hand, the GW source is located right behind the pulsar, then $t = t_p$ and the two metric perturbation components cancel exactly, again giving zero redshift. This is consistent with the transverse nature of GWs.

The actual quantity measured in PTA experiments is the timing residual $r(t)$. This is simply given by the integral over observing time of the redshift induced by the incident GW:

$$r(t) = \int_0^t dt' z(t', \hat{\Omega}),$$ (3.14)

where t is the time of a given pulsar observation, and the integral starts from the beginning of the timing experiment.

3.3.1.1   Generalization of the Residual Formula

Equation (3.13) describes the response of the pulsar-Earth detector to an incoming wave in the z direction. It is useful to generalize the formula in two ways. First, although for any given source-pulsar pair we can always define a frame in which the source is in the z direction and the pulsar lies in the (x, z) plane, PTAs combine observations of an ensemble of pulsars Foster and Backer 1990. It is therefore useful to write the pulsar response in a generic fixed frame which does not have a specific alignment with respect to the source-Earth-pulsar reference. Second, Eq. (3.13) is expressed in terms of the GW component along the direction defined by the projection of the pulsar location into a plane perpendicular to the incident wavefront (direction x in this case). It is however useful to write the response in terms of the two tensor polarizations of the GW wave h_+ and $h_\times$.

We consider a Cartesian reference frame (x, y, z) centred at the solar system barycentre.[1] The source $\hat{\Omega}$ and pulsar $\hat{p}$ locations are therefore defined in terms of the standard angles (θ, ϕ):

$$\hat{\Omega} = -(\sin\theta \cos\phi)\,\hat{x} - (\sin\theta \sin\phi)\,\hat{y} - \cos\theta\hat{z}$$ (3.15a)

---

[1] As discussed in Sect. 3.1, TOAs are computed by converting the pulse arrival time at the observatory to the pulse arrival time at the solar system barycentre. In fact, when we refer to 'TOAs measured on Earth', 'GW Earth term' etc., those have to be intended 'at the solar system barycentre'.

$$\hat{p} = (\sin \theta_p \cos \phi_p)\,\hat{x} + (\sin \theta_p \sin \phi_p)\,\hat{y} + \cos \theta_p \hat{z}. \tag{3.15b}$$

Note the minus sign in $\hat{\Omega}$, which is defined as the direction of the *incoming* wave.

The wave propagating from the $\hat{\Omega}$ direction consists of two polarization states h_+ and $h_\times$. The relation between those and the metric perturbation along a specific direction is given by

$$h_{ij}(t, \hat{\Omega}) = e_{ij}^{+}(\hat{\Omega})h_+(t, \hat{\Omega}) + e_{ij}^{\times}(\hat{\Omega})h_\times(t, \hat{\Omega}), \tag{3.16}$$

where the polarization tensors $e_{ij}^{A}(\hat{\Omega})$ (with $A = +, \times$) are defined as

$$e_{ij}^{+}(\hat{\Omega}) = \hat{m}_i \hat{m}_j - \hat{n}_i \hat{n}_j\,, \tag{3.17a}$$

$$e_{ij}^{\times}(\hat{\Omega}) = \hat{m}_i \hat{n}_j + \hat{n}_i \hat{m}_j\,. \tag{3.17b}$$

Here $\hat{m}, \hat{n}$ are the GW principal axes and define, together with the direction of the wave propagation $\hat{\Omega}$, an orthonormal basis. Note that $\hat{m}$ is aligned with the plus wave polarization. We therefore have two Cartesian coordinate systems: one is the 'detector frame' defined by $(\hat{x}, \hat{y}, \hat{z})$, and one is the 'wave propagation frame' defined by $(\hat{m}, \hat{n}, \hat{\Omega})$. To compute the response in the detector frame, one needs to project onto it the metric perturbation defined along the principal axes $\hat{m}, \hat{n}$ of the wave propagation frame. The principal axis $\vec{m}$ defines an angle ψ (counterclockwise about the wave propagation) with the line of nodes of the detector frame. We can thus perform a rotation by an angle ψ to express $\hat{m}, \hat{n}$ in the detector frame coordinates (Anderson et al. 2001):

$$\hat{m} = (\sin \phi \cos \psi - \sin \psi \cos \phi \cos \theta)\hat{x} - (\cos \phi \cos \psi + \sin \psi \sin \phi \cos \theta)\hat{y} + (\sin \psi \sin \theta)\hat{z}\,, \tag{3.18a}$$

$$\hat{n} = (-\sin \phi \sin \psi - \cos \psi \cos \phi \cos \theta)\hat{x} + (\cos \phi \sin \psi - \cos \psi \sin \phi \cos \theta)\hat{y} + (\cos \psi \sin \theta)\hat{z}. \tag{3.18b}$$

Now that we have defined all of the relevant quantities with respect to the detector frame, Eq. (3.13) can be generalized to

$$z(t, \hat{\Omega}) = \frac{1}{2}\frac{\hat{p}^i \hat{p}^j}{1 + \hat{p}^i \hat{\Omega}_i}\left\{e_{ij}^{+}(\hat{\Omega})\left[h_+(t_p, \hat{\Omega}) - h_+(t, \hat{\Omega})\right] - e_{ij}^{\times}(\hat{\Omega})\left[h_\times(t_p, \hat{\Omega}) - h_\times(t, \hat{\Omega})\right]\right\}, \tag{3.19}$$

which can be written in compact form as

$$z(t, \hat{\Omega}) = \sum_A F^A(\hat{\Omega})[h_A(t_p, \hat{\Omega}) - h_A(t, \hat{\Omega})] \tag{3.20}$$

where

$$F^A(\hat{\Omega}) = \frac{1}{2} \frac{\hat{p}^i \hat{p}^j}{1 + \hat{p}^i \hat{\Omega}_i} e_{ij}^A(\hat{\Omega}). \tag{3.21}$$

In practice, this notation separates the physics of GW emission, enclosed in the h_A terms, from all of the geometric factors arising from the transformation between the radiation and the detector frames, which are absorbed in the F^A response functions. Note that the latter are universal, i.e. they do not depend on the nature of the GW signal.[2] The explicit form of the response functions (or antenna beam patterns) is given by

$$F^+(\hat{\Omega}) = \frac{1}{2} \frac{(\hat{m} \cdot \hat{p})^2 - (\hat{n} \cdot \hat{p})^2}{1 + \hat{\Omega} \cdot \hat{p}}, \tag{3.22a}$$

$$F^\times(\hat{\Omega}) = \frac{(\hat{m} \cdot \hat{p})(\hat{n} \cdot \hat{p})}{1 + \hat{\Omega} \cdot \hat{p}}. \tag{3.22b}$$

Note that the response functions depend only upon the three direction cosines $\hat{m} \cdot \hat{p}$, $\hat{n} \cdot \hat{p}$ and $\hat{\Omega} \cdot \hat{p}$ and are independent of the specific choice of Cartesian detector frame, as expected.

3.3.1.2   Stochastic Background

The set of equations presented in Sect. 3.3.1.1 forms a useful method for computing the redshift (and the associated residual through Eq. (3.14)) induced by an incident deterministic GW with a generic form $h(t)$. We now generalize the derivation for a stochastic GWB generated by the incoherent superposition of uncorrelated sources randomly distributed in the sky. In this case, Eq. (3.20) is generalized to represent the incoming GWs in the Fourier domain as $\tilde{h}_A(f)$ and by integrating over all possible frequencies and incoming directions to obtain

$$z(t) = \sum_{A=+,\times} \int_{-\infty}^{\infty} df \int d^2\hat{\Omega}\, F^A(\hat{\Omega}) \tilde{h}_A(f, \hat{\Omega}) e^{-2\pi i f t} \left[1 - e^{-2\pi i f \tau} \right], \tag{3.23}$$

where $\tau = t - t_p$ has been defined in Sect. 3.3.1. The interesting quantity for a GWB is the ensemble average over the stochastic variable $\tilde{h}_A(f)$. Under the assumption of an isotropic stationary and unpolarized background, this ensemble average takes

---

[2]This is true so long as only GR tensor polarizations are considered. In alternative theories of gravity, scalar and vector polarizations might also arise, and require different response functions F (Yunes and Siemens 2013).

the form Maggiore (2000)

$$\langle \tilde{h}_A^*(f, \hat{\Omega}) \tilde{h}'_A(fi, \hat{\Omega}') \rangle = \delta(f - f') \frac{\delta^2(\hat{\Omega}, \hat{\Omega}')}{4\pi} \delta_{AA'} \frac{1}{2} S_h(f), \tag{3.24}$$

where $S_h(f)$ is the power spectral density of the GWB, and the $1/2$ factor comes from considering only positive frequencies, i.e. $0 < f < \infty$. The ensemble average of the timing residuals observed in a pair of MSPs denoted as a and b then becomes

$$\langle z_a(t) z_b(t) \rangle = \frac{1}{2} \int_{-\infty}^{\infty} df S_h(f) \int d^2\hat{\Omega} \frac{1}{4\pi} \sum_{A=+,\times} F_a^A(\hat{\Omega}) F_b^A(\hat{\Omega}). \tag{3.25}$$

The above result comes from substituting Eq. (3.24) into Eq. (3.23) and by noticing that all of the terms involving $e^{-2\pi i f \tau}$ can be neglected in the short wavelength limit, which is appropriate for PTAs. PTAs are in fact sensitive to nHz GWs, corresponding to parsec wavelengths, which is much shorter than the distance to the closest known MSP of about 150 pc (typical MSP distances are in the kpc range).

The integral over sky orientations in Eq. (3.25) was first computed by Hellings and Downs (1983) and takes the form

$$C(\zeta_{ab}) = \frac{1}{4} \left[1 + \frac{\cos \zeta_{ab}}{3} + 4(1 - \cos \zeta_{ab}) \ln \left(\sin \frac{\zeta_{ab}}{2} \right) \right]. \tag{3.26}$$

where ζ_{ab} is the angle between the pulsars a and b on the sky. Finally, the observable quantity in PTA observations is the ensemble average cross correlation in the timing residuals between two pulsars $r_{ab} = \langle r_a(t) r_b(t) \rangle$, where $r_x(t)$ is defined by Eq. (3.14). We can therefore integrate Eq. (3.25) over time to get the final form of the correlation in the timing residuals

$$r_{ab} = C(\zeta_{ab}) \int_0^{\infty} df \frac{S_h(f)}{4\pi^2 f^2} \tag{3.27}$$

Elaborating on Eq. (3.24), one can define a dimensionless characteristic strain $h_c(f)$ satisfying the relation (Maggiore 2000)

$$h_c^2(f) = 2f S(f). \tag{3.28}$$

Note that with this definition, h_c is connected to the energy density ρ_{GW} of the GWB via

$$\frac{2\pi^2}{3H_0^2} f^2 h_c^2(f) = \frac{1}{\rho_c} \frac{d\rho_{gw}}{d \ln f} \equiv \Omega_{gw}(f), \tag{3.29}$$

where $\rho_c = 3H_0^2/8\pi$ is the critical energy density of a flat Universe, and $H_0 = 100\, h\, \mathrm{km\, s^{-1}\, Mpc^{-1}}$ is the Hubble expansion rate. Substituting Eq. (3.28) into Eq. (3.27), we finally get

$$r_{ab} = \Gamma(\zeta_{ab}) \int_0^\infty df\, P_h(f), \qquad (3.30)$$

where we defined

$$P_h(f) = \frac{h_c^2(f)}{12\pi^2 f^3}, \qquad (3.31)$$

and we re-defined the Hellings and Downs (HD) correlation coefficients as

$$\Gamma(\zeta_{ab}) = \frac{3}{2} C(\zeta_{ab})(1 + \delta_{ab}). \qquad (3.32)$$

Note that $P_h(f)$ has dimensions of $[s^{-3}]$, which is appropriate for a spectral density of a time series. The $3/2$ renormalization and the δ_{ab} term ensure that the new correlation coefficient $\Gamma(\zeta_{ab}) = 1$ when $a = b$ (i.e., the GWB has perfect autocorrelation). Note that when $a \neq b$ and $\zeta_{ab} = 0$, $\Gamma_{ab} = 1/2$; this is because for pulsars in the same direction, but at different distances, only the Earth terms are phase correlated, whereas the pulsar terms act as an additional source of noise. We will see below that this has important implications for GWB detection with PTAs.

3.3.2 GW Sources Relevant to PTAs and Their Signals

In Sect. 3.3.1 we demonstrated that both deterministic and stochastic GW sources affect the pulse TOAs. Deterministic sources leave a distinctive fingerprint of the form $r(t)$ (cf Eq. (3.14)) that can be exactly determined once the waveform $h(t)$ is known. On the other hand, stochastic GWBs induce a correlated signal r_{ab} (cf Eq. (3.30)) that can be determined if the characteristic strain spectrum $h_c(f)$ of the GWB is known. We now discuss the GW sources relevant to PTAs and their signals.

3.3.2.1 Deterministic GW Signals

A signal is deterministic when its waveform can be univocally specified at any given time, pending the knowledge of the signal dependence on the physical parameters of its source. Expected deterministic signals in the PTA band are produced by individual SMBHBs (Begelman et al. 1980) inspiralling and merging along the cosmic history (see e.g., Volonteri et al. 2003), although more exotic sources have been proposed, such as (super)strings cusps and kinks (Damour and Vilenkin

2001). Deterministic GW signals can be either continuous or transient; we will see below that SMBHBs can produce either type of signals depending on their physical properties and in which stage of their evolution they are observed.

I: Inspiralling Supermassive Black Hole Binaries

The archetypal continuous deterministic GW source is a SMBHB adiabatically inspiralling in a quasi-circular orbit. PTAs are sensitive to systems with $M >$ $10^8 M_\odot$ at centi-parsec orbital separations (Sesana et al. 2009, 2012). For those systems, the inspiral time is typically much longer than the observation time T. In the circular orbit approximation, the system emits a monochromatic wave at twice its orbital frequency (i.e. $\omega = 2\omega_K$) of the form Sesana and Vecchio (2010):

$$h_+(t) = (1 + \cos^2 \iota)A \cos(\omega t + \Phi_0), \tag{3.33a}$$

$$h_\times(t) = -2 \cos \iota \, A \sin(\omega t + \Phi_0), \tag{3.33b}$$

where ι is the inclination of the SMBHB orbital plane with respect to the line-of-sight, Φ_0 is the initial GW phase, and

$$A = 2\frac{\mathcal{M}^{5/3}}{D_l} (\pi f)^{2/3} \approx 1.3 \times 10^{-15} \left(\frac{f}{10^{-7} \text{Hz}}\right)^{2/3} \left(\frac{\mathcal{M}}{10^9 M_\odot}\right)^{5/3} \left(\frac{D_l}{1 \text{Gpc}}\right)^{-1} \tag{3.34}$$

is the GW amplitude. In Eq. (3.34), the luminosity distance D_l is related to the standard comoving distance D_c as $D_l = (1 + z)D_c$, the chirp mass $\mathcal{M}$ is related to the masses M_1 and M_2 of the two as $\mathcal{M} = (M_1 M_2)^{3/5}/(M_1 + M_2)^{1/5}$, and $f = \omega/(2\pi)$ is the GW frequency, which is related to the phase as $\Phi(t) = 2\pi \int^t f(t')dt'$. Note that Eq. (3.34) is written in terms of redshifted quantities. Those are related to their binary-rest frame counterparts via $\mathcal{M} = (1 + z)\mathcal{M}_{\text{rf}}$, $f = f_{\text{rf}}/(1 + z)$, where z is the redshift of the GW source.[3] In general, the GW community prefers redshifted quantities because they are the direct observables of GW experiments, and because they absorb all $(1 + z)$ factors, simplifying the equation when dealing with sources at cosmological distances.

The associated redshift $z(t)$[4] can be computed by plugging Eqs. (3.33a), (3.33b) into Eq. (3.20). By integrating the redshift according to Eq. (3.14), the residual is found to be composed of a pulsar term and an Earth term:

$$r(t) = r^P(t) - r^e(t), \tag{3.35}$$

---

[3]Unless otherwise specified, we always use *redshifted* masses and frequencies to describe the GW signals.

[4]Note that here $z(t)$ is the induced redshift in the timing residual, not to be confused with the redshift of the source introduced above.

where

$$r^e(t) = \frac{A}{\omega} \left\{ (1 + \cos^2 \iota) F^+ \left[\sin(\omega t + \Phi_0) - \sin \Phi_0 \right] + \right.$$
$$\left. 2 \cos \iota F^\times \left[\cos(\omega t + \Phi_0) - \cos \Phi_0 \right] \right\},$$

$$r^p(t) = \frac{A_p}{\omega_p} \left\{ (1 + \cos^2 \iota) F^+ \left[\sin(\omega_p t + \Phi_p + \Phi_0) - \sin(\Phi_p + \Phi_0) \right] + \right.$$
$$\left. 2 \cos \iota F^\times \left[\cos(\omega_p t + \Phi_p + \Phi_0) - \cos(\Phi_p + \Phi_0) \right] \right\}.$$

$$(3.36)$$

We specify ω and $\omega_p = \omega(t - \tau)$, because the two angular frequencies might be different in the pulsar and Earth terms. Note that this also implies different amplitudes A and A_p, since those depend on $f = \omega/2\pi$ as per Eq. (3.34). In fact, in the quadrupole approximation, the evolution of the binary orbital frequency and GW phase can be written as

$$\omega_K(t) = \omega_K \left(1 - \frac{256}{5} \mathcal{M}^{5/3} \omega_K^{8/3} t \right)^{-3/8}, \tag{3.37}$$

$$\Phi(t) = \Phi_0 + \frac{1}{16\mathcal{M}^{5/3}} \left(\omega_K^{-5/3} - \omega_K(t)^{-5/3} \right). \tag{3.38}$$

Over the typical PTA experiment duration (decades), ω and ω_p can be approximated as constants, and we drop the time dependence accordingly. However, the delay τ between the pulsar and the Earth term—which is the pulsar-Earth light travel time multiplied by a geometric factor—is thousands of years, comparable with the evolution timescale of typical SMBHBs (Sesana and Vecchio 2010). ω_p depends on the pulsar distance and relative orientation with respect to the incoming GW source; it is therefore different among observed pulsars and is smaller than ω.

The nominal frequency resolution of a PTA experiment is $\Delta f \approx 1/T$, where T is the duration of the experiment:

- If $(\omega_p - \omega)/(2\pi) > \Delta f$ for most MSPs, then there is no interference between the pulsar and the Earth terms; the latter can be added coherently and the former can be considered either as separate components of the signal or as an extra incoherent source of noise.
- Conversely, if $(\omega_a - \omega)/(2\pi) < \Delta f$ for the majority of MSPs, then the pulsar terms add up to the respective Earth terms, affecting their phase coherency.

This distinction has an impact on the detection strategy; different techniques are better suited to either situation, and many different detection algorithms have been developed accordingly, as we will see in Sect. 3.3.4.1. Examples of timing residuals from a circular SMBHB are shown in the upper left panel of Fig. 3.3; note that the signals are not perfect sinusoids because of the effect of the lower frequency pulsar term.

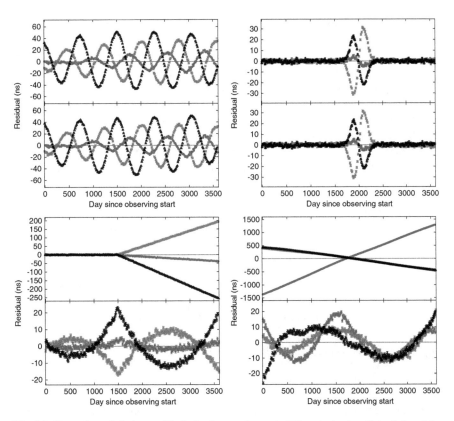

Fig. 3.3 Examples of timing residuals in three pulsars at different sky locations induced by selected GW signals plus noise. *Top left*: a continuous GW source generated by a circular SMBHB; *Top right*: a generic Gaussian burst; *Bottom left*: a burst with memory; *Bottom right*: a stochastic GWB from SMBHBs. In each panel top (bottom) plots show residuals before (after) fitting for the MSP spin and spindown. The fitting absorbs a large fraction of the signals with red spectra (from Burke-Spolaor (2015), courtesy of S. Burke-Spolaor)

II: Generic Bursts

Bursts are generally defined as signals that are well localized in time, i.e., lasting much shorter than the observation time T. Note that PTAs are sensitive to nHz-μHz frequencies, so that observable bursts will nevertheless last from weeks to several months. At such low frequencies and among the less exotic burst sources, we can expect defects to appear in a network of cosmic (super)strings when strings bend and reconnect, and which are known as cusps and kinks (Damour and Vilenkin 2001). For example, cusps have an extremely simple, linearly polarized waveform (Siemens and Olum 2003; Key and Cornish 2009)

$$h(t) = 2\pi A|t - t_*|^{1/3} \qquad A \approx \frac{G\mu L^{2/3}}{D_c} \tag{3.39}$$

where $G\mu$ is the string tension, D_c is the (comoving) distance to the cusp and L is its characteristic scale.

Another possible source of GW bursts consists of close encounters of SMBHs either on bound (elliptical) or unbound (parabolic, hyperbolic) orbits. Although the latter is extremely unlikely, the former might be a relatively common occurrence. It has, in fact, been shown that both three body scattering of ambient stars and torque exerted by a counter-rotating circumbinary disk can significantly increase the SMBHB eccentricity (see Dotti et al. (2012) and references therein). Another way to excite binary eccentricities is through the formation of a hierarchical SMBH triplet following two subsequent mergers (Hoffman and Loeb 2007; Bonetti et al. 2016). Bursts of GWs can therefore be emitted by highly eccentric SMBHBs with orbital frequencies $\ll 1/T$ at periastron passage (Amaro-Seoane et al. 2010). Eccentricity causes a 'split' of each polarization amplitude $h_+(t)$ and $h_\times(t)$ into harmonics according to (see, e.g., equations (5–6) in Willems et al. (2008) and references therein):

$$h_n^+(t) = A\left\{ -(1 + \cos^2 \iota)u_n(e) \cos\left[\frac{n}{2} \Phi(t) + 2\gamma(t)\right]\right.$$
$$\left. -(1 + \cos^2 \iota)v_n(e) \cos\left[\frac{n}{2} \Phi(t) - 2\gamma(t)\right] + \sin^2 \iota\, w_n(e) \cos\left[\frac{n}{2} \Phi(t)\right]\right\},$$
$$h_n^\times(t) = 2A \cos\iota\left\{u_n(e) \sin\left[\frac{n}{2} \Phi(t) + 2\gamma(t)\right] + v_n(e) \sin\left[\frac{n}{2} \Phi(t) - 2\gamma(t)\right]\right\}.$$

$$(3.40)$$

The coefficients $u_n(e)$, $v_n(e)$, and $w_n(e)$ are linear combinations of Bessel functions of the first kind $J_n(ne)$, $J_{n\pm1}(ne)$ and $J_{n\pm2}(ne)$, and $\gamma(t)$ is an additional precession term in the phase given by e. For $e \ll 1$, $|u_n(e)| \gg |v_n(e)|, |w_n(e)|$ and the expressions above reduce to the circular case of Eq. (3.33a), (3.33b). Waveforms for parabolic SMBHB encounters are given in Finn and Lommen (2010). In general, a GW burst is detected as a short duration distortion in the timing residuals. In the upper right panel of Fig. 3.3, we show an example of a burst with a generic Gaussian waveform.

III: Bursts with Memory

Besides the standard strain oscillation, GW bursts are also predicted to contain non-oscillatory components that result in a permanent deformation of spacetime. The final deformation depends on the radiation history of the source, and is therefore referred to as memory (Christodoulou 1991; Blanchet and Damour 1992; Favata 2009). Bursts displaying these features are known as bursts with memory (BWM). When the bursting source is a gravitationally-bound system, the memory arises from the fact that the radiated GW energy causes permanently non-vanishing second time derivatives in the source mass-energy quadrupole moments. Because of this, the spacetime metric relaxes to a configuration that differs from the pre-burst one. The merger of SMBHBs provides the most promising source of BWM for PTAs.

The permanent displacement in the spacetime metric for SMBHBs inspiralling in a quasi-circular orbit up to the merger was computed in Favata (2009). For the standard choice of polarization tensors—Eqs. (3.17a) and (3.17b)—, the $h_\times$ component of the strain vanishes and the remaining h_+ component takes the approximate form Madison et al. (2014)

$$h_+ \approx \frac{1 - \sqrt{8}/3}{24} \frac{\mu}{D_l} \sin^2 \iota (17 + \cos^2 \iota) \left[1 + \mathcal{O}(\mu^2/M^2) \right] \approx 1.5 \times 10^{-15} \frac{\mu}{10^9 M_\odot} \left(\frac{D_l}{1 \text{Gpc}} \right)^{-1},$$

(3.41)

where μ is the redshifted reduced mass, ι is the inclination angle just prior to the final merger, and the rightmost approximation has been obtained by averaging over source inclinations. This displacement quickly arises as a few % of the binary mass is radiated into GWs in the very last phase of the merger. The characteristic growth timescale is $t_r \approx 2\pi R_S/c \approx 1$ day M_9, where M_9 is the mass of the merger remnant in units of $10^9 M_\odot$ and R_s its Schwarzschild radius. After quickly ramping up, the perturbation simply settles to a constant value h. So for any practical purpose, the waveform is described by $h(t) = h_+ \Theta(t - t_0)$, where $\Theta(t - t_0)$ is the heaviside step function. The perturbation is therefore null until time t_0, and quickly jumps to the value given by Eq. (3.41), when the wave generated at the merger propagates through the detector. We can use Eq. (3.20) to get the associated redshift $z(t, \hat{\Omega})$ and integrate over time according to Eq. (3.14) to get the residual in the form

$$r(t) = \frac{1}{2} \cos(2\psi)(1 - \cos\theta)h_+[(t - t_e)\Theta(t - t_e) - (t - t_p)\Theta(t - t_p)]. \quad (3.42)$$

and $t_e - t_p \equiv \tau$ defined above. Since it is the integral of a constant, the residuals simply show a linear increase. Note that in this case we have a pulsar term, which is triggered at the time t_p at which the wave 'hits the pulsar', and an Earth term, which is triggered at the time t_e at which the wave 'hits the Earth'. As in the SMBHB case seen before, in a PTA, the Earth term will be correlated among all pulsars in the array, while the pulsar term will not. Contrary to the monochromatic waves, however, the two terms in general *do not* contribute to the detected signal at the same time. This is because $t_e - t_p$ is typically $\gg T$, the duration of the PTA experiment (unless the source is almost aligned with the considered pulsar). To imprint a signature onto the detected residuals, the 'trigger' time must occur within the duration T of the experiment. If this is not the case, then the signature is a continuous linear drift which is inevitably absorbed in a small correction to the pulsar frequency ν_0. Examples of BWM are shown in the lower left panel of Fig. 3.3. Contrary to the continuous wave case, the burst effect is largely absorbed by fitting for pulsar spin and spin derivative, which subtracts a quadratic function from the TOAs.

3.3.2.2 Stochastic Backgrounds

Stochastic GWBs in the PTA band can arise from a number of cosmological and astrophysical sources. As a first approximation, many calculations predict a characteristic strain spectrum with a single power-law shape

$$h_c = A \left(\frac{f}{\text{year}^{-1}} \right)^{-\alpha}, \tag{3.43}$$

where A is the strain amplitude at a reference frequency of 1 year^{-1}. The slope α differs depending on the specific background. On the cosmological side, cosmic string networks generate spectra with $\alpha = -5/3, -7/6, -1$ depending on several parameters defining the nature of the network (Damour and Vilenkin 2001; Ölmez et al. 2010). Standard inflation predicts $\alpha = -1$ with a signal amplitude that is well below foreseeable detection possibilities, even though several mechanisms can enhance the signal to detectable levels (see reviews in Chiara Guzzetti et al. 2016, Bartolo et al. 2016). Other inflationary relics can produce stronger GWBs, with $0.5 < \alpha < 2$ (Grishchuk 2005). Further cosmological GWBs include primordial BHs (Bugaev and Klimai 2011) or QCD phase transitions, and may have more complicated spectra (Caprini et al. 2010). The most promising signal for PTAs is, however, of astrophysical origin and stems from the cosmic population of SMBHBs (Rajagopal and Romani 1995; Jaffe and Backer 2003; Sesana et al. 2004).

Since galaxy mergers are common (Lacey and Cole 1993), we expect a large population of SMBHB to emit GWs in the PTA band *at any time* (Sesana et al. 2008). The superposition of many incoherent signals results in a GWB that is described by Eq. (3.43), with $\alpha = -2/3$ (Phinney 2001). The normalization A is affected by the poorly known SMBHB cosmic merger rate, but is predicted to be in the range of $10^{-16} < A < \text{few} \times 10^{-15}$ (Sesana 2013; McWilliams et al. 2014; Kulier et al. 2015; Ravi et al. 2015; Sesana et al. 2016). Following (Sesana et al. 2008) and assuming circular binaries, this stochastic GWB can be written in the form

$$h_c^2(f) = \int_0^\infty dz \int_0^\infty d\mathcal{M} \, \frac{d^3 N}{dz d\mathcal{M} d\ln f} \, h^2(f). \tag{3.44}$$

where h is the sky and polarisation averaged strain amplitude given by Thorne (1987)

$$h = \frac{8\pi^{2/3}}{10^{1/2}} \frac{\mathcal{M}^{5/3}}{D_L} f^{2/3}, \tag{3.45}$$

and the number of SMBHBs emitting per unit mass, redshift and log frequency is given by

$$\frac{d^3 N}{dz d\mathcal{M} d\ln f} = \frac{d^2 n}{dz d\mathcal{M}} \frac{dt_{\text{rf}}}{d\ln f_{\text{rf}}} \frac{dz}{dt_{\text{rf}}} \frac{dV_c}{dz}. \tag{3.46}$$

In Eq. (3.46), $d^2n/dz\,d\mathcal{M}$ is the cosmic merger rate density of SMBHBs, $dt_{\mathrm{rf}}/d\ln f_{\mathrm{rf}}$ is the time each binary spends in a given log frequency bin, and the other terms are standard cosmological relations. The level of the stochastic GWB therefore depends on the cosmic merger rate *and* on the mechanism driving the binary evolution through the $dt_{\mathrm{rf}}/d\ln f_{\mathrm{rf}}$ term. For GW driven binaries, $dt_{\mathrm{rf}}/d\ln f_{\mathrm{rf}} \propto f^{-8/3}$ and one recovers the $h_c \propto f^{-2/3}$ spectrum. However, since the GW emission efficiency has a steep f dependence, at the relatively large separations relevant to PTA (centiparsec), binaries might be still driven by the interaction with their stellar and gaseous environments (Kocsis and Sesana 2011; Ravi et al. 2014; Kelley et al. 2017). For typical astrophysical systems (see derivation in Sesana 2013), the transition frequency between gas/star and GW dominated evolution is:

$$f_{\mathrm{star/GW}} \approx 5 \times 10^{-9} M_8^{-7/10} q^{-3/10} \mathrm{Hz}$$

$$f_{\mathrm{gas/GW}} \approx 5 \times 10^{-9} M_8^{-37/49} q^{-69/98} \mathrm{Hz}, \tag{3.47}$$

which is potentially within the PTA range. If binaries are eccentric, things are further complicated by the fact that each system emits a series of harmonics. A full mathematical derivation including stellar coupling and eccentricity can be found in Rasskazov and Merritt (2016). The general effect of coupling with the environment is thus to produce a turnover of the spectrum at low frequencies, as shown in Fig. 3.4

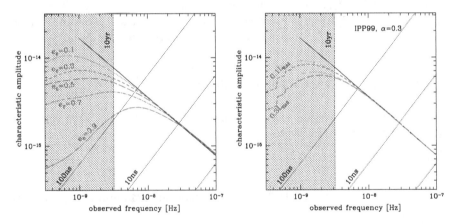

Fig. 3.4 Effect of the environment on h_c. *Left panel*: SMBHBs driven by stellar scattering; each blue line represents a population with a specific initial eccentricity, according to the model presented in Sesana (2010); *Right panel*: SMBHBs driven by interaction with a circumbinary disk modelled as in Ivanov et al. (1999). Green lines are for circular binaries, blue lines allow for a self-consistent eccentricity evolution (models from Kocsis and Sesana 2011). The upper and lower pairs of curves are for populations in which SMBHs accrete at 10% and 30% of the maximum allowed Eddington rate respectively, labelled in figure as $0.1L_{\mathrm{Edd}}$ and $0.3L_{\mathrm{Edd}}$. In both panels, the solid black lines represent the GW driven case $h_c \propto f^{-2/3}$, dashed lines mark residual levels according to $r = h/(2\pi f)$ to guide the eye, and the excluded region at $f < 3 \times 10^{-9}$ highlights the signal modification relevant to a PTA observation of $T \sim 10$ years. From Sesana (2015)

for selected models. Future detailed measurement of the GWB spectral shape and normalization with PTAs can therefore in principle constrain the cosmic merger rate of SMBHBs, their dynamical interaction with the environment and their eccentricity distribution (Chen et al. 2017).

3.3.3   PTA Sensitivity to Gravitational Waves

The major challenge of PTAs is to dig out a possible GW signal (whether deterministic or stochastic) from a plethora of noise sources, many of which are poorly understood. The output of a detector can in fact be written as

$$d(t) = s(t) + n(t), \tag{3.48}$$

where $d(t)$ is the recorded data, $s(t)$ is the putative GW signal and $n(t)$ is the detector noise. In Fourier space, for a Gaussian stationary noise, $n(t)$ satisfies the ensemble average condition (Maggiore 2000)

$$\langle \tilde{n}^*(f)\tilde{n}(f')\rangle = \delta(f - f')\frac{1}{2}P_n(f), \quad \langle n^2(t)\rangle = \int_0^\infty df\, P_n(f), \tag{3.49}$$

where $P_n(f)$ is the, noise spectral density[5] which has been defined for positive frequencies $0 < f < \infty$. In practice, the detectability of a signal depends on the noise spectral density $P_n(f)$ and on how it compares with the GW signal.

3.3.3.1   Sources of Noise in Pulsar Timing Arrays

An excellent review of the main noise sources relevant to PTAs is given in Cordes (2013). In practice, we can write

$$P_n(f) = P_{wn}(f) + P_{rn,ac}(f) + P_{rn,c}(f), \tag{3.50}$$

where $P_{wn}(f)$ is white noise, $P_{rn,ac}(f)$ is achromatic red noise and $P_{rn,c}(f)$ is chromatic red noise.

The power contributed by white noise takes the form

$$P_{wn}(f) = 2\sigma_{wn}^2 \Delta t, \tag{3.51}$$

---

[5]Note that the spectral density is usually referred to as $S(f)$. In our notation, $S(f)[s]$ is used in relation to the dimensionless GW strain. PTAs, however, measure TOAs and the associated power spectral density, denoted here as $P(f)$, has dimension of $[s^3]$. The relation between the two is $P(f) = S(f)/(12\pi^2 f^2)$.

where Δt is the cadence of observations (typically weeks) and σ_{wn} is the rms uncertainty in the TOA. The main sources of white noise in PTA observations are radiometer noise and jitter. Radiometer noise defines the maximum theoretical precision in measuring TOAs, and arises from the fact that folded pulses with finite S/N are matched to a theoretical template. Jitter is due to the intrinsic stochasticity of the phasing of individual pulses. Detail scaling for these noise sources is given in Cordes (2013); Wang (2015); typical figures of σ_{wn} are hundreds of ns (radiometer) and tens of ns (jitter).

Chromatic red noise, by definition, depends on the frequency of the observed radio photons and arises from frequency-dependent propagation effects in the ISM, in particular dispersion and scattering. Interaction of radio photons with the ISM's cold magnetized plasma yields a frequency-dependent delay in their group velocity. This causes a delay in TOAs that is proportional to the electron column density (referred to as dispersion measure, DM) travelled by the radio photons with a ν_γ^{-2} dependence, where ν_γ is the frequency of the observed photons (not to be mistaken with the spinning frequency of the MSP, ν). Scattering is the pulse broadening due to multiple paths travelled along the ISM and has a ν_γ^{-4} dependence. Note that, as both the Earth and the observed MSPs move in the Galaxy potential, the DM is typically time-dependent. Because of their frequency dependence, chromatic noise sources can be dealt with by using wideband receivers and fitting for the frequency dependence of the TOAs.[6]

Conversely, achromatic red noise is the same at all received radio frequencies and cannot be mitigated by means of wideband observations. This noise is intrinsic to the pulsar and is due to the complex torques arising by crust-superfluid interactions. Spin noise has been detected in several MSPs and has a very steep red spectrum $P_{sn}(f) \propto f^{-5\pm0.4}$ (Shannon and Cordes 2010). For comparison, a GWB with $h_c \propto f^{-2/3}$ results in $P_h(f) \propto f^{-13/3}$ according to Eq. (3.31). Spin noise can therefore be the most serious limiting factor for the detection of a stochastic GWB.

The noise sources that were considered thus far are supposedly uncorrelated among MSPs. There are however additional sources of noise that show specific correlation patterns. Clock offsets have the same effect on all pulsars, and therefore induce a monopole correlated signal. Errors in the solar system ephemeris (which are necessary for computing TOAs at the solar system barycentre) result in a dipole correlation pattern (Tiburzi et al. 2016). Fortunately, those are different from the quadrupole Γ_{ab} correlation diagnostic for a stochastic GWB, and advanced analysis methods can distinguish between them (Taylor et al. 2017). Note, however, that disentangling correlations with different patterns requires a sufficiently large number of high quality pulsars in the array. This is particularly problematic in current PTAs, which are dominated by a handful of good MSPs. In fact, the latest NANOGrav analysis (Arzoumanian et al. 2018), suggests that the current PTA

---

[6]Note that wideband observations entail other issues related to frequency dependence of the pulse profile. This can be dealt with, for example, by developing 2D (time-frequency) profile templates (Liu et al. 2014; Pennucci et al. 2014).

sensitivity to a stochastic GWB is limited by uncertainties in the solar system ephemeris measurements.

3.3.3.2 S/N Calculation and Scaling Relations

With an understanding of the GW signature imprinted on timing residuals and of the relevant sources of noise, we can estimate typical signal-to-noise ratio (S/N, ρ) of different GW signals as a function of the structural array parameters and assess prospects for their detectability. In the following, we make the distinction between deterministic signals and stochastic GWBs.

I: Deterministic Signals

For deterministic signals, the data can be matched-filtered with a template for $s(t)$. It can be shown (e.g. Sesana and Vecchio 2010) that in this case, the S/N of the GW signal is given by

$$\rho^2 = (r(t)|r(t)) \tag{3.52}$$

where $(\cdot|\cdot)$ is the weighted inner product defined as

$$(x|y) = 2 \int_0^{+\infty} \frac{\tilde{x}^*(f)\tilde{y}(f) + \tilde{x}(f)\tilde{y}^*(f)}{P_n(f)} df \simeq \frac{2}{P_0} \int_0^T x(t)y(t)dt . \tag{3.53}$$

and

$$\tilde{x}(f) = \int_{-\infty}^{+\infty} x(t)e^{-2\pi ift}. \tag{3.54}$$

Note that in the last step of Eq. (3.53), we implicitly assumed that the signal is monochromatic, and P_0 is the noise spectral density at the frequency of the signal. For an array with M pulsars identified by index a, we have

$$\rho_a^2 = \frac{2}{P_{0,a}} \int_0^{T_a} r_a^2(t)dt, \quad \rho^2 = \sum_a \rho_a^2. \tag{3.55}$$

The equation above also applies to a burst generated by very eccentric binaries by summing over all harmonics and considering the appropriate $P(f)$ at the observed frequency of each harmonic.

The integral in Eq. (3.55) can be easily computed using the residual formula for a circular SMBHB given in Eq. (3.36). For simplicity, we only consider the Earth term, and an array of M identical MSPs, dropping the a index. Individually-resolvable sources are usually expected to be observed at $f \equiv \omega/(2\pi) \gg 1/T$; we therefore assume white noise, $P_0 = 2\sigma^2 \Delta t$. Averaging over the antenna response functions F^+, $F^\times$ and orbital inclinations ι, and summing over all MSPs, we get

$$\rho^2 \approx \frac{M}{15\pi^2} \frac{T}{\Delta t} \frac{A^2}{\sigma^2 f^2}. \tag{3.56}$$

Noticing that $N = T/\Delta t$ is the number of observations and taking the square root we finally get

$$\rho \approx \frac{1}{\sqrt{15}\pi} \frac{A}{\sigma f} (NM)^{1/2}. \tag{3.57}$$

The S/N of a circular SMBHB is therefore proportional to the square root of the number of pulsars in the array and of the number of observations (i.e. the total observation time T, for a uniform observation cadence), and is inversely proportional to the rms residual σ. Equation (3.57) can be inverted to obtain the minimum amplitude A observable at a given S/N threshold:

$$A \approx \sqrt{15}\pi\rho\sigma f(NM)^{-1/2} = 9 \times 10^{-15} \frac{\rho}{5} \frac{\sigma}{100\,\text{ns}} \frac{f}{10^{-7}\text{Hz}} \left(\frac{N}{250}\right)^{-1/2} \left(\frac{M}{20}\right)^{-1/2}. \tag{3.58}$$

Although SMBHBs are abundant in the Universe (see figure 1 in Sesana et al. 2012), comparison between the above estimate and Eq. (3.34) shows that current PTAs are only sensitive to extremely massive SMBHBs, which are extremely rare. Equation (3.58) can be compared with the limits shown in Fig. 3.6. At 10^{-7} Hz, the EPTA upper limit is $\approx 3 \times 10^{-14}$, which is in line with the equation above, considering that the EPTA dataset is dominated by one pulsar with $\sigma = 130\,\text{ns}$ (Desvignes et al. 2016). We also note that the frequency dependence is somewhat flatter than f, indicating some red noise contribution. The turnover at $f < 10^{-8}$ is instead due to a combination of red noise and MSP spin and spindown fitting.

A similar derivation for BWM can be found in van Haasteren and Levin (2010), yielding

$$h_{\min} \approx 12\sqrt{3}\rho\sigma T^{-1}(NM)^{-1/2}\mathcal{F}(\chi) = 4.5 \times 10^{-16}\mathcal{F}(\chi) \frac{\rho}{5} \frac{\sigma}{100\,\text{ns}} \left(\frac{T}{10\text{years}}\right)^{-1} \left(\frac{N}{250}\right)^{-1/2} \left(\frac{M}{20}\right)^{-1/2}, \tag{3.59}$$

where $\mathcal{F}(\chi)$, given in van Haasteren and Levin (2010), is a function of $\chi = t_0/T$ (being t_0 the BWM arrival time) and has a minimum value of ≈ 1.4. When compared to Eq. (3.41), the above estimate suggests that PTAs can be sensitive to BWM out to much larger distances than inspiralling SMBHBs. Note however that while inspiralling SMBHBs are rather abundant, coalescences are extremely rare events. In fact the coalescence rate of SMBHBs with $M > 10^9 M_\odot$ throughout the Universe is $< 10^{-2}$ years (Blecha et al. 2016).

II: Stochastic Backgrounds

For stochastic signals, the strategy is to detect cross-correlated power in several detectors (i.e. in several pulsars). The S/N imprinted by a stochastic GWB in a PTA can be written as (Allen and Romano 1999; Maggiore 2000; Rosado et al. 2015)

$$\rho^2 = 2 \sum_{a=1,M} \sum_{b>a} T_{ab} \int \frac{\Gamma_{ab}^2 P_h^2}{P_{n,ab}^2} df, \tag{3.60}$$

where the sums run over all pulsar pairs, T_{ab} is the timespan for which both pulsars a and b are observed (note that, in general, MSPs have different time coverage, depending on when they were discovered, the schedule requirement at observatories, etc.) and Γ_{ab} is the HD correlation function defined by Eq. (3.32). The correlated noise term is given by

$$P_{n,ab}^2 = P_a P_b + P_h[P_a + P_b] + P_h^2(1 + \Gamma_{ab})^2, \tag{3.61}$$

where

$$P_{a,b} = 2\sigma_{a,b}^2 \Delta t + P_{rn,a,b}, \tag{3.62}$$

and P_h is related to the GWB characteristic strain via Eq. (3.31). In the following, we ignore the red noise component $P_{rn,a,b}$ for simplicity. Note that $P_{n,ab}$ has two distinct asymptotic trends. For $P_h \ll P_{a,b}$, i.e. in the weak signal limit, it reduces to $P_{n,ab}^2 \approx P_a P_b$. However, when $P_h > P_{a,b}$, i.e. in the strong signal limit, $P_{n,ab}^2 \approx P_h^2(1+\Gamma_{ab})^2$, meaning that the power of the noise is always comparable to the power of the signal, regardless of how strong the latter is. This has a profound impact for GWB observability with PTAs, as we now detail.

Let us again drop the a, b indexes by considering equal MSPs in the array. We further assume $M \gg 1$, so that we can substitute Γ_{ab} with the average value $\Gamma = 1/(4\sqrt{3})$ to get

$$\rho^2 \approx T\Gamma^2 M^2 \int \frac{P_h^2}{P_n^2} df. \tag{3.63}$$

In the limit $P_h \ll P_n$ and writing h_c according to Eq. (3.43), after some algebra and integrating in the range $1/T < f < \infty$ we get

$$\rho \approx 2 \times 10^{-13} N M \sigma^{-2} A^2 T^{10/3}. \tag{3.64}$$

The array sensitivity therefore increases quickly by improving timing precision and by extending the time baseline of the experiment (see Fig. 3.5). Observing more pulsars also help, but only linearly in S/N. The minimum detectable GWB therefore

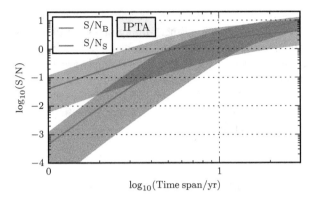

Fig. 3.5 S/N scaling with observation time of the loudest resolvable circular SMBHBs (red) and of the collective stochastic GWB (green) for 10^5 realizations of the cosmic SMBHB population. The coloured band shows the 90% confidence intervals, while the solid lines represent the median of all realizations. An IPTA-type array with 49 MSPs with σ between 500 ns and 9 μs was assumed. Note the two distinct scalings of the GWB S/N in the weak and strong signal regimes. We also note that the resolvable SMBHB S/N does not follow a simple $T^{1/2}$ scaling in this figure. This is because we do not plot the S/N of a specific source, but of the *loudest* source, which might change as lower frequencies (and better frequency resolution) are accessible with increasing T (from Rosado et al. 2015)

has a characteristic strain A given by

$$A \approx 7 \times 10^6 \rho^{1/2} \sigma T^{-5/3} (NM)^{-1/2}$$

$$\approx 10^{-16} \rho^{1/2} \frac{\sigma}{100\,\text{ns}} \left(\frac{T}{10\text{year}}\right)^{-5/3} \left(\frac{N}{250}\right)^{-1/2} \left(\frac{M}{20}\right)^{-1/2}. \qquad (3.65)$$

Note that although this lies at the lower end of the strain range predicted by cosmological models of SMBH assembly (Sesana 2013; McWilliams et al. 2014; Kulier et al. 2015; Ravi et al. 2015; Sesana et al. 2016), there are strong caveats. First, we ignored both red noise and MSP spin and spin-down fitting. The latter generally compromises the PTA sensitivity at $f < 2/T$ (see e.g. figure 1 in Cordes 2013) thus degrading the above estimate by a factor of a few. Second, when $\rho = 1$ we are already departing from the weak signal regime.

When $P_h > P_n$, things are drastically different. The integral in Eq. (3.63) reduces to $\int df$ giving

$$\rho \approx T^{1/2} \Gamma M (f_{\text{max}} - f_{\text{min}})^{1/2} \approx \Gamma M \Sigma^{1/2}. \qquad (3.66)$$

Here $f_{\text{min}} = 1/T$ and f_{max} is the maximum frequency for which the condition $P_h > P_n$ is satisfied. To get the rightmost expression in Eq. (3.66), we divided the frequency domain in resolution bins $\Delta f = 1/T$, and Σ is the number of frequency bins for which $P_h > P_n$ (usually a few). We now see that the S/N increases linearly with the number of pulsars in the array M and, as shown in Fig. 3.5, only with the square root of the observation time. Timing precision only plays a minor role by weakly affecting the f_{max} limit (or alternatively Σ). In practice, to make a confident detection of a stochastic GWB, it is absolutely necessary to include a large number of quality MSPs in the array. Note that, although we assumed $h_c \propto f^{-2/3}$, the derived scalings usually hold for any GWB shape, as shown in Vigeland and Siemens (2016) for broken power-law spectra approximating the GW signals from SMBHBs interacting with gas or stars.

3.3.4 Current Status of Gravitational Wave Searches

3.3.4.1 Analysis Methods

Whether the signal is deterministic or stochastic, the challenge of data analysis is to determine what the chances are that it is present in the data. The problem can be tackled either using a frequentist or a Bayesian philosophy. Reviewing principles and differences of those approaches is well beyond the scope of this contribution; we summarize here the main ideas and point the reader to the relevant PTA literature. The core aspect of all modern GW searches is the evaluation of the likelihood that some signal is present in the time series of the pulsar TOAs. Deferring technical details to e.g. van Haasteren and Levin (2013); Lentati et al. (2015), the likelihood function, marginalised over the timing parameters, can be written as

$$
\mathcal{L}(\vec{\theta}, \vec{\lambda}, \vec{\eta} | \vec{\delta t}) = \frac{1}{\sqrt{(2\pi)^{n-m} det(G^T C(\vec{\eta}, \vec{\theta})G)}} \times
$$

$$
\exp\left(-\frac{1}{2}(\vec{\delta t} - \vec{r}(\vec{\lambda}))^T G(G^T C(\vec{\eta}, \vec{\theta})G)^{-1} G^T(\vec{\delta t} - \vec{r}(\vec{\lambda}))\right). \quad (3.67)
$$

Here n is the length of the vector $\vec{\delta t} = \cup \delta t_a$ obtained by concatenating the individual pulsar TOA series δt_a, m is the number of parameters in the timing model, and the matrix G is related to the design matrix (see van Haasteren and Levin (2013) for details). $\vec{\theta}, \vec{\lambda}, \vec{\eta}$ are vectors of parameters describing the noise ($\vec{\theta}$), a deterministic GW signal ($\vec{\lambda}$) or the spectral shape of a stochastic GWB ($\vec{\eta}$). In general, the variance-covariance matrix C contains contributions from the putative GWB and from white and red noise: $C = C_{gw}(\vec{\eta}) + C_{wn}(\vec{\theta}) + C_{rn}(\vec{\theta})$. The detailed form of all the contributions to the variance-covariance matrix can be found in Lentati et al. (2015).

In frequentist searches, a detection statistic is constructed based on the likelihood function both in the null hypothesis and in the presence of a signal. Extensive Monte-Carlo simulations on synthetic data with injected signals are then performed to construct the detection probability as a function of the false alarm rate. Comparison with the value of the statistic obtained from the real dataset is then used to either claim a detection (with associated confidence) or to obtain an upper limit in case of no detection. This procedure has been detailed in Ellis et al. (2012) and has been used in several deterministic source searches (e.g. Zhu et al. 2014; Babak et al. 2016).

In Bayesian searches, the likelihood function is used to compute the odds ratio of the Bayesian evidence for the hypothesis that a signal is present in the data (model $\mathcal{H}_1$, with signal described by parameters $\vec{\lambda}$ and/or $\vec{\eta}$) versus the null hypothesis ($\mathcal{H}_0$). In the case of no prior preference of either model, the odds ratio reduces to the Bayes factor, $\mathcal{B}$:

$$\mathcal{B} = \frac{\int \mathcal{L}(\vec{\theta}, \vec{\lambda}|\vec{\delta t})\pi(\vec{\theta}, \vec{\lambda})d\vec{\theta}\,d\vec{\lambda}}{\int \mathcal{L}(\vec{\theta}|\vec{\delta t})\pi(\vec{\theta})d\vec{\theta}}, \qquad (3.68)$$

which is technically the ratio of the evidences for the hypothesis $\mathcal{H}_1$ and $\mathcal{H}_0$. The value of $\mathcal{B}$ is the statistic used to assess the presence of the signal (Kass and Raftery 1995). In the case of a detection, the shape of the likelihood function can be used to infer the parameters of the signal and their uncertainties, otherwise upper limits can be placed. Bayesian searches have been recently extensively applied to the search for both deterministic signals and stochastic GWBs.

3.3.4.2 Overview of Current Results

The search for deterministic signals in real data have so far focused on circular SMBHBs and BWM. In modern algorithms, a deterministic signal $r(\vec{\lambda}, t)$ is added to the model and a search is performed over the parameter space defined by $\vec{\lambda}$. For individual SMBHBs, $\vec{\lambda}$ includes the source amplitude, frequency, sky location, inclination phase and polarization (plus other parameters related to the pulsar term, when included in the search); for BWM, parameters include burst amplitude and trigger time t_0, sky location, and source inclination.

Yardley et al. (2010) obtained the first frequentist individual SMBHB upper limit on early PPTA data by looking for excess power as a function of frequency, placing a sky averaged upper limit on the source amplitude (cf Eq. (3.34)) of $A \approx 10^{-13}$ at 10 nHz. More recently, detection statistics for single SMBHBs have been calculated for circular systems either including the Earth term only (Babak and Sesana 2012), or adding the pulsar term (Ellis et al. 2012), as well as for eccentric binaries (Taylor et al. 2016). Alternative frequentist methods based on the construction of null streams have also been proposed (Zhu et al. 2015). In parallel, Taylor et al. (2014) developed a Bayesian pipeline that can handle generic circular SMBHBs. Searches on real data have been performed by the three major PTAs (Arzoumanian et al. 2014; Zhu et al. 2014; Babak et al. 2016), yielding null results. The EPTA placed the most stringent limit to date, shown in Fig. 3.6 (from Babak et al. 2016). Around 10 nHz, sources with $A > 10^{-14}$ can be confidently excluded. Compared to Eq. (3.34) this rules out the presence of centi-parsec SMBHBs of a few billion solar masses out

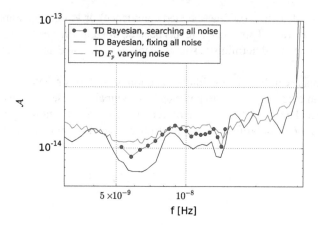

Fig. 3.6 Sky averaged 95% upper limit on the gravitational wave amplitude A of a circular SMBHB as a function of frequency, placed by three different searches (labelled in figure) performed on the EPTA dataset (from Babak et al. (2016), where the detailed descriptions of each method can be found)

to the distance of the Coma cluster. Note that those limits are consistent with our current understanding of SMBH assembly, as state-of-the-art models predict a mere 1% chance of making a detection at this sensitivity level (Babak et al. 2016).

Searches for BWM have been performed both with the PPTA (Wang et al. 2015) and NANOGrav (Arzoumanian et al. 2015) datasets. Both searches yielded comparable results, constraining the BWM rate to be less than ≈ 1 year^{-1} at $h = 10^{-13}$. To produce a strain of comparable amplitude, a $10^9 M_\odot$ SMBHB should merge in the Virgo cluster, which is an extremely unlikely event. In fact, Cordes and Jenet (2012), Ravi et al. (2015) estimated that the event rate for such a strong burst is $<10^{-6}$ year^{-1}, which makes these null results unsurprising.

All PTAs (including the IPTA) performed extensive searches for a stochastic GWB from SMBHBs, which is the most likely GW signal to be first detected by PTAs (Rosado et al. 2015). In isotropic GWB searches, the smoking gun of a detection is provided by the HD correlation pattern given by Eq. (3.32). The correlation can be used to construct an optimal correlation statistics based on the maximum likelihood estimator (Anholm et al. 2009; Chamberlin et al. 2015), which can be employed to obtain frequentist upper limits (Lentati et al. 2015). In advanced algorithms, the HD correlation is included in the analysis via the correlation matrix $C_{gw}(\vec{\eta})$, which, in the often used Fourier representation, takes the form (e.g. Lentati et al. 2015)

$$\Psi_{a,b,i,j} = \Gamma(\zeta_{ab})\varphi_i^{\text{GWB}}\delta_{ij}, \tag{3.69}$$

where indices i, j run over the difference frequency bins of the Fourier decomposition and φ_i is the power in the GW signal given by Eq. (3.31), which is evaluated at the central frequency f_i of the i-th bin. The matrix (3.69) is included in the appropriate Fourier representation of the likelihood function (3.67). Note that in the case of *anisotropic* GWB, the signal can be decomposed into spherical harmonics, and the power in different harmonics has different correlation patterns (since the HD curve is the correlation of the monopole component) that can also be included in the analysis (Gair et al. 2014; Taylor et al. 2015).

In the simplest searches, the $h_c(f)$ responsible for $P_h(f)$ is described by a single power law defined by the two parameters $\vec{\eta} = A, \alpha$ (cf Eq. (3.43)), however additional parameters can be (and have been) included in the search to describe a low frequency turnover. In the following, we always refer to the upper limit placed on A assuming an $f^{-2/3}$ GWB, appropriate for circular, GW-driven SMBHBs. Note that those can be easily converted into limits on Ω_{gw} by combining Eqs. (3.29) and (3.43). Systematic searches for GWBs in PTA data have been ongoing for more than a decade (Jenet et al. 2006; van Haasteren et al. 2011; Demorest et al. 2013), with early upper limits on A in the range $6 \times 10^{-15} - 1.1 \times 10^{-14}$, which is still higher than the range predicted by theoretical models. In recent years, improvements in the data quality and in the search algorithms made possible to push the sensitivity down to $A \approx 10^{-15}$, yielding null results. Using the first EPTA legacy data release, Lentati et al. (2015) placed an upper limit of $A = 3 \times 10^{-15}$ for a SMBHB GWB, while also providing upper limits on the cosmic string tension of $G\mu = 1.1 \times 10^{-7}$, which is competitive with CMB constraints (Planck Collaboration et al. 2016). Figure 3.7 (from Lentati et al. 2015) shows the measured correlation pattern of the six best EPTA MSPs as a function of their angular separation. It is clear that

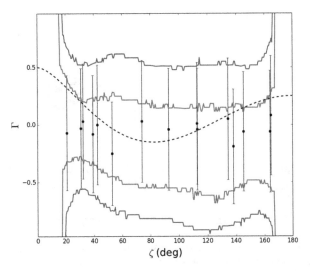

Fig. 3.7 The recovered correlation between pulsars as a function of angular separation on the sky in the EPTA analysis. The red and blue lines represent the 68% and 95% confidence intervals of the recovered correlation. Individual points represent the mean correlation coefficients with a 1σ uncertainty for each pulsar pair. The dashed line represents the HD correlation (from Lentati et al. 2015)

the measurement is still uninformative, and no detection can be claimed. Using the same dataset (Taylor et al. 2015) also performed the first search for an anisotropic GWB, constraining any amplitude in the higher multiples of the spherical harmonic decomposition to be less then 40% of the one in the monopole component (i.e. the isotropic part of the signal). Arzoumanian et al. (2016) used the NANOGrav 9 year dataset to place an upper limit of $A = 1.5 \times 10^{-15}$ and to further improve on EPTA limits on the string tension. They also studied spectra with a turnover and placed the first (weak) constraints on the astrophysical properties of the SMBHB population. The best upper limit to date was set to $A = 10^{-15}$ by Shannon et al. (2015) using Parkes data. The IPTA (Verbiest et al. 2016) also published its first upper limit at $A = 1.7 \times 10^{-15}$ using its first data release. This is not the most stringent limit, but it has been obtained by combining older individual PTA datasets. In fact, compared to the individual dataset used, the IPTA limit is a factor of ≈ 2 better, demonstrating the great potential of adding together more quality pulsars in a worldwide collaborative effort. Figure 3.8 shows a comparison of the individual PTA upper limits to current theoretical predictions from Sesana et al. (2016) assuming circular GW-driven SMBHBs, highlighting their uncertainty. In each panel, shaded areas represent the spread (larger than an order of magnitude) due to our poor knowledge of the cosmic

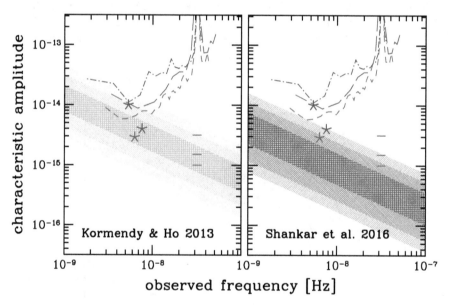

Fig. 3.8 Current upper limits on stochastic GWBs as a function of frequency for EPTA (dot-dashed green), NANOGrav (long-dashed blue), and PPTA (short-dashed red). For each curve, stars represent the integrated limits to an $f^{-2/3}$ background, and horizontal ticks represent their extrapolation at $f = 1 \, \text{year}^{-1}$, i.e. the upper limit on A defined by Eq. (3.43). Shaded areas represent the 68% 95% and 99.7% confidence intervals of $h_c(f)$ for selected SMBHB population models. The two panels show how uncertainties in the SMBH mass-host galaxy relation severely impact the expected signal level. See Sesana et al. (2016) for full details of the employed models (from Sesana et al. 2016)

galaxy merger rate. The difference between panels stems from different SMBH-galaxy relations (see Sesana et al. (2016) for model details), highlighting that even a single ingredient entering the computation can have a strong effect on the predicted signal. Moreover, note that possible SMBHB stalling (Simon and Burke-Spolaor 2016; Kelley et al. 2017), coupling with the environment (Kocsis and Sesana 2011; Ravi et al. 2014) and eccentricity (Enoki and Nagashima 2007; Sesana 2013; Huerta et al. 2015; Rasskazov and Merritt 2016) can all contribute to suppressing the signal at low frequencies, which makes any strong inference from GWB non-detection at a $A = 10^{-15}$ level problematic.

3.4   Future Prospects

In Sects. 3.2 and 3.3, we extensively described the state of the art in using pulsars as gravity probes. Precise timing of MSPs in particular, has already provided precise tests of GR and alternative theories of gravity in the strong-field regime, and the current PTA efforts are starting to place constraints on the cosmic population of SMBHBs, although no detection has yet been made. We now take a look at the near future, touching on several subjects that are relevant to improving the use of pulsars as tools for the study of gravity. With pulsars (at least for those that are not dominated by red timing noise, which is the case for most MSPs), the longer the data span, the more precise the TOAs. Therefore simply continuing to monitor them using the same radio telescopes will lead to more precise TOAs. This will in turn lead to more precise estimations of PK and PPN parameters for known pulsar binaries, and therefore more stringent tests of GR and alternative gravity theories. However the number of compact binaries that are useful for gravity tests (for which more than two PK parameters are determined) is limited. More precise TOAs will also naturally improve the constraints on a background of nanohertz GWs (see scaling relations in Sect. 3.3.3.2). However, to make a confident detection of a stochastic GWB, more pulsars with high precision are needed. The search for new pulsars is therefore critical to both tests of strong gravity and the search for nanohertz GWs.

Next-generation instruments such as the Large European Array for Pulsars (LEAP) (which is equivalent to a 200-m dish and has the sensitivity of SKA phase 1) (Bassa et al. 2016) already play a major role in improving the timing precision of known pulsars and the search for new ones. New instruments such as the South African MeerKAT (Booth et al. 2009) radio telescope are especially promising. With the Five hundred meter Aperture Spherical Telescope (FAST) (Nan et al. 2011) and the Square Kilometre Array (SKA) (Dewdney et al. 2009) telescopes, we will take a giant leap in sensitivity, providing both higher precision pulsar timing and surveys that will be able to find thousands of new pulsars (Smits et al. 2009,a). In particular, with its higher sensitivity, we expect major new scientific accomplishments with the arrival of the SKA.

3.4.1 High Precision Timing

The SKA (SKA1-MID and SKA2) will greatly improve the timing precision of known MSPs, including the Double Pulsar J0737–3039 (Burgay et al. 2003; Lyne et al. 2004) and triple systems such as PSR J0337+1715 (Ransom et al. 2014). The sensitivity of SKA1-MID should be comparable to that of both Arecibo (with an illuminated surface of about 200 m) and LEAP, but Arecibo has a limited range of observable declinations. For pulsars not visible with Arecibo, the increase in sensitivity will be remarkable. We expect SKA1-MID to improve the pulsar timing precision by one order of magnitude, while SKA2 should improve it by two orders of magnitude (Shao et al. 2015). This will provide PTAs with better TOAs and better constraints on a background of GWs, or better, the direct detection of a background or continuous source of GWs. This will also allow us to probe gravity in the strong-field regime, especially with the Double Pulsar PSR J0737–3039, which is visible from the southern sky and is particularly important for SKA. The TOA precision on the Double Pulsar is expected to reach 5 microseconds with SKA1-MID, while reaching sub-microsecond levels with SKA2, which will provide much tighter GR constraints. The eclipse of psr B (with its rapidly-changing flux density) will also be better measured by the SKA. Finally, the moment of inertia of psr A could also be measured and help constrain the equation of state of nuclear matter (Watts et al. 2015).

3.4.2 Finding More MSPs

The *SKA Galactic Census* will enable a search for new pulsars, with which we expect to expand the current pulsar population by a factor of three (Keane et al. 2015). In its first phase (SKA phase 1 or SKA1, which includes SKA1 LOW and SKA1 MID), it will have half of the total collecting area expected for SKA2. With its higher sensitivity, the telescope will take less time to achieve a particular S/N, therefore smearing due to varying accelerations will be less of an issue. We expect to find 10,000 normal pulsars with SKA2, including 1800 MSPs (Shao et al. 2015). The discovery of new, high-precision MSPs (whether solitary pulsars or binaries) will improve the PTAs' chances of detecting a background of nanohertz GWs or single continuous sources.

3.4.3 Finding More Highly-Relativistic Pulsars

The discovery of new compact binaries such as DNS binaries will lead to more stringent tests of GR (Kramer et al. 2006, 2004), while the discovery of new PSR—WD binaries will lead to more stringent tests of alternative theories of gravity, which usually predict the existence of gravitational dipole radiation, a varying gravitational constant and SEP violations (Ransom et al. 2014). In particular, we expect to find 100 DNS from SKA1 and 180 DNS from SKA2 (Keane et al. 2015). The SKA could also discover new triple systems (a pulsar with two compact objects) that can better constrain the SEP (specifically the Δ parameter). With a (NS-WD) inner binary and a NS as the outer star, the constraint on Δ would be many orders of magnitude tighter than with PSR J0337+1715 (Ransom et al. 2014) whose outer star is a WD. Additionally, the improved constraints on PK parameters thanks to the long-term monitoring of known pulsars such as the Double Pulsar, or to new pulsar discoveries could lead us to measure effects at 2PN, which could help constrain the equation of state of neutron stars (Watts et al. 2015). The SKA will also improve constraints on the Lense-Thirring effect. Discoveries with the SKA of new DNS binaries could enable a direct measurement of $\dot{x}$ as well as the Lense-Thirring contribution to $\dot{\omega}$ and their time derivatives $\ddot{\omega}$ and $\ddot{x}$. This is especially true for DNS with good timing precision, a close orbit, and a large angle between angular orbital momentum and pulsar spin (Shao et al. 2015).

3.4.4 Pulsar: Black Hole Systems

The discovery of pulsar—black hole (PSR-BH) binaries will open a completely new window on tests of strong gravity (Damour and Esposito-Farèse 1998; Wex and Kopeikin 1999). Indeed, possible PSR-BH systems are considered to be the *holy grail* for testing gravity in the strong-field regime, and better understanding the nature of black holes and their environments: the gravitational potential, the spacetime curvature, the compactness of the black hole. Pulsar timing allows for unique, high precision tests on the nature of black holes, including measuring the mass M_{BH}, angular momentum S_{BH} and charge Q_{BH} of the black hole. This in turn allows us to test of the *cosmic censorship conjecture* (which states that there cannot exist naked singularities) and the *no hair theorem* (stating that black holes are completely determined by M_{BH}, S_{BH} and Q_{BH}), which is violated in some alternative theories of gravity (Hawking and Penrose 1970; Kramer et al. 2004; Liu et al. 2012, 2014; Wex et al. 2013).

We expect to find the first PSR-BH binaries with SKA2 (Shao et al. 2015; Liu et al. 2014). There are three types of black hole environments that can be tested: pulsars could be found orbiting a stellar mass BH (Pfahl et al. 2005); an Intermediate Mass Black Hole (IMBH) in globular clusters (Devecchi et al. 2007; Clausen et al. 2014; Hessels et al. 2015); a supermassive BH such as SgrA* (Deneva 2010). In fact, dozens of radio pulsars are expected to be found in the Galactic Centre orbiting SgrA*. If the pulsars are close enough to the Galactic centre, the spacetime around the supermassive black hole (SgrA*) could be tested (Pfahl and Loeb 2004). The magnetar J1745–2900, detected in both X-rays and radio, was found to be near SgrA* (Mori et al. 2013; Kennea et al. 2013; Eatough et al. 2013). However it is not close enough to it to perform strong gravity tests. If pulsars can be found close enough to SgrA*, the determination of PK parameters through pulsar timing can help find a precise mass of SgrA* (through a measurement of $\dot{\omega}$, r, s and γ). Additionally, the determination of the spin-orbit coupling could help place an upper limit on the BH spin. We note that a recent paper suggested that, after analysing decades of pulsar timing data, a known MSP (PSR B1820–30A) is found to be orbiting an IMBH in the globular cluster NGC6624 (Perera et al. 2017). While a great result in demonstrating the existence of IMBHs, the orbital period of the system is too long to allow for tests of the spacetime around the black hole. Finally, it will be possible to probe the spacetime around black holes by combining pulsar timing data with the study of relativistic stellar orbits and the high-resolution imaging of the black hole horizon at sub-mm wavelengths, in particular with the Event Horizon Telescope. This will provide a highly-precise test of the no-hair theorem (Psaltis et al. 2016; Benkevitch et al. 2013).

3.4.5 Multi-Messenger Astronomy

The LIGO-Virgo detections of GWs from pairs of black holes and from a DNS (GW170817 Abbott et al. 2017) were truly historic and provide a new window on tests of strong gravity (Abbott et al. 2016). In the near future, we expect LIGO and Virgo to find other sources of GW, including more neutron stars in binary systems (DNS binaries or NS-BH systems) (Abadie et al. 2010), while we expect the space interferometer LISA (The LISA Consortium 2017) to find DNS and NS-BH systems with orbital periods of hours to minutes (see, e.g. Nelemans et al. 2001). The spectacular observational campaign following the GW detection of GW170817 opened the era of multi-messenger astronomy (Abbott et al. 2017): if the neutron star in the binary system is visible as a radio pulsar, the information from GW ground detectors can be combined with the electromagnetic radiation from the pulsar (through pulsar timing and the determination of PK parameters). This will enable us to learn about the environment of GW sources such as neutron stars in binary systems, which emit both radio waves (if the neutron star is visible as a radio pulsar) and GW waves. We might discover new physics, or at the very least, we will better constrain alternative theories of gravity Shao et al. (2017).

3.4.6 Detecting Nanohertz GWs

As already mentioned, in the FAST and SKA era, better timing together with the addition of new MSPs to PTAs will greatly increase the sensitivity of PTAs to a background of nanohertz GWs. The use of wideband receivers will help greatly in mitigating all chromatic effects related to pulse propagation in the ISM, allowing a better identification and removal of scattering and dispersion effects. Conversely, achromatic noise sources such as jitter and spin noise cannot be mitigated by extending the receiver band and might eventually dominate the rms uncertainty of the best timed MSPs (Cordes 2013). A promising avenue for improving PTA capabilities is to abandon the idea of analysing pre-determined TOAs, constructed by matching the profiles of individual measurements to a template, and instead directly use the pulse profiles of each individual observation. This type of profile domain analysis has been shown to potentially provide significant improvements in identifying scattering (Lentati et al. 2017) and spin noise (Shannon et al. 2016), and thanks to continuously improving algorithms, it can be also applied to wideband data (Lentati et al. 2017).

Despite the promise of great improvements, it is extremely difficult to forecast when PTAs will detect GWs. Theoretical exercises (Siemens et al. 2013; Taylor et al. 2016; Vigeland and Siemens 2016; Kelley et al. 2017) tend to predict a first detection within a decade. This depends on many unknowns, such as how many pulsars will be discovered and added to the PTAs each year, their intrinsic properties and stability, an optimization of the observing schedule (see e.g. Lee et al. 2012), and on the very uncertain GW predictions (cf Fig. 3.8). It is likely that the first glimpse of GWs in the data will come from the stochastic GWB rather than a deterministic source (Rosado et al. 2015). Given that current PTA sensitivities are dominated by a few, very stable systems, it is likely that this will not significantly change in the near future. The GWB will then show as additional red noise in the best timed pulsars, which will cause a saturation of PTA upper limits before any confident claim can be made through the detection of the HD correlation pattern.

Looking further ahead well within the SKA era, GW detection with PTAs will enable a series of scientific breakthrough including: (1) proving the existence of sub-pc SMBHBs, (2) understanding of the dynamics and cosmic history of SMBHBs, (3) identification and sky localization of individual sources, thus enabling multimessenger astronomy of massive objects in the low frequency regime, (4) tests of the existence of extra GW polarizations, (5) unprecedented constraints (or detection) on cosmic (super)string theories and other cosmological sources of GWs, such as first order phase transitions. A useful summary of the GW science enabled by future PTA detections can be found in Janssen et al. (2015).

3.5   Summary

MSPs are extremely precise clocks. Monitoring the TOAs of their radio pulses over many years (through the process of "pulsar timing") makes them rival atomic clocks on Earth. This process allows us to determine pulsar parameters with extremely high precision, including their rotation period, period derivative or the dispersion measure due to the interstellar medium. Through this process, we can learn about the strong-gravity environment around the neutron star, the interstellar medium, and possibly detect GWs from distant SMBHBs. For pulsars in binaries, we can constrain the strong-gravity environment around the neutron star through the fitting of PK parameters. The Double Pulsar PSR J0737−3039 provides the best test so far of GR, confirming GR within 0.05%. Constraints on violations of the Strong Equivalence Principle (SEP)—which would signal a breakdown of GR—as well as constraints on alternative theories of gravity, are however better achieved with pulsar—white dwarf (PSR-WD) binaries such PSR J1738+0333 and PSR J0348+0432. We also expect the triple system PSR J0337+1715 (one PSR with two WD) and future PSR-BH binaries to impose even stronger constraints. TOA precision naturally improves with longer datasets, therefore the long-term monitoring of known pulsars will yield better gravity constraints. However the timing precision of known pulsars such as the Double Pulsar will be largely improved with newer telescopes such as FAST or the SKA, which have much greater sensitivity. The number of interesting pulsar binaries that can constrain gravity theories is small (more than two PK parameters need to be determined in order to provide an extra consistency check on the theory of gravity). Therefore gravity constraints will be much improved with new pulsar searches such as with the SKA: an additional 1800 MSPs are expected to be found, and 10% of these could be pulsar binaries. In particular, we hope to find dozens of new DNS, which would lead to tighter constraints on GR, while PSR-WD binaries would provide interesting tests on alternative theories of gravity. Additionally, triple systems such as (PSR-WD-WD) systems but also (PSR-WD-NS) would be particularly interesting for constraining the SEP and alternative theories of gravity. The holy grail however would be a PSR-BH system, which would not only provide very tight constraints on gravity theories, but also uniquely probe the spacetime around black holes. Pulsar timing techniques can also be combined with other techniques to provide even more stringent tests: the no-hair theorem can be better constrained by combining data from pulsar timing with the imaging of the BH event horizon using the Event Horizon Telescope (Psaltis et al. 2016). In addition, pulsar timing data from pulsar binaries can be combined with future LIGO or LISA GW detections of neutron stars in binary systems, which may lead to the discovery of new physics or impose strong constraints on alternative theories of gravity (Shao et al. 2017). Cross-correlating the TOAs from an ensemble of MSPs thus forming a PTA, offers the unique possibility of detecting GWs in the nanohertz frequency range, which is inaccessible to both current ground and future space-based interferometers. The most promising GW sources in this frequency range are a cosmic population of SMBHBs, and the incoherent superposition of their signals will most likely be detected as a stochastic GWB with a red spectrum. Current PTAs already placed stringent limits on the amplitude of such a GWB, skimming the level at which theoretical predictions place the signal. The discovery of new MSPs with the SKA will greatly improve the sensitivity of current PTAs, most likely leading to a confident detection by the end of next decade. Together with LIGO/Virgo/Kagra in the kilohertz and LISA in the millihertz frequency range, PTAs will contribute to making gravitational wave astronomy possible across twelve orders of magnitude in frequency, which will allow us to probe the astrophysics of compact objects and strong gravity across ten orders of magnitudes in mass scale. With the prospects of great improvements ahead and the enormous potential for new discoveries that come with it, pulsar timing stands as a unique tool in the quest to understand the nature of gravity and the Cosmos.

Acknowledgements The authors wish to thank Michael Kramer and Norbert Wex for discussions and comments.

References

Abadie, J., Abbott, B.P., Abbott, R., Abernathy, M., Accadia, T., Acernese, F., Adams, C., Adhikari, R., Ajith, P., Allen, B., et al. Classical Quantum Gravity **27**, 173001 (2010). (Preprint, 1003.2480)

Abbott, B.P., Abbott, R., Abbott, T.D., Abernathy, M.R., Acernese, F., Ackley, K., Adams, C., Adams, T., Addesso, P., Adhikari, R.X., et al.: Phys. Rev. Lett. **116**, 061102 (2016). (Preprint, 1602.03837)

Abbott, B.P., Abbott, R., Abbott, T.D., Acernese, F., Ackley, K., Adams, C., Adams, T., Addesso, P., Adhikari, R.X., Adya, V.B., et al. Astron. J. **848**, L12 (2017). (Preprint, 1710.05833)

Abbott, B.P., Abbott, R., Abbott, T.D., Acernese, F., Ackley, K., Adams, C., Adams, T., Addesso, P., Adhikari, R.X., Adya, V.B., et al.: Phys. Rev. Lett. **119**, 161101 (2017). (Preprint, 1710.05832)

Allen, B., Romano, J.D.: Phys. Rev. D **59**, 102001 (1999). (Preprint, gr-qc/9710117)

Alsing, J., Berti, E., Will, C.M., Zaglauer, H.: Phys. Rev. D **85**, 064041 (2012). (Preprint, 1112.4903)

Amaro-Seoane, P., Sesana, A., Hoffman, L., Benacquista, M., Eichhorn, C., Makino, J., Spurzem, R.: Mon. Not. R. Astron. Soc. **402**, 2308–2320 (2010). (Preprint, 0910.1587)

Anderson, W.G., Brady, P.R., Creighton, J.D., Flanagan, É.É.: Phys. Rev. D **63**, 042003 (2001). (Preprint, gr-qc/0008066)

Anholm, M., Ballmer, S., Creighton, J.D.E., Price, L.R., Siemens, X.: Phys. Rev. D **79**, 084030 (2009). (Preprint, 0809.0701)

Antoniadis, J., van Kerkwijk, M.H., Koester, D., Freire, P.C.C., Wex, N., Tauris, T.M., Kramer, M., Bassa, C.G.: Mon. Not. R. Astron. Soc. **423**, 3316–3327 (2012). (Preprint, 1204.3948)

Antoniadis, J., Freire, P.C.C., Wex, N., Tauris, T.M., Lynch, R.S., van Kerkwijk, M.H., Kramer, M., Bassa, C., Dhillon, V.S., Driebe, T., Hessels, J.W.T., Kaspi, V.M., Kondratiev, V.I., Langer, N., Marsh, T.R., McLaughlin, M.A., Pennucci, T.T., Ransom, S.M., Stairs, I.H., van Leeuwen, J., Verbiest, J.P.W., Whelan, D.G.: Science **340**, 448 (2013). (Preprint, 1304.6875)

Arzoumanian, Z., et al.: Astrophys. J. **794**, 141 (2014). (Preprint, 1404.1267)

Arzoumanian, Z., et al.: Astrophys. J. **810**, 150 (2015). (Preprint, 1501.05343)

Arzoumanian, Z., et al.: Astrophys. J. **813**, 65 (2015). (Preprint, 1505.07540)

Arzoumanian, Z., et al.: Astrophys. J. **821**, 13 (2016). (Preprint, 1508.03024)

Arzoumanian, Z., Baker, P.T., Brazier, A., Burke-Spolaor, S., Chamberlin, S.J., Chatterjee, S., Christy, B., Cordes, J.M., Cornish, N.J., Crawford, F., Thankful Cromartie, H., Crowter, K., DeCesar, M., Demorest, P.B., Dolch, T., Ellis, J.A., Ferdman, R.D., Ferrara, E., Folkner, W.M., Fonseca, E., Garver-Daniels, N., Gentile, P.A., Haas, R., Hazboun, J.S., Huerta, E.A., Islo, K., Jenet, F., Jones, G., Jones, M.L., Kaplan, D.L., Kaspi, V.M., Lam M.T., Lazio, T.J.W., Levin, L., Lommen, A.N., Lorimer, D.R., Luo, J., Lynch R.S., Madison, D.R., McLaughlin, M.A., McWilliams, S.T., Mingarelli, C.M.F., Ng, C., Nice, D.J., Park, R.S., Pennucci, T.T., Pol, N.S., Ransom, S.M., Ray, P.S., Rasskazov, A., Siemens, X., Simon, J., Spiewak, R., Stairs, I.H., Stinebring, D.R., Stovall, K., Swiggum, J., Taylor, S.R., Vallisneri, M., Vigeland, S., Zhu, W.W.: ArXiv e-prints (2018). (Preprint, 1801.02617)

Babak, S., Sesana, A.: Phys. Rev. D **85**, 044034 (2012). (Preprint, 1112.1075)

Babak, S., et al.: Mon. Not. R. Astron. Soc. **455**, 1665–1679 (2016). (Preprint, 1509.02165)

Backer, D.C., Kulkarni, S.R., Heiles, C., Davis, M.M., Goss, W.M.: Nature **300**, 615–618 (1982)

Barker, B.M., O'Connell, R.F.: Phys. Rev. D **12**, 329–335 (1975)

Bartolo, N., Caprini, C., Domcke, V., Figueroa, D.G., Garcia-Bellido, J., Chiara Guzzetti, M., Liguori, M., Matarrese, S., Peloso, M., Petiteau, A., Ricciardone, A., Sakellariadou, M., Sorbo, L., Tasinato, G.: J. Cosmol. Astropart. Phys. **12**, 026 (2016). (Preprint, 1610.06481)

Bassa, C.G., Janssen, G.H., Karuppusamy, R., Kramer, M., Lee, K.J., Liu, K., McKee, J., Perrodin, D., Purver, M., Sanidas, S., Smits, R., Stappers, B.W.: Mon. Not. R. Astron. Soc. **456**, 2196–2209 (2016). (Preprint, 1511.06597)

Begelman, M.C., Blandford, R.D., Rees, M.J.: Nature **287**, 307–309 (1980)

Benkevitch, L., Fish, V.L., Johannsen, T., Akiyama, K., Broderick, A.E., Psaltis, D., Doeleman, S., Monnier, J.D., Baron, F.: American Astronomical Society Meeting Abstracts #221, vol. 221, p. 143.10 (2013)

Berti, E., et al.: Classical Quantum Gravity **32**, 243001 (2015). (*Preprint*, 1501.07274)

Bertotti, B., Carr, B.J., Rees, M.J.: Mon. Not. R. Astron. Soc. **203**, 945–954 (1983)

Bhat, N.D.R., Bailes, M., Verbiest, J.P.W.: Phys. Rev. D **77**, 124017 (2008). (Preprint, 0804.0956)

Bhattacharya, D., van den Heuvel, E.P.J.: Phys. Rep. **203**, 1–124 (1991)

Blanchet, L., Damour, T.: Phys. Rev. D **46**, 4304–4319 (1992)

Blecha, L., Sijacki, D., Kelley, L.Z., Torrey, P., Vogelsberger, M., Nelson, D, Springel, V., Snyder, G., Hernquist, L.: Mon. Not. R. Astron. Soc. **456**, 961–989 (2016). (Preprint, 1508.01524)

Boerner, G., Ehlers, J., Rudolph, E.: Astron. Astrophys. **44**, 417–420 (1975)

Bonetti, M., Haardt, F., Sesana, A., Barausse, E.: Mon. Not. R. Astron. Soc. **461**, 4419–4434 (2016). (Preprint, 1604.08770)

Booth, R.S., de Blok, W.J.G., Jonas, J.L., Fanaroff, B.: ArXiv e-prints (2009). (Preprint, 0910.2935)

Boyles, J., Lynch, R.S., Ransom, S.M., Stairs, I.H., Lorimer, D.R., McLaughlin, M.A., Hessels, J.W.T., Kaspi, V.M., Kondratiev, V.I., Archibald, A., Berndsen, A., Cardoso, R.F., Cherry, A., Epstein, C.R., Karako-Argaman, C., McPhee, C.A., Pennucci, T., Roberts, M.S.E., Stovall, K., van Leeuwen, J.: Astrophys. J. **763**, 80 (2013). (Preprint, 1209.4293)

Brans, C., Dicke, R.H.: Phys. Rev. **124**, 925–935 (1961)

Breton, R.P., Kaspi, V.M., Kramer, M., McLaughlin, M.A., Lyutikov, M., Ransom, S.M., Stairs, I.H., Ferdman, R.D., Camilo, F., Possenti, A. Science **321**, 104 (2008). (Preprint, 0807.2644)

Brumberg, V.A.: Essential Relativistic Celestial Mechanics, 271 p. Adam Hilger, Bristol (1991)

Bugaev, E., Klimai, P.: Phys. Rev. D **83**, 083521 (2011). (Preprint, 1012.4697)

Burgay, M., D'Amico, N., Possenti, A., Manchester, R.N., Lyne, A.G., Joshi, B.C., McLaughlin, M.A., Kramer, M., Sarkissian, J.M., Camilo, F., Kalogera, V., Kim, C., Lorimer, D.R.: Nature **426**, 531–533 (2003). (Preprint, astro-ph/0312071)

Burke-Spolaor, S.: ArXiv e-prints (2015). (Preprint, 1511.07869)

Caprini, C., Durrer, R., Siemens, X.: Phys. Rev. D **82**, 063511 (2010). (Preprint, 1007.1218)

Chamberlin, S.J., Creighton, J.D.E., Siemens, X., Demorest, P., Ellis, J., Price, L.R., Romano, J.D.: Phys. Rev. D **91**, 044048 (2015). (Preprint, 1410.8256)

Chen, S., Middleton, H., Sesana, A., Del Pozzo, W., Vecchio, A.: Mon. Not. R. Astron. Soc. **468**, 404–417 (2017). (Preprint, 1612.02826)

Chiara Guzzetti, M., Bartolo, N., Liguori, M., Matarrese, S.: ArXiv e-prints (2016). (Preprint, 1605.01615)

Christodoulou, D.: Phys. Rev. Lett. **67**, 1486–1489 (1991)

Clausen, D., Sigurdsson, S., Chernoff, D.F.: Mon. Not. R. Astron. Soc. **442**, 207–219 (2014). (Preprint, 1404.7502)

Clifton, T., Weisberg, J.M.: Astrophys. J. **679**, 687–696 (2008)

Cordes, J.M.: Classical Quantum Gravity **30**, 224002 (2013)

Cordes, J.M., Jenet, F.A.: Astrophys. J. **752**, 54 (2012)

Damour, T.: The problem of motion in Newtonian and Einsteinian gravity, pp. 128–198 (1987)

Damour, T.: Highlights in Gravitation and Cosmology. Iyer, B.R., Vishveshwara, C.V., Narlikar, J.V., Kembhavi, A.K., p. 393. Cambridge University Press, Cambridge (1988)

Damour, T., Deruelle, N.: Ann. Inst. Henri Poincaré Phys. Théor **43**(1), 107–132 (1985)

Damour, T., Deruelle, N.: Ann. Inst. Henri Poincaré Phys. Théor **44**(3), 263–292 (1986)

Damour, T., Esposito-Farèse, G.: Phys. Rev. D **53**, 5541–5578 (1996). (Preprint, gr-qc/9506063)

Damour, T., Esposito-Farèse, G.: Phys. Rev. D **54**, 1474–1491 (1996). (Preprint, gr-qc/9602056)
Damour, T., Esposito-Farèse, G.: Phys. Rev. D **58**, 042001 (1998). (Preprint, gr-qc/9803031)
Damour, T., Esposito-Farese, G.: Classical Quantum Gravity **9**, 2093–2176 (1992)
Damour, T., Esposito-Farese, G.: Phys. Rev. Lett. **70**, 2220–2223 (1993)
Damour, T., Ruffini, R.: Academie des Sciences Paris Comptes Rendus Serie Sciences Mathematiques **279**, 971–973 (1974)
Damour, T., Schafer, G.: Nuovo Cimento B Ser. **101**, 127–176 (1988)
Damour, T., Taylor, J.H.: Phys. Rev. D **45**, 1840–1868 (1992)
Damour, T., Vilenkin, A.: Phys. Rev. D **64**, 064008 (2001). (Preprint, gr-qc/0104026)
Deller, A.T., Bailes, M., Tingay, S.J.: Science **323**, 1327 (2009). (Preprint, 0902.0996)
Deller, A.T., Tingay, S.J., Bailes, M., Reynolds, J.E.: Astrophys. J. **701**, 1243–1257 (2009). (Preprint, 0906.3897)
Demorest, P.B., et al.: Astrophys. J. **762**, 94 (2013). (Preprint, 1201.6641)
Deneva, I.S.: Elusive neutron star populations: Galactic center and intermittent pulsars. Ph.D. thesis Cornell University (2010)
Desvignes, G., et al.: Mon. Not. R. Astron. Soc. **458**, 3341–3380 (2016). (Preprint, 1602.08511)
Detweiler, S.: Astrophys. J. **234**, 1100–1104 (1979)
Devecchi, B., Colpi, M., Mapelli, M., Possenti, A.: Mon. Not. R. Astron. Soc. **380**, 691–702 (2007). (Preprint, 0706.1656)
Dewdney, P.E., Hall, P.J., Schilizzi, R.T., Lazio, T.J.L.W.: IEEE Proc. **97**, 1482–1496 (2009)
Dotti, M., Sesana, A., Decarli, R.: Adv. Astron. **2012**, 940568 (2012). (Preprint, 1111.0664)
Eatough, R.P., Falcke, H., Karuppusamy, R., Lee, K.J., Champion, D.J., Keane, E.F., Desvignes, G., Schnitzeler, D.H.F.M., Spitler, L.G., Kramer, M., Klein, B., Bassa, C., Bower, G.C., Brunthaler, A., Cognard, I., Deller, A.T., Demorest, P.B., Freire, P.C.C., Kraus, A., Lyne, A.G., Noutsos, A., Stappers, B., Wex, N.: Nature **501**, 391–394 (2013). (Preprint, 1308.3147)
Edwards, R.T., Hobbs, G.B., Manchester, R.N.: Mon. Not. R. Astron. Soc. **372**, 1549–1574 (2006). (Preprint, astro-ph/0607664)
Ellis, J.A., Siemens, X., Creighton, J.D.E.: Astrophys. J. **756**, 175 (2012). (Preprint, 1204.4218)
Enoki, M., Nagashima, M.: Prog. Theor. Phys. **117**, 241–256 (2007). (Preprint, astro-ph/0609377)
Estabrook, F.B., Wahlquist, H.D.: Gen. Relativ. Gravit. **6**, 439–447 (1975)
Favata, M.: Phys. Rev. D **80**, 024002 (2009). (Preprint, 0812.0069)
Ferdman, R.D., Stairs, I.H., Kramer, M., Breton, R.P., McLaughlin, M.A., Freire, P.C.C., Possenti, A., Stappers, B.W., Kaspi, V.M., Manchester, R.N. Lyne, A.G.: Astrophys. J. **767**, 85 (2013). (Preprint, 1302.2914)
Finn, L.S., Lommen, A.N.: Astrophys. J. **718**, 1400–1415 (2010). (Preprint, 1004.3499)
Fonseca, E., Stairs, I.H., Thorsett, S.E.: Astrophys. J. **787**, 82 (2014). (Preprint, 1402.4836)
Foster, R.S., Backer, D.C.: Astrophys. J. **361**, 300–308 (1990)
Freire, P.C.C., Kramer, M., Wex, N.: Classical Quantum Gravity **29**, 184007 (2012). (Preprint, 1205.3751)
Freire, P.C.C., Wex, N., Esposito-Farèse, G., Verbiest, J.P.W., Bailes, M., Jacoby, B.A., Kramer, M., Stairs, I.H., Antoniadis, J., Janssen, G.H.: Mon. Not. R. Astron. Soc. **423**, 3328–3343 (2012). (Preprint, 1205.1450)
Gair, J., Romano, J.D., Taylor, S., Mingarelli, C.M.F.: Phys. Rev. D **90**, 082001 (2014). (Preprint, 1406.4664)
Gonzalez, M.E., Stairs, I.H., Ferdman, R.D., Freire, P.C.C., Nice, D.J., Demorest, P.B., Ransom, S.M., Kramer, M., Camilo, F., Hobbs, G., Manchester, R.N., Lyne, A.G.: Astrophys. J. **743**, 102 (2011). (Preprint, 1109.5638)
Grishchuk, L.P.: Phys. Usp. **48**, 1235–1247 (2005). (Preprint, gr-qc/0504018)
Hawking, S.W., Penrose, R.: Proc. R. Soc. London, Ser. A **314**, 529–548 (1970)
Hellings, R.W., Downs, G.S.: Astron. J. **265**, L39–L42 (1983)
Hessels, J., Possenti, A., Bailes, M., Bassa, C., Freire, P.C.C., Lorimer, D.R., Lynch, R., Ransom, S.M., Stairs, I.H.: Advancing Astrophysics with the Square Kilometre Array (AASKA14), vol. 47 (2015) (Preprint, 1501.00086)

Hoffman, L., Loeb, A.: Mon. Not. R. Astron. Soc. **377**, 957–976 (2007). (Preprint, astro-ph/0612517)

Hofmann, F., Müller, J., Biskupek, L.: Astron. Astrophys. **522**, L5 (2010)

Huerta, E.A., McWilliams, S.T., Gair, J.R., Taylor, S.R.: Phys. Rev. D **92**, 063010 (2015). (Preprint, 1504.00928)

Hulse, R.A., Taylor, J.H.: Astron. J. **195**, L51–L53 (1975)

Ivanov, P.B., Papaloizou, J.C.B., Polnarev, A.G.: Mon. Not. R. Astron. Soc. **307**, 79–90 (1999). (Preprint, astro-ph/9812198)

Jacoby, B.A.: Ph.D. thesis California Institute of Technology (2005)

Jaffe, A.H., Backer, D.C.: Astrophys. J. **583**, 616–631 (2003). (Preprint, astro-ph/0210148)

Janssen, G., Hobbs, G., McLaughlin, M., Bassa, C., Deller, A., Kramer, M., Lee, K., Mingarelli, C., Rosado, P., Sanidas, S., Sesana, A., Shao, L., Stairs, I., Stappers, B., Verbiest, J.P.W.: Advancing Astrophysics with the Square Kilometre Array (AASKA14), vol. 37 (2015). (Preprint, 1501.00127)

Jenet, F.A., Hobbs, G.B., van Straten, W., Manchester, R.N., Bailes, M., Verbiest, J.P.W., Edwards, R.T., Hotan, A.W., Sarkissian, J.M., Ord, S.M.: Astrophys. J. **653**, 1571–1576 (2006). (Preprint, astro-ph/0609013)

Johnston, S., Lorimer, D.R., Harrison, P.A., Bailes, M., Lyne, A.G., Bell, J.F., Kaspi, V.M., Manchester, R.N., D'Amico, N., Nicastro, L.: Nature **361**, 613–615 (1993)

Kaspi, V.M., Taylor, J.H., Ryba, M.F.: Astrophys. J. **428**, 713–728 (1994)

Kaspi, V.M., Lyne, A.G., Manchester, R.N., Crawford, F., Camilo, F., Bell, J.F., D'Amico, N., Stairs, I.H., McKay, N.P.F., Morris, D.J., Possenti, A.: Astrophys. J. **543**, 321–327 (2000). (Preprint, astro-ph/0005214)

Kass, R.E., Raftery, A.E.: J. Am. Stat. Assoc. **90**, 773–795 (1995)

Keane, E., Bhattacharyya, B., Kramer, M., Stappers, B., Keane, E.F., Bhattacharyya, B., Kramer, M., Stappers, B.W., Bates, S.D., Burgay, M., Chatterjee, S., Champion, D.J., Eatough, R.P., Hessels, J.W.T., Janssen, G., Lee, K.J., van Leeuwen, J., Margueron, J., Oertel, M., Possenti, A., Ransom, S., Theureau, G., Torne, P.: Advancing Astrophysics with the Square Kilometre Array (AASKA14), vol. 40 (2015) (Preprint, 1501.00056)

Kehl, M.S., Wex, N., Kramer, M., Liu, K.: ArXiv e-prints (2016). (Preprint, 1605.00408)

Kelley, L.Z., Blecha, L., Hernquist, L., Sesana, A.: ArXiv e-prints (2017). (Preprint, 1702.02180)

Kelley, L.Z., Blecha, L., Hernquist, L.: Mon. Not. R. Astron. Soc. **464**, 3131–3157 (2017). (Preprint, 1606.01900)

Kennea, J.A., Burrows, D.N., Kouveliotou, C., Palmer, D.M., Göğüş, E., Kaneko, Y., Evans, P.A., Degenaar, N., Reynolds, M.T., Miller, J.M., Wijnands, R., Mori, K., Gehrels, N.: Astron. J. **770**, L24 (2013). (Preprint, 1305.2128)

Key, J.S., Cornish, N.J.: Phys. Rev. D **79**, 043014 (2009). (Preprint, 0812.1590)

Kocsis, B., Sesana, A.: Mon. Not. R. Astron. Soc. **411**, 1467–1479 (2011). (Preprint, 1002.0584)

Kopeikin, S.M.: Astron. J. **439**, L5–L8 (1995)

Kopeikin, S.M.: Astron. J. **467**, L93 (1996)

Kramer, M.: Astrophys. J. **509**, 856–860 (1998). (Preprint, astro-ph/9808127)

Kramer, M.: Int. J. Mod. Phys. D **25**, 1630029–1630061 (2016). (Preprint, 1606.03843)

Kramer, M., Stairs, I.H.: Annu. Rev. Astron. Astrophys. **46**, 541–572 (2008)

Kramer, M., Backer, D.C., Cordes, J.M., Lazio, T.J.W., Stappers, B.W., Johnston, S.: New Astron. Rev. **48**, 993–1002 (2004). (Preprint, astro-ph/0409379)

Kramer, M., Stairs, I.H., Manchester, R.N., McLaughlin, M.A., Lyne, A.G., Ferdman, R.D., Burgay, M., Lorimer, D.R., Possenti, A., D'Amico, N., Sarkissian, J.M., Hobbs, G.B., Reynolds, J.E., Freire, P.C.C., Camilo, F.: Science **314**, 97–102 (2006). (Preprint, astro-ph/0609417)

Kulier, A., Ostriker, J.P., Natarajan, P., Lackner, C.N., Cen, R.: Astrophys. J. **799**, 178 (2015). (Preprint, 1307.3684)

Lacey, C., Cole, S.: Mon. Not. R. Astron. Soc. **262**, 627–649 (1993)

Lasky, P.D., et al.: Phys. Rev. X **6**, 011035 (2016). (Preprint, 1511.05994)

Lee, K., Jenet, F.A., Price, R.H., Wex, N., Kramer, M.: Astrophys. J. **722**, 1589–1597 (2010). (Preprint, 1008.2561)

Lee, K.J., Bassa, C.G., Janssen, G.H., Karuppusamy, R., Kramer, M., Smits, R., Stappers, B.W.:
Mon. Not. R. Astron. Soc. **423**, 2642–2655 (2012). (Preprint, 1204.4321)

Lentati, L., et al.: Mon. Not. R. Astron. Soc. **453**, 2576–2598 (2015). (Preprint, 1504.03692)

Lentati, L., Kerr, M., Dai, S., Hobson, M.P., Shannon, R.M., Hobbs, G., Bailes, M., Bhat, N.D.R.,
Burke-Spolaor, S., Coles, W., Dempsey, J., Lasky, P.D., Levin, Y., Manchester, R.N., Osłowski,
S., Ravi, V., Reardon, D.J., Rosado, P.A., Spiewak, R., van Straten, W., Toomey, L., Wang,
J., Wen, L., You, X., Zhu, X.: Mon. Not. R. Astron. Soc. **466**, 3706–3727 (2017). (Preprint,
1612.05258)

Lentati, L., Kerr, M., Dai, S., Shannon, R.M., Hobbs, G., Osłowski, S.: Mon. Not. R. Astron. Soc.
468, 1474–1485 (2017). (Preprint, 1703.02108)

Liu, K., Wex, N., Kramer, M., Cordes, J.M., Lazio, T.J.W.: Astrophys. J. **747**, 1 (2012). (Preprint,
1112.2151)

Liu, K., Desvignes, G., Cognard, I., Stappers, B.W., Verbiest, J.P.W., Lee K.J., Champion, D.J.,
Kramer, M., Freire, P.C.C., Karuppusamy, R.: Mon. Not. R. Astron. Soc. **443**, 3752–3760
(2014). (Preprint, 1407.3827)

Liu, K., Eatough, R.P., Wex, N., Kramer, M.: Mon. Not. R. Astron. Soc. **445**, 3115–3132 (2014).
(Preprint, 1409.3882)

Lorimer, D.R., Kramer, M.: Handbook of Pulsar Astronomy. Cambridge University Press,
Cambridge (2012)

Lorimer, D.R., Stairs, I.H., Freire, P.C., Cordes, J.M., Camilo, F., Faulkner, A.J., Lyne, A.G., Nice,
D.J., Ransom, S.M., Arzoumanian, Z., Manchester, R.N., Champion, D.J., van Leeuwen, J.,
Mclaughlin, M.A., Ramachandran, R., Hessels, J.W., Vlemmings, W., Deshpande, A.A., Bhat,
N.D., Chatterjee, S., Han, J.L., Gaensler, B.M., Kasian, L., Deneva, J.S., Reid, B., Lazio, T.J.,
Kaspi, V.M., Crawford, F., Lommen, A.N., Backer, D.C., Kramer, M., Stappers, B.W., Hobbs,
G.B., Possenti, A., D'Amico, N., Burgay, M.: Astrophys. J. **640**, 428–434 (2006). (Preprint,
astro-ph/0511523)

Lynch, R.S., Boyles, J., Ransom, S.M., Stairs, I.H., Lorimer, D.R., McLaughlin, M.A., Hessels,
J.W.T., Kaspi, V.M., Kondratiev, V.I., Archibald, A.M., Berndsen, A., Cardoso, R.F., Cherry,
A., Epstein, C.R., Karako-Argaman, C., McPhee, C.A., Pennucci, T., Roberts, M.S.E., Stovall,
K., van Leeuwen, J.: Astrophys. J. **763**, 81 (2013). (Preprint, 1209.4296)

Lyne, A.G., Burgay, M., Kramer, M., Possenti, A., Manchester, R.N., Camilo, F., McLaughlin,
M.A., Lorimer, D.R., D'Amico, N., Joshi, B.C., Reynolds, J., Freire, P.C.C.: Science **303**,
1153–1157 (2004). (Preprint, astro-ph/0401086)

Lyne, A., Hobbs, G., Kramer, M., Stairs, I., Stappers, B.: Science **329**, 408 (2010). (Preprint,
1006.5184)

Madison, D.R., Cordes, J.M., Chatterjee, S.: Astrophys. J. **788**, 141 (2014). (Preprint, 1404.5682)

Maggiore, M.: Phys. Rep. **331**, 283–367 (2000). (Preprint, gr-qc/9909001)

Maggiore, M.: Gravitational Waves, vol. 2. Oxford University Press, Oxford (2018)

Manchester, R.N., Hobbs, G.B., Teoh, A., Hobbs, M.: Astrophysics **129**, 1993–2006 (2005).
(Preprint, www.astro-ph/0412641)

Manchester, R.N., Kramer, M., Stairs, I.H., Burgay, M., Camilo, F., Hobbs, G.B., Lorimer, D.R.,
Lyne, A.G., McLaughlin, M.A., McPhee, C.A., Possenti, A., Reynolds, J.E., van Straten, W.:
Astrophys. J. **710**, 1694–1709 (2010). (Preprint, 1001.1483)

McWilliams, S.T., Ostriker, J.P., Pretorius, F.: Astrophys. J. **789**, 156 (2014). (Preprint, 1211.5377)

Mirshekari, S., Will, C.M.: Phys. Rev. D **87**, 084070 (2013). (Preprint, 1301.4680)

Mori, K., Gotthelf, E.V., Zhang, S., An, H., Baganoff, F.K., Barrière N.M., Beloborodov, A.M.,
Boggs, S.E., Christensen, F.E., Craig, W.W., Dufour F, Grefenstette, B.W., Hailey, C.J.,
Harrison, F.A., Hong, J., Kaspi, V.M., Kennea, J.A., Madsen, K.K., Markwardt, C.B., Nynka,
M., Stern, D., Tomsick, J.A., Zhang, W.W.: Astron. J. **770**, L23 (2013). (Preprint, 1305.1945)

Nan, R., Li, D., Jin, C., Wang, Q., Zhu, L., Zhu, W., Zhang, H., Yue, Y., Qian, L.: Int. J. Mod. Phys.
D **20**, 989–1024 (2011). (Preprint, 1105.3794)

Nelemans, G., Yungelson, L.R., Portegies Zwart, S.F.: Astron. Astrophys. **375**, 890–898 (2001).
(Preprint, astro-ph/0105221)

Nordtvedt, K.: Phys. Rev. **170**, 1186–1187 (1968)

Ölmez, S., Mandic, V., Siemens, X.: Phys. Rev. D **81**, 104028 (2010). (Preprint, 1004.0890)

Pennucci, T.T., Demorest, P.B., Ransom, S.M.: Astrophys. J. **790**, 93 (2014). (Preprint, 1402.1672)

Perera, B.B.P., McLaughlin, M.A., Kramer, M., Stairs, I.H., Ferdman, R.D., Freire, P.C.C., Possenti, A., Breton, R.P., Manchester, R.N., Burgay, M., Lyne, A.G., Camilo, F.: Astrophys. J. **721**, 1193–1205 (2010). (Preprint, 1008.1097)

Perera, B.B.P., Stappers, B.W., Lyne, A.G., Bassa, C.G., Cognard, I., Guillemot, L., Kramer, M., Theureau, G., Desvignes, G.: Mon. Not. R. Astron. Soc. **468**, 2114–2127 (2017). (Preprint, 1705.01612)

Peters, P.C., Mathews, J.: Phys. Rev. **131**, 435–440 (1963)

Pfahl, E., Loeb, A.: Astrophys. J. **615**, 253–258 (2004). (Preprint, astro-ph/0309744)

Pfahl, E., Podsiadlowski, P., Rappaport, S.: Astrophys. J. **628**, 343–352 (2005). (Preprint, astro-ph/0502122)

Phinney, E.S.: ArXiv Astrophysics e-prints (2001). (Preprint, astro-ph/0108028)

Planck Collaboration, Ade, P.A.R., Aghanim, N., Arnaud, M., Ashdown, M., Aumont, J., Baccigalupi, C., Banday, A.J., Barreiro, R.B., Bartlett, J.G., et al.: Astron. Astrophys. **594**, A13 (2016). (Preprint, 1502.01589)

Porto, R.A.: Phys. Rev. D **73**, 104031 (2006). (Preprint, gr-qc/0511061)

Possenti, A., Burgay, M.: The role of binary pulsars in testing gravity theories. In: Gravity: Where Do We Stand? p. 279. Springer, Cham (2016)

Psaltis, D., Wex, N., Kramer, M.: Astrophys. J. **818**, 121 (2016). (Preprint, 1510.00394)

Psaltis, D.: Living Rev. Relativ. **11**, 9 (2008). (Preprint, 0806.1531)

Rajagopal, M., Romani, R.W.: Astrophys. J. **446**, 543 (1995). (Preprint, astro-ph/9412038)

Ransom, S.M., Stairs, I.H., Archibald, A.M., Hessels, J.W.T., Kaplan, D.L., van Kerkwijk, M.H., Boyles, J., Deller, A.T., Chatterjee, S., Schechtman-Rook, A., Berndsen, A., Lynch, R.S., Lorimer, D.R., Karako-Argaman, C., Kaspi, V.M., Kondratiev, V.I., McLaughlin, M.A., van Leeuwen, J., Rosen, R., Roberts, M.S.E., Stovall, K.: Nature **505**, 520–524 (2014). (Preprint, 1401.0535)

Rasskazov, A., Merritt, D.: ArXiv e-prints (2016). (Preprint, 1606.07484)

Ravi, V., Wyithe, J.S.B., Shannon, R.M., Hobbs, G., Manchester, R.N.: Mon. Not. R. Astron. Soc. **442**, 56–68 (2014). (Preprint, 1404.5183)

Ravi, V., Wyithe, J.S.B., Shannon, R.M., Hobbs, G.: Mon. Not. R. Astron. Soc. **447**, 2772–2783 (2015). (Preprint, 1406.5297)

Reardon, D.J., et al.: Mon. Not. R. Astron. Soc. **455**, 1751–1769 (2016). (Preprint, 1510.04434)

Romani, R.W., Taylor, J.H.: Astron. J. **265**, L35–L37 (1983)

Rosado, P.A., Sesana, A., Gair, J.: Mon. Not. R. Astron. Soc. **451**, 2417–2433 (2015). (Preprint, 1503.04803)

Sazhin, M.V.: Sov. Astron. **22**, 36–38 (1978)

Sesana, A.: Astrophys. J. **719**, 851–864 (2010). (Preprint, 1006.0730)

Sesana, A.: Classical Quantum Gravity **30**, 224014 (2013). (Preprint, 1307.2600)

Sesana, A.: Mon. Not. R. Astron. Soc. **433**, L1–L5 (2013). (Preprint, 1211.5375)

Sesana, A.: Gravitational Wave Astrophysics. Astrophysics and Space Science Proceedings, Sopuerta, C.F. (ed.), vol. 40, p. 147. Springer, Cham (2015). (Preprint, 1407.5693)

Sesana, A., Vecchio, A.: Phys. Rev. D **81**, 104008 (2010). (Preprint, 1003.0677)

Sesana, A., Haardt, F., Madau, P., Volonteri, M.: Astrophys. J. **611**, 623–632 (2004). (Preprint, astro-ph/0401543)

Sesana, A., Vecchio, A., Colacino, C.N.: Mon. Not. R. Astron. Soc. **390**, 192–209 (2008). (Preprint, 0804.4476)

Sesana, A., Vecchio, A., Volonteri, M.: Mon. Not. R. Astron. Soc. **394**, 2255–2265 (2009). (Preprint, 0809.3412)

Sesana, A., Roedig, C., Reynolds, M.T., Dotti, M.: Mon. Not. R. Astron. Soc. **420**, 860–877 (2012). (Preprint, 1107.2927)

Sesana, A., Shankar, F., Bernardi, M., Sheth, R.K.: Mon. Not. R. Astron. Soc. **463**, L6–L11 (2016). (Preprint, 1603.09348)

Shannon, R.M., Cordes, J.M.: Astrophys. J. **725**, 1607–1619 (2010). (Preprint, 1010.4794)

Shannon, R.M., et al.: Science **349**, 1522–1525 (2015). (Preprint, 1509.07320)
Shannon, R.M., Lentati, L.T., Kerr, M., Bailes, M., Bhat, N.D.R., Coles, W.A., Dai, S., Dempsey, J., Hobbs, G., Keith, M.J., Lasky, P.D., Levin, Y., Manchester, R.N., Osłowski, S., Ravi, V., Reardon, D.J., Rosado, P.A., Spiewak, R., van Straten, W., Toomey, L., Wang, J.B., Wen, L., You, X.P., Zhu, X.J.: Astron. J. **828**, L1 (2016). (Preprint, 1608.02163)
Shao, L.: Phys. Rev. D **93**, 084023 (2016). (Preprint, 1602.05725)
Shao, L., Wex, N.: Classical Quantum Gravity **29**, (2012) (Preprint, 1209.4503)
Shao, L., Wex, N.: Classical Quantum Gravity **30**, 165020 (2013). (Preprint, 1307.2637)
Shao, L., Caballero, R.N., Kramer, M., Wex, N., Champion, D.J., Jessner, A.: Classical Quantum Gravity **30**, 165019 (2013). (Preprint, 1307.2552)
Shao, L., Stairs, I., Antoniadis, J., Deller, A., Freire, P., Hessels, J., Janssen, G., Kramer, M., Kunz, J., Laemmerzahl, C., Perlick, V., Possenti, A., Ransom, S., Stappers, B., van Straten, W.: Adv. Astrophys. Square Kilometre Array (AASKA14) 42 (2015). (Preprint 1501.00058)
Shao, L., Sennett, N., Buonanno, A., Kramer, M., Wex, N.: ArXiv e-prints (2017). (Preprint, 1704.07561)
Shapiro, I.I.: Phys. Rev. Lett. **13**, 789–791 (1964)
Siemens, X., Olum, K.D.: Phys. Rev. D **68**, 085017 (2003). (Preprint, gr-qc/0307113)
Siemens, X., Ellis, J., Jenet, F., Romano, J.D.: Classical Quantum Gravity **30**, 224015 (2013). (Preprint, 1305.3196)
Simon, J., Burke-Spolaor, S.: Astrophys. J. **826**, 11 (2016). (Preprint, 1603.06577)
Smits, R., Lorimer, D.R., Kramer, M., Manchester, R., Stappers, B., Jin, C.J., Nan, R.D., Li, D.: Astron. Astrophys. **505**, 919–926 (2009). (Preprint, 0908.1689)
Smits, R., Kramer, M., Stappers, B., Lorimer, D.R., Cordes, J., Faulkner, A.: Astron. Astrophys. **493**, 1161–1170 (2009). (Preprint, 0811.0211)
Stairs, I.H.: Living Rev. Relativ. **6**, 5 (2003). (Preprint, astro-ph/0307536)
Stairs, I.H., Thorsett, S.E., Arzoumanian, Z.: Phys. Rev. Lett. **93**, 141101 (2004). (Preprint, astro-ph/0408457)
Stairs, I.H., Faulkner, A.J., Lyne, A.G., Kramer, M., Lorimer, D.R., McLaughlin, M.A., Manchester, R.N., Hobbs, G.B., Camilo, F., Possenti, A., Burgay, M., D'Amico, N., Freire, P.C., Gregory, P.C.: Astrophys. J. **632**, 1060–1068 (2005). (Preprint, astro-ph/0506188)
Stella, L., Possenti, A.: Space Sci. Rev. **148**, 105–121 (2009)
Tauris, T.M., Kramer, M., Freire, P.C.C., Wex, N., Janka, H.T., Langer, N., Podsiadlowski, P., Bozzo, E., Chaty, S., Kruckow, M.U., van den Heuvel, E.P.J., Antoniadis, J., Breton, R.P., Champion, D.J.: ArXiv e-prints (2017). (Preprint, 1706.09438)
Taylor, J.H., Weisberg, J.M.: Astrophys. J. **345**, 434–450 (1989)
Taylor, J.H., Fowler, L.A., McCulloch, P.M.: Nature **277**, 437–440 (1979)
Taylor, S., Ellis, J., Gair, J.: Phys. Rev. D **90**, 104028 (2014). (Preprint, 1406.5224)
Taylor, S.R., et al.: Phys. Rev. Lett. **115**, 041101 (2015). (Preprint, 1506.08817)
Taylor, S.R., Huerta, E.A., Gair, J.R., McWilliams, S.T.: Astrophys. J. **817**, 70 (2016). (Preprint, 1505.06208)
Taylor, S.R., Vallisneri, M., Ellis, J.A., Mingarelli, C.M.F., Lazio, T.J.W., van Haasteren, R.: Astron. J. **819**, L6 (2016). (Preprint, 1511.05564)
Taylor, S.R., Lentati, L., Babak, S., Brem, P., Gair, J.R., Sesana, A., Vecchio, A.: Phys. Rev. D **95**, 042002 (2017). (Preprint, 1606.09180)
The LISA Consortium: ArXiv e-prints (2017). (Preprint, 1702.00786)
Thorne, K.S.: Gravitational Radiation, pp. 330–458 (1987)
Tiburzi, C., Hobbs, G., Kerr, M., Coles, W.A., Dai, S., Manchester, R.N., Possenti, A., Shannon, R.M., You, X.P.: Mon. Not. R. Astron. Soc. **455**, 4339–4350 (2016). (Preprint, 1510.02363)
van Haasteren, R., Levin, Y.: Mon. Not. R. Astron. Soc. **401**, 2372–2378 (2010). (Preprint, 0909.0954)
van Haasteren, R., Levin, Y.: Mon. Not. R. Astron. Soc. **428**, 1147–1159 (2013). (Preprint, 1202.5932)
van Haasteren, R., et al.: Mon. Not. R. Astron. Soc. **414**, 3117–3128 (2011). (Preprint, 1103.0576)

van Straten, W., Bailes, M., Britton, M., Kulkarni, S.R., Anderson, S.B., Manchester, R.N., Sarkissian, J.: Nature **412**, 158–160 (2001). (Preprint, astro-ph/0108254)

Verbiest, J.P.W., Bailes, M., van Straten, W., Hobbs, G.B., Edwards, R.T., Manchester, R.N., Bhat, N.D.R., Sarkissian, J.M., Jacoby, B.A., Kulkarni, S.R.: Astrophys. J. **679**, 675–680 (2008). (Preprint, 0801.2589)

Verbiest, J.P.W., et al.: Mon. Not. R. Astron. Soc. **458**, 1267–1288 (2016). (Preprint, 1602.03640)

Vigeland, S.J., Siemens, X.: Phys. Rev. D **94**, 123003 (2016). (Preprint, 1609.03656)

Volonteri, M., Haardt, F., Madau, P.: Astrophys. J. **582**, 559–573 (2003). (Preprint, astro-ph/0207276)

Wang, Y.: J. Phys. Conf. Ser. **610**, 012019 (2015). (Preprint, 1505.00402)

Wang, J.B., et al.: Mon. Not. R. Astron. Soc. **446**, 1657–1671 (2015). (Preprint, 1410.3323)

Watts, A., Espinoza, C.M., Xu, R., Andersson, N., Antoniadis, J., Antonopoulou, D., Buchner, S., Datta, S., Demorest, P., Freire, P., Hessels, J., Margueron, J., Oertel, M., Patruno, A., Possenti, A., Ransom, S., Stairs, I., Stappers, B.: Advancing Astrophysics with the Square Kilometre Array (AASKA14) **43** (2015). (Preprint, 1501.00042)

Weisberg, J.M., Taylor, J.H.: Gen. Relativ. Gravit. **13**, 1–6 (1981)

Weisberg, J.M., Taylor, J.H.: Astrophys. J. **576**, 942–949 (2002). (Preprint, astro-ph/0205280)

Weisberg, J.M., Nice, D.J., Taylor, J.H.: Astrophys. J. **722**, 1030–1034 (2010). (Preprint 1011.0718)

Wex, N., Kopeikin, S.M.: Astrophys. J. **514**, 388–401 (1999). (Preprint, astro-ph/9811052)

Wex, N., Liu, K., Eatough, R.P., Kramer, M., Cordes, J.M., Lazio, T.J.W.: Neutron Stars and Pulsars: Challenges and Opportunities after 80 years (IAU Symposium), van Leeuwen, J. (ed.), vol. 291, pp. 171–176 (2013). (Preprint, 1210.7518)

Wex, N.: ArXiv e-prints (2014). (Preprint, 1402.5594)

Will, C.M.: Astron. J. **393**, L59–L61 (1992)

Will, C.M.: Theory and Experiment in Gravitational Physics, p. 396. Cambridge University Press, Cambridge (1993)

Will, C.M.: *General Relativity and John Archibald Wheeler*. Astrophysics and Space Science Library, Ciufolini, I., Matzner, R.A.A. (eds.), vol. 367. Springer, New York (2010)

Will, C.M.: Living Rev. Relativ. **17**, 4 (2014). (Preprint, 1403.7377)

Will, C.M.: Living Rev. Relativ. **17**, 4 (2014). (Preprint, 1403.7377)

Will, C.M., Nordtvedt, K. Jr.: Astrophys. J. **177**, 757 (1972)

Will, C.M., Zaglauer, H.W.: Astrophys. J. **346**, 366–377 (1989)

Willems, B., Vecchio, A., Kalogera, V.: Phys. Rev. Lett. **100**, 041102 (2008). (Preprint, 0706.3700)

Yagi, K., Blas, D., Barausse, E., Yunes, N.: Phys. Rev. D **89**, 084067 (2014). (Preprint, 1311.7144)

Yardley, D.R.B., Hobbs, G.B., Jenet, F.A., Verbiest, J.P.W., Wen, Z.L., Manchester, R.N., Coles, W.A., van Straten W, Bailes, M., Bhat, N.D.R., Burke-Spolaor S, Champion, D.J., Hotan, A.W., Sarkissian, J.M.: Mon. Not. R. Astron. Soc. **407**, 669–680 (2010). (Preprint, 1005.1667)

Yunes, N., Siemens, X.: Living Rev. Relativ. **16**, 9 (2013). (Preprint, 1304.3473)

Zhu, X.J., et al.: Mon. Not. R. Astron. Soc. **444**, 3709–3720 (2014). (Preprint, 1408.5129)

Zhu, X.J., Wen, L., Hobbs, G., Zhang, Y., Wang, Y., Madison, D.R., Manchester R.N., Kerr, M., Rosado, P.A., Wang, J.B.: Mon. Not. R. Astron. Soc. **449**, 1650–1663 (2015). (Preprint, 1502.06001)

Zhu, W.W., Stairs, I.H., Demorest, P.B., Nice, D.J., Ellis, J.A., Ransom, S.M., Arzoumanian, Z., Crowter, K., Dolch, T., Ferdman, R.D., Fonseca, E., Gonzalez, M.E., Jones, G., Jones, M.L., Lam, M.T., Levin, L., McLaughlin, M.A., Pennucci, T., Stovall, K., Swiggum, J.: Astrophys. J. **809**, 41 (2015). (Preprint, 1504.00662)

Chapter 4
Accreting Pulsars: Mixing-up Accretion Phases in Transitional Systems

Sergio Campana and Tiziana Di Salvo

Abstract In the last 20 years our understanding of the millisecond pulsar population changed dramatically. Thanks to the large effective area and good time resolution of the NASA X-ray observatory Rossi X-ray Timing Explorer, we discovered that neutron stars in Low Mass X-ray Binaries (LMXBs) spins at frequencies between 200 and 750 Hz, and indirectly confirmed the recycling scenario, according to which neutron stars are spun up to millisecond periods during the LMXB-phase. In the meantime, the continuous discovery of rotation-powered millisecond pulsars in binary systems in the radio and gamma-ray band (mainly with the Fermi Large Area Telescope) allowed us to classify these sources into two "spiders" populations, depending on the mass of their companion stars: Black Widow pulsars, with very low-mass companion stars, and Redbacks, with larger mass companion stars possibly filling their Roche lobes without accretion of matter onto the neutron star. It was soon regained that millisecond pulsars in short orbital period LMXBs are the progenitors of the spider populations of rotation-powered millisecond pulsars, although a direct link between accretion-powered and rotation-powered millisecond pulsars was still missing. In 2013 the ESA X-ray observatory XMM-Newton spotted the X-ray outburst of a new accreting millisecond pulsar (IGR J18245$-$2452) in a source that was previously classified as a radio millisecond pulsar, probably of the Redback type. Follow up observations of the source when it went back to X-ray quiescence showed that it was able to swing between accretion-powered to rotation-powered pulsations in a relatively short timescale (few days), promoting this source as the direct link between the LMXB and the radio millisecond pulsar phases. Following discoveries showed that there exists a bunch of sources which alternates X-ray activity phases, showing X-ray coherent pulsations, to radio-loud phases,

S. Campana
INAF - Osservatorio Astronomico di Brera, Merate, LC, Italy
e-mail: sergio.campana@brera.inaf.it

T. Di Salvo (✉)
Dipartimento di Fisica e Chimica, Università di Palermo, Palermo, Italy
e-mail: tiziana.disalvo@unipa.it

© Springer Nature Switzerland AG 2018
L. Rezzolla et al. (eds.), *The Physics and Astrophysics of Neutron Stars*,
Astrophysics and Space Science Library 457,
https://doi.org/10.1007/978-3-319-97616-7_4

showing radio pulsations, establishing a new class of millisecond pulsars, the so-called transitional millisecond pulsars. In this review we describe these exciting discoveries and the properties of accreting and transitional millisecond pulsars, highlighting what we know and what we have still to learn about in order to fully understand the (sometime puzzling) behaviour of these systems and their evolutive connection.

4.1 The Links in the Chain: How a Neutron Star Becomes a Millisecond Pulsar

4.1.1 The Recycling Scenario: The Evolutionary Path Leading to the Formation of Millisecond Pulsars

A millisecond pulsar (hereafter MSP) is a fast rotating, weakly magnetised neutron star. A weak magnetic field for a neutron star means a magnetic field of $\sim 10^7$–10^8 G, that is several orders of magnitude higher than the strongest magnetic fields that can be produced on Earth laboratories (the highest magnetic field strength created on Earth is $\sim 9 \times 10^5$ G[1]), but weak with respect to the magnetic field of a newly born neutron star (which is $\gtrsim 10^{11}$–10^{12} G). MSPs have spin periods in the range 1–10 ms, corresponding to spin frequencies above 100 Hz. MSPs were first discovered in the radio band, with the detection of periodic radio pulses. The first discovered MSP is PSR B1937+21 (Backer et al. 1982), spinning roughly 641 times a second; this is to date the second fastest-spinning MSP among the ~ 300 that have been discovered so far. PSR J1748−2446ad, discovered in 2005 (Hessels et al. 2006), is the fastest-spinning pulsar currently known, spinning at 716 Hz. These millisecond spinning neutron stars are extreme physical objects: general and special relativity are fully in action, since their surfaces, attaining speeds close to one fifth of the speed of light, are located extremely close to their Schwarzschild radius. In addition electro-dynamical forces, caused by the presence of huge surface magnetic fields of several hundred million Gauss, display their spectacular properties accelerating electrons up to such energies to promote pair creation in a cascade process responsible for the emission in the radio and γ-ray bands. The rotational energy is swiftly converted and released into electromagnetic power which, in some cases, causes the neutron star to outshine with a luminosity of hundreds Suns.

Standard radio pulsars are usually isolated objects, with relatively high magnetic field strengths ($\gtrsim 10^{11}$ G) and relatively long spin periods ($\gtrsim 0.1$ s). Neutron star magnetic fields are probably the relic magnetic field of the progenitor star that is enhanced by "flux freezing", or conservation of the original magnetic flux, when the core of the progenitor star collapses to form a neutron star. Moreover, as the core of a massive star ($\geq 8 \, M_\odot$) is compressed and collapses into a neutron star, it

---

[1]https://phys.org/news/2011-06-world-strongest-magnetic-fields.html.

retains most of its angular momentum. But, because it has only a tiny fraction of the radius of its progenitor star, a neutron star is formed with very high rotation speed. An interesting example of a recently formed pulsar is the Crab pulsar, the central star in the Crab Nebula, a remnant of the supernova SN 1054, which exploded in the year 1054, less than a thousand years ago, that shows a spin period of 33 ms and a magnetic field strength of $B \gtrsim 4 \times 10^{12}$ G.

The strong magnetic field of the newly born neutron star and the high rotational velocity at its surface generate a strong Lorentz force resulting in the acceleration of protons and electrons on the star surface and the creation of an electromagnetic beam emanating from the poles of the magnetic field, which is responsible for the observed pulsed emission. In rotation-powered pulsars, the energy of the beam comes from the rotational energy of the pulsar, which therefore starts to spin-down. At zero-order the pulsar behaves as a rotating magnetic dipole which emits energy according to the Larmor formula (see Jackson's Classical Electrodynamics):

$$P_{\text{rad}} = \frac{2}{3} \frac{(\ddot{m}_{\perp})^2}{c^3} = \frac{2}{3} \frac{m_{\perp}^2 \Omega^4}{c^3} = \frac{2}{3c^3} (B R^3 \sin \alpha)^2 \left(\frac{2\pi}{P}\right)^4, \tag{4.1}$$

where $m_{\perp} = B R^3 \sin \alpha$ is the component of the magnetic dipole moment perpendicular to the rotation axis, B and R are the surface magnetic field and the neutron star radius, respectively, α is the angle between the rotation axis and the magnetic dipole axis, Ω is the spin angular frequency of the neutron star and P its spin period. The pulsar, therefore, gradually slows down at a rate that is higher for stronger magnetic fields and faster spins. If the magnetic field strength does not change significantly with time, we can estimate a pulsar's age from its spin period and the spin-down rate, by assuming that the pulsar's initial period P_0 was much shorter than the current period:

$$\tau \equiv \frac{P}{2\dot{P}}. \tag{4.2}$$

This is the timescale necessary to bring the pulsar from its initial spin period P_0 to its actual period at the observed spin-down rate, and is called characteristic age of the pulsar. Indeed the characteristic age of the Crab is ~ 2.5 kyr.

Soon after their discovery it became clear that MSPs are old neutron stars, with relatively weak magnetic fields, of characteristic ages comparable to the age of the Universe; these were therefore recognised as a different class of objects, called *recycled pulsars*. Binary systems that can host an old neutron star and may be responsible for the recycling are the so-called Low Mass X-ray Binaries (hereafter LMXBs), in which an old, weakly magnetised, neutron star accretes matter from a low-mass (less that 1 $M_{\odot}$) companion star. In these systems, matter transferred from the companion star enters the Roche lobe of the neutron star through the inner Lagrangian point and releases a large amount of gravitational energy before falling onto the neutron star. These systems may be up to five orders of magnitude more luminous than the Sun, and the temperature matter reaches close to the neutron star ranges from few keV up to a hundred keV, making these systems the brightest Galactic sources in the X-ray band. Because of the small size of the companion star, these systems are also quite compact, with orbital periods ranging from several minutes up to one day. For this reason, matter leaving the companion star, has a high specific angular momentum and cannot fall directly onto the neutron star. Besides an accretion disc is formed, where matter rotates with Keplerian velocities and looses energy until it reaches the innermost part of the system, close to the neutron star. When this matter accretes onto the neutron star surface it has relativistic velocities (up to half the velocity of light), and is able to efficiently accelerate the neutron star up to millisecond periods. Depending on the Equation of State (EoS) of ultra-dense matter, 0.1–$0.2\,M_\odot$ are sufficient to spin up a weakly magnetised neutron star to millisecond periods (Burderi et al. 1999). During this phase, because of the accretion of matter and angular momentum, the neutron star accumulates an extraordinary amount of mechanical rotational energy, up to 1% of its whole rest-mass energy.

4.1.2 Problems and Confirmation of the Recycling Scenario

The recycling scenario described above (see e.g. Bhattacharya and van den Heuvel 1991) establishes therefore a clear evolutive link between LMXBs and radio MSPs, with the former being the progenitors of the latter. However, till the end of nineties, there was no observational evidence confirming this evolutive scenario, since there was no evidence that LMXBs could host fast rotating neutron stars. In fact, despite thoroughly searched, no LMXB was found to show coherent pulsations, therefore unveiling its spin frequency. The fact that the large majority of LMXBs does not show coherent pulsations is still a problem. Several explanations have been invoked to interpret this fact, but none of them is fully satisfactory (see also Patruno and Watts 2012, and references therein, for further discussion of this issue). One possibility is that the magnetic axis is aligned with the rotation axis (e.g. Ruderman 1991), but this is excluded by the fact that MSPs, the descendants, show pulsations and therefore the magnetic and rotational axes are not aligned in these systems. Another possibility is that the magnetic field is not strong enough to channel the accreting matter to the neutron star magnetic poles or that optically thick matter around the neutron star may smear the coherent pulsations (e.g. Brainerd and Lamb 1987). However, the large majority of LMXBs are transient systems, showing large variation in luminosity, and going through soft (optically thick) X-ray spectra at high luminosity and hard (optically thin) X-ray spectra at low luminosity. During these stages the accretion rate decreases, the accreting matter becoming optically thin, allowing in principle the detection of X-ray pulsations at millisecond periods (e.g. Göğüş et al. 2007). Other two possibilities are that the neutron star magnetic field is buried by long phases of accretion (e.g. Romani 1990; Cumming et al. 2001) or that the coherent pulsations are very weak, beyond the sensitivity of current X-ray observatories.

Another problem of the recycling scenario was the lack of the observational link between LMXBs and MSPs, i.e. the lack of a system behaving like a LMXB during X-ray active phases and like a radio MSP during X-ray quiescence, when presumably the accretion rate goes down and the radio pulsar mechanism can switch on. However, lack of evidence does not mean evidence of lack, and both these problems were recently solved. In particular, in 1998 coherent pulsations were discovered for the first time in a transient LMXB (SAX J1808.4−3658, Wijnands and van der Klis 1998), and in 2013 the long-sought-for missing link between LMXBs and MSPs was finally found (IGR J18245−2452, a.k.a. M28I, Papitto et al. 2013a). In the next section we describe these discoveries and the related ones, and give the basic observational characteristics of these new classes of systems (see Table 4.1 for a summary of the main properties of these systems).

4.2 Millisecond Pulsars in LMXBs and Their Properties

4.2.1 The Discovery of a New Class of Fast Spinning Neutron Star

The situation dramatically changed in 1996, when the NASA observatory Rossi X-ray Timing Explorer (RXTE) was launched. RXTE was the first X-ray observatory coupling a large effective area (the Proportional Counter Array, PCA, had a total collecting area of $\sim$6500 cm^2) and good time resolution (up to 1 μs), hence suitable for the search of fast time variability. In 1996 quasi-coherent pulsations were detected in RXTE observations of the LMXB 4U 1728−34 at a frequency of $\sim$363 Hz, with amplitudes (rms) of 2.5–10% during six of the eight type-I bursts present in the observation (Strohmayer et al. 1996). The pulsations during these bursts showed frequency drifts of 1.5 Hz during the first few seconds but became effectively coherent during the burst decay. The 363 Hz pulsations were interpreted as rotationally induced modulations of inhomogeneous burst emission, and were considered the first compelling evidence for a millisecond spin period in a LMXB.

The direct evidence for the presence of a fast-spinning neutron star in a LMXB arrived 2 years later, in 1998, when observations performed with RXTE led to the discovery of the first millisecond pulsar in a LMXB, SAX J1808.4−3658. This transient LMXB, first observed by the Wide Field Camera (WFC) on board the X-ray satellite BeppoSAX, shows coherent pulsations with a period of 2.5 ms and an orbital period of 2.01 h (Wijnands and van der Klis 1998; Chakrabarty and Morgan 1998). For almost 4 years, SAX J1808.4−3658 was considered as a rare object in which some peculiarity of the system allowed for the detection of the neutron star spin. However, in the last 20 years, other 19 accreting millisecond pulsars have been discovered, the two most recent ones discovered in 2017 (Sanna et al. 2017a; Strohmayer and Keek 2017). All of them show coherent pulsations with periods in the range 1.7–6.0 ms (up to 60 ms if we also include the enigmatic

Table 4.1 Accreting X-ray pulsars in low mass X-ray binaries

Source	ν_s/P (Hz)/(ms)	P_{orb} (h)	f_x ($M_\odot$)	$M_{c,min}$ ($M_\odot$)	Companion type	Ref.
Accreting millisecond pulsars						
XSS J12270−4859	593 (1.7)	6.91	3.9×10^{-3}	0.27	MS	Roy et al. (2015) and de Martino et al. (2014)
PSR J1023+0038	592 (1.7)	4.75	1.1×10^{-3}	0.20	MS	Archibald et al. (2009) and Coti Zelati et al. (2014)
Aql X-1	550 (1.8)	18.95	1.4×10^{-2}	0.56	MS	Casella et al. (2008) and Mata Sánchez et al. (2017)
Swift J1749.4−2807	518 (1.9)	8.82	5.5×10^{-2}	0.59	MS	Altamirano et al. (2011) and D'Avanzo et al. (2011)
SAX J1748.9−2021	442 (2.3)	8.77	4.8×10^{-4}	0.1	MS	Altamirano et al. (2008) and Cadelano et al. (2017)
IGR J17498−2921	401 (2.5)	3.84	2.0×10^{-3}	0.17	MS	Papitto et al. (2011b)
XTE J1814−338	314 (3.2)	4.27	2.0×10^{-3}	0.17	MS	Markwardt and Swank (2003) and Wang et al. (2017)
IGR J18245−2452	254 (3.9)	11.03	2.3×10^{-3}	0.17	MS	Papitto et al. (2013a)
IGR J17511−3057	245 (4.1)	3.47	1.1×10^{-3}	0.13	MS	Papitto et al. (2010)
IGR J00291+5934	599 (1.7)	2.46	2.8×10^{-5}	0.039	BD	Galloway et al. (2005)
SAX J1808.4−3658	401 (2.5)	2.01	3.8×10^{-5}	0.043	BD	Wijnands and van der Klis (1998) and Wang et al. (2013)
HETE J1900.1−2455	377 (2.7)	1.39	2.0×10^{-6}	0.016	BD	Kaaret et al. (2006) and Elebert et al. (2008)
XTE J1751−305	435 (2.3)	0.71	1.3×10^{-6}	0.014	He WD	Markwardt et al. (2002) and D'Avanzo et al. (2009)
MAXI J0911−655	340 (2.9)	0.74	6.2×10^{-6}	0.024	He WD?	Sanna et al. (2017a)
NGC6440 X−2	206 (4.8)	0.95	1.6×10^{-7}	0.0067	He WD	Altamirano et al. (2010)
Swift J1756.9−2508	182 (5.5)	0.91	1.6×10^{-7}	0.007	He WD	Krimm et al. (2007)
IGR J16597−3704	105 (9.5)	0.77	1.2×10^{-7}	0.006	He WD	Sanna et al. (2018)
XTE J0929−314	185 (5.4)	0.73	2.9×10^{-7}	0.0083	C/O WD	Galloway et al. (2002) and Giles et al. (2005)
XTE J1807−294	190 (5.3)	0.67	1.5×10^{-7}	0.0066	C/O WD	Campana et al. (2003) and D'Avanzo et al. (2009)
IGR J17062−6143	164 (6.1)	> 0.28	–	–	–	Strohmayer and Keek (2017)

ν_s is the spin frequency, P_b the orbital period, f_x is the X-ray mass function, $M_{c,min}$ is the minimum companion mass for an assumed NS mass of 1.4 $M_\odot$. The companion types are: *WD* White Dwarf, *BD* Brown Dwarf, *MS* Main Sequence, *He Core* Helium Star
Adapted and updated from Patruno and Watts (2012)

LMXB IGR J17480−2446 recently discovered in the Globular Cluster Terzan 5; Papitto et al. 2011a), and all of them are found in compact systems, with orbital periods in the range 40 min to ∼10 h (with the exception is Aql X-1, one of the so-called intermittent millisecond pulsars, which has an orbital period of 19 h). Hence, very low mass donors, ≤0.2 $M_\odot$, are usually preferred. Another common feature of these systems is that all of them are transients, spending most of the time in a quiescent X-ray state; on occasions they show X-ray outbursts with moderate peak luminosities in the range 10^{36}–10^{37} erg s^{-1}. It was clear that we were facing a new class of astronomical objects, the so-called Accreting Millisecond X-ray Pulsars (hereafter AMSPs) that could constitute the bridge between the accretion-powered (LMXBs) and the rotation-powered (MSPs) neutron star sources.

4.2.2   Peculiar Behaviours and Intermittent Pulsations

Most (if not all) of the AMSPs are transient, as the vast majority of LMXBs in general. They spend most of time in quiescence with very low X-ray luminosity ($\lesssim 10^{31}$–10^{33} erg s^{-1}) and sometimes they show X-ray outbursts (reaching luminosities of ∼10^{36}–10^{37} erg s^{-1}) usually lasting from few days to ≤3 months. The shortest outburst recurrence time is 1 month for the globular cluster source NGC 6440 X-2, with an outburst duration of less than 4–5 days, whereas the longest outburst, from HETE J1900.1−2455, has lasted for ∼10 years (up to late 2015 when the source returned to quiescence, Degenaar et al. 2017). However, most of these systems have shown just one X-ray outburst during the last 20 years. The most regular among recurrent AMSPs, and therefore the best studied of these sources, is the first discovered AMSP, SAX J1808.4−3658, which has shown an X-ray outburst every 1.6–3.5 years, the latest one occurred in 2015 (Patruno et al. 2017; Sanna et al. 2017b). The outburst light curve is characterised by a fast rise (on a couple of days timescale), a slow exponential decay (with a timescale of ∼10 days) followed by a fast decay (with a timescale of ∼2 days). After the end of the main outburst, usually a flaring activity, called *reflares*, is observed, with a quasi-oscillatory behaviour and a variation in luminosity of up to three orders of magnitude on timescales ∼1–2 days. Moreover a strong ∼1 Hz oscillation is observed to modulate the reflares. A similar behaviour was also observed in the AMSP NGC 6440 X-2 (see e.g. Patruno and D'Angelo 2013). The reflaring behaviour has no clear explanation. Patruno et al. (2016) proposed a possible explanation in terms of either a strong propeller with a large amount of matter being expelled from the system or a trapped (dead) disc truncated at the co-rotation radius.

Another peculiar behaviour is the intermittency of the pulsations, important because it could bridge the gap between non-pulsating LMXBs and AMSPs. In 2005, the seventh discovered AMSP, HETE J1900.1−2455, went into X-ray outburst and showed X-ray pulsations at 377 Hz, with an orbital period of 1.39 h (Kaaret et al. 2006). Contrary to the usual behaviour of AMSPs, this outburst lasted for about 10 years. After the first 20 days of the outburst, the pulsations became intermittent for about 2.5 years. After that the pulsed fraction weakened with stringent upper limits ($\leq 0.07\%$, Patruno 2012). The most puzzling behaviour in this sense, was observed in the LMXB Aql X-1, which showed coherent X-ray pulsations, discovered in 1998 RXTE archival data, that appeared in only one ~150 s data segment out of a total exposure time of 1.5 Ms from more than 10 years of observations (Casella et al. 2008). The third intermittent pulsar is SAX J1748.9−2021, where pulsations were detected sporadically in several data segments and in three out of four outbursts observed by the source (Patruno et al. 2009a, see also Sanna et al. 2016 reporting on the 2015 outburst of the source). Interestingly, these AMSPs may have a long term average mass accretion rate higher with respect to the other AMSPs. To explain this behaviour, it has been proposed that a screening of the neutron star magnetic field by the accreting matter weakens its strength by orders of magnitude on timescale of few hundred days, so that it is less effective in truncating the accretion disc and channel matter to the magnetic poles (Patruno 2012). However, it is not clear if this hypothesis can explain all the phenomenology and more observations and theoretical efforts are needed to reach a satisfactory explanation.

A detailed review of most of the phenomenology of AMSPs can be found in Patruno and Watts (2012). In the following we will give an overview of the most debated issues on these systems, with particular attention to aspect regarding their evolution and their connection to rotation-powered MSPs.

4.2.3   Accretion Torques and Short-Term Spin Variations

Accretion torque theories can be tested studying the spin variations of AMSPs during accretion states. These studies can provide valuable information on the mass accretion rate and magnetic field of the neutron star in these systems, as well as their spin evolution. An open question is whether these accreting pulsars are spinning up during an outburst and spinning down in quiescence as predicted by the recycling scenario. Coherent timing has been performed on several sources of the sample, with controversial results. Although some AMSPs show pulse phase delays distributed along a second order polynomial, indicating an almost constant spin frequency derivative, other sources show strong timing noise which can hamper any clear measurement of the spin derivative. In fact, the phase delays behaviour as a function of time in these sources is sometimes quite complex and difficult to interpret, since phase shifts, most probably related to variations of the X-ray flux, are sometimes present.

The first AMSP for which a spin derivative has been measured is the fastest spinning (~599 Hz, in a 2.46 h orbit) among these sources, IGR J00291+5934. It is now generally accepted that this source shows spin up at a rate of ~(5–8) × 10^{-13} Hz s^{-1} (Falanga et al. 2005; Patruno 2010). Burderi et al. (2006) have attempted to fit the phase delays vs. time with physical models taking into account the observed decrease of the X-ray flux as a function of time during the X-ray outburst, with the aim to get a reliable estimate of the mass accretion rate onto the compact object. In the hypothesis that the spin-up of the source is caused by the accretion of matter and angular momentum from a Keplerian accretion disc, the mass accretion rate, $\dot{M}$, onto the neutron star can be calculated by the simple relation: $2\pi I \dot{\nu} = \dot{M}(GM_{NS}R)^{1/2}$, where I is the moment of inertia of the neutron star, $\dot{\nu}$ the spin frequency derivative, G the gravitational constant, M_{NS} the neutron star mass, R the accretion radius, and $(GM_{NS}R)^{1/2}$ the Keplerian specific angular momentum at the accretion radius. Because the X-ray flux, which is assumed to be a good tracer of the mass accretion rate, is observed to decrease along the outburst, this has to be included in the relation above in order to obtain the correct value of the mass accretion rate at the beginning of the outburst as well as its temporal evolution. Note that the accretion radius also depends on the mass accretion rate, $R \propto \dot{M}^{-\alpha}$, where α is usually assumed to be 2/7 (e.g. Ghosh and Lamb 1978), and therefore varies with time. Fitting the phase delays in this way, the spin frequency derivative at the beginning of the outburst results to be $\dot{\nu} \sim 1.2(2) \times 10^{-12}\,\mathrm{Hz\,s^{-1}}$, and the lower limit to the mass accretion rate at the beginning of the outburst, corresponding to $\alpha = 0$, is $\dot{M}_{-10} = 5.9\dot{\nu}_{-13}I_{45}m^{-2/3} = 70 \pm 10$, where $\dot{M}_{-10}$ is the mass accretion rate in units of $10^{-10}\,M_\odot\,\mathrm{year^{-1}}$, $\dot{\nu}_{-13}$ is the spin frequency derivative in units of $10^{-13}\,\mathrm{Hz\,s^{-1}}$, I_{45} the moment of inertia of the neutron star in units of $10^{45}\,\mathrm{g\,cm^2}$, and m is the neutron star mass in units of $M_\odot$. This would correspond to a bolometric luminosity of $\sim 7 \times 10^{37}\,\mathrm{erg\,s^{-1}}$, that is about an order of magnitude higher than the X-ray luminosity inferred from the observed X-ray flux, assuming a distance of 5 kpc. Once we will have a direct, independent, estimate of the distance to the source, we will have the possibility to test the $\dot{M}$ vs. X-ray luminosity relation, torque theories and/or the physical parameters of the neutron star.

Other AMSPs show clear parabolic trend of the pulse phase delays as a function of time during X-ray outburst, some of them showing spin-up (e.g. XTE J1807−294, $\dot{\nu} = 2.5(7) \times 10^{-14}\,\mathrm{Hz\,s^{-1}}$, Riggio et al. 2008; XTE J1751−305, $\dot{\nu} = 3.7(1.0) \times 10^{-13}\,\mathrm{Hz\,s^{-1}}$, Papitto et al. 2008; IGR J17511−3057, $\dot{\nu} = 1.6(2) \times 10^{-13}\,\mathrm{Hz\,s^{-1}}$, Riggio et al. 2011a) while others showing spin-down (e.g. XTE J0929−314, $\dot{\nu} = -9.2(4) \times 10^{-14}\,\mathrm{Hz\,s^{-1}}$, Galloway et al. 2002; XTE J1814−338, $\dot{\nu} = -6.7(7) \times 10^{-14}\,\mathrm{Hz\,s^{-1}}$, Papitto et al. 2007; IGR J17498−2921, $\dot{\nu} = -6.3(1.9) \times 10^{-14}\,\mathrm{Hz\,s^{-1}}$, Papitto et al. 2011b). Those sources showing spin-down suggest the possibility of an interaction of the neutron star magnetic field with the accretion disc outside the co-rotation radius (the radius in the disc where the Keplerian frequency equals the neutron star spin frequency). In fact the magnetic field lines can be threaded into the accretion disc and dragged by the high conductivity plasma so that an extra torque due to magnetic stresses has to be expected (see e.g. Wang 1987). In this case, an estimate of the magnetic field strength can be derived from the measured spin-down rate (see e.g. Di Salvo et al. 2007). However, there is not a general consensus on the interpretation of these phase residuals. Another possibility is that these are due to a timing noise caused by a pulse phase offset that varies in correlation with X-ray flux, such that noise in flux translates into timing noise (Patruno et al. 2009b). In this case, much stringent limits result on the spin derivatives in these sources (see Patruno and Watts 2012, and references therein). Although for some sources clear correlations are observed between abrupt jumps in the pulse phases and sharp variations in the X-ray flux, it is not clear yet how much of the phase variations can be ascribed to an accretion-rate-dependent hot spot location.

Certainly, the most debated case is SAX J1808.4−3658 whose phase variations are strongly dominated by timing noise. The pulse phase delays show a very puzzling behaviour, since a rather fast phase shift, by approximately 0.2 in phase, is present at day 14 from the beginning of the 2002 outburst (Burderi et al. 2006) (see Fig. 4.1, left panel). Interestingly, day 14 corresponds to a steepening of the exponential decay with time of the X-ray flux. However, analysing separately the phase delays of the fundamental and second harmonic of the pulse profile, Burderi et al. (2006) noted that the phase delays of the harmonic did not show any evidence of phase jumps (see Fig. 4.1, right panel). This is not an effect of the worse statistics of the phase delays derived from the harmonic, which of course show larger error bars.

This means that the phase jump in the fundamental is not related to an intrinsic spin variation (which would have affected the whole pulse profile), but is instead caused by a change of the shape of the pulse profile (perhaps related to the mechanism causing the increase of the steepness of the exponential decay of the X-ray flux). On the other hand, from the fitting of the phase delays of the second harmonic, under the hypothesis that these are a better trace of the spin of the neutron star, Burderi et al. (2006) find that the source shows a spin-up at the beginning of the outburst with $\dot{\nu}_0 = 4.4(8) \times 10^{-13}\,\mathrm{Hz\,s^{-1}}$, corresponding to a mass accretion rate of $\dot{M} \sim 1.8 \times 10^{-9}\,M_\odot\,\mathrm{year^{-1}}$, and a constant spin-down, with $\dot{\nu}_{sd} = 7.6(1.5) \times 10^{-14}\,\mathrm{Hz\,s^{-1}}$, dominating the phase delays at the end of the outburst. In this case, the mass accretion rate inferred from timing is only a factor of 2 larger than the observed X-ray luminosity at the beginning of the outburst, that is $\sim 10^{37}\,\mathrm{erg\,s^{-1}}$. The spin-down observed at the end of the outburst can be interpreted as due to a threading of the accretion disc by the neutron star magnetic field outside the co-rotation radius. Of course, in agreement with the expectation, the threading effect appears to be more relevant at the end of the outburst, when the mass accretion rate significantly decreases. In this case the magnetic moment, μ, of SAX J1808.43658 can be evaluated from the measured value of the spin-down, using the relation (see Rappaport et al. 2004):

$$2\pi I \dot{\nu}_{sd} \equiv \frac{\mu^2}{9r_{\mathrm{cor}}^3} \tag{4.3}$$

where r_{cor} is the co-rotation radius. The magnetic field found in this way is $B = (3.5 \pm 0.5) \times 10^8\,\mathrm{G}$, perfectly in agreement with other, independent, constraints (e.g. Burderi et al. 2006).

The fact that the second harmonic shows more regular phase residuals with respect to the fundamental has also been observed in other AMSPs (e.g. Riggio et al. 2008, 2011a; Papitto et al. 2012) and may indicate that the second harmonic is a good tracer of the neutron star spin frequency. A simple model proposed by Riggio et al. (2011b) (see also Papitto et al. 2012) may explain a similar behavior

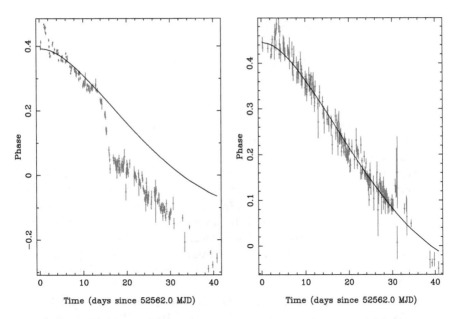

Fig. 4.1 Left: Phase vs. time for the fundamental of the pulse frequency of SAX J1808.4−3658. **Right:** Phase vs. time for the first harmonic of the pulse frequency of SAX J1808.4−3658. On top of the data, the best-fit function (including both the spin-up due to accretion and the spin-down at the end of the outburst) is plotted as a solid line (from Burderi et al. 2006)

in terms of modest variations of the relative intensity received by the two polar caps on to the neutron star surface, in the hypothesis that the two spots emit a signal of similar amplitude and with a similar harmonic content. The sum of the two signals (the total profile) will be the result of a destructive interference for what concerns the fundamental frequency, since it is the sum of two signals with a phase difference of $\sim\pi$. A constructive interference develops instead for the second harmonic of the total profile, since it is the sum of two signals with the same phase. The destructive interference regarding the fundamental frequency leads to large swings of the phase of the fundamental of the total profile due to modest variations of the relative intensity of the signals emitted by the two caps. In this case swings up to 0.5 phase cycles can be shown by the phase computed over the fundamental frequency of the observed profile, without correspondingly large variations of the phase of the second harmonic. Interestingly, IGR J00291+5934, showing a much more regular behaviour of the fundamental, shows a nearly sinusoidal pulse profile, with very little harmonic content (e.g. Galloway et al. 2005; Burderi et al. 2007).

4.2.4 Long-Term Spin Evolution

For AMSPs for which more than one outburst has been observed, it is possible to derive their long term spin evolution comparing the averaged spin frequency measured in each outburst. To date only six AMSPs have been monitored with high time resolution instruments in different outbursts: SAX J1808.4−3658, IGR J00291+5934, XTE J1751−305, Swift J1756.9−2508, NGC6440 X-2 and SAX J1748.9−2021 (although, with relatively low S/N and short outburst duration in the latter two sources). The best constrained is SAXJ1808.4−3658, for which secular spin evolution have now been measured over a 13 year baseline and shows a constant long-term spin-down at a rate of $\sim -1 \times 10^{-15}$ Hz s^{-1} (Hartman et al. 2008, 2009; Patruno et al. 2012; Sanna et al. 2017b). Because of the stability of the spin-down rate over the years, the most likely explanation appears to be loss of angular momentum via magnetic-dipole radiation, which is expected for a rapidly rotating neutron star with a magnetic field. The measured spin-down is consistent with a polar magnetic field of $(1.5–2.5) \times 10^8$ G. This is in agreement with the estimate above.

A spin down has also been measured for IGR J00291+5934 between the 2004 and 2008 outburst, at a rate of $-4.1(1.2) \times 10^{-15}$ Hz s^{-1} (Patruno 2010; Papitto et al. 2011c; Hartman et al. 2011), larger than that observed in SAX J1808.43658, as expected given that IGR J00291+5934 spins at a higher frequency. If interpreted in terms of magneto-dipole emission, the measured spin down translates into an estimate of the neutron star magnetic field of $(1.5–2) \times 10^8$ G. For the period between the 2008 and 2015 outbursts only an upper limit to the frequency evolution could be derived, $|\dot{\nu}| \leq 6 \times 10^{-15}$ Hz s^{-1} (3σ c.l., Sanna et al. 2017c), compatible with the previous estimate. Comparing the spin frequencies from 2002, 2005, 2007 and 2009 outbursts of XTE J1751−305, Riggio et al. (2011c) report a spin down at a rate of $\sim(1.2) \times 10^{-15}$ Hz s^{-1} and an inferred magnetic field of $\sim 4 \times 10^8$ G. Whereas for Swift J1756.9−2508 only an upper limit $(|\dot{\nu}_{sd}| \leq 2 \times 10^{-15}$ Hz s$^{-1})$, corresponding to a magnetic field $\leq 10^9$ G, has been reported (Patruno et al. 2010).

The fact that the spin-down during quiescent periods is probably due to magnetic-dipole radiation rises the interesting possibility that AMSPs may switch on as a radio pulsar during quiescence and ablate their donor. This is the so-called *hidden black widow* scenario proposed by Di Salvo et al. (2008) (see also Stella et al. 1994; Campana et al. 1998) to explain the long-term orbital evolution (see next section).

4.2.5 Orbital Evolution

The study of the orbital evolution in these systems is important to constrain the evolutionary path leading to the formation of radio MSPs. These studies, however, require a large timespan of data in order to constraint the orbital period derivative.

Hence, the main difficulty is given by the fact that most of the AMSPs rarely turn into X-ray outburst. For this reason, the best constraints on the orbital evolution in these systems come again from SAX J1808.4$-$3658, which has shown seven X-ray outburst to date, allowing to follow its orbital period over 17 years.

In the case of SAX J1808.4$-$3658 it is possible to see a clear parabolic trend of the time of passage to the ascending node versus time over the last 17 years (Di Salvo et al. 2008; Hartman et al. 2008; Burderi et al. 2009; Patruno et al. 2016; Sanna et al. 2017b). Interpreting this parabolic term as the orbital period derivative, gives orbital expansion at a quite high rate of $\dot{P}_{orb} = (3.4\text{--}3.9) \times 10^{-12}\,\text{s}\,\text{s}^{-1}$ (see also Sanna et al. 2016 who report a marginally significant, strong orbital expansion in the AMSP SAX J1748.9$-$2021). The observed orbital expansion implies a mass-radius index for the secondary $n < 1/3$ (see Di Salvo et al. 2008). In the reasonable hypothesis that the secondary star is a fully convective star out of thermal equilibrium and responds adiabatically to the mass transfer, a mass-radius index of $n = -1/3$ can be assumed for the secondary. However, this derivative is a factor $\sim$70 higher that the orbital derivative expected for conservative mass transfer, given the low averaged mass accretion rate onto the neutron star; since SAX J1808.4$-$3658 accretes for about 30 d every 2–4 year, the averaged X-ray luminosity from the source results to be $L_{obs} \sim 4 \times 10^{34}\,\text{erg}\,\text{s}^{-1}$.

A non-conservative mass transfer can explain the large orbital period derivative if we assume a mass transfer rate of $\dot{M} \sim 10^{-9}\,M_\odot\,\text{year}^{-1}$, and that this matter is expelled from the system with the specific angular momentum at the inner Lagrangian point (see Di Salvo et al. 2008; Burderi et al. 2009). In this case, the non-conservative mass transfer may be a consequence of the so-called *radio-ejection* model extensively discussed by Burderi et al. (2001). The basic idea is that a fraction of the transferred matter in the disc could be swept out by radiative pressure of the pulsar. In this case, the fast spinning neutron star may ablate the companion during quiescent periods (the so-called hidden black widow scenario proposed by Di Salvo et al. 2008). Alternatively, the large orbital period derivative observed in SAX J1808.4$-$3658 can be interpreted as the effect of short-term angular momentum exchange between the mass donor and the orbit (Hartman et al. 2009; Patruno et al. 2012), resulting from variations in the spin of the companion star (holding the star out of synchronous rotation) caused by intense magnetic activity driven by the pulsar irradiation. This mechanism has been invoked by Applegate and Shaham (1994) (hereafter A&S) to explain oscillating orbital residuals observed in some radio MSPs (see e.g. Arzoumanian et al. 1994). In this case, the energy flow in the companion needed to power the orbital period change mechanism can be supplied by tidal dissipation. However, the A&S mechanism envisages alternating epochs of orbital period increase and decrease, which is not yet observed from SAX J1808.4$-$3658. It also predicts that the system will evolve to longer orbital periods by mass and angular momentum loss on a timescale of 10^8 year (for a 2-h orbital period and a companion mass of 0.1–0.2 $M_\odot$), and thus requires a strong orbital period derivative, similar to that inferred from the quadratic trend observed in SAX J1808.4$-$3658. Therefore, also in the framework of the A&S mechanism, most of the orbital period variation observed in SAX J1808.4−3658 is probably caused by loss of matter and angular momentum, i.e. by a non-conservative mass transfer (see Sanna et al. 2017b for further discussions). The next outbursts from this source will tell us whether the orbital period increase will turn into a decrease or, instead, the orbital expansion will continue, and this will be crucial in order to discriminate between the two possibilities sketched above.

Another puzzling result comes from IGR J0029+5934, which has orbital parameters very similar to those of SAX J1808.4−3658, and is considered its *orbital twin*. IGR J0029+5934 has shown only four outbursts since its discovery, but tight upper limits could be derived on its orbital period derivative, $|\dot{P}_{orb}| < 5 \times 10^{-13}\,\mathrm{s\,s^{-1}}$ (90% confidence level Patruno 2017, see also Sanna et al. 2017c). This implies a much slower orbital evolution, on a timescale longer than ∼0.5 Gyr, as compared to the fast orbital evolution of its twin, ∼70 Myr. The orbital evolution observed in IGR J0029+5934 is compatible with the expected timescale of mass transfer driven by angular momentum loss via gravitational radiation, with no need of A&S mechanism or non-conservative mass transfer. In this case, it would be interesting to constrain the sign of the orbital period derivative in order to get information on the mass-radius index of the donor star and to infer whether it is in thermal equilibrium (implying orbital contraction) or not (implying orbital expansion).

4.2.6 Searches at Other Wavelengths

AMSPs are transient systems with inferred variations in the mass accretion rate onto the central source by a factor of ∼10^5 between outbursts and quiescence. Since the magnetospheric radius expands as the mass accretion rate decreases, it is easy to see that, while during an X-ray outburst the magnetospheric radius is expected to be very close to the neutron star surface, during X-ray quiescence the magnetospheric radius may expand beyond the light-cylinder radius (where an object co-rotating with the neutron star attains the speed of light). In this case, it is expected that any residual accretion is inhibited by the radiation pressure and, consequently, it is plausible to expect that the neutron star turns-on as a radio MSP until a new outburst episode pushes the magnetospheric radius back again, quenching radio emission and initiating a new accretion phase. The compelling possibility that these systems could swiftly switch from accretion-powered to rotation-powered magneto-dipole emitters during quiescence gives the opportunity to study a phase that could shed new light on the not yet cleared up radio pulsar emission mechanism. However, such a behaviour has been observed only recently, in the so-called transitional MSPs (see next sections), and in particular in the unique source IGR J18245−2452 (Papitto et al. 2013a), the long-sought-for "missing link" between LMXBs and radio MSPs; a binary system containing a neutron star alternating accretion-powered phases, in which it behaves like an AMSP, to rotation-powered phases, in which it behaves like a radio MSP. However, despite the huge observational effort made to catch, in a transient LMXB, the transition between the accretion-powered regime and the rotation-powered regime during quiescence, no pulsed radio emission has been found in other AMSPs. The most embarrassing problem is certainly the lack of pulsed radio emission from these systems during quiescence: many transient LMXBs and AMSPs have been thoroughly searched in radio during quiescence with disappointing negative results (Burgay et al. 2003; Iacolina et al. 2009, 2010; Patruno et al. 2017).

Burderi et al. (2001) (see also Burderi et al. 2002) have proposed a model, that naturally explains this non-detection, assuming that the radio pulsar mechanism switches on when a temporary significant reduction of the mass-transfer rate occurs. In some cases, even if the original mass transfer rate is restored, the accretion of matter onto the neutron star can be inhibited by the radiation pressure from the radio pulsar, which may be capable of ejecting out of the system most of the matter overflowing from the companion-star Roche lobe. In particular, Burderi et al. (2001) showed that in "wide systems", i.e. systems with orbital periods longer than a few hours, and with a sufficiently fast spinning neutron star, the switch-on of the radio pulsar can prevent any further accretion, even if the original mass transfer rate is restored. On the other hand, "compact" systems (orbital periods below a few hours) should show a cyclic behaviour since, once the temporary reduction of the accretion rate ends, the radiation pressure of the pulsar is unable to keep the matter outside the light cylinder radius and accretion resumes.

One of the strongest predictions of this model is the presence, during the radio-ejection phase, of a strong wind of matter emanating from the system: the mass released by the companion star swept away by the radiation pressure of the pulsar. This matter could cause strong free-free absorption in the radio band, hampering the detection of pulsed signals. A possible solution is then to observe at high radio frequencies (above 5–6 GHz, see Campana et al. 1998; Di Salvo and Burderi 2003), thus reducing the cross-section of free-free absorption which depends on ν^{-2}. Note that the pulsed radio flux also decreases with increasing frequency, although it is easy to see that this decrease is less steep with respect to the decrease of the effects caused by free-free absorption (see e.g. Burderi and King 1994). This may explain why some AMSPs indeed have radio counterparts (see Patruno and Watts 2012 for a review), although not pulsating. Interestingly, Iacolina et al. (2010) report a peak at 4σ significance for XTE J1751−305, obtained folding a radio observation of this source performed at Parkes radio telescope. This peak has a 40% probability of not being randomly generated over the 40,755 trial foldings of the dataset corresponding to one of the two observations performed at 8.5 GHz. This result is not confirmed in the other observation (at the same frequency) and thus deserves additional investigation in the future. Alternatively, pulsating radio emission should be searched in systems with long orbital periods, in which the matter transferred by the companion star is spread over a wider orbit.

Strong (indirect) evidences that a rotating magneto-dipole powers the quiescent emission of AMSPs, comes from observations of the quiescent emission from their identified optical counterpart. In the case of SAX J1808.4−3658, measures in the optical band show an unexpectedly large optical luminosity (Homer et al. 2001), inconsistent with both intrinsic emission from the companion star and X-ray reprocessing. The most probable explanation for the over-luminous optical counterpart of SAX J1808.4−3658 in quiescence, proposed by Burderi et al. (2003) and Campana et al. (2004), is that the magnetic dipole rotator is active during quiescence and its bolometric luminosity, given by the Larmor's formula (Eq. (4.1)), powers the reprocessed optical emission. Indeed, the optical luminosity and colours predicted by this model are perfectly in agreement with the observed values.

Similar results have been obtained for other AMSPs for which optical observations in quiescence have been performed. For XTE J0929−314 and XTE J1814−338, optical photometry in quiescence showed that the donor was irradiated by a source emitting in excess of the X-ray quiescent luminosity of these sources, requiring an energy source compatible with the spin-down luminosity of a MSP (D'Avanzo et al. 2009). IGR J00291+5934 showed evidences for a strongly irradiated companion in quiescence too (D'Avanzo et al. 2007).

The precise spin and orbital ephemerides of AMSPs are of fundamental importance to allow deep searches of their counterparts in the gamma-ray band, which has the advantage of not suffering the free-free absorption as in the radio band, but the disadvantage of the paucity of photons, which requires folding over years in order to reach the statistics needed for detecting a pulsed signal. Indeed, AMSPs, in analogy with MSPs (such as the so-called Black Widows and Red-Backs, detected in radio and most of them also in the gamma band; see Roberts 2013 as a review) are expected to show coherent pulsations in the gamma band during X-ray quiescence. The detection of a possible gamma-ray counterpart of the AMSP SAX J1808.4−3658 from 6 years of data from the Fermi/Large Area Telescope has been recently reported (de Oña Wilhelmi et al. 2016). The authors also searched for modulations of the flux at the known spin frequency or orbital period of the pulsar, but, taking into account all the trials, the modulation was not significant, preventing a firm identification via time variability. We expect that this result may be improved increasing the time span of the gamma-ray data.

4.3   The Missing Link

As we described in the previous sections a clear path has been delineated bringing an old, slowly spinning neutron star to become a millisecond radio pulsar. The last link of the chain was supposed to be disclosed with the discovery of the (transient) accreting millisecond X-ray pulsar SAX J1808.4−3658 (Wijnands and van der Klis 1998). But this source, as well as all the other of this class, do not show a radio pulsar soul during their quiescence (even if indirect hints for a turn on of the relativistic pulsar wind have been gathered, Burderi et al. 2003; Campana et al. 2004).

A new way was paved by radio surveys. The Faint Images of the Radio Sky at Twenty-cm (FIRST) radio survey carried out at the Very Large Array (VLA) covered ~10,000 square degrees discovering nearly one million of radio sources. Matching radio sources with optical surveys Bond et al. (2002) reported the discovery of a new magnetic cataclysmic variable with radio emission: FIRST J102347.6+003841 (hereafter J1023). Optical spectroscopic studies revealed the presence of an accretion disc in 2001 through the presence of double-horned emission lines (Szkody et al. 2003), which led Thorstensen and Armstrong (2005) to suggest that J1023 could be a neutron star low-mass X-ray binary. A bright millisecond radio pulsar (1.69 ms) was discovered in 2007 coincident with J1023 (Archibald et al. 2009). No signs of an accretion disc are discernible any more but the pulsed radio signal was eclipsed sporadically along the 4.8 h orbit. This was the first system testifying for the alternating presence of an accretion disc and of a millisecond radio pulsar, thus providing the first indirect evidence that the two souls can live within the same object.

Nature did even better. IGR J18245−2452 (J18245 hereafter) was discovered as a new X-ray transient in the globular cluster M28. XMM-Newton pointed observations revealed a pulsed signal at 3.93 ms making of J18245 an accretion X-ray pulsar, modulated at 11 h orbital period (Papitto et al. 2013a). Coincidentally, a radio pulsar was detected during a radio survey of M28 with the same spin period and binary orbital period (PSR 18245−2452I). J18245 closed definitely the chain: it is the first pulsar ever detected in X-rays and at radio wavelengths. Twenty days after the end of the outburst J18245 was detected again as a radio pulsar, closing the loop and demonstrating that the transition between the rotation-powered and the accretion-powered regime can occur on short timescales.

Incidentally, in mid 2013 the radio monitoring of J1023 failed to detect the radio pulsar any more. Simultaneously the optical, X-ray and gamma-ray flux increased by a factor of ~70 and ~5, respectively, even if the source did not enter in a proper outburst state (Patruno et al. 2014; Stappers et al. 2014). During this active state J1023 displayed a very peculiar behaviour with three different states: a high state during which X-ray pulsations are detected, a low state (a factor of ~7 dimmer) during which no pulsations were detected (and the upper limit on the pulsed fraction is lower than what observed during the high state) and a flaring state (with no pulsations too), outshining both states during which the source wildly varies. Transitions occur on a ~10 s timescale in X-rays (Archibald et al. 2015; Bogdanov et al. 2015). More details will be provided in the next section (see also Fig. 4.2). Another source distinctly showed this behaviour was XSS J12270−4859 (Papitto et al. 2015) (J12270 in the following), so that this puzzling X-ray light curve has become the archetypal way to identify a "transitional" X-ray pulsar.

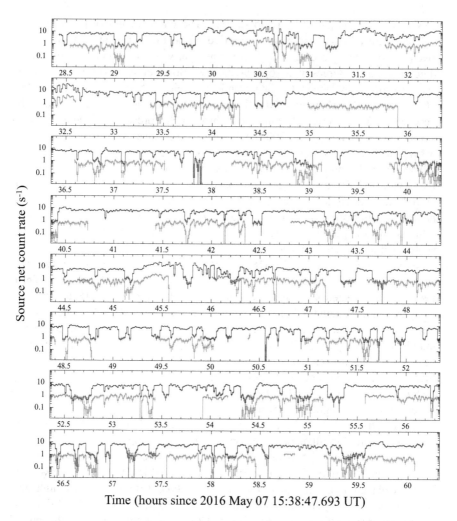

Source net count rate (s^{-1})

Time (hours since 2016 May 07 15:38:47.693 UT)

Fig. 4.2 Background-subtracted and exposure-corrected light curves of J1023 obtained with the XMM-Newton EPIC cameras (0.3–10 keV; black data), NuSTAR FPMA + FPMB (3–79 keV; red data) and Swift UVOT (UVM2 filter; blue data) during the time interval covered by XMM-Newton. For plotting purpose, light curves are shown with a binning time of 50 s and the vertical axis is plotted in logarithmic scale (from Coti Zelati et al. 2018)

4.4 Madamina, il catalogo è questo ("Don Giovanni", Mozart)

There are four transitional pulsars and a few candidates. At the moment of writing (October 2017) J1023 is in an active state. J18245 and J12270 are in quiescence, shining as radio pulsars. We summarise their main characteristics in Table 4.2.

Table 4.2 Parameters of transitional millisecond pulsars

Source	Spin period (ms)	Orbital period (h)	Distance (kpc)	Companion mass ($M_\odot$)	DM (pc cm^{-3})	Fermi detection	X-ray pulsations	Radio pulsations
J1023	1.69	4.75	1.37	~0.24	14.3	Y	Y	Y
J12270	1.69	6.91	1.4	~0.25	43.4	Y	Y	Y
J18245	3.93	11.03	5.5 (M28)	~0.2	119	N	Y	Y
J154439	–	~ 5.3	–	–	–	Y	N	N

4.4.1 PSR J1023+0038

J1023 was discovered as a peculiar magnetic cataclysmic variable with radio emission (Bond et al. 2002). Optical studies revealed signs for the presence of an accretion disc through double horned emission lines in 2001 (Szkody et al. 2003; Wang et al. 2009). Thorstensen and Armstrong (2005) performed photometric and spectroscopic optical observations campaign. The spectrum showed mid-G star features and the disappearance of the accretion disc signatures. Photometry showed a smooth orbital modulation at 4.75 h, with colour changes consistent with irradiation (but with no emission lines). Radial velocity studies and modelling of the light curves led them to show that the primary should be more massive than the Chandrasekhar mass, thus pointing to a LMXB, rather than a cataclysmic variable. Homer et al. (2006) used XMM-Newton data, lack of optical circular polarisation, and optical spectroscopic data to confirm this picture.

This changed with the discovery of a radio pulsar coincident with J1023. Archibald et al. (2009) discovered a bright, fast spinning (1.69 ms) MSP during a low-frequency pulsar survey carried out with Green Bank Telescope in 2007. The pulsar is eclipsed during a large fraction of the orbital period (at orbital phases 0.10–0.35). Due to its proximity and radio brightness it was possible to derive a very precise distance of 1.37 ± 0.04 kpc based on its radio parallax (Deller et al. 2012). XMM-Newton observations then revealed a $\sim 9 \times 10^{31}$ erg s^{-1} (0.5–10 keV) source showing pulsed emission at the radio period. The root-mean-squared pulsed fraction in the 0.25–2.5 keV energy range is $11 \pm 2\%$, whereas a 3σ upper limit of 20% is obtained at higher energies (Archibald et al. 2010; Bogdanov et al. 2011). The neutron star's parameters were determined thanks to the radio pulsar signal, with a spin period of 1.69 ms and a dipolar magnetic field of 9.7×10^7 G (Archibald et al. 2009; Deller et al. 2012).

Surprisingly, J1023 started not being detected in the radio band around June 2013 (Patruno et al. 2014; Stappers et al. 2014). Contemporaneously, the weak γ-ray flux detected by Fermi increased by a factor of ~ 5 (Stappers et al. 2014; Takata et al. 2014; Torres et al. 2017). This new state persisted in time and J1023 is still in this active state now (October 2017) with no signs of change. Together with the disappearance of the radio signal, J1023 brightened at all the other wavelengths. In the X-rays (0.5–10 keV) it brightened by a factor of ~ 30, with some flaring activity

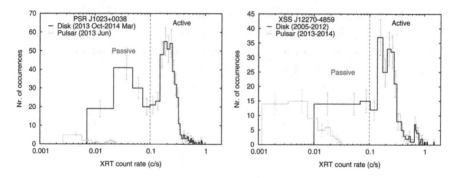

Fig. 4.3 Count rate distribution in the 0.3–10 keV XRT light curves of J1023 (left) and J12270 (right). The disc state (solid black histogram) and pulsar state (dashed green histogram) are shown separately on the same scale. The horizontal dotted line at $0.1\,\mathrm{c\,s^{-1}}$ for both sources marks the boundary between disc-active and disc-passive states (labeled in black and red, respectively, from Linares 2014)

reaching $10^{34}\,\mathrm{erg\,s^{-1}}$ (Patruno et al. 2014; Takata et al. 2014; Coti Zelati et al. 2014; Bogdanov et al. 2015). In the optical, J1023 brightened too by ∼1 mag and showed again the presence of several broad, double-horned emission lines typical of an accretion disc (Halpern et al. 2013; Coti Zelati et al. 2014).

At variance with any other LMXBs, J1023 shows a highly variable active state. A simple histogram of the observed count rates shows a bimodal distribution (Linares 2014) (see Fig. 4.3). A closer look (thanks to XMM-Newton observations) revealed the existence of three distinct states (Bogdanov et al. 2015; Archibald et al. 2015). These can be characterised as:

- a high state with a 0.3–80 keV luminosity $L_X \sim 7 \times 10^{33}\,\mathrm{erg\,s^{-1}}$ (Tendulkar et al. 2014) occurring for ∼80% of the time and during which X-ray pulsations at the neutron star spin period are detected with a r.m.s. pulsed fraction of $8.1 \pm 0.2\%$ (0.3–10 keV) (Archibald et al. 2015);
- a low state with a 0.3–80 keV luminosity $L_X \sim 10^{33}\,\mathrm{erg\,s^{-1}}$ occurring for ∼20% of the time and during which pulsations are not detected with a 95% r.m.s. upper limit $\lesssim 2.4\%$ (0.3–10 keV), much smaller than the detection during the high state;
- a flaring state during which sporadic bright flares occur reaching luminosities as high as of $\sim 10^{35}\,\mathrm{erg\,s^{-1}}$, with no pulsation too.

The transition between the high and the low states is very rapid, on a ∼10 s timescale and looks symmetric (ingress time equals to the egress time). It is not clear if similar variability has also been detected in the optical band, superimposed to the orbital modulation (Shahbaz et al. 2015; Jaodand et al. 2016). A phase-connected timing solution shows that the neutron star is spinning down at a rate ∼30% faster than the spin down due to rotational energy losses (Jaodand et al. 2016).

The 0.1–300 GeV flux of increased by a factor of ∼5 after the transition to the active state with a steep power law photon index of 2.5–3 (Stappers et al. 2014; Takata et al. 2014; Deller et al. 2015). Above 300 GeV VERITAS put instead an 95% upper limit of 7×10^{-13} erg cm^{-2} s^{-1} on the flux (Aliu et al. 2016).

Radio monitoring observations during the active state revealed a rapidly variable, flat spectrum persistent source. This emission is likely suggesting synchrotron emission as the origin. If this is in form of a jet or more generally of a propeller outflow is unknown (Deller et al. 2015). In addition, based on existing correlation in neutron star LMXBs between the radio luminosity and the X-ray luminosity, J1023 is brighter in radio, possibly suggesting a less efficient X-ray emission (Deller et al. 2015). Baglio et al. (2016) measured a linear polarisation of $0.90 \pm 0.17\%$ in the R band. In addition, the phase-resolved R-band curve shows a hint for a sinusoidal modulation. Lacking the Spectral Energy Distribution a red/nIR excess (characteristic of jet emission), the polarised emission likely comes from Thomson scattering with electrons in the disc.

Simultaneous X-ray (Chandra) and radio (VLA) monitoring showed a strong anti-correlated variability pattern, with radio emission strongly rising during X-ray low states (Bogdanov et al. 2018). A more articulated observing campaign involving XMM-Newton, NuSTAR, and Swift showed that X-rays al soft and hard (up to 80 keV) are strongly corrected with no lag, whereas X-rays and UV are not correlated (Coti Zelati et al. 2018).

Surprisingly, Ambrosino et al. (2017) discovered optical pulsations in J1023 during the active X-ray state. The pulsed fraction is at a level of ∼1%. Optical pulsation is present only in the high mode as X-ray pulsations (A. Papitto, private communication). This optical pulsed emission is puzzling. Ambrosino et al. (2017) convincingly showed that this emission cannot be explained by the cyclotron mechanism and cautiously favour a rotation-powered regime mechanism.

4.4.2 IGR J18245−2452

IGR J18245−2452 (J18245 in the following) was discovered by INTEGRAL/ISGRI during observations of the Galactic centre region (Eckert et al. 2013). J18245 lies in the globular cluster M28 at a distance of ∼5.5 kpc (Heinke et al. 2013; Romano et al. 2013; Homan and Pooley 2013). At this distance the peak outburst luminosity is ∼10^{37} erg s^{-1} (0.5–100 keV, e.g. De Falco et al. 2017). This luminosity led J18245 to be classified as a classical X-ray transient (i.e. not faint). A thermonuclear (type I) X-ray burst from J18245 was detected by Swift/XRT (Papitto et al. 2013b; Linares 2013). This marked the presence of a neutron star in the system. Further type I bursts were observed during the same outburst by MAXI (Serino et al. 2013) and INTEGRAL (De Falco et al. 2017). During an XMM-Newton observation, Papitto et al. (2013a) discovered a coherent periodicity in the X-ray flux at 3.9 ms. The pulsed signal is also modulated through Doppler shifts at the binary orbital period of 11.0 h, induced by a ∼0.2 $M_\odot$ companion star. Papitto et al. (2013a) were also able to associate the X-ray pulsar to a known radio pulsar previously discovered in M28, PSR J1824−2452I (Manchester et al. 2005), with the same spin and orbital periods. This provides the first direct evidence for a switch between an accretion-powered neutron star and a rotation-powered radio pulsar. The reactivation of the radio pulsar was very fast, with the detection of a pulsed signal less than 2 weeks after the end of the outburst.

A very peculiar state was observed when the source luminosity reached a mean level of a few 10^{36} erg s^{-1}. Rapid variation by a factor up to $\sim$100 were observed during two XMM-Newton observations (Ferrigno et al. 2014). In a hardness-intensity diagram two branches can be identified (see Fig. 4.4; Ferrigno et al. 2014). The brighter branch (blue branch) showed a tight correlation between hardness and intensity (as well as pulsed fraction). Below a threshold of $\sim$30 c s^{-1} in the pn instrument in addition to this branch a new one (magenta branch), appeared with scattered points at higher hardness. J18245 varied by a factor of $\sim$100 on a timescale as short as a few seconds. A spectral analysis at different spectral hardness shows a clear decrease in the power law spectral index from $\Gamma \sim 1.7$ to $\Gamma \sim 0.9$ and the disappearance of a black body component. The hardest spectrum is better described by partially covering power law model and it is achieved only occasionally at low count rates. A pulsed signal is always detected and the pulsed fraction is tightly correlated with the source count rate. In the magenta branch there is however no correlation of the hardness with the pulsed fraction, which is always at a level of 5–10% (Ferrigno et al. 2014).

M28 has been observed several times, however only thanks to the spatial resolution of the Chandra optics it is possible to gain information on the quiescence of J18245. Chandra observed M28 in 2002 finding J18245 in a quiescent state. The X-ray spectrum is well described by a simple power law model with a hard

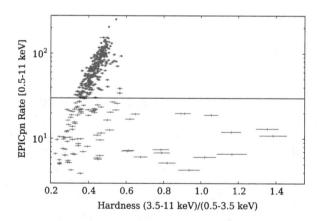

Fig. 4.4 Hardness-intensity diagram built by using two XMM-Newton observations with time bin of 200 s. The black solid line separates the different intensity states: the points represented in magenta and blue have a count rate lower and higher than 30 c s^{-1}, respectively (from Ferrigno et al. 2014)

photon index $\Gamma \sim 1.2$. The 0.5–10 keV unabsorbed luminosity is $\sim 2 \times 10^{32}\,\mathrm{erg\,s}^{-1}$ (Linares et al. 2014). No room for a soft component is left with an upper limit on the 0.1–10 keV luminosity of $\lesssim 7 \times 10^{31}\,\mathrm{erg\,s}^{-1}$. In a way similar to J1023 (but before J1023), also J18245 shows two different low-luminosity active states. They are readily apparent in a long Chandra observation taken in 2008 (Linares et al. 2014). The high and low state 0.5–10 keV luminosities are $\sim 4 \times 10^{33}\,\mathrm{erg\,s}^{-1}$ and $\sim 6 \times 10^{32}\,\mathrm{erg\,s}^{-1}$, respectively, with a factor of ~ 7 luminosity change. The spectra in the two states are fully compatible with a power law with photon index $\Gamma \sim 1.5$. Given the lower Chandra count rate, mode switching has been measure to occur on a timescale of $\lesssim 200\,\mathrm{s}$.

The optical counterpart has been identified thanks to HST images (Pallanca et al. 2013). The companion star has been detected during both quiescence and outburst, showing a two magnitude increase and the presence of the Hα line, indicating that accretion is taking place.

PSR J1824245 2I is known as a radio pulsar in M28 (detected during the quiescent period), but its observations were only sporadic and with large eclipses, variable from orbit to another as often happens in redbacks. In addition, the acceleration induced by the motion in the globular cluster prevents us from a firm measurement of the magnetic field.

4.4.3   XSS J12270−4859

XSS J12270−4859 (J12270 in the following) was discovered by the Rossi X-ray Timing Explorer during the high latitude slew survey (Sazonov and Revnivtsev 2004). Based on the presence of optical emission lines J12270 was initially classified as a cataclysmic variable hosting a magnetic white dwarf (Masetti et al. 2006; Butters et al. 2008), as for J1023. Unusual dipping and flaring behaviour led several authors to suggest a different classification involving a neutron star in a LMXB (Pretorius 2009; Saitou et al. 2009; de Martino et al. 2010, 2013; Hill et al. 2011; Papitto et al. 2015). de Martino et al. (2010) (see also Hill et al. 2011) were the first to associate J12270 to a relatively bright gamma-ray source detected by Fermi-LAT and emitting up to 10 GeV (3FGL J1227.9−4854). The source was puzzling and has been observed during this active state by XMM-Newton. An erratic behaviour typical of TMSP has been observed (de Martino et al. 2010). Despite flaring and dips, J12270 was stable at least over a 7 year period (de Martino et al. 2013). Dips correspond to the low state of J1023. A detailed analysis of the dips in J12270 disclosed three different types of dips: soft dips, dips with no spectral changes with respect to the active state and hard dips following flares. The pn spectrum can be modelled with a simple power law resulting in spectral indexes of: 1.64 ± 0.01 (active state); 1.65 ± 0.03 (flares); 1.71 ± 0.04 (dips), and 0.74 ± 0.08 (post-flare dips; de Martino et al. 2013). The 0.2–100 keV mean luminosity is $10^{34}\,\mathrm{erg\,s}^{-1}$ at a distance of 1.4 kpc. The ratio between the active and dip 0.2–10 keV luminosity is ~5, with the active 0.2–10 keV luminosity being 4×10^{33} erg s^{-1} (de Martino et al. 2013).

The XMM-Newton Optical Monitor (OM) was operated in fast timing mode (U and $UVM2$ filters) allowing for a strict comparison with X-ray data. The mean U magnitude was 16.6. Orbital modulation is readily apparent in the OM data. On top of this dips and flares are also present in the OM data, with a drop in the UV count rate by a factor of ~1.4 (de Martino et al. 2013). A lag analysis was also possible, showing no lag among soft (0.3–2 keV) and hard (2–10 keV) X-ray flux. The cross-correlation between U-band and X-rays show no lag, but indications that flares last longer in the optical-UV bands. A cross-correlation analysis on selected dips shows that UV and X-ray dips occur almost simultaneously, but the shape of UV dips is shallower, with a smoother decay and rise (de Martino et al. 2013). The optical spectrum shows prominent Hα, Hβ, He I, He II, and Bowen-blend N III/C III lines, as well as signs of irradiation (de Martino et al. 2014).

As J1023 during its active state also J12270 has been detected by Fermi in the 0.1–300 GeV, with a 0.1–10 GeV flux of 4×10^{-11} erg cm^{-2} s^{-1} and a steep power law spectrum with $\Gamma \sim 2.2$ and a cut-off energy of 8 GeV (Johnson et al. 2015; de Martino et al. 2010). Faint non-thermal radio emission was also detected with a flat spectral index (Masetti et al. 2006; Hill et al. 2011).

J12270 remained stable at all wavelengths for about a decade up to 2012 November/December, when a decline in flux at all bands was reported (Bassa et al. 2014; Bogdanov et al. 2014). This is in all respects similar to the transition of J1023 in the opposite sense: J12770 changed from an active state to a quiescent state, whereas the opposite transition was observed in J1023 in June 2013. J12270 decreased its optical brightness by 2 magnitudes, with all the optical emission lines disappearing. The X-ray spectrum hardened considerably with the spectrum described by power law with a photon index of $\Gamma = 1.1$ and a 0.3–10 keV unabsorbed luminosity of $\sim 2 \times 10^{32}$ erg s^{-1}, resulting in a factor of ~20 decrease with respect to the high state and ~4 with respect to the low state (de Martino et al. 2015; Bogdanov et al. 2014). At GeV energies the 0.1–100 GeV flux decreased by a factor of ~2 and the spectrum hardened to a power law with photon index 1.7 with a cut-off at 3 GeV (Johnson et al. 2015).

Radio observations revealed the presence of a millisecond radio pulsar at the position of J12270. The neutron star is spinning at 1.69 ms and has a magnetic field of 1.4×10^8 G (for a rotational energy of 9×10^{34} erg s^{-1}; Roy et al. 2015). The radio signal is absorbed for a large fraction of the orbit. After this discovery pulsations were searched at other frequencies and in older data. Pulsed emission was observed at GeV energies with a single peak emission nearly aligned with the radio main peak (Johnson et al. 2015). The pulsed signal was not detected in the X-ray band with a 3 σ upper limit on the rms pulsed fraction of ~7% (full band) or ~10% (0.5–2.5 keV; Papitto et al. 2015). Papitto et al. (2015) reanalysed XMM-Newton data during the high state and, knowing the pulse period, successfully detected a coherent pulsation. Pulsations were detected at an rms amplitude of ~8%, with a second harmonic stronger than the fundamental frequency. The amplitude is similar in the soft and hard X-ray bands. Polarimetric optical observation during the quiescent radio pulsar state, failed to detect any signal with a 3σ upper limit of 1.4% in the R band (Baglio et al. 2016).

4.4.4 1RXS J154439.4−112820

After the recognition of a peculiar variability pattern during the active state of J1023 and J12270, searches for new members of the transitional MSPs started, searching for rapid variability in the X-ray light curve of unidentified sources or cataclysmic variables and association with Fermi sources. With these characteristics Bogdanov and Halpern (2015) identified the forth member of the TMSP class in 1RXS J154439.4−112820 (J154439 in the following). J154439 is the only X-ray source within the 95% error circle of the Fermi source 3FL J1544.6−1125 (Stephen et al. 2010). J154439 has also been detected during the XMM Slew survey and by Swift/XRT, with flux variations by a factor of ∼3. Masetti et al. (2013) provided evidence for the presence of prominent emission lines (H and He) in the $R \sim 18.4$ optical counterpart, suggesting its identification as a cataclysmic variable.

During an XMM-Newton observation, J154439 showed characteristic rapid variations on a timescale of ∼10 s, passing randomly from ∼2 to ∼0.2 c s^{-1} and back. The overall spectrum is well fit with an absorbed power law model with $\Gamma = 1.7$ ($N_H = 1.4 \times 10^{21}$ cm^{-2}, consistent with the Galactic value). Spectral variations are only marginally evident with indexes of 1.67 ± 0.04 and 1.97 ± 0.28 (90% confidence level) for the high and low states, respectively (Bogdanov and Halpern 2015). Even if pn data were taken in timing mode, no X-ray periodicity was reported. A NuSTAR observation confirmed the spectral parameters over the 0.3–79 keV energy band, leading to a luminosity of 10^{33} erg s^{-1} at a scale distance of 1 kpc (Bogdanov 2016). Fast photometric data taken at the MDM Observatory showed fast variability also in the optical counterpart. MDM data and XMM-Newton/OM data revealed hints for an orbital modulation at a period of ∼5.2–5.4 h.

4.5 Once Upon a Time There Were a Magnetic Neutron Star Interacting with an Accretion Disc

The transition from and to the radio pulsar regime observed in the three well studied TMPS J1023, J18245, and J12270 with a good degree of certainty involves the presence of an accretion disc. In addition to an enhanced emission at all wavelengths, double horned emission lines were observed. This testifies that the switch on and off of the radio pulsar mechanism in binary systems can occur, at least, on timescales of tens of days and it is not a single occurrence in the neutron star lifetime.

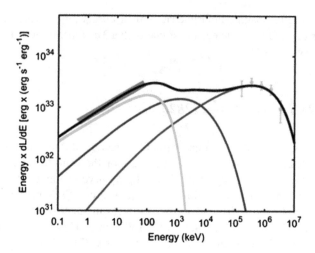

Fig. 4.5 Average spectral energy distribution (SED) observed from PSR J1023+0038 in X-rays (orange strip, from Tendulkar et al. 2014), and gamma-rays (cyan points, from Takata et al. 2014), evaluated for a distance of 1.37 kpc. The total SED evaluated with Papitto and Torres (2015) modelling is plotted as a black solid line. Synchrotron, Synchrotron-Self-Compton, and accretion flow (i.e. the sum of disc and neutron star emission) components are plotted as red, blue and green lines, respectively

The inference of the root cause of the sub-luminous disc state is more complicated. This sub-luminous state is dominated (~80% of the time) by the high mode during which X-ray pulsations are observed (e.g. in J1023). For this high mode Papitto and Torres (2015) (see also Bednarek 2015) showed that a scenario based on the propeller can explain the observed features at all wavelengths. The spectral energy distribution is characterised a broad emission from X-rays to gamma-rays (see Fig. 4.5). Papitto and Torres (2015) interpreted the gamma-ray part as due to the self-synchrotron Compton emission that originates at the turbulent boundary between the neutron star magnetosphere propelling the disc inflow at super-Keplerian speed. The X-ray emission is instead due to the sum of the synchrotron emission that originated from the same magnetospheric region and the luminosity emitted by the accretion flow, which however must be inefficient ($\lesssim 20\%$ of the conversion of gravitational energy) not to exceed the observed one.

Different suggestions have been put forward to explain the transitions among high and low mode (and flaring) and, more importantly, why only the three known TMSPs show this puzzling behaviour (and stability). The flux drops that bring the sources to the low mode are extremely peculiar and are very different from the dips currently observed in LMXBs, since there are no associated spectral changes and/or reduction of the soft X-ray flux, excluding absorption from intervening matter. Apparently, there is also no evidence for an X-ray luminosity dependence on the duration and frequency of flares and low flux mode intervals or any correlation between the separation between (and duration of) dips or flares (Bogdanov et al. 2015).

The fate of matter falling onto a magnetic neutron star is determined by the position of three different radii. Two are fixed and depends on the neutron star spin period: the corotation radius $r_{\mathrm{cor}} = (G\, M_{NS}\, P^2/(4\,\pi^2))^{1/3}$ (where magnetic field lines reach the Keplerian speed) and the light cylinder radius $r_{\mathrm{lc}} = c\, P/(2\,\pi)$ (with c the light velocity, where the magnetic field lines open, being unable to corotate with the neutron star). The truncation of the accretion disc by the magnetosphere occurs at the magnetospheric radius $r_{\mathrm{m}} = \xi\,(\mu^2/(2\,G\,M\,\dot{M}^2))^{1/7}$, with $\xi \sim 0.5$ (Campana et al. 2018) accounting for the disc geometry, $\mu = B\, R_{NS}^3$ the dipole magnetic moment, and $\dot{M}$ the mass accretion rate at the magnetospheric boundary. If the magnetospheric radius lies within the corotation radius, accretion proceeds unimpeded, however if $r_{\mathrm{m}} > r_{\mathrm{cor}}$ the incoming matter experiments a centrifugal force larger than gravity when it gets attached to the fields lines at the magnetospheric radius and (ideally) gets propelled out. If the mass accretion rate decreases further, r_{m} expands further, reaching at some point the light cylinder radius. At this point the magnetic field becomes radiative and matter is expelled further out by radiation pressure and relativistic particle wind (Campana et al. 1998; Burderi et al. 2001). A radio pulsar can in principle start working again. The observed luminosity in the sub-luminous state of J1023 and J12270 is relatively low and at the corresponding mass accretion rate (assuming that all the accreting material arrives at the neutron star surface) implies a magnetospheric radius well outside the corotation radius (Archibald et al. 2015; Bogdanov et al. 2015; Campana et al. 2016), so that the sources should be in the propeller regime. This poses a problem and suggests intriguing new insights for accretion physics at low luminosities.

The best and well studied example of propeller accretion is probably the accreting white dwarf AE Aquarii. In this system only $\sim$0.3% of the incoming material is able to reach the star surface. In AE Aqr most of the soft X-ray luminosity is produced in the inflow before ejection or accretion onto the surface (Oruru and Meintjes 2012). Correlation and lag analysis can shed light on the emission mechanisms in TMPS. J1023 has been well characterised. Hard (NuSTAR) and soft (XMM-Newton) X-ray emission appear well correlated (Coti Zelati et al. 2018). X-rays (XMM-Newton) instead do not correlate with UV (Swift/UVOT, Coti Zelati et al. 2018) nor with optical (B) emission (XMM-Newton/OM, Bogdanov et al. 2015), unless during the flaring state. This suggests that at least in J1023, as for AE Aqr, UV-optical emission comes from the accretion disc and the companion star.

Jaodand et al. (2016) were able to measure the overall period evolution of J1023. They found that J1023 is spinning down at a rate that is $\sim$30% larger than the pure dipole spin-down rate. This is consistent with the modelling of Parfrey et al. (2017) of AMSP magnetospheres in which the spin-down during propeller has the same functional form of the pulsar spin down and depends on how the disc inside the light cylinder opens some of the closed field lines, leading to an enhancement of the power extracted by the pulsar wind and the spin-down torque applied to the pulsar.

One class of scenarios involves a trapped disc. A propeller may not be able to eject matter so that the inflaming matter stays confined in the innermost part of the flow, trapping the magnetospheric radius close to the corotation radius (D'Angelo and Spruit 2010, 2012). Pure trapped disc models will not work being the oscillation timescale much shorter than what observed (Bogdanov et al. 2015). The switching between low and high mode might be the result of transitions between a non-accreting pure propeller mode and an accreting trapped-disc mode (Archibald et al. 2015; Bogdanov et al. 2015).

Alternatively, Campana et al. (2016) (see also Linares et al. 2014) proposed that high mode is connected to the propeller regime, whereas the low mode to the expulsion of the infalling matter by the neutron star pressure, with the neutron star in the radio pulsar state. During the high mode in the propeller regime, some matter leaks through the centrifugal barrier and accretes onto the neutron star, as shown in magnetohydrodynamical simulations (Romanova et al. 2005), and generates the observed X-ray pulsations. Detailed spectral modelling is consistent with a radiatively inefficient accretion disc close to the corotation radius during the high mode and receding beyond the light cylinder during the low mode. Contemporarily a shock emission sets in. Pulsed emission at a lower amplitude (~2–3%) and in the soft band only is predicted to occur. This scenario is in agreement with the broadband modelling by Papitto and Torres (2015) and by the observed optical polarisation coming from the disc (Baglio et al. 2016) and flat radio spectrum (Deller et al. 2015). The strong anti-correlation among X-ray and radio emission (Bogdanov et al. 2018) fits perfectly within this scenario: during the low state matter is expelled from the system by the relativistic pulsar wind generating in the shock strong radio emission. Parfrey and Tchekhovskoy (2018) showed through general-relativistic MHD simulations that this scenario is plausible.

A completely different scenario (Jaodand et al. 2016) is motivated by the observation of mode switching in radio pulsars, in which the pulse profile switches between two stable profiles. PSR B0943+10 showed that the mode switching in radio is accompanied by simultaneous switches in the X-ray band (profile and intensity; Mereghetti et al. 2013; Hermsen et al. 2013). Mode switching has never been observed in any X-ray pulsar so for the moment this might remain an intriguing suggestion.

Optical pulsations in the active X-ray state are puzzling. Ambrosino et al. (2017) showed that cyclotron emission can be ruled out and a rotation-powered mechanism might work. A pure (engulfed) radio pulsar however encounters problems in explaining the full phenomenology of J1023 (Campana et al. 2016). As noted in Bogdanov et al. (2018) J1023 and the other transitional ms pulsars are the only systems for which the neutron star rotational energy is comparable to the accretion energy. It might be that the neutron star always looses rotational energy, independently of the fate of the accreting matter. Accretion of matter is able to quench pulsed radio emission but pulsed optical emission might be generated. In this case we would expect pulsed optical emission to be present independently on the X-ray mode.

4.6   Open Questions

Among all the open questions that we described above, the most puzzling problem still regards the connection between LMXBs and AMSPs, or MSPs in general. It is still a mystery the reason why LMXBs usually do not show X-ray pulsations, not even at low mass accretion rates when presumably the magnetospheric radius is outside the neutron star surface and inside the co-rotation radius, thus allowing matter to be channelled by the magnetic field. Does it have to do with the magnetic field, that can be buried inside the neutron star surface by prolonged accretion phases? Another mystery is why AMSPs, although fast spinning and with a misaligned magnetic field do not show pulsed radio or gamma emission (as instead do other not accreting MSPs, including transitional MSPs) in quiescence, when accretion onto the neutron star decreases by orders of magnitude presumably pushing the magnetospheric radius outside the light cylinder radius. Does it have to do with the presence of matter around the system that is swept away by the radiation pressure of the pulsar? Is this enough to justify also non detections in the gamma-ray band? In other words, it would be important to understand why a system like J18245 (a.k.a. M28I), swinging between the rotation-powered and the accretion-powered phase in such short timescales is so rare. Is it because of its relatively large ($\sim$11 h) orbital period so that the matter outflowing the system is spread in a relatively wide orbit?

On the TMSP side, we have a number of open questions that will likely be answered in the next following years. Why do some sources (J1023 and XSS12270) showed a long (years) accretion-dominated state whereas J18245 showed a proper X-ray outburst? Why do the accretion-dominated state of these systems so stable (e.g. in J1023)? What is the root cause of the mode switching observed during the accretion-dominated state? Are X-ray pulsations detectable at a lower level (and/or in a softer energy band) also in the low mode of the active X-ray state? These are the basic questions that make these systems particularly attractive.

Recently optical pulsations detected in J1023 added a number of additional questions. What is the mechanism responsible for optical pulsation? Is the accretion-dominated state really powered by accretion? Can a rotation-powered mechanism remain active (i.e. not inhibited) during an accretion state?

Answering these questions will give important information on the still elusive radio pulsar and accretion emission mechanisms and on the evolutive path producing the different types of MSPs known today.

Acknowledgements   We would like to thank M.C. Baglio, L. Burderi, F. Coti Zelati, P. D'Avanzo, D. de Martino, R. Iaria, A. Papitto, N. Rea, A. Riggio, A. Sanna and L. Stella for useful discussions and comments over time.

References

Aliu, E., Archambault, S., Archer, A., et al.: A search for very high energy gamma rays from the missing link binary pulsar J1023+0038 with VERITAS. Astrophys. J. **831**, 193 (2016)

Altamirano, D., Casella, P., Patruno, A., Wijnands, R., van der Klis, M.: Intermittent millisecond X-ray pulsations from the neutron star X-ray transient SAX J1748.9–2021 in the globular cluster NGC 6440. Astrophys. J. **674**, L45 (2008)

Altamirano, D., Patruno, A., Heinke, C.O., et al.: Discovery of a 205.89 Hz accreting millisecond X-ray pulsar in the globular cluster NGC 6440. Astrophys. J. **712**, L58-L62 (2010)

Altamirano, D., Cavecchi, Y., Patruno, A., et al.: Discovery of an accreting millisecond pulsar in the eclipsing binary system SWIFT J1749.4–2807. Astrophys. J. **727**, L18 (2011)

Ambrosino, F., Papitto, A., Stella, L., et al.: Optical pulsations from a transitional millisecond pulsar. Nature Astron. **1**, 266 (2017)

Applegate, J.H., Shaham, J.: Orbital period variability in the eclipsing pulsar binary PSR B1957+20: evidence for a tidally powered star. Astrophys. J. **436**, 312–318 (1994)

Archibald, A.M., Stairs, I.H., Ransom, S.M., et al.: A radio pulsar/X-ray binary link. Science **324**, 1411 (2009)

Archibald, A.M., Kaspi, V.M., Bogdanov, S., Hessels, J.W.T., Stairs, I.H., Ransom, S.M., McLaughlin, M.A.: X-ray variability and evidence for pulsations from the unique radio pulsar/X-ray binary transition object FIRST J102347.6+003841. Astrophys. J. **722**, 88–95 (2010)

Archibald, A.M., Bogdanov, S., Patruno, A., et al.: Accretion-powered pulsations in an apparently quiescent neutron star binary. Astrophys. J. **807**, 62 (2015)

Arzoumanian, Z., Fruchter, A.S., Taylor, J.H.: Orbital variability in the eclipsing pulsar binary PSR B1957+20. Astrophys. J. **426**, 85–88 (1994)

Backer, D.C., Kulkarni, S.R., Heiles, C., Davis, M.M., Goss, W.M.: A millisecond pulsar. Nature **300**, 615–618 (1982)

Baglio, M.C., D'Avanzo, P., Campana, S., Coti Zelati, F., Covino, S., Russell, D.M.: Different twins in the millisecond pulsar recycling scenario: optical polarimetry of PSR J1023+0038 and XSS J12270–4859. Astron. Astrophys. **591**, A101 (2016)

Bassa, C.G., Patruno, A., Hessels, J.W.T., et al.: A state change in the low-mass X-ray binary XSS J12270–4859. Mon. Not. R. Astron. Soc. **441**, 1825–1830 (2014)

Bednarek, W.: γ-ray emission states in the redback millisecond pulsar binary system PSR J1227–4853. Mon. Not. R. Astron. Soc. **451**, L55–L59 (2015)

Bhattacharya, D., van den Heuvel, E.P.J.: Formation and evolution of binary and millisecond radio pulsars. Phys. Rep. **203**, 1–124 (1991)

Bogdanov, S.: A NuSTAR observation of the gamma-ray-emitting X-ray binary and transitional millisecond pulsar candidate 1RXS J154439.4–112820. Astrophys. J. **826**, 28 (2016)

Bogdanov, S., Halpern, J.P.: Identification of the high-energy gamma-ray source 3FGL J1544.6–1125 as a transitional millisecond pulsar binary in an accreting state. Astrophys. J. **803**, L27 (2015)

Bogdanov, S., Archibald, A.M., Hessels, J.W.T., et al.: A Chandra X-ray observation of the binary millisecond pulsar PSR J1023+0038. Astrophys. J. **742**, 97 (2011)

Bogdanov, S., Patruno, A., Archibald, A.M., Bassa, C., Hessels, J.W.T., Janssen, G.H., Stappers, B.W.: X-ray observations of XSS J12270–4859 in a new low state: a transformation to a disk-free rotation-powered pulsar binary. Astrophys. J. **789**, 40 (2014)

Bogdanov, S., Archibald, A.M., Bassa, C., et al.: Simultaneous Chandra and VLA observations of the transitional millisecond pulsar PSR J1023+0038: anti-correlated X-ray and radio variability. Astrophys. J. **806**, 148 (2015)

Bogdanov, S., Deller, A.T., Miller-Jones, J.C.A., et al.: Coordinated X-ray, ultraviolet, optical, and radio observations of the PSR J1023+0038 system in a low-mass X-ray binary state (2018). arXiv:1709.08574

Bond, H.E., White, R.L., Becker, R.H., O'Brien, M.S.: FIRST J102347.6+003841: the first radio-selected cataclysmic variable. Publ. Astron. Soc. Pac. **114**, 1359–1363 (2002)

Brainerd, J., Lamb, F.K.: Effect of an electron scattering cloud on X-ray oscillations produced by beaming. Astrophys. J. **317**, L33–L38 (1987)

Burderi, L., King, A.R.: The mass of the companion of PSR 1718–19. Astrophys. J. Lett. **430**, L57–L60 (1994)

Burderi, L., Possenti, A., Colpi, M., Di Salvo, T., D'Amico, N.: Neutron stars with submillisecond periods: a population of high-mass objects? Astrophys. J. **519**, 285–290 (1999)

Burderi, L., Possenti, A., D'Antona, F., et al.: Where may ultrafast rotating neutron stars be hidden? Astrophys. J. **560**, L71–L74 (2001)

Burderi, L., D'Antona, F., Burgay, M.: PSR J1740–5340: accretion inhibited by radio ejection in a binary millisecond pulsar in the globular cluster NGC 6397. Astrophys. J. **574**, 325–331 (2002)

Burderi, L., Di Salvo, T., D'Antona, F., Robba, N.R., Testa, V.: The optical counterpart to SAX J1808.4–3658 in quiescence: evidence of an active radio pulsar? Astron. Astrophys. **404**, L43–L46 (2003)

Burderi, L., Di Salvo, T., Menna, M.T., Riggio, A., Papitto, A.: Order in the chaos: spin-up and spin-down during the 2002 outburst of SAX J1808.4–3658. Astrophys. J. **653**, L133–L136 (2006)

Burderi, L., Di Salvo, T., Lavagetto, G., et al.: Timing an accreting millisecond pulsar: measuring the accretion torque in IGR J00291+5934. Astrophys. J. **657**, 961–966 (2007)

Burderi, L., Riggio, A., di Salvo, T., Papitto, A., Menna, M.T., D'Aì, A., Iaria, R.: Timing of the 2008 outburst of SAX J1808.4–3658 with XMM-Newton: a stable orbital-period derivative over ten years. Astron. Astrophys. **496**, L17–L20 (2009)

Burgay, M., Burderi, L., Possenti, A., et al.: A search for pulsars in quiescent soft X-ray transients. I. Astrophys. J. **589**, 902–910 (2003)

Butters, O.W., Norton, A.J., Hakala, P., Mukai, K., Barlow, E.J.: RXTE determination of the intermediate polar status of XSS J00564+4548, IGR J17195–4100, and XSS J12270–4859. Astron. Astrophys. **487**, 271–276 (2008)

Cadelano, M., Pallanca, C., Ferraro, F.R., Dalessandro, E., Lanzoni, B., Patruno, A.: The optical counterpart to the accreting millisecond X-ray pulsar SAX J1748.9–2021 in the globular cluster NGC 6440. Astrophys. J. **844**, 53 (2017)

Campana, S., Colpi, M., Mereghetti, S., Stella, L., Tavani, M.: The neutron stars of soft X-ray transients. Astron. Astrophys. Rev. **8**, 279–316 (1998)

Campana, S., Ravasio, M., Israel, G.L., Mangano, V., Belloni, T.: XMM-Newton observation of the 5.25 millisecond transient pulsar XTE J1807–294 in outburst. Astrophys. J. **594**, L39–L42 (2003)

Campana, S., D'Avanzo, P., Casares, J., et al.: Indirect evidence of an active radio pulsar in the quiescent state of the transient millisecond pulsar SAX J1808.4–3658. Astrophys. J. **614**, L49–L52 (2004)

Campana, S., Coti Zelati, F., Papitto, A., Rea, N., Torres, D.F., Baglio, M.C., D'Avanzo, P.: A physical scenario for the high and low X-ray luminosity states in the transitional pulsar PSR J1023+0038. Astron. Astrophys. **594**, A31 (2016)

Campana, S., Stella, L., Mereghetti, S., de Martino, D.: A universal relation for the propeller mechanisms in magnetic rotating stars at different scales. Astron. Astrophys. **610**, A46 (2018)

Casella, P., Altamirano, D., Patruno, A., Wijnands, R., van der Klis, M.: Discovery of coherent millisecond X-ray pulsations in Aquila X-1. Astrophys. J. **674**, L41 (2008)

Chakrabarty, D., Morgan, E.H.: The two-hour orbit of a binary millisecond X-ray pulsar. Nature **394**, 346–348 (1998)

Coti Zelati, F., Baglio, M.C., Campana, S., et al.: Engulfing a radio pulsar: the case of PSR J1023+0038. Mon. Not. R. Astron. Soc. **444**, 1783–1792 (2014)

Coti Zelati, F., Campana, S., Braito, V., et al.: Simultaneous broadband observations and high-resolution X-ray spectroscopy of the transitional millisecond pulsar PSR J1023+0038. Astron. Astrophys. **611**, A14 (2018)

Cumming, A., Zweibel, E., Bildsten, L.: Magnetic screening in accreting neutron stars. Astrophys. J. **557**, 958–966 (2001)

D'Angelo, C.R., Spruit, H.C.: Episodic accretion on to strongly magnetic stars. Mon. Not. R. Astron. Soc. **406**, 1208–1219 (2010)

D'Angelo, C.R., Spruit, H.C.: Accretion discs trapped near corotation. Mon. Not. R. Astron. Soc. **420**, 416-P (2012)

D'Avanzo, P., Campana, S., Covino, S., Israel, G.L., Stella, L., Andreuzzi, G.: The optical counterpart of IGR J00291+5934 in quiescence. Astron. Astrophys. **472**, 881–885 (2007)

D'Avanzo, P., Campana, S., Casares, J., Covino, S., Israel, G.L., Stella, L.: The optical counterparts of accreting millisecond X-ray pulsars during quiescence. Astron. Astrophys. **508**, 297–308 (2009)

D'Avanzo, P., Campana, S., Muñoz-Darias, T., Belloni, T., Bozzo, E., Falanga, M., Stella, L.: A search for the near-infrared counterpart of the eclipsing millisecond X-ray pulsar Swift J1749.4–2807. Astron. Astrophys. **534**, A92 (2011)

De Falco, V., Kuiper, L., Bozzo, E., Ferrigno, C., Poutanen, J., Stella, L., Falanga, M.: The transitional millisecond pulsar IGR J18245–2452 during its 2013 outburst at X-rays and soft gamma-rays. Astron. Astrophys. **603**, A16 (2017)

Degenaar, N., Ootes, L.S., Reynolds, M.T., Wijnands, R., Page, D.: A cold neutron star in the transient low-mass X-ray binary HETE J1900.1–2455 after 10 yr of active accretion. Mon. Not. R. Astron. Soc. **465**, L10–L14 (2017)

Deller, A.T., Archibald, A.M., Brisken, W.F., et al.: A parallax distance and mass estimate for the transitional millisecond pulsar system J1023+0038. Astrophys. J. **756**, L25 (2012)

Deller, A.T., Moldon, J., Miller-Jones, J.C.A., et al.: Radio imaging observations of PSR J1023+0038 in an LMXB state. Astrophys. J. **809**, 13 (2015)

de Martino, D., Falanga, M., Bonnet-Bidaud, J.-M., et al.: The intriguing nature of the high-energy gamma ray source XSS J12270–4859. Astron. Astrophys. **515**, A25 (2010)

de Martino, D., Belloni, T., Falanga, M., et al.: X-ray follow-ups of XSS J12270–4859: a low-mass X-ray binary with gamma-ray Fermi-LAT association. Astron. Astrophys. **550**, A89 (2013)

de Martino, D., Casares, J., Mason, E., et al.: Unveiling the redback nature of the low-mass X-ray binary XSS J1227.0–4859 through optical observations. Mon. Not. R. Astron. Soc. **444**, 3004–3014 (2014)

de Martino, D., Papitto, A., Belloni, T., et al.: Multiwavelength observations of the transitional millisecond pulsar binary XSS J12270–4859. Mon. Not. R. Astron. Soc. **454**, 2190–2198 (2015)

de Oña Wilhelmi, E., Papitto, A., Li, J., et al.: SAX J1808.4–3658, an accreting millisecond pulsar shining in gamma rays? Mon. Not. R. Astron. Soc. **456**, 2647–2653 (2016)

Di Salvo, T., Burderi, L.: Constraints on the neutron star magnetic field of the two X-ray transients SAX J1808.4–3658 and Aql X-1. Astron. Astrophys. **397**, 723–727 (2003)

Di Salvo, T., Burderi, L., Riggio, A., Papitto, A., Menna, M.T.: Order in the chaos? The strange case of accreting millisecond pulsars. In: The Multicolored Landscape of Compact Objects and Their Explosive Origins, vol. 924, pp. 613–622 (2007)

Di Salvo, T., Burderi, L., Riggio, A., Papitto, A., Menna, M.T.: Orbital evolution of an accreting millisecond pulsar: witnessing the banquet of a hidden black widow? Mon. Not. R. Astron. Soc. **389**, 1851–1857 (2008)

Eckert, D., Del Santo, M., Bazzano, A., et al.: IGR J18245–2452: a new hard X-ray transient discovered by INTEGRAL. The Astronomer's Telegram 4925 (2013)

Elebert, P., Callanan, P.J., Filippenko, A.V., Garnavich, P.M., Mackie, G., Hill, J.M., Burwitz, V.: Optical photometry and spectroscopy of the accretion-powered millisecond pulsar HETE J1900.1–2455. Mon. Not. R. Astron. Soc. **383**, 1581–1587 (2008)

Falanga, M., Kuiper, L., Poutanen, J., et al.: INTEGRAL and RXTE observations of accreting millisecond pulsar IGR J00291+5934 in outburst. Astron. Astrophys. **444**, 15–24 (2005)

Ferrigno, C., Bozzo, E., Papitto, A., et al.: Hiccup accretion in the swinging pulsar IGR J18245–2452. Astron. Astrophys. **567**, A77 (2014)

Galloway, D.K., Chakrabarty, D., Morgan, E.H., Remillard, R.A.: Discovery of a high-latitude accreting millisecond pulsar in an ultracompact binary. Astrophys. J. **576**, L137–L140 (2002)

Galloway, D.K., Markwardt, C.B., Morgan, E.H., Chakrabarty, D., Strohmayer, T.E.: Discovery of the accretion-powered millisecond X-ray pulsar IGR J00291+5934. Astrophys. J. **622**, L45–L48 (2005)

Ghosh, P., Lamb, F.K.: Disk accretion by magnetic neutron stars. Astrophys. J. **223**, L83–L87 (1978)

Giles, A.B., Greenhill, J.G., Hill, K.M., Sanders, E.: The optical counterpart of XTE J0929–314: the third transient millisecond X-ray pulsar. Mon. Not. R. Astron. Soc. **361**, 1180–1186 (2005)

Göğüş, E., Alpar, M.A., Gilfanov, M.: Is the lack of pulsations in low-mass X-ray binaries due to comptonizing coronae? Astrophys. J. **659**, 580–584 (2007)

Halpern, J.P., Gaidos, E., Sheffield, A., Price-Whelan, A.M., Bogdanov, S.: Optical observations of the binary MSP J1023+0038 in a new accreting state. The Astronomer's Telegram 5514 (2013)

Hartman, J.M., Patruno, A., Chakrabarty, D., et al.: The long-term evolution of the spin, pulse shape, and orbit of the accretion-powered millisecond pulsar SAX J1808.4–3658. Astrophys. J. **675**, 1468–1486 (2008)

Hartman, J.M., Patruno, A., Chakrabarty, D., Markwardt, C.B., Morgan, E.H., van der Klis, M., Wijnands, R.: A decade of timing an accretion-powered millisecond pulsar: the continuing spin down and orbital evolution of SAX J1808.4–3658. Astrophys. J. **702**, 1673–1678 (2009)

Hartman, J.M., Galloway, D.K., Chakrabarty, D.: A double outburst from IGR J00291+5934: implications for accretion disk instability theory. Astrophys. J. **726**, 26 (2011)

Heinke, C.O., Bahramian, A., Wijnands, R., Altamirano, D.: IGR J18245–2452 is a new transient located in the core of the globular cluster M28. The Astronomer's Telegram 4927 (2013)

Hermsen, W., Hessels, J.W.T., Kuiper, L., et al.: Synchronous X-ray and radio mode switches: a rapid global transformation of the pulsar magnetosphere. Science **339**, 436 (2013)

Hessels, J.W.T., Ransom, S.M., Stairs, I.H., Freire, P.C.C., Kaspi, V.M., Camilo, F.: A radio pulsar spinning at 716 Hz. Science **311**, 1901–1904 (2006)

Hill, A.B., Szostek, A., Corbel, S., et al.: The bright unidentified γ-ray source 1FGL J1227.9–4852: can it be associated with a low-mass X-ray binary? Mon. Not. R. Astron. Soc. **415**, 235–243 (2011)

Homan, J., Pooley, D.: A Chandra observation of the neutron-star transient IGR J18245–2452 in M28. The Astronomer's Telegram 5045 (2013)

Homer, L., Charles, P.A., Chakrabarty, D., van Zyl, L.: The optical counterpart to SAX J1808.4–3658: observations in quiescence. Mon. Not. R. Astron. Soc. **325**, 1471–1476 (2001)

Homer, L., Szkody, P., Henden, A., Chen, B., Schmidt, G.D., Fraser, O.J., West, A.A.: Characterizing three candidate magnetic cataclysmic variables from SDSS: XMM-Newton and optical follow-up observations. Astron. J. **132**, 2743–2754 (2006)

Iacolina, M.N., Burgay, M., Burderi, L., Possenti, A., Di Salvo, T.: Searching for pulsed emission from XTE J0929–314 at high radio frequencies. Astron. Astrophys. **497**, 445–450 (2009)

Iacolina, M.N., Burgay, M., Burderi, L., Possenti, A., Di Salvo, T.: Search for pulsations at high radio frequencies from accreting millisecond X-ray pulsars in quiescence. Astron. Astrophys. **519**, A13 (2010)

Jaodand, A., Archibald, A.M., Hessels, J.W.T., et al.: Timing observations of PSR J1023+0038 during a low-mass X-ray binary state. Astrophys. J. **830**, 122 (2016)

Johnson, T.J., Ray, P.S., Roy, J., et al.: Discovery of gamma-ray pulsations from the transitional redback PSR J1227–4853. Astrophys. J. **806**, 91 (2015)

Kaaret, P., Morgan, E.H., Vanderspek, R., Tomsick, J.A.: Discovery of the millisecond X-ray pulsar HETE J1900.1–2455. Astrophys. J. **638**, 963–967 (2006)

Krimm, H.A., Markwardt, C.B., Deloye, C.J., et al.: Discovery of the accretion-powered millisecond pulsar SWIFT J1756.9–2508 with a low-mass companion. Astrophys. J. **668**, L147–L150 (2007)

Linares, M.: IGR J18245–2452: an accreting neutron star and thermonuclear burster in M28. The Astronomer's Telegram 4960 (2013)

Linares, M.: X-ray states of redback millisecond pulsars. Astrophys. J. **795**, 72 (2014)

Linares, M., Bahramian, A., Heinke, C., et al.: The neutron star transient and millisecond pulsar in M28: from sub-luminous accretion to rotation-powered quiescence. Mon. Not. R. Astron. Soc. **438**, 251–261 (2014)

Manchester, R.N., Hobbs, G.B., Teoh, A., Hobbs, M.: The Australia telescope national facility pulsar catalogue. Astron. J. **129**, 1993–2006 (2005)

Markwardt, C.B., Swank, J.H.: XTE J1814–338. IAUC 8144 (2003)

Markwardt, C.B., Swank, J.H., Strohmayer, T.E., in 't Zand, J.J.M., Marshall, F.E.: Discovery of a second millisecond accreting pulsar: XTE J1751–305. Astrophys. J. **575**, L21–L24 (2002)

Masetti, N., Morelli, L., Palazzi, E., et al.: Unveiling the nature of INTEGRAL objects through optical spectroscopy. V. Identification and properties of 21 southern hard X-ray sources. Astron. Astrophys. **459**, 21–30 (2006)

Masetti, N., Sbarufatti, B., Parisi, P., et al.: BL Lacertae identifications in a ROSAT-selected sample of Fermi unidentified objects. Astron. Astrophys. **559**, A58 (2013)

Mata Sánchez, D., Muñoz-Darias, T., Casares, J., Jiménez-Ibarra, F.: The donor of Aquila X-1 revealed by high-angular resolution near-infrared spectroscopy. Mon. Not. R. Astron. Soc. **464**, L41–L45 (2017)

Mereghetti, S., Tiengo, A., Esposito, P., Turolla, R.: The variable X-ray emission of PSR B0943+10. Mon. Not. R. Astron. Soc. **435**, 2568–2573 (2013)

Oruru, B., Meintjes, P.J.: X-ray characteristics and the spectral energy distribution of AE Aquarii. Mon. Not. R. Astron. Soc. **421**, 1557–1568 (2012)

Pallanca, C., Dalessandro, E., Ferraro, F.R., Lanzoni, B., Beccari, G.: The optical counterpart to the X-ray transient IGR J1824–24525 in the globular cluster M28. Astrophys. J. **773**, 122 (2013)

Papitto, A., Torres, D.F.: A propeller model for the sub-luminous state of the transitional millisecond pulsar PSR J1023+0038. Astrophys. J. **807**, 33 (2015)

Papitto, A., di Salvo, T., Burderi, L., Menna, M.T., Lavagetto, G., Riggio, A.: Timing of the accreting millisecond pulsar XTE J1814–338. Mon. Not. R. Astron. Soc. **375**, 971–976 (2007)

Papitto, A., Menna, M.T., Burderi, L., di Salvo, T., Riggio, A.: Measuring the spin up of the accreting millisecond pulsar XTEJ1751–305. Mon. Not. R. Astron. Soc. **383**, 411–416 (2008)

Papitto, A., Riggio, A., di Salvo, T., et al.: The X-ray spectrum of the newly discovered accreting millisecond pulsar IGR J17511–3057. Mon. Not. R. Astron. Soc. **407**, 2575–2588 (2010)

Papitto, A., D'Aì, A., Motta, S., et al.: The spin and orbit of the newly discovered pulsar IGR J17480–2446. Astron. Astrophys. **526**, L3 (2011a)

Papitto, A., Bozzo, E., Ferrigno, C., et al.: The discovery of the 401 Hz accreting millisecond pulsar IGR J17498–2921 in a 3.8 h orbit. Astron. Astrophys. **535**, L4 (2011b)

Papitto, A., Riggio, A., Burderi, L., di Salvo, T., D'Aí, A., Iaria, R.: Spin down during quiescence of the fastest known accretion-powered pulsar. Astron. Astrophys. **528**, A55 (2011c)

Papitto, A., Di Salvo, T., Burderi, L., et al.: The pulse profile and spin evolution of the accreting pulsar in Terzan 5, IGR J17480–2446, during its 2010 outburst. Mon. Not. R. Astron. Soc. **423**, 1178–1193 (2012)

Papitto, A., Ferrigno, C., Bozzo, E., et al.: Swings between rotation and accretion power in a binary millisecond pulsar. Nature **501**, 517–520 (2013a)

Papitto, A., Bozzo, E., Ferrigno, C., Pavan, L., Romano, P., Campana, S.: A type-I X-ray burst detected by Swift/XRT from the direction of IGR J18245–2452. The Astronomer's Telegram 4959 (2013b)

Papitto, A., de Martino, D., Belloni, T.M., Burgay, M., Pellizzoni, A., Possenti, A., Torres, D.F.: X-ray coherent pulsations during a sub-luminous accretion disc state of the transitional millisecond pulsar XSS J12270–4859. Mon. Not. R. Astron. Soc. **449**, L26-L30 (2015)

Parfrey, K., Tchekhovskoy, A.: General-relativistic simulations of four states of accretion onto millisecond pulsars (2018). arXiv:1708.06362

Parfrey, K., Spitkovsky, A., Beloborodov, A.M.: Simulations of the magnetospheres of accreting millisecond pulsars. Mon. Not. R. Astron. Soc. **469**, 3656–3669 (2017)

Patruno, A.: The accreting millisecond X-ray pulsar IGR J00291+5934: evidence for a long timescale spin evolution. Astrophys. J. **722**, 909–918 (2010)

Patruno, A.: Evidence of fast magnetic field evolution in an accreting millisecond pulsar. Astrophys. J. **753**, L12 (2012)

Patruno, A.: The slow orbital evolution of the accreting millisecond pulsar IGR J0029+5934. Astrophys. J. **839**, 51 (2017)

Patruno, A., D'Angelo, C.: 1 Hz flaring in the accreting millisecond pulsar NGC 6440 X-2: disk trapping and accretion cycles. Astrophys. J. **771**, 94 (2013)

Patruno, A., Watts, A.L.: Accreting millisecond X-ray pulsars (2012). arXiv:1206.2727

Patruno, A., Altamirano, D., Hessels, J.W.T., Casella, P., Wijnands, R., van der Klis, M.: Phase-coherent timing of the accreting millisecond pulsar SAX J1748.9–2021. Astrophys. J. **690**, 1856–1865 (2009a)

Patruno, A., Wijnands, R., van der Klis, M.: An alternative interpretation of the timing noise in accreting millisecond pulsars. Astrophys. J. **698**, L60–L63 (2009b)

Patruno, A., Altamirano, D., Messenger, C.: The long-term evolution of the accreting millisecond X-ray pulsar Swift J1756.9–2508. Mon. Not. R. Astron. Soc. **403**, 1426–1432 (2010)

Patruno, A., Bult, P., Gopakumar, A., Hartman, J.M., Wijnands, R., van der Klis, M., Chakrabarty, D.: Accelerated orbital expansion and secular spin-down of the accreting millisecond pulsar SAX J1808.4–3658. Astrophys. J. **746**, L27 (2012)

Patruno, A., Archibald, A.M., Hessels, J.W.T., et al.: A new accretion disk around the missing link binary system PSR J1023+0038. Astrophys. J. **781**, L3 (2014)

Patruno, A., Maitra, D., Curran, P.A., et al.: The reflares and outburst evolution in the accreting millisecond pulsar SAX J1808.4–3658: a disk truncated near co-rotation? Astrophys. J. **817**, 100 (2016)

Patruno, A., Jaodand, A., Kuiper, L., et al.: Radio pulse search and X-ray monitoring of SAX J1808.4–3658: what causes its orbital evolution? Astrophys. J. **841**, 98 (2017)

Pretorius, M.L.: Time-resolved optical observations of five cataclysmic variables detected by INTEGRAL. Mon. Not. R. Astron. Soc. **395**, 386–393 (2009)

Rappaport, S.A., Fregeau, J.M., Spruit, H.: Accretion onto fast X-ray pulsars. Astrophys. J. **606**, 436–443 (2004)

Riggio, A., Di Salvo, T., Burderi, L., Menna, M.T., Papitto, A., Iaria, R., Lavagetto, G.: Spin-up and phase fluctuations in the timing of the accreting millisecond pulsar XTE J1807–294. Astrophys. J. **678**, 1273–1278 (2008)

Riggio, A., Papitto, A., Burderi, L., et al.: Timing of the accreting millisecond pulsar IGR J17511–3057. Astron. Astrophys. **526**, A95 (2011a)

Riggio, A., Papitto, A., Burderi, L., di Salvo, T.: A model to interpret pulse phase shifts in AMXPs: SAX J1808.4–3658 as a proof of concept. Am. Inst. Phys. Conf. Ser. **1357**, 151–154 (2011b)

Riggio, A., Burderi, L., di Salvo, T., Papitto, A., D'Aì, A., Iaria, R., Menna, M.T.: Secular spin-down of the AMP XTE J1751–305. Astron. Astrophys. **531**, A140 (2011c)

Roberts, M.R.E.: Surrounded by spiders! New black widows and redbacks in the galactic field. In: IAU Symposium, vol. 291, pp. 127–132 (2013)

Romani, R.W.: A unified model of neutron-star magnetic fields. Nature **347**, 741–743 (1990)

Romano, P., Barthelmy, S.D., Burrows, D.N., et al.: Swift observations of IGR J18245–2452. The Astronomer's Telegram 4929 (2013)

Romanova, M.M., Ustyugova, G.V., Koldoba, A.V., Lovelace, R.V.E.: Propeller-driven outflows and disk oscillations. Astrophys. J. **635**, L165–L168 (2005)

Roy, J., Ray, P.S., Bhattacharyya, B., et al.: Discovery of PSR J1227–4853: a transition from a low-mass X-ray binary to a redback millisecond pulsar. Astrophys. J. **800**, L12 (2015)

Ruderman, M.: Neutron star crustal plate tectonics. I - Magnetic dipole evolution in millisecond pulsars and low-mass X-ray binaries. Astrophys. J. **366**, 261–269 (1991)

Saitou, K., Tsujimoto, M., Ebisawa, K., Ishida, M.: Suzaku X-ray study of an anomalous source XSS J12270–4859. Publ. Astron. Soc. Jpn. **61**, L13–L16 (2009)

Sanna, A., Burderi, L., Riggio, A., et al.: Timing of the accreting millisecond pulsar SAX J1748.9–2021 during its 2015 outburst. Mon. Not. R. Astron. Soc. **459**, 1340–1349 (2016)

Sanna, A., Papitto, A., Burderi, L., et al.: Discovery of a new accreting millisecond X-ray pulsar in the globular cluster NGC 2808. Astron. Astrophys. **598**, A34 (2017a)

Sanna, A., Di Salvo, T., Burderi, L., et al.: On the timing properties of SAX J1808.4–3658 during its 2015 outburst. Mon. Not. R. Astron. Soc. **471**, 463–477 (2017b)

Sanna, A., Pintore, F., Bozzo, E., et al.: Spectral and timing properties of IGR J00291+5934 during its 2015 outburst. Mon. Not. R. Astron. Soc. **466**, 2910–2917 (2017c)

Sanna, A., Bahramian, A., Bozzo, E., et al. Discovery of 105 Hz coherent pulsations in the ultracompact binary IGR J16597–3704. Astron. Astrophys. **610**, L2 (2018)

Sazonov, S.Y., Revnivtsev, M.G.: Statistical properties of local active galactic nuclei inferred from the RXTE 3–20 keV all-sky survey. Astron. Astrophys. **423**, 469–480 (2004)

Serino, M., Takagi, T., Negoro, H., et al.: MAXI/GSC detected two X-ray bursts from IGR J18245–2452 in M28. The Astronomer's Telegram 4961 (2013)

Shahbaz, T., Linares, M., Nevado, S.P., et al.: The binary millisecond pulsar PSR J1023+0038 during its accretion state - I. Optical variability. Mon. Not. R. Astron. Soc. **453**, 3461–3473 (2015)

Stappers, B.W., Archibald, A.M., Hessels, J.W.T., et al.: A state change in the missing link binary pulsar system PSR J1023+0038. Astrophys. J. **790**, 39 (2014)

Stella, L., Campana, S., Colpi, M., Mereghetti, S., Tavani, M.: Do quiescent soft X-ray transients contain millisecond radio pulsars? Astrophys. J. **423**, L47–L50 (1994)

Stephen, J.B., Bassani, L., Landi, R., Malizia, A., Sguera, V., Bazzano, A., Masetti, N.: Using the ROSAT catalogue to find counterparts for unidentified objects in the first Fermi/LAT catalogue. Mon. Not. R. Astron. Soc. **408**, 422–429 (2010)

Strohmayer, T., Keek, L.: IGR J17062–6143 is an accreting millisecond X-ray pulsar. Astrophys. J. **836**, L23 (2017)

Strohmayer, T.E., Zhang, W., Swank, J.H., Smale, A., Titarchuk, L., Day, C., Lee, U.: Millisecond X-ray variability from an accreting neutron star system. Astrophys. J. **469**, L9 (1996)

Szkody, P., Fraser, O., Silvestri, N., et al.: Cataclysmic variables from the Sloan Digital Sky Survey. II. The second year. Astron. J. **126**, 1499–1514 (2003)

Takata, J., Li, K.L., Leung, G.C.K., et al.: Multi-wavelength emissions from the millisecond pulsar binary PSR J1023+0038 during an accretion active state. Astrophys. J. **785**, 131 (2014)

Tendulkar, S.P., Yang, C., An, H., et al.: NuSTAR observations of the state transition of millisecond pulsar binary PSR J1023+0038. Astrophys. J. **791**, 77 (2014)

Thorstensen, J.R., Armstrong, E.: Is first J102347.6+003841 really a cataclysmic binary? Astron. J. **130**, 759–766 (2005)

Torres, D.F., Ji, L., Li, J., Papitto, A., Rea, N., de Oña Wilhelmi, E., Zhang, S.: A search for transitions between states in redbacks and black widows using seven years of Fermi-LAT observations. Astrophys. J. **836**, 68-P (2017)

Wang, Y.-M.: Disc accretion by magnetized neutron stars - a reassessment of the torque. Astron. Astrophys. **183**, 257–264 (1987)

Wang, Z., Archibald, A.M., Thorstensen, J.R., Kaspi, V.M., Lorimer, D.R., Stairs, I., Ransom, S.M.: SDSS J102347.6+003841: A millisecond radio pulsar binary that had a hot disk during 2000–2001. Astrophys. J. **703**, 2017–2023 (2009)

Wang, Z., Breton, R.P., Heinke, C.O., Deloye, C.J., Zhong, J.: Multiband studies of the optical periodic modulation in the X-ray binary SAX J1808.4–3658 during its quiescence and 2008 outburst. Astrophys. J. **765**, 151 (2013)

Wang, L., Steeghs, D., Casares, J., et al.: System mass constraints for the accreting millisecond pulsar XTE J1814–338 using Bowen fluorescence. Mon. Not. R. Astron. Soc. **466**, 2261–2271 (2017)

Wijnands, R., van der Klis, M.: A millisecond pulsar in an X-ray binary system. Nature **394**, 344–346 (1998)

Chapter 5
Testing the Equation of State with Electromagnetic Observations

Nathalie Degenaar and Valery F. Suleimanov

Abstract Neutron stars are the densest, directly observable stellar objects in the universe and serve as unique astrophysical laboratories to study the behavior of matter under extreme physical conditions. This book chapter is devoted to describing how electromagnetic observations, particularly at X-ray, optical and radio wavelengths, can be used to measure the mass and radius of neutron stars and how this leads to constraints on the equation of state of ultra-dense matter. Having accurate theoretical models to describe the astrophysical data is essential in this effort. We will review different methods to constrain neutron star masses and radii, discuss the main observational results and theoretical developments achieved over the past decade, and provide an outlook of how further progress can be made with new and upcoming ground-based and space-based observatories.

5.1 Introduction to This Book Chapter

It is probably not an exaggeration to state that there are more reviews and book chapters written on the equation of state of neutron stars than there are actual constraining measurements (e.g. Lattimer 2012; Lattimer and Prakash 2016; Özel 2013; Özel and Freire 2016; Watts et al. 2015, 2016; Bogdanov 2016; Miller and Lamb 2016, to name a few in just the past 5 years). This is not surprising, however,

N. Degenaar (✉)
Anton Pannekoek Institute, University of Amsterdam, Amsterdam, The Netherlands
e-mail: degenaar@uva.nl

V. F. Suleimanov
Institut für Astronomie und Astrophysik, Kepler Center for Astro and Particle Physics, Universität Tübingen, Tübingen, Germany

Kazan (Volga region) Federal University, Kazan, Russia
e-mail: suleimanov@astro.uni-tuebingen.de

© Springer Nature Switzerland AG 2018
L. Rezzolla et al. (eds.), *The Physics and Astrophysics of Neutron Stars*,
Astrophysics and Space Science Library 457,
https://doi.org/10.1007/978-3-319-97616-7_5

as it merely underlines the immense interest in understanding the physics of ultra-dense matter that is not encountered on Earth. Such constraints can be obtained through electromagnetic observations of neutron stars, and this is therefore one of the prime pursuits in modern astrophysics.

Everything around us is constructed of atoms, which themselves consist of electrons and nucleons (i.e. protons and neutrons). This familiar structure of matter is, however, disrupted when matter is compressed to densities that reach beyond the *nuclear saturation density* of $\rho_0 \sim 2.8 \times 10^{14}\,\mathrm{g\,cm^{-3}}$. One of the main open questions in modern physics is how matter behaves at supra-nuclear densities, where particle interactions are governed by the strong force. For instance, does matter remain nucleonic or does it rather transition into more exotic forms of particles? What are the superfluid properties of matter compressed beyond the nuclear density? The detailed microphysics and particle interactions of dense matter result in a specific relation between the pressure and density of the bulk matter, which is called the *equation of state* (EOS) . Elucidating the behavior of matter at high densities thus implies understanding the dense-matter EOS (e.g. Lattimer 2012).

Mathematically, it is inherently difficult to describe multiple-particle interactions at high densities and therefore there are no unique theoretical predictions for the dense-matter EOS. Fortunately, the behavior of matter at supra-nuclear densities can be probed through different kinds of experiments. In Earth-based laboratories, particle acceleration experiments are conducted to probe matter near the nuclear saturation density at very high temperatures. However, probing the properties of matter beyond the nuclear density at (relatively) low temperatures solely relies on astrophysical observations of neutron stars.

Neutron stars are the remnants of once massive stars that ended their life in a supernova explosion. A defining property of neutron stars is that these objects are very compact; while being roughly a factor of 1.5 more massive than our Sun, their radius is almost a factor of $\sim 10^5$ smaller ($\sim 10\,\mathrm{km}$; Baade and Zwicky 1934; Wheeler 1966; Hewish et al. 1968). This extreme compactness implies that the density of neutron stars is incredibly high and must, in fact, reach beyond the nuclear density. Being the densest, directly observable stellar objects in our universe, neutron stars thus serve as natural laboratories to study the behavior of matter at supra-nuclear densities. Neutron stars come in different classes, which provides various angles to test and constrain the dense-matter EOS (Sect. 5.2.1).

Here, we aim to provide a student-level introduction into the many different ways in which we can obtain constraints on the neutron star EOS using observations of their electromagnetic radiation. This book chapter is organized as follows: after introducing some general concepts and background context in Sect. 5.2, we review the various ways in which the dense-matter EOS of neutron stars can be constrained from electromagnetic observations in Sect. 5.3, including a discussion on the uncertainties and systematic biases that the different techniques are subject to. In Sect. 5.4 we lay out what progress can be made in the future using new and upcoming facilities. A summary is provided in Sect. 5.5.

5.2 Basic Concepts: Neutron Stars and the Dense Matter Equation of State

5.2.1 The Plethora of Observable Neutron Stars

As is clear from the diverse content of this book, there are several different observational manifestations of neutron stars (e.g. Kaspi 2010, for a review). At the very basic level, we can distinguish neutron stars that are isolated and those that are part of a binary star system. Furthermore, neutron stars are often characterized by their magnetic field strength (B) and rotation period (P_s), which is often connected to their age and their environment.

Many neutron stars are located in a binary where they are accompanied by a main sequence or evolved star with a mass $M_c \lesssim 1\,M_\odot$. In such a configuration, neutron stars can manifest themselves as a *low-mass X-ray binary* (LMXB) , when accreting gas from their companion star, or as a non-accreting radio pulsar. In the former class of objects, the energy emitted by the neutron star (at X-ray wavelengths) is powered by the accretion process, whereas in the latter the rotational energy of the pulsar is tapped (producing mostly radio emission). Both types of neutron stars are typically spinning rapidly, at millisecond periods, and are assumed to be spun up to such high speeds by gaining angular momentum via accretion (e.g. Cook et al. 1994; Strohmayer et al. 1996; Wijnands and van der Klis 1998; Burderi et al. 1999; Archibald et al. 2009; Papitto et al. 2013a). These two classes of binary neutron stars are thus thought to be evolutionary linked, with millisecond radio pulsars descending from LMXBs (Alpar et al. 1982; Bhattacharya and van den Heuvel 1991). The magnetic field of these neutron stars is typically low ($B \lesssim 10^9\,G$), and is thought to have degraded by accretion (e.g. Radhakrishnan and Srinivasan 1982; Romani 1990; Bhattacharya and Srinivasan 1995; Cumming et al. 2001). Nevertheless, some neutron stars in LMXBs display coherent X-ray pulsations, which indicates that their magnetic field is strong enough to channel the accreted gas towards the magnetic poles (Wijnands and van der Klis 1998). This sub-class of pulsating LMXBs, of which 19 are known at the time of writing, are referred to as *accreting millisecond X-ray pulsars* (AMXPs; see Patruno et al. 2017, for a list).

Whereas accretion in LMXBs typically proceeds via an accretion disk that is fed by the Roche-lobe overflowing donor star, a modest number of neutron stars is known to accrete from the stellar wind of their low-mass companion star (Symbiotic X-ray binaries, SyXRBs; e.g. Chakrabarty and Roche 1997; Masetti et al. 2006; Lü et al. 2012). Neutron stars are also found to accrete from much more massive companions ($M_c \gtrsim 10\,M_\odot$), and are then referred to as high-mass X-ray binaries (HMXBs). Depending on the type of the companion and the particular process of mass transfer, neutron star HMXBs can be further divided into different sub-classes (e.g. Reig 2011, for a review). A small number of intermediate-mass X-ray binaries (IMXBs, companion mass $M_c \sim 1$–$10\,M_\odot$) are also known, but these are thought to be short-lived and quickly evolve into LMXBs (e.g. van den Heuvel 1975; Podsiadlowski et al. 2002; Pfahl et al. 2003; Tauris and van den Heuvel 2006).

The neutron stars in SyXRBs, HMXBs and IMXBs are typically spinning much more slowly (seconds) than those in LMXBs and also have stronger magnetic fields ($B \sim 10^{11}$–10^{13} G). A few slowly spinning radio pulsars with massive companion stars are known; these apparently have not been spun up yet by accretion (possibly due to their highly eccentric orbits) and might be the progenitors of HXMBs or IMXBs (e.g. Johnston et al. 1992; Kaspi et al. 1994; Stairs et al. 2001).

Isolated neutron stars also come in various flavors (e.g. Kaspi and Kramer 2016, for a recent review). For instance, one can distinguish young slowly spinning radio pulsars (including Rotating RAdio Transients, RRATs) and old millisecond radio pulsars (which likely originate from binaries but ablated their companion star). Furthermore, there are young and X-ray emitting (sometimes pulsating) neutron stars in the centers of supernova remnants (central compact objects, CCOs), nearby Dim Isolated Neutron Stars (DINS), and magnetars (originally further classified as Anomalous X-ray Pulsars, AXPs, and Soft Gamma Repeaters, SGRs). Several of these different classes of isolated neutron stars may be linked, possibly representing different stages in an overall thermal and magnetic field evolution (e.g. Faucher-Giguère and Kaspi 2006; Viganò et al. 2013; Mereghetti et al. 2015).

5.2.2   The Dense-Matter EOS and the Connection with Neutron Stars

The macroscopic properties of neutron stars, such as their maximum mass and corresponding radius, are determined by the details of the dense-matter EOS . Due to the complexity of the calculations and the inability to probe high densities via laboratory experiments, numerous EOSs with a wide range of different parameters have been constructed (e.g. Lattimer 2012, see also Fig. 5.1 left).

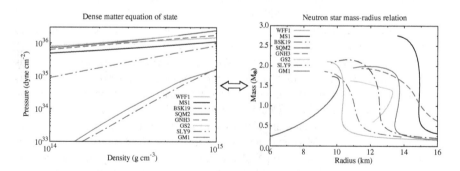

Fig. 5.1 Direct mapping between the EOS (left) and mass-radius relation for neutron stars (right). Shown are illustrative examples of different EOSs. Data for the following EOSs were taken from the compilation of Özel et al. (2016b): WFF1 = Wiringa et al. (1988), MS1 = Müller and Serot (1996), BSK19 = Potekhin et al. (2013), SQM2 = Prakash et al. (1995), GNH3 = Glendenning (1985), and GS2 = Glendenning and Schaffner-Bielich (1999). Added to this are two unified EOSs, SLY9 and GM1, from Fortin et al. (2016)

For each theoretical model describing the microphysics and strong-force inter-actions, i.e. for a given EOS of the bulk matter, the general relativistic structure equations (the Tolman-Oppenheimer-Volkoff equations) can be solved for an assumed central density, leading to a predicted mass (M) and radius (R) for the neutron star. Using a range of different central densities, the $M-R$ relation can be constructed for any given EOS, which is illustrated in Fig. 5.1. This unique mapping between the EOS and the basic properties of neutron stars implies that measuring a wide range of mass-radius pairs, with an accuracy of a few percent, can constrain the properties of ultra-dense matter through revise-engineering (Lindblom 1992). However, it is very challenging to determine both the mass and radius for a sizable sample of neutron stars with high accuracy. Some of the most constraining results have therefore instead come from measuring extrema.

For each EOS there is a maximum central density beyond which no stable configuration is possible, hence every EOS is characterized by a maximum neutron star mass (the Tolman-Oppenheimer-Volkoff limit). This is illustrated by Fig. 5.1, which displays the $M-R$ relation (right) for a selection of different EOSs (left). The EOS is effectively a measure for the compressibility of the matter, which is also referred to as its *softness*: The shallower the rise in pressure with increasing density, the more compressible the matter is, i.e. the softer the EOS. Since for soft EOSs the matter is more compressible, these generally predict neutron stars with smaller maximum masses than harder EOSs.

Looking at Fig. 5.1 (right), we can see that finding a neutron star with a high mass of $M \gtrsim 2\,M_\odot$ can put tight constraints on the EOS by eliminating entire families of soft EOS models. On the other hand, for many EOSs there is a regime in which the radius remains relatively constant over a fairly large range of mass (see Fig. 5.1), which implies that an accurate radius measurement can also provide important constraints. If in addition to the mass or the radius also the *compactness* (i.e. the ratio of mass and radius) can be determined, degeneracies can be broken so that valuable constraints on the EOS can be obtained. The quest of probing the dense-matter EOS through neutron stars can be approached in several different ways, exploiting electromagnetic observations of both isolated neutron stars and neutrons stars that are located in binaries (Sect. 5.3).

5.2.3 General Structure of a Neutron Star

Figure 5.2 shows a schematic drawing of the general structure of a neutron star. The density and pressure rise with increasing depth. The dense, liquid core lies beneath the solid crust, which is covered by a thin ocean/envelop and a very thin atmosphere. When a neutron star accretes, its structure and composition change (see Sect. 5.2.5). Below we describe the structure in a bit more detail.

The *core* makes up the largest part of the neutron star, containing approximately 99% of the total mass, and may be subdivided into an outer and an inner part. The outer core occupies the density range $\rho \sim (1-2)\rho_0$, where matter consists mainly

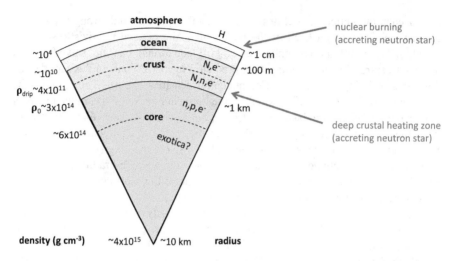

Fig. 5.2 Schematic drawing of the structure of a neutron star (not to scale). Some indicative numbers for the size and density are given, and the main particle constituents are indicated. *H* stands for hydrogen, *N* for nuclei, *n* for neutrons, e^- for electrons and *p* for protons. The dashed lines indicate dividing lines between the inner/outer crust and core. When a neutron star is accreting, nuclear burning (of the accreted gas) occurs on the surface of the neutron star and the bottom of the crust is severely heated due to density-driven fusion reactions. These zones are indicated with red text and arrows

of degenerate neutrons and merely a few percent of protons and electrons. Both protons and neutrons are expected to be superfluid in the outer core. In the inner core of the neutron star, the density may become as high as $\rho \sim (10\text{--}15)\rho_0$, depending on the EOS model. Due to the growing Fermi energies it may become energetically favorable for more exotic particles, rather than the standard composition of p, e^- and n, to occur at these high densities. For instance, neutrons may be replaced by hyperons, electrons may be replaced by pions or kaons and form a (superfluid) Boson-Einstein condensate, and perhaps the density becomes even so high that the attractive force between quarks can be neglected so that the quarks become unconfined. The occurrence of exotica in the core leads to a softer EOS, since it relaxes the Fermi surface and degeneracy pressure.

The *crust* typically covers about one tenth of the neutron star radius and can be subdivided into an inner and an outer part. The outer crust extends from the bottom of the atmosphere to the *neutron drip density*, $\rho_{\text{drip}} \sim 4.3 \times 10^{11}\,\text{g cm}^{-3}$, and matter consists of electrons and ions. Due to the rise in electron Fermi energy, the nuclei suffer inverse β-decay and become more neutron-rich with increasing density. The inner crust covers the region from the neutron drip density, where neutrons start to drip out of the nuclei, to the nuclear density and is composed of electrons, free superfluid neutrons and neutron-rich nuclei. With increasing density the nuclei grow heavier and the number of neutrons residing in the free neutron fluid (rather than in nuclei) increases. Nuclei begin to dissolve and merge together around the crust-core interface.

The *envelop or ocean* refers to a $\sim$100-m thick layer that lies on top of the crust and extends to a density of $\rho \sim 10^{10}\,\mathrm{g\,cm^{-3}}$. If one wants to study the interior temperature of neutron stars, which gives additional information on the EOS (see Sect. 5.3.8), it is essential to understand the properties of this layer: The envelop/ocean couples the sought-after temperature of the (isothermal) interior, T_B, to the observable effective surface temperature, T_{eff} (e.g. Gudmundsson et al. 1982, 1983; Potekhin et al. 1997; Heyl and Hernquist 1998; Geppert et al. 2004), which is then further modified by the atmosphere.

The *atmosphere* covers the neutron star and is very thin ($\sim$ cm). It is expected to be composed of pure H, since heavier elements should sink on very short timescales (e.g. Romani 1987; Bildsten et al. 1992; Özel et al. 2012b). However, for some isolated or quiescent neutron stars, observations suggest atmospheres composed of He or C (see Sects. 5.3.4.1 and 5.3.6.1). Moreover, when a neutron star is actively accreting, the atmosphere composition should be more complex (see Sect. 5.3.3.1). The atmosphere accounts for a negligible fraction of the total mass, but shapes the thermal photon spectrum emerging from the neutron star and is therefore a crucial ingredient in modeling their surface emission, which provides means to constrain their mass and radius (Sect. 5.2.6).

Due to the very steep temperature gradient in the atmosphere of a neutron star, its spectrum looks very different from a black body. Since the atmosphere consists of ions and free electrons, the opacity will be dominated by free-free absorption (absorption of a photon by a free electron in the Coulomb field of an ion) and is therefore proportional to ν^{-3}, where ν is the photon frequency. As a result of this strong dependence and the steep temperature gradient, high-energetic photons escape from deeper atmospheric layers (where the temperature is much higher), than low-energetic photons (e.g. Rajagopal and Romani 1996; Zavlin et al. 1996). For a particular temperature, the spectrum from a neutron star is thus expected to be harder than a pure black body spectrum. One important implication is that fitting black body models to such modified spectra results in an overestimation of the effective temperature and hence the size of the emission region is underestimated (e.g. Rutledge et al. 1999). Other parameters that shape the emerging spectrum are the chemical composition of the atmosphere, the surface magnetic field and the surface gravity (see e.g. Zavlin et al. 1996; Ho et al. 2008; Ho and Heinke 2009; Servillat et al. 2012; Potekhin et al. 2016, and references therein).

5.2.4   The Thermal Evolution of Neutron Stars: Cooling and Re-heating

Neutron stars are born in a supernova explosions, exhibiting very high temperatures of $T_B \sim 10^{12}$ K. In absence of an internal furnace, however, the neutron star will rapidly cool via neutrino emissions from its dense core and photon emissions from its surface (e.g. Yakovlev and Pethick 2004; Page et al. 2006, for reviews). Initially the core cools faster and is thermally decoupled from the hotter crust, but after ~ 1–100 year (depending on the neutron star structure), the crust has thermally relaxed. The crust and core of the neutron star will then be in thermal equilibrium, having a uniform temperature (e.g. Lattimer et al. 1994). In the ocean/envelop and atmosphere, there always continues to be a steep temperature gradient.

There are two factors that can alter the thermal evolution of a neutron star. Firstly, when a neutron star accretes gas from its surroundings, its composition changes and this provides a site for nuclear reactions that release energy and (temporarily) disrupt the thermal balance between the core and the crust (see Sect. 5.2.5). Secondly, in extreme cases the neutron star magnetic field can be a powerful source of energy that can alter the thermal balance. For instance, the decay of their magnetic field and fracturing of the crust by magnetic stresses can release significant amounts of energy that heat the outer layers of the neutron star (e.g. Arras et al. 2004; Pons et al. 2007, 2009; Aguilera et al. 2008; Cooper and Kaplan 2010). These processes are effective only for very strong magnetic fields ($B \gtrsim 10^{14}$ G), i.e. only relevant for magnetars. Indeed, the temperature of several magnetars have been observed to change in response to periods of magnetic activity (see Sect. 5.3.8).

5.2.5 Accretion: Effect on the Composition and Thermal Structure of Neutron Stars

When located in a binary, a neutron star may be able to accrete matter from a companion star. The transferred matter typically consists of H (or He), but as this gas accumulates on the surface of the neutron star, thermonuclear nuclear burning will transform it into heavier elements (e.g. Schatz et al. 1999). Under the weight of freshly accreted material, the burning ashes get pushed into the neutron star and eventually the original crust is fully replaced by an accreted one (e.g. Sato 1979; Haensel and Zdunik 1990). The temperatures involved are typically much lower ($T_B \sim 10^7$–10^8 K) than in a newborn neutron star ($T_B \sim 10^{12}$ K), and do not allow to overcome the nucleon Coulomb-barrier so thermonuclear fusion reactions cannot take place. As a result, an accreted crust is constructed of smaller nuclei and a larger number of free neutrons than a non-accreted crust (e.g. Haensel and Zdunik 1990). Due to the larger number of free neutrons, at a given density (at $\rho > \rho_{\mathrm{drip}} = 4.3 \times 10^{11}$ g cm^{-3}) the pressure in an accreted crust will be higher than for a crust made of cold catalyzed matter. This implies that the EOS of the inner crust will be harder when a neutron star has experienced accretion, i.e. the crust will be slightly thicker (e.g. Zdunik et al. 2017).

A non-equilibrium crust provides a site for nuclear reactions and hence a potential source of energy. An accreted matter element passes through a chain of non-equilibrium nuclear reactions as it is pushed deeper within the neutron star crust. This results in the release of a considerable amount of heat energy (~ 2 MeV per accreted baryon). Most of this energy is released in the pycnonuclear reaction chains that occur deep in the inner crust at densities of $\rho \sim 10^{12}$–10^{13} g cm^{-3} (see Fig. 5.2). The total energy deposited in the crustal reactions is rather similar for different initial compositions of the burning ashes (Haensel and Zdunik 2003).

The nuclear reactions induced in the crust due to accretion can reheat a neutron star and make it much hotter than an isolated neutron star of the same age. Indeed, Brown et al. (1998) showed that deep crustal heating can efficiently maintain the core of an accreting neutron star at a temperature $T_B \sim 5 \times 10^7$–10^8 K. Thermal surface radiation from accreting neutron stars is normally overwhelmed by X-ray emission from the hot accretion disk (except during X-ray bursts), but *transient* LMXBs exhibit periods of quiescence during which little or no accretion takes place and the thermal glow of the neutron star can be observed (see Sect. 5.3.4; e.g. Brown et al. 1998; Campana et al. 1998; Wijnands et al. 2013, 2017).

5.2.6 Surface Emission from Neutron Stars

Several techniques that are employed to constrain the neutron star EOS rely on detecting radiation directly from the stellar surface. The thermal radiation emitted by a stellar body depends on its temperature and radius. Therefore, if thermal radiation can be detected from (part of) the surface of a neutron star, this provides means to measure its radius. However, as gravitational effects come into play, the mass of the neutron star also enters the equations. Whereas in principle this complicates a simple measurement of the stellar radius, there may be ways to break degeneracies so that in fact both the mass and the radius can be determined.

The surface of a neutron star may not necessarily be emitting isotropically. Under certain circumstances, *hotspots* may be observable: regions that are hotter than the bulk of the surface and are offset from the rotational pole of the neutron star. As the hotter region rotates around the star, the observer will see its emission modulated (i.e. pulsed) at or near the stellar spin frequency. As we will see in this book chapter, both isotropic surface emission and hotspot emission can be employed to put constraints on the neutron star radius (and mass).

Typically, neutron stars have a temperature of $T_B \sim 10^5$ K (for radio pulsars) to $T_B \sim 10^8$ K (for LMXBs), which implies that the thermal surface radiation is emitted in the X-ray, (extreme) UV, or optical band. It is not obvious, however, that surface emission from a neutron star is detectable. In isolated neutron stars, for instance, the surface emission can be overpowered by non-thermal magnetospheric processes and in an X-ray binaries the gravitational energy released in the accretion process usually completely overwhelms the surface radiation from the neutron star. Fortunately there are situations in which thermal emission from the neutron star surface is directly observable. Firstly, during thermonuclear X-ray bursts, the neutron star surface briefly becomes brighter than the accretion flow due to the enormous amount of thermal energy liberated (Sect. 5.3.3). Secondly, many neutron star LMXBs are transient and the neutron star surface may be visible during their quiescent episodes (Sect. 5.3.4). Thirdly, in AMXPs the accretion flow is concentrated onto the magnetic poles, which creates observable hotspots (Sect. 5.3.5). Finally, thermal surface emission has also been detected for a number of (nearby) radio pulsars (Sect. 5.3.5), isolated neutron stars (Sect. 5.3.6), and magnetars (Sect. 5.3.8.8).

If the distance, D, is well known, the observed thermal flux density of the neutron star, F_ν, can be fitted to a model flux density $\mathcal{F}_{\nu,\infty}$:

$$F_\nu = \mathcal{F}_{\nu,\infty} K = \mathcal{F}_{\nu,\infty} \frac{R_\infty^2}{D^2} = \frac{\mathcal{F}_{\nu(1+z)}}{1+z} \frac{R^2}{D^2}, \qquad (5.1)$$

where $K = (R_\infty/D)^2$ is a normalization factor. Due to their extreme compactness, neutron stars gravitationally bend their own surface emission (e.g. Pechenick et al. 1983; Psaltis et al. 2000). Moreover, the emission is gravitationally lensed. The combined effect causes the observed angular size R_∞ to be related to the true physical radius of the neutron star as $R_\infty = R(1 + z)$, where z is the gravitational redshift defined as:

$$1 + z = (1 - 2GM/c^2 R)^{-1/2}, \qquad (5.2)$$

with G the gravitational constant and c the speed of light. The modelled flux density is thus related to the intrinsic flux density of the neutron star, $\mathcal{F}_{\nu(1+z)}$, as:

$$\mathcal{F}_{\nu,\infty} = \frac{\mathcal{F}_{\nu(1+z)}}{(1+z)^3}. \qquad (5.3)$$

Since z is significant for neutron stars, the stellar radius cannot be determined independently of its mass such that the outcome of the measurement is actually a strip along the mass-radius plane. Formally, further corrections need to be applied to account for the fast spin rates of neutron stars. For instance, due to Doppler boosting the approaching side of the neutron star will be brighter than the receding side. This asymmetry is further enhanced by frame dragging, as the space time differs from the simple Schwarzschild metric. Rapid rotation also causes oblateness (i.e. the radius at the equator is larger than at the poles). The combined effect implies that the surface emission is non-uniform (e.g. Miller et al. 1998a; Braje et al. 2000; Muno et al. 2003; Poutanen and Gierliński 2003; Cadeau et al. 2007; Morsink et al. 2007), and that any possible atomic lines from the atmosphere will be broadened (Özel and Psaltis 2003; Bhattacharyya et al. 2006; Chang et al. 2006; Bauböck et al. 2012). Nevertheless, even for the highest spin measured for neutron stars to date, the effects of rapid rotation are smaller than the typical systematic uncertainties involved in the EOS determinations (e.g. Steiner et al. 2013). Spin effects are often not taken into account.

If the neutron star magnetic field is sufficiently strong, the surface temperature may be inhomogeneous and this hampers inferring the true physical radius (Elshamouty et al. 2016a). Moreover, other emission processes (e.g. from magnetospheric processes, shocks from a pulsar wind, or an accretion flow; Campana et al. 1998) may contaminate the observed emission. This complicates inferring physical radii and introduces systematic effects.

Finally, and perhaps most importantly, the thermal radiation from the neutron star surface is shaped by the atmosphere. Using the observed surface emission as a tool to measure neutron star parameters and infer information about its interior properties thus requires a careful consideration of the atmospheric properties. Various classes of neutron stars are subject to different physical circumstances and therefore require their own mapping of the surface emission to the observed spectrum. Atmosphere models have thus been developed to cover a wide range of temperatures, ionization states, magnetic fields strengths, and compositions. This includes models for hydrogen atmospheres of cool, non-magnetic neutron stars ($B \lesssim 10^8$ G and $T_{eff} \sim 10^5$–10^7 K, applicable to quiescent LMXBs; e.g. Zavlin et al. 1996; Heinke et al. 2006; Haakonsen et al. 2012, Sect. 5.3.4.1), cool neutron stars with moderate magnetic field strengths ($B \sim 10^{11}$–10^{13} G, applicable to dim isolated neutron stars; e.g. Shibanov et al. 1992; Ho et al. 2008, Sect. 5.3.6.1), strongly magnetic neutron stars (i.e. magnetars; e.g. Özel 2003; van Adelsberg and Lai 2006, Sect. 5.3.8.8), C and He atmospheres (e.g. Ho and Heinke 2009, Sects. 5.3.4.1 and 5.3.6.1), hot bursting neutron stars with different metallicities (e.g. Suleimanov et al. 2011b, 2012a; Nättilä et al. 2015; Medin et al. 2016, Sect. 5.3.3.1), and weakly accreting neutron stars (Zampieri et al. 1995). Over the years, a large number of various grids of neutron star atmospheres and corresponding model spectra have been computed by many authors. Several of these are implemented into the X-ray spectral fitting package XSPEC, for instance *nsa* and *nsgrav* (Zavlin et al. 1996), *nsatmos* (Heinke et al. 2006), *nsx* (Ho and Heinke 2009), *carbatm* and *hatm* (Suleimanov et al. 2014, 2017b).

5.3 EOS Constraints from Electromagnetic Observations of Neutron Stars

As explained in Sect. 5.2.2, the observable properties that directly constrain the EOS are the neutron star mass and radius. Significant constraints on the EOS at supra-nuclear densities could be realized when both the mass and radius of a single neutron star are deduced from observations, but this is very challenging in practice. In the next Sections we review the many different approaches that can be taken to constrain the neutron star EOS using radio, optical and X-ray observations (Sects. 5.3.1–5.3.8), and how various methods can be combined for increased accuracy and checks for systematic biases (Sect. 5.3.9).

There are several radio pulsars in binaries for which very accurate mass estimates are available from measuring gravitational effects through radio pulsar timing (Sect. 5.3.1), combined with dynamical information on the donor star from optical observations (Sect. 5.3.2). However, so far there exists no accurate radius measurement for a radio pulsars that has its mass measured with high accuracy. Measuring neutron star radii relies on measuring radiation that comes directly from (part of) the neutron star surface (see Sect. 5.2.6). This can be achieved with observations of accreting neutron stars through modeling thermonuclear explosions that occur on their surface (Sect. 5.3.3) and studying their thermal glow during quiescent episodes (Sect. 5.3.4), or by modeling the pulse profiles of hotspots on the stellar surface (Sect. 5.3.5). Such hotspots may also occur on non-accreting binary radio pulsars, which allows to apply similar techniques to measure their radii. Furthermore, dim isolated neutron stars also emit thermal radiation that can be used to constrain their emission radii (Sect. 5.3.6).

Apart from putting direct constraints on the EOS through measuring M and R, a very high spin frequency can also provide stringent constraints on the dense matter EOS (Sect. 5.3.7). We also review a number of other approaches to obtain constraints on the neutron star core properties and EOS that are less developed at present (Sect. 5.3.8), but can perhaps in the future lead to better constraints, especially when combined with other methods (Sect. 5.3.9).

5.3.1 Mass Measurements of Radio Pulsars in Binaries

Radio observations can be used to detect every rotation of a neutron star, a technique that is known as *pulsar timing*. Such measurements provide very accurate constraints not only on the rotation period of the neutron star itself, but also on the orbital parameters of the binary. For dedicated reviews on neutron star masses obtained through pulsar timing techniques, and what advances will be brought by future instrumentation (see also Sect. 5.4.1.1), we refer to Kramer (2008); Watts et al. (2015), Özel and Freire (2016). Here, we briefly summarize the basic concepts and results that are most relevant for constraining the neutron star EOS.

By accurately timing every rotation of a radio pulsar, the binary mass function can be determined from the Keplerian orbital parameters:

$$f_{\rm ns} = \left(\frac{2\pi}{P_{\rm orb}} \right)^2 \frac{(a_{\rm ns} \sin i)^3}{G} = \frac{(M_{\rm c} \sin i)^3}{M_{\rm T}^2}, \tag{5.4}$$

where $P_{\rm orb}$ is the binary orbital period and $a_{\rm ns} \sin i$ the projection of the pulsar semi-major axis on the line of sight. This mass function has three unknown parameters: the angle i between the line of sight and the direction orthogonal to the orbital plane, the mass of the companion star $M_{\rm c}$, and the total mass of the binary $M_{\rm T} = M_{\rm c} + M$. The neutron star mass M can therefore not be obtained from timing its pulsations alone. However, in some occasions the projected semi-major axis of the companion's orbit ($a_{\rm c} \sin i$) can be measured: via radio pulsar timing in case of double pulsar binaries, or via phase-resolved optical spectroscopy if the companion is a white dwarf or main-sequence star. This provides a measurement of the mass ratio q of the two binary components:

$$q = \frac{M}{M_c} = \frac{(a_c \sin i)^3}{(a_{ns} \sin i)^3}. \tag{5.5}$$

Nevertheless, further information is required still to pinpoint the mass of the neutron star. For instance, some binaries that are viewed nearly edge-on allow to constrain the inclination from eclipses (where one of the binary components is periodically obscured by the other). Furthermore, optical studies can occasionally yield an independent measurement of the mass of the companion star (see Sect. 5.3.2). Finally, for some radio pulsars relativistic effects are measurable that allow to close the set of equations and obtain the masses of the binary components (e.g. Shao et al. 2015; Watts et al. 2015; Özel and Freire 2016, for details). For instance, for very eccentric binaries one can observe changes in the argument of periapsis, while for very compact binaries the orbital decay due to gravitational wave radiation can be measurable. In exceptional cases, when the binary is viewed nearly edge-on, it has been possible to measure a Shapiro delay arising from the pulses traveling through the gravitational potential of the companion star. If these various pieces of information are available, the masses of radio pulsars can be measured to very high precision (e.g. Thorsett et al. 1993; Thorsett and Chakrabarty 1999; Stairs 2004; Kiziltan et al. 2013).

Accurate mass measurements have so far been obtained for $\sim$40 radio pulsars and span a range of $M \sim 1.2$–$2.0\,M_\odot$ (see Özel and Freire 2016, for a recent overview). Assembling a firm statistical sample of mass measurements is of high interest for a variety of astrophysics questions, e.g. the dynamics of mass transfer and its role in binary evolution, the detailed physics of supernova explosions, and the birth-mass distribution of neutron stars (e.g. Özel et al. 2012b; Kiziltan et al. 2013). However, without radius determinations, these accurate mass determinations generally do not put any stringent constraints on the neutron star EOS. It is only the most extreme mass measurements that can provide interesting constraints by ruling out EOSs that predict lower maximum masses (see Sect. 5.2.2).

There are currently two radio pulsars with reliable mass measurements of $M \sim 2\,M_\odot$ that have allowed to put some interesting constraints on the neutron star EOS. The first is the millisecond radio pulsar PSR J1614$-$2230, for which the Shapiro delay can be measured and has led to a mass determination of $M = 1.928 \pm 0.017\,M_\odot$ (Demorest et al. 2010; Fonseca et al. 2016). The second is the millisecond radio pulsar PSR J0348+0432 for which the combination of radio pulse timing and optical studies of its white dwarf companion yielded a reliable mass estimate for both binary components, with $M = 2.01 \pm 0.03\,M_\odot$ for the neutron star (Antoniadis et al. 2013). A single extreme measurement can already provide very strong constraints on the neutron star EOS, primarily by constraining the nucleon interactions (Hebeler et al. 2013). However, the current extremes of $M \sim 2\,M_\odot$ do not necessarily rule out the occurrence of any exotic particles the core (e.g. Oertel et al. 2015; Fortin et al. 2016, 2017).

The above approach of measuring neutron star masses require the radio pulsars to be located in binaries and hence cannot be applied to isolated radio pulsars. However, some young (isolated) pulsars display sudden and temporary increases in the neutron star's spin. These *glitches* can be used to put some constraints on the superfluid properties and the masses of these neutron stars (see Sect. 5.3.8.1).

5.3.2 Optical Dynamical Mass Measurements of Neutron Stars in Binaries

If a neutron star is in a binary and periodic changes in the line-of-sight velocity of the companion star (K_2) can be measured from its optical emission, then the mass function $f(M, M_c)$ can be constructed from Newton's laws of Keplerian motion, which provides a lower limit on the mass of the neutron star:

$$M \geq f(M, M_c) \equiv \frac{M^3 \sin^3 i}{(M_T)^2} = \frac{K_2^3 P_{orb}}{2\pi G} \tag{5.6}$$

where $M_T = M_c + M$ is again the total binary mass. Furthermore, i is the inclination angle at which we view the binary orbit (with $i = 0°$ implying face-on and $i = 90°$ edge-on) and P_{orb} is the orbital period. When the orbital period can be measured along with the companion star's velocity variations, the mass function provides a lower limit on the neutron star mass; $M_c = 0$ and $i = 90°$ would yield $M = f$, but since $M_c > 0$, we have $M > f$ for any inclination angle.

Additional constraints from other types of measurements can turn this lower limit into an actual mass measurement. So far, this has been most successful in combination with accurate timing of radio pulsars (Sect. 5.3.1). For instance, for PSR J1614−2230 the Shapiro delay yielded i and M_c and hence allowed for measuring the pulsar mass of $M = 1.928 \pm 0.017$ M$_\odot$ via Eq. (5.6) since K_2 (from optical observations of the companion) and P_{orb} (from pulsar timing) can also be determined. Another way to remove degeneracies is through the detection of atmospheric lines of the companion star, which allowed for the accurate and constraining $M = 2.01 \pm 0.03$ M$_\odot$ mass measurement of PSR J0348+0432 (Antoniadis et al. 2013). The atmospheric lines detected from its white dwarf companion provide a second mass function, through the periodic variations in the central wavelength of these lines, and a mass measurement of the white dwarf via the gravitational redshift of the lines, hence allowing a determination of the mass of the pulsar.

Some binary radio pulsars are eclipsed, so-called *black widow pulsars*, which allows for a constraint on the inclination and can thus lead to a measurement of the neutron star. Some of these have similar or even higher mass estimates as PSR J1614−2230 and PSR J0348+0432, albeit with larger systematic uncertainties. For instance, for the black widow pulsar PSR B1757+20 a mass of $M \sim 2.40 \pm 0.12$ M$_\odot$

was estimated, but the binary inclination is highly uncertain and in fact allows for a mass as low as $M \sim 1.66\,M_\odot$ (van Kerkwijk et al. 2011). Furthermore, light curve fitting yielded an estimated mass of $M \sim 2.68 \pm 0.14\,M_\odot$ for the black widow PSR J1311−3430 (Romani et al. 2012), but the poor quality of the fit suggests the presence of un-modelled obscuration or emission; attempts to account for that result in a lower pulsar mass, down to $M \sim 1.8\,M_\odot$ (Romani et al. 2015). Whereas systematic uncertainties currently preclude from drawing firm conclusions, these methods can potentially be refined (e.g. Romani et al. 2015). Difficulties currently lie in the fact that the companions are oblated and even when irradiation is taken into account in the light curve modelling there is residual short time scale variability. Moreover, the surface is unevenly heated, and there is asymmetry in the light curves. All this significantly hinders a reliable determination of the binary inclination, which thus directly translates into uncertainties in the neutron star mass. Nevertheless the high mass estimates for these black widow pulsars are tantalising and foster the idea that neutron stars with $M > 2\,M_\odot$ may exist and will be found one day. If such high masses were to be confirmed that would firmly rule out several classes of EOSs. Note that in these studies timing of the radio pulses provides the binary parameters, but the mass measurements are obtained from modeling the companion's optical light curves (to constrain the orbital inclination) and optical spectroscopy (to measure the mass ratio).

Similar studies can be performed for X-ray binaries. So far, neutron star masses have been obtained from optical observations (combined with other techniques) for $\sim$10 HMXBs and $\sim$7 LMXBs. Together these span a range of $M \sim 1.1$–$1.9\,M_\odot$, but these measurements are much less accurate than those of millisecond radio pulsars (see Özel and Freire 2016, for an overview). The additional constraints required to obtain mass measurements for these objects can be obtained in different ways. For instance, eclipses are also observed for some high-inclination X-ray binaries, causing the X-ray emission near the neutron star to be periodically blocked by the donor star, in which case i can be accurately determined from the duration of X-ray eclipses (Horne 1985). Furthermore, X-ray pulse timing can directly constrain the orbit of the neutron star and hence its radial velocity semi-amplitude K_1. If the radial velocity semi-amplitude of the donor star (K_2) can also be measured from optical (or near-infrared) observations in quiescence, the mass ratio $q = M_c/M$ is obtained. Alternatively, this ratio can be determined from the broadening of absorption lines ($v \sin i$) from the companion star (e.g. Horne et al. 1986). A promising source to obtain this combined information is the only eclipsing millisecond X-ray pulsar, Swift J1749.4−2807, but crowding has so far prevented to identify the quiescent near-infrared counterpart (Jonker et al. 2013). In the future, X-ray polarization may be employed to constrain the inclination of some neutron star X-ray binaries via studies of their hotspot emission (see Sect. 5.4.3.5).

5.3.3 Radius Constraints from Thermonuclear X-ray Bursts

The X-ray luminosities of neutron star LMXBs in outburst vary in a wide range of $L_X \sim 10^{35}$–$10^{38}\,\mathrm{erg\,s^{-1}}$. This suggests that $\dot{M} = L_{acc}R/GM \sim 10^{15}$–$10^{18}\,\mathrm{g\,s^{-1}}$ of mass is accreted (where $L_{acc} \sim 2 \times L_X$ is the bolometric accretion flux; e.g. in 't Zand et al. 2007). This gas accumulates on the neutron star surface where it undergoes thermonuclear burning. Under certain conditions, mainly determined by the local mass-accretion rate and thermal properties of the ocean/envelop, the nuclear burning is unstable and results in runaway energy production that is observable as a thermonuclear X-ray burst (also called Type-I X-ray bursts; shortly X-ray bursts from here on). Currently, $\sim$110 X-ray bursting neutron star LMXBs are known. The majority of X-ray bursts have a duration of ten to hundreds of seconds and their spectra are generally well fitted with a black body model (Galloway et al. 2008), of which the temperature T_{BB} and normalization K_{BB} evolve in a characteristic way (see Fig. 5.3). We refer to Lewin et al. (1993) and Strohmayer and Bildsten (2006) for extensive reviews on X-ray bursts. Here, we focus on exploiting these events to constrain the neutron star properties.

Fig. 5.3 Results of fitting of a PRE burst of the LMXB SAX J1810.08−2609 with a black-body model, showing the evolution of the black-body flux (top), temperature (middle) and normalization (bottom) along the X-ray burst. The touchdown point where the expanding atmosphere rejoins the neutron star surface, is thought to occur at $t = 0$ (marked with the vertical dashed line). The part of the X-ray burst where the temperature is gradually and steadily decreasing (i.e. after the touchdown point), is referred to as the cooling tail

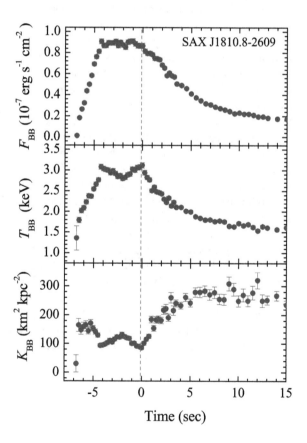

The X-ray bursting neutron stars are attractive candidates to constrain the EOS because during the X-ray bursts the neutron star surface becomes visible (Sect. 5.2.6). In the following, we assume that the stellar surface is uniformly emitting and that there is no other emission process that can alter the observed spectral flux distribution. It may seem that this last assumption cannot be fulfilled since the accretion flow is a prominent source of radiation, but we show below that for several X-ray bursting neutron stars the accretion rates are low enough to ignore its contribution to the burst spectrum.

There are two additional advantages that make X-ray bursts very suitable for constraining the neutron star parameters. Firstly, some X-ray bursts are so powerful that the radiation pressure causes the outer layers of the neutron star to expand, which is referred to as *photospheric radius expansion* (PRE). This implies that in these layers the radiation pressure force $g_{rad} = c^{-1} \int_0^\infty \kappa_\nu \mathcal{F}_\nu \, d\nu$ is larger than the local gravity $g = GM(1 + z)/R^2$ and that the luminosity exceeds the Eddington limit during the PRE burst phases. For every specific neutron star photosphere the value of this critical luminosity is unique, owing to the specific chemical composition and the difference in the opacities κ_ν. Usually, the Eddington luminosity is determined using the Thomson electron scattering opacity:

$$\kappa_e \approx 0.2 \, (1 + X) \, \text{cm}^2 \, \text{g}^{-1}, \tag{5.7}$$

where X is the hydrogen mass fraction. The expression for the observed Eddington luminosity is then:

$$L_{\text{Edd},\infty} = \frac{L_{\text{Edd}}}{(1+z)^2} = \frac{4\pi G M c}{\kappa_e}(1+z)^{-1}, \tag{5.8}$$

which corresponds to an observed bolometric Eddington flux of:

$$F_{\text{Edd}} = \frac{L_{\text{Edd},\infty}}{4\pi D^2}. \tag{5.9}$$

The Eddington flux is linked to the observable critical effective temperature $T_{\text{Edd},\infty}$, via:

$$F_{\text{Edd}} = \sigma_{\text{SB}} T^4_{\text{Edd},\infty}, \tag{5.10}$$

where σ_{SB} is the Stefan Boltzmann constant. We note that the parameters defined by Eqs. (5.8)–(5.10) are the prerequisite for measuring the neutron star radius and mass. Unfortunately, however, it is not possible to determine exactly at which burst phase the observed flux equals the Eddington flux. The flux at the touchdown point, where the black-body temperature T_{BB} reaches a maximum and the black-body normalization has a local minimum (see Fig. 5.3), is generally believed to be close to the Eddington flux. In one approach, the neutron star mass and radius are constrained by attempting to identify the Eddington flux of PRE bursts: this is often referred to as the *touchdown method*.

The spectral evolution of X-ray bursts allows us to obtain an additional constraints on the neutron star mass and radius. Since these parameters cannot change, an acceptable model must be able to describe the X-ray burst spectrum at every phase, from the Eddington flux near to peak down to the much lower fluxes in the X-ray burst tail, with constant values of M and R. Constraining the neutron star parameters via the spectral evolution of the X-ray burst is usually referred to as the *cooling-tail method*.

During X-ray bursts, in principle both the apparent surface area and the Eddington limit of the neutron star can be measured, hence the degeneracy between M and R can be broken using the equations laid out above. Although the approach is very promising, various systematic uncertainties (e.g. varying accretion emission during a burst, the emission radius, the detailed spectral properties, and emission anisotropies) need to be resolved before orthogonal constraints can reliably be obtained with high accuracy (e.g. Steiner et al. 2010; Galloway and Lampe 2012; Zamfir et al. 2012; in 't Zand et al. 2013; Degenaar et al. 2018, see Sect. 5.3.3.4).

5.3.3.1 Model Atmospheres for X-ray Bursts

The black-body fits give important initial information about an X-ray bursting neutron star, but this kind of fitting is not accurate enough to obtain reliable mass and radius measurements. Accurate model atmospheres of X-ray bursting neutron stars have to be computed for this aim to obtain model flux spectral distributions $\mathcal{F}_\nu$ (hereafter spectra). The model atmosphere parameters are the effective temperature T_{eff}, the surface gravity g, and the chemical composition of the atmosphere. The very first models of hot neutron star atmospheres (London et al. 1984, 1986; Lapidus et al. 1986; Ebisuzaki 1987; Madej 1991; Pavlov et al. 1991) demonstrated that Compton scattering is the main source of opacity and that the energy exchange between high energy photons escaping from the deep and hot atmospheric layers with the relatively cold surface electrons establishes the equilibrium spectra. These are close to the diluted black-body spectra (Fig. 5.4). Therefore, there are two parameters describing every model spectrum, the dilution factor w and the color correction factor f_c:

$$\mathcal{F}_\nu \approx w\,\pi\,B_\nu(f_c\,T_{\mathrm{eff}}), \qquad (5.11)$$

where B_ν is the Planck function. The color correction factor links the color temperature T_c, obtained from fitting a spectrum distorted by atmospheric effects to a Planck spectrum, to the effective temperature T_{eff} (i.e. $f_c = T_c/T_{\mathrm{eff}}$). We note that in first approximation, $w \approx f_c^{-1/4}$. However, approximation (5.11) actually doesn't conserve the bolometric flux

$$\int_0^\infty w\pi\,B_\nu(f_c\,T_{\mathrm{eff}})\,d\nu = w f_c^4\,\sigma_{\mathrm{SB}}\,T_{\mathrm{eff}}^4, \qquad (5.12)$$

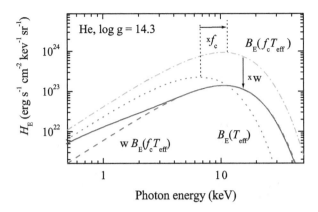

Fig. 5.4 Fits to model spectra (solid curve) with a diluted black body (dashed curve). Black-body spectra with temperatures T_{eff} (dotted curve) and $f_c T_{\text{eff}}$ (dot-dashed curve) are also shown. The model spectrum of a pure helium atmosphere with $\log g = 14.3$ and $\ell = 1.0$ is used. The fitting parameters are $f_c = 1.61$ and $w = 0.148$

and therefore the bolometric flux obtained from the black-body fit has to be multiplied by the bolometric correction $w f_c^4$.

At present, extended grids of hot neutron star atmospheres have been computed using the Kompaneets equation for the Compton scattering description (Suleimanov et al. 2011b), and the fully relativistic angle-depended redistribution function (Suleimanov et al. 2012a). The latest grids have been computed for different chemical compositions: pure hydrogen, pure helium and mixed H/He with a solar H/He ratio with different fractions of solar metallicities (i.e. heavy element abundances). The models are computed for nine surface gravities, with $\log g$ from 13.7 to 14.9 in steps of $\delta \log g = 0.15$. The relative luminosity ℓ is used instead of the effective temperature as a model parameter, $\ell = L/L_{\text{Edd}}$, or $T_{\text{eff}} = \ell^{1/4} T_{\text{Edd}}$. Here $L_{\text{Edd}} = L_{\text{Edd},\infty}(1+z)^2 = 4\pi R^2 gc/\kappa_e$ and $T_{\text{Edd}} = T_{\text{Edd},\infty}(1+z) = (gc/\sigma_{\text{SB}}\kappa_e)^{1/4}$ are the intrinsic Eddington luminosity and the intrinsic critical effective temperature, respectively. For every chemical composition and surface gravity, about 20 models with ℓ varying from 1.1 to 0.1 were computed. For three $\log g$ values (14.0, 14.3, and 14.6) the grids were extended down to $\ell = 0.001$. Some examples of the emergent model spectra are shown in Fig. 5.5.

To provide an anchor point to compare the theoretical behavior of X-ray bursts to actual data, the model atmosphere spectrum can be fitted by a diluted black body. Examples of the dependence on the relative luminosities are shown in Fig. 5.6. The values of the fit parameters depend significantly on the relative luminosity and the chemical composition, whereas the dependence on the surface gravity is less significant. The color correction factor f_c and the dilution factor w both evolve rapidly when the relative luminosity ℓ decreases from 1 to 0.5. Formally super-Eddington model atmospheres can occur because the effective electron scattering cross-section decreases with increasing temperature (Klein-Nishina reduction; see Paczynski 1983; Suleimanov et al. 2012a; Poutanen 2017).

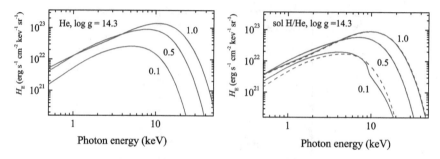

Fig. 5.5 Model spectra of pure helium (left) and solar H/He mix (right) atmospheres for $\log g = 14.3$ and relative luminosities $\ell = 0.1$, 0.5, and 1.0. The right panel shows the model spectra for solar (solid curves) and hundred times less heavy element abundances (dashed curves)

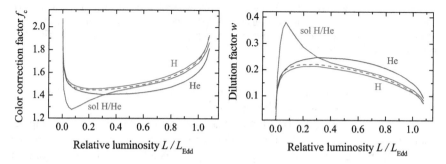

Fig. 5.6 The model dependences of color correction factors f_c (left) and dilution factors w (right), computed for pure hydrogen (red curves), pure helium (blue curves) and solar H/He mix (dark-yellow curves) atmospheres. The model curves for solar H/He mix are presented for two heavy element abundances: solar (solid curves) and reduced solar abundance (dashed curves). All these model curves were computed for $\log g = 14.3$

5.3.3.2 Methods for Obtaining Mass and Radius Measurements from X-ray Bursts

The potential of using X-ray bursts to obtain mass and radius measurements of neutron stars was recognized many years ago (Ebisuzaki 1987; van Paradijs et al. 1990; Damen et al. 1990; Lewin et al. 1993). However, the low quality of the early data and lack of sufficiently extended model atmosphere grids prevented to obtain meaningful results. With the availability of much better data (mainly from *RXTE*), the touchdown method has been applied to six different LMXBs in recent years (Table 5.1; see e.g. Özel 2006; Özel et al. 2009; Güver et al. 2010b). The results obtained via this approach were recently reviewed by Özel and Freire (2016), and yield a combined radius preference of $R = 9.8$–$11.0\,\mathrm{km}$ for a mass of $M = 1.4\,\mathrm{M_\odot}$. This method does not take into account the spectral evolution of X-ray bursts. The cooling-tail method has been laid out by Suleimanov et al. (2011a), Poutanen et al. (2014), Suleimanov et al. (2017c) and has been applied to 5 sources to date. Since

Table 5.1 Sources that have so far yielded the most constraining mass and/or radius limits

Approach	Sources	References
X-ray bursters	SAX J1748.9−2021	Güver and Özel (2013); Özel et al. (2016b)
	EXO 1745−248	Özel et al. (2009), Özel et al. (2016b)
	KS 1731−260	Özel et al. (2012a), Özel et al. (2016b)
	4U 1608−52	Güver et al. (2010a), Özel et al. (2016b), Poutanen et al. (2014)
	4U 1820−30	Güver et al. (2010b), Özel et al. (2016b), Suleimanov et al. (2017a)
	4U 1724−307	Suleimanov et al. (2011b), Nättilä et al. (2016), Özel et al. (2016b)
	4U 1702−429	Nättilä et al. (2016), Nättilä et al. (2017)
	SAX J1810.8−260	Nättilä et al. (2016), Suleimanov et al. (2017c)
	GS 1826−24	Zamfir et al. (2012)
Quiescent LMXBs	NGC 6397 U24	Guillot et al. (2011), Heinke et al. (2014), Özel et al. (2016b)
	47 Tuc X5	Heinke et al. (2003), Heinke et al. (2006), Bogdanov et al. (2016)
	47 Tuc X7	Heinke et al. (2003), Heinke et al. (2006), Bogdanov et al. (2016)
	M28 source 26	Becker et al. (2003), Servillat et al. (2012), Özel et al. (2016b)
	NGC 2808	Webb and Barret (2007), Servillat et al. (2008)
	M13	Gendre et al. (2003a), Webb and Barret (2007), Catuneanu et al. (2013), Özel et al. (2016b)
	ω Cen	Rutledge et al. (2002b), Gendre et al. (2003b), Heinke et al. (2014), Özel et al. (2016b)
	M30	Lugger et al. (2007), Guillot and Rutledge (2014), Özel et al. (2016b)
	NGC 6304	Özel et al. (2016b)
Radio pulsars	PSR J1614−2230	Demorest et al. (2010), Fonseca et al. (2016)
	PSR J0348+0432	Antoniadis et al. (2013)

this method is very promising but has not been described in a book review yet, we here focus on that method.

Before continuing to describe the cooling-tail method and its application in detail, we point out that there is a third approach where observable properties such as the recurrence time, peak flux and decay time of X-ray bursts are compared with theoretical light curve models to infer the surface redshift of the neutron star (e.g. Heger et al. 2007a). However, the only neutron star LMXB (out of ∼110 known X-ray bursters) that shows X-ray bursts with light curves that match the theoretical calculations, is GS 1826−24 (also called "the clocked burster"). Therefore, this method has only been applied to this particular source, yielding $R < 6.8$–11.3 km

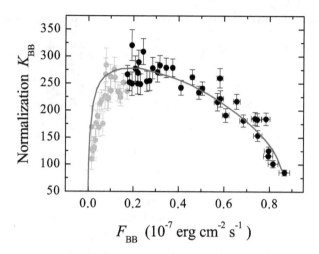

Fig. 5.7 The observed $K_{BB} - F_{BB}$ dependence (circles with error bars) for the cooling tail (after touchdown) of a PRE burst from SAX J1810.8−2609 (see Nättilä et al. 2016; Suleimanov et al. 2017c). The red solid curve is the best-fitting theoretical dependence $w - w f_c^4 \ell$ for solar H/He mix ($Z = 0.01 Z_\odot$) and $\log g = 14.3$, corresponding to the bright data points. The fitting parameters are $\Omega = 1261$ (km/10 kpc)^2 and $F_{Edd} = 0.776 \times 10^{-7}$ erg cm^{-2} s^{-1}

for $M < 1.2$–$1.7\,M_\odot$ (Zamfir et al. 2012). This is consistent with results obtained from the other X-ray burst analysis types as well as those obtained for quiescent neutron star LMXBs (see Sect. 5.3.4.2). For this approach the X-ray bursts do not have to show PRE (i.e. reach Eddington).

In the cooling-tail method, a black-body model is used to fit both the observed spectra and the model atmosphere spectra. This shows that the observed normalization K_{BB} depends only on the dilution factor when the late burst phases (after the touchdown point) are considered:

$$K_{BB} = w \frac{R^2 (1+z)^2}{D^2} = w\,\Omega. \qquad (5.13)$$

This means that all changes in the black-body normalization occurring after the touchdown point, which we can see in Fig. 5.3 (bottom), are connected to changes in the dilution factor. It is more convenient to demonstrate the evolution of the normalization as an observed $K_{BB} - F_{BB}$ dependence (see Fig. 5.7) and approximate it with the model curves $w - w f_c^4 \ell$ with two fit parameters[1]: F_{Edd} and the solid angle occupied by the neutron star on the sky Ω. Both fit parameters depend on the (poorly known) distance to the neutron star but their combination, the apparent

---

[1] Initially the suggested method was using the assumption $w = f_c^{-1/4}$ and the observed curves $K_{BB}^{-1/4} - F_{BB}$ were fitted with the model $f_c - \ell$ ones.

critical surface temperature $T_{Edd,\infty}$, is independent of distance:

$$T_{Edd,\infty} = \left(\frac{F_{Edd}}{\sigma_{SB}\Omega} \right)^{1/4} = \left(\frac{GMc}{\sigma_{SB}\, 0.2(1+X)\, R^2(1+z)^3} \right)^{1/4}. \tag{5.14}$$

The curve of equal $T_{Edd,\infty}$ on the neutron star M–R plane gives the possible values of the neutron star mass and radius, as shown in Fig. 5.8. However, this depends on the surface gravity of the model curves used and the assumed chemical composition. It is clear that for a given neutron star mass, the pure helium models ($X = 0$) will give a larger neutron star radius than pure hydrogen models ($X = 1$). Formally, the curve of constant $T_{Edd,\infty}$ gives the correct M and R at the point were $\log g$ is equal to the value used for the model curve computation. Therefore, we have to assume some chemical composition of the neutron star atmosphere and interpolate the model curves $w - w f_c^4\, \ell$ computed for nine surface gravities for every M–R pair. Then we can fit the observed curve $K_{BB} - F_{BB}$ by this specific model curve and obtain the χ^2 map on the M–R plane, allowing to define confidence regions (68, 90 and 99%). The distance to the source is the only free parameter for a fixed chemical composition, and its value is unique for every M–R pair. In Fig. 5.7 we also illustrate the importance of the chemical composition. The observed curve $K_{BB} - F_{BB}$ was fitted with the model curves $w - w f_c^4\, \ell$ computed for pure

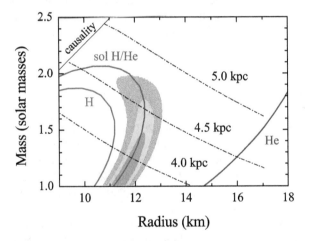

Fig. 5.8 The χ^2 confidence regions (68, 90 and 99% probabilities) in the M–R plane for SAX J1810.8–2609, obtained using the cooling-tail method for solar-mix H/He with reduced heavy element abundance ($Z = 0.01 Z_\odot$; Suleimanov et al. 2017c). The black dashed-dotted curves correspond to distances of 4.0, 4.5 and 5.0 kpc. The solid curves correspond to the best-fitting $T_{Edd,\infty}$ obtained for $\log g = 14.3$ and three chemical compositions: pure hydrogen (red curve), solar H/He mix with the reduced heavy element abundance (dark-yellow curve), and pure helium (blue curve). The fit parameters are $\Omega = 1310\,(\text{km}/10\,\text{kpc})^2$, $F_{Edd} = 0.774 \times 10^{-7}\,\text{erg cm}^{-2}\,\text{s}^{-1}$ for the pure hydrogen models, and $\Omega = 1088\,(\text{km}/10\,\text{kpc})^2$, $F_{Edd} = 0.757 \times 10^{-7}\,\text{erg cm}^{-2}\,\text{s}^{-1}$ for the pure helium models. The fit parameters for the solar H/He mix are the same as in Fig. 5.7

hydrogen and pure helium model atmospheres, and the corresponding curves of constant $T_{\text{Edd},\infty}$ are drawn. The pure hydrogen atmospheres are hard to sustain because of the significant mass accretion rate of bursting neutron stars. Realistically only two possibilities exist: solar H/He mix for normal LMXBs and pure helium atmospheres for ultracompact LMXBs in which the secondary star is a He white dwarf (such as 4U 1820−30). The derived neutron star radii are so different in these two cases that this allows us to distinguish helium-rich and solar-mix atmospheres using the cooling-tail method.

5.3.3.3 Obtained Results from the Cooling-Tail Method

Just like other approaches, the described cooling-tail method is only correct for an isolated, passively cooling neutron star. Of course, the X-ray bursting neutron stars accrete matter and the influence of the accretion flow could muddy with the neutron star mass and radius measurements. To reduce any bias from the presence of the accretion flow, the X-ray bursts for the analysis have to be carefully chosen and ideally should occur at low accretion rate. A detailed discussion of this issue was given by Poutanen et al. (2014); Kajava et al. (2014); Suleimanov et al. (2016). A careful inspection of the X-ray bursts detected by *RXTE* resulted in three LMXBs that show X-ray bursts at low mass-accretion rates: SAX J1810.8−260, 4U 1702−429, and 4U 1724−307. These were therefore selected to apply the cooling-tail method (Nättilä et al. 2016). In addition to the results from the burst analysis itself, limits from the nuclear physics were imposed (Steiner et al. 2015), and two different approximations for the model equation of state were applied (see Nättilä et al. 2016, for details). Moreover, the results from the three sources were combined by assuming that all these neutron stars have the same equation of state and hence lie on the same theoretical $M–R$ curve. This resulted in a predicted radius of $R = 10.5–12.8\,\text{km}$ for $M = 1.4\,M_\odot$, with some dependence on the assumed composition. Interestingly, the cooling-tail analysis of Nättilä et al. (2016) showed that SAX J1810.8−260 and 4U 1724−307 require atmospheres with solar mixed H/He abundances, whereas 4U 1702−429 needs to accrete helium-rich matter. Applying the same method to the UCXB 4U 1820−30 yielded $R = 10–12\,\text{km}$ for $M < 1.7\,M_\odot$ and $R = 8–12\,\text{km}$ for $M = 1.7–2.0\,M_\odot$ (using a pure-He atmosphere; Suleimanov et al. 2017a). The constraints obtained for this source are not very strict because the cooling tail phases that can be described by the theoretical atmosphere models is only very short (about a second). Finally, applying the cooling tail method to 4U 1608−52 yielded a lower limit on the radius of $R > 12\,\text{km}$ for $M = 1.0–2.4\,M_\odot$ and a constraint of $R = 13–16\,\text{km}$ for $M = 1.2–1.6\,M_\odot$ (Poutanen et al. 2014). The rather wide range for this source is mainly caused by the large distance uncertainty.

The most recent development regarding the cooling-tail method is that rather than fitting a diluted black body to the observed X-ray burst spectra and comparing that with similar fits to the model spectra, the model spectra are fitted directly to the observational data (Nättilä et al. 2017). This was applied for 4U 1702−429 and resulted in a preferred radius and mass of $R = 12.4 \pm 0.4\,\text{km}$ and $M = 1.9 \pm 0.3\,M_\odot$

for $5.1 < D < 6.2$ kpc and a hydrogen mass fraction of $X < 0.09$ (confirming previous suggestions of a H-poor atmosphere; Nättilä et al. 2016). An interesting aspect of this application is that it allows to put strong constraints on the atmosphere composition, due to the strong dependence of the spectral evolution on the element abundances. This first application is promising and the method can be further developed (Nättilä et al. 2017).

Comparing results from the cooling-tail and touchdown method suggests that the latter yields systematically lower radii (cf. Özel et al. 2016b). Indeed, both techniques have been applied to 4U 1608−52 (Güver et al. 2010a; Poutanen et al. 2014), 4U 1728−34 (Suleimanov et al. 2011b; Özel et al. 2016b), and 4U 1820−30 (Özel et al. 2016b; Suleimanov et al. 2017a) with inconsistent results. A concern for the X-ray bursts sample used in the touchdown analysis is that the accretion luminosity may be a significant source of contamination (Degenaar et al. 2018), which may be the reason that the spectral evolution of these bursts deviates significantly from the theoretically expected cooling tails (e.g. Poutanen et al. 2014; Kajava et al. 2014). On the other hand, the cooling-tail method received criticism for the poor quality of the black-body fits for some sources (Güver et al. 2012; Özel and Psaltis 2015).

In the past few years, results from X-ray burst studies have also been combined with those obtained from quiescent LMXBs in a Bayesian formalism. We further discuss that in Sect. 5.3.4.2.

5.3.3.4 Biases and Uncertainties in X-ray Burst Studies

Apart from the influence of the accretion flow mentioned in the previous section, there are other systematic uncertainties in the X-ray burst analysis that can bias the results.[2] For instance, the rapid rotation of the neutron star is typically neglected. We know that many neutron stars in LMBXs rotate at frequencies of several hundred Hz (Patruno et al. 2017). The resulting oblateness causes the equatorial radius to be larger than that of a non-rotating neutron star with the same mass (AlGendy and Morsink 2014). As a result the apparent area of the neutron star surface increases so that we obtain a larger neutron star radius. We refer to Bauböck et al. (2015a) for a discussion on this topic. Poutanen et al. (2014) applied the cooling-tail method to the fastest spinning neutron star LMXB 4U 1608−52 (620 Hz) and argued that the obtained radius limit would be $\sim$10% lower if its rapid rotation was taken into account.

Another important factor is the heavy element abundance in the atmospheres of X-ray bursters. Thermonuclear ashes could appear at the neutron star surface during early stages of the burst due to convection, and due to ejection of the surface layers during a strong PRE phase (e.g. Paczynski and Proszynski 1986; Weinberg et al. 2006; in't Zand and Weinberg 2010). Model atmospheres of hot neutron

---

[2]We note that the cooling-tail method is independent of distance: see Eq. (5.14).

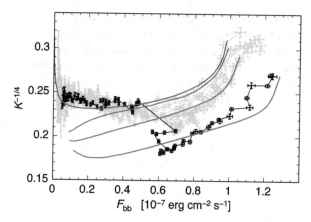

Fig. 5.9 Black-body normalization versus flux for X-ray bursts from HETE J1900.1−2455, together with the model curves $f_c - \ell$. The models were computed for $M = 1.4\,M_\odot$ and $R = 12\,km$ neutron star atmospheres with metal enhancement factors $\zeta = 0.01$ (blue curve), 1 (purple curve), 10 (red curve) and 40 (green curve). The track with the largest deviations is highlighted in black (see Kajava et al. 2017)

stars enriched with heavy elements were computed by Nättilä et al. (2015). This work showed that for increasing heavy element abundances, the color correction factor decreases and the dilution factor increases (see Fig. 5.9). Interestingly, a burst from HETE J1900.1−2455 showed a highly unusual cooling tail (Fig. 5.9). This odd behavior can be explained if an almost pure heavy metal atmosphere just after touchdown was covered by solar H/He mix matter at the later cooling phase, perhaps due to accretion (Kajava et al. 2017). The heavy metal enrichment could be less apparent if the heavy element abundance is lower, e.g. ten solar abundances, but this cannot change the result significantly (cf. the results of Suleimanov et al. (2011a) and Nättilä et al. (2016)).

Important and unsolved questions in X-ray burst analysis are whether the emitting radii measured during a single or during multiple bursts are identical and equal to the neutron star radius (e.g. Güver et al. 2012; Galloway and Lampe 2012), whether the bursts consistency reach the Eddington limit (e.g. Boutloukos et al. 2010; Güver et al. 2012), if the burst emission is by approximation isotropic (e.g. Zamfir et al. 2012), and whether the chemical composition varies (e.g. Bhattacharyya et al. 2010).

5.3.4 Radius Constrains from Thermally-Emitting Quiescent LMXBs

Using sensitive X-ray satellites, thermal surface radiation can also be detected from neutron star LMXBs during their quiescent episodes. If the neutron star is radiating uniformly, one can derive constraints on the mass and radius by fitting the thermal surface radiation (Sect. 5.2.6) with an appropriate neutron star atmosphere model (see Sect. 5.3.4.1). Neutron stars in LMXBs generally have low magnetic fields, and these are not expected to cause surface temperature inhomogeneities. However, even if their magnetic field may not cause temperature anisotropies in itself, when residual accretion takes place even a weak magnetic field may channel the accreted matter onto a fraction of the stellar surface and hence heating it non-uniformly (see Sect. 5.3.4.4). We know that at least in some LMXBs gas is still accreting onto the neutron star in quiescence. Evidence for such low-level accretion comes from irregular X-ray variability on a time scale of hours to days (e.g. Cackett et al. 2010; Degenaar and Wijnands 2012; Bernardini et al. 2013; Wijnands and Degenaar 2013; Coti Zelati et al. 2014). Moreover, the presence of a non-thermal emission component in the quiescent spectrum can be evidence for continued accretion (e.g. Fridriksson et al. 2011; Chakrabarty et al. 2014; D'Angelo et al. 2015).

A few dozen neutron star LMXBs have been observed in quiescence (e.g. Wijnands et al. 2017, for a recent overview). Most of these show a distinctive soft spectral component that is ascribed to thermal surface emission of the neutron star that was heated during previous accretion episodes (see Sect. 5.2.4). When fitted with a black body, temperatures of $kT_{BB} \sim 0.1$–0.2 keV are typically obtained, whereas neutron star atmosphere models yield $kT \sim 50$–150 eV. Many neutron star LMXBs also show a hard emission tail in their quiescent spectra that is typically modeled as a simple power-law with a photon index of ~ 1–2 (see Fig. 5.10).

The fractional contribution of the power-law component to the 0.5–10 keV flux varies widely between different sources (with some being fully dominated by the non-thermal emission, whereas it is absent in others; e.g. Jonker et al. 2004), but can also vary for a single source (e.g. Cackett et al. 2011; Fridriksson et al. 2011, see also Fig. 5.10). Although this non-thermal emission is often taken as evidence of ongoing accretion, it could perhaps also be associated with processes that involve the magnetic field of the neutron star (e.g. a pulsar wind or a shock from where the gas runs into the magnetosphere). This idea springs from the fact that the AMXPs, which show X-ray pulsations and hence a dynamically important magnetic field during outburst episodes, are often fully dominated by the non-thermal emission component (e.g. Degenaar et al. 2012, and references therein). A non-thermal component in the quiescent spectrum could thus imply that the surface temperature is not homogeneous, which is a problem for measuring the mass and radius. Moreover, such a high-energy tail increases the uncertainties on the fit parameters of the thermal component (Marino et al. 2018).

Since uncertainties in the distance translate directly into uncertainties in the measured masses and radii (see Sect. 5.3.4.4), neutron star LMXBs in globular clusters are often exploited for studying their quiescent thermal emission. There are numerous thermally-emitting quiescent neutron star LMXBs in globular clusters (e.g. Heinke et al. 2003; Guillot et al. 2009; Maxwell et al. 2012) and some of these are sufficiently bright to obtain good-quality spectra that allow for M and R constraints from detailed spectral fitting (see Sect. 5.3.4.2). Moreover opposed to neutron star LMXBs in the field, the quiescent X-rays of globular cluster sources are often strikingly non-variable and have strong constraints on the absence of any hard

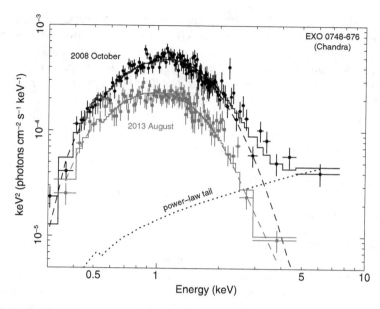

Fig. 5.10 Unfolded X-ray spectra of the quiescent neutron star LMXB EXO 0748−676. Shown are two observations taken ∼2 months (black, top) and ∼5 year (red, bottom) after the end of its ∼24-year long accretion outburst (Degenaar et al. 2014b). Whereas the 2013 spectrum only consisted of thermal emission (dotted line, fitted here with a neutron star atmosphere model NSATMOS; Heinke et al. 2006), the 2008 spectrum exhibited an additional hard power-law tail (dashed curve) that contributed ∼20% to the total unabsorbed 0.5–10 keV flux. A decrease in the thermal emission component is apparent and is thought to result from the thermal relaxation of the accretion-heated crust (see also Sect. 5.3.8.7)

X-ray emission component (e.g. Guillot et al. 2011; Heinke et al. 2014; Bahramian et al. 2015; Walsh et al. 2015; Bogdanov et al. 2016).

5.3.4.1 Model Atmospheres for Quiescent Neutron Stars

Since a strong stellar magnetic field can alter the ionization energies and opacities of the atmosphere, it is easier to study neutron stars that are expected to have a relatively low surface magnetic field (i.e. to reduce the number of free parameters involved in shaping the spectrum), like those in LMXBs (see Sect. 5.2.1). The atmosphere models usually applied to quiescent neutron star LMXBs assume a uniform and isotropic surface temperature, a negligible magnetic field strength ($B \lesssim 10^9$ G), and ignore the effect of rotation (see Sect. 5.2.6). Rutledge et al. (1999, 2001a,b) presented the first studies where a non-magnetic neutron star atmosphere model (Zavlin et al. 1996) was applied to thermal spectra of quiescent neutron star LMXBs. This delivered the first broad constraints on the stellar radii.

Due to the rapid gravitational settling of heavy elements (Romani 1987), for accretion rates of $\dot{M} \lesssim 10^{14}$ g s^{-1} ($\lesssim 10^{-12}$ M$_\odot$ yr^{-1}), the atmosphere of quiescent neutron stars is thought to consist of light elements (Bildsten et al. 1992). However, if mass is still supplied to the neutron star at a sufficiently high rate of $\dot{M} \gtrsim 10^{14}\,\mathrm{g\,s^{-1}}$, heavy elements may be dumped into the atmosphere fast enough not to stratify and hence a (potentially measurable) metal abundance may exist (e.g. Bildsten et al. 1992, see also Sect. 5.3.8). Moreover, some neutron star LMXBs are known to accrete from H-poor companions (e.g. Bildsten and Deloye 2004; Ivanova et al. 2008). In particular in globular clusters a significant fraction of the neutron star LMXBs have small orbits that can only fit small, H-deficit donor stars (e.g. Bahramian et al. 2014, for a recent list).

The impact of He-rich atmospheres on radius determinations has been investigated in a number of works (e.g. Servillat et al. 2012; Catuneanu et al. 2013; Heinke et al. 2014). This revealed that He-atmosphere models for low magnetic field neutron stars resulted in significantly larger masses and radii than fits with H atmospheres (see also Lattimer and Steiner 2014). This can be ascribed to the larger difference between the effective and color temperatures for He atmospheres, and underlines again the importance of the atmosphere composition in these studies. Nevertheless, as shown by Bogdanov et al. (2016), even minute traces of H in a He-rich donor can still lead to a H-dominated atmosphere. Furthermore, if residual accretion occurs and spallation is effective, H could also be produced in the atmosphere (Bildsten et al. 1992, 1993).

5.3.4.2   Observational Constraints from Globular Cluster Sources

Individual mass-radius measurements of quiescent LMXBs typically have uncertainties that are too large to provide meaningful constraints on the EOS (e.g. Heinke et al. 2006; Webb and Barret 2007; Marino et al. 2018). However, fairly tight constraints can be obtained when statistical techniques are applied to an ensemble of sources under the assumptions that these neutron stars exhibit the same radius. This approach is motivated by the fact that for many of the most plausible EOSs, the radius remains constant for different masses (Lattimer and Prakash 2001, see Fig. 5.1 right). A statistical Bayesian analysis technique was applied by Guillot et al. (2013) to a sample of five quiescent neutron star LMXBs in globular clusters. This resulted in a joint radius of $R = 9.1^{+1.3}_{-1.5}\,\mathrm{km}$ for a $M = 1.4\,\mathrm{M_\odot}$ neutron star. This work was extended by Guillot and Rutledge (2014) after adding one more globular cluster source and including new data for a previously analysed cluster, arriving at a similar measurement of $R = 9.4 \pm 1.2\,\mathrm{km}$, which would reject several EOSs. However, not all statistical uncertainties were explored in full (see Sect. 5.3.4.4; Heinke et al. 2014; Lattimer and Steiner 2014). Perhaps the most important source of bias in this work is the assumption that all neutron stars have pure-H atmospheres. As discussed in Sect. 5.3.4.1, it is likely that some globular cluster sources accrete from H-poor companions, which may have a big impact on the resulting radius measurements. Most recently, Steiner et al. (2018) analysed a sample of 8 globular cluster sources applying a Bayesian formalism and scrutinizing several uncertainties such as distances, atmosphere composition and surface temperature inhomogeneities (see Sect. 5.3.4.4). With this conservative treatment, it was found that a $M = 1.4\,M_\odot$ neutron star most likely has a radius of $R = 10$–14 km . Furthermore, this work showed that tighter constraints are only possible when stronger assumptions are made about the atmosphere composition of the neutron stars, the systematics of the observations, or the nature of dense matter (Steiner et al. 2018).

Statistical approaches have also been applied to obtain mass-radius constraints from combining samples of quiescent globular cluster sources with X-ray bursters. Steiner et al. (2010) used a Bayesian framework to combine the mass-radius results of three globular clusters with that of three X-ray bursters obtained from the touchdown method (see Sect. 5.3.3). Additional constraints implied by for instance causality and a theoretical minimum neutron star mass were included in this analysis, which led to preferred radii of $R = 11$–12 km for a $M = 1.4\,M_\odot$ neutron star. This analysis was followed-up by using an extended sample of six globular cluster sources (from Guillot and Rutledge 2014), accounting for the discovery of a $M = 2\,M_\odot$ neutron star (Demorest et al. 2010), and allowing for the possibility of He atmospheres. This led to a preferred radius of $R = 10.5$–12.7 km, ruling out a number of hard EOSs (Steiner et al. 2013).

An alternative Bayesian formalism was developed in which measured masses and radii are mapped to pressures at three fiducial radii (e.g. Read et al. 2009; Özel et al. 2009). This approach makes use of the fact that the radii of neutron stars are only sensitive to the EOS in a fairly narrow range of densities ($\rho \sim 2$–$7.5\,\rho_0$), so that the relevant equations can be mathematically described by only sampling a small number of points in this density range. This method was applied to a combined sample of globular cluster sources (the same as in Guillot and Rutledge 2014) and five X-ray bursters (touchdown method) by Özel et al. (2016b). Additional constraints from other types of analysis were also imposed, including the requirement to allow for a $M = 2\,M_\odot$ neutron star, and results of laboratory experiments in the vicinity of the nuclear saturation density (Tsang et al. 2012; Lattimer and Lim 2013, and references therein). Taken together, this led to a narrow preferred radius range of $R = 10$–11 km (assuming pure H atmospheres for all globular cluster sources). This approach was further extended by Bogdanov et al. (2016), who increased the globular cluster sample with new data from two more sources. The empirical EOS that is inferred from this most recent analysis is consistent with relatively small radii of $R = 9.9$–11.2 km around $M = 1.5\,M_\odot$. This would suggest a fairly soft EOS, with a lower pressure above $\rho = 2\rho_0$ than predicted by a number of hard, nucleonic EOSs (Bogdanov et al. 2016).

5.3.4.3 Observational Constraints from Field LMXBs

There have also been a few attempts to constrain masses and radii of neutron stars in field LMXBs. These efforts have concentrated on sources that are relatively bright in quiescence. One of such sources is EXO 0748−676 (shown in Fig. 5.10), which was recently studied by Cheng et al. (2017) in an attempt to constrain its mass and radius. However, this analysis exposed a worrisome dependence on the energy range over which the fits were performed with a best-fitting mass and radius of $M \sim 2\,M_\odot$ and $R \sim 11\,$km for 0.3–10 keV and $M \sim 1.5\,M_\odot$ and $R \sim 12\,$km for 0.3–10 keV. This was ascribed to the strong energy-dependence of the applied neutron star atmosphere model (Cheng et al. 2017). A similar type of analysis was recently attempted for the well-studied transient LMXB Aql X-1, leading to $M \sim 1.2\,M_\odot$ and $R \sim 10.5\,$km (Li et al. 2017). However, these measurements were only marginally consistent with those obtained from analysing its X-ray bursts (via both the touchdown and the cooling-tail method; Li et al. 2017). This is possibly related to the large number of systematic biases that come into play for the quiescent method, as described in Sect. 5.3.4.4. For instance, Aql X-1 exhibits a power-law emission component in its quiescent spectra (e.g. Cackett et al. 2011) and the neutron star likely continues to accrete at a low level (e.g. Coti Zelati et al. 2014).

5.3.4.4 Biases and Uncertainties in Studies of Quiescent LMXBs

Several works have discussed the different sources of systematic uncertainties for inferring masses and radii of quiescent LMXBs in detail (e.g. Heinke et al. 2014; Özel and Psaltis 2015; Bogdanov et al. 2016). We summarize these here:

Distance Since to first order we constrain the quantity R/D, uncertainties in the distance directly translate into uncertainties in R. For transient LMXBs in the field, distances are usually inferred from X-ray burst analysis (by assuming that the peak of an X-ray burst reaches the Eddington limit). However, the Eddington limit depends strongly on the atmosphere composition (see Sect. 5.3.3.1) and uncertainties can easily be as large as $\sim$20–50% (e.g. Kuulkers et al. 2003). For globular clusters distances can be more reliably measured through a number of techniques, yielding smaller uncertainties of $\sim$5–10%.

Atmosphere Composition The thermal emission observed from neutron stars is shaped by the atmosphere. In case of quiescent neutron star LMXBs, it was demonstrated that a different composition (H versus He) has a strong effect on the inferred masses and radii (see Sect. 5.3.4.4; e.g. Servillat et al. 2012; Catuneanu et al. 2013; Heinke et al. 2014). For neutron star LMXBs that display accretion outbursts, information on the chemical composition of the accreted matter can be obtained e.g. from their X-ray burst properties or from their optical spectra. However, the quiescent LMXBs in globular cluster typically have never shown an accretion outburst (e.g. Wijnands et al. 2013, for a discussion) and hence the exact composition of their atmospheres is unknown. The exceptions are ωCen, which has strong H features in its spectrum, and 47 Tuc X5, which has a long orbital period measured from X-ray eclipses and must thus contain a H-rich donor.

Modeling of the Interstellar Absorption The soft, thermal X-ray emission received from LMXBs is altered by interstellar extinction and a reliable estimate of the absorption column is therefore required for accurate $M-R$ determinations.

Particularly if the absorption column is high, absorption edges due to metals become prominent and modeling the X-ray spectrum is then sensitive to the assumed abundances for the ISM (e.g. Juett et al. 2004; Pinto et al. 2013; Schulz et al. 2016). The latter issue (explored by Heinke et al. 2014) can be circumvented by targeting objects with low interstellar absorption. Bogdanov et al. (2016) studied the effect of the changing absorption column in the high-inclination source 47 Tuc X5. It was found that the inclusion of episodes of strong absorption lowered the inferred neutron star radius.

CCD Pile-Up Even for dim quiescent LMXBs, the received count rate might be higher than the readout time of the CCD, causing multiple events to be recorded as single events with an artificially high energy. In general, pile-up causes a hardening of the X-ray spectrum. Spectra that suffer from pile-up can be corrected at the expense of loss of counts by ignoring the piled-up pixels in the data extraction, or by modeling the pile-up with a dedicated spectral model. The effect of CCD pile-up on mass and radius measurements of quiescent LMXBs was addressed specifically by Bogdanov et al. (2016). This analysis showed that even a pile-up fraction as low as $\sim 1\%$ can have a huge impact; not only are the confidence contours enlarged due to the statistical uncertainty in the pile-up model parameter, but there is also a displacement to lower M and R values if pile-up is not accounted for (caused by the artificial hardening of the spectrum). Some older works typically left a pile-up fraction up to a few percent uncorrected for, and hence may need to be revisited.

Instrument Calibration Uncertainties Since inferring the mass and radius requires an absolute determination of the thermal flux from the neutron star, it directly relies on the calibration of the instrument that is used for the measurement. To account for this, a systematic error of a few percent is often included in the analysis (e.g. Guillot et al. 2013).

Energy Range Considered for Spectral Fitting A study of the quiescent neutron star LMXB EXO 0748−676 recently highlighted that the energy range over which the thermal fits are performed (0.3–10 or 0.5–10 keV) can have a profound impact on the resulting mass and radius measurement (Cheng et al. 2017; Marino et al. 2018).

The Occurrence of Residual Accretion If accretion continues in quiescence, the assumptions of a steady state and passively cooling atmosphere, a purely thermal flux, or a uniform surface temperature may no longer be valid. Elshamouty et al. (2016a) performed a dedicated study of the impact of surface temperature inhomogeneities on radius measurements for quiescent LMXBs. For this study the sources X5 and X7 in the globular cluster 47 Tuc were used, as well as the field LMXB Cen X-4. Assuming that residual accretion would cause pulsed X-ray emission, current limits on the pulsed fraction of these quiescent neutron stars (on the order of $\sim 10\%$) allow the radii to be underestimated by $\sim 10\%$–30% (Elshamouty et al. 2016b). Based on this result, it has been argued that improving constraints on the presence of pulsations from quiescent LMXBs may be essential for progress in constraining their radii.

The Presence of a Hard Spectral Component The presence of a non-thermal emission component, even if it cannot be detected, can potentially bias the mass and radius measurements. In particular, the presence of a power-law emission tail would harden the spectrum and thus lead to a higher temperature measurement and underestimated radius. This issue was specifically addressed by Bogdanov et al. (2016) for X5 and X7 sources in 47 Tuc. It was found that for these two particular objects not accounting for the presence of an undetectable power-law emission component (with contributions to the total unabsorbed 0.5–10 keV flux constrained to be $\lesssim 0.2\%$ and $\lesssim 1.6\%$ for X7 and X5, respectively) would result in a $\sim 0.5\%$ change in the neutron star radius confidence limits.

5.3.5   X-ray Pulse Profile Modeling

In Sects. 5.3.3 and 5.3.4, we presented different ways to obtain radius (and mass) measurements by modeling the X-ray spectra of neutron star LMXBs during X-ray bursts and quiescent episodes. A critical assumption in those approaches is that it the neutron star is spherically symmetric and homogeneously emitting. However, there is a sizable number of neutron stars that display pulsed X-ray emission, modulated at their spin period, from surface hotspots (see Sect. 5.2.6). The exact shape of these *pulse profiles* (or waveforms) is affected by relativistic Doppler shifts, aberration, and light bending and hence depends on the compactness of the neutron star. Accurate modeling of the pulse profiles can thus provide M and R constraints (e.g. Lo et al. 2013; Miller and Lamb 2015).

There are different circumstances in which surface hotspots are produced. For instance, the radio pulsar mechanism is thought to produce energetic electrons and positrons that collide with the polar caps (e.g. Ruderman and Sutherland 1975; Arons 1981; Harding and Muslimov 2001) and thereby create hot spots that give rise to pulsed thermal emission. For accreting neutron stars, hotspots may occur when the stellar magnetic field concentrates the accretion flow onto the polar caps (e.g. Pringle and Rees 1972; Rappaport et al. 1977; Finger et al. 1996; Wijnands and van der Klis 1998). In addition, unstable thermonuclear burning is sometimes confined to specific parts of the neutron star and produces so-called *burst oscillations* at/near the spin of the neutron star (Strohmayer et al. 1996).[3] The temperatures of these various types of hotspots are such that these can be observed at X-ray wavelengths.

Pulsed X-ray emission is also observed for some isolated (slowly rotating) neutron stars, where the hotspots are possibly due to preferential leakage of heat from the crust/core along paths with a certain magnetic field orientation (e.g.

---

[3]The oscillations seen during the rise are thought to come from spreading of the burning front that is modulated by the neutron star spin period (e.g. Strohmayer et al. 1997), whereas the rapid variability seen during the cooling tails are thought to be associated with oscillatory behavior of the surface ("surface modes"; e.g. Muno et al. 2002; Piro and Bildsten 2005).

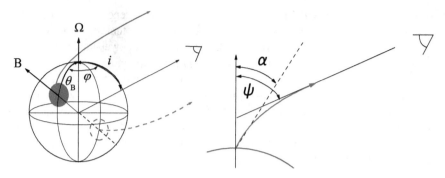

Fig. 5.11 Left: Geometry of a hotspot on the neutron star surface and relevant angles. B and Ω indicate the magnetic and rotational axis, respectively, i is the inclination angle of the rotation axis to the line of sight, θ_B is the hotspot co-latitude, and φ is the rotational phase angle. Right: The emitted α and the observed ψ angles of a light ray from the neutron star normal

Potekhin and Yakovlev 2001). However, the most accurate constraints from X-ray pulse timing can be obtained for neutron stars that spin rapidly, because Doppler boosting becomes more pronounced with increasing spin and this helps to reduce degeneracies in the data. Therefore, the technique of X-ray pulse profile modeling has mainly focussed on LMXBs and millisecond radio pulsars. The basic model and methodology of this technique was recently reviewed by Bogdanov et al. (2016). Here, we briefly discuss the concept, main results and its challenges.

There is a large number of parameters that shape the pulse profile. Some of these are geometrical factors, such as the angle between the rotation axis and our line of sight, the angle between the rotation axis and the center of the hotspots, as well as the geometry of the hotspots (see Fig. 5.11, left). Another important parameter is the angular distribution of the emergent radiation. While this is not important for a homogeneously emitting spherical star, hotspots are observed at different angles for different rotational phases. These parameters together set the observed pulse profile of slowly rotating stars in Newtonian gravity, where their influence scales with the stellar radius and does not depend on the mass. However, due the compactness and rapid spin of neutron stars, relativistic effects can become important and this introduces a mass dependence. The masses and radii of rapidly rotating neutron stars can thus be inferred from their X-ray pulse profiles. We consider the main relativistic effects separately.

5.3.5.1 Relativistic Effects: Light Bending

In general relativity, light rays do not travel in straight lines but rather along geodesic curves. The shape of these light trajectories depends on the geometry of space-time. We start with considering light bending in the Schwarzschild geometry (i.e. ignoring spin). In this case the light trajectory lies in one plane and only two angles need to be connected to the surface normal: the emitted angle α and the observed angle ψ (see

Fig. 5.11, right). The correct connection is given by the integral (Pechenick et al. 1983):

$$\psi = \int_R^\infty \frac{dr}{r^2} \left[\frac{1}{b^2} - \frac{1}{r^2} \left(1 - \frac{R_S}{r} \right) \right]^{-1/2}, \qquad (5.15)$$

where

$$b = R(1+z) \sin \alpha \qquad (5.16)$$

is the impact parameter, and $R_S = 2GM/c^2$ is the Schwarzschild radius. This integral allows us to compute the angle between the light ray and the surface normal at every distance r from the neutron star. A simple and useful analytical approximation of this integral was made by Beloborodov (2002):

$$1 - \cos \psi \approx (1 - \cos \alpha)(1+z)^2. \qquad (5.17)$$

This relation is sufficiently accurate for $R > 2R_S$ and is widely used for light bending computations.

5.3.5.2 Rapidly Rotating Spherical Neutron Stars

There are three principal effects caused by the rapid rotation of neutron stars: the Doppler effect, the time delay and the oblateness of a neutron star. The first two effects can be taken into account even assuming a spherical form of a rapidly rotating neutron star. In what follows, we use the description of Poutanen and Beloborodov (2006). The equation for the spectrum of the unit surface dS' is:

$$dF_E = \frac{\delta^4}{1+z} I'_{E'}(\alpha') \cos \alpha \, \frac{d \cos \alpha}{d \cos \psi} \frac{dS'}{D^2}, \qquad (5.18)$$

where D is the distance to the neutron star and δ is the Doppler factor, given by:

$$\delta = \frac{1}{\gamma \, (1 - \beta \cos \xi)}. \qquad (5.19)$$

Here $\gamma = 1/\sqrt{1 - \beta^2}$, $\beta = v/c$, and ξ is the angle between the direction of the velocity vector v and the line of sight. The velocity is determined as

$$v = 2\pi \nu_{rot} R \sin \theta \, (1+z), \qquad (5.20)$$

where ν_{rot} is the neutron star spin frequency, and θ is co-latitude of the given point. The observed photon energy E is shifted relative to the emitted energy E' both due to the gravitational redshift and the Doppler effect:

$$E = \frac{\delta}{1+z} E'. \tag{5.21}$$

The aberration, which changes the observed inclination of the surface unit to the line of sight, is taken into account using the Doppler factor:

$$\cos \alpha' = \delta \cos \alpha. \tag{5.22}$$

The transformation factor is $d \cos \alpha / d \cos \psi \approx (1 + z)^{-2}$, if Beloborodov's approximation (5.17) is used. The bolometric flux of the surface unit is:

$$dF = \frac{\delta^5}{(1+z)^2} I'(\alpha') \cos \alpha \, \frac{d \cos \alpha}{d \cos \psi} \frac{dS'}{D^2}, \tag{5.23}$$

where $I'(\alpha')$ is the emergent radiation intensity. Equations (5.18) and (5.23) allow us to compute the light trajectory of a small hotspot on a rapidly rotating spherical neutron star, if the connection between its spherical coordinates, θ and ϕ, and the angle ψ can be established:

$$\cos \psi = \cos i \cos \theta + \sin i \sin \theta \cos \phi, \tag{5.24}$$

where i is the inclination angle of the rotation axis to the line of sight. The angle ξ can be also computed:

$$\cos \xi = -\frac{\sin \alpha}{\sin \psi} \sin i \sin \phi, \tag{5.25}$$

where the coordinate ϕ is a rotational phase as well. The travel time of the spot emission to the observer depends on the spot position on the neutron star surface. The difference is small, but it could be significant if the neutron star rapidly rotates and turns to a large angle during a tiny time step. Therefore, the observed phase will differ from the intrinsic neutron star rotational phase

$$\phi \approx \phi_{obs} - \Delta \phi_{obs}, \tag{5.26}$$

where the phase delay depends on the time delay Δt

$$\Delta \phi_{obs} = 2\pi \nu_{rot} \Delta t. \tag{5.27}$$

This delay time depends on the impact parameter (5.16)

$$c \Delta t(b) = \int_R^\infty \frac{dr}{1 - R_S/r} \left\{ \left[1 - \frac{b^2}{r^2} \left(1 - \frac{R_S}{r} \right) \right]^{-1/2} - 1 \right\}, \tag{5.28}$$

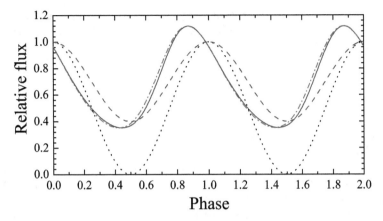

Fig. 5.12 The observed bolometric flux versus phase computed for a small hotspot on the surface of a rotating ($\nu_{rot} = 400$ Hz) neutron star with $M = 1.4$ M$_\odot$, $R = 2.5$ R_S, and angles $i = \theta_B = 45°$ (solid curve). Also shown are the light curves computed without any relativistic effects (dotted curve), with the light bending only (dashed curve), and without the time delay taken into account (dash-dotted curve). We created these curves using the same methods and parameters that produced figure 2 of Poutanen and Beloborodov (2006)

and there is a sufficiently accurate approximation for this expression:

$$c\Delta t = (1 - \cos\psi)\, R. \tag{5.29}$$

The influence of all described effects on the bolometric light curve of the rapidly rotating neutron star is shown in Fig. 5.12.

5.3.5.3 Rapidly Rotating Oblate Neutron Stars

Rapidly rotating neutron stars are oblate, i.e. their shape is not perfectly spherical. The theoretical models of rapidly rotating neutron stars were computed by many authors (see, e.g. Cook et al. 1994, and references therein). Their shape weakly depends on the details of their inner structure, and can be fitted using a few basic parameters only (see, e.g. Morsink et al. 2007). The most simple fit was suggested by AlGendy and Morsink (2014):

$$R(\theta) = R_{\text{eqv}}\left(1 - \bar\Omega^2(0.788 - 1.03x)\cos^2\theta\right), \tag{5.30}$$

where $x = GM/(c^2 R_{\text{eqv}})$ and $\bar\Omega = 2\pi\nu_{rot}\,(R_{\text{eqv}}^3/GM)^{1/2}$. Here R_{eqv} is the equatorial radius. These authors suggested approximate formulae for the neutron star moment of inertia, the quadruple momentum, and the surface gravity distribution with the centrifugal force taken into account.

The space-time in the vicinity of a rapidly rotating neutron star differs from the Schwarzschild metric, and is usually considered in the form suggested by Butterworth and Ipser (1976):

$$ds^2 = -e^{2\nu}c^2dt^2 + \bar{r}^2 \sin^2\theta \, B^2 \, e^{-2\nu}(d\phi - \varpi cdt)^2 \tag{5.31}$$
$$+e^{2(\zeta-\nu)}(d\bar{r}^2 + \bar{r}^2 \, d\theta^2),$$

where ϖ is an angular velocity of a local inertial frame at the neutron star surface. The radial coordinate $\bar{r}$ is connected with the circumferential coordinate r as $r = B e^{-\nu} \bar{r}$ (Friedman et al. 1986). AlGendy and Morsink (2014) have also suggested approximate formulae for the coordinate function B, ν and ζ, which depend on M, R_{eqv}, and ν_{rot} only. Another approach was taken by Bauböck et al. (2012), using the Kerr metric with a quadrupole correction. A correct consideration of the light travel paths in such complicate metrics is not easy. It can be done directly by applying a ray tracing method (see, e.g. Bauböck et al. 2012), which can be used to consider spectral changes (Bauböck et al. 2015a). For light curve computations a different, simplified approach was used in which the shape of the rapidly rotated neutron star is treated correctly, but the Schwarzschild metric is used to account for light bending (see, e.g. Morsink et al. 2007; Miller and Lamb 2015). Whereas this is much simpler than the ray tracing method, it gives acceptable results (Morsink et al. 2007). In this approach the main effect gives the angle between the radius-vector and the normal to the surface. Figure 5.13 shows a comparison of the light curves obtained by this method and those computed for spherical neutron stars. The effects of finite hotspot size were considered by Bauböck et al. (2015b).

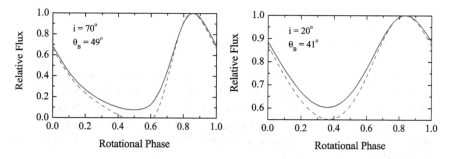

Fig. 5.13 The observed bolometric flux versus phase computed for one small hotspot on the surface of a rotating ($\nu_{rot} = 600$ Hz) oblate neutron star with $M = 1.4$ M$_\odot$, an equatorial radius $R_{eqv} = 16.4$ km, and angles $i = 70°$, $\theta_B = 49°$ (left), and $i = 20°$, $\theta_B = 41°$ (right). The profiles for spherical neutron stars with radii equal to the neutron star radius at the hotspot colatitudes (15.1 and 14.8 km) are also shown by dashed curves. We created these plots using the same methods and parameters that produced figures 3 and 4 of Morsink et al. (2007)

5.3.5.4 Angular Distribution of the Emergent Radiation

A hotspot on a neutron star surface is seen by a distant observer at different angles for different rotational phases. This implies that the angular distribution of the emergent radiation determines the observed flux variation of a rotating neutron star with a surface hotspot. It is not possible to observationally resolve the angular distribution of the surface radiation, hence theoretical models need to be employed to describe it. Existing neutron star atmosphere models give the required angular distributions of the emergent radiation.

The model angular distribution depends mainly on the temperature structure of the atmosphere and the dominant opacity sources at a given photon energy. The angular distribution of the emergent radiation computed for hot neutron star atmospheres ($kT_{eff} > 1$ keV), where the dominant source of electron scattering is close (Suleimanov et al. 2012a) to the Sobolev-Chandrasekhar angular distribution derived for a pure electron scattering atmosphere (Chandrasekhar 1960; Sobolev 1963). We note that this is correct only for sufficiently high photon energies, because the angular distribution at low photon energies, where free-free opacity dominates, becomes close to an isotropic distribution.

The angular distribution of the radiation emerging from relatively cold neutron stars is more complicated and cannot be described by any analytical relations. In this case only numerical distributions given by the model atmospheres have to be used (see, e.g. Bogdanov 2016, and references therein). Recently, the angular distributions for pure hydrogen model atmospheres in an effective temperature range of $T_{eff} = 0.5$–10 MK and for nine surface gravity values were computed and implemented into the X-ray spectral fitting package XSPEC (model HATM; Suleimanov et al. 2017b). The same was done for pure carbon atmospheres using an effective temperature range of $T_{eff} = 1$–4 MK (CARBATM in XSPEC).

The radiation angular distributions emerging from non-magnetized neutron star atmospheres are peaked relative to the atmosphere normal (*pencil beam*). This is not true for highly magnetized neutron star atmospheres. In that case the emergent radiation has a relatively narrow (a few degrees, depending on the magnetic field strength) peak near the normal, and a second broad smoothed peak at inclinations of $\alpha \approx 40$–60° (Pavlov et al. 1994). The total amount of energy radiated in the normal peak is relatively low, and the magnetized model atmospheres produce instead *fan-beamed* radiation. Examples of the angular distributions of non-magnetized and strongly-magnetized neutron star hydrogen atmospheres, as well as corresponding light curves, are shown in Fig. 5.14.

5.3.5.5 Application to Millisecond Pulsars

Millisecond radio pulsars with thermally emitting X-ray hotspots at their polar caps are attractive objects for constraining neutron star radii and masses via the pulse profile modeling technique. So far this has been attempted for three radio pulsars: PSR J0437−4715, PSR J0030+0451 and PSR J2124−3358. The procedure

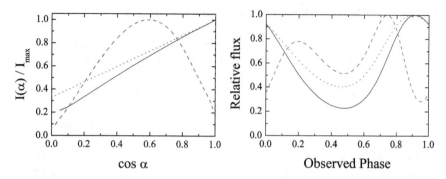

Fig. 5.14 Left: The model angular distributions computed at a photon energy of $E = 1$ keV for a pure hydrogen non-magnetized atmosphere ($T_{eff} = 2$ MK, solid curve; Suleimanov et al. 2017b) and a highly magnetized atmosphere ($T_{eff} = 1.2$ MK, $B = 1.2 \times 10^{13}$ G, dashed curve; Suleimanov et al. 2010a). The angular distribution for a pure electron scattering atmosphere is also shown by the dotted curve. Right: The observed flux at an energy of $E = 1$ keV versus the observed phase computed for a small bright spot on the surface of a rotating ($\nu_{rot}=600$ Hz) oblate neutron star with a mass $M = 1.4$ M$_\odot$, equatorial radius $R = 16.4$ km, and angles $i = 45°$ and $\theta_B = 45°$. The light curves computed using the model angular distributions for a pure hydrogen non-magnetized atmosphere (solid curve), pure hydrogen magnetized atmosphere (dashed curve), and for an isotropic angular distribution (dotted curve) are shown

was applied for the nearest (156.3 ± 1.3 pc; Deller et al. 2008) millisecond radio pulsar PSR J0437−4715 by Pavlov and Zavlin (1997). This pulsar has a relatively low spin period 5.75 ms ($\nu_{rot} = 174$ Hz), so that the effects of oblateness and Doppler boosting can be ignored. Pavlov and Zavlin (1997) evaluated the neutron star compactness $M/M_\odot \approx 1.4$–1.6 $R/10$ km using *ROSAT* observations and fixed angles $i = 40°$ and $\theta_B = 35°$ (Manchester and Johnston 1995). The obtained neutron star radius is $R \approx 10$–15 km taking into account the mass measured for this neutron star from radio pulse timing ($M = 1.76 \pm 0.2$ M$_\odot$; Verbiest et al. 2008). Studies of PSR J0437−4715 were continued by using new observations performed with *XMM-Newton* and applying hydrogen model atmospheres (Bogdanov et al. 2007; Bogdanov 2013). In these studies no angles were fixed and the resulting constraint on the neutron star radius is not very strict ($R > 11$ km). Similar studies were also performed for the next nearest radius pulsar, PSR J0030+0451, and another radio pulsar PSR J2124−3358 (Bogdanov et al. 2008; Bogdanov and Grindlay 2009). Unfortunately no mass measurement is available for these objects and only lower limits on their radii have been obtained so far ($R > 10.7$ and >7.8 km for PSR J0030+0451 and PSR J2124−3358, respectively).

The X-ray pulse profile modeling technique was also applied to three AMXPs: SAX J1808.4−3658, XTE J1814−338 and XTE J1807−294. Since these have higher spin frequencies (up to 600 Hz; e.g. Patruno et al. 2017), the relativistic Doppler effect has to be taken into account. Poutanen and Gierliński (2003) modeled the soft and hard X-ray pulse profiles of the first discovered and frequently active AMXP SAX J1808.4−3658 ($\nu_{rot}\approx400$ Hz) assuming a spherical shape and Schwarzschild spacetime, which led to a relatively small neutron star radius ($R \approx 8$–11 km for $M = 1.4$–1.6 $M_\odot$). The main source of uncertainties in these studies is the applied approximations for the angular distributions of the emergent radiation. Leahy et al. (2008) and Morsink and Leahy (2011) took into account the time delay effect as well as the neutron star oblateness in analysing the X-ray pulse profiles of this source. The obtained mass and radius limitations remain rather wide, $R \approx 5$–13 km for $M = 0.8$–1.7 $M_\odot$. Using the same technique, the masses and the radii of two other AMXPs were also constrained: XTE J1814−338 ($\nu_{rot} \approx 314$ Hz; Leahy et al. 2009) and XTE J1807−294 ($\nu_{rot} \approx 191$ Hz; Leahy et al. 2011). The derived confidence regions in the M–R plane are also large for these objects, with radii in the range of $R \approx 8$–24 km and masses of $M \approx 1 - 2.8$ $M_\odot$ for XTE J1807−294, and $R \approx 11$–20 km and $M \approx 1 - 2.6$ $M_\odot$ for XTE J1814−338. We note that neutron star radii of $R \approx 12$ km are compatible for all three found confidence regions. Finally, pulse profile modeling has been performed for two sources with burst oscillations, 4U 1636−536 (Nath et al. 2002) and XTE J1814−338 (Bhattacharyya et al. 2005), but these have not yielded strong constraints (e.g. Weinberg et al. 2001; Muno et al. 2002, 2003). Although current mass-radius constraints from X-ray pulse profile studies of different types of neutron stars are rather loose, there are very good prospects for improving this (Sects. 5.3.5.6 and 5.4).

5.3.5.6 Challenges of X-ray Pulse Profile Modeling

The reason why this modeling has so far not yielded stringent constraints is clear. Apart from the mass and radius, the pulse profiles depend on geometrical factors such as the size and location of the hotspots as well as the inclination angle between the observer's line of sight and the rotation axis of the neutron star. These geometrical parameters are difficult to determine and also introduce degeneracies with M and R. Modeling of AMXPs in particular is complicated due to the contaminating emission from the accretion disk, variations in their pulse profiles and Comptonization in the accretion column (e.g. Özel 2013; Miller and Lamb 2016). It is expected, however, that these dependencies can be resolved and M and R can be recovered from detailed modeling of the pulse profile (e.g. Psaltis et al. 2014; Lo et al. 2013; Miller and Lamb 2015).

Another important factor is the beaming pattern of the hotspot radiation, as high beaming can to some extent mimic the effects of decreased gravitational light bending and Doppler boosting. For X-ray bursts this is well understood from theoretical modeling (e.g. Madej 1991; Suleimanov et al. 2012a) and not a significant source of uncertainty (Miller 2013). The beaming pattern of the hotspots of accreting pulsars, however, is much less understood and diminishes the constraints on M and R that can be obtained through this method (e.g. Poutanen and Gierliński 2003; Leahy et al. 2009, 2011; Morsink and Leahy 2011). This is less an issue for the magnetic hotspots of rotation-powered pulsars, although there are some uncertainties about the beaming patterns from hydrogen atmospheres heated by the bombardment of relativistic particles in the magnetosphere (e.g. Bogdanov et al. 2008). Another complication of pulse profile modeling for accreting pulsars is the presence of an accretion disk that may block part of the hotspot radiation and introduce harmonic structure in the pulse profile (Poutanen 2008; Poutanen et al. 2009; Ibragimov and Poutanen 2009; Kajava et al. 2011).

Despite the existing challenges, the technique of X-ray pulse profile modeling is very promising and there are good prospects for obtaining much better constraints in the (near) future. Firstly, the recently launched *NICER* mission is expected to allow for detailed X-ray pulse profile modeling (see Sect. 5.4) and accurate methods of constraining neutron star masses and radii using *NICER* data have been developed (see, e.g. Miller and Lamb 2015; Miller 2016; Stevens et al. 2016; Özel et al. 2016a; Watts et al. 2016). Secondly, the relevant geometrical angles may be constrained using X-ray polarization, which could be achieved with several concept missions currently under investigation (Sect. 5.4). For resolving the problem with the angular distributions, extended theoretical computations have to be performed that allow to include the radiation-dominated shock and the radiation transport through it self-consistently, taking the energy balance into account as well.

As discussed in Sects. 5.3.3.4 and 5.3.4.4, X-ray spectroscopic measurements of accreting neutron stars are subject to a number of systematic uncertainties and biases. This does not appear to be the case, however, for X-ray pulse profile modeling. How to apply the pulse profile method in practice and what is required to obtain meaningful constraints on M and R was investigated by Lo et al. (2013); Psaltis et al. (2014); Miller and Lamb (2015). By calculating synthetic pulse profiles under various assumptions, e.g. for the hotspot size, stellar rotation frequency, inclination, the effect of the key pulse profile parameters on M, R, and other relevant parameters was studied. This showed that the uncertainties in M and R are most sensitive to the stellar rotation (with more rapid rotation resulting in smaller errors), spot inclination and observer inclination, and to a much lesser extent to various background components (instrumental, astrophysical, other emission components in the system). A key result of the parameter estimation studies of Lo et al. (2013); Miller and Lamb (2015) is that the mass and radius can be reliably obtained without any strong systematic errors. Currently this is the only method for determining radii for which this is the case.

5.3.6 Radius Constraints from Thermally-Emitting Isolated Neutron Stars

Part of the neutron stars located in the centers of SNRs are magnetars (e.g. Mereghetti et al. 2015, for a review), whereas others form the separate, small class of CCOs (e.g. Pavlov et al. 2002, 2004). These objects have thermal X-ray spectra that can be roughly described by black bodies with temperatures less than a few MK, or $kT_{BB} \sim 0.2$–0.6 keV. CCOs have typical X-ray luminosities of $L_X \sim 10^{33}$–

$10^{34}\,\mathrm{erg\,s^{-1}}$ and are not detected at other wavelengths. These slowly spinning neutron stars show no sign of radio pulsar activity (e.g. no radio pulsations, no pulsar wind nebula or no non-thermal X-ray emission; see short reviews of Gotthelf and Halpern 2007; de Luca 2008; Halpern and Gotthelf 2010a).

Individual CCOs are not completely identical (see e.g. table 1 of Gotthelf et al. 2013, for a list of currently known properties). For instance, three CCOs display X-ray pulsations with periods $P \sim 0.1$–0.5 s and their measured period derivatives $\dot{P}$ suggest dipole magnetic field strengths $B \lesssim 10^{11}$ G (e.g. Gotthelf et al. 2013, and references therein). The X-ray spectrum of 1E 1207.4−5209 shows absorption features associated with the cyclotron line and harmonics (Sanwal et al. 2002; Bignami et al. 2003; Suleimanov et al. 2010b, 2012b). The lowest absorption feature (~ 0.7 keV) is consistent with the magnetic field estimation $B \sim 10^{11}$ G if gravitational redshift is taken into account (Gotthelf et al. 2013).

Many of the CCOs have relatively good distance estimates from their associated SNRs, which make them attractive candidates for measuring the neutron star radii from their thermal X-ray emission (see Pavlov et al. 2002, 2004; Gotthelf et al. 2013, for reviews). The apparent radii obtained by fitting black bodies to the observed spectra are only a few km (Pavlov et al. 2004). For the pulsating CCOs the small sizes of the emitting area can be easily explained by the existence of relatively small hotspots on the neutron star surface. However, fitting the spectra with pure-hydrogen atmosphere models yields reasonable neutron star sizes in some cases. For example, the radius of 1E 1207.4−5209 obtained with a black-body fit is $R \approx 1$–3 km, while the radius derived from hydrogen atmosphere models is $R \approx 10$ km (Zavlin et al. 1996). This underlines the importance of considering appropriate neutron star model atmospheres for inferring stellar radii.

In this section we focus on the results of neutron star radii determinations of two CCOs, located in the SNRs Cas A and HESS J1731−347. These are interesting cases because there are indications that their atmospheres may be carbon rich, and their observed surface temperatures and thermal history give additional interesting information on the properties of the dense matter in their cores.

5.3.6.1 Model Atmospheres of CCOs

The technique of determining neutron star radii from the thermal X-ray spectra of CCOs again uses relation (5.1) of Sect. 5.2.6. The key uncertainty of distance is less an issue for nearby neutron stars (allowing for parallax measurements) and those associated with SNRs as their distance can usually be determined reliably and accurately (see, e.g. Pavlov et al. 2000; Rutledge et al. 2002a). However, the applied model spectrum also plays a significant role. Fortunately, the atmospheres of non-accreting neutron stars should be chemically homogeneous due to gravitational separation (Alcock and Illarionov 1980). As a result, the lightest chemical element of the neutron star envelop dominates the atmosphere chemical composition.

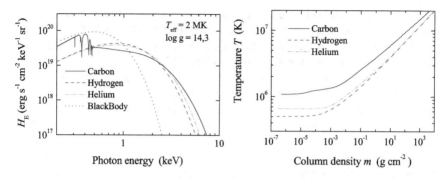

Fig. 5.15 Model spectra (left) of neutron star atmospheres with various chemical compositions and their temperature structures (right), all computed for $T_{eff} = 2\,MK$ and $\log g = 14.3$. The corresponding black-body spectrum is also shown in the left panel (red dotted curve)

As mentioned before, model spectra of fully ionized H and He atmospheres with relatively low effective temperatures (a few MK and less) are harder than black-body spectra of the same effective temperatures (see reviews by Zavlin and Pavlov 2002; Zavlin 2009, and Fig. 5.15). Moreover, neutron star atmosphere spectra are wider than black-body spectra; when high quality data are available, applying a simple black-body model often requires two temperature components (Fig. 5.15). When fitting H-atmosphere models instead of black bodies, this leads to larger neutron star radii (see details in Zavlin 2009). Helium model spectra are even harder and slightly more diluted than hydrogen model spectra (Fig. 5.15), and therefore result in larger radii still (e.g. Heinke et al. 2014). This is because the bremsstrahlung helium opacity is larger due to its Z^2 dependence, and in addition the temperature of the atmosphere is higher because of the increased opacity. Both factors lead to harder spectra than that of H-atmospheres.

The above mentioned effects are even stronger if an atmosphere dominated by C is considered. Carbon is not completely ionized at the neutron star typical temperatures, and the photo-ionization opacity is very strong near the photo-absorption edge of the hydrogen-like carbon ion ($\sim0.49\,keV$). The presence of this edge leads to a lower flux at the blue side of the edge, and energy prefers to escape even at higher photon energies than in H or He atmospheres (Ho and Heinke 2009). As shown in Fig. 5.15, the C-atmosphere spectra are harder and more diluted than the H ones. This leads to larger neutron star radii obtained from the spectral fit for a given distance (Ho and Heinke 2009).

5.3.6.2 Cooling of Isolated Neutron Stars

As mentioned in Sect. 5.2.4, neutron stars gradually cool down after being formed in a supernova explosion. The cooling trajectory after birth can be divided into two main stages. During the first $\sim10^5$–10^6 year a neutron star cools mainly due to neutrino emission from its dense core. Once the core temperature drops below $T_B \sim 10^8$ K, the neutrino emission processes become inefficient and in this second stage a neutron star cools mainly due to radiative losses from its surface. Young neutron stars, which are still in the neutrino cooling stage, are of particular interest because their surface temperatures are high enough (on the order of 1 MK) to be detected with sensitive X-ray satellites. There are a number of excellent works that detail neutron star cooling theory (e.g. Yakovlev and Pethick 2004; Page et al. 2006; Weisskopf et al. 2011). Here we briefly summarize the key elements of the thermal evolution of neutron stars.

Neutrinos can be produced in a variety of particle interactions in the dense neutron star core. The efficiency at which neutrinos are produced depends sensitively on its interior density and composition. In general, the more massive a neutron star is, the more rapidly it cools (e.g. Lattimer and Prakash 2001). Neutrons and protons in the core are likely in a superfluid state, which affects the efficiency of neutrino cooling (e.g. Gusakov et al. 2004; Yakovlev and Pethick 2004; Page et al. 2009; Wijnands et al. 2013). On the one hand, the pairing of nucleons in Cooper pairs decreases the primary neutrino emission process, but on the other hand the forming and breaking of Cooper pairs itself can generate a strong splash of neutrino emission. This can lead to short-lived, fast cooling episodes in the thermal history of neutron stars, which possibly have been observed in the CCO Cas A (e.g. Shternin et al. 2011, and references therein).

The temperature of neutron star cores cannot be directly measured, but need to be inferred from the observable surface temperature. This is not an easy task. For instance, during the first ~ 100 year, neutron stars are not isothermal because the core cools faster than the crust. As a result the surface temperature stays relatively constant, tracing the temperature of the hot crust, and then drops fast when the crust starts to cool and reaches thermal equilibrium with the core. This phase is called the initial thermal relaxation (e.g. Lattimer et al. 1994). During the following neutrino cooling stage, the connection between the core temperature and the surface temperature is governed by the electron thermal conductivity in the neutron star ocean/envelop (see e.g. Potekhin et al. 1997). The efficiency of the thermal conductivity depends particularly on the envelop chemical composition: for the same interior temperature, a light element envelop will yield a higher surface temperature than a metal-rich envelop due to their different thermal conductivities. Figure 5.16 (left) displays examples of neutron star thermal evolution curves for different interior properties, illustrating the different stages of cooling.

Although the core temperature of neutron stars does not provide direct constraints on the EOS, the efficiency of neutrino cooling gives some insight into its density. Furthermore, studying neutron star cooling can give interesting additional information about the behavior of ultra-dense matter, particularly the superfluidity of neutrons and protons. Very strong magnetic fields such as encountered in magnetars can change the thermal evolution significantly, but this is not relevant for the moderate magnetic field strengths of CCOs (e.g. Pons et al. 2009).

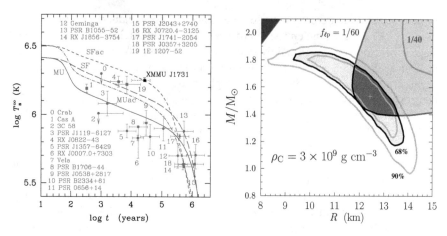

Fig. 5.16 Left: Surface temperatures for a number of cooling isolated neutron stars, including the CCOs in Cas A and HESS J1731−347 (XMMU J1731), versus their ages compared with theoretical cooling curves for a $M = 1.5\,M_\odot$ neutron star. Here, MU refers to a non-superfluid star with slow core cooling via modified Urca processes and an envelop of iron, SF is for strong proton superfluidity in the core and a similar envelop, while MUac and SFac refer to the same models as MU and SF but with a pure-C envelop. Right: M and R constraints for the CCO in HESS J1731−247 from fitting the observed spectra and applying cooling theory. It was assumed that the C envelop extents to $\rho = 3 \cdot 10^9\,\mathrm{g\,cm^{-3}}$ and that the suppression factor for the superfluidity of protons is $f_{\ell p} = 1/60$. The dark shaded area is the resulting confidence M–R region. For further details we refer to Ofengeim et al. (2015)

5.3.6.3   The Neutron Star in Cas A

The CCO in Cas A was discovered by *Chandra* during its first-light observations (Tananbaum 1999). Black-body fits to its featureless X-ray spectrum yielded a high temperature ($T_{BB} = 6$–$8\,$MK) and a small emitting radius ($R_{BB} = 0.2$–$0.5\,$km) for a distance of $3.4\,$kpc (Pavlov et al. 2000). A one-component black-body fit did not provided a statistically acceptable description of a later obtained *Chandra* spectrum. The higher-quality data required at least two thermal components, both fitted with H atmospheres, with significantly different temperatures and sizes ($T_1 \approx 4.5\,$MK, $R_1 \approx 0.4\,$km and $T_2 \approx 1.6\,$MK, $R_2 \approx 12\,$km; Pavlov and Luna 2009). Interpreting the smallest and hottest of the two components as a hotspot may not be consistent with the lack of pulsed X-ray emission. The 3σ upper limit on the pulsed fraction is ∼16%, assuming a sinusoidal pulse shape, and a homogeneous H atmosphere model gives an emitting size of $R \sim 4$–$5.5\,$km.[4] A possible solution for this discrepancy was provided by Ho and Heinke (2009), who found that a pure-C atmosphere model provides a good fit to the data and yields a reasonable neutron star size of $R \sim 10$–

---

[4] For reference, the pulsed fractions of other CCOs are ∼11% for RX J0822.0−4300 in SNR Puppis A, ∼9% for 1E 1207.4−5209 in SNR PKS 1209−51/52, and ∼64% for CXOU J185238.6+004020 in SNR Kes 79 (Gotthelf et al. 2013).

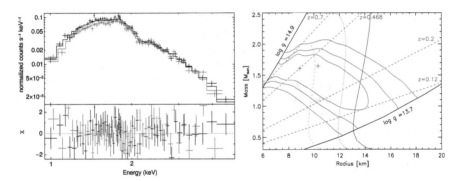

Fig. 5.17 Left: Spectra of the CCO in Cas A along with fits to a C-atmosphere model (log g = 14.45 and $z = 0.375$) for the *Chandra* observations of 2006 (black) and 2012 (red). It is assumed that the entire neutron star surface is emitting and that D =3.4 kpc (from Posselt et al. 2013). Right: M–R confidence contours (68%, 90%, 99%) for the CCO in Cas A obtained using the C-atmosphere model fit. The two black curves indicate the range of the considered log g values, while the dashed blue lines indicate different gravitational redshift parameters. The over-plotted curves are the M–R relations for three possible EOSs according to Hebeler et al. (2013). The best-fit model is marked with a red cross and the best-fit model obtained by Heinke and Ho (2010) is marked with a blue cross. This plot was kindly provided by B. Posselt as adapted from Posselt et al. (2013)

14 km. The observed X-ray spectra of the CCO in Cas A are shown in Fig. 5.17 (left) together with the C-atmosphere model fits. The resulting confidence contours on the M–R plane are shown in Fig. 5.17 (right).

What also makes Cas A interesting is that it was found to display a significant temperature decrease over ∼10 year time, which would point to unusually fast cooling of the neutron star (Heinke and Ho 2010). The presence of a significant temperature evolution was confirmed in subsequent studies (Shternin et al. 2011; Elshamouty et al. 2013), but has been questioned by Posselt et al. (2013). If the observed rapid cooling in Cas A is real, it provides very important insight into the physics of neutron star cores. In particular, such a fast cooling stage can be accounted for by a neutrino emission splash that results from the transition to a neutron superfluidity phase (Page et al. 2011; Shternin et al. 2011), and would be direct evidence that the neutrons in the core are superfluid.

5.3.6.4 The Neutron Star in HESS J1731−347

The CCO at the center of the TeV-emitting supernova remnant HESS J1731−347, also known as G 353.6−0.7, was discovered with *XMM-Newton* in 2007 (Acero et al. 2009; Tian et al. 2010; Abramowski et al. 2011). This neutron star emits a black-body-like X-ray spectrum with $kT \sim 0.5$ keV and is strongly absorbed at low energies (hydrogen column density of $N_H \sim 1.5 \times 10^{22}$ cm^{-2}; Acero et al. 2009; Halpern and Gotthelf 2010b; Bamba et al. 2012). X-ray timing studies yield an upper limit on the pulsed fraction of $\sim$7–8% (for sinusoidal pulsations; Klochkov et al. 2015). The host SNR is most likely located either in the Scutum-Crux arm at $D \sim$3 kpc or in the Norma-Cygnus arm at $D \sim 4.5$ kpc, whereas the measured X-ray absorption and ^{12}CO emission suggest $D > 3.2$ kpc (Abramowski et al. 2011).

Fitting the X-ray spectrum of the CCO in HESS J1731$-$347 with an absorbed black-body model for $D = 3.2$ kpc leads to an unrealistic neutron star radius of $R \sim$0.5 km (Klochkov et al. 2013, 2015). Fitting the spectra with H-model atmospheres yield too small radii for this distance as well, but C atmospheres give acceptable radii (Suleimanov et al. 2014). This results in $M = 1.55^{+0.28}_{-0.24}$ M$_\odot$ and $R = 12.4^{+0.9}_{-2.2}$ km for $D = 3.2$ kpc (see Fig. 5.16; Klochkov et al. 2015).

The neutron star in HESS J1731$-$347 also stands out in its thermal properties. The inferred temperature is much higher than that of other CCOs ($T_{\rm eff,\infty} = 1.78^{+0.04}_{-0.02}$ MK), and unusual for its estimated age of $\sim$27 kyr (see Fig. 5.16; Tian et al. 2008). Modern neutron star cooling theory (e.g. Weisskopf et al. 2011) limits the radius to $R > 12$ km, because smaller radii cannot produce the observed neutron star temperature for this age (Klochkov et al. 2015). We note that agreement between the observed temperature and measured age of the CCO can only be achieved if the neutron star is covered by a thick C envelop ($\Delta M \sim 10^{-8}$ M$_\odot$) and if the protons in the core are superfluid. Detailed consideration of neutron star cooling theory leads to even stronger constraints of the neutron star parameters (see Fig. 5.16 right; Ofengeim et al. 2015).

5.3.7 EOS Constraints from Measuring Fast Spin Rates

Apart from mass and radius measurements, the spin of neutron stars can potentially also provide interesting constraints on the EOS. This is because very fast spin rates constrain the maximum neutron star radius (e.g. Cook et al. 1994). The spin frequency of the neutron star must be lower than the Keplerian frequency, otherwise it would shed mass at its equator due to centrifugal forces. The mass-shredding limit depends on the EOS, M and R (Haensel et al. 2009):

$$f_{\rm max} = C_{\rm dev} \left(\frac{M}{\rm M_\odot} \right)^{1/2} \left(\frac{R}{10\,{\rm km}} \right)^{-3/2} {\rm kHz}. \qquad (5.32)$$

Since the deviation factor $C_{\rm dev}$ depends on the EOS, the above equation can provide a limit on R. The more compact the neutron star (i.e. the smaller R for a given M), the higher the supported rotation can be. Therefore, softer EOSs allow for a higher spin rate (Haensel et al. 2009).

The fastest spinning radio pulsar currently known is MSP J1748$-$2446ad, which has a rotation frequency of 716 Hz (1.396 ms; Hessels et al. 2006). The fastest known AMXP in an LMXB is IGR J00291+5938, which spins at 599 Hz (Galloway et al. 2005), whereas the highest rotation rate for a neutron star LMXB inferred from X-ray burst oscillations is 620 Hz for 4U 1608−52 (Galloway et al. 2008). The centrifugal break-up frequency predicted by most EOSs is $f_{max} \sim 1.5$–2 kHz (e.g. Lattimer 2011). Rotation speeds <1 ms would rule out certain families of hard EOSs, and would be particularly constraining in combination with a large mass.

The lack of more rapidly spinning neutron stars has been taken as evidence that there is some mechanism that limits their spin-up (e.g. spin-equilibrium, spin-down by magnetic dipole radiation, gravitational wave emission; Papaloizou and Pringle 1978; Wagoner 1984; Bildsten 1998; Melatos and Payne 2005; di Salvo et al. 2008; Burderi et al. 2009; Haskell and Patruno 2011; Patruno et al. 2012, 2017). However, there are also physical reasons that might make it difficult to find very rapidly spinning neutron stars, even if they exist. For instance, since accretion is though to spin up neutron stars, high spin may be naively expected in sources that accrete at high rates. However, high accretion rates are also thought to suppress the magnetic field and in absence of channeled accretion no pulsations are expected to be produced (e.g. Cumming et al. 2001; Romanova et al. 2008). Moreover, it has been suggested that strong accretion promotes spin alignment, which would weaken the pulsations (e.g. Ruderman 1991; Lamb et al. 2009). Furthermore, in case of X-ray bursters, very rapid spin may suppress flame spreading and make bursts shorter and weaker (e.g. Spitkovsky et al. 2002; Cavecchi et al. 2013).

5.3.8 Constraints from Other Types of Electromagnetic Observations

The radio, optical, X-ray observations and analysis techniques discussed in Sects. 5.3.1–5.3.7 provide the most direct constraints on neutron star masses and radii to date. However, there are several other observational phenomena that can also put some interesting constraints on the dense matter EOS, particularly with upcoming facilities (see Sect. 5.4). These are briefly discussed below.

5.3.8.1 Mass Measurements from Glitches in Young Radio Pulsars

The most accurate mass measurements have been obtained for rapidly spinning radio pulsars in binary systems (Sect. 5.3.1). However, some constraints on the masses of young, slowly spinning radio pulsars can be obtained from observing *glitches* . These are a sudden increase in the spin period of young, slowly spinning radio pulsars (e.g. Espinoza et al. 2011, for a review). This is thought to be caused by the fact that unlike the normal matter in the neutron star crust, the superfluid component does not slow down due to the electromagnetic energy losses (Anderson and Itoh 1975). Superfluids rotate by forming vortices that are usually "pinned" to the normal matter and the area density of these vortices determines the spin rate of the superfluid. The superfluid therefore acts as a reservoir of angular momentum. As the neutron star spins down, an increasing lag develops between the normal matter in the crust (rotating at the stellar spin rate) and the superfluid (rotating faster); once it reaches a critical value, the superfluid vortices will suddenly "unpin" and transfer angular momentum to the normal matter, explaining the observed jump in spin frequency that is referred to as a glitch.

Recently, two different approaches have been developed to measure neutron star masses from the angular momentum reservoir inferred from glitches (e.g. Ho et al. 2015; Pizzochero et al. 2017). Firstly, Ho et al. (2015) uses observable quantities inferred from X-ray and radio data such as the pulsar spin and its time derivative, the glitching activity and the temperature of the neutron star, and couples these to theoretical models for (temperature sensitive) superfluidity and the EOS to infer neutron star masses for about a dozen glitching radio pulsars. Secondly, Pizzochero et al. (2017) show how the maximum observed amplitude and recurrence time of glitches can constrain the mass of nearly two dozen glitching radio pulsars when combined with microphysical models of the interactions between the normal and superfluid matter. This analysis showed that lower-mass neutron stars produce larger-amplitude glitches.

The mass measurements obtained via these means depend on the assumed EOS and are subject to a number of systematic uncertainties. However, future advances in theoretical modeling and radio/X-ray observing (see Sect. 5.4) allow to further develop these methods. In particular, if an independent mass measurement for a glitching pulsar can be obtained, that can be used to tightly constrain the EOS even if the neutron star mass is not extreme. This seems particular promising with the discovery of young radio pulsars in binaries (e.g. Lyne et al. 2015). For neutron stars that have a smaller mass, the moment of inertia in the crust will be larger and hence stronger amplitude glitches can be produced.

5.3.8.2   Gravitationally Redshifted Lines and Edges During X-ray Bursts

As discussed in Sect. 5.3.3, the neutron star surface is visible during X-ray bursts. As the radiation from the X-ray burst, ignited in the accreted ocean/envelop, passes through a metal-rich atmosphere, this can potentially create absorption lines or edges (e.g. Rajagopal and Romani 1996; Brown et al. 2002). The rotational broadening of such a line depends on R, whereas its centroid energy depends on the ratio M/R. If the inclination is known, this leads to an independent measure of M and R (e.g. Özel and Psaltis 2003; Özel 2006).

Narrow atomic features can be detected with high-resolution X-ray spectrographs such as e.g. the gratings aboard the currently active missions *Chandra* and *XMM-Newton* (see Sect. 5.4.3.3 for prospects with future missions). Since rapid rotation will further broaden narrow spectral features through Doppler smearing, this would be most promising for slowly spinning neutron stars. Several attempts in this direction have been made (e.g. Kong et al. 2007; Galloway et al. 2010b; in 't Zand et al. 2013, 2017), but only in one case narrow spectral features were claimed to be seen (Cottam et al. 2002). However, the later discovery of a high spin rate for this particular neutron star (552 Hz; Galloway et al. 2010a) rules out that the putative lines originated from the stellar atmosphere (Lin et al. 2010). Moreover, attempts to solidify the result by performing new observations failed to detect the features claimed in the initial study (Cottam et al. 2008).

5.3.8.3   Radius Lower Limits from mHz QPOs

Quasi-Periodic Oscillations (QPOs) at mHz frequencies are detected for a handful of LMXBs and are thought to be associated with quasi-stable burning on the neutron star surface (e.g. Revnivtsev et al. 2001; Yu and van der Klis 2002; Heger et al. 2007b; Altamirano et al. 2008; Linares et al. 2012; Keek et al. 2014; Lyu et al. 2014, 2016). As recently argued by Stiele et al. (2016), the maximum black-body emitting radius measured during a QPO cycle provides a lower limit on the radius of the neutron star (it is uncertain whether the entire surface should be emitting). Applying this approach to 4U 1636−536 using *RXTE* data resulted in a lower limit of $R > 11$ km, after accounting for various uncertainties. Better constraints could be obtained using data from new missions such as *NICER* and *HXMT* (Sect. 5.4). A possible advantage is that the mHz QPOs are not expected to cause significant changes in the accretion flow that can complicate the radius measurements, which is the case for X-ray bursts (e.g. van Paradijs and Lewin 1986; Ballantyne and Everett 2005; in 't Zand et al. 2013; Ji et al. 2014; Worpel et al. 2015; Degenaar et al. 2018).

5.3.8.4   Radius Upper Limits from Accretion Disk Reflection

The X-ray spectra of many neutron star LMXBs show broad emission lines near 6.5 keV that are interpreted as radiation that is reflected off the inner edge of the accretion disk (e.g. George and Fabian 1991; Matt et al. 1991; Fabian and Ross 2010). Accurate modeling of this Fe-K line, as well as the corresponding Compton hump near 20–30 keV, is widely used to infer the location of the inner edge of the accretion disk. In principle this provides an upper limit on the radius of the neutron star, since the disk must truncate at the stellar surface if not before (e.g. Bhattacharyya and Strohmayer 2007; Cackett et al. 2008). In several LMXBs the inner disk radii measured from reflection analysis appear to be significantly truncated away from the neutron star, which is attributed for instance to the presence of a geometrically thick boundary layer where the accretion flow impacts the stellar surface (e.g. D'Aí et al. 2014; Ludlam et al. 2017a), the magnetic field of the neutron star (e.g. Degenaar et al. 2014a; van den Eijnden et al. 2017), or evaporation of the inner accretion disk at low mass-accretion rates (e.g. Papitto et al. 2013b). However, in about a dozen sources the inferred inner disk radii are small and may suggest that the disk is running into the neutron star surface, hence providing an upper limit on the stellar radius (e.g. Miller et al. 2013; Degenaar et al. 2015; Chiang et al. 2016; Ludlam et al. 2016, 2017b). This information can be useful when combined results obtained from other techniques (see Sect. 5.3.9).

5.3.8.5 Radius and Mass Upper Limits from kHz QPOs

Like disk reflection, another way to obtain an upper limit on the neutron star radius
from studying the accretion flow properties are kHz QPOs, which are detected for
about two dozen neutron star LMXBs. Although the origin of this rapid variability
(which can be as fast as $\sim 10^3$ Hz) is not well understood, it is typically linked to the
very inner accretion disk. Requiring an observed kHz QPO frequency to be lower
than the Keplerian frequency at the neutron star surface places an upper limit on the
neutron star radius (Miller et al. 1998b). For most EOSs, however, the neutron star
radii lie within the innermost stable circular orbit (ISCO); the Keplerian frequency
at the ISCO is set by the neutron star mass, and therefore requiring that the kHz QPO
frequency is lower than that at the ISCO (i.e. assuming that the QPO is produced
at or outside the ISCO) places an upper bound on the neutron star mass as well
(Kluzniak et al. 1990; Miller et al. 1998b). However, the limits obtained in this way
are not particularly constraining (e.g. van Straaten et al. 2000). More precise mass
measurements would in principle be possible (e.g. Stella et al. 1999; Psaltis et al.
1999; Barret et al. 2006), but heavily relies on the specific interpretation of the kHz
QPO, of which there is no consensus.

5.3.8.6 Core Cooling of Transiently Accreting Neutron Stars

As discussed in Sect. 5.3.6 in the context of thermally-emitting isolated neutron
stars, the temperature of the neutron star core is set by the rate at which it is cooling
through neutrino emissions. This is related to its mass since more massive neutron
stars should have higher interior densities that lead to more efficient neutrino cooling
(rendering more massive neutron stars colder; e.g. Yakovlev and Pethick 2004; Page
et al. 2006). Comparing the inferred temperatures and ages of a number of isolated
neutron stars with theoretical calculations of their thermal evolution shows that their
cores are likely not dense enough to allow for very efficient cooling mechanisms
and hence does not point to particularly massive neutron stars (e.g. Yakovlev and
Pethick 2004; Page et al. 2009, see also Fig. 5.16). This is in sharp contrast with
neutron stars in LMXBs for which similar types of tests (see below) suggest that
efficient core cooling is taking place and would thus point to more massive neutron
stars. It is possible that the lack of very cool objects among the isolated neutron
stars is a selection effect, since relatively high temperatures are required to detect
and identify them as neutron stars.

When located in LMXBs, neutron stars can be reheated through the nuclear
reactions that are induced in the crust during accretion episodes (Sect. 5.2.5) and
can heat the core on a thermal timescale of $\sim 10^4$ year (Brown et al. 1998; Colpi
et al. 2001; Wijnands et al. 2013). In such systems, knowledge or estimates of
the accretion history can then be compared to their core temperature (e.g. Potekhin
et al. 1997; Brown et al. 2002; Yakovlev and Pethick 2004), to determine the rate
of neutrino cooling. This suggests that enhanced cooling mechanisms should be
operating in several neutron stars and hence that these objects should be relatively

massive (e.g. Yakovlev and Pethick 2004; Heinke et al. 2009; Wijnands et al. 2013; Han and Steiner 2017). Although this approach does not allow for accurate mass measurements and there are many systematic uncertainties both in the observations and the models (see Wijnands et al. 2013; Han and Steiner 2017, for discussions), it does provide means to pick out potentially massive neutron stars that could be interesting objects for other types of studies (e.g. optical dynamical mass measurements; see Sect. 5.4.2.2).

5.3.8.7 The Potential of Crust Cooling of Transiently Accreting Neutron Stars

As discussed in Sect. 5.2.5, the crust of a neutron star is heated during accretion phases due to nuclear reactions. Dedicated X-ray monitoring of ~10 transient LMXBs in quiescence following accretion outbursts have revealed a steady decrease in the thermal X-ray flux and inferred neutron star temperature over the course of about a decade. This has been ascribed to the thermal relaxation of the accretion-heated crust (e.g. Wijnands et al. 2013, 2017, for reviews). Such a cooling trajectory depends on the properties of the outburst (which determines how long and intense the crust was heated), the microphysics of the crust such as its structure and composition (which set the thermal conductivity and the nuclear reactions), and the thickness of the crust. The latter is of particular interest, since it is determined by the surface gravity. Therefore, if all the microphysics of neutron star crusts were understood, and the outburst is closely monitored (as is often the case nowadays), the only free parameter determining the cooling trajectory is the compactness of the neutron star (e.g. Wijnands et al. 2013; Deibel et al. 2015).

Although there are still many uncertainties about the thermal and transport properties of neutron star crusts (e.g. Page and Reddy 2013, for a review), studies to improve this are well under way and are alongside providing interesting constraints on other physical properties of neutron stars. For instance, these studies can also reveal the presence of non-spherical shapes for the nuclei at the bottom of the crust (referred to as nuclear pasta; e.g. Horowitz et al. 2015; Deibel et al. 2017). Moreover, some interesting constraints on the core heat capacity have recently been obtained: this approach can potentially lead to more stringent constraints in the future and limit the number of baryons in the core that can be bound in a superfluid (Cumming et al. 2017; Degenaar et al. 2017). Crust cooling studies are thus also interesting avenue to learn more about neutron star crusts as well as their cores, and in principle also have the potential to lead to constraints on the compactness of these neutron stars as well as the dense matter EOS.

5.3.8.8 Constraints from Magnetars

Although the heating mechanism is likely different, it appears that crust cooling is also observed in transient magnetars (of which about two dozen are known; e.g. Coti

Zelati et al. 2017, for a recent overview). Such studies may provide similar prospects as the crust cooling studies of transiently accreting neutron stars (Sect. 5.3.8.7). Furthermore, magnetars show glitches (e.g. Dib et al. 2008; Dib and Kaspi 2014), which can potentially be used to measure their masses in a similar fashion as done for young radio pulsars (Sect. 5.3.8.1). Finally, on rare occasions QPOs have been detected during active flaring episodes of magnetars and ascribed to seismic vibrations (e.g. Israel et al. 2005; Watts and Strohmayer 2006; Huppenkothen et al. 2013). It has been proposed that these magnetar QPOs offer a view into the neutron star M and R (e.g. Strohmayer and Watts 2005; Watts and Reddy 2007; Steiner and Watts 2009; Gabler et al. 2012), but the mode frequencies also depend on unknown factors such as the magnetic field strength, superfluid properties, and crust composition. Moreover, there is some controversy over the interpretation of magnetar QPOs (see e.g. Watts et al. 2016, for a discussion).

5.3.9 Combining Different Methods for Improved Constraints

Radio pulsar timing has been combined with optical observations to obtain very accurate mass constraints for a number of radio pulsars (see Sect. 5.3.2). Moreover, constraints from samples of X-ray bursters and quiescent LMXBs have been combined in statistical frameworks to obtain accurate radius constraints (see Sect. 5.3.4.2). This underlines the power of combining different, complementary techniques for mass and radius measurements. In this section we briefly explore which other methods can be combined to obtain improved EOS constraints. A cross-comparison between techniques is also very important to identify and better understand the systematic uncertainties subject to each approach.

It would be valuable to obtain M and R for a single neutron star both from its quiescent thermal emission and its X-ray bursts. However, for the quiescent method mostly globular cluster sources have been used that have never been seen to exhibit an accretion outburst (and hence X-ray bursts). Conversely, the quiescent emission of X-ray bursters is often either contaminated by a hard spectral component, or so dim that no accurate constraints can be obtained. Nevertheless, a few attempts have been made. For instance for the prolific transiently accreting neutron star Aql X-1, which has been observed in quiescence numerous times (e.g. Cackett et al. 2011; Campana et al. 2014) and is fairly bright. Analysis of its quiescent spectra were compared to that of a number of PRE bursts (Li et al. 2017). Both methods led to reasonably constrained $M-R$ confidence intervals, but these overlap only marginally: the constraints from the quiescent analysis are shifted to lower radius and mass compared to the constraints inferred from the burst analysis. This can likely be attributed to systematic effects (see Sect. 5.3.4.4). Another source for which quiescent and X-ray burst measurements can potentially be combined is 4U 1608−52. It is relatively bright in quiescence (albeit exhibiting a power-law spectral component) and displays an accretion outburst every few years during which it shows (PRE) X-ray bursts (e.g. Poutanen et al. 2014). Moreover, 4U 1608−52 also exhibits a number of other phenomena that can potentially lead to M and R constraints such as disk reflection and mHz QPOs (see below).

It can also be interesting to combine lower limits on the neutron star radius obtained from the mHz QPOs (Sect. 5.3.8.3) with upper limits inferred from disk reflection modeling (Sect. 5.3.8.4) or kHz QPOs (Sect. 5.3.8.5). For instance, reflection studies constrain the radius of the neutron star in 4U 1636−536 to $R \lesssim 11$ km (e.g. Ludlam et al. 2017b), whereas the lower limit inferred from its mHz QPOs is $R \gtrsim 11$ km (Stiele et al. 2016). Without scrutinizing the systematic errors and assumptions of both methods, it is striking that these two independent approaches come together at the same value for the radius. It is conceivable that during the mHz QPO the entire surface was radiating and that the disk in 4U 1636−536 is truly truncating at the neutron star surface, hence that the stellar radius is $R \sim 11$ km. Another example of a neutron star that displays mHz QPOs (Yu and van der Klis 2002) and possibly has the disk running in the neutron star surface (Degenaar et al. 2015) is 4U 1608−52. Combining these different X-ray techniques can thus possibly bracket the radius measurement for this neutron star. It is also X-ray bright in quiescence, displays PRE bursts (Poutanen et al. 2014) and burst oscillations (at 620 Hz; Galloway et al. 2008). It could thus also be a good target for pulse profile modeling with the new mission *NICER* (Sect. 5.4.3.4) and X-ray polarization studies (Sect. 5.4.3.5). Finally, the companion stars of some transient neutron star LMXBs may also be bright enough to be studied in quiescence at optical wavelengths with the future generation of instruments (Sect. 5.4.2.2). This can provide independent and complimentary constraints on the neutron star mass.

For accreting neutron stars, X-ray pulse profile modeling can be performed in two different ways. For the AMXPs, we observe emission from hotspots at the magnetic poles (in addition to emission from a shock that forms just above the surface as the rapidly in-falling material is abruptly decelerated). For burst oscillations, on the other hand, hotspots arise due to unstable thermonuclear burning zones on the stellar surface (i.e. not confined to the magnetic poles). Several neutron stars show both coherent X-ray pulsations and burst oscillations; ideally, one would want to use both types of hotspots to model the resulting pulse profile and obtain $M-R$ constraints, to check for consistency and to calibrate both methods (Watts et al. 2016). An important breakthrough could also be provided by the detection of surface atomic lines (Sect. 5.3.8.2) in the hotspot emission of a neutron star for which the pulse profile (Sect. 5.3.5) can also be accurately modeled (e.g. Rauch et al. 2008). Combining these two pieces of information yields complementary and independent measurements of M and R. Lastly, some radio pulsars with accurate M measurements show thermal emission that may allow for a R measurement through X-ray pulse profile modeling with *NICER* (see Sect. 5.4.3.4).

5.4 Future Prospects with New and Upcoming Instrumentation

Over the past decade we have witnessed significant developments in inferring the dense matter EOS from electromagnetic observations of neutron stars. We have started to gain significant constraints on the pressure-density relation and our knowledge of the superfluid properties of their interiors is steadily growing. However, it is at present not yet possible to infer the composition of the dense core of neutron stars (i.e. nuclear versus exotic matter). The most important challenges that presently limit tighter constraints are systematic uncertainties, limited data quality, and small number statistics. Below we give an overview of the exciting prospects of new and future instrumentation to continue and improve determinations of neutron star masses and radii in the next decade and beyond.

5.4.1 The Future Generation of Radio Telescopes

5.4.1.1 Significantly Increasing the Number of Mass Measurements of Radio Pulsars

The Square Kilometre Array (*SKA*) will be the world's largest radio telescope and is expected to begin science operations in the early 2020s. The *SKA* can provide major breakthroughs in neutron star research, including EOS constrains (Watts et al. 2015). In particular, it is expected to discover significant numbers of radio pulsars. This allows for an increased number of mass measurements, including those with extreme properties that put the most stringent constraints on the EOS (e.g. Keane et al. 2015; Hessels et al. 2015, see also Sect. 5.4.4). Furthermore, the precise timing techniques that are being developed for pulsar timing arrays (PTAs) have started to yield accurate masses for more pulsars and may eventually lead to finding more extreme ones (e.g. Reardon et al. 2016; Fonseca et al. 2016).

5.4.1.2 Measuring the Moment of Inertia of Radio Pulsars

Radio and gamma-ray observations of radio pulsars have the potential to measure their moment of inertia . Since this is a function of both M and R, it allows the radius to be measured if the mass can be determined independently (e.g. via radio pulsar timing and/or optical studies of the companion star). Attempts have been made for PSR J1614−2230, one of the two radio pulsars with an accurate high mass measurement of $M \sim 2\ M_\odot$ (Demorest et al. 2010; Fonseca et al. 2016). Unfortunately, the spin-down luminosity inferred from its gamma-ray emission does not provide very tight constraints and this is not likely to improve with new instrumentation (Watts et al. 2015). However, long-term observations of the double pulsar system PSR J0737−3039 will eventually lead to measurement of the moment of inertia of one of the pulsars, resulting in a radius constraint with ∼5% accuracy (Lyne et al. 2004; Lattimer and Schutz 2005; Kramer and Wex 2009). As mentioned by Watts et al. (2015), the *SKA* (Sect. 5.4.1.1) can possibly discover more pulsar systems for which the moment of inertia will be measurable, but this is likely challenging due to the highly restrictive requirements for the system geometry.

5.4.2 The New and Future Generation of Optical Telescopes

5.4.2.1 Accurate Distance Determinations with *Gaia*

As discussed in Sect. 5.3.4.4, one of the most important systematic uncertainties in inferring radii from neutron stars in LMXBs, is that their distance is usually not accurately known. *Gaia*, launched in 2013, is an astrometry mission dedicated to measure the parallaxes of stars with unprecedented precision. In a few years, this may provide accurate distances for a number of neutron star LMXBs, either located in the field or in globular clusters. This will strongly reduce the systematic uncertainties in the radii determined for these objects from X-ray observations.

5.4.2.2 Dynamical Mass Measurements with the Next Generation Optical Telescopes

The highly improved sensitivity of upcoming optical facilities such as the European Extremely Large Telescope (*E-ELT*), the Thirty Meter Telescope (*TMT*) and the Large Synoptic Survey Telescope (*LSST*) will be transformable in obtaining dynamical mass measurements for neutron stars in binary systems. Synergies with the *SKA* are expected to be particularly promising (e.g. Antoniadis et al. 2015).

5.4.3 Advances from X-ray Astronomy

5.4.3.1 Radius Measurements from X-ray Bursts with New X-ray Telescopes

Radius constraints from X-ray bursting neutron star LMXBs have all been obtained with the wealth of data provided by NASA's *RXTE*, which was decommissioned in 2012. However, in 2015 the Indian satellite *ASTROSAT* was launched (Singh et al. 2014) and in 2017 the Chinese Hard X-ray Modulation Telescope (*HXMT*) was brought into orbit (Zhang et al. 2014). Their X-ray detectors have similar capabilities to those of *RXTE*, and both missions thus provide continued opportunities to measure neutron star radii from X-ray bursts In addition, the *NICER* mission installed in 2017 is a very promising tool for X-ray burst studies (e.g. Keek et al. 2016), and can potentially also lead to more accurate radius constraints from bursting neutron stars.

5.4.3.2 Radius Constraints of dim Thermally-Emitting Neutron Stars with *Athena*

Further in the future, currently planned for launch in the late 2020s, we will have access to the ESA mission *Athena* (Barcons et al. 2017). Although dense matter is not a core science goal of *Athena*, its very high collective area, soft X-ray coverage and good spectral resolution does allow for more accurate modeling of the thermal spectra of quiescent neutron stars, studying the cooling tails of X-ray bursts,[5] modeling the pulse profiles of different kinds of pulsars, and searching for gravitationally redshifted lines in the surface emission of neutron stars (Motch et al. 2013). Furthermore, *Athena* might provide more stringent limits on the pulsed fraction of quiescent neutron star LMXBs, which is of high importance to assess the possible presence of temperature inhomogeneities (Elshamouty et al. 2016b).

5.4.3.3 High-Resolution X-ray Spectroscopy to Search for Atomic Spectral Features

Detecting gravitationally redshifted atomic features during X-ray bursts or low-level accretion activity, which would constrain the compactness of a neutron star (Sect. 5.3.8.2), requires high-resolution X-ray spectrographs. This is incorporated in the design of *Athena* (the X-ray Integral Field Unit, X-IFU) and was provided by short-lived mission *Hitomi*, which may find follow-up with the X-ray Astronomy Recovery Mission (*XARM*). There are thus some future prospects for searching for narrow, gravitationally redshifted features from neutron star atmospheres.

5.4.3.4 Accurate X-ray Pulse Profile Modeling with *NICER*

The Neutron Star Interior Composition ExploreR (*NICER*) is a NASA mission that was successfully installed on the International Space Station in 2017 June (Gendreau et al. 2012; Arzoumanian et al. 2014). It covers the energy range from 0.2–12 keV, has a very high effective area and unprecedented timing precision (absolute time-tagging of <300 ns). *NICER* is dedicated to achieve precise (~5%) mass and radius measurements for a few selected neutron stars through highly accurate pulse-profile modeling of pulsars (e.g. Bogdanov 2016; Miller 2016; Özel et al. 2016a, see also Sect. 5.3.5). Moreover, neutron stars that display rapid

---

[5]Although pile-up can be an issue when studying bursts with *Athena* (e.g. Keek et al. 2016).

oscillations during the rise of an X-ray burst are very promising targets to obtain EOS constraints with *NICER* (e.g. Watts et al. 2016). Obtaining independent constraints on geometrical factors significantly improves the constraints obtained from such pulse profile modeling. This can uniquely be achieved from X-ray polarization studies, of which several concepts are currently being investigated (Sect. 5.4.3.5).

5.4.3.5 Geometrical Constraints from X-ray Polarization

As discussed in Sect. 5.3.5, pulse profile modeling is a very promising technique to obtain mass and radius constraints without systematic errors, if a number of geometrical factors can be determined. Fortunately, the hotspot emission that is used by this technique is expected to be polarized (e.g. Rees 1975; Meszaros et al. 1988; Viironen and Poutanen 2004). Phase-resolved measurements of the angle and degree of polarization can then constrain both the inclination angle of the observer and the hotspot, thereby breaking degeneracies that current limit pulse profile modeling. AMXPs and X-ray burst oscillation sources are prime targets for such polarimetry studies. The concept missions *eXTP* (China/Europe; Zhang et al. 2016), *IXPE* (NASA; Weisskopf et al. 2016), and *XIPE* (ESA; Soffitta et al. 2016) have X-ray polarimeters in their design concepts that would facilitate such studies.

5.4.4   Searching for the Most Rapidly Spinning Radio and X-ray Pulsars

As discussed in Sect. 5.3.7, very high spin rates also have the potential to put interesting constraints on the neutron star EOS. The future holds great prospects for continuing and enhancing searches for rapidly spinning neutron stars. At radio wavelengths, such advances will be brought through timing studies with *SKA* and pulsar timing arrays, whereas at X-ray wavelengths such studies are currently facilitated by searching for coherent X-ray pulsations and X-ray burst oscillations with *ASTROSAT*, *HXMT* and *NICER*. Several mission concepts currently under investigation would also facilitate such studies (e.g. *eXTP* and *STROBE-X*; for both the current design has a higher effective area than *RXTE*, which allows for detecting weaker pulsations; e.g. Zhang et al. 2016; Wilson-Hodge et al. 2016).

5.5   Conclusions

Neutron stars are unique, natural laboratories to constrain the EOS of ultra-dense matter. In particular, measuring the mass and radius of several neutron stars with <10% errors can place strong constraints on the EOS (e.g. Özel et al. 2010; Steiner et al. 2010). Such measurements are facilitated by observations at radio, optical and X-ray wavelengths. Precise and reliable mass measurements have so far only been obtained for radio pulsars ($\sim$40 objects with masses ranging between $M \sim 1.2$–2.0 $M_\odot$), whereas radius measurements have only been obtained for about two dozen neutron stars in LMXBs ($\sim$15 objects leading to an overall estimate of $R \sim 10$–12 km for an assumed mass of $M \sim 1.4$–1.5 $M_\odot$).

Whereas the masses of radio pulsars are determined with high accuracy, the radius measurements of neutron star LMXBs are more strongly model dependent and subject to a number of systematic uncertainties. Nevertheless, over the past decade much progress was made in obtaining masses and radii from the thermal X-ray emission of neutron star LMXBs. Alongside, there have been important theoretical developments, including a detailed assessment of spin effects (e.g. Bauböck et al. 2012, 2015a), advanced calculations of atmosphere models (e.g. Ho and Heinke 2009; Suleimanov et al. 2011b, 2012a; Nättilä et al. 2015) and the development of statistical analysis methods (e.g. Özel et al. 2010; Özel and Psaltis 2015; Steiner et al. 2010, 2013). Moreover, there are many different techniques that can ultimately lead to constraints on the neutron star EOS. Several approaches are still under development and can be improved with new and upcoming observatories. There lies great power in applying different techniques to individual neutron stars: orthogonal constraints can be obtained or cross-checks can be performed that allows us to address the systematic uncertainties of different methods.

Finally, it is important to note that although probing the behavior of ultra-dense matter is typically the main scientific driver to try and constrain the neutron star EOS, there are much wider implications. The behavior of matter near and beyond the nuclear density is governed through the strong interactions and understanding this fundamental force is important for several other areas of astrophysics. For instance, it plays an important role in core-collapse supernova explosions, the dynamics of compact object mergers that involve a neutron star and the formation timescale of black holes, as well as the precise gravitational wave and neutrino signals produced in these processes, the resulting mass loss and nucleosynthesis, and associated γ-ray bursts and hypernovae. The recent discovery of gravitational wave signals and electromagnetic counterparts of the merger of two neutron stars has ever more increased interest in constraining the neutron star EOS.

Acknowledgements The authors acknowledge support from NewCompStar COST Action MP1304. ND is supported by a Vidi grant from the Netherlands Organization for Scientific Research (NWO). VS is supported by Deutsche Forschungsgemeinschaft (DFG) grant WE 1312/51-1.

For the creation of Figure 5.1, the authors thank Morgane Fortin for providing unified EOSs and Feryal Ö for making her compilation of EOSs publicly available online.[6]

---

[6]http://xtreme.as.arizona.edu/NeutronStars

References

Abramowski, A., et al.: Astron. Astrophys, **531**, A81 (2011)
Acero, F., et al.: Proceedings of the 31th ICRC 2009, Lodz, arXiv:0907.0642 (2009)
Aguilera, D.N., Pons, J.A., Miralles, J.A.: Astrophys. J. Lett. **673**, L167 (2008)
Alcock, C., Illarionov, A.: Astrophys. J. **235**, 534 (1980)
AlGendy, M., Morsink, S.M.: Astrophys. J. **791**, 78 (2014)
Alpar, M.A., et al.: Nature **300**, 728 (1982)
Altamirano, D., et al.: Astrophys. J. Lett. **673**, L35 (2008)
Anderson, P.W., Itoh, N.: Nature **256**, 25 (1975)
Antoniadis, J.: et al.: Science **340**, 448 (2013)
Antoniadis, J., et al.: Advancing astrophysics with the square kilometre array. In: Proceedings of
    Science, vol. 157 (2015)
Archibald, A.M., et al.: Science **324**, 1411 (2009)
Arons, J.: Astrophys. J. **248**, 1099 (1981)
Arras, P., Cumming, A., Thompson, C.: Astrophys. J. Lett. **608**, L49 (2004)
Arzoumanian, Z., et al.: Space telescopes and instrumentation 2014: ultraviolet to gamma ray. In:
    Proceedings of SPIE, vol. 9144, 914420 (2014)
Baade, W., Zwicky, F.: Proc. Natl. Acad. Sci. **20**, 259 (1934)
Bahramian, A., et al.: Astrophys. J. **780**, 127 (2014)
Bahramian, A., et al.: Mon. Not. R. Astron. Soc. **452**, 3475 (2015)
Ballantyne, D.R., Everett, J.E.: Astrophys. J. **626**, 364 (2005)
Bamba, A., et al.: Astrophys. J. **756**, 149 (2012)
Barcons, X., et al.: Astron. Nachr. **338**, 153 (2017)
Barret, D., Olive, J.-F., Miller, M.C.: Mon. Not. R. Astron. Soc. **370**, 1140 (2006)
Bauböck, M., et al.: Astrophys. J. **753**, 175 (2012)
Bauböck, M., et al.: Astrophys. J. **799**, 22 (2015a)
Bauböck, M., Psaltis, D., Özel, F.: Astrophys. J. **811**, 144 (2015b)
Becker, W., et al.: Astrophys. J. **594**, 798 (2003)
Beloborodov, A.M.: Astrophys. J. Lett. **566**, L85 (2002)
Bernardini, F., et al.: Mon. Not. R. Astron. Soc. **436**, 2465 (2013)
Bhattacharya, D., Srinivasan, G.: X-ray binaries **1995**, 495 (1995)
Bhattacharya, D., van den Heuvel, E.P.J.: Phys. Rev. **203**, 1 (1991)
Bhattacharyya, S., Strohmayer, T.E.: Astrophys. J. Lett. **664**, L103 (2007)
Bhattacharyya, S., et al.: Astrophys. J. **619**, 483 (2005)
Bhattacharyya, S., Miller, M.C., Lamb, F.K.: Astrophys. J. **644**, 1085 (2006)
Bhattacharyya, S., Miller, M.C., Galloway, D.K.: Mon. Not. R. Astron. Soc. **401**, 2 (2010)
Bignami, G.F., et al.: Nature **423**, 725 (2003)
Bildsten: Astrophys. J. Lett. **501**, L89 (1998)
Bildsten, L., Deloye, C.J.: Astrophys. J. Lett. **607**, L119 (2004)
Bildsten, L., Salpeter, E.E., Wasserman, I.: Astrophys. J. **384**, 143 (1992)
Bildsten, L., Salpeter, E.E., Wasserman, I.: Astrophys. J. **408**, 615 (1993)
Bogdanov, S.: Astrophys. J. **762**, 96 (2013)
Bogdanov, S.: Eur. Phys. J. A **52**, 37 (2016)
Bogdanov, S., Grindlay, J.E.: Astrophys. J. **703**, 1557 (2009)
Bogdanov, S., Rybicki, G.B., Grindlay, J.E.: Astrophys. J. **670**, 668 (2007)
Bogdanov, S., Grindlay, J.E., Rybicki, G.B.: Astrophys. J. **689**, 407 (2008)
Bogdanov, S., et al.: Astrophys. J. **831**, 184 (2016)
Boutloukos, S., Miller, M.C., Lamb, F.K.: Astrophys. J. Lett. **720**, L15 (2010)
Braje, T.M., Romani, R.W., Rauch, K.P.: Astrophys. J. **531**, 447 (2000)
Brown, E.F., Bildsten, L., Rutledge, R.E: Astrophys. J. Lett. **504**, L95 (1998)
Brown, E.F., Bildsten, L., Chang, P.: Astrophys. J. **574**, 920 (2002)
Burderi, L., et al.: Astrophys. J. **519**, 285 (1999)

Burderi, L., et al.: Astron. Astrophys. **496**, L17 (2009)
Butterworth, E.M., Ipser, J.R.: Astrophys. J. **204**, 200 (1976)
Cackett, E.M., et al.: Astrophys. J. **674**, 415 (2008)
Cackett, E.M., et al.: Astrophys. J. **720**, 1325 (2010)
Cackett, E.M., et al.: Mon. Not. R. Astron. Soc. **414**, 3006 (2011)
Cadeau, C., et al.: Astrophys. J. **654**, 458 (2007)
Campana, S., et al.: Astron. Astrophys. Rev. **8**, 279 (1998)
Campana, S., et al.: Mon. Not. R. Astron. Soc. **441**, 1984 (2014)
Catuneanu, A., et al.: Astrophys. J. **764**, 145 (2013)
Cavecchi, Y., et al.: Mon. Not. R. Astron. Soc. **434**, 3526 (2013)
Chakrabarty, D., Roche, P.: Astrophys. J. **489**, 254 (1997)
Chakrabarty, D., et al.: Astrophys. J. **797**, 92 (2014)
Chandrasekhar, S.: Radiative Transfer. Dover Publication, New York (1960)
Chang, P., et al.: Astrophys. J. Lett. **636**, L117 (2006)
Cheng, Z., et al.: Mon. Not. R. Astron. Soc. **471**, 2605 (2017)
Chiang, C.-Y., et al.: Astrophys. J. **821**, 105 (2016)
Colpi, M., et al.: Astrophys. J. Lett. **548**, L175 (2001)
Cook, G.B., Shapiro, S.L., Teukolsky, S.A.: Astrophys. J. **424**, 823 (1994)
Cooper, R.L., Kaplan, D.L.: Astrophys. J. Lett. **708**, L80 (2010)
Coti Zelati, F., et al.: Mon. Not. R. Astron. Soc. **438**, 2634 (2014)
Coti Zelati, F., et al.: Mon. Not. R. Astron. Soc. **471**, 1819 (2017)
Cottam, J., Paerels, F., Mendez, M.: Nature **420**, 51 (2002)
Cottam, J., et al.: Astrophys. J. **672**, 504 (2008)
Cumming, A., Zweibel, E., Bildsten, L.: Astrophys. J. **557**, 958 (2001)
Cumming, A., et al.: Phys. Rev. C **95**, 025806 (2017)
D'Aí, A., et al.: Eur. Phys. J. Web Conf. **64**, 06006 (2014)
Damen, E., et al.: Astron. Astrophys. **237**, 103 (1990)
D'Angelo, C.R., et al.: Mon. Not. R. Astron. Soc. **449**, 2803 (2015)
de Luca, A.: In: Bassa, C., Wang, Z., Cumming, A., Kaspi, V.M. (eds.) AIPC, vol. 983, 40 Years
    of Pulsars: Millisecond Pulsars, Magnetars and More, pp. 311–319 (2008)
Degenaar, N., Wijnands, R.: Mon. Not. R. Astron. Soc. **422**, 581 (2012)
Degenaar, N., Patruno, A., Wijnands, R.: Astrophys. J. **756**, 148 (2012)
Degenaar, N., et al.: Astrophys. J. Lett. **796**, L9 (2014a)
Degenaar, N., et al.: Astrophys. J. **791**, 47 (2014b)
Degenaar, N., et al.: Mon. Not. R. Astron. Soc. **451**, L85 (2015)
Degenaar, N., et al.: Mon. Not. R. Astron. Soc. **465**, L10 (2017)
Degenaar, N., et al.: Space Sci. Rev. **214**, 15 (2018)
Deibel, A., et al.: Astrophys. J. Lett. **809**, L31 (2015)
Deibel, A., et al. Astrophys. J. **839**, 95 (2017)
Deller, A.T., et al.: Astrophys. J. Lett. **685**, L67 (2008)
Demorest, P.B., et al.: Nature **467**, 1081 (2010)
di Salvo, T., et al.: Mon. Not. R. Astron. Soc. **389**, 1851 (2008)
Dib, R., Kaspi, V.M.: Astrophys. J. **784**, 37 (2014)
Dib, R., Kaspi, V.M., Gavriil, F.P.: Astrophys. J. **673**, 1044 (2008)
Ebisuzaki, T.: Publ. Astron. Soc. Jpn. **39**, 287 (1987)
Elshamouty, K.G., et al.: Astrophys. J. **777**, 22 (2013)
Elshamouty, K.G., Heinke, C.O., Chouinard, R.: Mon. Not. R. Astron. Soc. **463**, 78 (2016a)
Elshamouty, K.G., et al.: Astrophys. J. **826**, 162 (2016b)
Espinoza, C.M., et al.: Mon. Not. R. Astron. Soc. **414**, 1679 (2011)
Fabian, A.C., Ross, R.R.: Space Sci. Rev. **157**, 167 (2010)
Faucher-Giguère, C.-A., Kaspi, V.M.: Astrophys. J. **643**, 332 (2006)
Finger, M.H., et al.: Nature **381**, 291 (1996)
Fonseca, E., et al.: Astrophys. J. **832**, 167 (2016)
Fortin, M., et al.: Phys. Rev. C **94**, 035804 (2016)

Fortin, M., et al.: Phys. Rev. C **95**, 065803 (2017)
Fridriksson, J.K., et al.: Astrophys. J. **736**, 162 (2011)
Friedman, J.L., Ipser, J.R., Parker, L.: Astrophys. J. **304**, 115 (1986)
Gabler, M., et al.: Mon. Not. R. Astron. Soc. **421**, 2054 (2012)
Galloway, D.K., Lampe, N.: Astrophys. J. **747**, 75 (2012)
Galloway, D.K., et al.: Astrophys. J. Lett. **622**, L45 (2005)
Galloway, D.K., et al.: Astrophys. J. Suppl. Ser. **179**, 360 (2008)
Galloway, D.K., et al.: Astrophys. J. Lett. **711**, L148 (2010a)
Galloway, D.K., et al.: Astrophys. J. **724**, 417 (2010b)
Gendre, B., Barret, D., Webb, N.: Astron. Astrophys. **403**, L11 (2003a)
Gendre, B., Barret, D., Webb, N.: Astron. Astrophys. **400**, 521 (2003b)
Gendreau, K.C., Arzoumanian, Z., Okajima, T.: Space telescopes and instrumentation 2012: ultraviolet to gamma ray. In: Proceedings of SPIE, vol. 8443, 844313 (2012)
George, I.M., Fabian, A.C.: Mon. Not. R. Astron. Soc. **249**, 352 (1991)
Geppert, U., Küker, M., Page, D.: Astron. Astrophys. **426**, 267 (2004)
Glendenning, N.K.: Astrophys. J. **293**, 470 (1985)
Glendenning, N.K., Schaffner-Bielich, J.: Phys. Rev. C **60**, 025803 (1999)
Gotthelf, E.V., Halpern, J.P.: Astrophys. J. Lett. **664**, L35 (2007)
Gotthelf, E.V., Halpern, J.P., Alford, J.: Astrophys. J. **765**, 58 (2013)
Gudmundsson, E.H., Pethick, C.J., Epstein, R.I.: Astrophys. J. Lett. **259**, L19 (1982)
Gudmundsson, E.H., Pethick, C.J., Epstein, R.I.: Astrophys. J. **272**, 286 (1983)
Guillot, S., Rutledge, R.E.: Astrophys. J. Lett. **796**, L3 (2014)
Guillot, S., et al.: Astrophys. J. **699**, 1418 (2009)
Guillot, S., Rutledge, R.E., Brown, E.F.: Astrophys. J. **732**, 88 (2011)
Guillot, S., et al.: Astrophys. J. **772**, 7 (2013)
Gusakov, M.E., et al.: Astron. Astrophys. **423**, 1063 (2004)
Güver, T., Özel, F.: Astrophys. J. Lett. **765**, L1 (2013)
Güver, T., et al.: Astrophys. J. **712**, 964 (2010a)
Güver, T., et al.: Astrophys. J. **719**, 1807 (2010b)
Güver, T., Psaltis, D., Özel, F.: Astrophys. J. **747**, 76 (2012)
Haakonsen, C.B., et al.: Astrophys. J. **749**, 52 (2012)
Haensel, P., Zdunik, J.L.: Astron. Astrophys. **229**, 117 (1990)
Haensel, P., Zdunik, J.L.: Astron. Astrophys. **404**, L33 (2003)
Haensel, P., et al.: Astron. Astrophys. **502**, 605 (2009)
Halpern, J.P., Gotthelf, E.V.: Astrophys. J. **709**, 436 (2010a)
Halpern, J.P., Gotthelf, E.V.: Astrophys. J. **710**, 941 (2010b)
Han, S., Steiner, A.W.: Phys. Rev. C **96**, 035802 (2017)
Harding, A.K., Muslimov, A.G.: Astrophys. J. **556**, 987 (2001)
Haskell, B., Patruno, A.: Astrophys. J. Lett. **738**, L14 (2011)
Hebeler, K., et al.: Astrophys. J. **773**, 11 (2013)
Heger, A., et al.: Astrophys. J. Lett. **671**, L141 (2007a)
Heger, A., Cumming, A., Woosley, S.E.: Astrophys. J. **665**, 1311 (2007b)
Heinke, C.O., Ho, W.C.G.: Astrophys. J. Lett. **719**, L167 (2010)
Heinke, C.O., et al.: Astrophys. J. **588**, 452 (2003)
Heinke, C.O., et al.: Astrophys. J. **644**, 1090 (2006)
Heinke, C.O., et al.: Astrophys. J. **691**, 1035 (2009)
Heinke, C.O., et al.: Mon. Not. R. Astron. Soc. **444**, 443 (2014)
Hessels, J., et al.: Advancing astrophysics with the square kilometre array. In: Proceedings of Science, vol. 47 (2015)
Hessels, J., et al.: Science **311**, 1901 (2006)
Hewish, A., et al.: Nature **217**, 709 (1968)
Heyl, J.S., Hernquist, L.: Mon. Not. R. Astron. Soc. **300**, 599 (1998)
Ho, W.C.G., Heinke, C.O.: Nature **462**, 71 (2009)
Ho, W.C.G., Potekhin, A.Y., Chabrier, G.: Astrophys. J. Suppl. Ser. **178**, 102 (2008)

Ho, W.C.G., et al.: Sci. Adv. **1**, e1500578 (2015)

Horne, K.: Mon. Not. R. Astron. Soc. **213**, 129 (1985)

Horne, K., Wade, R.A., Szkody, P.: Mon. Not. R. Astron. Soc. **219**, 791 (1986)

Horowitz, C.J., et al.: Phys. Rev. Lett. **114**, 031102 (2015)

Huppenkothen, D., et al.: Astrophys. J. **768**, 87 (2013)

Ibragimov, A., Poutanen, J.: Mon. Not. R. Astron. Soc. **400**, 492 (2009)

in't Zand, J.J.M., Weinberg, N.N.: Astron. Astrophys. **520**, A81 (2010)

in't Zand, J.J.M., Jonker, P.G., Markwardt, C.B.: Astron. Astrophys. **465**, 953 (2007)

in't Zand, J.J.M., et al.: Astron. Astrophys. **553**, A83 (2013)

in't Zand, J.J.M., et al.: 11th Integral conference. In: PoS. ArXiv:1703.07221 (2017, to be published)

Israel, G.L., et al.: Astrophys. J. Lett. **628**, L53 (2005)

Ivanova, N., et al.: Mon. Not. R. Astron. Soc. **386**, 553 (2008)

Ji, L., et al.: Astrophys. J. Lett. **791**, L39 (2014)

Johnston, S., et al.: Astrophys. J. Lett. **387**, L37 (1992)

Jonker, P.G., et al.: Mon. Not. R. Astron. Soc. **354**, 666 (2004)

Jonker, P.G., et al.: Mon. Not. R. Astron. Soc. **429**, 523 (2013)

Juett, A.M., Schulz, N.S., Chakrabarty, D.: Astrophys. J. **612**, 308 (2004)

Kajava, J.J.E., et al.: Mon. Not. R. Astron. Soc. **417**, 1454 (2011)

Kajava, J.J.E., et al.: Mon. Not. R. Astron. Soc. **445**, 4218 (2014)

Kajava, J.J.E., et al.: Mon. Not. R. Astron. Soc. **464**, L6 (2017)

Kaspi, V.M.: Proc. Natl. Acad. Sci. **107**, 7147 (2010)

Kaspi, V.M., Kramer, M.: In: Blandford, R., Sevrin, A. (eds.) Rapporteur Talks in the Proceedings of the 26th Solvay Conference on Physics on Astrophysics and Cosmology, pp 22-61, World Scientific, Singapore (2015/2016). ArXiv:1602.07738

Kaspi, V.M., et al.: Astrophys. J. Lett. **423**, L43 (1994)

Keane, E., et al.: Advancing astrophysics with the square kilometre array, In: PoS, vol. 40 (2015)

Keek, L., et al.: Astrophys. J. **789**, 121 (2014)

Keek, L., Wolf, Z., Ballantyne, D.R.: Astrophys. J. **826**, L12 (2016)

Kiziltan, B., et al.: Astrophys. J. **778**, 66 (2013)

Klochkov, D., et al.: Astron. Astrophys. **556**, A41 (2013)

Klochkov, D., et al.: Astron. Astrophys. **573**, A53 (2015)

Kluzniak, W., Michelson, P., Wagoner, R.V.: Astrophys. J. **358**, 538 (1990)

Kong, A.K.H., et al.: Astrophys. J. Lett. **670**, L17 (2007)

Kramer, M.: In: Röser, S. (ed.) Reviews in Modern Astronomy, vol. 20, p. 255 (2008)

Kramer, M., Wex, N.: Classical Quantum Gravity **26**, 073001 (2009)

Kuulkers, E., et al.: Astron. Astrophys. **399**, 663 (2003)

Lamb, F.K., et al.: Astrophys. J. Lett. **705**, L36 (2009)

Lapidus, I.I., Sunyaev, R.A., Titarchuk, L.G.: Sov. Astron. Lett. **12**, 383 (1986)

Lattimer, J.M.: Astrophys. Space Sci. **336**, 67 (2011)

Lattimer, J.M.: Annu. Rev. Nucl. Part. Sci. **62**, 485 (2012)

Lattimer, J.M., Lim, Y.: Astrophys. J. **771**, 51 (2013)

Lattimer, J.M., Prakash, M.: Astrophys. J. **550**, 426 (2001)

Lattimer, J.M., Prakash, M.: Phys. Rev. **621**, 127 (2016)

Lattimer, J.M., Schutz, B.F.: Astrophys. J. **629**, 979 (2005)

Lattimer, J.M., Steiner, A.W.: Astrophys. J. **784**, 123 (2014)

Lattimer, J.M., et al.: Astrophys. J. **425**, 802 (1994)

Leahy, D.A., Morsink, S.M., Cadeau, C.: Astrophys. J. **672**, 1119 (2008)

Leahy, D.A., et al.: Astrophys. J. **691**, 1235 (2009)

Leahy, D.A., Morsink, S.M., Chou, Y.: Astrophys. J. **742**, 17 (2011)

Lewin, W.H.G., van Paradijs, J., Taam, R.E.: Space Sci. Rev. **62**, 223 (1993)

Li, Z., et al.: Astrophys. J. **845**, 8 (2017)

Lin, J., et al.: Astrophys. J. **723**, 1053 (2010)

Linares, M., et al.: Astrophys. J. **748**, 82 (2012)

Lindblom, L.: Astrophys. J. **398**, 569 (1992)
Lo, K.H., et al.: Astrophys. J. **776**, 19 (2013)
London, R.A., Howard, W.M., Taam, R.E.: Astrophys. J. Lett. **287**, L27 (1984)
London, R.A., Taam, R.E., Howard, W.M.: Astrophys. J. **306**, 170 (1986)
Lü, G.-L., et al.: Mon. Not. R. Astron. Soc. **424**, 2265 (2012)
Ludlam, R.M., et al.: Astrophys. J. **824**, 37 (2016)
Ludlam, R.M., et al.: Astrophys. J. **838**, 79 (2017a)
Ludlam, R.M., et al.: Astrophys. J. **836**, 140 (2017b)
Lugger, P.M., et al.: Astrophys. J. **657**, 286 (2007)
Lyne, A.G., et al.: Science **303**, 1153 (2004)
Lyne, A.G., et al.: Mon. Not. R. Astron. Soc. **451**, 581 (2015)
Lyu, M., Méndez, M., Altamirano, D.: Mon. Not. R. Astron. Soc. **445**, 3659 (2014)
Lyu, M., et al.: Mon. Not. R. Astron. Soc. **463**, 2358 (2016)
Madej, J.: Astrophys. J. **376**, 161 (1991)
Manchester, R.N., Johnston, S.: Astrophys. J. Lett., **441**, L65 (1995)
Marino, A., et al.: Mon. Not. R. Astron. Soc. **479**, 3634 (2018)
Masetti, N., et al.: Astron. Astrophys. **453**, 295 (2006)
Matt, G., Perola, G.C., Piro, L.: Astron. Astrophys. **247**, 25 (1991)
Maxwell, J.E., et al.: Astrophys. J. **756**, 147 (2012)
Medin, Z., et al.: Astrophys. J. **832**, 102 (2016)
Melatos, A., Payne, D.J.B.: Astrophys. J. **623**, 1044 (2005)
Mereghetti, S., Pons, J.A., Melatos, A.: Space Sci. Rev. **191**, 315 (2015)
Meszaros, P., et al.: Astrophys. J. **324**, 1056 (1988)
Miller, M.C.: Astrophys. J. **822**, 27 (2016)
Miller, M.C., et al.: Astrophys. J. Lett. **779**, L2 (2013)
Miller, M.C.: Timing neutron stars: pulsations, oscillations and explosions. In: Belloni, T., Mendez,
    M., Zhang, C.M. (eds.) ASSL, Springer, Berlin (2013, to appear). ArXiv:1312.0029
Miller, M.C., Lamb, F.K.: Astrophys. J. **808**, 31 (2015)
Miller, M.C., Lamb, F.K.: Eur. Phys. J. A **52**, 63 (2016)
Miller, M.C., Lamb, F.K., Cook, G.B.: Astrophys. J. **509**, 793 (1998a)
Miller, M.C., Lamb, F.K., Psaltis, D.: Astrophys. J. **508**, 791 (1998b)
Morsink, S.M., Leahy, D.A.: Astrophys. J. **726**, 56 (2011)
Morsink, S.M., et al.: Astrophys. J. **663**, 1244 (2007)
Motch, C., et al.: Athena white paper. (2013). ArXiv:1306.2334
Müller, H., Serot, B.D.: Nucl. Phys. A **606**, 508 (1996)
Muno, M.P., Özel, F., Chakrabarty, D.: Astrophys. J. **581**, 550 (2002)
Muno, M.P., Özel, F., Chakrabarty, D.: Astrophys. J. **595**, 1066 (2003)
Nath, N.R., Strohmayer, T.E., Swank, J.H.: Astrophys. J. **564**, 353 (2002)
Nättilä, J., et al.: Astron. Astrophys. **581**, A83 (2015)
Nättilä, J., et al.: Astron. Astrophys. **591**, A25 (2016)
Nättilä, J., et al.: Astron. Astrophys. **608**, A31 (2017)
Oertel, M., et al.: J. Phys. G Nucl. Phys. **42**, 075202 (2015)
Ofengeim, D.D., et al.: Mon. Not. R. Astron. Soc. **454**, 2668 (2015)
Özel, F.: Astrophys. J. **583**, 402 (2003)
Özel, F.: Nature **441**, 1115 (2006)
Özel, F.: Rep. Prog. Phys. **76**, 016901 (2013)
Özel, F., Freire, P.: Annu. Rev. Astron. Astrophys. **54**, 401 (2016)
Özel, F., Psaltis, D.: Astrophys. J. Lett. **582**, L31 (2003)
Özel, F., Psaltis, D.: Astrophys. J. **810**, 135 (2015)
Özel, F., Güver, T., Psaltis, D.: Astrophys. J. **693**, 1775 (2009)
Özel, F., Baym, G., Güver, T.: Phys. Rev. D **82**, 101301 (2010)
Özel, F., Gould, A., Güver, T.: Astrophys. J. **748**, 5 (2012a)
Özel, F., et al.: Astrophys. J. **757**, 55 (2012b)
Özel, F., et al.: Astrophys. J. **832**, 92 (2016a)

Özel, F., et al.: Astrophys. J. **820**, 28 (2016b)
Paczynski, B.: Astrophys. J. **267**, 315 (1983)
Paczynski, B., Proszynski, M.: Astrophys. J. **302**, 519 (1986)
Page, D., Reddy, S.: In: Bertulani, C.A., Piekarewicz, J. (eds.) Neutron Star Crust. Nova Science
    Publishers Inc., London (2013). arXiv:1201.5602
Page, D., Geppert, U., Weber, F.: Nucl. Phys. A **777**, 497 (2006)
Page, D., et al.: Astrophys. J. **707**, 1131 (2009)
Page, D., et al.: Phys. Rev. Lett. **106**, 081101 (2011)
Papaloizou, J., Pringle, J.E.: Mon. Not. R. Astron. Soc. **184**, 501 (1978)
Papitto, A., et al.: Nature **501**, 517 (2013a)
Papitto, A., et al.: Mon. Not. R. Astron. Soc. **429**, 3411 (2013b)
Patruno, A., Haskell, B., D'Angelo, C.: Astrophys. J. **746**, 9 (2012)
Patruno, A., Haskell, B., Andersson, N.: Astrophys. J. ArXiv:1705.07669 (2017, submitted)
Pavlov, G.G., Luna, G.J.M.: Astrophys. J. **703**, 910 (2009)
Pavlov, G.G., Zavlin, V.E.: Astrophys. J. Lett. **490**, L91 (1997)
Pavlov, G.G., Shibanov, I.A., Zavlin, V.E.: Mon. Not. R. Astron. Soc. **253**, 193 (1991)
Pavlov, G.G., et al.: Astron. Astrophys. **289**, 837 (1994)
Pavlov, G.G., et al.: Astrophys. J. Lett. **531**, L53 (2000)
Pavlov, G.G., et al.: Neutron stars in supernova remnants. In: Slane, P.O., Gaensler, B.M. (eds.)
    Astronomical Society of the Pacific Conference Series, vol. 271, p. 247 (2002)
Pavlov, G.G., Sanwal, D., Teter, M.A.: Young neutron stars and their environments. In: Camilo, F.,
    Gaensler, B.M. (eds.) IAU Symposium, vol. 218, p. 239 (2004)
Pechenick, K.R., Ftaclas, C., Cohen, J.M.: Astrophys. J. **274**, 846 (1983)
Pfahl, E., Rappaport, S., Podsiadlowski, P.: Astrophys. J. **597**, 1036 (2003)
Pinto, C., et al.: Astron. Astrophys. **551**, A25 (2013)
Piro, A. L., Bildsten, L.: Astrophys. J. **629**, 438 (2005)
Pizzochero, P.M., et al.: Nat. Astron. **1**, 0134 (2017)
Podsiadlowski, P., Rappaport, S., Pfahl, E.D.: Astrophys. J. **565**, 1107 (2002)
Pons, J.A., et al.: Phys. Rev. Lett. **98**, 071101 (2007)
Pons, J.A., Miralles, J.A., Geppert, U.: Astron. Astrophys. **496**, 207 (2009)
Posselt, B., et al.: Astrophys. J. **779**, 186 (2013)
Potekhin, A.Y., Yakovlev, D.G.: Astron. Astrophys. **374**, 213 (2001)
Potekhin, A.Y., Chabrier, G., Yakovlev, D.G.: Astron. Astrophys. **323**, 415 (1997)
Potekhin, A.Y., et al.: Astron. Astrophys. **560**, A48 (2013)
Potekhin, A.Y., Ho, W.C.G., Chabrier, G.: (2016). ArXiv:1605.01281
Poutanen, J.: In: Wijnands, R., Altamirano, D., Soleri, P., Degenaar, N., Rea, N., Casella, P.,
    Patruno, A. Linares, M. (eds.) AIPC, vol. 1068, pp. 77–86 (2008)
Poutanen, J.: Astrophys. J. **835**, 119 (2017)
Poutanen, J., Beloborodov, A.M.: Mon. Not. R. Astron. Soc. **373**, 836 (2006)
Poutanen, J., Gierliński, M.: Mon. Not. R. Astron. Soc. **343**, 1301 (2003)
Poutanen, J., Ibragimov, A., Annala, M.: Astrophys. J. Lett. **706**, L129 (2009)
Poutanen, J., et al.: Mon. Not. R. Astron. Soc. **442**, 3777 (2014)
Prakash, M., Cooke, J.R., Lattimer, J.M.: Phys. Rev. D **52**, 661 (1995)
Pringle, J.E., Rees, M.J.: Astron. Astrophys. **21**, 1 (1972)
Psaltis, D., Belloni, T., van der Klis, M.: Astrophys. J. **520**, 262 (1999)
Psaltis, D., Özel, F., DeDeo, S.: Astrophys. J. **544**, 390 (2000)
Psaltis, D., Özel, F., Chakrabarty, D.: Astrophys. J. **787**, 136 (2014)
Radhakrishnan, V., Srinivasan, G.: Current Sci. **51**, 1096 (1982)
Rajagopal, M., Romani, R.W.: Astrophys. J. **461**, 327 (1996)
Rappaport, S., et al.: Astrophys. J. Lett. **217**, L29 (1977)
Rauch, T., Suleimanov, V., Werner, K.: Astron. Astrophys. **490**, 1127 (2008)
Read, J.S., et al.: Phys. Rev. D **79**, 124032 (2009)
Reardon, D.J., et al.: Mon. Not. R. Astron. Soc. **455**, 1751 (2016)

Rees, M.J.: in Annals of the New York academy of sciences. In: Bergman, P.G., Fenyves, E.J., Motz, L. (eds.) Seventh Texas Symposium on Relativistic Astrophysics, vol. 262, pp. 367–375 (1975)

Reig, P.: Astrophys. Space Sci. **332**, 1 (2011)

Revnivtsev, M., et al.: Astron. Astrophys. **372**, 138 (2001)

Romani, R.W.: Astrophys. J. **313**, 718 (1987)

Romani, R.W.: Nature **347**, 741 (1990)

Romani, R.W., et al.: Astrophys. J. Lett. **760**, L36 (2012)

Romani, R.W., Filippenko, A.V., Cenko, S.B.: Astrophys. J. **804**, 115 (2015)

Romanova, M.M., Kulkarni, A.K., Lovelace, R.V.E.: Astrophys. J. Lett. **673**, L171 (2008)

Ruderman, M.: Astrophys. J. **366**, 261 (1991)

Ruderman, M.A., Sutherland, P.G.: Astrophys. J. **196**, 51 (1975)

Rutledge, R.E., et al.: Astrophys. J. **514**, 945 (1999)

Rutledge, R.E., et al.: Astrophys. J. **559**, 1054 (2001a)

Rutledge, R.E., et al.: Astrophys. J. **551**, 921 (2001b)

Rutledge, R.E., et al.: Astrophys. J. **578**, 405 (2002a)

Rutledge, R.E., et al.: Astrophys. J. **577**, 346 (2002b)

Sanwal, D., et al.: Astrophys. J. Lett. **574**, L61 (2002)

Sato, K.: Progress Theor. Phys. **62**, 957 (1979)

Schatz, H., et al.: Astrophys. J. **524**, 1014 (1999)

Schulz, N.S., Corrales, L., Canizares, C.R.: Astrophys. J. **827**, 49 (2016)

Servillat, M., Webb, N.A., Barret, D.: Astron. Astrophys. **480**, 397 (2008)

Servillat, M., et al.: Mon. Not. R. Astron. Soc. **423**, 1556 (2012)

Shao, L., et al.: Advancing astrophysics with the square kilometre array. In: PoS, vol. 42 (2015)

Shibanov, I.A., et al.: Astron. Astrophys. **266**, 313 (1992)

Shternin, P.S., et al.: Mon. Not. R. Astron. Soc. **412**, L108 (2011)

Singh, K.P., et al.: Space telescopes and instrumentation 2014: ultraviolet to gamma ray. In: Proceedings of SPIE, vol. 9144, 91441S (2014)

Sobolev, V.V.: A Treatise on Radiative Transfer. Princeton, New York (1963)

Soffitta, P., et al.: Space telescopes and instrumentation 2016: ultraviolet to gamma ray. In: Proceedings of SPIE, vol. 9905, 990515 (2016)

Spitkovsky, A., Levin, Y., Ushomirsky, G.: Astrophys. J. **566**, 1018 (2002)

Stairs, I.H.: Science **304**, 547 (2004)

Stairs, I.H., et al.: Mon. Not. R. Astron. Soc. **325**, 979 (2001)

Steiner, A.W., Watts, A.L.: Phys. Rev. Lett. **103**, 181101 (2009)

Steiner, A.W., Lattimer, J.M., Brown, E.F.: Astrophys. J. **722**, 33 (2010)

Steiner, A.W., Lattimer, J.M., Brown, E.F.: Astrophys. J. Lett. **765**, L5 (2013)

Steiner, A.W., et al.: Phys. Rev. C **91**, 015804 (2015)

Steiner, A.W., et al.: Mon. Not. R. Astron. Soc. **476**, 421 (2018)

Stella, L., Vietri, M., Morsink, S.M.: Astrophys. J. Lett. **524**, L63 (1999)

Stevens, A.L., et al.: Astrophys. J. **833**, 244 (2016)

Stiele, H., Yu, W., Kong, A.K.H.: Astrophys. J. **831**, 34 (2016)

Strohmayer, T., Bildsten, L.: In: Lewin, W., van der Klis, M. (eds.) Compact stellar X-ray Sources. Cambridge Astrophysics Series, vol. 39. Cambridge University Press, Cambridge (2006), pp. 113–156

Strohmayer, T.E., Watts, A.L.: Astrophys. J. Lett. **632**, L111 (2005)

Strohmayer, T.E., et al.: Astrophys. J. Lett. **469**, L9 (1996)

Strohmayer, T.E., Zhang, W., Swank, J.H.: Astrophys. J. Lett. **487**, L77 (1997)

Suleimanov, V., et al.: Astron. Astrophys. **522**, A111 (2010a)

Suleimanov, V.F., Pavlov, G.G., Werner, K.: Astrophys. J. **714**, 630 (2010b)

Suleimanov, V., et al.: Astrophys. J. **742**, 122 (2011a)

Suleimanov, V., Poutanen, J., Werner, K.: Astron. Astrophys. **527**, A139 (2011b)

Suleimanov, V.F., Poutanen, J., Werner, K.: Astron. Astrophys. **545**, A120 (2012a)

Suleimanov, V.F., Pavlov, G.G., Werner, K.: Astrophys. J. **751**, 15 (2012b)

Suleimanov, V.F, et al.: Astrophys. J. Suppl. Ser. **210**, 13 (2014)
Suleimanov, V.F., et al.: Eur. Phys. J. A **52**, 20 (2016)
Suleimanov, V.F., et al.: Mon. Not. R. Astron. Soc. **472**, 3905 (2017a)
Suleimanov, V.F., et al.: Astron. Astrophys. **600**, A43 (2017b)
Suleimanov, V.F., et al.: Mon. Not. R. Astron. Soc. **466**, 906 (2017c)
Tananbaum, H.: IAU Circ. 7246 (1999)
Tauris, T.M., van den Heuvel, E.P.J.: Formation and evolution of compact stellar X-ray sources. In: Lewin, W.H.G., van der Klis, M. (eds.) In: Compact Stellar X-ray Sources. Cambridge University Press, Cambridge, pp. 623–665 (2006)
Thorsett, S.E., Chakrabarty, D.: Astrophys. J. **512**, 288 (1999)
Thorsett, S.E., et al.: Astrophys. J. Lett. **405**, L29 (1993)
Tian, W.W., et al.: Astrophys. J. Lett. **679**, L85 (2008)
Tian, W.W., et al.: Astrophys. J. **712**, 790 (2010)
Tsang, M.B., et al.: Phys. Rev. C **86**, 015803 (2012)
van Adelsberg, M., Lai, D.: Mon. Not. R. Astron. Soc. **373**, 1495 (2006)
van den Eijnden, J., et al.: Mon. Not. R. Astron. Soc. **466**, L98 (2017)
van den Heuvel, E.P.J.: Astrophys. J. Lett. **198**, L109 (1975)
van Kerkwijk, M.H., Breton, R.P., Kulkarni, S.R.: Astrophys. J. **728**, 95 (2011)
van Paradijs, J., Lewin, H.G.: Astron. Astrophys. **157**, L10 (1986)
van Paradijs, J., et al.: Publ. Astron. Soc. Jpn. **42**, 633 (1990)
van Straaten, S., et al.: Astrophys. J. **540**, 1049 (2000)
Verbiest, J.P.W., et al.: Astrophys. J. **679**, 675 (2008)
Viganò, D., et al.: Mon. Not. R. Astron. Soc. **434**, 123 (2013)
Viironen, K., Poutanen, J.: Astron. Astrophys. **426**, 985 (2004)
Wagoner, R.V.: Astrophys. J. **278**, 345 (1984)
Walsh, A.R., Cackett, E.M., Bernardini, F.: Mon. Not. R. Astron. Soc. **449**, 1238 (2015)
Watts, A.L., Reddy, S.: Mon. Not. R. Astron. Soc. **379**, L63 (2007)
Watts, A.L., Strohmayer, T.E.: Astrophys. J. Lett. **637**, L117 (2006)
Watts, A., et al.: Advancing astrophysics with the square kilometre array. In: PoS, vol. 43 (2015)
Watts, A.L., et al.: Rev. Mod. Phys. **88**, 021001 (2016)
Webb, N.A., Barret, D.: Astrophys. J. **671**, 727 (2007)
Weinberg, N., Miller, M. C., Lamb, D.Q.: Astrophys. J. **546**, 1098 (2001)
Weinberg, N., Bildsten, L., Schatz, H.: Astrophys. J. **639**, 1018 (2006)
Weisskopf, M.C., et al.: Astrophys. J. **743**, 139 (2011)
Weisskopf, M.C., et al.: Space telescopes and instrumentation 2016: ultraviolet to gamma ray, 2016. In: Proceedings of SPIE, vol. 9905, 990517 (2016)
Wheeler, J.A.: Annu. Rev. Astron. Astrophys. **4**, 393 (1966)
Wijnands, R., Degenaar, N.: Mon. Not. R. Astron. Soc. **434**, 1599 (1966)
Wijnands, R., van der Klis, M.: Nature **394**, 344 (1998)
Wijnands, R., Degenaar, N., Page, D.: Mon. Not. R. Astron. Soc. **432**, 2366 (2013)
Wijnands, R., Degenaar, N., Page, D.: J. Astrophys. Astron. **38**, 49 (2017)
Wilson-Hodge, C.A., et al.: Proceedings of SPIE, vol. 9905. Society of Photo-Optical Instrumentation Engineers (SPIE) Conference Series, 99054Y (2016)
Wiringa, R.B., Fiks, V., Fabrocini, A.: Phys. Rev. C **38**, 1010 (1988)
Worpel, H., Galloway, D.K., Price, D.J.: Astrophys. J. **801**, 60 (2015)
Yakovlev, D.G., Pethick, C.J.: Annu. Rev. Astron. Astrophys. **42**, 169 (2004)
Yu, W., van der Klis, M.: Astrophys. J. Lett. **567**, L67 (2002)
Zamfir, M., Cumming, A., Galloway, D.K.: Astrophys. J. **749**, 69 (2012)
Zampieri, L., et al.: Astrophys. J. **439**, 849 (1995)
Zhang, S., et al.: Space telescopes and instrumentation 2014: ultraviolet to gamma ray. In: Proceedings of SPIE, vol. 9144, 914421 (2014)
Zavlin, V.E.: In: Becker, W. (ed.) Astrophysics and Space Science Library. vol. 357, p. 181. Springer, Berlin (2009)

Zavlin, V.E., Pavlov, G.G.: In: Becker,W., Lesch, H., Trümper, J. (eds.) Neutron stars, pulsars, and
    supernova remnants, vol. 263 (2002)
Zavlin, V.E., Pavlov, G.G., Shibanov, Y.A.: Astron. Astrophys. **315**, 141 (1996)
Zdunik, J.L., Fortin, M., Haensel, P.: Astron. Astrophys. **599**, A119 (2017)
Zhang, S.N., et al.: SPIE **9905**, E1 (2016)

Chapter 6
Nuclear Equation of State for Compact Stars and Supernovae

G. Fiorella Burgio and Anthea F. Fantina

Abstract The equation of state (EoS) of hot and dense matter is a fundamental input to describe static and dynamical properties of neutron stars, core-collapse supernovae and binary compact-star mergers. We review the current status of the EoS for compact objects, that have been studied with both ab-initio many-body approaches and phenomenological models. We limit ourselves to the description of EoSs with purely nucleonic degrees of freedom, disregarding the appearance of strange baryonic matter and/or quark matter. We compare the theoretical predictions with different data coming from both nuclear physics experiments and astrophysical observations. Combining the complementary information thus obtained greatly enriches our insight into the dense nuclear matter properties. Current challenges in the description of the EoS are also discussed, mainly focusing on the model dependence of the constraints extracted from either experimental or observational data, the lack of a consistent and rigorous many-body treatment at zero and finite temperature of the matter encountered in compact stars (e.g. problem of cluster formation and extension of the EoS to very high temperatures), the role of nucleonic three-body forces, and the dependence of the direct URCA processes on the EoS.

6.1 Introduction

An equation of state (EoS) is a relation between thermodynamic variables describing the state of matter under given physical conditions. Independent variables are usually the particle numbers N_i, the temperature T, and the volume V; alternatively, one can use the particle number densities $n_i = N_i/V$, the corresponding particle

G. Fiorella Burgio
Istituto Nazionale di Fisica Nucleare, Sez. di Catania, Catania, Italy
e-mail: fiorella.burgio@ct.infn.it

A. F. Fantina (✉)
Grand Accélérateur National d'Ions Lourds (GANIL), CEA/DRF - CNRS/IN2P3, Caen, France
e-mail: anthea.fantina@ganil.fr

© Springer Nature Switzerland AG 2018
L. Rezzolla et al. (eds.), *The Physics and Astrophysics of Neutron Stars*,
Astrophysics and Space Science Library 457,
https://doi.org/10.1007/978-3-319-97616-7_6

number fractions being $Y_i = n_i/n_B = N_i/N_B$ (e.g. $Y_e = N_e/N_B$ for the electron fraction, N_B and N_e being the baryon and electron number, respectively). In addition, conservation laws hold, so that there are conserved quantities such as the total baryon number, the total electric charge number, and the total lepton number.

In astrophysics, EoSs are usually implemented in hydrodynamic (or hydrostatic) models that describe the evolution (or the static structure) of the macroscopic system. An EoS can be determined if the system is in thermodynamic equilibrium, i.e. if thermal, mechanical, and chemical equilibrium are achieved. In particular, the latter generally is not attained in main sequence stars or in explosive nucleosynthesis. In these scenarios, a full reaction network that takes into account the reaction cross sections of the species present in the medium has to be considered. Otherwise, if the timescale of the nuclear reactions is much shorter than the dynamic evolution timescales, nuclear statistical equilibrium (NSE) can be assumed. This situation is typically achieved for temperatures $T \gtrsim 0.5\,\text{MeV}$ (Iliadis 2007). On the other hand, weak interactions are not generally in equilibrium. Specifically, in core-collapse supernovae (CCSNe), the electron-capture reaction, $p + e^- \rightarrow n + \nu_e$, is not in equilibrium for baryon densities below $n_B \approx 10^{-3} - 10^{-4}\,\text{fm}^{-3}$ (or equivalently, for mass-energy densities below $\rho_B \approx$ few $10^{11}\,\text{g cm}^{-3}$). In the first stages of CCSNe, neutrinos are not in equilibrium and are not included in the EoS, but treated in transport schemes. In later stages of CCSNe, and in (proto-) neutron stars ((P)NSs), neutrinos are trapped and weak interactions are in equilibrium. They can thus be included in the EoS, and a lepton (or neutrino) fraction can be introduced. For (mature) cold NSs, beta equilibrium without neutrinos is usually achieved since neutrinos become untrapped, and the electron fraction is fixed by charge neutrality together with the beta-equilibrium condition.

The determination of an EoS for compact objects is one of the main challenges in nuclear astrophysics, because of the wide range of densities, temperatures, and isospin asymmetries encountered in these astrophysical objects. Moreover, current nuclear physics experiments cannot probe all the physical conditions found in compact stars, thus theoretical models are required to extrapolate to unknown regions. The set of independent thermodynamic variables for the most general EoS are usually the baryon density n_B, the temperature T, and the charge fraction, e.g. the electron fraction Y_e, or alternatively the charge density (see also the CompOSE manual (Typel et al. 2015) for an explanation of the thermodynamic variables and potentials). However, for cold NSs, the temperature (below 1 MeV) is lower than typical nuclear energies and the zero-temperature approximation can be adopted, thus making, together with the beta-equilibrium condition, the independent variables of the EoS reduced to the density only. On the other hand, in CCSNe, in compact-star mergers, and in black-hole (BH) formation, the temperature can rise to a few tens or even above a hundred MeV. Therefore, the approximate range of thermodynamic variables over which the most general EoS has to be computed is: $10^{-11} \lesssim n_B \lesssim$ few fm^{-3}, $0 \leq T \lesssim 150\,\text{MeV}$, and $0 < Y_e < 0.6$ (see, e.g., Figs. 1–2 in Oertel et al. 2017 and O'Connor and Ott 2011).

In this Chapter, we aim to give an overview of the present status of the EoSs for compact-star modelling, with particular focus on the underlying many-body methods, and to discuss some of the current challenges in the field. After a brief introduction on the nucleon-nucleon interaction in Sect. 6.2.1, we will review the theoretical many-body methods in Sect. 6.2.2, both microscopic (Sect. 6.2.2.1) and phenomenological (Sect. 6.2.2.2). In Sect. 6.2.3 we will discuss the constraints on the EoS obtained in both nuclear physics experiments (Sect. 6.2.3.1) and astrophysical observations (Sect. 6.2.3.2). We will present in Sect. 6.2.4 the application of the EoSs in compact-object modelling: we will first discuss the zero-temperature NS case (Sect. 6.2.4.1), then we will introduce some widely used general purpose EoSs and discuss their impact in CCSNe, BH formation, and in binary mergers (Sect. 6.2.4.2). A brief description of the available online databases on the EoSs is given in Sect. 6.2.5. Finally, in Sect. 6.3, we will discuss some of the current challenges for the EoS modelling and in Sect. 6.4 we will draw our conclusions.

6.2   Current Status of Many-Body Methods and Equation of State

6.2.1   The Nucleon-Nucleon Interaction: A Brief Survey

The properties of the nuclear medium are strongly determined by the features of the nucleon-nucleon (NN) interaction, in particular the presence of a hard repulsive core. The nuclear Hamiltonian should in principle be derived from the quantum chromodynamics (QCD), but this is a very difficult task which presently cannot be realised. There are three basic classes of bare nucleonic interactions:

- Phenomenological interactions mediated by meson exchanges;
- Chiral expansion approach;
- Models that include explicitly the quark-gluon degrees of freedom.

In the phenomenological approaches, quark degrees of freedom are not treated explicitly but are replaced by hadrons—baryons and mesons—in which quarks are confined. Very refined and complete phenomenological models have been constructed for the NN interactions, e.g. the Paris potential (Lacombe et al. 1980), the Bonn potential (Machleidt and Slaus 2001), the Nijmegen potentials (Nagels et al. 1977, 1978) also with hyperon-nucleon (YN) (Maessen et al. 1989) and hyperon-hyperon (YY) potentials (Rijken et al. 1999). Those phenomenological models have been tested using thousands of experimental data on NN scattering cross sections, from which the phase shifts in different two-body channels are extracted with high precision up to an energy of about 300 MeV in the laboratory, even if discrepancies between the results of different groups still persist (Machleidt and Slaus 2001).

The most widely known potential models are the Urbana (Lagaris and Pand-haripande 1981) and Argonne potentials, the latest version called the $v18$ potential (Wiringa et al. 1995). The structure of the NN potential is very complex and depends on many quantities characterising a two-nucleon system. These quantities enter via operator invariants consistent with the symmetries of the strong interactions, and involve spin, isospin, and orbital angular momentum. The NN potential acting between a nucleon pair ij is a Hermitian operator $\hat{v}_{ij}$ in coordinate, spin, and isospin spaces. A sufficiently generic form of $\hat{v}_{ij}$ able to reproduce the abundance of NN scattering data is

$$\hat{v}_{ij} = \sum_{u=1}^{18} v_u(r_{ij})\hat{O}_{ij}^u, \tag{6.1}$$

where the first fourteen operators are charge-independent, i.e., invariant with respect to rotation in the isospin space:

$$\begin{aligned}
\hat{O}_{ij}^{u=1,\dots 14} = {}& 1,\ \boldsymbol{\tau}_i \cdot \boldsymbol{\tau}_j,\ \boldsymbol{\sigma}_i \cdot \boldsymbol{\sigma}_j,\ (\boldsymbol{\sigma}_i \cdot \boldsymbol{\sigma}_j)(\boldsymbol{\tau}_i \cdot \boldsymbol{\tau}_j),\ \hat{S}_{ij},\ \hat{S}_{ij}(\boldsymbol{\tau}_i \cdot \boldsymbol{\tau}_j), \\
& \hat{\mathbf{L}} \cdot \hat{\mathbf{S}},\ \hat{\mathbf{L}} \cdot \hat{\mathbf{S}}(\boldsymbol{\tau}_i \cdot \boldsymbol{\tau}_j),\ \hat{L}^2,\ \hat{L}^2(\boldsymbol{\tau}_i \cdot \boldsymbol{\tau}_j),\ \hat{L}^2(\boldsymbol{\sigma}_i \cdot \boldsymbol{\sigma}_j), \\
& \hat{L}^2(\boldsymbol{\sigma}_i \cdot \boldsymbol{\sigma}_j)(\boldsymbol{\tau}_i \cdot \boldsymbol{\tau}_j),\ (\hat{\mathbf{L}} \cdot \hat{\mathbf{S}})^2,\ (\hat{\mathbf{L}} \cdot \hat{\mathbf{S}})^2(\boldsymbol{\tau}_i \cdot \boldsymbol{\tau}_j). \tag{6.2}
\end{aligned}$$

The notation $\mathbf{r}_{ij} = \mathbf{r}_i - \mathbf{r}_j$ indicates the relative position vector, whereas $\boldsymbol{\sigma}_i$ and $\boldsymbol{\sigma}_j$ are spins (in units of $\hbar/2$), and $\boldsymbol{\tau}_i$ and $\boldsymbol{\tau}_j$ are isospins (in units of $\hbar/2$). The relative momentum is denoted by $\hat{\mathbf{p}}_{ij} = \hat{\mathbf{p}}_i - \hat{\mathbf{p}}_j$; $\hat{\mathbf{L}} = \mathbf{r}_{ij} \times \hat{\mathbf{p}}_{ij}$ is the total orbital angular momentum, and $\hat{L}^2$ its square in the centre-of-mass system. The spin-orbit coupling enters via $\hat{\mathbf{L}} \cdot \hat{\mathbf{S}}$, being $\hat{\mathbf{S}} = (\boldsymbol{\sigma}_i + \boldsymbol{\sigma}_j)/2$ the total spin (in units of $\hbar$). Analogously, we define the total isospin $\hat{\mathbf{T}} = (\boldsymbol{\tau}_i + \boldsymbol{\tau}_j)/2$. The tensor coupling enters via the tensor operator

$$\hat{S}_{ij} = 3(\boldsymbol{\sigma}_i \cdot \mathbf{n}_{ij})(\boldsymbol{\sigma}_j \cdot \mathbf{n}_{ij}) - \boldsymbol{\sigma}_i \cdot \boldsymbol{\sigma}_j, \tag{6.3}$$

where $\mathbf{n}_{ij} = \mathbf{r}_{ij}/r_{ij}$. Both the spin-orbit and tensor couplings are necessary for explaining experimental data. The terms with $\hat{O}_{ij}^{u=15,\dots 18}$ are small and break charge independence, and they correspond to $v_{np}(T = 1) = v_{nn} = v_{pp}$, while the charge symmetry implies only that $v_{nn} = v_{pp}$. Modern fits to very precise nucleon scattering data indicate the existence of charge-independence breaking. However, the effect of such forces on the energy of nucleonic matter is much smaller than the uncertainties of many-body calculations and therefore can be neglected while constructing the EoS.

A different approach to the study of the NN interaction is the one based on quark and gluon degrees of freedom, thus connecting the low energy nuclear physics phenomena with the underlying QCD structure of the nucleons. This is quite difficult because the whole hadron sector is in the non-perturbative regime, due to confinement. A possible strategy is based on the systematic use of the symmetries embodied in the hadronic QCD structure. The main symmetry which is explicitly broken in the confined matter is the chiral symmetry, since the bare u and d quark masses in non-strange matter are just a few MeV. According to the general Goldstone theorem, this results in the physical mass of the pion, which suggests to treat the pion degrees of freedom explicitly and to describe the short range part by structureless contact terms. Along this line, Weinberg (1990, 1991) proposed a scheme for including in the interaction a series of operators which reflect the partially broken chiral symmetry of QCD. The strength parameters associated to each operator are then determined by fitting the NN phase shifts, the properties of deuteron and of few-body nuclear systems. The method is then implemented in the framework of the Effective Field Theory (EFT), i.e. by ordering the terms according to their dependence on the physical parameter q/m, where m is the nucleon mass and q a generic momentum that appears in the Feynman diagram for the considered process. This parameter is assumed to be small and each term is dependent on a given power of this parameter thus fixing its relevance. In this way a hierarchy of the different terms of the forces is established. In particular, the pion exchange term is treated explicitly and is considered the lowest order (LO) term of the expansion. Moreover, it is found that the three-body forces (TBFs) so introduced are of higher order than the simplest two-body forces and they are treated on an equal footing. They arise first at next-to-next leading order (N^2LO) and, as a consequence, because of the hierarchy intrinsic in the chiral expansion, TBFs are expected to be smaller than two-body forces, at least within the range of validity of the expansion, whereas four-body forces appear only at next-to-next-to-next leading order (N^3LO) level, and so on. It has to be stressed that in the TBFs the same couplings that fix the two-body forces have to be used and, in general, only a few additional parameters must be introduced as the order increases. Therefore the TBFs are automatically consistent with the two-body forces, and so on for the higher order many-nucleon forces. At present the nucleonic interaction has been calculated up to N^4LO (Hu et al. 2017). An exhaustive list of higher order diagrams up to N^3LO can be found in review papers (Meißner 2005; Epelbaum et al. 2009). This Chiral Perturbation Expansion (ChPE) can be used to construct NN interactions that are of reasonably good quality in reproducing the two-body data (Entem and Machleidt 2002; Holt et al. 2010). The assumption of a small q/m parameter in principle restricts the applications of these forces to not too large momenta, and therefore to a not too large density of nuclear matter. It turns out that the safe maximum density is around the saturation value, n_0. This method has been refined along the years and many applications can be found in the literature.

Another approach inspired by the QCD theory of strong interaction has been developed in Fujiwara et al. (2007), Oka et al. (1987), Oka and Yazaki (1980), Shimizu (1989), Valcarce et al. (2005). In this approach, based on the resonating-group method (RGM), the quark degree of freedom is explicitly introduced and the NN interaction is constructed from gluon and meson exchange between quarks, the latters being confined inside the nucleons. The resulting interaction is highly non-local due to the RGM formalism and contains a natural cut-off in momentum. The most recent model, named fss2 (Fujiwara et al. 2001, 2007), reproduces closely the experimental phase shifts, and fairly well the data on the few-body systems, e.g. the triton binding energy is reproduced within 300 keV. Recently, it has been shown that the fss2 interaction is able to reproduce correctly the nuclear matter saturation point without the TBF contribution (Baldo and Fukukawa 2014).

Recently, a further possibility of constructing the NN interaction based on lattice QCD has been explored, see Aoki et al. (2012), Beane et al. (2011) for a review. This tool, from which one should be able in principle to calculate the hadron properties directly from the QCD Lagrangian, is extremely expensive from the numerical point of view and current simulations can be performed only with large quark masses. In fact, an accurate simulation has to be made on a fine grid spacing and large volumes, thus requiring high performance computers. Hopefully in the next few years high precision calculations will be possible, especially for those channels where scarce experimental data are available, e.g. the nucleon-hyperon interaction.

A further class of NN interactions is based on renormalization group (RG) methods (see, e.g., Entem and Machleidt (2002), Bogner et al. (2010) for a complete review). The main effect of the hard core in the NN interaction is to produce scattering to high momenta of the interacting particles. A possible way to soften the hard core from the beginning is by integrating out all the momenta larger than a certain cut-off Λ and "renormalize" the interaction to an effective interaction V_{low} in such a way that it is equivalent to the original interaction for momenta $q < \Lambda$. The V_{low} interaction turns out to be much softer, since no high momentum components are present and, as a consequence, three- and many-body forces emerge automatically from a pure two-body force. The short range repulsion is replaced by the non local structure of the interaction. The cut-off Λ is taken above 300 MeV in the laboratory, corresponding to relative momentum $q \approx 2.1\,\mathrm{fm}^{-1}$, that is the largest energy where the experimental data are established. The fact that V_{low} is soft has the advantage to be much more manageable than a hard core interaction, in particular it can be used in perturbation expansion and in nuclear structure calculations in a more efficient way (Bogner et al. 2010; Furnstahl and Hebeler 2013).

6.2.2 Theoretical Many-Body Methods

The theoretical description of matter in extreme conditions is a very challenging task. Moreover, current nuclear physics experiments cannot probe all the physical conditions encountered in compact stars. Therefore, theoretical models are required to extrapolate to unknown regions. The undertaken theoretical approaches also depend on the relevant degrees of freedom of the problem, from nuclei and nucleons at lower densities and temperature, to additional particles, such as hyperons and quarks, at high densities and temperature. The current theoretical many-body approaches to describe a nuclear system can be divided into two main categories:

1. **Ab-initio (microscopic) approaches**, that start from "realistic" two-body inter-actions fitted to experimental NN scattering data and to the properties of bound few-nucleon systems. Examples of these kinds of models are Green's function methods, (Dirac-)Brueckner Hartree-Fock, variational, coupled cluster, and Monte Carlo methods. Despite the tremendous progress that has been done in the last years, these methods cannot yet be applied to large finite nuclear systems. Nevertheless, recent developments allow ab-initio methods to reach medium to "heavy" nuclei, see e.g. Cipollone et al. (2013), Somà et al. (2013), Binder et al. (2014), Cipollone et al. (2015). Therefore, in the description of dense matter, the ab-initio models are usually restricted to homogeneous matter; thus, they are not applied to describe clustered matter (like in SN cores or NS crusts).
2. **Phenomenological approaches**, that rely on effective interactions which depend on a certain number of parameters fitted to reproduce properties of finite nuclei and nuclear matter. This class of methods are widely used in nuclear structure and astrophysical applications. Among them, there are self-consistent mean-field models and shell-model approaches. In astrophysics, the latters have been employed, for example, to study electron-capture rates on nuclei relevant for SN simulations (see e.g. Langanke and Martínez-Pinedo (2014) and references therein). Alternatively, models based on self-consistent mean-field approaches are widely used, in particular, to build EoSs of dense matter. These methods, based on the nuclear energy-density functional (EDF) theory, can be either non-relativistic (e.g. using Skyrme or Gogny interactions) or relativistic (based on an effective Lagrangian with baryon and meson fields). A more macroscopic approach to treat the many-body system is the (finite-range) liquid-drop model, which parameterizes the energy of the system in terms of global properties such as volume energy, asymmetry energy, surface energy, etc. and whose parameters are fitted phenomenologically. The liquid-drop model usually describes well the trend of nuclear binding energies and has been largely applied to construct EoSs for compact stars.

In the following, we will not aim at giving a complete review on the different theoretical many-body approaches (see, e.g., Ring and Schuck 2004; Müther and Polls 2000; Bender et al. 2003; Baldo and Burgio 2012; Duguet 2014; Carlson et al. 2015), but we will give an overview of the two kinds of approaches, focusing on the latest advances.

6.2.2.1 Ab-Initio Approaches

A microscopic many-body method is characterised mainly by two basic elements: the realistic bare interaction among nucleons and the many-body scheme followed in the calculation of the EoS. The many-body methods can be enumerated as follows:

- The Bethe-Brueckner-Goldstone (BBG) diagrammatic method and the corresponding hole-line expansion,
- The relativistic Dirac-Brueckner Hartree-Fock (DBHF) approach,
- The variational method,
- The coupled cluster expansion,
- The self-consistent Green's function (SCGF),
- The renormalization group (RG) method,
- Different methods based on Monte Carlo (MC) techniques.

A brief survey of all those methods is given below. For further details the reader is left to the quoted references.

- *The Bethe-Brueckner-Goldstone expansion.*
  The BBG many-body theory is based on the re-summation of the perturbation expansion of the ground-state energy of nuclear matter (Baldo 1999; Baldo and Maieron 2007). The original bare NN interaction is systematically replaced by an effective interaction that describes the in-medium scattering processes, the so-called G-matrix, that takes into account the effect of the Pauli principle on the scattered particles, and the in-medium potential $U(k)$ felt by each nucleon, k being the momentum. The corresponding integral equation for the G-matrix can be written as

$$\langle k_1 k_2 | G(\omega) | k_3 k_4 \rangle = \langle k_1 k_2 | v | k_3 k_4 \rangle + \sum_{k_3' k_4'} \langle k_1 k_2 | v | k_3' k_4' \rangle$$

$$\times \frac{\left(1 - \Theta_F(k_3')\right)\left(1 - \Theta_F(k_4')\right)}{\omega - e_{k_3'} + e_{k_4'} + i\eta} \langle k_3' k_4' | G(\omega) | k_3 k_4 \rangle \ , \quad (6.4)$$

where v is the bare NN interaction, ω is the starting energy, the two factors $(1 - \Theta_F(k))$ force the intermediate momenta to be above the Fermi momentum ("particle states"), the single-particle energy being $e_k = \hbar^2 k^2 / 2m + U(k)$, with m the particle mass, and the summation includes spin-isospin variables. The main feature of the G-matrix is that it is defined even for bare interactions with an infinite hard core, thus making the perturbation expansion more manageable. The introduction and choice of the in-medium single-particle potential are essential to make the re-summed expansion convergent. The resulting nuclear EoS can be calculated with good accuracy in the Brueckner two hole-line approximation with the continuous choice for the single-particle potential, the results in this scheme being quite close to the calculations which include also the three hole-line contribution (Song et al. 1998).

One of the well known results of all non-relativistic many-body approaches is the need of introducing TBFs in order to reproduce correctly the saturation point in symmetric nuclear matter. For this purpose, TBFs are reduced to a density dependent two-body force by averaging over the generalised coordinates (position, spin, and isospin) of the third particle, assuming that the probability of having two particles at a given distance is reduced according to the two-body correlation function. In the BBG calculations for nuclear matter, a phenomenological approach to the TBF was adopted, based on the so-called Urbana model for finite nuclei, which consists of an attractive two-pion exchange contribution between two nucleons via the excitation of a third nucleon, e.g. a Δ-baryon (Fujita and Miyazawa 1957), supplemented by a parameterized repulsive part (Carlson et al. 1983; Pieper 2008; Pieper et al. 2001; Pudliner et al. 1995), adjusted to the properties of light nuclei. In the nuclear matter case, the two parameters contained in the Urbana TBF (Baldo et al. 1997; Zhou et al. 2004; Li et al. 2006) were accurately tuned in order to get an optimal nuclear matter saturation point. In symmetric nuclear matter, this TBF produces a shift in the binding energy of about $+1\,\mathrm{MeV}$ and of $-0.01\,\mathrm{fm}^{-3}$ in density. The problem of such a procedure is that the TBF is dependent on the two-body force. The connection between two-body and TBFs within the meson-nucleon theory of nuclear interaction is extensively discussed and developed in Zuo et al. (2002a,b). At present the theoretical status of microscopically derived TBFs is still quite rudimentary; however, a tentative approach has been proposed using the same meson-exchange parameters as the underlying NN potential. Results have been obtained with the Argonne $v18$ (Wiringa et al. 1995), the Bonn B (Brockmann and Machleidt 1990), and the Nijmegen 93 potentials (Li et al. 2008; Li and Schulze 2008). Alternatively, latest nuclear matter calculations (Logoteta et al. 2015) used a new class of chiral inspired TBF, showing that the considered TBF models are not able to reproduce simultaneously the correct saturation point and the properties of three- and four-nucleon systems.

Recently, it has been shown that the role of TBF is greatly reduced if the NN potential is based on a realistic constituent quark model (Baldo and Fukukawa 2014) which can explain at the same time few-nucleon systems and nuclear matter, including the observational data on NSs and the experimental data on heavy-ion collisions (HICs) (Fukukawa et al. 2015). An extensive comparison among several EoSs obtained using different two-body and TBFs is illustrated afterwards.

- *The Dirac-Brueckner Hartree-Fock approach.*
The relativistic approach is the framework on which the nuclear EoS should be ultimately based. The best relativistic treatment developed so far is the Dirac-Brueckner approach, about which excellent review papers can be found in the literature (see, e.g., Machleidt 1989). In the relativistic context, the only two-body forces that have been used are the ones based on meson exchange models. The DBHF method has been developed in analogy with the non-relativistic case, where the two-body correlations are described by introducing the in-medium relativistic G-matrix. This is a difficult task, and in general one keeps the interaction as instantaneous (static limit) and a reduction to a three-dimensional formulation from a four-dimensional one. The main relativistic effect is due to the use of the spinor formalism which has been shown (Brown et al. 1987) to be equivalent to introducing a particular TBF, the so-called Z-diagram. This TBF

turns out to be repulsive and consequently produces a saturating effect. In fact the DBHF gives a better saturation point than the BHF. In this way, a definite link between DBHF and BHF + TBF is established. Indeed, including in BHF only these particular TBFs, one gets results close to DBHF calculations, see e.g. Li et al. (2006). Generally speaking, the EoS calculated within the DBHF method turns out to be stiffer above saturation than the ones calculated from the BHF + TBF method. Currently, some features of this method are still controversial and the results depend strongly on the method used to determine the covariant structure of the in-medium G-matrix.

- *The variational method.*

In the variational method one assumes that the ground-state trial wave function Ψ can be written as

$$\Psi_{trial}(r_1, r_2, \ldots \ldots) = \prod_{i<j} f(r_{ij}) \Phi(r_1, r_2, \ldots .) , \qquad (6.5)$$

where Φ is the unperturbed ground-state wave function, properly antisymmetrised, and the product runs over all possible distinct pairs of particles. The correlation factors f are determined by the Ritz-Raleigh variational principle, i.e. by imposing that the mean value of the Hamiltonian gets a minimum

$$\frac{\delta}{\delta f} \frac{\langle \Psi_{trial} | H | \Psi_{trial} \rangle}{\langle \Psi_{trial} | \Psi_{trial} \rangle} = 0 . \qquad (6.6)$$

In principle this is a functional equation for f and it is intended to transform the uncorrelated wave function $\Phi(r_1, r_2, \ldots .)$ to the correlated one, and can be written explicitly in a closed form only if additional suitable approximations are introduced. Once the trial wave function is determined, all the expectation values of other operators can be calculated. Therefore the main task in the variational method is to find a suitable ansatz for the correlation factors f. Several different methods exist for the calculation of f, e.g. in the nuclear context the Fermi-Hyper-Netted-Chain (FHNC) (Fantoni and Rosati 1975; Pandharipande and Wiringa 1979) calculations have been proved to be efficient.

For nuclear matter at low densities, two-body correlations play an essential role, and this justifies the assumption that f is actually a two-body operator $\hat{F}_{ij}$. Generally one assumes that $\hat{F}$ can be expanded in the same spin-isospin, spin-orbit, and tensor operators appearing in the NN interaction (Fantoni and Fabrocini 1998; Carlson et al. 2015). Due to the formal structure of the Argonne NN forces, most variational calculations have been performed with this class of NN interactions, often supplemented by the Urbana TBFs. Many excellent review papers exist in the literature on the variational method and its extensive use for the determination of nuclear matter EoS, e.g. Pandharipande and Wiringa (1979), Navarro et al. (2002). The best known and most used variational nuclear matter EoS is the Akmal-Pandharipande-Ravenhall (APR) (Akmal et al. 1998).

A detailed discussion on the connection between variational method and BBG expansion can be found in Baldo and Maieron (2007).

Other methods based on the variational principle are widely used in nuclear physics to evaluate expectation values. Among those, we mention the coupled-cluster theory, proposed in Coester (1958), Coester and Kümmel (1960), in which the correlation operator is represented in terms of the cluster operator. The method has been proved to be successful in recent nuclear matter calculations with chiral NN interactions (Hagen et al. 2007, 2014) and also in nuclear structure calculations (Hagen et al. 2012, 2010). The variational Monte Carlo (VMC) approach is also widely used in nuclear physics to evaluate expectation values. Several calculations have been performed for light nuclei, including two and three-body correlations (Wiringa et al. 2014), but the EoS of homogeneous nuclear matter is hard to obtain, due to the increasingly large computational effort with the number of nucleons (see Navarro et al. (2002), Carlson et al. (2015) for complete reviews).

- *Chiral effective field theory (χ EFT) approach.*
  High-precision nuclear potentials based on chiral perturbation theory (ChPT) (Entem and Machleidt 2003; Machleidt and Entem 2011) are nowadays widely employed to link QCD, the fundamental theory of strong interactions, to nuclear many-body phenomena. In particular, for nuclear matter, many-nucleon forces are of course relevant. In this case another scale appears, k_F/m, k_F being the Fermi momentum, which is of the same order of the pion mass m_π at saturation and it is smaller than a typical hadron scale. In the chiral limit it is then natural to expand in k_F/m, and this expansion can be obtained from the vacuum ChPT expansion (Kaiser et al. 2002). For nuclear matter the correction thus obtained with respect to the vacuum diagrams gives a direct contribution to the EoS of nuclear matter, and this correction is clearly proportional to a power of k_F/m. Also in this case a cut-off must be introduced, and its tuning allows to obtain a saturation point and compressibility in fair agreement with phenomenology. Along the same lines more sophisticated expansions can be developed, including a power counting modified for finite density systems, where the small scale is fixed by both k_F and m_π/m. The results thus obtained are in good agreement with the most advanced non-relativistic many-body calculations (Lacour et al. 2011). A different approach can be developed, where the many-nucleon interactions built in vacuum are directly used in nuclear matter calculations. In this case the ChPT is used in conjunction with the EFT scheme. In recent years, χEFT has been used for studying nuclear matter within various theoretical frameworks like many-body perturbation theory (Hebeler et al. 2011; Wellenhofer et al. 2014; Coraggio et al. 2014; Drischler et al. 2016), SCGF framework (Carbone et al. 2013b), in-medium chiral perturbation theory (Holt et al. 2013), the BHF approach (Kohno 2013; Li and Schulze 2012), and quantum Monte Carlo methods (Gezerlis et al. 2013; Roggero et al. 2014; Lynn et al. 2016). Several reliable calculations have been performed up to twice the saturation density n_0, beyond which uncertainties were estimated by analysing the order-by-order convergence in the chiral expansion and the many-body perturbation theory

(Coraggio et al. 2014; Holt and Kaiser 2017). Variations in the resolution scale (Bogner et al. 2005) and low-energy constants appearing in the two-nucleon and three-nucleon forces were systematically explored (Hebeler and Schwenk 2010). It has been found that the theoretical uncertainty band grows rapidly with the density beyond n_0, due to the missing third-order terms at low densities and higher-order contributions in the chiral expansion. This has consequences not only for the EoS, but also for the symmetry energy at saturation density, S_0, and the slope parameter L, as discussed in Baldo and Burgio (2016).

- *Self-consistent Green's function.*

Another way to approach the many-body problem is through the many-body Green's functions formalism (Dickhoff and Van Neck 2008). In this approach one performs a diagrammatic analysis of the many-body propagators in terms of free one-body Green's functions and two-body interactions. The perturbative expansion results in an infinite series of diagrams, among which one has to choose those which are relevant for the considered physical problem. Depending on the approximation, one can either choose a given number of diagrams or sum an infinite series of them, in analogy with the BHF approach. In the description of nuclear matter, the method is conventionally applied at the ladder approximation level, which encompasses at once particle-particle and hole-hole propagation, and this represents the main difference with respect to the G-matrix, where only particle-particle propagators are included. At a formal level, the comparison between the BHF and the SCGF approaches is not straightforward. Even though both approaches arise from a diagrammatic expansion, the infinite subsets of diagrams considered in the two approaches are not the same, and the summation procedures are also somewhat different. Whereas the BHF formalism in the continuous choice can be derived from the ladder SCGF formalism after a series of approximations, this is not the case for the full BBG expansion. In principle, if both BBG and SCGF were carried out to all orders, they should yield identical results. BBG theory, however, is an expansion in powers of density (or hole-lines), and the three-hole line results seem to indicate that it converges quickly. The error in the SCGF expansion is more difficult to quantify, as one cannot directly compute (or even estimate) which diagrams have to be included in the expansion. Reviews on the applications of the method to nuclear problems can be found in Müther and Polls (2000), Dickhoff and Barbieri (2004).

Also for the SCGF method the inclusion of TBFs is essential. So far TBFs were not included in the ladder approximation, however a method has been developed recently in Carbone et al. (2013b), and applied to symmetric nuclear matter using chiral nuclear interactions. TBFs are included via effective one-body and two-body interactions, and are found to improve substantially the saturation point (Carbone et al. 2013a).

One has to notice that because of the well-known Cooper instability (Cooper 1956), through which a fermionic many-body system with an attractive interaction tends to form pairs at the Fermi surface, low-temperature nuclear matter is unstable with respect to the formation of a superfluid or superconducting state. The Cooper instability shows up as a pole in the T-matrix when the temperature falls below the critical temperature for the transition to the superfluid/supercon-ducting state. Therefore current calculations are often performed at temperatures above the critical temperature and extrapolated to zero temperature, see Frick (2004) for details.

- *Quantum Monte Carlo methods.*

  Quantum Monte Carlo (QMC) methods are very successful in describing the ground state of fermionic systems, like liquid ^{3}He, or bosons, like atomic liquid ^{4}He. Modern computer technology has allowed the extension of the QMC method to nuclear systems, which have more complicated interactions and correlation structures. The mostly used versions are the auxiliary field diffusion Monte Carlo (AFDMC) (Gandolfi et al. 2009) and the Green's function Monte Carlo (GFMC) (Carlson et al. 2003) methods, which differ in the treatment of the spin and isospin degrees of freedom. It has to be noticed that the computing time increases exponentially with the number of particles, which limits the number of nucleons considered by GFMC up to 16 neutrons. The largest nucleus considered is ^{12}C. The AFDMC strategy allows to efficiently sample spin-isospin correlations in systems with a sufficient number of nucleons (N = 114). A recent comparison has demonstrated that both methods give very close results for neutron drops with $N \leq 16$ (Gandolfi et al. 2011). However, the accuracy of the different QMC versions is limited by the fermion sign problem (Schmidt and Kalos 1987), for which different approximations are adopted (Carlson et al. 2012, 2015). This seriously limits the potentiality of the QMC approach.

  In spite of its recent progress, it is not yet possible to perform GFMC and AFDMC calculations with the Argonne $v18$ potential, mainly due to technical problems associated with the spin-orbit structure of the interaction and the trial wave function, which induce very large statistical errors. In order to overcome this problem, the full operatorial structure of current high-quality NN potentials has been simplified and more manageable NN potentials have been developed containing less operators with readjusted parameters. In particular, we mention the V8', V6', and V4' potentials (Pudliner et al. 1997; Wiringa and Pieper 2002), eventually supplemented with the Urbana TBFs. Recently, a local chiral potential has been developed (Gezerlis et al. 2013) which is well suited for QMC techniques.

Finite-Temperature Equation of State

In the latest stage of the SN collapse the EoS of asymmetric nuclear matter at finite temperature plays a major role in determining the final evolution. Microscopic calculations of the nuclear EoS at finite temperature are quite few. The variational calculation by Friedman and Pandharipande (1981) was one of the first few semi-microscopic investigations. In the resulting EoS for symmetric nuclear matter, one recognizes the familiar Van der Waals shape, which entails a liquid-gas phase transition, with a definite critical temperature T_c, i.e. the temperature at which the minimum in the Van der Waals isotherm disappears. In the Friedman and Pandharipande work, the critical temperature turns out to be around $T_c = 18 - 20$ MeV. The values of the critical temperature, however, depend on the theoretical scheme, as well as on the particular NN interaction adopted. In particular, non-relativistic Brueckner-like calculations at finite temperature (Baldo and Ferreira 1999), where the formalism by Bloch and De Dominicis (1958, 1959a,b) was followed, confirmed the Friedman and Pandharipande findings with very similar values of T_c. The main difficulty in this approach is the lack of thermodynamic consistency. In fact the thermodynamic relation $P = -\mathscr{F} + \mu n$, which connects the pressure P with the free energy density $\mathscr{F}$, the chemical potential μ, and the number density n and usually referred to as the Hughenoltz-Van Hove theorem, is not satisfied. In other words, the pressure calculated in such a way does not coincide with the pressure calculated from $P = -\Omega/V$ (Ω being the grand potential and V the volume). In Baldo and Ferreira (1999), a procedure was proposed in order to overcome this problem: the pressure is calculated from the derivative of the free energy per particle so that the Hughenoltz-Van Hove theorem is automatically satisfied. The difficulty is that the chemical potential determined by fixing the density in the Fermi distribution is not strictly the one extracted from the derivative of $\mathscr{F}$, as it should be. In any case, the procedure looks most reliable within the Brueckner scheme (see Baldo and Ferreira (1999) for details). For completeness, we remind the reader that the Brueckner approximation, both at zero and finite temperature, violates the Hugenoltz-Van Hove theorem. On the contrary, the Hughenoltz-Van Hove theorem is strictly fulfilled within the SCGF method (Baym 1962; Rios et al. 2006, 2008). The results at the two-body correlation level, when only two-body forces are used, in some cases are similar to the Brueckner ones, in some others they differ appreciably according to the forces used. The main difference with the Brueckner scheme is the introduction in the ladder summation of the hole-hole propagation, which gives a repulsive contribution. As a result, the critical temperature in the SCGF approach with Argonne v_{18} potential is found to be about $T_c \approx 11.6$ MeV, whereas in the BHF approach $T_c \approx 18.1$ MeV (Rios et al. 2008), depending on the adopted NN interaction. As far as the DBHF is concerned, it turns out that the critical temperature within this scheme is definitely smaller than in the non-relativistic scheme, about 10 MeV against 18–20 MeV (Ter Haar and Malfliet 1986; Haar and Malfliet 1987; Huber et al. 1998). This cannot be due to relativistic effects, since the critical density is about $1/3$ of the saturation density, but to a different behaviour of the Dirac-Brueckner EoS at low density. This point remains to be clarified. Indeed, there are experimental data from heavy-ion reactions that point towards a value of $T_c > 15$ MeV (Borderie and Rivet 2008; Karnaukhov et al. 2008).

Results and Discussion

We will discuss here the results obtained with some of the widely used many-body methods illustrated above. The simplest constraint that has to be considered is the reproduction of the phenomenological saturation point; we will see that this condition is not trivially fulfilled. Other constraints will be analysed afterwards.

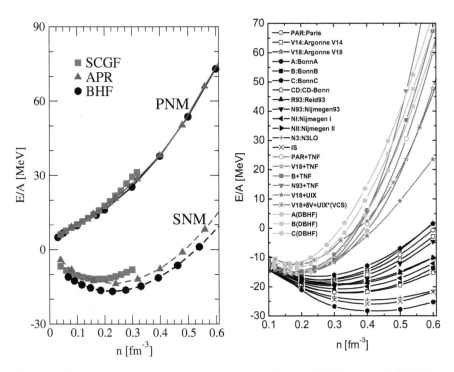

Fig. 6.1 Left panel: Symmetric and pure neutron matter EoS from BHF (black circles), SCGF (red squares), and APR (blue triangles) schemes including only two-body forces. Right panel: Energy per nucleon of symmetric nuclear matter obtained with different NN and TBF interactions, for different theoretical approaches; courtesy of H.-J. Schulze. See text for details

We begin by discussing a comparison between the BHF, SCGF, and APR EoS with only two-body forces. Results are displayed for the binding energy per nucleon, E/A, of symmetric nuclear matter (SNM) and pure neutron matter (PNM) in Fig. 6.1, left panel, where the Argonne $v18$ NN potential is adopted. We notice a substantial agreement between all methods for PNM calculations, whereas for SNM some differences show up. It is well known that the discrepancies between SCGF and BHF result in an overall repulsive effect in the binding energy (Dewulf et al. 2003), which is mainly due to the inclusion in the SCGF expansion of the hole-hole propagation. Those effects are quite sizeable in SNM. For instance, the saturation point shifts from $n_0 = 0.25 \, \text{fm}^{-3}$, $E(n_0)/A = -16.8 \, \text{MeV}$ for BHF to $n_0 = 0.17 \, \text{fm}^{-3}$, $E(n_0)/A = -11.9 \, \text{MeV}$ for SCGF. While the shift seems to go towards the right saturation density, the value of the SCGF saturation energy is quite high.

In the right panel of Fig. 6.1, we display the energy per particle in SNM obtained with a set of NN potentials and with different TBFs (TNF in the legend). On the standard BHF level (black curves) one obtains in general too strong binding, varying between the results with the Paris (Lacombe et al. 1980), $v18$ (Wiringa et al. 1995), and Bonn C potentials (Machleidt 1989; Brockmann and Machleidt 1990) (less binding), and those with the Bonn A (Machleidt and Slaus 2001), N³LO (Entem and Machleidt 2002, 2003), and IS (Doleschall et al. 2003) potentials (very strong binding). Including TBFs, with the Paris, Bonn B, $v18$, and Njimegen 93 (Stoks et al. 1994) potentials, adds considerable repulsion and yields results slightly less repulsive than the DBHF ones with the Bonn potentials (green curves). This is not surprising, because it is well known that the major effect of the DBHF approach amounts to include the TBF corresponding to nucleon-antinucleon excitation by 2σ exchange within the BHF calculation. In those BHF calculations microscopic TBFs have been included and those turn out to be more repulsive at high density than the phenomenological TBF, i.e. the one derived from the Urbana UIX model (full red symbols). This is a clear sign of uncertainty in the role of TBF at large density. The blue curve with asterisks represents the results of the APR EoS obtained with the Urbana UIX TBF, which was adjusted to reproduce the saturation point, by varying mainly one parameter, as it has been done in the BHF approach. We notice that the effect of the TBF is quite moderate around saturation ($\delta n = -0.01\,\mathrm{fm}^{-3}$, $\delta E = +1\,\mathrm{MeV}$), but they are essential to get the correct saturation point.

The contribution of the TBF to the saturation mechanism is quite relevant when chiral forces are used. In fact, as illustrated in Fig. 6.2 (left panel), without TBF the EoS does not display an apparent saturation, and anyhow close to saturation density the TBF contribution is quite large, of several MeV, in contrast to the case of meson exchange interactions, where the TBF contribution around saturation is of

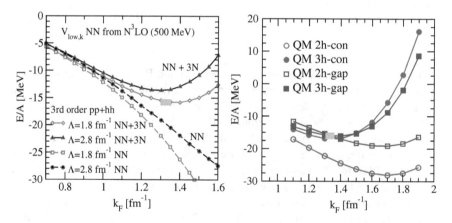

Fig. 6.2 Left panel: EoS of symmetric nuclear matter including third-order perturbative corrections based on chiral interaction at N³LO level. The cut-off for TBFs is fixed at $\Lambda_{3N} = 2.0\,\mathrm{fm}^{-1}$, while the cut-off Λ of the (evolved) chiral two-body interaction is set equal to the two indicated values. Figure adapted from Hebeler et al. (2011). Right panel: EoS of symmetric nuclear matter from the quark model (QM) interaction fss2. The open (full) symbols correspond to the two (three) hole-line calculations, respectively. The circles (squares) indicate the calculation with the continuous (gap) choice for the single-particle potential. In both panels, the green box indicates the saturation point

the order of 1 MeV. These calculations are perturbative in character, as indicated in the labels, but this feature holds true also in more refined calculations. We notice the relevance of the momentum cut-off Λ, that is introduced in order to control the point interaction forces, that otherwise would produce a divergent contribution. In general the chiral two-body forces are evolved according to the RG method before they are employed in the many-body calculations, as in Hebeler et al. (2011). The same procedure has been followed in the BHF calculations of Sammarruca et al. (2015), where a similar relevance of the TBF was found. The most sophisticated many-body calculation with chiral forces is probably the one of Hagen et al. (2014), where the coupled cluster method was employed up to a selected set of three-body clusters. In the latter paper it was also found that it is difficult with the same chiral forces to fit both the binding energy of few nucleon systems (H, ^{3}He) and the saturation point. This feature is common to the meson exchange forces discussed above, for which the same difficulty was found. A similar conclusion was found in the BHF calculations of Logoteta et al. (2016), where it is suggested to fit simultaneously the few-body binding energy and the saturation point.

More recently it has been shown (Baldo and Fukukawa 2014) that the fss2 interaction is able to reproduce also the correct nuclear matter saturation point without any additional parameter or need to introduce TBFs. This is illustrated in Fig. 6.2 (right panel), where the EoS for symmetric matter is reported. The open symbols correspond to the EoS calculated at the BHF level of approximation with the gap (GC, squares) and the continuous (CC, circles) choices, while the full symbols correspond to the EoS calculated by including the three hole-line contribution. One can see that also in this case the final EoS is insensitive to the choice of the single-particle potential. The main result of this calculation is that the saturation point is reproduced without the introduction of TBFs. Note that this is the only two-body interaction that is able to reproduce with a fair accuracy both the binding energy of few nucleon systems and the saturation point of nuclear matter, without the need of TBFs.

The conclusion one can draw from this rapid review of results with different forces is that the relevance of the TBFs is model dependent and that the explicit introduction of the quark degrees of freedom reduces strongly the relevance of the TBFs and allows to connect few-body systems to nuclear matter.

In the following, we will only consider microscopic EoSs which fit correctly the saturation point. The considered set of EoS includes variational calculations (APR) (Akmal et al. 1998), BHF calculations with TBFs, both phenomenological (Baldo et al. 1997; Taranto et al. 2013) and microscopically derived (Grangé et al. 1989; Zhou et al. 2004), and relativistic Dirac-Brueckner calculations (Fuchs 2006). A comparison among these EoSs, along with some phenomenological EoSs, will be discussed in Sect. 6.2.2.2.

6.2.2.2   Phenomenological Approaches

Phenomenological approaches make use of effective interactions instead of bare ones to treat dense matter, either homogeneous or clustered. Most of these approaches rely on the (nuclear) EDF theory, that has proved to be successful in reproducing the properties of medium-mass and heavy nuclei (Bender et al. 2003; Stone and Reinhard 2007) but can also be applied to describe infinite systems, either inhomogeneous (like SN cores ot NS crusts) or homogeneous (like NS cores). Indeed, nuclear EDFs presently provide a complete and accurate description of ground-state properties and collective excitations over the whole nuclear chart (e.g. Bender et al. 2003; Lalazissis et al. 2004). Non-uniform (nucleonic) clustered matter, that is present at subsaturation density at relatively low temperatures, can be treated using various models, like the NSE model, liquid-drop type models, (semi-classical) Thomas-Fermi models, etc. On the other hand, a different approach to construct the (phenomenological) EoS is to use purely parameterized EoSs, that do not rely on any description of the NN interaction. An example is given by the piecewise polytropic EoS for nuclear matter of Read et al. (2009), while a metamodel for the nucleonic EoS inspired from a Taylor expansion around the saturation density of symmetric nuclear matter is proposed and parameterized in terms of the empirical parameters in Margueron et al. (2018a) and employed to analyse global properties of NSs in Margueron et al. (2018b).

Nuclear EDF/Mean-Field Approaches

The density functional theory has been very successfully applied in various fields of physics and chemistry. The advantage of this method is to recast the complex many-body problem of interacting particles (like nucleons) into an effective independent particle approach (see, e.g., Bender et al. (2003), Lacroix (2010), Duguet (2014) for a review). The total energy of the system is thus expressed as a functional of the nucleon number densities, the kinetic energy densities, and the spin-current densities, which are functions of the three spatial coordinates. It has been proved that the exact ground state of the system can be obtained from an energy minimisation procedure (see Hohenberg and Kohn (1964), Kohn and Sham (1965) for the case of electron systems). The issue lies in the fact that the exact form of the functional itself is not known a priori. Therefore, one has to rely on phenomenological functionals, either relativistic, usually derived from a Lagrangian, or non-relativistic, traditionally derived from effective forces of Skyrme or Gogny type. In the nuclear context, this approach has been often referred to as the self-consistent (relativistic) mean-field Hartree-Fock method, or the Hartree-Fock+BCS and Hartree-Fock-Bogoliubov (HFB) methods if pairing is included (see, e.g., Ring and Schuck (2004), Brink and Broglia (2005), and Chap. 8 in this book for details on pairing). The EDFs depend on a certain number of parameters fitted to reproduce some properties of known nuclei and nuclear matter, as well as ab-initio calculations of infinite nuclear matter. The non-uniqueness of the fitting procedure and the choice of the experimental data used to fit the parameters have led to several different functionals, that may give very different predictions when applied outside the domain where they were fitted (see, e.g., Goriely and Capote 2014). The situation is particularly critical for astrophysical applications, where extrapolations of nuclear masses are required for the description of the deepest regions of the NS crust, in SN cores, and in nucleosynthesis calculations. However, the reliability of these EDFs for very neutron-rich systems can be partially tested by comparing their predictions for the properties of pure neutron matter with results obtained from microscopic ab-initio calculations. Moreover, another question arises as whether the EDF parameters determined by fitting nuclear data at zero temperature can be reliably used when applying the EDFs at finite temperature. Different studies have shown that the temperature dependence of the couplings is rather weak up to a few tens of MeV (e.g., Moustakidis and Panos 2009; Fantina et al. 2012b; Fedoseew and Lenske 2015), but it remains to be clarified whether these conclusions still hold at higher temperatures ($\gtrsim 100$ MeV) that can be reached in CCSNe or NS mergers (see Sect. 6.3.2).

- *Non-relativistic EDFs.*

  Non-relativistic approaches usually start from an Hamiltonian $\hat{H}$ for the many-body system, $\hat{H} = \hat{T} + \hat{V}$, where $\hat{T} = \sum_i \hat{p}^2/2m_i$ is the kinetic term ($\hat{p}$ being the momentum operator and m_i the mass of the species i) and $\hat{V}$ is the potential term. The latter accounts for the two-body (pseudo)potential, that allows one to incorporate physical properties like effective masses. Three-body interactions were included explicitly in the seminal work by Vautherin and Brink (1972), while most recent EDFs rather employ density-dependent terms that include in an effective way higher-order correlations. However, these terms can generate some issues when implemented beyond mean field (e.g., Bender et al. 2003). The total energy of the system E can also be written in terms of only the EDF without knowing explicitly the underlying Hamiltonian, $E = \int d^3r \, \mathscr{E}_{\text{EDF}} + E_{\text{Coul}}$, where $\mathscr{E}_{\text{EDF}}$ is the energy functional that includes the kinetic energy density and the interaction term modelling the effective interaction among particles, and E_{Coul} is the Coulomb energy. In calculations including pairing, the pair energy, E_{pair}, has to be accounted for, and in finite nuclei the corrections for spurious motion, E_{corr}, have to be subtracted (e.g., Bender et al. 2003).

  The Skyrme-type effective interactions are zero-range density-dependent interactions and they are widely used in nuclear structure and in astrophysical applications since they allow for fast numerical computations. Since the pioneer work of Skyrme (1956), several extensions have been proposed (see, e.g., Lesinski et al. 2007; Bender et al. 2009; Chamel et al. 2009; Margueron et al. 2009; Margueron and Sagawa 2009; Zalewski et al. 2010; Fantina et al. 2011; Hellemans et al. 2012; Margueron et al. 2012; Davesne et al. 2015), allowing to include and study, for example, the tensor part of the EDF, the spin-density-dependent terms, as well as a surface-peaked effective-mass term. The accuracy in reproducing experimentally measured properties of finite nuclei has been greatly increased in recent well-calibrated Skyrme-type EDFs (see, e.g., Goriely et al. (2010, 2013a,b, 2016a), Chamel et al. (2015), and Washiyama et al. 2012; Kortelainen et al. 2014). Even though in some cases Skyrme forces may exhibit some instabilities and self-interaction errors (see, e.g., the discussions in Cao et al. 2010; Chamel 2010; Chamel and Goriely 2010; Hellemans et al. 2013; Navarro and Polls 2013; Pastore et al. 2014, 2015), these can be cured with appropriate modifications of the EDF. In Skyrme forces, usually the pairing interaction is specified separately, even if attempts to construct the pairing force starting from the same Skyrme interaction exist (e.g., Dobaczewski et al. 1984), although this results in more involving calculations. Many Skyrme models have been recently compared against several nuclear matter constraints in Dutra et al. (2012). However, most of the criteria chosen by the authors to discriminate among the different parameterizations are still matter of debate, particularly regarding the symmetry energy coefficients (see e.g. Li et al. 2014), and most of the constraints are known with large error bars (see also Sects. 6.2.3.1 and 6.3.1). Therefore, it might be premature to rule out some models on those basis (see e.g. Stevenson et al. 2013).

On the other hand, finite-range (density-dependent) interactions are generally derived from the Gogny interaction (Dechargé and Gogny 1980). For these EDFs, the same finite-range interaction has been generally employed for the pairing term. However, this kind of EDFs are less widely used in astrophysics with respect to the Skyrme ones, because of the more involving numerical computations (see, e.g., Goriely et al. (2009, 2016b), Hilaire et al. (2016); see also Sellahewa and Rios (2014) for an analysis of different Gogny interactions and their predictions of the homogeneous-matter properties).

In addition to the Skyrme and Gogny effective interactions, other non-relativistic approaches have been developed. The two-body separable monopole (SMO) interaction has been designed to be an effective interaction whose terms are separable in the space (and isospin) coordinates with parameters fitted to the properties of finite nuclei (Stevenson et al. 2001; Rikovska Stone et al. 2002). Other approaches include the three-range Yukawa (M3Y) type interactions (Nakada 2003) and the local EDF developed, e.g., in Fayans et al. (2000), Fayans et al. (2001) in which the self-consistent Gor'kov equations are solved to study nuclear ground-state properties. More recently, new EDFs have been constructed within an approach inspired by the Kohn-Sham density functional theory (Baldo et al. 2010, 2013). These Barcelona-Catania-Paris(-Madrid) (BCP and BCPM) EDFs have been derived by introducing in the functional results from microscopic nuclear and neutron-matter BHF calculations, and by adding appropriate surface, Coulomb, and spin-orbit contributions. With a reduced number of parameters, these EDFs yield a very good description of properties of finite nuclei.

Nevertheless, a particular attention has to be paid when applying non-relativistic EDFs at high densities, where the EoSs based on these EDFs may become superluminal.

- *Relativistic mean-field (RMF) and relativistic Hartree-Fock (RHF) models.*
  RMF models have been successfully employed in nuclear structure, to describe both nuclei close to the valley of stability and exotic nuclei (see, e.g., Nikšić et al. (2011) for a review, and Dutra et al. (2014, 2016a,b) for a recent comparison of different RMF parameterizations). RMF models have been constructed based on the framework of quantum hadrodynamics (see, e.g., Fetter and Walecka 1971; Walecka 1974; Serot 1992). The basic idea of these models is the same as for non-relativistic mean-field approaches: the many-body state is built up as an independent particle or quasiparticle state from the single-particle wave functions, which are, in this framework, four-component Dirac spinors. A nucleus is thus described as a system of Dirac nucleons whose motion is governed by the Dirac equation. The NN interaction can be described as zero-range (point coupling), where the single-particle potentials entering the Dirac equations are functions of the various relativistic densities, or as finite-range interaction, in terms of an exchange of mesons through an effective Lagrangian $\mathscr{L} = \mathscr{L}_{\mathrm{nuc}} + \mathscr{L}_{\mathrm{mes}} + \mathscr{L}_{\mathrm{int}}$, where the different terms account for the nucleon, the free meson, and the interaction contribution, respectively. The isoscalar scalar σ meson and the isoscalar vector ω meson mediate the long and short-range part of the interaction, respectively, in symmetric nuclear matter, while isovector mesons (like the isovector vector ρ meson and the isovector scalar δ meson) need to be included as well to treat isospin-asymmetric matter. It is also possible to reformulate the model in terms of the corresponding EDF. The RMF total energy is then given by $E = \int d^3r \, \mathscr{E}_{\mathrm{RMF}} + E_{\mathrm{Coul}}$, where $\mathscr{E}_{\mathrm{RMF}}$ includes the nucleon, meson, and interaction contributions. As in non-relativistic EDFs, E_{pair} has to be included when accounting for pairing and the centre-of-mass correction has to be subtracted (Bender et al. 2003) in finite nuclei. The interaction term depends on the nucleon-meson coupling constants that are usually determined by fitting nuclei or nuclear-matter properties. In particular, coupling to scalar mesons is needed to obtain a correct spin-orbit interaction in finite nuclei. However, in the RMF, spin-orbit splitting occurs without the recourse to an assumed spin-orbit interaction. The Klein-Gordon equations for the meson fields, coupled to the Dirac equations for the nucleons, are solved self-consistently in the RMF approximation, where the meson-field operators are replaced by their expectation values in the nuclear ground state. However, for a quantitative description of nuclear matter and finite nuclei, one needs to include a medium dependence of the effective mean-field interactions accounting for higher-order many-body effects, analogously to non-relativistic EDFs. A medium dependence can either be introduced by including non-linear (NL) meson self-interaction terms in the Lagrangian, or by assuming an explicit density dependence (DD) for the meson-nucleon couplings. The former approach has been employed in constructing several phenomenological RMF interactions, like the popular NL3 (Lalazissis et al. 1997), PK1, PK1R (Long et al. 2004), and FSUGold (Todd-Rutel and Piekarewicz 2005) (see, e.g., Maslov et al. (2015) for a recent study with a NL Walecka model, see Boguta and Bodmer 1977; Serot and Walecka 1986). In the second approach, the functional form of the density dependence

of the coupling can be derived by comparing results with microscopic Dirac-Brueckner calculations of symmetric and asymmetric nuclear matter or it can be fully phenomenological, with parameters adjusted to experimental data (see, e.g. the DD-RMF models of Nikšić et al. 2002; Long et al. 2004; Typel 2005; Gögelein et al. 2008; Roca-Maza et al. 2011; Antić and Typel 2015). The density dependence gives rise to the so-called rearrangement contributions which are essential for the thermodynamic consistency of the model. Generalised (g)RMF models, which are an extension of the DD-RMF models, where the degrees of freedom of nucleons and (light) clusters are included in the Lagrangian, have been also formulated (see Sect. 6.2.2.3).

On the other hand, point-coupling models have been developed (see, e.g., Nikolaus et al. 1992; Rusnak and Furnstahl 1997; Bürvenich et al. 2002; Nikšić et al. 2008; Zhao et al. 2010), recently reaching a level of accuracy comparable to that of standard meson-exchange effective interactions when applied for the description of finite nuclei. Parameters of these models can also be constrained by χEFTs (Finelli et al. 2003, 2004, 2006).

However, these models do not explicitly take into account the antisymmetrisation of the many-body wave function. Despite the computational more involving character of the finite-range interaction mediated by meson exchange, relativistic Hartree-Fock including exchange terms and relativistic HFB accounting for pairing have also been implemented (see, e.g., Meng et al. (2006) for a review and the more recent Long et al. 2007, 2010).

A RMF model incorporating the internal quark structure of baryons is the *quark-meson coupling* model. This approach treats nucleons as bound states of three quarks and interacting via meson exchange. In addition to standard mesons, pions are also included. This model has been applied to study NS properties, e.g., in Thomas et al. (2013), Whittenbury et al. (2014).

As an illustrative example, in Fig. 6.3, the energy per particle for symmetric (SNM, upper panels) and pure neutron matter (PNM, lower panels) is plotted as a function of baryon density, both for microscopic models (left panels) and for different phenomenological functionals (right panels). Among the latters, we show the non-relativistic SLy4 (Chabanat et al. 1998) and BSk21 (Goriely et al. 2010) Skyrme-type EDFs, the D1M (Goriely et al. 2009) Gogny EDF, and the BCPM (Baldo et al. 2010, 2013) functional, and the RMF NL3 (Lalazissis et al. 1997) and DD-MEδ (Roca-Maza et al. 2011) models. In both SNM and PNM, up to about twice the saturation density, all approaches, except very stiff EoSs, yield similar results. Microscopic EoSs diverge at higher densities because of the different treatment of TBFs and three-body correlations. Since microscopic calculations of PNM are very accurate, they can serve as benchmark calculations to constrain more phenomenological models. The spread in these results can thus provide an estimate of the current theoretical uncertainties (see also Gandolfi et al. 2015). For phenomenological approaches, a similar spread at high density can be noticed. Indeed, these models have parameters that are fitted on experimental data known with some uncertainties around saturation (see also Sect. 6.2.3.1), thus their

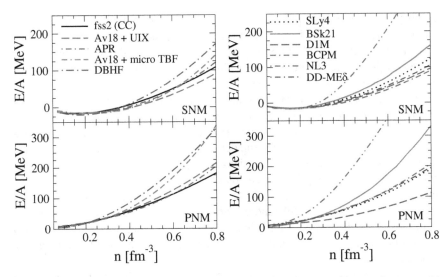

Fig. 6.3 Energy per particle in symmetric (upper panels) and pure neutron matter (lower panels) as a function of baryon number density for different models, both microscopic (left panels) and phenomenological (right panels). See text for details

behaviour at larger densities where no experimental data are available can be very different. It has to be mentioned that, at low density, the appearance of clusters has to be considered in the EoS (Röpke et al. 2013); the treatment of clustered matter will be discussed in the next Section.

6.2.2.3 Approaches to Treat Non-uniform Matter

Non-uniform nuclear matter (either nuclei or clusters) is expected to be present at low densities (below saturation) and relatively low temperatures, thus in the crust of NSs and in SN cores. At present, the best ab-initio many-body calculations employing realistic interactions are not affordable to describe inhomogeneous matter. Therefore, one has to rely on different approximations based on phenomenological effective interactions. These approaches either (1) use the so-called single-nucleus approximation, i.e. the composition of matter is assumed to be made of one representative heavy nucleus (the one that is energetically favoured), possibly together with light nuclei (often represented by alpha particles) and unbound nucleons, or (2) consider the distribution of an ensemble of nuclei. It has been shown that employing the single-nucleus approximation instead of considering a full distribution of nuclei has a small impact on thermodynamic quantities (Burrows and Lattimer 1984). However, differences might be significant if the composition is dominated by light nuclei, or in the treatment of nuclear processes like electron captures in CCSNe. Indeed, the nucleus that is energetically favoured from thermodynamic arguments might not be the one with the highest reaction rate. There are different ways to identify the onset of instability with respect to cluster formation, thus the transition from uniform to non-uniform matter (see, e.g., Landau and Liftshitz's textbook (Landau and Lifshitz 1980) and Chap. 7 of this book), although currently there exists no rigorous treatment to describe cluster formation beyond the single-nucleus approximation (see also Sect. 6.3.2).

As for electrons, in stellar environments like compact stars, they are usually treated as a non-interacting degenerate background gas (see, e.g., Lattimer 1996; Haensel et al. 2007). In cold NS crusts, electron-charge screening (spatial polarisation) effects are small and the electron density is essentially uniform (Haensel et al. 2007; Chamel and Fantina 2016); at densities $\rho_B \gg 10\,AZ\,\mathrm{g\,cm^{-3}}$ ($\sim 10^4\,\mathrm{g\,cm^{-3}}$ for iron, A and Z being the nucleus mass and proton number, respectively), the electrons can be treated as a quasi-ideal Fermi gas (Chamel and Haensel 2008). For temperatures $T \gtrsim 1\,\mathrm{MeV}$ and densities $\gtrsim 10^6\,\mathrm{g\,cm^{-3}}$, leptons (electrons and neutrinos) are relativistic, in particle-antiparticle pair equilibrium and in thermal equilibrium with nuclear matter (see, e.g., Lattimer et al. 1985; Lattimer and Douglas Swesty 1991).

Nucleons can be either treated as a uniform system of interacting particles, or distributed within a defined shaped and sized cell. In the latter case, often the Wigner-Seitz (WS) approximation is used: matter is divided in cells, each one charged neutral. While at lower densities the cell is usually assumed spherical, centred around the positive charged ion surrounded by an essentially uniform electron and eventually free (unbound) nucleon (neutron and, at finite temperature, free proton) gas, at higher densities nuclei can be non-spherical and other geometries of the cell are considered. The standard way to calculate the EoS is then, for each thermodynamic condition, to minimise the (free) energy of the system with respect to the variational variables, e.g. the nucleus atomic and mass number, the volume (or radius) of the cell, and the free nucleon densities, under baryon number and charge conservation[1] (see, e.g., the pioneer work of Baym et al. 1971a). If additional structures, like the so-called "pasta" phases, are included, the minimisation is also performed on the shape of the cell (see, e.g., the pioneer works of Ravenhall et al. 1983; Hashimoto et al. 1984).

Within the single-nucleus approximation, different models have been developed:

- *(Compressible) Liquid-Drop Models.*
  Liquid-drop models parameterize the energy of the system in terms of global properties such as volume, asymmetry, surface, and Coulomb energy; their

---

[1]Note that in the outer crust of cold catalysed NSs, the classical way to determine the EoS is to use the so-called BPS model (Baym et al. 1971b). In this model, the outer crust is supposed to be made of fully ionised atoms arranged in a body-centred cubic lattice at $T = 0$ and to contain homogeneous crystalline structures made of one type of nuclides, coexisting with a degenerate electron gas (no free nucleons are present). The EoS in each layer of pressure P is found by minimising the Gibbs free energy per nucleon, the only microscopic input being nuclear masses (see, e.g., Haensel et al. 2007).

parameters are fitted phenomenologically. In these models, nucleons inside neutron-proton clusters and free neutrons outside are assumed to be uniformly distributed, and are treated separately. Moreover, clusters have a sharp surface, and quantum shell effects, despite playing a critical role in determining the equilibrium composition (particularly in the NS outer crust), are neglected. This approach has been among the earliest to be used in astrophysical applications to treat non-uniform matter at zero and finite temperature, because of its applicability and reduced computational cost (see, e.g., Baym et al. 1971a; Lattimer 1981; Lattimer et al. 1985; Lattimer and Douglas Swesty 1991; Lorenz et al. 1993; Watanabe et al. 2000; Douchin and Haensel 2000, 2001; Oyamatsu and Iida 2007; Nakazato et al. 2011).

- *(Extended) Thomas-Fermi ((E)TF) models.*
  These models allow for a consistent treatment of nucleons "inside" and "outside" clusters and are a computationally very fast approximation to the full Hartree-Fock equations. The total energy of the system is written as a functional of the density of each species and their gradients. Indeed, the (E)TF approximation to the energy density derived from a given nuclear EDF consists in expressing the kinetic-energy densities and the spin-current densities upon which the EDF depends as a function of the nucleon number densities (and their derivatives). As a consequence, shell effects in the energy density are lost, but can be restored perturbatively using the Strutinski Integral (SI) theorem (Brack et al. 1985; Onsi et al. 2008; Pearson et al. 2012, 2015). The density of nucleons in the cell can be either parameterized (e.g. using a Fermi-like profile) or obtained self-consistently. These approaches have been developed in both non-relativistic and relativistic framework, at zero and finite temperature (see, e.g., Sumiyoshi et al. 1995; Cheng et al. 1997; Shen et al. 1998a; Onsi et al. 2008; Avancini et al. 2010; Shen et al. 2011c; Okamoto et al. 2012; Pearson et al. 2012; Zhang and Shen 2014; Sharma et al. 2015).

- *Self-consistent mean-field models.*
  Hartree-Fock models are fully quantum mechanical (see, e.g., Negele and Vautherin 1973; Baldo et al. 2007; Grill et al. 2011). As a result, shell effects, which are found to disappear at temperatures above 2–3 MeV, and pairing (in the Hartree-Fock+BCS and in the HFB approaches), which needs to be considered at temperatures below 1 MeV (Brack and Quentin 1974), are naturally included. However, these models are computationally very expensive and their current implementation is plagued by the occurrence of spurious neutron shell effects (Chamel et al. 2007). Non-relativistic interactions are usually employed (e.g., Negele and Vautherin 1973; Bonche and Vautherin 1982; Magierski et al. 2003; Gögelein and Müther 2007; Newton and Stone 2009; Pais and Stone 2012; Papakonstantinou et al. 2013; Pais et al. 2014; Sagert et al. 2016), but RMF models have been also used (Maruyama et al. 2005).

At finite temperature, different configurations are expected to be realised. A way to find these configurations is to solve the equations of motion and eventually exploit the ergodicity of the dynamics. This can be done using time-dependent Hartree-Fock models and dynamical extended time-dependent Hartree-Fock approaches based on a wavelet representation (see, e.g., Schuetrumpf et al. (2013) for the former models applied at finite temperature using Skyrme functionals—i.e. applicable to SN matter, and Sébille et al. (2009, 2011) for the latter approach applied in the zero-temperature approximation—i.e. for NSs). Another method is the *classical molecular dynamics*, where nucleons are represented by point-like particles instead of single-particle wave functions. So-called *quantum molecular dynamics*, where nucleons are treated as wave packets, have also been developed. However, in both cases particles move according to classical equations of motion and quantum effects (like shell effects) are not taken into account (see, e.g. Watanabe et al. (2009), Dorso et al. (2012), Maruyama et al. (2012), López et al. (2014), Schneider et al. (2014), Caplan et al. (2015), Giménez Molinelli and Dorso (2015), Horowitz et al. (2015), see also Chap. 7 in this book). However, these approaches are very time consuming. On the other hand, one can assume that the system is in thermodynamic equilibrium and use NSE (or "statistical") models, where cluster degrees of freedom are introduced explicitly. These models suppose that the system is composed of a statistical ensemble of nuclei and nucleons in thermal, mechanical, and chemical equilibrium. The NSE is achieved when the characteristic time for nuclear processes is much shorter than the timescales associated to the hydrodynamic evolution of the system, and typically above $T \gtrsim 0.5\,\mathrm{MeV}$ (Iliadis 2007). Approaches considering an ensemble of nuclei are:

- *(Extended) NSE.*

  In the simplest version, NSE approaches treat the matter constituents as a mixture of non-interacting ideal gases governed by the Saha equation, where a Maxwell-Boltzmann statistics is employed, although quantum statistics (e.g. Fermi-Dirac for nucleons) can be incorporated. The nuclear binding energies required as input of NSE calculations can be either experimental, whenever available (Audi et al. 2003; Wang et al. 2012), or theoretical (e.g. obtained from liquid-drop like models, or from more microscopic EDF-based mass models). A limitation of standard NSE-based models is that they neglect interactions and in-medium effects, that are known to be very important in nuclear matter, especially at high densities. For this reason, homogeneous matter expected to be present in NS cores, as well as the crust-core boundary in NSs, or matter at densities close to saturation density, cannot be correctly described by this kind of approaches, and microscopic or phenomenological models have to be applied instead. Therefore, extended NSE models, where the distribution of clusters is obtained self-consistently under conditions of statistical equilibrium and interactions are taken into account, are developed. For example, in-medium corrections of nuclear binding energies, either due to temperature or to the presence of unbound nucleons, have been calculated for Skyrme interactions in Papakonstantinou et al. (2013), Aymard et al. (2014) within a local-density and ETF approximation, respectively. Some NSE models neglect the screening of the Coulomb interaction due to the electron background, while it is accounted for in other models, usually in the WS approximation (see, e.g., Raduta and

Gulminelli 2009, 2010; Blinnikov et al. 2011). The interactions between the cluster and the surrounding gas are often treated with an excluded-volume method (e.g., Hempel and Schaffner-Bielich 2010; Raduta and Gulminelli 2010; Hempel et al. 2012; Furusawa et al. 2013b; Raduta et al. 2014). However, the difference between the excluded-volume approach and the quantal picture proper of microscopic calculations leads to two different definitions of clusters in dense matter which in turn give differences in the observables (Papakonstantinou et al. 2013). Excited states, that are populated at finite temperature, can be incorporated, either employing temperature-dependent degeneracy factors (as, e.g., in Ishizuka et al. 2003; Raduta and Gulminelli 2010; Gulminelli and Raduta 2015), or using temperature-dependent coefficients in the mass formula (as, e.g., in Buyukcizmeci et al. 2014). Despite thermodynamic quantities are not very much affected by the presence of the ensemble of nuclei with respect to the single-nucleus approximation picture, quantitative differences arise in the matter composition, in particular concerning the contribution of light and intermediate mass nuclei (see, e.g., Gulminelli and Raduta 2015; Raduta et al. 2016). Among the first applications of a NSE model for the EoS of SN cores at low densities is that of Hillebrandt et al. (1984), Hillebrandt and Wolff (1985). NSE models have been subsequently employed for conditions encountered in CCSN, e.g., in Botvina and Mishustin (2010), Hempel and Schaffner-Bielich (2010), Raduta and Gulminelli (2010), Blinnikov et al. (2011), Hempel et al. (2012), Furusawa et al. (2013b), Buyukcizmeci et al. (2014), Raduta et al. (2014), Gulminelli and Raduta (2015), Furusawa et al. (2017b) (see also Buyukcizmeci et al. (2013) for a comparison of methods). As shown in Fig. 6.4 for a typical condition encountered in the SN collapse, the NSE approach predicts a broad bi-modal distribution, centred around magic numbers. This behaviour cannot be reproduced within the single-nucleus approximation, widely employed in SN simulations (see also Sect. 6.2.4).

Most of the aforementioned works make use of the excluded-volume approximation, which is less reliable for light nuclei, thus other approaches to treat interactions have been developed, as discussed below.

- *Virial EoS.*
The virial expansion, originally formulated by Uhlenbeck and Beth (1936), Beth and Uhlenbeck (1937), is based on an expansion of the grand canonical potential in powers of the particle fugacities $z_i = \exp[(\mu_i - m_i c^2)/T]$, μ_i being the chemical potential of the particle i and m_i its mass. It can thus be seen as an extension of NSE models to account for correlations between particles at low density and finite temperature. This method relies on two assumptions: (1) the system is in a gas phase and has not undergone phase transition with decreasing temperature or increasing density, and (2) the fugacity is small, so that the partition function can be expanded in powers of z. The virial coefficients in the expansion are functions of temperature, and they are related to the two-, three-, and N-body correlations (see, e.g., Bedaque and Rupak (2003), Liu et al. (2009) for a discussion on coefficients beyond the second order). In particular, the second virial coefficient is directly related to the two-body

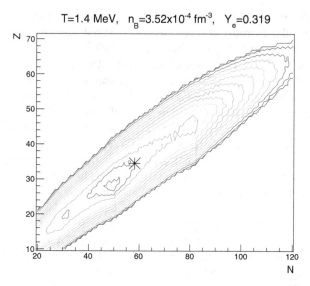

Fig. 6.4 Distribution of nuclei (blue to red: less to more abundant) for $n_B = 3.52 \times 10^{-4}\,\mathrm{fm}^{-3}$, $T = 1.4\,\mathrm{MeV}$, $Y_e = 0.319$, calculated within the NSE model of Gulminelli and Raduta (2015). The star corresponds to the result obtained in the single-nucleus approximation. Courtesy of Ad. R. Raduta

scattering phase shifts; thus, one can derive a model-independent approach up to $n_B \approx 10^{-5} - 10^{-4}\,\mathrm{fm}^{-3}$. In the context of nuclear matter relevant for compact stars, the virial EoS has been applied, e.g., in Horowitz and Schwenk (2006a,b) to describe neutron matter and matter composed of nucleons and alpha particles, and in Shen et al. (2010b, 2011a) to model non-uniform matter at low densities. However, this treatment is limited to light particles.

- *Models with in-medium mass shifts.*
  In-medium mass shifts are a way to account for correlation effects in the medium, avoiding the use of an excluded volume. Nucleons and bound states (clusters) are treated on the same footing, as different constituent particles. This approach also points out the appearance of the Mott effect due to Pauli blocking that prevents the formation of clusters at sufficiently high densities, and allows one to obtain the medium (density- and temperature-dependent) modification of cluster binding energies that enter into the EoS. This in-medium modification approach has been included in different NSE-based models, like the *quantum statistical model*, based on the thermodynamic Green's function method and developed, e.g., in Röpke et al. (1982, 1983); for recent applications of this approach to the description of light nuclei in nuclear matter at subsaturation densities, see, e.g., Typel et al. (2010), Röpke (2011), Röpke et al. (2013), Röpke (2015). It has also been incorporated in the *generalised (g)RMF models*, which are an extension of the DD-RMF models where nucleon and (light) cluster degrees of freedom are included in the Lagrangian (see, e.g. Typel et al. 2010; Voskresenskaya

and Typel 2012; Typel 2013; Typel et al. 2014). It turns out that the gRMF smoothly interpolates between the low-density virial EoS and the high-density limit of nucleonic matter, while the precise form of the transition depends, among other factors, on the choice of the coupling strength of the clusters to the meson fields. A comparison of models using quantum statistical and gRMF models and the excluded-volume approach shows a rather good agreement at temperatures greater than a few MeV (Hempel et al. 2011). The effect of light clusters in RMF models in nuclear matter and in the pasta phase has been also investigated, e.g., in Avancini et al. (2012), Pais et al. (2015).

6.2.3 Constraints on the Equation of State

Theoretical models for the EoS can be constrained by both nuclear physics and astrophysical observations (see, e.g., Page and Reddy (2006), Klähn et al. (2006), Lattimer and Prakash (2007), Trümper (2011), Tsang et al. (2012), Li and Han (2013), Lattimer and Lim (2013), Lattimer and Steiner (2014), Stone et al. (2014), Horowitz et al. (2014) for a discussion). However, in many cases, constraints on the EoS are not directly obtained from the raw data, but theoretical modelling is required to infer the constraints, or to extrapolate them in a region of the phase diagram not accessible by experiments or observations, thus making the constraints model dependent.

6.2.3.1 Constraints from Nuclear Physics Experiments

Valuable information on the many-body theories of nuclear matter is given by available data coming mainly from nuclear structure studies and heavy-ion collisions (HICs). Nuclear matter is an idealised infinite uniform system of nucleons, where the Coulomb interaction is switched off. Within the liquid-drop model of nuclei, if we put $E_{Coul} = 0$ and in the limit of $A \to \infty$, the energy per nucleon, E/A, depends only on the neutron and proton densities, and because of charge symmetry of nuclear forces, it does not change if protons are replaced by neutrons and vice versa. Symmetric nuclear matter, with an equal number of neutrons and protons, is the simplest approximation to the bulk nuclear matter in heavy atomic nuclei. On the other hand, pure neutron matter is the simplest approximation to the matter as found in NS cores.

As in the droplet model functional (Myers and Swiatecki 1969), it is convenient to express the binding energy in terms of the baryon density n_B and the asymmetry parameter $\delta = (N - Z)/A$, N (Z) being the neutron (proton) number and $A = N + Z$. Usually, this energy is written as

$$E(n_B, \delta) = E(n_B, 0) + S(n_B)\delta^2 , \qquad (6.7)$$

$E(n_B, 0)$ being the energy of symmetric nuclear matter ($\delta = 0$) and $S(n_B)$ the symmetry energy. Both these terms can be expanded around the saturation density for symmetric matter, n_0, as

$$E(n_B, 0) = E(n_0) + \frac{1}{18} K_0 \epsilon^2 , \tag{6.8}$$

$$S(n_B) = S_0 + \frac{1}{3} L\epsilon + \frac{1}{18} K_{sym}\epsilon^2 . \tag{6.9}$$

where $E(n_0)$ characterises the binding energy in symmetric nuclear matter at saturation, $\epsilon = (n_B - n_0)/n_0$, K_0 is the incompressibility at the saturation point, $S(n_0) \equiv S_0$ is the symmetry energy coefficient at saturation, and the parameters L and K_{sym} characterise the density dependence of the symmetry energy around saturation. The value of n_0 and of the binding energy per nucleon for symmetric nuclear matter at saturation, $E(n_0)/A \equiv E_0/A$, can be extracted from experimentally measured nuclear masses, yielding $n_0 = 0.16 \pm 0.01 \, \text{fm}^{-3}$ and $E_0/A = -16.0 \pm 1.0 \, \text{MeV}$ (Wang et al. 2012; Angeli and Marinova 2013). The uncertainties in these parameters result from the uncertainties in the experimental measurements and from the non-uniqueness of the fit of mass formulae used to reproduce thousands of nuclear masses. The coefficient S_0 determines the increase in the energy per nucleon due to a small asymmetry δ, whereas the incompressibility K_0 gives the curvature of $E(n_B)$ at $n_B = n_0$ and the associated increase of the energy per nucleon of symmetric nuclear matter due to a small compression or rarefaction. These parameters are defined as :

$$K_0 \equiv 9n_0^2 \left(\frac{\partial^2 E}{\partial n_B^2}\right)_{n_B=n_0,\delta=0} , \quad S_0 \equiv \frac{1}{2}\left(\frac{\partial^2 E}{\partial \delta^2}\right)_{n_B=n_0,\delta=0} , \tag{6.10}$$

and the higher-order symmetry energy coefficients, L and K_{sym}, are defined as

$$L \equiv 3n_0 \left(\frac{\partial S(n_B)}{\partial n_B}\right)_{n_B=n_0} , \quad K_{sym} \equiv 9n_0^2 \left(\frac{\partial^2 S(n_B)}{\partial n_B^2}\right)_{n_B=n_0} . \tag{6.11}$$

The extraction of K_0 from experimental data is complicated and not unambiguous. Roughly speaking, RMF models predict larger values of K_0 with respect to non-relativistic EDFs (a large list of theoretical calculations of K_0 is given, e.g., in Stone et al. 2014). Analysis of isoscalar giant monopole resonance in heavy nuclei suggests $K_0 = 240 \pm 10 \, \text{MeV}$ (Colò et al. 2004) (a tighter constraint is reported in Piekarewicz (2004), $K_0 = 248 \pm 8 \, \text{MeV}$, while Blaizot (1980) gives $K_0 = 210 \pm 30 \, \text{MeV}$). However, it has been argued that data actually give information on the density dependence of the incompressibility around $0.1 \, \text{fm}^{-3}$ (Khan and Margueron 2013). HIC experiments (either flow experiments or kaon production experiments) would point to a rather "soft" EoS (e.g., Fuchs et al. 2001; Sturm et al. 2001; Danielewicz et al. 2002; Hartnack et al. 2006; Le Fèvre

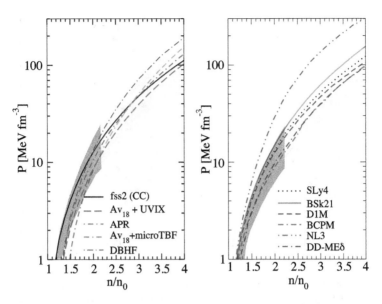

Fig. 6.5 Pressure versus baryon density (in units of saturation density) of symmetric nuclear matter for different microscopic (left panel) and phenomenological (right panel) models. The shaded area at lower (higher) density corresponds to constraints inferred from KaoS (flow) experiment (Danielewicz et al. 2002; Lynch et al. 2009)

et al. 2016). However, the inferred constraints remain model dependent since the data interpretation requires complex theoretical simulations (see also Sect. 6.3.1). A discussion on the HIC constraints in relation with compact stars has been done, e.g., in Sagert et al. (2012), Hempel et al. (2015). A comparison of the pressure versus density predicted by microscopic and phenomenological models with the results of the analysis on the flow and the kaon-production experiments (Danielewicz et al. 2002; Lynch et al. 2009) is shown in Fig. 6.5. Although not obvious, one can see that most of the EoSs, except very stiff ones, agree with these constraints (shaded area); only marginal deviations occur at the highest densities where the analysis is less reliable due to the possible presence of additional degrees of freedom or a phase transition. A refinement of the boundary could in principle put a more stringent constraint on the EoS, or even rule out some of them.

Of particular importance for compact-star physics is the symmetry energy and its density dependence, which has been shown to affect the composition of NS crusts, the crust-core transition, and the neutron drip (see, e.g., Li et al. (2014), Baldo and Burgio (2016) and references therein).[2] The value of the symmetry energy at

---

[2]It has also to be mentioned that not only the density dependence but also the temperature dependence of the symmetry energy, although not discussed here, can be important and may potentially have an impact in the CCSN dynamics (e.g., Donati et al. 1994; Dean et al. 2002; Fantina et al. 2012a; Agrawal et al. 2014).

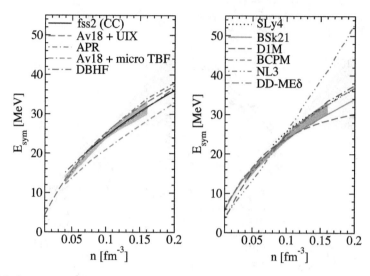

Fig. 6.6 Symmetry energy as a function of baryon number density calculated with different microscopic (left panel) and phenomenological (right panel) models. The smaller (larger) shaded area corresponds to constraints inferred from the analysis of IAS (and corresponding extrapolation) of Danielewicz and Lee (2014)

saturation, S_0, can be extracted, e.g., from nuclear masses, isobaric analog state (IAS) phenomenology, skin width data, and HICs; additional constraints come also from the NS data analysis. Constraints on L can be obtained, e.g., from the study of dipole resonances, electric dipole polarizability, and neutron skin thickness (see, e.g., Lattimer (2012), Tsang et al. (2012), Paar et al. (2014) and Sect. 6.3.1). While S_0 is fairly well constrained to lie around 30 MeV, the values of the slope of the symmetry energy, L, and of higher order coefficients like K_{sym}, at saturation, are still very uncertain and poorly constrained. For example, combining different data, Lattimer and Lim (2013) give $29.0 < S_0 < 32.7$ MeV, $40.5 < L < 61.9$ MeV, while a more recent work suggests $30.2 < S_0 < 33.7$ MeV, $35 < L < 70$ MeV (Danielewicz and Lee 2014). In Fig. 6.6, we display the symmetry energy versus baryon number density for different microscopic (left panel) and phenomenological (right panel) models. Note that for microscopic models, the curves of E_{sym} are given by the difference between the energy of pure neutron matter and that of symmetric matter, while for phenomenological models E_{sym} is calculated from the definition $E_{\text{sym}}(n) = 1/2(\partial^2 E/\partial \delta^2)|_{\delta=0}$. Shaded areas represent constraints inferred from a study of the IAS and its extrapolation at lower and higher densities (see Fig. 15 in Danielewicz and Lee 2014). All the considered models, except very soft or very stiff ones, agree with these constraints. This means that the latters cannot be used to extract a simple functional parameterization of the density dependence of the symmetry energy. In particular, if one assumes a power law dependence, i.e. $E_{\text{sym}} \approx n^\alpha$, the exponential index α cannot be constrained within a meaningful accuracy. Additional constraints, not reported here, have also been inferred at higher

Table 6.1 Nuclear parameters at saturation (saturation density n_0, energy per baryon E_0/A, incompressibility K_0, symmetry energy coefficient S_0, and slope of the symmetry energy L), for different microscopic and phenomenological models discussed in the text

EoS	n_0 [fm^{-3}]	E_0/A [MeV]	K_0 [MeV]	S_0 [MeV]	L [MeV]
fss2 (CC)	0.157	-16.3	219.0	31.8	52.0
Av$_{18}$+ UVIX	0.16	-15.98	212.4	31.9	52.9
APR	0.16	-16.0	247.3	33.9	53.8
Av$_{18}$+ micro TBF	0.17	-16.0	254.0	30.3	59.2
DBHF	0.18	-16.15	230.0	34.4	69.4
SLy4	0.16	-15.97	229.9	32.0	45.9
BSk21	0.158	-16.05	245.8	30.0	46.6
D1M	0.165	-16.03	225.0	28.55	24.83
BCPM	0.16	-16.0	213.75	31.92	52.96
NL3	0.148	-16.3	271.76	37.4	118.3
DD-MEδ	0.152	-16.12	219.1	32.35	52.85
Empirical values	0.16 ± 0.01	-16.0 ± 1.0	240 ± 10	30–34	35–70

Empirical values are taken from Wang et al. (2012), Angeli and Marinova (2013), Colò et al. (2004), Danielewicz and Lee (2014)

density, around and even beyond twice saturation density, from ASY-EOS and FOPI-LAND data (e.g., Russotto et al. 2016).

Finally, in Table 6.1, we list the nuclear parameters at saturation for the different microscopic and phenomenological models considered in the text, showing that all these models, except very stiff ones, agree reasonably well with the empirical values.

6.2.3.2 Constraints from Astrophysics

Astrophysical observations can provide complementary information on the region of the dense-matter EoS which is not experimentally accessible in the laboratory. These constraints mainly come from observations of NSs, either isolated or in binary systems (see also Chap. 5 of this book).

- **NS masses and radii**. The most precise and stringent astrophysical constraints on the EoS come, at the present time, from the measurement of NS masses (see, e.g., Özel and Freire (2016) and the nsmasses website[3] for a recent compilation). Indeed, the maximum mass of a NS is a direct consequence of general relativity and depends to a large extent on the high-density part of the EoS (above nuclear saturation density), where the EoS remains at present very uncertain. According

---

[3]https://stellarcollapse.org/nsmasses.

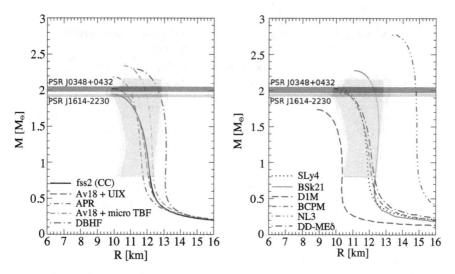

Fig. 6.7 Gravitational mass versus radius of non-rotating NSs for different nucleonic EoSs based on both microscopic (left panel) and phenomenological (right panel) models (results for the latters are taken from Fantina et al. 2013; Gonzalez-Boquera et al. 2017; Sharma et al. 2015; Fattoyev et al. 2010; Wang et al. 2014). Horizontal bands correspond to the measured masses of PSR J0348+0432 (Antoniadis et al. 2013) and PSR J1614 − 2230 (Fonseca et al. 2016). Shaded areas correspond to the 68% and 95% confidences derived in Steiner et al. (2013b)

to different calculations, the maximum mass of spherical non-rotating NSs is predicted to lie in the range $1.5 M_\odot \lesssim M_{max} \lesssim 2.5 M_\odot$, $M_\odot$ being the mass of the Sun (see, e.g. Chamel et al. (2013b,c) for a review). Recently, the mass of two NSs in binary systems have been precisely measured using the Shapiro delay: PSR J1614−2230 (Demorest et al. 2010), with a mass $M = 1.928 \pm 0.017\ M_\odot$ (Fonseca et al. 2016), and PSR J0348+0432, with a mass $M = 2.01 \pm 0.04\ M_\odot$ (Antoniadis et al. 2013). The latter mass is sufficiently high to put quite strong constraints on the EoS at densities four times larger than nuclear saturation, but it still remains compatible with a large class of models (see Fig. 6.7). However, this measured mass, which is about three times larger than the maximum mass of a star made of an ideal neutron Fermi gas, is a clear observational indication of the dominating role of strong interactions in NSs. There also exist several less precise measurements of NS masses with values around and even above $2\ M_\odot$. These measurements mainly refer to NSs in X-ray binaries or millisecond pulsar systems, where accretion, stellar wind, possible filling of Roche lobe by the companion, light-curve modelling, and other uncertainties could all play an important role. For this reason, the error of these mass measurements is quite large (see, e.g., Haensel et al. (2007), van Leeuwen (2013) for a discussion).

On the other hand, constraints on the EoS have been proposed using low-mass NSs. Podsiadlowski et al. (2005) suggested to probe the EoS using the observations of J0737−3039, a double pulsar system whose pulsar B has a mass $M = 1.2489 \pm 0.0007 M_\odot$ (Kramer et al. 2006). The characteristics of the system suggest that pulsar B was formed after the accretion-induced collapse of an oxygen-neon-magnesium core that becomes unstable against electron capture. Taking into account the uncertainties in the conditions of the pre-collapse core, it has been estimated that the critical baryonic mass of pulsar B for the onset of electron capture should be $1.366 < M_b < 1.375 M_\odot$ (Podsiadlowski et al. 2005). If this scenario is correct, the knowledge of both the gravitational and the baryonic mass of pulsar B leads to a constraint on the EoS. However, there are various caveats in this analysis (e.g. neglect of mass loss during the SN, variation of the critical mass due to carbon flashes, formation history of this system), which can considerably change the constraint on the EoS (Podsiadlowski et al. 2005; Tauris et al. 2013). In particular, Kitaura et al. (2006) carried out hydrodynamical simulations of stellar collapse taking mass loss into account and found $M_b = 1.360 \pm 0.002 M_\odot$. A similar system has been observed recently, J1756-2251, where the mass of the pulsar is $1.230 \pm 0.007 M_\odot$ (one of the lowest NS mass measured with high accuracy) (Ferdman et al. 2014). However, the constraint inferred from the M versus M_b relation strongly depends on the assumptions of the model and cannot therefore definitely rule out EoSs that do not satisfy it.

A definite and stringent constraint on the EoS via the mass-radius relation would be the measurement of both mass and radius of the same object (see, e.g., Read et al. 2009; Steiner et al. 2013b; Özel et al. 2016). However, precise estimations of NS radii are very difficult because more model dependent than those of masses, mostly because observations of NS radii are indirect and the determination of the radius from observations is affected by large uncertainties (e.g., composition of the atmosphere, distance of the source, magnetic field, accretion; see, e.g., Potekhin 2014; Fortin et al. 2015). Observations of the thermal emission from NSs can provide valuable constraints on their masses and radii. The most reliable constraints are expected to be inferred from observations of transient low-mass X-ray binaries (LMXBs) in globular clusters because their distances can be accurately determined and their atmospheres, most presumably weakly-magnetised and primarily composed of hydrogen, can be reliably modelled. Constraints can also come from observations of type I X-ray bursts, the manifestations of explosive thermonuclear fusion reactions triggered by the accretion of matter onto the NS surface. Recently, Steiner et al. (2010, 2013b) determined a probability distribution of masses and radii by analysing observations of type I X-ray bursters and transient LMXBs in globular clusters (see Fig. 6.7). However, this kind of analysis is still a matter of debate (see, e.g., Lattimer 2012; Galloway and Lampe 2012; Guillot et al. 2013; Güver et al. 2013; Poutanen et al. 2014). Additional information on radii could also be inferred from X-ray pulsation in millisecond pulsars (see, e.g., Bogdanov 2016b).

Table 6.2 Properties of non-rotating NSs (maximum mass M_{max} and corresponding radius R, and central density n_{cen}), for different microscopic and phenomenological models

EoS	M_{max} [$M_\odot$]	R [km]	n_{cen} [fm^{-3}]
fss2 (CC)	1.94	9.9	1.87
Av$_{18}$+ UVIX	2.03	9.8	1.24
APR	2.19	9.9	1.15
Av$_{18}$+ micro TBF	2.34	10.6	1.01
DBHF	2.30	11.2	0.97
SLy4	2.05	10.0	1.21
BSk21	2.28	11.1	0.97
D1M	1.74	8.9	1.57
BCPM	1.98	10.0	1.25
NL3	2.78	13.4	0.67
DD-MEδ	1.97	10.2	1.20

Future high-precision telescopes and missions like NICER, ATHENA+, and SKA are expected to improve our knowledge on the NS mass-radius relation (see, e.g., Watts et al. 2015; Bogdanov 2016a).

In Fig. 6.7, we display the gravitational mass M versus radius R for non-rotating NSs,[4] obtained with different microscopic (left panel) and phenomenological (right panel) EoSs of asymmetric and beta-stable matter whose underlying models have been discussed in Sect. 6.2.2. Note that only the EoSs based on the SLy4, BSk21, and BCPM EDFs are unified, while the others have been supplemented with an EoS for the crust (the BPS EoS (Baym et al. 1971b), except for the D1M model where the EoS of Douchin and Haensel (2001) was used). Properties of NSs calculated with these EoSs are also reported in Table 6.2. These calculations assume that only nucleonic and leptonic (electrons and eventually muons) degrees of freedom are present inside the NS. Horizontal bands correspond to the precise measurements of NS masses (Antoniadis et al. 2013; Fonseca et al. 2016), while shaded areas correspond to the $M - R$ constraint inferred in Steiner et al. (2013b) (see their Fig. 1). Except very "soft" or "stiff" EoSs, the other considered EoSs are at least marginally compatible with the latter (model-dependent) constraint. Instead, EoSs that predict a maximum NS mass below the observed ones have to be ruled out.[5] However, it would be premature to discard the underlying nuclear interaction as well. Indeed, analyses of HIC experiments (e.g., Fuchs et al. 2001; Sturm et al. 2001; Hartnack et al. 2006; Xiao et al. 2009) seem to favour soft (hadronic) EoSs. This apparent discrepancy could be resolved by considering the occurrence of a transition to an "exotic" phase in NS cores (see, e.g., the discussion in Chamel et al. 2013a). On the other hand, BHF microscopic calculations that include also hyperon degrees of freedom (e.g., Baldo et al. 2000; Schulze and Rijken 2011; Vidaña et al. 2011)

---

[4] It will be explained in Sect. 6.2.4.1 how to construct a $M - R$ relation for a given EoS.

[5] Although rotation can increase the predicted maximum mass, this increase amounts to a few % only even for the fastest spinning pulsar known (e.g., Chamel et al. 2013b,c; Fantina et al. 2013).

show that the NS EoS becomes softer, and the value of the NS maximum mass is substantially reduced, well below the observational limit of 2 $M_\odot$. This poses a serious problem for the microscopic theory of NS interior (see the discussion in Chap. 7 of this book).

Another possibility to get a constraint on the mass-radius relation is to use observations of the gravitational redshift of photons emitted from the NS surface, z_{surf}, a quantity related to the compactness ratio (proportional to M/R, see Haensel et al. 2007). Spectroscopic study of the gamma-ray burst GRB 790305 from the soft gamma-ray repeater SGR 0526−066 (Higdon et al. 1990) suggested a value $z_{\mathrm{surf}} = 0.23 \pm 0.07$ for this object. Cottam et al. (2002) also reported the detection of absorption lines in the spectra of several X-ray bursts from the LMXB EXO 0748−676, but this detection has not been confirmed by subsequent observations (Cottam et al. 2008). Moreover, it has been argued that the widths of these absorption lines are incompatible with the measured rotational frequency of this NS (Lin and Özel 2010), but this point remains to be clarified (Bauböck et al. 2013).

Finally, the detection of neutrinos from SN1987A allows an estimation of the energy released during the SN core collapse. Indeed, since 99% of the energy of the SN was carried away by neutrinos of all flavours, the energy of the neutrino can be considered a measurement of the binding energy of the newly born NS. Defining the binding energy as the mass defect with respect to a cloud of iron dust (Douchin and Haensel 2001) leads to a constraint on the NS gravitational mass. However, EoSs are usually found to be compatible with this constraint, thus making hard to rule out an EoS from it.

- **NS rotation**. Rotation of pulsars can be accurately measured. The spin frequency of a NS must be lower than the Keplerian frequency, i.e. the frequency beyond which the star will be disrupted as a result of mass shedding. Since the value of the Keplerian frequency obtained from numerical simulations of rotating NSs depends on the EoS (see, e.g., Stergioulas (2003) for a review), an observed frequency above the Keplerian one predicted for a given EoS would rule the EoS out. Depending on the stiffness of the EoS, the highest possible rotational frequency for the maximum mass configuration has been found to lie in the range between ∼1.6 and ∼2 kHz (Krastev et al. 2008; Haensel et al. 2009; Fantina et al. 2013). Even the observation of PSR J1748−2446ad, the fastest spinning pulsar known (Hessels et al. 2006), with a frequency of 716 Hz, cannot put stringent constraints on existing EoSs, because its rotational frequency still remains small compared to the Keplerian frequency. Only observations of NSs with spin frequencies larger than about 1 kHz (see, e.g., Bejger et al. 2007; Krastev et al. 2008; Haensel et al. 2009), could change the picture.

- **NS cooling**. Cooling observations are a promising way to probe the NS interior. Indeed, NS cooling depends on the composition and on the occurrence of superfluidity that determine heat transport properties (see, e.g., Weber (1999), Yakovlev and Pethick (2004), Page et al. (2006), Potekhin et al. (2015), and Chaps. 8 and 9 in this book), thus potentially giving complementary information on the EoS. For example, constraints on the mass-radius relation have been derived, using cooling models, from the observation of the central compact object in the SN remnant HESS J1731−347, that appears to be the hottest observed isolated cooling NS (Klochkov et al. 2015; Ofengeim et al. 2015). Recently, the impact of the stiffness of the EoS and in-medium effects on the cooling have also been studied (Grigorian et al. 2016). A prominent role in the NS cooling is played by the neutrino emission due to the so-called direct URCA processes (e.g. Taranto et al. 2016), which set in only if the proton fraction is larger than a certain threshold value. The proton fraction depends on the nuclear symmetry energy, and hence on the EoS. We will discuss this open question more in details in Sect. 6.3.4.

- **NS moment of inertia**. The mass and radius of a rotating NS can be constrained by measuring its moment of inertia I (see e.g. Lattimer and Schutz 2005; Worley et al. 2008; Lattimer and Prakash 2016). Indeed, I can be expressed as a function of the NS mass and radius (see e.g. the empirical formula for a slowly rotating NS proposed in Lattimer and Schutz (2005) and that holds for a wide class of EoSs, except for the very soft ones). Therefore, the radius could be determined if the mass and the moment of inertia of the NS is known. However, the moment of inertia of a rotating NS has not yet been measured. A lower bound can be inferred from the timing observations of the Crab pulsar, assuming that the loss of the pulsar spin energy goes mainly into accelerating the nebula (see, e.g., Bejger and Haensel 2003; Haensel et al. 2007): only the EoSs predicting a value of I higher than that estimated for Crab ($I \approx 1.4 - 3.1 \times 10^{45}$ g cm^2, see Haensel et al. 2007; Fantina et al. 2013) are acceptable. However, the main uncertainty in this lower limit lies in the mass of the nebula, thus this constraint remains approximate.

- **Gravitational waves and oscillations**. The merger of compact binary stars is expected to be the main source of the gravitational-wave signals observed with gravitational-wave detectors (see e.g. Baiotti and Rezzolla (2017) and also Chaps. 3, 10, 12 in this book). It has been argued that the detection of gravitational waves from the post-merger phase of binary NSs could discriminate among a set of candidate EoSs (see e.g. Bauswein et al. 2012, 2014; Takami et al. 2014, 2015; Bauswein et al. 2016; Rezzolla and Takami 2016). Also, quasi-periodic oscillations in soft gamma-ray repeaters could be used to derive constraints on the EoS (see e.g. Steiner and Watts 2009; Sotani et al. 2012; Gabler et al. 2013). This research area thus might be a promising way for constraining the EoS in the future (see e.g. Bose et al. (2018); Rezzolla et al. (2018) and Chap. 10 for a discussion on the effect of the EoS on the gravitational-wave signal from binary mergers).[6]

---

[6]During the refereeing process of this Chapter, the gravitational-wave signal from a binary NS merger, GW170817, has been observed in the galaxy NGC 4993 (Abbott et al. 2017a), in association with the detection of a gamma-ray burst (GRB 170817A) and electromagnetic counterparts (Abbott et al. 2017a,b).

6.2.4 Applications to Compact Objects

In this Section we discuss some applications of the EoSs for compact objects, starting with the zero-temperature case relevant for NSs, then presenting some finite-temperature general purpose EoSs and their impact in compact-object simulations.

6.2.4.1 Applications to Neutron Stars

The NS physics and EoS have been extensively discussed, e.g., in Haensel et al. (2007), Chamel and Haensel (2008), Lattimer and Prakash (2016) (see also Chap. 7 in this book). Since the temperature in cold isolated NSs is below ~ 1 MeV, lower than characteristic nuclear Fermi energy, the zero-temperature approximation can be used in constructing the EoS. The EoS is the necessary microphysics ingredient to determine the NS macroscopic properties, e.g. the mass-radius relation. Indeed, the structure of stationary, non-rotating, and unmagnetised NSs is determined by integrating the Tolman-Oppenheimer-Volkoff (TOV) equations for hydrostatic equilibrium in general relativity (Tolman 1939; Oppenheimer and Volkoff 1939) (see Haensel et al. (2007) for details),

$$\frac{dP}{dr} = -\frac{G\rho\mathcal{M}}{r^2}\left(1 + \frac{P}{\rho c^2}\right)\left(1 + \frac{4\pi Pr^3}{\mathcal{M}c^2}\right)\left(1 - \frac{2G\mathcal{M}}{rc^2}\right)^{-1}, \qquad (6.12)$$

where the function $\mathcal{M}(r)$ is defined by

$$\frac{d\mathcal{M}}{dr} = 4\pi r^2 \rho, \qquad (6.13)$$

with the boundary condition $\mathcal{M}(0) = 0$. The gravitational mass of the NS is given by $M = \mathcal{M}(R)$, R being the circumferential radius of the star where $P(R) = 0$. In order to solve these equations, an EoS, $P(\rho)$, must be specified. The latter depends on the properties of dense matter which still remain very uncertain, especially in the core of NSs. A number of EoS for NSs are available, either with only nucleonic degrees of freedom, or with hyperonic and quark matter. The majority of them are non-unified, i.e. they are built piecewise, matching different models, each one applied to a specific region of the NS. On the other hand, in *unified* EoSs, all the regions of the NS (outer crust, inner crust, and the core) are calculated using the same nuclear interaction (e.g., Douchin and Haensel 2001; Fantina et al. 2013; Miyatsu et al. 2013; Sharma et al. 2015). A unified and thermodynamically consistent treatment is important to properly locate the NS boundaries (such as the crust-core interface), that are important for the NS dynamics and which may leave imprints on astrophysical observables. The use of non-unified EoSs may also lead to considerable uncertainties in the NS radius determination (Fortin et al. 2016). In Fig. 6.8, we show the NS pressure as a function of the baryon number density

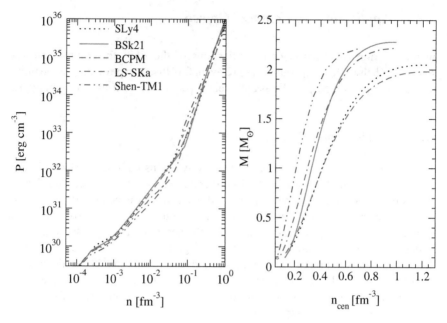

Fig. 6.8 Pressure versus baryon number density inside the NS (left panel) and NS gravitational mass versus central baryon density (right panel), for different unified EoSs (data are taken from Fantina et al. (2013), Sharma et al. (2015))

(left panel) and the NS mass versus the central density (right panel) resulting from the resolution of the TOV equations, Eqs. (6.12)–(6.13), for some unified EoSs. For those based on the SLy4 (Chabanat et al. 1998; Douchin and Haensel 2001), BSk21 (Goriely et al. 2010; Fantina et al. 2013), and BCPM (Baldo et al. 2010, 2013; Sharma et al. 2015) EDFs, the EoS of the outer crust is calculated in the standard BPS model (Baym et al. 1971b); for the former (SLy4), the outer-crust EoS is that of Haensel and Pichon (1994), while for the latters nuclear masses are taken from the experimental data in Wang et al. (2012) whenever available, complemented with theoretical mass models from HFB calculations with the corresponding functional. For the inner crust, clusters are described within the compressible liquid-drop model in the EoS based on SLy4, thus no shell effects are included; however, different shapes of the WS cells (spheres, cylinders) are considered. In the EoS based on the BSk21 EDF, the EoS for the inner crust has been calculated within the ETF model using a parameterized nucleon density distribution and with proton shell corrections included using the Strutinski Integral method, while in the BCPM EoS the self-consistent TF approach is employed and nuclear pasta is accounted for. For the nucleonic liquid core, the EoSs are computed with the same functionals applied in the inner crust; note that for the BCPM EoS, the core EoS is derived in the framework of the BBG theory. One can also obtain a unified EoS from a general purpose EoS, if applied at zero (or very small) temperature and in beta equilibrium. For comparison, we display in Fig. 6.8 the results for two interactions of such EoSs, that will be discussed in the next Section: the LS EoS in its SKa version[7] and the Shen EoS (Shen et al. 1998a) with the TM1 parameter set. All the considered EoSs predict a maximum mass around or above 2 $M_\odot$. However, significant differences arise for the prediction of the central density and the radius of lower mass NSs. Indeed, for a 1.5 $M_\odot$ NS, radii vary from ~11.6 km for the EoS based on the SLy4 EDF to ~14.4 km for the Shen-TM1 EoS (see, e.g., Fig. 16 in Sharma et al. 2015).

6.2.4.2   General Purpose Equations of State

Except for the case of "cold" (catalysed) NSs, for which the zero-temperature approximation can be used (see Sect. 6.2.4.1), for PNSs, CCSNe, and binary mergers, finite-temperature EoSs are crucially needed. A detailed analysis of the finite-temperature properties of the bulk EoS relevant for CCSNe, PNSs, and binary mergers, has been done, e.g., in Constantinou et al. (2014, 2015). A wide range of densities, temperatures, and charge fractions, describing both clustered and homogeneous matter, is covered by the so-called "general purpose" EoSs. These EoSs are therefore suitable for applications to SNe and mergers. However, at present, only a few of them are available and direct applicable to simulations. Moreover, for several of them, the underlying nuclear models are in disagreement with current constraints from either astrophysics (e.g. measured mass of NSs) or nuclear physics (experimental and/or theoretical constraints); see, e.g., the discussion in Oertel et al. (2017) and Sect. 6.2.3. We list below the general purpose EoSs with only nucleonic degrees of freedom currently used in astrophysical applications.

- **H&W**. The Hillebrandt and Wolff (H&W) EoS (Hillebrandt et al. 1984; Hillebrandt and Wolff 1985) has been calculated using a NSE-network based on the model of El Eid and Hillebrandt (1980), including 470 nuclei in the density range 10^9–3×10^{12} g cm^{-3}. At higher densities, the EoS is computed in the single-nucleus approximation (Hillebrandt et al. 1984); the nuclear interaction employed is the Skyrme interaction with the SKa parameter set (Köhler 1976). This EoS is still used in recent numerical simulations (Janka 2012).

- **LS**. The Lattimer and Swesty (LS) EoS (Lattimer and Douglas Swesty 1991) is a very widely used EoS in numerical simulations.[8] It models matter as a mixture of heavy nuclei (treated in the single-nucleus approximation), α particles, free neutrons and protons, immersed in a uniform gas of leptons and photons. Nuclei are described within a medium-dependent liquid-drop model, and a simplified NN interaction of Skyrme type is employed for nucleons. Alpha particles are described as hard spheres obeying an ideal Boltzmann gas statistics. Interactions

---

[7]http://www.astro.sunysb.edu/lattimer/EOS/main.html.

[8]The original EoS routine is available for three different parameterizations, according to the value of the incompressibility of the underlying nuclear interaction ($K_0 = 180, 220,$ and 375 MeV), at http://www.astro.sunysb.edu/dswesty/lseos.html. More recent tables are given at http://www.astro.sunysb.edu/lattimer/EOS/main.html.

between heavy nuclei and the gas of α particles and nucleons are treated in an excluded-volume approach. With increasing density, shape deformations of nuclei (non-spherical nuclei and bubble phases) are taken into account by modifying the Coulomb and surface energies, and the transition to uniform matter is described by a Maxwell construction.

- **STOS**. The Shen et al. (STOS) EoS (Shen et al. 1998a,b, 2011c) is another widely used EoS. As the LS EoS, matter is described as a mixture of heavy nuclei (treated in the single-nucleus approximation), α particles, and free neutrons and protons, immersed in a homogeneous lepton gas. For nucleons, a RMF model with the TM1 interaction (Sugahara and Toki 1994) is used; α particles are described as an ideal Boltzmann gas with excluded-volume corrections. The properties of the heavy nucleus are determined by WS-cell calculations within the TF approach employing parameterized density distributions of nucleons and α particles. The translational energy and entropy contribution of heavy nuclei, as well as the presence of a bubble phase, are neglected (see Zhang and Shen (2014) for a study of the effect of a possible bubble phase and a comparison between self-consistent TF calculations and those using a parameterized density distribution).

- **FYSS**. The EoS of Furusawa et al. (2011, 2013b, 2017b) is based on a NSE model, including light and heavy nuclei up to $Z \sim 1000$. For nuclei, the liquid-drop model is employed, including temperature-dependent bulk energies and shell effects (Furusawa et al. 2013b, 2017b). For light nuclei, Pauli- and self-energy shifts (Typel et al. 2010) are incorporated (Furusawa et al. 2013b, 2017b). The nuclear interaction used is the RMF parameterization TM1 (Sugahara and Toki 1994). The pasta phases for heavy nuclei are also taken into account. The FYSS EoS has been applied in CCSN simulations to study the effect of light nuclei in Furusawa et al. (2013a), and to investigate the dependence of weak-interaction rates on the nuclear composition during stellar core collapse in Furusawa et al. (2017a).

- **HS**. The Hempel and Schaffner-Bielich (HS) (Hempel and Schaffner-Bielich 2010) EoS is based on the extended NSE model, taking into account an ensemble of nuclei (several thousands, including light ones) and interacting nucleons. Nuclei are described as classical Maxwell-Boltzmann particles, and nucleons are described within the RMF model employing different parameterizations. Binding energies are taken from experimental data whenever available (Audi et al. 2003), or from theoretical nuclear mass tables. Coulomb energies and screening due to the electron gas are calculated in the WS approximation, while excited states of the nuclei are treated with an internal partition function, as in Ishizuka et al. (2003). Excluded-volume effects are implemented in a thermodynamic consistent way so that it is possible to describe the transition to uniform matter. At present, EoS tables are available for the following parameterizations: TMA (Toki et al. 1995; Hempel and Schaffner-Bielich 2010; Hempel et al. 2012), TM1 (Sugahara and Toki 1994; Hempel et al. 2012), FSUgold (Todd-Rutel and Piekarewicz 2005; Hempel et al. 2012), NL3 (Lalazissis et al. 1997; Fischer et al. 2014b),

DD2 (Typel et al. 2010; Fischer et al. 2014b), and IU-FSU (Fattoyev et al. 2010; Fischer et al. 2014b).

- **SFHo, SFHx**. The SFHo and the SFHx EoSs (Steiner et al. 2013a) are based on the HS EoS, using two new RMF parameterizations fitted to some NS radius determinations. These parameterizations have rather low values of the slope of the symmetry energy, L, with respect to those used in the HS EoS.

- **SHT, SHO**. The EoSs of G. Shen et al., SHT (Shen et al. 2011b) and SHO (Shen et al. 2011a), are computed using different methods in different density-temperature domains.[9] At high densities, uniform matter is described within a RMF model. For non-uniform matter at intermediate densities, calculations are performed in the (spherical) WS approximation, incorporating nuclear shell effects (Shen et al. 2010a). The same RMF parameterization is employed. In this regime, matter is modelled as a mixture of one average nucleus and nucleons, but no α particles (Shen et al. 2010a). At lower densities, a virial EoS for a non-ideal gas consisting of neutrons, protons, α particles, and 8980 heavy nuclei ($A \geq 12$) from a finite-range droplet model mass table is employed (Shen et al. 2010b). Second-order virial corrections are included among nucleons and α particles, Coulomb screening is included for heavy nuclei, and no excluded-volume effects are considered. In the SHT (SHO) model, the RMF NL3 (FSUGold) parameterization is used. Since the original FSUGold EoS has a maximum NS mass of $1.7 M_\odot$, a modification in the pressure has been introduced at high density (above $0.2 \, \text{fm}^{-3}$), in order to increase the maximum NS mass to $2.1 M_\odot$. In order to produce a full table on a fine grid that is thermodynamically consistent, a smoothing and interpolation scheme is used (Shen et al. 2011b,a).

Recently, EoSs at finite temperature incorporating additional degrees of freedom have been also developed. Indeed, the appearance of additional particles such as hyperons, pions, or even a transition to quark matter, cannot be excluded in the density-temperature regime encountered during CCSNe or mergers.

- *EoSs with hyperons and/or pions*. In the RMF framework, Ishizuka et al. (2008) extended the STOS EoS (Shen et al. 1998a), including hyperons and pions. Scalar coupling constants of hyperons to nucleons are chosen to reproduce hyperonic potential extracted from hypernuclear data, while vector couplings are fixed based on symmetries. These authors also investigate the impact of the EoS in NSs and in a spherical, adiabatic collapse of a $15 M_\odot$ star without neutrino transfer: hyperon effects are found to be small for the density and temperature encountered. Pions are treated as an ideal free Bose gas. Although, without interactions, charged pions condensate below some critical temperature, Ishizuka et al. (2008) mention that pion condensation is suppressed when considering a πN repulsive interaction. Moreover, the effect of pions on the EoS is expected to become non-negligible at high temperature, where the free gas approximation should be valid. The EoS of Ishizuka et al. (2008) has been applied, e.g., in

---

[9]The SHT EoS is also available at the website http://cecelia.physics.indiana.edu/gang_shen_eos/.

Nakazato et al. (2012) to investigate hyperons in BH-forming failed SNe. An
extension of the STOS EoS including pions is discussed, e.g., in Nakazato et al.
(2008). In the non-relativistic framework, Oertel et al. (2012) added hyperons
extending the model of Balberg and Gal (1997), that is based on a non-relativistic
potential similar to that used in the LS EoS (Lattimer and Douglas Swesty 1991)
for nucleons. The hyperon couplings are chosen to be compatible with the single-
particle hyperonic potentials in nuclear matter and with the measured NS mass
of Demorest et al. (2010). Pions are also included in this model, as an ideal free
Bose gas. A version of this EoS, including only pions, has been employed to
study BH formation (Peres et al. 2013).

Other models, including only Λ hyperons, have been also developed, e.g.
extending the STOS EoS (Shen et al. 2011c), the LS EoS (Gulminelli et al.
2013; Peres et al. 2013), or the HS model (Banik et al. 2014) (an extended HS
model with the DD2 interaction including hyperons and quarks with a constant
speed of sound, $c_s^2 = 1/3$, has been considered in Heinimann et al. 2016). The
possibility of a phase transition at the onset of hyperons has been discussed, e.g.,
in Schaffner-Bielich et al. (2000, 2002), Gulminelli et al. (2012), Oertel et al.
(2016). At low temperatures, the onset of hyperons occurs between about 2 and 3
times the saturation density. The impact of additional particles on thermodynamic
quantities (especially on the pressure) may be important for high temperature and
densities (see, e.g., Oertel et al. 2017). The role of hyperons in the dynamical
collapse of a non-rotating massive star to a BH and in the formation and evolution
of a PNS has been studied in Banik (2014) using the hyperonic STOS EoS (Shen
et al. 2011c).

- *EoSs with quarks.* Some EoSs also consider a phase transition to quark matter.
  The MIT bag model is applied, e.g., in Nakazato et al. (2008, 2013), Sagert et al.
  (2009, 2010), Fischer et al. (2011, 2014a), and the transition from hadronic to
  quark phase is modelled with a Gibbs construction. The parameters of the model,
  the bag constant B and the strange quark mass, impact the onset of the appearance
  of the quark matter. The inclusion of a gas of pions raises the density of the
  transition to the quark phase due to the softening of the hadronic part of the EoS
  (Nakazato et al. 2008). The possible impact of a quark phase in the core-collapse
  dynamics will be briefly discussed below.

Applications to Core-Collapse Supernovae and Black-Hole Formation

The main microphysics ingredients playing a crucial role in the CCSN dynamics
are the EoS, the electro-weak processes (specifically, the electron capture on free
protons and nuclei), and the neutrino transport (see also Chap. 1 in this book).
It has been shown that these inputs can have an important effect on the collapse
dynamics and the shock propagation (see, e.g. Bethe (1990), Mezzacappa (2005),
Janka (2012), Janka et al. (2007, 2016), Burrows (2013) for a review). However,
their complex interplay and strong feedback with hydrodynamics make difficult
to predict a priori whether a small modification of one of these inputs can have a

considerable effect on the explosion (usually, effects are expected to be moderated, according to the Mazurek's law; see Lattimer and Prakash 2000; Janka 2012). In particular, the impact of the EoS is twofold: (1) it determines the thermodynamic quantities acting on the hydrodynamics (e.g. the pressure and entropy) and (2) it determines the composition of matter thus affecting the electron-capture rates. Concerning the latters, it has been shown that the single-nucleus approximation is not adequate to properly describe electron-capture rates during collapse. Indeed, the most probable nucleus is not necessarily the one for which the rate is higher and this may have an impact on the Y_e evolution thus on the collapse dynamics (see, e.g., Langanke and Martínez-Pinedo (2003, 2014) for a review, and Hix et al. 2003; Sullivan et al. 2016; Raduta et al. 2016, 2017; Furusawa et al. 2017a).

Several studies have been carried out on the impact of the EoS on the infall and post-bounce phase, most in spherical symmetry, employing either EoSs in the single-nucleus approximation or based on a NSE approach (see, e.g. Sumiyoshi et al. 2005; Hempel et al. 2012; Janka 2012; Steiner et al. 2013a; Fischer et al. 2014b; Togashi et al. 2014). Roughly speaking, a "softer" EoS would lead to a more compact and faster contracting PNS producing higher neutrino luminosities (Marek et al. 2009), and to larger shock radii (Janka 2012; Suwa et al. 2013) in multi-dimensional simulations, resulting in a more favourable situation for explosion. Different ("soft" versus "stiff") EoSs may also potentially impact the gravitational-wave signal from SN (see, e.g. Marek et al. 2009; Scheidegger et al. 2010; Richers et al. 2017). However, it is not straightforward to correlate single nuclear parameters to the collapse dynamics, because different EoSs usually differ in many properties predicted by the underlying nuclear model and because spurious correlations between nuclear parameters can exist for a given model. Moreover, other input parameters like the progenitor structure can impact the outcome of the simulations (see also Chap. 1 in this book). Therefore, systematic investigations are difficult to perform, also because of computational costs of multi-dimensional simulations, and no strong conclusive statements can be drawn.

Since the EoS determines the maximum mass that the hot PNS can support, it also impacts the time from bounce until BH formation (t_{BH}). The sensitivity of t_{BH} to the EoS has been investigated, e.g., in Sumiyoshi et al. (2007), Sumiyoshi (2007), Fischer et al. (2009), O'Connor and Ott (2011), Ott et al. (2011), using the LS and the STOS EoSs, and in Nakazato et al. (2012), Peres et al. (2013), Banik (2014), Char et al. (2015), where EoSs with additional degrees of freedom (hyperons, quarks, or pions) have been employed. Especially in failed CCSNe, high temperatures and densities can be reached, so additional particles are expected to be more abundant. It is generally found that the softening of the EoS thus induced reduces t_{BH}, because the EoS supports less massive PNS with respect to the nucleonic EoS (see, e.g. Nakazato et al. 2008; Sumiyoshi et al. 2009; Nakazato et al. 2012; Peres et al. 2013; Banik 2014; Char et al. 2015).

Some works have also claimed that a transition to a quark phase could have a non-negligible impact on the core-collapse dynamics (see, e.g., Gentile et al. 1993; Drago and Tambini 1999; Sagert et al. 2009; Fischer et al. 2011). In particular, it has been found that this phase transition can lead to a second shock wave triggering the explosion (Sagert et al. 2009). Conditions for heavy-element nucleosynthesis in the explosion of massive stars triggered by a quark-hadron phase transition have also been investigated (e.g., Fischer et al. 2011; Nishimura et al. 2012). However, the EoS applied by Sagert et al. (2009), based on the MIT bag model for the quark phase, was found to be in disagreement with the $2M_\odot$ maximum mass constraint, and subsequent works could not systematically confirm the aforementioned scenario, leaving the question still open (see, e.g. Sagert et al. 2010, 2012; Nakazato et al. 2013; Fischer et al. 2014a).

Applications to Binary Mergers

Binary compact objects, either NSs or BHs, may also provide valuable information on the EoS of dense matter. Indeed, they are promising sources of gravitational waves, they may produce short gamma-ray bursts (GRBs), and they are thought to be one of the main astrophysical scenarios for r-process nucleosynthesis (see, e.g., Shibata and Taniguchi (2011), Faber and Rasio (2012); Rosswog (2015) for a review); all these scenarios are sensitive to the EoS.[10]

Several studies show that the gravitational-wave frequency is related to the tidal deformability during the late inspiral phase of compact binary systems, and thus depends on the EoS (see, e.g., Shibata and Taniguchi 2011; Faber and Rasio 2012; Read et al. 2013; Maselli et al. 2013; Kumar et al. 2017). Moreover, the frequencies of the gravitational waves emitted during the post-merger phase are also sensitive to the NS EoS (see, e.g., Sekiguchi (2011), Bauswein et al. (2014), Takami et al. (2014), Bauswein and Stergioulas (2015), Palenzuela et al. (2015), Takami et al. (2015), Rezzolla and Takami (2016), Baiotti and Rezzolla (2017); see also Chap. 10 in this book).

It has also been proposed to probe the EoS using the analysis of short GRBs that are thought to be associated to binary-merger events (see, e.g., Fan et al. 2013; Lasky et al. 2014; Fryer et al. 2015; Lawrence et al. 2015).

Finally, the conditions and characteristics of r-process nucleosynthesis and the amount of ejected material depend on the thermodynamic conditions and matter composition of the ejecta thus on the EoS (see, e.g., Goriely et al. (2011), Bauswein et al. (2013), Wanajo et al. (2014); see also Chap. 11 in this book).

---

[10]Several such studies have been conducted very recently, after the detection of the GW170817 event (Abbott et al. 2017a). The associated observations of the gamma-ray burst GRB 170817A and electromagnetic counterparts for this event suggest indeed that GW170817 was produced by the coalescence of two NSs followed by a short gamma-ray burst and a kilonova powered by the radioactive decay of r-process nuclei synthesised in the ejecta (Abbott et al. 2017a,b).

6.2.5   CompOSE and Other Online EoS Databases

CompOSE is an online database that has been developed within the European Science Foundation (ESF) funded "CompStar" network and the European Cooperation in Science and Technology (COST) Action MP1304 "NewCompStar". The database is hosted at the website http://compose.obspm.fr. A manual describing how to use the database and how to include one's own EoS into it is also provided. As stated on the main page of the website, "The online service CompOSE provides data tables for different state of the art equations of state (EoS) ready for further usage in astrophysical applications, nuclear physics and beyond." It is not only a repository of EoS tables, but also provides a set of tools to manage the tables, such as interpolation schemes and data handling softwares. At the time being, CompOSE hosts several one-parameter EoSs, suitable for application to NSs, and general purpose EoSs, applicable to SN matter. More details and extensive explanations are given in Typel et al. (2015).

Other online EoS databases that collect different available EoSs exist. STELLARCOLLAPSE.ORG,[11] provides tabulated EoSs, as well as other resources for stellar collapse applications. EOSBD[12] aims "to summarize and share the current information on nuclear EoS which is available today from theoretical/experimental/observational studies of nuclei and dense matter". The Ioffe website[13] provides EoSs of fully ionised electron-ion plasma, EoSs and opacities for partially ionised hydrogen in strong magnetic fields, unified EoSs for NS crust and core, and some hyperonic EoSs; references to the original works are also given. Relativistic EoS tables for SN are also provided online.[14]

6.3   Challenges and Future Prospects

6.3.1   Model Dependence of Data Extrapolations

One of the big issues of obtaining constraints on the EoS from experimental or observational data resides in the extrapolation of the raw data. Indeed, the majority of the constraints result from combining raw data with theoretical models, thus making the constraints model dependent. A typical example among astrophysical observations is the determination of NS radii (see Sect. 6.2.3.2 and Chap. 5 in this book). Concerning constraints coming from nuclear physics experiments, issues arise since the state of matter in SNe and NSs is different compared to that

---

[11] http://www.stellarcollapse.org.

[12] http://aspht1.ph.noda.tus.ac.jp/eos/index.html.

[13] http://www.ioffe.ru/astro/NSG/nseoslist.html.

[14] http://user.numazu-ct.ac.jp/~sumi/eos/; http://phys-merger.physik.unibas.ch/~hempel/eos.html.

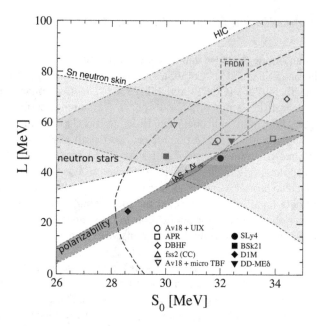

Fig. 6.9 Slope of the symmetry energy L versus the symmetry energy coefficient S_0. Shaded areas correspond to different experimental constraints. Symbols correspond to S_0 and L predicted by different microscopic and phenomenological models. See the text for details

in HICs: matter in SNe can be more isospin asymmetric and has to be charge neutral, while there is a net charge in HICs. For example, the extraction of the pressure versus density constraint in symmetric nuclear matter shown in Fig. 6.5 is subject to uncertainties of the transport models, which depend on a number of parameters that are not fully constrained. Another important example is given by the inferred constraints on the symmetry energy. The latters are abundant at saturation density (see, e.g., Tsang et al. 2012; Lattimer and Lim 2013; Lattimer and Steiner 2014). A (non complete) compilation of different experimental constraints is collected in Fig. 6.9, together with the values of (S_0, L) predicted by different theoretical models, both microscopic (empty symbols) and phenomenological (filled symbols).

1. The green shaded area marked as "HIC" corresponds to the constraints inferred from study of isospin diffusion in HICs (Tsang et al. 2009);
2. The turquoise shaded area labelled "Sn neutron skin" reports the constraints inferred from the analysis of neutron skin thickness in Sn isotopes (Chen et al. 2010);
3. The blue shaded area labelled "polarizability" represents the constraints on the electric dipole polarizability deduced in Roca-Maza et al. (2015). In the latter work, available experimental data on the electric dipole polarizability, α_D, of ^{68}Ni, ^{120}Sn, and ^{208}Pb are compared with the predictions of random-phase

approximation calculations, using a representative set of nuclear EDFs. From the correlation between the neutron skin thickness of a neutron-rich nucleus and L, and between $\alpha_D S_0$ and the neutron skin thickness, Roca Maza et al. extracted a relation between S_0 and L for the three nuclei under study (see Eqs. (12)–(14) in Roca-Maza et al. (2015), and their Fig. 5). The overlap of these constraints is shown in Fig. 6.9;

4. The "FRDM" rectangle corresponds to the values of S_0 and L inferred from finite-range droplet mass model calculations (Möller et al. 2012). These boundaries were derived by varying the considered sets of data along with different refinements of the model. Therefore, they can be biased by the uncertainties of the approach, and probably the constraints turn out to be too severe;

5. The isobaric analog state (IAS) phenomenology and the skin width data can put tight constraints on the density dependence of the symmetry energy up to saturation. These constraints give a range of possible values for S_0 and L, which are displayed in the "IAS $+\Delta r_{np}$" diagonal region, which represents simultaneous constraints of Skyrme-Hartree-Fock calculations of IAS and the ^{208}Pb neutron-skin thickness (Danielewicz and Lee 2014).

Finally, the horizontal band labelled "neutron stars" is obtained by considering the 68% confidence values for L obtained from a Bayesian analysis of mass and radius measurements of NSs (Steiner et al. 2013b), while the dashed curve is the unitary gas bound on symmetry energy parameters of Tews et al. (2017) (see their Eqs. (24)–(25) with $Q_n = 0$): values of (S_0, L) to the right of the curve are permitted. Constraints have also been derived from measurements of collective excitations, like giant dipole resonances (see, e.g., Trippa et al. 2008; Lattimer and Lim 2013; Lattimer and Steiner 2014) and pigmy dipole resonances (see, e.g., the discussion in Daoutidis and Goriely 2011; Reinhard and Nazarewicz 2013). However, we do not display the former constraint, whose band would largely superpose with the other constraints for $S_0 > 30\,\mathrm{MeV}$, and the latter, because of the large theoretical and experimental uncertainties. Note that there is no area of the parameter space where all the considered constraints are simultaneously fulfilled. This is likely to be due to the current uncertainties in the experimental measurements and to the model dependencies that plague the extraction of the constraints from the raw data. Although combining different constraints reduces the uncertainties in the (S_0, L) parameter space, no definitive conclusion can be drawn and, except for models predicting a too high (or low) value of the symmetry energy parameters, no theoretical models can be ruled out a priori on this basis. Finally, it has also to be clarified whether the derived correlations among different parameters and observables have a physical origin or are due to spurious correlations between the model parameters.

6.3.2 Many-Body Treatment at Finite Temperature, Cluster Formation

A unified and consistent treatment of the different phases of matter, both at zero and finite temperature, is extremely challenging (see also Chap. 7 in this book). While either microscopic or phenomenological approaches are suitable to describe homogeneous matter, the correct description of cluster formation, and more generally of phase transitions, both at low and high densities and temperatures, is far from being a trivial task. Indeed, at present, there exist no consistent and rigorous treatment at zero and finite temperature of cluster formation beyond the single-nucleus approximation. In extended NSE models, interactions between a cluster and the surrounding gas are often treated in the excluded-volume approach, but from virial and quantal approaches it is found that cluster properties themselves are modified by the presence of a gas (e.g., Horowitz and Schwenk 2006a; Typel et al. 2010; Hempel et al. 2011) and interactions among clusters should be also considered (e.g., Typel 2014). Moreover, (1) these in-medium effects are density and temperature dependent and (2) with increasing temperature, excited states of nuclei become populated and need to be incorporated in the model. The way of implementing them is not unique, and the different treatments lead to a considerable spread in the predictions of extended NSE models (see, e.g., Buyukcizmeci et al. 2013).

Another issue concerns the extension of the many-body methods and the extrapolation of their predictions, particularly at high density and temperature. For example, the non-uniqueness of the fitting procedure of the EDF parameters and the choice of the experimental data used to fit the parameters have led to different EDFs, thus yielding a large spread in their predictions outside of the domain where the EDFs were fitted (e.g., Goriely and Capote 2014). Especially for compact-object applications, this question can be critical, since extrapolations of nuclear masses are needed to describe the deepest regions of the NS crust and SN cores. On the other hand, the nuclear interaction itself can be temperature dependent. The temperature dependence of the EoS is very important for the physics of CCSNe, PNSs, and compact-star mergers, where densities larger than the saturation density and temperatures up to hundreds of MeV can be reached. In particular, the stiffness of the EoS and the temperature dependence of the pressure can be crucial in determining the final fate of the CCSNe. Therefore, efforts should be devoted on the study of the EoS in the high-temperature regime. In Sect. 6.2.2.1, the current state of the art of microscopic calculations of the finite-temperature EoS has been already discussed. The extension of those calculations at large temperatures is not a trivial task. For instance, in the Bloch-De Dominicis theoretical framework (Bloch and De Dominicis 1959b), on which the finite-temperature BHF approach is based, several higher order diagrams have to be included in the expansion, both at two and three hole-line level. These additional contributions could have sizeable effects that are not straightforward to predict a priori. For phenomenological models, the question arises as whether the EDF parameters determined by fitting nuclear data at zero temperature can be reliably used when applying the EDFs at finite temperature. Thermal properties of asymmetric nuclear matter have been investigated within a relativistic model, showing that the couplings are weakly dependent on temperature, up to a few tens of MeV (Fedoseew and Lenske 2015). Similar conclusions can be deduced, e.g., from Moustakidis and Panos (2009), Fantina et al. (2012b). In the former work, where an extra term has been added to a Skyrme-type interaction, it has been shown that the temperature dependence of the couplings is weak up to about 30 MeV. Also, a good agreement is obtained when comparing the free energy and pressure of nuclear matter for Brussels-Montreal Skyrme models with ab-initio calculations at finite temperature, up to 20 MeV (see Fig. 1 in Fantina et al. 2012b). It remains to be determined whether these conclusions still hold at the highest temperatures ($\gtrsim 100$ MeV) that can be reached in CCSNe or binary mergers.

6.3.3 Role of Three-Body Forces

In Sect. 6.2.2.1 we have shown that a NN interaction based on quark degrees of freedom (Baldo and Fukukawa 2014; Fukukawa et al. 2015) is able to reproduce at the same time the three-body properties, and the saturation point of nuclear matter without introducing TBFs, just using some parameters fitted on the NN phase shifts and deuteron properties. These results were obtained by including in the many-body calculation the three-body correlations within the hole-line expansion of the BBG formalism, indicating that the explicit introduction of the quark structure of the nucleons is relevant for the NN interaction. Additional interactions based on quark degrees of freedom should be considered, in order to understand if they have similar properties and eventually to pinpoint the key reasons of their performance, which is comparable with that of the best NN interaction based on meson exchange processes or on the chiral symmetry of QCD.

6.3.4 Composition and URCA Process

During the first 10^5–10^6 year, a NS cools down mainly via neutrino emission. In the absence of superfluidity, three main processes are usually taken into account: the direct URCA (DU), the modified URCA (MU), and the NN bremsstrahlung (BNN) processes. The most efficient neutrino emission is the DU process, a sequence of neutron decays, $n \rightarrow p + e^- + \bar{\nu}_e$, and electron captures, $p + e^- \rightarrow n + \nu_e$. For this process and for npe NS matter, the neutrino emissivity is given by Yakovlev et al. (2001)

$$Q^{(DU)} \approx 4.0 \times 10^{27} \left(\frac{Y_e \, n_B}{n_0} \right)^{1/3} \frac{m_n^\star m_p^\star}{m_n^2} T_9^6 \Theta(k_{F_p} + k_{F_e} - k_{F_n}) \, \text{erg cm}^{-2} \, \text{s}^{-1} ,$$

$$(6.14)$$

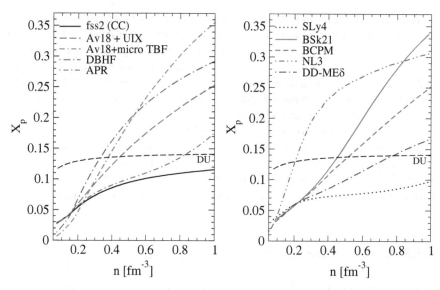

Fig. 6.10 Proton fraction versus baryon number density for different microscopic (left panel) and phenomenological (right panel) models. The dashed black lines labelled "DU" mark the threshold for the DU process to occur (Klähn et al. 2006)

where m_n is the neutron mass, $m_n^\star$ ($m_p^\star$) is the neutron (proton) effective mass, T_9 is the temperature in units of 10^9 K, Θ is the Fermi function, and $k_{F,p}$, $k_{F,e}$, and $k_{F,n}$ are the proton, electron, and neutron Fermi momenta, respectively. It thus considerably depends on the temperature and on the nucleon effective masses. If muons are present, then the corresponding DU process may also become possible, in which case the neutrino emissivity is increased by a factor of 2. If it takes place, the DU process enhances neutrino emission and NS cooling rates by a large factor compared to MU and BNN processes. The role of the DU processes has been long questioned in the past years, since it depends on the adopted EoS and the values of the superfluidity gaps, on which, at present, there is no consensus. Concerning the EoS, the energy and momentum conservation imposes a proton fraction threshold for this process to occur (Lattimer et al. 1991; Klähn et al. 2006), $X_p \approx 11$–15%, that is mainly determined by the symmetry energy. In Fig. 6.10, we display the proton fraction versus the baryon density in NS matter for different microscopic (left panel) and phenomenological (right panel) models. For the former models, we observe that, except fss2 (CC), all the considered microscopic approaches are characterised by a quite low value of the threshold density. For instance, for the BHF with Av18+ UIX, the DU process sets in at $0.44\,\mathrm{fm}^{-3}$, thus the DU process operates in NS with masses $M > 1.10\,M_\odot$, while for the APR EoS the onset of DU is shifted to larger density, $0.82\,\mathrm{fm}^{-3}$, due to the lower values of the symmetry energy, hence X_p. Among the phenomenological models considered here, only the EoS based on the SLy4 EDF forbids the DU process, while the EoS based on NL3 (DD-MEδ) has the lowest (Cavagnoli et al. 2011) and highest (Wang et al. 2014) threshold density. For the EoS based on the BSk21 (BCPM) EDF, the threshold density is $0.45\,\text{fm}^{-3}$ (Fantina et al. 2013) and $0.53\,\text{fm}^{-3}$ (Sharma et al. 2015), thus the DU process occurs for NS with $M > 1.59 M_\odot$ ($M > 1.35 M_\odot$). Incidentally, Klähn et al. (2006) argued that no DU process should occur in NSs with typical masses in the range $M \sim 1 - 1.5 M_\odot$. From the observational point of view, the pulsar in CTA1, the transiently accreting millisecond pulsar SAX J1808.4−3658, and the soft X-ray transient 1H 1905+000 appear to be very cold, thus suggesting that these NSs may cool very fast via the DU process (Jonker et al. 2007; Heinke et al. 2009; Page et al. 2009; Abdo et al. 2012). Moreover, the low luminosity from several young SN remnants likely to contain a still unobserved NS (Kaplan et al. 2004, 2006) could suggest further evidence for a DU process (Shternin and Yakovlev 2008; Page et al. 2009). If DU processes actually occur in those objects and the NS masses were known, they could put constraints on the EoSs unfavouring those that forbid DU for those masses. In fact, a key parameter that could discriminate whether the DU occurs in a NS is its mass. Unfortunately, the masses of these cooling objects are not precisely measured, if not known at all. An object of particular interest is Cassiopeia A (Heinke and Ho 2010), that can potentially give information on the interior of the NS (see, e.g., Page et al. 2011; Shternin et al. 2011; Blaschke et al. 2012, 2013; Sedrakian 2013; Taranto et al. 2016) and on the nuclear symmetry energy and the nuclear pasta (Newton et al. 2013). Its fast cooling was claimed to be a direct proof of superfluidity in NSs, even if more recent analyses put a word of caution on the initial data (Posselt et al. 2013; Elshamouty et al. 2013; Ho et al. 2015).

A further critical point of most current cooling simulations is the fact that a given EoS is combined with pairing gaps obtained within a different theoretical framework and using different input interactions, thus resulting in an inconsistent analysis. Recently, some progress has been made along this direction (Taranto et al. 2016), concluding that the possibility of strong DU processes cannot be excluded from the cooling analysis.

The current results confirm the extreme difficulty to draw quantitative conclusions from the current NS cooling data. In particular, the present substantial theoretical uncertainty regarding superfluidity gaps and thermal conductivity (see also Chap. 8 in this book) calls for a renewed effort in the theoretical activity of the next few years.

6.4 Conclusions

The EoS of hot and dense matter is a crucial input to describe static and dynamical properties of compact objects. However, constructing such a (unified) EoS is a very challenging task. The physical conditions prevailing in these astrophysical objects are so extreme that it is currently impossible to reproduce them in terrestrial laboratories. Therefore, theoretical models are required. Nevertheless, ab-initio calculations cannot be at present applied to determine the EoS in all the regions of NSs and SNe, mainly because of computational cost, thus more phenomenological models have to be employed. In this Chapter, we have reviewed the current status of the EoS for compact objects. We have presented the different underlying many-body methods, both microscopic and phenomenological, for homogeneous and inhomogeneous matter, considering only nucleonic degrees of freedom. We have discussed these models with respect to constraints coming from both nuclear physics experiments and astrophysical observations: apart from the precise measurements of the $2M_\odot$ NSs, other constraints are less strict since often model dependent. New terrestrial experiments and facilities such as RIKEN, FAIR, HIE-ISOLDE, SPIRAL 2, FRIB, and TRIUMF, and new-generation telescopes and projects such as ATHENA+, NICER, and SKA, and gravitational-wave detectors such as Advanced Virgo and LIGO, and LISA promise to provide more and more precise data that can significantly contribute to probe the internal structure of compact objects, allowing unprecedented comparisons with theoretical predictions. Finally, we have discussed some of the present challenges in the EoS modelling. Indeed, despite many recent advances in the many-body treatment, still issues have to be faced in the description of the EoS. These include (1) the model dependence of the constraints inferred from experimental nuclear physics and astrophysical data, (2) the lack of a consistent and rigorous many-body treatment both at zero and finite temperature of cluster formation beyond the single-nucleus approximation, (3) the treatment and role of nucleonic TBFs, and (4) the description of the cooling in NSs. The current theoretical uncertainties require significant efforts to be undertaken in these directions in the next few years.

Acknowledgements This work has been partially supported by the COST action MP1304 "New-CompStar". The authors would like to thank Ad. R. Raduta for providing Fig. 6.4, H. J. Schulze for providing Fig. 6.1, M. Baldo for providing us with data for Fig. 6.2 and for valuable comments on the manuscript, N. Chamel and S. Goriely for fruitful discussions and providing us with data for Figs. 6.3, 6.5, and 6.6, C. Providência for data for Figs. 6.3 and 6.5, F. Gulminelli for valuable comments on the manuscript, and X. Roca Maza for insightful discussions.

References

Abbott, B.P., Abbott, R., Abbott, T.D., Acernese, F., Ackley, K., Adams, C., Adams, T., Addesso, P., Adhikari, R.X., Adya, V.B., et al.: Gravitational waves and gamma-rays from a binary neutron star merger: GW170817 and GRB 170817A. Astrophys. J. Lett. **848**, L13 (2017a). https://doi.org/10.3847/2041-8213/aa920c

Abbott, B.P., Abbott, R., Abbott, T.D., Acernese, F., Ackley, K., Adams, C., Adams, T., Addesso, P., Adhikari, R.X., Adya, V.B., et al.: GW170817: observation of gravitational waves from a binary neutron star inspiral. Phys. Rev. Lett. **119**(16), 161101 (2017b). https://doi.org/10.1103/PhysRevLett.119.161101

Abdo, A.A., Wood, K.S., DeCesar, M.E., Gargano, F., Giordano, F., Ray, P.S., Parent, D., Harding, A.K., Miller, M.C., Wood, D.L., Wolff, M.T.: PSR J0007+7303 in the CTA1 supernova remnant: new gamma-ray results from two years of fermi large area telescope observations. Astrophys. J. **744**, 146 (2012). https://doi.org/10.1088/0004-637X/744/2/146

Agrawal, B.K., De, J.N., Samaddar, S.K., Centelles, M., Viñas, X.: Symmetry energy of warm nuclear systems. Eur. Phys. J. A **50**, 19 (2014). https://doi.org/10.1140/epja/i2014-14019-8

Akmal, A., Pandharipande, V.R., Ravenhall, D.G.: Equation of state of nucleon matter and neutron star structure. Phys. Rev. C **58**, 1804–1828 (1998). https://journals.aps.org/prc/abstract/10. 1103/PhysRevC.58.1804

Angeli, I., Marinova, K.P.: Table of experimental nuclear ground state charge radii: an update. At. Data Nucl. Data Tables **99**, 69–95 (2013). https://doi.org/10.1016/j.adt.2011.12.006

Antić, S., Typel, S.: Neutron star equations of state with optical potential constraint. Nucl. Phys. A **938**, 92–108 (2015). https://doi.org/10.1016/j.nuclphysa.2015.03.004

Antoniadis, J., Freire, P.C.C., Wex, N., Tauris, T.M., Lynch, R.S., van Kerkwijk, M.H., Kramer, M., Bassa, C., Dhillon, V.S., Driebe, T., Hessels, J.W.T., Kaspi, V.M., Kondratiev, V.I., Langer, N., Marsh, T.R., McLaughlin, M.A., Pennucci, T.T., Ransom, S.M., Stairs, I.H., van Leeuwen, J., Verbiest, J.P.W., Whelan, D.G.: A massive pulsar in a compact relativistic binary. Science **340**, 1233232 (2013). https://doi.org/10.1126/science.1233232

Aoki, S., Doi, T., Hatsuda, T., Ikeda, Y., Inoue, T., Ishii, N., Murano, K., Nemura, H., Sasaki, K., HAL QCD Collaboration: Lattice quantum chromodynamical approach to nuclear physics. Prog. Theor. Exp. Phys. **2012**(1), 01A105 (2012). https://doi.org/10.1093/ptep/pts010

Audi, G., Wapstra, A.H., Thibault, C.: The Ame2003 atomic mass evaluation (II). Tables, graphs and references. Nucl. Phys. A **729**, 337–676 (2003). https://doi.org/10.1016/j.nuclphysa.2003. 11.003

Avancini, S.S., Chiacchiera, S., Menezes, D.P., Providência, C.: Warm "pasta" phase in the Thomas-Fermi approximation. Phys. Rev. C **82**(5), 055807 (2010). https://doi.org/10.1103/ PhysRevC.82.055807

Avancini, S.S., Barros Jr., C.C., Brito, L., Chiacchiera, S., Menezes, D.P., Providência, C.: Light clusters in nuclear matter and the "pasta" phase. Phys. Rev. C **85**(3), 035806 (2012). https:// doi.org/10.1103/PhysRevC.85.035806

Aymard, F., Gulminelli, F., Margueron, J.: In-medium nuclear cluster energies within the extended Thomas-Fermi approach. Phys. Rev. C **89**(6), 065807 (2014). https://doi.org/10.1103/ PhysRevC.89.065807

Baiotti, L., Rezzolla, L.: Binary neutron star mergers: a review of Einstein's richest laboratory. Rep. Prog. Phys. **80**(9), 096901 (2017). https://doi.org/10.1088/1361-6633/aa67bb

Balberg, S., Gal, A.: An effective equation of state for dense matter with strangeness. Nucl. Phys. A **625**, 435–472 (1997). https://doi.org/10.1016/S0375-9474(97)81465-0

Baldo, M. (ed.): Nuclear Methods and The Nuclear Equation of State. World Scientific, Singapore (1999). https://doi.org/10.1142/2657

Baldo, M., Burgio, G.F.: Properties of the nuclear medium. Rep. Prog. Phys. **75**(2), 026301 (2012). https://doi.org/10.1088/0034-4885/75/2/026301

Baldo, M., Burgio, G.F.: The nuclear symmetry energy. Prog. Part. Nucl. Phys. **91**, 203–258 (2016). https://doi.org/10.1016/j.ppnp.2016.06.006

Baldo, M., Ferreira, L.S.: Nuclear liquid-gas phase transition. Phys. Rev. C **59**, 682–703 (1999). https://doi.org/10.1103/PhysRevC.59.682

Baldo, M., Fukukawa, K.: Nuclear matter from effective quark-quark interaction. Phys. Rev. Lett. **113**(24), 242501 (2014). https://doi.org/10.1103/PhysRevLett.113.242501

Baldo, M., Maieron, C.: Equation of state of nuclear matter at high baryon density. J. Phys. G: Nucl. Part. Phys. **34**, R243 (2007). https://doi.org/10.1088/0954-3899/34/5/R01

Baldo, M., Bombaci, I., Burgio, G.F.: Microscopic nuclear equation of state with three-body forces and neutron star structure. Astron. Astrophys. **328**, 274–282 (1997)

Baldo, M., Burgio, G.F., Schulze, H.J.: Hyperon stars in the Brueckner-Bethe-Goldstone theory. Phys. Rev. C **61**(5), 055801 (2000). https://doi.org/10.1103/PhysRevC.61.055801

Baldo, M., Saperstein, E.E., Tolokonnikov, S.V.: A realistic model of superfluidity in the neutron star inner crust. Eur. Phys. J. A **32**, 97–108 (2007). https://doi.org/10.1140/epja/i2006-10356-5

Baldo, M., Robledo, L., Schuck, P., Viñas, X.: Energy density functional on a microscopic basis. J. Phys. G: Nucl. Part. Phys. **37**(6), 064015 (2010). https://doi.org/10.1088/0954-3899/37/6/ 064015

Baldo, M., Robledo, L.M., Schuck, P., Viñas, X.: New Kohn-Sham density functional based on microscopic nuclear and neutron matter equations of state. Phys. Rev. C **87**(6), 064305 (2013). https://doi.org/10.1103/PhysRevC.87.064305

Banik, S.: Probing the metastability of a protoneutron star with hyperons in a core-collapse supernova. Phys. Rev. C **89**(3), 035807 (2014). https://doi.org/10.1103/PhysRevC.89.035807

Banik, S., Hempel, M., Bandyopadhyay, D.: New hyperon equations of state for supernovae and neutron stars in density-dependent hadron field theory. Astrophys. J. Suppl. Ser. **214**, 22 (2014). https://doi.org/10.1088/0067-0049/214/2/22

Bauböck, M., Psaltis, D., Özel, F.: Narrow atomic features from rapidly spinning neutron stars. Astrophys. J. **766**, 87 (2013). https://doi.org/10.1088/0004-637X/766/2/87

Bauswein, A., Stergioulas, N.: Unified picture of the post-merger dynamics and gravitational wave emission in neutron star mergers. Phys. Rev. D **91**(12), 124056 (2015). https://doi.org/10.1103/PhysRevD.91.124056

Bauswein, A., Janka, H.T., Hebeler, K., Schwenk, A.: Equation-of-state dependence of the gravitational-wave signal from the ring-down phase of neutron-star mergers. Phys. Rev. D **86**(6), 063001 (2012). https://doi.org/10.1103/PhysRevD.86.063001

Bauswein, A., Goriely, S., Janka, H.T.: Systematics of dynamical mass ejection, nucleosynthesis, and radioactively powered electromagnetic signals from neutron-star mergers. Astrophys. J. **773**, 78 (2013). https://doi.org/10.1088/0004-637X/773/1/78

Bauswein, A., Stergioulas, N., Janka, H.T.: Revealing the high-density equation of state through binary neutron star mergers. Phys. Rev. D **90**(2), 023002 (2014). https://doi.org/10.1103/PhysRevD.90.023002

Bauswein, A., Stergioulas, N., Janka, H.T.: Exploring properties of high-density matter through remnants of neutron-star mergers. Eur. Phys. J. A **52**, 56 (2016). https://doi.org/10.1140/epja/i2016-16056-7

Baym, G.: Self-consistent approximations in many-body systems. Phys. Rev. **127**, 1391–1401 (1962). https://doi.org/10.1103/PhysRev.127.1391

Baym, G., Bethe, H.A., Pethick, C.J.: Neutron star matter. Nucl. Phys. A **175**, 225–271 (1971a). https://doi.org/10.1016/0375-9474(71)90281-8

Baym, G., Pethick, C., Sutherland, P.: The ground state of matter at high densities: equation of state and stellar models. Astrophys. J. **170**, 299 (1971b). https://doi.org/10.1086/151216

Beane, S.R., Detmold, W., Orginos, K., Savage, M.J.: Nuclear physics from lattice QCD. Prog. Part. Nucl. Phys. **66**, 1–40 (2011). https://doi.org/10.1016/j.ppnp.2010.08.002

Bedaque, P.F., Rupak, G.: Dilute resonating gases and the third virial coefficient. Phys. Rev. B **67**(17), 174513 (2003). https://doi.org/10.1103/PhysRevB.67.174513

Bejger, M., Haensel, P.: Accelerated expansion of the Crab Nebula and evaluation of its neutron-star parameters. Astron. Astrophys. **405**, 747–751 (2003). https://doi.org/10.1051/0004-6361:20030642

Bejger, M., Haensel, P., Zdunik, J.L.: Rotation at 1122 Hz and the neutron star structure. Astron. Astrophys. **464**, L49–L52 (2007). https://doi.org/10.1051/0004-6361:20066902

Bender, M., Heenen, P.H., Reinhard, P.G.: Self-consistent mean-field models for nuclear structure. Rev. Mod. Phys. **75**, 121–180 (2003). https://doi.org/10.1103/RevModPhys.75.121

Bender, M., Bennaceur, K., Duguet, T., Heenen, P.H., Lesinski, T., Meyer, J.: Tensor part of the Skyrme energy density functional. II. Deformation properties of magic and semi-magic nuclei. Phys. Rev. C **80**(6), 064302 (2009). https://doi.org/10.1103/PhysRevC.80.064302

Beth, E., Uhlenbeck, G.E.: The quantum theory of the non-ideal gas. II. Behaviour at low temperatures. Physica **4**, 915–924 (1937). https://doi.org/10.1016/S0031-8914(37)80189-5

Bethe, H.A.: Supernova mechanisms. Rev. Mod. Phys. **62**, 801–866 (1990). https://doi.org/10.1103/RevModPhys.62.801

Binder, S., Langhammer, J., Calci, A., Roth, R.: Ab initio path to heavy nuclei. Phys. Lett. B **736**, 119–123 (2014). http://dx.doi.org/10.1016/j.physletb.2014.07.010; //www.sciencedirect.com/science/article/pii/S0370269314004961

Blaizot, J.P.: Nuclear compressibilities. Phys. Rep. **64**, 171–248 (1980). https://doi.org/10.1016/0370-1573(80)90001-0

Blaschke, D., Grigorian, H., Voskresensky, D.N., Weber, F.: Cooling of the neutron star in Cassiopeia A. Phys. Rev. C **85**(2), 022802 (2012). https://doi.org/10.1103/PhysRevC.85. 022802

Blaschke, D., Grigorian, H., Voskresensky, D.N.: Nuclear medium cooling scenario in light of new Cas A cooling data and the $2M_\odot$ pulsar mass measurements. Phys. Rev. C **88**(6), 065805 (2013). https://doi.org/10.1103/PhysRevC.88.065805

Blinnikov, S.I., Panov, I.V., Rudzsky, M.A., Sumiyoshi, K.: The equation of state and composition of hot, dense matter in core-collapse supernovae. Astron. Astrophys. **535**, A37 (2011). https:// doi.org/10.1051/0004-6361/201117225

Bloch, C., De Dominicis, C.: Un développement du potentiel de gibbs d'un système quantique composé d'un grand nombre de particules. Nucl. Phys. **7**, 459–479 (1958). https://doi.org/10. 1016/0029-5582(58)90285-2

Bloch, C., De Dominicis, C.: Un développement du potentiel de Gibbs d'un système composé d'un grand nombre de particules (II). Nucl. Phys. **10**, 181–196 (1959a). https://doi.org/10.1016/ 0029-5582(59)90203-2

Bloch, C., De Dominicis, C.: Un développement du potentiel de Gibbs d'un système quantique composé d'un grand nombre de particules III: La contribution des collisions binaires. Nucl. Phys. **10**, 509–526 (1959b). https://doi.org/10.1016/0029-5582(59)90241-X

Bogdanov, S.: Constraining the state of ultra-dense matter with the neutron star interior composition explorer. In: AAS/High Energy Astrophysics Division, vol. 15, p. 105.05 (2016a)

Bogdanov, S.: Prospects for neutron star equation of state constraints using "recycled" millisecond pulsars. Eur. Phys. J. A **52**, 37 (2016b). https://doi.org/10.1140/epja/i2016-16037-x

Bogner, S.K., Schwenk, A., Furnstahl, R.J., Nogga, A.: Is nuclear matter perturbative with low-momentum interactions? Nucl. Phys. A **763**, 59–79 (2005). https://doi.org/10.1016/j.nuclphysa. 2005.08.024

Bogner, S.K., Furnstahl, R.J., Schwenk, A.: From low-momentum interactions to nuclear structure. Prog. Part. Nucl. Phys. **65**, 94–147 (2010). https://doi.org/10.1016/j.ppnp.2010.03.001

Boguta, J., Bodmer, A.R.: Relativistic calculation of nuclear matter and the nuclear surface. Nucl. Phys. A **292**, 413–428 (1977). https://doi.org/10.1016/0375-9474(77)90626-1

Bonche, P., Vautherin, D.: Mean-field calculations of the equation of state of supernova matter II. Astron. Astrophys. **112**, 268–272 (1982)

Borderie, B., Rivet, M.F.: Nuclear multifragmentation and phase transition for hot nuclei. Prog. Part. Nucl. Phys. **61**, 551–601 (2008). https://doi.org/10.1016/j.ppnp.2008.01.003

Bose, S., Chakravarti, K., Rezzolla, L., Sathyaprakash, B.S., Takami, K.: Neutron-star radius from a population of binary neutron star mergers. Phys. Rev. Lett. **120**(3), 031102 (2018). https:// doi.org/10.1103/PhysRevLett.120.031102

Botvina, A.S., Mishustin, I.N.: Statistical approach for supernova matter. Nucl. Phys. A **843**(1), 98–132 (2010). http://dx.doi.org/10.1016/j.nuclphysa.2010.05.052; http://www.sciencedirect.com/ science/article/pii/S0375947410005099

Brack, M., Quentin, P.: Selfconsistent calculations of highly excited nuclei. Phys. Lett. B **52**, 159–162 (1974). https://doi.org/10.1016/0370-2693(74)90077-X

Brack, M., Guet, C., Håkansson, H.B.: Selfconsistent semiclassical description of average nuclear properties—a link between microscopic and macroscopic models. Phys. Rep. **123**, 275–364 (1985). https://doi.org/10.1016/0370-1573(86)90078-5

Brink, D., Broglia, R.: Nuclear Superfluidity: Pairing in Finite Systems. Cambridge University Press, Cambridge (2005)

Brockmann, R., Machleidt, R.: Relativistic nuclear structure. I. Nuclear matter. Phys. Rev. C **42**, 1965–1980 (1990). https://doi.org/10.1103/PhysRevC.42.1965

Brown, G.E., Weise, W., Baym, G., Speth, J.: Relativistic effects in nuclear physics. Comment. Nucl. Part. Phys. **17**, 39–62 (1987)

Burrows, A.: Colloquium: perspectives on core-collapse supernova theory. Rev. Mod. Phys. **85**, 245–261 (2013). https://doi.org/10.1103/RevModPhys.85.245

Burrows, A., Lattimer, J.M.: On the accuracy of the single-nucleus approximation in the equation of state of hot, dense matter. Astrophys. J. **285**, 294–303 (1984). https://doi.org/10.1086/162505

Bürvenich, T., Madland, D.G., Maruhn, J.A., Reinhard, P.G.: Nuclear ground state observables and QCD scaling in a refined relativistic point coupling model. Phys. Rev. C **65**(4), 044308 (2002). https://doi.org/10.1103/PhysRevC.65.044308

Buyukcizmeci, N., Botvina, A.S., Mishustin, I.N., Ogul, R., Hempel, M., Schaffner-Bielich, J., Thielemann, F.K., Furusawa, S., Sumiyoshi, K., Yamada, S., Suzuki, H.: A comparative study of statistical models for nuclear equation of state of stellar matter. Nucl. Phys. A **907**, 13–54 (2013). https://doi.org/10.1016/j.nuclphysa.2013.03.010

Buyukcizmeci, N., Botvina, A.S., Mishustin, I.N.: Tabulated equation of state for supernova matter including full nuclear ensemble. Astrophys. J. **789**, 33 (2014). https://doi.org/10.1088/0004-637X/789/1/33

Cao, L.G., Colò, G., Sagawa, H.: Spin and spin-isospin instabilities and Landau parameters of Skyrme interactions with tensor correlations. Phys. Rev. C **81**(4), 044302 (2010). https://doi.org/10.1103/PhysRevC.81.044302

Caplan, M.E., Schneider, A.S., Horowitz, C.J., Berry, D.K.: Pasta nucleosynthesis: molecular dynamics simulations of nuclear statistical equilibrium. Phys. Rev. C **91**(6), 065802 (2015). https://doi.org/10.1103/PhysRevC.91.065802

Carbone, A., Cipollone, A., Barbieri, C., Rios, A., Polls, A.: Self-consistent Green's functions formalism with three-body interactions. Phys. Rev. C **88**(5), 054326 (2013a). https://doi.org/10.1103/PhysRevC.88.054326

Carbone, A., Polls, A., Rios, A.: Symmetric nuclear matter with chiral three-nucleon forces in the self-consistent Green's functions approach. Phys. Rev. C **88**(4), 044302 (2013b). https://doi.org/10.1103/PhysRevC.88.044302

Carlson, J., Pandharipande, V.R., Wiringa, R.B.: Three-nucleon interaction in 3-, 4- and ∞-body systems. Nucl. Phys. A **401**, 59–85 (1983). https://doi.org/10.1016/0375-9474(83)90336-6

Carlson, J., Morales, J., Pandharipande, V.R., Ravenhall, D.G.: Quantum Monte Carlo calculations of neutron matter. Phys. Rev. C **68**(2), 025802 (2003). https://doi.org/10.1103/PhysRevC.68.025802

Carlson, J., Gandolfi, S., Gezerlis, A.: Quantum Monte Carlo approaches to nuclear and atomic physics. Prog. Theor. Exp. Phys. **2012**(1), 01A209 (2012). https://doi.org/10.1093/ptep/pts031

Carlson, J., Gandolfi, S., Pederiva, F., Pieper, S.C., Schiavilla, R., Schmidt, K.E., Wiringa, R.B.: Quantum Monte Carlo methods for nuclear physics. Rev. Mod. Phys. **87**, 1067–1118 (2015). https://doi.org/10.1103/RevModPhys.87.1067

Cavagnoli, R., Menezes, D.P., Providência, C.: Neutron star properties and the symmetry energy. Phys. Rev. C **84**(6), 065810 (2011). https://doi.org/10.1103/PhysRevC.84.065810

Chabanat, E., Bonche, P., Haensel, P., Meyer, J., Schaeffer, R.: A Skyrme parametrization from subnuclear to neutron star densities Part II. Nuclei far from stabilities. Nucl. Phys. A **635**, 231–256 (1998). https://doi.org/10.1016/S0375-9474(98)00180-8

Chamel, N.: Self-interaction errors in nuclear energy density functionals. Phys. Rev. C **82**(6), 061307 (2010). https://doi.org/10.1103/PhysRevC.82.061307

Chamel, N., Fantina, A.F.: Electron exchange and polarization effects on electron captures and neutron emissions by nuclei in white dwarfs and neutron stars. Phys. Rev. D **93**(6), 063001 (2016). https://doi.org/10.1103/PhysRevD.93.063001

Chamel, N., Goriely, S.: Spin and spin-isospin instabilities in asymmetric nuclear matter at zero and finite temperatures using Skyrme functionals. Phys. Rev. C **82**(4), 045804 (2010). https://doi.org/10.1103/PhysRevC.82.045804

Chamel, N., Haensel, P.: Physics of neutron star crusts. Living Rev. Relativ. **11**, 10 (2008). https://doi.org/10.12942/lrr-2008-10

Chamel, N., Naimi, S., Khan, E., Margueron, J.: Validity of the Wigner-Seitz approximation in neutron star crust. Phys. Rev. C **75**(5), 055806 (2007). https://doi.org/10.1103/PhysRevC.75.055806

Chamel, N., Goriely, S., Pearson, J.M.: Further explorations of Skyrme-Hartree-Fock-Bogoliubov mass formulas. XI. Stabilizing neutron stars against a ferromagnetic collapse. Phys. Rev. C **80**(6), 065804 (2009). https://doi.org/10.1103/PhysRevC.80.065804

Chamel, N., Fantina, A.F., Pearson, J.M., Goriely, S.: Phase transitions in dense matter and the maximum mass of neutron stars. Astron. Astrophys. **553**, A22 (2013a). https://doi.org/10.1051/0004-6361/201220986

Chamel, N., Haensel, P., Zdunik, J.L., Fantina, A.F.: Erratum: on the maximum mass of neutron stars. Int. J. Mod. Phys. E **22**(11), 1392004 (2013b). https://doi.org/10.1142/S0218301313920043; http://www.worldscientific.com/doi/abs/10.1142/S0218301313920043

Chamel, N., Haensel, P., Zdunik, J.L., Fantina, A.F.: On the maximum mass of neutron stars. Int. J. Mod. Phys. E **22**, 1330018 (2013c). https://doi.org/10.1142/S021830131330018X

Chamel, N., Pearson, J.M., Fantina, A.F., Ducoin, C., Goriely, S., Pastore, A.: Brussels–Montreal nuclear energy density functionals, from atomic masses to neutron stars. Acta Phys. Pol. B **46**, 349 (2015). https://doi.org/10.5506/APhysPolB.46.349

Char, P., Banik, S., Bandyopadhyay, D.: A comparative study of hyperon equations of state in supernova simulations. Astrophys. J. **809**, 116 (2015). https://doi.org/10.1088/0004-637X/809/2/116

Chen, L.W., Ko, C.M., Li, B.A., Xu, J.: Density slope of the nuclear symmetry energy from the neutron skin thickness of heavy nuclei. Phys. Rev. C **82**(2), 024321 (2010). https://doi.org/10.1103/PhysRevC.82.024321

Cheng, K.S., Yao, C.C., Dai, Z.G.: Properties of nuclei in the inner crusts of neutron stars in the relativistic mean-field theory. Phys. Rev. C **55**, 2092–2100 (1997). https://doi.org/10.1103/PhysRevC.55.2092

Cipollone, A., Barbieri, C., Navrátil, P.: Isotopic chains around oxygen from evolved chiral two- and three-nucleon interactions. Phys. Rev. Lett. **111**(6), 062501 (2013). https://doi.org/10.1103/PhysRevLett.111.062501

Cipollone, A., Barbieri, C., Navrátil, P.: Chiral three-nucleon forces and the evolution of correlations along the oxygen isotopic chain. Phys. Rev. C **92**(1), 014306 (2015). https://doi.org/10.1103/PhysRevC.92.014306

Coester, F.: Bound states of a many-particle system. Nucl. Phys. **7**, 421–424 (1958). https://doi.org/10.1016/0029-5582(58)90280-3

Coester, F., Kümmel, H.: Short-range correlations in nuclear wave functions. Nucl. Phys. **17**, 477–485 (1960). https://doi.org/10.1016/0029-5582(60)90140-1

Colò, G., van Giai, N., Meyer, J., Bennaceur, K., Bonche, P.: Microscopic determination of the nuclear incompressibility within the nonrelativistic framework. Phys. Rev. C **70**(2), 024307 (2004). https://doi.org/10.1103/PhysRevC.70.024307

Constantinou, C., Muccioli, B., Prakash, M., Lattimer, J.M.: Thermal properties of supernova matter: the bulk homogeneous phase. Phys. Rev. C **89**(6), 065802 (2014). https://doi.org/10.1103/PhysRevC.89.065802

Constantinou, C., Muccioli, B., Prakash, M., Lattimer, J.M.: Thermal properties of hot and dense matter with finite range interactions. Phys. Rev. C **92**(2), 025801 (2015). https://doi.org/10.1103/PhysRevC.92.025801

Cooper, L.N.: Bound electron pairs in a degenerate fermi gas. Phys. Rev. **104**, 1189–1190 (1956). https://doi.org/10.1103/PhysRev.104.1189

Coraggio, L., Holt, J.W., Itaco, N., Machleidt, R., Marcucci, L.E., Sammarruca, F.: Nuclear-matter equation of state with consistent two- and three-body perturbative chiral interactions. Phys. Rev. C **89**(4), 044321 (2014). https://doi.org/10.1103/PhysRevC.89.044321

Cottam, J., Paerels, F., Mendez, M.: Gravitationally redshifted absorption lines in the X-ray burst spectra of a neutron star. Nature **420**, 51–54 (2002). https://doi.org/10.1038/nature01159

Cottam, J., Paerels, F., Méndez, M., Boirin, L., Lewin, W.H.G., Kuulkers, E., Miller, J.M.: The burst spectra of EXO 0748-676 during a long 2003 XMM-Newton observation. Astrophys. J. **672**, 504–509 (2008). https://doi.org/10.1086/524186

Danielewicz, P., Lacey, R., Lynch, W.G.: Determination of the equation of state of dense matter. Science **298**, 1592–1596 (2002). https://doi.org/10.1126/science.1078070

Danielewicz, P., Lee, J.: Symmetry energy II: isobaric analog states. Nucl. Phys. A **922**, 1–70 (2014). https://doi.org/10.1016/j.nuclphysa.2013.11.005

Daoutidis, I., Goriely, S.: Impact of the nuclear symmetry energy on the pygmy dipole resonance. Phys. Rev. C **84**(2), 027301 (2011). https://doi.org/10.1103/PhysRevC.84.027301

Davesne, D., Navarro, J., Becker, P., Jodon, R., Meyer, J., Pastore, A.: Extended Skyrme pseudopotential deduced from infinite nuclear matter properties. Phys. Rev. C **91**(6), 064303 (2015). https://doi.org/10.1103/PhysRevC.91.064303

Dean, D.J., Langanke, K., Sampaio, J.M.: Temperature dependence of the symmetry energy. Phys. Rev. C **66**(4), 045802 (2002). https://doi.org/10.1103/PhysRevC.66.045802

Dechargé, J., Gogny, D.: Hartree-Fock-Bogolyubov calculations with the D 1 effective interaction on spherical nuclei. Phys. Rev. C **21**, 1568–1593 (1980). https://doi.org/10.1103/PhysRevC.21.1568

Demorest, P.B., Pennucci, T., Ransom, S.M., Roberts, M.S.E., Hessels, J.W.T.: A two-solar-mass neutron star measured using Shapiro delay. Nature **467**, 1081–1083 (2010). https://doi.org/10.1038/nature09466

Dewulf, Y., Dickhoff, W.H., van Neck, D., Stoddard, E.R., Waroquier, M.: Saturation of nuclear matter and short-range correlations. Phys. Rev. Lett. **90**(15), 152501 (2003). https://doi.org/10.1103/PhysRevLett.90.152501

Dickhoff, W.H., Barbieri, C.: Self-consistent Green's function method for nuclei and nuclear matter. Prog. Part. Nucl. Phys. **52**, 377–496 (2004). https://doi.org/10.1016/j.ppnp.2004.02.038

Dickhoff, W.H., Van Neck, D.: Many-Body Theory Exposed! Propagator Description of Quantum Mechanics in Many-Body Systems, 2nd edn. World Scientific, Singapore (2008). https://doi.org/10.1142/6821

Dobaczewski, J., Flocard, H., Treiner, J.: Hartree-Fock-Bogolyubov description of nuclei near the neutron-drip line. Nucl. Phys. A **422**, 103–139 (1984). https://doi.org/10.1016/0375-9474(84)90433-0

Doleschall, P., Borbély, I., Papp, Z., Plessas, W.: Nonlocality in the nucleon-nucleon interaction and three-nucleon bound states. Phys. Rev. C **67**(6), 064005 (2003). https://doi.org/10.1103/PhysRevC.67.064005

Donati, P., Pizzochero, P.M., Bortignon, P.F., Broglia, R.A.: Temperature dependence of the nucleon effective mass and the physics of stellar collapse. Phys. Rev. Lett. **72**, 2835–2838 (1994). https://doi.org/10.1103/PhysRevLett.72.2835

Dorso, C.O., Giménez Molinelli, P.A., López, J.A.: Topological characterization of neutron star crusts. Phys. Rev. C **86**(5), 055805 (2012). https://doi.org/10.1103/PhysRevC.86.055805

Douchin, F., Haensel, P.: Inner edge of neutron-star crust with SLy effective nucleon-nucleon interactions. Phys. Lett. B **485**, 107–114 (2000). https://doi.org/10.1016/S0370-2693(00)00672-9

Douchin, F., Haensel, P.: A unified equation of state of dense matter and neutron star structure. Astron. Astrophys. **380**, 151–167 (2001). https://doi.org/10.1051/0004-6361:20011402

Drago, A., Tambini, U.: Finite temperature quark matter and supernova explosion. J. Phys. G: Nucl. Part. Phys. **25**, 971–979 (1999). https://doi.org/10.1088/0954-3899/25/5/302

Drischler, C., Hebeler, K., Schwenk, A.: Asymmetric nuclear matter based on chiral two- and three-nucleon interactions. Phys. Rev. C **93**(5), 054314 (2016). https://doi.org/10.1103/PhysRevC.93.054314

Duguet, T.: The nuclear energy density functional formalism. In: Scheidenberger, C., Pfützner, M. (eds.) The Euroschool on Exotic Beams, Volume IV. Lecture Notes in Physics, vol. 879, p. 293. Springer, Berlin (2014). https://doi.org/10.1007/978-3-642-45141-6_7

Dutra, M., Lourenço, O., Sá Martins, J.S., Delfino, A., Stone, J.R., Stevenson, P.D.: Skyrme interaction and nuclear matter constraints. Phys. Rev. C **85**(3), 035201 (2012). https://doi.org/10.1103/PhysRevC.85.035201

Dutra, M., Lourenço, O., Avancini, S.S., Carlson, B.V., Delfino, A., Menezes, D.P., Providência, C., Typel, S., Stone, J.R.: Relativistic mean-field hadronic models under nuclear matter constraints. Phys. Rev. C **90**(5), 055203 (2014). https://doi.org/10.1103/PhysRevC.90.055203

Dutra, M., Lourenço, O., Menezes, D.P.: Erratum: stellar properties and nuclear matter constraints [Phys. Rev. C93, 025806 (2016)]. Phys. Rev. C **94**(4), 049901 (2016a). https://doi.org/10.1103/PhysRevC.94.049901

Dutra, M., Lourenço, O., Menezes, D.P.: Stellar properties and nuclear matter constraints. Phys. Rev. C **93**(2), 025806 (2016b). https://doi.org/10.1103/PhysRevC.93.025806

El Eid, M.F., Hillebrandt, W.: A new equation of state of supernova matter. Astron. Astrophys., Suppl. Ser. **42**, 215–226 (1980)

Elshamouty, K.G., Heinke, C.O., Sivakoff, G.R., Ho, W.C.G., Shternin, P.S., Yakovlev, D.G., Patnaude, D.J., David, L.: Measuring the cooling of the neutron star in cassiopeia a with all chandra X-ray observatory detectors. Astrophys. J. **777**, 22 (2013). https://doi.org/10.1088/0004-637X/777/1/22

Entem, D.R., Machleidt, R.: Accurate nucleon-nucleon potential based upon chiral perturbation theory. Phys. Lett. B **524**, 93–98 (2002). https://doi.org/10.1016/S0370-2693(01)01363-6

Entem, D.R., Machleidt, R.: Accurate charge-dependent nucleon-nucleon potential at fourth order of chiral perturbation theory. Phys. Rev. C **68**(4), 041001 (2003). https://doi.org/10.1103/PhysRevC.68.041001

Epelbaum, E., Hammer, H.W., Meißner, U.G.: Modern theory of nuclear forces. Rev. Mod. Phys. **81**, 1773–1825 (2009). https://doi.org/10.1103/RevModPhys.81.1773

Faber, J.A., Rasio, F.A.: Binary neutron star mergers. Living Rev. Relativ. **15**, 8 (2012). https://doi.org/10.12942/lrr-2012-8

Fan, Y.Z., Wu, X.F., Wei, D.M.: Signature of gravitational wave radiation in afterglows of short gamma-ray bursts? Phys. Rev. D **88**(6), 067304 (2013). https://doi.org/10.1103/PhysRevD.88.067304

Fantina, A.F., Margueron, J., Donati, P., Pizzochero, P.M.: Nuclear energy functional with a surface-peaked effective mass: global properties. J. Phys. G: Nucl. Part. Phys. **38**(2), 025101 (2011). https://doi.org/10.1088/0954-3899/38/2/025101

Fantina, A.F., Blottiau, P., Margueron, J., Mellor, P., Pizzochero, P.M.: Effects of the temperature dependence of the in-medium nucleon mass on core-collapse supernovae. Astron. Astrophys. **541**, A30 (2012a). https://doi.org/10.1051/0004-6361/201118187

Fantina, A.F., Chamel, N., Pearson, J.M., Goriely, S.: Unified equation of state for neutron stars and supernova cores using the nuclear energy-density functional theory. J. Phys. Conf. Ser. **342**, 012003 (2012b). https://doi.org/10.1088/1742-6596/342/1/012003

Fantina, A.F., Chamel, N., Pearson, J.M., Goriely, S.: Neutron star properties with unified equations of state of dense matter. Astron. Astrophys. **559**, A128 (2013). https://doi.org/10.1051/0004-6361/201321884

Fantoni, S., Fabrocini, A.: Correlated basis function theory for fermion systems. In: Navarro, J., Polls, A. (eds.) Microscopic Quantum Many-Body Theories and Their Applications. Lecture Notes in Physics, vol. 510, p. 119. Springer, Berlin (1998). https://doi.org/10.1007/BFb0104526

Fantoni, S., Rosati, S.: The hypernetted-chain approximation for a fermion system. Nuovo Cim. A **25**, 593–615 (1975). https://doi.org/10.1007/BF02729302

Fattoyev, F.J., Horowitz, C.J., Piekarewicz, J., Shen, G.: Relativistic effective interaction for nuclei, giant resonances, and neutron stars. Phys. Rev. C **82**(5), 055803 (2010). https://doi.org/10.1103/PhysRevC.82.055803

Fayans, S.A., Tolokonnikov, S.V., Trykov, E.L., Zawischa, D.: Nuclear isotope shifts within the local energy-density functional approach. Nucl. Phys. A **676**, 49–119 (2000). https://doi.org/10.1016/S0375-9474(00)00192-5

Fayans, S.A., Zawischa, D.: Local energy-density functional approach to many-body nuclear systems with S-wave pairing. Int. J. Mod. Phys. B **15**, 1684–1702 (2001). https://doi.org/10.1142/S0217979201006203

Fedoseew, A., Lenske, H.: Thermal properties of asymmetric nuclear matter. Phys. Rev. C **91**(3), 034307 (2015). https://doi.org/10.1103/PhysRevC.91.034307

Ferdman, R.D., Stairs, I.H., Kramer, M., Janssen, G.H., Bassa, C.G., Stappers, B.W., Demorest, P.B., Cognard, I., Desvignes, G., Theureau, G., Burgay, M., Lyne, A.G., Manchester, R.N., Possenti, A.: PSR J1756-2251: a pulsar with a low-mass neutron star companion. Mon. Not. R. Astron. Soc. **443**, 2183–2196 (2014). https://doi.org/10.1093/mnras/stu1223

Fetter, A.L., Walecka, J.D.: Quantum Theory of Many-Particle Systems. International Series in Pure and Applied Physics. McGraw-Hill, San Francisco (1971)

Finelli, P., Kaiser, N., Vretenar, D., Weise, W.: Nuclear many-body dynamics constrained by QCD and chiral symmetry. Eur. Phys. J. A **17**, 573–578 (2003). https://doi.org/10.1140/epja/i2003-10004-8

Finelli, P., Kaiser, N., Vretenar, D., Weise, W.: Relativistic nuclear model with point-couplings constrained by QCD and chiral symmetry. Nucl. Phys. A **735**, 449–481 (2004). https://doi.org/10.1016/j.nuclphysa.2004.02.001

Finelli, P., Kaiser, N., Vretenar, D., Weise, W.: Relativistic nuclear energy density functional constrained by low-energy QCD. Nucl. Phys. A **770**, 1–31 (2006). https://doi.org/10.1016/j.nuclphysa.2006.02.007

Fischer, T., Whitehouse, S.C., Mezzacappa, A., Thielemann, F.K., Liebendörfer, M.: The neutrino signal from protoneutron star accretion and black hole formation. Astron. Astrophys. **499**, 1–15 (2009). https://doi.org/10.1051/0004-6361/200811055

Fischer, T., Sagert, I., Pagliara, G., Hempel, M., Schaffner-Bielich, J., Rauscher, T., Thielemann, F.K., Käppeli, R., Martínez-Pinedo, G., Liebendörfer, M.: Core-collapse supernova explosions triggered by a quark-hadron phase transition during the early post-bounce phase. Astrophys. J. Suppl. Ser. **194**, 39 (2011). https://doi.org/10.1088/0067-0049/194/2/39

Fischer, T., Klähn, T., Sagert, I., Hempel, M., Blaschke, D.: Quark Matter in Core Collpase Supernova Simulations. Acta Phys. Pol. B Proc. Suppl. **7**, 153 (2014a). https://doi.org/10.5506/APhysPolBSupp.7.153

Fischer, T., Hempel, M., Sagert, I., Suwa, Y., Schaffner-Bielich, J.: Symmetry energy impact in simulations of core-collapse supernovae. Eur. Phys. J. A **50**, 46 (2014b). https://doi.org/10.1140/epja/i2014-14046-5

Fonseca, E., Pennucci, T.T., Ellis, J.A., Stairs, I.H., Nice, D.J., Ransom, S.M., Demorest, P.B., Arzoumanian, Z., Crowter, K., Dolch, T., Ferdman, R.D., Gonzalez, M.E., Jones, G., Jones, M.L., Lam, M.T., Levin, L., McLaughlin, M.A., Stovall, K., Swiggum, J.K., Zhu, W.: The NANOGrav nine-year data set: mass and geometric measurements of binary millisecond pulsars. Astrophys. J. **832**, 167 (2016). https://doi.org/10.3847/0004-637X/832/2/167

Fortin, M., Zdunik, J.L., Haensel, P., Bejger, M.: Neutron stars with hyperon cores: stellar radii and equation of state near nuclear density. Astron. Astrophys. **576**, A68 (2015). https://doi.org/10.1051/0004-6361/201424800

Fortin, M., Providência, C., Raduta, A.R., Gulminelli, F., Zdunik, J.L., Haensel, P., Bejger, M.: Neutron star radii and crusts: uncertainties and unified equations of state. Phys. Rev. C **94**(3), 035804 (2016). https://doi.org/10.1103/PhysRevC.94.035804

Frick, T.: Self-consistent Green's function in nuclear matter at finite temperature. Ph.D.thesis, Universitaet Tuebingen (2004)

Friedman, B., Pandharipande, V.R.: Hot and cold, nuclear and neutron matter. Nucl. Phys. A **361**, 502–520 (1981)

Fryer, C.L., Belczynski, K., Ramirez-Ruiz, E., Rosswog, S., Shen, G., Steiner, A.W.: The fate of the compact remnant in neutron star mergers. Astrophys. J. **812**, 24 (2015). https://doi.org/10.1088/0004-637X/812/1/24

Fuchs, C.: Kaon production in heavy ion reactions at intermediate energies. Prog. Part. Nucl. Phys. **56**, 1–103 (2006). https://doi.org/10.1016/j.ppnp.2005.07.004

Fuchs, C., Faessler, A., Zabrodin, E., Zheng, Y.M.: Probing the nuclear equation of state by K^+ production in heavy-ion collisions. Phys. Rev. Lett. **86**, 1974–1977 (2001). https://doi.org/10.1103/PhysRevLett.86.1974

Fujita, J., Miyazawa, H.: Pion theory of three-body forces. Prog. Theor. Phys. **17**, 360–365 (1957). https://doi.org/10.1143/PTP.17.360

Fujiwara, Y., Fujita, T., Kohno, M., Nakamoto, C., Suzuki, Y.: Resonating group study of baryon baryon interactions for the complete baryon octet: NN interaction. Phys. Rev. C **65**, 014002 (2001). https://doi.org/10.1103/PhysRevC.65.014002

Fujiwara, Y., Suzuki, Y., Nakamoto, C.: Baryon-baryon interactions in the SU(6) quark model and their applications to light nuclear systems. Prog. Part. Nucl. Phys. **58**, 439–520 (2007). https://doi.org/10.1016/j.ppnp.2006.08.001

Fukukawa, K., Baldo, M., Burgio, G.F., Lo Monaco, L., Schulze, H.J.: Nuclear matter equation of state from a quark-model nucleon-nucleon interaction. Phys. Rev. C **92**, 065802 (2015). https://doi.org/10.1103/PhysRevC.92.065802

Furnstahl, R.J., Hebeler, K.: New applications of renormalization group methods in nuclear physics. Rep. Prog. Phys. **76**(12), 126301 (2013). https://doi.org/10.1088/0034-4885/76/12/126301

Furusawa, S., Yamada, S., Sumiyoshi, K., Suzuki, H.: A new baryonic equation of state at sub-nuclear densities for core-collapse simulations. Astrophys. J. **738**, 178 (2011). https://doi.org/10.1088/0004-637X/738/2/178

Furusawa, S., Nagakura, H., Sumiyoshi, K., Yamada, S.: The influence of inelastic neutrino reactions with light nuclei on the standing accretion shock instability in core-collapse supernovae. Astrophys. J. **774**, 78 (2013a). https://doi.org/10.1088/0004-637X/774/1/78

Furusawa, S., Sumiyoshi, K., Yamada, S., Suzuki, H.: New equations of state based on the liquid drop model of heavy nuclei and quantum approach to light nuclei for core-collapse supernova simulations. Astrophys. J. **772**, 95 (2013b). https://doi.org/10.1088/0004-637X/772/2/95

Furusawa, S., Nagakura, H., Sumiyoshi, K., Kato, C., Yamada, S.: Dependence of weak interaction rates on the nuclear composition during stellar core collapse. Phys. Rev. C **95**(2), 025809 (2017a). https://doi.org/10.1103/PhysRevC.95.025809

Furusawa, S., Sumiyoshi, K., Yamada, S., Suzuki, H.: Supernova equations of state including full nuclear ensemble with in-medium effects. Nucl. Phys. A **957**, 188–207 (2017b). https://doi.org/10.1016/j.nuclphysa.2016.09.002

Gabler, M., Cerdá-Durán, P., Stergioulas, N., Font, J.A., Müller, E.: Imprints of superfluidity on magnetoelastic quasiperiodic oscillations of soft gamma-ray repeaters. Phys. Rev. Lett. **111**(21), 211102 (2013). https://doi.org/10.1103/PhysRevLett.111.211102

Galloway, D.K., Lampe, N.: On the consistency of neutron-star radius measurements from thermonuclear bursts. Astrophys. J. **747**, 75 (2012). https://doi.org/10.1088/0004-637X/747/1/75

Gandolfi, S., Illarionov, A.Y., Schmidt, K.E., Pederiva, F., Fantoni, S.: Quantum Monte Carlo calculation of the equation of state of neutron matter. Phys. Rev. C **79**(5), 054005 (2009). https://doi.org/10.1103/PhysRevC.79.054005

Gandolfi, S., Carlson, J., Pieper, S.C.: Cold neutrons trapped in external fields. Phys. Rev. Lett. **106**(1), 012501 (2011). https://doi.org/10.1103/PhysRevLett.106.012501

Gandolfi, S., Gezerlis, A., Carlson, J.: Neutron matter from low to high density. Annu. Rev. Nucl. Part. Sci. **65**, 303–328 (2015). https://doi.org/10.1146/annurev-nucl-102014-021957

Gentile, N.A., Aufderheide, M.B., Mathews, G.J., Swesty, F.D., Fuller, G.M.: The QCD phase transition and supernova core collapse. Astrophys. J. **414**, 701–711 (1993). https://doi.org/10.1086/173116

Gezerlis, A., Tews, I., Epelbaum, E., Gandolfi, S., Hebeler, K., Nogga, A., Schwenk, A.: Quantum Monte Carlo calculations with chiral effective field theory interactions. Phys. Rev. Lett. **111**(3), 032501 (2013). https://doi.org/10.1103/PhysRevLett.111.032501

Giménez Molinelli, P.A., Dorso, C.O.: Finite size effects in neutron star and nuclear matter simulations. Nucl. Phys. A **933**, 306–324 (2015). https://doi.org/10.1016/j.nuclphysa.2014.11.005

Gögelein, P., Müther, H.: Nuclear matter in the crust of neutron stars. Phys. Rev. C **76**(2), 024312 (2007). https://doi.org/10.1103/PhysRevC.76.024312

Gögelein, P., van Dalen, E.N.E., Fuchs, C., Müther, H.: Nuclear matter in the crust of neutron stars derived from realistic NN interactions. Phys. Rev. C **77**(2), 025802 (2008). https://doi.org/10.1103/PhysRevC.77.025802

Gonzalez-Boquera, C., Centelles, M., Viñas, X., Rios, A.: Higher-order symmetry energy and neutron star core-crust transition with Gogny forces. Phys. Rev. C **96**(6), 065806 (2017). https://doi.org/10.1103/PhysRevC.96.065806

Goriely, S., Capote, R.: Uncertainties of mass extrapolations in Hartree-Fock-Bogoliubov mass models. Phys. Rev. C **89**(5), 054318 (2014). https://doi.org/10.1103/PhysRevC.89.054318

Goriely, S., Hilaire, S., Girod, M., Péru, S.: First Gogny-Hartree-Fock-Bogoliubov nuclear mass model. Phys. Rev. Lett. **102**(24), 242501 (2009). https://doi.org/10.1103/PhysRevLett.102. 242501

Goriely, S., Chamel, N., Pearson, J.M.: Further explorations of Skyrme-Hartree-Fock-Bogoliubov mass formulas. XII. Stiffness and stability of neutron-star matter. Phys. Rev. C **82**(3), 035804 (2010). https://doi.org/10.1103/PhysRevC.82.035804

Goriely, S., Bauswein, A., Janka, H.T.: r-process nucleosynthesis in dynamically ejected matter of neutron star mergers. Astrophys. J. Lett. **738**, L32 (2011). https://doi.org/10.1088/2041-8205/ 738/2/L32

Goriely, S., Chamel, N., Pearson, J.M.: Further explorations of Skyrme-Hartree-Fock-Bogoliubov mass formulas. XIII. The 2012 atomic mass evaluation and the symmetry coefficient. Phys. Rev. C **88**(2), 024308 (2013a). https://doi.org/10.1103/PhysRevC.88.024308

Goriely, S., Chamel, N., Pearson, J.M.: Hartree-Fock-Bogoliubov nuclear mass model with 0.50 MeV accuracy based on standard forms of Skyrme and pairing functionals. Phys. Rev. C **88**(6), 061302 (2013b). https://doi.org/10.1103/PhysRevC.88.061302

Goriely, S., Chamel, N., Pearson, J.M.: Further explorations of Skyrme-Hartree-Fock-Bogoliubov mass formulas. XVI. Inclusion of self-energy effects in pairing. Phys. Rev. C **93**(3), 034337 (2016a). https://doi.org/10.1103/PhysRevC.93.034337

Goriely, S., Hilaire, S., Girod, M., Péru, S.: The Gogny-Hartree-Fock-Bogoliubov nuclear-mass model. Eur. Phys. J. A **52**, 202 (2016b). https://doi.org/10.1140/epja/i2016-16202-3

Grangé, P., Lejeune, A., Martzolff, M., Mathiot, J.F.: Consistent three-nucleon forces in the nuclear many-body problem. Phys. Rev. C **40**, 1040–1060 (1989). https://doi.org/10.1103/PhysRevC. 40.1040

Grigorian, H., Voskresensky, D.N., Blaschke, D.: Influence of the stiffness of the equation of state and in-medium effects on the cooling of compact stars. Eur. Phys. J. A **52**, 67 (2016). https:// doi.org/10.1140/epja/i2016-16067-4

Grill, F., Margueron, J., Sandulescu, N.: Cluster structure of the inner crust of neutron stars in the Hartree-Fock-Bogoliubov approach. Phys. Rev. C **84**(6), 065801 (2011). https://doi.org/10. 1103/PhysRevC.84.065801

Guillot, S., Servillat, M., Webb, N.A., Rutledge, R.E.: Measurement of the radius of neutron stars with high signal-to-noise quiescent low-mass X-ray binaries in globular clusters. Astrophys. J. **772**, 7 (2013). https://doi.org/10.1088/0004-637X/772/1/7

Gulminelli, F., Raduta, A.R.: Unified treatment of subsaturation stellar matter at zero and finite temperature. Phys. Rev. C **92**(5), 055803 (2015). https://doi.org/10.1103/PhysRevC.92.055803

Gulminelli, F., Raduta, A.R., Oertel, M.: Phase transition toward strange matter. Phys. Rev. C **86**(2), 025805 (2012). https://doi.org/10.1103/PhysRevC.86.025805

Gulminelli, F., Raduta, A.R., Oertel, M., Margueron, J.: Strangeness-driven phase transition in (proto-)neutron star matter. Phys. Rev. C **87**(5), 055809 (2013). https://doi.org/10.1103/ PhysRevC.87.055809

Güver, T., Özel, F.: The mass and the radius of the neutron star in the transient low-mass X-ray binary SAX J1748.9-2021. Astrophys. J. Lett. **765**, L1 (2013). https://doi.org/10.1088/2041-8205/765/1/L1

Haar, B.T., Malfliet, R.: Nucleons, mesons and deltas in nuclear matter a relativistic Dirac-Brueckner approach. Phys. Rep. **149**, 207–286 (1987). https://doi.org/10.1016/0370-1573(87)90085-8

Haensel, P., Pichon, B.: Experimental nuclear masses and the ground state of cold dense matter. Astron. Astrophys. **283**, 313–318 (1994)

Haensel, P., Potekhin, A.Y., Yakovlev, D.G.: Neutron Stars 1: Equation of State and Structure. Springer, New York (2007). https://doi.org/10.1007/978-0-387-47301-7

Haensel, P., Zdunik, J.L., Bejger, M., Lattimer, J.M.: Keplerian frequency of uniformly rotating neutron stars and strange stars. Astron. Astrophys. **502**, 605–610 (2009). https://doi.org/10. 1051/0004-6361/200811605

Hagen, G., Papenbrock, T., Dean, D.J., Schwenk, A., Nogga, A., Włoch, M., Piecuch, P.: Coupled-cluster theory for three-body Hamiltonians. Phys. Rev. C 76(3), 034302 (2007). https://doi.org/10.1103/PhysRevC.76.034302

Hagen, G., Papenbrock, T., Dean, D.J., Hjorth-Jensen, M.: Ab initio coupled-cluster approach to nuclear structure with modern nucleon-nucleon interactions. Phys. Rev. C 82(3), 034330 (2010). https://doi.org/10.1103/PhysRevC.82.034330

Hagen, G., Hjorth-Jensen, M., Jansen, G.R., Machleidt, R., Papenbrock, T.: Evolution of shell structure in neutron-rich calcium isotopes. Phys. Rev. Lett. 109(3), 032502 (2012). https://doi.org/10.1103/PhysRevLett.109.032502

Hagen, G., Papenbrock, T., Ekström, A., Wendt, K.A., Baardsen, G., Gandolfi, S., Hjorth-Jensen, M., Horowitz, C.J.: Coupled-cluster calculations of nucleonic matter. Phys. Rev. C 89(1), 014319 (2014). https://doi.org/10.1103/PhysRevC.89.014319

Hartnack, C., Oeschler, H., Aichelin, J.: Hadronic matter is soft. Phys. Rev. Lett. 96(1), 012302 (2006). https://doi.org/10.1103/PhysRevLett.96.012302

Hashimoto, M., Seki, H., Yamada, M.: Shape of nuclei in the crust of a neutron star. Prog. Theor. Phys. 71, 320–326 (1984). https://doi.org/10.1143/PTP.71.320

Hebeler, K., Schwenk, A.: Chiral three-nucleon forces and neutron matter. Phys. Rev. C 82(1), 014314 (2010). https://doi.org/10.1103/PhysRevC.82.014314

Hebeler, K., Bogner, S.K., Furnstahl, R.J., Nogga, A., Schwenk, A.: Improved nuclear matter calculations from chiral low-momentum interactions. Phys. Rev. C 83(3), 031301 (2011). https://doi.org/10.1103/PhysRevC.83.031301

Heinimann, O., Hempel, M., Thielemann, F.K.: Towards generating a new supernova equation of state: a systematic analysis of cold hybrid stars. Phys. Rev. D 94(10), 103008 (2016). https://doi.org/10.1103/PhysRevD.94.103008

Heinke, C.O., Ho, W.C.G.: Direct observation of the cooling of the Cassiopeia A neutron star. Astrophys. J. Lett. 719, L167–L171 (2010). https://doi.org/10.1088/2041-8205/719/2/L167

Heinke, C.O., Jonker, P.G., Wijnands, R., Deloye, C.J., Taam, R.E.: Further constraints on thermal quiescent X-ray emission from SAX J1808.4-3658. Astrophys. J. 691, 1035–1041 (2009). https://doi.org/10.1088/0004-637X/691/2/1035

Hellemans, V., Heenen, P.H., Bender, M.: Tensor part of the Skyrme energy density functional. III. Time-odd terms at high spin. Phys. Rev. C 85(1), 014326 (2012). https://doi.org/10.1103/PhysRevC.85.014326

Hellemans, V., Pastore, A., Duguet, T., Bennaceur, K., Davesne, D., Meyer, J., Bender, M., Heenen, P.H.: Spurious finite-size instabilities in nuclear energy density functionals. Phys. Rev. C 88(6), 064323 (2013). https://doi.org/10.1103/PhysRevC.88.064323

Hempel, M., Schaffner-Bielich, J.: A statistical model for a complete supernova equation of state. Nucl. Phys. A 837, 210–254 (2010). https://doi.org/10.1016/j.nuclphysa.2010.02.010

Hempel, M., Schaffner-Bielich, J., Typel, S., Röpke, G.: Light clusters in nuclear matter: excluded volume versus quantum many-body approaches. Phys. Rev. C 84(5), 055804 (2011). https://doi.org/10.1103/PhysRevC.84.055804

Hempel, M., Fischer, T., Schaffner-Bielich, J., Liebendörfer, M.: New equations of state in simulations of core-collapse supernovae. Astrophys. J. 748, 70 (2012). https://doi.org/10.1088/0004-637X/748/1/70

Hempel, M., Hagel, K., Natowitz, J., Röpke, G., Typel, S.: Constraining supernova equations of state with equilibrium constants from heavy-ion collisions. Phys. Rev. C 91(4), 045805 (2015). https://doi.org/10.1103/PhysRevC.91.045805

Hessels, J.W.T., Ransom, S.M., Stairs, I.H., Freire, P.C.C., Kaspi, V.M., Camilo, F.: A radio pulsar spinning at 716 Hz. Science 311, 1901–1904 (2006). https://doi.org/10.1126/science.1123430

Higdon, J.C., Lingenfelter, R.E.: Gamma-ray bursts. Annu. Rev. Astron. Astrophys. 28, 401–436 (1990). https://doi.org/10.1146/annurev.aa.28.090190.002153

Hilaire, S., Goriely, S., Péru, S., Dubray, N., Dupuis, M., Bauge, E.: Nuclear reaction inputs based on effective interactions. Eur. Phys. J. A 52, 336 (2016). https://doi.org/10.1140/epja/i2016-16336-2

Hillebrandt, W., Wolff, R.G.: Models of type II supernova explosions. In: Arnett, W.D., Truran, J.W. (eds.) Nucleosynthesis: Challenges and New Developments, p. 131. University of Chicago Press, Chicago (1985)

Hillebrandt, W., Nomoto, K., Wolff, R.G.: Supernova explosions of massive stars - the mass range 8 to 10 solar masses. Astron. Astrophys. **133**, 175–184 (1984)

Hix, W.R., Messer, O.E.B., Mezzacappa, A., Liebendörfer, M., Sampaio, J., Langanke, K., Dean, D.J., Martínez-Pinedo, G.: Consequences of nuclear electron capture in core collapse supernovae. Phys. Rev. Lett. **91**(20), 201102 (2003). https://doi.org/10.1103/PhysRevLett.91.201102

Ho, W.C.G., Elshamouty, K.G., Heinke, C.O., Potekhin, A.Y.: Tests of the nuclear equation of state and superfluid and superconducting gaps using the Cassiopeia A neutron star. Phys. Rev. C **91**(1), 015806 (2015). https://doi.org/10.1103/PhysRevC.91.015806

Hohenberg, P., Kohn, W.: Inhomogeneous electron gas. Phys. Rev. **136**, 864–871 (1964). https://doi.org/10.1103/PhysRev.136.B864

Holt, J.W., Kaiser, N.: Equation of state of nuclear and neutron matter at third-order in perturbation theory from chiral effective field theory. Phys. Rev. C **95**(3), 034326 (2017). https://doi.org/10.1103/PhysRevC.95.034326

Holt, J.W., Kaiser, N., Weise, W.: Density-dependent effective nucleon-nucleon interaction from chiral three-nucleon forces. Phys. Rev. C **81**(2), 024002 (2010). https://doi.org/10.1103/PhysRevC.81.024002

Holt, J.W., Kaiser, N., Weise, W.: Nuclear chiral dynamics and thermodynamics. Prog. Part. Nucl. Phys. **73**, 35–83 (2013). https://doi.org/10.1016/j.ppnp.2013.08.001

Horowitz, C.J., Schwenk, A.: Cluster formation and the virial equation of state of low-density nuclear matter. Nucl. Phys. A **776**, 55–79 (2006a). https://doi.org/10.1016/j.nuclphysa.2006.05.009

Horowitz, C.J., Schwenk, A.: The virial equation of state of low-density neutron matter. Phys. Lett. B **638**, 153–159 (2006b). https://doi.org/10.1016/j.physletb.2006.05.055

Horowitz, C.J., Brown, E.F., Kim, Y., Lynch, W.G., Michaels, R., Ono, A., Piekarewicz, J., Tsang, M.B., Wolter, H.H.: A way forward in the study of the symmetry energy: experiment, theory, and observation. J. Phys. G: Nucl. Part. Phys. **41**(9), 093001 (2014). https://doi.org/10.1088/0954-3899/41/9/093001

Horowitz, C.J., Berry, D.K., Briggs, C.M., Caplan, M.E., Cumming, A., Schneider, A.S.: Disordered nuclear pasta, magnetic field decay, and crust cooling in neutron stars. Phys. Rev. Lett. **114**(3), 031102 (2015). https://doi.org/10.1103/PhysRevLett.114.031102

Hu, J., Zhang, Y., Epelbaum, E., Meißner, U.G., Meng, J.: Nuclear matter properties with nucleon-nucleon forces up to fifth order in the chiral expansion. Phys. Rev. C **96**(3), 034307 (2017). https://doi.org/10.1103/PhysRevC.96.034307

Huber, H., Weber, F., Weigel, M.K.: Symmetric and asymmetric nuclear matter in the relativistic approach at finite temperatures. Phys. Rev. C **57**, 3484–3487 (1998). https://doi.org/10.1103/PhysRevC.57.3484

Iliadis, C.I.: Nuclear Physics of Stars. Wiley-VCH, Weinheim (2007). https://doi.org/10.1002/9783527692668

Ishizuka, C., Ohnishi, A., Sumiyoshi, K.: Liquid-gas phase transition of supernova matter and its relation to nucleosynthesis. Nucl. Phys. A **723**, 517–543 (2003). https://doi.org/10.1016/S0375-9474(03)01324-1

Ishizuka, C., Ohnishi, A., Tsubakihara, K., Sumiyoshi, K., Yamada, S.: Tables of hyperonic matter equation of state for core-collapse supernovae. J. Phys. G: Nucl. Part. Phys. **35**(8), 085201 (2008). https://doi.org/10.1088/0954-3899/35/8/085201

Janka, H.T.: Explosion mechanisms of core-collapse supernovae. Annu. Rev. Nucl. Part. Sci. **62**, 407–451 (2012). https://doi.org/10.1146/annurev-nucl-102711-094901

Janka, H.T., Langanke, K., Marek, A., Martínez-Pinedo, G., Müller, B.: Theory of core-collapse supernovae. Phys. Rep. **442**, 38–74 (2007). https://doi.org/10.1016/j.physrep.2007.02.002

Janka, H.T., Melson, T., Summa, A.: Physics of core-collapse supernovae in three dimensions: a sneak preview. Annu. Rev. Nucl. Part. Sci. **66**, 341–375 (2016). https://doi.org/10.1146/annurev-nucl-102115-044747

Jonker, P.G., Steeghs, D., Chakrabarty, D., Juett, A.M.: The cold neutron star in the soft X-ray transient 1H 1905+000. Astrophys. J. Lett. **665**, L147–L150 (2007). https://doi.org/10.1086/521079

Kaiser, N., Fritsch, S., Weise, W.: Chiral dynamics and nuclear matter. Nucl. Phys. A **697**, 255–276 (2002). https://doi.org/10.1016/S0375-9474(01)01231-3

Kaplan, D.L., Frail, D.A., Gaensler, B.M., Gotthelf, E.V., Kulkarni, S.R., Slane, P.O., Nechita, A.: An X-ray search for compact central sources in supernova remnants. I. SNRS G093.3+6.9, G315.4-2.3, G084.2+0.8, and G127.1+0.5. Astrophys. J. Suppl. Ser. **153**, 269–315 (2004). https://doi.org/10.1086/421065

Kaplan, D.L., Gaensler, B.M., Kulkarni, S.R., Slane, P.O.: An X-ray search for compact central sources in supernova remnants. II. Six large-diameter SNRs. Astrophys. J. Suppl. Ser. **163**, 344–371 (2006). https://doi.org/10.1086/501441

Karnaukhov, V.A., Oeschler, H., Budzanowski, A., Avdeyev, S.P., Botvina, A.S., Cherepanov, E.A., Karcz, W., Kirakosyan, V.V., Rukoyatkin, P.A., Skwirczyńska, I., Norbeck, E.: Critical temperature for the nuclear liquid-gas phase transition (from multifragmentation and fission). Phys. At. Nucl. **71**, 2067–2073 (2008). https://doi.org/10.1134/S1063778808120077

Khan, E., Margueron, J.: Determination of the density dependence of the nuclear incompressibility. Phys. Rev. C **88**(3), 034319 (2013). https://doi.org/10.1103/PhysRevC.88.034319

Kitaura, F.S., Janka, H.T., Hillebrandt, W.: Explosions of O-Ne-Mg cores, the Crab supernova, and subluminous type II-P supernovae. Astron. Astrophys. **450**, 345–350 (2006). https://doi.org/10.1051/0004-6361:20054703

Klähn, T., Blaschke, D., Typel, S., van Dalen, E.N.E., Faessler, A., Fuchs, C., Gaitanos, T., Grigorian, H., Ho, A., Kolomeitsev, E.E., Miller, M.C., Röpke, G., Trümper, J., Voskresensky, D.N., Weber, F., Wolter, H.H.: Constraints on the high-density nuclear equation of state from the phenomenology of compact stars and heavy-ion collisions. Phys. Rev. C **74**(3), 035802 (2006). https://doi.org/10.1103/PhysRevC.74.035802

Klochkov, D., Suleimanov, V., Pühlhofer, G., Yakovlev, D.G., Santangelo, A., Werner, K.: The neutron star in HESS J1731-347: central compact objects as laboratories to study the equation of state of superdense matter. Astron. Astrophys. **573**, A53 (2015). https://doi.org/10.1051/0004-6361/201424683

Köhler, H.S.: Skyrme force and the mass formula. Nucl. Phys. A **258**, 301–316 (1976). https://doi.org/10.1016/0375-9474(76)90008-7

Kohn, W., Sham, L.J.: Self-consistent equations including exchange and correlation effects. Phys. Rev. **140**, 1133–1138 (1965). https://doi.org/10.1103/PhysRev.140.A1133

Kohno, M.: Nuclear and neutron matter G-matrix calculations with a chiral effective field theory potential including effects of three-nucleon interactions. Phys. Rev. C **88**(6), 064005 (2013). https://doi.org/10.1103/PhysRevC.88.064005

Kortelainen, M., McDonnell, J., Nazarewicz, W., Olsen, E., Reinhard, P.G., Sarich, J., Schunck, N., Wild, S.M., Davesne, D., Erler, J., Pastore, A.: Nuclear energy density optimization: shell structure. Phys. Rev. C **89**(5), 054314 (2014). https://doi.org/10.1103/PhysRevC.89.054314

Kramer, M., Stairs, I.H., Manchester, R.N., McLaughlin, M.A., Lyne, A.G., Ferdman, R.D., Burgay, M., Lorimer, D.R., Possenti, A., D'Amico, N., Sarkissian, J.M., Hobbs, G.B., Reynolds, J.E., Freire, P.C.C., Camilo, F.: Tests of general relativity from timing the double pulsar. Science **314**, 97–102 (2006). https://doi.org/10.1126/science.1132305

Krastev, P.G., Li, B.A., Worley, A.: Constraining properties of rapidly rotating neutron stars using data from heavy-ion collisions. Astrophys. J. **676**, 1170–1177 (2008). https://doi.org/10.1086/528736

Kumar, B., Biswal, S.K., Patra, S.K.: Tidal deformability of neutron and hyperon stars within relativistic mean field equations of state. Phys. Rev. C **95**(1), 015801 (2017). https://doi.org/10.1103/PhysRevC.95.015801

Lacombe, M., Loiseau, B., Richard, J.M., Mau, R.V., Côté, J., Pirès, P., de Tourreil, R.: Parametrization of the Paris N-N potential. Phys. Rev. C **21**, 861–873 (1980). https://doi.org/10.1103/PhysRevC.21.861

Lacour, A., Oller, J.A., Meißner, U.G.: Non-perturbative methods for a chiral effective field theory of finite density nuclear systems. Ann. Phys. **326**, 241–306 (2011). https://doi.org/10.1016/j.aop.2010.06.012

Lacroix, D.: Introduction - strong interaction in the nuclear medium: new trends. ArXiv e-prints (2010). arxiv:1001.5001

Lagaris, I.E., Pandharipande, V.R.: Phenomenological two-nucleon interaction operator. Nucl. Phys. A **359**, 331–348 (1981). https://doi.org/10.1016/0375-9474(81)90240-2

Lalazissis, G.A., König, J., Ring, P.: New parametrization for the Lagrangian density of relativistic mean field theory. Phys. Rev. C **55**, 540–543 (1997). https://doi.org/10.1103/PhysRevC.55.540

Lalazissis, G.A., Ring, P., Vretenar, D. (eds.): Extended Density Functionals in Nuclear Structure Physics. Lecture Notes in Physics, vol. 641. Springer, Berlin (2004). https://doi.org/10.1007/b95720

Landau, L.D., Lifshitz, E.M.: Statistical Physics. Pt.1, Pt.2. Pergamon, Oxford (1980)

Langanke, K., Martínez-Pinedo, G.: Nuclear weak-interaction processes in stars. Rev. Mod. Phys. **75**, 819–862 (2003). https://doi.org/10.1103/RevModPhys.75.819

Langanke, K., Martínez-Pinedo, G.: The role of electron capture in core-collapse supernovae. Nucl. Phys. A **928**, 305–312 (2014). https://doi.org/10.1016/j.nuclphysa.2014.04.015

Lasky, P.D., Haskell, B., Ravi, V., Howell, E.J., Coward, D.M.: Nuclear equation of state from observations of short gamma-ray burst remnants. Phys. Rev. D **89**(4), 047302 (2014). https://doi.org/10.1103/PhysRevD.89.047302

Lattimer, J.M.: The equation of state of hot dense matter and supernovae. Annu. Rev. Nucl. Part. Sci. **31**, 337–374 (1981). https://doi.org/10.1146/annurev.ns.31.120181.002005

Lattimer, J.M.: The nuclear equation of state and supernovae. In: Nuclear Equation of State, pp. 83–208. World Scientific, Singapore (1996)

Lattimer, J.M.: The nuclear equation of state and neutron star masses. Annu. Rev. Nucl. Part. Sci. **62**, 485–515 (2012). https://doi.org/10.1146/annurev-nucl-102711-095018

Lattimer, J.M., Douglas Swesty, F.: A generalized equation of state for hot, dense matter. Nucl. Phys. A **535**, 331–376 (1991). https://doi.org/10.1016/0375-9474(91)90452-C

Lattimer, J.M., Lim, Y.: Constraining the symmetry parameters of the nuclear interaction. Astrophys. J. **771**, 51 (2013). https://doi.org/10.1088/0004-637X/771/1/51

Lattimer, J.M., Prakash, M.: Nuclear matter and its role in supernovae, neutron stars and compact object binary mergers. Phys. Rep. **333–334**, 121–146 (2000). https://doi.org/10.1016/S0370-1573(00)00019-3

Lattimer, J.M., Prakash, M.: Neutron star observations: prognosis for equation of state constraints. Phys. Rep. **442**, 109–165 (2007). https://doi.org/10.1016/j.physrep.2007.02.003

Lattimer, J.M., Prakash, M.: The equation of state of hot, dense matter and neutron stars. Phys. Rep. **621**, 127–164 (2016). https://doi.org/10.1016/j.physrep.2015.12.005

Lattimer, J.M., Schutz, B.F.: Constraining the equation of state with moment of inertia measurements. Astrophys. J. **629**, 979–984 (2005). https://doi.org/10.1086/431543

Lattimer, J.M., Steiner, A.W.: Constraints on the symmetry energy using the mass-radius relation of neutron stars. Eur. Phys. J. A **50**, 40 (2014). https://doi.org/10.1140/epja/i2014-14040-y

Lattimer, J.M., Pethick, C.J., Ravenhall, D.G., Lamb, D.Q.: Physical properties of hot, dense matter: the general case. Nucl. Phys. A **432**, 646–742 (1985). https://doi.org/10.1016/0375-9474(85)90006-5

Lattimer, J.M., Pethick, C.J., Prakash, M., Haensel, P.: Direct URCA process in neutron stars. Phys. Rev. Lett. **66**, 2701–2704 (1991). https://doi.org/10.1103/PhysRevLett.66.2701

Lawrence, S., Tervala, J.G., Bedaque, P.F., Miller, M.C.: An upper bound on neutron star masses from models of short gamma-ray bursts. Astrophys. J. **808**, 186 (2015). https://doi.org/10.1088/0004-637X/808/2/186

Le Fèvre, A., Leifels, Y., Reisdorf, W., Aichelin, J., Hartnack, C.: Constraining the nuclear matter equation of state around twice saturation density. Nucl. Phys. A **945**, 112–133 (2016). http://dx.doi.org/10.1016/j.nuclphysa.2015.09.015; http://www.sciencedirect.com/science/article/pii/S0375947415002225

Lesinski, T., Bender, M., Bennaceur, K., Duguet, T., Meyer, J.: Tensor part of the Skyrme energy density functional: spherical nuclei. Phys. Rev. C **76**(1), 014312 (2007). https://doi.org/10.1103/PhysRevC.76.014312

Li, B.A., Han, X.: Constraining the neutron-proton effective mass splitting using empirical constraints on the density dependence of nuclear symmetry energy around normal density. Phys. Lett. B **727**(1–3), 276–281 (2013). http://dx.doi.org/10.1016/j.physletb.2013.10.006; https://www.sciencedirect.com/science/article/pii/S0370269313007995

Li, Z.H., Schulze, H.J.: Neutron star structure with modern nucleonic three-body forces. Phys. Rev. C **78**, 028801 (2008). https://doi.org/10.1103/PhysRevC.78.028801

Li, Z.H., Schulze, H.J.: Nuclear matter with chiral forces in Brueckner-Hartree-Fock approximation. Phys. Rev. C **85**(6), 064002 (2012). https://doi.org/10.1103/PhysRevC.85.064002

Li, Z.H., Lombardo, U., Schulze, H.J., Zuo, W., Chen, L.W., Ma, H.R.: Nuclear matter saturation point and symmetry energy with modern nucleon-nucleon potentials. Phys. Rev. C **74**(4), 047304 (2006). https://doi.org/10.1103/PhysRevC.74.047304

Li, Z.H., Lombardo, U., Schulze, H.J., Zuo, W.: Consistent nucleon-nucleon potentials and three-body forces. Phys. Rev. C **77**, 034316 (2008). https://doi.org/10.1103/PhysRevC.77.034316

Li, B.A., Ramos, A., Verde, G., Vidaña, I.: Topical issue on nuclear symmetry energy. Eur. Phys. J. A **50**, 9 (2014). http://dx.doi.org/10.1140/epja/i2014-14009-x

Lin, J., Özel, F., Chakrabarty, D., Psaltis, D.: The incompatibility of rapid rotation with narrow photospheric X-ray lines in EXO 0748-676. Astrophys. J. **723**, 1053–1056 (2010). https://doi.org/10.1088/0004-637X/723/2/1053

Liu, X.J., Hu, H., Drummond, P.D.: Virial expansion for a strongly correlated fermi gas. Phys. Rev. Lett. **102**(16), 160401 (2009). https://doi.org/10.1103/PhysRevLett.102.160401

Logoteta, D., Vidaña, I., Bombaci, I., Kievsky, A.: Comparative study of three-nucleon force models in nuclear matter. Phys. Rev. C **91**(6), 064001 (2015). https://doi.org/10.1103/PhysRevC.91.064001

Logoteta, D., Bombaci, I., Kievsky, A.: Nuclear matter saturation with chiral three-nucleon interactions fitted to light nuclei properties. Phys. Lett. B **758**, 449–454 (2016). https://doi.org/10.1016/j.physletb.2016.05.042

Long, W., Meng, J., Giai, N.V., Zhou, S.G.: New effective interactions in relativistic mean field theory with nonlinear terms and density-dependent meson-nucleon coupling. Phys. Rev. C **69**(3), 034319 (2004). https://doi.org/10.1103/PhysRevC.69.034319

Long, W., Sagawa, H., Giai, N.V., Meng, J.: Shell structure and ρ-tensor correlations in density dependent relativistic Hartree-Fock theory. Phys. Rev. C **76**(3), 034314 (2007). https://doi.org/10.1103/PhysRevC.76.034314

Long, W.H., Ring, P., Giai, N.V., Meng, J.: Relativistic Hartree-Fock-Bogoliubov theory with density dependent meson-nucleon couplings. Phys. Rev. C **81**(2), 024308 (2010). https://doi.org/10.1103/PhysRevC.81.024308

López, J.A., Ramírez-Homs, E., González, R., Ravelo, R.: Isospin-asymmetric nuclear matter. Phys. Rev. C **89**(2), 024611 (2014). https://doi.org/10.1103/PhysRevC.89.024611

Lorenz, C.P., Ravenhall, D.G., Pethick, C.J.: Neutron star crusts. Phys. Rev. Lett. **70**, 379–382 (1993). https://doi.org/10.1103/PhysRevLett.70.379

Lynch, W.G., Tsang, M.B., Zhang, Y., Danielewicz, P., Famiano, M., Li, Z., Steiner, A.W.: Probing the symmetry energy with heavy ions. Prog. Part. Nucl. Phys. **62**, 427–432 (2009). https://doi.org/10.1016/j.ppnp.2009.01.001

Lynn, J.E., Tews, I., Carlson, J., Gandolfi, S., Gezerlis, A., Schmidt, K.E., Schwenk, A.: Chiral three-nucleon interactions in light nuclei, neutron-α scattering, and neutron matter. Phys. Rev. Lett. **116**(6), 062501 (2016). https://doi.org/10.1103/PhysRevLett.116.062501

Machleidt, R.: The meson theory of nuclear forces and nuclear structure. In: Advances in Nuclear Physics, pp. 189–376. Springer, Boston (1989)

Machleidt, R., Entem, D.R.: Chiral effective field theory and nuclear forces. Phys. Rep. **503**, 1–75 (2011). https://doi.org/10.1016/j.physrep.2011.02.001

Machleidt, R., Slaus, I.: Topical review: the nucleon-nucleon interaction. J. Phys. G: Nucl. Part. Phys. **27**, R69–R108 (2001). https://doi.org/10.1088/0954-3899/27/5/201

Maessen, P.M.M., Rijken, T.A., de Swart, J.J.: Soft-core baryon-baryon one-boson-exchange models. II. Hyperon-nucleon potential. Phys. Rev. C **40**, 2226–2245 (1989). https://doi.org/10.1103/PhysRevC.40.2226

Magierski, P., Bulgac, A., Heenen, P.H.: Exotic nuclear phases in the inner crust of neutron stars in the light of Skyrme-Hartree-Fock theory. Nucl. Phys. A **719**, C217–C220 (2003). https://doi.org/10.1016/S0375-9474(03)00921-7

Marek, A., Janka, H.T., Müller, E.: Equation-of-state dependent features in shock-oscillation modulated neutrino and gravitational-wave signals from supernovae. Astron. Astrophys. **496**, 475–494 (2009). https://doi.org/10.1051/0004-6361/200810883

Margueron, J., Sagawa, H.: Extended Skyrme interaction: I. Spin fluctuations in dense matter. J. Phys. G: Nucl. Part. Phys. **36**(12), 125102 (2009). https://doi.org/10.1088/0954-3899/36/12/125102

Margueron, J., Goriely, S., Grasso, M., Colò, G., Sagawa, H.: Extended Skyrme interaction: II. Ground state of nuclei and of nuclear matter. J. Phys. G: Nucl. Part. Phys. **36**(12), 125103 (2009). https://doi.org/10.1088/0954-3899/36/12/125103

Margueron, J., Grasso, M., Goriely, S., Colò, G., Sagawa, H.: Extended Skyrme interaction in the spin channel. Prog. Theor. Phys. Suppl. **196**, 172–175 (2012). https://doi.org/10.1143/PTPS.196.172

Margueron, J., Hoffmann Casali, R., Gulminelli, F.: Equation of state for dense nucleonic matter from metamodeling. I. Foundational aspects. Phys. Rev. C **97**(2), 025805 (2018a). https://doi.org/10.1103/PhysRevC.97.025805

Margueron, J., Hoffmann Casali, R., Gulminelli, F.: Equation of state for dense nucleonic matter from metamodeling. II. Predictions for neutron star properties. Phys. Rev. C **97**(2), 025806 (2018b). https://doi.org/10.1103/PhysRevC.97.025806

Maruyama, T., Tatsumi, T., Voskresensky, D.N., Tanigawa, T., Chiba, S.: Nuclear "pasta" structures and the charge screening effect. Phys. Rev. C **72**(1), 015802 (2005). https://doi.org/10.1103/PhysRevC.72.015802

Maruyama, T., Watanabe, G., Chiba, S.: Molecular dynamics for dense matter. Prog. Theor. Exp. Phys. **2012**(1), 01A201 (2012). https://doi.org/10.1093/ptep/pts013

Maselli, A., Gualtieri, L., Ferrari, V.: Constraining the equation of state of nuclear matter with gravitational wave observations: tidal deformability and tidal disruption. Phys. Rev. D **88**(10), 104040 (2013). https://doi.org/10.1103/PhysRevD.88.104040

Maslov, K.A., Kolomeitsev, E.E., Voskresensky, D.N.: Making a soft relativistic mean-field equation of state stiffer at high density. Phys. Rev. C **92**(5), 052801 (2015). https://doi.org/10.1103/PhysRevC.92.052801

Meißner, U.G.: Modern theory of nuclear forces. Nucl. Phys. A **751**, 149–166 (2005). https://doi.org/10.1016/j.nuclphysa.2005.02.023

Meng, J., Toki, H., Zhou, S.G., Zhang, S.Q., Long, W.H., Geng, L.S.: Relativistic continuum Hartree Bogoliubov theory for ground-state properties of exotic nuclei. Prog. Part. Nucl. Phys. **57**, 470–563 (2006). https://doi.org/10.1016/j.ppnp.2005.06.001

Mezzacappa, A.: Ascertaining the core collapse supernova mechanism: the state of the art and the road ahead. Annu. Rev. Nucl. Part. Sci. **55**, 467–515 (2005). https://doi.org/10.1146/annurev.nucl.55.090704.151608

Miyatsu, T., Yamamuro, S., Nakazato, K.: A new equation of state for neutron star matter with nuclei in the crust and hyperons in the core. Astrophys. J. **777**, 4 (2013). https://doi.org/10.1088/0004-637X/777/1/4

Möller, P., Myers, W.D., Sagawa, H., Yoshida, S.: New finite-range droplet mass model and equation-of-state parameters. Phys. Rev. Lett. **108**(5), 052501 (2012). https://doi.org/10.1103/PhysRevLett.108.052501

Moustakidis, C.C., Panos, C.P.: Equation of state for β-stable hot nuclear matter. Phys. Rev. C **79**(4), 045806 (2009). https://doi.org/10.1103/PhysRevC.79.045806

Müther, H., Polls, A.: Two-body correlations in nuclear systems. Prog. Part. Nucl. Phys. **45**, 243–334 (2000). https://doi.org/10.1016/S0146-6410(00)00105-8

Myers, W.D., Swiatecki, W.J.: Average nuclear properties. Ann. Phys. **55**, 395–505 (1969). https://doi.org/10.1016/0003-4916(69)90202-4

Nagels, M.M., Rijken, T.A., de Swart, J.J.: Baryon-baryon scattering in a one-boson-exchange-potential approach. II. Hyperon-nucleon scattering. Phys. Rev. D **15**, 2547–2564 (1977). https://doi.org/10.1103/PhysRevD.15.2547

Nagels, M.M., Rijken, T.A., de Swart, J.J.: Low-energy nucleon-nucleon potential from Regge-pole theory. Phys. Rev. D **17**, 768–776 (1978). https://doi.org/10.1103/PhysRevD.17.768

Nakada, H.: Hartree-Fock approach to nuclear matter and finite nuclei with M3Y-type nucleon-nucleon interactions. Phys. Rev. C **68**(1), 014316 (2003). https://doi.org/10.1103/PhysRevC.68.014316

Nakazato, K., Sumiyoshi, K., Yamada, S.: Astrophysical implications of equation of state for hadron-quark mixed phase: compact stars and stellar collapses. Phys. Rev. D **77**(10), 103006 (2008). https://doi.org/10.1103/PhysRevD.77.103006

Nakazato, K., Iida, K., Oyamatsu, K.: Curvature effect on nuclear "pasta": is it helpful for gyroid appearance? Phys. Rev. C **83**(6), 065811 (2011). https://doi.org/10.1103/PhysRevC.83.065811

Nakazato, K., Furusawa, S., Sumiyoshi, K., Ohnishi, A., Yamada, S., Suzuki, H.: Hyperon matter and black hole formation in failed supernovae. Astrophys. J. **745**, 197 (2012). https://doi.org/10.1088/0004-637X/745/2/197

Nakazato, K., Sumiyoshi, K., Yamada, S.: Stellar core collapse with hadron-quark phase transition. Astron. Astrophys. **558**, A50 (2013). https://doi.org/10.1051/0004-6361/201322231

Navarro, J., Polls, A.: Spin instabilities of infinite nuclear matter and effective tensor interactions. Phys. Rev. C **87**(4), 044329 (2013). https://doi.org/10.1103/PhysRevC.87.044329

Navarro, J., Guardiola, R., Moliner, I.: The Coupled Cluster Method and Its Applications, pp. 121–178. World Scientific, Singapore (2002). https://doi.org/10.1142/9789812777072_0003

Negele, J.W., Vautherin, D.: Neutron star matter at sub-nuclear densities. Nucl. Phys. A **207**, 298–320 (1973). https://doi.org/10.1016/0375-9474(73)90349-7

Newton, W.G., Stone, J.R.: Modeling nuclear "pasta" and the transition to uniform nuclear matter with the 3D Skyrme-Hartree-Fock method at finite temperature: core-collapse supernovae. Phys. Rev. C **79**(5), 055801 (2009). https://doi.org/10.1103/PhysRevC.79.055801

Newton, W.G., Murphy, K., Hooker, J., Li, B.A.: The cooling of the Cassiopeia A neutron star as a probe of the nuclear symmetry energy and nuclear pasta. Astrophys. J. Lett. **779**, L4 (2013). https://doi.org/10.1088/2041-8205/779/1/L4

Nikolaus, B.A., Hoch, T., Madland, D.G.: Nuclear ground state properties in a relativistic point coupling model. Phys. Rev. C **46**, 1757–1781 (1992). https://doi.org/10.1103/PhysRevC.46.1757

Nikšić, T., Vretenar, D., Finelli, P., Ring, P.: Relativistic Hartree-Bogoliubov model with density-dependent meson-nucleon couplings. Phys. Rev. C **66**(2), 024306 (2002). https://doi.org/10.1103/PhysRevC.66.024306

Nikšić, T., Vretenar, D., Lalazissis, G.A., Ring, P.: Finite- to zero-range relativistic mean-field interactions. Phys. Rev. C **77**(3), 034302 (2008). https://doi.org/10.1103/PhysRevC.77.034302

Nikšić, T., Vretenar, D., Ring, P.: Relativistic nuclear energy density functionals: mean-field and beyond. Prog. Part. Nucl. Phys. **66**, 519–548 (2011). https://doi.org/10.1016/j.ppnp.2011.01.055

Nishimura, N., Fischer, T., Thielemann, F.K., Fröhlich, C., Hempel, M., Käppeli, R., Martínez-Pinedo, G., Rauscher, T., Sagert, I., Winteler, C.: Nucleosynthesis in core-collapse supernova explosions triggered by a quark-hadron phase transition. Astrophys. J. **758**, 9 (2012). https://doi.org/10.1088/0004-637X/758/1/9

O'Connor, E., Ott, C.D.: Black hole formation in failing core-collapse supernovae. Astrophys. J. **730**, 70 (2011). https://doi.org/10.1088/0004-637X/730/2/70

Oertel, M., Fantina, A.F., Novak, J.: Extended equation of state for core-collapse simulations. Phys. Rev. C **85**(5), 055806 (2012). https://doi.org/10.1103/PhysRevC.85.055806

Oertel, M., Gulminelli, F., Providência, C., Raduta, A.R.: Hyperons in neutron stars and supernova cores. Eur. Phys. J. A **52**, 50 (2016). https://doi.org/10.1140/epja/i2016-16050-1

Oertel, M., Hempel, M., Klähn, T., Typel, S.: Equations of state for supernovae and compact stars. Rev. Mod. Phys. **89**(1), 015007 (2017). https://doi.org/10.1103/RevModPhys.89.015007

Ofengeim, D.D., Kaminker, A.D., Klochkov, D., Suleimanov, V., Yakovlev, D.G.: Analysing neutron star in HESS J1731-347 from thermal emission and cooling theory. Mon. Not. R. Astron. Soc. **454**, 2668–2676 (2015). https://doi.org/10.1093/mnras/stv2204

Oka, M., Yazaki, K.: Nuclear force in a quark model. Phys. Lett. B **90**, 41–44 (1980). https://doi. org/10.1016/0370-2693(80)90046-5

Oka, M., Shimizu, K., Yazaki, K.: Hyperon-nucleon and hyperon-hyperon interaction in a quark model. Nucl. Phys. A **464**, 700–716 (1987). https://doi.org/10.1016/0375-9474(87)90371-X

Okamoto, M., Maruyama, T., Yabana, K., Tatsumi, T.: Three-dimensional structure of low-density nuclear matter. Phys. Lett. B **713**, 284–288 (2012). https://doi.org/10.1016/j.physletb.2012.05. 046

Onsi, M., Dutta, A.K., Chatri, H., Goriely, S., Chamel, N., Pearson, J.M.: Semi-classical equation of state and specific-heat expressions with proton shell corrections for the inner crust of a neutron star. Phys. Rev. C **77**(6), 065805 (2008). https://doi.org/10.1103/PhysRevC.77.065805

Oppenheimer, J.R., Volkoff, G.M.: On massive neutron cores. Phys. Rev. **55**, 374–381 (1939). https://doi.org/10.1103/PhysRev.55.374

Ott, C.D., O'Connor, E., Peng, F., Reisswig, C., Sperhake, U., Schnetter, E., Abdikamalov, E., Diener, P., Löffler, F., Hawke, I., Meakin, C.A., Burrows, A.: New open-source approaches to the modeling of stellar collapse and the formation of black holes. Astrophys. Space Sci. **336**, 151–156 (2011). https://doi.org/10.1007/s10509-010-0553-1

Oyamatsu, K., Iida, K.: Symmetry energy at subnuclear densities and nuclei in neutron star crusts. Phys. Rev. C **75**(1), 015801 (2007). https://doi.org/10.1103/PhysRevC.75.015801

Özel, F., Freire, P.: Masses, radii, and the equation of state of neutron stars. Annu. Rev. Astron. Astrophys. **54**, 401–440 (2016). https://doi.org/10.1146/annurev-astro-081915-023322

Özel, F., Psaltis, D., Güver, T., Baym, G., Heinke, C., Guillot, S.: The dense matter equation of state from neutron star radius and mass measurements. Astrophys. J. **820**, 28 (2016). https:// doi.org/10.3847/0004-637X/820/1/28

Paar, N., Moustakidis, C.C., Marketin, T., Vretenar, D., Lalazissis, G.A.: Neutron star structure and collective excitations of finite nuclei. Phys. Rev. C **90**(1), 011304 (2014). https://doi.org/ 10.1103/PhysRevC.90.011304

Page, D., Reddy, S.: Dense matter in compact stars: theoretical developments and observational constraints. Annu. Rev. Nucl. Part. Sci. **56**, 327–374 (2006). https://doi.org/10.1146/annurev. nucl.56.080805.140600

Page, D., Geppert, U., Weber, F.: The cooling of compact stars. Nucl. Phys. A **777**, 497–530 (2006). https://doi.org/10.1016/j.nuclphysa.2005.09.019

Page, D., Lattimer, J.M., Prakash, M., Steiner, A.W.: Neutrino emission from cooper pairs and minimal cooling of neutron stars. Astrophys. J. **707**, 1131–1140 (2009). https://doi.org/10. 1088/0004-637X/707/2/1131

Page, D., Prakash, M., Lattimer, J.M., Steiner, A.W.: Rapid cooling of the neutron star in Cassiopeia A triggered by neutron superfluidity in dense matter. Phys. Rev. Lett. **106**(8), 081101 (2011). https://doi.org/10.1103/PhysRevLett.106.081101

Pais, H., Stone, J.R.: Exploring the nuclear pasta phase in core-collapse supernova matter. Phys. Rev. Lett. **109**(15), 151101 (2012). https://doi.org/10.1103/PhysRevLett.109.151101

Pais, H., Newton, W.G., Stone, J.R.: Phase transitions in core-collapse supernova matter at subsaturation densities. Phys. Rev. C **90**(6), 065802 (2014). https://doi.org/10.1103/PhysRevC.90. 065802

Pais, H., Chiacchiera, S., Providência, C.: Light clusters, pasta phases, and phase transitions in core-collapse supernova matter. Phys. Rev. C **91**(5), 055801 (2015). https://doi.org/10.1103/ PhysRevC.91.055801

Palenzuela, C., Liebling, S.L., Neilsen, D., Lehner, L., Caballero, O.L., O'Connor, E., Anderson, M.: Effects of the microphysical equation of state in the mergers of magnetized neutron stars with neutrino cooling. Phys. Rev. D **92**(4), 044045 (2015). https://doi.org/10.1103/PhysRevD. 92.044045

Pandharipande, V.R., Wiringa, R.B.: Variations on a theme of nuclear matter. Rev. Mod. Phys. **51**, 821–860 (1979). https://doi.org/10.1103/RevModPhys.51.821

Papakonstantinou, P., Margueron, J., Gulminelli, F., Raduta, A.R.: Densities and energies of nuclei in dilute matter at zero temperature. Phys. Rev. C **88**(4), 045805 (2013). https://doi.org/10.1103/PhysRevC.88.045805

Pastore, A., Martini, M., Davesne, D., Navarro, J., Goriely, S., Chamel, N.: Linear response theory and neutrino mean free path using Brussels-Montreal Skyrme functionals. Phys. Rev. C **90**(2), 025804 (2014). https://doi.org/10.1103/PhysRevC.90.025804

Pastore, A., Tarpanov, D., Davesne, D., Navarro, J.: Spurious finite-size instabilities in nuclear energy density functionals: spin channel. Phys. Rev. C **92**(2), 024305 (2015). https://doi.org/10.1103/PhysRevC.92.024305

Pearson, J.M., Chamel, N., Goriely, S., Ducoin, C.: Inner crust of neutron stars with mass-fitted Skyrme functionals. Phys. Rev. C **85**(6), 065803 (2012). https://doi.org/10.1103/PhysRevC.85.065803

Pearson, J.M., Chamel, N., Pastore, A., Goriely, S.: Role of proton pairing in a semimicroscopic treatment of the inner crust of neutron stars. Phys. Rev. C **91**(1), 018801 (2015). https://doi.org/10.1103/PhysRevC.91.018801

Peres, B., Oertel, M., Novak, J.: Influence of pions and hyperons on stellar black hole formation. Phys. Rev. D **87**(4), 043006 (2013). https://doi.org/10.1103/PhysRevD.87.043006

Piekarewicz, J.: Unmasking the nuclear matter equation of state. Phys. Rev. C **69**(4), 041301 (2004). https://doi.org/10.1103/PhysRevC.69.041301

Pieper, S.C.: The Illinois extension to the Fujita-Miyazawa three-nucleon force. In: Sakai, H., Sekiguchi, K., Gibson, B.F. (eds.) New Facet of Three Nucleon Force - 50 Years of Fujita Miyazawa. American Institute of Physics Conference Series, vol. 1011, pp. 143–152. American Institute of Physics, College Park (2008). https://doi.org/10.1063/1.2932280

Pieper, S.C., Pandharipande, V.R., Wiringa, R.B., Carlson, J.: Realistic models of pion-exchange three-nucleon interactions. Phys. Rev. C **64**(1), 014001 (2001). https://doi.org/10.1103/PhysRevC.64.014001

Podsiadlowski, P., Dewi, J.D.M., Lesaffre, P., Miller, J.C., Newton, W.G., Stone, J.R.: The double pulsar J0737-3039: testing the neutron star equation of state. Mon. Not. R. Astron. Soc. **361**, 1243–1249 (2005). https://doi.org/10.1111/j.1365-2966.2005.09253.x

Posselt, B., Pavlov, G.G., Suleimanov, V., Kargaltsev, O.: New constraints on the cooling of the Central Compact Object in Cas A. Astrophys. J. **779**, 186 (2013). https://doi.org/10.1088/0004-637X/779/2/186

Potekhin, A.Y.: Atmospheres and radiating surfaces of neutron stars. Phys. Usp. **57**, 735–770 (2014). https://doi.org/10.3367/UFNe.0184.201408a.0793

Potekhin, A.Y., Pons, J.A., Page, D.: Neutron stars-cooling and transport. Space Sci. Rev. **191**, 239–291 (2015). https://doi.org/10.1007/s11214-015-0180-9

Poutanen, J., Nättilä, J., Kajava, J.J.E., Latvala, O.M., Galloway, D.K., Kuulkers, E., Suleimanov, V.F.: The effect of accretion on the measurement of neutron star mass and radius in the low-mass X-ray binary 4U 1608-52. Mon. Not. R. Astron. Soc. **442**, 3777–3790 (2014). https://doi.org/10.1093/mnras/stu1139

Pudliner, B.S., Pandharipande, V.R., Carlson, J., Wiringa, R.B.: Quantum Monte Carlo calculations of A ≤ 6 nuclei. Phys. Rev. Lett. **74**, 4396–4399 (1995). https://doi.org/10.1103/PhysRevLett.74.4396

Pudliner, B.S., Pandharipande, V.R., Carlson, J., Pieper, S.C., Wiringa, R.B.: Quantum Monte Carlo calculations of nuclei with A ≤ 7. Phys. Rev. C **56**, 1720–1750 (1997). https://doi.org/10.1103/PhysRevC.56.1720

Raduta, A.R., Gulminelli, F.: Thermodynamics of clusterized matter. Phys. Rev. C **80**(2), 024606 (2009). https://doi.org/10.1103/PhysRevC.80.024606

Raduta, A.R., Gulminelli, F.: Statistical description of complex nuclear phases in supernovae and proto-neutron stars. Phys. Rev. C **82**(6), 065801 (2010). https://doi.org/10.1103/PhysRevC.82.065801

Raduta, A.R., Aymard, F., Gulminelli, F.: Clusterized nuclear matter in the (proto-)neutron star crust and the symmetry energy. Eur. Phys. J. A **50**, 24 (2014). http://dx.doi.org/10.1140/epja/i2014-14024-y

Raduta, A.R., Gulminelli, F., Oertel, M.: Modification of magicity toward the dripline and its impact on electron-capture rates for stellar core collapse. Phys. Rev. C **93**(2), 025803 (2016). https://doi.org/10.1103/PhysRevC.93.025803

Raduta, A.R., Gulminelli, F., Oertel, M.: Stellar electron capture rates on neutron-rich nuclei and their impact on stellar core collapse. Phys. Rev. C **95**(2), 025805 (2017). https://doi.org/10.1103/PhysRevC.95.025805

Ravenhall, D.G., Pethick, C.J., Wilson, J.R.: Structure of matter below nuclear saturation density. Phys. Rev. Lett. **50**, 2066–2069 (1983). https://doi.org/10.1103/PhysRevLett.50.2066

Read, J.S., Lackey, B.D., Owen, B.J., Friedman, J.L.: Constraints on a phenomenologically parametrized neutron-star equation of state. Phys. Rev. D **79**(12), 124032 (2009). https://doi.org/10.1103/PhysRevD.79.124032

Read, J.S., Baiotti, L., Creighton, J.D.E., Friedman, J.L., Giacomazzo, B., Kyutoku, K., Markakis, C., Rezzolla, L., Shibata, M., Taniguchi, K.: Matter effects on binary neutron star waveforms. Phys. Rev. D **88**(4), 044042 (2013). https://doi.org/10.1103/PhysRevD.88.044042

Reinhard, P.G., Nazarewicz, W.: Information content of the low-energy electric dipole strength: Correlation analysis. Phys. Rev. C **87**(1), 014324 (2013). https://doi.org/10.1103/PhysRevC.87.014324

Rezzolla, L., Takami, K.: Gravitational-wave signal from binary neutron stars: a systematic analysis of the spectral properties. Phys. Rev. D **93**(12), 124051 (2016). https://doi.org/10.1103/PhysRevD.93.124051

Rezzolla, L., Most, E.R., Weih, L.R.: Using gravitational-wave observations and quasi-universal relations to constrain the maximum mass of neutron stars. Astrophys. J. Lett. **852**, L25 (2018). https://doi.org/10.3847/2041-8213/aaa401

Richers, S., Ott, C.D., Abdikamalov, E., O'Connor, E., Sullivan, C.: Equation of state effects on gravitational waves from rotating core collapse. Phys. Rev. D **95**(6), 063019 (2017). https://doi.org/10.1103/PhysRevD.95.063019

Rijken, T.A., Stoks, V.G.J., Yamamoto, Y.: Soft-core hyperon-nucleon potentials. Phys. Rev. C **59**, 21–40 (1999). https://doi.org/10.1103/PhysRevC.59.21

Rikovska Stone, J., Stevenson, P.D., Miller, J.C., Strayer, M.R.: Nuclear matter and neutron star properties calculated with a separable monopole interaction. Phys. Rev. C **65**(6), 064312 (2002). https://doi.org/10.1103/PhysRevC.65.064312

Ring, P., Schuck, P.: The Nuclear Many-Body Problem. Physics and Astronomy Online Library. Springer, Berlin (2004). https://www.springer.com/fr/book/9783540212065

Rios, A., Polls, A., Ramos, A., Müther, H.: Entropy of a correlated system of nucleons. Phys. Rev. C **74**(5), 054317 (2006). https://doi.org/10.1103/PhysRevC.74.054317

Rios, A., Polls, A., Ramos, A., Müther, H.: Liquid-gas phase transition in nuclear matter from realistic many-body approaches. Phys. Rev. C **78**(4), 044314 (2008). https://doi.org/10.1103/PhysRevC.78.044314

Roca-Maza, X., Viñas, X., Centelles, M., Ring, P., Schuck, P.: Relativistic mean-field interaction with density-dependent meson-nucleon vertices based on microscopical calculations. Phys. Rev. C **84**(5), 054309 (2011). https://doi.org/10.1103/PhysRevC.84.054309

Roca-Maza, X., Viñas, X., Centelles, M., Agrawal, B.K., Colò, G., Paar, N., Piekarewicz, J., Vretenar, D.: Neutron skin thickness from the measured electric dipole polarizability in ^{68}Ni ^{120}Sn and ^{208}Pb. Phys. Rev. C **92**(6), 064304 (2015). https://doi.org/10.1103/PhysRevC.92.064304

Roggero, A., Mukherjee, A., Pederiva, F.: Quantum Monte Carlo calculations of neutron matter with nonlocal chiral interactions. Phys. Rev. Lett. **112**(22), 221103 (2014). https://doi.org/10.1103/PhysRevLett.112.221103

Röpke, G.: Parametrization of light nuclei quasiparticle energy shifts and composition of warm and dense nuclear matter. Nucl. Phys. A **867**, 66–80 (2011). https://doi.org/10.1016/j.nuclphysa.2011.07.010

Röpke, G.: Nuclear matter equation of state including two-, three-, and four-nucleon correlations. Phys. Rev. C **92**(5), 054001 (2015). https://doi.org/10.1103/PhysRevC.92.054001

Röpke, G., Münchow, L., Schulz, H.: Particle clustering and Mott transitions in nuclear matter at finite temperature (I). Method and general aspects. Nucl. Phys. A **379**, 536–552 (1982). https://doi.org/10.1016/0375-9474(82)90013-6

Röpke, G., Schmidt, M., Münchow, L., Schulz, H.: Particle clustering and Mott transition in nuclear matter at finite temperature (II) Self-consistent ladder Hartree-Fock approximation and model calculations for cluster abundances and the phase diagram. Nucl. Phys. A **399**, 587–602 (1983). https://doi.org/10.1016/0375-9474(83)90265-8

Röpke, G., Bastian, N.U., Blaschke, D., Klähn, T., Typel, S., Wolter, H.H.: Cluster-virial expansion for nuclear matter within a quasiparticle statistical approach. Nucl. Phys. A **897**, 70–92 (2013). https://doi.org/10.1016/j.nuclphysa.2012.10.005

Rosswog, S.: The multi-messenger picture of compact binary mergers. Int. J. Mod. Phys. D **24**, 1530012 (2015). https://doi.org/10.1142/S0218271815300128

Rusnak, J.J., Furnstahl, R.J.: Relativistic point-coupling models as effective theories of nuclei. Nucl. Phys. A **627**, 495–521 (1997). https://doi.org/10.1016/S0375-9474(97)00598-8

Russotto, P., Gannon, S., Kupny, S., Lasko, P., Acosta, L., Adamczyk, M., Al-Ajlan, A., Al-Garawi, M., Al-Homaidhi, S., Amorini, F., et al.: Results of the ASY-EOS experiment at GSI: the symmetry energy at suprasaturation density. Phys. Rev. C **94**(3), 034608 (2016). https://doi.org/10.1103/PhysRevC.94.034608

Sagert, I., Fischer, T., Hempel, M., Pagliara, G., Schaffner-Bielich, J., Mezzacappa, A., Thielemann, F.K., Liebendörfer, M.: Signals of the QCD phase transition in core-collapse supernovae. Phys. Rev. Lett. **102**(8), 081101 (2009). https://doi.org/10.1103/PhysRevLett.102.081101

Sagert, I., Fischer, T., Hempel, M., Pagliara, G., Schaffner-Bielich, J., Thielemann, F.K., Liebendörfer, M.: Strange quark matter in explosive astrophysical systems. J. Phys. G: Nucl. Part. Phys. **37**(9), 094064 (2010). https://doi.org/10.1088/0954-3899/37/9/094064

Sagert, I., Tolos, L., Chatterjee, D., Schaffner-Bielich, J., Sturm, C.: Soft nuclear equation-of-state from heavy-ion data and implications for compact stars. Phys. Rev. C **86**(4), 045802 (2012). https://doi.org/10.1103/PhysRevC.86.045802

Sagert, I., Fann, G.I., Fattoyev, F.J., Postnikov, S., Horowitz, C.J.: Quantum simulations of nuclei and nuclear pasta with the multiresolution adaptive numerical environment for scientific simulations. Phys. Rev. C **93**(5), 055801 (2016). https://doi.org/10.1103/PhysRevC.93.055801

Sammarruca, F., Coraggio, L., Holt, J.W., Itaco, N., Machleidt, R., Marcucci, L.E.: Toward order-by-order calculations of the nuclear and neutron matter equations of state in chiral effective field theory. Phys. Rev. C **91**(5), 054311 (2015). https://doi.org/10.1103/PhysRevC.91.054311

Schaffner-Bielich, J., Gal, A.: Properties of strange hadronic matter in bulk and in finite systems. Phys. Rev. C **62**(3), 034311 (2000). https://doi.org/10.1103/PhysRevC.62.034311

Schaffner-Bielich, J., Hanauske, M., Stöcker, H., Greiner, W.: Phase transition to hyperon matter in neutron stars. Phys. Rev. Lett. **89**(17), 171101 (2002). https://doi.org/10.1103/PhysRevLett.89.171101

Scheidegger, S., Käppeli, R., Whitehouse, S.C., Fischer, T., Liebendörfer, M.: The influence of model parameters on the prediction of gravitational wave signals from stellar core collapse. Astron. Astrophys. **514**, A51 (2010). https://doi.org/10.1051/0004-6361/200913220

Schmidt, K.E., Kalos, M.H.: Few- and many-fermion problems. In: Binder, K. (ed.) Applications of the Monte Carlo Method in Statistical Physics. Topics in Current Physics, vol. 36, pp. 125–143. Springer, Berlin (1987). ISBN: 978-3-540-17650-3; https://doi.org/10.1007/978-3-642-51703-7_4

Schneider, A.S., Berry, D.K., Briggs, C.M., Caplan, M.E., Horowitz, C.J.: Nuclear "waffles". Phys. Rev. C **90**(5), 055805 (2014). https://doi.org/10.1103/PhysRevC.90.055805

Schuetrumpf, B., Klatt, M.A., Iida, K., Maruhn, J.A., Mecke, K., Reinhard, P.G.: Time-dependent Hartree-Fock approach to nuclear "pasta" at finite temperature. Phys. Rev. C **87**(5), 055805 (2013). https://doi.org/10.1103/PhysRevC.87.055805

Schulze, H.J., Rijken, T.: Maximum mass of hyperon stars with the Nijmegen ESC08 model. Phys. Rev. C **84**(3), 035801 (2011). https://doi.org/10.1103/PhysRevC.84.035801

Sébille, F., Figerou, S., de la Mota, V.: Self-consistent dynamical mean-field investigation of exotic structures in isospin-asymmetric nuclear matter. Nucl. Phys. A **822**, 51–73 (2009). https://doi.org/10.1016/j.nuclphysa.2009.02.013

Sébille, F., de La Mota, V., Figerou, S.: Probing the microscopic nuclear matter self-organization processes in the neutron star crust. Phys. Rev. C **84**(5), 055801 (2011). https://doi.org/10.1103/PhysRevC.84.055801

Sedrakian, A.: Rapid cooling of Cassiopeia A as a phase transition in dense QCD. Astron. Astrophys. **555**, L10 (2013). https://doi.org/10.1051/0004-6361/201321541

Sekiguchi, Y., Kiuchi, K., Kyutoku, K., Shibata, M.: Effects of hyperons in binary neutron star mergers. Phys. Rev. Lett. **107**(21), 211101 (2011). https://doi.org/10.1103/PhysRevLett.107.211101

Sellahewa, R., Rios, A.: Isovector properties of the Gogny interaction. Phys. Rev. C **90**(5), 054327 (2014). https://doi.org/10.1103/PhysRevC.90.054327

Serot, B.D.: Quantum hadrodynamics. Rep. Prog. Phys. **55**, 1855–1946 (1992). https://doi.org/10.1088/0034-4885/55/11/001

Serot, B.D., Walecka, J.D.: The relativistic nuclear many body problem. Adv. Nucl. Phys. **16**, 1–327 (1986)

Sharma, B.K., Centelles, M., Viñas, X., Baldo, M., Burgio, G.F.: Unified equation of state for neutron stars on a microscopic basis. Astron. Astrophys. **584**, A103 (2015). https://doi.org/10.1051/0004-6361/201526642

Shen, H., Toki, H., Oyamatsu, K., Sumiyoshi, K.: Relativistic equation of state of nuclear matter for supernova and neutron star. Nucl. Phys. A **637**, 435–450 (1998a). https://doi.org/10.1016/S0375-9474(98)00236-X

Shen, H., Toki, H., Oyamatsu, K., Sumiyoshi, K.: Relativistic equation of state of nuclear matter for supernova explosion. Prog. Theor. Phys. **100**, 1013–1031 (1998b). https://doi.org/10.1143/PTP.100.1013

Shen, G., Horowitz, C.J., Teige, S.: Equation of state of dense matter from a density dependent relativistic mean field model. Phys. Rev. C **82**(1), 015806 (2010a). https://doi.org/10.1103/PhysRevC.82.015806

Shen, G., Horowitz, C.J., Teige, S.: Equation of state of nuclear matter in a virial expansion of nucleons and nuclei. Phys. Rev. C **82**(4), 045802 (2010b). https://doi.org/10.1103/PhysRevC.82.045802

Shen, G., Horowitz, C.J., O'Connor, E.: Second relativistic mean field and virial equation of state for astrophysical simulations. Phys. Rev. C **83**(6), 065808 (2011a). https://doi.org/10.1103/PhysRevC.83.065808

Shen, G., Horowitz, C.J., Teige, S.: New equation of state for astrophysical simulations. Phys. Rev. C **83**(3), 035802 (2011b). https://doi.org/10.1103/PhysRevC.83.035802

Shen, H., Toki, H., Oyamatsu, K., Sumiyoshi, K.: Relativistic equation of state for core-collapse supernova simulations. Astrophys. J. Suppl. Ser. **197**, 20 (2011c). https://doi.org/10.1088/0067-0049/197/2/20

Shibata, M., Taniguchi, K.: Coalescence of black hole-neutron star binaries. Living Rev. Relativ. **14**, 6 (2011). https://doi.org/10.12942/lrr-2011-6

Shimizu, K.: Study of baryon baryon interactions and nuclear properties in the quark cluster model. Rep. Prog. Phys. **52**, 1–56 (1989). https://doi.org/10.1088/0034-4885/52/1/001

Shternin, P.S., Yakovlev, D.G.: A young cooling neutron star in the remnant of Supernova 1987A. Astron. Lett. **34**, 675–685 (2008). https://doi.org/10.1134/S1063773708100034

Shternin, P.S., Yakovlev, D.G., Heinke, C.O., Ho, W.C.G., Patnaude, D.J.: Cooling neutron star in the Cassiopeia A supernova remnant: evidence for superfluidity in the core. Mon. Not. R. Astron. Soc. **412**, L108–L112 (2011). https://doi.org/10.1111/j.1745-3933.2011.01015.x

Skyrme, T.H.R.: CVII. The nuclear surface. Philos. Mag. **1**, 1043–1054 (1956). https://doi.org/10.1080/14786435608238186

Somà, V., Barbieri, C., Duguet, T.: Ab initio Gorkov-Green's function calculations of open-shell nuclei. Phys. Rev. C **87**(1), 011303 (2013). https://doi.org/10.1103/PhysRevC.87.011303

Song, H.Q., Baldo, M., Giansiracusa, G., Lombardo, U.: Bethe-Brueckner-Goldstone expansion in nuclear matter. Phys. Rev. Lett. **81**, 1584–1587 (1998). https://doi.org/10.1103/PhysRevLett.81.1584

Sotani, H., Nakazato, K., Iida, K., Oyamatsu, K.: Probing the equation of state of nuclear matter via neutron star asteroseismology. Phys. Rev. Lett. **108**(20), 201101 (2012). https://doi.org/10.1103/PhysRevLett.108.201101

Steiner, A.W., Watts, A.L.: Constraints on neutron star crusts from oscillations in giant flares. Phys. Rev. Lett. **103**(18), 181101 (2009). https://doi.org/10.1103/PhysRevLett.103.181101

Steiner, A.W., Lattimer, J.M., Brown, E.F.: The equation of state from observed masses and radii of neutron stars. Astrophys. J. **722**, 33–54 (2010). https://doi.org/10.1088/0004-637X/722/1/33

Steiner, A.W., Hempel, M., Fischer, T.: Core-collapse supernova equations of state based on neutron star observations. Astrophys. J. **774**, 17 (2013a). https://doi.org/10.1088/0004-637X/774/1/17

Steiner, A.W., Lattimer, J.M., Brown, E.F.: The neutron star mass-radius relation and the equation of state of dense matter. Astrophys. J. Lett. **765**, L5 (2013b). https://doi.org/10.1088/2041-8205/765/1/L5

Stergioulas, N.: Rotating stars in relativity. Living Rev. Relativ. **6**, 3 (2003). https://doi.org/10.12942/lrr-2003-3

Stevenson, P., Strayer, M.R., Rikovska Stone, J.: Many-body perturbation calculation of spherical nuclei with a separable monopole interaction. Phys. Rev. C **63**(5), 054309 (2001). https://doi.org/10.1103/PhysRevC.63.054309

Stevenson, P.D., Goddard, P.M., Stone, J.R., Dutra, M.: Do Skyrme forces that fit nuclear matter work well in finite nuclei? In: American Institute of Physics Conference Series, vol. 1529, pp. 262–268 (2013). https://doi.org/10.1063/1.4807465

Stoks, V.G.J., Klomp, R.A.M., Terheggen, C.P.F., de Swart, J.J.: Construction of high-quality NN potential models. Phys. Rev. C **49**, 2950–2962 (1994). https://doi.org/10.1103/PhysRevC.49.2950

Stone, J.R., Reinhard, P.G.: The Skyrme interaction in finite nuclei and nuclear matter. Prog. Part. Nucl. Phys. **58**, 587–657 (2007). https://doi.org/10.1016/j.ppnp.2006.07.001

Stone, J.R., Stone, N.J., Moszkowski, S.A.: Incompressibility in finite nuclei and nuclear matter. Phys. Rev. C **89**(4), 044316 (2014). https://doi.org/10.1103/PhysRevC.89.044316

Sturm, C., Böttcher, I., Dębowski, M., Förster, A., Grosse, E., Koczoń, P., Kohlmeyer, B., Laue, F., Mang, M., Naumann, L., Oeschler, H., Pühlhofer, F., Schwab, E., Senger, P., Shin, Y., Speer, J., Ströbele, H., Surówka, G., Uhlig, F., Wagner, A., Waluś, W.: Evidence for a soft nuclear equation-of-state from kaon production in heavy-ion collisions. Phys. Rev. Lett. **86**, 39–42 (2001). https://doi.org/10.1103/PhysRevLett.86.39

Sugahara, Y., Toki, H.: Relativistic mean-field theory for unstable nuclei with non-linear σ and ω terms. Nucl. Phys. A **579**, 557–572 (1994). https://doi.org/10.1016/0375-9474(94)90923-7

Sullivan, C., O'Connor, E., Zegers, R.G.T., Grubb, T., Austin, S.M.: The sensitivity of core-collapse supernovae to nuclear electron capture. Astrophys. J. **816**, 44 (2016). https://doi.org/10.3847/0004-637X/816/1/44

Sumiyoshi, K.: Influence of equation of state in supernova simulations: neutrinos from proto-neutron star and black hole formation. Prog. Theor. Phys. Suppl. **168**, 610–613 (2007). https://doi.org/10.1143/PTPS.168.610

Sumiyoshi, K., Oyamatsu, K., Toki, H.: Neutron star profiles in the relativistic Brueckner-Hartree-Fock theory. Nucl. Phys. A **595**, 327–345 (1995). https://doi.org/10.1016/0375-9474(95)00388-5

Sumiyoshi, K., Yamada, S., Suzuki, H., Shen, H., Chiba, S., Toki, H.: Postbounce evolution of core-collapse supernovae: long-term effects of the equation of state. Astrophys. J. **629**, 922–932 (2005). https://doi.org/10.1086/431788

Sumiyoshi, K., Yamada, S., Suzuki, H.: Dynamics and neutrino signal of black hole formation in nonrotating failed supernovae. I. Equation of state dependence. Astrophys. J. **667**, 382–394 (2007). https://doi.org/10.1086/520876

Sumiyoshi, K., Ishizuka, C., Ohnishi, A., Yamada, S., Suzuki, H.: Emergence of hyperons in failed supernovae: trigger of the black hole formation. Astrophys. J. Lett. **690**, L43–L46 (2009). https://doi.org/10.1088/0004-637X/690/1/L43

Suwa, Y., Takiwaki, T., Kotake, K., Fischer, T., Liebendörfer, M., Sato, K.: On the importance of the equation of state for the neutrino-driven supernova explosion mechanism. Astrophys. J. **764**, 99 (2013). https://doi.org/10.1088/0004-637X/764/1/99

Takami, K., Rezzolla, L., Baiotti, L.: Constraining the equation of state of neutron stars from binary mergers. Phys. Rev. Lett. **113**(9), 091104 (2014). https://doi.org/10.1103/PhysRevLett.113.091104

Takami, K., Rezzolla, L., Baiotti, L.: Spectral properties of the post-merger gravitational-wave signal from binary neutron stars. Phys. Rev. D **91**(6), 064001 (2015). https://doi.org/10.1103/PhysRevD.91.064001

Taranto, G., Baldo, M., Burgio, G.F.: Selecting microscopic equations of state. Phys. Rev. C **87**, 045803 (2013). https://journals.aps.org/prc/abstract/10.1103/PhysRevC.87.045803

Taranto, G., Burgio, G.F., Schulze, H.J.: Cassiopeia A and direct Urca cooling. Mon. Not. R. Astron. Soc. **456**, 1451–1458 (2016). https://doi.org/10.1093/mnras/stv2756

Tauris, T.M., Langer, N., Moriya, T.J., Podsiadlowski, P., Yoon, S.C., Blinnikov, S.I.: Ultra-stripped type IC supernovae from close binary evolution. Astrophys. J. Lett. **778**, L23 (2013). https://doi.org/10.1088/2041-8205/778/2/L23

Ter Haar, B., Malfliet, R.: Equation of state of nuclear matter in the relativistic Dirac-Brueckner approach. Phys. Rev. Lett. **56**, 1237–1240 (1986). https://doi.org/10.1103/PhysRevLett.56.1237

Tews, I., Lattimer, J.M., Ohnishi, A., Kolomeitsev, E.E.: Symmetry parameter constraints from a lower bound on neutron-matter energy. Astrophys. J. **848**, 105 (2017). https://doi.org/10.3847/1538-4357/aa8db9

Thomas, A.W., Whittenbury, D.L., Carroll, J.D., Tsushima, K., Stone, J.R.: Equation of state of dense matter and consequences for neutron stars. In: European Physical Journal Web of Conferences, vol. 63, p. 03004 (2013). https://doi.org/10.1051/epjconf/20136303004

Todd-Rutel, B.G., Piekarewicz, J.: Neutron-rich nuclei and neutron stars: a new accurately calibrated interaction for the study of neutron-rich matter. Phys. Rev. Lett. **95**(12), 122501 (2005). https://doi.org/10.1103/PhysRevLett.95.122501

Togashi, H., Takano, M., Sumiyoshi, K., Nakazato, K.: Application of the nuclear equation of state obtained by the variational method to core-collapse supernovae. Prog. Theor. Exp. Phys. **2014**(2), 023D05 (2014). https://doi.org/10.1093/ptep/ptu020

Toki, H., Hirata, D., Sugahara, Y., Sumiyoshi, K., Tanihata, I.: Relativistic many body approach for unstable nuclei and supernova. Nucl. Phys. A **588**, c357–c363 (1995). https://doi.org/10.1016/0375-9474(95)00161-S

Tolman, R.C.: Static solutions of einstein's field equations for spheres of fluid. Phys. Rev. **55**, 364–373 (1939). https://doi.org/10.1103/PhysRev.55.364

Trippa, L., Colò, G., Vigezzi, E.: Giant dipole resonance as a quantitative constraint on the symmetry energy. Phys. Rev. C **77**(6), 061304 (2008). https://doi.org/10.1103/PhysRevC.77.061304

Trümper, J.E.: Observations of neutron stars and the equation of state of matter at high densities. Prog. Part. Nucl. Phys. **66**, 674–680 (2011). https://doi.org/10.1016/j.ppnp.2011.01.018

Tsang, M.B., Zhang, Y., Danielewicz, P., Famiano, M., Li, Z., Lynch, W.G., Steiner, A.W.: Constraints on the density dependence of the symmetry energy. Phys. Rev. Lett. **102**(12), 122701 (2009). https://doi.org/10.1103/PhysRevLett.102.122701

Tsang, M.B., Stone, J.R., Camera, F., Danielewicz, P., Gandolfi, S., Hebeler, K., Horowitz, C.J., Lee, J., Lynch, W.G., Kohley, Z., Lemmon, R., Möller, P., Murakami, T., Riordan, S., Roca-Maza, X., Sammarruca, F., Steiner, A.W., Vidaña, I., Yennello, S.J.: Constraints on the symmetry energy and neutron skins from experiments and theory. Phys. Rev. C **86**(1), 015803 (2012). https://doi.org/10.1103/PhysRevC.86.015803

Typel, S.: Relativistic model for nuclear matter and atomic nuclei with momentum-dependent self-energies. Phys. Rev. C **71**(6), 064301 (2005). https://doi.org/10.1103/PhysRevC.71.064301

Typel, S.: Clusters in nuclear matter and the equation of state for astrophysical applications. In: Gonçalves, V.P., da Silva, M.L.L., Thiago de Santana Amaral, J., Machado, M.V.T. (eds.) American Institute of Physics Conference Series, vol. 1520, pp. 68–102 (2013). https://doi.org/10.1063/1.4795946

Typel, S.: Neutron skin thickness of heavy nuclei with α-particle correlations and the slope of the nuclear symmetry energy. Phys. Rev. C **89**(6), 064321 (2014). https://doi.org/10.1103/PhysRevC.89.064321

Typel, S., Röpke, G., Klähn, T., Blaschke, D., Wolter, H.H.: Composition and thermodynamics of nuclear matter with light clusters. Phys. Rev. C **81**(1), 015803 (2010). https://doi.org/10.1103/PhysRevC.81.015803

Typel, S., Wolter, H.H., Röpke, G., Blaschke, D.: Effects of the liquid-gas phase transition and cluster formation on the symmetry energy. Eur. Phys. J. A **50**, 17 (2014). https://doi.org/10.1140/epja/i2014-14017-x

Typel, S., Oertel, M., Klähn, T.: CompOSE CompStar online supernova equations of state harmonising the concert of nuclear physics and astrophysics compose.obspm.fr. Phys. Part. Nucl. **46**, 633–664 (2015). https://doi.org/10.1134/S1063779615040061

Uhlenbeck, G.E., Beth, E.: The quantum theory of the non-ideal gas I. Deviations from the classical theory. Physica **3**, 729–745 (1936). https://doi.org/10.1016/S0031-8914(36)80346-2

Valcarce, A., Garcilazo, H., Fernandez, F., Gonzalez, P.: Quark-model study of few-baryon systems. Rep. Prog. Phys. **68**, 965–1042 (2005). https://doi.org/10.1088/0034-4885/68/5/R01

van Leeuwen, J. (ed.): Neutron Stars and Pulsars (IAU S291): Challenges and Opportunities after 80 Years. Proceedings of the International Astronomical Union Symposia and Colloquia. Cambridge University Press, Cambridge (2013). https://books.google.fr/books?id=jHX3mgEACAAJ

Vautherin, D., Brink, D.M.: Hartree-fock calculations with Skyrme's interaction. I. Spherical nuclei. Phys. Rev. C **5**, 626–647 (1972). https://journals.aps.org/prc/abstract/10.1103/PhysRevC.5.626

Vidaña, I., Logoteta, D., Providência, C., Polls, A., Bombaci, I.: Estimation of the effect of hyperonic three-body forces on the maximum mass of neutron stars. Europhys. Lett. **94**, 11002 (2011). https://doi.org/10.1209/0295-5075/94/11002

Voskresenskaya, M.D., Typel, S.: Constraining mean-field models of the nuclear matter equation of state at low densities. Nucl. Phys. A **887**, 42–76 (2012). https://doi.org/10.1016/j.nuclphysa.2012.05.006

Walecka, J.D.: A theory of highly condensed matter. Ann. Phys. **83**, 491–529 (1974). https://doi.org/10.1016/0003-4916(74)90208-5

Wanajo, S., Sekiguchi, Y., Nishimura, N., Kiuchi, K., Kyutoku, K., Shibata, M.: Production of all the r-process nuclides in the dynamical ejecta of neutron star mergers. Astrophys. J. **789**, L39 (2014). https://doi.org/10.1088/2041-8205/789/2/L39

Wang, M., Audi, G., Wapstra, A., Kondev, F., MacCormick, M., Xu, X., Pfeiffer, B.: The ame2012 atomic mass evaluation. Chin. Phys. C **36**(12), 1603 (2012). http://stacks.iop.org/1674-1137/36/i=12/a=003

Wang, S., Zhang, H.F., Dong, J.M.: Neutron star properties in density-dependent relativistic mean field theory with consideration of an isovector scalar meson. Phys. Rev. C **90**(5), 055801 (2014). https://doi.org/10.1103/PhysRevC.90.055801

Washiyama, K., Bennaceur, K., Avez, B., Bender, M., Heenen, P.H., Hellemans, V.: New parametrization of Skyrme's interaction for regularized multireference energy density functional calculations. Phys. Rev. C **86**(5), 054309 (2012). https://doi.org/10.1103/PhysRevC.86. 054309

Watanabe, G., Iida, K., Sato, K.: Thermodynamic properties of nuclear "pasta" in neutron star crusts. Nucl. Phys. A **676**, 455–473 (2000). https://doi.org/10.1016/S0375-9474(00)00197-4

Watanabe, G., Sonoda, H., Maruyama, T., Sato, K., Yasuoka, K., Ebisuzaki, T.: Formation of nuclear "Pasta" in supernovae. Phys. Rev. Lett. **103**(12), 121101 (2009). https://doi.org/10. 1103/PhysRevLett.103.121101

Watts, A., Espinoza, C.M., Xu, R., Andersson, N., Antoniadis, J., Antonopoulou, D., Buchner, S., Datta, S., Demorest, P., Freire, P., Hessels, J., Margueron, J., Oertel, M., Patruno, A., Possenti, A., Ransom, S., Stairs, I., Stappers, B.: Probing the neutron star interior and the Equation of State of cold dense matter with the SKA. In: Advancing Astrophysics with the Square Kilometre Array (AASKA14) 43 (2015). https://doi.org/10.22323/1.215.0043

Weber, F.: Pulsars as Astrophysical Laboratories for Nuclear and Particle Physics. Series in High Energy Physics, Cosmology and Gravitation. Taylor & Francis, Milton Park (1999). https:// books.google.fr/books?id=uxHBcb2XHaAC

Weinberg, S.: Nuclear forces from chiral lagrangians. Phys. Lett. B **251**, 288–292 (1990). https:// doi.org/10.1016/0370-2693(90)90938-3

Weinberg, S.: Effective chiral lagrangians for nucleon-pion interactions and nuclear forces. Nucl. Phys. B **363**, 3–18 (1991). https://doi.org/10.1016/0550-3213(91)90231-L

Wellenhofer, C., Holt, J.W., Kaiser, N., Weise, W.: Nuclear thermodynamics from chiral low-momentum interactions. Phys. Rev. C **89**(6), 064009 (2014). https://doi.org/10.1103/ PhysRevC.89.064009

Whittenbury, D.L., Carroll, J.D., Thomas, A.W., Tsushima, K., Stone, J.R.: Quark-meson coupling model, nuclear matter constraints, and neutron star properties. Phys. Rev. C **89**(6), 065801 (2014). https://doi.org/10.1103/PhysRevC.89.065801

Wiringa, R.B., Pieper, S.C.: Evolution of nuclear spectra with nuclear forces. Phys. Rev. Lett. **89**(18), 182501 (2002). https://doi.org/10.1103/PhysRevLett.89.182501

Wiringa, R.B., Stoks, V.G.J., Schiavilla, R.: Accurate nucleon-nucleon potential with charge-independence breaking. Phys. Rev. C **51**, 38–51 (1995). https://journals.aps.org/prc/abstract/ 10.1103/PhysRevC.51.38

Wiringa, R.B., Schiavilla, R., Pieper, S.C., Carlson, J.: Nucleon and nucleon-pair momentum distributions in A ≤ 12 nuclei. Phys. Rev. C **89**(2), 024305 (2014). https://doi.org/10.1103/ PhysRevC.89.024305

Worley, A., Krastev, P.G., Li, B.A.: Nuclear constraints on the moments of inertia of neutron stars. Astrophys. J. **685**, 390–399 (2008). https://doi.org/10.1086/589823

Xiao, Z., Li, B.A., Chen, L.W., Yong, G.C., Zhang, M.: Circumstantial evidence for a soft nuclear symmetry energy at suprasaturation densities. Phys. Rev. Lett. **102**(6), 062502 (2009). https:// doi.org/10.1103/PhysRevLett.102.062502

Yakovlev, D.G., Pethick, C.J.: Neutron star cooling. Annu. Rev. Astron. Astrophys. **42**, 169–210 (2004). https://doi.org/10.1146/annurev.astro.42.053102.134013

Yakovlev, D.G., Kaminker, A.D., Gnedin, O.Y., Haensel, P.: Neutrino emission from neutron stars. Phys. Rep. **354**, 1–155 (2001). https://doi.org/10.1016/S0370-1573(00)00131-9

Zalewski, M., Olbratowski, P., Satuła, W.: Surface-peaked effective mass in the nuclear energy density functional and its influence on single-particle spectra. Phys. Rev. C **81**(4), 044314 (2010). https://doi.org/10.1103/PhysRevC.81.044314

Zhang, Z.W., Shen, H.: Relativistic equation of state at subnuclear densities in the thomas-fermi approximation. Astrophys. J. **788**, 185 (2014). https://doi.org/10.1088/0004-637X/788/2/185

Zhao, P.W., Li, Z.P., Yao, J.M., Meng, J.: New parametrization for the nuclear covariant energy density functional with a point-coupling interaction. Phys. Rev. C **82**(5), 054319 (2010). https:// doi.org/10.1103/PhysRevC.82.054319

Zhou, X.R., Burgio, G.F., Lombardo, U., Schulze, H.J., Zuo, W.: Three-body forces and neutron star structure. Phys. Rev. C **69**(1), 018801 (2004). https://doi.org/10.1103/PhysRevC.69.018801

Zuo, W., Lejeune, A., Lombardo, U., Mathiot, J.F.: Interplay of three-body interactions in the EOS of nuclear matter. Nucl. Phys. A **706**, 418–430 (2002a). https://doi.org/10.1016/S0375-9474(02)00750-9

Zuo, W., Lejeune, A., Lombardo, U., Mathiot, J.F.: Microscopic three-body force for asymmetric nuclear matter. Eur. Phys. J. A **14**, 469–475 (2002b). https://doi.org/10.1140/epja/i2002-10031-y

Chapter 7
Phases of Dense Matter in Compact Stars

David Blaschke and Nicolas Chamel

Abstract Formed in the aftermath of gravitational core-collapse supernova explosions, neutron stars are unique cosmic laboratories for probing the properties of matter under extreme conditions that cannot be reproduced in terrestrial laboratories. The interior of a neutron star, endowed with the highest magnetic fields known and with densities spanning about ten orders of magnitude from the surface to the centre, is predicted to exhibit various phases of dense strongly interacting matter, whose physics is reviewed in this chapter. The outer layers of a neutron star consist of a solid nuclear crust, permeated by a neutron ocean in its densest region, possibly on top of a nuclear "pasta" mantle. The properties of these layers and of the homogeneous isospin asymmetric nuclear matter beneath constituting the outer core may still be constrained by terrestrial experiments. The inner core of highly degenerate, strongly interacting matter poses a few puzzles and questions which are reviewed here together with perspectives for their resolution. Consequences of the dense-matter phases for observables such as the neutron-star mass-radius relationship and the prospects to uncover their structure with modern observational programmes are touched upon.

D. Blaschke (✉)
Institute of Theoretical Physics, University of Wroclaw, Wroclaw, Poland

Bogoliubov Laboratory of Theoretical Physics, Joint Institute for Nuclear Research, Dubna, Russia

National Research Nuclear University (MEPhI), Moscow, Russia
e-mail: david.blaschke@ift.uni.wroc.pl

N. Chamel
Institute of Astronomy and Astrophysics, Université Libre de Bruxelles, Brussels, Belgium
e-mail: nchamel@ulb.ac.be

© Springer Nature Switzerland AG 2018
L. Rezzolla et al. (eds.), *The Physics and Astrophysics of Neutron Stars*,
Astrophysics and Space Science Library 457,
https://doi.org/10.1007/978-3-319-97616-7_7

7.1 Introduction

7.1.1 Cosmic Laboratories

Neutron stars are the stellar remnants of massive stars at the end point of their evolution (see, e.g., Haensel et al. 2007). Neutron stars have a mass between one and two times that of the Sun, but packed into a space only 20 km across (100,000 times smaller than the Sun's diameter). The average density of a neutron star can thus exceed a few hundred-thousand billion grams per cubic centimeter - a density higher than that found inside the heaviest atomic nuclei. Neutron stars are not only the most compact observed stars in the Universe, but they are also endowed with the strongest magnetic fields known, which could reach millions of billions times that of the Earth. Neutron-star observations thus offer the unique opportunity to explore the properties of matter under extreme conditions, which cannot be reproduced in the laboratory.

7.1.2 Cold Catalysed Matter Hypothesis

During the formation of a neutron star, the hot compressed matter in the collapsed stellar core is assumed to undergo all kinds of nuclear and electroweak processes such that the compact stellar remnant cools down by following a sequence of full thermodynamic quasi equilibrium states. The resulting neutron star thus eventually consists of "cold catalysed matter", i.e., electrically charge neutral matter in its absolute ground state, after all internal heat has been released (Harrison and Wheeler 1958; Harrison et al. 1965). This scenario supposes that the reaction rates are much higher than the cooling rate. In reality, the composition of a neutron star may not only depend on the particular internal conditions prevailing during its formation, but also on its subsequent evolution, as well as on its environment. In particular, the constitution of the collapsed core may become "frozen in" as it cools down (see, e.g. Goriely et al. (2012) for a recent discussion). Moreover, the accretion of matter from a companion star or a high enough magnetic field may notably change the composition of the neutron star. These different situations will be separately discussed.

7.2 Surface Layers of a Neutron Star

Neutron stars are expected to be surrounded by a very thin atmosphere consisting of a plasma of electrons and light elements (mainly hydrogen and helium though heavier elements like carbon may also be present (Heinke and Heinke 2009)). Its properties such as the effective temperature, the composition, and the magnetic field configuration, can be inferred by analyzing the thermal X-ray emission from neutron stars (see, e.g. Potekhin et al. 2015). The region beneath consists of a solid crust (see, e.g., Chamel and Haensel 2008).

7.2.1 Nonaccreted Neutron-Star Crusts

The outermost region of a nonaccreting neutron star is expected to be made of iron ^{56}Fe, the end-product of stellar nucleosynthesis. The properties of compressed iron can be probed in terrestrial laboratories up to pressures of order 10^{14} dyn cm^{-2} with nuclear explosions and laser-driven shock-wave experiments (see, e.g., Batani et al. 2002; Fortov and Lomonosov 2010; Ping 2013). Under these conditions, iron has an hexagonal close-packed structure (Ping 2013; Stixrude 2012). Transitions to a face-centred cubic lattice and a body-centred cubic lattice are expected at pressures of about 6×10^{13} dyn cm^{-2} and 4×10^{14} dyn cm^{-2} respectively for temperatures below $\sim 10^4$–10^5 K (Stixrude 2012). Although such pressures are tremendous according to terrestrial standards, they still remain negligibly small compared to those prevailing in a neutron star (note also that observed middle-aged neutron stars have typical surface temperatures of order 10^6 K, see e.g. Potekhin and Chabrier (2018)). In particular, the highest density of iron that has been experimentally attained is only about three times the density at the surface of a cold neutron star, and 14 orders of magnitude lower than the density at the stellar center. This corresponds to a depth of about 0.1 mm for a star with a mass $\mathcal{M} = 1.4\mathcal{M}_\odot$ and a radius $R = 12$ km.[1] Deeper in the star, recourse must therefore be made to theoretical models (see, e.g. Lai et al. (1991) and references therein). At a density $\rho_{\rm eip} \approx 2 \times 10^4$ g cm^{-3} (about 22 cm below the surface[2]), the interatomic spacing $a_N = [3/(4\pi n_N)]^{1/3}$ (with n_N the number density of atomic nuclei) becomes comparable with the atomic radius $\sim a_0/Z^{1/3}$ (with a_0 the Bohr radius and Z the atomic number - here $Z = 26$). At densities $\rho \gg \rho_{\rm eip}$, atoms are crushed into a dense plasma of nuclei and free electrons. This ionization is not triggered by thermal excitations but by the prodigious gravitational pressure (see, e.g., Haensel et al. (2007)). Beyond this point, electrons are very weakly perturbed by ions, and thus behave as an essentially ideal Fermi gas. Because electrons are highly degenerate, they provide the necessary pressure to counterbalance the weight of the layers above. The attractive Coulomb interactions between electrons and ions reduce the electron Fermi gas pressure, but their relative contribution is only of order $Z^{2/3}e^2/(\hbar v_F)$ (Haensel et al. 2007) (with e the elementary electric charge, $\hbar$

---

[1]The depth z_0 below the surface of a cold and fully catalysed neutron star at a pressure P_0 was estimated as $z_0 \approx \int_0^{P_0} \frac{dP}{\rho g_s}$, with $g_s = \frac{G\mathcal{M}}{R^2}\left(1 - \frac{2G\mathcal{M}}{Rc^2}\right)^{-1/2}$, see, e.g., Chamel and Haensel (2008), and we have made use of an interpolation of the "QEOS" equation of state from More et al. (1988), as tabulated in Lai et al. (1991).

[2]The depth was calculated combining the equations of state labeled "QEOS" from More et al. (1988) and "TFD" in Table 5 of Lai et al. (1991).

the Planck-Dirac constant, and v_F the electron Fermi velocity). The (electric charge) polarization of electrons around ions (also referred to as electron screening) leads to a correction of order $Z^{4/3}e^4/(\hbar v_F)^2$ (Haensel et al. 2007). For the nuclides present in the crust, the electron exchange correction of order $e^2/(\hbar v_F)$ turns out be of the same magnitude. For a discussion of higher-order corrections due to electron correlations, finite size of nuclei, and quantum-zero point motion of nuclei, see e.g. Guo et al. (2007), Pearson et al. (2011). Although electron-ion interactions are very small, they still play a major role in the equilibrium structure of the crust. As the density reaches about $7 \times 10^6 \, \mathrm{g \, cm^{-3}}$ (about 11 m below the surface[3]), electrons become relativistic since the interelectron spacing $a_e = [3/(4\pi n_e)]^{1/3}$ (with n_e the electron number density) is comparable with the electron Compton wavelength $\lambda_e = \hbar/(m_e c)$ (m_e is the electron mass, and c is the speed of light).

7.2.2 Accreted Neutron-Star Crusts

The composition of the surface layers of a neutron star may be changed by the fallback of material from the envelope ejected during the supernova explosion, and more importantly by the accretion of matter from a stellar companion (see, e.g., Chamel and Haensel (2008)). The accreted material forms an hydrogen rich envelope around the star. Stable hydrogen burning produces helium, which accumulates in a layer beneath. These helium ashes ignite under specific conditions of density (typically $\sim 10^6$–$10^7 \, \mathrm{g \, cm^{-3}}$) and temperature. For some range of accretion rates, helium burning is unstable, converting within seconds all the envelope into nuclides in the nickel-cadmium range (Schatz et al. 2001). These thermonuclear explosions are observed as X-ray bursts, with luminosity up to about $10^{38} \, \mathrm{erg \, s^{-1}}$ ($\approx$ Eddington limit for neutron stars), and with a typical decay lasting a few tens of seconds. Multiplying the burst luminosity by its duration we get an estimate of the total burst energy $\sim 10^{39}$–10^{40} erg. X-ray bursts are quasiperiodic, with typical recurrence time of about hours to days. Less frequent but more energetic ($\sim 10^{42}$ erg) are superbursts lasting for a few hours, with recurrence times of several years. These superbusts are presumably triggered by the unstable burning of carbon at densities $\sim 10^8$–$10^9 \, \mathrm{g \, cm^{-3}}$, and their ashes are predicted to consist mainly of ^{66}Ni, ^{64}Ni, ^{60}Fe, and ^{54}Cr (Schatz et al. 2003).

Because the equilibrium structure of such multicomponent plasmas is highly uncertain, various properties of accreted neutron-star crusts such as their breaking strain remain poorly known. According to classical molecular dynamics simulations, ashes tend to arrange on a regular body-centred cubic lattice as in catalysed crusts (Horowitz and Berry 2009). However, the system may have not fully relaxed to its true equilibrium state due to the very slow dynamics of crystal growth.

---

[3] The depth was calculated combining the equations of state labeled "QEOS" and "TFD" in Table 5 of Lai et al. (1991), and the equation of state calculated in Pearson et al. (2011).

Alternatively, valuable insight can be gained from analyses of the dynamical stability of given crystal structures (but with arbitrary composition) (Kozhberov and Baiko 2015). The reduced computational cost allows to take into account quantum effects, which can play a key role at low temperatures. Recently, a different approach has been followed using genetic algorithms (Engstrom et al. 2016). Although limited to a few ternary crystals, unconstrained global searches of the equilibrium structure have revealed a very rich phase diagram with noncubic lattices. In view of the importance of the crust structure, these theoretical studies should be pursued. Astrophysical observations can provide complementary information. In particular, the possibility of an amorphous crust has been ruled out by cooling simulations of the observed thermal relaxation of transiently accreting neutron stars (Shternin et al. 2007). Gravitational-wave observations (or lack thereof) may provide additional information on the structure of neutron-star crusts (Hoffman and Heyl 2012).

7.2.3 Highly Magnetised Neutron Stars

Neutron stars are not only the most compact observed stars in the universe, but are also among the strongest magnets known, with typical surface magnetic fields of order 10^{12} G (Seiradakis and Wielebinski 2004). A few radio pulsars have been recently found to have significantly higher surface magnetic fields of order 10^{13}–10^{14} G (Ng and Kaspi 2011). Surface magnetic fields of order 10^{14}–10^{15} G have been inferred in soft-gamma ray repeaters (SGRs) and anomalous x-ray pulsars (AXPs) from both spin-down and spectroscopic studies (Olausen and Kaspi 2014; Tiengo et al. 2013; An et al. 2014). Various observations suggest that the interior magnetic field may be even higher (Stella et al. 2005; Kaminker et al. 2007; Vietri et al. 2007; Rea et al. 2010; Makishima et al. 2014). At the time of this writing, 11 SGRs and 12 AXPs have been already identified (Olausen and Kaspi 2014). It is now widely believed that these objects belong to a different class of neutron stars called *magnetars* (see e.g. Woods and Thompson (2006) for a review), as proposed by Duncan and Thomson in 1992 (Thompson and Duncan 1992). Numerical simulations confirmed that magnetic fields of order $\sim 10^{15}$–10^{16} G can be produced during supernovae explosions due to the magnetorotational instability (Ardeljan et al. 2005). Theoretical considerations corroborated by numerical simulations suggest that neutron stars may potentially possess internal magnetic fields as high as 10^{18} G (see, e.g. Kiuchi and Yoshida 2008; Frieben and Rezzolla 2012; Pili et al. 2014; Chatterjee et al. 2015 and references therein).

The properties of the surface layers of a neutron star can be drastically different in the presence of a high magnetic field. The electron motion perpendicular to the magnetic field lines is quantised into Landau orbitals with a characteristic magnetic length scale (see, e.g., Haensel et al. (2007)) $a_m = a_0\sqrt{B_{\mathrm{at}}/B}$, with

$$B_{\mathrm{at}} = \frac{m_e^2 e^3 c}{\hbar^3} \simeq 2.35 \times 10^9 \text{ G}. \tag{7.1}$$

For magnetic fields $B \gg B_{at}$, atoms are expected to adopt a very elongated shape along the magnetic field lines and to form linear chains. The attractive interaction between these chains could lead to a phase transition into a magnetically condensed phase with a surface density estimated as (Lai and Shapiro 1991)

$$\rho_s \simeq 560 A Z^{-3/5} B_{12}^{6/5} \, \text{g cm}^{-3}, \tag{7.2}$$

where $B_{12} \equiv B/(10^{12}) \, \text{G}$ (assuming $B \ll 10^{18}$ G, see e.g. Chamel et al. (2012)). For iron with $B = 10^{15}$ G, we obtain $\rho_s \simeq 1.8 \times 10^7 \, \text{g cm}^{-3}$, as compared to $7.86 \, \text{g cm}^{-3}$ in the absence of magnetic fields. In deeper regions of the crust, the density ρ at pressure P is approximately given by (Chamel et al. 2012)

$$\rho \approx \rho_s \left(1 + \sqrt{\frac{P}{P_0}} \right), \tag{7.3}$$

where

$$P_0 \simeq 1.45 \times 10^{20} B_{12}^{7/5} \left(\frac{Z}{A} \right)^2 \, \text{dyn cm}^{-2}. \tag{7.4}$$

The presence of a high magnetic field does not change the equilibrium structure of a body-centred cubic crystal of ions embedded in a uniform charge neutralizing electron background (Kozhberov 2016). However, it has been found that in the strongly quantising regime, the long-range part of the ion-ion potential exhibits Friedel oscillations that may lead to the formation of strongly coupled filaments aligned with the magnetic field (Bedaque et al. 2013). This possibility should be further examined.

The absence of spectral features in the thermal emission from seven radio-quiet isolated neutron stars (usually referred to as the 'Magnificent Seven') could be the consequence of the magnetic condensation of their surface (Turolla et al. 2004), with magnetic fields of order 10^{13}–10^{14} G as estimated from X-ray timing data. The presence of high magnetic fields has recently found additional support from optical polarimetry measurements (Mignani et al. 2017). Future X-ray polarimetry measurements could potentially allow to discriminate between the case of a gaseous atmosphere and a condensed magnetic surface.

7.3 General Considerations on Dense Stellar Plasmas

In this section, we consider fully ionised Coulomb plasmas at densities ρ above $10^7 \, \text{g cm}^{-3}$ and below the onset of neutron emission by nuclei.

7.3.1 Gravitational Stratification

Because the pressure has to vary continuously throughout the star and the nucleon number is conserved, the suitable thermodynamic potential for determining the composition is the Gibbs free energy per nucleon g (Tondeur 1971; Baym et al. 1971). As shown in the appendix of Chamel and Fantina (2015), g remains the suitable thermodynamic potential in the presence of a high magnetic field. At a given pressure, hot dense matter in full thermodynamic equilibrium generally consists of an admixture of various nuclear species (Hempel et al. 2012; Furusawa1 et al. 2013; Buyukcizmeci et al. 2014; Gulminelli and Raduta 2015; Grams et al. 2018). As the temperature decreases, the distribution of nuclei becomes very narrow. The crust of a cold nonaccreted neutron star is thus expected to be stratified into different layers, each of which consists of a body-centred cubic crystal made of a single nuclear species. Substitutional binary ionic compounds with cesium chloride structure can only possibly exist at the interface between two adjacent strata (Chamel and Fantina 2016b). Multinary ionic compounds may however be present in the crust of accreted neutron stars (Chamel 2017a). We shall not discuss this possibility here. In the single-nucleus approximation, each layer is described as a one-component crystal of pointlike nuclei in a uniform charge neutralizing electron background. Retaining only the electrostatic correction to the ideal electron Fermi gas model, and expanding g to first order in $\alpha = e^2/(\hbar c)$, the transition between two adjacent strata made of nuclei (A_1, Z_1) and (A_2, Z_2) respectively, is determined by the condition (Chamel and Fantina 2016b)

$$\mu_e + C\,\alpha\hbar c n_e^{1/3}\left(\frac{4}{3}\frac{Z_1^{5/3}}{A_1} - \frac{1}{3}\frac{Z_1^{2/3}Z_2}{A_2} - \frac{Z_2^{5/3}}{A_2}\right)\left(\frac{Z_1}{A_1} - \frac{Z_2}{A_2}\right)^{-1} = \mu_e^{1\to 2}, \quad (7.5)$$

$$\mu_e^{1\to 2} \equiv \left[\frac{M'(A_2, Z_2)c^2}{A_2} - \frac{M'(A_1, Z_1)c^2}{A_1}\right]\left(\frac{Z_1}{A_1} - \frac{Z_2}{A_2}\right)^{-1} + m_e c^2, \quad (7.6)$$

where μ_e is the electron Fermi energy, $M'(A_1, Z_1)$ and $M'(A_2, Z_2)$ are the nuclear masses, C is a dimensionless structure constant (see discussion below).

According to the Bohr-van Leeuwen theorem (Van Vleck 1932), the electrostatic correction is independent of the magnetic field apart from a negligibly small contribution due to quantum fluctuations of ion motion (Baiko 2009). Equations (7.5) and (7.6) thus remain valid for highly magnetised neutron stars (provided the electrostatic correction remains small[4]), but nuclear masses hence $\mu_e^{1\to 2}$ could change substantially (especially if the magnetic field strength exceeds 10^{17} G)

---

[4]Because the electron chemical potential scales as $\mu_e \sim 2pi^2 m_e c^2 \lambda_e^3 n_e B_{\mathrm{rel}}/B$ with $B_{\mathrm{rel}} \simeq 4.4 \times 10^{13}$ G in strongly quantising magnetic fields, the expansion of g to first order in α eventually breaks down in high enough magnetic fields. See also Sect. 7.4.2.

compared to their values in the absence of a magnetic field (Peña Arteaga et al. 2011; Stein et al. 2016a). As shown in Sect. 7.3.2, mechanical stability requires $Z_1/A_1 > Z_2/A_2$ so that Eqs. (7.5) and (7.6) are well defined. We adopt here the convention to include the rest mass of Z electrons in $M'(A, Z)$. The reason is that experimental *atomic* masses $M(A, Z)$ are generally tabulated rather than *nuclear* masses $M'(A, Z)$. The latter can be obtained from the former after subtracting out the binding energy of the atomic electrons (see, e.g., Eq. (A4) of Lunney et al. (2003)). The threshold condition (7.5) still holds in hot dense matter whether the Coulomb plasma is in a solid or a liquid state, provided the temperature T is well below the electron Fermi temperature defined by

$$T_{Fe} = \frac{\mu_e - m_e c^2}{k_B} \approx 5.93 \times 10^9 \frac{\mu_e}{m_e c^2} \text{ K}. \tag{7.7}$$

In principle, nuclei in hot dense matter may coexist with a gas of nucleons and light particles, but their fraction is negligibly small at $T \ll T_{Fe}$ (see, e.g., Haensel et al. 2007). The structure constant is very-well approximated by the ion-sphere model (Salpeter 1954)

$$C = -\frac{9}{10} \left(\frac{4\pi}{3} \right)^{1/3}, \tag{7.8}$$

which also provides a lower bound (Lieb and Narnhofer 1975). The structure constant of the solid phase (assuming a perfect body-centred cubic crystal) is $C_{bcc} \simeq -1.4442$, whereas in the liquid phase $C_{liq} \simeq -1.4621$ (see, e.g., Haensel et al. (2007)). Crystallization of a one-component plasma of ions with charge Z occurs at the temperature given by (see, e.g., Haensel et al. 2007)

$$T_m = \frac{e^2}{a_e k_B \Gamma_m} Z^{5/3}, \tag{7.9}$$

where k_B is Boltzmann's constant, and Γ_m is the Coulomb coupling parameter at melting. In the absence of magnetic fields, $\Gamma_m \simeq 175$ (Haensel et al. 2007). In the presence of a high magnetic field, Coulomb crystals are expected to be more stable so that $\Gamma_m \lesssim 175$ (Potekhin and Chabrier 2013). Transitions between multicomponent phases as in accreted neutron-star crusts have been addressed in Chamel and Fantina (2016b).

Because the transition between two adjacent layers occurs at a fixed pressure, it is accompanied by a density discontinuity given by (Chamel and Fantina 2016b)

$$\frac{\bar{n}_2^{min} - \bar{n}_1^{max}}{\bar{n}_1^{max}} = \frac{A_2}{Z_2} \frac{Z_1}{A_2} \left[1 + \frac{1}{3} C\alpha\hbar c \left(n_e^{2/3} \frac{d\mu_e}{dn_e} \right)^{-1} \left(Z_1^{2/3} - Z_2^{2/3} \right) \right] - 1, \tag{7.10}$$

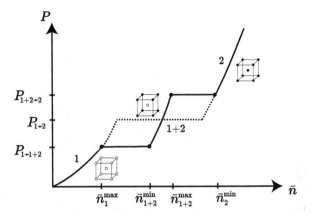

Fig. 7.1 Schematic representation of the pressure P versus mean baryon number density $\bar{n}$ for a transition between two pure body-centred cubic solid phases of nuclei (A_1, Z_1) and (A_2, Z_2) accompanied by the formation of a substitutional binary compound. For comparison, the transition leading to the coexistence of pure phases is indicated by the dotted line. The figure is not to scale. Taken from Chamel and Fantina (2016b)

as schematically illustrated in Fig. 7.1. It is to be understood that the electron density is obtained from Eq. (7.5). In principle, pure phases of nuclei (A_1, Z_1) and (A_2, Z_2) can coexist at intermediate densities. However, since inside a self-gravitating body in hydrostatic equilibrium the pressure must increase monotonically with depth (see, e.g., Haensel et al. 2007), such coexisting phases cannot be present in the crust of a neutron star. Instead, binary compounds could form at the boundary but only over a small range of pressures $(P_{1+2\rightarrow 2} - P_{1\rightarrow 1+2})/P_{1\rightarrow 2} \ll 1$, and provided $Z_1 \neq Z_2$ (Chamel and Fantina 2016b). As a matter of fact, transitions from a pure phase to a compound phase is still accompanied by density jumps. Very accurate analytical expressions for the pressure at the transition between two adjacent strata as well as the average baryon number densities of each layer in the absence of magnetic field can be found in Chamel and Fantina (2016b). The errors were found to lie below 0.2% for the pressures, and below 0.07% for the densities. Approximate expressions in the presence of a strongly quantising magnetic field are given in Chamel et al. (2017a).

7.3.2 Matter Neutronization and Onset of Neutron Emission

As a consequence of Le Chatelier's principle, the bulk modulus $K = \bar{n}\,dP/d\bar{n}$ of matter in equilibrium must be positive: the density must therefore increase with pressure. Using Eq. (7.10) and ignoring the small electrostatic correction, we conclude that nuclei become progressively more neutron rich with increasing depth $(Z_2/A_2 < Z_1/A_1)$ (see also De Blasio (2000) where the restoring force acting on displaced ions was explicitly calculated). This neutronization is achieved by electron captures, and various nuclear processes (e.g. fissions, fusions) depending on the local conditions. In turn, assuming that the electrostatic contribution in the left-hand side of Eq. (7.5) is small,[5] $\mu_e^{1\rightarrow2}$ must be positive implying that $M'(A_2, Z_2)/A_2 > M'(A_1, Z_1)/A_1$. Combining these two inequalities, we find that as the composition changes nucleons become less bound $B(A_2, Z_2)/A_2 < B(A_1, Z_1)/A_1$, where $B(A, Z) = Zm_pc^2 + (A - Z)m_nc^2 - M'(A, Z)c^2$ denotes the total binding energy of the nucleus (A, Z). At some point, $\Delta N > 0$ neutrons will become unbound and start to drip out of nuclei. This transition marks the boundary between the outer and inner regions of the crust. Due to baryon number conservation, the daughter nuclei will be of the form $(A - \Delta N, Z - \Delta Z)$, with $Z \geq \Delta Z \geq \Delta N + Z - A$ (this follows from the requirement that the daughter nuclei must contain positive numbers of neutrons and protons). Assuming $\Delta Z \neq 0$, and ignoring neutron band-structure effects, the onset of this transition is determined by the condition (Chamel et al. 2015a)

$$\mu_e + C\alpha\hbar cn_e^{1/3}\left[\frac{Z^{5/3} - (Z - \Delta Z)^{5/3}}{\Delta Z} + \frac{1}{3}Z^{2/3}\right] = \mu_e^{\text{drip}}, \qquad (7.11)$$

where

$$\mu_e^{\text{drip}} \equiv \frac{M'(A - \Delta N, Z - \Delta Z)c^2 - M'(A, Z)c^2 + m_nc^2\Delta N}{\Delta Z} + m_ec^2. \qquad (7.12)$$

In the case $\Delta Z = 0$ and $\Delta N > 0$, the threshold condition reads

$$M'(A, Z) - M'(A - \Delta N, Z) = \Delta Nm_n, \qquad (7.13)$$

independently of the electron background: this shows that a nucleus unstable against neutron emission in vacuum is also unstable in a stellar environment at *any* density. This means in particular that the neutron-drip transition in the crust of a neutron star must be necessarily triggered by reactions such that $\Delta Z \neq 0$. Since $\mu_e^{\text{drip}} > m_ec^2$, the nucleus must satisfy the following constraint

$$M'(A - \Delta N, Z - \Delta Z) - M'(A, Z) + m_n\Delta N > 0. \qquad (7.14)$$

Equations (7.5), (7.10), and (7.11) are applicable to both weakly and highly magnetised neutron stars (with suitable values for the nuclear masses), accreted and nonaccreted (to the extent that the crust consists of pure layers). Very accurate analytical expressions for the pressure and the density at the neutron-drip transition can be found in Chamel et al. (2015a,c).

---

[5]This assumption may be violated in the presence of a very high magnetic field, as discussed in Sect. 7.3.1.

7.4 Outer Crust of a Neutron Star

7.4.1 Nonaccreted Neutron-Star Crusts

The internal constitution of the outermost layers of a nonaccreted (fully catalysed) neutron star crust is completely determined by experimental atomic mass measurements. In particular, the strata of iron ^{56}Fe is predicted to extend up to a density of about 8×10^6 g cm^{-3} (about 12 m below the surface[6]). Using the data from the 2016 Atomic Mass Evaluation (AME) (Wang et al. 2017) and applying the model of Pearson et al. (2011), the region beneath is found to consist of a succession of layers made of ^{62}Ni, ^{64}Ni, ^{66}Ni, ^{86}Kr, ^{84}Se, ^{82}Ge, and ^{80}Zn with increasing depth. Although electron charge polarisation effects are very small, they may change the composition (see, e.g., Chamel and Fantina 2016a). For instance, ignoring this correction leads to the appearance of ^{58}Fe (whose mass is experimentally known) in between the layers of ^{62}Ni and ^{64}Ni. This nuclide is still present if the electron polarisation is implemented using the interpolating formula of Potekhin and Chabrier (2000) instead of the limiting Thomas-Fermi expression for $Z \to +\infty$ (Salpeter 1961) employed in Pearson et al. (2011). The existence or not of this nuclide in the crust is also found to depend on the precision of the nuclear mass measurements: using the data from the 2012 AME (Audi et al. 2012), which differs by less than 1 keV/c^2 compared to the more recent value (Wang et al. 2017), ^{58}Fe thus disappears independently on how the electron polarisation correction is calculated.

The composition of the innermost regions is more uncertain due to the lack of experimental data, and can only be explored using nuclear mass models (see, e.g., Guo et al. 2007; Pearson et al. 2011; Rüster et al. 2006; Roca-Maza and Piekarewicz 2008; Chamel et al. 2015b; Sharma et al. 2015; Utama et al. 2016; Fantina et al. 2017; Chamel et al. 2017b for recent calculations). The most accurate nuclear mass models achieve to fit all known experimental masses with a root-mean square deviation of a few hundred keV/c^2 (Pearson et al. 2013; Sobiczewski and Litvinov 2014), whereas errors of the order of a keV/c^2 can change the composition as previously discussed. Therefore, the main sources of uncertainties are nuclear masses. Because of nuclear pairing effects, nuclei with even numbers of neutrons and protons are more tightly bound that neighboring nuclei in the nuclear chart. For this reason, odd nuclei were generally not considered in earlier calculations of neutron-star crusts, and the search for the equilibrium nuclides was limited to a restricted set of even-even nuclei (130 even–even nuclei in the seminal work of Baym et al. (1971)). However, the crustal composition is determined by the full thermodynamic equilibrium with respect to all processes, not only strong nuclear reactions. Therefore, there is no fundamental reason to rule out odd nuclei. As a

---

[6]The depth was calculated combining the equations of state labeled "QEOS" and "TFD" in Table 5 of Lai et al. (1991), and the equation of state calculated in Pearson et al. (2011).

matter of fact, the microscopic nuclear mass model HFB-21 predicts the presence of ^{79}Cu and ^{121}Y in the outer crust (Pearson et al. 2011). The nuclear shell structure has a profound influence on the crustal composition, with a predominance of nuclides with neutron magic numbers $N = 50$ and $N = 82$. Because of the requirement of β equilibrium and electric charge neutrality, the proton number Z is more tightly constrained and for this reason, only a few crustal layers are made of nuclei with proton magic number $Z = 28$, and the only doubly magic nucleus predicted by various nuclear mass models to be present in the crust is ^{78}Ni (Xu et al. 2014; Hagen et al. 2016).

Progress in experimental techniques over the past decades have allowed to probe deeper the interior of a neutron star crust, up to a density $\sim 6 \times 10^{10}$ g cm^{-3}. For comparison, the most exotic experimentally measured nuclide predicted to exist in the crust of a neutron star in 1971 was ^{84}Se ($Z/A \simeq 0.405$) at densities up to about 8.2×10^9 g cm^{-3} (Baym et al. 1971). More recently, the presence of the neutron-rich zinc isotope ^{82}Zn ($Z/A \simeq 0.366$) that was predicted by some nuclear mass models has been ruled out by experiments at the ISOLDE-CERN facility (Wolf et al. 2013; Kreim et al. 2013). In contrast to the modeling of gravitational core-collapse supernova explosions that require the knowledge of the properties of a very large ensemble of nuclei, significant advances in the understanding of the crust of a nonaccreted neutron star could thus be achieved in the near future by measurements of a few exotic nuclei. Experiments could also provide crucial information on the evolution of the nuclear-shell structure towards the neutron-drip line (see, e.g. Steppenbeck et al. 2013 and references therein). As discussed in Sect. 7.3.2, the nuclei (A, Z) that could possibly exist in the outer crust of a neutron star beneath the layer of ^{80}Zn must satisfy the *experimental* constraints

$$Z/A < 0.375, \qquad M'(A, Z)/A > 930.848 \, \text{MeV}/c^2 \,, \qquad (7.15)$$

where we have made use of the latest mass data from Wang et al. (2017). These conditions apply to *all* regions of the outer crust below that containing ^{80}Zn. The conditions (7.15) rule out the doubly magic nuclei ^{48}Ca, ^{48}Ni, and ^{56}Ni since $Z/A \simeq 0.417, 0.583$ and 0.5 respectively. On the other hand, the doubly magic nucleus ^{78}Ni ($Z/A \simeq 0.359$) is not necessarily excluded.

Under the assumption of cold catalysed matter, the neutron drip transition is determined considering all possible electron capture and neutron emission processes with all possible values of ΔZ and ΔN. The lowest threshold pressure is reached for $\Delta Z = Z$ and $\Delta N = A$ (see e.g. Chamel et al. 2015a). In this case, Eq. (7.14) reduces to $M'(A, Z)/A < m_n$: this condition is satisfied by any nucleus and therefore does not provide any additional constraint. Using microscopic nuclear mass models, the neutron-drip density and pressure are predicted to lie in the range $\rho_{\text{drip}} \sim 4.2\text{--}4.5 \times 10^{11}$ g cm^{-3}, and $P_{\text{drip}} \sim 7.7\text{--}8.0 \times 10^{29}$ dyn cm^{-2} respectively (Chamel et al. 2015a; Fantina et al. 2016). As discussed in Sect. 7.3.2, the equilibrium nucleus found by minimising the Gibbs free energy per nucleon must be stable against neutron emission. Still, it may lie beyond the "neutron-drip line" in the chart of nuclides. Let us recall that this line is generally defined at each value of the proton number Z by the lightest isotope for which the neutron separation energy S_n, defined by $S_n(A, Z) \equiv M'(A-1, Z)c^2 - M'(A, Z)c^2 + m_n c^2$ is negative, i.e., the lightest isotope that is unstable with respect to the emission of one neutron. Because of pairing and shell effects, many nuclei beyond the neutron-rich side of the neutron drip line are actually stable. The outer crust of a neutron star may actually contain ultradrip nuclei. For instance, the HFB-21 mass model predicts the presence of ^{124}Sr with $S_n = 0.83\,$MeV; the isotope at the neutron-drip line is ^{121}Sr with $S_n = -0.33\,$MeV (Pearson et al. 2011).

7.4.2 Highly Magnetised Neutron Stars

According to the magnetar theory, neutron stars are born with very high magnetic fields of order $B \sim 10^{16}$–$10^{17}\,$G. The presence of such high magnetic fields may alter the formation of the crust.

As discussed in Sect. 7.2.3, the motion of electrons is quantised into Landau orbitals in the presence of a high magnetic field. If the magnetic field strength exceeds the value

$$B_{\mathrm{rel}} = \frac{m_e^2 c^3}{e\hbar} \simeq 4.41 \times 10^{13}\,\mathrm{G}\,, \qquad (7.16)$$

the orbital size a_m is lower than the electron Compton wavelength λ_e, so that the electron motion is relativistic. This situation is encountered in SGRs, AXPs, as well as in some radio pulsars. The energy levels of a relativistic electron gas in a magnetic field were first calculated by Rabi (1928). The quantising effects of the magnetic field on the properties of the outer crust of a neutron star are most important when only the first Rabi level is occupied. In such case, the magnetic field is usually referred to as strongly quantising. This situation arises when $\rho < \rho_B$ and $T < T_B$ with

$$\rho_B = \frac{A}{Z} m \frac{B_\star^{3/2}}{\sqrt{2}\pi^2 \lambda_e^3} \simeq 2.07 \times 10^6 \frac{A}{Z} B_\star^{3/2}\,\mathrm{g\,cm^{-3}}\,, \qquad (7.17)$$

$$T_B = \frac{m_e c^2}{k_{\mathrm{B}}} B_\star \simeq 5.93 \times 10^9 B_\star\,\mathrm{K}\,, \qquad (7.18)$$

where $B_\star \equiv B/B_{\mathrm{rel}}$ and $m = M'(A, Z)/A$ is the mean mass per nucleon.

For the "low" magnetic fields $B_\star \lesssim 1$ prevailing in most neutron stars, the internal constitution of their outer crust is essentially the same as in the absence of magnetic fields except possibly near the stellar surface, as discussed in Sect. 7.2.3. For higher magnetic fields $B_\star \gg 1$ as measured in SGRs, AXPs, and some radio pulsars, the composition is found to depend on the magnetic field (Lai and Shapiro 1991; Chamel et al. 2012; Nandi and Bandyopadhyay 2011; Chamel et al. 2013b;

Nandi and Bandyopadhyay 2013). In particular, the maximum baryon number density $\bar{n}^{\mathrm{max}}$ up to which a nuclide (A_1, Z_1) is present exhibit typical quantum oscillations as a function of the magnetic field strength (Chamel et al. 2017a, 2015c, 2016). Beyond some magnetic field strength $B_\star^{1\to2} \approx 1/2(\mu_e^{1\to2}/(m_ec^2))^2$, where $\mu_e^{1\to2}$ is defined by Eq. (7.6), electrons are confined to the lowest Rabi level thus leading to an essentially linear increase of $\bar{n}^{\mathrm{max}}$ with $B_\star$ (small nonlinearities are introduced by the electrostatic interactions). Since the electron Fermi energy increases with density, the magnetic field is strongly quantising in any layer of the outer crust if $B_\star > B_\star^{\mathrm{drip}} \approx 1/2(\mu_e^{\mathrm{drip}}/(m_ec^2))^2$, where μ_e^{drip} is the electron Fermi energy at the neutron-drip transition defined by Eq. (7.12). Typical values for $B_\star^{\mathrm{drip}}$ are around 1300 (Chamel et al. 2012, 2017a, 2015c, 2016). Depending on the value of the magnetic field strength, some nuclides may disappear and others appear. For example, the nickel isotopes ^{66}Ni and ^{64}Ni are no longer present in the crust for $B_\star > 67$ and $B_\star > 1668$ respectively, whereas ^{88}Sr and ^{132}Sn appear at $B_\star = 859$ and $B_\star = 1989$ respectively (see, e.g. Chamel et al. (2017a) for the detailed composition). All in all, the crust of a neutron star becomes less neutron-rich in the presence of a high magnetic field. Moreover, the neutron-drip transition is shifted to either higher or lower densities depending on the magnetic field strength. The lowest density is reached for $B_\star = B_\star^{\mathrm{drip}}$ and is given by $\rho_{\mathrm{drip}}^{\mathrm{min}} \approx (3/4)\rho_{\mathrm{drip}}(B_\star = 0)$ (Chamel et al. 2015c). In the strongly quantising regime, the neutron-drip density increases almost linearly with $B_\star$. In all these calculations, the same nuclear masses as in the absence of magnetic fields were employed. However, high enough magnetic fields can also influence the structure of nuclei (Peña Arteaga et al. 2011; Stein et al. 2016a), inducing additional changes in the crustal composition (Basilico et al. 2015). However, complete nuclear mass tables corrected for the presence of a magnetic field are not yet available.

7.4.3   Accreted Neutron-Star Crusts

The composition of the outer crust of an accreting neutron star in a low-mass X-ray binary can be very different from that of a nonaccreting neutron star depending on the duration of accretion. In particular, the original outer crust, containing a mass $\sim 10^{-5}\,M_\odot$ (with $M_\odot$ the mass of the Sun), is pushed down by the accreted material and is molten into the liquid core in about 10^4 years assuming an accretion rate of $10^{-9}\,M_\odot$ per year.

The constitution of the outer crust of an accreted neutron star depends on the composition of the X-ray burst ashes, produced in the outermost region of the star at densities $\rho \lesssim 10^7\,\mathrm{g\,cm}^{-3}$. As this material sinks into deeper layers due to accretion, it may undergo electroweak and nuclear reactions. The evolution of a matter element was followed in Gupta et al. (2007, 2008), Schatz et al. (2014), Lau et al. (2018) considering a reaction network of many nuclei. Such calculations are computationally very expensive, and require the knowledge of a very large number of nuclear inputs (e.g. nuclear masses, reaction rates), some of which have not been experimentally measured and must therefore be estimated using nuclear models. For this reason, the composition of accreted crusts remains more uncertain than that of catalysed crusts despite recent progress in nuclear mass measurements (Estradé et al. 2011; Meisel et al. 2015, 2016). A numerically more tractable approach consists of assuming that the rate of an energetically allowed reaction is much faster than the accretion rate. At densities $\rho > 10^8 \, \mathrm{g \, cm^{-3}}$, matter is strongly degenerate, and is "relatively cold" ($T \lesssim 10^8$ K), so that thermonuclear processes are strongly suppressed. Under these conditions, the most important reactions are single electron captures. Multiple electron captures are very rare and can thus be ignored. For instance, the double electron capture by ^{56}Fe occurs on a timescale of about 10^{20} years (Blaes et al. 1990). The first electron capture by a nucleus (A, Z)

$$(A, Z) + e^- \longrightarrow (A, Z - 1) + \nu_e \,, \tag{7.19}$$

proceeds in *quasi-equilibrium*: this reaction occurs as soon as the electron Fermi energy μ_e exceeds some threshold value μ_e^β, as determined by Eqs. (7.11) and (7.12) with $\Delta Z = 1$ and $\Delta N = 0$ (in the case of a transition to an excited state of the daughter nucleus with energy E_{ex}, the mass $M'(A, Z - 1)$ must be replaced by $M'(A, Z - 1) + E_{\mathrm{ex}}/c^2$). In other words, this reaction is allowed if the Gibbs free energy per nucleon is lowered. The daughter nucleus is generally highly unstable, and captures a second electron *off-equilibrium* with an energy release Q:

$$(A, Z - 1) + e^- \longrightarrow (A, Z - 2) + \nu_e + Q \,. \tag{7.20}$$

With these assumptions, the final composition of accreted neutron-star crusts is independent of the details of the reaction rates. It is only determined by the initial composition of X-ray bursts ashes and by nuclear masses (as well as the energies E_{ex} for transitions to excited states). Such calculations have been carried out in Steiner (2012) using a liquid droplet model with empirical nuclear-shell corrections. A further simplification consists of approximating the distribution of nuclides by a single nucleus (Haensel and Zdunik 1990, 2003, 2008). This computationally very fast treatment was shown to provide a fairly accurate estimate for the total heat released as compared to reaction network calculations (Haensel and Zdunik 2008). The composition of the outer crust of accreted neutron stars have been recently determined using microscopic nuclear mass models (Fantina et al. 2018).

As the X-ray burst ashes sink into the crust, their proton number decreases due to electron captures whereas their mass number remains unchanged. At some point, the daughter nuclei will be so neutron rich that free neutrons will be emitted. This transition, which marks the boundary between the outer and inner regions of the crust, will occur when the threshold electron chemical potential μ_e^{drip} for neutron emission (i.e. $\Delta N > 0$ and $\Delta Z = 1$) will become lower than the threshold electron chemical potential μ_e^β for electron capture alone. This condition can be equivalently expressed as (Chamel et al. 2015a)

$$M'(A - \Delta N, Z - 1) - M'(A, Z - 1) + \Delta N m_n < 0 \,. \tag{7.21}$$

Depending on the composition of ashes, and employing microscopic nuclear mass models, the neutron-drip density and pressure are expected to lie in the range $\rho_{drip} \sim 2.6 - 6.5 \times 10^{11}\,\mathrm{g\,cm^{-3}}$, and $P_{drip} \sim 4.4 - 13 \times 10^{29}\,\mathrm{dyn\,cm^{-2}}$ respectively (Fantina et al. 2016).

7.5 Inner Crust

The neutron-saturated clusters constituting the inner crust of a neutron star owe their existence to the presence of a highly degenerate surrounding neutron liquid: neutron emission processes, which would lead to the immediate disintegration of these clusters in vacuum, are energetically forbidden in a neutron star due to the Pauli exclusion principle since neutron continuum states are already occupied. As a newly formed neutron star cools down, free neutrons in the inner crust are expected to become superfluid by forming Cooper pairs analogously to electrons in conventional superconductors[7] (see, e.g. Chamel and Haensel 2008; Sedrakian and Clark 2006; Margueron et al. 2012; Page et al. 2014; Graber et al. 2017; Chamel 2017b for recent reviews; see also Sedrakian and Haskell in this volume).

For all these reasons, the inner crust of a neutron star is a unique environment, whose extreme conditions imposed by the huge gravitational field of the star cannot be reproduced in terrestrial laboratories. The description of the inner crust of a neutron star must therefore rely on theoretical models. Because clusters are inseparable from their neutron environment, both should be consistently treated. Despite considerable progress in the development of many-body methods for the description of light and medium mass nuclei on the one hand, and infinite homogeneous neutron matter on the other hand (see, e.g. Hagen et al. 2016; Dickhoff and Barbieri 2004; Baldo and Biurgio 2012; Carlson et al. 2015; Hagen et al. 2016; Holt et al. 2016; Hergert et al. 2016; Oertel et al. 2017), ab initio calculations of the inner crust of a neutron star still remain out of reach. Various phenomenological approaches have thus been followed (see, e.g. Haensel et al. 2007; Chamel and Haensel 2008 for comprehensive reviews), and will be briefly discussed below.

7.5.1 Nonaccreted Neutron-Star Crusts

State-of-the-art calculations pioneered by Negele and Vautherin (1973) rely on the self-consistent nuclear energy density functional theory, traditionally formulated in

---

[7]The high temperatures $\sim 10^7$ K prevailing in neutron stars prevent the formation of electron pairs recalling that the highest critical temperatures of terrestrial superconductors do not exceed ~ 200 K (Drozdov et al. 2015). See also Ginzburg (1969).

terms of effective nucleon-nucleon interactions in the "mean-field approximation" (see, e.g. Duguet (2014) for a recent review). It should be stressed that in principle, this theory could predict the *exact* ground state of any nuclear system. In practice, however, this theory remains phenomenological because the exact nuclear energy density functional is not known. Still, the accuracy of any given functional can be tested against experimental nuclear data, as well as results from microscopic calculations following different many-body approaches. The most accurate existing functionals are able to fit essentially all measured nuclear masses and charge radii with a root-mean square deviation of about $0.5–0.6 \, \mathrm{MeV}/c^2$ and $0.03 \, \mathrm{fm}$ respectively, while reproducing at the same time properties of homogeneous infinite nuclear matter (equation of state, pairing gaps, effective masses) obtained from ab initio calculations (Pearson et al. 2013; Goriely et al. 2016). In the nuclear energy density functional theory, the ground-state energy is obtained by solving the self-consistent Hartree-Fock-Bogoliubov (HFB) equations describing independent quasiparticles in an average field induced by the underlying particles (see, e.g. Chamel et al. 2013a). In the absence of nuclear pairing, the HFB equations reduce to the Hartree-Fock (HF) equations. As recently shown in Pastore et al. (2017), the HFB energy can be estimated with an error of a few keV per nucleon by applying the decoupling approximation, according to which the nuclear pairing phenomenon is described by the Bardeen-Cooper-Schrieffer (BCS) equations (Bardeen et al. 1957). A relativistic version of the nuclear energy density functional theory has been developed, whereby nucleons interact with various effective meson fields (see, e.g., Bender et al. (2003) for a discussion of both relativistic and nonrelativistic approaches). This theory has been usually referred to as relativistic "mean-field" (RMF) model, although it can account for many-body correlations beyond the mean-field approximation as its nonrelativistic version. In this approach, causality is guaranteed at high density. It has thus been widely employed in studies of dense-matter in the core of neutron stars (Glendenning 2000).

Nuclear-energy density functional calculations of neutron-star crusts (see, e.g. Margueron et al. 2012) have been traditionally performed using an approximation originally introduced by Eugene Wigner and Frederick Seitz in 1933 in solid-state physics (Wigner and Seitz 1933), whereby clusters with their surrounding neutrons are confined inside spherical cells independent from each other. The Wigner-Seitz approximation allows for relatively fast numerical computations. The crustal composition obtained in this way was found to be very sensitive to nuclear pairing effects (Baldo et al. 2007; Grill et al. 2011), but also to the choice of boundary conditions (Baldo et al. 2007, 2006). This stems from the artificial quantisation of unbound neutron states (Chamel et al. 2007; Margueron et al. 2008; Pastore et al. 2016). Over the past years, a few three-dimensional calculations of the ground state of cold dense matter have been undertaken in cubic cells with strictly periodic boundary conditions (Magierski and Heenen 2002; Gögelein and Müther 2007). However, the reliability of such calculations is still limited by the appearance of spurious neutron shell effects (see, e.g. Newton and Stone 2009; Fattoyev et al. 2017) that can only be completely alleviated either by choosing a large enough cell (much larger than the screening length) (Sébille et al. 2009, 2011) or by considering a single Wigner-cell of the crystal lattice (a truncated octahedron in case of a body-centred cubic lattice) with Bloch boundary conditions (Chamel 2012; Schuetrumpf and Nazarewicz 2015). In the former case, the results could also be biased by the arbitrary choice of the shape of the "supercell" (see, e.g., Giménez Molinelli and Dorso 2015). The latter approach is computationally more tractable due to the reduced size of the cell and the smaller number of particles. On the other hand, it requires the prior knowledge of the crystal structure. This, however, does not appear as a serious limitation, except for describing the densest part of the crust (see Sect. 7.6). Indeed, clusters are generally assumed to form a body-centred cubic lattice (in some layers, however, a face-centred cubic lattice might be energetically favoured, as found, e.g. in Okamoto et al. (2013)). Although the presence of the neutron liquid leads to an induced interaction between clusters that could potentially affect the structure of the inner crust (Kobyakov and Pethick 2014), refined estimates of the induced interaction have dismissed this possibility (Kobyakov and Pethick 2016). Whether the crust is described by a supercell with periodic boundary conditions or by a single Wigner-Seitz cell with Bloch boundary conditions, nuclear-energy density functional calculations are computationally extremely costly.

For this reason, many studies of the inner crust of a neutron star rely on compressible liquid drop models (see, e.g. Douchin and Haensel 2001; Newton et al. 2013; Deibel et al. 2014; Gulminelli and Raduta 2015; Bao and Shen 2016; Fortin et al. 2016; Tews 2017; Lim and Holt 2017 for recent calculations). In this approach, the nuclear clusters and the neutron liquid are considered as two distinct homogeneous phases. The bulk and surface properties of these two phases can in principle be determined using the nuclear energy density functional theory. However, empirical parametrisations of the surface properties are often employed. Despite its simplicity, the liquid drop model has shed light on various aspects of dense matter. In particular, the formation of neutron-proton clusters was shown to arise from a detailed balance between Coulomb and surface effects (Baym et al. 1971), leading to the prediction of complex configurations commonly referred to as "pastas" in the densest layers of the crust (see Sect. 7.6). Moreover, the liquid drop model has been employed to study the role of nuclear parameters, such as the symmetry energy, on the crustal composition and the crust-core boundary (see, e.g., Newton et al. 2013; Bao and Shen 2016).

A more realistic treatment of dense matter in neutron-star crusts is to employ semi-classical methods such as the Thomas-Fermi (TF) approximation (see, e.g. Sharma et al. 2015; Gögelein and Müther 2007; Okamoto et al. 2013; Fortin et al. 2016; Lim and Holt 2017; Oyamatsu and Iida 2007; Avancini et al. 2008; Miyatsu et al. 2013; Grill et al. 2014; Iida and Oyamatsu 2014), or its higher-order extensions (Onsi et al. 1997; Goriely et al. 2008; Martin and Urban 2015). The neutron liquid and the clusters are not treated separately, but are described in terms of continuous neutron and proton distributions. This method is based on a systematic expansion of the smooth part of the single-particle quantum density of states in powers of $\hbar$ (see, e.g., Brack et al. (1985) for a review; see Centelles et al. (1993) for the relativistic version). The accuracy of this development has been recently studied in Papakonstantinou et al. (2013), Aymard et al. (2014). The shell correction to the total energy of a Wigner-Seitz cell (responsible for the oscillatory part of the single-particle quantum density of states) can be added perturbatively via the Strutinsky integral (Dutta et al. 2004; Onsi et al. 2008; Pearson et al. 2012, 2015, 2018). This so called ETFSI method is not only a computationally high-speed approximation to the full HF+BCS equations, but it also avoids the pitfalls discussed above that plague the current numerical implementations of the nuclear energy density functional theory. These calculations show that proton-shell effects still play an important role in determining the ground-state composition of the inner crust, with clusters containing predominantly $Z = 40$ protons (Pearson et al. 2012). The appearance of this magic number can be understood from the fact that the strength of the spin-orbit coupling tends to decrease with the neutron excess, as measured in ordinary nuclei (Schiffer et al. 2004). However, the importance of shell effects is mitigated by pairing (Pearson et al. 2015). Depending on the symmetry energy, clusters with $Z = 20, 50, 58$ or 92 can also be favored (Pearson et al. 2018). In the densest region of the crust, configurations with different cluster sizes may differ by a few keV per nucleon at most, suggesting that this region could be very heterogeneous. Further studies beyond the mean-nucleus approximation should thus be pursued. At high enough density, protons may drip out of clusters (see, e.g. Pearson et al. 2018; Pethick et al. 1995), and are expected to become superconducting at low enough temperatures.

With further compression, the crust dissolves into an homogeneous mixture of nucleons and electrons. The crust-core transition appears to be essentially continuous from the thermodynamic point of view, and occurs at a density between about one third and one half of the nuclear saturation density depending on the symmetry energy and on the nuclear model employed (see, e.g. Newton et al. 2013; Bao and Shen 2016; Ducoin et al. 2011; Iida and Oyamatsu 2014; Sulaksono et al. 2014; Seif and Basu 2014; Gonzalez-Boquera et al. 2017; Fang et al. 2017 for recent studies).

7.5.2 Highly Magnetised Neutron Stars

As discussed in Sects. 7.2.3 and 7.4.2, the presence of a high magnetic field may have a profound influence on the composition and on the properties of dense matter. However, few studies have been devoted so far to the role of a high magnetic field on the inner crust of a neutron star. Rabi quantisation of electron motions was studied in Nandi and Bandyopadhyay (2011, 2013), Nandi et al. (2011) within the TF approximation. It was found that the inner crust of highly magnetised neutron stars is more proton rich, similarly to the results obtained in the outer crust (see Sect. 7.4.2). As a consequence, magnetised crusts contain larger clusters separated by a smaller distance, whereas the neutron liquid is more dilute. However, these effects were found to be negligible for magnetic field strengths below $\sim 10^{16}$–10^{17} G. In addition to the modification of the electron motion, the effects of the magnetic field on nucleons was studied in de Lima et al. (2013) within the TF approximation. The transition density between different cluster types was found to exhibit typical oscillations as a function of the magnetic field strength, as also found in the outer crust for the density at the interface between different layers. Studies of spatial instabilities in highly magnetised homogeneous stellar matter within the RMF approach and taking into account the anomalous magnetic moment of nucleons suggest that the densest part of the crust could consist of alternating layers of homogeneous and inhomogeneous regions (Fang et al. 2016, 2017; Chen 2017). However, quantisation effects of the magnetic field on the crust-core transition are washed out at temperatures of order 10^9 K (Fang et al. 2017). The spin polarisation of matter adds to the complexity of these phases (see, e.g. Rabhi et al. (2015) and references therein). Moreover, the formation of Cooper pairs in the spin-singlet channel thus becomes disfavoured in the presence of a high magnetic field. Calculations in pure neutron matter suggest that neutron superfluidity in the crust of a neutron star is destroyed if the magnetic field strength exceeds 10^{17} G (Stein et al. 2016b). However, calculations of inhomogeneous neutron superfluidity in magnetised crusts are still lacking. This warrants further studies.

7.5.3 Accreted Neutron-Star Crusts

Comparatively very few studies have been devoted to the inner regions of accreted neutron-star crusts. This stems from the fact that the rates of electroweak and strong nuclear reactions that could possibly occur in this environment where neutron-proton clusters coexist with free neutrons are poorly known. Moreover, a very large number of different nuclear species are expected to exist in any given layer of the accreted crust. Because these clusters owe their existence to the surrounding neutrons, their properties cannot be experimentally measured. A consistent treatment of a distribution of clusters with free neutrons in the framework of the nuclear energy-density functional theory is computationally extremely challenging as this would require solving the HFB equations in considerably larger cells than in nonaccreted crusts.

For all these reasons, the composition of the inner crust of accreted neutron stars has been mainly studied so far using compressible liquid-drop models, in the single-nucleus approximation (see, Haensel and Zdunik (1990, 2003, 2008) and references therein) and taking into account the distribution of different clusters (Steiner 2012). The composition was determined by following the sequence of electron captures, neutron emissions, and pycnonuclear reactions that occur as the elements present at the bottom of the outer crust sink into deeper layers. These calculations indicate that the accretion of matter onto a neutron star can radically change the internal constitution of the inner crust. Very recently, the role of nuclear shell effects on the properties of accreted crusts has been studied in the framework of the nuclear energy density functional theory using the ETFSI method (Fantina et al. 2018).

7.6 Nuclear Pasta Mantle

In the outer crust of a neutron star, nucleons are bound inside nuclei, whose size (typically 5–6 fm) is negligibly small compared to the ion spacing (a_N varies from about 10^5 fm at the surface to 10^2 fm at the neutron-drip point). Consequently, the composition of the outer crust (determined by nuclear physics) and its structure (determined by the physics of dense Coulomb plasmas) can be studied separately. Whereas the spatial arrangement of nuclei in the outer crust is governed by the long-range Coulomb interactions, the shape and size of an individual nucleus are controlled by the short-range nuclear and Coulomb interactions. Nuclei expected to be present in the outer crust of a neutron star are almost spherical. Considering axially symmetric nuclei, the nuclear surface can be parametrised by the radius $R_N(\theta) = R_0(1 + \beta_2 Y_2^0(\theta) + \beta_4 Y_4^0(\theta))$, where θ is the polar angle, Y_ℓ^m are spherical harmonics, β_2 and β_4 are dimensionless parameters characterising quadrupole and hexadecapole deformations. Positive (respectively negative) values for β_2 lead to prolate (respectively oblate) spheroid nuclei. Typical orders of magnitude for these latter parameters are $|\beta_2| \sim 10^{-1}$ and $|\beta_4| \sim 10^{-2}$. In most regions of the inner crust, nuclear clusters remain sufficiently far apart that their structure is not influenced by the presence of neighbouring clusters. Therefore, clusters are expected to be quasi spherical as in the outer crust. However, because the average distance between clusters decreases with increasing density, clusters eventually fuse and connect into herringbone structures (Watanabe et al. 2009) similarly to percolating networks as speculated earlier by Ogasawara & Sato (Ogasawara and Sato 1982).

Matter in the densest layers of the crust is frustrated and various exotic configurations collectively referred to as "nuclear pastas" might appear, as first shown in Ravenhall et al. (1983), Hashimoto et al. (1984), Oyamatsu et al. (1984) (see, e.g., Chamel and Haensel 2008; Watanabe and Maruyama 2012 for a review). Frustrated nuclear systems are also encountered in heavy-ion collisions although under different physical conditions (see, e.g. Botvina and Mishustin 2005). If they exist, nuclear pastas would behave like liquid crystals and would thus form a liquid mantle below the crust (Pethick and Potekhin 1998). Nuclear pastas have been studied following the same treatments as described in Sect. 7.5 for the inner crust: compressible liquid drop models (see, e.g. Newton et al. 2013; Lim and Holt 2017; Nakazato et al. 2011; Gupta and Arumugam 2013; Viñas et al. 2017 for recent calculations), semi-classical methods (see, e.g. Sharma et al. 2015; Gögelein and Müther 2007; Okamoto et al. 2013; Fortin et al. 2016; Lim and Holt 2017; Oyamatsu and Iida 2007; Avancini et al. 2008; Grill et al. 2014; Iida and Oyamatsu 2014; Martin and Urban 2015; Viñas et al. 2017, and the nuclear energy density functional theory (see, e.g. Magierski and Heenen 2002; Gögelein and Müther 2007; Newton and Stone 2009; Fattoyev et al. 2017; Sébille et al. 2009, 2011; Schuetrumpf and Nazarewicz 2015; Sagert et al. 2016; Pais and Stone 2012). The existence of nuclear pastas have been also investigated using the quark-meson coupling model (Grams et al. 2017), in which nucleons in dense matter are described by non-overlapping static spherical "bags" of quarks interacting through the self-consistent exchange of mesons in the mean-field approximation. In all these approaches, a few specific nuclear shapes are generally considered, or some symmetries are assumed.

The formation of nuclear pastas has been explored using molecular dynamics (see, e.g. Dorso et al. 2012 for a review). By treating nucleons as classical pointlike particles interacting through a two-body potential, classical molecular dynamics allow for large-scale simulations with $\sim 10^3$–10^5 particles in a box with periodic boundary conditions (Dorso et al. 2012; Schneider et al. 2013, 2014; Horowitz et al. 2015; Schneider et al. 2016; Berry et al. 2016). Although no other constraints are imposed, the nuclear pasta configurations obtained from these simulations could still be influenced by the geometry of the box (Giménez Molinelli and Dorso 2015) and the treatment of Coulomb interactions (Alcain et al. 2014). Quantum molecular dynamics simulations (see, e.g. Watanabe and Maruyama 2012; Maruyama et al. 2012 for a review) rely on a semi-classical description in which the state of a nucleon is represented by a Gaussian wave packet moving (classically) in a mean field. The antisymmetrisation of the many-body wave function is implemented in an effective way through a phenomenological repulsive potential, so called Pauli potential. Calculations have been performed with $\sim 10^3$ nucleons (Watanabe and Maruyama 2012; Maruyama et al. 2012; Nandi and Schramm 2016). Fermi statistics is properly taken into account in fermionic molecular dynamics, in which the many-body wavefunction is expressed as a Slater determinant. However, such simulations are computationally extremely costly (the computing time scales as N^4 as compared to N^2 for classical and quantum molecular dynamics, where N is the number of particles). For this reason, only a few fermionic molecular dynamics simulations have been carried out so far (Vantournhout et al. 2011; Vantournhout and Feldmeier 2012). Finally, nuclear pastas have been studied using a nuclear analog of the Ising model, in which the nucleon positions are discretized in a three-dimensional cubic lattice, and assuming that each lattice site can be occupied by only one nucleon at most (Hasnaoui and Piekarewicz 2013).

The presence of nuclear pastas in neutron stars remains uncertain and model-dependent. Still, all models supporting their existence predict the following sequence of configurations with increasing density: polpette/gnocchi (spherical clusters), spaghetti (cylindrical clusters), lasagne (slabs), bucatini/penne/maccheroni (cylindrical holes), Swiss cheese (spherical holes). Intermediate phases have been found by some models such as "nuclear waffles" (slabs with holes) (Sébille et al. 2011; Schneider et al. 2014; Williams and Koonin 1985), cross-rods (Pais and Stone 2012; Schneider et al. 2014; Lassaut et al. 1987), or slabs connected by helical ramps (Berry et al. 2016). Because the nuclear pasta mantle is very dense, it may represent a sizable fraction of the crustal mass (Lorenz et al. 1993), and thus may have important astrophysical consequences (see, e.g., Chamel and Haensel (2008)).

As recently discussed in Fortin et al. (2016) (see also the contribution from Fantina and Burgio in this volume), a thermodynamical inconsistent treatment of the crust-core boundary can lead to significant errors on the predicted masses and radii of neutron stars. Moreover, thermodynamical inconsistencies could trigger spurious hydrodynamical instabilities. A unified description based on the same model is therefore of utmost importance. We shall come back to this issue when addressing the quark-hadron phase transition in the core.

7.7 Nuclear and Hypernuclear Matter

A recent review about the equation of state (EoS) for supernovae, neutron stars and their mergers has been given in Oertel et al. (2017), where also the very different theoretical approaches to the nuclear matter problem are shortly reviewed (see also Fantina and Burgio in this volume). Very roughly one can distinguish two classes of systematic approaches to nuclear and hypernuclear matter: the ones based on (realistic) baryon-baryon interactions that can describe also the phase shift data as well as properties of nuclear and hypernuclear clusters on the one hand and the approaches based on relativistic density functionals with elaborated (density-dependent) couplings to mesonic mean fields that are fitted to describe observable properties of large nuclei and predict the behaviour of the EoS at saturation densities and beyond. While the nonrelativistic approaches based on potentials (Brueckner-Hartree-Fock, Brueckner-Bethe-Goldstone, Variational approach) require three-body forces in order to obtain acceptable ground state properties of nuclear matter and the $2M_\odot$ constraint, the relativistic Dirac-Brueckner-Hartree-Fock (DBHF) approach fulfills constraints from compact star physics and heavy-ion collisions without them (Klähn et al. 2006). A modern constraint for theories of nuclear matter is provided by the ab-initio chiral effective field theory (χEFT) approach that provides a link between QCD (in its low-energy limit as chiral perturbation theory) and nuclear matter properties with nucleons, pions and a nucleon-nucleon contact interaction as the elements of a systematic perturbation theory. The χEFT shall be considered as a benchmark for the pure neutron matter case (no bound states) at subnuclear densities where errors due to unaccounted higher orders are sufficiently small. See Fig. 7.2 for a comparison of approaches.

We would like to note that the density-dependent relativistic mean-field theory "DD2" lies perfectly within the bounds provided by the grey band of the χEFT results. Unfortunately, this band broadens quickly with density so that at supersaturation densities it does no longer provide reliable constraints.

In order to explore the nuclear EoS beyond the saturation density of $n_0 = 0.15\,\text{fm}^{-3}$, it has therefore been suggested (Hebeler et al. 2013; Read et al. 2009; Raithel et al. 2016) to consider sequences of polytropic EoS

$$P(n) = \kappa_i (n/n_0)^{\Gamma_i}, \; n_i < n < n_{i+1}, \; i = 1 \ldots N, \tag{7.22}$$

and by varying the polytropic indices Γ_i (and eventually also the switch densities n_i) to obtain a most generic class of high-density EoS from which the one(s) can be chosen that perform best with respect to constraints from compact star observations. Such a multi-polytropic ansatz is also capable of describing a hadron-to-quark matter phase transition, with a constraint pressure region (as in the Maxwell

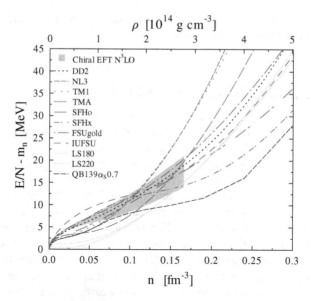

Fig. 7.2 Energy per baryon in pure neutron matter for different supernova EoS, compared to results of χEFT (grey band (Krüger et al. 2013)), from Fischer et al. (2014)

construction case) or with changing pressure as for the case of an extended mixed (pasta) phase in the compact star (Alvarez-Castillo and Blaschke 2017; Alvarez-Castillo et al. 2017; Zdunik et al. 2006). We shall come back to this in Sect. 7.10 below.

7.7.1 Hyperons and Hyperon Puzzle

At densities of two to three times saturation density the hyperon threshold is expected to be crossed. It has been shown within Brueckner-Bethe-Goldstone calculations (Baldo et al. 2000) that under conservative assumptions for the forces involving hyperons the appearance of these additional degrees of freedom softens the EoS and hence lowers the maximum mass of a neutron star so that even the well-constrained binary radio pulsar masses of typically $\sim 1.4\,M_\odot$ cannot be reached. Without additional repulsive hyperon interactions this is exactly what all neutron star calculations based on nuclear hyperon model EoSs predict. This is called the "hyperon puzzle" (Zdunik and Haensel 2013). In general, the EoS beyond saturation density is not well constrained (Oertel et al. 2017). For hyperons and their interaction with nucleons and themselves the situation is even worse. However, from the observed existence of hyperons and massive neutron stars it seems evident that nucleon-hyperon and likely hyperon-hyperon repulsion (Rijken and Schulze 2016) plays an important role (unless hyperons themselves play no decisive role for neutron star structure). Meanwhile, different approaches have been successfully applied which allow for stable neutron stars with up to two solar masses. Repulsion stiffens the hyperon matter EoS sufficiently to account for massive neutron stars. At the same time the stiffening results in an onset of hyperon degrees of freedom at higher densities. Thus, the fraction of hyperons is reduced. In another scenario this idea is taken to the limit where the hyperon onset density is larger than the densities one would find in a neutron star. Generally, in RMF models for hypernuclear matter (Hofmann et al. 2001) there is no severe hyperon puzzle, because the unknown density dependent scaling of the meson masses and couplings of the can be defined in the spirit of a relativistic density functional theory such that all known constraints on the stiffness of the EoS from heavy-ion collisions and compact star observations can be fulfilled (Maslov et al. 2015, 2016). A particular role plays the sufficiently repulsive effect of the ϕ-meson mean field (Weissenborn et al. 2012). This solution of the hyperon puzzle holds also for hypermatter stars with (color superconducting) quark matter cores (Bonanno and Sedrakian 2012; Lastowiecki et al. 2012). A third solution to the hyperon problem is a transition to sufficiently stiff quark matter (Baldo et al. 2003) which can happen before or after the transition to hyperon matter.

7.7.2 Δ Isobars and Δ Puzzle

While a large amount of literature is devoted to the discussion of hyperons and the hyperon puzzle, only little work has been done to discuss the presence of $\Delta(1232)$ isobars which, once they appear in neutron star matter, have a similar softening effect on the EoS as the hyperons. That Δ isobars have so far been largely neglected may be due to the outcome of the seminal paper (Glendenning 1985) which concluded that they may appear only at too high densities to be relevant for the structure of compact star interiors. In Drago et al. (2014a,b) it has been shown that the threshold density strictly correlates with the L parameter, the derivative of the symmetry energy in nuclear matter, which according to the compilation of constraints in Lattimer and Lim (2013) lies in the range of $40.5\,\text{MeV} \leq L \leq 61.9\,\text{MeV}$. For the relativistic mean field EoS SFHo (Steiner et al. 2013) in this range the Δ^- isobar occurs even at lower densities than the Λ hyperon in neutron star matter and leads to a further softening of the EoS. Thus the Δ isobars too limit the maximum mass of a hadronic compact star to $\sim 1.5–1.6\,M_\odot$ (Drago et al. 2014b) and entail a Δ puzzle .

A solution of this puzzle may be similar in spirit to those possibilities discussed as solutions of the hyperon puzzle, namely, that there could be stiffening effects in the hadronic EoS at supernuclear densities (due to, e.g., multi-pomeron exchange (Yamamoto et al. 2016) in the baryon-baryon interaction or the density dependence of meson masses and their couplings in RMF models (Kolomeitsev et al. 2017)) or a sufficiently early phase transition to stiff quark matter could occur. As an equilibrium phase transition, the latter possibility would require a rapid stiffening of the quark matter EoS after the transition so that a maximum mass above the

$2\,M_\odot$ constraint could be reached. This would entail the "reconfinement" problem as in the case of the corresponding solution of the hyperon puzzle. In order to circumvent problems with unphysical crossing of EoS an interpolation scheme has been suggested which we discuss more in detail in Sect. 7.9.2 below. This scheme realises a crossover transition between two phases which leads to an intermediate phase that is stiffer than those it is bridging. The mass-radius relation obtained with such a construction applied to several soft hadronic EoS shows a sufficiently large maximum mass and an increase in radius at the transition.

Another solution of the Δ puzzle has been suggested within the scenario of coexistence of *two families* of compact stars (Berezhiani et al. 2003; Drago et al. 2004; Bombaci et al. 2004), where the second family may belong to the class of strange quark matter stars with a sufficiently large maximum mass to fulfill the $2\,M_\odot$ constraint. The hadronic branch of compact star configurations would be metastable against the "decay" to the strange quark matter branch by a nucleation of strange quark matter droplets in the stars interior. The nucleation time scales could be sufficiently long to populate the branches of both families of solutions. The compact stars with smaller maximum mass would then be hadronic stars while the more massive stars with larger radii would be strange quark matter stars. For details, see Drago et al. (2014a) The main caveat of the two-family scenario is that the timescale for the nucleation of a strange quark matter droplet in the metastable hadronic phase by quantum tunneling (Iida and Sato 1998; Bombaci et al. 2007) sensitively depends on the value of the bag constant employed in those models (and also on other parameters, if present). This makes any predictions for the lifetime of the metastable state rather arbitrary. Once a strange quark matter droplet is formed, the conversion of the hadronic star to a strange quark star proceeds on a millisecond timescale as has been demonstrated in simulations of the turbulent combustion process (Herzog and Röpke 2011; Pagliara et al. 2013).

7.7.3   The Case of the d^*(2380) Dibaryon

The recent discovery of the d^*(2380) dibaryon (see (Clement 2017) for a recent review) has triggered an investigation of its possible role in compact star interiors (Vidaña et al. 2018). It has been found that the d^*(2380) would appear at densities around three times nuclear saturation density and comprise about 20% of the matter in the core of massive compact stars and even higher fractions possible in the course of compact star merger events. The formation of the d^*(2380) dibaryon in compact star matter can be understood as the conversion of pairs of neutrons and protons, thus effectively reducing their number density and introducing a rather heavy degree of freedom instead. This leads to a strong softening of the equation of state, an effect like the occurrence of a mixed quark-hadron matter phase. The presence of the dibaryon in the composition of a high-mass neutron star has an effect on the cooling of the star since it adds fast direct Urca type cooling processes, similar to the pair-breaking and pair-formation processes in superfluid neutron star matter.

7.7.4 Chirally Symmetric Nuclear Matter Phase

There is evidence from recent finite-temperature lattice QCD simulations of the
FASTSUM collaboration that the restoration of chiral symmetry takes place already
within the hadronic phase by parity doubling (Aarts et al. 2017), which is then
signalled by a mass degeneracy of the baryonic chiral partner states (the nucleon
and the $N^*(1535)$) in such a way that the lower-mass state is almost independent
of the medium while the higher mass state comes down (as in the well-known case
of pion and the sigma meson described by the NJL model, see Klevansky (1992)
and references therein). A theoretical description can be based on the parity doublet
model (Detar and Kunihiro 1989; Jido et al. 2001a,b) which has been applied to
the phenomenology of hot dense matter, including neutron stars (Dexheimer et al.
2008; Mukherjee et al. 2017) A recent extension of the model to the quark level of
description can address a deconfinement transition (Benic et al. 2015a) and has been
applied to the discussion of compact star phenomenology (Marczenko et al. 2018)
In this work it was shown that the account for the quark substructure of the baryons
entails a stiffening of the asymmetric nuclear matter just above the saturation density
before the actual chiral restoration occurs which results in the doubling of nucleonic
degrees of freedom (parity doubling). This transition proceeds as a rather strong
first-order phase transition that entails an almost horizontal branch at $\sim 2\,M_\odot$ in
the corresponding $M - R$ diagram, still before quark deconfinement. A pattern
like this had been suggested earlier as the signature of a strong deconfinement
transition (Alvarez-Castillo et al. 2016) In view of this possible strong effect of
the combination of quark substructure and parity doubling on the hadronic EoS this
suggestion should be revisited. Another prediction of the parity doubling model is
a new formula for the threshold value of the proton fraction above which the fast
direct URCA cooling process would become operative in compact stars. Thus the
chiral symmetry restoration by parity doubling in the nuclear EoS should be taken
into the consideration of modern compact star cooling theories that are employed
when interpreting cooling data.

7.7.5 Stiffness from Quark Pauli Blocking

We have just discussed the importance of repulsive interactions for solving the
hyperon puzzle. But how generic is the presence of repulsive interactions in hadron-
hadron scattering and in dense hadronic matter and what is their origin? Is it just the
exchange of vector mesons ($\omega, \phi, \varrho, \dots$)? There is also the concept of the excluded
volume, based on the fact that baryons (hadrons in general) are finite size objects
due to their quark substructure, that gives rise to an increase in the pressure of the
system, like a repulsive interaction. For standard excluded volume approaches the
pressure even diverges when the available volume goes to zero (closest packing of
hard-sphere hadrons). How to avoid such a divergence? In simple terms, a hadron

can be seen as an MIT bag of quarks with a radius. When at high densities the bags merge the individuality of hadrons gets lost and a big bag filled with quark matter emerges, described by the thermodynamical bag model. This solution, the phase transition to quark matter, was already suggested for the Hagedorn resonance gas model by Hagedorn and Rafelski (Hagedorn and Rafelski 1980; Rafelski and Hagedorn 1980).

Another solution has been suggested as a reformulation of the concept of the excluded volume so that the available volume is a function of the density which deviates from the system volume as a function of the density so that below a certain density there is no effect, but for high densities it goes to zero asymptotically, for instance as a Gaussian function. For details of the thermodynamically consistent formulation of such an approach, see Typel (2016). This concept can be carried even further by allowing the available volume to be a function not only of the density, but also of the temperature and to exceed the system volume so that attractive interactions can be modeled too. Recently, with this generalization of the concept a QCD phase diagram has been obtained that resembles first order and crossover transitions, separated by a critical endpoint (Typel and Blaschke 2018), as expected for full QCD.

Despite this phenomenological success and its thermodynamic consistency, the excluded volume model does not provide a microphysical understanding for the origin of the repulsion between hadrons and its medium dependence which determines the stiffness of high-density matter. A satisfactory explanation can be given by accounting for the Pauli exclusion principle on the quark level which leads to a repulsive quark exchange interaction between hadrons, very similar to molecular systems where the apparent hard sphere behavior of atoms and molecules originates from the Pauli blocking among electrons in the atomic orbitals. A quantitative estimate for the Pauli blocking effect in nuclear matter has been given on the basis of a simple string model for quark confinement and the baryon spectrum in Röpke et al. (1986a), where it was found that the result very well corresponds to the repulsive part of the density dependent repulsion in the Skyrme-type model by Vautherin and Brink (Vautherin and Brink 1972). The Pauli quenching effect is not only density- but also temperature and momentum dependent so that one can deduce from it a temperature and density dependent effective nucleon mass (Röpke et al. 1986b). For the applications to compact star physics it is very important that the Pauli blocking is flavor dependent. It gives a major contribution to the symmetry energy and allows, in principle, to determine the contribution to the pressure of hypernuclear matter.

Quark Pauli blocking leads to a positive energy shift of baryons, so that the occurrence of hyperons and Δ isobars in dense matter may become inhibited as for $T = 0$ matter the critical densities for the occurrence of these species are shifted to higher values. Pauli blocking is due to an admixture of 6-quark or even higher multiquark wave functions in dense baryonic matter. It may on the other hand provide a mechanism for binding two Δ isobars in the $d^*(2380)$ dibaryon by quark exchange forces since a $\Delta - \Delta$ molecule has overlap with a three-diquark state.

As in the simple-minded picture of merging bags the quark exchange effect between hadrons is a precursor of quark deconfinement! The additional pressure drives the possible phase transition to lower densities, the effect of the antisymmetrisation among quarks of overlapping hadron wave functions generates multiquark components of the many-particle wave function in dense nuclear matter. With such a microphysical background an early calculation of stable hybrid stars with interacting quark matter cores at high-mass could be presented (Blaschke et al. 1990). The effect of partial quark delocalisation leads to the emergence of a quark Fermi sea. Such effects have been described as typical for quarkyonic matter (Kojo et al. 2010a,b). The question arises how partial chiral symmetry restoration in dense nuclear matter could modify the Pauli blocking effect. In a recent exploratory calculation that used the Pauli quenching result of Röpke et al. (1986a), Blaschke et al. (1990) with density-dependent quark masses, a strong enhancement of the effect has been found with results for the equation of state (Blaschke et al. 2018) that correspond to those from the modified excluded volume model by Typel (Typel 2016).

7.8 Quark Matter

Deconfined quark matter shall be described in terms of quarks and gluons, the degrees of freedom of QCD, the gauge field theory of strong interactions. For applications in neutron stars, i.e. at high densities and zero temperature, the ab initio approach to solve QCD numerically by lattice QCD simulations , is not available due to the sign problem. At zero baryon density the lattice QCD simulations have reached a quality where they reproduce at low temperatures the hadron mass spectrum and the hadron resonance gas thermodynamics which can be directly compared with phenomenological approaches. It goes so far that up to temperatures including the pseudocritical temperature of the hadron-to-quark matter crossover transition the thermodynamics of QCD can be described as a hadron resonance gas. The lesson for the $T = 0$ domain of neutron star physics therefore can be that it has to develop effective field theory approaches which are nonperturbative so that they can address the formation of hadronic (baryonic) bound states as well as collective phenomena like dynamical chiral symmetry breaking and color superconductivity. At the same time, the effective models shall try to capture as many aspects of QCD as possible, like the (approximate) chiral symmetry of its Lagrangian in the light quark sector and its dynamical breaking. See Fig. 7.3.

In this section we want to recapitulate a few challenging questions in this context and discuss the progress which has been made in these directions, in particular with the help of the COST Action "NewCompStar".

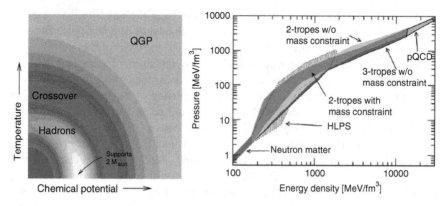

Fig. 7.3 Left panel: Schematic view of the speed of sound in the QCD phase diagram [This figure was kindly provided by M.G. Alford and C.J. Horowitz who created it during the INT-16-2b program on "The Phases of Dense Matter" at the INT Seattle (2016)]. Dark color stands for low c_s, for asymptotic values of temperature and chemical potential $c_s^2 = 1/3$. At low temperatures, in-between nuclear saturation density and the perturbative QCD region, c_s must come close to 1 (speed of light), for the EoS to fulfill the $2\,M_\odot$ constraint on the pulsar mass. Right panel: Multi-polytrope interpolations of the compact star EoS between the well established limits of pure neutron matter (in red) and perturbative QCD (in yellow), from Kurkela et al. (2014). Compact star phenomenology can constrain the EoS for energy densities up to about $2\,\mathrm{GeV/fm}^3$, see Hebeler et al. (2013) (HLPS) and Alvarez-Castillo and Blaschke (2017)

7.8.1 Asymptotically High Densities

Although one can hope that perturbative results will constrain effective models at large densities the perturbative domain does not overlap with the densities one expects in a neutron star (Kurkela et al. 2014). Despite the progress that has been made (Kurkela et al. 2010) in bringing perturbative calculations of the properties of hot and dense QCD closer to the hadronic domain, the QCD phase transition from confined quarks in dense hadronic matter to deconfined quarks in a quark-gluon plasma is characterized by features which are not accessible at any finite order of perturbative QCD. A standard example for this statement is the description of chiral symmetry breaking which in the light quark sector is a distinctively non-perturbative feature.

The most common approach to avoid these difficulties is to apply effective quark matter models which are not derived from QCD but account for certain characteristic features, most notably again the dynamical breaking of chiral symmetry. Prominent examples for effective models are the thermodynamic bag model as the high density limit of the MIT bag model, and effective relativistic mean field models, typically of the Nambu–Jona-Lasinio type (Buballa 2005). A next approach which is well developed for the study of hadrons is the non-perturbative Dyson-Schwinger formalism which starts from the QCD action and derives gap equations to determine QCD's n-point Green-functions in a medium. As these typically couple to higher order Green-functions, truncation schemes are introduced which, if chosen wisely, preserve key features of QCD (Cloet and Roberts 2014).

7.8.2 Dynamical Chiral Symmetry Breaking

As described for hyperon matter, quark matter would be too soft to account for massive neutron stars if repulsive interactions would not be taken into account. However, vector repulsion arises as naturally as the breaking of chiral symmetry in relativistic effective models and in the Dyson-Schwinger approach as vector and scalar part of the dressing (self-energy) of the fermion propagator.

The dynamical mass generation mechanism is common to all modern EoS models for quark matter in neutron stars. It can be obtained already with the simplest dynamical models that contain a 4-fermion interaction with sufficiently strong coupling in the scalar channel, such as the celebrated Nambu–Jona-Lasinio (NJL) model of low-energy QCD that became a "workhorse" for quark matter phenomenology. There are a few excellent reviews on the NJL model and its application to finite temperature and density, e.g., Klevansky (1992), Hatsuda and Kunihiro (1994), Buballa (2005), Fukushima and Hatsuda (2011), Fukushima and Sasaki (2013) to which we refer the reader for details.

For definiteness, we like to show the typical form of an NJL model Lagrangian for applications to compact star physics as it is discussed in Baym et al. (2018), Kojo et al. (2015). The Lagrangian of the three-flavor NJL model is

$$\mathcal{L} = \overline{q}(\gamma^\mu p_\mu - \hat{m}_q + \mu_q \gamma^0)q + \mathcal{L}^{(4)} + \mathcal{L}^{(6)}, \qquad (7.23)$$

where q is the quark field operator with color, flavor, and Dirac indices, $\overline{q} = q^\dagger \gamma^0$, $\hat{m}_q$ the quark current mass matrix, μ_q the (flavor dependent) quark chemical potential and $\mathcal{L}^{(4)} = \mathcal{L}_\sigma^{(4)} + \mathcal{L}_d^{(4)} + \mathcal{L}_V^{(4)}$ and $\mathcal{L}^{(6)} = \mathcal{L}_\sigma^{(6)} + \mathcal{L}_{\sigma d}^{(6)}$ are four and six-quark interaction terms, chosen to reflect the symmetries of QCD.

The first of the four-quark interactions, a contact interaction with coupling constant $G > 0$,

$$\mathcal{L}_\sigma^{(4)} = G \sum_{j=0}^{8} \left[(\overline{q}\tau_j q)^2 + (\overline{q} i \gamma_5 \tau_j q)^2 \right] = 8G \text{tr}(\phi^\dagger \phi), \qquad (7.24)$$

describes spontaneous chiral symmetry breaking, where τ_j ($j = 0, \ldots, 8$) are the generators of the flavor-U(3) symmetries, and in Eq. (7.24),

$$\phi_{ij} = (\overline{q}_R)_a^j (q_L)_a^i \qquad (7.25)$$

is the chiral operator with flavor indices i, j (with summation over the color index a); the right and left quark chirality components are defined by $q_{R,L} = \frac{1}{2}(1 \pm \gamma_5)q$.

The second of the four-quark terms describes the scattering of a pair of quarks in the s-wave, spin-singlet, flavor- and color-antitriplet channel; this interaction leads to BCS pairing of quarks:

$$
\mathscr{L}_d^{(4)} = H \sum_{A,A'=2,5,7} \left[\left(\bar{q} i \gamma_5 \tau_A \lambda_{A'} C \bar{q}^T \right) \left(q^T C i \gamma_5 \tau_A \lambda_{A'} q \right) \right.
$$

$$
\left. + \left(\bar{q} \tau_A \lambda_{A'} C \bar{q}^T \right) \left(q^T C \tau_A \lambda_{A'} q \right) \right],
$$

$$
= 2H \mathrm{tr}(d_L^\dagger d_L + d_R^\dagger d_R), \tag{7.26}
$$

with $H > 0$. Here τ_A and $\lambda_{A'}$ (A, $A' = 2, 5, 7$) are the antisymmetric generators of U(3) flavor and SU(3) color, respectively, and

$$
(d_{L,R})_{ai} = \epsilon_{abc} \epsilon_{ijk} (q_{L,R})_b^j \mathscr{C} (q_{L,R})_c^k \tag{7.27}
$$

are diquark operators of left- and right-handed chirality, with $\mathscr{C} = i \gamma^0 \gamma^2$ the charge conjugation operator. The diquark pairing interaction leads as well to an attractive correlation between two quarks inside confined hadrons and, in constituent quark models, plays a role in the observed mass splittings of hadrons (De Rujula et al. 1975; Anselmino et al. 1993; Jaffe 2005). This interaction, in weak coupling, arises from single gluon exchange; however at the densities of interest in neutron stars, the non-linearities of QCD prevent direct calculation of this interaction, and so one must treat it phenomenologically.

In addition

$$
\mathscr{L}_V^{(4)} = -g_V (\bar{q} \gamma^\mu q)^2, \tag{7.28}
$$

with $g_V > 0$, is the Lagrangian for the phenomenological vector interaction, which produces universal repulsion between quarks (Kunihiro 1991).

The six-quark interactions represent the effects of the instanton-induced QCD axial anomaly, which breaks the $U(1)_A$ axial symmetry of the QCD Lagrangian. The resulting Kobayashi-Maskawa-'t Hooft (KMT) interaction leads to an effective coupling between the chiral and diquark condensates of the form (Kobayashi and Maskawa 1970; 't Hooft 1986):

$$
\mathscr{L}_\sigma^{(6)} = -8K (\det \phi + \text{h.c.}), \tag{7.29}
$$

$$
\mathscr{L}_{\sigma d}^{(6)} = K' (\mathrm{tr}[(d_R^\dagger d_L) \phi] + \text{h.c.}), \tag{7.30}
$$

where K and K' are positive constants. Provided that $K' \simeq K$ (which one expects on the basis of the Fierz transformation connecting the corresponding interaction vertices) the six-quark interactions encourage the coexistence of the chiral and diquark condensates.

Before discussing the hadron-to-quark matter phase transition under neutron star constraints in Sect. 7.9 and the phenomenology of hybrid compact stars with quark matter interior in Sect. 7.10 we briefly discuss some of the challenges in describing deconfined quark matter in the nonperturbative domain of the QCD phase diagram that borders the hadronic matter phases.

7.8.3 Color Superconductivity

Quark matter at finite densities and zero or small temperature can exhibit an extremely rich phase structure due to different pairing mechanisms which arise from the coupling of color, flavor and spin degrees of freedom and result in a variety of different possible condensates (Alford 2001; Alford et al. 2008). The importance of condensates is illustrated by the color-flavor locked phase which can appear in three flavor (up, down, strange) matter and is shown to be the asymptotic ground state of quark matter at low temperature (Schäfer and Wilczek 1999a). This zoo of color superconducting phases bears a rich potential for the phenomenology, in particular for the cooling of compact stars with color superconducting quark matter cores. While for applications in hybrid star cooling it is the smallest diquark pairing gaps that matter most (Grigorian et al. 2005), for direct effects on the EoS the largest gaps are most important. Those are usually found in the scalar diquark channels where they can be of the order of the dynamically generated quark masses and thus result in effects on the EoS as important as chiral symmetry breaking. For the discussion of the QCD phase diagram it is therefore crucial to solve the coupled gap equations for masses and diquark gaps simultaneously. For the three-flavor case this has been accomplished first in Ruester et al. (2005), Blaschke et al. (2005), Abuki and Kunihiro (2006). For usual NJL model parametrisations critical temperatures for onset of color superconductivity around 50 MeV arise. When quarks get coupled to the Polyakov-loop and thus their excitation is strongly suppressed in the confined phase and still moderately suppressed in the deconfined one, the critical temperature for the two-flavor superconducting (2SC) phase can become as high as 150 MeV and thus be of the order of the chiral restoration temperature and temperatures for chemical freeze-out in heavy-ion collisions, see Blaschke et al. (2010), Abuki et al. (2008). It is worth to highlight that the restoration of chiral symmetry is flavor dependent. This results in a sequential appearance of quark flavors at different densities (Blaschke et al. 2009; Alford and Sedrakian 2017), with the almost simultaneous occurrence of the corresponding color superconducting phase - 2SC for two-flavor quark matter and color-flavor-locking (CFL) for three-flavor quark matter. With a neutralizing surrounding of positively charged hadrons even a single flavor phase phase is possible due to an effect analogous to the neutron drip line - the down-quark dripline occurs and these down quarks can obey a single flavor pairing pattern resulting in the color-spin-locking (d-CSL) phase (Blaschke et al. 2009).

In the case of a strongly asymmetric occupation of the Fermi levels of different quark flavors, as in the case of β-equilibrated compact star matter, an energetically

Fig. 7.4 Schematic phase
diagram of dense nuclear
matter with two critical
endpoints of first order phase
transitions, from Kojo et al.
(2015). The crossover at low
temperatures is due to the
BEC-BCS crossover
(Hatsuda et al. 2006; Abuki
et al. 2010)

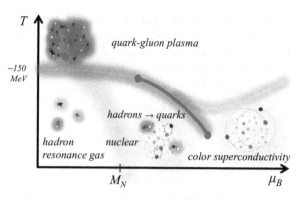

favorable ground state is found when the diquark Cooper pairs have finite momentum. One speaks of crystalline phases of QCD or, in analogy to such a situation
in the electronic superconductors in a magnetic field, of the Larkin-Ovchinnikov-
Fulde-Ferrell (LOFF) state of matter. An example of (rotating) hybrid star sequence
with crystalline color superconducting quark matter cores fulfilling also the modern
maximum mass constraint has been presented in Ippolito et al. (2008). For a review
on crystalline phases of QCD, see Anglani et al. (2014).

Usually the quark mass gap and the diquark pairing gap repel each other, but
for strong coupling it is possible that a coexistence of chiral symmetry breaking
and color superconductivity occurs which changes the order of the chiral restoration
transition from first order to crossover. A particularly interesting scenario is the
crossover transition of this kind which results from the coupling of chiral and
diquark gaps via the six-quark interaction term (7.30) that stems from the Fierz
transformed KMT determinant interaction (Hatsuda et al. 2006; Abuki et al. 2010),
see Fig. 7.4 for an illustration of this case which is in accordance with the idea
of quark-hadron continuity (Schäfer and Wilczek 1999b) or more general parton-
hadron duality (Wetterich 1999).

The question arises whether such a picture of a crossover transition at zero
temperature in compact stars would exclude characteristic signatures of a strong
first order phase transition in the mass-radius diagram such as mass twin stars and a
third family sequence. We come back to this question in Sect. 7.10 below.

7.8.4 Stiffness

The stiffness of quark matter in NJL type models is determined by the coupling
of quarks to a vector mean-field which has a self-interaction resulting from the 4-
fermion coupling (7.28) in the model Lagrangian, analogous to the linear Walecka
model for nucleons. The inclusion of this coupling was essential for obtaining stable
branches of hybrid stars with masses reaching up to and above $2 M_\odot$ when a NJL
type model was used for the quark matter EoS, see Klähn et al. (2007) for a first

paper fulfilling this constraint. While increasing the vector coupling g_V raises the maximum mass of the hybrid star sequence, increasing the scalar diquark coupling H decreases it, together with a decrease in the onset mass for the deconfinement transition. A systematic study of the parameter space has been performed in Klähn et al. (2013) which showed that stable hybrid stars which reach up to $2\,M_\odot$ could be obtained with a NJL model quark matter core and a hadronic shell described by the DBHF EoS with the Bonn-A interaction only when both, the vector coupling and the diquark coupling were nonvanishing and sufficiently strong.

Recently, an eight-quark interaction term in the vector channel,

$$\mathscr{L}_V^{(8)} = -G\eta_4(\bar{q}\gamma_\mu q)^4, \tag{7.31}$$

has been introduced in Benic (2014) in order to describe a stiffening of quark matter at high densities which improves the stability of compact hybrid star configurations against gravitational collapse (Benic et al. 2015b). Such an interaction is indispensable when one wants to describe within a microscopically motivated EoS model the occurrence of a third family of hybrid stars fulfilling the $2\,M_\odot$ mass constraint. Such a higher order quark interaction can be motivated by the existence of vector boson couplings of fourth order. Such boson self-coupling terms occur in the nonlinear Walecka model, but they are also important for explaining a sufficient repulsion in the high-energy nucleon-nucleon scattering by multi-pomeron exchange (Rijken and Schulze 2016).

7.8.5 Confinement

The NJL model finds wide application as a microscopic model for quark matter in compact stars since in this case at $T = 0$ the dynamical chiral symmetry breaking mechanism also mimics confinement (see, e.g., Klähn et al. 2013; Hell and Weise 2014 and references therein). This is because at the mean field level the value of the constituent quark mass is independent of the chemical potential μ (as well as the pressure P) so that the quark density $n = \partial P/\partial\mu$ vanishes as long as μ stays below its critical value μ_c. For $\mu > \mu_c$ the quark mass jumps down and the pressure evolves with μ so that there is a finite quark density. However, the NJL model has no true confinement and this becomes apparent at finite temperatures, where the partial pressure and density of the quarks becomes of the same order as the hadronic one already for temperatures below the pseudocritical temperature $T_c = 154 \pm 9\,\mathrm{MeV}$ known from lattice QCD simulations. This problem is severe when applications for finite temperature systems are to be considered like protoneutron stars in supernova simulations or hot, hypermassive neutron stars in neutron star merger events. Unfortunately, the minimal coupling of the Dirac quarks to a homogeneous gluon background field within the Polyakov-loop improved NJL (PNJL) model would not be an appropriate approach to account for confining effects in $T = 0$ quark matter since we have no possibility to calibrate a chemical potential dependence of the Polyakov-loop potential[8] with lattice QCD simulations, unlike the case at vanishing baryon density and finite temperatures where the PNJL model is an acceptable model for low-energy QCD matter (Ratti et al. 2006; Roessner et al. 2007).

A standard minimal way do account for quark confinement in a thermodynamic description is the thermodynamic bag model which consists in adding to the pressure of a relativistic Fermi gas of quarks (plus Bose gas of gluons, eventually improved by perturbative corrections of $\mathcal{O}(\alpha_S)$) a negative pressure contribution $P_{bag} = -B$, with $B > 0$ being the bag constant. This makes sure that at low temperatures and chemical potentials always the hadronic matter dominates the thermodynamics because it has a nonnegative pressure. The nonperturbative bag pressure can be understood to originate from quark and gluon condensates, $B = B_q + B_g$. The NJL model allows to calculate the medium-dependent value of a chiral condensate $B_q \propto \langle \bar{q}q \rangle$ (Buballa 2005) and fits for the μ-dependent quark bag function $B_q(\mu)$ from a nonlocal color superconducting have been provided for hybrid neutron star calculations (Grigorian et al. 2004). Qualitatively, the density dependent bag function starts at a finite value and gets lowered with increasing density. Such functions have been modelled on heuristic grounds also in Burgio et al. (2002a,b) and may be regarded as a density functional. Since typical NJL-type models have no gluon dynamics, an understanding of the gluonic contribution to the bag function and its medium dependence is lacking. Again qualitatively, one can expect that the gluon condensate should melt at increasing temperature and density such that a positive pressure contribution should result. Such bag-like contributions to the pressure of NJL-model quark matter have been introduced, e.g. in Pagliara and Bielich (2008), Bonanno and Sedrakian (2012), Blaschke et al. (2010), where due to this addition a stable hybrid star with a CFL quark matter core could be obtained. In NJL quark matter models without such a contribution the hybrid stars turn gravitationally unstable as soon as the strange quark flavor appears (Klähn et al. 2007; Alford et al. 2007; Klähn et al. 2013). Another argument to add a (medium dependent) bag constant to an NJL-type model is to enforce coincidence of the chiral restoration transition in the QCD phase diagram with the hadron-to-quark matter transition, see Klähn and Fischer (2015). In these models an appropriately chosen vector interaction (7.28) has to be added in order to achieve a sufficient stiffening of the resulting hybrid star EoS in order to fulfill the $2\,M_\odot$ constraint.

However, since bag type models, even if improved by coupling to vector and scalar mean fields, have no dynamical confinement we should therefore advance to models that make use of a confining interquark potential, like the field correlator method (FCM), see Logoteta and Bombaci (2013) and references therein. This model constructs from color electric and magnetic correlators a mean vector field that contributes an energy shift to the quark and gluon distribution functions and

---

[8]An ansatz for the Polyakov-loop potential which is applicable also at $T = 0$ has been suggested by Dexheimer and Schramm (2010). For an application to hybrid stars, see Blaschke et al. (2010).

strongly suppresses them at low densities. This model reproduces rather well a constant speed of sound (CSS) model for quark matter with, however, a low squared speed of sound $c_s^2 \sim 1/3$ (Alford et al. 2015). This model has the strength that at finite temperatures its input parameters (gluon condensate G_2 and vector field V_I) can be fixed with lattice QCD. However, when these values are used for the construction of hybrid stars there is a problem to fulfill the $2\,M_\odot$ constraint (Pereira 2011). Moreover, no absolutely stable strange quark matter is obtained (Pereira 2013).

Another possibility to include confining effects into a field-theoretic description of quark matter is the color dielectric model (CDM) (Alberico et al. 2002). As opposed to the FCM the color dielectric field χ is a scalar and determines the quark masses. At low densities and thus low values of χ the quark masses diverge and this suppresses the excitation of these degrees of freedom. Interestingly in the CDM, like in the bag model, absolutely stable strange quark matter is possible. Maieron et al. (2004) showed that for the CDM and a bag model with density-dependent B the maximum masses of hybrid stars are never larger than $1.6\,M_\odot$.

Very recently, a relativistic density functional approach to quark matter has been developed (Kaltenborn et al. 2017) which implements a confining mechanism in the spirit of the CDM in its density functional which in this case is motivated by the string-flip model of quark matter (Röpke et al. 1986a) and is a generalization of Li et al. (2015) and an earlier density functional approach developed for heavy-ion collision applications (Khvorostukhin et al. 2006). We will refer to this approach as SFM below. Its density functional subsumes all aspects required for the definition of a class of hybrid star EoS which may fulfill all known constraints from the phenomenology of compact stars.

7.8.6   Strange Quark Matter in Neutron Star Cores

We first discuss the situation where normal nuclear or hyperon matter deconfines at a critical density and forms a quark-gluon plasma. This problem is still far from being solved consistently in a way where hadrons would be taken into account as actually confined quarks which then deconfine dynamically. The standard way to circumvent a detailed description of deconfinement is to choose a two phase approach where nuclear and quark matter are modeled independently and the transition is constructed thermodynamically consistently, viz. in terms of a Maxwell (or similar) construction. In case of the Maxwell construction one simply determines at which baryon-chemical potential the pressure of the nuclear and quark phase are equal and thus defines the transition point. By construction, this implies a first order phase transition which 'switches' from a given nuclear equation of state to a softer quark matter equation of state. Therefore, this procedure requires a nuclear EoS which is stiff enough to support at least a two solar mass neutron star, and a quark matter EoS which is softer but stiff enough to do the same. How exactly this happens can vary. The quark matter EoS can mimic the nuclear EoS, be generally softer or at some density can turn even stiffer than the underlying nuclear EoS so that a second crossing of the pressure curves occurs ("reconfinement problem" (Zdunik and Haensel 2013)). The latter scenario is justified if one assumes that at densities far enough beyond the transition a comparison of both phases is meaningless as the nuclear EoS does not describe any physical reality anymore. In any of these scenarios, repulsion is a crucial feature to account for the existence of massive neutron stars and, as stated earlier, a natural property of relativistic models. For the onset density of hyperon matter we discussed how it is pushed to increasingly high densities with increasing stiffness or repulsion. The same would be true for quark matter if condensates are not taken into account. However, condensates couple colors and flavors and can lower the transition density. Therefore, quark models can account for transition densities far below the hyperon threshold.

For the appearance of strange matter in neutron stars, a couple of scenarios emerge which we divide into two groups. First, nuclear matter can deconfine directly into (three flavor) strange matter as one would find it for the thermodynamical bag model, as described earlier. For a dynamical treatment of this transition as a nucleation process see, e.g., Bombaci et al. (2016), Bhattacharyya et al. (2017) and references therein. Second, a sequential transition from nuclear to two-flavor followed by a transition to three-flavor matter takes place. Similar to this scenario, there are two more cases which would not result in a strange matter core but are not less realistic. The sequential transition results in neutron star configurations which are stable at all densities below the strange quark threshold but unstable beyond due to the softening of the EoS with this new degree of freedom. This situation has been described for an NJL model with diquark couplings (Klähn et al. 2007). However, choosing a finite constant as offset to the quark pressure can alter this result (Bonanno and Sedrakian 2012) and render neutron stars with strange matter core stable (the thermodynamic properties of matter are described in terms of pressure derivatives, hence a constant offset keeps them intact). Another way to stabilize strange core configurations is a strongly density dependent stiffening of the strange matter EoS following the transition (Benic et al. 2015b; Kaltenborn et al. 2017). This can result in situations where stable two flavor and three flavor quark core neutron star configurations are separated by a population gap at intermediate central neutron star densities. In extreme scenarios, this can generate separated mass twin configurations, viz. two neutron star families with similar masses but very different radii (Blaschke et al. 2013; Alvarez-Castillo and Blaschke 2017). If members of both families could be observed, this would be a strong indicator for a first order QCD-like phase transition where (the smaller) compact stars can carry a core made of strange matter (Bastian et al. 2018). Note, however, that according to the two-families scenario of Drago et al. (2014a) the smaller stars would be the hadronic ones (with a maximum mass of 1.5–$1.6\,M_\odot$) while the very massive ones would be strange stars with a larger radius. They can coexist in this scenario because the hadronic stars are metastable with respect to the nucleation of strange quark matter.

Another indicator for a first order phase transition in dense matter would be the observation of a delayed second neutrino signal after a supernova. This has been suggested based on simulations which applied a bag model (Sagert et al. 2009;

Fischer et al. 2011). Although such a measurement would be very exciting it is not clear how it would address the question whether the transition involved strange matter, or quark matter at all.

7.8.7 Absolutely Stable Strange Matter?

The hypothesis that strange matter could be absolutely stable, bases on the observation that the appearance of strange quarks lowers the energy per baryon. As two flavor quark matter is evidently less stable than Fe (otherwise Fe would decay into it's quark components and so would we) this leaves a window where two-flavor matter is less and strange matter more stable than iron (Bodmer 1971; Terazawa 1979; Witten 1984; Farhi and Jaffe 1984). This hypothesis has been supported by certain parametrisations of the thermodynamic bag model. Choosing the proper bag constant one can indeed find exactly the proposed situation. If this scenario is reality it would have a number of interesting consequences. Therefore, the search for stable strange matter inspired a multitude of experiments and has born many new ideas. Strange matter could form strange nuggets of extreme density with rather small atomic numbers and hence extremely low cross sections. It could form objects very similar to a neutron star, almost entirely made of strange matter, see Bombaci (2001), Weber (2005) for reviews of this and other scenarios involving strangeness in compact stars. A seed of strange matter in a neutron star could destabilize the surrounding matter and thus trigger a conversion of the neutron star interior into strange matter. Recently, it has been proposed that muonic bundles that have been observed at ALICE (CERN) are produced by strangelets (Kankiewicz et al. 2018). Currently, other ways to detect strangelets are actively investigated (VanDevender et al. 2017). The idea of absolutely stable strange matter is certainly appealing.

Theoretically, as mentioned before, the hypothesis has been based on the thermodynamic bag model. It should be noticed, that this model lacks a key feature of QCD, namely chiral symmetry breaking, viz. quarks in the thermodynamic bag model are assumed to have bare masses and therefore extremely small threshold densities (the critical chemical potential scales with the effective mass). NJL-type models, which do generate dressed quark masses, do not confirm that strange matter is absolutely stable. The reason for this is easily found: In the (low) density domain where the bag model predicts absolutely stable strange matter, NJL type models find chiral symmetry to be broken, hence significantly larger quark masses and consequently a higher energy per particle, too high to render strange matter absolutely stable, see Fig. 7.5. An appealing and sometimes confusing feature of the thermodynamic bag model is that it originates from the MIT bag model which has been developed to describe hadron properties. This evidently seems possible even though the model assumes bare quark masses and makes it easier to believe that massless quarks could form stable strangelets. This confusion can be cleared if one realizes that the MIT bag model does not only assume bare quark masses but a bag constant which is introduced as synonym for 'all we don't know about confinement

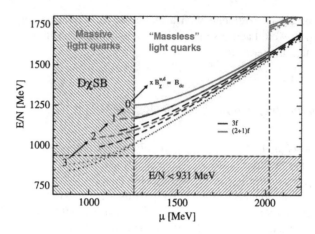

Fig. 7.5 Energy per baryon vs. baryon chemical potential for the vBag quark matter model (Klähn and Fischer 2015). Absolutely stable strange quark matter could be obtained if dynamical chiral symmetry breaking (DχSB) is neglected. However: The vertical band marks the approximate region where chiral symmetry is broken. For the shown curves DχSB is ignored for all (black) or all but the s-quark (red). At densities below the s-quark threshold this allows to effectively compare two(red)- and three-flavor(black) quark matter for different effective bag constants. In none of the cases strange matter is more stable than iron ($E/N < 931$ MeV) in a density domain where chiral symmetry is restored. See also Klähn and Blaschke (2018)

and further interactions'. The fathers of the MIT bag model stated this explicitly. In Klähn and Fischer (2015) we illustrated how to translate a model with chiral symmetry breaking into a bag type model and where this model would break, which is when chiral symmetry is broken. For an earlier investigation of this kind, see Buballa (2005) for the NJL model case and Gocke et al. (2001) for a nonlocal chiral quark model. Effects of color superconductivity in this context have been studied and parametrized in Grigorian et al. (2004).

The bag constant mostly originates from chiral symmetry breaking. It is subtracted because the difference between the vacuum pressure of massive quarks and effectively massless quarks is negative. The absolute value is reduced by confinement. Thinking of the absolute value of the bag constant as the energy of a hadron, this makes sense as confinement reduces the energy of the system of chirally broken quarks by the binding energy. The MIT bag model is not consistent in the sense that it ignores any relation between quark mass and bag constant. In vacuum this is a reasonable approximation for two reasons: First, the model describes hadrons and does not attempt to predict any dynamical property of individual constituent quarks. Second, it can be fitted to observables and thus repairs the inherent shortcomings. The thermodynamic bag model addresses both effects, chiral symmetry breaking and confinement, in terms of one parameter - the bag constant. Statements regarding quark matter based on the thermodynamic bag model, in particular at low densities where the bag constant affects the total pressure significantly, should be considered with considerable caution as it is all but clear, that the model actually describes deconfined matter.

It should be noticed, that effective quark models with density dependent quark masses have been suggested which indeed would predict absolutely stable strange matter (Dondi et al. 2017; N.-U. Bastian, 2017, private communication). A distinct feature of these models is a steep *concave* decrease of the strange quark mass at comparably low density opposing to the typically *convex* behavior. This reduces the effective quark mass drastically already at low densities which makes it very similar to to the thermodynamic bag model, evidently with similar results regarding the stability of strange matter. It would be interesting to see, how a microscopic approach which generates this kind of density behavior would perform describing hadron properties.

7.9 Hadron-to-Quark Matter Phase Transition

7.9.1 Maxwell Construction and Beyond

When the hadronic and quark matter phases are described with EoS given by relations between the pressure and chemical potential (for $T = 0$, which is relevant for the NS modeling) $P_H(\mu)$ and $P_Q(\mu)$ correspondingly, one can find the critical value of the baryochemical potential μ_c from the condition of equal pressures,

$$P_Q(\mu_c) = P_H(\mu_c) = P_c, \tag{7.32}$$

for which the phases are in mechanical equilibrium with each other. The value P_c defines the Maxwell construction. Quantities characterizing the quark-, hadron-, and mixed phases are denoted by the subscripts Q, H, and M, respectively.

Assuming the surface tension to be smaller than the critical value σ_c, a mixed phase could have an influence on the compact stars structure. The adequate description of the physics of pasta phases is a complicated problem which requires to take into account sizes and shapes of structures as well as transitions between them. It has been dealt with in the literature within different methods and approximations (Voskresensky et al. 2003; Yasutake et al. 2014; Maruyama et al. 2008; Watanabe et al. 2003; Horowitz et al. 2005, 2015; Newton and Stone 2009).

Here, instead of the full solution we would like to use a simple modification of the Maxwell construction which mimics the result of pasta matter studies (Ayriyan and Grigorian 2018). This means that the pressure as a function of the baryon density in the mixed phase is not constant but rather a monotonously rising function. We are assuming that close to the phase transition point (that would be obtained using the Maxwell construction) the EoS of both phases are changing due to finite size and Coulomb effects (see Fig. 7.6), so that the effective mixed phase EoS $P_M(\mu)$ could be described in the parabolic form

$$P_M(\mu) = a(\mu - \mu_c)^2 + b(\mu - \mu_c) + P_c + \Delta P. \tag{7.33}$$

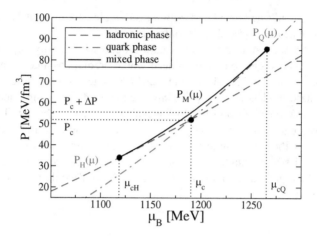

Fig. 7.6 Maxwell construction of a first-order phase transition between a hadronic EoS $P_H(\mu)$ and a quark matter EoS $P_Q(\mu)$ occurring at the critical pressure $P_c = P(\mu_c)$. The modification by a parabolic function $P_M(\mu)$ mimicks pasta phases. Adapted from Ayriyan et al. (2018)

Here we have introduced the pressure shift ΔP at μ_c as a free parameter of the model which determines the mixed phase pressure at this point

$$P_M(\mu_c) = P_c + \Delta P = P_M. \tag{7.34}$$

The parameter ΔP shall be related to the largely unknown surface tension between the hadronic and quark matter phases of the strongly interacting system at the phase transition. Infinite surface tension corresponds to the Maxwell construction and thus to $\Delta P = 0$, while a vanishing surface tension is the case of a Gibbs construction under global charge conservation (also called Glendenning construction (Glendenning 1992)) that for known examples looks similar to the results of our mixed phase construction for the largest values considered here, $\Delta_P \sim 0.07 \ldots 0.10$. A more quantitative relation between ΔP and the surface tension would require a fit of the mixed phase parameter to a pasta phase calculation for given surface tensions. Such a calculation can be done along the lines of Yasutake et al. (2014).

According to the mixed phase construction shown in Fig. 7.6 we have two critical chemical potentials, μ_{cH} for transition from H-phase to M-phase and μ_{cQ} for the transition from the M-phase to Q-phase. Together with the coefficients a and b from Eq. (7.33) this are four unknowns which shall be determined from the four equations for the continuity of the pressure

$$P_M(\mu_{cH}) = P_H(\mu_{cH}) = P_H, \tag{7.35}$$

$$P_M(\mu_{cQ}) = P_Q(\mu_{cQ}) = P_Q, \tag{7.36}$$

and of the baryon number density $n(\mu) = dP(\mu)/d\mu$

$$n_M(\mu_{cH}) = n_H(\mu_{cH}), \tag{7.37}$$

$$n_M(\mu_{cQ}) = n_Q(\mu_{cQ}). \tag{7.38}$$

From the Eqs. (7.35) and (7.36) for the pressure a and b are found and can be eliminated from the set of equations for the densities. By solving the remaining Eqs. (7.37) and (7.38) for the densities numerically one can find the values for the critical chemical potentials μ_{cH} and μ_{cQ}.

In Fig 7.7 we show results of this mixed phase construction for the example of EoS investigated in Benic et al. (2015b) for different values of the parameter Δ_P (upper panel) and the corresponding solutions of the TOV equations (lower panel). With this modification of the Maxwell construction one can examine the robustness of the third family solutions against pasta phase effects. Further details and other EoS combinations, see Ayriyan et al. (2018), Alvarez-Castillo et al. (2017), Ayriyan and Grigorian (2018).

The procedure for "mimicking pasta phases" described in this subsection is general and can be superimposed to any hybrid EoS which was obtained using the Maxwell construction of a first-order phase transition from a low-density to a high-density phase with their given EoS. A certain limitation occurs in the case of sequential phase transitions (see, e.g., Blaschke et al. 2009; Alford and Sedrakian 2017) when the broadening of the phase transition region as shown in Fig. 7.6 would affect the adjacent transition.

7.9.2 Interpolation

The method of interpolating hadron and quark matter EoS has been suggested for the neutron star EoS by Masuda, Hatsuda and Takatsuka in Masuda et al. (2013a,b) in analogy to an earlier developed technique for vanishing baryochemical potential along the temperature axis of the QCD phase diagram (Asakawa and Hatsuda 1997). While the first construction was flawed because it was not performed in natural thermodynamic variables, the revised work corrected for this and the method has found extensive use (Kojo et al. 2015; Blaschke et al. 2012) as a tool to bridge the gap in our knowledge about the nonperturbative domain of the hadron-to-quark matter transition before a reliable unified approach based on quark degrees of freedom will be developed. The situation is illustrated in Fig. 7.8, taken from Kojo et al. (2015).

This method can be contrasted to the Maxwell construction and its modification which was described before. In applying the Maxwell construction we tacitly assume that the two EoS to be matched have a common domain of validity in the space of the thermodynamic variables so that their direct matching makes sense. However, this may not be the case when, the low-density and high-density forms

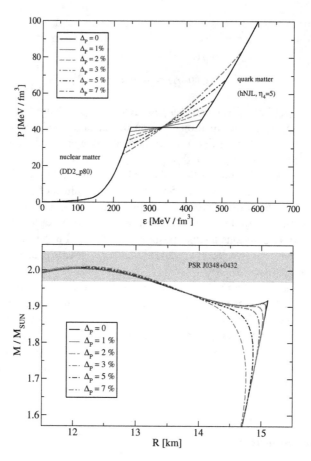

Fig. 7.7 Two-phase model for the hadron-to-quark matter transition obtained by a Maxwell construction ($\Delta_P = \Delta P/P_c = 0$) and its modifications ($\Delta_P = 1, 2, 3, 5, 7$ %) which mimic mixed phases with pasta structures. Upper panel: P vs. ε for the EoS from Benic et al. (2015b) where the nuclear matter phase is DD2 with excluded volume and the quark matter phase is a NJL model with 8-quark interactions. Lower panel: Solutions of the TOV equations for this EoS show that for $\Delta_P > 1\%$ the sequence of hybrid star solutions does not form a third family separated from that of the neutron star ones. Adapted from Ayriyan et al. (2018)

of the equation of state become not trustworthy before they would meet in the $P - \mu$ plane. For example when a soft nuclear matter EoS gets extrapolated to densities where already quark exchange effects should play a role and a perturbative QCD EoS is drawn into the low density region around the nuclear matter density where it is lacking confinement, chiral symmetry breaking etc. This is illustrated in Fig. 7.9. Note by comparing with the Maxwell construction case of Fig. 7.6 that

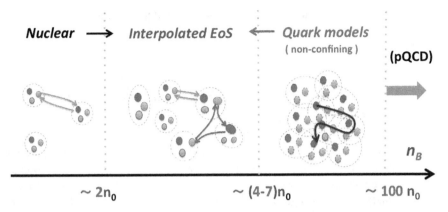

Fig. 7.8 The idea of interpolating between known EoS of nuclear matter at low densities and deconfined quark matter at high densities is to bridge the gap in our quantitative knowledge of the physical mechanisms that govern quark deconfinement and its interplay with chiral symmetry restoration, color superconductivity and other collective phenomena in the strongly coupled quark-gluon plasma. Quark exchange processes between baryons are illustrated as well as the emergence of dense quark matter with itinerant, delocalised quarks. From Kojo et al. (2015)

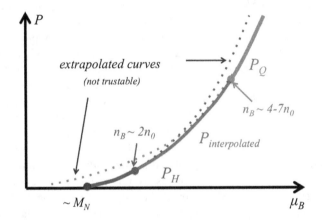

Fig. 7.9 Illustration of the interpolation scheme in the pressure vs. baryochemical potential diagram. This method shall be applied when a Maxwell construction makes no sense because the low-density and high-density forms of the equation of state loose their applicability before they would meet. In this case the curves with the higher pressure which should be the physical ones are the not trustable parts of the given EoS. From Kojo et al. (2015)

the interpolated part of the EoS lies now below the (unphysical) matching of the asymptotic EoS.[9]

---

[9]If one would not pay attention and apply a Maxwell construction to this example one would have to let the physical pressure go along the larger pressure, which would mean to join the two not trustable parts of the model EoS. A recent example for matching unphysical EoS is Annala et al.

For a systematic discussion of interpolation techniques in obtaining hybrid star EoS from given hadronic and quark matter model inputs, see Alvarez-Castillo et al. (2018).

7.9.3   Metastable Hadronic Stars Coexisting with Strange Quark Stars

So far we have considered hadron-to-quark matter phase transitions in compact stars in equilibrium for which the Gibbs conditions of phase equilibrium apply. However, as it has been discussed above in Sect. 7.7.2, the nucleation of strange quark matter from a metastable hadronic phase by quantum tunneling (Iida and Sato 1998; Bombaci et al. 2007) could take sufficiently long time for the metastable hadronic star branches of compact stars to become populated and observable, together with the final state of the conversion process, the branch of stars consisting of absolutely stable strange quark matter. This is called the two-family scenario (Berezhiani et al. 2003; Drago et al. 2004; Bombaci et al. 2004) the simultaneous existence of very compact ($R < 10$ km) hadronic stars with a maximum mass of only 1.5–$16 M_\odot$ with less compact ($R > 10.5$ km) strange quark matter stars fulfilling the $2 M_\odot$ mass constraint could be explained. While the conversion of a hadronic star to a strange quark star starting from a seed droplet of strange quark matter is completed within milliseconds by turbulent combustion (Herzog and Röpke 2011; Pagliara et al. 2013), the timescale for the seed formation by quantum tunneling varies between those timescales and irrelevant periods beyond the age of the Universe. That such a huge variation results from small changes in parameters of the model, like the bag constant of the strange quark matter EoS limits the predictive power of such a scenario.

Nevertheless, the two-family scenario has recently been applied also in interpreting the merger event GW170817 that was first observed by the LIGO detectors as a chirped gravitational wave signal and subsequently as a gamma-ray burst by the FERMI and INTEGRAL satellites as well as in all other wavelengths of the electromagnetic spectrum (Abbott et al. 2017b). In Drago and Pagliara (2018), Drago et al. (2018) it has been suggested that between the one-family and two-family scenarios could be discriminated once the gravitational wave signal at the moment of the merger could be detected.

7.9.4   Towards a Unified Quark-Nuclear EoS

We have promised to come back to the issue of a unified EoS. This time not for crust and core matter, but rather a unified description of quark and nuclear matter

---

(2017), where the authors in such grounds suggest a new type of holographic hybrid stars, with nuclear matter core surrounded by a quark matter shell.

on the basis of a microscopic quark model where nucleons appear as bound states of quarks. As in the case of the unified EoS for nuclear matter and clustered nuclear matter, the point is to describe strongly coupled deconfined quark matter and the quark bound states in medium on the basis of the same model for the quark interaction.

In an early, nonrelativistic potential model approach (Röpke et al. 1986a) this problem has been solved in these steps: (i) solve the three-quark Schrödinger equation for the nucleon ground state with a confining potential, (ii) determine the density-dependent nucleon self-energy (Pauli blocking) shift from the quark exchange interaction in the two-nucleon system (antisymmetrisation of the 6-quark wave function) in the nucleonic matter phase, (iii) determine the self-consistent Hartree shift in the quark matter phase when the confining interaction is saturated within the range of the nearest neighbors (string-flip model), (iv) Maxwell construction of the phase transition between the Fermi gas of nucleons with repulsive (quark exchange) interactions and the self-consistent Hartree approximation for the quark plasma with saturated confinement interaction. The result is a unified EoS for quark-nuclear matter with a first-order deconfinement phase transition where the description of both phases is based on the same confining interaction potential.

The main flaw of the old string-flip model is that it works with constituent quarks and thus does not account for the chiral symmetry restoration which we expect to occur in the vicinity of the deconfinement one.

An alternative route to the unified EoS is to start from the NJL model as a paradigmatic field theoretic model for dynamical chiral symmetry breaking and its restoration in a hot dense medium. This model must then be considered beyond the mean-field level in order to describe baryons as bound states of three quarks. Technically, a bosonization of the NJL model (7.23) in the meson and diquark channels is performed and a Faddeev equation for the nucleon as a quark-diquark bound state can be derived from a partial resummation of a subclass of diagrams in the one-quark loop expansion, see Cahill (1989), Reinhardt (1990). This scheme has been carried out for $T = 0$ and finite baryochemical potential in Bentz and Thomas (2001) to arrive at an EoS for nuclear matter with composite nucleons coupled to the scalar and vector meson mean fields, thus being reminiscent of a Walecka model. The saturation was obtained at too high densities but could be brought to the phenomenological value by adopting asd hoc an 8-quark interaction or by imposing an infrared cutoff to the quark propagators which should mimic confinement. Subsequently, the phase transition to NJL model quark matter was obtained from a Maxwell construction (Bentz et al. 2003) thus fulfilling the aim to obtain a unified quark-nuclear matter EoS within one microscopic model for the quark interactions. This NJL model based approach was extended in Lawley et al. (2006) to the case of isospin-asymmetric hybrid star matter in β-equilibrium with a phase transition to color superconducting quark matter where also mass-radius diagrams for hybrid compact star sequences have been obtained. Unfortunately, Sarah Lawley left Physics and this promising development was discontinued.

Another promising step has been done in Wang et al. (2011) where in a slightly simplified quark-diquark interaction scheme of the NJL model type the dissociation of the baryon in the QCD phase diagram with 2SC phase has been described in terms of the baryon spectral function. Interestingly, the baryon in the medium is a Borromean state: as a three-quark state it is bound while the two-quark state (diquark) is unbound, see also the discussion in Blaschke et al. (2015). Unfortunately, this development did not lead to an EoS of quark-nuclear matter yet. This step can be expected from a cluster virial expansion for quark matter on the basis of the Φ-derivable approach which is equivalent to a generalized Beth-Uhlenbeck approach when the choice of diagrams in the Φ-functional is restricted to all two-loop diagrams that can be drawn with quark cluster Green's functions (Blaschke 2015; Bastian et al. 2018).

As drawback of the NJL model based approach remains the lack of confinement. This becomes particularly severe when the EoS is required not only at zero temperature, as in the case of supernova collapse and neutron star merger simulations. In these situations with temperatures of the order of the Fermi energy there are free quarks contributing to the thermodynamics of the system in the hadronic phase. The coupling to the Polyakov-loop suppresses quarks at finite temperatures and provides an improvement but is not applicable at low and vanishing temperatures for high baryon densities. In this situation a relativistic density functional approach to quark matter can provide a solution which suppresses colored states in the confinement domain at low densities by diverging scalar self-energies (masses). An EoS for the finite-temperature applications of this concept to heavy-ion collisions has been developed in Khvorostukin et al. (2006). The finite-temperature generalization of the SFM EoS has been applied to supernova simulations in Fischer et al. (2017), where it was demonstrated that with this EoS the QCD phase transition can trigger a successful explosion for massive progenitors such as blue supergiant stars as massive as 50 $M_\odot$. These very promising developments at the mean-field level (Kaltenborn et al. 2017; Khvorostukin et al. 2006) have now to be followed by developing a relativistic cluster virial expansion on this basis (Bastian et al. 2018).

7.10 Hybrid Stars in the Mass-Radius Diagram

It is known that there is a one-to-one relationship between the mass-radius $(M - R)$ relationship of (cold) compact stars and their EoS given, e.g., as pressure-energy density $(P(\varepsilon))$ relation via the Tolman–Oppenheimer–Volkoff (TOV) equations. On this basis, measurements of mass and radius of pulsars can be used to determine the EoS of strongly interacting matter in β-equilibrium in a domain of densities and temperatures that is not accessible to terrestrial laboratory experiments. Naturally, the question appeared whether neutron star observations could help to settle the question whether there is at least one critical point in the QCD phase diagram. To this end one would have to prove that the deconfinement phase transition that eventually occurs in neutron star interiors is of first order. Exploiting the above relation between EoS and M-R diagram the task can be rephrased: Is there a feature in the possible M-R sequences of (hybrid) compact stars that would signal a first order phase transition in the corresponding EoS? If this question could be answered affirmatively, what are then the prospects to identify such a feature by observations of compact stars? The following subsections will discuss answers to these two questions.

7.10.1 Phase Diagram of Hybrid Stars

In Alford et al. (2013) Alford et al. have introduced the notion of a phase diagram for hybrid stars, i.e. stars where the core matter undergoes a phase transition of first order with a jump in the energy density $\Delta\varepsilon$ (latent heat) occurring when a critical pressure P_{trans} is reached at a critical energy density ε_{trans}trans. This phase diagram is spanned by latent heat and critical pressure, both in units of the energy density at the onset of the transition, see Fig. 7.10. It exhibits a universal dividing line (shown as solid line in red color) which corresponds to the criterion for a gravitational instability at the onset of the phase transition and is given by (Seidov 1971)

$$\frac{\Delta\varepsilon}{\varepsilon_{crit}} \geq \frac{1}{2} + \frac{3}{2}\frac{P_{crit}}{\varepsilon_{crit}}, \qquad (7.39)$$

so that above this line stable hybrid stars are absent (A) or, if the high-density phase of matter in the inner core is stiff enough, a disconnected (D) "third family" (Gerlach 1968) sequence of stable hybrid stars emerges. The position of the (almost vertical)

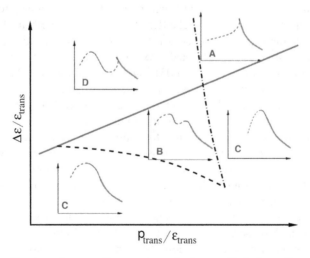

Fig. 7.10 Classification of mass-radius sequences in the so-called phase diagram for hybrid compact stars, spanned by the latent heat $\Delta\varepsilon$ and the critical pressure p_{trans} of their underlying equation of state, normalized to the critical energy density ε_{trans} at the onset of the phase transition. Adapted from Alford et al. (2013)

dividing line is not universal, it depends on the EoS. Below the Seidov line, the onset of the phase transition in the inner core is followed by a connected (C) branch of stable hybrid stars. In this domain there is a special, triangular-shaped region where after the connected branch of hybrid stars an instability occurs which divides it from a third family sequence of compact hybrid stars, so that in this region the sequence is characterized by both (B): a connected and a disconnected branch of stable hybrid stars.

The existence of a disconnected hybrid star branch in the $M - R$ diagram can thus be identified as the distinctive and observationally accessible feature for detecting a (strong) first-order phase transition in compact star interiors. Due to the mass defect which accompanies the compactification because of the gain in gravitational binding energy, there is a certain range in masses for which pairs of compact stars exist that have the same mass but different radii and different internal composition, the so-called "mass twins" (Glendenning and Kettner 2000). Their existence is equivalent to the existence of a third family and therefore to the strong first order phase transition in the stellar core.

7.10.2   The Story of the Twin Stars

The possibility of a third family of compact stars and its relation to a phase transition in stellar matter was recognized by Gerlach (1968) already in 1968, and after a while discussed by Kämpfer (1981) in the context of pion condensation and quark matter in compact stars. After another 17 years, the phenomenon was reconsidered by Glendenning and Kettner (2000) who coined the term "twins" for the effect accompanying the third family. Schertler et al. (2000) and Bhattacharyya et al. (2004) took up the idea that twin stars may signal a phase transition in a compact star, but all these solutions predicted maximum masses which were well below the 2.1 $M_\odot$ constraint (Nice et al. 2005) which was published in 2005 (and taken back in 2007), so that mass twins were put aside, also under the impression of the "masquerade" effect (Alford et al. 2005), to be discussed below. But the idea of mass twins as an observable signature for a phase transition was too good to be forgotten. In order to revive it one had to demonstrate that it could be reconciled with the meanwhile discovered very precise constraints on the maximum mass of a compact star EoS, i.e., 1.97±0.04 $M_\odot$ (Demorest et al. 2010) for PSR J1614-2230[10] and 2.01 ± 0.04 $M_\odot$ (Antoniadis et al. 2013) for PSR J0438+432. Early in 2013, it could be demonstrated that high-mass twins were theoretically possible, when a realistic nuclear EoS model was combined with a constant speed of sound (CSS) model for the high-density phase (Alvarez-Castillo and Blaschke 2013). With such a hybrid star model EoS one could describe twin stars at any mass, while fulfilling

---

[10]Note that this mass for PSR J1614-2230 reported in 2010 was down-corrected in 2016 to 1.947± 0.018 $M_\odot$ (Fonseca et al. 2016) and most recently to 1.908±0.016 $M_\odot$ (Arzoumanian et al. 2018).

the 2 $M_\odot$ mass constraint, see the recent work on the classification of twin star solutions (Christian et al. 2018). Even a fourth family and mass triples of compact stars are possible (Alford and Sedrakian 2017). What are the astrophysical chances to discover twin stars, at high or low mass? We discuss two possibilities.

7.10.2.1   High-Mass Twin Stars: J1614-2230 & J0438+432

Historically, as soon as the second 2 $M_\odot$ pulsar was discovered (Antoniadis et al. 2013) in 2013, it was suggested that radius measurements of these stars should have the best chance to discover the case of high-mass twins (Blaschke et al. 2013) and thus to prove the existence of a critical endpoint in the QCD phase diagram (Benic et al. 2015b; Alvarez-Castillo et al. 2016). Since the deconfinement transition, if possible in compact star interiors at all, should have the best chances to occur in the most massive stars, i.e., the stars with the highest possible densities in their core.[11] However, while PSR J1614-2230 is a millisecond pulsar for which the neutron star interior composition explorer (NICER) (https://heasarc.gsfc.nasa.gov/docs/nicer) in principle could provide a radius measurement (given a sufficient duration of this experiment on board of the international space station (ISS)), such a radius measurement is not possible for PSR J0438+432. Then one could hope for the discovery of another high-mass pulsar, e.g., by the square kilometer array (SKA) (http://www.ska.ac.za). The case is illustrated in the left panel of Fig. 7.11 for the hybrid star EoS of Kaltenborn et al. (2017) with the DD2_p40 hadronic EoS and the SFM with screening parameter $\alpha = 0.2$, where the radius difference of the twins at 2 $M_\odot$ is about 3 km. In that figure we show also the most recent result 1.947 ± 0.018 $M_\odot$ for the mass measurement of PSR J1614-2230 (Arzoumanian et al. 2018).

Note that the case of two compact star families with an overlapping range in masses but different radii has also been considered in another context, namely for the conversion of metastable hadronic stars to strange stars (Bombaci et al. 2016; Bhattacharyya et al. 2017; Dondi et al. 2017), where the branch of strange stars can reach or exceed the 2 $M_\odot$ mass constraint at larger radii than the smaller and less massive hadronic stars have. The testable consequences of this two families scenario in the context of gravitational wave signals from compact star merger events have been recently discussed in Drago and Pagliara (2018), Drago et al. (2018).

---

[11] Note that the transition from a configuration at the endpoint of the neutron star branch in the $M - R$ diagram to the third family branch of hybrid stars (triggered, e.g., by mass accretion or spin-down) occurs under simultaneous conservation of baryon mass and angular momentum, as has been demonstrated in Bejger et al. (2017) for the case of the high-mass twin EoS (Benic et al. 2015b). It could therefore occur at the free-fall timescale, accompanied by a burst-type phenomenon (Alvarez-Castillo et al. 2015) (such as a fast radio burst (Falcke and Rezzolla 2014)) unlike the case discussed earlier in Glendenning et al. (1997) where an angular momentum mismatch had to be compensated by, e.g., dipole radiation an estimated timescale for the transition period of 10^5 years.

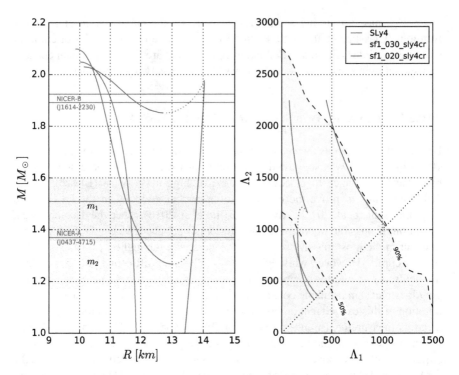

Fig. 7.11 Left panel: $M - R$ relations for a stiff (DD2_p40, blue line) and a soft (Sly4, red line) hadronic EoS. The result for hybrid stars binaries with hybrid stars from the third family branch obtained by a phase transition to quark core matter (SFM with screening parameters $\alpha = 0.2$, and 0.3, green lines) together with the mass ranges m_1 and m_2 of the stars in the binary merger GW170817. The red horizontal lines indicate the the mass bands for the nearest millisecond pulsar J0437-4715 (Reardon et al. 2016) and for the high-mass pulsar J1614-2230 (Arzoumanian et al. 2018). Both pulsars are target for the radius measurement by NICER (https://heasarc.gsfc.nasa. gov/docs/nicer). Right panel: LVC constraint (Abbott et al. 2017b) on tidal deformabilities Λ_2 vs. Λ_1 for the low-spin prior with a 50% (grey region, bordered by dashed line) and a 90% (just ashed line) confidence level region from the binary compact star merger GW170817, compared to results shown in the left panel. A merger of two hybrid stars from the third family branch is equivalent to that of a soft hadronic EoS. From Bejger et al. (2018), see also Paschalidis et al. (2018)

7.10.2.2 Low-Mass Twin Stars: GW170817 and NICER

With the detection of gravitational waves from the inspiral phase of the neutron star merger GW170817 by the LIGO Scientific and Virgo Collaboration (Abbott et al. 2017b) and the subsequent detection of the associated kilonova event in the galaxy NGC4993 in all ranges of the electromagnetic spectrum (Abbott et al. 2017a), the window for multi-messenger astronomy has been opened. It is possible that already this merger event with stars from the mass ranges $M_1 = 1.36$–$1.60\ M_\odot$ and $M_2 = 1.16$–$1.36\ M_\odot$ could have been a binary not of two regular neutron stars but rather a neutron star and a hybrid star or even a pair two hybrid stars, see the right panel of Fig. 7.11. With the same hybrid EoS as for the high-mass twin case we can obtain also typical mass (low-mass) twins, just by slightly varying the unknown screening parameter to be $\alpha = 0.3$. From examining the LIGO constraint on tidal deformabilities in the right panel of Fig. 7.11, it is clear that a neutron star - neutron star scenario for a soft hadronic EoS (SLy4) can not be distinguished from a hybrid star - hybrid star scenario. A hybrid star - neutron star scenario is also possible, but a neutron star - neutron star scenario with a stiff hadronic EoS is only marginally compatible with GW170817, see Paschalidis et al. (2018). In Annala et al. (2018) the constraint on tidal deformability has been translated to a limiting radius for a 1.4 $M_\odot$ neutron star (within the neutron star - neutron star scenario) of $R_{1.4,\mathrm{max}} = 13.6\,\mathrm{km}$.

In Fig. 7.11 we show also the mass range $1.44 \pm 0.07\ M_\odot$ of PSR J0437-4715 (Reardon et al. 2016) which is the primary target of the radius measurement by NICER (https://heasarc.gsfc.nasa.gov/docs/nicer). While at this moment the low-mass twin case is only an option, competing with the case of an ordinary neutron star merger with soft hadronic EoS, the soft hadronic EoS could be ruled out should NICER (https://heasarc.gsfc.nasa.gov/docs/nicer) announce a radius for the nearest millisecond pulsar PSR J0437-4715 in excess of $\sim$13 km (Paschalidis et al. 2018).

7.10.3 Masquerade?

It may well be that the deconfinement phase transition in the dense matter EoS does not leave a recognizable imprint on the mass-radius diagram of compact stars. This could have different reasons, for instance: (i) the critical density for the onset is too high to be reached in compact star interiors, even at highest masses, (ii) the quark matter EoS is too soft to carry a hadronic mantle so that the onset of quark matter results in gravitational instability and thus determines the value of the maximum mass of neutron stars, (iii) the transition is too weak (the latent heat $\Delta\varepsilon$ is too small) to lead to a recognizable (D)isconnected third family branch but rather results in a hybrid star branch (C)onnected to the neutron star one and not distinguishable from the pure neutron star case.

This latter case has been dubbed "masquerade" in Alford et al. (2005). It would require other methods of detecting the deconfinement transition in neutron star interiors than to measure the mass-radius dependence. For instance the cooling behaviour could be qualitatively different for stars above the threshold mass. The best case, however, would be a galactic supernova event which would emit a detectable second neutrino burst that has to be associated with the conversion of nuclear matter to quark matter.

7.11   Conclusions

With their tremendous gravity and their very high magnetic fields, the interior of neutron stars exhibits novel phases that cannot be reproduced in terrestrial laboratories. The atmosphere and the surface of a neutron star can be probed by spectroscopic measurements, although the inferred composition may still depend on the adopted atmospheric model. The interior of a neutron star is not directly observable. The outer most region is predicted to consist of a solid crust made of fully ionised nuclei embedded in a highly degenerate electron gas. Under the assumption of cold catalysed matter, nuclei are arranged on a perfect body-centred. cubic lattice, and their composition is completely determined by experimental atomic mass measurements, currently up to a density of about $6 \times 10^{10} \, \mathrm{g\,cm}^{-3}$. Future measurements will allow us to drill deeper into the crust. However, the cold catalysed matter hypothesis may not be very realistic. The composition of a newly formed hot neutron star is likely to become frozen when the crust solidifies, or at even earlier times. Which nuclides would thus be produced and how would they arrange? Could the crust consist of a disordered solid? Following the composition of a matter element as it cools down is very challenging as this requires the knowledge of all the relevant reaction rates, in addition to the dense-matter properties that govern the thermal evolution. Moreover, the presence of a high magnetic field or the accretion of matter from a companion star may radically change the constitution and the structure of the crust. Different neutron stars may thus have different crusts depending on their history. With increasing pressure, nuclei become progressively more neutron rich until some neutrons become unbound. The onset of neutron emission by nuclei, at densities of order $10^{11} \, \mathrm{g\,cm}^{-3}$ marks the transition to the inner crust, where neutron-proton clusters are immersed in a neutron ocean (superfluid at low enough temperatures). With further compression, clusters might fuse and form a liquid mantle of nuclear "pastas" that eventually dissolve into a uniform mixture of nucleons and electrons at a density of order $10^{14} \, \mathrm{g\,cm}^{-3}$.

At supersaturation densities, heavier baryons such as hyperons and deltas may be excited before the deconfined quark matter phases appear, most likely in a superconducting state. The details of this transition are largely unknown because no benchmark exist. So we describe the main challenging questions that our present understanding faces and try some answers. What is the confinement/deconfinement mechanism in cold degenerate matter? How is it intertwined with chiral symmetry breaking/restoration in the presence of strong diquark correlations (color super-conductivity)? Which mechanisms determine the stiffness of nuclear matter and quark matter? How could eventually a unified description of quark and nuclear matter be achieved? Do hyperons occur in neutron stars or does nature choose to deconfine strongly interacting matter before heavy baryons would get excited? Can hybrid stars with strange quark matter content be stable? Does strangeness occur in compact star interiors at all? How far is the asymptotic perturbative QCD description from the density range probed by the phenomenology of compact stars?

We discussed the existence of strange matter in compact stars in hadronic and quark matter phases. The appearance of hyperons leads to a hyperon puzzle in approaches based on effective baryon-baryon potentials but is not a severe problem in relativistic mean field models. The puzzle is resolved for a stiffening of hadronic matter at supersaturation densities, an effect based on the quark Pauli quenching between hadrons. We further outlined the conflict between the necessity to implement dynamical chiral symmetry breaking for a realistic quark matter model and the condition of undressed, approximate massless quarks for the appearance of absolutely stable strange quark matter. The existence of absolutely stable strange quark matter cannot be excluded on theoretical grounds only. However, we outlined the problems of the reasoning that lead to this hypothesis. In general, the role of strangeness in compact stars in hadronic or quark matter realizations remains unsettled.

As a workhorse for the theory chiral quark models of the NJL type have been discussed and their extension to a relativistic density-functional theory has been advertised because it allows to implement features of quark confinement simultaneously with dynamical chiral symmetry breaking, color superconductivity and stiffening at high densities which all are required for the phenomenological description of compact star physics. Further unknown white spots in the theory of superdense neutron star matter can be circumvented with interpolating models and Bayesian techniques. One of the modern questions with a chance to be answered by next generations of observational campaigns is the nature of the deconfinement transition. Is it a strong first order transition which would give rise to the phenomenon of mass twin stars?

Acknowledgements The work of N.C. was supported by Fonds de la Recherche Scientifique - FNRS (Belgium) under grants n° CDR-J.0187.16 and CDR-J.0115.18. D.B. received support from Narodowe Centrum Nauki - NCN (Poland) under contract No. UMO-2014/13/B/ST9/02621 (Opus7) and by the MEPhI Academic Excellence programme under contract no. 02.a03.21.0005. This work was also partially supported by the European Cooperation in Science and Technology (COST) Action MP1304 *NewCompStar*.

References

Aarts, G., Allton, C., De Boni, D., Hands, S., Jäger, B., Praki, C., Skullerud, J.I.: J. High Energy Phys. **1706**, 034 (2017)

Abbott, B.P., et al., [LIGO Scientific and Virgo and Fermi-GBM and INTEGRAL Collaborations]: Astrophys. J. **848**(2), L13 (2017a)

Abbott, B.P. et al., [LIGO Scientific and Virgo Collaborations]: Phys. Rev. Lett. **119**(16), 161101 (2017b)

Abuki, H., Kunihiro, T.: Nucl. Phys. A **768**, 118 (2006)

Abuki, H., Anglani, R., Gatto, R., Nardulli, G., Ruggieri, M.: Phys. Rev. D **78**, 034034 (2008)

Abuki, H., Baym, G., Hatsuda, T., Yamamoto, N.: Phys. Rev. D **81**, 125010 (2010)

Alberico, W.M., Drago, A., Ratti, C.: Nucl. Phys. A **706**, 143 (2002)

Alcain, P.N., Giménez Molinelli, P.A., Nichols, J.I., Dorso, C.O.: Phys. Rev. C **89**, 055801 (2014)

Alford, M.G.: Ann. Rev. Nucl. Part. Sci. **51**, 131 (2001)

Alford, M.G., Sedrakian, A.: Phys. Rev. Lett. **119**(16), 161104 (2017)

Alford, M., Braby, M., Paris, M.W., Reddy, S.: Astrophys. J. **629**, 969 (2005)

Alford, M., Blaschke, D., Drago, A., Klähn, T., Pagliara, G., Schaffner-Bielich, J.: Nature **445**, E7 (2007)

Alford, M.G., Schmitt, A., Rajagopal, K., Schäfer, T.: Rev. Mod. Phys. **80**, 1455 (2008)

Alford, M.G., Han, S., Prakash, M.: Phys. Rev. D **88**(8), 083013 (2013)

Alford, M.G., Burgio, G.F., Han, S., Taranto, G., Zappala, D.: Phys. Rev. D **92**(8), 083002 (2015)

Alvarez-Castillo, D.E., Blaschke, D. (2013). Arxiv:1304.7758 [astro-ph.HE]

Alvarez-Castillo, D.E., Blaschke, D.B.: Phys. Rev. C **96**(4), 045809 (2017)

Alvarez-Castillo, D.E., Bejger, M., Blaschke, D., Haensel, P., Zdunik, L. (2015). Arxiv:1506.08645 [astro-ph.HE]

Alvarez-Castillo, D., Benic, S., Blaschke, D., Han, S., Typel, S.: Eur. Phys. J. A **52**(8), 232 (2016)

Alvarez-Castillo, D., Blaschke, D., Typel, S.: Astron. Nachr. **338**(9–10), 1048 (2017)

Alvarez-Castillo, D.E., Blaschke, D.B., Grunfeld, A.G., Pagura, V.P. (2018). Arxiv:1805.04105 [hep-ph]

An, H., et al.: Astrophys. J. **790**, 60 (2014)

Anglani, R., Casalbuoni, R., Ciminale, M., Ippolito, N., Gatto, R., Mannarelli, M., Ruggieri, M.: Rev. Mod. Phys. **86**, 509 (2014)

Annala, E., Ecker, C., Hoyos, C., Jokela, N., Rodríguez-Fernández, D., Vuorinen, A. (2017). Arxiv:1711.06244 [astro-ph.HE]

Annala, E., Gorda, T., Kurkela, A., Vuorinen, A.: Phys. Rev. Lett. **120**(17), 172703 (2018)

Anselmino, M., Predazzi, E., Ekelin, S., Fredriksson, S., Lichtenberg, D.B.: Rev. Mod. Phys. **65**, 1199 (1993)

Antoniadis, J., et al.: Science **340**, 1233232 (2013)

Ardeljan, N.V., Bisnovatyi-Kogan, G.S., Moiseenko, S.G.: Mon. Not. R. Astron. Soc. **359**, 333 (2005)

Arzoumanian, Z., et al. ,[NANOGrav Collaboration]: Astrophys. J. Suppl. **235**(2), 37 (2018)

Asakawa, M., Hatsuda, T.: Phys. Rev. D **55**, 4488 (1997)

Audi, G., Wang, M., Wapstra, A.H., Kondev, F.G., MacCormick, M., Xu, X., Pfeiffer, B.: Chin. Phys. C **36**, 1287 (2012)

Avancini, S.S., Menezes, D.P., Alloy, M.D., Marinelli, J.R., Moraes, M.M.W., Providência, C.: Phys. Rev. C **78**, 015802 (2008)

Aymard, F., Gulminelli, F., Margueron, J.: Phys. Rev. C **89**, 065807 (2014)

Ayriyan, A., Grigorian, H.: EPJ Web Conf. **173**, 03003 (2018)

Ayriyan, A., Bastian, N.-U., Blaschke, D., Grigorian, H., Maslov, K., Voskresensky, D.N.: Phys. Rev. C **97**(4), 045802 (2018)

Baiko, D.A.: Phys. Rev. E **80**, 046405 (2009)

Baldo, M., Biurgio, F.: Rep. Prog. Phys. **75**, 026301 (2012)

Baldo, M., Burgio, G.F., Schulze, H.J.: Phys. Rev. C **61**, 055801 (2000)

Baldo, M., Burgio, G.F., Schulze, H.-J. (2003). Astro-ph/0312446

Baldo, M., Saperstein, E.E., Tolokonnikov, S.V.: Nucl. Phys. A **775**, 235 (2006)

Baldo, M., Saperstein, E.E., Tolokonnikov, S.V.: Eur. Phys. J. A **32**, 97 (2007)

Bao, S.S., Shen, H.: Phys. Rev. C **93**, 025807 (2016)

Bardeen, J., Cooper, L.N., Schrieffer, J.R.: Phys. Rev. **108**, 1175 (1957)

Basilico, D., Peña Arteaga, D., Roca-Maza, X., Coló, G.: Phys. Rev. C **92**, 035802 (2015)

Bastian, N.U.F., Blaschke, D., Fischer, T., Ropke, G.: Universe **4**, 67 (2018)

Bastian, N.U.F., Blaschke, D.B., Cierniak, M., Fischer, T., Kaltenborn, M.A.R., Marczenko, M., Typel, S.: EPJ Web Conf. **171**, 20002 (2018)

Batani, D., Morelli, A., Tomasini, M., Benuzzi-Mounaix, A., Philippe, F., Koenig, M., Marchet,B., Masclet, I., Rabec, M., Reverdin, Ch., Cauble, R., Celliers, P., Collins, G., Da Silva, L., Hall, T., Moret, M., Sacchi, B., Baclet, P., Cathala, B.: Phys. Rev. Lett. **88**, 235502 (2002)

Baym, G., Pethick, C., Sutherland, P.: Astrophys. J. **170**, 299 (1971)

Baym, G., Bethe, H.A., Pethick, C.J.: Nucl. Phys. A **175**, 225 (1971)

Baym, G., Hatsuda, T., Kojo, T., Powell, P.D., Song, Y., Takatsuka, T.: Rep. Prog. Phys. **81**(5), 056902 (2018)

Bedaque, P.F., Mahmoodifar, S., Sen, S.: Phys. Rev. C **88**, 055801 (2013)

Bejger, M., Blaschke, D., Haensel, P., Zdunik, J.L., Fortin, M.: Astron. Astrophys. **600**, A39 (2017)

Bejger, M., Blaschke, D., Bastian, N.-U.: (2018, in preparation)

Bender, M., Heenen, P.-H., Reinhard, P.-G.: Rev. Mod. Phys. **75**, 121 (2003)

Benic, S.: Eur. Phys. J. A **50**, 111 (2014)

Benic, S., Mishustin, I., Sasaki, C.: Phys. Rev. D **91**(12), 125034 (2015a)

Benic, S., Blaschke, D., Alvarez-Castillo, D.E., Fischer, T., Typel, S.: Astron. Astrophys. **577**, A40 (2015b)

Bentz, W., Thomas, A.W.: Nucl. Phys. A **696**, 138 (2001)

Bentz, W., Horikawa, T., Ishii, N., Thomas, A.W.: Nucl. Phys. A **720**, 95 (2003)

Berezhiani, Z., Bombaci, I., Drago, A., Frontera, F., Lavagno, A.: Astrophys. J. **586**, 1250 (2003)

Berry, D.K., Caplan, M.E., Horowitz, C.J., Huber, G., Schneider, A.S.: Phys. Rev. C **94**, 055801 (2016)

Bhattacharyya, A., Ghosh, S.K., Hanauske, M., Raha, S.: Astron. Astrophys. **418**, 795 (2004)

Bhattacharyya, S., Bombaci, I., Logoteta, D., Thampan, A.V.: Astrophys. J. **848**(1), 65 (2017)

Blaes, O., Blanford, R., Madau, P., Koonin, S.: Astrophys. J. **363**, 612 (1990)

Blaschke, D.: PoS **BaldinISHEPPXXII**, 113 (2015). [Arxiv:1502.06279 [nucl-th]]

Blaschke, D., Tovmasian, T., Kämpfer, B.: Sov. J. Nucl. Phys. **52**, 675 (1990) [Yad. Fiz. **52**, 1059 (1990)]

Blaschke, D., Fredriksson, S., Grigorian, H., Öztas, A.M., Sandin, F.: Phys. Rev. D **72**, 065020 (2005)

Blaschke, D., Sandin, F., Klähn, T., Berdermann, J.: Phys. Rev. C **80**, 065807 (2009)

Blaschke, D.B., Sandin, F., Skokov, V.V., Typel, S.: Acta Phys. Polon. Suppl. **3**, 741 (2010)

Blaschke, D., Berdermann, J., Lastowiecki, R.: Prog. Theor. Phys. Suppl. **186**, 81 (2010)

Blaschke, D., Alvarez Castillo, D.E., Benic, S., Contrera, G., Lastowiecki, R.: PoS ConfinementX **249** (2012). [arXiv:1302.6275 [hep-ph]]

Blaschke, D., Alvarez-Castillo, D.E., Benic, S.: PoS CPOD **2013**, 063 (2013). [Arxiv:1310.3803 [nucl-th]]

Blaschke, D., Dubinin, A.S., Zablocki, D.: PoS **BaldinISHEPPXXII**, 083 (2015)

Blaschke, D., Grigorian, H., Röpke, G.: (2018, in preparation)

Bodmer, A.R.: Phys. Rev. D **4**, 1601 (1971)

Bombaci, I.: Lect. Notes Phys. **578**, 253 (2001)

Bombaci, I., Parenti, I., Vidana, I.: Astrophys. J. **614**, 314 (2004)

Bombaci, I., Lugones, G., Vidana, I.: Astron. Astrophys. **462**, 1017 (2007)

Bombaci, I., Logoteta, D., Vidaña, I., Providência, C.: Eur. Phys. J. A **52**(3), 58 (2016)

Bonanno, L., Sedrakian, A.: Astron. Astrophys. **539**, A16 (2012)

Botvina, A.S., Mishustin, I.N.: Phys. Rev. C **72**, 048801 (2005)

Brack, M., Guet, C., Håkansson, H.-B.: Phys. Rep. **123**, 275 (1985)

Buballa, M.: Phys. Rep. **407**, 205 (2005)

Burgio, G.F., Baldo, M., Sahu, P.K., Santra, A.B., Schulze, H.J.: Phys. Lett. B **526**, 19 (2002a)

Burgio, G.F., Baldo, M., Sahu, P.K., Schulze, H.J.: Phys. Rev. C **66**, 025802 (2002b)

Buyukcizmeci, N., Botvina, A.S., Mishustin, I.N.: Astrophys. J. **789**, 33 (2014)

Cahill, R.T.: Aust. J. Phys. **42**, 171 (1989)

Carlson, J., Gandolfi, S., Pederiva, F., Pieper, S.C., Schiavilla, R., Schmidt, K.E., Wiringa, R.B.: Rev. Mod. Phys. **87**, 1067 (2015)

Centelles, M., Viñas, X., Barranco, M., Schuck, P.: Ann. Phys. **221**, 165 (1993)

Chamel, N.: Phys. Rev. C **85**, 035801 (2012)

Chamel, N.: J. Phys. Conf. Ser. **932**, 012039 (2017a)

Chamel, N.: J. Astrophys. Astron. **38**, 43 (2017b)

Chamel, N., Fantina, A.F.: Phys. Rev. D **92**, 023008 (2015)

Chamel, N., Fantina, A.F.: Phys. Rev. D **93**, 063001 (2016a)

Chamel, N., Fantina, A.F.: Phys. Rev. C **94**, 065802 (2016b)

Chamel, N., Haensel, P.: Living Rev. Relativ. **11**, 10 (2008). https://link.springer.com/article/10. 12942/lrr-2008-10

Chamel, N., Naimi, S., Khan, E., Margueron, J.: Phys. Rev. C **75**, 055806 (2007)

Chamel, N., Pavlov, R.L., Mihailov, L.M., Velchev, Ch.J., Stoyanov, Zh.K., Mutafchieva, Y.D., Ivanovich, M.D., Pearson, J.M., Goriely, S.: Phys. Rev. C **86**, 055804 (2012)

Chamel, N., Goriely, S., Pearson, J.M.: In: Broglia, R.A., Zelevinsky, V. (eds.) Fifty Years of Nuclear BCS: Pairing in Finite Systems, pp. 284–296. World Scientific Publishing Co. Pte. Ltd., Singapore (2013a)

Chamel, N., Pavlov, R.L., Mihailov, L.M., Velchev, Ch.J., Stoyanov, Zh.K., Mutafchieva, Y.D., Ivanovich, M.D.: Bulg. J. Phys. **40**, 275 (2013b)

Chamel, N., Fantina, A.F., Zdunik, J.L., Haensel, P.: Phys .Rev. C **91**, 055803 (2015a)

Chamel, N., Pearson, J.M., Fantina, A.F., Ducoin, C., Goriely, S., Pastore, A.: Acta Phys. Polon. B **46**, 349 (2015b)

Chamel, N., Stoyanov, Zh.K., Mihailov, L.M., Mutafchieva, Y.D., Pavlov, R.L., Velchev, Ch.J.: Phys. Rev. C **91**, 065801 (2015c)

Chatterjee, D., Elghozi, T., Novak, J., Oertel, M.: Mon. Rot. R. Astron. Soc. **447**, 3785 (2015)

Chamel, N., Mutafchieva, Y.D., Stoyanov, Zh.K., Mihailov, L.M., Pavlov, R.L.: J. Phys. Conf. Ser. **724**, 012034 (2016)

Chamel, N., Mutafchieva, Y.D., Stoyanov, Zh.K., Mihailov, L.M., Pavlov, R.L.: In: Tadjer, A. et al. (eds.) Quantum Systems in Physics, Chemistry, and Biology. Progress in Theoretical Chemistry and Physics, vol. 30, pp. 181–191. Springer, Berlin (2017a)

Chamel, N., Fantina1, A.F., Pearson, J.M., Goriely, S.: EPJ Web Conf. **137**, 09001 (2017b)

Chen, Y.J., Phys. Rev. C **95**, 035807 (2017)

Christian, J.E., Zacchi, A., Schaffner-Bielich, J.: Eur. Phys. J. A **54**(2), 28 (2018)

Clement, H.: Prog. Part. Nucl. Phys. **93**, 195 (2017)

Cloet, I.C., Roberts, C.D.: Prog. Part. Nucl. Phys. **77**, 1 (2014)

De Blasio, F.V.: Astron. Astrophys. **353**, 1129 (2000)

de Lima, R.C.R., Avancini, S.S., Providência, C.: Phys. Rev. C **88**, 035804 (2013)

De Rujula, A., Georgi, H., Glashow, S.L.: Phys. Rev. D **12**, 147 (1975)

Deibel, A.T., Steiner, A.W., Brown, E.F.: Phys. Rev. C **90**, 025802 (2014)

Demorest, P., Pennucci, T., Ransom, S., Roberts, M., Hessels, J.: Nature **467**, 1081 (2010)

Detar, C.E., Kunihiro, T.: Phys. Rev. D **39**, 2805 (1989)

Dexheimer, V., Schramm, S., Zschiesche, D.: Phys. Rev. C **77**, 025803 (2008)

Dexheimer, V.A., Schramm, S.: Nucl. Phys. Proc. Suppl. **199**, 319 (2010)

Dickhoff, W.H., Barbieri, C.: Prog. Part. Nucl. Phys. **52**, 377 (2004)

Dondi, N.A., Drago, A., Pagliara, G.: EPJ Web Conf. **137**, 09004 (2017)

Dorso, C.O., Giménez Molinelli, P.A., Lópoez, J.A.: In: Bertulani, C., Piekarewicz, J. (eds.) Neutron Star Crust, pp. 151–169. Nova Science Publishers, New York (2012)

Dorso, C.O., Giménez Molinelli, P.A., López, J.A.: Phys. Rev. C **86**, 055805 (2012)

Douchin, F., Haensel, P.: Astron. Astrophys. **380**, 151 (2001)

Drago, A., Pagliara, G.: Astrophys. J. **852**(2), L32 (2018)

Drago, A., Lavagno, A., Pagliara, G.: Phys. Rev. D **69**, 057505 (2004)

Drago, A., Lavagno, A., Pagliara, G.: Phys. Rev. D **89**(4), 043014 (2014)

Drago, A., Lavagno, A., Pagliara, G., Pigato, D.: Phys. Rev. C **90**(6), 065809 (2014)

Drago, A., Pagliara, G., Popov, S.B., Traversi, S., Wiktorowicz, G.: Universe **4**(3), 50 (2018)

Drozdov, A.P., Eremets, M.I., Troyan, I.A., Ksenofontov, V., Shylin, S.I.: Nature **525**, 73 (2015)

Ducoin, C., Margueron, J., Providência, C., Vidaña, I.: Phys. Rev. C **83**, 045810 (2011)

Duguet, T.: Lecture Notes in Physics, vol. 879, pp. 293–350. Springer, Berlin (2014)

Dutta, A.K., Onsi, M., Pearson, J.M.: Phys. Rev. C **69**, 052801 (R) (2004)

Engstrom, T.A., Yoder, N.C., Crespi, V.H.: Astrophys. J. **818**, 183 (2016)

Estradé, A., et al.: Phys. Rev. Lett. **107**, 172503 (2011)

Falcke, H., Rezzolla, L.: Astron. Astrophys. **562**, A137 (2014)

Fang, J., Pais, H., Avancini, S., Providência, C.: Phys. Rev. C **94**, 062801(R) (2016)

Fang, J., Pais, H., Pratapsi, S., Avancini, S., Li, J., Providência, C.: Phys. Rev. C **95**, 045802 (2017)

Fang, J., Pais, H., Pratapsi, S., Providência, C.: Phys. Rev. C **95**, 062801 (R) (2017)
Fantina, A.F., Chamel, N., Mutafchieva, Y.D., Stoyanov, Zh.K., Mihailov, L.M., Pavlov, R.L.: Phys. Rev. C **93**, 015801 (2016)
Fantina, A.F., Chamel, N., Pearson, J.M., Goriely, S.: Nuovo Cimento C **39**, 400 (2017)
Fantina, A.F., Zdunik, J.L., Chamel, N., Pearson, J.M., Haensel, P., Goriely, S. (2018). Arxiv:1806.03861
Farhi, E., Jaffe, R.L.: Phys. Rev. D **30**, 2379 (1984)
Fattoyev, F.J., Horowitz, C.J., Schuetrumpf, B.: Phys. Rev. C **95**, 055804 (2017)
Fischer, T., et al.: Astrophys. J. Suppl. **194**, 39 (2011)
Fischer, T., Hempel, M., Sagert, I., Suwa, Y., Schaffner-Bielich, J.: Eur. Phys. J. A **50**, 46 (2014)
Fischer, T., et al.: Publ. Astron. Soc. Austral. **34**, 67 (2017)
Fischer, T., Bastian, N.U.F., Wu, M.R., Typel, S., Klähn, T., Blaschke, D.B. (2017). Arxiv:1712.08788 [astro-ph.HE]
Fonseca, E., et al.: Astrophys. J. **832**, 167 (2016)
Fortin, M., Providência, C., Raduta, Ad.R., Gulminelli, F., Zdunik, J.L., Haensel, P., Bejger, M.: Phys. Rev. C **94**, 035804 (2016)
Fortov, V.E., Lomonosov, I.V.: Shock Waves **20**, 53 (2010)
Frieben, J., Rezzolla, L.: Mon. Not. R. Astron. Soc. **427**, 3406 (2012)
Fukushima, K., Hatsuda, T.: Rep. Prog. Phys. **74**, 014001 (2011)
Fukushima, K., Sasaki, C.: Prog. Part. Nucl. Phys. **72**, 99 (2013)
Furusawa1, S., Sumiyoshi, K., Yamada, S., Suzuki, H.: Astrophys. J. **772**, 95 (2013)
Gerlach, U.H.: Phys. Rev. **172**, 1325 (1968)
Giménez Molinelli, P.A., Dorso, C.O.: Nucl. Phys. **A933**, 306 (2015)
Ginzburg, V.L.: J. Stat. Phys. **1**, 3 (1969)
Glendenning, N.K.: Astrophys. J. **293**, 470 (1985)
Glendenning, N.K.: Phys. Rev. D **46**, 1274 (1992)
Glendenning, N.K.: Compact Stars: Nuclear Physics, Particle Physics, and General Relativity. Springer, Berlin (2000)
Glendenning, N.K., Kettner, C.: Astron. Astrophys. **353**, L9 (2000)
Glendenning, N.K., Pei, S., Weber, F.: Phys. Rev. Lett. **79**, 1603 (1997)
Gocke, C., Blaschke, D., Khalatyan, A., Grigorian, H. (2001). [hep-ph/0104183]
Gögelein, P., Müther, H.: Phys. Rev. C **76**, 024312 (2007)
Gonzalez-Boquera, C., Centelles, M., Viñas, X., Rios, A.: Phys. Rev. C **96**, 065806 (2017)
Goriely, S., Samyn, M., Pearson, J.M., Onsi, M.: Nucl. Phys. **A750**, 425 (2008)
Goriely, S., Pearson, J.M., Chamel, N.: In: Bertulani, C., Piekarewicz, J. (eds.) Neutron Star Crust, p. 213. Nova Science Publishers, New York (2012)
Goriely, S., Chamel, N., Pearson, J.M.: Phys. Rev. C **93**, 034337 (2016)
Graber, V., Andersson, N., Hogg, M.: Int. J. Mod. Phys. D **26**, 1730015 (2017)
Grams, G., Santos, A.M., Panda, P.K., Providência, C., Menezes, D.P.: Phys. Rev. C **95**, 055807 (2017)
Grams, G., Giraud, S., Fantina, A.F., Gulminelli, F.: Phys. Rev. C **97**, 035807 (2018)
Grigorian, H., Blaschke, D., Aguilera, D.N.: Phys. Rev. C **69**, 065802 (2004)
Grigorian, H., Blaschke, D., Voskresensky, D.: Phys. Rev. C **71**, 045801 (2005)
Grill, F., Margueron, J., Sandulescu, N.: Phys. Rev. C **84**, 065801 (2011)
Grill, F., Pais, H., Providência, C., Vidaña, I., Avancini, S.S.: Phys. Rev. C **90**, 045803 (2014)
Guo, L., et al.: Phys. Rev. C **76**, 065801 (2007)
Gulminelli, F., Raduta, Ad.R.: Phys. Rev. C **92**, 055803 (2015)
Gupta, N., Arumugam, P.: Phys. Rev. C **87**, 028801 (2013)
Gupta, S., Brown, E.F., Schatz, H., Möller, P., Kratz, K.-L.: Astrophys. J. **662**, 1188 (2007)
Gupta, S., Kawano, T., Möller, P.: Phys. Rev. Lett. **101**, 231101 (2008)
Haensel, P., Zdunik, J.L.: Astron. Astrophys. **227**, 431 (1990)
Haensel, P., Zdunik, J.L.: Astron. Astrophys. **404**, L33 (2003)
Haensel, P., Zdunik, J.L.: Astron. Astrophys. **480**, 459 (2008)

Haensel, P., Potekhin, A.Y., Yakovlev, D.G.: Neutron Stars 1: Equation of State and Structure. Springer, Berlin (2007)
Hagedorn, R., Rafelski, J. (1980). CERN-TH-2947, C80-08-24-1
Hagen, G., Jansen, G.R., Papenbrock, T.: Phys. Rev. Lett. **117**, 172501 (2016)
Hagen, G. et al.: Nat. Phys. **12**, 186 (2016)
Harrison, B.K., Wheeler, J.A.: Onzième Conseil de Physique Solvay. Stoops, Bruxelles (1958)
Harrison, B.K., Thorne, K.S., Wakano, M., Wheeler, J.A.: Gravitation Theory and Gravitational Collapse. The University of Chicago Press, Chicago (1965)
Hashimoto, M., Seki, H., Yamada, M.: Prog. Theor. Phys. **71**, 320 (1984)
Hasnaoui, K.H.O., Piekarewicz, J.: Phys. Rev. C **88**, 025807 (2013)
Hatsuda, T., Kunihiro, T.: Phys. Rep. **247**, 221 (1994)
Hatsuda, T., Tachibana, M., Yamamoto, N., Baym, G.: Phys. Rev. Lett. **97**, 122001 (2006)
Hebeler, K., Lattimer, J.M., Pethick, C.J., Schwenk, A.: Astrophys. J. **773**, 11 (2013)
Hell, T., Weise, W.: Phys. Rev. C **90**(4), 045801 (2014)
Hempel, M., Fischer, T., Schaffner-Bielich, J., Liebendörfer, M.: Astrophys. J. **748**, 70 (2012)
Hergert, H., Bogner, S.K., Morris, T.D., Schwenk, A., Tsukiyama, K.: Phys. Rep. **621**, 165 (2016)
Herzog, M., Röpke, F.K.: Phys. Rev. D **84**, 083002 (2011)
Ho, W.C.G., Heinke, C.O.: Nature **462**, 71 (2009)
Hoffman, K., Heyl, J.: Mon. Not. R. Astron. Soc.**426**, 2404 (2012)
Hofmann, F., Keil, C.M., Lenske, H.: Phys. Rev. C **64**, 025804 (2001)
Holt, J.W., Rho, M., Weise, W.: Phys. Rep. **621**, 2 (2016)
Horowitz, C.J., Berry, D.K.: Phys. Rev. C **79**, 065803 (2009)
Horowitz, C.J., Perez-Garcia, M.A., Berry, D.K., Piekarewicz, J.: Phys. Rev. C **72**, 035801 (2005)
Horowitz, C.J., Berry, D.K., Briggs, C.M., Caplan, M.E., Cumming, A., Schneider, A.S.: Phys. Rev. Lett. **114**(3), 031102 (2015)
Iida, K., Oyamatsu, K.: Eur. Phys. J. **A50**, 42 (2014)
Iida, K., Sato, K.: Phys. Rev. C **58**, 2538 (1998)
Ippolito, N., Ruggieri, M., Rischke, D., Sedrakian, A., Weber, F.: Phys. Rev. D **77**, 023004 (2008)
Jaffe, R.L.: Phys. Rep. **409**, 1 (2005)
Jido, D., Hatsuda, T., Kunihiro, T.: Phys. Rev. D **63**, 011901 (2001a)
Jido, D., Oka, M. Hosaka, A.: Prog. Theor. Phys. **106**, 873 (2001b)
Kaltenborn, M.A.R., Bastian, N.U.F., Blaschke, D.B.: Phys. Rev. D **96**(5), 056024 (2017)
Kaminker, A.D., Yakovlev, D.G., Potekhin, A.Y., Shibazaki, N., Shternin, P.S., Gnedin, O.Y.: Astrophys. Space Sci. **308**, 423 (2007)
Kämpfer, B.: J. Phys. A **14**, L471 (1981)
Kankiewicz, P., Rybczynski, M., Wlodarczyk, Z., Wilk, G.: EPJ Web Conf. **171**, 20003 (2018)
Khvorostukin, A.S., Skokov, V.V., Toneev, V.D., Redlich, K.: Eur. Phys. J. C **48**, 531 (2006)
Kiuchi, K., Yoshida, S.: Phys.Rev.D **78**, 044045 (2008)
Kobayashi, M., Maskawa, T.: Prog. Theor. Phys. **44**, 1422–1424 (1970)
Kobyakov, D., Pethick, C.J.: Phys. Rev. Lett. **112**, 112504 (2014)
Kobyakov, D., Pethick, C.J.: Phys. Rev. C **94**, 055806 (2016)
Kojo, T., Hidaka, Y., McLerran, L., Pisarski, R.D.: Nucl. Phys. A **843**, 37 (2010a)
Kojo, T., Pisarski, R.D., Tsvelik, A.M.: Phys. Rev. D **82**, 074015 (2010b)
Kojo, T., Powell, P.D., Song, Y., Baym, G.: Phys. Rev. D **91**(4), 045003 (2015)
Kolomeitsev, E.E., Maslov, K.A., Voskresensky, D.N.: Nucl. Phys. A **961**, 106 (2017)
Kozhberov, A.A.: Astrophys. Space Sci. **361**, 256 (2016)
Kozhberov, A., Baiko, D.: Phys. Plasmas **22**, 092903 (2015)
Klähn, T., Blaschke, D.: EPJ Web Conf. **171**, 08001 (2018)
Klähn, T., Fischer, T.: Astrophys. J. **810**(2), 134 (2015)
Klähn, T., et al.: Phys. Rev. C **74**, 035802 (2006)
Klähn, T., Blaschke, D., Sandin, F., Fuchs, C., Faessler, A., Grigorian, H., Röpke, G., Trümper, J.: Phys. Lett. B **654**, 170 (2007)
Klähn, T., Lastowiecki, R., Blaschke, D.B.: Phys. Rev. D **88**(8), 085001 (2013)
Klevansky, S.P.: Rev. Mod. Phys. **64**, 649 (1992)

Kreim, S., Hempel, M., Lunney, D., Schaffner-Bielich, J.: Int. J. Mass Spectrom. **349–350**, 63 (2013)

Krüger, T., Tews, I., Hebeler, K., Schwenk, A.: Phys. Rev. C **88**, 025802 (2013)

Kunihiro, T.: Phys. Lett. B **271**, 395 (1991)

Kurkela, A., Fraga, E.S., Schaffner-Bielich, J., Vuorinen, A.: Astrophys. J. **789**, 127 (2014)

Kurkela, A., Romatschke, P., Vuorinen, A.: Phys. Rev. D **81**, 105021 (2010)

Lassaut, M., Flocard, H., Bonche, P., Heenen, P.H., Suraud, E.: Astron. Astrophys. **183**, L3 (1987)

Lastowiecki, R., Blaschke, D., Grigorian, H., Typel, S.: Acta Phys. Polon. Supp. **5**, 535 (2012)

Lattimer, J.M., Lim, Y.: Astrophys. J. **771**, 51 (2013)

Lawley, S., Bentz, W., Thomas, A.W.: J. Phys. G **32**, 667 (2006)

Lai, D., Shapiro, S.L.: Astrophys. J. **383**, 745 (1991)

Lai, D., Abrahams, A.M., Shapiro, S.L.: Astrophys. J. **377**, 612 (1991)

Lau, R., et al.: Astrophys. J. **859**, 62 (2018)

Li, A., Zuo, W., Peng, G.X.: Phys. Rev. C **91**(3), 035803 (2015)

Lieb, E.H., Narnhofer, H.: J. Stat. Phys. **12**, 291 (1975)

Lim, Y., Holt, J.W.: Phys. Rev. C **95**, 034326 (2017)

Logoteta, D., Bombaci, I.: Phys. Rev. D **88**, 063001 (2013)

Lorenz, C.P., Ravenhall, D.G., Pethick, C.J.: Phys. Rev. Lett. **70**, 379 (1993)

Lunney, D., Pearson, J.M., Thibault, C.: Rev. Mod. Phys. **75**, 1021 (2003)

Magierski, P., Heenen, P.-H.: Phys. Rev. C **65**, 045804 (2002)

Maieron, C., Baldo, M., Burgio, G.F., Schulze, H.J.: Phys. Rev. D **70**, 043010 (2004)

Makishima, K., Enoto, T., Hiraga, J.S., Nakano, T., Nakazawa, K., Sakurai, S., Sasano, M., Murakami, H.: Phys. Rev. Lett. **112**, 171102 (2014)

Marczenko, M.L., Blaschke, D., Redlich, K., Sasaki, C. (2018). Arxiv:1805.06886 [nucl-th]

Maruyama, T., Chiba, S., Schulze, H.J., Tatsumi, T.: Phys. Lett. B **659**, 192 (2008)

Maslov, K.A., Kolomeitsev, E.E., Voskresensky, D.N.: Phys. Lett. B **748**, 369 (2015)

Maslov, K.A., Kolomeitsev, E.E., Voskresensky, D.N.: Nucl. Phys. A **950**, 64 (2016)

Masuda, K., Hatsuda, T., Takatsuka, T.: Astrophys. J. **764**, 12 (2013a)

Masuda, K., Hatsuda, T., Takatsuka, T.: Prog. Theor. Exp. Phys. **2013**(7), 073D01 (2013b)

Margueron, J., Sandulescu, N.: In: Bertulani, C., Piekarewicz, J. (eds.) Neutron Star Crust, p. 65. Nova Science Publishers, New York (2012)

Margueron, J., Van Giai, N., Sandulescu, N.: In: Lombardo, U., Baldo, M., Burgio, F., Schulze H.-J. (eds.) Exotic States of Nuclear Matter. Proceedings of the International Symposium EXOCT07, p. 362. World Scientific Publishing, Singapore (2008)

Martin, N., Urban, M.: Phys. Rev. C **92**, 015803 (2015)

Maruyama, T., Watanabe, G., Chiba, S.: Prog. Theor. Exp. Phys. **2012**, 01A201 (2012)

Meisel, Z., et al.: Phys. Rev. Lett. **115**, 162501 (2015)

Meisel, Z., et al.: Phys. Rev. C **93**, 035805 (2016)

Mignani, R.P., Testa, V., González Caniulef, D., Taverna, R., Turolla, R., Zane, S., Wu, K.: Mon. Not. R. Astron. Soc. **465**, 492 (2017)

Miyatsu, T., Yamamuro, S., Nakazato, K.: Astrophys. J. **777**, 4 (2013)

More, R.M., Warren, K.H., Young, D.A., Zimmerman, G.B.: Phys. Fluids **31**, 3059 (1988)

Mukherjee, A., Schramm, S., Steinheimer, J., Dexheimer, V.: Astron. Astrophys. **608**, A110 (2017)

Nakazato, K., Iida, K., Oyamatsu, K.: Phys. Rev. C **83**, 065811 (2011)

Nandi, R., Bandyopadhyay, D.: J. Phys. Conf. Ser. **312**, 042016 (2011)

Nandi, R., Bandyopadhyay, D.: J. Phys. Conf. Ser. **420**, 012144 (2013)

Nandi, R., Schramm, S.: Phys. Rev. C **94**, 025806 (2016)

Nandi, R., Bandyopadhyay, D., Mishustin, I.N., Greiner, W.: Astrophys. J. **736**, 156 (2011)

Negele, J.W., Vautherin, D.: Nucl. Phys. **A 207**, 298 (1973)

Newton, W.G., Stone, J.R.: Phys. Rev. C **79**, 055801 (2009)

Newton, W.G., Gearheart, M., Li, B.-A.: Astrophys. J. Suppl. Ser. **204**, 9 (2013)

Ng, C.Y., Kaspi, V.M.: AIP Conf. Proc. **1379**, 60 (2011)

Nice, D.J., Splaver, E.M., Stairs, I.H., Loehmer, O., Jessner, A., Kramer, M., Cordes, J.M.: Astrophys. J. **634**, 1242 (2005)

Oertel, M., Hempel, M., Klähn, T., Typel, S.: Rev. Mod. Phys. **89**(1), 015007 (2017)

Ogasawara, R., Sato, K.: Prog. Theor. Phys. **68**, 222 (1982)

Okamoto, M., Maruyama, T., Yabana, K., Tatsumi, T.: Phys. Rev. C **88**, 025801 (2013)

Olausen, S.A., Kaspi, V.M.: Astrophys. J. Suppl. Ser. **212**, 6 (2014)

Onsi, M., Przysiezniak, H., Pearson, J.M.: Phys. Rev. C **55**, 3139 (1997)

Onsi, M., Dutta, A.K., Chatri, H., Goriely, S., Chamel, N., Pearson, J.M.: Phys. Rev. C **77**, 065805 (2008)

Oyamatsu, K., Iida, K.: Phys. Rev. C **75**, 015801 (2007)

Oyamatsu, K., Hashimoto, M., Yamada, M.: Prog. Theor. Phys. **72**, 373 (1984)

Page, D., Lattimer, J.M., Prakash, M.: In: Bennemann, K.H., Ketterson, J.B. (eds.) Novel Superfluids: Volume 2, p. 505. Oxford University Press, Oxford (2014)

Pagliara, G., Schaffner-Bielich, J.: Phys. Rev. D **77**, 063004 (2008)

Pagliara, G., Herzog, M., Rpke, F.K.: Phys. Rev. D **87**(10), 103007 (2013)

Pais, H., Stone, J.R.: Phys. Rev. Lett. **109**, 151101 (2012)

Papakonstantinou, P., Margueron, J., Gulminelli, F., Raduta, Ad.R.: Phys. Rev. C **88**, 045805 (2013)

Paschalidis, V., Yagi, K., Alvarez-Castillo, D., Blaschke, D.B., Sedrakian, A.: Phys. Rev. D **97**(8), 084038 (2018)

Pastore, A., Shelley, M., Baroni, S., Diget, C.A.: J. Phys. G: Nucl. Part. Phys. **44**, 094003 (2017)

Pastore, A., Shelley, M., Diget, C.A.: In: Proceedings of 26th International Nuclear Physics Conference. PoS(INPC2016)145 (2016)

Pearson, J.M., Chamel, N., Goriely, S.: Phys. Rev. C **83**, 065810 (2011)

Pearson, J.M., Chamel, N., Goriely, S., Ducoin, C.: Phys. Rev. C **85**, 065803 (2012)

Pearson, J.M., Goriely, S., Chamel, N.: Int. J. Mass Spectrom. **349–350**, 57 (2013)

Pearson, J.M., Chamel, N., Pastore, A., Goriely, S.: Phys. Rev. C **91**, 018801 (2015)

Pearson, J.M., Chamel, N., Potekhin, A.Y., Fantina, A.F., Ducoin, C., Dutta, A.K., Goriely, S.: Mon. Not. R. Astron. Soc. (2018). https://doi.org/10.1093/mnras/sty2413

Peña Arteaga, D., Grasso, M., Khan, E., Ring, P.: Phys. Rev. C **84**, 045806 (2011)

Pereira, F.I.M.: Nucl. Phys. A **860**, 102 (2011)

Pereira, F.I.M.: Nucl. Phys. A **897**, 151 (2013)

Pethick, C.J., Potekhin, A.Y.: Phys. Lett. **B 427**, 7 (1998)

Pethick, C.J., Ravenhall, D.G., Lorenz, C.P.: Nucl. Phys. **A 584**, 675 (1995)

Pili, A.G., Bucciantini, N., Del Zanna, L.: Mon. Not. R. Astron. Soc. **439**, 3541 (2014)

Ping, Y.: Phys. Rev. Lett. **111**, 065501 (2013)

Potekhin, A.Y., Chabrier, G.: Phys. Rev. E **62**, 8554 (2000)

Potekhin, A.Y., Chabrier, G.: Astron. Astrophys. **550**, A43 (2013)

Potekhin, A.Y., Chabrier, G.: Astron. Astrophys. **609**, A74 (2018)

Potekhin, A.Y., De Luca, A., Pons, J.A.: Space Sci. Rev. **191**, 171 (2015)

Rabhi, A., Pérez-García, M.A., Providência, C., Vidaña, I.: Phys. Rev. C **91**, 045803 (2015)

Rabi, I.I.: Z. Phys. **49**, 507 (1928)

Rafelski, J., Hagedorn, R. (1980). CERN-TH-2969

Raithel, C.A., Özel, F., Psaltis, D.: Astrophys. J. **831**(1), 44 (2016)

Ratti, C., Thaler, M.A., Weise, W.: Phys. Rev. D **73**, 014019 (2006)

Ravenhall, D.G., Pethick, C.J., Wilson, J.R.: Phys. Rev. Lett. **50**, 2066 (1983)

Rea, N., Esposito, P., Turolla, R., Israel, G.L., Zane, S., Stella, L., Mereghetti, S., Tiengo, A., Götz, D., Göğüş, E., Kouveliotou, C.: Science **330**, 944 (2010)

Read, J.S., Lackey, B.D., Owen, B.J., Friedman, J.L.: Phys. Rev. D **79**, 124032 (2009)

Reardon, D.J., et al.: Mon. Not. R. Astron. Soc. **455**(2), 1751 (2016)

Reinhardt, H.: Phys. Lett. B **244**, 316 (1990)

Rijken, T.A., Schulze, H.J.: Eur. Phys. J. A **52**(2), 21 (2016)

Roca-Maza, X., Piekarewicz, J.: Phys. Rev. C **78**, 025807 (2008)

Roessner, S., Ratti, C., Weise, W.: Phys. Rev. D **75**, 034007 (2007)

Röpke, G., Blaschke, D., Schulz, H.: Phys. Rev. D **34**, 3499 (1986a)

Röpke, G., Blaschke, D., Schulz, H.: Phys. Lett. B **174**, 5 (1986b)

Ruester, S.B., Werth, V., Buballa, M., Shovkovy, I.A., Rischke, D.H.: Phys. Rev. D **72**, 034004 (2005)

Rüster, S.B., Hempel, M., Schaffner-Bielich, J.: Phys. Rev. C **73**, 035804 (2006)

Sagert, I., Fischer, T., Hempel, M., Pagliara, G., Schaffner-Bielich, J., Mezzacappa, A., Thielemann, F.-K., Liebendörfer, M.: Phys. Rev. Lett. **102**, 081101 (2009)

Sagert, I., Fann, G.I., Fattoyev, F.J., Postnikov, S., Horowitz, C.J.: Phys. Rev. C **93**, 055801 (2016)

Salpeter, E.E.: Aust. J. Phys. **7**, 373 (1954)

Salpeter, E.E.: Astrophys. J. **134**, 669 (1961)

Sébille, F., Figerou, S., de la Mota, V.: Nucl. Phys. **A822**, 51 (2009)

Sébille, F., de la Mota, V., Figerou, S.: Phys. Rev. C **84**, 055801 (2011)

Schäfer, T., Wilczek, F.: Phys. Rev. D **60**, 114033 (1999a)

Schäfer, T., Wilczek, F.: Phys. Rev. Lett. **82**, 3956 (1999b)

Schatz, H., Aprahamian, A., Barnard, V., Bildsten, L., Cumming, A., Ouellette, M., Rauscher, T., Thielemann, F.-K., Wiescher, M.: Phys. Rev. Lett. **86**, 3471 (2001)

Schatz, H., Bildsten, L., Cumming, A.: Astrophys. J. **583**, L87 (2003)

Schatz, H., Gupta, S., Möller, P., Beard, M., Brown, E.F., Deibel, A.T., Gasques, L.R., Hix, W.R., Keek, L., Lau, R., Steiner, A.W., Wiescher, M.: Nature **505**, 62 (2014)

Schertler, K., Greiner, C., Schaffner-Bielich, J., Thoma, M.H.: Nucl. Phys. A **677**, 463 (2000)

Schiffer, J.P., Freeman, S.J., Caggiano, J.A., Deibel, C., Heinz, A., Jiang, C.-L., Lewis, R., Parikh, A., Parker, P.D., Rehm, K.E., Sinha, S., Thomas, J.S.: Phys. Rev. Lett. **92**, 162501 (2004)

Schneider, A.S., Horowitz, C.J., Hughto, J., Berry, D.K.: Phys. Rev. C **88**, 065807 (2013)

Schneider, A.S., Berry, D.K., Briggs, C.M., Caplan, M.E., Horowitz, C.J.: Phys. Rev. C **90**, 055805 (2014)

Schneider, A.S., Berry, D.K., Caplan, M.E., Horowitz, C.J., Lin, Z.: Phys. Rev. C **93**, 065806 (2016)

Schuetrumpf, B., Nazarewicz, W.: Phys. Rev. C **92**, 045806 (2015)

Sedrakian, A., Clark, J.W.: Pairing in Fermionic Systems, p. 135. World Scientific Publishing, Singapore (2006)

Seidov, Z.F.: Sov. Astron. **15**, 347 (1971)

Seif, W.M., Basu, D.N.: Phys. Rev. C **89**, 028801 (2014)

Seiradakis, J.H., Wielebinski, R.: Astron. Astrophys. Rev. **12**, 239 (2004)

Sharma, B.K., Centelles, M., Viñas, X., Baldo, M., Burgio, G.F.: Astron. Astrophys. **584**, A103 (2015)

Shternin, P.S., Yakovlev, D.G., Haensel, P., Potekhin, A.Y.: Mon. Not. R. Astron. Soc. **382**, L43 (2007)

Sobiczewski, A., Litvinov, Y.A.: Phys. Rev. C **89**, 024311 (2014)

Stein, M., Maruhn, J., Sedrakian, A., Reinhard, P.-G.: Phys. Rev. C **94**, 035802 (2016a)

Stein, M., Sedrakian, A., Huang, X.-G., Clark, J.W.: Phys. Rev. C **93**, 015802 (2016b)

Steiner, A.W.: Phys. Rev. C **85**, 055804 (2012)

Steiner, A.W., Hempel, M., Fischer, T.: Astrophys. J. **774**, 17 (2013)

Stella, L., Dall'Osso, S., Israel, G.L.: Astrophys. J. **634**, L165 (2005)

Steppenbeck, D. et al.: Nature **502**, 207 (2013)

Stixrude, L.: Phys. Rev. Lett. **108**, 055505 (2012)

Sulaksono, A., Alam, N., Agrawal, B.K.: Int. J. Mod. Phys. **E23**, 1450072 (2014)

't Hooft, G.: Phys. Rep. **142**, 357 (1986)

Terazawa, H.: Tokyo University Report No. INS-336 (1979)

Tews, I.: Phys. Rev. C **95**, 015803 (2017)

Thompson, C., Duncan, R.C.: Astrophys. J. **392**, L9 (1992)

Tiengo, A., et al.: Nature **500**, 312 (2013)

Tondeur, F.: Astron. Astrophys. **14**, 451 (1971)

Turolla, R., Zane, S., Drake, J.J.: Astrophys. J. **603**, 265 (2004)

Typel, S.: Eur. Phys. J. A **52**(1), 16 (2016)

Typel, S., Blaschke, D.: Universe **4**, 32 (2018)

Utama, R., Piekarewicz, J., Prosper, H.B.: Phys. Rev. C **93**, 014311 (2016)

VanDevender, J.P., et al.: Sci. Rep. **7**, 8758 (2017)

Vantournhout, K., Feldmeier, H.: J. Phys. Conf. Ser. **342**, 012011 (2012)

Vantournhout, K., Neff, T., Feldmeier, H., Jachowicz, N., Ryckebusch, J.: Prog. Part. Nucl. Phys. **66**, 271 (2011)

Van Vleck, J.H.: The Theory of Electric and Magnetic Susceptibilities. Oxford University Press, London (1932)

Vautherin, D., Brink, D.M.: Phys. Rev. C **5**, 626 (1972)

Vietri, M., Stella, L., Israel, G.L.: Astrophys. J. **661**, 1089 (2007)

Vidaña, I., Bashkanov, M., Watts, D.P., Pastore, A.: Phys. Lett. B **781**, 112 (2018)

Viñas, X., Gonzalez-Boquera, C., Sharma, B.K., Centelles, M.: Acta Phys. Polon. **B 10**, 259 (2017)

Voskresensky, D.N., Yasuhira, M., Tatsumi, T.: Nucl. Phys. A **723**, 291 (2003)

Wang, J.C., Wang, Q., Rischke, D.H.: Phys. Lett. B **704**, 347 (2011)

Wang, M., et al.: Chin. Phys. C **41**, 030002 (2017)

Watanabe, G., Maruyama, T.: In: Bertulani, C. Piekarewicz, J. (eds.) Neutron Star Crust, p. 23. Nova Science Publishers, New York (2012)

Watanabe, G., Sonoda, H., Maruyama, T., Sato, K., Yasuoka, K., Ebisuzaki, T.: Phys. Rev. Lett. **103**, 121101 (2009)

Watanabe, G., Sato, K., Yasuoka, K., Ebisuzaki, T.: Phys. Rev. C **68**, 035806 (2003)

Weber, F.: Prog. Part. Nucl. Phys. **54**, 193 (2005)

Weissenborn, S., Chatterjee, D., Schaffner-Bielich, J. Phys. Rev. C **85**(6), 065802 (2012); Erratum: [Phys. Rev. C **90**(1), 019904 (2014)]

Wetterich, C.: Phys. Lett. B **462**, 164 (1999)

Wigner, E.P., Seitz, F.: Phys. Rev. **43**, 804 (1933)

Witten, E.: Phys. Rev. D **30**, 272 (1984)

Williams, R.D., Koonin, S.E.: Nucl. Phys. A **435**, 844 (1985)

Wolf, R.N. et al.: Phys. Rev. Lett. **110**, 041101 (2013)

Woods, P.M., Thompson, C.: In: Lewin, W.H.G., van der Klis, M. (eds.), Compact Stellar X-ray Sources, p. 547. Cambridge University Press, Cambridge (2006)

Xu, Z.Y., et al.: Phys. Rev. Lett. **113**, 032505 (2014)

Yamamoto, Y., Furumoto, T., Yasutake, N., Rijken, T.A.: Eur. Phys. J. A **52**(2), 19 (2016)

Yasutake, N., Lastowiecki, R., Benic, S., Blaschke, D., Maruyama, T., Tatsumi, T.: Phys. Rev. C **89**, 065803 (2014)

Zdunik, J.L., Haensel, P.: Astron. Astrophys. **551**, A61 (2013)

Zdunik, J.L., Bejger, M., Haensel, P., Gourgoulhon, E.: Astron. Astrophys. **450**, 747 (2006)

Chapter 8
Superfluidity and Superconductivity in Neutron Stars

Brynmor Haskell and Armen Sedrakian

Abstract This review focuses on applications of the ideas of superfluidity and superconductivity in neutron stars in a broader context, ranging from the microphysics of pairing in nucleonic superfluids to macroscopic manifestations of superfluidity in pulsars. The exposition of the basics of pairing, vorticity and mutual friction can serve as an introduction to the subject. We also review some topics of recent interest, including the various types of pinning of vortices, glitches, and oscillations in neutron stars containing superfluid phases of baryonic matter.

8.1 Introduction

Neutron stars are one of the most extreme astrophysical laboratories in the universe. They allow us to probe physics in strong gravitational fields in the regime where general-relativistic corrections can be as large as 20%, the magnetic fields deduced at their surfaces $B \leq 10^{15}$ G are the largest measured in Nature, and their interiors are expected to contain the densest forms of matter. For typical neutron star masses $\simeq 1.4 - 2.0 \, M_\odot$ ($M_\odot$ being the solar mass) and radii $R \simeq 10$–14 km, the central densities of neutron stars can easily exceed the nuclear saturation density $n_s = 0.16 \, \mathrm{fm}^{-3}$ by factors of a few up to ten.

At the same time neutron stars are extreme low-temperature laboratories: the high densities of their interiors imply large Fermi energies of fermions $\varepsilon_F \simeq$ 10–100 MeV, which turn out to be much higher than the characteristic interior temperatures of mature neutron stars $T \sim 10^8$ K $\simeq 0.01$ MeV. Because of the attractive long-range component of the nuclear force and the high degeneracy $\varepsilon_F \gg$

B. Haskell (✉)
Nicolaus Copernicus Astronomical Center, Polish Academy of Sciences, Warszawa, Poland
e-mail: bhaskell@camk.edu.pl

A. Sedrakian
Frankfurt Institute for Advanced Studies, Frankfurt-Main, Germany
e-mail: sedrakian@fias.uni-frankfurt.de

© Springer Nature Switzerland AG 2018
L. Rezzolla et al. (eds.), *The Physics and Astrophysics of Neutron Stars*,
Astrophysics and Space Science Library 457,
https://doi.org/10.1007/978-3-319-97616-7_8

T the neutrons and protons (and presumably some hyperons) become superfluid and superconducting at critical temperatures of the order of $T_c \simeq 10^9\text{--}10^{10}$ K.

Neutron superfluidity in a neutron star crust and its core, as well as proton superconductivity in the core, profoundly alter its dynamics, just as the emergence of these phenomena does in terrestrial experiments. For example, superfluid neutrons can now flow relative to the 'normal' component of the star with little or no viscosity, as standard reactions and scattering processes giving rise to bulk and shear viscosity are strongly suppressed. An important factor in neutron star dynamics is the appearance of an array of vortices in the neutron condensate. In analogy to a laboratory superfluid in a rotating container, the neutron superfluid mimics large scale rotation by creating an array of quantised vortices each carrying a quantum of circulation. Interactions between vortices and the normal component open a new dissipative channel, known as mutual friction. These interactions may be strong enough to 'pin' the vortices and freeze the rotation rate of the superfluid neutrons. The angular momentum thus stored is then released catastrophically during discrete events, which are thought to be the cause of the observed 'glitches', i.e., sudden spin-up episodes observed in pulsars. These phenomena reflect the interior dynamics of neutron stars and thus can potentially provide an insight into the physics of superfluids in their interiors.

This chapter provides an educational introduction and an overview of the field of superfluidity and superconductivity in neutron stars. The first part of the chapter reviews the microphysics of nuclear pairing in neutron stars by providing an elementary introduction to the microscopic theory of nuclear pairing and a review of current issues such as medium polarization corrections to the pairing, pairing in higher partial waves and in strong magnetic fields (Sect. 8.2). The interaction of vortices with the ambient fluid at the microphysical level, which leads to the phenomenon of mutual friction between the superfluid and the normal fluid, is reviewed in Sect. 8.3. This is followed by a discussion of hydrodynamics of superfluids in neutron stars (Sect. 8.4). Section 8.5 is devoted to the interactions of vortices with flux-tubes and nuclear clusters, i.e., their pinning to various structures. This is followed by a discussion of the macrophysics of rotational anomalies in neutron stars in Sect. 8.6. We provide our concluding remarks in Sect. 8.7.

8.2 Microscopic Pairing Patterns in Neutron Stars

8.2.1 General Ideas

The microscopic understanding of the pairing mechanism in nucleonic matter in neutron stars is based on the theory advanced by Bardeen, Cooper, and Schrieffer (BCS) in 1957 to explain the superconductivity of some metals at low temperatures (Bardeen et al. 1957). The key ingredient of this theory is the notion of an attractive interaction between two electrons which is mediated by lattice phonon exchange. According to the Cooper theorem (Cooper 1956) low-temperature fermions which fill a Fermi sphere can bind to form Cooper pairs if there is an attractive interaction between them. The bound states of electrons (typically with total spin 0) form a coherent many-body state which carries an electric current without any resistivity below a certain critical temperature T_c. The overwhelming success of the BCS theory in explaining the wealth of experimental data encouraged applications of the key ideas of this theory in other fields of physics, including nuclear physics. In contrast to electronic materials, where the direct interaction between the electrons is repulsive due to the Coulomb force between same-charge particles, in nuclear systems the dominant long-range piece of the interaction between the nucleons (neutrons and protons) is attractive. It is not surprising then that superconductivity and superfluidity in nuclear systems - finite nuclei and neutron stars - were conjectured shortly after the advent of the BCS theory by Bohr et al. (1958), Migdal (1959) and others.

Fully microscopic calculations of the pairing properties of neutron and proton matter in neutron stars were carried out following the discovery of pulsars in 1967 and their identification with neutron stars. Although at the time nuclear interactions were not known as precisely as nowadays, the first computations of the pairing gaps of about 1 MeV in neutron and proton matter are consistent with present day calculations (see the reviews by Lombardo and Schulze (2001), Dean and Hjorth-Jensen (2003), Sedrakian and Clark (2006), Page et al. (2013) and Gezerlis et al. (2014) and references therein).

A useful reference for the understanding of the patterns of pairing in neutron stars is the partial wave analysis of the nuclear interaction. In fact, the experimental measurements of the nuclear scattering are given per partial wave. At low energies the nucleon-nucleon (nn) scattering is dominated by two S-wave interactions, specifically the $^3S_1-^3D_1$ coupled partial wave and the $^1S_0$ partial wave; here we use the standard spectroscopic notations to specify the scattering channels, i.e., $^{2S+1}L_J$, with $L = 0, 1, 2$ mapped to S, P, D, where L is the orbital angular momentum, S is the total spin and J is the total angular momentum, which is the sum of the former two vectors. Thus, at low energies the $L = 0$ states, which have symmetrical wave-function in the coordinate space, dominate. The total wave function contains however spin and isospin components which must be selected in a manner to satisfy the Pauli principle, which dictates that the total wave function must be anti-symmetrical. As a consequence, the neutron-neutron and proton-proton scattering which always has a total isospin $T = 1$ symmetrical component, cannot occur in the spin symmetrical $S = 1$ and spatially symmetrical $L = 0$ state. Therefore, the strongly attractive interaction in the $^3S_1-^3D_1$ channel which binds the deuteron (binding energy $E_d = -2.2\,\text{MeV}$) does not lead to pairing in neutron dominated matter, where neutrons and protons form vastly separate Fermi surfaces. Then, for large differences in the numbers of protons and neutrons (isospin asymmetry) only same isospin Cooper pairs can arise in the remaining $^1S_0$ partial wave channel.

At laboratory energies of nn scattering larger that $E_L = 250\,\text{MeV}$ the measured scattering phase-shift in the $^1S_0$-wave interaction channel becomes negative, i.e., the interaction becomes repulsive, see Fig. 8.1. However, already at $E_L \simeq 160\,\text{MeV}$ the $^3P_2 -^3 F_2$ tensor interaction becomes the most attractive channel for $T = 1$

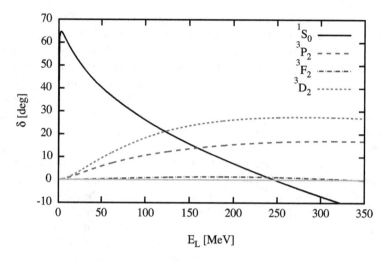

Fig. 8.1 The nucleon-nucleon scattering phase shifts as a function of laboratory energy for the channels where pairing in neutron stars matter appears. The S and P wave scattering is responsible for neutron-neutron and proton-proton pairing, whereas the $^3D_2$ wave scattering can occur only between neutrons and protons. The P and F waves are coupled by the non-central tensor component of the nuclear interaction

(neutron-neutron and proton-proton) pairs. The corresponding density in neutron star matter is obtained by noting that the center of mass energy of two scattering nucleons is $E_L/2$, which should be of the order of the Fermi energy of neutrons or protons. (Here we specialize the discussion to the high-density and low-temperature regime of interest to superfluidity in neutron stars). Neutron Fermi energies become of the order of $\varepsilon_{Fn} \simeq 60\,\mathrm{MeV}$ at the nuclear saturation density $n_s = 0.16\,\mathrm{fm}^{-3}$. Thus, we anticipate that neutron pairing in the $^1S_0$-wave vanishes at densities slightly above the saturation density and that the core of the star contains superfluid featuring neutron pairs in the $^3P_2$–$^3F_2$ partial wave. The spatial component of the wave-function of these Cooper pairs is anti-symmetrical whereas the spin ($S = 1$) and isospin ($T = 1$) components are symmetrical. Clearly, the pairing in this so-called triplet spin-1 channel is consistent with the Pauli principle for two neutrons. Because the proton fraction in a neutron star core is small, about 5-10% of the net number density, their Fermi energies, and consequently the center of mass scattering energies, remain low. Therefore, proton pairs arise in the $^1S_0$-wave up to quite high densities. It is conceivable that, at densities higher than a few times the nuclear saturation density, higher partial waves can contribute to the pairing in neutron star matter. For example, if the partial densities of neutrons and protons are forced to be close to each other by some mechanism, i.e., matter is isospin symmetrical, then neutron-proton pairs can be formed in the most attractive $^3D_2$ partial wave with a wave function which is symmetrical in space, antisymmetrical in isospace ($T = 0$) and symmetrical in spin ($S = 1$). A mechanism that can enforce equal numbers of neutrons and protons is, e.g., meson condensation (Weber 1999).

The BCS theory was originally formulated in terms of a variational wave function of a coherent state which minimized the energy of an ensemble of electrons interacting via contact (attractive) interaction (Bardeen et al. 1957). Here we will outline an alternative formulation based on the method of *canonical transformations* due to Bogoljubov (1958).

Consider a macroscopic number of N fermions which are described by the *pairing Hamiltonian*

$$H - \mu N = \sum_{p,\sigma} \varepsilon_p a_{p,\sigma}^\dagger a_{p,\sigma} - \sum_{p_1+p_2=p_3+p_4} V_{\text{eff}}(p_1, p_2; p_3, p_4) a_{p_3}^\dagger a_{p_4}^\dagger a_{p_1} a_{p_2}. \quad (8.1)$$

The first term is the kinetic energy, the second term is the attractive interaction energy, where $V_{\text{eff}}(p_1, p_2; p_3, p_4)$ is an effective pairing interaction. Here $a_{p,\sigma}^\dagger$ and $a_{p,\sigma}$ are the particle creation and annihilation operators for particles with spin $\sigma = \uparrow \downarrow$ and momentum p. Note that we work in the grand-canonical ensemble, so that instead of fixing the number of particles we assume that our system is connected to a reservoir of particles; μ is the chemical potential - the energy needed to add or remove a particle to the system. The application of the method of canonical transformations to the Hamiltonian (8.1) requires new creation and annihilation operators defined as

$$a_{p,\uparrow} = u_p \alpha_{p\uparrow} + v_p \alpha_{-p\downarrow}^\dagger, \quad (8.2)$$

$$a_{p,\downarrow} = u_p \alpha_{p\downarrow} - v_p \alpha_{-p\uparrow}^\dagger. \quad (8.3)$$

The requirement that the anti-commutation relations obeyed by the new operators are the same as those obeyed by the original fermionic ones leads us to

$$\{\alpha_{p,\sigma}, \alpha_{p',\sigma'}^\dagger\} = \alpha_{p,\sigma}^\dagger \alpha_{p',\sigma'} + \alpha_{p',\sigma'} \alpha_{p,\sigma}^\dagger = \delta_{pp'} \delta_{\sigma,\sigma'}. \quad (8.4)$$

It follows then that the functions u_p and v_p are not independent, but $u_p^2 + v_p^2 = 1$, i.e., there is a single *independent* function, say v_p. This parameter is found from the minimization of the statistical average of the Hamiltonian (8.1)

$$E - \mu N = \langle H - \mu N \rangle, \quad (8.5)$$

where $\langle \ldots \rangle$ stands for mean value with the occupation numbers defined as $\langle \alpha_{p,\downarrow}^\dagger \alpha_{p,\downarrow} \rangle = n_{p,\downarrow}$ and $\langle \alpha_{p,\uparrow}^\dagger \alpha_{p,\uparrow} \rangle = n_{p,\uparrow}$. The energy is then given by

$$E - \mu N = \sum_p \varepsilon(p) \left[u_p^2 (n_{p,\uparrow} + n_{p,\downarrow}) + v_p^2 (2 - n_{p,\uparrow} - n_{p,\downarrow}) \right]$$

$$- \sum_{pp'} V_{\text{eff}}(p, p') u_p v_p u_{p'} v_{p'} Q(p) Q(p'), \quad (8.6)$$

where $Q(\boldsymbol{p}) \equiv (1 - n_{\boldsymbol{p},\uparrow} - n_{\boldsymbol{p},\downarrow})$. Eliminating u_p from (8.6) via $u_p^2 = 1 - v_p^2$ and performing variations with respect to v_p we find then

$$2\varepsilon_p = \frac{\Delta(\boldsymbol{p})(1 - 2v_p^2)}{u_p v_p}, \tag{8.7}$$

where

$$\Delta(\boldsymbol{p}) = \sum_{\boldsymbol{p}'} V_{\text{eff}}(\boldsymbol{p}, \boldsymbol{p}') u_p v_p (1 - n_{\boldsymbol{p},\downarrow} - n_{\boldsymbol{p},\uparrow}), \tag{8.8}$$

is the so-called *gap equation*. From Eq. (8.7) we find for the *Bogolyubov amplitudes*

$$u_p^2 = \frac{1}{2}\left(1 + \frac{\varepsilon_p}{E_p}\right), \qquad v_p^2 = \frac{1}{2}\left(1 - \frac{\varepsilon_p}{E_p}\right), \tag{8.9}$$

where the quasiparticle spectrum in the superconductor is defined as

$$E_p = \sqrt{\varepsilon_p^2 + \Delta_p^2}. \tag{8.10}$$

On substituting $u_p v_p = \Delta(\boldsymbol{p})/2E_p$ in Eq. (8.8), we find a non-linear integral equation for the gap function $\Delta(\boldsymbol{p})$ which can be solved for any given effective interaction $V_{\text{eff}}(\boldsymbol{p}, \boldsymbol{p}')$. It is easy to verify that E_p is indeed the quasiparticle energy, by taking the variation of the total energy with respect to the occupation numbers, i.e., by computing the variation[1]

$$\frac{\delta(E - \mu N)}{\delta n_{\boldsymbol{p},\uparrow}} = \varepsilon_p(u_p^2 - v_p^2) + 2u_p v_p \Delta(\boldsymbol{p}) = E_p. \tag{8.11}$$

Equation (8.10) demonstrates the fundamental property of the superconductors: *the spectrum of a superconductor contains an energy gap* Δ. As a consequence the excitations can be created in the system if a Cooper pair breaks, which means that energy of the order of 2Δ must be supplied to the superconductor. The main property of superconductors - the absence of dissipation of current - follows from the existence of the gap in their spectrum. In the case of uncharged fermionic superfluids (e.g. neutron matter) the same property is referred to as superfluidity (i.e., fluid motions without dissipation). In equilibrium, the occupation numbers of fermions are given by the Fermi function $f(p) = (e^{E_p/T} + 1)^{-1}$, where T is the temperature. At low temperatures the fermionic momenta are restricted to the vicinity of the Fermi surface; then, assuming an isotropic (S-wave) interaction, we can simplify the

---

[1]When computing the variation with respect to the occupation numbers one needs to assume that the Bogolyubov coefficients are constant, because they are determined from the condition $\delta(E - \mu N)/\delta v_p = 0$.

gap equation by changing the integration measure $\sum_p = m^* p_F \int d\varepsilon_p \int d\Omega/(2\pi)^3$ to find

$$1 = Gv \int_0^\Lambda \frac{d\varepsilon_p}{2\sqrt{\varepsilon_p^2 + \Delta^2}} \tanh\left(\frac{\sqrt{\varepsilon_p^2 + \Delta^2}}{2T}\right), \tag{8.12}$$

where $v = p_F m^*/\pi^2$ is the *density of states*, m^* is the effective mass, p_F is the Fermi momentum, and for the sake of illustrations below we assume a momentum-independent contact interaction $V_{\text{eff}}(p, p') = G$, which in turn requires a cut-off Λ to regularize the integral in the ultraviolet. The latter cut-off is physically well-motivated, as the effective pairing interactions are typically localized close to the Fermi surface.[2]

Consider now analytical solutions of the gap equation in the limiting cases $T \to T_c$ and $T \to 0$, where T_c is the critical temperature of phase transition. For $T = 0$, the tanh function is unity and a straightforward integration gives

$$1 = \frac{Gv}{2} \text{arcsinh}\left(\frac{\Lambda}{\Delta(0)}\right) \simeq \ln\left(\frac{2\Lambda}{\Delta(0)}\right), \tag{8.13}$$

where in the last step we assumed *weak coupling*, i.e., $\Delta \ll \Lambda$ to expand $\lim_{x\to\infty} \text{arcsinh} \, x = \ln(2x) + O(x^2)$; this can be written in a more familiar form

$$\Delta(0) = 2\Lambda \exp\left(-\frac{2}{Gv}\right), \tag{8.14}$$

which is the famous *gap equation* of the BCS theory. It demonstrates the exponential sensitivity of the pairing gap to the effective attractive interaction G. In the limit $T \to T_c$, we can set in the integrand of the gap equation $\Delta = 0$. An elementary integration then gives $T_c = (2\Lambda\gamma/\pi) \exp(-2/Gv)$, where $\gamma \equiv e^C$ and $C = 0.57$ is the Euler constant. Combining the results for T_c and $\Delta(0)$ we obtain a relation between them: $\Delta(0) = \pi T_c/\gamma = 1.76 \, T_c$.

The limiting expressions for the gap function can be easily extended to include the next-to-leading order terms in the two limiting cases discussed above:

$$\Delta(T) = \Delta(0) - \sqrt{2\pi\Delta(0)T} \exp\left(-\frac{\Delta(0)}{T}\right) \tag{8.15}$$

for $T \to 0$ and

$$\Delta(T) = \pi\sqrt{\frac{8}{7\zeta(3)}} [T_c(T_c - T)]^{1/2} = 3.06 [T_c(T_c - T)]^{1/2} \tag{8.16}$$

---

[2]The full nuclear interaction can be renormalized via resummations of infinite series such as to contain only components close to the Fermi surface.

for $T \to T_c$, where $\zeta(x)$ is the Riemann's ζ-function with $\zeta(3) = 1.20205$. The temperature dependence of the gap function in the whole temperature regime can be obtained numerically. An analytical fit to the $\Delta(T)$ function, useful for practical applications, is given in Mühlschlegel (1959). Having the temperature dependence of the gap one still needs fit formulae to the gap function at zero temperature $\Delta(0)$ (or equivalently the critical temperature T_c) as a function of density or Fermi momentum. We point out that accurate fits can be obtained with the functional form

$$\Delta(k_F) = a \exp(-k_F^2) + \sum_{n=0}^{4} c_n k_F^n, \tag{8.17}$$

where k_F is the Fermi wave-number. The fit coefficients to the S- and P-wave neutron and S-wave proton gaps can be found in Sinha and Sedrakian (2015).

8.2.2 Effective Interactions

An important issue in computations of the gaps in neutron star matter is the proper determination of the effective pairing interaction $V_{\text{eff}}(p, p')$. As a first approximation one may use the bare nuclear interaction in the gap equation, which provides us with a useful reference result. The most important correction to this interaction arises from *polarization effects* or *screening*, which in diagrammatic language can be understood as a summation of infinite series of particle-hole diagrams. We will review these effects on the basis of the *Landau Fermi liquid theory* and will compare to the alternatives thereafter. In our discussion we will follow the original Landau approach; for computations which are based on this type approach, but include more advanced diagrammatic treatments of the problem see Wambach et al. (1993), Schulze et al. (1996), Schwenk et al. (2003), Lombardo et al. (2004) and Shen et al. (2005).

We now suppose that the bare interaction depends only on the momentum transfer in the collision $q = p_1 - p_3$ and write it in terms of all possible combinations of the spin and isospin components:

$$V_{\text{bare}}(q) = \frac{1}{\nu} \left\{ F_q + G_q(\boldsymbol{\sigma} \cdot \boldsymbol{\sigma}') + \left[F_q' + G_q'(\boldsymbol{\sigma} \cdot \boldsymbol{\sigma}') \right] (\boldsymbol{\tau} \cdot \boldsymbol{\tau}') \right\}, \tag{8.18}$$

where $\boldsymbol{\sigma}$ and $\boldsymbol{\tau}$ are the vectors formed from the Pauli matrices in the spin and isospin spaces and F_q, F_q', G_q and G_q' are the so-called *Landau parameters*. For simplicity the tensor and spin-orbit interactions are neglected as they are small in neutron matter. The summation of geometrical series of particle-hole diagrams then gives for the effective interaction

$$V_{\text{eff}} = \frac{1}{\nu} \left\{ \tilde{F}_q + \tilde{G}_q(\boldsymbol{\sigma} \cdot \boldsymbol{\sigma}') + \left[\tilde{F}_q' + \tilde{G}_q'(\boldsymbol{\sigma} \cdot \boldsymbol{\sigma}') \right] (\boldsymbol{\tau} \cdot \boldsymbol{\tau}') \right\}, \tag{8.19}$$

where $\tilde{A}_q = A_q[1 + L(q)A_q]^{-1}$, A stands for any of the Landau parameters. It is seen that the screening leads to a renormalization of the Landau parameters $A \rightarrow \tilde{A}$, where the function $L(q)$, which represents the single particle-hole loop, is the *Lindhard function*. In the low-temperature regime of interest the momenta of fermions are restricted to lie on their Fermi surface, therefore the Landau parameters will depend only on the relative orientation of the momenta of particles. Then, it is useful to expand these into spherical harmonics

$$A(q) = \sum_l A_l P_l(\cos\theta), \tag{8.20}$$

where P_l are the Legendre polynomials, A again stands for any of the Landau parameters above, θ is the scattering angle which is related to the magnitude of the momentum transfer via $q = 2p_F \sin\theta/2$, where p_F is the Fermi momentum.

In pure neutron matter $\tau \cdot \tau' = 1$. Then, keeping the lowest-order harmonics in the expansion (8.20) and for scattering with total $S = 0$ (i.e. $\sigma \cdot \sigma' = -3$) the effective pairing interaction is given by

$$\nu(p_F)V_{\text{eff}}(q) = F_0^n \left[1 - \mathcal{L}(q)F_0^n\right] - 3G_0^n \left[1 - \mathcal{M}(q)G^n\right], \tag{8.21}$$

where $F^n = F + F'$ and $G^n = G + G'$ describe the effective interaction in the density and spin channels respectively, $\mathcal{L}(q) = L(q)[1 + L(q)F_0^n]^{-1}$ and $\mathcal{M}(q) = L(q)[1 + L(q)G_0^n]^{-1}$ are screening corrections to the direct part of the effective interaction $\propto 1$. In general the Lindhard function is complex; in the low-temperature regime of interest its imaginary part (which is related to the damping of particle-hole excitations) can be neglected. It assumes a simple form for zero energy transfer at fixed momentum: $L(x) = -1 + (2x)^{-1}(1 - x^2)\ln|(1 - x)/(1 + x)|$, with $x = q/2p_F$ (Fetter and Walecka 1971).

The Landau parameters and the effective interaction in Eq. (8.21) have been computed extensively over the past several decades within various approximations. There is a general agreement that the density fluctuations $\propto F_0^n$ enhance, whereas the spin-fluctuations $\propto G_0^n$ reduce the attraction in the pairing interaction. The overall effect of the spin fluctuations at subnuclear densities is numerically larger and, consequently, these fluctuations reduce the pairing gap in neutron matter. As can be seen from Eq. (8.14) the dimensionless quantity (coupling) determining the magnitude of the gap involves, apart from the effective interaction, also the density of states. In general the quasiparticle spectrum can be written as

$$\varepsilon(p) = \frac{p^2}{2m} + U(p), \tag{8.22}$$

where $U(p)$ is the single particle potential felt by a particle in the nuclear environment. The full momentum dependence in Eq. (8.22) can be approximated by introducing an effective quasiparticle mass. An expansion of the potential $U(p)$ around the Fermi momentum leads to

$$\varepsilon(p) = \frac{p_F}{m^*}(p - p_F) - \mu^*, \quad \frac{m^*}{m} = \left(1 + \frac{m}{p_F}\frac{\partial U(p)}{\partial p}\bigg|_{p=p_F}\right)^{-1}, \qquad (8.23)$$

where $\mu^* \equiv -\epsilon(p_F) + \mu - U(p_F)$. Because the effective mass $m^*/m < 1$ in neutron matter, the modifications of the single-particle energies lead to a reduction of the gap. The effective mass also plays a significant role in the multifluid hydrodynamics of superfluids in neutron stars, as discussed in Sect. 8.4.

Correlations in neutron matter can be studied in a number of alternative theories, for example, Monte Carlo sampling of systems with odd and even numbers of neutrons, whereby the gap is defined as the energy difference between odd and even states (Gandolfi et al. 2008; Gezerlis and Carlson 2010). Wave-function based approaches minimize the energy of the BCS state using correlated basis functions (CBF), which are built to account for the operator structure of the nuclear interaction (Chen et al. 1993).

In Fig. 8.2 we show selected benchmark results for pairing gaps in different theories of pairing. The BCS result with bare single particle energies was obtained using a low-momentum interaction, which leads to slightly larger gaps compared

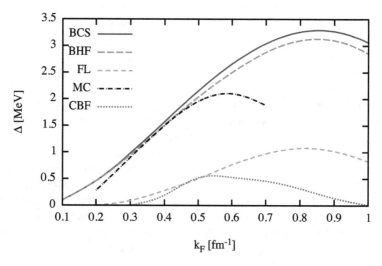

Fig. 8.2 The pairing gap in the low-density neutron matter relevant for neutron star crusts as a function of Fermi momentum. The curves labeled "BCS" and "BHF" show the result for low-momentum interaction using bare and medium modified single particle spectra, whereby the renormalization of the particle spectrum is taken into account in the Bruckner-Hartree-Fock theory (Sedrakian et al. 2003). The screening effects, which strongly suppress the gap, are shown on the examples based on Fermi-liquid "FL" (Wambach et al. 1993) and correlated basis functions "CBF" theories (Chen et al. 1993). We also show the results of auxiliary field Monte Carlo "MC" simulations (Gandolfi et al. 2008), which are closer to the results without screening corrections

to full (realistic) interaction which contains a hard core. The modifications of the single particle spectrum, computed within the Bruckner-Hartree-Fock theory (without invoking effective mass approximation) leads to a moderate reduction of the gap (Sedrakian et al. 2003). If one includes the screening corrections in the effective interaction, then the gap is reduced by a factor of three. There is consensus among the Fermi-liquid (Wambach et al. 1993; Schulze et al. 1996; Schwenk et al. 2003; Lombardo et al. 2004; Shen et al. 2005) and CBF approaches (Chen et al. 1993; Pavlou et al. 2016) on the magnitude of the reduction. Auxiliary field and Green's functions Monte Carlo calculations on the other hand predict pairing gaps at low densities which are closer to the BCS result (Gandolfi et al. 2008; Gezerlis and Carlson 2010).

8.2.3 Higher Partial Wave Pairing

As seen from Fig. 8.1 at high densities (energies) the dominant pairing interaction between neutrons is in the $^3P_2 - ^3F_2$ partial wave. The pairing pattern for spin-1 condensates is more complex than for spin-0 S-wave condensates because of competition between states with various projections of the orbital angular momentum and complications due to the tensor coupling of the $^3P_2$ partial wave to the $^3F_2$ one, see Baldo et al. (1998); Zverev et al. (2003); Schwenk and Friman (2004); Maurizio et al. (2014). To solve the P-wave gap equation (8.8) one starts with an expansion of the pairing interaction in partial waves

$$V_{\text{eff.}}(\boldsymbol{p}, \boldsymbol{p'}) = 4\pi \sum_L (2L + 1) P_L(\hat{\boldsymbol{p}} \cdot \hat{\boldsymbol{p'}}) V_L(p, p'), \tag{8.24}$$

where P_L are the Legendre polynomial, and an associated expansion of the gap function in spherical harmonics Y_{LM}

$$\Delta(\boldsymbol{p}) = \sum_{L,M} \sqrt{\frac{4\pi}{2l+1}} Y_{LM}(\hat{\boldsymbol{p}}) \Delta_{LM}(p), \tag{8.25}$$

where L and M are the total orbital momentum and its z-component. The nonlinearity of the gap equation prevents a straightforward solution for its components $\Delta_{LM}(p)$; a common approximation is to perform an angle average in the denominator of the kernel of the gap equation by replacing the factor $\sqrt{\varepsilon(p)^2 + |\Delta(\boldsymbol{p})|^2} \to \sqrt{\varepsilon(p)^2 + D(p)^2}$ where *the angle averaged gap* is given by

$$D(p)^2 \equiv \int \frac{d\Omega}{4\pi} |\Delta(\boldsymbol{p})|^2 = \sum_{L,M} \frac{1}{2L+1} |\Delta_{LM}(k)|^2. \tag{8.26}$$

With this approximation the angular integrals are trivial and we obtain a one-dimensional gap equation for the L-th component of the gap

$$\Delta_L(p) = -\int_0^\infty \frac{dp'\,p'}{\pi} \frac{V_L(p, p')}{\sqrt{\varepsilon(p')^2 + \left[\sum_{L'} \Delta_{L'}(p')^2\right]}} \Delta_L(p'). \qquad (8.27)$$

Although the denominator contains a sum over gap components with different values of L, the contributions from channels other than the $^3P_2$ can be neglected as they are numerically insignificant in the range of densities (energies) where P-wave pairing is dominant.

The non-central tensor part of the nuclear force couples the $^3P_2$ wave to the $^3F_2$ wave; this coupling affects the value of the gap. The modification of the gap equation which takes into account this tensor coupling requires a simple doubling of the components of the gap equation. The coupled channel gap equation reads

$$\begin{pmatrix} \Delta_L \\ \Delta_{L'} \end{pmatrix} = \int_0^\infty \frac{dp'\,p'^2}{\pi E(p')} \begin{pmatrix} -V_{LL} & V_{LL'} \\ V_{L'L} & -V_{L'L'} \end{pmatrix} \begin{pmatrix} \Delta_L \\ \Delta_{L'} \end{pmatrix}, \qquad (8.28)$$

where now $D(k)^2 = \Delta_L(k)^2 + \Delta_{L'}(k)^2$.[3]

Note that the angle averaged approximation provides a numerically accurate value of the angle averaged gap on the Fermi surface. However, in a number of problems, such as neutrino and axion emission from P-wave superfluids the angle dependence of the gap equation is an important factor and should be taken into account by solving the gap equation without the angle averaged approximation. Such solutions show that the angle dependence of the gap function can lead to nodes on the Fermi surface, as for example in the case of solutions of the form $\Delta(\theta) = \Delta_0 \sin\theta$. However, "stretched" solutions with fully gaped Fermi surface $\Delta(\theta) = \Delta_0(1 + \cos^2\theta)$ are viable candidates for angle dependence of the pairing gap and it is difficult to distinguish between these options from energy minimization arguments alone.

Figure 8.1 shows that the interaction is attractive among neutrons and protons in the $^3D_2$ channel and it is stronger than the attraction in the $^3P_2$ channel. Thus, in a hypothetical high-density isospin-symmetrical phase of nuclear matter one would expect D-wave pairing instead of the P-wave pairing, which is the dominant channel in the high density neutron matter. Consequently, there should be a transition from D- to P-wave pairing discussed above. Thus, as the imbalance between neutrons and protons (isospin asymmetry) changes from zero to larger values one would experience a transition from D- to P-wave pairing. Computations show that already small asymmetries destroy the D-wave pairing in nuclear matter (Alm et al.

---

[3]Note that the tensor coupling arises also in the case of pairing in the $^3S_1$-$^3D_1$ channel which acts only between neutrons and protons and is relevant when the isospin asymmetry between neutrons and protons in not too large (Lombardo et al. 2001).

1996), therefore for its realization one needs nearly symmetrical nuclear matter. Above nuclear saturation density such situations can arise in some special cases as, for example, when a mesonic condensate forms (Weber 1999).

8.2.4 Effects of Strong Magnetic Fields on Pairing

Neutron stars are highly magnetized objects and the magnetic field in the stellar interior is modified by the presence of superconductivity. The topology and properties of the magnetic field depend strongly on the type of superconductivity, which depends on the Ginzburg-Landau parameter $\kappa_{GL} = \lambda/\xi_p$, where ξ_p is the proton coherence length, which roughly determines the size of the Cooper pairs, and λ is the penetration depth of the magnetic field in superconducting matter. In most of the neutron star core one has $\kappa_{GL} > 1/\sqrt{2}$ and type II superconductivity is expected, in which case the magnetic field penetrates the superconductor by forming an array of quantized flux tubes. In laboratory type II superconductors the field can only penetrate the superconductor for field strengths between a lower critical field H_{c1} and an upper critical field H_{c2}. In neutron stars the situation is somewhat different, as one still has the upper critical field H_{c2} (essentially the field at which flux tubes are so densely packed that their cores touch), but magnetic fields can still penetrate the core below the lower critical field H_{c1}. This is the case because the magnetic flux cannot be expelled effectively from the superconducting core due to its high electric conductivity; the time-scale for such expulsion is of the order of the secular timescales (Baym et al. 1969). In the deep core of the neutron star, on the other hand, it is possible to have $\kappa_{GL} < 1/\sqrt{2}$, and in this case we expect type I superconductivity, in which fluxtubes are not energetically favourable and the field is arranged in domains of unpaired proton matter of much larger spatial dimensions (Bruk 1973; Sedrakian et al. 1997; Link 2003; Buckley et al. 2004; Sedrakian 2005).

A class of neutron stars known as magnetars have surface magnetic fields of the order of 10^{15} G and it has been conjectured that their interior fields could be significantly larger (Turolla et al. 2015). While modifications of the equation of state of matter require fields which are close to the limiting fields 10^{18} G compatible with gravitational stability, pairing phenomena which occur near the Fermi surface are affected by much lower fields, of the order $B \sim 10^{16} - 10^{17}$ G. The interaction energy of the magnetic field with the nucleon spin is $\mu_N B$, where $\mu_N = e\hbar/2m_p$ is the nuclear magneton. A numerical estimate gives $\mu_N B \simeq \pi(B/10^{18} \text{ Gauss})$ MeV. Therefore, fields of the order of 10^{16} G would substantially affect the pairing gaps of the order of 1 MeV via the spin–B-field interaction. The interaction of the magnetic field with the neutron or proton spin induces an imbalance in the number of spin-up and spin-down particles, which implies that the Cooper pairing in the S wave will be suppressed. Indeed in this case the number of Cooper pairs will be limited by the number of spin-down particles, with the excess spin-up particles remaining unpaired (Stein et al. 2016). This *Pauli paramagnetic suppression* acts for both proton and neutron condensates and the associated critical field is within the range $H_c^{Para} \sim 10^{16}\text{-}10^{17}$ G. Note that similar suppression of the S-wave pairing arises in the condensed matter context and the associated magnetic field is known as the *Chandrasekhar-Clogston limiting field*.

In the case of the proton condensate the upper critical field H_{c2} turns out to be smaller than the field associated with the Pauli paramagnetic ordering (Sinha and Sedrakian 2015), therefore the proton condensate is destroyed for even lower magnetic fields $H_{c2} \simeq 10^{15}$ G. We have seen that the S-wave gaps and in particular the proton gap depends substantially on the density. If the magnetic field strength can be assumed approximately constant in the core of a magnetar the size of the superconducting region will depend on the magnitude of the field B and will be limited by the condition $B \leq H_{c2}$.

The role of the magnetic field in the P-wave pairing is not well understood from the microscopic point of view. Because the Cooper pairs in this case form spin-1 objects, the spin-magnetic field interaction will not be destructive. The consequences of the suppression of the nucleonic pairing on the phenomenology of magnetars are discussed elsewhere (Sedrakian et al. 2017).

8.3 Microphysics of Mutual Friction

In this section we concentrate on the interaction of vortex lines in neutron stars with ambient components. The discussion includes both the neutron vortex lines which are formed in response to the rotation of the star and the magnetic flux-tubes which, as discussed above, form if the proton superconductor is type-II. This interaction is known under the general name *mutual friction* and appears naturally in the superfluid hydrodynamics including vortices, which we will discuss in the following Sect. 8.4. The mutual friction is an important ingredient of the description of superfluid dynamics as it determines the dynamical time-scales of coupling of superfluid to normal (non-superfluid) matter and rotational anomalies in neutron stars, such as glitches, time-noise, precession etc., which are in part discussed in the subsequent Sect. 8.6. The mutual friction is quantified in terms of a dimensionfull drag parameter η defined via the force exerted by ambient fluids on the vortex $f_d = \eta(v_v - v_e)$, where v_v is the vortex velocity, v_e is the velocity of the normal component. In superfluids it is balanced by the Magnus (lifting) force $f_M = \rho_n[(v_s - v_v) \times v]$, acting on a vortex with circulation vector v placed in a superfluid flow with velocity v_s. It is, therefore, convenient to use the dimensionless drag-to-lift ratio $\mathcal{R} = \eta/\rho_n\kappa$, where ρ_n is the mass density of the (neutron) superfluid component and $\kappa = \pi\hbar/m_n$ is the quantum of circulation, m_n being the neutron mass. For massless vortices $f_M + f_d = 0$, which is known as *the force balance equation*. We shall discuss these quantities in more detail in the context of superfluid hydrodynamics in Sect. 8.4.

The neutron vortices which carry the angular momentum of neutron star interiors arise in the S and P wave superfluids where the Cooper pairs have total spin-0 and spin-1, as discussed in Sect. 8.2. A vortex in a neutral fermionic superfluid has a core of the order of the coherence length ξ, where quasiparticle pairing is quenched and, therefore, scattering centers are available for interactions. Consider first a vortex in a neutron superfluid with an S-wave symmetry of the condensate. The core of the vortex contains fermionic states which are given by Caroli and Matricon (1965)

$$
\begin{pmatrix} u_{q_\parallel,\mu}(r_\perp) \\ v_{p_\parallel,\mu}(r_\perp) \end{pmatrix} = e^{ip_\parallel z} \left(e^{i\theta(\mu-\frac{1}{2})} \, e^{i\theta(\mu+\frac{1}{2})} \right) \begin{pmatrix} u'_\mu(r) \\ v'_\mu(r) \end{pmatrix},
\tag{8.29}
$$

where r, θ, z are cylindrical coordinates with the axis of symmetry along the vortex circulation (here $\parallel$ and $\perp$ are the parallel and perpendicular to vortex components), and μ is the azimuthal quantum number, which assumes half-integer positive values. It is seen that the states are plain waves along the vortex circulation and are quantized in the orthogonal direction. The radial part of the wave-function is given by

$$
\begin{pmatrix} u'_\mu(r) \\ v'_\mu(r) \end{pmatrix} = 2 \left(\frac{2}{\pi p_\perp r} \right)^{1/2} e^{-K(r)} \begin{pmatrix} \cos\left(p_\perp r - \frac{\pi\mu}{2}\right) \\ \sin\left(p_\perp r - \frac{\pi\mu}{2}\right) \end{pmatrix},
\tag{8.30}
$$

where $p_\perp = \sqrt{p^2 - p_F^2}$, p_F being the neutron Fermi momentum, and the function in the exponent is given by

$$
K(r) = \frac{p_F}{\pi p_\perp \Delta_n} \int_0^r \Delta(r')dr' \simeq \frac{p_F r}{\pi p_\perp \xi} \left(1 + \frac{\xi e^{-r/\xi}}{r} \right).
\tag{8.31}
$$

The eigenstates of neutrons in the core of a vortex are given by

$$
\varepsilon_\mu(p) \simeq \frac{\pi\mu\Delta_n^2}{2\varepsilon_{Fn}} \left(1 + \frac{p^2}{2p_F^2} \right),
\tag{8.32}
$$

where ε_{Fn} is the Fermi energy of neutrons, Δ_n is the asymptotic value of the gap far from the vortex core and we used an expansion in the small quantity p/p_F.

Electrons will couple to the core quasiparticles of the neutron vortex via the interaction of the electron charge e with the magnetic moment of neutrons $\mu_n = -1.913\mu_N$, where $\mu_N = e\hbar/2m_p$ is the nuclear magneton (Feibelman 1971). The momentum relaxation time scale for electrons off neutron vortices is given by Bildsten and Epstein (1989)

$$
\tau_{eV}[^1S_0] = \frac{1.6 \times 10^3}{\Omega_s} \frac{\Delta_n}{T} \left(\frac{\varepsilon_{Fe}}{\varepsilon_{Fn}} \right)^2 \left(\frac{\varepsilon_{Fn}}{2m_n} \right)^{1/2} \exp\left(\frac{\varepsilon_{1/2}^0}{T} \right),
\tag{8.33}
$$

where ε_{Fe} is the Fermi energy of electrons, Δ_n is the S-wave neutron pairing gap, $\varepsilon_{1/2}^0$ is given by Eq. (8.32) with $\mu = 1/2$ and Ω_s is the superfluid's angular velocity. The relaxation time is strongly temperature dependent because of the Boltzmann exponential factor involving the eigenstates of the vortex core quasiparticles.

The vortex structure of the P-wave superfluid was studied by Sauls et al. (1982) using a tensor order parameter $A_{\mu\nu}(r)$, $\mu, \nu = 1, 2, 3$ which is traceless and symmetric. It can be decomposed in cylindrical coordinates (r, ϕ, z) as

$$A_{\mu\nu} = \frac{\Delta}{\sqrt{2}} e^{i\phi} \left\{ [f_1 \hat{r}_\mu \hat{r}_\nu + f_2 \hat{\phi}_\mu \hat{\phi}_\nu - (f_1 + f_2) \hat{z}_\mu \hat{z}_\nu + i g(r_\mu \hat{\phi}_\nu + r_\nu \hat{\phi}_\mu)] \right\},$$

(8.34)

where $g(r)$ and $f_{1,2}(r)$ are the radial functions describing the vortex profile and Δ is the average value of the gap in the $^3P_2$ channel. Vortices in a P-wave superfluid are intrinsically magnetized, with the magnetization given by $M_V(r) = (\gamma_n \hbar) \sigma(r)/2 = g_n \mu_N \sigma(r)/2$, where $\gamma_n = g_n \mu_N \hbar^{-1}$ is the gyromagnetic ratio of the neutron and $g_n = -3.826$, $\hbar \sigma/2$ is the spin density which can be estimated for the P-wave vortex as (Sauls et al. 1982)

$$\sigma(r) = \frac{\nu_n \Delta_n^2}{3} \ln\left(\frac{\Lambda}{T_c}\right) g(r)[f_1(r) - f_2(r)],$$

(8.35)

where Λ is the BCS cut-off, ν_n - the neutron density of states at the Fermi surface. The magnitude of the vortex magnetization that follows from Eq. (8.35) is estimated as (Sauls et al. 1982)

$$|M_V(^3P_2)| = \frac{g_n \mu_N}{2} n_n \left(\frac{\Delta}{\varepsilon_{Fn}}\right)^2 \simeq 10^{11} \text{ G.}$$

(8.36)

Ambient electrons which coexist with the P-wave superfluid because of approximate β-equilibrium among neutrons, protons and electrons, will scatter off the magnetized vortices via the QED interaction term $-e\boldsymbol{\gamma} \cdot A(r)$, where the vector potential associated with the vortex is given by $A(r) = A(r)\hat{\phi}$

$$A(r) = \frac{1}{r} \int_0^r |M_V(r')| r' dr'.$$

(8.37)

The relaxation time for the electron-vortex scattering is given by

$$\tau_{eV}[^3P_2] \simeq \frac{7.91 \times 10^8}{\Omega_s} \left(\frac{k_{Fn}}{\text{fm}}\right) \left(\frac{\text{MeV}}{\Delta_n}\right) \left(\frac{n_e}{n_n}\right)^{2/3}.$$

(8.38)

An important feature of this relaxation time is its near independence of the temperature (a weak temperature dependence arises because of the temperature dependence of the gap). Therefore, it provides an asymptotic lower limit on the scattering rate at low temperatures ($T \ll \Delta_n$) where the relaxation time $\tau_{eV}[^1S_0]$ given by Eq. (8.33) is exponentially suppressed. Numerically $\tau_{eV}[^3P_2]$ is of the order of ten days for the period of the Vela pulsar and varies weakly with the density within the core region where P-wave superfluid resides.

So far, for simplicity, we neglected the proton component of the core of a neutron star. However, as we describe below, the proton fluid can dramatically modify the mutual friction in the core of the star. Consider first a normal (non-superconducting) fluid of protons. At high densities the proton energies can indeed become large enough so that the $^1S_0$-wave interaction becomes negative and pairing vanishes. As described in Sect. 8.2.4, strong magnetic fields $B \geq H_{c2} \simeq 10^{15}$ G will also unpair the proton fluid. Non-superconducting protons will couple to electrons on short plasma time-scale, the relevant scale being set by the plasma frequency. Therefore, protons will compete with electrons in providing the most efficient interaction with the neutron vortices and, eventually, the coupling between the charged plasma component and the neutron superfluid. The key advantage of protons over electrons is that they couple to neutrons via the strong nuclear force, instead of much weaker electromagnetism. The relaxation time for the proton scattering off the quasiparticles in the cores of S-wave neutron vortices is given by Sedrakian (1998)

$$\tau_{pV}[^1S_0] = \frac{0.71}{\Omega_s} \frac{m_n^* m_p^*}{m_n \mu_{pn}^*} \left(\frac{\varepsilon_{Fp}}{\varepsilon_{Fn}}\right)^2 \frac{\varepsilon_{1/2}^0}{T} \exp\left(\frac{\varepsilon_{1/2}^0}{T}\right) \frac{\xi_n^2}{\langle \sigma_{np} \rangle}, \qquad (8.39)$$

where $\mu_{pn}^* = m_p^* m_n^* / (m_n^* + m_p^*)$ is the reduced mass of the neutron-proton system (entering the relation between the cross-section and the scattering amplitude squared), $\varepsilon_{1/2}^0$ is the lowest eigenstate of vortex core excitations Eq. (8.32), and $\langle \sigma_{np} \rangle$ is an average neutron-proton cross-section (for a more precise definition see Eq. (20) of Sedrakian (1998)). The bare neutron mass m_n stems from the quantum of circulation $\kappa = \pi \hbar / m_n$ defining the number of vortices in terms of spin frequency Ω_s. Numerical evaluation of Eq. (8.39) shows that in the relevant range of temperatures $T \simeq 10^7\text{-}10^8$ K the relaxation time is of the order $\tau_{pV}[^1S_0] \simeq 10^{-2}$ sec for the period of the Vela pulsar, which is much shorter than the time-scales for electromagnetic coupling of electrons given by Eqs. (8.33) and (8.38). Only at temperatures of the order of several 10^6 K the process (8.38) takes over; however such low temperatures are unlikely to be achieved in neutron star cores. One potentially important consequence of the shortness of the relaxation time (8.39) is that the superfluid cores of magnetars will couple to the remaining stellar plasma on short dynamical time-scales once proton superconductivity is suppressed by the unpairing mechanisms discussed in Sect. 8.2.4. This will limit the possibilities of explaining glitches and long-timescale variability in terms of superfluid dynamics of magnetar cores (Sedrakian 2016).

Next let us turn to the case where protons are superconducting. A fundamentally new aspect in this case is the *entrainment* of the proton condensate by the neutron condensate which leads to magnetization of a neutron vortex by protonic entrainment currents (Vardanyan and Sedrakyan 1981; Alpar et al. 1984b). (The entrainment effect was first proposed by Andreev and Bashkin (1976) for charge-neutral mixtures of superfluid phases of helium, but they did not discuss vortices). The effective flux of the neutron vortex is given by

$$\phi^* = k\phi_0, \quad k = \frac{m_p^*}{m_p}, \tag{8.40}$$

where $\phi_0 = \pi\hbar c/e$ is the quantum of flux. Numerically, the magnitude of the field in this case is by four orders larger than due to the spontaneous magnetization (Muzikar et al. 1980), therefore much shorter relaxation times are expected. The calculation of the electron relaxation on a neutron vortex discussed in the case of P-wave super-fluid can be repeated in the case of the magnetization by entrainment currents (Alpar et al. 1984b). It is convenient to define the zero-radius scattering rate as

$$\tau_0^{-1} = \frac{2cn_v}{k_{eF}} \left(\frac{\pi^2 \phi_*^2}{4\phi_0^2} \right), \tag{8.41}$$

where the term separated in the bracket is an approximation to the exact Aharonov-Bohm scattering result where $\sin^2(\pi/2)(\phi_*/\phi_0)$ appears instead. The full finite-range scattering rate is then given by Alpar et al. (1984b)

$$\tau_{e\phi}^{-1} = \frac{3\pi}{32} \left(\frac{\varepsilon_{Fe}}{m_p c^2} \right) \frac{\tau_0^{-1}}{k_{eF}\lambda}, \tag{8.42}$$

where λ is the penetration depth. As was the case with the relaxation time (8.38) the scattering rate from magnetized neutron vortices is temperature independent in the first approximation. Numerically, the relaxation time is within the range of seconds and is about four order of magnitude shorter than the one given by (8.38), as a direct consequence of the induced field being larger by the same amount.

As discussed in Sect. 8.2.4, the magnetic field of a neutron star will penetrate the superconducting proton fluid by either forming quantized vortices (in the case where the proton superconductor is type-II) or domains of unpaired proton matter (in the case where it is type-I). Consider first a type-II superconductor. For fields of the order of 10^{12} G and typical rotation frequencies of neutron stars ($\Omega \simeq 100$ Hz), the number of proton vortices (or flux tubes) per area of a single neutron vortex (assuming for the sake of argument colinear neutron and proton vortex lines) is of the order of 10^{13}. Therefore, one may anticipate that proton vortices will strongly affect neutron vortex dynamics. There exist several scenarios for how neutron and proton vortex systems interact and we review them in turn.

One possible scenario assumes that the proton vortex network continues into the crust via magnetic field lines and therefore is frozen into the crustal electron-ion plasma. Neutron vortices might then become pinned on or between these vortices because of the long-range hydrodynamical interaction between them on characteristic scales of the order of λ. Microscopically, crossing the vortices may lead to an additional pinning force on the scale of ξ where the condensate is quenched (Muslimov and Tsygan 1985b; Sauls 1989), but the long-range $\sim \lambda$ hydrodynamical force is the dominant component. The pinning of neutron vortices to proton flux tubes may thus substantially affect the dynamics of neutron vorticity and to some extent can be viewed as mutual friction. The magnitude of the pinning force strongly depends on the relative orientation of vortices and flux tubes, which does not permit to draw model independent conclusions on the relative importance of the pinning force. In some models there are flux tubes associated with the different components of the magnetic fields (poloidal, toroidal, etc), therefore the geometry of the flux tube network itself is a complicated problem. These models will be discussed in more detail in Sect. 8.5 below.

In the vortex cluster model (Sedrakyan and Shakhabasyan 1991) a neutron vortex carries a cluster of proton vortices colinear with the neutron vortex, which are arranged within the region where the entrainment induced field exceeds the lower critical field of the proton superconductor H_{c1}. Such a cluster may substantially enhance the scattering rate estimate given in Alpar et al. (1984b). Larger scattering rate and large forces on the neutron vortex from the electron fluid can lead (counterintuitively) to longer relaxation times for the neutron superfluid than predicted in Alpar et al. (1984b); as a consequence post-glich relaxation time-scales are compatible with the vortex cluster model (Sedrakian and Sedrakian 1995).

Understanding of mutual friction in the case of type-I superconducting protons is difficult because of the lack of model-independent predictions for the domain structure and size of type-I superconductors. A tractable case is where a neutron vortex carries a colinear normal proton domain; in this case it can be shown that the neutron vortex motion induces an electric current within the domain which leads in turn to Ohmic dissipation of electron current (Sedrakian 2005). The dimensionless drag to lift ratio for this process was found to be of the order of 10^{-4}, which makes precession in neutron stars compatible with the type-I superconductivity of protons (Link 2003).

We now turn to the discussion of the mutual friction in the S-wave superfluid within the neutron star's crust. Here the lattice of crustal nuclei is the physically distinct new component which was absent in our discussion of the core physics. Because in the crust the protons are confined to the nuclei and the superfluid is S-wave, the only process that can be carried over from the discussion above is the one give by Eq. (8.33). However, according to our current understanding it is not the dominant process of mutual friction in the crust.

The stationary state of a neutron vortex might require its pinning on a nucleus or in the space between nuclei. As we shall see in Sect. 8.5 there are substantial differences in the estimates of the pinning force in the crust, however the most advanced treatments of pinning based on the density functional theory of superfluid matter indicate that the pinning occurs between the nuclei (Wlazłowski et al. 2016). The sign of the pinning force makes, however, little difference; if the pinning of vortices is strong, then these can respond to the changes in the rotation rate of the crust via thermally activate creep (see Link (2014) and references therein). Vortex creep theory postulates a form of the radial velocity of a vortex which depends exponentially on the ratio of the pinning potential to the temperature. If pinning is strong the associated drag-to-lift ratio could be large and in the range that can account for long time-scale relaxations of glitches (Link 2014).

The strong pinning regime may not arise when the vortex lattice is oriented randomly with respect to the basis vectors of the nuclear lattice. A neutron vortex moving in the crust will couple to the phonon modes of the nuclear lattice (Jones 1990). The one-phonon processes lead to a weak coupling of the superfluid to the crust with $\eta \simeq 10$ g cm^{-1} s^{-1} which implies small dimensionless drag-to-lift ratio $\eta/\rho_n \kappa \ll 1$. The interaction of a neutron vortex moving relative to the nuclei in the crust will generate oscillation modes (Kelvin modes or kelvons) on the vortex and will thus dissipate the kinetic energy of vortex motion into the energy of oscillations (Epstein and Baym 1992). This dissipation can be viewed as mutual friction, because energy and momentum is transferred between the superfluid and the crust. Epstein and Baym (1992) express the drag-to-lift ratio in terms of a dissipation angle $\eta/\rho_n \kappa = \tan \theta_d$ and find an upper limit on this angle $\theta_d \leq 0.7$. This implies that throughout the crust this dissipation mechanism leads to dissipation angles and spin-up time-scales which are close to the maximal one $t_{s,\max} \approx (2\Omega)^{-1}$. It has been argued that the randomness of nuclear potentials may suppress the kelvon-excitation mechanism (Jones 1992) with two-phonon processes being important in a certain range of parameters.

Because the relative orientation of the circulation vector of the vortex lattice and the crustal lattice basis vectors may be random or dependent on the history of solidification of the crust and nucleation of the superfluid phase, it remains an open question whether the pinned regime is realized in the crust of a neutron stars. Numerical simulations (Wlazłowski et al. 2016) and/or astrophysical constraints coming from glitch observations may eventually distinguish between the various models. The problem of pinning in the crust of a neutron star is discussed in more detail in Sect. 8.5 below.

8.4 Superfluid Hydrodynamics

Let us now discuss how to model a superfluid neutron star on the macroscopic, hydrodynamical scale. In order to do hydrodynamics it is necessary to average over length-scales that are large enough for the constituents to be considered as fluids. In the case of superfluids this means coarse graining not only over length-scales much larger than the mean free path of the particles, but also over length scales much larger than the characteristic scale of vortices or flux tubes.

A superfluid forms a macroscopic coherent state, therefore it can be described by a *macroscopic* wave function $\psi(r) = |\psi(r)| \exp[i \chi(r)]$, which implies that the number density of the superfluid is given by $|\psi(r)|^2$ and the momentum is just the gradient of the phase $p = \hbar \nabla \psi(r)$. This implies immediately that

$$\nabla \times p = \nabla \times \nabla \chi(r) = 0, \tag{8.43}$$

i.e., the superfluid is irrotational. However, rotating superfluids support rotation by forming quantized vortices, above a certain critical angular velocity Ω_{c1}. Indeed, the minimization of the free energy of the fluid in the rotating frame (Khalatnikov 1989), i.e.,

$$F_r = F - \boldsymbol{\Omega} \cdot \boldsymbol{J}, \tag{8.44}$$

where $\boldsymbol{J}$ the angular momentum of the fluid and F is its free energy in the laboratory frame, leads to a solution predicting rigid rotation, which is supported by the condensate through the formation of an array of superfluid vortices. The circulation of a single vortex is quantised as

$$\oint \boldsymbol{p} \cdot d\boldsymbol{l} = m_n \, n \, \kappa, \quad n = 1, 2, \ldots \tag{8.45}$$

where $\kappa = \pi \hbar / m_n \approx 2 \times 10^{-3}$ cm^2 s^{-1} is the quantum of circulation and m_n is the neutron mass.

In neutron stars the hydrodynamical description of the neutron fluid requires thus averaging over length-scales larger than the inter-vortex separation, in order to define the rotation rate by averaging over many vortices. This means that in general a fluid element must be small compared to the radius of the star ($R \simeq 10$ km), but large compared to the inter-vortex separation $d_n \simeq 10^{-3} \, (P/10 \text{ ms})^{1/2}$ cm, with $P = 2\pi / \Omega$ being the spin period of the star.

In a realistic neutron star one has to account for multiple superfluid/super-conducting fluids, and a minimal model, such as the one we now present, must account at least for a neutron-proton conglomerate with the background of electrons. Consider first the simpler Newtonian case (Vardanyan and Sedrakyan 1981; Mendell and Lindblom 1991; Mendell 1991; Sedrakian and Sedrakian 1995; Mendell 1998; Glampedakis et al. 2011; Graber et al. 2015; Passamonti et al. 2017b; Antonelli and Pizzochero 2017). In the case of interest to us, the indices x, y label either protons (p) or neutrons (n) and electrons are assumed to form a charge neutralizing background moving with proton fluid on timescales much shorter than those of interest here, such as dynamical and oscillation timescales of milliseconds, or glitch timescales of minutes or more (Mendell 1991).

We will not list the complete set of magneto-hydrodynamics equations here and will concentrate on the ingredients that are needed in the modeling the dynamics of a rotating and magnetized neutron star following Andersson and Comer (2006). Corrections due to coupling of the superfluid to its excitations are ignored. These can be accounted for by including an additional fluid of excitations to the analysis, which also allows to recover the standard hydrodynamical equations used for superfluid helium (Prix 2004; Haskell et al. 2012c). Each constituent x conserves its mass density ρ_x (in the following a summation over latin indices is assumed)

$$\partial_t \rho_x + \nabla_i (\rho_x v_x^i) = 0, \tag{8.46}$$

and Euler equations for their velocities v_x^i are given by

$$(\partial_t + v_x^j \nabla_j)(v_i^x + \varepsilon_x w_i^{yx}) + \nabla_i(\tilde{\mu}_x + \Phi) + \varepsilon_x w_{yx}^j \nabla_i v_j^x = (f_i^{x,mf} + f_i^{x,pin} + f_i^{x,mag})/\rho_x,$$
(8.47)

where $w_i^{yx} = v_i^y - v_i^x$, $\tilde{\mu}_x = \mu_x/m_x$ is the chemical potential per unit mass. The key new aspect of these treatments is the *entrainment* effect which causes the momentum and the velocities of the components to not be parallel, but rather related by

$$p_i^x = m_x(v_i^x + \varepsilon_x w_i^{yx}).$$
(8.48)

This observation was first made by Andreev and Bashkin in the context of charge neutral ^{3}He-^{4}He mixtures (Andreev and Bashkin 1976) as mentioned above. Here $\varepsilon_x = 1 - m_x^*/m_x$ is the entrainment coefficient which accounts for the non-dissipative coupling between the components. (It is related to quantity k defined in (8.40) by $\varepsilon_x = 1 - k$). Note that the microphysical calculations predict for the effective masses $m_x^* \lesssim m_x$ in the core, in which case the entrainment coefficient is positive. The opposite relation holds in the crust of a neutron stars, i.e. $m_x^* \gg m_x$, due to the band structure of the nuclear lattice and Bragg scattering (Chamel 2012).

The gravitational potential Φ in Eq. (8.47) obeys the Poisson equation

$$\nabla^2 \Phi = 4\pi G \sum_x \rho_x.$$
(8.49)

On the right-hand side of Eq. (8.47) the forces are as follows. The first term $f_i^{x,pin}$ is the force due to pinned vortices, $f_i^{x,mf}$ is the mutual friction force mediated by free vortices, and $f_i^{x,mag}$ is the force due to the magnetic field, which as we shall see depends strongly on whether the protons are superconducting or not. We will discuss the contributions from the mutual friction and magnetic forces in detail below in this section, whereas the pinning force is discussed in Sect. 8.5.

For laminar flows and straight vortices, the mutual friction force has the standard Hall-Vienen-Bekarevich-Khalatnikov form (Hall and Vinen 1956; Khalatnikov 1989)

$$f_i^{x,mf} = \kappa n_v \rho_n \mathcal{B}' \epsilon_{ijk} \hat{\Omega}_n^i w_{xy}^k + \kappa n_v \rho_n \mathcal{B} \epsilon_{ijk} \hat{\Omega}_n^j \epsilon^{klm} \hat{\Omega}_l^n w_m^{xy},$$
(8.50)

where Ω_n^j is the angular velocity of the neutrons (a hat represents a unit vector) and $\mathcal{B}$ and $\mathcal{B}'$ are the mutual friction coefficients. The neutron vortex density per unit area is n_v, and is linked to the rotation rate (at a cylindrical radius ϖ) by the relations

$$\kappa n_v(\varpi) = 2\tilde{\Omega} + \varpi \frac{\partial}{\partial \varpi}\tilde{\Omega}, \qquad \tilde{\Omega} \equiv \Omega_n + \varepsilon_n(\Omega_p - \Omega_n),$$
(8.51)

obtained by imposing that the circulation derived by integrating over a contour the smoothed average momentum is the sum of the quantised circulations of the $\mathcal{N}(\varpi)$ vortices enclosed, i.e.

$$\oint \epsilon^{ijk} \nabla_j p_k^{\mathrm{n}} = 2\pi \int_0^\varpi m_n \tilde{\Omega} r \, dr = \mathcal{N}(\varpi) m_{\mathrm{n}} \kappa, \tag{8.52}$$

and we assume here and below singly quantized vortices. The parameters $\mathcal{B}$ and $\mathcal{B}'$ depend on the microphysical processes giving rise to the mutual friction, as described in Sect. 8.3, and can be expressed in terms of a dimensionless drag-to-lift ratio parameter $\mathcal{R}$ (see also Sect. 8.3) related to the dimensionfull drag parameter η [g cm^{-1} s^{-1}] as

$$\mathcal{R} = \frac{\eta}{\kappa \rho_{\mathrm{n}}}, \tag{8.53}$$

according to

$$\mathcal{B} = \frac{\mathcal{R}}{1 + \mathcal{R}^2}, \qquad \mathcal{B}' = \frac{\mathcal{R}^2}{1 + \mathcal{R}^2}. \tag{8.54}$$

To connect to the discussion of the relaxation times computed in Sect. 8.3, we now express η, or equivalently $\mathcal{B}$ and $\mathcal{B}'$ in terms of these microscopic time-scales. We distinguish two cases of non-relativistic and ultra-relativistic unpaired excitations, which we assume to be protons (p) or electrons (e). The force exerted by non-superconducting quasiparticles per single vortex is given in general by: (Bildsten and Epstein 1989)

$$\boldsymbol{f}_d = \frac{2}{\tau n_{\mathrm{v}}} \int f(\boldsymbol{p}, \boldsymbol{v}_{\mathrm{v}}) \boldsymbol{p} \frac{d^3 p}{(2\pi\hbar)^3} = -\eta \boldsymbol{v}_{\mathrm{v}}, \tag{8.55}$$

where $\boldsymbol{v}_{\mathrm{v}}$ is the velocity of the vortex in a frame co-moving with the normal component and $f(\boldsymbol{p}, \boldsymbol{v}_{\mathrm{v}})$ is the non-equilibrium distribution function, which we expand assuming small perturbation about the equilibrium distribution function f_0, that is, $f(\boldsymbol{p}, \boldsymbol{v}_{\mathrm{v}}) = f_0(\boldsymbol{p}) + (\partial f_0/\partial \epsilon)(\boldsymbol{p} \cdot \boldsymbol{v}_{\mathrm{v}})$. In the low-temperature limit $\partial f_0/\partial \epsilon \simeq -\delta(\epsilon - \epsilon_F)$, where ε_F is the corresponding Fermi energy. After integration one finds

$$\eta_{\mathrm{p}} = m_{\mathrm{p}}^* \frac{n_{\mathrm{p}}}{\tau_{\mathrm{p}} n_{\mathrm{v}}}, \qquad \eta_{\mathrm{e}} = \frac{\hbar k_{\mathrm{e}}}{c} \frac{n_{\mathrm{e}}}{\tau_{\mathrm{e}} n_{\mathrm{v}}}, \tag{8.56}$$

where $n_{\mathrm{p,e}}$ are the proton/electron number densities. Note that if both electron and proton quasiparticles are present then the contributions from η_{e} and η_{p} need to be summed, just as in the case of the ordinary transport coefficients.

The parameters $\mathcal{B}$ or $\mathcal{R}$ can be extracted from the timescales obtained in Sect. 8.3, using

$$\tau_{mf} = \frac{1}{2\Omega_s(0)\mathcal{B}} , \qquad (8.57)$$

where $\Omega_s(0)$ is the spin frequency of the superfluid at $t = 0$.

From the equations in (8.47) we can also see that in the case of two constant-density rigidly rotating fluids, with moments of inertia I_p and I_n and frequencies Ω_n and Ω_p, an initial difference in rotation rate $\Delta\Omega = \Omega_p - \Omega_n$ will be erased by mutual friction according to $\Delta\Omega(t) = \Delta\Omega(0)\exp(-t/\tau_{su})$ (Bildsten and Epstein 1989; Andersson et al. 2006), with

$$\tau_{su} \approx \left(\frac{I_p}{I_n + I_p}\right)\frac{1}{2\Omega_n(0)\mathcal{B}} , \qquad (8.58)$$

where $\Omega_n(0)$ is the spin frequency of the neutron fluid at $t = 0$. Here for the sake of argument we neglected the external torques and the entrainment.

The expression for the mutual friction in Eq. (8.51) is appropriate for straight vortices in a triangular array which corresponds to the minimum of the free energy of a rotating superfluid. However, vortices are likely to bend due to their finite rigidity and this effect can easily be included in the expression for the mutual friction, see Khalatnikov (1989) for a discussion in single component fluids and its extension to multi-component fluids in Mendell (1991); Sedrakian and Sedrakian (1995); Andersson et al. (2007). Furthermore, it is well known from laboratory experiments with superfluid ^{4}He that a counterflow along the vortex axis can trigger the Glaberson-Donnelly instability (Glaberson et al. 1974; Donnelly 1991) and destabilise the vortex lattice, creating a turbulent tangle. In the case of an isotropic tangle a phenomenological form for the mutual friction, due to Gorter and Mellink (1949), is

$$f_i^{GM} = \frac{8\pi^2\rho_n}{3\kappa}\left(\frac{\xi_1}{\xi_2}\right)^2 \mathcal{B}^3 w_{pn}^2 w_i^{pn}, \qquad (8.59)$$

where the phenomenological parameters are set to $\xi_1 \approx 0.3$ and $\xi_2 \approx 1$. In neutron star interiors the presence of large relative flows between the 'normal' and superfluid components and large Reynolds numbers, of the order of Re$\geq 10^7$ are likely to lead to superfluid turbulence and the presence of a vortex tangle. According to Peralta et al. (2006); Melatos and Peralta (2007); Andersson et al. (2007) a polarized tangle is expected in a rotating pulsar.

Let us shift our attention to the magnetic force. The equations of magneto-hydrodynamics for a superfluid and superconducting neutron star have been initially considered by Vardanyan and Sedrakyan (1981); Mendell and Lindblom (1991); Mendell (1991); Sedrakian and Sedrakian (1995); Mendell (1998). More recently detailed studies were carried out by Glampedakis et al. (2011); Graber et al. (2015);

Passamonti et al. (2017b) which to various degree also include discussion of the evolution of the magnetic field in a superconducting neutron star. For the current discussion let us restrict out attention to the simplified case of a two component neutron star, in which the electrons are assumed to move with the protons. In this case one finds

$$f^i_{p,mag} = \frac{1}{4\pi} \left[B^j \nabla_j (H_{c1} \hat{B}^i) - B \nabla^i H_{c1} \right] - \frac{\rho_p}{4\pi} \nabla^i \left(B \frac{\partial H_{c1}}{\partial \rho_p} \right), \qquad (8.60)$$

$$f^i_{n,mag} = \frac{1}{4\pi} \left[\mathcal{W}^j_n \nabla_j (H_{vn} \hat{\mathcal{W}}^i_n) - \mathcal{W}_n \nabla^i H_{vn} \right] - \frac{\rho_n}{4\pi} \nabla^i \left(B \frac{\partial H_{c1}}{\partial \rho_n} \right), \qquad (8.61)$$

where a hat indicates a unit vector and we have defined

$$\mathcal{W}^i_p = \epsilon^{ijk} \nabla_j (v^p_k + \varepsilon_p w^{np}_k) + a_p B^i = n_{vp} k^i_p \qquad (8.62)$$

with $k^i = \kappa \hat{k}^i$ pointing along the local vortex direction, $a_p = e/mc \simeq 9.6 \times 10^3$ $G^{-1} s^{-1}$ and n_{vp} the surface density of proton vortices. The total magnetic induction B^i is the sum of three terms

$$B^i = B^i_p + B^i_n + b^i_L, \qquad (8.63)$$

where B^i_p is the contribution due to the proton vortices, B^i_n is the contribution due to the neutron vortices and b^i_L is the London filed. The modulus of the induction is $B = \sqrt{B_i B^i}$. Generally $|B_p| \gg |B_n| \approx |b_L|$, i.e., the strengths of both the average field due to neutron vortices and the London field is negligible compared to neutron star interior magnetic fields ($|B_n| \approx |b_L| \approx 10^{-2}$ G). The term $H_{vn} = 4\pi a_p \varepsilon_{vn}/\kappa \approx 10 \times H_{c1}$ plays the role of an effective magnetic field (Glampedakis et al. 2011) and depends on the energy per unit length of a neutron vortex

$$\varepsilon_{vn} \approx \frac{\kappa^2}{4\pi} \frac{\rho_n}{1 - \varepsilon_n} \log \left(\frac{l_v}{\xi_n} \right), \qquad (8.64)$$

where ξ_n is the coherence length of the vortex and l_v the inter-vortex separation, and we can approximate $\log(l_v/\xi_n) \approx 20 - 1/2 \log(\Omega/100 \text{ rad/s})$ (Andersson et al. 2007). To study the coupled evolution of the fluid and magnetic field these equations need to be coupled to the induction equations for the magnetic field:

$$\partial_t B^i = \epsilon^{ijk} \epsilon_{klm} \nabla_j (v^l_e B^m) - \frac{1}{a_p} \epsilon^{ijk} \nabla_j f^e_k, \qquad (8.65)$$

where, under the assumption that $n_{vp} \gg n_v$ one finds

$$f_e^i \approx \frac{1}{4\pi\rho_p} \frac{\mathcal{R}_p}{1+\mathcal{R}_p^2} \left[\mathcal{R}_p B^j \nabla_j (H_{c1}\hat{B}^i) - B\nabla^i H_{c1} - \epsilon^{ijk} B_j \nabla_k H_{c1} \epsilon^{ijk} \hat{B}_j B^l \nabla_l \hat{B}_k \right],$$
$$(8.66)$$

where $\mathcal{R}_p$ is the drag parameter describing the scattering of electrons on proton vortices.

One may also expect an additional contribution to the mutual friction, due to the flux-tube and neutron vortex interaction, of the form

$$f_{pn,mf}^i = \rho_n \kappa n_v \frac{\mathcal{C}_v}{1+\mathcal{R}_p^2} [\mathcal{R}_p f_*^i + \epsilon^{ijk} \hat{\mathcal{W}}_j^p f_{*k}], \qquad (8.67)$$

$$f_*^i \approx \frac{1}{4\pi a_p \rho_p} [\hat{B}^j \nabla_j (H_{c1}\hat{B}^i) - \nabla^i H_{c1}], \qquad (8.68)$$

where $\mathcal{C}_v$ is a phenomenological coefficient that parameterises the strength of the resistive interaction between the proton and neutron vortex arrays, which may also drive the evolution of the magnetic field (Ruderman et al. 1998).

The magnetic field configuration of superfluid and superconducting neutron stars has been analysed in detail in recent years both by studying equilibrium models (Lander 2013; Henriksson and Wasserman 2013; Lander 2014) and, more recently, by studying the evolution of the coupled core and crust magnetic fields (Elfritz et al. 2016). In general superfluidity and superconductivity have a strong impact on the timescales for the evolution of the core magnetic field (Passamonti et al. 2017a), and for strongly magnetized neutron stars ($B \geq 10^{14}$ G) could lead to the expulsion of the toroidal field from the core, and significant rearrangement of the crustal magnetic field on timescales comparable to, or shorter than, the age of the star.

8.4.1 Relativistic Fluids

What has been presented up to now is the Newtonian framework for describing superfluid and superconducting neutron stars. One can develop a similar framework in general relativity (Carter and Khalatnikov 1992, 1994; Carter and Langlois 1998), for a review see Andersson and Comer (2007). First define the number density four-currents of each component

$$n_x^\mu = n_x u_x^\mu, \qquad (8.69)$$

with normalization $u_\mu^x u_x^\mu = -1$ and Greek letters representing four dimensional space-time indices; summation is implicit over repeated Greek indices. A master function Λ is then defined which is a function of the scalars of the system, in particular the number densities n_x and $n_{xy}^2 = n_{xy}^2 = -g_{\mu\nu}n_x^\mu n_y^\nu$, where $g_{\mu\nu}$ is the metric (Carter and Khalatnikov 1994). In the case of co-moving fluids Λ is (up to the sign) simply the local thermodynamical energy density. In the general case Λ includes relative flows of the fluids by definition.

Having at our disposal the master function we can proceed to define the conjugate momenta in the standard fashion of the Lagrangian theory (Carter and Khalatnikov 1992, 1994; Carter and Langlois 1998)

$$\pi_\mu^x = g_{\mu\nu}(\mathcal{A}^x n_x^\nu + \mathcal{A}^{xy} n_y^\nu), \tag{8.70}$$

$$\mathcal{A}^x = -2\frac{\partial \Lambda}{\partial n_x^2}, \tag{8.71}$$

$$\mathcal{A}^{xy} = \mathcal{A}^{yx} = -\frac{\partial \Lambda}{\partial n_{xy}^2}, \tag{8.72}$$

where the effect of entrainment is encoded in the coefficients $\mathcal{A}^{xy}$. The stress-energy tensor is defined as

$$T_\nu^\mu = \Psi \delta_\nu^\mu - \sum_x n_x^\mu \pi_\nu^x, \tag{8.73}$$

with the generalised pressure Ψ defined as

$$\Psi = \Lambda - \sum_x n_x^\mu \pi_\mu^x. \tag{8.74}$$

The equations of motion for the fluid can then be written as a set of Euler equations

$$\sum_x n_x^\mu \nabla_{[\mu}\pi_{\nu]} = 0, \tag{8.75}$$

to be solved together with the Einstein's equations of general relativity and conservation equations for the individual four-currents of the components

$$\nabla_\mu n_x^\mu = 0. \tag{8.76}$$

Note that the solution to the equations above automatically satisfies the energy-momentum conservation $\nabla_\mu T_\nu^\mu = 0$.

Relativistic superfluid hydrodynamics of the type described above was formulated initially by Carter and Khalatnikov (1992, 1994) and Carter and Langlois (1998). The corresponding equations where adapted to differentially rotating relativistic superfluid neutron stars in Langlois et al. (1998) in the case of cold equations of state. Equilibrium configurations of two-fluid (superfluid and normal component featuring) neutron stars were constructed in Refs. Prix et al. (2005); Sourie et al. (2016). Accounting for heat transport, dissipation and in particular vortex-mediated mutual friction is more challenging in general relativity than in Newtonian physics (Gusakov et al. 2013), as standard approaches by Eckart (1940); Landau and Lifshitz (1959) lead to causality and stability problems. While Israel and Stewart (1979); Carter (1991) resolve some of these issues, and progress has been made making maximal use of the variational approach (Andersson and Comer 2015) the general relativistic formulation is, however, not complete. Some recent advances in the problem of mutual friction and vortex motion in a relativistic framework can be found in Andersson et al. (2016); Gusakov (2016).

8.5   Pinning Effects

In the previous discussion we have considered mainly vortices that are free to move with respect to the fluid components and experience a standard drag force, linear in the difference in velocity between said component and the vortices themselves. (A brief discussion of pinning in the crust was given at the end of Sect. 8.3 to complete the discussion of microphysics of mutual friction). However the interaction between vortices and ions in the crust, or flux-tubes in the core, can be strong enough to balance the Magnus force, and 'pin' the vortices, preventing them from moving, similarly to static friction. We now turn to the detailed discussion of the pinning effects in the superfluid core and in the crust of the neutron star.

8.5.1   Vortex-Flux Tube Pinning and Interactions

In the outer core of the neutron star, where neutrons are superfluid and protons are expected to form a type II superconductor, the magnetic field is confined to flux-tubes with flux quantum $\phi_0 = \pi \hbar c / e \approx 2 \times 10^{-7}$ G cm^2. The neutron vortices thus co-exist with an array of far more numerous flux tubes, with average spacing

$$l_\phi = \left(\frac{B}{\phi_0} \right)^{-1/2} \approx 4 \times 10^3 \left(\frac{B}{10^{12} \text{ G}} \right)^{-1/2} \text{ fm.} \qquad (8.77)$$

Proton flux tubes are also less rigid than neutron vortices. Indeed the tension of a neutron vortex is given by

$$T_v = \frac{\rho_n \kappa^2}{4\pi} \ln \left(\frac{l_v}{\xi_n} \right) \approx 10^9 \left(\frac{\rho_n}{2 \times 10^{14} \text{ g cm}^{-3}} \right) \text{ erg cm}^{-1} \qquad (8.78)$$

with the typical inter-vortex separation $l_v \approx 10^{-3}$ (P/10 ms)$^{1/2}$ cm. The flux-tube tension is given by Harvey et al. (1986)

$$T_\Phi = \left(\frac{\phi_0}{4\pi\lambda} \right)^2 \ln \left(\frac{\lambda}{\xi_p} \right) \approx 10^7 \left(\frac{m_p^*/m_p}{0.5} \right)^{-1} \left(\frac{x_p}{0.05} \right) \left(\frac{\rho_n}{2 \times 10^{14} \text{g cm}^{-3}} \right) \text{ erg cm}^{-1},$$
$$\qquad (8.79)$$

where m_p^* is the effective mass of the protons, $\xi_p \approx 20$ fm the coherence length of a proton vortex and $\lambda \approx 100$ fm is the London penetration depth of the magnetic field. Neutron vortices will thus be immersed in a tangle of far more numerous flux tubes.

The interaction between neutron vortices and flux tubes changes the energy of the system in two ways: there will be a change in condensation energies, as the superfluid and superconducting cores overlap, and also a contribution from the interaction between the magnetic fields of the two vortices, as described in Sect. 8.3. If the overlap reduces the energy of the overall configuration neutron vortices are effectively 'pinned' to flux tubes, to some extent in the same way as they can be pinned to ions in the crust.

The contribution due to the change in condensation is given by Muslimov and Tsygan (1985a) and Sauls (1989)

$$\Delta E_c \approx 0.13 \, \text{MeV} \left(\frac{\Delta_p}{1 \text{MeV}} \right) \left(\frac{x_p}{0.05} \right)^{-1} \left(\frac{m_n^*/m_n}{1} \right)^{-2} \left(\frac{m_p^*/m_p}{0.5} \right)^{-1} \tag{8.80}$$

with m_n^* the neutron effective mass, Δ_p the proton pairing gap, while the change in energy due to the magnetic interaction (Jones 1991a; Chau et al. 1992):

$$\Delta E_{\text{mag}} = 2 \frac{B_n^i B_i^{\Phi}}{8\pi} (\pi \lambda^2 l_\lambda), \tag{8.81}$$

with B_n^i the magnetic field along the neutron vortex, $B_{\Phi}^i \approx 10^{15}$ G that along the proton vortex, and l_λ is the overlap length between vortex and flux-tube. Keeping in mind that neutron vortices can be considered rigid on lengthscales approximately 100 times larger than those over which a fluxtube can bend, we can average the expression in (8.81) to obtain (Link 2012):

$$\Delta E_{\text{mag}} \approx 10 \left(\frac{m_p^*/m_p}{0.5} \right)^{-1/2} \left(\frac{|m_p - m_p^*|/m_p^*}{0.5} \right) \left(\frac{x_p}{0.05} \right)^{1/2} \left(\frac{\rho_n}{2 \times 10^{14} \text{ g/cm}^3} \right)^{1/2} \text{MeV}. \tag{8.82}$$

In general one may expect also a dependence on the inclination angle of the global magnetic field with the rotation axis, and on vortex tension, see Gügercinoğlu and Alpar (2016); Sedrakian et al. (2011) for a discussion of toroidal flux tubes and the effect of bending and pinning in this case. Nevertheless the above averaged expression illustrates that the pinning force will be sizeable. Pinned vortices can thus 'push' magnetic flux tubes, possibly winding up a strong toroidal component of the magnetic field in magnetars (Glampedakis and Andersson 2011) and leading to the long term expulsion of magnetic flux from the star as the vortex array expands while the star spins down (Srinivasan et al. 1990; Ruderman et al. 1998; Jahan-Miri 2000; Jones 2006).

If, on the other hand, the pinning force cannot balance the Magnus force, vortices are forced to cut through flux tubes. This process will excite Kelvin waves along the vortex, leading to strong dissipation and mutual friction (Epstein and Baym 1992; Link 2003). The energy released at every vortex/flux tube intersection is: (Link 2003)

$$\Delta E_{vf} = \frac{2}{\pi} \frac{\Delta E_{mag}^2}{\rho_n \kappa \lambda} (v_\lambda w_{np})^{-1/2}, \tag{8.83}$$

where $v_\lambda = \bar{h}/2m_K \lambda \approx 10^9$ cm/s is the characteristic velocity of a kelvon of effective mass m_K. The energy loss rate per unit volume is thus

$$\dot{\mathcal{E}} = \frac{n_v w_{pn}}{l_\phi^2} \Delta E_{vf}. \tag{8.84}$$

By equating the expression in (8.84) to the work done by the mutual friction force we can derive the drag coefficient for vortex/flux tube cutting (Haskell et al. 2014)

$$\mathcal{R} = \mathcal{R}_0 \left(\frac{v_\lambda}{w_{pn}}\right)^{3/2},$$

$$\mathcal{R}_0 = \frac{2}{\pi} \left(\frac{\Delta E_{mag}}{\rho_n \kappa \lambda v_\lambda}\right)^2 \approx 1.3 \times 10^{-10} \left(\frac{B}{10^{12}G}\right). \tag{8.85}$$

Note that the mutual friction coefficient is now velocity dependent, and the lower the relative velocity the larger the friction. In practice as soon as vortices start moving they are unlikely to be able to continue and the system will move back towards the pinned state (Haskell et al. 2014).

8.5.2 Pinning-Repinning of Vortices

Let us consider the forces acting on a massless vortex segment. If the vortex is free the force balance equation, averaged over a number n_v of vortices per unit area, is

$$\kappa n_v \epsilon_{ijk} \hat{\Omega}^j (v_n^k - v_v^k) + \kappa n_v \mathcal{R}(v_i^p - v_i^v) = 0. \tag{8.86}$$

One can solve Eq. (8.86) for the vortex velocity v_v^i (the direction of which will depend on $\mathcal{R}$) and obtain the standard form of mutual friction in Eq. (8.50). Pinning to the ions in the crust or flux tubes in the core modifies the force balance equation, because now some of the vortices may be immobilized by the pinning. When averaging over a large number of vortices we can assume that only a fraction γ of them is free, leading to a force balance equation of the form:

$$\gamma \kappa n_v \epsilon_{ijk} \hat{\Omega}^j (v_n^k - v_v^k) + \gamma \kappa n_v \mathcal{R}(v_i^p - v_i^v) + (1 - \gamma)\kappa n_v \epsilon_{ijk} \hat{\Omega}^j (v_n^k - v_p^k) + f_i^{pin} = 0 \tag{8.87}$$

where f_i^{pin} now balances the Magnus force on the $(1 - \gamma)n_v$ pinned vortices, for which we have assumed that $v_v^i|_{\text{pinned}} = v_p^i$. The force acting on the fluids is thus $f_i^{\text{n,pin}} = -f_i^{\text{p,pin}} = f_i^{\text{pin}}$.

The quantity that is needed for the equations of motion in Eqs. (8.47) and (8.87) is thus the pinning force per unit length acting on a vortex, which is highly uncertain. The pinning force per pinning site can, in fact, be quite readily obtained theoretically, as it depends only on the difference in energy between the configuration where the vortex overlaps with an individual pinning site, and that in which it is outside. Nevertheless even in this case significant uncertainties remain, with different results in the literature disagreeing also on whether the interaction is attractive or repulsive, i.e. on whether one has pinning to nuclei or interstitial pinning (Pizzochero et al. 2002; Donati and Pizzochero 2004, 2006). Note, however, that to understand the dynamics of the fluid we are mainly interested in the magnitude of the pinning force, and not in its sign.

For the case of pinning of neutron vortices to nuclei in the crust the maximum pinning force acting on a vortex can be estimated as (Alpar 1977)

$$|F_{\text{pin}}| \approx \left(n_{\text{out}} E_{\text{out}}^{\text{cond.}} - n_{\text{in}} E_{\text{in}}^{\text{cond.}} \right) \frac{V}{\xi_n}, \tag{8.88}$$

where V is the volume of a nuclear cluster and ξ_n is the coherence length of the neutron vortex, which defines the scale of the interaction, 'in' and 'out' refer to quantities taken within the nuclear cluster and in the free neutron gas, $E^{\text{cond.}} \simeq 3\Delta_n^2/8\varepsilon_{Fn}$ is the condensation energy of neutron fluid per unit volume. The pinning force per unit length depends however on the difference in energy between different configurations of a vortex that encounters several pinning sites and may bend to reduce its energy. It is thus, generally a function of the orientation of the vortex with respect to the lattice

$$f_{\text{pin}}(\theta, \phi) = |F_{\text{pin}}| \frac{\Delta n(\theta, \phi)}{l_T}, \tag{8.89}$$

where the angles (θ, ϕ) are taken with respect to a reference axis and $l_T \approx 10^2 - 10^3 R_{WS}$ is the length scale over which a vortex can bend, determined by the tension in Eq. (8.78), and R_{WS} is the radius of the Wigner-Seitz cell. It was pointed out early on in Jones (1991b) that vortex rigidity plays an important role, as for an infinite vortex all configurations would be energetically equivalent (i.e. they would intercept the same number of pinning sites) and there would thus be no pinning at all.

Recently calculations have been carried out for a realistic setup by Seveso et al. (2016), who averaged the expression in Eq. (8.89) over all orientations of a vortex with respect to a BCC lattice. The authors find nuclear pinning, i.e. vortices are pinned to the nuclei. Wlazłowski et al. (2016) studied the interactions of a vortex with a pinning site in the time dependent local density approximation, i.e.,

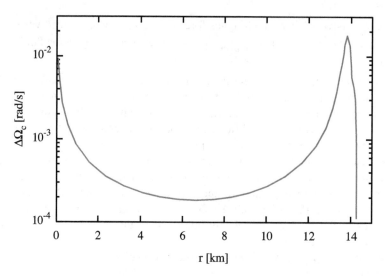

Fig. 8.3 Dependence of the critical lag $\Delta\Omega_c$ for unpinning on the internal cylindrical radius of the star for a $1.4M_\odot$ neutron star model described by the GM1 equation of state as described in Seveso et al. (2016) and Seveso et al. (2012)

it includes superfluid dynamics in addition to static interactions. In this case, the pinning is interstitial, i.e., the vortices are localized in-between the nuclear clusters.

The angular momentum reservoir is, however, independent of the sign of the pinning force, and both works (Wlazłowski et al. 2016; Seveso et al. 2016) obtain pinning forces that are dynamically significant and can explain the observed glitching activity of the Vela pulsar. By simply balancing the Magnus force with the pinning force, Seveso et al. (2016) find that velocity differences up to $|w^{pn}| \approx 10^4$ cm s^{-1} can be sustained in the crust, which can explain the observed glitching activity of the Vela and other pulsars, also in the presence of strong entrainment. An example of the critical lag profile in a neutron star, obtained for the pinning of Seveso et al. (2016), is shown in Fig. 8.3.

The results of dynamical simulations can also be used to study the problem of repinning of free vortices. This is a fundamental issue, as while estimating when a pinned vortex will unpin allows us to estimate how much angular momentum can be stored and released in a glitch, calculating when a vortex will re-pin allows us to understand whether vortices will 'creep' out, gradually spinning down the star, or expel vorticity in 'avalanches'.

Macroscopic vortex dynamics in a spinning down container was studied on the basis of Gross-Pitaevskii equations by Warszawski and Melatos (2011); Warszawski et al. (2012), who have shown that the main unpinning trigger for vortices is the proximity effect, i.e. the change in Magnus force due to the motion of neighbouring vortices, which can lead to forward or backward propagating vortex avalanches. These can, in turn, trigger a glitch. Computational limitations, however, constrain these simulations to a small number of vortices (typically of the order of hundreds) separated by at the most tens of pinning sites. This is in contrast with the situation encountered in neutron stars where a large number $N_v \gtrsim 10^{12}$ of vortices must move together in a glitch. These are on average separated by a large number (of the order of 10^{10}) pinning sites. Nevertheless, the external spin-down drives the system by increasing the lag between the superfluid and normal fluid to the critical value for unpinning. It is thus crucial to understand whether the system can self-adjust and hover close enough to the critical lag that vortices can unpin and skip over many pinning sites in order to knock on neighbouring vortices and allow an avalanche to propagate.

The problem of vortex re-pinning was investigated by Sedrakian (1995); Haskell and Melatos (2016), by considering vortex motion in a parabolic pinning potential. Pinning of a moving vortex in a random potential was also studied in Link (2009). In particular Haskell and Melatos (2016) calculated the mean-free path of a straight vortex for scattering off cylindrical pinning sites, and found that the main parameters that control repinning are the strength of the mutual friction and, crucially, how close the system is to the critical threshold lag for unpinning. From Fig. 8.4 we can deduce that if the system is within 5% of the critical lag for unpinning, a vortex can move a distance comparable to the inter-vortex separation and knock on other vortices, causing an avalanche, for realistic values of the mutual friction. Studies by Chauve et al. (2001); Marchetti et al. (2003); Fily et al. (2010) have also shown that the geometry of the lattice can play a crucial role, with a more disordered lattice behaving like a plastic system, in which unpinned and pinned vortices coexist, and an ordered lattice behaving like an elastic system in which there is a sharp transition to mass unpinning.

Before moving on, let us note that if protons form a type II superconductor in the core, as was discussed in Sect. 8.5.1, the energy cost of vortex/flux-tube cutting can lead to pinning. A simple estimate of the pinning force per unit length $f^{\mathrm{pin},\phi}$ can be obtained from the expression for the overlap energy in (8.81)

$$f^{\mathrm{pin},\phi} \approx \frac{E_{\mathrm{mag}}}{\lambda l_\phi}, \tag{8.90}$$

where λ is the London penetration length, l_ϕ the distance between fluxtubes and E_{mag} is defined in Eq. (8.82). The force (8.90) is balanced by the Magnus force for a critical velocity $|w^{\mathrm{pn}}| \approx 5 \times 10^3 (B/10^{12}\mathrm{G})^{1/2}$ cm s^{-1}, which indicates that this force could play an important role in a glitching model based on unpinning of vortices in the core. Note, however, that the estimate (8.90) does not account for the effect of averaging over different orientations of the vortex lattice, and is thus an upper limit on the pinning force.

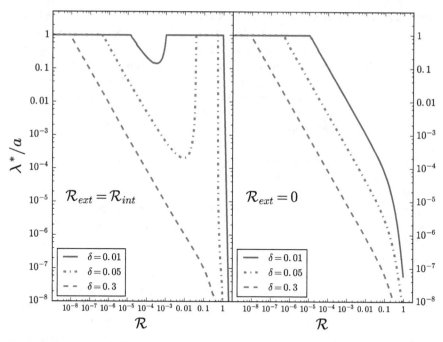

Fig. 8.4 Mean free path λ^* of a vortex, normalized to the intervortex spacing a, for different values of $\delta = (\Delta\Omega_c - \Delta\Omega)/\Delta\Omega_c$, with $\Delta\Omega_c$ the critical lag for unpinning, as described in Haskell and Melatos (2016). The left panel shows the case in which mutual friction is described by the same parameter $\mathcal{R}$ inside and outside the pinning potential, while the right panel is for the case in which there is no mutual friction outside the pinning potential. In general avalanches can propagate if mutual friction is weak, and especially in the case in which mutual friction is the same everywhere, if $\delta \lesssim 0.05$, i.e. the system is close to the critical lag for unpinning

8.6 Macrophysics of Superfluidity in Neutron Stars

8.6.1 Glitches and Post-glitch Relaxations

Pulsar glitches are sudden spin up events in the otherwise steadily decreasing rotational frequency of pulsars. These were observed in the Vela pulsar soon after the discovery of radio pulsars (Reichley and Downs 1969; Radhakrishnan and Manchester 1969). The initial jump in frequency in the case of the Vela pulsar is instantaneous to the accuracy of the data (with the best upper limit of $\tau_r \lesssim 40$ s coming from the Vela 2000 glitch (Dodson et al. 2002)), but is often accompanied by an increase in spin-down rate that relaxes back towards the pre-glitch values on longer timescales (ranging from minutes to months). The long-time scales of relaxations following glitches were taken as an evidence for the presence of a loosely coupled superfluid component in the star (Baym et al. 1969).

Initial studies attributed the long relaxation times to the slow coupling of the core of the star due to the weak coupling of vortices to the electron fluid according to Eq. (8.39), see Baym et al. (1969); Feibelman (1971). Anderson and Itoh put forward the hypothesis that glitches are linked to a pinned superfluid in the star, that is decoupled from the observable 'normal' component, and whose sudden re-coupling (due to unpinning) leads to an exchange of angular momentum and a glitch (Anderson and Itoh 1975).

Following the idea of a pinned superfluid in the crust of neutron stars (Anderson and Itoh 1975) the initial models of glitches and post-glitch relaxation concentrated on the details of the physics of vortex pinning and unpinning mainly in the crust and the fits of the models to the observed behaviour of the Vela pulsar (Alpar et al. 1984a,c, 1985; Cheng et al. 1988). The glitch in these models was generated by a sudden unpinning of neutron vortices within the crust of the neutron star. The observed subsequent slow relaxation of the rotation frequency and its derivative was modelled in terms of thermal creep of neutron vortices against pinning barriers. The 'creeping' velocity of neutron vortices is given in these models by

$$v_r \approx v_0 \exp(-E_a/k_B T), \tag{8.91}$$

where $v_0 \approx 10^7$ cm s^{-1} (Alpar et al. 1984a), k_B is Boltzmann's constant and E_a is the activation energy for unpinning (Link and Epstein 1991). This latter quantity in a first approximation can be taken as $E_a \approx E_p (1 - \Delta\Omega/\Delta\Omega_c)$, with E_p the pinning energy, $\Delta\Omega$ the lag between the superfluid neutrons and the crust, and $\Delta\Omega_c$ the critical lag for unpinning. The equations of motion for the frequency of the observable 'normal' component of the star Ω_p are

$$I_p \dot{\Omega}_p = N_{ext} + \sum_i I_n^i \frac{2\Omega_n^i}{\varpi} v_r^i, \tag{8.92}$$

where I_p is the moment of inertia of the crust and all components tightly coupled to it, ϖ the cylindrical radius, N_{ext} is the external torque, and the superfluid is divided into a number i of different regions, with associated moment of inertia I_n^i, angular velocity Ω_n^i and vortex velocity v_r^i, calculated from (8.91). The solutions to Eq. (8.92) admit two regimes. If the steady state lag $\Delta\Omega$ is much smaller that $\Delta\Omega_c$ then the response of the system is linear and the region contributes to the frequency relaxation exponentially after a glitch (Alpar et al. 1993). This is also the regime that can be modelled in terms of mutual friction coupling due to a small number of free vortices (van Eysden and Melatos 2010; Haskell et al. 2012a; van Eysden 2014; Howitt et al. 2016). If $\Delta\Omega \approx \Delta\Omega_c$ the response will be nonlinear, and the contribution of the region to the relaxation takes the form of a Fermi function (Alpar et al. 1984a; Link 2014). Recent work has also shown that the non-linear response of creep to glitches can be used to interpret the inter-glitch behaviour of the Vela pulsar, and predict the occurrence of the next glitch (Akbal et al. 2016). The creep model has been more recently extended to the scenario where neutron vortices creep against the core flux-tubes (Link 2014; Gügercinoğlu and Alpar 2016), with essentially the same physical picture of post-jump relaxation involved. Pinning of superfluid neutron vortices to proton vortices in the core naturally extends the reservoir of angular momentum available for a glitch thus potentially explaining the observed activity of the Vela pulsar (Gügercinoğlu and Alpar 2014), although a large amount of pinned vorticity in the core is not consistent with linear models for the recovery of Vela glitches (Haskell et al. 2013). The original crustal vortex creep model relied on the assumption, derived from the short relaxation times found in Alpar et al. (1984b), that the core is coupled to the crust on short dynamical time-scales, which are unobservable in glitches and their relaxations.

An alternative to crust-based models is the vortex cluster model of dynamics of superfluid neutron vortices in the core of a neutron star, where the coupling between the superfluid and normal component occurs on much longer time-scales (Sedrakian and Sedrakian 1995). Models of pulsar glitches and post-glitch relaxations based on the superfluid core rotation in the absence of pinning between the neutron vortices and flux-tubes were developed within the vortex cluster model and applied to Vela glitches in Sedrakian et al. (1995). The glitch itself can arise in these models through the interaction of vortex clusters with the crust-core interface (Sedrakian and Cordes 1999); the core moment of inertia alone was estimated to be sufficient to account for both glitches and post-glitch relaxations (Sedrakian et al. 1995; Sedrakian and Cordes 1999).

If the system hovers close to the critical unpinning threshold vortex, avalanches may propagate in the neutron star interior (Haskell and Melatos 2016), making them a viable mechanism for triggering a glitch (Cheng et al. 1988; Warszawski et al. 2012). A glitching pulsar would thus behave as a self organised critical system, in which slowly increasing global stresses (due to the external spin-down torque that drives the increase in lag and thus Magnus force) are released rapidly and locally via nearest neighbour interactions between vortices. Such a system is scale invariant and one expects the distribution of sizes of the avalanches to be a power-law, and the distribution of waiting times an exponential. This is generally what is observed in the pulsar population (Melatos et al. 2008), with the notable exception of the Vela pulsar and PSR J0537-6910, which exhibit a quasi-periodicity in their glitching behaviour and for which the glitches can be predicted (Middleditch et al. 2006; Akbal et al. 2016). This behaviour in the case of PSR J0537 is generally attributed to the presence of crust-quakes, but may also be the consequence of the timescale of the external driving (the spin-down) being short compared to the time-scale on which stress is released locally by the vortices. Another system in which the distribution of glitch sizes appears to deviate from a power-law is the Crab pulsar (Espinoza et al. 2014) for which there appears to be a cut off for small glitch sizes. This behaviour is, however, natural if one considers not only the exchange of angular momentum due to vortex motion, but also the coupling timescale due to mutual friction. Haskell (2016) investigated this by considering a multifluid system described by Eq. (8.47), in which however only a number γn_v of vortices is free at any given time, with $\gamma \leq 1$. The value of γ is randomly drawn from a power-law distribution after a waiting time t_w (randomly drawn from an exponential distribution) and the results of these simulations show that small values of γ not only correspond to a small amount of angular momentum, but also to an effective reduction in the average

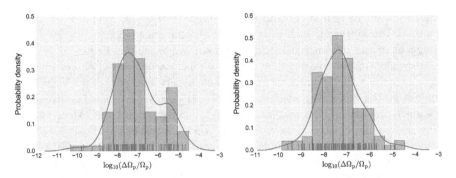

Fig. 8.5 Probability distribution function for glitch sizes $\Delta\Omega_p/\Omega_p$ for a microscopic waiting time $t_w = 0.1$ days and two different microscopic power law indices: $n = -1.05$ (left) and $n = -1.5$, as described in Haskell (2016). In both cases there is a clear deviation from a power-law, with the appearance of a cutoff at low sizes, as observed in the Crab pulsar (Espinoza et al. 2014)

mutual friction and an increased coupling timescale. In this case the event does not appear as a sudden jump in frequency, but is more gradual and closer to timing noise, thus not being recognised as a glitch by detection algorithms and producing a cutoff in the size distribution for small glitches, see Fig. 8.5.

In addition to the mechanisms mentioned above there exist a number of other candidate mechanism for glitches. As already mentioned crust quakes may drive glitches (Ruderman 1969, 1976; Carter et al. 2000) as has been suggested for PSR J0537-6910 and also the Crab pulsar (Middleditch et al. 2006), and hydrodynamical instabilities may also lead to a glitch (Mastrano and Melatos 2005; Glampedakis and Andersson 2009).

In the above discussion of the pinning-based trigger mechanisms for a glitches we tacitly assumed that pinning occurs between vortices and ions in the crust or vortices and flux-tubes in the core. The extent and location of the pinning region can, however, be studied more quantitatively by examining both the size of the maximum glitch recorded in a pulsar (Pizzochero 2011; Seveso et al. 2012), and its 'activity' $\mathcal{A}$, i.e the amount of spin-down that is reversed by glitches during observations

$$\mathcal{A} = \frac{1}{t_{\text{obs}}} \sum_i \frac{\Delta\Omega_p^i}{\Omega_p}, \tag{8.93}$$

where the sum is performed over all recorded glitches in a time t_{obs} and $\Delta\Omega_p^i$ is the recorded size of glitch i. From angular momentum conservation over a glitch one has that the ratio between the moment of inertia of the superfluid reservoir I_n and that of the 'normal' component I_p is

$$\frac{I_n}{I_p} \approx -\frac{\Omega_p}{\dot\Omega_p}\mathcal{A}(1 - \varepsilon_n). \tag{8.94}$$

Andersson et al. (2012); Chamel (2013) noted that in the presence of strong neutron entrainment ε_n, such as is predicted in the crust where $\varepsilon_n \approx 10$ due to Bragg scattering (Chamel 2012) (although see Watanabe and Pethick (2017) for a description of how pairing may lead to weaker entrainment), the crust cannot store enough angular momentum to explain the observed activity of the Vela pulsar (unless the star has a very small mass $M \lesssim 1M_\odot$). The core must be involved in the glitch mechanism. The neutron superfluid is, in fact, expected to extend into the core and models that extend the reservoir beyond the crust can predict the observed activity of the Vela and other pulsars (Newton et al. 2015; Ho et al. 2015). The observed activity of a pulsar, together with the size of its maximum glitch, can then potentially be used to determine the mass of a glitching pulsar and constrain the equation of state of dense matter (Ho et al. 2015; Pizzochero et al. 2017; Antonelli et al. 2018).

Finally let us note that the effect of both classical and superfluid turbulence have been ignored in the above discussion, but my have an impact on glitch physics. Transitions between turbulent and laminar regimes could explain the short spin up timescales and long inter-glitch timescales (Peralta et al. 2006; Melatos and Peralta 2007) and shear-driven turbulence can contribute to low frequency fluctuations in the spin of the star, i.e. so-called 'timing noise' (Melatos and Link 2014).

8.6.2 Oscillations in Superfluid Stars

Neutron stars are expected to be prolific emitters of gravitational waves (Haskell et al. 2015; Lasky 2015) and in particular there are several modes of oscillation of the star that could lead to detectable emission. The most promising modes for ground based detection are the f-mode, or fundamental mode, and the r-mode, analogous to Rossby waves in the ocean, which can be driven unstable due to gravitational wave emission and grow to large amplitudes (Andersson 1998; Friedman and Morsink 1998). In order to assess the detectability of these signals it is thus crucial to understand in which region of parameter space the modes can be driven unstable by gravitational wave emission, and in which region, on the other hand, they are rapidly damped by viscosity. At high temperatures ($T \gtrsim 10^9$ K) bulk viscosity is the main damping mechanisms and matter is not expected to be superfluid. At lower temperatures, however, superfluidity has a strong impact on the damping. On the one side superfluidity leads to a suppression in the neutron-neutron and neutron-proton scattering events that give rise to shear viscosity, which in this case is mainly due to electron-electron scattering and is reduced with respect to the case in which neutrons are normal (Andersson et al. 2005). On the other hand superfluidity opens a new dissipative channel by allowing for vortex mediated mutual friction.

It has been established early on that the doubling of the degrees of freedom in the superfluid component doubles the number of oscillations modes; these have been studied in Newtonian theory (Lindblom and Mendell 1994; Lee 1995; Sedrakian and

Wasserman 2001a) and general relativistic setting (Comer et al. 1999). The $l \leq 2$ modes of incompressible superfluid self-gravitating fluids (both axially symmetric and tri-axial) were studied by Sedrakian and Wasserman (2001a,b) in the presence of mutual friction and viscosity using the tensor virial method. In the absence of shear viscosity the mutual friction can be eliminated from the equations describing the center-of-mass motions of the two fluids and it acts to damp only the relative motions of the two fluids. Shear viscosity which acts only in the normal fluid breaks the symmetry of the Euler equations for the normal fluid and superfluid and, therefore, couples these two sets of modes (Sedrakian and Wasserman 2001a,b). These initial studies were followed by studies which obtained the analogues of the f and p modes in superfluid neutron stars (Prix and Rieutord 2002) and included the general relativity in the mode description in the case of non-rotating stars (Andersson et al. 2002; Gualtieri et al. 2014). Furthermore, much work has been concentrated on the r-modes of the superfluid neutron stars (Lee and Yoshida 2003; Yoshida and Lee 2003a,b; Prix et al. 2004; Haskell et al. 2009; Andersson et al. 2010) which may play a key role in the dynamics and stability of rapidly rotating neutron stars. Progress has been made in understanding the oscillations modes at finite temperature (Gusakov and Andersson 2006; Kantor and Gusakov 2011; Andersson et al. 2013b) as well as the influence of crust elasticity (Passamonti and Andersson 2012).

Let us examine this problem in detail. The linearised version of the equations of motion in (8.47) can be written in terms of two sets of degrees of freedom, one that represents the 'total' motion of the fluid (and would be present also in a normal, non-superfluid, star), and one that represents the counter-moving motion. Let us consider the case in which mutual friction is the only dissipative mechanism acting on the system. By introducing the total mass flux

$$\rho \delta v^k = \rho_{\mathrm{n}} \delta v_{\mathrm{n}}^j + \rho_{\mathrm{p}} \delta v_{\mathrm{p}}^j \tag{8.95}$$

with $\rho = \rho_{\mathrm{n}} + \rho_{\mathrm{p}}$ and where δ represents Eulerian perturbations, we can write a 'total' Euler equation for the total velocity v^i, in a frame rotating with angular velocity Ω,

$$\partial_t \delta v_i + \nabla_i \delta \Phi + \frac{1}{\rho} \nabla_i \delta p - \frac{1}{\rho^2} \delta \rho \nabla_i p + 2\epsilon_{ijk} \Omega^j \delta v^k = 0 \tag{8.96}$$

where the pressure p is obtained from

$$\nabla_i p = \rho_{\mathrm{n}} \nabla_i \tilde{\mu}_{\mathrm{n}} + \rho_{\mathrm{p}} \nabla_i \tilde{\mu}_{\mathrm{p}}. \tag{8.97}$$

We also have the standard continuity equation

$$\partial_t \delta \rho + \nabla_j (\rho \delta v^j) = 0. \tag{8.98}$$

As already mentioned these are identical to the perturbed equations of motion for a single fluid system, and quite notably the mutual friction term drops out of the equations. The mutual friction naturally appears in the equations of motion for the second degree of freedom, which we can write in terms of the perturbed relative velocity $\delta w^j = \delta v_p^j - \delta v_n^j$. The 'difference' Euler equation is

$$(1 - \varepsilon_n x_p^{-1})\partial_t \delta w_i + \nabla_i \delta\beta + 2\tilde{\mathcal{B}}'\epsilon_{ijk}\Omega^j \delta w^k - \tilde{\mathcal{B}}\epsilon_{ijk}\hat{\Omega}^j \epsilon^{klm}\Omega_l \delta w_m = 0, \qquad (8.99)$$

where we have defined the local deviation from chemical equilibrium

$$\delta\beta = \delta\tilde{\mu}_p - \delta\tilde{\mu}_n \qquad (8.100)$$

and $\tilde{\mathcal{B}}' = 1 - \mathcal{B}'/x_p$ and $\tilde{\mathcal{B}} = \mathcal{B}/x_p$ with $x_p = \rho_p/\rho$ the proton fraction. The continuity equation for the proton fraction is:

$$\partial_t \delta x_p + \frac{1}{\rho}[x_p(1 - x_p)\rho \delta w^j] + \delta v^j \nabla_j x_p = 0 \qquad (8.101)$$

The degrees of freedom are thus explicitly coupled unless x_p is constant. In general we do not expect to find any mode of oscillation in a realistic neutron star that is purely co-moving, and thus all modes are affected, to some extent, by mutual friction.

We can define a conserved energy for the system by first defining a 'kinetic' term as an integral over a volume V

$$E_k = \frac{1}{2}\int \left[|\delta v|^2 + (1 - \varepsilon_n/x_p)x_p(1 - x_p)|\delta w|^2\right]\rho \, dV \qquad (8.102)$$

and a 'potential' term

$$E_p = \frac{1}{2}\int \left\{\rho\left(\frac{\partial\rho}{\partial p}\right)_\beta |\delta h|^2 + \left(\frac{\partial\rho}{\partial\beta}\right)_p [2\mathrm{Re}(\delta h\delta\beta^*) + |\delta\beta|^2] - \frac{1}{4\pi G}|\nabla_i\delta\Phi|^2\right\} dV, \qquad (8.103)$$

where a star represents complex conjugation. The perturbed equations of motion we have written down explicitly include dissipative terms due to mutual friction. It is, however, instructive to consider the non dissipative case and ignore the contribution due to mutual friction. In this case $\partial_t(E_k + E_p) = 0$ and one can solve the problem for a mode with time dependence $e^{i\omega t}$. If damping is weak, and procedes on a timescale τ much longer than the period of the mode, i.e. one has $\omega = \omega_r + i/\tau$, with $\tau \gg 1/\omega_r$, we can use the solution of the non-dissipative problem to estimate the damping timescale as

$$\tau = \left|\frac{2(E_p + E_k)}{\partial_t E}\right|, \qquad (8.104)$$

where $\partial_t E$ is obtained from a dissipation integral, in which the dissipative terms due to viscosity are evaluated using the non dissipative solution. In the case of mutual friction one can see that this takes the form (Andersson et al. 2009)

$$\partial_t E_{\mathcal{B}} = -2 \int \rho_n \mathcal{B} \Omega [\delta_i^m - \hat{\Omega}^m \hat{\Omega}^i] \delta w^{i*} \delta w_m \, dV. \tag{8.105}$$

If the damping timescale is sufficiently short, the estimate of τ in (8.104) matches that obtained from the full mode solution. Often, however, the full solution to the dissipative problem is not available, and (8.104) is the only way to asses the impact of viscosity.

Finally one must calculate the timescale for gravitational waves to drive the mode. The energy lost to gravitational waves can be obtained from a multipole expansion (Thorne 1980)

$$\partial_t E_{gw} = -\omega_r \sum_l N_l \omega_i^{2l+1} (|\delta D_{lm}|^2 + |\delta J_{lm}|^2), \tag{8.106}$$

where ω_i is the frequency of the mode in the intertial frame $N_l = (4\pi G)(l+1)(l+2)/\{c^{2l+1}l(l-1)[(2l+1)!!]^2\}$ and the mass multipoles are

$$\delta D_{lm} \approx \int \delta T_{00} Y_{lm}^* r^l dV \tag{8.107}$$

and the current multipoles are:

$$\delta J_{lm} \approx -\int \delta T_{0j} Y_{j,lm}^{B*} dV, \tag{8.108}$$

with $Y_{j,lm}^B$ the magnetic multipoles (Thorne 1980) and $T_{\alpha\beta}$ is the two-fluid stress energy tensor defined in Sect. 8.4.1.

We refer the interested reader to Andersson and Kokkotas (2001); Andersson et al. (2009) for a detailed discussion of the calculation, and simply point out that in general

$$T_{00} \approx \delta\rho, \tag{8.109}$$

$$T_{0j} \approx \rho \delta v_j + \delta\rho v_j, \tag{8.110}$$

i.e., only the co-moving degree of freedom radiates gravitationally to leading order in rotation.

We are now potentially equipped to calculate the driving timescale τ_{gw} and compare it to the mutual friction damping timescale $\tau_{\mathcal{B}}$ for realistic neutron star modes. Clearly if $\tau_{\mathcal{B}} \lesssim \tau_{gw}$ mutual friction will damp the mode faster than gravitational radiation can drive it, and suppress the instability, while in the opposite case a mode can grow to large amplitudes and radiate gravitationally.

Let us first of all examine the fundamental, or f-mode. This is essentially a 'surface' mode for which the frequency $\omega_f \sim \bar{\rho}^{1/2}$, with $\bar{\rho}$ the average density of the star. In the case of the f-mode both Newtonian and relativistic studies have shown that mutual friction completely suppresses the gravitational-wave driven instability below the superfluid transition temperature (Andersson et al. 2009; Gaertig et al. 2011). It may thus play a role in hot newly born neutron stars (Pnigouras and Kokkotas 2016; Surace et al. 2016) but is unlikely to be active in older, colder stars.

The situation is different for the r-mode. To first order in rotation this mode is purely axial and comoving, leading to a single multipole solution of the form:

$$\delta v^j = \left[\sum_l \left(\frac{m}{r^2 \sin\theta} U_l Y_l^m \hat{e}_\theta^j + \frac{i}{r^2 \sin\theta} U_l \partial_\theta Y_l^m \hat{e}_\phi^j \right) \right] \exp\left(i\omega_0 t\right) \qquad (8.111)$$

where $U_l = Ar^{l+1}$ with A a constant, $\omega_0 = 2m/l(l+1)$ is the frequency of the mode in the rotating frame, and we are using spherical coordinates, with $\hat{e}_\theta^j$ and $\hat{e}_\phi^j$ unit vectors and Y_l^m are spherical harmonics. The $l = m = 2$ r-mode thus provides the strongest contribution to gravitational wave emission, as in a single fluid star. However, as we have mentioned, the co-moving motion couples to the counter-moving degrees of freedom at higher orders in rotation, leading to mutual friction damping (Passamonti et al. 2009).

For the r-mode standard mutual friction due to electron scattering off vortex cores has little effect on the instability (Lindblom and Mendell 2000; Haskell et al. 2009). Strong mutual friction due to vortex/flux-tube cutting in the core can, however, have a strong impact on the instability and damp it in a large section of parameter space. Furthermore strong dissipation due to vortex-flux tube interactions limits the growth of the mode and sets an effective saturation amplitude for it that may be smaller than the saturation amplitude due to non-linear couplings to other modes (Haskell et al. 2014).

A number of authors have also suggested that for specific temperatures there can be avoided crossings between the superfluid r-modes and other inertial modes, leading to enhanced mutual friction damping (Lee and Yoshida 2003; Gusakov et al. 2014a,b; Kantor and Gusakov 2017). This is in an interesting scenario as it would reconcile our understanding of the r-mode instability in superfluid neutron star with observational data on spins and temperatures of neutron stars in Low Mass X-ray Binaries (Ho et al. 2011; Haskell et al. 2012b; Mahmoodifar and Strohmayer 2013). This scenario also predicts the existence of hot, rapidly rotating neutron stars (Chugunov et al. 2014; Kantor et al. 2016).

Finally it has been pointed out that in the presence of pinned vorticity the r-mode can grow unstable if a large lag develops between the superfluid and the crust, also possibly triggering a glitch (Glampedakis and Andersson 2009; Andersson et al. 2013a).

8.6.3 Long-Term Variabilities

In addition of the phenomena discussed above neutron stars exhibit long-term variability, which has been attributed to free precession (Stairs et al. 2000; Kerr et al. 2016). Theoretical studies of precession in neutron stars containing a superfluid component indicate that the analogue of free precession in classical systems will be damped if the superfluid is coupled strongly to the normal component (Shaham 1977; Sedrakian et al. 1999). In addition to this analogue of the classical precession, a fast precession mode appears, which is associated with the doubling of the degrees of freedom. To leading order it is independent of the deformation of the star and scales as the ratio of the moments of inertia of the superfluid and normal components. Nevertheless, free precession of neutron stars has been modelled and applied to the available data with some success (Jones and Andersson 2001; Wasserman 2003; Akgün et al. 2006; Jones 2012; Ashton et al. 2017), which might be evidence for weak coupling of the superfluid to the normal component. Long term variabilities in neutron stars can also be understood in terms of Tkachenko waves - oscillations of the vortex lattice in the neutron superfluid (Ruderman 1970). The corresponding modes have been studied in a number of setting, including the damping by mutual friction and shear viscosity (Noronha and Sedrakian 2008; Shahabasyan 2009; Haskell 2011; Shahabasyan and Shahabasyan 2011), and have been shown to be in the range relevant for the long-term periodicities observed in pulsars.

8.7 Conclusions and Future Directions

This chapter provided an educational introduction to the physics of superfluidity and superconductivity as well as a discussion of selected subjects of current interest. Some basic aspects of the physics of superfluidity in neutron stars are now well established: the rough magnitudes of the pairing gaps, the existence of vortices and flux tubes, the basic channels of mutual friction and mechanisms of vortex pinning. The basic contours of the macro-physics behaviour in glitches, their relaxations and other anomalies such as precession and oscillations are also emerging. However, there are significant uncertainties related to the details of the theory. For example, it is a matter of debate whether glitches occur in the crust, in the core or in both components of the star. The same applies to the post-glitch response of various superfluid shells that respond to a glitch on observable dynamical time-scales. Finally, there are phenomena that are not understood well at the basic level, such as for example, the possibility of free precession in neutron stars.

One of the main difficulties in modelling these phenomena lies in the large separation in scales that exists in neutron stars between the interactions of vortices, flux tubes and clusters on the Fermi scale, vortex-vortex interactions on the scale of millimeters and the large scale hydrodynamics of the star. Future efforts must thus focus on bridging the gap between these scales, both from a theoretical and a computational point of view.

Future theoretical developments in the field will definitely obtain impetus from observational programs in radio, X-ray and gravitational wave astronomy. The SKA radio observatory, to become operational in the upcoming decade, has the potential of significant discoveries in pulsars astrophysics through the strong increase of the number (up to 30.000) of observed pulsars and its high sensitivity. The currently operating gravitational wave observatories are sensitive probes of continuous gravitational wave radiation from pulsars and current upper limits on such radiation are sensitive enough to constraint some physics of neutron star interiors (e.g., the rigidity of the crust). In the future, some of the transients seen in the electromagnetic spectrum may become observable through gravitational waves, thus providing complementary information on their dynamics. Finally, current and future X-ray observations of pulsars are expected to put stronger constraints on the thermal evolution models of neutron stars and their gross parameters (in particular, neutron star radii will be measured by the NICER experiment to a high precision). Combined these 'multi-messenger' observations of neutron stars will provide us with a deeper theoretical understanding of the workings of superfluids in neutron star interiors.

Acknowledgements B.H. has received funding from the European Union's Horizon 2020 research and innovation programme under grant agreement No. 702713. A.S. is supported by the Deutsche Forschungsgemeinschaft (Grant No. SE 1836/3-2). We acknowledge the support by the NewCompStar COST Action MP1304 and by the PHAROS COST Action CA 16214.

References

Akbal, O., Alpar, M.A., Buchner, S., Pines, D.: Nonlinear interglitch dynamics, the braking index of the Vela pulsar and the time to the next glitch. ArXiv e-prints (2016). 1612.03805

Akgün, T., Link, B., Wasserman, I.: Precession of the isolated neutron star PSR B1828-11. MNRAS **365**, 653–672 (2006). https://doi.org/10.1111/j.1365-2966.2005.09745.x. astro-ph/0506606

Alm, T., Röpke, G., Sedrakian, A., Weber, F.: $^3D_2$ pairing in asymmetric nuclear matter. Nucl. Phys. A **604**, 491–504 (1996). https://doi.org/10.1016/0375-9474(96)00153-4

Alpar, M.A.: Pinning and threading of quantized vortices in the pulsar crust superfluid. Astrophys. J. **213**, 527–530 (1977) . https://doi.org/10.1086/155183

Alpar, M.A., Pines, D., Anderson, P.W., Shaham, J.: Vortex creep and the internal temperature of neutron stars. I - general theory. Astrophys. J. **276**, 325–334 (1984a). https://doi.org/10.1086/161616

Alpar, M.A., Langer, S.A., Sauls, J.A.: Rapid postglitch spin-up of the superfluid core in pulsars. Astrophys. J. **282**, 533–541 (1984b). https://doi.org/10.1086/162232

Alpar, M.A., Anderson, P.W., Pines, D., Shaham, J.: Vortex creep and the internal temperature of neutron stars. II - VELA pulsar. Astrophys. J. **278**, 791–805 (1984c). https://doi.org/10.1086/161849

Alpar, M.A., Nandkumar, R., Pines, D.: Vortex creep and the internal temperature of neutron stars The Crab pulsar and PSR 0525 + 21. Astrophys. J. **288**, 191–195 (1985). https://doi.org/10.1086/162780

Alpar, M.A., Chau, H.F., Cheng, K.S., Pines, D.: Postglitch relaxation of the VELA pulsar after its first eight large glitches - a reevaluation with the vortex creep model. Astrophys. J. **409**, 345–359 (1993). https://doi.org/10.1086/172668

Andersson, N.: A new class of unstable modes of rotating relativistic stars. Astrophys. J. **502**, 708–713 (1998). https://doi.org/10.1086/305919.gr-qc/9706075

Andersson, N., Comer, G.L.: A flux-conservative formalism for convective and dissipative multi-fluid systems, with application to Newtonian superfluid neutron stars. Classical Quantum Gravity **23**, 5505–5529 (2006). https://doi.org/10.1088/0264-9381/23/18/003. physics/0509241

Andersson, N., Comer, G.L.: Relativistic fluid dynamics: physics for many different scales. Living Rev. Relativ. **10**, 1 (2007). https://doi.org/10.12942/lrr-2007-1. gr-qc/0605010

Andersson, N., Comer, G.L.: A covariant action principle for dissipative fluid dynamics: from formalism to fundamental physics. Classical Quantum Gravity **32**, 075008 (2015). https://doi.org/10.1088/0264-9381/32/7/075008. 1306.3345

Anderson, P.W., Itoh, N.: Pulsar glitches and restlessness as a hard superfluidity phenomenon. Nature **256**, 25–27 (1975). https://doi.org/10.1038/256025a0

Andersson, N., Kokkotas, K.D.: The R-mode instability in rotating neutron stars. Int. J. Modern Phys. D **10**, 381–441 (2001). https://doi.org/10.1142/S0218271801001062. gr-qc/0010102

Andersson, N., Comer, G.L., Langlois, D.: Oscillations of general relativistic superfluid neutron stars. Phys. Rev. D **66**, 104002 (2002). https://doi.org/10.1103/PhysRevD.66.104002. gr-qc/0203039

Andersson, N., Comer, G.L., Glampedakis, K.: How viscous is a superfluid neutron star core?. Nucl. Phys. A **763**, 212–229 (2005). https://doi.org/10.1016/j.nuclphysa.2005.08.012. astro-ph/0411748

Andersson, N., Sidery, T., Comer, G.L.: Mutual friction in superfluid neutron stars. Mon. Not. R. Astron. Soc. **368**, 162–170 (2006). https://doi.org/10.1111/j.1365-2966.2006.10147.x. astro-ph/0510057

Andersson, N., Sidery, T., Comer, G.L.: Superfluid neutron star turbulence. Mon. Not. R. Astron. Soc. **381**, 747–756 (2007). https://doi.org/10.1111/j.1365-2966.2007.12251.x. astro-ph/0703257

Andersson, N., Glampedakis, K., Haskell, B.: Oscillations of dissipative superfluid neutron stars. Phys. Rev. D **79**, 103009 (2009). https://doi.org/10.1103/PhysRevD.79.103009. 0812.3023

Andersson, N., Haskell, B., Comer, G.L.: r-modes in low temperature color-flavor-locked superconducting quark stars. Phys. Rev. D **82**, 023007 (2010). https://doi.org/10.1103/PhysRevD.82.023007. 1005.1163

Andersson, N., Glampedakis, K., Ho, W.C.G., Espinoza, C.M.: Pulsar glitches: the crust is not enough. Phys. Rev. Lett. **109**, 241103 (2012). https://doi.org/10.1103/PhysRevLett.109.241103. 1207.0633

Andersson, N., Glampedakis, K., Hogg, M.: Superfluid instability of r-modes in "differentially rotating" neutron stars. Phys. Rev. D **87**, 063007 (2013a). https://doi.org/10.1103/PhysRevD.87.063007. 1208.1852

Andersson, N., Krüger, C., Comer, G.L., Samuelsson, L.: A minimal model for finite temperature superfluid dynamics. Classical Quantum Gravity **30**, 235025 (2013b). https://doi.org/10.1088/0264-9381/30/23/235025. 1212.3987

Andersson, N., Wells, S., Vickers, J.A.: Quantised vortices and mutual friction in relativistic superfluids. Classical Quantum Gravity **33**, 245010 (2016). https://doi.org/10.1088/0264-9381/33/24/245010. 1601.07395

Andreev, A.F., Bashkin, E.P.: Three-velocity hydrodynamics of superfluid solutions. Sov. J. Exp. Theor. Phys. **42** 164 (1976)

Antonelli, M., Pizzochero, P.M.: Axially symmetric equations for differential pulsar rotation with superfluid entrainment. Mon. Not. R. Astron. Soc. **464** (2017), 721–733. https://doi.org/10.1093/mnras/stw2376. 1603.02838

Antonelli, M., Montoli, A., Pizzochero, P.M.: Effects of general relativity on glitch amplitudes and pulsar mass upper bounds. Mon. Not. R. Astron. Soc. (2018) https://doi.org/10.1093/mnras/sty130. 1710.05879

Ashton, G., Jones, D.I., Prix, R.: On the free-precession candidate PSR B1828-11: evidence for increasing deformation. Mon. Not. R. Astron. Soc. **467**, 164–178 (2017). https://doi.org/10. 1093/mnras/stx060. 1610.03508

Baldo, M., Elgarøy, Ø., Engvik, L., Hjorth-Jensen, M., Schulze, H.-J.: $^3P_2$-$^3F_2$ pairing in neutron matter with modern nucleon-nucleon potentials. Phys. Rev. C **58**, 1921–1928 (1998). https:// doi.org/10.1103/PhysRevC.58.1921. nucl-th/9806097

Bardeen, J., Cooper, L.N., Schrieffer, J.R.: Theory of superconductivity. Phys. Rev. **108**, 1175–1204 (1957). https://doi.org/10.1103/PhysRev.108.1175

Baym, G., Pethick, C., Pines, D., Ruderman, M.: Spin up in neutron stars : the future of the vela pulsar. Nature **224**, 872–874 (1969). https://doi.org/10.1038/224872a0

Bildsten, L., Epstein, R.I.: Superfluid dissipation time scales in neutron star crusts. ApJ **342**, 951–957 (1989). https://doi.org/10.1086/167652

Bogoljubov, N.N.: On a new method in the theory of superconductivity. Il Nuovo Cimento (1955–1965) **7**, 794–805 (1958). https://doi.org/10.1007/BF02745585

Bohr, A., Mottelson, B.R., Pines, D.: Possible analogy between the excitation spectra of nuclei and those of the superconducting metallic state. Phys. Rev. **110**, 936–938 (1958). https://doi.org/10. 1103/PhysRev.110.936

Bruk, Y.M.: Intermediate state in superconducting neutron star. Astrophysics **9**, 134–143 (1973). https://doi.org/10.1007/BF01011420

Buckley, K.B., Metlitski, M.A., Zhitnitsky, A.R.: Vortices and type-I superconductivity in neutron stars. Phys. Rev. C **69**, 055803 (2004). https://doi.org/10.1103/PhysRevC.69.055803. hep-ph/0403230

Caroli, C., Matricon, J.: Excitations électroniques dans les supraconducteurs purs de 2ème espèce. Physik der Kondensierten Materie **3**, 380–401 (1965). https://doi.org/10.1007/BF02422806

Carter, B.: Convective variational approach to relativistic thermodynamics of dissipative fluids. Proc. R. Soc. London, Ser. A **433**, 45–62 (1991). https://doi.org/10.1098/rspa.1991.0034

Carter, B., Khalatnikov, I.M.: Momentum, vorticity, and helicity in covariant superfluid dynamics. Ann. Phys. **219**, 243–265 (1992). https://doi.org/10.1016/0003-4916(92)90348-P

Carter, B., Khalatnikov, I.M.: Canonically covariant formulation of Landau's Newtonian superfluid dynamics. Rev. Math. Phys. **6**, 277–304 (1994). https://doi.org/10.1142/S0129055X94000134

Carter, B., Langlois, D.: Relativistic models for superconducting-superfluid mixtures. Nucl. Phys. B **531**, 478–504 (1998). https://doi.org/10.1016/S0550-3213(98)00430-1. gr-qc/9806024

Carter, B., Langlois, D., Sedrakian, D.M.: Centrifugal buoyancy as a mechanism for neutron star glitches. Astron. Astrophys. **361**, 795–802 (2000). astro-ph/0004121

Chamel, N.: Neutron conduction in the inner crust of a neutron star in the framework of the band theory of solids. Phys. Rev. C **85**, 035801 (2012). https://doi.org/10.1103/PhysRevC.85.035801

Chamel, N.: Crustal entrainment and pulsar glitches. Phys. Rev. Lett. **110**, 011101 (2013). https:// doi.org/10.1103/PhysRevLett.110.011101. 1210.8177

Chau, H.F., Cheng, K.S., Ding, K.Y.: Implications of 3P2 superfluidity in the interior of neutron stars. Astrophys. J. **399**, 213–217 (1992). https://doi.org/10.1086/171917

Chauve, P., Le Doussal, P., Jörg Wiese, K.: Renormalization of pinned elastic systems: how does it work beyond one loop? Phys. Rev. Lett. **86**, 1785–1788 (2001). https://doi.org/10.1103/ PhysRevLett.86.1785. cond-mat/0006056

Chen, J.M.C., Clark, J.W., Davé, R.D., Khodel, V.V.: Pairing gaps in nucleonic superfluids. Nucl. Phys. A **555**, 59–89 (1993). https://doi.org/10.1016/0375-9474(93)90314-N

Cheng, K.S., Pines, D., Alpar, M.A., Shaham, J.: Spontaneous superfluid unpinning and the inhomogeneous distribution of vortex lines in neutron stars. Astrophys. J. **330**, 835–846 (1988). https://doi.org/10.1086/166517

Chugunov, A.I., Gusakov, M.E., Kantor, E.M.: New possible class of neutron stars: hot and fast non-accreting rotators. Mon. Not. R. Astron. Soc. **445** (2014), 385–391. https://doi.org/10. 1093/mnras/stu1772. 1408.6770

Comer, G.L., Langlois, D., Lin, L.M.: Quasinormal modes of general relativistic superfluid neutron stars. Phys. Rev. D **60**, 104025 (1999). https://doi.org/10.1103/PhysRevD.60.104025. gr-qc/9908040

Cooper, L.N.: Bound electron pairs in a degenerate Fermi gas. Phys. Rev. **104**, 1189–1190 (1956). https://doi.org/10.1103/PhysRev.104.1189

Dean, D.J., Hjorth-Jensen, M.: Pairing in nuclear systems: from neutron stars to finite nuclei. Rev. Mod. Phys. **75**, 607–656 (2003). https://doi.org/10.1103/RevModPhys.75.607. nucl-th/0210033

Dodson, R.G., McCulloch, P.M., Lewis, D.R.: High time resolution observations of the January 2000 glitch in the vela pulsar. Astrophys. J. **564**, L85–L88 (2002). https://doi.org/10.1086/339068. astro-ph/0201005

Donati, P., Pizzochero, P.M.: Fully consistent semi-classical treatment of vortex-nucleus interaction in rotating neutron stars. Nucl. Phys. A **742**, 363–379 (2004). https://doi.org/10.1016/j.nuclphysa.2004.07.002

Donati, P., Pizzochero, P.M.: Realistic energies for vortex pinning in intermediate-density neutron star matter. Phys. Lett. B **640**, 74–81 (2006). https://doi.org/10.1016/j.physletb.2006.07.047

Donnelly, R.J.: Quantized Vortices in Helium II. University of Massachusetts, Amherst (1991)

Eckart, C.: The thermodynamics of irreversible processes. III. Relativistic theory of the simple fluid. Phys. Rev. **58**, 919–924 (1940). https://doi.org/10.1103/PhysRev.58.919

Elfritz, J.G., Pons, J.A., Rea, N., Glampedakis, K., Viganò, D.: Simulated magnetic field expulsion in neutron star cores. Mon. Not. R. Astron. Soc. **456**, 4461–4474 (2016). https://doi.org/10.1093/mnras/stv2963. 1512.07151

Epstein, R.I., Baym, G.: Vortex drag and the spin-up time scale for pulsar glitches. Astrophys. J. **387**, 276–287 (1992). https://doi.org/10.1086/171079

Espinoza, C.M., Antonopoulou, D., Stappers, B.W., Watts, A., Lyne, A.G.: Neutron star glitches have a substantial minimum size. Mon. Not. R. Astron. Soc. **440**, 2755–2762 (2014). https://doi.org/10.1093/mnras/stu395. 1402.7219

Feibelman, P.J.: Relaxation of electron velocity in a rotating neutron superfluid: application to the relaxation of a pulsar's slowdown rate. Phys. Rev. D **4**, 1589–1597 (1971). https://doi.org/10.1103/PhysRevD.4.1589

Fetter, A.L., Walecka, J.D.: Quantum Theory of Many-Particle Systems. McGraw-Hill, Boston (1971)

Fily, Y., Olive, E., di Scala, N., Soret, J.C.: Critical behavior of plastic depinning of vortex lattices in two dimensions: molecular dynamics simulations. Phys. Rev. B **82**, 134519 (2010). https://doi.org/10.1103/PhysRevB.82.134519. 0912.3835

Friedman, J.L., Morsink, S.M.: Axial Instability of rotating relativistic stars. Astrophys. J. **502** (1998), 714–720. https://doi.org/10.1086/305920. gr-qc/9706073

Gaertig, E., Glampedakis, K., Kokkotas, K.D., Zink, B.: f-mode instability in relativistic neutron stars. Phys. Rev. Lett. **107**, 101102 (2011). https://doi.org/10.1103/PhysRevLett.107.101102. 1106.5512

Gandolfi, S., Illarionov, A.Y., Fantoni, S., Pederiva, F., Schmidt, K.E.: Equation of state of superfluid neutron matter and the calculation of the $^1S_0$ pairing gap. Phys. Rev. Lett. **101**, 132501 (2008). https://doi.org/10.1103/PhysRevLett.101.132501. 0805.2513

Gezerlis, A., Carlson, J.: Low-density neutron matter. Phys. Rev. C **81**, 025803 (2010). https://doi.org/10.1103/PhysRevC.81.025803. 0911.3907

Gezerlis, A., Pethick, C.J., Schwenk, A.: Pairing and superfluidity of nucleons in neutron stars. ArXiv e-prints (2014). 1406.6109

Glaberson, W.I., Johnson, W.W., Ostermeier, R.M.: Instability of a vortex array in He II. Phys. Rev. Lett. **33**, 1197–1200 (1974). https://doi.org/10.1103/PhysRevLett.33.1197

Glampedakis, K., Andersson, N.: Hydrodynamical trigger mechanism for pulsar glitches. Phys. Rev. Lett. **102**, 141101 (2009). https://doi.org/10.1103/PhysRevLett.102.141101. 0806.3664

Glampedakis, K., Andersson, N.: Magneto-rotational neutron star evolution: the role of core vortex pinning. Astrophys. J. **740**, L35 (2011). https://doi.org/10.1088/2041-8205/740/2/L35. 1106.5997

Glampedakis, K., Andersson, N., Samuelsson, L.: Magnetohydrodynamics of superfluid and superconducting neutron star cores. Mon. Not. R. Astron. Soc. **410**, 805–829 (2011). https://doi.org/10.1111/j.1365-2966.2010.17484.x. 1001.4046

Gorter, C.J., Mellink, J.H.: On the irreversible processes in liquid helium II. Physica 15, 285–304 (1949). https://doi.org/10.1016/0031-8914(49)90105-6

Graber, V., Andersson, N., Glampedakis, K., Lander, S.K.: Magnetic field evolution in superconducting neutron stars. Mon. Not. R. Astron. Soc. 453, 671–681 (2015). https://doi.org/10.1093/mnras/stv1648. 1505.00124

Gualtieri, L., Kantor, E.M., Gusakov, M.E., Chugunov, A.I.: Quasinormal modes of superfluid neutron stars. Phys. Rev. D 90, 024010 (2014). https://doi.org/10.1103/PhysRevD.90.024010. 1404.7512

Gügercinoğlu, E., Alpar, M.A.: Vortex creep against Toroidal flux lines, crustal entrainment, and pulsar glitches. Astrophys. J. 788, L11 (2014). https://doi.org/10.1088/2041-8205/788/1/L11

Gügercinoğlu, E., Alpar ,M.A.: Microscopic vortex velocity in the inner crust and outer core of neutron stars. Mon. Not. R. Astron. Soc. 462, 1453–1460 (2016). https://doi.org/10.1093/mnras/stw1758. 1607.05092

Gusakov, M.E.: Relativistic formulation of the Hall-Vinen-Bekarevich-Khalatnikov superfluid hydrodynamics. Phys. Rev. D 93, 064033 (2016). https://doi.org/10.1103/PhysRevD.93.064033. 1601.07732

Gusakov, M.E., Andersson, N.: Temperature-dependent pulsations of superfluid neutron stars. Mon. Not. R. Astron. Soc. 372, 1776–1790 (2006). https://doi.org/10.1111/j.1365-2966.2006.10982.x. astro-ph/0602282

Gusakov, M.E., Kantor, E.M., Chugunov, A.I., Gualtieri, L.: Dissipation in relativistic superfluid neutron stars. Mon. Not. R. Astron. Soc. 428, 1518–1536 (2013). https://doi.org/10.1093/mnras/sts129. 1211.2452

Gusakov, M.E., Chugunov, A.I., Kantor, E.M.: Instability windows and evolution of rapidly rotating neutron stars. Phys. Rev. Lett. 112, 151101 (2014). https://doi.org/10.1103/PhysRevLett.112.151101. 1310.8103

Gusakov, M.E., Chugunov, A.I., Kantor, E.M.: Explaining observations of rapidly rotating neutron stars in low-mass X-ray binaries. Phys. Rev. D 90, 063001 (2014). https://doi.org/10.1103/PhysRevD.90.063001. 1305.3825

Hall, H.E., Vinen, W.F.: The rotation of liquid helium II. II. The theory of mutual friction in uniformly rotating helium II. Proc. R. Soc. London, Ser. A 238, 215–234 (1956). https://doi.org/10.1098/rspa.1956.0215

Harvey, J.A., Ruderman, M.A., Shaham, J.: Effects of neutron-star superconductivity on magnetic monopoles and core field decay. Phys. Rev. D 33, 2084–2091 (1986). https://doi.org/10.1103/PhysRevD.33.2084

Haskell, B.: Tkachenko modes in rotating neutron stars: the effect of compressibility and implications for pulsar timing noise. Phys. Rev. D 83, 043006 (2011). https://doi.org/10.1103/PhysRevD.83.043006. 1011.1180

Haskell, B.: The effect of superfluid hydrodynamics on pulsar glitch sizes and waiting times. Mon. Not. R. Astron. Soc. 461, L77–L81 (2016). https://doi.org/10.1093/mnrasl/slw103. 1603.04304

Haskell, B., Melatos, A.: Pinned vortex hopping in a neutron star crust. Mon. Not. R. Astron. Soc. 461, 2200–2211 (2016). https://doi.org/10.1093/mnras/stw1334. 1510.03136

Haskell, B., Andersson, N., Passamonti, A.: r modes and mutual friction in rapidly rotating superfluid neutron stars. Mon. Not. R. Astron. Soc. 397 (2009), 1464–1485. https://doi.org/10.1111/j.1365-2966.2009.14963.x. 0902.1149

Haskell, B., Pizzochero, P.M., Sidery, T.: Modelling pulsar glitches with realistic pinning forces: a hydrodynamical approach. Mon. Not. R. Astron. Soc. 420, 658–671 (2012a). https://doi.org/10.1111/j.1365-2966.2011.20080.x. 1107.5295

Haskell, B., Degenaar, N., Ho, W.C.G.: Constraining the physics of the r-mode instability in neutron stars with X-ray and ultraviolet observations. Mon. Not. R. Astron. Soc. 424, 93–103 (2012b). https://doi.org/10.1111/j.1365-2966.2012.21171.x. 1201.2101

Haskell, B., Andersson, N., Comer, G.L.: Dynamics of dissipative multifluid neutron star cores. Phys. Rev. D 86, 063002 (2012c). https://doi.org/10.1103/PhysRevD.86.063002. 1204.2894

Haskell, B., Pizzochero, P.M., Seveso, S.: Investigating superconductivity in neutron star interiors with glitch models. Astrophys. J. **764**, L25 (2013). https://doi.org/10.1088/2041-8205/764/2/L25. 1209.6260

Haskell, B., Glampedakis, K., Andersson, N.: A new mechanism for saturating unstable r modes in neutron stars. Mon. Not. R. Astron. Soc. **441**, 1662–1668 (2014). https://doi.org/10.1093/mnras/stu535. 1307.0985

Haskell, B., Andersson, N., D'Angelo, C., Degenaar, N., Glampedakis, K., Ho, W.C.G. et al.: Gravitational waves from rapidly rotating neutron stars. In: Sopuerta, C.F. (ed.) Gravitational Wave Astrophysics. Astrophysics and Space Science Proceedings, vol. 40, p. 85 (2015). 1407.8254. DOI

Henriksson, K.T., Wasserman, I.: Poloidal magnetic fields in superconducting neutron stars. Mon. Not. R. Astron. Soc. **431**, 2986–3002 (2013). https://doi.org/10.1093/mnras/stt338. 1212.5842

Ho, W.C.G., Andersson, N., Haskell, B.: Revealing the physics of r modes in low-mass X-ray binaries. Phys. Rev. Lett. **107**, 101101 (2011). https://doi.org/10.1103/PhysRevLett.107. 101101. 1107.5064

Ho, W.C.G., Espinoza, C.M., Antonopoulou, D., Andersson, N.: Pinning down the superfluid and measuring masses using pulsar glitches. Sci. Adv. **1**, e1500578–e1500578 (2015). https://doi.org/10.1126/sciadv.1500578. 1510.00395

Howitt, G., Haskell, B., Melatos, A.: Hydrodynamic simulations of pulsar glitch recovery. Mon. Not. R. Astron. Soc. **460**, 1201–1213 (2016). https://doi.org/10.1093/mnras/stw1043. 1512.07903

Israel, W., Stewart, J.M.: Transient relativistic thermodynamics and kinetic theory. Ann. Phys. **118**, 341–372 (1979) . https://doi.org/10.1016/0003-4916(79)90130-1

Jahan-Miri, M.: Flux expulsion and field evolution in neutron stars. Astrophys. J. **532**, 514–529 (2000). https://doi.org/10.1086/308528. astro-ph/9910490

Jones, P.B.: Rotation of the neutron-drip superfluid in pulsars - the resistive force. Mon. Not. R. Astron. Soc. **243**, 257–262 (1990)

Jones, P.B.: Neutron superfluid spin-down and magnetic field decay in pulsars. Mon. Not. R. Astron. Soc. **253**, 279–286 (1991a). https://doi.org/10.1093/mnras/253.2.279

Jones, P.B.: Rotation of the neutron-drip superfluid in pulsars - the interaction and pinning of vortices. Astrophys. J. **373**, 208–212 (1991b). https://doi.org/10.1086/170038

Jones, P.B.: Rotation of the neutron-drip superfluid in pulsars - the Kelvin phonon contribution to dissipation. Mon. Not. R. Astron. Soc. **257**, 501–506 (1992). https://doi.org/10.1093/mnras/257.3.501

Jones, P.B.: Type II superconductivity and magnetic flux transport in neutron stars. Mon. Not. R. Astron. Soc. **365**, 339–344 (2006). https://doi.org/10.1111/j.1365-2966.2005.09724.x. astro-ph/0510396

Jones, D.I.: Pulsar state switching, timing noise and free precession. MNRAS **420**, 2325–2338 (2012). https://doi.org/10.1111/j.1365-2966.2011.20238.x. 1107.3503

Jones, D.I., Andersson, N.: Freely precessing neutron stars: model and observations. Mon. Not. R. Astron. Soc. **324**, 811–824 (2001). https://doi.org/10.1046/j.1365-8711.2001.04251.x. astro-ph/0011063

Kantor, E.M., Gusakov, M.E.: Temperature effects in pulsating superfluid neutron stars. Phys. Rev. D **83**, 103008 (2011). https://doi.org/10.1103/PhysRevD.83.103008. 1105.4040

Kantor, E.M., Gusakov, M.E.: Temperature-dependent r-modes in superfluid neutron stars stratified by muons. ArXiv e-prints (2017). 1705.06027

Kantor, E.M., Gusakov, M.E., Chugunov, A.I.: Observational signatures of neutron stars in low-mass X-ray binaries climbing a stability peak. Mon. Not. R. Astron. Soc. **455**, 739–753 (2016). https://doi.org/10.1093/mnras/stv2352. 1512.02428

Kerr, M., Hobbs, G., Johnston, S., Shannon, R.M.: Periodic modulation in pulse arrival times from young pulsars: a renewed case for neutron star precession. MNRAS **455**, 1845–1854 (2016). https://doi.org/10.1093/mnras/stv2457. 1510.06078

Khalatnikov, I.: An introduction to the theory of superfluidity. Advanced Books Classics Series. Addison-Wesley Publishing Company, Boston (1989)

Landau, L.D., Lifshitz, E.M.: Fluid Mechanics. Pergamon Press, Oxford (1959)

Lander, S.K.: Magnetic fields in superconducting neutron stars. Phys. Rev. Lett. **110**, 071101 (2013). https://doi.org/10.1103/PhysRevLett.110.071101. 1211.3912

Lander, S.K.: The contrasting magnetic fields of superconducting pulsars and magnetars. Mon. Not. R. Astron. Soc. **437**, 424–436 (2014). https://doi.org/10.1093/mnras/stt1894. 1307.7020

Langlois, D., Sedrakian, D.M., Carter, B.: Differential rotation of relativistic superfluid in neutron stars. Mon. Not. R. Astron. Soc. **297**, 1189–1201 (1998). https://doi.org/10.1046/j.1365-8711.1998.01575.x. astro-ph/9711042

Lasky, P.D.: Gravitational waves from neutron stars: a review. Publications of the Astronomical Society of Australia **32**, e034 (2015). https://doi.org/10.1017/pasa.2015.35. 1508.06643

Lee, U., Yoshida, S.: r-modes of neutron stars with superfluid cores. Astrophys. J. **586**, 403–418 (2003). https://doi.org/10.1086/367617. astro-ph/0211580

Lindblom, L., Mendell, G.: The oscillations of superfluid neutron stars. Astrophys. J. **421** (1994), 689–704. https://doi.org/10.1086/173682

Lindblom, L., Mendell, G.: r-modes in superfluid neutron stars. Phys. Rev. D **61**, 104003 (2000). https://doi.org/10.1103/PhysRevD.61.104003. gr-qc/9909084

Link, B.: Constraining hadronic superfluidity with neutron star precession. Phys. Rev. Lett. **91**, 101101 (2003). https://doi.org/10.1103/PhysRevLett.91.101101. astro-ph/0302441

Link, B.: Dynamics of quantum vorticity in a random potential. Phys. Rev. Lett. **102**, 131101 (2009). https://doi.org/10.1103/PhysRevLett.102.131101. 0807.1945

Link, B.: Instability of superfluid flow in the neutron star core. Mon. Not. R. Astron. Soc. **421**, 2682–2691 (2012). https://doi.org/10.1111/j.1365-2966.2012.20498.x. 1111.0696

Link, B.: Thermally activated post-glitch response of the neutron star inner crust and core. I. Theory. Astrophys. J. **789**, 141 (2014). https://doi.org/10.1088/0004-637X/789/2/141. 1311.2499

Link, B.K., Epstein, R.I.: Mechanics and energetics of vortex unpinning in neutron stars. Astrophys. J. **373** (1991), 592–603. https://doi.org/10.1086/170078

Lee, U.: Nonradial oscillations of neutron stars with the superfluid core. Astron. Astrophys. **303**, 515 (1995)

Lombardo, U., Schulze, H.-J.: Superfluidity in neutron star matter. In: Blaschke, D., Glendenning, N.K., Sedrakian, A. (eds.) Physics of Neutron Star Interiors. Lecture Notes in Physics, vol. 578, p. 30. Springer, Berlin (2001). astro-ph/0012209

Lombardo, U., Nozières, P., Schuck, P., Schulze, H.-J., Sedrakian, A.: Transition from BCS pairing to Bose-Einstein condensation in low-density asymmetric nuclear matter. Phys. Rev. C **64**, 064314 (2001). https://doi.org/10.1103/PhysRevC.64.064314. nucl-th/0109024

Lombardo, U., Schuck, P., Shen, C.: Screening effects on $^1S_0$ pairing in nuclear matter. Nucl. Phys. A **731**, 392–400 (2004). https://doi.org/10.1016/j.nuclphysa.2003.11.051

Mahmoodifar, S., Strohmayer, T.: Upper bounds on r-mode amplitudes from observations of low-mass X-ray binary neutron stars. Astrophys. J. **773**, 140 (2013). https://doi.org/10.1088/0004-637X/773/2/140. 1302.1204

Marchetti, M.C., Middleton, A.A., Saunders, K., Schwarz, J.M.: Driven depinning of strongly disordered media and anisotropic mean-field limits. Phys. Rev. Lett. **91**, 107002 (2003). https://doi.org/10.1103/PhysRevLett.91.107002. cond-mat/0302275

Mastrano, A., Melatos, A.: Kelvin-Helmholtz instability and circulation transfer at an isotropic-anisotropic superfluid interface in a neutron star. Mon. Not. R. Astron. Soc. **361**, 927–941 (2005). https://doi.org/10.1111/j.1365-2966.2005.09219.x. astro-ph/0505529

Maurizio, S., Holt, J.W., Finelli, P.: Nuclear pairing from microscopic forces: singlet channels and higher-partial waves. Phys. Rev. C **90**, 044003 (2014). https://doi.org/10.1103/PhysRevC.90.044003. 1408.6281

Melatos, A., Link, B.: Pulsar timing noise from superfluid turbulence. Mon. Not. R. Astron. Soc. **437**, 21–31 (2014). https://doi.org/10.1093/mnras/stt1828. 1310.3108

Melatos, A., Peralta, C.: Superfluid turbulence and pulsar glitch statistics. Astrophys. J. **662**, L99–L102 (2007). https://doi.org/10.1086/518598. 0710.2455

Melatos, A., Peralta, C., Wyithe, J.S.B.: Avalanche dynamics of radio pulsar glitches. Astrophys. J. **672**, 1103–1118 (2008). https://doi.org/10.1086/523349. 0710.1021

Mendell, G.: Superfluid hydrodynamics in rotating neutron stars. I - Nondissipative equations. II - Dissipative effects. ApJ **380**, 515–540 (1991). https://doi.org/10.1086/170609

Mendell, G.: Magnetohydrodynamics in superconducting-superfluid neutron stars. MNRAS **296**, 903–912 (1998). https://doi.org/10.1046/j.1365-8711.1998.01451.x. astro-ph/9702032

Mendell, G., Lindblom, L.: The coupling of charged superfluid mixtures to the electromagnetic field. Ann. Phys. **205**, 110–129 (1991). https://doi.org/10.1016/0003-4916(91)90239-5

Middleditch, J., Marshall, F.E., Wang, Q.D., Gotthelf, E.V., Zhang, W.L Predicting the starquakes in PSR J0537-6910. Astrophys. J. **652**, 1531–1546 (2006). https://doi.org/10.1086/508736. astro-ph/0605007

Migdal, A.B.: Superfluidity and the moments of inertia of nuclei. Nucl. Phys. A **13**, 655–674 (1959). https://doi.org/10.1016/0029-5582(59)90264-0

Mühlschlegel, B.: Die thermodynamischen Funktionen des Supraleiters. Zeitschrift fur Physik **155**, 313–327 (1959). https://doi.org/10.1007/BF01332932

Muslimov, A.G., Tsygan, A.I.: Neutron star superconductivity and superfluidity, and the decay of pulsar magnetic fields. Pisma v Astronomicheskii Zhurnal **11**, 196–202 (1985a)

Muslimov, A.G., Tsygan, A.I.: Vortex lines in neutron star superfluids and decay of pulsar magnetic fields. Astrophys. Space Sci. **115**, 43–49 (1985b). https://doi.org/10.1007/BF00653825

Muzikar, P., Sauls, J.A., Serene, J.W.: $^3P_2$ pairing in neutron-star matter: magnetic field effects and vortices. Phys. Rev. D **21**, 1494–1502 (1980). https://doi.org/10.1103/PhysRevD.21.1494

Newton, W.G., Berger, S., Haskell, B.: Observational constraints on neutron star crust-core coupling during glitches. Mon. Not. R. Astron. Soc. **454**, 4400–4410 (2015). https://doi.org/10.1093/mnras/stv2285. 1506.01445

Noronha, J., Sedrakian, A.: Tkachenko modes as sources of quasiperiodic pulsar spin variations. Phys. Rev. D **77**, 023008 (2008). https://doi.org/10.1103/PhysRevD.77.023008. 0708.2876

Page, D., Lattimer, J.M., Prakash, M., Steiner, A.W.: Stellar Superfluids. ArXiv e-prints (2013). 1302.6626

Pavlou, G.E., Mavrommatis, E., Moustakidis, C., Clark, J.W.: Microscopic study of $^1S_0$ superfluidity in dilute neutron matter. ArXiv e-prints (2016). 1612.02188

Passamonti, A., Andersson, N.: Towards real neutron star seismology: accounting for elasticity and superfluidity. Mon. Not. R. Astron. Soc. **419** (2012), 638–655. https://doi.org/10.1111/j.1365-2966.2011.19725.x. 1105.4787

Passamonti, A., Haskell, B., Andersson, N.: Oscillations of rapidly rotating superfluid stars. Mon. Not. R. Astron. Soc. **396**, 951–963 (2009). https://doi.org/10.1111/j.1365-2966.2009.14751.x. 0812.3569

Passamonti, A., Akgün, T., Pons, J., Miralles, J.A.: On the magnetic field evolution timescale in superconducting neutron star cores. ArXiv e-prints (2017a). 1704.02016

Passamonti, A., Akgün, T., Pons, J.A., Miralles, J.A.: The relevance of ambipolar diffusion for neutron star evolution. Mon. Not. R. Astron. Soc. **465** (2017b), 3416–3428. https://doi.org/10.1093/mnras/stw2936. 1608.00001

Peralta, C., Melatos, A., Giacobello, M., Ooi, A.: Transitions between turbulent and laminar superfluid vorticity states in the outer core of a neutron star. Astrophys. Space Sci. **651**, 1079–1091 (2006). https://doi.org/10.1086/507576. astro-ph/0607161

Pizzochero, P.M.: Angular momentum transfer in vela-like pulsar glitches. Astrophys. J. **743**, L20 (2011). https://doi.org/10.1088/2041-8205/743/1/L20. 1105.0156

Pizzochero, P.M., Barranco, F., Vigezzi, E., Broglia, R.A.: Nuclear impurities in the superfluid crust of neutron stars: quantum calculation and observable effects on the cooling. Astrophys. J. **569**, 381–394 (2002). https://doi.org/10.1086/339284

Pizzochero, P.M., Antonelli, M., Haskell, B., Seveso, S.: Constraints on pulsar masses from the maximum observed glitch. Nat. Astron. **1**, 0134 (2017). https://doi.org/10.1038/s41550-017-0134. 1611.10223

Pnigouras, P., Kokkotas, K.D.: Saturation of the f-mode instability in neutron stars. II. Applications and results. Phys. Rev. D **94**, 024053 (2016). https://doi.org/10.1103/PhysRevD.94.024053. 1607.03059

Prix, R.: Variational description of multifluid hydrodynamics: uncharged fluids. Phys. Rev. D **69**, 043001 (2004). https://doi.org/10.1103/PhysRevD.69.043001. physics/0209024

Prix, R., Rieutord, M.: Adiabatic oscillations of non-rotating superfluid neutron stars. Astron. Astrophys. **393**, 949–963 (2002). https://doi.org/10.1051/0004-6361:20021049. astro-ph/0204520

Prix, R., Comer, G.L., Andersson, N.: Inertial modes of non-stratified superfluid neutron stars. Mon. Not. R. Astron. Soc. **348**, 625–637 (2004). https://doi.org/10.1111/j.1365-2966.2004. 07399.x. astro-ph/0308507

Prix, R., Novak, J., Comer, G.L.: Relativistic numerical models for stationary superfluid neutron stars. Phys. Rev. D **71**, 043005 (2005). https://doi.org/10.1103/PhysRevD.71.043005. gr-qc/0410023

Radhakrishnan, V., Manchester, R.N.: Detection of a change of state in the pulsar PSR 0833-45. Nature **222**, 228–229 (1969). https://doi.org/10.1038/222228a0

Reichley, P.E., Downs, G.S.: Observed decrease in the periods of pulsar PSR 0833-45. . Nature **222**, 229–230 (1969). https://doi.org/10.1038/222229a0

Ruderman, M.: Neutron starquakes and pulsar periods. Nature **223**, 597–598 (1969). https://doi.org/10.1038/223597b0

Ruderman, M.: Long period oscillations in rotating neutron stars. Nature **225**, 619–620 (1970). https://doi.org/10.1038/225619a0

Ruderman, M.: Crust-breaking by neutron superfluids and the VELA pulsar glitches. Astrophys. J. **203**, 213–222 (1976). https://doi.org/10.1086/154069

Ruderman, M., Zhu, T., Chen, K.: Neutron star magnetic field evolution, crust movement, and glitches. Astrophys. J. **492**, 267–280 (1998). https://doi.org/10.1086/305026. astro-ph/9709008

Sauls, J.: Superfluidity in the interiors of neutron stars. In: Ögelman, H., van den Heuvel, E.P.J. (eds.) NATO Advanced Science Institutes (ASI) Series C, vol. 262, p. 457 (1989)

Sauls, J.A., Stein, D.L., Serene, J.W.: Magnetic vortices in a rotating $^3P_2$ neutron superfluid. Phys. Rev. D **25**, 967–975 (1982). https://doi.org/10.1103/PhysRevD.25.967

Schulze, H.-J., Cugnon, J., Lejeune, A., Baldo, M., Lombardo, U.: Medium polarization effects on neutron matter superfluidity. Phys. Lett. B **375**, 1–8 (1996). https://doi.org/10.1016/0370-2693(96)00213-4

Schwenk, A., Friman, B.: Polarization contributions to the spin dependence of the effective interaction in neutron matter. Phys. Rev. Lett. **92**, 082501 (2004). https://doi.org/10.1103/PhysRevLett.92.082501. nucl-th/0307089

Schwenk, A., Friman, B., Brown, G.E.: Renormalization group approach to neutron matter: quasiparticle interactions, superfluid gaps and the equation of state. Nucl. Phys. A **713**, 191–216 (2003). https://doi.org/10.1016/S0375-9474(02)01290-3. nucl-th/0207004

Sedrakian, A.D.: Vortex repinning in neutron star crusts. Mon. Not. R. Astron. Soc. **277**, 225–234 (1995). https://dx.doi.org/10.1093/mnras/277.1.225

Sedrakian, A.: Damping of differential rotation in neutron stars. Phys. Rev. D **58**, 021301 (1998). https://doi.org/10.1103/PhysRevD.58.021301. astro-ph/9806156

Sedrakian, A.: Type-I superconductivity and neutron star precession. Phys. Rev. D **71**, 083003 (2005). https://doi.org/10.1103/PhysRevD.71.083003. astro-ph/0408467

Sedrakian, A.: Rapid rotational crust-core relaxation in magnetars. Astron. Astrophys. **587**, L2 (2016). https://doi.org/10.1051/0004-6361/201628068. 1601.00056

Sedrakian, A., Clark, J.W.: Nuclear Superconductivity in Compact Stars: BCS Theory and Beyond, p. 135. World Scientific Publishing Co, Singapore (2006)

Sedrakian, A., Cordes, J.M.: Vortex-interface interactions and generation of glitches in pulsars. Mon. Not. R. Astron. Soc. **307**, 365–375 (1999). https://doi.org/10.1046/j.1365-8711.1999. 02638.x. astro-ph/9806042

Sedrakian, A.D., Sedrakian, D.M.: Superfluid core rotation in pulsars. I. Vortex cluster dynamics. Astrophys. J. **447**, 305 (1995). https://doi.org/10.1086/175876

Sedrakian, A., Wasserman, I.: Perturbations of self-gravitating, ellipsoidal superfluid-normal fluid mixtures. Phys. Rev. D **63**, 024016 (2001). https://doi.org/10.1103/PhysRevD.63.024016. astro-ph/0004331

Sedrakian, A., Wasserman, I.: The tensor virial method and its applications to self-gravitating superfluids. In: Blaschke, D., Glendenning, N.K., Sedrakian, A. (eds.) Physics of Neutron Star Interiors. Lecture Notes in Physics, vol. 578, p. 97. Springer, Berlin (2001). astro-ph/0101527

Sedrakian, A.D., Sedrakian, D.M., Cordes, J.M., Terzian, Y.: Superfluid core rotation in pulsars. II. Postjump relaxations. Astrophys. J. **447**, 324 (1995). https://doi.org/10.1086/175877

Sedrakian, D.M., Sedrakian, A.D., Zharkov, G.F.: Type I superconductivity of protons in neutron stars. MNRAS **290**, 203–207 (1997). https://doi.org/10.1093/mnras/290.1.203. astro-ph/9710280

Sedrakian, A., Wasserman, I., Cordes, J.M.: Precession of isolated neutron stars. I. effects of imperfect pinning. Astrophys. J. **524**, 341–360 (1999). https://doi.org/10.1086/307777. astro-ph/9801188

Sedrakian, A., Kuo, T.T.S., Müther, H., Schuck, P.: Pairing in nuclear systems with effective Gogny and V $_{low-k}$ interactions. Phys. Lett. B **576**, 68–74 (2003). https://doi.org/10.1016/j.physletb. 2003.09.090. nucl-th/0308068

Sedrakian, D.M., Hayrapetyan, M.V., Malekian, A.: Toroidal magnetic field of a superfluid neutron star in a postnewtonian approximation. Astrophysics **54**, 100–110 (2011). https://doi.org/10. 1007/s10511-011-9161-1

Sedrakian, A., Huang, X.-G., Sinha, M., Clark, J.W.: From microphysics to dynamics of magnetars. ArXiv e-prints (2017). 1701.00895

Sedrakyan, D.M., Shakhabasyan, K.M.: Superfluidity and the magnetic field of pulsars. Sov. Phys. Usp. **34**, 555–571 (1991). https://doi.org/10.1070/PU1991v034n07ABEH002446

Seveso, S., Pizzochero, P.M., Haskell, B.: The effect of realistic equations of state and general relativity on the 'snowplough' model for pulsar glitches. Mon. Not. R. Astron. Soc. **427**, 1089–1101 (2012). https://doi.org/10.1111/j.1365-2966.2012.21906.x. 1205.6647

Seveso, S., Pizzochero, P.M. , Grill, F., Haskell, B.: Mesoscopic pinning forces in neutron star crusts. Mon. Not. R. Astron. Soc. **455**, 3952–3967 (2016). https://doi.org/10.1093/mnras/ stv2579

Sinha, M., Sedrakian, A.: Magnetar superconductivity versus magnetism: neutrino cooling processes. Phys. Rev. C **91**, 035805 (2015). https://doi.org/10.1103/PhysRevC.91.035805. 1502.02979

Shahabasyan, M.K.: Vortex lattice oscillations in rotating neutron stars with quark "CFL" cores. Astrophysics **52**, 151–155 (2009). https://doi.org/10.1007/s10511-009-9044-x

Shahabasyan, K.M., Shahabasyan, M.K.: Oscillations in the angular velocity of pulsars. Astrophysics **54**, 111–116 (2011). https://doi.org/10.1007/s10511-011-9162-0

Shaham, J.: Free precession of neutron stars - role of possible vortex pinning. Astrophys. J. **214** (1977), 251–260. https://doi.org/10.1086/155249

Shen, C., Lombardo, U., Schuck, P.: Screening of nuclear pairing in nuclear and neutron matter. Phys. Rev. C **71**, 054301 (2005). https://doi.org/10.1103/PhysRevC.71.054301. nucl-th/0503008

Sourie, A., Oertel, M., Novak, J.: Numerical models for stationary superfluid neutron stars in general relativity with realistic equations of state. Phys. Rev. D **93**, 083004 (2016). https:// doi.org/10.1103/PhysRevD.93.083004. 1602.06228

Srinivasan, G., Bhattacharya, D., Muslimov, A.G., Tsygan, A.J.: A novel mechanism for the decay of neutron star magnetic fields. Curr. Sci. **59**, 31–38 (1990)

Stairs, I.H., Lyne, A.G., Shemar, S.L.: Evidence for free precession in a pulsar. Nature **406**, 484–486 (2000). https://doi.org/10.1038/35020010

Stein, M., Sedrakian, A., Huang, X.-G., Clark, J.W.: Spin-polarized neutron matter: critical unpairing and BCS-BEC precursor. Phys. Rev. C **93**, 015802 (2016). https://doi.org/10.1103/ PhysRevC.93.015802. 1510.06000

Surace, M., Kokkotas, K.D., Pnigouras, P.: The stochastic background of gravitational waves due to the f-mode instability in neutron stars. Astron. Astrophys. **586** (2016), A86. https://doi.org/ 10.1051/0004-6361/201527197. 1512.02502

Thorne, K.S.: Multiple expansions of gravitational radiation. Rev. Modern Phys. **52**, 299–340 (1980). https://doi.org/10.1103/RevModPhys.52.299

Turolla, R., Zane, S., Watts, A.L.: Magnetars: the physics behind observations. A review. Rep. Prog. Phys. **78**, 116901 (2015). https://doi.org/10.1088/0034-4885/78/11/116901. 1507.02924

van Eysden, C.A.: Short-period pulsar oscillations following a glitch. Astrophys. J. **789**, 142 (2014). https://doi.org/10.1088/0004-637X/789/2/142. 1403.1046

van Eysden, C.A., Melatos, A.: Pulsar glitch recovery and the superfluidity coefficients of bulk nuclear matter. Mon. Not. R. Astron. Soc. **409**, 1253–1268 (2010). https://doi.org/10.1111/j.1365-2966.2010.17387.x. 1007.4360

Vardanyan, G.A., Sedrakyan, D.M.: Magnetohydrodynamics of superfluid superconducting mixtures. Sov. Phys. JETP **54**, 919 (1981)

Wambach, J., Ainsworth, T.L., Pines, D.: Quasiparticle interactions in neutron matter for applications in neutron stars. Nucl. Phys. A **555**, 128–150 (1993). https://doi.org/10.1016/0375-9474(93)90317-Q

Wasserman, I.: Precession of isolated neutron stars - II. Magnetic fields and type II superconductivity. Mon. Not. R. Astron. Soc. **341**, 1020–1040 (2003). https://doi.org/10.1046/j.1365-8711.2003.06495.x. astro-ph/0208378

Warszawski, L., Melatos, A.: Gross-Pitaevskii model of pulsar glitches. Mon. Not. R. Astron. Soc. **415**, 1611–1630 (2011) . https://doi.org/10.1111/j.1365-2966.2011.18803.x. 1103.6090

Warszawski, L., Melatos, A., Berloff, N.G.: Unpinning triggers for superfluid vortex avalanches. Phys. Rev. B **85**, 104503 (2012). https://doi.org/10.1103/PhysRevB.85.104503. 1203.5133

Watanabe, G., Pethick, C.J.: Superfluid density of neutrons in the inner crust of neutron stars: new life for pulsar glitch models. Phys. Rev. Lett. **119**, 062701 (2017). https://doi.org/10.1103/PhysRevLett.119.062701. 1704.08859

Weber, F. (ed.): Pulsars as astrophysical laboratories for nuclear and particle physics (1999)

Wlazłowski, G., Sekizawa, K., Magierski, P., Bulgac, A., Forbes, M.M.: Vortex pinning and dynamics in the neutron star crust. Phys. Rev. Lett. **117**, 232701 (2016). https://doi.org/10.1103/PhysRevLett.117.232701. 1606.04847

Yoshida, S., Lee, U.: r-modes in relativistic superfluid stars. Phys. Rev. D **67**, 124019 (2003a). https://doi.org/10.1103/PhysRevD.67.124019. gr-qc/0304073

Yoshida, S., Lee, U.: Inertial modes of neutron stars with a superfluid core. MNRAS **344**, 207–222 (2003b). https://doi.org/10.1046/j.1365-8711.2003.06816.x. astro-ph/0302313

Zverev, M.V., Clark, J. W., Khodel, V.A.: $^3P_2$-$^3F_2$ pairing in dense neutron matter: the spectrum of solutions. Nucl. Phys. A **720**, 20–42 (2003). https://doi.org/10.1016/S0375-9474(03)00653-5. nucl-th/0301028

Chapter 9
Reaction Rates and Transport in Neutron Stars

Andreas Schmitt and Peter Shternin

Abstract Understanding signals from neutron stars requires knowledge about the transport inside the star. We review the transport properties and the underlying reaction rates of dense hadronic and quark matter in the crust and the core of neutron stars and point out open problems and future directions.

9.1 Introduction

9.1.1 Context

Transport describes how conserved quantities such as energy, momentum, particle number, or electric charge are transferred from one region to another. Such a transfer occurs if the system is out of equilibrium, for instance through a temperature gradient or a non-uniform chemical composition. Different theoretical methods are used to understand transport, depending on how far the system is away from its equilibrium state. If the system is close to equilibrium locally and perturbations are on large scales in space and time, hydrodynamics is a powerful technique. Further away from equilibrium other techniques are required, for example kinetic theory, which can also be used to provide the transport coefficients needed in the hydrodynamic equations. In any case, transport is determined by interactions on a microscopic level, and it is the resulting transport properties that we are concerned with in this review.

Signals from neutron stars are sensitive to equilibrium properties such as the equation of state but also, to a large extent, to transport properties – here, by

A. Schmitt (✉)
Mathematical Sciences and STAG Research Centre, University of Southampton, Southampton, UK
e-mail: a.schmitt@soton.ac.uk

P. Shternin
Ioffe Institute, St. Petersburg, Russia

© Springer Nature Switzerland AG 2018
L. Rezzolla et al. (eds.), *The Physics and Astrophysics of Neutron Stars*,
Astrophysics and Space Science Library 457,
https://doi.org/10.1007/978-3-319-97616-7_9

neutron stars we mean all objects with a radius of about 10 km and a mass of about 1–2 solar masses, including the possibilities of hybrid stars, which have a quark matter core, and pure quark stars. Therefore, understanding transport is crucial to interpret astrophysical observations, and, turning the argument around, we can use astrophysical observations to improve our understanding of transport in dense matter and thus ultimately our understanding of the microscopic interactions.

Transport properties are most commonly computed from particle collisions. (Although, in strongly coupled systems, the picture of well-defined particles scattering off each other has to be taken with care.) These can be scattering processes in which energy and momentum is exchanged without changing the chemical composition of the system, or these can be flavor-changing processes from the electroweak interaction. Understanding transport thus amounts to understanding the rates of these processes, as a function of temperature and density. Electroweak processes are well understood, but large uncertainties arise if the strong interaction is involved in a reaction that contributes to transport. Therefore, approximations such as weak-coupling techniques or one-pion exchange for nucleon-nucleon collisions are being used, and efforts in current research aim at improving these approximations.

In a neutron star, most of the particles involved in these processes are fermions: electrons, muons, neutrinos, neutrons, protons, hyperons, and quarks. Since the Fermi momenta of these fermions are typically much larger than the temperature (neutrinos are an exception), transport probes the excitations in small vicinities of the corresponding Fermi surfaces. (It can also probe the values of the Fermi momenta themselves since momentum conservation of a given reaction imposes a constraint on them.) Some of the processes we discuss involve bosonic excitations, for instance the lattice phonons in the crust, the superfluid mode, or mesons such as pions and kaons. Typically, their contribution is smaller because, well, they do not have a Fermi surface and thus the rates and transport coefficients contain higher powers of temperature. Therefore, purely bosonic contributions are usually only relevant if the fermionic ones are suppressed, for example through an energy gap from Cooper pairing.

9.1.2 Phenomenological and Theoretical Motivations

Computing transport properties of matter inside neutron stars is motivated by phenomenological and theoretical considerations. The phenomenological motivation is of course to understand astrophysical data that are sensitive to transport. Our focus in the main part of the review is on the transport properties themselves, and we discuss observations only in passing. Therefore, let us now list some of the relevant phenomena which are intimately connected with transport. (Here we only include very few selected references, which we think are useful for further reading; many more references will be given in the main part.)

- Oscillatory modes of the star, most importantly r-modes, become unstable with respect to the emission of gravitational waves (Andersson 1998; Friedman and Morsink 1998). We know that these instabilities must be damped because otherwise we would not observe fast rotating stars. Viscous damping plays a major role, and knowledge of both bulk and shear viscosity (which are important in different temperature regimes) is required (Haskell 2015).

- Pulsar glitches, sudden jumps in the rotation frequency of the star, are commonly explained through pinning and un-pinning of superfluid vortices in the inner crust of the star (Haskell and Melatos 2015). A quantitative treatment requires the understanding of superfluid transport, including entrainment effects of the superfluid in the crust, and possibly hydrodynamical instabilities.

- The interpretation of thermal radiation of neutron stars depends on knowledge about heat transport in the outermost layers of the star, the atmosphere and the ocean (Potekhin 2014; Potekhin et al. 2007; Potekhin et al. 2015b).

- Cooling of neutron stars, for instance isolated neutron stars and quiescent X-ray transients, requires understanding of the microscopic neutrino emission processes. Together with thermodynamic properties such as the specific heat and other transport properties such as heat conductivity, the cooling process can be modeled (Yakovlev and Pethick 2004; Page et al. 2006).

- Understanding the time evolution of magnetic fields in neutron stars and its coupling to the thermal evolution requires magnetohydrodynamical simulations. As an input from microscopic physics electrical and thermal conductivities are needed (Viganò et al. 2013). Additional complications may arise from superconductivity and magnetic flux tubes in the core.

- In accreting neutron stars, the crust is forced out of equilibrium by the accreted matter, and in some cases, for instance 'quasi-persistent' sources, the subsequent relaxation process can be observed in real time. Nuclear reactions, including pyco-nuclear fusion, contribute to the so-called 'deep crustal heating' (Haensel and Zdunik 2008), and transport properties of the crust such as thermal conductivity are needed to understand the relaxation process (Degenaar et al. 2014). An important role is possibly played by transport properties of inhomogeneous phases in the crust/core transition region ('nuclear pasta') (Horowitz et al. 2015). Deep crustal heating also plays a pivotal role in maintaining high observed temperatures of X-ray transients (Brown et al. 1998).

- Crust relaxation is also important for magnetar flares. Similar to accretion, the crust is disrupted, now by a catastrophic rearrangement of the magnetic field. Crustal transport properties in the presence of a magnetic field become important (Turolla et al. 2015).

- The neutron star in the Cassiopeia A supernova remnant has undergone unusually rapid cooling in the past decade (Heinke and Ho 2010; Shternin et al. 2011; Elshamouty et al. 2013; Ho et al. 2015). If true (the reliability of this data is under discussion (Posselt et al. 2013); Posselt and Pavlov 2018) this indicates an unusual neutrino emission process, for instance Cooper pair breaking and formation at the critical temperature for neutron superfluidity (Shternin et al. 2011); Page et al. (2011).

- Core-collapse supernovae and the evolution of the resulting proto-neutron star are sensitive to neutrino transport and neutrino-nucleus reactions. The phenomenological implications include direct neutrino signals (Janka 2017), nucleosynthesis, the mechanism of the supernova explosion itself, cooling of proto-neutron stars, and pulsar kicks (Janka et al. 2012).
- Neutron star mergers have proved to be multi-messenger events, emitting detectable gravitational waves and electromagnetic signals (Abbott et al. 2017a,b). Simulations of the merger process within general relativity are being performed, using (magneto)hydrodynamics, where viscous effects may be important (Alford et al. 2018). Merger events explore transport at larger temperatures than neutron stars in (near-)equilibrium. Similar to proto-neutron stars from supernovae, the evolution of merger remnants requires understanding of neutrino reactions and transport.

The theoretical motivation for understanding transport in neutron star matter can—at least for the ultra-dense regions in the interior of the star—be put in the wider context of understanding transport in matter underlying the theory of Quantum Chromodynamics (QCD) or, even more generally speaking, of understanding transport in relativistic, strongly interacting theories. This perspective connects some of the results in this review with questions about the correct formulation of relativistic, dissipative (superfluid) hydrodynamics, about the validity of the quasiparticle picture and thus of kinetic theory, about non-perturbative effects in QCD scattering processes, about universal results and bounds for shear viscosity and other transport coefficients and so forth. These questions are being discussed extensively in the recent and current literature, be it from an abstract theoretical perspective, e.g., within the gauge/gravity correspondence, or in a more applied context such as relativistic heavy-ion collisions or cold atomic gases. Neutron stars may appear to be too specific and too complicated to be viewed as a clean laboratory for these questions, but we think it is worth pointing out these connections, and they will be touched in some sections of this review.

9.1.3 Purpose and Structure of this Review

We intend to collect and comment on recent results in the literature, pointing out open problems and future directions, with an emphasis on the theoretical, rather than the observational, questions. We include pedagogical derivations and explanations in most parts, making this review accessible for non-experts in transport theory and neutron star physics. In particular, we start in Sect. 9.2 by introducing some basic concepts of transport theory and explain how the basic approach must be extended and adjusted to the extreme conditions inside a neutron star. After this introductory section, we have structured the review by moving from the outer layers of the star into the central regions. Since we thereby move from low densities to ultra-high densities, we encounter various distinct phases with very distinct transport properties. We start from the crust in Sect. 9.3, where the matter composition is rather well known: a lattice or a strongly coupled liquid of ions coexists with an electron gas, and, in the inner crust, with a neutron (super)fluid. As we move through the crust/core interface, we encounter the so-called nuclear pasta phases, and eventually end up in a region of nuclear matter, composed of neutrons and protons, with electrons and muons accounting for charge neutrality. Additionally, hyperons may be present, and possibly meson condensates. We discuss transport of hadronic matter in the core in Sect. 9.4. At sufficiently large densities, matter becomes deconfined and we enter the quark matter phase. Since the density at which this transition happens is unknown, we do not know whether quark matter exists in the core of neutron stars (or whether there are pure quark stars). Transport properties of quark matter, which we discuss in Sect. 9.5, are one important ingredient to answer that question. For readers unfamiliar with quark matter and its possible phases, we have included an introductory section and overview in Sects. 9.5.1 and 9.5.2. At the end of Sect. 9.5—although being a somewhat decoupled topic— we briefly discuss possible effects of quantum anomalies on transport in neutron stars. In all sections, our main goal was, besides some introductory and pedagogical discussions, to focus on the most recent results and their impact for future research. In some parts, for instance in Sect. 9.5 about quark matter, we have tried to give a more complete overview, including older results, which is possible because of the smaller amount of existing literature compared to nuclear matter. Reaction rates in the core from the weak interaction are discussed in Sects. 9.4.2 and 9.5.3. The rates for these processes are interesting by themselves since they directly feed into the cooling behavior of the star. They are also interesting for the bulk viscosity because bulk viscosity in a neutron star is dominated by chemical re-equilibration and thus by flavor-changing processes. We discuss bulk viscosity, including the rates for other leptonic and non-leptonic flavor-changing processes, for hadronic matter in Sect. 9.4.3 and for quark matter in Sect. 9.5.4. Shear viscosity, thermal and electrical conductivity, are discussed together since they are determined by similar processes, some of which rely on the strong interaction, and we discuss them in Sects. 9.3.1, 9.4.1, and 9.5.5.

9.1.4 Related Reviews

There are a number of reviews that (partially) deal with transport properties in neutron stars, having some overlap with our work, and which we recommend for further reading. Page and Reddy (2012) review transport in the inner crust of the star. A more exhaustive overview of the crust is given by Chamel and Haensel (2008), discussing transport as well as details of the structure and connections to observations. Potekhin et al. (2015b) review cooling of isolated neutron stars and discuss transport and thermodynamic properties that are needed to understand the cooling process, including the effect of strong magnetic fields. Cooling in proto-neutron stars just after a core collapse supernova explosion has recently been reviewed by Roberts and Reddy (2017). Many of the currently used results for neutrino emissivity in hadronic matter, including superfluid phases, can already be found in the review by Yakovlev et al. (2001). Superfluidity in neutron stars and some of its effects on transport and reaction rates are reviewed by Page et al. (2014). For a detailed discussion of many-body techniques for hadronic matter inside neutron stars, including neutrino emission processes, see the review by Sedrakian (2007). Transport properties of quark matter are discussed in chapter VII of the review about color superconductivity by Alford et al. (2008a), for a pedagogical discussion of neutrino emissivity in quark matter see Schmitt (2010). Our review serves as an update to some of these earlier reviews and has a somewhat different focus than most of them, bringing together theoretical results for transport properties from the crust through nuclear matter in the core up to ultra-dense deconfined quark matter.

There are several aspects of transport and reaction rates in neutron stars which we do not discuss or only touch very briefly: we will not elaborate on reactions relevant for neutrino transport in supernovae (Burrows et al. 2006) and neutrino-nucleus reactions relevant for supernovae nucleosynthesis (Balasi et al. 2015). Nuclear astrophysics in a broader context is discussed by Wiescher et al. (2012) and Schatz (2016), and we refer the reader to the review by Meisel et al. (2018) and more specific literature regarding nuclear reactions in accreting crusts (Yakovlev et al. 2006; Gupta et al. 2007; Gupta et al. 2008; Haensel and Zdunik 2008; Steiner 2012; Afanasjev et al. 2012; Schatz et al. 2014). Neutrino emission reactions in the crust are summarized by Yakovlev et al. (2001) with more recent updates by Chamel and Haensel (2008) and Potekhin et al. (2015b), and we have nothing to add to these reviews. Finally, we will not discuss transport in the outer layers of the star, including the atmosphere and the heat blanketing envelopes, where radiative transfer, transport of non-degenerate electrons (Potekhin et al. 2015a; Potekhin et al. 2015b), and diffusion processes (Beznogov and Yakovlev 2013; Beznogov et al. 2016), among others, are important.

9.2 Basic Concepts of Transport Theory

9.2.1 Basic Equations and Transport Coefficients

We start with a brief introduction to the basic concepts that will be used throughout this review. The goal of this section is to provide the definition of the most important transport coefficients, to show how they appear in the hydrodynamical framework and how they are computed from kinetic theory. In the present section, we shall present a general setup for a dilute gas of one non-relativistic fermionic species. Further assumptions and specifications will be made in the subsequent sections. Our starting point is the Boltzmann equation for the non-equilibrium fermionic distribution function $f(x, p, t)$,

$$\frac{\partial f}{\partial t} + u \cdot \frac{\partial f}{\partial x} + R \cdot \frac{\partial f}{\partial p} = I[f], \tag{9.1}$$

where the particle velocity u is related to momentum via $p = mu$ with the particle mass m, and R is the external force which we do not specify for now, except for assuming that $\nabla_p \cdot R = 0$. For instance, it can include the gravitational force or, if the particles carry electric charge, the Lorentz force. The collision term is

$$I[f] = - \int_{p_1} \int_{p'} \int_{p'_1} W(p, p_1; p', p'_1)[f f_1 (1 - f')(1 - f'_1) - (1 - f)(1 - f_1) f' f'_1], \tag{9.2}$$

with the abbreviations $f_1 = f(x_1, p_1, t)$, $f' = f(x', p', t)$, $f'_1 = f(x'_1, p'_1, t)$, and

$$\int_p \equiv \int \frac{d^3 p}{(2\pi\hbar)^3}. \tag{9.3}$$

The collision integral gives the number of collisions per unit time in which a particle with a given momentum p is lost in a scattering process with another ingoing particle with momentum p_1 to produce two outgoing particles with momenta p' and p'_1, plus the number of collisions of the inverse process, in which a particle with momentum p is created. The transition rates $W(p, p_1; p', p'_1)$ depend on the details of the collision process and contain energy and momentum conservation of the process. Their specific form is not needed for now; we shall see later how the Boltzmann equation is solved approximately in specific cases. For notational convenience, we have omitted the spin variable. One may think of the momentum to actually be a pair of momentum and spin and the momentum integral to include the sum over spin. We have written the collision term in the simplified form that only contains scattering of a given, single particle species with itself. Later, we shall discuss approximate solutions to the Boltzmann equation for more than one particle species, for instance electrons and ions in the neutron star crust.

The Boltzmann equation allows us to derive an equation for the transport of any dynamical variable $\psi(x, p, t)$. To this end, we introduce the average value of ψ per particle as

$$\langle \psi \rangle = \frac{1}{n} \int_p \psi f, \qquad n = \int_p f, \tag{9.4}$$

where n is the number density. Multiplying the Boltzmann equation with ψ and integrating over momentum then yields

$$\frac{\partial n \langle \psi \rangle}{\partial t} + \nabla \cdot (n \langle \psi u \rangle) = n \left(\left\langle \frac{\partial \psi}{\partial t} \right\rangle + \langle u \cdot \nabla \psi \rangle + \left\langle R \cdot \frac{\partial \psi}{\partial p} \right\rangle \right) + \int_p \psi I[f], \tag{9.5}$$

where ∇ is the spatial gradient. The first two terms in the parentheses on the right-hand side account for the production of ψ due to its space and time variations, the third therm gives the supply from forces, and the last term gives the production rate from collisions.

From the transport equation (9.5) we derive the hydrodynamic equations by choosing ψ to be a quantity that is conserved in a collision, such that the momentum integral over ψ times the collision term vanishes. These invariants are $\psi = 1$, which corresponds to particle number conservation, energy $\psi = p^2/(2m)$, and momentum components $\psi = p_i$. Thus we obtain three equations (two scalar equations, one vector equation) that do not depend on the collision term explicitly (but contain the non-equilibrium distribution function, which in principle has to be determined from the full Boltzmann equation). These equations can be written as

$$\frac{\partial \rho}{\partial t} + \nabla \cdot \boldsymbol{g} = 0, \tag{9.6a}$$

$$\frac{\partial \mathcal{E}}{\partial t} + \nabla \cdot \boldsymbol{j}_{\mathcal{E}} = n\boldsymbol{R} \cdot \boldsymbol{v}, \tag{9.6b}$$

$$\frac{\partial g_i}{\partial t} + \partial_j (\Pi_{ji} + \pi_{ji}) = nR_i. \tag{9.6c}$$

Here we have introduced the center-of-mass velocity $\boldsymbol{v}$. In the present case of a single fluid, this velocity is identical to the drift velocity of the (single) fluid $\langle \boldsymbol{u} \rangle$. For multi-fluid mixtures, there is a drift velocity for each fluid, which of course does not have to be identical to the total velocity $\boldsymbol{v}$ of the mixture. This case will become important in the next section, where we discuss electrons in an ion background with a nonzero $\boldsymbol{\vartheta} \equiv \langle \boldsymbol{u} \rangle - \boldsymbol{v}$. In Eq. (9.6) we have also introduced mass density $\rho = mn$, momentum density $\boldsymbol{g} = \rho \boldsymbol{v}$, energy density $\mathcal{E} = \mathcal{E}_0 + \rho v^2/2$, and stress tensor $\Pi_{ij} = \rho v_i v_j + \delta_{ij} P$, where the energy density in the co-moving frame of the fluid $\mathcal{E}_0$ and the pressure P are given by

$$\mathcal{E}_0 = n\langle \varepsilon \rangle = \frac{\rho}{2} \langle w^2 \rangle, \qquad P = \frac{\rho}{3} \langle w^2 \rangle = \frac{2}{3} \mathcal{E}_0, \tag{9.7}$$

where $\boldsymbol{w} \equiv \boldsymbol{u} - \boldsymbol{v}$ is the difference between the single-particle velocity and the macroscopic center-of-mass velocity, and

$$\varepsilon = \frac{m w^2}{2} \tag{9.8}$$

is the single-particle energy in the co-moving frame of the fluid. The flux terms in the energy conservation (9.6b) and momentum conservation (9.6c) equations are

$$j_{\mathcal{E},i} = (\mathcal{E} + P)v_i + \pi_{ij}v_j + j_{T,i} = \frac{m}{2} \int_p u^2 u_i f, \tag{9.9a}$$

$$\Pi_{ij} + \pi_{ij} = m \int_p u_i u_j f, \tag{9.9b}$$

which include the dissipative contributions, which vanish in equilibrium,

$$\boldsymbol{j}_T \equiv n\langle \varepsilon \boldsymbol{w} \rangle \,, \qquad \pi_{ij} \equiv \rho \langle w_i w_j \rangle - \delta_{ij} P \,. \tag{9.10}$$

We assume that close to equilibrium we can apply the thermodynamic relations $\mathcal{E}_0 + P = \mu n + T s$ and $d\mathcal{E}_0 = \mu dn + T ds$ locally, with the t and $\boldsymbol{x}$ dependent chemical potential μ, entropy density s, and temperature T. Using these relations, together with Eqs. (9.6a) and (9.6c), the energy conservation (9.6b) can be written as an equation for entropy production. And, using Eq. (9.6a), the momentum conservation (9.6c) can be written in the form of the Navier-Stokes equation. Hence, Eqs. (9.6b) and (9.6c) become

$$\frac{\partial s}{\partial t} + \nabla \cdot \left(s\boldsymbol{v} + \frac{\boldsymbol{j}_T}{T} \right) = -\frac{\pi_{ji}\partial_j v_i + \boldsymbol{j}_T \cdot \nabla T/T}{T} \equiv \varsigma \,, \tag{9.11a}$$

$$\frac{\partial v_i}{\partial t} + (\boldsymbol{v} \cdot \nabla)v_i = -\frac{\partial_i P}{\rho} + \frac{R_i}{m} - \frac{\partial_j \pi_{ji}}{\rho} \,, \tag{9.11b}$$

where we have defined the entropy production rate ς. Instead of deriving the hydrodynamical equations from the Boltzmann equation, we can also view them as an effective theory where dissipative terms can be added systematically with certain transport coefficients. These transport coefficients are then an input to hydrodynamics, for instance computed from a kinetic approach. From Eq. (9.11a) we see that the dissipative part is composed of products of the thermodynamic forces $\nabla T/T$ and $\partial_i v_j$ and the corresponding thermodynamic fluxes $\boldsymbol{j}_T$ and π_{ij}. The usual transport coefficients are then introduced by assuming linear relations between them with the coefficients being thermal conductivity κ, shear viscosity η, and bulk viscosity ζ,

$$\boldsymbol{j}_T = -\kappa \nabla T, \tag{9.12a}$$

$$\pi_{ij} = -2\eta \left(v_{ij} - \frac{\delta_{ij}}{3} \nabla \cdot \boldsymbol{v} \right) - \zeta \delta_{ij} \nabla \cdot \boldsymbol{v} \,, \tag{9.12b}$$

where we have abbreviated

$$v_{ij} \equiv \frac{\partial_i v_j + \partial_j v_i}{2} \,. \tag{9.13}$$

In principle, one can systematically expand the fluxes in powers of derivatives and thus create terms beyond linear order (Burnett 1935, 1936; García-Colín et al. 2008). Higher-order hydrodynamical coefficients are rarely used in the non-relativistic context (see however Chao and Schäfer (2012) and Schäfer (2014) for a discussion of second-order hydrodynamics, motivated by applications to unitary Fermi gases). In contrast, second-order *relativistic* hydrodynamics has been studied much more extensively, motivated by the acausality of the first-order equations and by applications to relativistic heavy-ion collisions (Israel and Stewart 1979; Romatschke 2010; Denicol et al. 2012). Here we will not go beyond first order.

The simple one-component monatomic gas discussed above does not have a bulk viscosity ζ because ζ is proportional to the trace of π_{ij}, as we see from Eq. (9.12b), and the trace of π_{ij} vanishes in our simple example, as Eq. (9.10) shows, due to the relation between energy density and pressure in Eq. (9.7). In more general cases, the hydrostatic pressure is not given by (9.7), and the bulk viscosity is nonzero. Notice that the three terms in Eqs. (9.12) have different spatial symmetry and do not couple. We can compute the rate of the total entropy change $\dot{S}$ of the system by integrating Eq. (9.11a) over the volume V of the system. Making use of Eqs. (9.12), we obtain

$$\dot{S} = \int_V \frac{d^3x}{T} \left[2\eta \left(v_{ij} - \frac{\delta_{ij}}{3} \nabla \cdot \boldsymbol{v} \right)^2 + \zeta (\nabla \cdot \boldsymbol{v})^2 + \frac{\kappa (\nabla T)^2}{T} \right] - \int_{\partial V} d\boldsymbol{\sigma} \cdot \frac{\boldsymbol{j}_T}{T} .$$

$$(9.14)$$

The first integral gives the total entropy production by the dissipative processes inside the system, while the surface integral corresponds to the heat exchange with the external thermostat. Due to the second law of thermodynamics, all phenomenological coefficients κ, η, and ζ have to be non-negative.

In more general cases, the entropy production equation (9.11a) contains more terms, for instance related to diffusion in multi-component mixtures. Some of these terms will be discussed in the following sections. When additional dissipative processes are considered, the equations become more cumbersome, but the principal scheme is the same.

9.2.2 Calculating Transport Coefficients in the Chapman-Enskog Approach

Kinetic theory allows us to compute the transport coefficients on microscopic grounds. The basic idea is to expand the distribution function around the local equilibrium distribution function. The kinetic equation then describes the evolution of the system towards local equilibrium. There exist two elaborate methods for this expansion, namely Grad's moment method (Grad 1949) and the Chapman-Enskog method (Chapman et al. 1999), see also the textbooks by Kremer (2010) and Zhdanov (2002) for extensive discussions of both methods. Here we give a brief sketch of the Chapman-Enskog method. We write the distribution function as

$$f(\boldsymbol{x}, \boldsymbol{u}, t) \approx f^{(0)} + \delta f , \qquad \delta f = -\frac{\partial f^{(0)}}{\partial \varepsilon} \Phi + \mathcal{O}(\Phi^2) \approx \frac{f^{(0)}(1 - f^{(0)})}{k_B T} \Phi ,$$

$$(9.15)$$

with a small correction $\Phi(x, t)$ to the Fermi-Dirac function in local equilibrium

$$f^{(0)}(x, u, t) = \left\{ \exp\left[\frac{\varepsilon(x, u, t) - \mu(x, t)}{k_B T(x, t)}\right] + 1 \right\}^{-1} , \qquad (9.16)$$

where k_B is the Boltzmann constant and where ε from Eq. (9.8) is a function of $w(x, u, t) = u - v(x, t)$. The idea of the following approximation is to only keep the lowest order in Φ and also drop higher-order terms in the derivatives of T, μ, and v. Inserting the ansatz (9.15) into the Boltzmann equation (9.1) yields the following lowest order equation

$$\frac{\partial f^{(0)}}{\partial t} + u \cdot \frac{\partial f^{(0)}}{\partial x} + R \cdot \frac{\partial f^{(0)}}{\partial p} \approx I_{\text{lin}}[\Phi] , \qquad (9.17)$$

where $I_{\text{lin}}[\Phi]$ is the linearized collision term. To be more general than in the previous section we do not specify its expression for now. [Linearizing the collision term (9.2) yields Eq. (9.31).] Note that on the left-hand side the terms proportional to Φ are counted as higher order since they are multiplied by derivatives of T, μ, and v. Certain integral constraints on the deviation functions Φ can be obtained from the condition that number density, momentum, and energy in a gas volume element must be the same if calculated with the local equilibrium distribution (9.16) and with the full function f (Pitaevskii and Lifshitz 2008).

Let us for now assume the system to be incompressible, which is a good approximation for instance for the neutron star crust. On account of the continuity equation (9.6a), this is equivalent to $\nabla \cdot v = 0$. (In an incompressible fluid, the density of a fluid element is constant in time, $\partial_t \rho + v \cdot \nabla \rho = 0$.) As a consequence, there is no dissipation through bulk viscosity. We shall come back to bulk viscosity later when we address the core of the star. There, bulk viscosity *is* an important source of dissipation. We also focus on static systems, i.e., we shall neglect all time derivatives. Extending the results of the previous section, we will include the electrical conductivity. To this end, we set $R = -eE$, where E is the electric field and e is the elementary charge. For now, we do not include a magnetic field and keep the assumption of a single particle species. This assumption deserves a comment. The expression (9.11a) does not contain the external force R, indicating that the force does not create dissipation. Of course, the work done by the force R affects the energy conservation (9.6b), but this only enters the bulk motion, as Eq. (9.11b) shows. Dissipation from the electric field emerges if there exists a friction force which opposes the diffusive motion. This is not described by the collision integral (9.2), but is realized in a multi-component system such as the electron-ion plasma in the neutron star crust or nuclear matter in the core made of neutrons, protons, and leptons. In this case, as already mentioned below Eq. (9.6), the average velocity of the constituents $\langle u \rangle$ is different from the center-of-mass velocity v of the mixture. This gives rise to an electric (and diffusive) current $j = -en\vartheta = -en(\langle u \rangle - v)$. In the neutron star crust (liquid or solid), due to the small mass ratio m_e/m_i of electron and ion masses, the contribution of the ion diffusion to the electric current can be neglected. Therefore, the rest frame of the ions is, to a good approximation, identical to the center-of mass frame and we can keep working with a single particle species (the electrons).

With these assumptions, we find for the left-hand side of Eq. (9.17),

$$
u \cdot \frac{\partial f^{(0)}}{\partial x} - e E \cdot \frac{\partial f^{(0)}}{\partial p} = -\frac{\partial f^{(0)}}{\partial \varepsilon} \left[\frac{\varepsilon - h}{T} w \cdot \nabla T + e w \cdot E^* + p_i w_j \left(v_{ij} - \frac{\delta_{ij}}{3} \nabla \cdot v \right) \right].
$$
(9.18)

Here we work in the co-moving frame of the total fluid, i.e., we have set $v = 0$ after taking the derivatives, such that from now on we have $w = u = p/m$. We have added a term proportional to $\nabla \cdot v$ (which is zero in our approximation) in order to reproduce the structure needed for the shear viscosity, defined the enthalpy per particle $h = \mu + sT/n$, and the effective electric field

$$
E^* = E + \frac{\nabla \mu}{e} + \frac{s}{n} \frac{\nabla T}{e}.
$$
(9.19)

The enthalpy is included in the thermal conduction term (proportional to ∇T) to eliminate the convective heat flux [cf. first term in Eq. (9.9a)].

In order to express the dissipative currents in terms of the deviation function Φ, we re-derive the entropy production equation (9.11a) as follows. We assume the entropy density of the system close to equilibrium to be given by the usual statistical expression

$$
s = -k_B \int_p [f \ln f + (1 - f) \ln(1 - f)].
$$
(9.20)

This suggests to set $\psi = \ln f + (f^{-1} - 1) \ln(1 - f)$ in the general transport equation (9.5). The right-hand side of that equation, including the collision term as well as the terms from the explicit (x, p, t)-dependence of ψ, yields the entropy production

$$
T_\varsigma = k_B \int_p [\ln f - \ln(1 - f)] I[f] = -\int_p \Phi I[f] = j \cdot E^* - j_T \cdot \frac{\nabla T}{T} - \pi_{ij} \partial_j v_i.
$$
(9.21)

In the second step we have performed the linearization according to Eq. (9.15), taking into account that ς vanishes for the local equilibrium function $f^{(0)}$. In the third step, we have used that, according to the Boltzmann equation (9.17), we can replace the collision integral by Eq. (9.18), and we have expressed the fluxes in terms of Φ,

$$
j = e \int_p \Phi \frac{\partial f^{(0)}}{\partial \varepsilon} w , \qquad j_T = -\int_p \Phi \frac{\partial f^{(0)}}{\partial \varepsilon} (\varepsilon - h) w , \qquad \pi_{ij} = -\int_p \Phi \frac{\partial f^{(0)}}{\partial \varepsilon} p_i w_j .
$$
(9.22)

Now, generalizing Eq. (9.12a), we introduce the transport coefficients associated with the electric and heat fluxes,

$$
\begin{pmatrix} E^* \\ j_T \end{pmatrix} = \begin{pmatrix} \frac{1}{\sigma} & -Q_T \\ -Q_T T & -\kappa \end{pmatrix} \begin{pmatrix} j \\ \nabla T \end{pmatrix} ,
\tag{9.23}
$$

where σ is the electrical conductivity and Q_T is the thermopower. The form of the non-diagonal terms is a consequence of Onsager's symmetry principle (Pitaevskii and Lifshitz 2008). Notice that due to the same spatial rank-one tensor structure of the thermodynamic forces $\nabla T/T$ and E^*, their linear response laws are coupled. The perturbation that drives the shear viscosity is the second-rank tensor (9.12b), hence the corresponding response law decouples. In terms of the transport coefficients, the local entropy production rate (9.21) becomes

$$
T\varsigma = \kappa \frac{(\nabla T)^2}{T} + \frac{j^2}{\sigma} + 2\eta \left(v_{ij} - \frac{\delta_{ij}}{3} \nabla \cdot v \right)^2 ,
\tag{9.24}
$$

implying the non-negativeness of κ, η, and σ.

The transport coefficients η, κ, σ, Q_T can now be computed as follows. To compute the shear viscosity, we make the ansatz

$$
\Phi = -A_\eta(\varepsilon) \left(p_i w_j - \frac{\delta_{ij}}{3} p \cdot w \right) \left(v_{ij} - \frac{\delta_{ij}}{3} \nabla \cdot v \right) ,
\tag{9.25}
$$

where, in an isotropic system, the unknown function A_η only depends on the particle energy. This function has to be determined by inserting the ansatz for Φ into the linearized Boltzmann equation (9.17). We can express the shear viscosity through A_η as

$$
\eta = -\frac{2}{15} \int_p p^2 w^2 A_\eta(\varepsilon) \frac{\partial f^{(0)}}{\partial \varepsilon} .
\tag{9.26}
$$

This relation is obtained by inserting the ansatz (9.25) into π_{ij} from Eq. (9.22), using the form of the viscous stress tensor (9.12b) and the angular integral in velocity (or momentum) space (remember that $p = mw$ in the frame we are working in)

$$
\int \frac{d\Omega_p}{4\pi} w_i w_j \left(w_k w_\ell - \frac{\delta_{k\ell}}{3} w^2 \right) = \frac{w^4}{15} \left(\delta_{ik}\delta_{j\ell} + \delta_{i\ell}\delta_{jk} - \frac{2}{3}\delta_{ij}\delta_{k\ell} \right) .
\tag{9.27}
$$

In Eq. (9.26) we have multiplied the result by a factor 2 from the sum over the 2 spin degrees of freedom of a spin-$\frac{1}{2}$ fermion (such that now the integral does not implicitly include the spin sum anymore).

To compute electrical and thermal conductivities and the thermopower, we use the ansatz

$$\Phi = -A_\kappa(\varepsilon)\frac{\varepsilon - h}{T}\boldsymbol{w} \cdot \nabla T - A_\sigma(\varepsilon)e\,\boldsymbol{w} \cdot \boldsymbol{E}^*, \tag{9.28}$$

with A_κ and A_σ computed from the linearized Boltzmann equation, and the transport coefficients are found in an analogous way as just demonstrated for the shear viscosity: we insert the ansatz (9.28) into $\boldsymbol{j}$ and $\boldsymbol{j}_T$ from Eq. (9.22), perform the angular integral,

$$\int \frac{d\Omega_p}{4\pi} w_i w_j = \frac{w^2 \delta_{ij}}{3}, \tag{9.29}$$

and compare the result with Eq. (9.23) to obtain (again taking into account the 2 spin degrees of freedom)

$$\sigma = -\frac{2e^2}{3}\int_p w^2 A_\sigma(\varepsilon)\frac{\partial f^{(0)}}{\partial \varepsilon}, \tag{9.30a}$$

$$\sigma Q_T = -\frac{2e}{3}\int_p w^2 A_{\kappa,\sigma}(\varepsilon)\frac{\varepsilon - h}{T}\frac{\partial f^{(0)}}{\partial \varepsilon}, \tag{9.30b}$$

$$\kappa + \sigma Q_T^2 T = -\frac{2}{3}\int_p w^2 A_\kappa(\varepsilon)\frac{(\varepsilon - h)^2}{T}\frac{\partial f^{(0)}}{\partial \varepsilon}, \tag{9.30c}$$

from which σ, Q_T, and κ can be computed. As a conseqence of Onsager's symmetry principle, we have obtained two expressions for σQ_T, using either A_σ or A_κ in the integral.

In general, even the solution of the linearized Boltzmann equation is not an easy task and various methods and approximations are used. First, one needs to specify the explicit expression for the collision integral. For instance, the linearization of the collision integral (9.2) gives

$$I_{\text{lin}}[\Phi] = -\frac{1}{k_B T}\int_{p_1}\int_{p'}\int_{p_1'} W(\boldsymbol{p}, \boldsymbol{p}_1, \boldsymbol{p}', \boldsymbol{p}_1')f^{(0)}f_1^{(0)}(1 - f'^{(0)})(1 - f_1'^{(0)})(\Phi + \Phi_1 - \Phi' - \Phi_1'), \tag{9.31}$$

where we have used $f^{(0)}f_1^{(0)}(1 - f'^{(0)})(1 - f_1'^{(0)}) = (1 - f^{(0)})(1 - f_1^{(0)})f'^{(0)}f_1'^{(0)}$ due to energy conservation.

One of the simplest cases is realized when the collision integral can be written in the form of the (energy-dependent) relaxation-time approximation,

$$I = -\sum_{lm}\frac{\delta f^{lm}}{\tau^l(\varepsilon)}Y_{lm}(\Omega_p), \tag{9.32}$$

which takes into account the angular dependence of the deviation to the equilibrium distribution function by expanding it in spherical harmonics Y_{lm}. Here $\tau^l(\varepsilon)$ is the relaxation time for the perturbation of multiplicity l. The solution of the Boltzmann equation is then

$$A_\sigma(\varepsilon) = A_\kappa(\varepsilon) = \tau^1(\varepsilon) , \qquad A_\eta(\varepsilon) = \tau^2(\varepsilon) . \tag{9.33}$$

When the relaxation time approximation is not available, one usually represents the functions $A(\varepsilon)$ in the form of a series expansion in some basis functions. This basis has to be chosen carefully for a satisfactory convergence of the expansion. In some cases the infinite chain of equations for the coefficients can be solved analytically and the exact solution for the transport coefficients is obtained from (9.30) (in form of an infinite series). In practice, the chain of equations is truncated at a finite number of coefficients. The truncation procedure is justified on the basis of the variational principle of kinetic theory (Ziman 2001). The variational principle uses the fact that the entropy production rate calculated from (9.21) with the linearized collision integral $I[\Phi]$ is a semi-positive definite functional of Φ. This is readily seen for the binary collision integral (9.31) since the probability W is positive, but it holds in general. Suppose that the arbitrary function $\tilde{\Phi}$ is subject to the constraint

$$\int_p X\tilde{\Phi} = \int_p I_{\mathrm{lin}}[\tilde{\Phi}]\tilde{\Phi} = -T\varsigma[\tilde{\Phi}] , \tag{9.34}$$

where we have abbreviated (9.18) by X. The variational principle states that over the class of such functions, the entropy production is maximal for the solution of the Boltzmann equation $X = I_{\mathrm{lin}}[\Phi]$, in other words $\varsigma[\Phi] \geq \varsigma[\tilde{\Phi}]$. Increasing the number of terms in the functional expansion and maximizing the functional $\varsigma[\tilde{\Phi}]$ under the constraint (9.34) one approaches the exact solution. This principle can be reformulated to give a direct limit on the diagonal coefficients in the Onsager relations. For instance, setting the thermodynamic forces to zero, $\nabla T/T = 0$ and $v_{ij} = 0$, and keeping only E^*, one obtains the electrical conductivity by minimizing

$$\frac{1}{\sigma} \leq \frac{E^{*2}}{T\varsigma[\tilde{\Phi}]} \tag{9.35}$$

over the functions subject to (9.34). Notice that the off-diagonal coefficient Q_T cannot be constrained in this way.

The variational principle discussed here applies for the stationary case in the absence of a magnetic field. The extension of the variational principle beyond this approximation is non-trivial and is outside the scope of the present section.

9.2.3   Towards Neutron Star Conditions

In this section we briefly comment on some modifications and extensions of the
kinetic theory laid out in the previous sections due to the specific conditions inside
neutron stars. We mention plasma effects, transport in Fermi liquids, relativistic
effects, and effects from Cooper pairing.

9.2.3.1   Plasma Effects

Electrically charged particles, for instance electrons in the crust and in the core,
interact via the long-range Coulomb potential. This seems to be at odds with
the concept of instant binary collisions, which forms the basis of the Boltzmann
approach to compute transport properties of dilute gases. However, the interaction
between charged particles in a plasma is screened and thus is effectively damped
on length scales $r > r_D$, where r_D is the Debye screening length. Therefore,
the Boltzmann equation becomes appropriate to describe the processes occurring
on large scales, provided the screened interaction potential is used in the collision
integral (Pitaevskii and Lifshitz 2008). The screening itself depends on the distribution functions of the plasma components, which severely complicates the solution.
However, for weak deviations from equilibrium, when the linearized Boltzmann
equation is used, the screening which enters the collision integral in Eq. (9.17) can
be calculated from the equilibrium distribution functions (i.e., in the collisionless
limit). Additional justification comes from the degeneracy conditions, which are
appropriate for electrons in most parts of the star (and other charged particles in
the core). In this case, only a small fraction of the thermal excitations contribute to
transport phenomena. Moreover, the kinetic energy of the particles increases with
density stronger than the Coulomb interaction energy. In other words, the denser
the gas is, the closer it is to the ideal Fermi gas (Landau and Lifshitz 1980).
All these properties allow us to use the formalism of the linearized Boltzmann
equation discussed above. Note that the force term R should contain the Lorentz
force with the self-consistent electromagnetic field. The generalized Ohm law (9.23)
then is written in the co-moving frame of the plasma and contains the electric field
measured in this frame, $E' = E + \frac{1}{c} v \times B$. We will return to this aspect in more
details in Sect. 9.4.1.5.

The ions in the neutron star crust are non-degenerate and non-ideal. The
discussion of their transport phenomena is more involved. Fortunately, the ion
contribution is usually negligible, see Sect. 9.3.

9.2.3.2   Transport in Fermi Liquids

Nuclear matter in the core of a neutron star is a strongly interacting, non-ideal,
multi-component fluid. The kinetic theory of rarefied gases described above cannot

be applied directly. However, the relevant temperatures are low and the matter is highly degenerate. In this case, the framework of Landau's Fermi-liquid theory (Baym and Pethick 1991) can be used to describe the low-energy excitations of the system. The excitations are considered as a dilute gas of quasiparticles which obey the Fermi-Dirac distribution (9.16) in momentum space, normalized to give the total local number density n of the real particles. The single-quasiparticle energy $\varepsilon(p)$ is a functional of the distribution function f, the quasiparticle Fermi momentum is $p_F = \hbar(3\pi^2 n)^{1/3}$, and in equilibrium the spectrum of quasiparticles in the vicinity of the Fermi surface is described by the effective mass on the Fermi surface $m^* = p_F/v_F$, where

$$\varepsilon^{(0)} - \mu = v_F(p - p_F), \qquad v_F = \left(\frac{\partial \varepsilon^{(0)}}{\partial p}\right)_{p=p_F}, \qquad (9.36)$$

with the Fermi velocity v_F, and the superscript (0) indicates equilibrium.

The evolution of the quasiparticle distribution function is described by the Landau transport equation

$$\frac{\partial f}{\partial t} + u \cdot \frac{\partial f}{\partial x} - \nabla \varepsilon \cdot \frac{\partial f}{\partial p} = I[f], \qquad (9.37)$$

where now $u = \nabla_p \varepsilon$. The Eq. (9.37) is different from the Boltzmann equation (9.1) since the term $\nabla \varepsilon$ is present even in the absence of external forces R. This is because the energy spectrum—being a functional of f—changes from one coordinate point to another. Thus, $\nabla \varepsilon$ contains the combined effects of the external forces and the effective field resulting from interactions between quasiparticles. In addition, the quasiparticle velocity is coordinate-dependent for the same reason.

Transport coefficients of the Fermi-liquid are computed by considering a small deviation from local equilibrium and performing the linearization of the Landau equation in a way similar to Sect. 9.2.2 (Baym and Pethick 1991; Pitaevskii and Lifshitz 2008). However, there is an important difference. The local equilibrium distribution function is $f^{(0)}(\varepsilon^{(0)})$, but the conservation laws from the collision integral employ the true quasiparticle energies ε. Hence the collision integral vanishes for the functions $f^{(0)}(\varepsilon)$ instead of true local distribution function. As a consequence, the linearized collision integral depends not on $\delta f = f - f^{(0)}(\varepsilon^{(0)})$ but on $\delta \tilde{f} = f - f^{(0)}(\varepsilon)$, and the definition of the function Φ (9.15) is modified to

$$\delta \tilde{f} = -\frac{\partial f^{(0)}}{\partial \varepsilon} \Phi. \qquad (9.38)$$

Since the definitions of the fluxes also contain the true quasiparticle energies and velocities, they are given by the expressions (9.22) with Φ redefined according to (9.38). Therefore, in the stationary case, we obtain formally identical equations as in the above derivation. Fermi-liquid effects do not appear explicitly. The same is true if a magnetic field is taken into account (Pitaevskii and Lifshitz 2008). In more general cases, terms containing δf can appear on the left-hand side of the linearized Boltzmann equation. This situation is realized for instance when the bulk viscosity of the Fermi liquid is considered (Baym and Pethick 1991; Sykes and Brooker 1970).

9.2.3.3 Relativistic Effects

Neutron stars are ultra-dense objects, and thus relativistic effects are important for the transport in the star. They manifest themselves in various forms, and we have to distinguish between effects on a microscopic level (e.g., calculations of transport coefficients) and a macroscopic level (e.g., simulations based on hydrodynamic equations), as well as between effects from special relativity (large velocities) and general relativity (spacetime curvature on scales of interest). In this review, we are almost exclusively concerned with microscopic calculations, where we can usually ignore effects from general relativity. The reason is the large separation of the scale on which the gravitational field changes inside the star from the microscopic scales on which the equilibration processes (collisions or reactions) operate (Tauber and Weinberg 1961; Thorne 1966). If the mean free paths of the particles are microscopic in this sense, one can study transport processes in the local Lorentz frame, and gravity effectively does not appear in the analysis. If the mean free path, however, becomes comparable to the macroscopic scale of gravity, one has to consider the full general relativistic transport equation (Vereshchagin and Aksenov 2017)

$$p^\mu \frac{\partial f}{\partial x^\mu} - \Gamma^\mu_{\ \nu\rho} p^\nu p^\rho \frac{\partial f}{\partial p^\mu} = I[f], \tag{9.39}$$

where we have omitted external forces, where $\Gamma^\mu_{\ \nu\rho}$ are the Christoffel symbols, x^μ is the spacetime four-vector, p^μ the four-momentum, and $I[f]$ is the collision integral (specified in the local reference frame). This situation occurs for instance for neutrino transport in supernovae and proto-neutron stars (Pons et al. 1999). In neutron stars, this general approach may be important for example in superfluid phases if the only available excitations are the Goldstone modes, whose mean free path can become of the order of the size of the star, see Sects. 9.4.1.4 and 9.5.5.2. Effects from general relativity are also important when transport coefficients— computed from a microscopic approach—are used as an input for hydrodynamic equations. These equations, when they concern the structure of the whole star or a significant fraction of it, must be formulated within general relativity. An example is the equation for the radial component of the heat flux in a cooling star (Thorne 1966, 1977),

$$F_r = -\kappa \, e^{-\lambda - \phi} \frac{\partial \tilde{T}}{\partial r}, \tag{9.40}$$

where κ is the thermal conductivity, λ and ϕ appear in the parametrization of the metric,

$$ds^2 = e^{2\phi} d(ct)^2 - e^{2\lambda} dr^2 - r^2(d\theta^2 + \sin^2\theta \, d\varphi^2) , \qquad (9.41)$$

and $\tilde{T} \equiv Te^{\phi}$ is the redshifted temperature. (It is the redshifted temperature, not the temperature T, which is constant in equilibrium.)

To connect the non-relativistic hydrodynamic equations of Sect. 9.2.1 to a covariant formalism, one introduces the (special) relativistic stress-energy tensor,

$$T^{\mu\nu} = T^{\mu\nu}_{\text{ideal}} + T^{\mu\nu}_{\text{diss}} , \qquad (9.42)$$

where we have separated the ideal part $T^{\mu\nu}_{\text{ideal}}$ from the dissipative contribution $T^{\mu\nu}_{\text{diss}}$,

$$T^{\mu\nu}_{\text{ideal}} = (\epsilon + P)v^{\mu} v^{\nu} - g^{\mu\nu} P , \qquad (9.43a)$$

$$T^{\mu\nu}_{\text{diss}} = \kappa(\Delta^{\mu\gamma} v^{\nu} + \Delta^{\nu\gamma} v^{\mu})[\partial_{\gamma} T + T(v \cdot \partial)v_{\gamma}]$$

$$+ \eta \Delta^{\mu\gamma} \Delta^{\nu\delta} \left(\partial_{\delta} v_{\gamma} + \partial_{\gamma} v_{\delta} - \frac{2}{3} g_{\gamma\delta} \partial \cdot v \right) + \zeta \Delta^{\mu\nu} \partial \cdot v . \quad (9.43b)$$

Here, ϵ and P are energy density and pressure measured in the rest frame of the fluid, $g^{\mu\nu} = (1, -1, -1, -1)$ is the metric tensor in flat space, $v^{\mu} = \gamma(1, \boldsymbol{v})$ is the four-velocity with the Lorentz factor γ and the three-velocity $\boldsymbol{v}$ used in Sects. 9.2.1 and 9.2.2. We have abbreviated $\Delta^{\mu\nu} = g^{\mu\nu} - v^{\mu}v^{\nu}$, and the transport coefficients κ, η, ζ are heat conductivity, shear and bulk viscosity, as in the non-relativistic formulation (9.12). In the non-relativistic limit, using the notation from Sect. 9.2.1, $T^{00}_{\text{ideal}} \to \mathcal{E}$ is the energy density, $T^{0i}_{\text{ideal}} \to g_i$ is the momentum density, $T^{i0}_{\text{ideal}} \to (\mathcal{E} + P)v_i$ is the non-dissipative part of the energy flux $\boldsymbol{j}_{\mathcal{E}}$, and $T^{ij}_{\text{ideal}} \to \Pi_{ij}$ is the non-relativistic stress tensor. The dissipative terms are formulated in the so-called Eckart frame (Eckart 1940), where—in contrast to the Landau frame (Landau and Lifshitz 1987)—the conserved four-current $j^{\mu} = nv^{\mu}$ does not receive dissipative corrections (Kovtun 2012). The hydrodynamic equations are then obtained from the conservation laws for the stress-energy tensor and the current,

$$\partial_{\mu} T^{\mu\nu} = \partial_{\mu} j^{\mu} = 0 . \qquad (9.44)$$

They reduce to Eqs. (9.6) in the non-relativistic limit. We will briefly return to this relativistic formulation in Sect. 9.5.4.2, but otherwise we will not discuss any of the effects illustrated by Eqs. (9.39), (9.40), and (9.43). In particular, since we do not discuss neutrino transport in supernovae, no effects from general relativity will be further discussed. Therefore, when we use 'relativistic' in the rest of the review, we mean effects from special relativity in the following simple sense: relativistic effects are important if the rest mass (times the speed of light) of a given particle species is not overwhelmingly larger than its Fermi momentum. (In this case, the Fermi velocity introduced in the previous section, i.e., the slope of the dispersion relation at the Fermi surface, becomes a sizable fraction of the speed of light.) With this criterion, the ions in the crust and the nucleons in the core are often treated non-relativistically (for ultra-high densities in the core, this treatment becomes questionable), while the lighter electrons and quarks are relativistic (except for electrons at very low densities in the outer crust).

9.2.3.4   Transport with Cooper Pairing

The effect of Cooper pairing on reaction rates and transport will be discussed specifically in various sections throughout the review. As a preparation and a simple overview, we now give some general remarks that may be helpful to understand and put into perspective the more detailed discussions and results. For a pedagogical introduction, bringing together elements from non-relativistic and relativistic approaches to Cooper pairing in superfluids and superconductors see Schmitt (2015).

Cooper pairing in neutron stars is expected to occur in the inner crust for neutrons and in the core for neutrons, protons, and, if present, for hyperons and quarks. The critical temperatures of these systems vary over several orders of magnitude, depending on the form of matter, on density, and on the particular pairing channel. Moreover, it is prone to large uncertainties because the attractive force needed for Cooper pairing originates from the strong interaction. Nevertheless, a rough benchmark to keep in mind is $T_c \sim 1\,\mathrm{MeV}$, which is the maximal critical temperature reached for nuclear matter[1] (with significantly smaller values for neutron triplet pairing) and which is exceeded by about an order of magnitude, maybe even two, by quark matter, where $T_c \sim (10-100)\,\mathrm{MeV}$ (also in quark matter, there are pairing patterns with significantly lower critical temperatures). In any case, we conclude that the temperatures inside the star—except for very young neutron stars—are sufficiently low to allow for Cooper pairing. The resulting stellar superfluids and superconductors (Alford et al. 2008a; Page et al. 2014; Haskell and Sedrakian 2018, this volume) are similar to their relatives in the laboratory, but the situation in the star is typically more complicated. For instance, the neutron superfluid in the inner crust coexists with a lattice of ions, the core might be a superconductor and a superfluid at the same time, and quark matter might introduce effects of color superconductivity. In addition, the star rotates and has a magnetic field, which suggests the presence of superfluid vortices and possibly magnetic flux tubes, which may coexist and interact with each other. Therefore, understanding superfluid transport in the environment of a neutron star is a difficult task, and some care is required in using results from ordinary superfluids.

---

[1]In units where $k_B = 1$, temperature and energy have the same units, 1 MeV corresponds to $1.160 \times 10^{10}\,\mathrm{K}$.

One obvious effect of Cooper pairing is the suppression of reaction rates and scattering processes of the fermions that pair. This effect is very easy to understand. Cooper pairing induces an energy gap Δ in the quasiparticle dispersion relation (one needs a finite amount of energy to break up a pair), and thus, for temperatures much smaller than the gap, quasiparticles are not available for a given process. As a consequence, if at least one of the participating fermions is gapped, the rate is exponentially suppressed by a factor $\exp(-\Delta/T)$ for $T \ll \Delta$. The suppression is milder if the pairing is not isotropic and certain directions in momentum space are left ungapped. This is conceivable for some forms of neutron pairing and in certain color-superconducting quark matter phases. In this case, if for instance only one- or zero-dimensional regions of the Fermi surface contribute (as opposed to the full two-dimensional Fermi surface in the unpaired case), the rate is suppressed by a power of the small parameter T/Δ. Except for these special cases, at low temperatures we can usually neglect the processes suppressed by Cooper pairing and can restrict ourselves to contributions from ungapped fermions or other low-energy excitations, if present.

At larger temperatures, as we move towards the critical temperature T_c, the form of the exponential suppression no longer holds and the rate in the Cooper-paired phase has to be evaluated numerically. Since particle number conservation is broken spontaneously, particles can be deposited into or created from the Cooper pair condensate. This effect induces subprocesses that are called Cooper pair breaking and formation processes. They are particularly interesting in nuclear matter, where more efficient processes, such as the direct Urca process, are suppressed. Then, somewhat counterintuitively, an enhancement of the neutrino emission is possible as the system cools through the critical temperature for neutron superfluidity.

While Cooper pairing removes fermionic degrees of freedom from transport at low temperatures, it introduces one or several massless bosonic excitations if a *global* symmetry is spontaneously broken by the formation of a Cooper pair condensate. This is due to the Goldstone theorem, and the corresponding Goldstone mode for superfluidity is, following the terminology of superfluid helium, usually called phonon (or 'superfluid mode', or 'superfluid phonon' to distinguish it from the lattice phonons in the neutron star crust). In this case, the broken global symmetry is the $U(1)$ associated with particle number conservation. Superfluid neutron matter and the color-flavor locked (CFL) quark matter phase both have a phonon. Transport through phonons is mostly computed with the help of an effective theory, and we will quote some of the resulting transport properties in hadronic and quark matter. If Cooper pairing breaks additional global symmetries, such as rotational symmetry, additional Goldstone modes appear. This is possible in $^3P_2$ neutron pairing (Bedaque and Nicholson 2013; Bedaque and Reddy 2014) and in spin-one color superconductivity (Pang et al. 2011).

If instead a *local* symmetry is spontaneously broken, there is no Goldstone mode. This is the case for Cooper pairing of protons and for quark matter phases other than CFL such as the so-called 2SC phase (although, due to the presence of electrons and the resulting screening effects, the Goldstone mode in a proton superconductor can be 'resurrected' (Baldo and Ducoin 2011)). As in ordinary superconductivity, the would-be Goldstone boson is replaced by an additional degree of freedom of the gauge field, which acquires a magnetic mass. One obvious consequence is the well-known Meissner effect, which is of relevance for the magnetic field evolution in neutron stars. Magnetic screening can also indirectly affect transport properties if a certain transport property is dominated by one (unpaired) particle species that is charged under the gauge symmetry which is spontaneously broken by Cooper pairing of a different species (even though the species that pairs does not contribute to transport itself because it is gapped). This situation occurs in nuclear matter when electrons experience a modified electromagnetic interaction due to pairing of protons, and in the 2SC phase of quark matter, where the different particle species are electrons and the different colors and flavors of quarks, which are not all paired in this specific phase, and the relevant gauge bosons are the gluons and the photon.

As we know from some of the earliest experiments with superfluid helium, a superfluid at nonzero temperature (below T_c) behaves as a two-fluid system (Tisza 1938; Landau 1941) (for the connection of the two-fluid picture to an underlying microscopic theory see for instance Alford et al. 2013). This means that, in a hydrodynamic approach, there are two independent velocity fields: one for the superfluid component, which is the Cooper pair condensate in a fermionic superfluid (or the Bose-Einstein condensate in a bosonic superfluid such as ^{4}He), and one for the so-called normal component, which corresponds to the phonons and possibly a fraction of the fermions which have remained unpaired. Since only the normal component carries entropy, the two-fluid nature has obvious consequences for heat transport, which now can occur through a counterflow of the two fluid components. While this mechanism proves extremely efficient in laboratory experiments with superfluid helium, it may be less effective in the more complicated situation in a neutron star. For instance, in the inner crust of the star the counterflow of the normal and superfluid components becomes dissipative due to the presence of electrons which damp the motion of the normal fluid through induced electron-phonon interactions (Page and Reddy 2012). Another consequence of the two-fluid behavior is the existence of second sound. (The phonon, first and second sound are in general three different excitations. At low temperatures, the phonon excitation is identical to first sound, while close to the critical temperature it is identical to second sound (Alford et al. 2014a).) In superfluid helium, first and second sound are predominantly density and temperature oscillations, respectively, for all temperatures $T < T_c$. This is not necessarily true for other superfluids and it has been shown that first and second sound may exchange their roles (Alford et al. 2014a).

Two-fluid systems allow for additional transport coefficients. For instance, in the hydrodynamics of a superfluid, usually three independent bulk viscosity coefficients are taken into account (Khalatnikov 1965). In a neutron star, the situation might become even more complicated due to the presence of additional fluid components, e.g., a nonzero-temperature neutron superfluid coexisting with electrons and protons, such that we have to deal with an involved multi-fluid system. One interesting feature of multi-fluids with relevance for the physics of neutron stars is the possibility of hydrodynamical instabilities due to a counterflow between the fluids. Such an instability may occur for the neutron superfluid in the inner crust, if it moves (locally) with a sufficiently large nonzero velocity relative to the ion lattice. In this review, we shall not further discuss multi-fluid transport in detail (except for the transport coefficients of a single superfluid at nonzero T) and refer the reader to the recent literature and references therein (Gusakov et al. 2009a,b; Glampedakis et al. 2012a; Chamel 2013; Andersson et al. 2013; Haber et al. 2016; Andersson et al. 2017).

Finally, let us mention another very important consequence of Cooper pairing, which has been related to various astrophysical observations such as pulsar glitches (Haskell and Melatos 2015), namely the formation of rotational vortices in a superfluid and of magnetic flux tubes in a superconductor. (A magnetic field enters a type-II superconductor through quantized magnetic flux tubes if its magnitude lies between the upper and lower critical magnetic fields. The presence of a superfluid, to which the superconductor couples, may change the textbook-like behavior of type-II superconductors qualitatively (Haber and Schmitt 2017a,b).) Besides ordinary vortices in hadronic matter, quark matter in the core of neutron stars may contain so-called semi-superfluid vortices (Balachandran et al. 2006; Alford et al. 2016) in the CFL phase and/or color magnetic flux tubes (Alford and Sedrakian 2010; Glampedakis et al. 2012b) in the CFL or 2SC phases (the latter are not protected by topological arguments and it is unknown if they are energetically stable objects in the neutron star environment). As for most of the multi-fluid aspects, we will not review the transport properties of superfluids in the presence of vortices. For various aspects of the hydrodynamics of these systems, including the possibility of superfluid turbulence and possible boundaries between phases with and without (or with a different kind of) vortices, see Hall and Vinen (1956), Khalatnikov (1965), Donnelly (1999), Gusakov (2016), Gusakov and Dommes (2016), and Graber et al. (2017).

9.3 Transport in the Crust and the Crust/Core Transition Region

9.3.1 Thermal and Electrical Conductivity and Shear Viscosity

The main carriers which determine the transport processes in the neutron star crust are electrons. The electrons in the crust form an almost ideal, degenerate gas. The degeneracy temperature T_F for electrons is

$$T_{Fe} = \frac{\mu_e - m_e c^2}{k_B} = 5.9 \times 10^9 \text{K} \left(\sqrt{1 + x_r^2} - 1 \right), \tag{9.45}$$

where $x_r = p_{Fe}/(m_e c)$ is the electron relativistic parameter, with the electron Fermi momentum p_{Fe}, the electron rest mass m_e, and the electron chemical potential (including the rest mass) $\mu_e = m_e c^2 \sqrt{1 + x_r^2} \equiv m_e^* c^2$. In a one-component plasma with ion charge number Z and total nucleon number per ion[2] A, $x_r \approx (\rho_6 Z/A)^{1/3}$, where ρ_6 is the mass density ρ in units of $10^6 \, \mathrm{g\,cm^{-3}}$. In most of the crust, $\rho_6 \gg 1$ and the electrons are ultra-relativistic. We will not discuss electrons in non-degenerate or partially degenerate conditions $T \gtrsim T_{Fe}$. The effects of non-degenerate electrons are important when the thermal structure of the stellar heat blanket is calculated. In non-degenerate regions the radiative contribution to heat transport is relevant, which we also do not discuss here, for details see Potekhin et al. (2015a; 2015b).

For degenerate electrons ($T \ll T_{Fe}$) the analysis of the Boltzmann equation is simplified since the transport is mainly provided by those electrons whose energies lie in a narrow thermal band near the Fermi surface $|\varepsilon - \mu_e| \lesssim k_B T$. When using Eqs. (9.25) and (9.30), it is safe to set $h = \mu_e$ and neglect the thermopower correction in Eq. (9.30c). As a result, it is convenient to present the transport coefficients of interest in the form

$$\sigma = \frac{e^2 n_e \tau_\sigma}{m_e^*}, \tag{9.46a}$$

$$\kappa = \frac{\pi^2 k_B^2 T n_e \tau_\kappa}{3 m_e^*}, \tag{9.46b}$$

$$Q_T = \frac{\pi^2 k_B^2 T m_e^*}{3 e p_{Fe}^2}(3 + \xi), \tag{9.46c}$$

$$\eta = \frac{n_e p_{Fe}^2 \tau_\eta}{5 m_e^*}, \tag{9.46d}$$

where τ_σ, τ_κ, and τ_η are the effective relaxation times, and $\xi \sim 1$ is a dimensionless factor which can change sign depending on the electron scattering mechanism. For brevity, we will not consider the thermopower coefficient further. The inverse quantities $\nu_{\sigma,\kappa,\eta} = \tau_{\sigma,\kappa,\eta}^{-1}$ are called the effective collision frequencies. If the relaxation time approximation (9.32) is applicable, the effective relaxation times become the actual relaxation times evaluated at the Fermi surface, $\tau_\sigma = \tau_\kappa = \tau_e^1(\mu)$, $\tau_\eta = \tau_e^2(\mu)$, cf. Eq. (9.33), since one approximates $\frac{\partial f^{(0)}}{\partial \varepsilon} \approx -\delta(\varepsilon - \mu_e)$. In this case, we obtain the standard Wiedemann-Franz rule for conductivities,

$$\frac{\kappa}{\sigma} = \frac{\pi^2 k_B^2 T}{3 e^2}. \tag{9.47}$$

The relaxation time approximation holds when electron-ion collisions are the dominant scattering mechanism and the energy ω transferred in these collision is

---

[2]In the inner crust, unbound neutrons exist and the ion mass number A_{nuc} is less than A. The ion mass is then $m_i = A_{\mathrm{nuc}} m_u$, with m_u being the atomic unit mass (Chamel and Haensel 2008).

small $\omega \ll k_B T$. When this is not the case, the variational calculations outlined in Sect. 9.2.2 are usually employed. It turns out that already the simplest variational approximation gives a satisfactory estimate for astrophysical conditions. Moreover, the violation of the Wiedemann-Franz rule is not as dramatic as in ordinary metals at low temperature (Yakovlev and Urpin 1980).

When there are different relaxation mechanisms for the electron distribution function, for instance collisions with different particle species, the respective collision integrals must be added on the right-hand side of the Boltzmann equation. In practice, one usually considers different mechanisms separately to obtain the effective collision frequency ν_{ej} for each scattering process. Due to the strong degeneracy of electrons, the cumulated collision frequency $\nu_{tot} = \sum_j \nu_{ej}$ obtained in this way is a good approximation to the solution of the Boltzmann equation with all mechanisms included. This is known as Matthiessen's rule (Ziman 2001). The variational principle of kinetic theory allows us to estimate the error introduced by this approximation (Ziman 2001), see also Potekhin et al. (2015b). Below we consider the most important processes that determine the electron transport.

9.3.1.1 Electron-Ion Collisions

The main process for electron transport is their scattering off ions. The ions in the neutron star crust form a strongly coupled non-ideal plasma, whose state is defined by an ion coupling parameter Γ. For a one-component plasma (in the sense that only one sort of ions is present)

$$\Gamma = \frac{Z^2 e^2}{a_{WZ} k_B T} \approx 153 \, x_r \left(\frac{Z}{50}\right)^{5/3} \left(\frac{T}{10^8 \, \text{K}}\right)^{-1}, \tag{9.48}$$

where the ion Wigner-Seitz cell radius a_{WZ} is defined by the relation

$$\frac{4\pi}{3} a_{WZ}^3 n_i = 1. \tag{9.49}$$

When $\Gamma \ll 1$, ions are in the gaseous phase, at $\Gamma \gtrsim 1$ in the liquid phase, and at $\Gamma = \Gamma_m \approx 175$ (Potekhin and Chabrier 2000) the ion liquid crystallizes and is thought to form a body-centered cubic lattice (Chamel and Haensel 2008). This condition and Eq. (9.48) define the (density-dependent) melting temperature T_m. Notice that the melting point can shift substantially if the electron polarization or magnetic field effects are taken into account (Potekhin and Chabrier 2000, 2013). Another important parameter is the ion plasma temperature

$$T_{pi} = \frac{\hbar}{k_B} \left(\frac{4\pi Z^2 e^2 n_i}{m_i}\right)^{1/2}, \tag{9.50}$$

above which the thermodynamic properties are classical, and below which quantum effects should be taken into account. In the context of electron transport, the important point is that at $T < T_{pi}$ the typical energy transferred in the electron-ion collisions is $\omega \sim k_B T$ and the relaxation time approximation cannot be used (Yakovlev and Urpin 1980). If $T_{pi} < T_m$, quantum effects are only important in the crystalline phase. A temperature regime where quantum effects are relevant in the liquid phase can in principle be realized for light elements and high densities. In this case, the properties of the liquid—including transport properties—are modified, but also the crystallization point itself (the value $\Gamma_m \approx 175$ is obtained from a classical estimate, not taking into account zero-point vibrations). Calculations show that at some density the crystallization temperature starts to decrease and reaches zero at a certain critical density, above which no crystallization occurs (Chabrier 1993; Jones and Ceperley 1996). However, the importance of a quantum liquid regime for neutron star envelopes is questionable since nuclear reactions (electron captures and pycnonuclear burning) would not allow light elements to exist at sufficiently large densities, see Sect. 2.3.5 of Haensel et al. (2007) for more details. Therefore, here we discuss quantum corrections only for the solid phase (see footnote 4 for a brief remark about results for the quantum liquid regime).

For any phase state of the ions, the effective electron-ion collision frequency, to be used in (9.46), is usually written in terms of the effective Coulomb logarithm Λ_{ei},

$$
\nu_{ei} = \frac{4\pi Z^2 e^4 n_i}{p_{Fe}^2 \nu_{Fe}} \Lambda_{ei} \approx 8.8 \times 10^{17} \frac{Z}{50} \sqrt{1 + x_r^2} \, \Lambda_{ei} \, \text{s}^{-1} , \tag{9.51}
$$

where $\nu_{Fe} = p_{Fe}/m_e^*$ and we have omitted the transport indices σ, κ, η for brevity. The Coulomb logarithm is a central quantity in the transport theory of electromagnetic plasmas. In the (classical) liquid regime, $1 \lesssim \Gamma < \Gamma_m$, we have $\Lambda_{ei} \sim 1$, while in the solid regime $\Lambda_{ei} \propto T/T_m$ at $T_m > T \gtrsim 0.15 T_{pi}$ and $\Lambda_{ei} \propto T^2/(T_m T_{pi})$ at $T \lesssim 0.15 T_{pi}$ (Potekhin et al. 1999; Chugunov and Yakovlev 2005; Potekhin et al. 2015b). For a one-component plasma it was calculated by Potekhin et al. (1999) and Chugunov and Yakovlev (2005), including various effects such as electron screening, non-Born and relativistic corrections, ion-ion correlations in the liquid regime, and multi-phonon processes in the solid regime. The main complication in the calculation of the Coulomb logarithm is to properly take into account the ion-ion correlations that are important in a strongly non-ideal Coulomb liquid. In the conditions of the neutron star crust, the typical electron kinetic energy is much larger than the electron-ion interaction energy (as mentioned in Sect. 9.2.3.1), and electrons can be treated as quasi-free particles scattering off the static electric potential created by charge density fluctuations in the ion system. The resulting expression in the first-order Born approximation, which is equally applicable in liquid and solid states can be written as (Baym 1964)

$$
\Lambda_{ei} = \int_{q_0}^{2k_{Fe}} \frac{dk}{k} |k^2 U(k)|^2 \left[1 - \beta_r^2 \frac{k^2}{4k_{Fe}^2} \right] R(k) \int_{-\infty}^{+\infty} d\omega \frac{z}{e^z - 1} G(k, z) S(\omega, k) ,
$$

$$
\tag{9.52}
$$

where $z = \hbar\omega/(k_B T)$, $k_{Fe} = p_{Fe}/\hbar$, $\beta_r = v_{Fe}/c$, $U(k)$ is the Fourier transform of the effective potential describing single electron-ion scattering[3]

$$U(k) = \frac{F(k)}{k^2 \epsilon(k)}, \tag{9.53}$$

which includes electron screening via the static dielectric function $\epsilon(k)$ and finite-size corrections for nuclei through the form-factor term $F(k)$, and the term in square brackets describes the relativistic suppression of the backward scattering. In the liquid phase, $q_0 = 0$, while in the solid phase, $q_0 = q_{BZ} = (6\pi^2 n_i)^{1/3}$, see below. The functions $R(k)$ and $G(k, z)$ are kinematic factors depending on the transport property that is calculated, namely $R_{\sigma,\kappa}(k) = 1$, $R_\eta(k) = 3[1 - k^2/(4k_{Fe}^2)]$, $G_{\sigma,\eta}(k, z) = 1$, and

$$G_\kappa(k, z) = 1 + \frac{z^2}{\pi^2}\left(3\frac{k_{Fe}^2}{k^2} - \frac{1}{2}\right). \tag{9.54}$$

Finally, $S(\omega, k)$ is the dynamical structure factor which describes the ion density fluctuations,

$$S(\omega, k) = \frac{1}{2\pi N_i} \int_{-\infty}^{+\infty} dt \int d^3x\, d^3x'\, e^{ik\cdot(x-x')-i\omega t} \left\langle \delta\hat{n}^\dagger(x, t)\delta\hat{n}(x', 0)\right\rangle_{eq}, \tag{9.55}$$

where $\langle \ldots \rangle_{eq}$ stands for average over the Gibbs ensemble of ions (thermal average), N_i is the total number of ions, and

$$\delta\hat{n}(x, t) = \hat{n}_i(x, t) - \langle \hat{n}_i(x, t)\rangle_{eq}, \tag{9.56}$$

with the ion number density operator $\hat{n}_i(x, t)$.

Let us first consider a liquid with a temperature reasonably far above the melting temperature, $T > T_m$. Then $\langle \hat{n}_i(x, t)\rangle_{eq} = n_i$ takes into account the uniform compensating background. Ignoring quantum effects in the liquid, as argued above, the $z \to 0$ limit can be used in the integrand of the ω-integration in Eq. (9.52), and one is left with the static structure factor $S(k)$. This case corresponds to the relaxation time approximation, and one obtains the Ziman formula known from transport theory of liquid metals (Ziman 1961). The Wiedemann-Franz rule (9.47) is also fulfilled. The static structure factor can be calculated from numerical simulations of the Coulomb plasma. In the absence of correlations, $S(k) \to 1$.

---

[3]The long-range nature of the Coulomb interaction leads to a logarithmic divergence of the integral in (9.52) since $U(k) \propto k^{-2}$ at small k, which is regularized by plasma screening, see Sect. 9.2.3.1. Therefore, very roughly, $\Lambda_{ei} \sim \log[2k_{Fe}/\max(q_0, r_D^{-1})]$, and hence the name 'Coulomb logarithm'.

Potekhin et al. (1999) used static structure factors obtained by Young et al. (1991) and provided a useful analytical fit for the Coulomb logarithm that can be readily used in simulations.

Now consider the case $T < T_m$, when ions are assumed to form a perfect one-component body-centered cubic (bcc) crystal. The high symmetry of the cubic lattice implies that the transport properties are isotropic (Harrison 1980). In this case, the electrons are scattered off phonons, i.e., lattice vibrations. The Coulomb logarithm is still given by Eq. (9.52), where an expression for the structure factor can now be obtained using a multi-phonon expansion. For temperatures not too close to the melting temperature the single-phonon contribution to the structure factor is sufficient (Flowers and Itoh 1976; Yakovlev and Urpin 1980). In this regime, useful approximate expressions for the collision frequencies (that however do not include various corrections already mentioned above) are (Yakovlev and Urpin 1980; Baiko and Yakovlev 1995; Chugunov and Yakovlev 2005)

$$
v_{\mathrm{ei}}^{\kappa,\sigma} = \alpha_f u_{-2}\beta_r^{-1}\frac{k_B T}{\hbar}\left(2 - \beta_r^2\right) F_{\kappa,\sigma}\left(\frac{T}{T_{\mathrm{pi}}}\right), \qquad v_{\mathrm{ei}}^{\eta} = \alpha_f u_{-2}\beta_r^{-1}\frac{k_B T}{\hbar}\left(3 - \beta_r^2\right) F_\eta\left(\frac{T}{T_{\mathrm{pi}}}\right),
$$
(9.57)

where α_f is the fine structure constant, $u_{-2} = 13.0$ is one of the frequency moments of the bcc lattice, and the functions $F(T/T_{\mathrm{pi}})$ describe quantum corrections,

$$
F_\sigma(t) = F_\eta(t) = \frac{t}{\sqrt{t^2 + a_0^2}},
$$
(9.58a)

$$
F_\kappa(t) = F_\sigma(t) + \frac{t}{\pi^2 u_{-2}\left(t^2 + a_2^2\right)^{3/2}}\frac{\ln(4Z) - 1 - \beta_r^2}{2 - \beta_r^2},
$$
(9.58b)

where $a_0 = 0.13$ and $a_2 = 0.11$. Accordingly, when $T \gtrsim 0.15 T_{\mathrm{pi}}$ one can set $F_{\sigma,\kappa,\eta} = 1$ in Eq. (9.57). In this classical limit, the relaxation time approximation still works fairly well and the Wiedemann-Franz rule $v_{\mathrm{ei}}^\sigma = v_{\mathrm{ei}}^\kappa$ applies. The difference between v_{ei}^η and $v_{\mathrm{ei}}^{\kappa,\sigma}$ is due to the difference in the kinematic factor R in Eq. (9.52). At low temperatures, $T \lesssim 0.15 T_{\mathrm{pi}}$, the relaxation time approximation breaks down and quantum effects are important. Since $F_{\kappa,\sigma,\eta}(t) \propto t$, the quantum corrections suppress the electron-ion collisions in this limit. Because of the second term in Eq. (9.58b), which is a consequence of the factor (9.54), $v_{\mathrm{ei}}^\sigma \neq v_{\mathrm{ei}}^\kappa$ and the Wiedemann-Franz rule is violated. This violation is, however, not as dramatic as for terrestrial solids (Yakovlev and Urpin 1980).

It is important to stress that the electron-phonon interaction in Coulomb crystals in the astrophysical environment is very different from that in terrestrial metals. For the latter, normal processes within one Brillouin zone are dominant, $k \lesssim q_{\mathrm{BZ}}$, while in the astrophysical context, since electrons are quasi-free, with $k_{Fe} \gg q_{\mathrm{BZ}}$, the typical momentum transfer is large compared to q_{BZ}, and Umklapp processes, which transfer an electron from one Brillouin zone to another, play the major role.

At very low temperatures, the picture of quasi-free electrons is modified, since the distortion of the quasi-spherical Fermi surface by band gaps becomes important. This suppresses the Umklapp processes. However, Chugunov (2012) has shown that this 'freezing' of the Umklapp processes is only important at $T \lesssim 10^{-2} T_{pi}$ and is relatively slow, see also Page and Reddy (2012). In practice, at these temperatures the transport is dominated by other processes (see below), and the freezing of Umklapp processes can be safely neglected in practical calculations.

As the temperature of the Coulomb solid approaches the melting temperature, $T \rightarrow T_m$, the single-phonon picture is no longer valid. Baiko et al. (1998) calculated the multi-phonon contribution to the structure factor $S(\omega, k)$ in the harmonic approximation; these results were later incorporated in analytical fits by Potekhin et al. (1999). Recent quantum Monte Carlo simulations have shown that the harmonic approximation works well up to the vicinity of the melting temperature (Abbar et al. 2015). Note that in a pure perfect lattice, only the inelastic part $S'(\omega, k)$ of the total structure factor $S(\omega, k) = S'(\omega, k) + S''(k)\delta(\omega)$ contributes to transport properties. The elastic term $S''(k)$ describes Bragg diffraction (zero-phonon process). It does not contribute to scattering, but it leads to a renormalization of the electron ground state (which are the Bloch waves) and the appearance of the electron band structure. Notice that the elastic component is automatically taken out by $\langle \hat{n}_i(\mathbf{x}, t) \rangle_{eq}$ in Eq. (9.56) (Baym 1964; Rosenfeld and Stott 1990). The Bragg elastic contribution to an unmodified density (charge) correlator $\langle \hat{n}^\dagger \hat{n} \rangle$ is

$$S''(\mathbf{k}) = e^{-2W(k)} (2\pi)^3 n_i \sum_{\mathbf{G}} \delta(\mathbf{k} - \mathbf{G}), \qquad (9.59)$$

where the summation is taken over the reciprocal lattice vectors $\mathbf{G}$ and the exponent $W(k)$ is the Debye-Waller factor (Harrison 1980), which describes thermal damping of the Bragg peaks. In addition, Baiko et al. (1998) have proposed that in the liquid regime, sufficiently close to the melting point, an incipient long-range order exists, which is preserved during the typical electron scattering time. Solid-like features such as a shear mode are observed in a strongly coupled system in the liquid regime both in numerical experiments and in laboratory. Thus, Baiko et al. (1998) suggested that the electrons obey the local band structure which is preserved during the electron relaxation. As a consequence, in order to account for this ion local ordering in the electron transport, they proposed to subtract an 'elastic' contribution given by Eq. (9.59) averaged over the orientations of $\mathbf{k}$ from the total liquid structure factor. This procedure removes the large jumps of the Coulomb logarithm and hence of the transport coefficients at the melting point. This prescription allowed Potekhin et al. (1999) and Chugunov and Yakovlev (2005) to construct a single fit for Λ_{ei} valid in both liquid and solid regimes. An interesting feature of the approach by Potekhin et al. (1999) is that they do not fit the numerical results for the Coulomb logarithms. Instead, they introduce a fitting expression for the effective potential which encapsulates the contributions from non-Born terms, electron screening, ion correlations, the Debye-Waller factor, and the structure factor. The Coulomb logarithms are then found by analytical integration in Eq. (9.52).[4]

This approach is attractive but it was criticized in Itoh et al. (2008) and Daligault and Gupta (2009). The main argument is that in the simple terrestrial metals the jump in resistivity at the melting point is a well-established indication of a solid-liquid transition (e.g., Schaeffer et al. 2012). It seems that a convincing way to describe electron transport in the disordered state of the strongly coupled Coulomb melt is missing. It is, in principle, possible to extract the behavior of the crustal thermal conductivity from studies of the crustal cooling in X-ray transients after the outburst stages (Brown and Cumming 2009; Page and Reddy 2013; Meisel et al. 2018). However, in this case, effects related to the multi-component composition of the accreted crust will probably dominate (Mckinven et al. 2016).

9.3.1.2 Impurities and Mixtures

The crustal lattice is not expected to be strictly perfect. Like terrestrial crystalline solids, it can possess various defects, which are jointly called impurities. One usually considers impurities in the form of charge fluctuations and introduces the impurity parameter

$$Q = \sum_j Y_j (Z_j - \langle Z \rangle)^2, \tag{9.60}$$

where the summation is taken over the different ion species, Y_j and Z_j are number fraction and charge number of each species, respectively, and $\langle Z \rangle$ is the mean charge. If the impurities are relatively rare and weakly correlated, electron-phonon interactions and electron-impurity scatterings can be considered as different transport relaxation mechanisms. Employing Matthiessen's rule, the total electron-ion collision frequency is expressed as $\nu_{ei} = \nu_{e-ph} + \nu_{e-imp}$. The electron-impurity effective collision frequency ν_{e-imp} is calculated form Eq. (9.51) by substituting $Z^2 \rightarrow Q$ and using the Coulomb logarithm from Eq. (9.52) with the elastic structure factor $S(k) = 1$. Since the elastic scattering is temperature-independent, it limits the collision frequencies at low temperatures. In the simplest model of Debye screening,

---

[4]This fit has also been applied to transport coefficients in a liquid at $T \lesssim T_{pi}$, where quantum effects become important. It is supposed (Potekhin et al. 1999; Chugunov and Yakovlev 2005) to give a more reliable estimate than the use of direct numerical calculations based on the classical structure factors. This is reasonable since a unified analytical expression in both liquid and solid phase is used and in the latter phase quantum effects are properly included, see Potekhin et al. (1999) for a detailed argumentation. Robust results for transport coefficients in the quantum liquid domain are not present in the literature up to our knowledge since the structure factors in the quantum liquid regime are unknown.

$U(k) \propto (k^2 + k_D^2)^{-1}$, and the integration in Eq. (9.52) gives

$$\Lambda_{imp}^{\kappa,\sigma} = \frac{1}{2}\left[1 + 4\beta_r^2 \xi_S^2\right] \ln\left(1 + \xi_S^{-2}\right) - \frac{\beta_r^2}{2} - \frac{1 + \beta_r^2 \xi_S^2}{2 + 2\xi_S^2}, \tag{9.61a}$$

$$\Lambda_{imp}^{\eta} = \frac{3}{2}\left[1 + 3\beta_r^2 \xi_S^4 + 2\xi_S^2(1 + \beta_r^2)\right] \ln\left(1 + \xi_S^{-2}\right) - \frac{9}{2}\beta_r^2 \xi_S^2 - \frac{3}{4}\beta_r^2 - 3, \tag{9.61b}$$

where $\xi_S = k_D/(2k_{Fe})$. The screening wavenumber k_D in principle acquires contributions from Thomas-Fermi screening of degenerate electrons and impurity screening, $k_D^2 = k_{TF}^2 + k_{imp}^2$, however k_{imp} can usually be neglected (e.g., Chugunov and Yakovlev 2005).

In the opposite case, when no crystal is formed in a multi-component plasma (in a liquid, or in a glassy solid), the so-called plasma additivity rule can be used (Potekhin et al. 1999), and $Z^2 n_i \Lambda_{ei}$ is replaced by $\sum_j Z_j^2 n_j \Lambda_{ei}^j$, where Λ_{ei}^j is the Coulomb logarithm for scattering off the ion species j. A modification of this rule was proposed by Daligault and Gupta (2009) based on large scale molecular dynamical simulations. They suggest that it is more accurate to use $\langle Z \rangle^{1/3} Z_j^{5/3}$ instead of Z_j^2.

The intermediate case is more complicated. Molecular dynamics simulations strongly suggest that the crystallization of the multi-component Coulomb plasma occurs even in the case of large impurity parameter Q (Horowitz et al. 2009; Horowitz and Berry 2009). An amorphous crust structure was also proposed, see for instance Daligault and Gupta (2009). Some studies show that the diffusion in the solid phase is relatively rapid and quickly relaxes amorphous structures to a regular lattice (Hughto et al. 2011). In addition, an amorphous crustal structure is in contradiction with observations (Shternin et al. 2007; Brown and Cumming 2009). Already in the case of a moderate impurity parameter, $Q \sim 1$, the simple prescription of electron scattering as a sum of phonon contribution and uncorrelated impurity scattering is questionable. In fact, all information about electron-ion scattering (from lattice vibrations or impurities) is encoded in the structure factor, which naturally takes into account correlations in the minority species on the same footing as the correlations in the majority species. The structure factor of a multi-component solid can be obtained from numerical simulations. To calculate the transport properties it is necessary to correctly separate the Bragg contribution, which does not contribute to scattering, from the total structure factor. This is not as simple as in case of one-component plasma (Horowitz et al. 2009). The remaining part of the structure factor is then used to calculate the Coulomb logarithms. As a result, both classical molecular dynamics simulations (Horowitz et al. 2009; Horowitz and Berry 2009) and recent quantum path integral Monte Carlo approach (Abbar et al. 2015; Roggero and Reddy 2016) show that the simple impurity expression based on the parameter Q underestimates the Coulomb logarithm and hence overestimates the corresponding values of transport coefficients. Moreover, Roggero and Reddy (2016) found that their results for a broad range of Q can be approximated by the standard lattice + impurity formalism, where the effective impurity parameter $\tilde{Q} = L(\Gamma)Q$ is used.[5] The factor $L(\Gamma)$ is generally larger than one and increases with Γ. Roggero and Reddy (2016) find $L(\Gamma) \approx 2 - 4$ for the conditions they consider. Note that classical simulations can treat only the high-temperature case $T > T_{pi}$, while the quantum simulations of Roggero and Reddy (2016) were the first to investigate the multi-component solid for $T < T_{pi}$, where the dynamical effects in Eq. (9.52) are important.

9.3.1.3 Other Processes

Let us briefly describe other processes which contribute to transport in neutron star crusts. Electrons in the crust can scatter off electrons, not only off ions. For degenerate electrons, Matthiessen's rule is a good approximation, and the electron-electron collision frequency ν_{ee} is simply added to the electron-ion collision frequency ν_{ei}. The impact of the contribution from electron-electron scattering on thermal conductivity κ and shear viscosity η was analyzed in Shternin and Yakovlev (2006) and Shternin (2008a). Note that in this approximation electron-electron scattering does not change the charge current and therefore does not contribute to the electrical conductivity.[6]

In most part of the neutron star crust, electrons are relativistic and their collisions are mediated by the current-current (magnetic) interaction, in contrast to the electron-ion Coulomb interaction. The current-current interaction occurs through exchange of transverse plasmons, which leads to a peculiar temperature and density dependence of the transport coefficients, as we describe in detail in the context of lepton and quark transport in the core of the star, see Sects. 9.4.1.2 and 9.5.5.1. However, except for a very low-temperature, pure one-component plasma, the electron-electron collisions are found to be unimportant. (They can be important in a low-Z plasma, i.e., in white dwarfs and degenerate cores of the red giants. In fact, the correct inclusion of the electron-electron collisions have important consequences for the position of the red giant branch tip in the Hertzsprung-Russell diagram (Cassisi et al. 2007).)

Ions in the liquid phase (or phonons in the crystalline solid phase) can also contribute to transport properties of neutron star crusts. The ion contribution to shear viscosity was considered by Caballero et al. (2008) and is found to be negligible. A similar conclusion for the thermal conductivity was reached by Chugunov and Haensel (2007), see also Pérez-Azorín et al. (2006). However, in a certain parameter region, the ion contribution can be significant in the magnetized crust for the heat

---

[5] An appropriate average of individual species Γ's calculated from the first equality in Eq. (9.48) is used as the mixture Γ parameter.

[6] This is not the case in the non-degenerate plasma, where Matthiessen's rule does not hold, and both *ee* and *ei* collisions need to be considered on the right-hand side of the Boltzmann equation. The impact of *ee* collisions is then especially pronounced at small Z (Braginskii 1958).

conduction across the field lines. Still, simulations suggest that its importance is limited also in this case, see for example Potekhin et al. (2015b).

9.3.1.4   Inner Crust: Free Neutron Transport

In the inner crust of a neutron star the density becomes sufficiently high for neutrons to detach from nuclei. The structure of the inner crust then consists of a lattice of nuclear clusters (where charged protons are localized) alongside with the gas of unbound (or 'free') neutrons, see, e.g., Chamel and Haensel (2008). In addition, the neutrons are believed to form Cooper pairs in the $^1S_0$ channel. The charge distribution in the nuclear clusters in the inner crust differs from the point-like nuclei in the outer crust. This is taken into account by introducing nuclear form factors in the electron-nuclei scattering potential. These corrections have been included by Gnedin et al. (2001) for Coulomb logarithms relevant to thermal and electrical conductivities and by Chugunov and Yakovlev (2005) for shear viscosity. The finite size of the charge distribution generally reduces the collision frequencies and hence increases the values of electron transport coefficients.

In the presence of a large amount of free neutrons, electron-neutron scattering can become important. The relativistic electrons interact with the neutron spins (magnetic moments). This contribution was analyzed by Flowers and Itoh (1976). Recently, an induced interaction between electrons and neutrons was proposed (Bertoni et al. 2015), which can be effectively understood as occurring via exchange of lattice phonons. However, Bertoni et al. (2015) found that the contribution from this interaction is never relevant when calculating kinetic coefficients in the inner crust. In contrast, a similar interaction can be important in the core (see Sect. 9.4.1.4). If neutrons are superfluid, both these contributions are further suppressed.

Since the gas of unbound neutrons is present in the inner crust, they can also contribute themselves to the transport properties. For instance, the thermal conductivity becomes a sum of electron and neutron contributions, $\kappa = \kappa_e + \kappa_n$. The neutron contribution for normal neutrons was discussed by Bisnovatyi-Kogan and Romanova (1982) and more recently by Deibel et al. (2017), and it was found to be negligible compared to the electron contribution, except probably the region near the crust-core boundary (Deibel et al. 2017). We are not aware of any calculations for the shear viscosity of the neutron fluid in the inner crust. The potential importance of the free neutron transport is further reduced if one takes into account that the unbound neutrons move in the periodic potential of the nuclear lattice, hence their spectrum shows a band structure. Chamel (2012) has argued that due to Bragg scattering of neutrons the actual density of conducting neutrons that participate in transport is much smaller than the total density of unbound neutrons, which further reduces the role of neutrons.

When neutrons are superfluid, a collective superfluid mode ('superfluid phonons') can contribute to transport, as explained in Sect. 9.2.3.4. Initial estimates suggested that the superfluid phonon contribution to the thermal conductivity can be important in magnetized stars (Aguilera et al. 2009). However, more detailed considerations which include the neutron band structure have shown that this contribution is always less than the contribution of lattice phonons (Chamel et al. 2013, 2016). We will come back to collective modes in the discussion of the core, see Sects. 9.4.1.4 and 9.4.2.3 for superfluid phonons in nuclear matter, and Sects. 9.5.4.2 and 9.5.5.2 for superfluid phonons in quark matter.

9.3.1.5 Transport in a Magnetic Field

The magnetic field $\boldsymbol{B}$ in the crust modifies the motion of charged particles in the directions perpendicular to the direction of the magnetic field $\boldsymbol{b} \equiv \boldsymbol{B}/B$. It can be strong enough to have an influence on the transport properties. Electrons are light and thus lower fields affect their transport (compared to the fields needed to affect ions). We start from the situation where the electron motion across the magnetic field is not quantized. In this case, magnetic field effects are characterized by the Hall magnetization parameter

$$\omega_g \tau = 1760 \frac{B_{12}}{\sqrt{1+x_r^2}} \frac{\tau}{10^{-16}\,\text{s}}, \tag{9.62}$$

where τ is a characteristic relaxation time, $B_{12} \equiv B/(10^{12}\,\text{G})$, and $\omega_g = |e|B/(m_e^*c)$ is the electron gyrofrequency, which is related to the electron cyclotron frequency $\omega_c = |e|B/(m_e c)$ by $\omega_g = \omega_c/\sqrt{1+x_r^2}$. If $\omega_g \tau \gtrsim 1$, the electron transport becomes anisotropic. Let us first consider the electrical and thermal conductivities (the perturbation of multiplicity $l = 1$, see Sect. 9.2.2). The general expressions for the currents (9.23) are modified such that the kinetic coefficients become tensors instead of scalars κ, σ, $Q_T \rightarrow \hat{\kappa}$, $\hat{\sigma}$, $\hat{Q}_T$. Accordingly, one introduces the effective relaxation time tensors $\hat{\tau}^{\sigma,\kappa}$ via Eq. (9.46). The symmetry relations for the kinetic coefficients in isotropic media suggest that these tensors have only three independent components. If one aligns the z-axis along $\boldsymbol{b}$, these are longitudinal $\hat{\tau}_{zz} \equiv \tau_\parallel$, transverse $\hat{\tau}_{yy} = \hat{\tau}_{xx} \equiv \tau_\perp$, and Hall terms $\hat{\tau}_{xy} = -\hat{\tau}_{yx} \equiv \tau_\Lambda$. These tensors are found from the solution of the linearized Boltzmann equation in an external magnetic field. The procedure is similar as described in Sect. 9.2.2. However, now the force $\boldsymbol{R}$ contains the magnetic field contribution, and the term $\frac{e}{c}\boldsymbol{w} \times \boldsymbol{B}\frac{\partial \delta f}{\partial \boldsymbol{p}}$ must be retained in Eq. (9.18) (Ziman 2001). One usually adopts the relaxation time approximation (9.32), where the relaxation time $\tau(\varepsilon)$ is taken from the non-magnetic problem. In this approximation, the solution to the linearized Boltzmann equation gives

$$\tau_\parallel = \tau, \qquad \tau_\perp = \frac{\tau}{1+(\omega_g \tau)^2}, \qquad \tau_\Lambda = \frac{\omega_g \tau^2}{1+(\omega_g \tau)^2}. \tag{9.63}$$

In fact, an averaging of these relaxation times should be performed following Eq. (9.30). However, in degenerate matter it is sufficient to set $\tau = \tau(\mu)$, like in the non-magnetized case. In the limit of weak magnetization, $\omega_g \tau \ll 1$, one has $\tau_{\|} = \tau_{\perp} = \tau$ and $\tau_{\wedge} = 0$. In the opposite case of a large Hall magnetization parameter, the electron transport across the magnetic field becomes strongly suppressed and ion or neutron contributions can become important.

If the magnetic field is sufficiently strong, the quantization of the transverse electron motion can no longer be neglected. This happens when $\hbar\omega_g \gtrsim k_B T$. The electrons then occupy several Landau levels (weakly quantizing field) or only the lowest Landau level (strongly quantizing field). In either case, the magnetic field also modifies the thermodynamic properties of the system. Transport along and across the magnetic field must be considered separately. The thermal and electrical conductivities in a quantizing magnetic field in different regimes were investigated by many authors (Kaminker and Yakovlev 1981; Yakovlev 1984; Hernquist 1984; Potekhin 1996; Potekhin and Yakovlev 1996). The results for both quantizing and non-quantizing fields in the relaxation time approximation were reconsidered and summarized by Potekhin (1999). He suggested that in the case of strongly degenerate electrons the form of Eq. (9.63) holds, but two different relaxation times τ must be used in the expressions for the parallel component $\tau_{\|}$ and the transverse components $\tau_{\perp}$ and $\tau_{\wedge}$. In the weakly quantizing limit, these two relaxation times oscillate around τ, approaching it in the non-quantizing limit. Based on the model of the effective electron-ion potential (Potekhin et al. 1999) (see Sect. 9.3.1.1), Potekhin (1999) constructed useful fitting expressions to calculate Coulomb logarithms appropriate for thermal and electrical conductivities of magnetized electrons in both quantizing and non-quantizing limits in liquid or solid neutron star crusts, as long as quantum effects on the ion motion can be ignored, i.e., at $T \gtrsim T_{pi}$. By construction, these expressions provide transport coefficients which behave smoothly across the liquid-solid phase transition (recall the discussion in Sect. 9.3.1.1).

Recently, finite-temperature effects on the electrical conductivity of warm magnetized matter in the neutron star crust were discussed by Harutyunyan and Sedrakian (2016). These authors used the relaxation time approximation, but included also the transverse plasmon exchange channel when calculating the electron-ion transport cross-section. This channel was found to be suppressed by a small factor $k_B T / (m_i c^2)$ and does not contribute to the relaxation time.

All results for transport coefficients in magnetized matter described above were based on various sorts of the relaxation time approximation. This approach is justified if the scattering probability does not depend on $\boldsymbol{B}$ and if the scattering is elastic. Both these approximations fail in general at low temperatures, when the crust is solid (e.g., Baiko 2016; Chugunov and Haensel 2007). In this case, a more general expression for the collision integral must be used, and the solution of the Boltzmann equation becomes more complicated. Unfortunately, the construction of the variational principle in the magnetized case is challenging (Ziman 2001). In the standard approaches, the solution of the Boltzmann equation corresponds only to a stationary point of the variational functional among the class of the trial functions, not to its maximum.[7] However, for degenerate matter, relying on the experience from the non-magnetized case, one expects the lowest-order expansion of the deviation function Φ to give appropriate results. Based on this expectation, Baiko (2016) studied electron electrical and thermal conductivities in the magnetized, solid crust employing the Ziman (2001) approach. He used the one-phonon approximation for the electron-lattice interaction and took into account the phonon spectra distortion due to the magnetic field. The magnetic field leads to the appearance of a soft phonon mode, with quadratic dispersion at small wavenumbers. This mode is easier to excite than the usual non-magnetized acoustic phonon, therefore the electrical and thermal resistivities increase. Employing the lowest order of the variational method, and aligning the magnetic field along one of the symmetry axes of the crystal, Baiko (2016) found that the thermal and electrical conductivity tensors are expressed via effective relaxation times as in Eq. (9.63), but like for a quantizing magnetic field, two different relaxation times enter the longitudinal and transverse parts. The difference between these effective relaxation times increases with magnetic field. At low temperatures, $T \lesssim T_{\mathrm{pi}}$, both relaxation times are appreciably larger than in the field-free case. The results of Baiko (2016) are strictly valid in the non-quantizing case. In this case, phonons are weakly magnetized. However, the results are also relevant for weakly quantized fields, when electrons populate several Landau levels, and yield estimates of the transport coefficients averaged over the quantum oscillations. The most relevant case of highly magnetized phonons, where the influence of the magnetic field on κ and σ is largest, corresponds to the strongly quantizing magnetic field, where electrons populate only the lowest Landau level and the approach used by Baiko (2016) is inappropriate. An accurate analysis of the transport properties of quantized electrons in strongly magnetized Coulomb crystals has yet do be done.

The effects of a magnetic field on the shear viscosity of the crust has not received as much attention as the thermal and electrical conductivities. The electron shear viscosity was considered by Ofengeim and Yakovlev (2015) for the non-quantizing magnetic field, taking into account only electron-ion collisions in the relaxation-time approximation. In an anisotropic medium, where the anisotropy is for instance caused by an external magnetic field, the viscous stress tensor contains the fourth-rank tensor $\eta_{\alpha\beta\gamma\delta}$ instead of the scalar coefficient η. Symmetry constraints leave five independent shear viscosity coefficients $\eta_0 \ldots \eta_4$ (Pitaevskii and Lifshitz 2008). For degenerate electrons, the expressions for the five shear viscosity coefficients in the relaxation time approximation are rather simple (Ofengeim and Yakovlev 2015). The coefficients η_0, η_2, and η_4 are given by Eq. (9.46d) where $\tau_\parallel$, $\tau_\perp$, and τ_Λ from Eq. (9.63) are used, respectively. The coefficient η_0 is independent of B and can be called longitudinal viscosity in analogy to longitudinal conductivities. The two remaining coefficients can be found from the relations $\eta_1(B) = \eta_2(2B)$ and $\eta_3(B) = \eta_4(2B)$. The 'Hall' viscosity coefficients η_3 and η_4 do not enter

---

[7]A promising variant of the variational principle was recently suggested by Reinholz and Röpke (2012), where the positive-definite variational functional was proposed.

the expression for the energy dissipation rate. More accurate calculations of the shear viscosity of the magnetized neutron star crust should deal with the various effects outlined in the previous discussion on conductivities. This remains for future studies.

9.3.2 Transport in the Pasta Phase

As the density in the inner crust increases, the size of the nuclei—or better: the size of the nuclear clusters—increases until the clusters start to overlap. The density at the crust-core interface ρ_{cc}, where nuclei are fully dissolved in uniform nuclear matter, is about $\rho_{cc} \approx \rho_0/2 = 1.4 \times 10^{14}$ g cm^{-3}, where ρ_0 is the mass density at nuclear saturation. It is now generally believed that the transition region hosts several phases that are characterized by peculiar shapes of the nuclear clusters, reminiscent of various shapes of pasta. Hence the term 'nuclear pasta' for these phases. Loosely speaking, when the spherical nuclear clusters start to touch, as a result of the competition between the nuclear attraction and Coulomb repulsion of protons, it may become energetically favorable for them to rearrange and form elongated structures like rods, or two-dimensional slabs. This was first pointed out by Ravenhall et al. (1983) and Hashimoto et al. (1984). In a simple picture, five subsequent phases appear as we increase density, i.e., as we move from the crust into the core of the star: first, usual large spherically shaped clusters ('gnocchi'), then cylindrical rods ('spaghetti'), then plane-parallel slabs ('lasagna'), followed by the inverted phases, with rod-like voids, then spherical voids ('anti-gnocchi' or 'swiss cheese') in nuclear matter. At high temperature, the pasta is in the liquid state, but at low temperatures it is thought to freeze in ordered or disordered structures, for example in a regular lattice of slabs. The pasta region is estimated to exist between densities of about 10^{14} g cm^{-3} and ρ_{cc}, being about 100 m thick; the total mass of the pasta layer can be as large as the mass of the rest of the crust. The appearance of the pasta phases in simulations depends on the details of the interaction and implementation, and there are models that predict less pasta phases, mixtures of different phases, or do not predict the pasta phases at all (Douchin and Haensel 2000; Oyamatsu and Iida 2007). In modern models, where large-scale simulations are employed, there is a rich variety of possibilities for pasta phases, see for instance Alcain et al. (2014), Schneider et al. (2014), Horowitz et al. (2015), Berry et al. (2016), and Caplan and Horowitz (2017) and Yakovlev (2015) for more detailed reviews.

The complexity of the nuclear pasta naturally suggests that its transport properties can be very different from the rest of the crust. The main contribution to the conductivities and shear viscosity comes from electrons which now scatter off the non-trivial charge density fluctuations of the pasta phase. In applications, it is not unreasonable to treat the transport coefficients for the pasta phase as phenomenological quantities. In analogy to the treatment of disorder in the crust, one can introduce an effective impurity parameter $\tilde{Q}$ to parametrize the transport coefficients. For instance, assuming that the pasta layer has much lower electrical conductivity (with $\tilde{\mathcal{Q}} \approx 100$) than the rest of the crust, Pons et al. (2013) were able to explain the existence of the maximal spin period of X-ray pulsars (see also Viganò et al. (2013) for the effect of the resistive pasta layer on the magnetic field evolution of isolated neutron stars). In a similar way, assuming that the pasta is a thermal insulator with $\tilde{\mathcal{Q}} \approx 40$, Horowitz et al. (2015) were able to explain the late-time crustal cooling in the quasi-persistent X-ray transient MXB 1659–29 (Cackett et al. 2013). Notice that the effective impurity parameter, of course, does not have to be the same when different transport coefficients κ, σ, or η are considered.

Horowitz and Berry (2008) computed shear viscosity and thermal conductivity of the pasta phase based on classical molecular dynamics simulations. The electrons scatter off the charged protons, whose correlated dynamics is described by the proton structure factor S_p. Thus, one can use the expressions given in the previous section for electron-ion scattering with S_p replacing S, and using the proton charge $Z = 1$. As pointed out by Horowitz and Berry (2008), this approach applies also if nuclei form spherical clusters (ions) of a charge Z, which will be reflected in the proton structure factor. The results show that the transport coefficients obtained in this way do not change dramatically when non-spherical pasta phases are considered. In fact, Horowitz and Berry (2008) obtained the same order-of-the magnitude values as can be inferred from Chugunov and Yakovlev (2005), where spherical nuclei were considered in the same density range. Since classical molecular dynamics simulations are used, these results are applicable only for high temperatures. Horowitz and Berry (2008) set $T = 1$ MeV and the proton fraction $Y_p = 0.2$. These values do not apply directly to neutron star crusts, and the authors discuss how smaller proton fractions and smaller temperatures might modify their conclusions. Similar conclusions were reached recently by Nandi and Schramm (2018) based on quantum molecular dynamics simulations for a wider range of parameters than in Horowitz and Berry (2008).

Horowitz and Berry (2008) and Nandi and Schramm (2018) used the expression (9.52) for calculating the Coulomb logarithm. This expression is based on the angular-averaged structure factor $S_p(q)$, which assumes isotropic, or nearly isotropic, scattering. It is clear that this is not the case in nuclear pasta. One can imagine that electron scattering should be much stronger in directions across the pasta clusters than in directions along them. This is indeed reflected in the strong dependence of the structure factor $S_p(\boldsymbol{q})$ on the direction of the vector $\boldsymbol{q}$ (Schneider et al. 2014, 2016). Transport in nuclear pasta is essentially anisotropic, and the transport theory of anisotropic solids must be applied. The solution of the Boltzmann equation becomes more complicated since the collision integral involves anisotropic scatterings. The transport coefficients in anisotropic materials become tensor quantities, just like for the magnetized case considered above, where the anisotropy (gyrotropy) was induced by the magnetic field.

If the scattering is still elastic, but anisotropic, the relaxation-time approximation (9.32) generalizes to

$$I_e = - \sum_{lml'm'} \delta f^{lm}(\varepsilon) \left[\hat{\nu}_e(\varepsilon) \right]_{lm}^{l'm'} Y_{l'm'}(\Omega_p), \tag{9.64}$$

where $\hat{\nu}_e(\varepsilon)$ is the inverse relaxation time (collision frequency) matrix. In the isotropic case, one has $\left[\hat{\nu}_e(\varepsilon) \right]_{lm}^{l'm'} = \left[\tau_e^l(\varepsilon) \right]^{-1} \delta_{ll'} \delta_{mm'}$, and we recover Eq. (9.32). In principle, the expression for the matrix elements $\left[\hat{\nu}_e(\varepsilon) \right]_{lm}^{l'm'}$ can be expressed in integral form employing the proton structure factor $S_p(q)$ in a similar way to Eqs. (9.51)–(9.52).

An essential property of the general anisotropic case is that the perturbations of different multiplicities l can mix. It is customary to assume, however, that this mixing is small and can be neglected, so that $l = l'$ in Eq. (9.64), see for instance Askerov (1981). Yakovlev (2015) employed this approximation and considered electrical and thermal conductivities (i.e., $l = 1$) of the anisotropic pasta, including also a magnetic field, and assuming that the pasta phase has a symmetry axis (not necessarily aligned with the magnetic field). The $l = 1$ perturbation of the distribution function can be written as $\Phi = -\boldsymbol{w} \cdot \boldsymbol{\vartheta}$, where the vector $\boldsymbol{\vartheta}$ has to be determined. If we orient the z-axis of the laboratory system along the pasta symmetry axis, the generalized relaxation time approximation can be written as (Yakovlev 2015)

$$I_e = -\frac{\partial f^{(0)}}{\partial \varepsilon} \left[\nu_a(\varepsilon) w_z \vartheta_z + \nu_p(\varepsilon) \boldsymbol{w}_p \cdot \boldsymbol{\vartheta}_p \right], \tag{9.65}$$

where the collision frequencies ν_a and ν_p describe relaxation along and across the symmetry axis, respectively, where $\boldsymbol{w}_p$ is the electron velocity component transverse to the symmetry axis, and $\boldsymbol{\vartheta}_p$ is the corresponding component of $\boldsymbol{\vartheta}$. Using this expression for the collision integral, one solves the Boltzmann equation containing electric field, temperature gradient, and external magnetic field to find the vector $\boldsymbol{\vartheta}$. Then, the thermal and electrical conductivity tensors (and the thermopower) can be found from the expressions for the currents. They remain tensor quantities even in the absence of an external magnetic field. Since the relaxation time approximation is used, the thermal and electrical conductivity tensors are still related via the Wiedemann-Franz law (9.47). Yakovlev (2015) discussed the general structure of the solutions and their qualitative properties for the case where one of the principal collision frequencies is much larger than the other, say $\nu_p \gg \nu_a$. That means that the heat or charge transport is much more efficient along the symmetry direction of the pasta phase than across it. The net effect of this anisotropy on the transport in the inner crust of the neutron star will depend on the predominant orientation of the nuclear clusters (they can be aligned with the radius, be predominantly perpendicular, or form a disordered domain-like structure). We refer the reader to the original work (Yakovlev 2015) for a discussion of the rich variety of possibilities.

Microscopic calculations of the relaxation time tensor for nuclear pasta remain a task for future studies. Schneider et al. (2016) made a step towards this goal by running a large classical molecular dynamics simulation. They find a pasta slab phase with a number of topological defects and calculate the static proton structure factor $S_p(q)$, including the full angular dependence. While they do not present a full calculation of the transport properties, they perform simple estimates for the angular dependence of the kinetic coefficients. They find that the relaxation along two symmetry axes can differ by an order of magnitude, thus supporting the assumptions of Yakovlev (2015). As before, the molecular dynamics simulations were performed at high temperatures and proton fractions and thus cannot be directly applied to the neutron star crust. However, they found that topological defects present in the pasta decrease the values of transport coefficients, and this decrease can be described by the effective impurity parameter $\tilde{Q} \sim 30$, a value having the same order of magnitude as inferred from astrophysical observations (Pons et al. 2013; Horowitz et al. 2015). This suggests that detailed investigations of the transport properties in the pasta phase along these lines are promising directions for the future.

9.4 Transport in the Core: Hadronic Matter

At densities above ρ_{cc}, the nuclear clusters dissolve completely and the matter in neutron stars is uniform and neutron-rich. The simplest composition is $npe\mu$ matter, where muons (μ) appear when the difference between neutron and proton chemical potentials becomes larger than the muon mass, which occurs at densities around ρ_0. The matter is usually thought to be in (or close to) equilibrium with respect to weak processes. The condition of beta-decay and the inverse process of lepton capture to proceed at the same rate then imposes the following relation between the chemical potentials,

$$\mu_n = \mu_p + \mu_\ell, \tag{9.66}$$

where ℓ stands for electrons or muons, such that $\mu_e = \mu_\mu$. We have omitted the neutrino chemical potential μ_ν because at typical neutron star temperatures neutrinos leave the system once they are created (exceptions are the hot cores of proto-neutron stars, binary neutron star mergers, and supernovae interiors, where neutrinos can be trapped and thus $\mu_\nu \neq 0$). This beta-equilibrated matter, together with the condition of electric charge neutrality, is highly asymmetric, or neutron-rich: the typical proton fraction in neutron star cores is $x_p \lesssim 15\%$ (e.g., Haensel et al. 2007). Electrons and muons form almost ideal degenerate gases (electrons are ultra-relativistic, while muons become relativistic soon after their threshold). The nucleons, however, form a highly non-ideal, strongly-interacting liquid, where nuclear many-body effects are of utmost importance. In this Sect. we discuss the transport properties of this high-density nuclear matter, starting from the non-superfluid case, and including effects of superfluidity later on. We will also briefly discuss some of the effects of hyperons, in particular in Sect. 9.4.3, where we address the bulk viscosity of hadronic matter. At even larger densities, it is conceivable that a transition to deconfined quark matter occurs. The transport properties of various possible phases of quark matter are discussed separately in Sect. 9.5.

9.4.1 Shear Viscosity, Thermal and Electrical Conductivity

9.4.1.1 General Formalism

Transport coefficients of nuclear matter in neutron star cores are calculated within the transport theory for Landau Fermi liquids, outlined in Sect. 9.2.3.2, adapted to multi-component systems. The response to external perturbations is described by a system of Landau transport equations for quasiparticles (9.37), whose solution, as discussed in Sect. 9.2.3.2, is equivalent to the solution of the system of linearized Boltzmann equations for the transport coefficients we are interested in. For a given quasiparticle species 'c', the collision term of the linearized Boltzmann equation $I_c = \sum_i I_{ci}$ contains a sum of collision integrals for collisions with other species 'i', each of the form (9.31), with binary transition probabilities W_{ci}. Depending on the antisymmetrization of the particle states in the calculation of W_{ci}, the symmetry factor $(1 + \delta_{ci})^{-1}$ must be included in (9.31) to avoid double counting of collisions within the same particle species.

Quasiparticle scattering occurs within the thermal width of the Fermi surface, and thus the typical energy transfer in the collision event is of the order of temperature. Therefore, the collisions cannot be considered elastic, and the relaxation time approximation—frequently used in the previous section, where electron-ion collisions were considered—is generally not applicable. The system of transport equations must be solved retaining the full form of the collision integrals on the right-hand side of the Boltzmann equation, for instance with variational methods.

Some general properties of the transport coefficients can be deduced immediately from Eq. (9.31). Due to the strong degeneracy, the Pauli blocking factors $f^{(0)} f_1^{(0)} (1 - f'^{(0)})(1 - f_1'^{(0)})$ effectively place all quasiparticles on their respective Fermi surfaces, and for each momentum integration in Eq. (9.31) we can write

$$d^3 p \approx p_F m^* d\varepsilon \, d\Omega_p \,, \tag{9.67}$$

where the change from momentum to energy integration has produced the density of states on the Fermi surface $\propto p_F^2/v_F = m^* p_F$. It is customary to describe the deviation function $\Phi(\varepsilon)$ in terms of a series expansion over the dimensionless excitation energy $x = (\varepsilon - \mu)/(k_B T)$. Moreover, in traditional Fermi liquids, the rate W_{ci} is considered to be independent of the energy transfer in the collisions. This leads to a T^2 behavior of the collision integral (9.31) irrespective of the details of collisions, which only reorient the quasiparticle momenta, leaving their absolute values intact.

In the simplest variational solution, the deviation functions are assumed to have the form (9.25) or (9.28), appropriate for the perturbation in question, with a constant effective relaxation time (9.33) for each quasiparticle species. Then the Boltzmann equation reduces to a system of algebraic equation for the effective relaxation times

$$1 = \sum_i v_{ci}\tau_c + \sum_{i \neq c} v'_{ci}\tau_i, \tag{9.68}$$

where, according to the discussion above, the effective collision frequencies are $v_{ci} \propto T^2 m_c^* m_i^{*2} \langle W_{ci} \rangle_{\text{tr}}$, with a slightly different effective mass dependence for the mixing terms, $v'_{ci} \propto m_c^{*2} m_i^*$. Here, $\langle W_{ci} \rangle_{\text{tr}}$ is an effective transport scattering cross-section, which is the angular average of W_{ci} at the Fermi surface with appropriate kinematic factors [cf. Eqs. (9.51)–(9.54)]. Hence, the effective relaxation times $\tau_c \propto T^{-2}$, and this temperature dependence is reflected in the transport coefficients. Thus in a normal Fermi liquid one obtains

$$\eta \propto T^{-2}, \quad \sigma \propto T^{-2}, \quad \kappa \propto T^{-1}. \tag{9.69}$$

Notice that the effective collision frequencies are not the same for κ, η, and σ.

The exact result obeys the same general properties as the variational solution. The correction to the variational solution for any transport coefficient, say κ, can be written as $\kappa = C_\kappa \kappa_{\text{var}}$ where C_κ is a temperature-independent correction factor. This factor is found from the solution of a system of dimensionless integral equations for $\Phi_c(x)$. For a one-component Fermi liquid, the exact solution was constructed in Brooker and Sykes (1968), Sykes and Brooker (1970), Højgård Jensen et al. (1968) in the form of a rapidly converging series, see Baym and Pethick (1991) for details. The integral equation for $\Phi_c(x)$ can be also solved numerically by iterative methods. In any case, the correction constants $C_{\kappa,\sigma,\eta}$ were found to be in the range 1–1.4, which is unimportant for practical purposes in astrophysical applications. An exact analytic expression for the spin response of the Fermi liquid in an external (oscillating) magnetic field was recently constructed from the transport equation by Pethick and Schwenk (2009). The exact solution of the transport equation was generalized to multi-component Fermi liquids in the neutron star context by Flowers and Itoh (1979) and then analyzed in a general form by Anderson et al. (1987). Owing to large uncertainties present in the various parameters describing the neutron star matter, the simplest variational result seems to be a sufficient approximation in all cases.

For $npe\mu$ matter in neutron star cores, the collision frequencies in Eq. (9.68) are determined by electromagnetic interactions between charged particles, and by strong interactions between baryons. It turns out that the lepton and nucleon subsystems in Eq. (9.68) decouple and can be considered separately (Flowers and Itoh 1979). Then, the thermal conductivity (or shear viscosity) can be written as

$\kappa = \kappa_{e\mu} + \kappa_{np}$. The situation is different for the electrical conductivity, which is relevant in the presence of a magnetic field, see Sect. 9.4.1.5.

9.4.1.2 Lepton Sector

The lepton (electron and muon) transport coefficients are mediated by the collisions within themselves and with charged protons, which now can be considered as passive scatterers. Since the electromagnetic collisions are long-range (and hence small-angle), the corresponding collision frequencies are determined by the character of plasma screening (see Sect. 9.2.3.1). Explicitly, the differential transition rate W_{ci} is proportional to the squared matrix element for electromagnetic interaction, which, in an isotropic plasma, can be written as

$$M_{ci} \propto \frac{J_1^{(0)} J_2^{(0)}}{q^2 + \Pi_l(\omega, q)} - \frac{\boldsymbol{J}_{1t} \cdot \boldsymbol{J}_{2t}}{q^2 - \omega^2 + \Pi_t(\omega, q)}, \tag{9.70}$$

where ω and q are the energy and momentum transferred in the collision, respectively, $J^{(0)}$ and $\boldsymbol{J}_t$ are time-like and transverse (with respect to $\boldsymbol{q}$) space-like components of the transition current, respectively, and Π_l and Π_t are the longitudinal and transverse polarization functions. In conditions present in neutron star cores, the long-wavelength $q \ll p_F$ and static $\omega v_F \ll q$ limits are appropriate since the transferred energy is of the order of the temperature, $\omega \sim T$. The first term in Eq. (9.70) corresponds to the electric (Coulomb) interaction, while the second term corresponds to the magnetic (Ampère) part of the interaction. The second term is essentially relativistic and is suppressed for non-relativistic particles by the ratio $J_t/J^{(0)} \propto u/c$. Therefore, the magnetic term is not that important in the crust (see Sect. 9.3.1), where the dominant contribution to transport coefficients comes from electron collisions with heavy non-relativistic ions, but it becomes significant in the core. In the context of plasma physics, the relativistic collision integral (9.31), taking into account longitudinal and transverse screening as in Eq. (9.70), was first derived by Silin (1961). Alternatively, the two terms in Eq. (9.70) can be viewed as resulting from interaction via longitudinal and transverse virtual plasmon exchange.

The dominance of the transverse plasmon exchange in the transport properties of relativistic plasmas was realized by Heiselberg et al. (1992) and worked out by Heiselberg and Pethick (1993) in the context of unpaired quark matter, see Sect. 9.5.5.1. The reason is as follows. To lowest order, the longitudinal screening is static, $\Pi_l = q_l^2$, where q_l^2 is the Thomas-Fermi screening wavenumber. Therefore, the dominant contribution to the part of the collision frequency that is mediated by the longitudinal interaction comes from $q \lesssim q_l$. In contrast, the transverse plasmon (photon) screening is essentially dynamical in the from of Landau damping, so that

$$\Pi_t = i \frac{\pi}{4} \frac{\omega}{qc} q_t^2, \tag{9.71}$$

where $q_t \sim q_l$ is a characteristic transverse wavenumber ($q_l = q_t$ if all charged particles are ultra-relativistic). Hence, the dominant contribution to the 'transverse' part of the collision frequency comes from $q \lesssim [\pi\omega/(4cq_t)]^{1/3}q_t \ll q_l$. The latter inequality is due to the low temperature ($\omega \sim T$) and has two important consequences. First, the transverse plasmon exchange dominates the collisions between the relativistic particles in a degenerate plasma, and second, the scattering probability depends on the energy transfer of the collision. As a consequence, the temperature behavior of lepton transport coefficients in neutron star cores is essentially non-Fermi liquid (in contrast to the general theory outlined in the previous section). The modification of the temperature behavior depends on the kinematics of the problem in question and on the relation between the 'longitudinal' and 'transverse' contributions. For the thermal conductivity, the effective collision frequency becomes $\nu_{ci} \propto T$ if transverse plasmon exchange fully dominates the interaction, while for shear viscosity and electrical conductivity in the same limit, $\nu_{ci} \propto T^{5/3}$ (Heiselberg and Pethick 1993). Note that the energy dependence of the scattering rate does not change the conclusion that the simple variational solution described by Eq. (9.68) remains a sufficient approximation. It can be shown that the correction to the variational solution for the transverse-dominated collisions does not exceed 10% (Shternin and Yakovlev 2007, 2008a).

The lepton transport coefficients for $npe\mu$ matter with correct account for the transverse plasmon exchange were analyzed in Shternin and Yakovlev (2007, 2008a), Shternin (2008b). The low-temperature result (when the transverse plasmon exchange dominates) for the thermal conductivity is (Shternin and Yakovlev 2007)

$$\kappa_{e\mu} = \kappa_e + \kappa_\mu = \frac{\pi^2}{54\zeta(3)} \frac{k_B c (p_{Fe}^2 + p_{F\mu}^2)}{\hbar^2 \alpha_f}$$

$$= 2.43 \times 10^{22} \left(\frac{n_B}{n_0}\right)^{2/3} \left(x_e^{2/3} + x_\mu^{2/3}\right) \frac{\text{erg}}{\text{cm s K}}, \qquad (9.72)$$

where $n_0 = 0.16 \text{ fm}^{-3}$ is the number density at nuclear saturation, n_B is the total baryon number density, x_e and x_μ are the electron and muon number density fractions, respectively. The expression for the electron and muon contributions to the shear viscosity (Shternin and Yakovlev 2008a) $\eta_{e\mu} = \eta_e + \eta_\mu$ is more cumbersome in analytical form, and we only give the numerical result

$$\eta_{e\mu} = 8.43 \times 10^{20} \left(\frac{n_B}{n_0}\right)^{14/9} \left(\frac{T}{10^8 \text{ K}}\right)^{-5/3} \frac{x_e^2 + x_\mu^2}{(x_e^{2/3} + x_\mu^{2/3} + x_p^{2/3})^{2/3}} \frac{\text{g}}{\text{cm s}}. \qquad (9.73)$$

We also give the expression for the electrical conductivity of non-magnetized $npe\mu$ matter (Shternin 2008b),

$$\sigma = 1.86 \times 10^{30} \left(\frac{n_B}{n_0}\right)^{8/9} \left(\frac{T}{10^8 \text{ K}}\right)^{-5/3} \frac{x_e^{1/3} + x_\mu^{1/3} + x_p^{1/3}}{(x_e^{2/3} + x_\mu^{2/3} + x_p^{2/3})^{2/3}} \text{ s}^{-1}. \qquad (9.74)$$

This result for the electrical conductivity is already the full result for $npe\mu$ matter (and thus we have not added the subscript '$e\mu$'), because the baryon sector (neutrons) does not contribute to the electrical conductivity in the non-magnetized case. The result (9.74) is of the same (very large) order of magnitude as the classical estimate of Baym et al. (1969), rendering Ohmic dissipation in neutron star cores insignificant. In magnetized matter, the situation changes dramatically, as we will discuss in Sect. 9.4.1.5.

The thermal conductivity is temperature-independent and depends only on the carrier Fermi momentum. The result (9.72) is valid for all practically relevant temperatures and densities in (non-superfluid) neutron star cores (Shternin and Yakovlev 2007). In contrast, the result (9.73) can significantly overestimate the shear viscosity since the dominance of transverse collisions is not always strict (especially at lower densities), for details see Shternin and Yakovlev (2008a) and Kolomeitsev and Voskresensky (2015). The same is true for the electrical conductivity in Eq. (9.74). A relatively compact expression obtained from a fit for $\eta_{e\mu}$ that is valid in a broad temperature and density range can be found in Kolomeitsev and Voskresensky (2015).

We conclude this subsection by noting the advantage of the lepton kinetic coefficients. Since they are mediated by electromagnetic collisions, the final analytical expressions can be used for any equation of state (since they depend only on the effective masses of charged particles and their Fermi momenta). In addition, they can easily be updated to include other charged particles acting as passive scatterers, for instance hyperons.

9.4.1.3 Baryon Sector

The nucleon transport coefficients in $npe\mu$ cores are governed by collisions between neutrons and protons mediated by the strong interaction. Since nuclear matter in the core of neutron stars is highly asymmetric (in other words, the proton fraction x_p is small), the proton contribution to transport coefficients is small, and it is enough to treat them only as a passive scatterers for neutrons. In this case, the system of equations (9.68) reduces to one equation for the effective neutron relaxation time τ_n (e.g., Baiko et al. 2001). It is sometimes assumed that due to x_p being small the results for pure neutron matter are appropriate for neutron star cores, at least at low densities. However, pure neutron matter turns out to be a bad approximation for assessing the transport coefficients, since the protons cannot be ignored even if $x_p \approx 0.01$ (Baiko et al. 2001; Shternin et al. 2013). The reason is that the effective transport cross-section for neutron-proton collisions is larger than that for neutron-neutron collisions due to inclusion of the $T_z = 0$ isospin channel in scattering, and different kinematics of these collisions (Shternin et al. 2013). Applying the general theory outlined in Sect. 9.4.1.1 for neutrons scattering off neutrons and protons, one obtains the following results for thermal conductivity and shear viscosity,

$$\kappa_n = 1.03 \times 10^{22} \frac{n_n}{n_0} \left(\frac{m_n^*}{m_N}\right)^{-2} \left(\frac{T}{10^8 \text{ K}}\right)^{-1}$$

$$\times \left[\left(\frac{m_n^*}{m_N}\right)^2 m_\pi^2 S_{\kappa nn} + \left(\frac{m_p^*}{m_N}\right)^2 m_\pi^2 S_{\kappa np}\right]^{-1} \frac{\text{erg}}{\text{cm s K}}, \qquad (9.75a)$$

$$\eta_n = 2.15 \times 10^{17} \left(\frac{n_n}{n_0}\right)^{5/3} \left(\frac{m_n^*}{m_N}\right)^{-2} \left(\frac{T}{10^8 \text{ K}}\right)^{-2}$$

$$\times \left[\left(\frac{m_n^*}{m_N}\right)^2 m_\pi^2 S_{\eta nn} + \left(\frac{m_p^*}{m_N}\right)^2 m_\pi^2 S_{\eta np}\right]^{-1} \frac{\text{g}}{\text{cm s}}, \qquad (9.75b)$$

where $m_N = 939$ MeV$/c^2$ is the nucleon mass (neglecting the mass difference between neutron and proton), and we have used the same notations as in Baiko et al. (2001), Shternin and Yakovlev (2008a), Shternin et al. (2013), and Kolomeitsev and Voskresensky (2015). The quantities $S_{\kappa/\eta NN}$ ($N = n, p$) are the quasiparticle scattering rates W_{ci} averaged with certain phase factors, for details see for example Shternin et al. (2013), Shternin et al. (2017). They have the meaning of effective transport cross-sections and are normalized by the relevant nuclear force scale—the inverse pion mass squared, $m_\pi^{-2} \approx 20$ mb in natural units. Note that the numerical prefactors in Eqs. (9.75) include the correction constants $C_\kappa \approx 1.2$ and $C_\eta \approx 1.05$, as discussed at the end of Sect. 9.4.1.1 (Shternin et al. 2013). These corrections are of course irrelevant for most applications in neutron star physics.

The main ingredients for the calculation of the nucleon transport coefficients are the effective masses of the nucleons m^* on the Fermi surface and the quasiparticle scattering rates W_{ci}. Both quantities are strongly affected by in-medium effects and should be calculated using a microscopic many-body approach. Thus, the results for the nucleon transport coefficients are model-dependent and, in principle, their calculation should be based on the same microscopic model as the calculation of the equation of state. From Eqs. (9.75) we see that the effective mass enters the expressions for the transport coefficients in fourth power (because it describes the density of states and four quasiparticle states are involved in binary collisions). A moderate modification of the effective masses thus results in a strong modification of the transport coefficients. Therefore, the simplest way to include in-medium effects is to compute the effective mass modification, but use the free-space scattering rate which is well-known from experiment. This approach is particularly appealing because of its universality. The resulting expressions can be used for any equation of state of dense nuclear matter. Convenient fitting expressions for effective collision frequencies within this approach can be found in Baiko et al. (2001), Shternin and Yakovlev (2008a). Typical values of transport cross-sections at $n_B = n_0$ in Eqs. (9.75) are $S_{\kappa nn} \sim S_{\eta nn} \sim S_{\eta np} \approx 0.2 m_\pi^{-2}$, while $S_{\kappa np} \approx 0.4 m_\pi^{-2}$. Note that older results by Flowers and Itoh (1979), obtained via the same approach, turned out to be incorrect. Unfortunately, the in-medium modifications of the scattering rates themselves can be substantial. At present, the theoretical uncertainties are rather large and can result in order of magnitude differences in the final results.

Several many-body approaches have been employed in the calculation of the transport coefficients. The main problem is to properly take into account the particle correlations appearing in the strongly interacting liquid. Additional complications arise from the need to include three-body nucleon forces, which are necessary to reproduce the empirical saturation point of symmetric nuclear matter (see for instance Baldo 1999). Results have been obtained within the Brueckner-Hartree-Fock (BHF) scheme, where the in-medium G-matrix is used in place of the quasiparticle interaction (Benhar et al. 2010; Zhang et al. 2010; Shternin et al. 2013, 2017), within the effective quasiparticle interaction constructed on top of the G-matrix (Wambach et al. 1993) (for neutron matter only), within the in-medium T-matrix approach (also for pure neutron matter) (Sedrakian et al. 1994), and using the Correlated Basis Function and the cluster expansion technique (Benhar et al. 2010; Benhar and Valli 2007; Carbone and Benhar 2011), which employs the variationally constructed effective interaction. All these approaches start from 'realistic' nuclear potentials, which are designed to fit the data on the free-space scattering phase shifts and properties of bound few-body systems.

A somewhat different approach is based on the Landau-Migdal Fermi-liquid theory for nuclear matter. In this approach, the long-range pion-exchange part of the nucleon-nucleon interaction is considered explicitly, while the short-range part of the potential is absorbed into a number of phenomenological constants. The key point of the theory is an in-medium modification of the pion propagator (Migdal et al. 1990; Migdal 1978) leading to the softening of the pion mode. This softening is strongly density-dependent and becomes important at $n_B \gtrsim n_0$. At larger densities, this can lead to pion condensation. It is assumed that above the saturation density n_0, the nucleon quasiparticle scattering is fully determined by the medium-modified one-pion exchange (MOPE), where also the interaction vertices are modified due to short-range nuclear correlations. Since the pion mode is soft, the effective range of the nucleon interaction increases, which leads to a strong enhancement of the scattering rates, especially at high densities. As a consequence, the effective collision frequencies in Eq. (9.68) become larger and the transport coefficients reduce substantially. The calculations in this model (for pure neutron mater) were performed by Blaschke et al. (2013) for thermal conductivity and by Kolomeitsev and Voskresensky (2015) for shear viscosity.

Let us compare the lepton and nucleon contributions to the thermal conductivity and shear viscosity of non-superfluid nuclear matter. The strong non-Fermi-liquid behavior of the lepton thermal conductivity (remember that $\kappa_{e\mu}$ is constant in T) makes it smaller than the baryon contribution $\kappa_n \propto T^{-1}$ regardless of the microscopic model used to calculate the latter quantity. However, the key result is that the thermal conductivity is large such that the neutron star core is isothermal (more precisely, accounting for effects of general relativity, the redshifted temperature is spatially constant), and the precise value of κ is not important. This value is only interesting for the cooling of young neutron stars, as it regulates the duration of the thermal relaxation in the newly-born star (Gnedin et al. 2001; Shternin and Yakovlev 2008b; Blaschke et al. 2013). For instance, delaying the thermal relaxation due to the decrease of κ_n in the MOPE model allowed Blaschke et al. (2013) to fit the cooling data of the Cas A neutron star. The situation is different for the shear viscosity: here, the leptonic contribution is proportional to $T^{-5/3}$ to leading order and is not damped at low temperatures. The calculations reported in the literature show that $\eta_{e\mu}$ can be either larger or smaller than the nucleonic contribution η_n, see, however, the discussion in Shternin et al. (2013).

All considerations above assume a uniform Fermi liquid. Let us briefly address the possibility of proton localization, originally proposed by Kutschera and Wójcik (1989, 1990). In this scenario, for small proton fractions, the protons in neutron star cores can be localized in a potential well produced by neutron density fluctuations induced by the protons themselves. The protons occupy some bound ground state and do not form a Fermi sea. This model has its analogy in the polaron problem in solids (Kutschera and Wójcik 1993). Recently, proton localization was reconsidered for some realistic equations of state (Szmaglinski et al. 2006; Kubis and Wójcik 2015). The authors find that protons can localize at densities $n_B \gtrsim (0.5 - 1)\,\mathrm{fm}^{-3}$, i.e., well in the range that can occur in the interior of neutron stars. The transport properties of nuclear matter with localized protons were studied by Baiko and Haensel (1999). They considered npe matter, where the localized protons are uncorrelated. Then the problem has much in common with transport properties of the neutron star crust with charged impurities, see Sect. 9.3.1.2. The electrons and neutrons now scatter off themselves and off the localized protons. Unless the protons can be excited in their sites, the scattering is elastic and the collision frequencies to be used in Eq. (9.68) become temperature-independent and dominate over temperature-dependent collision frequencies for other scatterings (cf. Sect. 9.3.1.2). Clearly, this leads to a strong decrease of the transport coefficients at low temperatures compared to the results without localization (Baiko and Haensel 1999). The consequences of proton localization on neutron star cooling was investigated in Baiko and Haensel (2000), but other possible astrophysical implications are largely unexplored. Baiko and Haensel (1999) considered a completely disordered system of localized protons, and it was proposed that these impurities can form a lattice (Kutschera and Wójcik 1995).

9.4.1.4 Effects of Cooper Pairing

So far we have neglected the effects of neutron and proton pairing on the transport coefficients. Pairing directly affects the dissipation in the neutron and proton subsystem since the structure of the excitations in superfluid matter is changed. However, as we will see immediately, proton pairing also affects the leptonic transport coefficients that were discussed in Sect. 9.4.1.2 without pairing.

We start by considering the electron and muon transport in the presence of proton pairing in the $^1S_0$ channel. The effect of proton superconductivity is twofold. First, it modifies the scattering rates of leptons off the protons (the protonic excitations) and second, it modifies the screening properties of the plasma which regulates the electromagnetic interaction in Eq. (9.70). In the static limit, the longitudinal part of the polarization operator Π_l, which describes longitudinal plasmon screening, remains unaffected (e.g. Arseev et al. 2006; Gusakov 2010). In contrast, the character of the transverse screening changes dramatically. The most important difference from the non-superconducting case is that now the transverse screening is predominantly static ($\Pi_t \neq 0$ for $\omega \to 0$). In this case, the collision probability becomes ω-independent, which restores the standard Fermi-liquid behavior of the collision frequencies between the electrons and muons, $\nu_{ci} \propto T^2$.

The long-wavelength ($q \to 0$) transverse plasmon (photon) in superconductor acquires a Meissner mass. However, at larger q the screening mass gradually drops and in the Pippard limit, $q\xi \gg 1$, it becomes inversely proportional to q. Here, $\xi \sim v_{Fp}/\Delta_p$ is the coherence length, with the energy gap Δ_p in the proton quasiparticle spectrum from Cooper pairing. It turns out that both the long wavelength and Pippard limits may be applicable to the neutron star cores (Shternin 2018). Here we review the situation which is realized at high densities or low Δ_p, where the Pippard limit for the transverse screening is relevant and refer to Shternin (2018) for a more detailed discussion. At low temperatures $T \ll T_{cp}$, the protons give the dominant contribution to the polarization function Π_t, and one obtains [cf. Eq. (9.71)]

$$\Pi_t \approx \pi \alpha_f p_{Fp}^2 \frac{\Delta_p}{q}, \qquad \text{(for } q\xi \gg 1) . \qquad (9.76)$$

Thus the characteristic screening wavenumber in this case is $\propto \Delta_p^{1/3}$ instead of $\propto \omega^{1/3}$ in Sect. 9.4.1.2. The detailed behavior of screening at intermediate temperatures and a crossover from static to dynamical screening was discussed by Shternin and Yakovlev (2007, 2008a). They found that the 'transverse' part of the collision frequencies dominates in this case as well.

Taking into account lepton collisions with protons is more involved. The main low-energy excitations of the proton system are the single-particle excitations, namely the Bogoliubov quasiparticles. In addition to the presence of the energy gap in the quasiparticle spectrum, one needs to take into account that the number of quasiparticles is not conserved. They can be excited from the Cooper pair condensate, or coalesce into it. As a consequence, the collision integral describing electron-proton scattering is more complicated than in Eq. (9.31). However, at low temperatures the main effect of pairing is the exponential reduction of the number of quasiparticles and thus the exponential reduction in the collision frequencies, $\nu_{cp} \propto \exp(-\Delta_p/T)$ (Shternin and Yakovlev 2007, 2008a). Therefore, for temperatures much lower than the critical temperature for proton pairing, $T \ll T_{cp}$, the details of the lepton-proton collisions are not important, since they are suppressed. The transport coefficients are dominated in this case by collisions in the lepton subsystem, and, taking into account the screening modification, one derives the following compact leading-order expression for the thermal conductivity instead of Eq. (9.72) (Shternin and Yakovlev 2007),

$$
\kappa_{e\mu}^{SF} = \frac{5}{24} \frac{k_B c p_{Fp}^2}{\alpha_f \hbar^2} \frac{\Delta_p}{k_B T} = 3.87 \times 10^{24} \left(\frac{n_B}{n_0}\right)^{2/3} x_p^{2/3} \frac{\Delta_p}{1\,\text{MeV}} \left(\frac{T}{10^8\,\text{K}}\right)^{-1} \frac{\text{erg}}{\text{cm s K}}.
$$
(9.77)

This result shows the standard Fermi-liquid dependence $\kappa \propto T^{-1}$ and is several orders of magnitude larger than the non-superfluid result (9.72). This is not really important in practice since the star becomes isothermal in a short time in both cases.

Similarly, one can compute the low-temperature expression for the leptonic shear viscosity (Shternin and Yakovlev 2008a),

$$
\eta_{e\mu}^{SF} = 7.60 \times 10^{21} \left(\frac{n_B}{n_0}\right)^{14/9} \frac{(x_e^2 + x_\mu^2) x_p^{2/9}}{x_e^{2/3} + x_\mu^{2/3}} \left(\frac{\Delta_p}{1\,\text{MeV}}\right)^{1/3} \left(\frac{T}{10^8\,\text{K}}\right)^{-2} \frac{\text{g}}{\text{cm s}}.
$$
(9.78)

Comparing Eq. (9.78) with Eq. (9.73), we see that the effect of proton Cooper pairing on the lepton shear viscosity is less dramatic than in the case of thermal conductivity. This is due to a weaker dependence of η on the screening momentum than κ. In turns out that Eq. (9.78) is a better approximation to the full result for $\eta_{e\mu}^{SF}$ than the non-superfluid expression (9.73). The electrical conductivity in the presence of proton pairing cannot be treated in similar simple way, see Sect. 9.4.1.5.

Recently, an effective lepton-neutron interaction was proposed by Bertoni et al. (2015). The idea is that the neutron quasiparticle in the neutron star core is in fact a neutron dressed by a neutron-proton cloud. Thus it possesses an effective electric charge and interacts with charged leptons on the same ground as the protons. Within field-theoretical language, this lepton-neutron interaction is induced by a proton particle-hole excitation which is coupled to neutrons (Bertoni et al. 2015). Estimates show that this effective interaction can be relevant when the protons are in the superconducting state. Moreover, at $T \ll T_{cp}$, the effective lepton-neutron collisions can dominate over the inter-lepton collisions, thus providing a dominant contribution to lepton transport coefficients. A detailed rigorous treatment of this interaction is yet to be done and would be highly desired. Notice that such a coupling can also modify the screening properties of the photons in the nuclear medium and therefore other collisions mediated by electromagnetic interactions. This is currently under investigation (Stetina et al. 2018).

Let us turn now to the nuclear (hadronic) sector in the presence of pairing. Recall that the main carriers are neutrons. If they are unpaired, but the protons are gapped, only neutron-proton collisions are affected. This situation can be treated in a similar way as the lepton-proton collisions above. The result has not yet been computed in detail for $T \lesssim T_{cp}$, but at low temperatures the main effect is the exponential suppression of the collision frequency (Baiko et al. 2001; Shternin and Yakovlev 2008a). The damping of neutron-proton scattering leads to increase of the neutron effective relaxation times and, as a consequence, of neutron transport coefficients, which are now governed by the neutron-neutron scattering only.

A more interesting situation occurs if neutrons pair. In general, the transport equations in the superfluid are complicated, as one needs to account for anomalous contributions (the response of the condensate), just like in terrestrial fermionic superfluids such as liquid ^{3}He (Vollhardt and Wölfle 1990). However, the situation simplifies greatly when the temporal and spatial scales of the external perturbation are large compared to $\hbar\Delta^{-1} \sim 10^{-22}$ s and $\xi \sim 10^{-11}$ cm, which is the appropriate limit for the transport coefficients. In this case, the response of the condensate is instantaneous, such that it can be considered to be in local equilibrium. Then, the kinetics of the system is described by the transport equation for Bogoliubov quasiparticles, whose streaming (left-hand side) term reduces to a standard streaming term of the Boltzmann equation (Vollhardt and Wölfle 1990), see also Gusakov (2010) for the case of superfluid mixtures. The transport coefficients can then, in principle, be calculated along the lines laid out in Sect. 9.4.1.1, provided the collision integral is specified. The latter can be derived from the normal-state collision integral by applying the Bogoliubov transformations. The resulting expression takes into account non-conservation of quasiparticles. Since neutron pairing in the core of a neutron star is expected to occur in the anisotropic ^3P_j state, additional complications arise. This is analogous to certain phases of superfluid ^{3}He, which break rotational symmetry as well, resulting in anisotropic transport properties (Vollhardt and Wölfle 1990). It is also comparable to anisotropic transport in nuclear pasta phases discussed in Sect. 9.3.2. In the context of neutron stars, this was not studied in detail (see however Shahzamanian and Yavary 2005 and Juri and Hernández 1988). A thorough investigation of the Bogoliubov quasiparticle contribution to transport coefficients remains an open problem, but the key features at low temperatures can be worked out, neglecting all the modifications to the quasiparticle collision integral except the spectrum modification by the gap (for the neutron thermal conductivity this was done by Baiko et al. 2001).

Naively, one might think that since the number of available quasiparticles is exponentially suppressed at low temperatures, their contribution to transport coefficients is exponentially suppressed as well. This is true if there exists a scattering mechanism which effectively limits the quasiparticle mean free path. If, however, the main contribution to the collision probability comes from collisions between the Bogoliubov quasiparticles, it is suppressed roughly by the same factor. As a result, the exponential factors cancel each other, and one is left with the standard Fermi-liquid dependence of the thermal conductivity, $\kappa \propto T^{-1}$, as shown in the context of the so-called B phase of superfluid ^{3}He (Pethick et al. 1977). Similar arguments show that the shear viscosity tends to a constant value which is not far from its value at T_c. Moreover, it can be shown that a relation analogous to the Wiedemann-Franz rule (9.47) applies (Pethick et al. 1977),

$$\frac{\kappa T}{\eta} = \frac{5\Delta^2}{p_F^2}. \tag{9.79}$$

In neutron star cores, this situation can be realized when neutron pairing occurs at lower temperatures than proton pairing, such that $T < T_{cn} < T_{cp}$. Then, neutron-proton collisions are suppressed exponentially stronger than neutron-neutron collisions (there are much less proton excitations than neutron ones) and do not participate in neutron transport (Baiko et al. 2001). Nevertheless, with lowering temperature, the transport coefficients stay large until other neutron relaxation mechanisms start to dominate over neutron-neutron scattering, for instance neutron-lepton scattering due to the neutron magnetic moment or interactions with collective excitations. This will lead to a strong suppression of the transport coefficients compared to the limiting values discussed by Pethick et al. (1977). Evidently, a similar analysis in the opposite case, $T_{cn} > T_{cp}$, leads to analogous results.

Let us note that the quasiparticle mean free path increases exponentially with decreasing temperature. Eventually, at about $0.1\,T_c$, it becomes of the order of the size of the superfluid region. In this case, the bulk hydrodynamical picture is inappropriate to describe the transport since the quasiparticles move 'ballistically'. Moreover, one needs to take into account the spatial structure of the superfluid region (baryon density, the gap value Δ, and the gravitational potential change on the mean free path scale) and the interaction of the quasiparticles with the boundaries of the superfluid region (more precisely, with the edges of the critical temperature profile, where $T_{cn}(n_B)$ is lower and the macroscopic hydrodynamical picture is restored). If the spatial scale of the external perturbations is smaller than the quasiparticle mean free path (or the frequency of the perturbation is larger than the collision frequency), the response of the quasiparticle system cannot be considered in the local equilibrium approximation. To our knowledge, these effects have not been studied yet. Nevertheless, one expects that leptons dominate the transport in this regime.

At low temperatures, since the number of single-particle excitations is suppressed exponentially, low-energy collective modes become the relevant degrees of freedom, if they exist. As the superfluid condensate spontaneously breaks the $U(1)$ internal symmetry related to baryon number conservation, at least one gapless collective mode must exist in the superfluid system according to the Goldstone theorem (see Sect. 9.2.3.4). This fundamental mode is called superfluid phonon because of its acoustic dispersion relation. When this is the only low-lying collective excitation, it fully defines the transport properties of the system. This situation is realized, for instance, in superfluid ^{4}He or in cold atomic gases. In this case, an effective theory can be constructed on general grounds that describes the phonon dispersion and the interactions between phonons (e.g., Son 2002; Son and Wingate 2006; Greiter et al. 1989). In the context of neutron star cores, the phase that comes closest to this scenario (in the sense that there are no additional low-energy excitations such as leptons or other unpaired fermions) is the color-flavor locked phase of quark matter, and we give some details on the field-theoretical description of superfluid phonons in Sect. 9.5.5.2. The phonon transport coefficients are calculated from the solution of the appropriate kinetic equation that includes phonon scatterings in the collision term (Khalatnikov 1965). In the case of neutron pairing, shear viscosity and thermal conductivity mediated by phonon-phonon interactions were investigated in Manuel and Tolos (2011), Manuel and Tolos (2013), Manuel et al. (2014), Tolos et al. (2016). These studies suggest that the phonon contribution is important in a narrow range of temperatures, 10^9 K $\lesssim T < T_{cn}$, where, in fact, the validity of the effective theory is questionable. In reality, the excitation spectra and in neutron star cores is richer and various scattering mechanisms can be important (Bedaque and Reddy 2014; Kolomeitsev and Voskresensky 2015). Superfluid phonons of the neutron component couple to leptons indirectly via the neutron-proton interaction; this provides an efficient scattering mechanism for phonons, decreasing their mean free path. According to Bedaque and Reddy (2014), this makes the superfluid phonon contribution to transport coefficients negligible (cf. discussion in Sect. 9.3.1.4). The phonon coupling with the Bogoliubov quasiparticles was investigated by Kolomeitsev and Voskresensky (2015) with a similar conclusion. Other low-energy excitations which can exist in neutron star cores in the presence of nucleon Cooper pairing are as follows. In metallic superconductors, the collective mode is massive due to presence of Coulomb interaction. However, it was proposed that the efficient plasma screening in nuclear matter 'resurrects' the Goldstone mode of the proton condensate (Baldo and Ducoin 2011) (it corresponds to the oscillation of a charge-neutral mixture of proton pair condensate and leptons). This mode was found to effectively scatter on leptons and does not contribute to transport (Bedaque and Reddy 2014). Since neutrons in the core form Cooper pairs in the anisotropic $^3P_2$ state, the condensate spontaneously breaks rotational symmetry, and one expects the appearance of corresponding Goldstone modes. These modes were termed 'angulons'[8] by Bedaque et al. (2003). The properties of angulons were studied by Bedaque and Nicholson (2013), and their contribution to transport properties of neutron star cores by Bedaque and Reddy (2014). Leinson (2012) analyzed the collective modes of the order parameter on a microscopic level for all temperatures and did not find a gapless mode. Instead, he found modes similar to 'normal-flapping' modes of the A-phase of superfluid ^{3}He (Vollhardt and Wölfle 1990), which are not massless at finite temperatures. This, however, contradicts the recent study by Bedaque et al. (2015), who find that angulons have zero mass at any temperature. The reason for the contradiction is unknown, and a consistent picture of the low-lying excitations in the superfluid phases of neutron star cores is yet to be developed. Nevertheless, even the massive mode can contribute to the transport properties provided its mass is sufficiently small (of the order of T). Note also that Fermi-liquid effects can strongly modify the properties of the collective modes at nonzero temperatures. It is possible that such modes become purely diffusive, not being well-defined quasiparticle excitations at finite q. These issues were discussed in detail for a general Fermi liquid that becomes superfluid by Leggett (1966a,b). Finally, we note that the warnings stated above regarding transport in the 'ballistic' regime apply also for the superfluid phonons (and other collective modes). The phonon mean free path grows by powers of T in comparison to the exponential growth of the mean free path for Bogoliubov quasiparticles, but it can still easily become of the order of the size of the superfluid region in the star. The dissipation in this situation must be treated with caution, see also remarks and references below Eq. (9.158) in the context of quark matter.

---

[8]Note that the same term was recently used in a different context (Schmidt and Lemeshko 2015).

9.4.1.5 Magnetic Field Effects

Transport properties of the magnetized neutron star core can be addressed using similar methods as for the crust, discussed in Sect. 9.3.1.5, but generalized to the case of multi-component mixtures. As in the crust, the anisotropy induced by the magnetic field renders the transport coefficients anisotropic. In contrast to the crust, the fermionic excitations are not expected to become strongly quantized by the magnetic field due to the larger effective masses.[9] Still, the collision probabilities can depend on the magnetic field, for instance due to a modification of the plasma screening, although this effect has never been investigated, to the best of our knowledge. The anisotropic thermal conductivity has never been considered since it is assumed to be very large anyhow. Similarly, the shear viscosity in magnetized star cores has not been calculated (except for the attempt to study the collisionless problem by Banik and Nandi (2013)).

However, the problem of electrical conductivity has gained considerable attention because this quantity is one of the key ingredients in the magnetic field evolution in neutron stars, see for instance Graber et al. (2015) and references therein. The response of multi-component mixtures to an external electromagnetic field differs qualitatively from the case of the electron conductivity described in Sect. 9.3.1.5 because of the relative motion between all components in the plasma. This is especially pronounced if neutral species are present in the mixture, like in the case of a partially ionized plasma or in neutron star matter. We briefly review the details of the calculations following Yakovlev and Shalybkov (1991a).

Let us assume that the plasma as a whole moves with a non-relativistic velocity $\boldsymbol{v}$ under the influence of an external force (electric field, but we will be a bit more general at this point). This force induces an $l = 1$ perturbation to the distribution function that is given by $\Phi_i = -\boldsymbol{w}_i \cdot \boldsymbol{\vartheta}_i$, where $i = 1, \ldots, N$ labels the constituents of the multi-component system. The unknown vectors $\boldsymbol{\vartheta}_i$ are energy-dependent, but, as a first approximation, can be assumed to be constant (as we saw above, this is a fairly good approximation in degenerate matter). Then, these vectors are exactly the diffusion velocities of the constituents of the mixture (relative to the total velocity $\boldsymbol{v}$). If the co-moving frame is defined to ensure zero total momentum of the fluid element, the diffusion velocities obey the linear constraints (Yakovlev and Shalybkov 1991a)

$$\sum_i \frac{\mu_i n_i}{c^2} \boldsymbol{\vartheta}_i = 0. \tag{9.80}$$

---

[9] Remember that the effective mass of electrons becomes larger as we increase the density. For a rough estimate, let us assume $n_B = n_0$, an electron density $n_e = n_B/10$, and $v_{Fe} \approx c$. Then, the effective mass $m_e^* = p_{Fe}/v_{Fe}$ squared can be translated into a magnetic field $B \sim 4 \times 10^{18}$ G, at which quantizing effects would become important. This field is larger than what is typically expected for neutron star cores, given that the largest measured surface magnetic fields are of the order of 10^{15} G (According to the data in the ATNF pulsar catalogue on the time of writing (version 1.58) http://www.atnf.csiro.au/research/pulsar/psrcat, Manchester et al. 2005).

Here, in a generalization of the center-of-mass velocity to a degenerate fluid of relativistic particles, the mass density $m_i n_i$ has been replaced by $\mu_i n_i / c^2$, where the chemical potentials μ_i include the rest mass.

Instead of the single Eq. (9.17), one then obtains a system of linearized kinetic equations. In order to determine the vectors $\boldsymbol{\vartheta}_i$, one multiplies these equations by $\boldsymbol{p}_i$ and integrates them over momenta to arrive at (neglecting temperature gradients for simplicity, i.e., there are no thermo-diffusion and thermo-electric effects)

$$-\frac{\mu_i n_i}{c^2}\dot{\boldsymbol{v}} + \boldsymbol{F}_i + \frac{q_i n_i}{c}\boldsymbol{\vartheta}_i \times \boldsymbol{B} = \sum_j J_{ij}(\boldsymbol{\vartheta}_i - \boldsymbol{\vartheta}_j), \tag{9.81}$$

where the friction term on the right-hand side is given by the symmetric matrix J_{ij}, which is related to the effective collision frequencies [see Eq. (9.68)] by $J_{ij} = n_i m_i^* \nu_{ij}$. The driving term on the left-hand side contains the body forces $\boldsymbol{F}_i = n_i \boldsymbol{R}_i$, which do not depend on $\boldsymbol{\vartheta}_i$, and the magnetic part of the Lorentz forces. The system (9.81) contains more unknowns than equations since it also determines the plasma acceleration $\dot{\boldsymbol{v}}$, but is closed by Eq. (9.80). After $\dot{\boldsymbol{v}}$ is eliminated, the solution can be written as (Yakovlev and Shalybkov 1991b)

$$\boldsymbol{\vartheta}_i = -\hat{\mathcal{D}}_{ij}\left(\boldsymbol{F}_j - X_j \boldsymbol{F}\right), \tag{9.82}$$

where $\boldsymbol{F}$ is the total force and $X_j = n_j \mu_j \left(\sum_k n_k \mu_k\right)^{-1}$ is the mass fraction of species j. The auxiliary tensor $\hat{\mathcal{D}}_{ij}$ has rank $N-1$ and can be expressed in different ways (Lam 2006). The resulting drift velocities are of course the same for any representation of $\hat{\mathcal{D}}_{ij}$. Due to the linear constraint (9.80), one can use any $N-1$ independent linear combinations of diffusion velocities for forming the current terms in the final hydrodynamical expressions. A natural choice for one of these combinations is the electric (charge) current $\boldsymbol{j} = \sum_i q_i n_i \boldsymbol{\vartheta}_i$. In the three-component npe plasma containing a neutral (n) component and a charged (pe) fluid, the natural choice of the second independent velocity is the 'ambipolar drift' velocity, i.e., the relative velocity of the charged fluid with respect to neutrons. This description has become standard, starting from the work by Goldreich and Reisenegger (1992). However, in the more general case where several charged and neutral species coexist (for instance when muons and/or hyperons are included), this description becomes less convenient. We thus prefer to keep the more symmetric choice, that is to work with the drift velocities $\boldsymbol{\vartheta}_i$ subject to (9.80) and use the full system of multi-fluid hydrodynamical equations. Of course any representation leads to the same physical results. For the generalized Ohm law and the induction equation in npe matter this was discussed in detail by Shalybkov and Urpin (1995). It is instructive to write the expression for the entropy generation rate in collisions with the help of the general Eq. (9.21),

$$T\varsigma|_{\text{coll}} = \frac{1}{2}\sum_{ij} J_{ij}(\boldsymbol{\vartheta}_i - \boldsymbol{\vartheta}_j)^2. \tag{9.83}$$

Now consider the electrical conductivity problem, where the force terms in Eq. (9.81) are solely given by the electromagnetic field (Yakovlev and Shalybkov 1991a),

$$F_i = q_i n_i E' \equiv q_i n_i \left(E + \frac{1}{c} v \times B \right).$$ (9.84)

Then, the electric current and electric field E' in the co-moving system are related via the (generalized) Ohm law $j = \hat{\sigma} E'$, where the electrical conductivity tensor $\hat{\sigma}$ is expressed through the tensor $\hat{D}_{ij}$ as

$$\hat{\sigma} = -\sum_{i,j} q_i q_j n_i n_j \hat{D}_{ij}.$$ (9.85)

It is convenient to introduce also the resistivity tensor $\hat{R} = \hat{\sigma}^{-1}$. The final result for non-superfluid npe matter assuming charge neutrality and neglecting electron-neutron collisions reads

$$R_\parallel = [\sigma(B = 0)]^{-1}, \qquad R_\perp = R_\parallel + \frac{X_n^2}{J_{pn} c^2} B^2, \qquad R_\Lambda = \frac{1 - 2X_e}{n_e e c} B,$$ (9.86)

where $R_\parallel$ and $R_\perp$ are the resistivities parallel and perpendicular to the magnetic field, respectively, and R_Λ is the Hall resistivity. The second term in the expression for $R_\perp$ is proportional to B^2 and thus is responsible for a considerable increase of the resistivity in strong magnetic fields. As we see from its form, it originates from the friction of the neutron fluid with the charged components, governed by the strong forces. In this sense, it can be viewed as the result of ambipolar diffusion (Goldreich and Reisenegger 1992). Note that the increase of the transverse resistivity in a magnetized plasma containing neutral species is a well-known effect in physics of space plasmas (Bykov and Toptygin 2007). This has important consequence on the dissipation of the magnetic field energy. Indeed, the field energy dissipation rate per unit volume is

$$\dot{W}_B = -j \cdot E' = -j_\parallel^2 R_\parallel - j_\perp^2 R_\perp,$$ (9.87)

where $j_\parallel$ and $j_\perp$ are the components of the electric current along and transverse to B, respectively. Therefore, as first noted by Haensel et al. (1990), the increase of $R_\perp$ in Eq. (9.86) can lead to accelerated dissipation. Since the neutron-proton collision frequencies scale as $J_{np} \propto T^2$, the increase in $R_\perp$ becomes pronounced at lower temperatures. The microscopic calculation of the friction coefficients J_{ij} (or effective collision frequencies) were performed by Yakovlev and Shalybkov (1991b) and updated in Shternin (2008b) (as described in Sect. 9.4.1.2 and 9.4.1.3). Recall that the recent results by Bertoni et al. (2015) indicate the possible importance of the electron-neutron collisions. The influence of this effect on the electrical conductivity in neutron star cores has not been studied yet.

This simple picture is modified if other driving forces (in addition to the electromagnetic field) are present on the left-hand side of Eq. (9.81). If gradients of temperature or chemical potentials are present, one deals with thermo-electric and electro-diffusion effects. The Eqs. (9.80)–(9.83) still hold, but the expressions for the hydrodynamical currents are different (Landau and Lifshitz 1987; Landau et al. 1984). For instance, the electric current is not proportional to E', but also depends on thermal and chemical gradients. In the following we do not consider any thermal gradient and focus on diffusion effects, which were found to be important in the problem of the magnetic field evolution in neutron stars (Goldreich and Reisenegger 1992; Pethick 1992). The reason is that the timescale of the field evolution can be comparable to the typical times of the reactions that are responsible for chemical equilibration (see Sect. 9.4.2). The driving forces for diffusion are the gradients of the chemical potentials, which are added to the force term in the kinetic equation[10]

$$F_i = q_i n_i E' - n_i \nabla \mu_i. \tag{9.88}$$

In a one-component fluid (Sect. 9.2.2), the chemical potential gradient can be absorbed into an effective electric field, see Eq. (9.19). Multiple $\nabla \mu_i$'s do not allow this simple prescription. The diffusion velocities are still found by Eq. (9.82). They now receive contributions from the electric field and from the gradients. The microscopic equation for the entropy generated in the collisions still has the form given in Eq. (9.83), but the macroscopic expression now reads

$$T \varsigma|_{\text{coll}} = \sum_i F_i \cdot \vartheta_i = j \cdot E' - \sum_i n_i \vartheta_i \cdot \nabla \mu_i. \tag{9.89}$$

The first term on the right-hand side corresponds to Ohmic heating, while the second term describes entropy generation due to the irreversible diffusion process. The heat source for the latter is the particle (chemical) energy $\mu_i d n_i$. If chemical reactions are taken into account, $dn_i = -\nu_i d\xi$, with the stoichiometric coefficients ν_i and the reaction extent ξ, such that $\Gamma \equiv \dot{\xi}$ is the reaction rate. The thermodynamic force that drives the relaxation to chemical equilibrium is[11]

$$\delta \mu = \sum_i \nu_i \mu_i. \tag{9.90}$$

---

[10]The driving force terms should also contain gravitational acceleration to ensure the proper equilibrium state (Shalybkov and Urpin 1995; Lam 2006); we omit the corresponding terms for brevity.

[11]In the chemistry literature, (the negative of) $\delta \mu$ is called reaction affinity and usually denoted by $\mathcal{A}$. All reactions with nonzero reaction affinity we discuss later (Sects. 9.4.2, 9.4.3, and 9.5.4) have stoichiometric coefficients $\nu_i = \pm 1$, and the notation $\delta \mu$ is used mostly (but not exclusively) in the neutron star literature we refer to.

In equilibrium, $\delta\mu = 0$, and the reactions tend to move to this point. The thermodynamic flux conjugate to $\delta\mu$ is Γ. In the linear regime, $\Gamma = \lambda\,\delta\mu$, where λ is the corresponding transport coefficient, which, to a first approximation, only depends on the equilibrium state. The entropy generation from the chemical reactions is

$$T\varsigma|_{\mathrm{react}} = \Gamma\,\delta\mu = \lambda\,\delta\mu^2 , \tag{9.91}$$

thus the second law of thermodynamics requires $\lambda > 0$.

If we allow for chemical reactions, the conservation laws are modified: source terms appear in the continuity equations for the constituent species, recoil terms emerge in the momentum conservation equations (usually neglected as second-order), and the energy conservation law is also modified. For instance, using the continuity equation including the source term,

$$\frac{\partial n_i}{\partial t} + \nabla \cdot (n_i \boldsymbol{v}) + \nabla \cdot (n_i \boldsymbol{\vartheta}_i) = -\Gamma \nu_i , \tag{9.92}$$

one rewrites Eq. (9.89) as

$$-\dot{W}_B - \sum_i \mu_i \left(\frac{\partial}{\partial t} + \boldsymbol{v} \cdot \nabla \right) n_i = T\varsigma|_{\mathrm{react}} + T\varsigma|_{\mathrm{coll}} = \lambda\,\delta\mu^2 + \frac{1}{2}\sum_{ij} J_{ij}(\boldsymbol{\vartheta}_i - \boldsymbol{\vartheta}_j)^2 , \tag{9.93}$$

where we have used Eq. (9.80) and assumed div $\boldsymbol{v} = 0$. (The interplay between fluid compression, i.e., div $\boldsymbol{v} \neq 0$, and the chemical reactions is discussed in detail in Sect. 9.4.3.) It is frequently assumed that the magnetic field evolution is quasistationary, such that the second ('chemical') term on the left-hand side of Eq. (9.93) can be neglected. Then, Eq. (9.93) describes the transfer of energy of the magnetic field to heat via binary collisions and reactions (Gusakov et al. 2017; Castillo et al. 2017).

The problem of the magnetic field evolution in neutron star cores attracts persistent attention (Castillo et al. 2017; Gusakov et al. 2017; Passamonti et al. 2017; Glampedakis et al. 2011; Beloborodov and Li 2016; Graber et al. 2015; Hoyos et al. 2008; Elfritz et al. 2016; Bransgrove et al. 2018), and a more complete list of references can be found in the cited works. We did not discuss general relativistic effects and the influence of Cooper pairing, which can both be important. A full dynamical analysis that includes magnetic, thermal, and chemical evolution is highly demanded but has not been performed yet.

9.4.2 Reaction Rates from the Weak Interaction

9.4.2.1 General Treatment

Neutrino emissivity is a key ingredient in the study of the neutron star evolution. Minutes after the birth of a neutron star, neutrinos escape the star freely, taking away energy and thus cooling the star. The core is the main source of neutrinos, which are produced in various reactions involving the weak interaction. In the exhaustive review by Yakovlev et al. (2001) the wealth of neutrino-producing reactions possible in neutron star cores are described systematically. Naturally, the most important processes among them involve baryons. In this section we mainly focus on the recent results for these reactions, and mention others—less efficient ones—only briefly. As for the case of the transport coefficients discussed in Sect. 9.4.1.3, the main recent efforts have been focused on improving the treatment of in-medium effects.

The problem of calculating the neutrino emissivity is closely related to the more general problem of neutrino transport in dense matter, which is of utmost importance in supernovae and proto-neutron star studies. The emissivity can be calculated from the gain term in the corresponding neutrino transport equation. If this is done in the framework of non-equilibrium transport theory, one is in principle able to study the weak response of dense matter in a systematic way, including situations far from equilibrium. For a pedagogical discussion of the real-time Green's function approach see Sedrakian (2008, 2007) and references therein. Alternatively, emissivities can be found using the optical theorem without employing the neutrino transport equation (see, e.g., Voskresensky 2001). In any case, the rates can be expressed through the contraction of the weak currents with the polarization of the medium. The latter accounts for all many-body processes which exist in dense matter, and its microscopic calculation is not straightforward. Fortunately, in neutron star cores the quasiparticle approximation is well-justified. In this approximation, the reaction rates can be equivalently calculated using Fermi's Golden Rule based on the squared matrix element of the process. Due to its transparency, this approach is most commonly used for the weak reactions in hadronic cores of neutron stars. In this section we will follow this prescription and discuss the results beyond the quasiparticle picture at the end. This also allows us to make a close connection to the previous section. We note, however, that the approach based on Green's functions is particularly advantageous in the case of pairing, where some difficulties can arise with the use of Fermi's Golden Rule. We illustrate this approach in Sect. 9.5.3, where neutrino emission from quark matter is considered.

The weak reactions are naturally classified by the number of quasiparticles involved, since each fermion generally adds a phase factor T/μ (cf. Sect. 9.4.1.1), and by the type of the weak current (neutral or charged) responsible for the process.[12] Both types contribute to the neutrino emissivity, but only the flavor-changing reactions (that go via the charged weak current) are responsible for

---

[12]In this section we set $k_B = 1$ for brevity, such that T/μ is dimensionless.

establishing beta-equilibrium. The first kinematically allowed processes with the lowest number of involved quasiparticles give the dominant contribution to the reaction rates. Therefore, the most powerful neutrino emission mechanism is the so-called baryon direct Urca process, which consists of a pair of reactions going via the charged weak current

$$\mathcal{B}_1 \rightarrow \mathcal{B}_2 + \ell + \bar{\nu}_\ell, \tag{9.94a}$$

$$\mathcal{B}_2 + \ell \rightarrow \mathcal{B}_1 + \nu_\ell, \tag{9.94b}$$

where $\mathcal{B}_{1,2}$ stand for baryons [for instance $(\mathcal{B}_1, \mathcal{B}_2) = (n, p)$ in the case of nuclear matter], ℓ for leptons, and ν_ℓ for the corresponding neutrino. In equilibrium, the rates of the reactions (9.94a) and (9.94b) are equal. They scale as T^5 and the neutrino emissivity as T^6, see below. It is essential that direct Urca processes in strongly degenerate matter have a threshold: in the quasiparticle approximation all fermions in Eqs. (9.94) are placed on their Fermi surfaces [see Eq. (9.67)]. Momentum conservation implies that the Fermi momenta of the triple $(\mathcal{B}_1, \mathcal{B}_2, \ell)$ satisfy the triangle condition (if one neglects the small neutrino momentum which is of the order T). In the highly isospin-asymmetric cores of neutron stars [$(\mathcal{B}_1, \mathcal{B}_2) = (n, p)$] this strongly limits both the electron ($\ell = e$) and muon ($\ell = \mu$) direct Urca processes. Only in matter with a sufficiently large proton fraction (and as a consequence of the electric neutrality, lepton fraction) the triangle condition for the Fermi momenta of the triple (n, p, ℓ) can be satisfied. In electrically neutral npe matter, the threshold for the proton fraction x_p is 11%. Therefore, not all equations of state allow for the direct Urca process to operate. Depending on the $x_p(n_B)$ profile, some equations of state can allow direct Urca processes for massive stars, and for some equations of state the proton fraction never exceeds the direct Urca threshold.

The counterpart to (9.94) via the neutral weak current is

$$\mathcal{B}_1 \rightarrow \mathcal{B}_1 + \bar{\nu}_\ell + \nu_\ell. \tag{9.95}$$

In the quasiparticle approximation, this reaction is kinematically forbidden, unless the baryons $\mathcal{B}_1$ form a Cooper pair condensate, see Sect. 9.4.2.3.

When the direct Urca processes are not allowed, the next-order processes in the number of quasiparticles take the lead. They include an additional baryon $\mathcal{C}$ which couples to the emitting baryons via the strong force.[13] The presence of a spectator relieves the triangle condition, but the price to pay is the reduced phase space of the process by a factor of $(T/\mu)^2$. In the presence of the spectator, the neutral-current reaction

$$\mathcal{B}_1 + \mathcal{C} \rightarrow \mathcal{B}_1 + \mathcal{C} + \bar{\nu}_\ell + \nu_\ell \tag{9.96}$$

---

[13]The spectator particle $\mathcal{C}$ can also be a lepton coupling via the electromagnetic forces, but this process is negligible.

becomes possible, and it is the familiar bremsstrahlung emission. The most important processes are, however, the charged-current reactions

$$B_1 + C \rightarrow B_2 + C + \ell + \bar{\nu}_\ell, \tag{9.97a}$$

$$B_2 + C + \ell \rightarrow B_1 + C + \nu_\ell, \tag{9.97b}$$

called modified Urca reactions (see Yakovlev et al. 2001 for the origin of the nomenclature). The reactions (9.96)–(9.97) are sometimes jointly called the electroweak bremsstrahlung of the lepton pairs. Depending on the relations between the Fermi momenta of the five degenerate fermions involved in the modified Urca reactions (9.97), these reactions can also have thresholds. However, in practice, this is never really important in neutron star conditions. For the complete classification of all phase-space restrictions for reactions (9.94)–(9.97) see Kaminker et al. (2016).

The expression for the rate Γ and the neutrino emissivity ϵ_ν of any of the reactions (9.94)–(9.97) can be expressed via Fermi's Golden Rule as follows (e.g., Yakovlev et al. 2001; Shapiro and Teukolsky 1983)

$$\begin{pmatrix} \Gamma \\ \epsilon_\nu \end{pmatrix} = \int \prod_{j=i,f} \frac{d^3 p_j}{(2\pi)^3} \mathcal{F}_{fi} (2\pi)^4 \delta^{(4)} \left(P_f - P_i \right) \begin{pmatrix} 1 \\ \omega_\nu \end{pmatrix} s \left| M_{fi} \right|^2, \tag{9.98}$$

where M_{fi} is the transition amplitude for the reaction (summed over initial and averaged over final polarizations), P_i and P_f are total four-momenta of initial and final particles, respectively, integration is done over the whole phase-space of reacting quasiparticles (including neutrinos),[14] and the symmetry factor s corrects the phase volume in case of indistinguishable collisions. The quantity $\mathcal{F}_{fi}$ is the Pauli blocking factor

$$\mathcal{F}_{fi} = \prod_i f_F(\varepsilon_i - \mu_i) \prod_{f \neq \nu} \left[1 - f_F(\varepsilon_f - \mu_f) \right], \tag{9.99}$$

where $f_F(y) \equiv (e^{y/T} + 1)^{-1}$. It contains products of the distribution functions for all fermions except neutrinos (which escape the star). At small temperatures, it effectively puts all the quasiparticles on the respective Fermi surfaces when the matrix element is calculated [cf. Eq. (9.31)]. Finally, the energy ω_ν, which enters the expression for the emissivity ϵ_ν, is the neutrino/antineutrino energy in the reactions (9.94), (9.97) and the total energy of the neutrino pair in the case of the bremsstrahlung reaction (9.96).

The calculation of Γ and ϵ_ν is similar to the calculation of collision frequencies described in Sect. 9.4.1.1. First, we assume that the matrix element M_{fi} does not depend on the neutrino energy. Then, the integrals over the absolute values

---

[14]We keep the same normalizations of the baryon and lepton wave functions for brevity, while traditionally one puts 2ε in the denominator for relativistic particles.

of the momenta are rewritten as energy integrals—for the degenerate particles according to Eq. (9.67) and for the neutrino $d^3 p_\nu = c^{-3} \varepsilon_\nu^2 d\varepsilon_\nu d\Omega_\nu$. Introducing the dimensionless energy variables $x = (\varepsilon - \mu)/T$ as in Sect. 9.4.1.1, one can express the energy-conserving delta-function as

$$\delta(E_f - E_i) = T^{-1} \delta \left(\sum_f x_f - \sum_i x_i + \frac{\delta\mu}{T} \right), \tag{9.100}$$

where $\delta\mu$ is the chemical potential difference between the final and initial particles in the reaction, as introduced in Eq. (9.90). Recall that the freely escaping neutrinos have zero chemical potential. Finally, we can write both Γ and ϵ_ν in the following generic form,

$$\begin{pmatrix} \Gamma \\ \epsilon_\nu \end{pmatrix} = \frac{s}{(2\pi)^{3n-4}} \hat{\Omega} T^k I_\epsilon \left(\frac{\delta\mu}{T} \right) \langle |M_{fi}|^2 \rangle \prod_{j \neq \nu} p_{Fj} m_j^*, \tag{9.101}$$

where $\hat{\Omega}$ is the angular integral over quasiparticle momenta orientations for fixed absolute values of momenta, so that the possible relative orientations are restricted by momentum conservation (Kaminker et al. 2016), $\langle |M_{fi}|^2 \rangle$ stands for the angular-averaged matrix element, n is the number of reacting quasiparticles, and the last product comes from the quasiparticle densities of states on the respective Fermi surfaces. The factor $I_\epsilon(\delta\mu/T)$ in Eq. (9.101) is the energy integral over the dimensionless variables $x_{i,f}$. Due to Eq. (9.100), it depends on the ratio $\delta\mu/T$. In beta-equilibrium and for bremsstrahlung reactions, $\delta\mu = 0$. The temperature dependence is given by the factor T^k in Eq. (9.101), where the exponent k depends on the specific reaction and on whether we compute Γ or ϵ_ν: each degenerate fermion on either side of the reaction gives one power of T, the neutrino contributes a factor T^3, and one power of T is subtracted due to the energy-conserving delta function. Therefore, the rate for the direct Urca process (three fermions) is proportional to T^5 and the rate for the modified Urca process (five fermions) is proportional to T^7. The corresponding emissivities have an extra factor T due to the neutrino energy ω_ν (Shapiro and Teukolsky 1983). Slightly different considerations should be carried out for the bremsstrahlung reactions, where the assumption of an energy-independent matrix element no longer holds. Instead, as we discuss below, the leading energy-dependence of the bremsstrahlung matrix elements is $M_{fi} \propto \omega_\nu^{-1}$ and can be factored out from the angular integration. After the factorization, the decomposition (9.101) still holds where the energy-independent part of the angular-averaged squared matrix element stays in place of $\langle |M_{fi}|^2 \rangle$. The temperature dependence for the bremsstrahlung reaction (9.96) is the same as for the modified Urca reactions (9.97) since the appearance of the squared neutrino pair energy in the denominator is compensated by the fact that the integration is now performed over the momenta of both neutrino and antineutrino in the outgoing channel of the reaction (see for instance Yakovlev et al. 2001).

Let us return to the Urca reactions, which are responsible for the processes of beta-equilibration. We denote quantities related to the forward and backward reactions by superscripts '+' and '−', respectively. Then, the net neutrino emissivity from a pair of forward and backward reactions is $\epsilon_\nu(\delta\mu) = \epsilon_\nu^+ + \epsilon_\nu^-$ and the rate of the composition change is $\Delta\Gamma(\delta\mu) = \Gamma^+ - \Gamma^-$. In equilibrium, $\delta\mu = 0$ and $\Delta\Gamma = 0$, while $\epsilon_\nu = 2\epsilon_\nu^+$. Beta-equilibration processes increase the entropy of the system, while the neutrino emission takes away energy. Thus the total heat release of the Urca reactions in non-equilibrium is

$$T\varsigma = \Delta\Gamma\,\delta\mu - \epsilon_\nu. \tag{9.102}$$

This means that the pair of the Urca reactions can either cool or heat the star depending on the relation of two terms in Eq (9.102), which in turn depends on the degree of departure from equilibrium.

Both $\Delta\Gamma$ and ϵ_ν can be expressed as a product of the equilibrium reaction rate times a function which depends solely on $\delta\mu/T$. The analytical expressions for these functions in non-superfluid matter can be found, for example, in Reisenegger (1995). When the deviation from beta-equilibrium is small ($|\delta\mu| \ll T$), the response to this deviation is linear, $\Delta\Gamma \propto \delta\mu$, while $\epsilon_\nu \approx \epsilon_\nu^{eq}$. In the suprathermal regime, when $|\delta\mu| \gg T$, the phase space available for the reacting particles is determined by $\delta\mu$ instead of T. Then, $I_\epsilon \propto (\delta\mu/T)^k$, such that the Urca reaction rates become temperature-independent, and $\delta\mu$ enters the final expressions in place of T, see for instance Yakovlev et al. (2001) for details. Moreover, Flores-Tulián and Reisenegger (2006) proved the general expression

$$\frac{\partial\epsilon_\nu}{\partial\delta\mu} = 3\Delta\Gamma. \tag{9.103}$$

Thus the reactions generate heat if $\epsilon_\nu(\delta\mu)$ is steeper than $\delta\mu^3$ and cool the star via neutrino emission otherwise. The advantage of the relation (9.103) is that it also works in case of pairing (Flores-Tulián and Reisenegger 2006).

9.4.2.2 Neutrino Emission of Nuclear Matter

We now use the general formalism outlined above to quantitatively discuss the main neutrino reactions relevant in neutron star cores. We start with the most efficient one, the direct Urca reaction (9.94). The possibility of its occurrence and importance for neutron stars was first pointed out by Boguta (1981), who found that the threshold conditions for the direct Urca process to operate can be fulfilled in some relativistic mean field models. This paper was unnoticed for a decade until Lattimer et al. (1991) rediscovered this possibility and argued that the sufficiently large proton fraction can be achieved for many realistic density dependencies of the symmetry energy of nuclear matter. In the limit of small momentum q transferred from leptons to nucleons, the weak charged current of nucleons contains vector (V) and axial-vector (A) contributions. For non-relativistic nucleons, when nucleon recoil can be neglected, the direct Urca matrix element averaged over the directions of neutrino momenta is (Lattimer et al. 1991; Yakovlev et al. 2001)

$$|M_{\rm DU}|^2 = 2G_F^2 \cos^2\theta_C (g_V^2 + 3g_A^2), \qquad (9.104)$$

where $G_F = 1.17 \times 10^{-5}$ GeV^{-2} is the Fermi coupling constant, θ_C is the Cabibbo angle with $\sin\theta_C \approx 0.22$, and $g_V = 1$ and $g_A \approx 1.26$ are the nucleon weak vector and axial vector coupling constants.

Inserting the expression (9.104) into Eq. (9.101), one obtains for the rate of the $np\ell$ reaction (in equilibrium) in physical units

$$\epsilon_\nu^{\rm DU} = 4 \times 10^{21} \left(\frac{n_{\rm B}}{n_0}\right)^{1/3} x_\ell^{1/3} \frac{m_p^* m_n^*}{m_N^2} \left(\frac{T}{10^8\,K}\right)^6 \Theta_{np\ell}\ {\rm erg\ cm}^{-3}\,{\rm s}^{-1}, \qquad (9.105)$$

where $\Theta_{np\ell}$ is the step function accounting for the threshold. The reaction rates $\Gamma = \Gamma^+ = \Gamma^-$ in equilibrium and the composition change rate in the subthermal regime are (Yakovlev et al. 2001)

$$\Gamma^{\rm DU} = \frac{0.118}{T}\epsilon_\nu^{\rm DU}, \qquad \Delta\Gamma^{\rm DU} = \frac{0.158}{T}\frac{\delta\mu}{T}\epsilon_\nu^{\rm DU} \qquad ({\rm for}\ \delta\mu \ll T). \qquad (9.106)$$

We note that the characteristic time for a nucleon to participate in a direct Urca reaction is quite large, $\tau^{\rm DU} \sim n_{\rm B}/\Gamma^{\rm DU} \approx 500$ yr (for $n_{\rm B} = n_0$, $T = 10^8$ K, and $x_\ell = 0.1$). Nevertheless, the direct Urca reaction is the strongest neutrino emission process and—if it operates—cools the star very fast.

The classical result (9.105) changes when relativistic corrections are taken into account. Apart from the nuclear recoil effect (Leinson and Pérez 2001), such corrections arise from additional terms in the nucleon charged weak current, corresponding to a non-trivial spatial structure of nucleons. In addition to the vector and axial vector contributions, the weak currents contain tensor (T) and induced pseudoscalar (P) terms (see, for instance, Timmermans et al. (2002) for the discussion of the weak hadron currents). The T-terms are of the order q/m_p and describe weak magnetism. According to Leinson (2002) the weak magnetism contribution can be as large as 50% of the rate (9.105). In the closely related context of the neutrino opacity due to charged weak current reactions, these corrections were taken into account in Horowitz and Pérez-García (2003), see also Roberts and Reddy (2017b) and references therein for a recent discussion on this subject. The contribution from the P-terms turns out to be proportional to the lepton mass and thus is found to be unimportant for the direct Urca processes with electrons (Leinson 2012). However, for the muonic direct Urca ($\ell = \mu$), this contribution can be substantial (Timmermans et al. 2002). For modifications of the direct Urca rate in various parametrizations in the framework of relativistic mean field theories see also Ding et al. (2016).

Another modification of the rate (9.105) results from in-medium effects (cf. Sect. 9.4.1). The simplest manifestation of the in-medium effects is through the modifications of the effective nucleon masses m_p^* and m_n^* (see, e.g., Baldo et al. (2014) for an illustration of the range of uncertainty). Dong et al. (2016) investigated the suppression of the emissivities due to Fermi surface depletion. The depletion is quantified in terms of the 'quasiparticle strength' $z_F < 1$, which appears in the numerator of the single-particle propagator in the interacting system (Dickhoff and Van Neck 2008). The reaction rates are basically multiplied by powers of z_F depending of the number of quasiparticle involved. In this case it is important to use the effective masses calculated in the same order of the theory. Corrections leading to $z_F < 1$ are counterbalanced by an increase of the effective mass (Schwenk et al. 2003; Zuo et al. 1998; Baldo and Grasso 2000; Schwenk et al. 2004), thus the overall corrections are not dramatic. Also, the couplings g_V and g_A (and, in principle, the tensor and pseudoscalar couplings g_T and g_P) are renormalized in a dense medium due to nucleon correlations (Voskresensky 2001). This effect is assumed to be not very important for direct Urca processes (Kolomeitsev and Voskresensky 2015).

In any case, all these corrections (relativistic, weak magnetism, in-medium) modify the estimate (9.105) at most by a factor of a few, leaving all the principal astrophysical consequences based on its high rate intact.

Now we turn to the situation where the proton fraction is small and the triangle condition forbids the direct Urca process. Then the next order processes come into play, which all involve the strong interaction. The corresponding reaction rates are subject to uncertainties in the description of nucleon-nucleon interactions in medium. In this regard, the situation is similar to the problem of kinetic coefficients discussed in Sect. 9.4.1 but is more complicated even if in-medium effects are not considered. This can be understood by looking at the relevant diagrams for the bremsstrahlung reaction (9.96) presented in Fig. 9.1. The dashed line represents the emitted lepton pair, while the block T_{NN} represents the nucleon strong interaction amplitude. With respect to the strong interaction vertex, the emission of the lepton pair occurs from 'external legs' in diagrams (a) and (b), but there are also rescattering contributions (c) and emission from internal meson exchange lines (d) (Hanhart et al. 2001). Moreover, the amplitude T_{NN} is half on-shell in contrast to the on-shell amplitude involved in the kinetic coefficients calculations.

Despite the considerable progress achieved in recent decades in the treatment of the nucleon interaction, in practice the standard benchmark for the electroweak bremsstrahlung reactions follows the work of Friman and Maxwell (1979), who used the lowest-order one-pion exchange (OPE) model to describe the nucleon interaction. The neutron star cooling simulations which employ neutrino emissivities calculated in the Friman and Maxwell (1979) model are called 'standard cooling scenarios' (e.g., Yakovlev and Pethick 2004). However, it is well-known that at energies relevant to neutron stars, OPE in the Born approximation overpredicts the cross-section by a factor of a few. Friman and Maxwell (1979) also estimated the in-medium effects of long-range and short-range correlations by considering a special form of the correlated potential, utilizing the set of the Landau-Migdal parameters, and investigating the role of the one-ρ exchange. In their calculations, Friman and

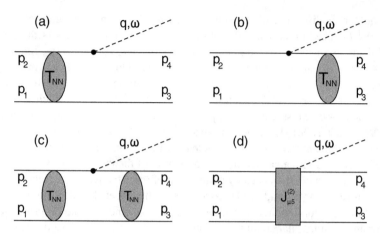

Fig. 9.1 Feynman diagrams contributing to the NN electroweak bremsstrahlung reactions. Here $p_1 \ldots p_4$ label initial and final nucleon momenta, q, ω are the emitted lepton pair [$\nu\bar{\nu}$ for bremsstrahlung reactions (9.96) and $\bar{\nu}_\ell \ell$ for modified Urca reactions (9.97)] momentum and energy, respectively. T_{NN} is the quasiparticle scattering amplitude. In diagram (d), the $J^{(2)}_{\mu 5}$ block represents the two-body axial vector current. The figure is redrawn from Hanhart et al. (2001)

Maxwell considered diagrams (a) and (b) and used the non-relativistic V-A model for the weak vertices. They found that in the limit of small q, the vector current contributions of diagrams (a) and (b) cancels exactly for both OPE and Landau interactions. This is true for both neutral current and charged current (modified Urca) reactions (although the cancellation in latter case is nontrivial and involves exchange contributions). Therefore they concluded that the neutrino emission is dominated by the axial vector current. We will see below that this result survives in a more elaborate treatment.

One may use a more universal approach based on the soft electroweak bremsstrahlung theorem. This method is similar to the soft-photon theorems for the electromagnetic emission (Low 1958; Burnett and Kroll 1968) which relates the cross-sections of the radiative processes in the leading and sub-leading orders to the corresponding cross-sections of the non-radiative processes. The soft photon theorem was extended to the axial vector currents by Adler and Dothan (1966) and first applied to neutrino emission in Hanhart et al. (2001) and Timmermans et al. (2002). Hanhart et al. (2001) employed the dominant term in the soft expansion, while Timmermans et al. (2002) proved the general soft electroweak bremsstrahlung theorem. They also analyzed the full relativistic structure of the weak currents and the strong interaction amplitude. It was found that in the extreme non-relativistic limit the vector current contribution vanishes irrespectively of the details of the structure of the amplitude T_{NN}, which generalizes the Friman & Maxwell result. The basis of the soft emission theorem is the requirement of the vector current conservation and the partial axial vector current conservation. In the soft limit, the dominant contribution comes from the diagrams (a)–(b) in Fig. 9.1, while the diagrams (c)–(d) are of higher order in q, ω.

The hadronic part of the matrix element corresponding to diagram (b) in Fig. 9.1 can be written as

$$M_{fi} \propto \langle f | \hat{\Gamma}^{Z/W} \mathcal{G}_N(p_2 - q; \varepsilon_2 - \omega) T_{NN} | i \rangle, \qquad (9.107)$$

where $\hat{\Gamma}^{Z/W}$ is the neutral or charged current weak vertex, and $\mathcal{G}_N$ is the nucleon propagator in the intermediate nucleon line. Non-relativistically, the quasiparticle propagator can be written in the form

$$\mathcal{G}_N(p; \varepsilon) = \left[\frac{p^2}{2m_N^*} - \frac{p_{FN}^2}{2m_N^*} - (\varepsilon - \mu_N) \right]^{-1}. \qquad (9.108)$$

When the lepton pair energy and momentum are small, $\omega \ll \varepsilon_2$, $q \ll p_2$, one obtains $\mathcal{G}_N(p_2 - q; \varepsilon_2 - \omega) \approx \omega^{-1}$. Therefore, in the soft limit, $M_{fi} \propto \omega^{-1}$.

For the bremsstrahlung reactions (9.96) the emitted energy is equal to the energy of the neutrino pair $\omega = \omega_\nu$, that is of the order of the temperature $T \lesssim 10$ MeV. Therefore, the soft limit is directly applicable. In contrast, for the modified Urca processes (9.94), the emitted energy is basically the degenerate lepton Fermi energy $\mu_\ell \sim 100$ MeV, which is not small. We thus first discuss the reactions (9.96), although they are generally less important for neutron stars than the modified Urca processes. For nn (or pp) bremsstrahlung, the soft limit matrix element of the axial vector nucleon current in the non-relativistic limit can be expressed via the on-shell scattering amplitude as (Hanhart et al. 2001; Timmermans et al. 2002)

$$J_{fi}^A \propto \frac{g_A}{\omega_\nu} [T_{NN}, S]_{fi}, \qquad (9.109)$$

where S is the operator of the total spin of the nucleon pair and the square brackets denote the commutator. The denominator ω_ν comes from the virtual nucleon propagator in the dominant 'external legs' diagrams [diagrams (a) or (b) in Fig. 9.1]. This allows us to calculate the reaction rates and emissivities in a model-independent way based on the experimentally measured phase shifts (in other words, T_{NN}). A similar approach was used in Baiko et al. (2001) for transport coefficients (see Sect. 9.4.1). An analogous (but different) expression can be written for the np bremsstrahlung (Timmermans et al. 2002).

It is customary to write the expression for the neutrino emissivity in the Friman & Maxwell form (Friman and Maxwell 1979; Yakovlev et al. 2001)

$$\epsilon_\nu^{nn} = 7.5 \times 10^{11} \left(\frac{m_n^*}{m_N} \right)^4 \left(\frac{n_n}{n_0} \right)^{1/3} \left(\frac{T}{10^8 \text{ K}} \right)^8 \alpha_{nn}^{\text{ex}} \beta_{nn} \mathcal{N}_\nu \text{ erg cm}^{-3} \text{ s}^{-1}, \qquad (9.110)$$

where $\mathcal{N}_\nu = 3$ is the number of neutrino flavors and $\alpha_{nn}^{\mathrm{ex}} \approx 0.8$ is the dimensionless factor coming from the angular averaged OPE matrix element; its density dependence is very mild.[15] (The superscript 'ex' indicates that the exchange contribution is included.) Note that Friman and Maxwell (1979) overestimated the exchange contribution by a factor of two because they used incorrect symmetry factor $s = 1/2$ [see Eq. (9.98)] instead of $s = 1/4$ which should be used when working with the antisymmetrized amplitudes[16] (Lykasov et al. 2008; Hannestad and Raffelt 1998; Yakovlev et al. 2001; Hanhart et al. 2001). The same incorrect factor seems to have been used in Maxwell (1987); Schwenk et al. (2004); van Dalen et al. (2003). The factor β_{nn} takes into account all other corrections beyond OPE, see below. Similar expressions can be written down for np and pp bremsstrahlung, resulting in $\epsilon_\nu^{pp} < \epsilon_\nu^{np} < \epsilon_\nu^{nn}$ (see for instance Yakovlev et al. 2001). The calculations show that the use of the realistic T_{NN} matrix instead of the OPE leads to the suppression of the neutrino emissivity approximately by a factor of four [so $\alpha_{nn}^{\mathrm{ex}}(T_{NN}) \approx 0.2$] (Hanhart et al. 2001; Timmermans et al. 2002; van Dalen et al. 2003; Li et al. 2009, 2015). Note that the inclusion of additional meson exchange terms (Friman and Maxwell 1979; van Dalen et al. 2003) results in a better agreement with T-matrix calculations. Li et al. (2015) quantified the limits of applicability for the soft bremsstrahlung theorem for a certain realistic model of the nucleon interactions. They found that the approximation (9.109) is accurate within 10% up to $\omega_\nu = 60$ MeV. Relativistic corrections for the densities of interest are within 5–15% (van Dalen et al. 2003).

Up to now we have not considered in-medium effects. Since the bremsstrahlung reactions are based on nucleon-nucleon collisions, one deals with similar complications as in Sect. 9.4.1.3, where we discussed transport coefficients. As usual, one medium effect is in the renormalization of the effective masses. Since m^* enters the bremsstrahlung emissivities in the fourth power, bremsstrahlung is more sensitive to the effective mass than the direct Urca processes. Apart from the effective masses and possible renormalization of weak coupling constants, correlations in the medium modify the quasiparticle scattering rates. It was found by van Dalen et al. (2003) that the use of the G-matrix of BHF theory (three-body forces not included) instead of the T-matrix results in some 30% increase of the bremsstrahlung rate. This is in contrast to estimates by Blaschke et al. (1995), who found a decrease of the in-medium emissivity by a factor of 10–20. A possible source of this discrepancy may be the omission of the tensor $^3P_2 - {}^3F_2$ coupling in the latter work (van Dalen et al. 2003). Schwenk et al. (2004) used quasiparticle scattering amplitudes which are constructed from renormalization group methods based on the low momentum

---

[15]The numerical prefactor in Eq. (9.110) is calculated assuming a charge-independent value of the pion-nucleon coupling constant $f_{NN\pi}^2 = 0.08$, as used in Friman and Maxwell (1979); Yakovlev et al. (2001). Using $f_{NN\pi}^2 = 0.075$, as in Timmermans et al. (2002), results in a prefactor 6.6 instead of 7.5. This difference plays no role in practical applications.

[16]This gives $1/2$, and another $1/2$ is needed to account for double counting of collisions when integrating over distributions of initial particles.

universal potential V_{low-k} (Schwenk and Friman 2004). Their results computed to second-order give an overall reduction of the emissivity by a density-dependent factor of $4 - 10$. This reduction is higher at lower densities and thus more relevant to supernova studies (remember that their values must be divided by two (Lykasov et al. 2008)).

More general calculations can be performed in the framework of linear response theory by studying the response of nuclear matter to a weak probe. As it is clear from Eq. (9.109) and the discussion above, essential information is contained in the dynamical spin response function $S_\sigma(\omega, q)$, see for example Lykasov et al. (2008), Hannestad and Raffelt (1998) and references therein. For results within Landau's Fermi liquid theory see Lykasov et al. (2008) and a general overview of the correlated basis function approach to the weak response is given in Benhar and Lovato (2015). In any case, at present the medium modifications of the bremsstrahlung rates deep in the interior of neutron stars are quite uncertain. One can easily imagine a modification by a factor of two in any direction.

In the medium-modified OPE (MOPE) model of Migdal et al. (1990), Voskresensky (2001), the emissivity receives a strong density-dependent correction due to the softening of the pion mode. This correction results in a factor that can be written as (Kolomeitsev and Voskresensky 2015)

$$\beta_{nn}^{MOPE} = 3 \left(\frac{n}{n_0} \right)^{4/3} \frac{[\Gamma(n)/\Gamma(n_0)]^6}{(\tilde{\omega}_\pi/m_\pi)^3}, \tag{9.111}$$

where $\Gamma(n) \approx [1 + 1.6(n/n_0)^{1/3}]^{-1}$ and $\tilde{\omega}_\pi$ is the effective pion gap in the medium. The adopted density dependence of the pion gap results in suppression at $n \lesssim n_0$ and in significant enhancement at $n \gtrsim n_0$ up to a factor of $\beta_{nn}^{MOPE} \approx 100$ depending on a model adopted for the pion gap.

Finally, in the exotic case of proton localization, also discussed in Sect. 9.4.1, the neutrino emissivity due to scattering of neutrons off the localized protons was considered by Baiko and Haensel (1999). Its interesting feature is the T^6 temperature dependence of the emissivity, compared to T^8 for the ordinary bremsstrahlung processes. This contribution could thus be very important, provided that the phenomenon of proton localization is realized in neutron stars. According to the results of Baiko and Haensel (1999), the ratio $\epsilon_\nu^{n-loc.p}/\epsilon_\nu^{MU}$ can be as large as $2 \times 10^3 \, T_8^{-2}$, dominating the neutrino emission in the neutron star core.

Unfortunately, the situation for the modified Urca reactions (9.97) is even more cumbersome. As stated above, the applicability of soft electroweak theorems is not justified since the energy of the lepton pair is not small, $\omega \sim p_{F\ell}$. Therefore it is not immediately clear that the 'leg' diagrams (a) and (b) in Fig. 9.1 give the dominant contribution. Moreover, off-shell amplitudes should be used. In $npe\mu$ matter one considers two branches of the modified Urca processes, namely neutron and proton branches [$\mathcal{C} = n$ and $\mathcal{C} = p$ in Eq. (9.97)]. We focus on the neutron branch, for the proton branch see Yakovlev et al. (2001).[17] In the OPE approximation, the emissivity of the modified Urca process from the leg diagrams is (Friman and Maxwell 1979; Yakovlev et al. 2001)

$$\epsilon_\nu^{MU} = 8.1 \times 10^{13} \left(\frac{m_n^*}{m_n}\right)^3 \left(\frac{m_p^*}{m_p}\right)^4 \left(\frac{p_{F\ell}c}{\mu_\ell}\right) \left(\frac{n_p}{n_0}\right)^{1/3} \left(\frac{T}{10^8 \ K}\right)^8 \alpha_{MU}^{n\ell} \ \text{erg cm}^{-3} \ s^{-1},$$

$$(9.112)$$

where $\alpha_{MU}^{n\ell} \approx 1$ comes from the averaged matrix element (Yakovlev et al. 2001). Comparing with Eq. (9.110), one sees that the neutrino emissivity from the modified Urca process is more than 50 times stronger than that of the bremsstrahlung process. The reaction rate is given by the equation analogous to Eq. (9.113), but with different numerical constants in the prefactors

$$\Gamma^{MU} = \frac{0.106}{T}\epsilon_\nu^{MU}, \qquad \Delta\Gamma^{MU} = \frac{0.129}{T}\frac{\delta\mu}{T}\epsilon_\nu^{MU} \qquad (\text{for } \delta\mu \ll T). \qquad (9.113)$$

The correlation effects considered by Friman and Maxwell (1979) reduce the rate in Eq. (9.112) approximately by a factor 1/2. From the above discussion of the nn bremsstrahlung, one expects further reduction of the emissivity when going beyond the OPE approximation towards the full scattering amplitude. Indeed, according to estimates by Blaschke et al. (1995), the expected reduction is about 1/4 with respect to the OPE result and the use of the in-medium T-matrix leads to further reduction by an additional factor of $0.6 - 0.9$.

In the recent work by Dehghan Niri et al. (2016), the in-medium modified Urca emissivity was calculated starting from the correlated nucleon pair states (see also Sawyer and Soni 1979 and Haensel and Jerzak 1987). The pair correlation functions are determined in the lowest-order constrained variational (LOCV) procedure. The LOCV functions turn out to be similar to those obtained in the BHF scheme (Baldo and Moshfegh 2012). The modified Urca emissivity calculated in this way by construction effectively includes rescattering diagrams of type (c) in Fig. 9.1 (as pointed out already by Friman and Maxwell 1979). It was found that the LOCV result at two-body level shows a reduction of the emissivity from the Friman & Maxwell result. The reduction becomes more pronounced with increasing density, reaching a factor of 4 at $n = 3n_0$. However, the inclusion of phenomenological three-body forces (in the Urbana IX model (Carlson et al. 1983)) eliminates this reduction, and the LOCV result with three-body forces turns out to be close to that of Friman and Maxwell (1979).

One might expect that in the MOPE model of in-medium nuclear interactions (Migdal et al. 1990; Voskresensky 2001) the correction of the emissivity of the modified Urca process is similar to the bremsstrahlung correction given in

---

[17]Note that the angular factor $\hat{\Omega}$ [see Eq. (9.101)] for the proton branch is slightly incorrect in Yakovlev et al. (2001), see Gusakov (2002), Kaminker et al. (2016) for details.

Eq. (9.111). This is not the case. According to the analysis of Voskresensky (2001), at $n \gtrsim n_0$, diagrams of type (d) dominate, which describe in-medium conversion of a virtual charged pion to a neutral pion with the emission of a real lepton pair. The modification factor for the medium modified Urca reaction with respect to the free one-pion exchange result is

$$\beta_{\mathrm{MMU}} = 3 \left(\frac{n}{n_0} \right)^{10/3} \frac{[\Gamma(n)/\Gamma(n_0)]^6}{(\tilde{\omega}_\pi/m_\pi)^8}. \tag{9.114}$$

Comparing with Eq. (9.111), one finds a higher power of the pion gap $\tilde{\omega}_\pi$ in the denominator and a stronger density dependence in the prefactor. With typical parameters, one obtains enhancement by a factor of $\beta_{\mathrm{MMU}} \sim 3$ at $n \sim n_0$ and up to $\beta_{\mathrm{MMU}} \approx 5 \times 10^3$ at $n = 3n_0$. Note, however, that this enhancement strongly depends on the uncertain values of the pion gap $\tilde{\omega}_\pi$ and the vertex correction Γ entering Eqs. (9.111) and (9.114).

The modified Urca process is dominant when the direct Urca process is forbidden. When the density is sufficiently close to (but still below) the direct Urca threshold, one needs to take into account the softening of the nucleon propagation in the virtual lines when examining the 'leg' contributions (a) and (b) in Fig. 9.1. This can increase the modified Urca rates significantly in comparison to the standard result (Shternin et al. 2018). For example, consider diagram (b) for the neutron branch of the modified Urca reaction, i.e., p_2 corresponds to a neutron line. After emitting the lepton pair with momentum $q \approx p_{F\ell}$ and energy $\omega \approx \mu_\ell + \omega_\nu$, the neutron transforms to a virtual proton with energy $\varepsilon = \mu_n - \omega$ and momentum $k = p_2 - q$ well above the Fermi surface ($k > p_{Fp}$). The standard practice (e.g., Friman and Maxwell 1979) is to set the proton propagator to $\mathcal{G}_0^{-1} = -\omega \approx -\mu_\ell$ in Eq. (9.108). However, in the case of backward emission $q \uparrow\downarrow p_2$ ($k = p_{Fn} - p_{F\ell}$), the intermediate momentum k can be close to k_{Fp} and in beta-stable matter $\varepsilon \approx \mu_p$ (we neglect ω_ν here). When $\rho > \rho_{\mathrm{DU}}$ this results in a pole on the real-axis ($\mathcal{G}^{-1} = 0$ for some values of q), manifesting opening of the direct Urca process, while for $\rho \to \rho_{\mathrm{DU}}$, the intermediate proton line softens in a certain (backward) part of the phase space. In other words, $\mathcal{G}^{-1}$ can be much smaller than $\mathcal{G}_0^{-1}$ when $k \to p_{Fp}$, leading to a strong enhancement of the neutrino emissivity. As a consequence, only the vicinity of $k \approx p_2 - q$ is important when calculating the matrix element in (9.107), i.e., only weakly off-shell values of T_{NN} are needed. In this sense, one reinstalls the soft bremsstrahlung theorem in a certain way. A crude estimate of the effect of the nucleon softening can be obtained by neglecting all momentum dependence in (9.107) except for $\mathcal{G}$. For the contribution of diagram (b) in Fig. 9.1 one gets (Shternin et al. 2018)

$$\frac{\left\langle \left| M_{fi}^{(b)} \right|^2 \right\rangle_\Omega}{\left\langle \left| M_{fi}^{(b)0} \right|^2 \right\rangle_\Omega} \approx \frac{m_p^{*2} \mu_\ell}{2 p_{Fp}^2 p_{F\ell}\, \delta p}, \qquad \delta p \ll p_{Fn}, \tag{9.115}$$

where $M_{fi}^{(b)}$ and $M_{fi}^{(b)0}$ are calculated using $\mathcal{G}$ and $\mathcal{G}_0$, respectively, and $\delta p = p_{Fn} - p_{Fp} - p_{F\ell}$ measures the distance from the direct Urca threshold in terms of momenta. A slightly different result is found for the (a) diagram[18] but the δp^{-1} asymptotic behavior is the same. The exchange contributions somewhat complicate this picture, however they are of next order in δp. The correction (9.115) to the modified Urca rates leads to a more pronounced density dependence of Q^{MU} than given in Eq. (9.112) and a significant rate enhancement at $\rho \to \rho_{DU}$. Moreover, calculations show that the rates are enhanced by a factor of several for all relevant densities in neutron star cores (even far from ρ_{DU}). Notice that this result is universal in a sense that it does not depend on the particular model employed for the strong interaction. A more detailed study of this effect would be interesting.

Let us note that the softening of the intermediate nucleon, which results in the enhancement of the modified Urca rate given by Eq. (9.115), has similarity with the MOPE result (9.114) where the enhancement is due to softening of the intermediate pion. We note, however, that the dominance of the diagram (d) contribution over diagrams (a)-(c) in MOPE calculations was found without taking into account enhancement of the latter by the effects described above. This can alter the MOPE result.[19]

At high temperatures, one needs to go beyond the quasiparticle approximation and take into account coherence effects such as the Landau-Pomeranchuk-Migdal (LPM) effect. When the quasiparticle lifetime becomes small [the spectral width $\gamma(\omega)$ of the quasiparticle becomes large], it undergoes multiple scattering during the formation time of the radiation. In this case, the picture of well-defined quasiparticles fails and the nuclear medium basically plays the role of the spectator in reactions (9.96)–(9.97), such that the process (9.95) essentially becomes allowed. Calculations of the reaction rates and emissivities become more involved (Knoll and Voskresensky 1995). The finite spectral width in the nucleon propagators regularize the infrared divergence (9.109), leading to the LPM suppression of the reaction rates. This becomes important when $\omega \sim T \lesssim \gamma(\omega)$. According to various calculations, the threshold temperature is rather large, $T \gtrsim 5$ MeV (van Dalen et al. 2003; Sedrakian and Dieperink 1999; Lykasov et al. 2008; Shen et al. 2013).

9.4.2.3 Neutrino Emission in the Presence of Cooper Pairing of Nucleons

The onset of the pairing instability has a strong effect on the reaction rates and the neutrino emission, as already mentioned in Sect. 9.2.3.4. As discussed in Sect. 9.4.1.4, the presence of the gap in the quasiparticle spectrum reduces the available phase space for the reaction to proceed and the reaction rates become

---

[18]One should substitute m_p^* by m_n^* and one of the factors p_{Fp} in the denominator by p_{Fn} in Eq. (9.115).

[19]Notice that it is not sufficient simply to compare Eqs. (9.114) and (9.115), since the 'leg' diagrams (a)-(b) also possess MOPE enhancement, c.f. Eq. (9.111).

strongly suppressed (in the close-to-equilibrium situation). In addition, the number of quasiparticles is not conserved now, which opens new reaction channels, namely those corresponding to Cooper pair breaking and formation.

The modifications due to Cooper pairing are usually described by superfluid 'reduction factors',

$$\epsilon_\nu^{\mathrm{SF}} = \epsilon_\nu^N R_\epsilon \left(\left\{ \frac{\Delta_i}{T} \right\}, \frac{\delta\mu}{T} \right), \tag{9.116a}$$

$$\Delta\Gamma^{\mathrm{SF}} = \Delta\Gamma^N R_\Gamma \left(\left\{ \frac{\Delta_i}{T} \right\}, \frac{\delta\mu}{T} \right), \tag{9.116b}$$

where SF refers to superfluid and N to non-superfluid. The factors R_ϵ and R_Γ describe the superfluid modifications of the total emissivity and the equilibration rate (for the composition-changing reactions), respectively, and depend on the superfluid gaps Δ_i, where i labels the superfluid species,[20] and on the chemical potential imbalance $\delta\mu$. Calculating these factors accurately is a complicated task. Effects of superfluidity enter the original expressions for the rate and the emissivity (9.98) through the superfluid quasiparticle distribution functions in (9.99) and the energy spectra in the delta-function (9.100). They also affect the matrix element M_{fi}, allowing for the quasiparticle number non-conservation, and, moreover, the weak interaction vertices can be affected by the response of the condensate [this has crucial consequences when the emission due to Cooper pairing (9.95) is considered, see below]. All effects of pairing can be taken into account consistently by starting from the full propagators in the so-called Nambu-Gorkov space. We shall briefly discuss this approach in Sect. 9.5.3 for the direct Urca process in color-superconducting quark matter, see Eq. (9.135) and discussion below that equation. In almost all calculations of the reduction factors $R_{\epsilon/\Gamma}$ we are aware of for nuclear matter, the modifications of the reaction cross-sections are not considered, and only the phase-space modifications are included. For the direct Urca processes (9.94) this approach is well-justified, see section 4.3.1 in Yakovlev et al. (2001). The relative contribution of the number-conserving channels of the reaction and channels which include breaking and formation of Cooper pairs are considered in Sedrakian (2005), Sedrakian (2007). At high temperatures, the scattering contribution dominates, while at $T \rightarrow 0$ its contribution decreases to one half of the total rate. In practice, there is no need to consider these contributions separately and one can use M_{DU} (9.104) without superfluid modifications (Yakovlev et al. 2001). The effects of the superfluid coherence factors on the matrix element of the electroweak bremsstrahlung reactions have, to our knowledge, not been explored. Therefore, in the following, we briefly discuss the results for the superfluid reduction factor obtained without superfluid effects on M_{fi}. Such factors are universal since they

---

[20]For simplicity, here we only use the term superfluidity, including proton Cooper pairing. The distinction between superfluidity and superconductivity is not important in the present context.

do not depend on the details of the strong interaction and are assumed to reflect the main properties of the correct results.

Let us first consider beta-equilibrated matter, $\delta\mu = 0$ (Yakovlev et al. 1999, 2001; Gusakov 2002). Even with the above simplifications, the calculation of $R_{\epsilon/\Gamma}$ require laborious efforts because it has to account for the possible coexistence of proton pairing and (anisotropic) neutron pairing. Recall that protons are assumed to pair in the isotropic $^1S_0$ state, while neutrons in the neutron star core are paired in the $^3P_2(m_J = 0, \pm1, \pm2)$ channel with a possible admixture from the $^3F_2$ channel. Only the cases $m_J = 0$ and $|m_J| = 2$ are considered in detail since they do not include integration over the azimuthal angle of the quasiparticle momentum about the quantization axis. The angular integration over the polar angle must be carried out, which precludes using the angular-energy decomposition in the form of Eq. (9.101).[21] At very low temperatures, $T \ll T_c$, where T_c is the critical temperature for pairing, participation in the reaction of a paired fermion species in the $^1S_0$ or $^3P_2(m_J = 0)$ channels leads to an exponential suppression of the rates. The case $|m_J| = 2$ is qualitatively different since the gap contains nodes on the Fermi surface. Then the suppression is given by a power law in T/T_c. At intermediate temperatures, most interesting in practice, the suppression factors show an approximate power-law dependence for any superfluidity type, even for a fully gapped spectrum. Numerical results and fitting expressions for direct Urca, modified Urca, and bremsstrahlung reactions for various combinations of pairing types, isotropic and anisotropic, can be found in Gusakov (2002), which also contains a review of other works.

The beta-equilibrium conditions can be perturbed by various processes, for instance by compression. In the superfluid case, since the reaction rates are suppressed, the system cannot counterbalance a growing perturbation $\delta\mu$. If $\delta\mu$ becomes larger than the pairing gap Δ the reactions become unblocked. If the direct Urca process is allowed by momentum conservation, the threshold value is $\delta\mu_{th} = \Delta_n + \Delta_p$. Otherwise, if the modified Urca process is responsible for beta-equilibration, $\delta\mu_{th} = \Delta_n + \Delta_p + 2\min(\Delta_n, \Delta_p)$ (Reisenegger 1997). When $\delta\mu > \delta\mu_{th}$, the beta-equilibration reaction which decreases $\delta\mu$ is allowed and is no longer suppressed by the presence of gaps. The value of $\delta\mu$ determines the phase space for the reaction, like in case of normal matter in the supra-thermal regime. Reisenegger (1997) first suggested this effect[22] and qualitatively described it by introducing step-like suppression factors $R_{\epsilon,\Gamma} = \Theta(\delta\mu - \delta\mu_{th})$ [it is understood

---

[21] In the case of bremsstrahlung or modified Urca, which can include two neutrons with anisotropic pairing, the matrix element in principle cannot be taken out of the integration since it depends on the relative orientation of the scattered particles even without superfluid modifications. This is always neglected. Note, however, that in this case the region of the momenta orientation that imply lowest gaps will be extracted from M_{fi}. One can expect considerable modifications if the angular dependence of M_{fi} is not flat (this is the case for n–p scattering, which contributes to modified Urca and n-p bremsstrahlung rates).

[22] This idea was re-discovered recently under the name of 'gap-bridging' in Alford et al. (2012a), Alford and Pangeni (2017).

that the 'N' quantities in Eqs. (9.116) are calculated including $\delta\mu$]. Later these results were improved in Villain and Haensel (2005), Pi et al. (2010), Petrovich and Reisenegger (2010), González-Jiménez et al. (2015), where discussions of the behavior of the R-factors and complications of the numerical scheme can be found. In the recent study by González-Jiménez et al. (2015), the most general case of anisotropic pairing in considered, but unfortunately no analytical approximation for the reduction factors are given. It would be nice (but not easy) to obtain approximations similar to those presented in Gusakov (2002), but for the non-equilibrium case. In fact, according to Eq. (9.103), it is enough to find one of the factors R_ϵ or R_Γ (Flores-Tulián and Reisenegger 2006).

Now let us turn to the neutral weak current emission associated with the Cooper pair breaking and formation (CPF) processes in the reaction given in Eq. (9.95). These processes were already mentioned in Sect. 9.2.3.4. The process (9.95) is a first-order process in the number of quasiparticles and therefore does not explicitly depend on the strong interaction details (although strong interactions determine, for instance, the value of the gap). This process is kinematically forbidden in the normal matter but becomes allowed if the nucleons pair. It was proposed by Flowers et al. (1976) and later rediscovered by Voskresensky and Senatorov (1987) The expression for the emissivity can be written as (Yakovlev et al. 2001)

$$\epsilon_\nu^{CP} = 1.17 \times 10^{14} \left(\frac{m_N^*}{m_N}\right) \left(\frac{p_{FN}}{m_N c}\right) \left(\frac{T}{10^8 \text{ K}}\right)^7 \alpha^{CP} F(\Delta_N/T) \mathcal{N}_\nu \text{ erg cm}^{-3} \text{ s}^{-1},$$

(9.117)

where, as usual, $\alpha^{CP} = \alpha_V^{CP} + \alpha_A^{CP}$ is a dimensionless number that arises from the matrix element of the process containing vector α_V^{CP} and axial-vector α_A^{CP} contributions and $F(\Delta_N/T)$ comes from the energy integration (and angular integration in the anisotropic case). Near the critical temperature $T \to T_{cN}$, the function F approaches zero linearly, $F \propto (1 - T/T_{cN})$, and for low temperatures F behaves qualitatively like the reduction factors $R_{\epsilon/\Gamma}$, i.e., the emissivity is exponentially suppressed, unless there are nodes of the gap on the Fermi surface, in which case F behaves according to a power-law in temperature (Yakovlev et al. 2001). Thus, at low temperatures, the CPF emission is strongly suppressed. However, the function F has a maximum at $T \sim 0.8 T_{cN}$ and in the vicinity of this temperature the CPF emission can be the dominant neutrino emission mechanism in the superfluid neutron star core. Therefore, the CPF process is an important ingredient in the so-called minimal cooling scenario of the thermal evolution of isolated neutron stars (Page et al. 2004; Gusakov et al. 2004; Page et al. 2009).

During the last decade, significant improvements in CPF emission studies were made. Crucially, one has to take into account consistently the response of the condensate (collective modes), which enters the emissivity through the anomalous part of the weak vertices. This is achieved by a proper renormalization of the vertices, which ensures vector current conservation (Kundu and Reddy 2004). As a consequence, the CPF emission for singlet-paired matter is strongly suppressed in the non-relativistic limit, as pointed out by Leinson and Pérez (2006), and later elaborated in Sedrakian et al. (2007), Leinson (2008), Kolomeitsev and Voskresensky (2008), Steiner and Reddy (2009), Leinson (2009), Kolomeitsev and Voskresensky (2010), Kolomeitsev and Voskresensky (2011), Sedrakian (2012). Although some controversy about the results from different approaches still exists, the main conclusion is that $\alpha_V^{CP} \propto g_V^2 v_{FN}^4$, being small in the non-relativistic limit. Recall that the singlet nucleon pairing is present in the low-density, hence non-relativistic ($v_{FN} \ll 1$), domain. The axial-vector contribution does not receive any vertex correction because there is no spin response from the condensate in the case of singlet pairing. However, this contribution itself is a relativistic effect, $\alpha^{CP} \approx \alpha_A^{CP} = \tilde{\alpha}_A^{CP} g_A^2 v_{FN}^2$, where $\tilde{\alpha}_A^{CP}$ is a numerical factor of order unity whose precise value is subject to debates. The final conclusion is that the CPF emission from the singlet pairing is not important, being much smaller than the bremsstrahlung contribution even when the latter is suppressed by the superfluid reduction factor R_ϵ.

The situation is different for the case of triplet pairing of neutrons, which is thought to occur in a large fraction of the neutron star core. Without taking into account the condensate response, the CPF emission from the triplet superfluid is given by Eq. (9.117) with $\alpha^{CP} = g_V^2 + 2g_A^2 \approx 4.17$ (Yakovlev et al. 2001). Up to very high densities, where the triplet pairing is usually found to vanish, the neutrons can still be considered non-relativistic. By analogy with the case of singlet pairing, one expects that the vector current conservation would suppress the vector contribution to the emissivity also in the triplet case. This leads to the approximation $\alpha^{CP} = 2g_A^2$, by a factor of about 0.78 less than the initial result (Page et al. 2009). However, in contrast to the singlet case, the order parameter of the triplet superfluid varies under the action of the axial-vector field (Leinson 2010). As a consequence, this modifies the axial-vector contribution to the emissivity. This was considered by Leinson (2010), using an angular-averaged gap as an approximation. He indeed found the suppression of the vector contribution in the non-relativistic limit, while for the axial-vector contribution the result is $\alpha^{CP} \approx \alpha_A^{CP} = \frac{1}{2} g_A^2 \approx 0.8$. Thus taking into account the condensate response reduces the emissivity by a factor of 0.19 compared values given by Yakovlev et al. (2001). This quenching has observational consequences if the real-time thermal evolution of the superfluid neutron star can be observed, see for instance Shternin and Yakovlev (2015). The actual angular dependence of the gap and Fermi-liquid effects modify this result only slightly (within 10% according to Leinson 2013). Thus, even with the more elaborate treatment of the superfluid response to weak perturbations, the neutrino emissivity due to CPF processes from triplet neutron superfluidity can be the dominant neutrino emission mechanism at $T \sim 0.8 \, T_{cn}$.

Collective modes in the superfluid (see Sect. 9.4.1.4) can also contribute to the neutrino emissivity. The emission related to the collisions of the Goldstone modes—angulons—in the triplet superfluid was considered in Bedaque et al. (2003) and found to be always negligible. However, Bedaque and Sen (2014) recently considered the case of a strong magnetic field, to which the neutron fluid couples through the neutron magnetic moment. Since the magnetic field breaks rotational symmetry explicitly, one of the angulon modes acquires a gap of the order of $eB/(m_n^*c)$ and its decay to a neutrino pair becomes kinematically allowed. The resulting neutrino emissivity can be written in a form similar to Eq. (9.117), where the function $F(\Delta_N/T)$ is replaced by the B-dependent function $h(g_n B/(aT))$, where g_n is the neutron magnetic moment, and $a = 4.81$. This function $h(x)$ peaks at $x \sim 7$ and is exponentially suppressed at large x (small T). According to the numerical estimates in Bedaque and Sen (2014), the neutrino emissivity due to the 'magnetized angulon' decay can be larger than that of the CPF process at $T \approx 10^7$ K provided the interior magnetic field is as large as $B \sim 10^{15}$ G (the situation where the magnetic field is confined in flux tubes of the proton superconductor is also discussed).

9.4.2.4 Electromagnetic Bremsstrahlung

The preceding sections do not contain all neutrino emission processes in the cores of neutron stars. Other possibly relevant processes are discussed in Yakovlev et al. (2001), see also Potekhin et al. (2015b). Here we briefly discuss new results for the electromagnetic bremsstrahlung emission, obtained after, and thus not included in, Yakovlev et al. (2001). The emission from the electromagnetic bremsstrahlung

$$\ell + C \rightarrow \ell + C + \nu + \bar{\nu}, \qquad (9.118)$$

where $\ell = e, \mu$ is a lepton and C is some electrically charged particle, is thought to be several orders of magnitude smaller than those from collisions mediated by strong interactions (Yakovlev et al. 2001). Still, the lepton-lepton bremsstrahlung may be the dominant process for low-temperature superfluid matter (with both neutrons and protons in the paired state), where the neutrino emission processes involving baryons are suppressed. The studies reviewed in Yakovlev et al. (2001) underestimated the significance of the bremsstrahlung in electromagnetic collisions. The reason is the same as discussed in Sect. 9.4.1.2—correctly taking into account screening of the transverse part of the interaction makes these collisions much more efficient. The proper transverse screening was considered for the electron-electron bremsstrahlung in Jaikumar et al. (2005), with the result

$$Q^{ee} = 1.7 \times 10^{12} \left(\frac{T}{10^8 \text{ K}}\right)^7 \left(\frac{n_e}{n_0}\right)^{2/3} \tilde{N} \left(\frac{m_D c^2}{2T}\right) \mathcal{N}_\nu \text{ erg cm}^{-3} \text{ s}^{-1}, \qquad (9.119)$$

where m_D is the electric (Debye) screening mass (corresponding to $\hbar q_l/c$ in the notation of Sect. 9.4.1.2) and $\tilde{N} \leqslant 1$ is a slowly varying function, approaching unity in the strongly degenerate limit. Like in the case of the thermal conductivity (Sect. 9.4.1.2), dynamical screening borrows one power of the temperature from the expression for emissivity, so it behaves like T^7 instead of T^8 for standard bremsstrahlung reactions (here we neglect the temperature dependence of $\tilde{N}$, which becomes important if the temperature approaches the plasma temperature). According to Eq. (9.119), the emissivity in *e-e* collisions becomes increasingly important with lowering the temperature and was underestimated by more than five orders of magnitude before that work. The bremsstrahlung emission from electron-proton (or other charged baryons) collisions should obey a similar enhancement, although the transverse channel is suppressed by the relativistic factor v_{Fp}^2 (see Sect. 9.4.1.2).

The domain of immediate importance of Eq. (9.119) is in the possible region of the inner core where the singlet proton pairing is absent, but the neutron triplet pairing exists. Then the neutrino emission due to the process in question can compete with the proton-proton bremsstrahlung due to strong forces. In the case of proton pairing, the expression (9.119) is expected to modify in the same way as the transport coefficients discussed in Sect. 9.4.1.4. Detailed studies of these effects for realistic conditions in neutron star cores are desirable but not performed yet.

9.4.3   Bulk Viscosity

As we can see from Eq. (9.14), the bulk viscosity ζ is responsible for dissipation in the presence of a nonzero divergence $\nabla \cdot \boldsymbol{v}$. Via the continuity equation (9.6a), this divergence is identical to compression and expansion of a fluid element. In a neutron star, certain oscillations lead to local, periodic compression and expansion. Therefore, bulk viscosity is an important transport property of the matter inside the star if we are interested in the damping of these oscillations. The dominant contribution to bulk viscosity is given by electroweak reactions because their time scale becomes comparable to the period of the oscillations of the star, which are typically of the order of the rotation period. Since rotation periods are of the order of a millisecond or larger, re-equilibration processes from the strong interaction play no role for bulk viscosity. The 'resonance' between the weak interaction and the oscillation frequency occurs in a certain temperature regime, usually for relatively high temperatures of about 1 MeV or higher. Bulk viscosity is thus particularly important for young neutron stars or in neutron star mergers.

To explain the interplay between the reaction rates of the weak processes and an externally given volume oscillation, let us briefly review how the bulk viscosity of dense hadronic matter is computed (Sawyer 1989a; Haensel et al. 2000). We denote the angular frequency of the volume oscillation by ω, such that we can write the volume as $V(t) = V_0[1 + \delta v(t)]$, with a (dimensionless) volume perturbation $\delta v(t) = \delta v_0 \cos \omega t \ll 1$. Then, on the one hand, we can write the dissipated energy density, averaged over one oscillation period $\tau = 2\pi/\omega$, as

$$\langle \dot{\mathcal{E}} \rangle_\tau = -\frac{\zeta}{\tau} \int_0^\tau dt \, (\nabla \cdot \boldsymbol{v})^2 \approx -\frac{\zeta \omega^2 \delta v_0^2}{2}, \tag{9.120}$$

where we have used the continuity equation (at zero velocity $v = 0$) to relate the divergence of the velocity field to the change in the total particle number density, which, in turn, is directly related to the change in volume if the total particle number is fixed. On the other hand, the dissipated energy density can be expressed in terms of the mechanical work done by the induced pressure oscillations,

$$\langle \dot{\mathcal{E}} \rangle_\tau = \frac{1}{\tau} \int_0^\tau dt \, P(t) \frac{d\delta v}{dt} , \tag{9.121}$$

where the pressure is

$$P(t) = P_0 + \frac{\partial P}{\partial V} V_0 \delta v + \sum_{x=n,p,e} \frac{\partial P}{\partial n_x} \delta n_x . \tag{9.122}$$

The oscillation in the pressure is in general out of phase compared to the volume oscillation because of the microscopic re-equilibration processes which induce changes in the number densities of the particle species δn_x. For this derivation, we consider the simplest form of hadronic matter, made of neutrons, protons and electrons. We discuss extensions to more complicated forms of matter and their bulk viscosity below.

From Eqs. (9.120) and (9.121) we compute the bulk viscosity. Let us for now assume the electroweak re-equilibration process is the direct Urca process, given by

$$p + e \rightarrow n + \nu_e , \qquad n \rightarrow p + e + \bar{\nu}_e . \tag{9.123}$$

In chemical equilibrium, the reactions (9.123) do not change the various densities because they occur with the same rate, and the sum of the chemical potentials of the ingoing particles is the same as the sum of the chemical potentials of the outgoing particles, $\delta\mu \equiv \mu_p + \mu_e - \mu_n = 0$. We assume that neutrinos and anti-neutrinos leave the system once they are produced. They can thus only be outgoing particles and we set their chemical potential to zero, $\mu_\nu \approx 0$. This assumption has to be dropped for very young (proto-)neutron stars where the temperature is large and the mean free path of neutrinos becomes much smaller than the size of the star. Then, neutrino absorption processes need to be taken into account in the calculation of the bulk viscosity, as discussed by Lai (2001). A non-equilibrium situation occurs if the equality of chemical potentials is disrupted, $\delta\mu \neq 0$. Such a disruption can be induced by the volume oscillation if the various particle species respond differently to compression and expansion. The situation considered here is particularly simple because there is a single process and a single $\delta\mu$. In general, there can be multiple processes related to the same $\delta\mu$, for instance if we include modified Urca processes (whose contribution, if the more efficient direct Urca process is allowed, can be neglected). A more complicated situation occurs if multiple processes are related to multiple $\delta\mu$'s, for instance if we include strangeness in the form of hyperons. We shall sketch the derivation of the bulk viscosity for such a case in the context of quark matter, see Sect. 9.5.4.1. Here we proceed with the single process (9.123).

In this case, the changes in the densities are all locked together,

$$\frac{dn_n}{dt} = -\frac{dn_e}{dt} = -\frac{dn_p}{dt} = \Gamma[\delta\mu(t)] \approx \lambda\,\delta\mu(t)\,, \tag{9.124}$$

where $\Gamma[\delta\mu(t)]$ is the number of neutrons produced per unit time and volume in the process $p + e \to n + \nu_e$. Using the general terminology employed at the end of Sect. 9.4.1.5, the stoichiometric coefficients of the reaction $p + e \to n + \nu_e$ are -1 for n and $+1$ for e and p, counting how many particles of a given species are created and annihilated in the given process. These numbers (in this case simply plus or minus signs) appear in Eq. (9.124). On the right-hand side of Eq. (9.124) we have applied the linear approximation for small $\delta\mu$, such that now all information about the reaction rate is included in λ, using the same notation as in Eq. (9.91). According to our definition of $\delta\mu$, a net production of neutrons sets in for $\delta\mu > 0$, from which we conclude that $\lambda > 0$. The difference in chemical potentials $\delta\mu$ oscillates due to the volume oscillation and due to the weak reactions,

$$\frac{d\delta\mu}{dt} = \frac{\partial\delta\mu}{\partial V}\frac{dV}{dt} + \sum_{x=n,p,e}\frac{\partial\delta\mu}{\partial n_x}\frac{dn_x}{dt}$$

$$= -B\frac{d\delta v}{dt} - \lambda C\delta\mu(t)\,, \tag{9.125}$$

where we have used Eq. (9.124) and abbreviated

$$B \equiv \frac{\partial P}{\partial n_p} + \frac{\partial P}{\partial n_e} - \frac{\partial P}{\partial n_n}\,, \qquad C \equiv \frac{\partial\delta\mu}{\partial n_p} + \frac{\partial\delta\mu}{\partial n_e} - \frac{\partial\delta\mu}{\partial n_n}\,. \tag{9.126}$$

These quantities are evaluated in equilibrium, i.e., they only depend on the equation of state, not on the electroweak reaction rate. We can also express the pressure (9.122) with the help of B,

$$P(t) = P_0 + \frac{\partial P}{\partial V}V_0\delta v + B\delta n_e\,. \tag{9.127}$$

In general, $\delta\mu(t)$ oscillates out of phase with the volume $\delta v(t)$, and we make the ansatz $\delta\mu(t) = \text{Re}[\delta\mu_0 e^{i\omega t}]$, with the complex amplitude $\delta\mu_0$. The differential equation (9.125) then yields algebraic equations for real and imaginary parts of $\delta\mu_0$, which can easily be solved. We compute δn_e by integrating Eq. (9.124), then insert the result into the pressure (9.127) and the result into the expression for the bulk viscosity, which is obtained from Eqs. (9.120) and (9.121). This yields

$$\zeta = -\frac{\lambda B\,\text{Re}(\delta\mu_0)}{\omega^2\delta v_0} = \frac{\lambda B^2}{(\lambda C)^2 + \omega^2}\,. \tag{9.128}$$

This is the basic form of the bulk viscosity of nuclear matter, as a function of the thermodynamic quantities B and C, the reaction rate λ, and the external angular frequency ω. It shows that bulk viscosity is a resonance phenomenon: the viscosity is maximal when the time scales set by the external oscillation frequency and the microscopic reaction rate match. Since the microscopic reaction rate typically increases with temperature T, the bulk viscosity as a function of T at fixed ω is a function with a maximum at T given by $C\lambda(T) = \omega$. It is now obvious that the strong processes, which are responsible for *thermal* equilibrium, do not contribute to the bulk viscosity because they operate on much shorter time scales. Bulk viscosity in a neutron star is utterly dominated by weak processes, which are responsible for *chemical* re-equilibration. It has been pointed out by Alford and Schmitt (2007) that the dissipation due to the out-of-phase oscillations of volume (externally given) and chemical potentials (response of the system) is completely analogous to an electric circuit with alternating voltage (externally given) and electric current (response of the system). In this analogy, which is mathematically exact and physically very plausible, the analogue of the resistance is B^{-1} and the analogue of the capacitance is $B/(C\lambda)$, while the inductance is zero.

We show the bulk viscosity of hadronic matter in Fig. 9.2. If the direct Urca process is allowed, the conversion of neutrons into protons and vice versa is faster and thus the maximum of the bulk viscosity occurs at a smaller temperature compared to the case where only the modified Urca process is at work. Since the strong interaction is needed for the modified Urca process, the corresponding rates are prone to large uncertainties, as discussed in the previous section. The bulk viscosity due to the modified Urca process used by Alford et al. (2010a), from which Fig. 9.2 is taken, is based on free one-pion exchange interaction (Friman and Maxwell 1979; Haensel 1992; Reisenegger 1995). Kolomeitsev and Voskresensky (2015) showed that medium modifications in the MOPE model can enhance the rate of the modified Urca process and thus may shift the maximum of the bulk viscosity to lower temperatures. Figure 9.2 also includes a comparison with quark matter, whose bulk viscosity we discuss in Sect. 9.5.4. We see that the bulk viscosity peaks at even lower temperatures than that of hadronic matter with the direct Urca process. The reason is that in (unpaired) quark matter the more efficient non-leptonic, strangeness-changing, process $u + d \leftrightarrow u + s$ is the dominant chemical re-equilibration process. The figure also shows that the equation of state, through the susceptibilities B and C, has a sizable effect on the bulk viscosity. This effect has also been studied by Vidaña (2012), with an emphasis on the role of the symmetry energy for the bulk viscosity.

The bulk viscosity also receives contribution from muons. Muons appear in the direct (or modified) Urca processes (9.123) with electron and electron neutrino replaced by muon and muon neutrino (Haensel et al. 2000, 2001). One can also consider the purely leptonic processes that convert an electron into a muon and vice versa,

$$e + e \leftrightarrow \mu + e + \nu + \bar{\nu}, \qquad e + \mu \leftrightarrow \mu + \mu + \nu + \bar{\nu}. \tag{9.129}$$

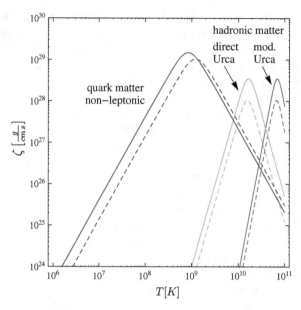

Fig. 9.2 Bulk viscosity of unpaired nuclear matter from Urca processes with angular frequency $\omega = 8.4\,\mathrm{kHz}$ and baryon density $n = 2n_0$. If the direct Urca process is allowed, the reaction rate is faster and the maximum of the bulk viscosity is at a lower temperature. The dashed lines are obtained by using non-interacting matter for the susceptibilities, while the solid curves use the equation of state of Akmal et al. (1998). The plot also shows the result for unpaired strange quark matter from the non-leptonic process $u + d \leftrightarrow u + s$, to be discussed in Sect. 9.5.4.1, see also Fig. 9.4. Again, the dashed curve represents non-interacting quark matter, while the solid curve includes effects of the interaction on the thermodynamics. Figure from Alford et al. (2010a). https://doi.org/10.1088/0954-3899/37/12/125202. © IOP Publishing. Reproduced with permission. All rights reserved

These processes are the dominant contribution to the bulk viscosity for temperatures well below the critical temperature for hadronic superfluidity (Alford and Good 2010).

As the result for quark matter in Fig. 9.2 suggests, the presence of strangeness has a significant effect on the bulk viscosity. The reason is that the phase space for a non-leptonic (strangeness-changing) process is typically much larger than that for a semi-leptonic process because the leptons have a negligibly small Fermi momentum. In hadronic matter, the presence of hyperons thus leads to a very different result for the bulk viscosity, with a maximum typically at smaller temperatures that for ordinary nuclear matter. The bulk viscosity based on the strangeness-changing processes

$$n + n \leftrightarrow p + \Sigma^-,\tag{9.130a}$$

$$n + p \leftrightarrow p + \Lambda,\tag{9.130b}$$

$$n + n \leftrightarrow n + \Lambda,\tag{9.130c}$$

has been computed by Jones (2001); Haensel et al. (2002); Lindblom and Owen (2002); van Dalen and Dieperink (2004); Chatterjee and Bandyopadhyay (2006). Effects of a large magnetic field were taken into account by Sinha and Bandyopadhyay (2009), and the bulk viscosity in quark/hadron mixed phases was computed by Drago et al. (2005).

The curves in Fig. 9.2 show the result for unpaired matter. Cooper pairing of nucleons and/or hyperons change the underlying reaction rates dramatically and (to a much smaller extent) the susceptibilities that enter the bulk viscosity. Therefore, the energy gap in the nucleon dispersions has to be taken into account, leading to a suppression of the reaction rates. This suppression is exponential for temperatures much smaller than the critical temperature if at least one of the participating particles [say the neutron or the proton for the direct Urca process (9.123)] is gapped with an isotropic gap. A power law suppression occurs if the pairing leaves a node at the Fermi sphere where excitations with infinitesimally small energy are possible. This is conceivable for certain phases of $^3P_2$ pairing of neutrons (the milder suppression is of course only possible if at the same time there are unpaired protons). As a consequence of the suppression of the reaction rate, the bulk viscosity is suppressed for small temperatures $T \ll T_c$. The effect of pairing on the bulk viscosity of hadronic matter was calculated by Haensel et al. (2000, 2001, 2002). If neutrons form a superfluid, the corresponding Goldstone mode may contribute to the bulk viscosity and, depending on the equation of state, there may be a temperature regime where its contribution is dominant (Manuel et al. 2013). Superfluidity also has an effect on the hydrodynamics of the system. Since a superfluid at finite temperature is effectively a two-fluid system, there is more than a single bulk viscosity coefficient. The additional coefficients have been computed for superfluid nuclear matter from Urca processes by Gusakov (2007), from phonons by Manuel et al. (2013), and for superfluid nucleon-hyperon matter by Gusakov and Kantor (2008). The effect of these additional coefficients on the instability window for r-mode oscillations appears to be small (Haskell and Andersson 2010).

9.5 Transport in the Core: Quark Matter

9.5.1 General Remarks

Matter at sufficiently large baryon density is deconfined and quarks and gluons rather than baryons and mesons become the relevant degrees of freedom. This phase of matter is called quark matter or, especially at large temperatures where the gluons contribute to the thermodynamics, quark-gluon plasma. In the context of neutron stars, by quark matter we always mean three-flavor quark matter (or 'strange quark matter') made of up, down, and strange quarks. The reason is that the charm, bottom, and top quarks are too heavy to exist at the densities and temperatures typical for a neutron star. Therefore, even when we use perturbative methods which can only be trusted at extremely large densities, we ignore the heavy flavors because eventually we are interested in extrapolating our results down to neutron star densities. It is uncertain whether quark matter exists in the interior of neutron stars because we do not precisely know the central density of the star and, more importantly, we do not know the critical density at which nuclear matter turns into quark matter. It is conceivable that this transition is a crossover (Schäfer and Wilczek 1999; Alford et al. 1999a; Hatsuda et al. 2006; Schmitt et al. 2011), such that there is no well-defined transition density, just like the transition from the hadronic phase to the quark-gluon plasma at large temperatures and small baryon densities (Aoki et al. 2006). Astrophysical data may provide important clues for the question of the location and nature of the deconfinement transition at large densities. Connecting observations from neutron stars to properties of ultra-dense matter is an intriguing example of probing our understanding of fundamental theories such as QCD with the help of astrophysics. In the case of quark matter (and also ultra-dense nuclear matter), the interplay between astrophysics and theory is particularly important because currently there is no rigorous first-principle calculation of dense QCD, unless we go to even larger densities where perturbative methods become reliable (Kurkela et al. 2010a,b, 2014). The reason is that even brute force methods on the lattice fail due to the so-called sign problem, although there has been recent progress towards evading and/or mitigating the sign problem (Aarts 2016; Glesaaen et al. 2016; Gattringer and Langfeld 2016). Ideally, we would like a given phase of dense matter to be identifiable in an unambiguous way from a set of astrophysical observations. Of course, in reality, several distinct phases may show very similar behavior with respect to the observables that are accessible to us. For instance, many of the quark matter phases that we discuss in the following are basically indistinguishable from each other through bulk properties such as the equation of state and thus mass and radius of the star. But they do differ from each other in their low-energy properties, for instance because of different Cooper pairing patterns. Therefore, it is mostly the transport, less the thermodynamics, that differs from phase to phase.

When we compute transport properties of quark matter, many aspects are similar to what we have discussed for hadronic matter in the previous sections: we are obviously interested in the same quantities, i.e., neutrino emissivity, viscosity coefficients, etc., and the methods we use are often the same, even though the formulations in the literature may sometimes look different. Nevertheless, there are some general differences which are useful to keep in mind before we go into more details. Firstly, quark matter is relativistic because the quark masses are small compared to the quark chemical potential and thus compared to the Fermi momentum.[23] (For the up and down quarks, the Fermi velocity v_F introduced in Eq. (9.36) is very close to the speed of light, while the strange quark is heavy enough to induce sizable corrections to this ultra-relativistic limit.) Therefore,

---

[23]In this section, we work in natural units, $c = \hbar = k_B = 1$, such that mass, energy, momentum, and temperature have the same units.

all microscopic calculations are performed in a relativistic framework, which for nuclear matter is only necessary at very large densities where the nucleon rest mass becomes comparable to the Fermi energy. Secondly, when we want to treat quark matter rigorously with currently available methods, we need to approach neutron star densities 'from above', i.e., we often assume quarks to be weakly interacting to be able to apply perturbation theory and then extrapolate the results down in density. This becomes relevant for some of the results discussed here, but not for all since, as we know from the previous section, not all transport properties rely on a precise knowledge of the strong interaction and are rather dominated by the electroweak interaction. Thirdly, quark matter has a larger variety of candidate phases for neutron star cores than nuclear matter because there are 9 fermion species in three-flavor, three-color quark matter. As a result, there is a multitude of different possible patterns of Cooper pairing (Alford et al. 2008a), which is particularly interesting with regard to transport properties.

We will summarize the current state of the art of reaction rates and transport properties in quark matter that are relevant for neutron stars. We attempt to give a comprehensive account of the current knowledge, which is possible because there are considerably fewer studies about quark matter transport than about nuclear matter transport. The results about quark matter we present here were obtained starting from a few works in the early eighties through a peak period around 2005–2008 and including very recent progress that is still ongoing, with interesting ideas and prospectives for future work.

9.5.2 Phases of Quark Matter: Overview

As a preparation, especially for readers unfamiliar with dense QCD, it is useful to start with a brief overview about the relevant quark matter phases and their basic properties. In many cases, these basic properties already give us a rough idea about the behavior of the transport properties which we shall then discuss in more detail.

Just as we know the properties of low-density 'ordinary' nuclear matter, we have solid, albeit only theoretical, knowledge about quark matter at extremely high densities. If the density is sufficiently large to apply weak-coupling methods and to neglect all three quark masses compared to the quark chemical potential, the ground state is the color-flavor locked (CFL) phase (Alford et al. 1998, 1999b). While in nuclear matter more complicated phases including meson condensates and hyperons may occur as we move away from ordinary nuclear matter to *larger* densities (towards the center of the neutron star), more complicated phases of quark matter occur as we move towards *lower* densities (starting from the asymptotically dense regime, which is beyond neutron star densities).

In the CFL phase, all quarks participate in Cooper pairing,[24] and as a consequence the dispersions of all fermionic quasiparticles are gapped. At zero temperature, the number densities of all quark species are identical, and therefore the CFL phase is 'automatically' neutral, without any electrons or muons (recall that the electric charges of up, down and strange quarks happen to add up to zero). This makes CFL very special with respect to transport because at low energies all fermionic degrees of freedom are suppressed and can be neglected. The CFL phase breaks chiral symmetry spontaneously, and thus there is a set of (pseudo-)Goldstone modes, very similar to the mesons that arise from 'usual' chiral symmetry breaking through a chiral condensate. At low temperatures, these light bosons dominate (some of) the transport properties of the CFL phase. The CFL mesons appear with the same quantum numbers as the mesons from quark-antiquark condensation, which is a consequence of the identical symmetry breaking pattern. However, their masses are different: in CFL, the kaons, not the pions, are the lightest mesons.

The CFL phase is a superfluid because it spontaneously breaks the $U(1)$ symmetry associated with baryon number conservation, and thus, as discussed in Sect. 9.2.3.4, the usual complications of superfluid transport arise, such as the two-fluid picture at nonzero temperature, or the existence of quantized vortices in rotating CFL. Moreover, superfluidity implies the existence of an exactly massless Goldstone mode, which yields the dominant contribution for instance to the shear viscosity of CFL. The transport properties of CFL are determined by an effective theory for the pseudo-Goldstone modes and the superfluid mode. The form of this effective theory, in turn, is entirely given by the symmetry breaking pattern of CFL, just like usual chiral perturbation theory. Therefore, if CFL persists down to neutron star densities, we have a very solid knowledge of the low-energy physics of quark matter, although the numerical coefficients of the effective theory can only be determined reliably at weak coupling and become uncertain as we move towards lower densities.

The opposite of CFL, in a way, is unpaired quark matter, where none of the quark species forms Cooper pairs. Unpaired quark matter probably exists only at high temperatures $T \gtrsim 10\,\text{MeV}$, because at lower temperatures Cooper pairing in some form seems unavoidable (Alford et al. 2008a). Nevertheless, unpaired quark matter is an important concept and its transport properties, even at low temperature, are relevant. The reason is that almost all quark matter phases except for CFL have some unpaired quarks or quarks with a very small pairing gap, which dominate transport. Thus, up to numerical prefactors, the result for unpaired quark matter is a good approximation for these phases in many instances. The calculation of transport properties for unpaired quark matter is, in a sense, more difficult than for CFL because we need to know the interaction via gluons in a strongly coupled

---

[24]Cooper pairing in quark matter always implies color superconductivity because at least some of the gluons acquire a Meissner mass, in CFL all eight of them. Whether a color superconductor is also an electromagnetic superconductor and a superfluid is more subtle and will not be discussed in full detail here.

regime (unless the transport property of interest is dominated by the electroweak interactions). Therefore, shear viscosity, electrical and thermal conductivities of unpaired quark matter are typically based on perturbative calculations, assuming the strong coupling constant α_s to be small.

Between the two extreme cases of CFL (present at asymptotically large densities and sufficiently small temperatures) and completely unpaired quark matter (strictly speaking only present at temperatures higher than in a neutron star but important conceptually), there are many possible quark matter phases where quarks 'partially' pair. These phases are likely to be relevant for neutron stars and thus their transport properties have been studied extensively. 'Partial' pairing means that certain quark colors or flavors remain unpaired and/or that Cooper pairing does not occur in all directions in momentum space and even may vary spatially. Such phases necessarily appear at moderate densities because going down in density means a decrease in quark chemical potential and an increase in the strange quark mass, somewhere between the current mass of about 100 MeV and the vacuum constituent mass of about 500 MeV. As a consequence, the strange quark mass cannot be neglected at densities in the cores of neutron stars, and the particularly symmetric situation at asymptotically large densities is disrupted. Why does the less symmetric situation of different quark masses eventually lead to a breakdown of CFL? The reason is that the gain in free energy from Cooper pairing is maximized if the two participating fermion species have the same Fermi surface and pairing occurs over the entire surface in momentum space. If the fermions that 'want' to pair have different Fermi surfaces, an energy cost is involved in Cooper pairing, and this cost may be too large to create a paired state. Different masses, together with the conditions of beta-equilibrium and charge neutrality, provide such a difference in Fermi surfaces because, at least for the most favorable spin-0 channel, pairing takes place between quarks of different color and flavor. Therefore, CFL is under stress if we move away from asymptotically large densities. The system is expected to react in a series of phase transitions, producing more complicated quark matter phases. The exact sequence of these phases can be determined in a controlled way at very large densities and weak coupling, but as we move to lower densities, we have less rigorous knowledge of the phase structure and rely mostly on model calculations or bold extrapolations from ultra-high densities. In particular, it is not known where in this sequence of phases nuclear matter takes over. It is conceivable that CFL persists down to densities where the transition to hadronic matter occurs, possibly leading to a nuclear/CFL interface inside a neutron star. It is also possible that other color-superconducting phases exist in the core of neutron stars, possibly breaking rotational and/or translational invariance. Also, since the QCD coupling increases with lower energies, $\alpha_s \gtrsim 1$ at densities relevant for astrophysics, the color-superconducting phases may be replaced by something qualitatively different, possibly involving elements from the Bardeen-Cooper-Schrieffer–Bose-Einstein-condensation (BCS-BEC) crossover (Deng et al. 2007) seen in atomic gases or possibly showing features of the quarkyonic phase that is predicted in QCD in the limit of infinite number of colors N_c (McLerran and Pisarski 2007).

9.5.3   Neutrino Emissivity

As for hadronic matter, neutrino emissivity is interesting in itself because it is the main cooling mechanism of the star, and the rates for the neutrino processes can be relevant for the bulk viscosity of quark matter inside a neutron star. We consider the processes

$$d \rightarrow u + e + \bar{\nu}_e, \qquad u + e \rightarrow d + \nu_e. \tag{9.131}$$

These are the analogues of the direct Urca processes in nuclear matter (9.94). In quark matter, the triangle inequality from momentum conservation does not pose a severe constraint on this process because the Fermi surfaces between up and down quark are not as different as the ones for neutrons and protons in nuclear matter. Therefore, second-order neutrino processes such as bremsstrahlung are usually negligible in quark matter. We first discuss the processes (9.131) and later summarize the results that involve strangeness. Generalizing the definition (9.98), where Fermi's Golden Rule was applied directly, the neutrino emissivity is

$$\epsilon_\nu \equiv 2 \frac{\partial}{\partial t} \int \frac{d^3 p_\nu}{(2\pi)^3} \, p_\nu \, f_\nu(t, p_\nu), \tag{9.132}$$

where p_ν is the neutrino momentum, and the factor 2 accounts for neutrinos and anti-neutrinos. The change of the neutrino distribution function f_ν is computed from the gain term in the neutrino transport equation, which takes the form

$$\frac{\partial}{\partial t} f_\nu(t, p_\nu) = -\frac{\cos^2 \theta_C G_F^2}{8} \int \frac{d^3 p_e}{(2\pi)^3 p_\nu p_e} L_{\lambda\sigma} f_F(p_e - \mu_e)$$
$$\times f_B(p_\nu + \mu_e - p_e) \text{Im} \, \Pi_R^{\lambda\sigma}(Q), \tag{9.133}$$

with the Fermi and Bose distribution functions $f_{F,B}(x) = (e^{x/T} \pm 1)^{-1}$ for the electron with energy p_e and chemical potential μ_e and the W-boson with four-momentum $Q = (q_0, \boldsymbol{q}) = (p_e - p_\nu - \mu_e, \boldsymbol{p}_e - \boldsymbol{p}_\nu)$. We have abbreviated $L^{\lambda\sigma} \equiv \text{Tr}\left[(\gamma_0 p_e - \boldsymbol{\gamma} \cdot \boldsymbol{p}_e) \gamma^\sigma (1 - \gamma^5)(\gamma_0 p_\nu - \boldsymbol{\gamma} \cdot \boldsymbol{p}_\nu) \gamma^\lambda (1 - \gamma^5)\right]$, with the Dirac matrices γ^σ ($\sigma = 0, 1, 2, 3$) and $\gamma^5 = i\gamma^0\gamma^1\gamma^2\gamma^3$. (Note that the subscript ν stands for neutrino and is thus not used as a Lorentz index.) Finally, $\text{Im} \, \Pi_R^{\lambda\sigma}$ is the imaginary part of the retarded W-boson self-energy

$$\Pi^{\lambda\sigma}(Q) = \frac{T}{V} \sum_K \text{Tr}[\Gamma_-^\lambda S(K) \Gamma_+^\sigma S(P)], \tag{9.134}$$

with the quark propagator S, which is a matrix in color, flavor, and Dirac space and—in the case of Cooper pairing—in Nambu-Gorkov space. The electroweak vertices are diagonal in Nambu-Gorkov space, $\Gamma_\pm^\lambda = \text{diag}[\gamma^\lambda(1 - \gamma^5)\tau_\pm, -\gamma^\lambda(1 +$

$\gamma^5)\tau_\mp]$, where $\tau_\pm = (\tau_1 \pm i\tau_2)/2$ with the Pauli matrices τ_1, τ_2 are matrices in flavor space which ensure that an up and a down quark interact at the vertex, i.e., K and P correspond to the up and down quark four-momenta, respectively.

As an instructive example for the neutrino emissivity in quark matter due to the Urca processes, let us discuss the 2SC phase (Bailin and Love 1984). In the 2SC phase all strange quarks and all quarks of one color, say blue, are ungapped. This phase is an example for the less symmetrically paired phases mentioned above. The up and down quarks participating in the processes (9.131) can be either gapped or ungapped. Since the weak interaction does not change the color of the quarks, they are either both gapped (when they are red or green) or both ungapped (when they are blue). The contribution of the gapped sector, here shown for a single color, is (Schmitt 2010)

$$\frac{\partial}{\partial t} f_\nu(t, \boldsymbol{p}_\nu) = \frac{\pi \cos^2 \theta_C G_F^2}{4} \sum_{e_1, e_2 = \pm} \int \frac{d^3 \boldsymbol{p}_e d^3 \boldsymbol{k}}{(2\pi)^3 (2\pi)^3 p_\nu p_e} L_{\lambda\sigma} \left(T^{\lambda\sigma} B_k^{e_1} B_p^{e_2} + U^{\lambda\sigma} \frac{\Delta^2}{4\varepsilon_k \varepsilon_p} \right)$$

$$\times f_F(p_e - \mu_e) f_F(-e_1 \varepsilon_k) f_F(e_2 \varepsilon_p) \delta(q_0 - e_1 \varepsilon_k + e_2 \varepsilon_p), \qquad (9.135)$$

where $\boldsymbol{k}$ and $\boldsymbol{p}$ are the up and down quark three-momenta, respectively. We have denoted the Bogoliubov coefficients by $B_k^e = (1 - e\xi_k/\varepsilon_k)/2$ with $\xi_k = k - \mu_u$ and the quasiparticle dispersion $\varepsilon_k^2 = \xi_k^2 + \Delta^2$, where μ_u is the up quark chemical potential, and analogously for the down quark with chemical potential μ_d. Moreover, we have abbreviated $T^{\lambda\sigma} \equiv \text{Tr}\left[\gamma^\lambda(1 - \gamma^5)\gamma^0 \Lambda_k^- \gamma^\sigma(1 - \gamma^5)\gamma^0 \Lambda_p^-\right]$, and $U^{\lambda\sigma} \equiv \text{Tr}\left[\gamma^\lambda(1 - \gamma^5)\gamma^5 \Lambda_k^+ \gamma^\sigma(1 + \gamma^5)\gamma^5 \Lambda_p^-\right]$, with the energy projectors $\Lambda_k^\pm = (1 \pm \gamma^0 \boldsymbol{\gamma} \cdot \hat{\boldsymbol{k}})/2$. The contribution proportional to Δ^2 comes from the so-called anomalous propagators, the off-diagonal components of the quark propagator $S(K)$ in Nambu-Gorkov space. Their effect was discussed in detail and evaluated numerically for the 2SC phase by Jaikumar et al. (2006), see Fig. 9.3 for a diagrammatic representation of normal and anomalous contributions to the W-boson self-energy.

The expression in Eq. (9.135) is instructive for the neutrino emissivity in the presence of Cooper pairing because it shows 4 potential subprocesses that arise from summing over e_1 and e_2. Naively, one would expect the distribution functions for the process $u + e \to d + \nu_e$ to appear in the form $f_e f_u (1 - f_d)$ (we have neglected f_ν since the neutrinos leave the star once they are created). We see that the combinations $f_e f_u f_d$, $f_e (1 - f_u)(1 - f_d)$, $f_e (1 - f_u) f_d$ appear as well [note that for the Fermi distribution $f(-x) = 1 - f(x)$]. The reason is that quasiparticles in the superconductor are mixtures of particles and holes (this momentum-dependent mixture is quantified by the Bogoliubov coefficients) and are thus allowed to appear on either side of the reaction. Since we have started from a general form of the reaction rate that is based on the full structure of the propagator, all four reactions are included automatically and we do not have to set up a separate calculation of these Cooper pair breaking and formation processes. This is analogous to nuclear matter with superfluid neutrons, see Sect. 9.4.2.3. In that case, since the direct Urca

Fig. 9.3 Normal and anomalous contributions to the W-boson self-energy needed to compute the neutrino emissivity in the presence of Cooper pairing, here shown for pairing between up and down quarks, for instance in the 2SC phase. The first diagram contains two 'normal' propagators (diagonal elements in Nambu-Gorkov space), the double line indicating that they include the effect of pairing through a modified dispersion relation. The second diagram contains two 'anomalous' propagators (off-diagonal elements in Nambu-Gorkov space), whose diagrammatic structure indicates that they describe a propagating fermion that is absorbed by the condensate $\langle ud \rangle$ and continues to propagate as a charge-conjugate fermion. The two contributions are obtained naturally in the Nambu-Gorkov formalism by performing the trace over Nambu-Gorkov space in Eq. (9.134)

process is usually suppressed, the Cooper pair breaking and formation processes are discussed for the neutral current process $n + n \rightarrow n + n + \nu + \bar{\nu}$, which, in the presence of Cooper pairing, allows for the processes $\{nn\} \rightarrow n + n + \nu + \bar{\nu}$ and $n + n \rightarrow \{nn\} + \nu + \bar{\nu}$, where $\{\ldots\}$ denotes the Cooper pair condensate. The quark version of these processes is $\{uu\} \rightarrow u + u + \nu + \bar{\nu}$ and $u + u \rightarrow \{uu\} + \nu + \bar{\nu}$ (assuming single-flavor quark Cooper pairing), which yields a neutrino emissivity $\epsilon_\nu \propto T^7$ (Jaikumar and Prakash 2001), just like in nuclear matter.

If we are only interested in small temperatures compared to the energy gap Δ, the Cooper pair breaking and formation processes are irrelevant, and the contribution from the gapped quarks is exponentially suppressed, $\epsilon_\nu \propto e^{-\Delta/T}$. As a consequence, the neutrino emissivity of the 2SC phase is, at small temperatures, utterly dominated by the unpaired blue quarks. At higher temperatures, as we approach the critical temperature T_c (for the 2SC phase, $T_c \sim 10\,\text{MeV}$), Eq. (9.135) has to be evaluated numerically.

To compute the neutrino emissivity for unpaired quarks, we may set $\Delta = 0$ in Eq. (9.135). As a result, the dispersions ε_k of the quarks (assumed to be massless) become dispersions of free fermions. However, it is crucial to include the effect of the strong interaction, i.e., to treat the system as a Fermi liquid rather than a non-interacting system of quarks. Otherwise, the phase space for the Urca process is zero and the neutrino emissivity vanishes. Fermi liquid corrections are included by writing the Fermi momenta of up and down quarks as $\mu_u[1 - \mathcal{O}(\alpha_s)]$ and $\mu_d[1 - \mathcal{O}(\alpha_s)]$. The result for 2-flavor unpaired quark matter is (reinstating all color degrees of freedom, $N_c = 3$)

$$\epsilon_\nu^{\text{unp.}} \approx \frac{457}{630} \cos^2 \theta_C G_F^2 \alpha_s \mu_e \mu_u \mu_d T^6 \left(1 + \frac{4\alpha_s}{9\pi} \ln \frac{\Lambda}{T} \right)^2 , \qquad (9.136)$$

where the electron chemical potential is related to the quark chemical potentials via $\mu_u + \mu_e = \mu_d$ in β-equilibrium. The logarithmic correction of Schäfer and Schwenzer (2004) to the standard result by Iwamoto (1980, 1982) arises if non-Fermi liquid effects are included for the quarks, and can lead to an enhancement of the neutrino emissivity at low temperatures (Schäfer and Schwenzer 2004). The energy scale that appears in the logarithm is of the order of the screening scale, $\Lambda \propto g\mu$, where g is the strong coupling constant related to α_s by $\alpha_s = g^2/(4\pi)$, see Gerhold and Rebhan (2005) for a calculation of Λ. Higher order corrections to this result have been computed by Adhya et al. (2012). The strange quark mass has to be included if the result is generalized to strange quark matter. A mass term can easily be added in the quark dispersion, but the result for the emissivity becomes more complicated and is best evaluated numerically. One effect of the mass is to open up the phase space such that the emissivity would be nonzero even if the Fermi liquid corrections were neglected, as discussed by Iwamoto (1980); Wang et al. (2006). A dynamical quark mass from a chiral density wave has a similar effect. The chiral density wave is an anisotropic phase in which the chiral condensate oscillates between between scalar and pseudoscalar components, and the neutrino emissivity depends on the dynamical mass and the wave vector that determines this oscillation (Tatsumi and Muto 2014). This phase, possibly in coexistence with quark Cooper pairing is a candidate phase in the vicinity of a potential first-order chiral phase transition between the hadronic matter and quark matter.

A similar calculation as outlined here for the unpaired and 2SC phases applies to the so-called Larkin-Ovchinnikov-Fulde-Ferrell (LOFF) phases and to color superconductors where Cooper pairs have total spin one. These two classes of phases are further important examples of the less symmetric phases that are expected to arise for a large mismatch in Fermi surfaces. An estimate of this mismatch, based on an expansion for small strange quark masses m_s, is given by comparing m_s^2/μ to the energy gap Δ, where μ is the quark chemical potential (baryon chemical potential divided by $N_c = 3$). In neutron stars, exotic phases like LOFF or spin-one pairing thus occur if the attractive interaction (for which Δ is a measure) is not strong enough to overcome the mismatch m_s^2/μ (which increases with decreasing density because m_s increases and μ decreases). In the LOFF phase, the system reacts to the mismatch in Fermi surfaces by forming Cooper pairs only in certain directions in momentum space, resulting in Cooper pairs with nonzero momentum (Alford et al. 2001; Anglani et al. 2014). In general, a finite number of different Cooper pair momenta will be realized in a given phase, resulting in counter-propagating currents and in a crystalline structure with periodically varying gap function. Since there are directions in momentum space where the quasiparticle dispersion is ungapped, the neutrino emissivity of the LOFF phase is qualitatively very similar to unpaired quark matter, as shown by Anglani et al. (2006). Spin-one color superconductors arise unavoidably in single-flavor Cooper pairing. This form of pairing is the only possible one if the mismatch in Fermi momenta of quarks of different flavor is sufficiently large to prevent any form of cross-flavor pairing. Spin-one color superconductors break rotational symmetry and typically exhibit ungapped directions in momentum space as well. Therefore, as for the

LOFF phase, their neutrino emissivity has the same T^6 behavior as unpaired quark matter. A possible exception is the color-spin locked phase (CSL), where, in a certain variant, all quarks are gapped. However, weak coupling calculations suggest that another variant of CSL, where there *are* unpaired quasifermions, is energetically preferred (Schmitt 2005) (although there are fewer paired quarks, the larger value of the gap function overcomes this lack of pairing). If we only consider the gapped branches, there are striking similarities of the neutrino emissivity in spin-one color superconductors, computed by Schmitt et al. (2006); Wang et al. (2006); Berdermann et al. (2016), to the neutrino emissivity of $^3P_2$ phases in nuclear matter (Yakovlev et al. 2001), which we have briefly discussed in Sect. 9.4.2.3.

The neutrino emissivity of CFL is qualitatively different from the phases with ungapped fermions. In CFL, neutrino emissivity is dominated by the Goldstone modes, and the relevant processes are

$$\pi^{\pm}, K^{\pm} \rightarrow e^{\pm} + \bar{\nu}_e , \qquad \pi^0 \rightarrow \nu_e + \bar{\nu}_e , \qquad \phi + \phi \rightarrow \phi + \nu_e + \bar{\nu}_e . \qquad (9.137)$$

Here, $\pi^{\pm}$ and $K^{\pm}$ are the CFL mesons mentioned above, which have the same quantum numbers as, but different masses than, their counterparts from usual chiral symmetry breaking. In particular, the kaons are the lightest mesons in CFL, with masses of a few MeV. Since these masses are larger than typical temperatures of neutron stars, the resulting neutrino emissivities are exponentially suppressed. The superfluid mode ϕ is massless and thus does not show this exponential suppression. However, the emissivity is proportional to a large power of T, which makes this result very small as well (Jaikumar et al. 2002),

$$\epsilon_\nu^{\mathrm{CFL}} \sim \frac{G_F^2 T^{15}}{f_\phi \mu^4} , \qquad (9.138)$$

where f_ϕ is the analogue of the pion decay constant for the spontaneous breaking of baryon number. We conclude that the CFL phase basically does not contribute to the neutrino emissivity.

Neutrino emissivities of quark matter have been included in cooling calculations for hybrid stars (Grigorian et al. 2005; Popov et al. 2006; Hess and Sedrakian 2011; Noda et al. 2013), and quark matter may provide an explanation for the rapid cooling of the neutron star in Cassiopeia A (Sedrakian 2013, 2016a). In this scenario, the star cools through a transition from the 2SC phase with very inefficient cooling to a crystalline color superconductor, where there are unpaired fermions. This explanation assumes that there is no contribution of the strange quarks and—on purely phenomenological grounds—that there is some residual pairing mechanism for the blue quarks in 2SC. While the explanation of the rapid cooling in nuclear matter is based on the transition from an unpaired phase to the superfluid phase, quark matter may thus potentially show a similar behavior via a transition from one paired phase to another.

9.5.4 Bulk Viscosity

9.5.4.1 Unpaired Quark Matter

We have already discussed the definition and physical meaning of the bulk viscosity ζ in Sect. 9.4.3 and can immediately start from the expression

$$\zeta = -\frac{2}{\omega^2 \delta v_0^2} \frac{1}{\tau} \int_0^\tau dt\, P(t) \frac{d\delta v}{dt}\,, \tag{9.139}$$

which follows from Eqs. (9.120) and (9.121). The pressure $P(t)$ is given by Eq. (9.122), where now, for unpaired quark matter, the oscillations in density occur for the three quark flavors and the electron, δn_u, δn_d, δn_s, δn_e. We consider the processes

$$u + d \leftrightarrow u + s\,, \tag{9.140a}$$

$$u + e \to d + \nu_e\,, \qquad d \to u + e + \bar{\nu}_e \tag{9.140b}$$

$$u + e \to s + \nu_e\,, \qquad s \to u + e + \bar{\nu}_e\,. \tag{9.140c}$$

The non-leptonic process $u + d \leftrightarrow u + s$ will turn out to be the dominant one, but it is instructive to keep the leptonic processes. This allows us to sketch the calculation of the bulk viscosity for a more complicated scenario as outlined at the beginning of Sect. 9.4.3. Namely, we now have two out-of-equilibrium chemical potentials $\delta\mu_1 \equiv \mu_s - \mu_d$ and $\delta\mu_2 \equiv \mu_d - \mu_u - \mu_e$, relevant for the reactions (9.140a) and (9.140b). The relevant difference in chemical potentials for the reaction (9.140c) is then $\delta\mu_1 + \delta\mu_2$, and thus not an independent quantity. As independent changes in densities we keep δn_d, δn_e. The changes in up and strange quark densities then are $\delta n_u = \delta n_d - \delta n_e$ and $\delta n_s = -\delta n_d - \delta n_e$. The change in the electron density comes from the processes (9.140b) and (9.140c), and the change in the down quark number density comes from the processes (9.140a) and (9.140b), and in analogy to Eq. (9.124) we write in the linear approximation

$$\frac{dn_e}{dt} \approx (\lambda_2 + \lambda_3)\delta\mu_2(t) + \lambda_3\delta\mu_1(t)\,, \qquad \frac{dn_d}{dt} \approx \lambda_1\delta\mu_1(t) - \lambda_2\delta\mu_2(t)\,, \tag{9.141}$$

where λ_1, λ_2, λ_3 have to be computed from the microscopic processes. The result for the non-leptonic process (9.140a) is (Wang and Lu 1984; Sawyer 1989b; Madsen 1993)

$$\lambda_1 \approx \frac{64 \sin^2\theta_C \cos^2\theta_C G_F^2}{5\pi^3}\mu_d^5 T^2\,, \tag{9.142}$$

while the leptonic processes (9.140b) and (9.140c) yield (Iwamoto 1980, 1982)

$$\lambda_2 \approx \frac{17\cos^2\theta_C G_F^2}{15\pi^2}\alpha_s\mu_u\mu_d\mu_e T^4 , \qquad (9.143a)$$

$$\lambda_3 \approx \frac{17\sin^2\theta_C G_F^2}{40\pi^2}\mu_s m_s^2 T^4 , \qquad (9.143b)$$

where λ_2 is obtained from the same calculation that leads to the neutrino emissivity (9.136), and the leptonic process including the strange quark is computed to lowest order in the strange quark mass m_s. Generalizing Eq. (9.125), we have two differential equations for $\delta\mu_1$ and $\delta\mu_2$,

$$\frac{d\delta\mu_i}{dt} = \frac{\partial\delta\mu_i}{\partial V}V_0\frac{d\delta v}{dt} + \sum_{x=u,d,s,e}\frac{\partial\delta\mu_i}{\partial n_x}\frac{dn_x}{dt}$$

$$= -B_i\frac{d\delta v}{dt} - \alpha_i\delta\mu_1(t) - \beta_i\delta\mu_2(t) , \qquad i = 1, 2 , \qquad (9.144)$$

where B_i are combinations of thermodynamic functions in equilibrium, and α_i, β_i contain thermodynamic functions and the reaction rates $\lambda_1, \lambda_2, \lambda_3$. As in Sect. 9.4.3, we use the ansatz $\delta\mu_i(t) = \mathrm{Re}(\delta\mu_{i0}e^{i\omega t})$ with complex amplitudes $\delta\mu_{i0}$, such that (9.144) can be solved for real and imaginary parts of $\delta\mu_{10}$ and $\delta\mu_{20}$. The bulk viscosity (9.139) then becomes

$$\zeta = \frac{a_1\mathrm{Re}(\delta\mu_{10}) + a_2\mathrm{Re}(\delta\mu_{20})}{\omega^2\delta v_0} , \qquad (9.145)$$

where a_1 and a_2 are combinations of $B_1, B_2, \lambda_1, \lambda_2, \lambda_3$. Computing $\mathrm{Re}(\delta\mu_{10})$ and $\mathrm{Re}(\delta\mu_{10})$ from Eq. (9.144) yields the final expression in terms of thermodynamic functions in equilibrium, the reaction rates, and the externally given frequency ω. This result is very lengthy and entangles all reaction rates in a complicated way with the thermodynamic functions (Alford and Schmitt 2007; Sa'd et al. 2007a). For a qualitative discussion we introduce the inverse time scales $\gamma_{nl} = \lambda_1/\mu_s^2$ for the non-leptonic process (9.140a) and $\gamma_l = \lambda_2/\mu_s^2 \approx \lambda_3/\mu_s^2$ for the leptonic processes (9.140b) and (9.140c) and assume $\gamma_{nl} \gg \gamma_l$. Then, with some simple estimates of the thermodynamic functions, and ignoring numerical prefactors, we find (Alford and Schmitt 2007)

$$\zeta \propto \gamma_{nl}\frac{\gamma_{nl}\gamma_l + \omega^2}{\gamma_{nl}^2\gamma_l^2 + \gamma_{nl}^2\omega^2 + \omega^4} . \qquad (9.146)$$

From this result, various limit cases can be derived, depending on whether the external frequency ω is of the order of the leptonic rate, the nonleptonic rate, in between these rates etc. The most relevant case turns out to be $\omega \approx \gamma_{nl} \gg \gamma_l$, in which the slower leptonic processes can be completely neglected. Reinstating the thermodynamic functions, we obtain (Madsen 1992)

$$\zeta \approx \frac{\lambda_1 B^2}{(\lambda_1 C)^2 + \omega^2} , \qquad (9.147)$$

where

$$B \equiv n_d \frac{\partial \mu_d}{\partial n_d} - n_s \frac{\partial \mu_s}{\partial n_s} , \qquad C \equiv \frac{\partial \mu_d}{\partial n_d} + \frac{\partial \mu_s}{\partial n_s} . \qquad (9.148)$$

We have recovered the result (9.128) derived in the context of nuclear matter for a single reaction rate. The result for unpaired quark matter as a function of temperature for a fixed frequency ω is plotted in Fig. 9.4.

The calculation of the bulk viscosity in unpaired quark matter outlined here has been improved and extended in the literature in several ways. Firstly, Cooper pairing needs to be taken into account, and we shall discuss the results for various phases in the following subsection (Fig. 9.4 collects most of these results). Secondly, the supra-thermal regime, where the amplitude of the oscillations in chemical potential become large compared to the temperature, has been studied by Alford et al. (2010a), who have generalized earlier numerical results for strange quark matter by Madsen (1992). Shovkovy and Wang (2011) studied this regime together with the interplay of leptonic and non-leptonic processes. The bulk viscosity in the presence of large amplitudes is important if the time evolution and in particular the saturation of unstable r-modes is studied (Alford et al. 2012b). Thirdly, as for the neutrino emissivity, see Eq. (9.136), non-Fermi liquid effects can be included in

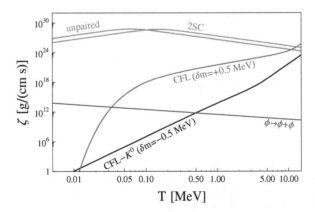

Fig. 9.4 Bulk viscosity for unpaired quark matter, the 2SC phase, and the CFL phase. For the CFL phase, the result is shown from the process that only involves the superfluid mode ϕ and from processes that involve kaons and the superfluid mode, for thermal kaons $\delta m = m_{K^0} - \mu_{K^0} > 0$ and condensed kaons $\delta m < 0$. The parameters chosen here are the quark chemical potential $\mu = 400\,\text{MeV}$, the frequency $\omega/(2\pi) = 1\,\text{ms}^{-1}$, and the kaon mass $m_{K^0} = 10\,\text{MeV}$. The figure is reproduced with modifications from Alford et al. (2008b)

the calculation of unpaired quark matter. Most importantly, they modify the result for the dominant non-leptonic process $u + d \leftrightarrow u + s$ (Schwenzer 2012)

$$\lambda_1 \approx \frac{64 \sin^2 \theta_C \cos^2 \theta_C G_F^2}{5\pi^3} \mu_d^5 T^2 \left(1 + \frac{4\alpha_s}{9\pi} \ln \frac{\Lambda}{T}\right)^4 . \tag{9.149}$$

The correction factor has a higher power compared to the leptonic process that leads to the neutrino emissivity (9.136) because now 4, not 2 quarks participate in the process. If this result is extrapolated to realistic values of the strong coupling, $\alpha_s \sim 1$, the enhancement due to the long-range interactions is larger than for the emissivity. As a consequence, the maximum of the bulk viscosity—at a fixed frequency ω—is shifted to smaller temperatures. Alford and Schwenzer (2014) pointed out that this may have interesting consequences for the r-mode instability, see Fig. 9.5. We have to keep in mind, however, firstly, that completely unpaired quark matter is unlikely to exist in this (sufficiently cold) temperature regime

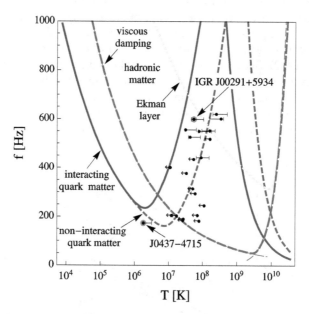

Fig. 9.5 R-mode instability window, computed from bulk viscosity (high T) and shear viscosity (low T) in nuclear and quark matter. Stars are unstable with respect to the emission of gravitational waves through the r-mode instability if they rotate with higher frequencies than given by the critical curves shown here. The observed pulsars shown as data points are in the stable region only for interacting quark matter with the non-Fermi-liquid corrections from Eq. (9.149). These corrections enhance the reaction rate for the conversion of down into strange quarks and thus shift the maximum of the bulk viscosity towards lower temperatures, leading to a shifted stability region compared to noninteracting quark matter. Hadronic matter is consistent with the data only by including additional processes such as dissipation from boundary layer rubbing at the crust of the star. Reprinted figure with permission from Alford and Schwenzer, Phys. Rev. Lett. 113, 251102 (2014). Copyright (2014) by the American Physical Society

because some form of Cooper pairing is expected to occur, and, secondly, that unpaired quark matter provides extremely efficient cooling due to the presence of direct Urca processes (9.136), which is difficult to reconcile with observations (Chugunov et al. 2017). Finally, Huang et al. (2010) have computed the bulk viscosity for unpaired strange quark matter in the presence of a magnetic field, which induces an anisotropic bulk viscosity and, for very large fields, $B \gtrsim 10^{18}$ G, a hydrodynamical instability.

9.5.4.2 Color-Superconducting Quark Matter

If Cooper pairing between the quarks is taken into account, the derivation outlined above is still valid, but the thermodynamic functions B and C in Eq. (9.147) become different, and of course the reaction rates (9.142) and (9.143) have to be recomputed. Let us first discuss the 2SC phase and its bulk viscosity from the process $u + d \leftrightarrow u + s$, which was computed by Alford and Schmitt (2007). Since the weak interaction does not change color, there are $N_c \times N_c = 9$ subprocesses from the $N_c = 3$ possible colors at each of the two electroweak vertices of the process $u + d \leftrightarrow u + s$. In the 2SC phase, only if the color at both vertices is blue, all participating quarks are ungapped. Therefore, at temperatures much smaller than the critical temperature T_c, where all processes with at least one gapped quark are exponentially suppressed, the reaction rate in the 2SC phase is 1/9 times the rate of unpaired quark matter. This does not necessarily mean that the bulk viscosity in the 2SC phase is smaller. It rather means that the maximum of the bulk viscosity, where the rate is in resonance with the frequency ω, is assumed at a larger temperature (if the approximation $T \ll T_c$ is still valid at this temperature). The peak value of the bulk viscosity in the 2SC phase is different from the unpaired phase because the peak frequency is different and the thermodynamic functions B and C are different. It turns out, however, that for typical values of the strange quark mass and the superconducting gap, the peak values are very similar, as we can see in Fig. 9.4.

The bulk viscosity has also been computed in spin-one color superconductors, from the process $u + e \leftrightarrow d + \nu_e$ (Sa'd et al. 2007b; Berdermann et al. 2016), which is the dominant one if strange quarks are ignored, and taking into account the non-leptonic process $u + d \leftrightarrow u + s$ (Wang and Shovkovy 2010), whose reaction rate for four different spin-one color superconductors was computed by Wang et al. (2010). The conclusion is very similar as for the 2SC phase: since there are unpaired quarks in all possible spin-one phases (with the exception mentioned above in the context of neutrino emissivity), one can, for a rough estimate, neglect the contributions of the gapped branches, and the reaction rates become, up to a numerical prefactor, the same as for unpaired quark matter. The bulk viscosity thus behaves qualitatively similar to 2SC quark matter.

The CFL phase behaves differently because there are no ungapped quarks that can contribute to chemical re-equilibration processes, and the bulk viscosity is dominated from bosonic low-energy degrees of freedom, such as the kaon and the superfluid mode via the processes

$$K^0 \leftrightarrow \phi + \phi, \tag{9.150a}$$

$$\phi \leftrightarrow \phi + \phi. \tag{9.150b}$$

The process involving the neutral kaon has been computed for thermal kaons by Alford et al. (2007). As mentioned above, the kaons are the lightest pseudo-Goldstone modes from chiral symmetry breaking in CFL. If the strange quark mass is taken into account, kaon condensation occurs on top of the Cooper pair condensation, giving rise to the so-called CFL-K^0 phase (Bedaque and Schäfer 2002; Kaplan and Reddy 2002). This phase is the next phase down in density if we start from the CFL phase at asymptotically large densities and include the effects of the strange quark mass in a systematic way. It is therefore a very important phase and a viable candidate for the interior of neutron stars. The bulk viscosity of the CFL-K^0 phase is also dominated by the process (9.150a), where K^0 now denotes the Goldstone mode from kaon condensation (Alford et al. 2008b; Schmitt 2008). This Goldstone mode would be exactly massless if strangeness was an exact symmetry. Taking into account the effect of the weak interactions, one finds a mass of about 50 keV for this mode (Son 2001), smaller than the temperatures at which the bulk viscosity of the CFL-K^0 phase becomes sizable. Since all the above arguments about the bulk viscosity remain valid, the bulk viscosity of CFL also peaks at a certain temperature. In Fig. 9.4 we see that this maximum is reached only at temperatures larger than 10 MeV. This is due to the less efficient reaction (9.150a) compared to contributions from quarks. Therefore, inside neutron stars, the CFL bulk viscosity, with or without kaon condensation, is very small compared to other quark matter phases. One might think that the potentially large result for the bulk viscosity at high temperatures is relevant for proto-neutron stars or neutron star mergers, where temperatures may well reach 10 MeV or more. However, we expect the critical temperatures for kaon condensation (Alford et al. 2008c) and the critical temperature of CFL itself to be of the order of 10 MeV, and thus the results beyond this temperature have to be taken with care.

At much lower temperatures, the process (9.150b), which only involves the exactly massless superfluid mode ϕ, is expected to be dominant (Manuel and Llanes-Estrada 2007). The result shown in Fig. 9.4 for this process should be taken seriously only for temperatures larger than about 50 keV, because for smaller temperatures the mean free path is of the order of or larger than the size of the star, indicating that we are no longer in the hydrodynamic regime. Since the CFL phase is a superfluid, there is more than one bulk viscosity coefficient because a superfluid at nonzero temperature can be viewed as a two-fluid system, as mentioned in Sect. 9.2.3.4. Let us denote the full relativistic stress-energy tensor of a superfluid by $T^{\mu\nu}_{\text{ideal}} + T^{\mu\nu}_{\text{diss}}$, as we did in Sect. 9.2.3.3 for a normal fluid. Then, the dissipative terms in first-order hydrodynamics that are usually considered are

$$T_{\text{diss}}^{\mu\nu} = \eta \Delta^{\mu\gamma} \Delta^{\nu\delta} \left(\partial_\delta v_\gamma + \partial_\gamma v_\delta - \frac{2}{3} g_{\gamma\delta} \partial \cdot v \right) + \Delta^{\mu\nu} \left[\zeta_1 \partial_\gamma \left(\frac{n_s}{\sigma} w^\gamma \right) + \zeta_2 \partial \cdot v \right]$$

$$+\kappa (\Delta^{\mu\gamma} v^\nu + \Delta^{\nu\gamma} v^\mu)[\partial_\gamma T + T(v \cdot \partial) v_\gamma] . \tag{9.151}$$

We have denoted $\Delta^{\mu\nu} = g^{\mu\nu} - v^\mu v^\nu$ with the metric tensor $g^{\mu\nu} = \text{diag}(1, -1, -1, -1)$ and the four-velocity of the normal fluid v^μ. Moreover, n_s is the superfluid density, $w^\mu \equiv \partial^\mu \psi - \mu v^\mu$ with the chemical potential μ measured in the rest frame of the normal fluid and $\partial^\mu \psi / \sigma$ the four-velocity of the superfluid with the phase of the condensate ψ and the chemical potential measured in the rest frame of the superfluid $\sigma = (\partial_\mu \psi \partial^\mu \psi)^{1/2}$. For a single fluid, $n_s = 0$, Eq. (9.151) reduces to the normal-fluid expression (9.43b) with $\zeta = \zeta_2$. The Josephson equation, which relates the chemical potential μ to the phase of the condensate, is also modified by dissipative corrections,

$$v \cdot \partial \psi = \mu + \zeta_3 \partial_\mu \left(\frac{n_s}{\sigma} w^\mu \right) + \zeta_4 \partial \cdot v . \tag{9.152}$$

In general, there are even more possible dissipative terms and thus more coefficients in a two-fluid system (Gusakov 2007), which are usually neglected. As mentioned in Sect. 9.2.3.3, the form given here corresponds to the Eckart frame, where, in contrast to the Landau frame, there is no explicit dissipative correction to the conserved current. With $\zeta_4 = \zeta_1$ due to the Onsager symmetry principle, there are three independent bulk viscosity coefficients ζ_1, ζ_2, ζ_3 (which have different units), and ζ_2 corresponds to ζ discussed above for a single fluid. The bulk viscosity coefficients have been estimated for the process (9.150b) in the zero-frequency limit by Mannarelli and Manuel (2010) with the result

$$\zeta_1 \sim \frac{m_s^2}{T\mu} , \qquad \zeta_2 \sim \frac{m_s^4}{T} , \qquad \zeta_3 \sim \frac{1}{T\mu^2} , \tag{9.153}$$

and for the process (9.150a) by Bierkandt and Manuel (2011). The bulk viscosity coefficients of CFL have been applied to the damping of r-modes by Andersson et al. (2010), but, as argued above, the dissipative effects from bulk viscosity in CFL are very small. This is not changed by the additional bulk viscosity coefficients from superfluidity.

9.5.5 Shear Viscosity, Thermal and Electrical Conductivity

9.5.5.1 Unpaired Quark Matter

As we have already seen in Sect. 9.4, the physics behind shear viscosity of dense matter in neutron stars is different from the physics behind bulk viscosity. In the case of shear viscosity, it is thermal, not chemical, re-equilibration and thus for the baryonic and the quark contributions the strong, not the electroweak, interaction becomes relevant. (Recall, however, that even in the calculation of electroweak processes and chemical re-equilibration the strong interaction plays a role because the participating quarks interact strongly with each other.) Electrically neutral unpaired quark matter must contain electrons because the strange quark mass induces an imbalance between the number densities of up, down, and strange quarks, and electrons are needed to neutralize the system. We shall discuss the electron contribution in the context of the 2SC phase below, but first focus on the contribution from quarks alone. If we assume the QCD coupling to be weak, the quasiparticle picture is valid and we can use a kinetic approach to compute the quark-quark scattering rate from one-gluon exchange. This calculation proceeds along the same lines as outlined in Sect. 9.4.1. Here we simply give the final results for shear viscosity η, thermal conductivity κ, and electrical conductivity σ for unpaired quark matter at low temperatures $T \ll \mu$, computed by Heiselberg and Pethick (1993),

$$\eta^{\text{unp.}} \approx 4.4 \times 10^{-3} \frac{\mu^4 m_D^{2/3}}{\alpha_s^2 T^{5/3}} = 5.5 \times 10^{-3} \frac{\mu^4}{\alpha_s^{5/3} T (T/\mu)^{2/3}}$$

$$\approx 2.97 \times 10^{15} \left(\frac{\mu}{500\,\text{MeV}}\right)^{14/3} \left(\frac{T}{1\,\text{MeV}}\right)^{-5/3} \frac{\text{g}}{\text{cm}\,\text{s}}, \tag{9.154a}$$

$$\kappa^{\text{unp.}} \approx 0.5 \frac{m_D^2}{\alpha_s^2} \approx 2.53 \times 10^{21} \left(\frac{\mu}{500\,\text{MeV}}\right)^2 \frac{\text{erg}}{\text{cm}\,\text{s}\,\text{K}} \tag{9.154b}$$

$$\sigma^{\text{unp.}} \approx 0.01 \frac{e^2 \mu^2 m_D^{2/3}}{\alpha_s T^{5/3}} \approx 2.72 \times 10^{25} \left(\frac{\mu}{500\,\text{MeV}}\right)^{8/3} \left(\frac{T}{1\,\text{MeV}}\right)^{-5/3} \text{s}^{-1}, \tag{9.154c}$$

where $m_D^2 = N_f g^2 \mu^2/(2\pi^2)$ is the gluon electric screening mass (squared). The results show a similar non-Fermi-liquid behavior as the leptonic results discussed in Sect. 9.4.1.2 for the same reason: the magnetic interaction that governs the quasiparticles collisions is screened dynamically. For the estimates given here, we have set $\alpha_s \approx 1$ and $N_f = 3$. Jaccarino et al. (2012) have performed a numerical comparison of the quark matter shear viscosity to that of nuclear matter.

It is interesting to compare these results, in particular the shear viscosity, with other QCD calculations and general expectations from strongly coupled systems. The weak-coupling result for the QCD shear viscosity in the opposite limit, $T \gg \mu$, is $\eta \approx aT^3/[\alpha_s^2 \ln(b/\alpha_s)]$, with numerical coefficients a and b (for massless quarks and $N_c = N_f = 3$, $a \approx 1.35$, $b \approx 0.46$) (Huot et al. 2007). This result, together with the entropy density $s \propto T^3$ yields a prediction for the dimensionless ratio η/s. This ratio, in turn, is $\eta/s = 1/(4\pi)$ for a large class of (infinitely) strongly coupled theories, which can be shown with the help of holographic methods based on the gauge/gravity duality (Kovtun et al. 2003). Experimental data suggests that the shear viscosity in a quark-gluon plasma created in a heavy-ion collision is remarkably close to that value, which is difficult to explain by a naive extrapolation of the weak-coupling result to large values of α_s. This has led to the conclusion that heavy-ion collisions produce a quark-gluon plasma which is strongly coupled. We may ask the same question in the context of dense QCD: to which extent are we allowed to extrapolate the weak-coupling result to more moderate densities present in neutron stars and should we rather be using non-perturbative methods? We know that Cooper pairing is one non-perturbative effect, which partially answers this question. If we ignore Cooper pairing for now, the entropy density of $N_f = N_c = 3$ quark matter at low temperatures is $s \approx 2\mu^2 T$, and thus with Eq. (9.154a) we find $\eta/s \approx 2.7 \times 10^{-3}(\mu/T)^{8/3}\alpha_s^{-5/3}$. Interestingly, this result is qualitatively different from the $T \gg \mu$ result because of the appearance of the dimensionless ratio μ/T. In particular, η/s appears to become large for low temperatures and fixed α_s, having no chance to approach $1/(4\pi)$, even when we boldly extrapolate to large values of α_s. Holographic strong-coupling calculations at large N_c by Mateos et al. (2007) suggest that $\eta/s = 1/(4\pi)$ does not receive corrections from a baryon chemical potential, although Myers et al. (2009) found a μ/T dependence for more exotic theories, which were compared to and contrasted by Fermi-liquid theory by Davison et al. (2014). Putting these more exotic theories aside, one might be tempted to conclude that weak-coupling transport in dense QCD is even more different from strong-coupling transport than it is in hot QCD. It is, however, conceivable that $\eta/s = 1/(4\pi)$ is not a good benchmark for dense QCD, for instance because of the large-N_c limit that underlies this holographic result. In any case, it would be very interesting to go beyond the weak-coupling calculation of high-density transport properties of quark matter. Since the quasiparticle picture is no longer valid at strong coupling, the shear viscosity can then no longer been calculated from a collision integral, and the more general Kubo formalism should be employed, which allows for a general spectral density. First steps in this direction have been made by Iwasaki et al. (2008); Iwasaki and Fukutome (2009); Lang et al. (2015); Harutyunyan et al. (2017), who used this formalism to compute shear viscosity and thermal conductivity of quark matter at finite T and μ. These calculations where performed within the Nambu–Jona-Lasinio model and for temperatures larger than relevant for neutron stars (a calculation within the same model, but using the Boltzmann approach, was performed recently by Deb et al. (2016)).

9.5.5.2 Color-Superconducting Quark Matter

The results (9.154) were extended to the 2SC phase (without strange quarks) by Alford et al. (2014b), also in the weak-coupling regime. In the 2SC phase, no global symmetry is broken and thus there are no Goldstone modes. Therefore, we expect the main contribution to come from ungapped fermionic modes. Besides the effect on the fermionic modes, we now have to take into account the effect of pairing on the gauge bosons, as discussed in general terms in Sect. 9.2.3.4 and for lepton transport in nuclear matter in Sect. 9.4.1.4. Remember that screening determines the range of the interaction, and weaker screening results in a more efficient relaxation mechanism. In the 2SC phase, different gluons are screened differently, depending on whether they couple to the unpaired blue quarks or to the paired red and green quarks. Let us first discuss the static screening of the gauge bosons. The three gluons which only see red and green quarks—corresponding to the three generators T_{1-3} of the $SU(3)$ gauge group—are neither magnetically screened (as in unpaired quark matter) nor electrically screened (unlike in unpaired quark matter). This is because the red and green charges are all confined in Cooper pairs and the Cooper pairs themselves carry color charge anti-blue. The gluons corresponding to T_{4-7} acquire a Meissner mass and also are electrically screened. Since the relevant interactions with gluons T_{1-7} involve at least one red or green quark, they do not matter for the shear viscosity at temperatures much smaller than the gap, $T \lesssim 10\,\mathrm{MeV}$. It remains the 8th gluon and the photon. Their behavior is complicated because they mix, and in the 2SC phase they do so differently in electric and magnetic sectors (Schmitt et al. 2004). In the electric sector, there is a screened gluon T_8, and a screened photon corresponding to the generator Q of the electromagnetic gauge group $U(1)$. In the magnetic sector, there is a screened gluon $\tilde{T}_8$ (with a small admixture of the photon), and an *unscreened* photon $\tilde{Q}$ (with a small admixture of the 8th gluon). Because of this rotated photon with zero magnetic screening mass, the 2SC phase is not an electromagnetic superconductor, i.e., does not show an electromagnetic Meissner effect. It does show a *color* Meissner effect for 5 out of the 8 gluons. (The CFL phase also has such a magnetically unscreened rotated photon and is thus no electromagnetic superconductor either; in CFL quark matter all 8 gluons—one of them having a small admixture of the photon—acquire a Meissner mass.) The rotated photon is screened dynamically in the form of Landau damping like the ordinary transverse photon in nuclear matter (Sect. 9.4.1.2) or the transverse gluons in unpaired quark matter of the previous subsection.

The dominant contribution to the shear viscosity comes from unpaired fermions: blue quarks and electrons. We collect their charges with respect to the 8th gluon and the photon (and their rotated versions in the magnetic sector) in Table 9.1. Since the rotated photon is weakly screened, it provides the dominant contribution to the collision frequencies, effectively suppressing the relaxation times and hence the transport coefficients for the species which interact via $\tilde{Q}$. The only unpaired fermion that does not couple to $\tilde{Q}$ is the blue down (bd) quark. Therefore—and although it interacts via the strong interaction—it has the longest relaxation time and gives the dominant contribution to the shear viscosity at sufficiently small temperatures (Alford et al. 2014b),

Table 9.1 Static 2SC screening masses (squared) of the eighth gluon T_8 and the photon Q, in units of $\mu^2/(3\pi^2)$ (from Schmitt et al. (2004), with $N_f = 2$), and charges of the unpaired fermions in the 2SC phase, assuming the strong coupling constant to be much larger than the electromagnetic coupling $g \gg e$

	T_8	Q	$\tilde{T}_8$	$\tilde{Q}$
	Electric screening		Magnetic screening	
Screening mass	$3g^2$	$2e^2$	$g^2/3$	0
Blue up	$-g/\sqrt{3}$	$2e/3$	$-g/\sqrt{3}$	e
Blue down	$-g/\sqrt{3}$	$-e/3$	$-g/\sqrt{3}$	0
Electron	0	$-e$	$-e^2/(g\sqrt{3})$	$-e$

The unpaired fermions dominate the transport properties, and at sufficiently small temperatures shear viscosity, thermal and electrical conductivities are dominated by the blue down quark because it does not couple to the only unscreened gauge boson, the rotated photon $\tilde{Q}$. (Here the electric charge is given in Heaviside-Lorentz units, such that $e^2 = 4\pi\alpha_f$)

$$\eta_{bd}^{2SC} \approx 2.3 \times 10^{-3} \frac{\mu^4}{\alpha_s^{3/2} T(T/\mu)} . \tag{9.155}$$

This result is qualitatively different from the unpaired result (9.154a) because in unpaired quark matter all quarks experience unscreened magnetic interactions. In that case, the electron contribution becomes important as well, for a short discussion see Alford et al. (2014b).

At larger temperatures, dynamical screening of $\tilde{Q}$ becomes stronger [recall Eq. (9.71)] and the interaction via the rotated magnetic photon no longer dominates over the interaction via the screened gauge bosons. As a consequence, electrons become dominant and the result (9.155) is no longer valid. In fact, it is only valid at very small temperatures, $T/\mu \sim 10^{-5}$. (The contribution of blue up quarks is never important since they have smaller Fermi momentum that blue down quarks.) For the numerical results for all temperatures, including thermal and electrical conductivities, see Alford et al. (2014b). These results show in particular that for the thermal conductivity the transition from quark-dominated to electron-dominated regime occurs at a much higher temperature than for shear viscosity, not unlike the competition of lepton and nucleon contributions to κ and η in nuclear matter. The electrical conductivity has also been computed close to the critical temperature and taking into account an external magnetic field by Kerbikov and Andreichikov (2015).

If the mismatch between the up and down quark Fermi momenta is large, isotropic pairing is no longer possible. The red and green quarks that participate in Cooper pairing then develop ungapped quasiparticle excitations in certain regions in momentum space. Their contribution to the shear viscosity, which is dominated by transverse, Landau damped gluons T_{1-3}, has been computed for the (anisotropic, but not crystalline) Fulde-Ferrell phase by Sarkar and Sharma (2017). The result is small compared to the contribution of the completely unpaired blue quarks, but elements of the calculation may be transferred in future studies to LOFF phases in

CFL, where there are no completely unpaired quarks, only few electrons, and all non-abelian gauge bosons have nonzero electric and magnetic screening masses. While this calculation has not been done yet, we briefly review the results from the Goldstone modes in the pure (isotropic) CFL phase. The calculation of the contribution from the superfluid mode is based on the effective Lagrangian

$$\mathcal{L} = \frac{3(D_\mu \psi D^\mu \psi)^2}{4\pi^2} \tag{9.156}$$

$$= \frac{1}{2}[(\partial_0 \phi)^2 - v^2(\nabla \phi)^2] - \frac{\pi}{9\mu^2}\partial_0\phi(\partial_\mu\phi\partial^\mu\phi) + \frac{\pi^2}{108\mu^4}(\partial_\mu\phi\partial^\mu\phi)^2 + \dots, \tag{9.157}$$

where ψ is the phase of the condensate introduced below Eq. (9.151), the covariant derivative acting on this phase is $D_\mu\psi = \partial_\mu\psi - A_\mu$ with $A_\mu = (\mu, 0)$, the rescaled field of the superfluid mode is $\phi = 3\mu\psi/\pi$, the velocity of the Goldstone mode is $v = 1/\sqrt{3}$, and we have dropped the terms linear and constant in ϕ in the second line. The result from $\phi + \phi \leftrightarrow \phi + \phi$ scattering for the shear viscosity was computed by Manuel et al. (2005),

$$\eta_\phi^{\text{CFL}} \approx 1.3 \times 10^{-4} \frac{\mu^4}{T(T/\mu)^4} \approx 6.96 \times 10^{22} \left(\frac{\mu}{500\,\text{MeV}}\right)^8 \left(\frac{T}{1\,\text{MeV}}\right)^{-5} \frac{\text{g}}{\text{cm}\,\text{s}}. \tag{9.158}$$

Alford et al. (2010b) calculated the contribution of kaon scattering $K^0 + K^0 \leftrightarrow K^0 + K^0$ to shear viscosity in the CFL-K^0 phase, where the relevant excitation is the Goldstone mode K^0 from kaon condensation. It was found that this contribution is smaller than that of the superfluid mode ϕ. However, the relevant mean free path (the 'shear mean free path') of the phonons becomes of the order of or larger than the radius of the star at temperatures lower than about 1 MeV. (In this ballistic regime, an 'effective shear viscosity' can be induced from shear stresses at the boundary of the system, for instance in superfluid cold atoms in an optical trap (Mannarelli et al. 2013; Mannarelli 2013).) This is not the case for the kaons, which therefore may provide the dominant contribution to shear viscosity in this regime. The thermal conductivity of CFL due to phonons was obtained from a simple mean free path estimate by Shovkovy and Ellis (2002). Later, Braby et al. (2010) made this estimate more precise by a calculation within kinetic theory, and it was found

$$\kappa_\phi^{\text{CFL}} \gtrsim 4.01 \times 10^{-2} \frac{\mu^8}{\Delta^6} \approx 1.04 \times 10^{26} \left(\frac{\mu}{500\,\text{MeV}}\right)^8 \left(\frac{\Delta}{50\,\text{MeV}}\right)^{-6} \frac{\text{erg}}{\text{cm}\,\text{s}\,\text{K}}. \tag{9.159}$$

This large thermal conductivity suggests that a CFL quark matter core of a neutron star becomes isothermal within a few seconds (Braby et al. 2010). In addition, Braby et al. (2010) also computed the kaon contribution. This was done in the CFL, not the CFL-K^0, phase, i.e., from a massive kaon $m_{K^0} \sim 10\,\text{MeV}$ instead of the (approximately) massless Goldstone kaon in the CFL-K^0 phase, which was used in the calculation for the shear viscosity we just mentioned. It was found that the contribution from the kaons for typical parameter values is much smaller than the phonon contribution (9.159). Neither for the shear viscosity nor for the thermal conductivity, scattering processes due to interactions between the superfluid mode and the kaon have been taken into account so far.

The shear viscosity of spin-1 color superconductors has not yet been computed. In most phases, the dominant contribution can be expected to come from unpaired quarks, and the calculation would be similar as for instance in the 2SC phase, with possible complications from anisotropies and ungapped directions in momentum space, like in the case of the Fulde-Ferrell calculation mentioned above. Only in the fully gapped version of the CSL phase (which seems to be disfavored, at least at weak coupling, as mentioned above) Goldstone modes would become important. An effective theory for the massless modes has been worked out by Pang et al. (2011), which can be used to compute the shear viscosity in CSL quark matter.

9.5.6   Axial Anomaly in Neutron Stars

9.5.6.1   Anomaly-Induced Transport

Transport in the presence of a chiral imbalance, i.e., in systems where there are more left-handed than right-handed fermions or vice versa, is qualitatively different from 'usual' transport. The reason is the chiral anomaly, which leads to the non-conservation of the axial current due to quantum effects. Anomaly-induced transport (or short: anomalous transport[25]) has been discussed extensively in the recent literature, with applications in a multitude of different systems, reviewed recently in a pedagogical article by Landsteiner (2016). One prominent manifestation of anomalous transport is the 'chiral magnetic effect', where a dissipationless electric current is induced in the direction of a background magnetic field. This effect has been predicted to occur in non-central heavy-ion collisions (Kharzeev et al. 2008), where large magnetic fields are created and where a chiral imbalance can be generated by fluctuations of the gluon fields through the QCD anomaly (while the anomaly of Quantum Electrodynamics (QED) then provides the mechanism for the creation of the electric current). Signatures of the chiral magnetic effect have been seen in the data, although the interpretation still leaves room for alternative explanations (Kharzeev et al. 2016). An unambiguous manifestation of the chiral magnetic effect has been observed in so-called Weyl semi-metals, which exhibit

---

[25]The term 'anomalous transport' is used in various contexts with different meaning, for instance in plasma physics, where it refers to unusual diffusion behavior and has nothing to do with the quantum anomaly. Confusion can be avoided by using the more cumbersome, but less ambiguous, 'anomaly-induced transport'. Also 'chiral transport' or 'anomalous chiral transport' is sometimes used.

chiral quasiparticles (Li et al. 2016). The chiral magnetic effect is one example among various anomaly-induced phenomena. Others are the 'chiral vortical effect', where the role of the magnetic field is played by a nonzero vorticity, and the 'chiral separation effect', where the role of the difference in left- and right-handed fermion densities is played by their sum and an axial current, not a vector current, is generated. In a hydrodynamic formulation, anomalous effects generate additional terms with new— 'anomalous'—transport coefficients (Son and Surowka 2009). Also an anomalous version of kinetic theory has been formulated (Stephanov and Yin 2012; Son and Yamamoto 2013).

It is natural to ask whether anomalous transport plays a role in dense matter and whether it has observable consequences for neutron stars. Sizable effects can only come from massless or very light particles because a mass breaks chiral symmetry explicitly and thus tends to suppress any effects from the chiral anomaly. It has been suggested by Ohnishi and Yamamoto (2014) that a dynamical instability ('chiral plasma instability') due to the chiral magnetic effect for electrons occurs in core collapse supernovae, possibly producing the very strong magnetic fields in magnetars. However, the chiral imbalance for electrons created from the electron capture process is completely washed out by the nonzero electron mass (Grabowska et al. 2015; Kaplan et al. 2017), although one might naively think that this mass is negligible in the astrophysical context. The instability may nevertheless be realized if the electrons experience instead an effective chiral chemical potential from the fluid helicity generated in the neutrino gas through the chiral vortical effect (Yamamoto 2016). It has also been suggested that pulsar kicks originate from chiral imbalance in leptons, either from electrons, which however would require a very small crust (possibly in quark stars) (Charbonneau and Zhitnitsky 2010; Charbonneau et al. 2010), or from neutrinos due to the chiral separation effect from the magnetic field, treating electrons and neutrinos as a single fluid (Kaminski et al. 2016). One may also ask whether anomalous transport of neutrinos has an effect on the dynamics or even the very existence of core-collapse supernova explosions. This question is motivated by the different behavior of a chiral fluid with respect to magnetohydrodynamic turbulence, pointed out by Yamamoto (2016) and Pavlović et al. (2017). These studies have only begun recently, and it remains to be seen whether (proto-)neutron stars or supernova explosions, maybe also neutron star mergers and the hyper-massive neutron stars resulting from them, provide yet another system where effects of the quantum anomaly become manifest on macroscopic scales.

9.5.6.2 Axions

Another anomaly-related effect with relevance to neutron stars, now specifically from the QCD anomaly, is the existence of axions. Axions, which are a promising hypothetical candidate for cold dark matter, arise from the most natural solution to the so-called strong CP problem: the axial anomaly effectively—via axial rotations—induces a CP-violating term proportional to $G_{\mu\nu}\tilde{G}^{\mu\nu}$ to the QCD

Lagrangian, where $G_{\mu\nu}$ and $\tilde{G}^{\mu\nu}$ are the gluon field strength tensor and its dual. We know that the prefactor of this term, which is an angle $\theta \in [-\pi, \pi]$, must be extremely small because of very tight experimental constraints on the electric dipole moment of the neutron. Rather than viewing θ as a parameter, whose smallness then would be very difficult to understand, Peccei and Quinn (1977) have suggested a dynamical mechanism that leads to extremely small values for θ. This mechanism is based on the spontaneous breaking of a global anomalous $U(1)$ symmetry, and the axion is the corresponding (not exactly massless) Goldstone mode (Weinberg 1978; Wilczek 1978). The exact implementation of this mechanism in the Standard Model leaves room for different models, which essentially fall into two classes, introduced by Kim (1979) and Shifman et al. (1980) one the one hand and Dine et al. (1981) and Zhitnitsky (1980) on the other hand. Although axions are expected to couple to electrons, photons, and nucleons, so far a positive signal for the axion or any axion-like particle has remained elusive in experimental searches. Constraints on the coupling strengths (and thus on the axion mass) are obtained for instance from cooling of white dwarfs, from cosmology, and from solar physics (Raffelt 2008). In addition, supernova explosions (Keil et al. 1997) and the cooling of neutron stars (Umeda et al. 1997) can potentially contribute to these constraints. To this end, the reaction rates for axions in a nuclear medium have to be calculated. Many of these calculations are analogous to the calculations of the neutrino reaction rates reviewed in Sect. 9.4.2 and share the same problems and uncertainties. Axions can be emitted from bremsstrahlung in electron scattering processes from ions in the crust (Iwamoto 1984), or from bremsstrahlung in nucleon-nucleon collisions $N + N \rightarrow N + N + a$ in the core, where N can be a neutron or a proton and a is the axion. The calculation of the latter process involves knowledge of the strong interaction between the nucleons, just like the analogous neutrino-emitting process $N + N \rightarrow N + N + \nu + \bar{\nu}$ and like the modified Urca process. Therefore, the rate contains significant uncertainties. It was first computed using the one-pion exchange interaction for neutrons by Iwamoto (1984) and later extended to the cases that involve protons (Iwamoto 2001; Stoica et al. 2009). These results are expected to present upper limits since medium corrections to the interactions are likely to reduce the rates (Keil et al. 1997; Hanhart et al. 2001; Fischer et al. 2016). Recently, Keller and Sedrakian (2013) computed the axion emissivity for superfluid nuclear matter from pair breaking and formation processes. In unpaired quark matter, the rate from the analogous process $q + q \rightarrow q + q + a$, where q is an u, d, or s quark has been computed by Anand et al. (1990), using one-gluon exchange for the quark interaction. The axion emissivities can be used to study numerically the axion contribution to the cooling of neutron stars. Such a simulation is naturally prone to large uncertainties, but conservative estimates yield an upper bound for the axion mass of the order of $0.1\,\text{eV}$ (Umeda et al. 1997; Sedrakian 2016b), consistent with limits set by the direct neutrino detection from supernova SN 1987A.

9.6 Outlook

We have seen that understanding transport in neutron star matter requires a variety of different techniques and theoretical results—sometimes even if we ask for the explanation of a single, specific astrophysical observation. Current efforts combine nuclear and particle physics with elements of condensed matter and solid state physics, using and developing methods from hydrodynamics, kinetic theory, many-body physics, quantum field theory, and general relativity. In many ways, neutron stars are a unique laboratory, with matter under more extreme conditions than anywhere else. This laboratory is far away from us and we seem to have very limited access to the matter deep inside the star. It is thus easy to get discouraged regarding precise tests of the transport properties that we predict theoretically. Nevertheless, as we have pointed out, transport properties do provide us with an important tool to interpret astrophysical data and eventually answer the question about what the interior of the neutron star is made of. And, most importantly for future studies, the current exciting results from gravitational-wave astronomy promise more, and more precise, data for the near future, especially if combined with electromagnetic signals as for the recently observed neutron star merger event (Abbott et al. 2017a,b). Neutron star mergers are sensitive to both the equation of state and transport of (relatively hot) dense matter. Moreover, a possible future detection of gravitational waves from isolated neutron stars (Glampedakis and Gualtieri 2018, this volume) would be another spectacular testing ground for transport in ultra-dense matter. The reason is that potential sources such as the r-mode instability and a sustained ellipticity of the star are intricately linked to transport properties, such as viscous effects and the formation and evolution of magnetic flux tube arrays.

Throughout the review we have pointed out open questions and unsolved problems. Many of them are inevitably related to our limited quantitative grip on the strong interaction, i.e., on QCD at baryon densities significantly larger than nuclear saturation density. This concerns for example the modified Urca process or shear viscosity of ultra-dense matter, be it nuclear or quark matter. First-principle QCD calculations on the lattice exist for thermodynamic quantities at zero baryon density, and there are some promising attempts to extend these calculations, firstly, to finite baryon densities and, secondly, to transport properties. Nevertheless, both extensions are extremely difficult, let alone implementing them simultaneously. Therefore, in the foreseeable future, the input from the strong interaction to transport properties of dense matter will most likely not go beyond the use of effective theories, phenomenological models, or extrapolations from perturbative calculations.

Other open problems that we have mentioned are related to transport in a magnetic field and transport in the presence of Cooper pairing. This concerns for instance microscopic calculations of transport properties in the crust and the inhomogeneous nuclear pasta phases, which obviously become very cumbersome through the anisotropy induced by a magnetic field. It also concerns more macroscopic magnetohydrodynamic studies (which we did not discuss in detail), which currently do not yield a satisfactory picture of the magnetic field evolution if compared to observational data. Also in the case of Cooper pairing, microscopic calculations become much more complicated, and we have pointed out various approaches and approximations used for that case. Transport properties of many possible phases, in particular a large part of the multitude of possible color-superconducting phases, have already been discussed in the literature and significant progress has been made. Challenges for future studies are for instance the nature of interfaces between superfluid and superconducting phases, especially if there are rotational vortices and/or magnetic flux tubes, possibly even color-magnetic flux tubes in quark matter. Again, also more macroscopic studies are difficult, and many things remain to be understood, for instance multi-fluid effects due to nonzero temperatures, or the time evolution of rotating superfluids, say the neutron superfluid in the crust or the color-flavor locked phase in a possible quark matter core.

Finally, let us emphasize the need of cross-disciplinary approaches for future efforts in the field of transport theory of dense matter. It is obvious that theoretical studies from nuclear and particle physics have to be combined with observational astrophysics. Maybe less obvious are parallels to other fields that deal with strongly coupled systems where transport properties can be measured. For instance, transport in unitary atomic Fermi gases has been studied in detail, including effects of superfluidity. One example is the study of critical velocities in two-component superfluids (Delehaye et al. 2015), which is of possible relevance to superfluid neutron star matter. It is even conceivable that future experiments with cold atoms can be 'designed' to mimic, at least qualitatively, effects that we expect in neutron stars, such as unpinning of vortices from a lattice structure. Also experiments with more traditional superfluids such as liquid helium might shed some light on questions we encounter in neutron stars (Graber et al. 2017). Transport also plays a prominent role in relativistic heavy-ion collisions, which provide a laboratory for strongly interacting matter at larger temperatures and lower baryon densities. Future experiments aim, in fact, at increasing the densities in these collisions, possibly reaching beyond nuclear saturation density (Friman et al. 2011; Blaschke et al. 2016). In any case, heavy-ion collisions raise various interesting fundamental questions about (relativistic) hydrodynamics and its regime of applicability, and we can imagine that insights gained in these studies might, even if not being directly applicable, give interesting input and pose relevant questions also in the context of neutron stars.

Acknowledgements We thank M. Alford, A. Chugunov, A. Kaminker, A. Rebhan, S. Reddy, A. Sedrakian, I. Shovkovy, and D. Yakovlev for useful comments and discussions and acknowledge support from the NewCompStar network, COST Action MP1304. A.S. is supported by the Science & Technology Facilities Council (STFC) in the form of an Ernest Rutherford Fellowship. P.S. is supported by the Foundation for the Advancement of Theoretical Physics and Mathematics "BASIS" and the Russian Foundation for Basic Research grant # 16-32-00507-mol-a.

References

Aarts, G.: J. Phys. Conf. Ser. **706**, 022004 (2016). 1512.05145

Abbar, S., Carlson, J., Duan, H., Reddy, S.: Phys. Rev. C **92**, 045809 (2015). 1503.01696

Abbott, B.P. et al. (LIGO Scientific Collaboration and Virgo Collaboration): Phys. Rev. Lett. **119**, 161101 (2017a). 1710.05832

Abbott, B.P., et al.: Astrophys. J. **848**, L12 (2017b). 1710.05833

Adhya, S.P., Roy, P.K., Dutt-Mazumder, A.K.: Phys. Rev. D **86**, 034012 (2012). 1204.2684

Adler, S.L., Dothan, Y.: Phys. Rev. **151**, 1267 (1966)

Afanasjev, A.V., Beard, M., Chugunov, A.I., Wiescher, M., Yakovlev, D.G.: Phys. Rev. C **85**, 054615 (2012). 1204.3174

Aguilera, D. N., Cirigliano, V., Pons, J.A., Reddy, S., Sharma, R.: Phys. Rev. Lett. **102**, 091101 (2009). 0807.4754

Akmal, A., Pandharipande, V.R., Ravenhall, D.G.: Phys. Rev. C **58**, 1804 (1998). nucl-th/9804027

Alcain, P.N., Giménez Molinelli, P.A., Dorso, C.O.: Phys. Rev. C **90**, 065803 (2014). 1406.1550

Alford, M.G., Good, G.: Phys. Rev. C **82**, 055805 (2010). 1003.1093

Alford, M.G., Pangeni, K.: Phys. Rev. C **95**, 015802 (2017). 1610.08617

Alford, M.G., Schmitt, A.: J. Phys. G **34**, 67 (2007). nucl-th/0608019

Alford, M.G., Sedrakian, A.: J. Phys. G **37**, 075202 (2010). 1001.3346

Alford, M.G., Schwenzer, K.: Phys. Rev. Lett. **113**, 251102 (2014). 1310.3524

Alford, M.G., Rajagopal, K., Wilczek, F.: Phys. Lett. B **422**, 247 (1998). hep-ph/9711395

Alford, M.G., Berges, J., Rajagopal, K.: Nucl. Phys. B **558**, 219 (1999a). hep-ph/9903502

Alford, M.G., Rajagopal, K., Wilczek, F.: Nucl. Phys. B **537**, 443 (1999b). hep-ph/9804403

Alford, M.G., Bowers, J.A., Rajagopal, K.: Phys. Rev. D **63**, 074016 (2001). hep-ph/0008208

Alford, M.G., Braby, M., Reddy, S., Schäfer, T.: Phys. Rev. C **75**, 055209 (2007). nucl-th/0701067

Alford, M.G., Schmitt, A., Rajagopal, K., Schäfer, T.: Rev. Mod. Phys. **80**, 1455 (2008a). 0709.4635

Alford, M.G., Braby, M., Schmitt, A.: J. Phys. G **35**, 115007 (2008b). 0806.0285

Alford, M.G., Braby, M., Schmitt, A.: J. Phys. G **35**, 025002 (2008c). 0707.2389

Alford, M.G., Mahmoodifar, S., Schwenzer, K.: J. Phys. G **37**, 125202 (2010a). 1005.3769

Alford, M.G., Braby, M., Mahmoodifar, S.: Phys. Rev. C **81**, 025202 (2010b). 0910.2180

Alford, M.G., Reddy, S., Schwenzer, K.: Phys. Rev. Lett. **108**, 111102 (2012a). 1110.6213

Alford, M.G., Mahmoodifar, S., Schwenzer, K.: Phys. Rev. D **85**, 044051 (2012b). 1103.3521

Alford, M.G., Mallavarapu, S.K., Schmitt, A., Stetina, S.: Phys. Rev. D **87**, 065001 (2013). 1212.0670

Alford, M.G., Mallavarapu, S.K., Schmitt, A., Stetina, S.: Phys. Rev. D **89**, 085005 (2014a). 1310.5953

Alford, M.G., Nishimura, H., Sedrakian, A.: Phys. Rev. C **90**, 055205 (2014b). 1408.4999

Alford, M.G., Mallavarapu, S.K., Vachaspati, T., Windisch, A.: Phys. Rev. C **93**, 045801 (2016). 1601.04656

Alford, M.G., Bovard, L., Hanauske, M., Rezzolla, L., Schwenzer, K.: Phys. Rev. Lett. **120**, 041101 (2018). 1707.09475

Anand, J.D., Goyal, A., Jha, R.N.: Phys. Rev. D **42**, 996 (1990)

Andersson, N. Astrophys. J.: **502**, 708 (1998). gr-qc/9706075

Anderson, R.H., Pethick, C.J., Quader, K.F.: Phys. Rev. B **35**, 1620 (1987)

Andersson, N., Haskell, B., Comer, G.L.: Phys. Rev. D **82**, 023007 (2010). 1005.1163

Andersson, N., Krüger, C., Comer, G.L., Samuelsson, L.: Class. Quant. Grav. **30**, 235025 (2013). 1212.3987

Andersson, N., Comer, G.L., Hawke, I.: Class. Quant. Grav. **34**, 125001 (2017). 1610.00445

Anglani, R., Nardulli, G., Ruggieri, M., Mannarelli, M.: Phys. Rev. D **74**, 074005 (2006). hep-ph/0607341

Anglani, R., Casalbuoni, R., Ciminale, M., Ippolito, N., Gatto, R., Mannarelli, M., Ruggieri, M.: Rev. Mod. Phys. **86**, 509 (2014). 1302.4264

Aoki, Y., Endrődi, G., Fodor, Z., Katz, S.D., Szabo, K.K.: Nature **443**, 675 (2006). hep-lat/0611014

Arseev, P.I., Loiko, S.O., Fedorov, N.K.: Phys. Usp. **49**, 1 (2006)

Askerov, B.M.: Electron Transport Phenomena in Semiconductors. World Scientific, Singapore (1981)

Benhar, O., Valli, M.: Phys. Rev. Lett. **99**, 232501 (2007). 0707.2681

Baiko, D.A., Haensel, P.: Acta Phys. Pol. B **30**, 1097 (1999). astro-ph/9906312

Baiko, D.A., Haensel, P.: Astron. Astrophys. **356**, 171 (2000). astro-ph/0004185

Baiko, D.A.: Mon. Not. R. Astron. Soc. **458**, 2840 (2016). 1602.08969

Baiko, D.A., Yakovlev, D.G.: Astron. Lett. **21**, 702 (1995). astro-ph/9604164

Baiko, D.A., Kaminker, A.D., Potekhin, A.Y., Yakovlev, D.G.: Phys. Rev. Lett. **81**, 5556 (1998). physics/9811052

Baiko, D.A., Haensel, P., Yakovlev, D.G.: Astron. Astrophys. **374**, 151 (2001). astro-ph/0105105

Bailin, D., Love, A.: Phys. Rep. **107**, 325 (1984).

Balachandran, A.P., Digal, S., Matsuura, T.: Phys. Rev. D **73**, 074009 (2006). hep-ph/0509276

Balasi, K.G., Langanke, K., Martínez-Pinedo, G.: Prog. Part. Nucl. Phys. **85**, 33 (2015). 1503.08095

Baldo, M. (ed.): Nuclear Methods and the Nuclear Equation of State. International Review of Nuclear Physics, vol. 8. World Scientific, Singapore (1999)

Baldo, M., Ducoin, C.: Phys. Rev. C **84**, 035806 (2011). 1105.1311

Baldo, M., Grasso, A.: Phys. Lett. B **485**, 115 (2000). nucl-th/0003039

Baldo, M., Moshfegh, H.R.: Phys. Rev. C **86**, 024306 (2012). 1209.2270

Baldo, M., Burgio, G. F., Schulze, H.-J., Taranto, G.: Phys. Rev. C **89**, 048801 (2014). 1404.7031

Banik, S., Nandi, R.: In: Jain, S.R., Thomas, R.G., Datar, V.M.: Proceedings of the DAE Symposium on Nuclear Physics, vol. 58, p. 820. Prudent Arts & Fab., Mumbai (2013)

Baym, G.: Phys. Rev. **135**, 1691 (1964)

Baym, G., Pethick, C.: Landau Fermi-Liquid Theory: Concepts and Applications. Wiley, New York (1991)

Bedaque, P.F., Schäfer, T.: Nucl. Phys. A **697**, 802 (2002). hep-ph/0105150

Baym, G., Pethick, C., Pines, D.: Nature **224**, 674 (1969)

Bedaque, P.F., Nicholson, A.N.: Phys. Rev. C **87**, 055807 (2013). [Erratum: Phys. Rev. C **89**, 029902 (2014)], 1212.1122

Bedaque, P.F., Reddy, S.: Phys. Lett. B **735**, 340 (2014). 1307.8183

Bedaque, P.F., Sen, S.: Phys. Rev. C **89**, 035808 (2014). 1312.6632

Bedaque, P.F., Rupak, G., Savage, M.J.: Phys. Rev. C **68**, 065802 (2003). nucl-th/0305032

Bedaque, P.F., Nicholson, A.N., Sen, S.: Phys. Rev. C **92**, 035809 (2015). 1408.5145

Beloborodov, A.M., Li, X.: Astrophys. J. **833**, 261 (2016). 1605.09077

Benhar, O., Lovato, A.: Int. J. Mod. Phys. E **24**, 1530006 (2015). 1506.05225

Benhar, O., Polls, A., Valli, M., Vidaña, I.: Phys. Rev. C **81**, 024305 (2010). 0911.5097

Berdermann, J., Blaschke, D., Fischer, T., Kachanovich, A.: Phys. Rev. D **94**, 123010 (2016). 1609.05201

Berry, D.K., Caplan, M.E., Horowitz, C.J., Huber, G., Schneider, A.S.: Phys. Rev. C **94**, 055801 (2016)

Bertoni, B., Reddy, S., Rrapaj, E.: Phys. Rev. C **91**, 025806 (2015). 1409.7750

Beznogov, M.V., Yakovlev, D.G.: Phys. Rev. Lett. **111**, 161101 (2013). 1307.6060

Beznogov, M.V., Potekhin, A.Y., Yakovlev, D.G.: Mon. Not. R. Astron. Soc. **459**, 1569 (2016). 1604.00538

Bierkandt, R., Manuel, C.: Phys. Rev. D **84**, 023004 (2011). 1104.5624

Bisnovatyi-Kogan, G.S., Romanova, M.M.: Sov. Phys. JETP **56**, 243 (1982)

Blaschke, D., Ropke, G., Schulz, H., Sedrakian, A.D., Voskresensky, D.N.: Mon. Not. R. Astron. Soc. **273**, 596 (1995)

Blaschke, D., Grigorian, H., Voskresensky, D.N.: Phys. Rev. C **88**, 065805 (2013). 1308.4093

Blaschke, D., Schaffner-Bielich, J., Schulze, H.-J.: Eur. Phys. J. A **52**, 267 (2016)

Boguta, J.: Phys. Lett. B **106**, 255 (1981)

Braby, M., Chao, J., Schäfer, T.: Phys. Rev. C **81**, 045205 (2010). 0909.4236

Braginskii, S.I.: Sov. Phys. JETP **6(33)**, 358 (1958)

Bransgrove, A., Levin, Y., Beloborodov, A.: Mon. Not. R. Astron. Soc. **473**, 2771 (2018). 1709.09167

Brooker, G.A., Sykes, J.: Phys. Rev. Lett. **21**, 279 (1968)

Brown, E.F., Cumming, A.: Astrophys. J. **698**, 1020 (2009). 0901.3115

Brown, E.F., Bildsten, L., Rutledge, R.E.: Astrophys. J. Lett. **504**, L95 (1998). astro-ph/9807179

Burnett, D.: Proc. Lond. Math. Soc. **39**, 385 (1935)

Burnett, D.: Proc. Lond. Math. Soc. **40**, 382 (1936)

Burnett, T.H., Kroll, N M.: Phys. Rev. Lett. **20**, 86 (1968)

Burrows, A., Reddy, S., Thompson, T.A.: Nucl. Phys. A **777**, 356 (2006). astro-ph/0404432

Bykov, A.M., Toptygin, I.: Phys. Usp. **50**, 141 (2007)

Caballero, O.L., Postnikov, S., Horowitz, C.J., Prakash, M.: Phys. Rev. C **78**, 045805 (2008). 0807.4353

Cackett, E.M., Brown, E.F., Cumming, A., Degenaar, N., Fridriksson, J.K., Homan, J., Miller, J.M., Wijnands, R.: Astrophys. J. **774**, 131 (2013). 1306.1776

Caplan, M.E., Horowitz, C.J.: Rev. Mod. Phys. **89**, 041002 (2017). 1606.03646

Carbone, A., Benhar, O.: J. Phys. Conf. Ser. **336**, 012015 (2011)

Carlson, J., Pandharipande, V.R., Wiringa, R.B.: Nucl. Phys. A **401**, 59 (1983)

Cassisi, S., Potekhin, A.Y., Pietrinferni, A., Catelan, M., Salaris, M.: Astroph. J. **661**, 1094 (2007). astro-ph/0703011

Castillo, F., Reisenegger, A., Valdivia, J.A.: Mon. Not. R. Astron. Soc. **471**, 507 (2017). 1705.10020

Chabrier, G.: Astroph. J. **414**, 695 (1993)

Chamel, N.: Phys. Rev. C **85**, 035801 (2012)

Chamel, N.: Phys. Rev. Lett. **110**, 011101 (2013). 1210.8177

Chamel, N., Haensel, P.: Living Rev. Relativ. **11**, 10 (2008). 0812.3955

Chamel, N., Page, D., Reddy, S.: Phys. Rev. C **87**, 035803 (2013). 1210.5169

Chamel, N., Page, D., Reddy, S.: J. Phys. Conf. Ser. **665**, 012065 (2016)

Chao, J., Schäfer, T.: Ann. Phys. **327**, 1852 (2012). 1108.4979

Chapman, S., Cowling, T.G., Burnett, D.: The Mathematical Theory of Non-Uniform Gases: An Account of the Kinetic Theory of Viscosity, Thermal Conduction and Diffusion in Gases, 3rd edn. Cambridge University Press, Cambridge (1999)

Charbonneau, J., Zhitnitsky, A.: J. Cosmol. Astropart. Phys. **1008**, 010 (2010). 0903.4450

Charbonneau, J., Hoffman, K., Heyl, J.: Mon. Not. R. Astron. Soc. **404**, L119 (2010). 0912.3822

Chatterjee, D., Bandyopadhyay, D.: Phys. Rev. D **74**, 023003 (2006). astro-ph/0602538

Chugunov, A.I.: Astron. Lett. **38**, 25 (2012)

Chugunov, A.I., Haensel, P.: Mon. Not. R. Astron. Soc. **381**, 1143 (2007). 0707.4614

Chugunov, A.I., Yakovlev, D.G.: Astron. Rep. **49**, 724 (2005). astro-ph/0511300

Chugunov, A.I., Gusakov, M.E., Kantor, E.M.: Mon. Not. R. Astron. Soc. **468**, 291 (2017). 1610.06380

Davison, R.A., Goykhman, M., Parnachev, A.: J. High Energy Phys. **07**, 109 (2014). 1312.0463

Daligault, J., Gupta, S.: Astrophys. J. **703**, 994 (2009). 0905.0027

Deb, P., Kadam, G.P., Mishra, H.: Phys. Rev. D **94**, 094002 (2016). 1603.01952

Degenaar, N., Medin, Z., Cumming, A., Wijnands, R., Wolff, M.T, Cackett, E.M., Miller, J.M., Jonker, P.G., Homan, J., Brown, E.F.: Astroph. J. **791**, 47 (2014). 1403.2385

Dehghan Niri, A., Moshfegh, H. R., Haensel, P.: Phys. Rev. C **93**, 045806 (2016)

Deibel, A., Cumming, A., Brown, E.F., Reddy, S.: Astrophys. J. **839**, 95 (2017). 1609.07155

Delehaye, M., Laurent, S., Ferrier-Barbut, I., Jin, S., Chevy, F., Salomon, C.: Phys. Rev. Lett. **115**, 265303 (2015). 1510.06709

Deng, J., Schmitt, A., Wang, Q.: Phys. Rev. D **76**, 034013 (2007). nucl-th/0611097

Denicol, G.S., Niemi, H., Molnar, E., Rischke, D.H.: Phys. Rev. D **85**, 114047 (2012). [Erratum: Phys. Rev. D **91**, 039902 (2015)]. 1202.4551

Dickhoff, W.H., Van Neck, D.: Many-Body Theory Exposed: Propagator Description of Quantum Mechanics in Many-Body Systems, 2nd edn. World Scientific, Singapore (2008)

Dine, M., Fischler, W., Srednicki, M.: Phys. Lett. B **104**, 199 (1981)

Ding, W.-B., Qi, Z.-Q., Hou, J.-W., Mi, G., Bao, T., Yu, Z., Liu, G.-Z., Zhao, E.-G.: Commun. Theor. Phys. **66**, 474 (2016)

Dong, J.M., Lombardo, U., Zhang, H.F., Zuo, W.: Astrophys. J. **817**, 6 (2016). 1512.02746

Donnelly, R.J.: J. Phys. Condens. Matter **11**, 7783 (1999)

Douchin, F., Haensel, P.: Phys. Lett. B **485**, 107 (2000). astro-ph/0006135

Drago, A., Lavagno, A., Pagliara, G.: Phys. Rev. D **71**, 103004 (2005). astro-ph/0312009

Eckart, C.: Phys. Rev. **58**, 919 (1940)

Elfritz, J.G., Pons, J.A., Rea, N., Glampedakis, K., Viganò, D.: Mon. Not. R. Astron. Soc. **456**, 4461 (2016). 1512.07151

Elshamouty, K.G., Heinke, C.O., Sivakoff, G.R., Ho, W.C.G., Shternin, P.S., Yakovlev, D.G., Patnaude, D.J., David, L.: Astroph. J. **777**, 22 (2013). 1306.3387

Fischer, T., Chakraborty, S., Giannotti, M., Mirizzi, A., Payez, A., Ringwald, A.: Phys. Rev. D **94**, 085012 (2016). 1605.08780

Flores-Tulián, S., Reisenegger, A.: Mon. Not. R. Astron. Soc. **372**, 276 (2006). astro-ph/0606412

Flowers, E., Itoh, N.: Astrophys. J. **206**, 218 (1976)

Flowers, E., Itoh, N.: Astrophys. J. **230**, 847 (1979)

Flowers, E., Ruderman, M., Sutherland, P.: Astrophys. J. **205**, 541 (1976)

Friedman, J.L., Morsink, S.M.: Astrophys. J. **502**, 714 (1998). gr-qc/9706073

Friman, B.L., Maxwell, O.V.: Astrophys. J. **232**, 541 (1979)

Friman, B., Hohne, C., Knoll, J., Leupold, S., Randrup, J., Rapp, R., Senger, P.: Lect. Notes Phys. **814**, 1 (2011)

García-Colín, L., Velasco, R., Uribe, F.: Phys. Rep. **465**, 149 (2008)

Gattringer, C., Langfeld, K.: Int. J. Mod. Phys. A **31**, 1643007 (2016). 1603.09517

Gerhold, A., Rebhan, A.: Phys. Rev. D **71**, 085010 (2005). hep-ph/0501089

Glampedakis, K., Gualtieri, L.: This volume (2018). 1709.07049

Glampedakis, K., Jones, D.I., Samuelsson, L.: Mon. Not. R. Astron. Soc. **413**, 2021 (2011). 1010.1153

Glampedakis, K., Andersson, N., Lander, S.K.: Mon. Not. R. Astron. Soc. **420**, 1263 (2012a). 1106.6330

Glampedakis, K., Jones, D.I., Samuelsson, L.: Phys. Rev. Lett. **109**, 081103 (2012b). 1204.3781

Glesaaen, J., Neuman, M., Philipsen, O.: J. High Energy Phys. **03**, 100 (2016). 1512.05195

Gnedin, O.Y., Yakovlev, D.G., Potekhin, A.Y.: Mon. Not. R. Astron. Soc. **324**, 725 (2001). astro-ph/0012306

Goldreich, P., Reisenegger, A.: Astrophys. J. **395**, 250 (1992)

González-Jiménez, N., Petrovich, C., Reisenegger, A.: Mon. Not. R. Astron. Soc. **447**, 2073 (2015). 1411.6500

Graber, V., Andersson, N., Glampedakis, K., Lander, S.K.: Mon. Not. R. Astron. Soc. **453**, 671 (2015). 1505.00124

Graber, V., Andersson, N., Hogg, M.: Int. J. Mod. Phys. D **26**, 1730015 (2017). 1610.06882

Grabowska, D., Kaplan, D.B., Reddy, S.: Phys. Rev. D **91**, 085035 (2015). 1409.3602

Grad, H.: Commun. Pure Appl. Math. **2**, 331 (1949). ISSN 1097-0312

Greiter, M., Wilczek, F., Witten, E.: Mod. Phys. Lett. B **3**, 903 (1989)

Grigorian, H., Blaschke, D., Voskresensky, D.: Phys. Rev. C **71**, 045801 (2005). astro-ph/0411619

Gupta, S., Brown, E.F., Schatz, H., Moeller, P., Kratz, K.-L.: Astrophys. J. **662**, 1188 (2007). astro-ph/0609828

Gupta, S.S., Kawano, T., Möller, P.: Phys. Rev. Lett. **101**, 231101 (2008). 0811.1791

Gusakov, M.E.: Astron. Astrophys. **389**, 702 (2002). astro-ph/0204334

Gusakov, M.E.: Phys. Rev. D **76**, 083001 (2007). 0704.1071

Gusakov, M.E.: Phys. Rev. C **81**, 025804 (2010). 1001.4452

Gusakov, M.E.: Phys. Rev. D **93**, 064033 (2016). 1601.07732

Gusakov, M., Dommes, V.: Phys. Rev. D **94**, 083006 (2016). 1607.01629

Gusakov, M.E., Kantor, E.M.: Phys. Rev. D **78**, 083006 (2008). 0806.4914

Gusakov, M.E., Kaminker, A.D., Yakovlev, D.G., Gnedin, O.Y.: Astron. Astrophys. **423**, 1063 (2004). astro-ph/0404002

Gusakov, M.E., Kantor, E.M., Haensel, P.: Phys. Rev. C **79**, 055806 (2009a). 0904.3467

Gusakov, M.E., Kantor, E.M., Haensel, P.: Phys. Rev. C **80**, 015803 (2009b). 0907.0010

Gusakov, M.E., Kantor, E.M., Ofengeim, D.D.: Phys. Rev. D **96**, 103012 (2017). 1705.00508

Haber, A., Schmitt, A.: EPJ Web Conf. **137**, 09003 (2017a). 1612.01865

Haber, A., Schmitt, A.: Phys. Rev. D **95**, 116016 (2017b). 1704.01575

Haber, A., Schmitt, A., Stetina, S.: Phys. Rev. D **93**, 025011 (2016). 1510.01982

Haensel, P., Jerzak, A.J.: Astron. Astrophys. **179**, 127 (1987)

Haensel, P., Zdunik, J.L.: Astron. Astrophys. **480**, 459 (2008). 0708.3996

Haensel, P.: Astron. Astrophys. **262**, 131 (1992)

Haensel, P., Urpin, V.A., Yakovlev, D.G.: Astron. Astrophys. **229**, 133 (1990)

Haensel, P., Levenfish, K.P., Yakovlev, D.G.: Astron. Astrophys. **357**, 1157 (2000). astro-ph/0004183

Haensel, P., Levenfish, K.P., Yakovlev, D.G.: Astron. Astrophys. **327**, 130 (2001). astro-ph/0103290

Haensel, P., Levenfish, K.P., Yakovlev, D.G.: Astron. Astrophys. **381**, 1080 (2002). astro-ph/0110575

Haensel, P., Potekhin, A.Y., Yakovlev, D.G., Neutron Stars 1: Equation of State and Structure. Astrophysics and Space Science Library, vol. 326. Springer Science+Buisness Media, New York (2007)

Hall, H., Vinen, W.: Proceedings of the Royal Society of London A: Mathematical, Physical and Engineering Sciences, vol. 238. The Royal Society, (1956), pp. 215–234

Hanhart, C., Phillips, D.R., Reddy, S.: Phys. Lett. B **499**, 9 (2001). astro-ph/0003445

Hannestad, S., Raffelt, G.: Astrophys. J. **507**, 339 (1998). astro-ph/9711132

Harrison, W.A.: Solid State Theory. Dover Publications, New York (1980)

Harutyunyan, A., Sedrakian, A.: Phys. Rev. C **94**, 025805 (2016). 1605.07612

Harutyunyan, A., Rischke, D.H., Sedrakian, A.: Phys. Rev. D **95**, 114021 (2017). 1702.04291

Hashimoto, M., Seki, H., Yamada, M.: Prog. Theor. Phys. **71**, 320 (1984)

Haskell, B.: Int. J. Mod. Phys. E **24**, 1541007 (2015)

Haskell, B., Andersson, N.: Mon. Not. R. Astron. Soc. **408**, 1897 (2010). 1003.5849

Haskell, B., Melatos, A.: Int. J. Mod. Phys. D **24**, 1530008 (2015). 1502.07062

Haskell, B., Sedrakian, A.: This volume (2018). 1709.10340

Hatsuda, T., Tachibana, M., Yamamoto, N., Baym, G.: Phys. Rev. Lett. **97**, 122001 (2006). hep-ph/0605018

Heinke, C.O., Ho, W.C.G.: Astrophys. J. **719**, L167 (2010). 1007.4719

Heiselberg, H., Pethick, C. J.: Phys. Rev. D **48**, 2916 (1993)

Heiselberg, H., Baym, G., Pethick, C.J., Popp, J.: Nucl. Phys. A **544**, 569 (1992)

Hess, D., Sedrakian, A.: Phys. Rev. D **84**, 063015 (2011). 1104.1706

Ho, W.C.G., Elshamouty, K.G., Heinke, C.O., Potekhin, A.Y.: Phys. Rev. C **91**, 015806 (2015). 1412.7759

Horowitz, C.J., Berry, D.K.: Phys. Rev. C **78**, 035806 (2008). 0807.2603

Horowitz, C.J., Pérez-García, M.A.: Phys. Rev. C **68**, 025803 (2003). astro-ph/0305138

Hoyos, J., Reisenegger, A., Valdivia, J.A.: Astron. Astrophys. **487**, 789 (2008). 0801.4372

Hernquist, L.: Astrophys. J. Suppl. **56**, 325 (1984)

Højgård Jensen, H., Smith, H., Wilkins, J.W.: Phys. Lett. A **27**, 532 (1968)

Horowitz, C.J., Berry, D.K.: Phys. Rev. C **79**, 065803 (2009). 0904.4076

Horowitz, C.J., Caballero, O.L., Berry, D.K.: Phys. Rev. E. **79**, 026103 (2009). 0804.4409

Horowitz, C.J., Berry, D.K., Briggs, C.M., Caplan, M.E., Cumming, A., Schneider, A.S.: Phys. Rev. Lett. **114**, 031102 (2015). 1410.2197

Huang, X.-G., Huang, M., Rischke, D.H., Sedrakian, A.: Phys. Rev. D **81**, 045015 (2010). 0910.3633

Hughto, J., Schneider, A.S., Horowitz, C.J., Berry, D.K.: Phys. Rev. E **84**, 016401 (2011). 1104.4822

Huot, S.C., Jeon, S., Moore, G. D.: Phys. Rev. Lett. **98**, 172303 (2007). hep-ph/0608062
Israel, W., Stewart, J.M.: Ann. Phys. **118**, 341 (1979)
Itoh, N., Uchida, S., Sakamoto, Y., Kohyama, Y., Nozawa, S.: Astrophys. J. **677**, 495 (2008). 0708.2967
Iwamoto, N.: Phys. Rev. Lett. **44**, 1637 (1980)
Iwamoto, N.: Ann. Phys. **141**, 1 (1982)
Iwamoto, N.: Phys. Rev. Lett. **53**, 1198 (1984)
Iwamoto, N.: Phys. Rev. D **64**, 043002 (2001)
Iwasaki, M., Fukutome, T.: J. Phys. G **36**, 115012 (2009)
Iwasaki, M., Ohnishi, H., Fukutome, T.: J. Phys. G **35**, 035003 (2008). hep-ph/0703271
Jaccarino, D., Plumari, S., Greco, V., Lombardo, U., Santra, A.B.: Phys. Rev. D **85**, 103001 (2012)
Jaikumar, P., Prakash, M.: Phys. Lett. B **516**, 345 (2001). astro-ph/0105225
Jaikumar, P., Prakash, M., Schäfer, T.: Phys. Rev. D **66**, 063003 (2002). astro-ph/0203088
Jaikumar, P., Gale, C., Page, D.: Phys. Rev. D **72**, 123004 (2005). hep-ph/0508245
Jaikumar, P., Roberts, C.D., Sedrakian, A.: Phys. Rev. C **73**, 042801 (2006). nucl-th/0509093
Janka, H.-T.: In: Alsabti, A.W., Murdin, P. (eds.) Handbook of Supernovae, pp. 1575–1604. Springer International Publishing, Berlin (2017). 1702.08713
Janka, H.-T., Hanke, F., Hüdepohl, L., Marek, A., Müller, B., Obergaulinger, M.: Prog. Theor. Exp. Phys. **2012**, 01A309 (2012). 1211.1378
Jones, P.B.: Phys. Rev. D **64**, 084003 (2001)
Jones, M.D., Ceperley, D.M.: Phys. Rev. Lett. **76**, 4572 (1996)
Juri, L.O., Hernández, E.S.: Phys. Rev. C **37**, 376 (1988)
Kaminker, A.D., Yakovlev, D.G.: Theor. Math. Phys. **49**, 1012 (1981)
Kaminker, A.D., Yakovlev, D.G., Haensel, P.: Astrophys. Space Sci. **361**, 267 (2016). 1607.05265
Kaminski, M., Uhlemann, C.F., Bleicher, M.: Schaffner-Bielich, J.: Phys. Lett. B **760**, 170 (2016). 1410.3833
Kaplan, D.B., Reddy, S.: Phys. Rev. D **65**, 054042 (2002). hep-ph/0107265
Kaplan, D.B., Reddy, S., Sen, S.: Phys. Rev. D **96**, 016008 (2017). 1612.00032
Keller, J., Sedrakian, A.: Nucl. Phys. A **897**, 62 (2013). 1205.6940
Keil, W., Janka, H.-T., Schramm, D.N., Sigl, G., Turner, M.S., Ellis, J.R.: Phys. Rev. D **56**, 2419 (1997). astro-ph/9612222
Kerbikov, B.O., Andreichikov, M.A.: Phys. Rev. D **91**, 074010 (2015). 1410.3413
Khalatnikov, I.M.: Introduction to the Theory of Superfluidity. Benjamin, New York (1965)
Kharzeev, D.E., McLerran, L.D., Warringa, H.J.: Nucl. Phys. A **803**, 227 (2008). 0711.0950
Kharzeev, D.E., Liao, J., Voloshin, S.A., Wang, G.: Prog. Part. Nucl. Phys. **88**, 1 (2016). 1511.04050
Kim, J.E.: Phys. Rev. Lett. **43**, 103 (1979)
Knoll, J., Voskresensky, D.N.: Phys. Lett. B **351**, 43 (1995). hep-ph/9503222
Kolomeitsev, E.E., Voskresensky, D.N.: Phys. Rev. C **77**, 065808 (2008). 0802.1404
Kolomeitsev, E.E., Voskresensky, D.N.: Phys. Rev. C **81**, 065801 (2010). 1003.2741
Kolomeitsev, E.E., Voskresensky, D.N.: Phys. At. Nucl. **74**, 1316 (2011). 1012.1273
Kolomeitsev, E.E., Voskresensky, D.N.: Phys. Rev. C **91**, 025805 (2015). 1412.0314
Kovtun, P.: J. Phys. A **45**, 473001 (2012). 1205.5040
Kovtun, P., Son, D.T., Starinets, A.O.: J. High Energy Phys. **10**, 064 (2003). hep-th/0309213
Kubis, S., Wójcik, W.: Phys. Rev. C **92**, 055801 (2015). 1602.07511
Kundu, J., Reddy, S.: Phys. Rev. C **70**, 055803 (2004). nucl-th/0405055
Kurkela, A., Romatschke, P., Vuorinen, A.: Phys. Rev. D **81**, 105021 (2010a). 0912.1856
Kurkela, A., Romatschke, P., Vuorinen, A., Wu, B.: ArXiv e-prints (2010b). 1006.4062
Kurkela, A., Fraga, E.S., Schaffner-Bielich, J., Vuorinen, A.: Astrophys. J. **789**, 127 (2014). 1402.6618
Kutschera, M., Wójcik, W.: Phys. Lett. B **223**, 11 (1989)
Kutschera, M., Wójcik, W.: Acta Phys. Pol. B **21**, 823 (1990)
Kutschera, M., Wójcik, W. Ł.: Phys. Rev. C **47**, 1077 (1993)
Kutschera, M., Wójcik, W. Ł.: Nucl. Phys. A **581**, 706 (1995)

Kremer, G.: An Introduction to the Boltzmann Equation and Transport Processes in Gases. Interaction of Mechanics and Mathematics. Springer, Berlin (2010)

Lai, D.: AIP Conf. Proc. **575**, 246 (2001). astro-ph/0101042

Lam, S.H.: Phys. Fluids **18**, 073101 (2006)

Landau, L.: Phys. Rev. **60**, 356 (1941)

Landau, L.D., Lifshitz, E.M.: Statistical Physics, Part 1. Course of Theoretical Physics, vol. 5, 3rd edn. Elsevier, Oxford (1980)

Landau, L.D., Lifshitz, E.M.: Fluid Mechanics, Second Edition: Volume 6 (Course of Theoretical Physics), 2nd edn. Butterworth-Heinemann, London, (1987)

Landau, L.D., Pitaevskii, L.P., Lifshitz, E.M.: Electrodynamics of Continuous Media, Second Edition: Volume 8 (Course of Theoretical Physics), 2nd edn. Butterworth-Heinemann, London (1984)

Landsteiner, K.: Acta Physiol. Pol. B **47**, 2617 (2016). 1610.04413

Lang, R., Kaiser, N., Weise, W.: Eur. Phys. J. A **51**, 127 (2015). 1506.02459

Lattimer, J.M., Prakash, M., Pethick, C.J., Haensel, P.: Phys. Rev. Lett. **66**, 2701 (1991)

Leggett, A.J.: Phys. Rev. **147**, 119 (1966a)

Leggett, A.J.: Prog. Theor. Phys. **36**, 417 (1966b)

Leinson, L.B.: Nucl. Phys. A **707**, 543 (2002). hep-ph/0207116

Leinson, L.B.: Phys. Rev. C **78**, 015502 (2008). 0804.0841

Leinson, L.B.: Phys. Rev. C **79**, 045502 (2009). 0904.0320

Leinson, L.B.: Phys. Rev. C **81**, 025501 (2010). 0912.2164

Leinson, L.B.: Phys. Rev. C **85**, 065502 (2012). 1206.3648

Leinson, L.B.: Phys. Rev. C **87**, 025501 (2013). 1301.5439

Leinson, L.B., Pérez, A.: Phys. Lett. B **518**, 15 (2001). hep-ph/0110207

Leinson, L.B., Pérez, A.: Phys. Lett. B **638**, 114 (2006). astro-ph/0606651

Li, Y., Liou, M. K., Schreiber, W.M.: Phys. Rev. C **80**, 035505 (2009)

Li, Y., Liou, M.K., Schreiber, W.M., Gibson, B.F.: Phys. Rev. C **92**, 015504 (2015)

Li, Q., Kharzeev, D.E., Zhang, C., Huang, Y., Pletikosić, I., Fedorov, A.V., Zhong, R.D., Schneeloch, J.A., Gu, G.D., Valla, T.: Nat. Phys. **12**, 550 (2016). 1412.6543

Lindblom, L., Owen, B.J.: Phys. Rev. D **65**, 063006 (2002). astro-ph/0110558

Low, F.E.: Phys. Rev. **110**, 974 (1958)

Lykasov, G.I., Pethick, C.J., Schwenk, A.: Phys. Rev. C **78**, 045803 (2008). 0808.0330

Madsen, J.: Phys. Rev. D **46**, 3290 (1992)

Madsen, J.: Phys. Rev. **D47**, 325 (1993)

Manchester, R.N., Hobbs, G.B., Teoh, A., Hobbs, M.: Astron. J. **129**, 1993 (2005). astro-ph/0412641

Mannarelli, M.: PoS Confinement X **261** (2013). 1301.6074

Mannarelli, M., Manuel, C.: Phys. Rev. D **81**, 043002 (2010). 0909.4486

Mannarelli, M., Manuel, C., Tolos, L.: Ann. Phys. **336**, 12 (2013). 1212.5152

Manuel, C., Llanes-Estrada, F.J.: J. Cosmol. Astropart. Phys. **0708**, 001 (2007). 0705.3909

Manuel, C., Tolos, L.: Phys. Rev. D **84**, 123007 (2011). 1110.0669

Manuel, C., Tolos, L.: Phys. Rev. D **88**, 043001 (2013). 1212.2075

Manuel, C., Dobado, A., Llanes-Estrada, F.J.: J. High Energy Phys. **09**, 076 (2005). hep-ph/0406058

Manuel, C., Tarrus, J., Tolos, L.: J. Cosmol. Astropart. Phys. **1307**, 003 (2013). 1302.5447

Manuel, C., Sarkar, S., Tolos, L.: Phys. Rev. C **90**, 055803 (2014). 1407.7431

Mateos, D., Myers, R.C., Thomson, R.M.: Phys. Rev. Lett. **98**, 101601 (2007). hep-th/0610184

Maxwell, O.V.: Astrophys. J. **316**, 691 (1987)

Mckinven, R., Cumming, A., Medin, Z., Schatz, H.: Astrophys. J. **823**, 117 (2016). 1603.08644

McLerran, L., Pisarski, R.D.: Nucl. Phys. A **796**, 83 (2007). 0706.2191

Meisel, Z., Deibel., A., Keek, L., Shternin, P., Elfritz, J.: J. Phys. G **45**, 093001 (2018). 1807.01150

Migdal, A.B.: Rev. Mod. Phys. **50**, 107 (1978)

Migdal, A.B., Saperstein, E.E., Troitsky, M.A., Voskresensky, D.N.: Phys. Rep. **192**, 179 (1990)

Myers, R.C., Paulos, M.F., Sinha, A.: J. High Energy Phys. **06**, 006 (2009). 0903.2834

Nandi, R., Schramm, S.: Astrophys. J. **852**, 135 (2018). 1709.09793

Noda, T., Hashimoto, M.-A., Yasutake, N., Maruyama, T., Tatsumi, T., Fujimoto, M.: Astrophys. J. **765**, 1 (2013). 1109.1080

Ofengeim, D.D., Yakovlev, D.G.: Europhys. Lett. **112**, 59001 (2015). 1512.03915

Ohnishi, A., Yamamoto, N.: ArXiv e-prints (2014). 1402.4760

Oyamatsu, K., Iida, K.: Phys. Rev. C **75**, 015801 (2007). nucl-th/0609040

Page, D., Reddy, S.: Phys. Rev. Lett. **111**, 241102 (2013). 1307.4455

Page, D., Lattimer, J.M., Prakash, M., Steiner, A.W.: Astrophys. J. **707**, 1131 (2009). 0906.1621

Page, D., Lattimer, J.M., Prakash, M., Steiner, A.W.: Astrophys. J. Suppl. **155**, 623 (2004). astro-ph/0403657

Page, D., Geppert, U., Weber, F.: Nucl. Phys. A **777**, 497 (2006). astro-ph/0508056

Page, D., Prakash, M., Lattimer, J.M., Steiner, A. W.: Phys. Rev. Lett. **106**, 081101 (2011). 1011.6142

Page, D., Reddy, S.: In: Bertulani, C.A., Piekarewicz, J. (eds.) Neutron Star Crust. Nova Science, Hauppauge (2012). 1201.5602

Page, D., Lattimer, J.M., Prakash, M., Steiner, A.W.: In: Bennemann, K.H., Ketterson, J.B. (eds.) Novel Superfluids: Volume 2, p. 505. Oxford University Press, Oxford (2014). 1302.6626

Pang, J.-Y., Brauner, T., Wang, Q.: Nucl. Phys. A **852**, 175 (2011). 1010.1986

Passamonti, A., Akgün, T., Pons, J.A., Miralles, J.A.: Mon. Not. R. Astron. Soc. **465**, 3416 (2017). 1608.00001

Pavlović, P., Leite, N., Sigl, G.: Phys. Rev. D **96**, 023504 (2017). 1612.07382

Peccei, R.D., Quinn, H.R.: Phys. Rev. Lett. **38**, 1440 (1977)

Pérez-Azorín, J.F., Miralles, J.A., Pons, J.A.: Astron. Astrophys. **451**, 1009 (2006). astro-ph/0510684

Pethick, C.J.: In: Pines, D., Tamagaki, R., Tsuruta, S. (eds.) Structure and Evolution of Neutron Stars, p. 115. Basic Books, New York (1992)

Pethick, C.J., Schwenk, A.: Phys. Rev. C **80**, 055805 (2009). 0908.0950

Pethick, C.J., Smith, H., Bhattacharyya, P.: Phys. Rev. B **15**, 3384 (1977)

Petrovich, C., Reisenegger, A.: Astron. Astrophys. **521**, A77 (2010). 0912.2564

Pi, C.-M., Zheng, X.-P., Yang, S.-H.: Phys. Rev. C **81**, 045802 (2010). 0912.2884

Pitaevskii, L., Lifshitz, E.: In: Landau, L.D., Lifshitz, E.M. (eds.) Physical Kinetics. Course of Theoretical Physics, vol. 10. Butterworth-Heinemann, London (2008)

Pons, J.A., Reddy, S., Prakash, M., Lattimer, J.M., Miralles, J.A.: Astrophys. J. **513**, 780 (1999). astro-ph/9807040

Pons, J.A., Viganò, D., Rea, N.: Nat. Phys. **9**, 431 (2013). 1304.6546

Popov, S., Grigorian, H., Blaschke, D.: Phys. Rev. C **74**, 025803 (2006). nucl-th/0512098

Posselt, B., Pavlov, G.G.: ArXiv e-prints (2018). 1808.00531

Posselt, B., Pavlov, G.G., Suleimanov, V., Kargaltsev, O.: Astroph. J. **779**, 186 (2013). 1311.0888

Potekhin, A.Y.: Astron. Astrophys. **306**, 999 (1996). astro-ph/9603133

Potekhin, A.Y.: Astron. Astrophys. **351**, 787 (1999). astro-ph/9909100

Potekhin, A.Y.: Phys. Usp. **57**, 735 (2014)

Potekhin, A.Y., Chabrier, G.: Phys. Rev. E. **62**, 8554 (2000). astro-ph/0009261

Potekhin, A.Y., Chabrier, G.: Astron. Astrophys. **550**, A43 (2013). 1212.3405

Potekhin, A.Y., Yakovlev, D.G.: Astron. Astrophys. **314**, 341 (1996). astro-ph/9604130

Potekhin, A.Y., Baiko, D.A., Haensel, P., Yakovlev, D.G.: Astron. Astrophys. **346**, 345 (1999). astro-ph/9903127

Potekhin, A.Y., Chabrier, G., Yakovlev, D.G.: Astroph. Space Sci. **308**, 353 (2007). astro-ph/0611014

Potekhin, A.Y., De Luca, A., Pons, J.A.: Space Sci. Rev. **191**, 171 (2015a). 1409.7666

Potekhin, A.Y., Pons, J.A., Page, D.: Space Sci. Rev. **191**, 239 (2015b). 1507.06186

Ravenhall, D.G., Pethick, C.J., Wilson, J.R.: Phys. Rev. Lett. **50**, 2066 (1983)

Raffelt, G.G.: Lect. Notes Phys. **741**, 51 (2008), hep-ph/0611350.

Reinholz, H., Röpke, G.: Phys. Rev. E. **85**, 036401 (2012)

Reisenegger, A.: Astrophys. J. **442**, 749 (1995). astro-ph/9410035

Reisenegger, A.: Astrophys. J. **485**, 313 (1997). astro-ph/9612179
Roberts, L.F., Reddy, S.: In: Alsabti, A.W., Murdin, P. (eds.) Handbook of Supernovae, pp. 1605–1635. Springer International Publishing, Berlin (2017a). 1612.03860
Roberts, L.F., Reddy, S.: Phys. Rev. C **95**, 045807 (2017b). 1612.02764
Roggero, A., Reddy, S.: Phys. Rev. C **94**, 015803 (2016). 1602.01831
Romatschke, P.: Int. J. Mod. Phys. E **19**, 1 (2010). 0902.3663
Rosenfeld, A.M., Stott, M.J.: Phys. Rev. B **42**, 3406 (1990)
Sa'd, B.A., Shovkovy, I.A., Rischke, D.H.: Phys. Rev. D **75**, 125004 (2007a). astro-ph/0703016
Sa'd, B.A., Shovkovy, I.A., Rischke, D.H.: Phys. Rev. D **75**, 065016 (2007b). astro-ph/0607643
Sarkar, S., Sharma, R.: Phys. Rev. D **96**, 094025 (2017). 1701.00010
Sawyer, R.F., Soni, A.: Astrophys. J. **230**, 859 (1979)
Sawyer, R.F.: Phys. Rev. D **39**, 3804 (1989a)
Sawyer, R.F.: Phys. Lett. B **233**, 412 (1989b). [Erratum: Phys. Lett. B **347**, 467 (1995)]
Schaeffer, A.M.J., Talmadge, W. B., Temple, S.R., Deemyad, S.: Phys. Rev. Lett. **109**, 185702 (2012)
Schäfer, T.: Phys. Rev. A **90**, 043633 (2014). 1404.6843
Schäfer, T., Schwenzer, K.: Phys. Rev. D **70**, 114037 (2004). astro-ph/0410395
Schäfer, T., Wilczek, F.: Phys. Rev. Lett. **82**, 3956 (1999). hep-ph/9811473
Schatz, H.: J. Phys. G **43**, 064001 (2016). 1606.00485
Schatz, H., Gupta, S., Möller, P., Beard, M., Brown, E.F., Deibel, A.T., Gasques, L.R., Hix, W.R., Keek, L., Lau, R., et al.: Nature **505**, 62 (2014). 1312.2513
Schmidt, R., Lemeshko, M.: Phys. Rev. Lett. **114**, 203001 (2015). 1502.03447
Schmitt, A.: Phys. Rev. D **71**, 054016 (2005). nucl-th/0412033
Schmitt, A.: Prog. Theor. Phys. Suppl. **174**, 14 (2008)
Schmitt, A.: Lect. Notes Phys. **811**, 1 (2010). 1001.3294
Schmitt, A.: Lect. Notes Phys. **888**, 1 (2015). 1404.1284
Schmitt, A., Wang, Q., Rischke, D.H.: Phys. Rev. D **69**, 094017 (2004). nucl-th/0311006
Schmitt, A., Shovkovy, I.A., Wang, Q.: Phys. Rev. D **73**, 034012 (2006). hep-ph/0510347
Schmitt, A., Stetina, S., Tachibana, M.: Phys. Rev. D **83**, 045008 (2011). 1010.4243
Shifman, M.A., Vainshtein, A., Zakharov, V.I.: Nucl. Phys. B **166**, 493 (1980)
Schneider, A.S., Berry, D.K., Briggs, C.M., Caplan, M.E., Horowitz, C.J.: Phys. Rev. C **90**, 055805 (2014). 1409.2551
Schneider, A.S., Berry, D.K., Caplan, M.E., Horowitz, C.J., Lin, Z.: Phys. Rev. C **93**, 065806 (2016). 1602.03215
Schwenk, A., Friman, B.: Phys. Rev. Lett. **92**, 082501 (2004). nucl-th/0307089
Schwenk, A., Friman, B., Brown, G.E.: Nucl. Phys. A **713**, 191 (2003). nucl-th/0207004
Schwenk, A., Jaikumar, P., Gale, C.: Phys. Lett. B **584**, 241 (2004). nucl-th/0309072
Schwenzer, K.: ArXiv e-prints (2012). 1212.5242
Sedrakian, A.: Phys. Lett. B **607**, 27 (2005). nucl-th/0411061
Sedrakian, A.: Prog. Part. Nucl. Phys. **58**, 168 (2007). nucl-th/0601086
Sedrakian, A.: Phys. Particles Nucl. **39**, 1155 (2008). astro-ph/0701017
Sedrakian, A.: Phys. Rev. C **86**, 025803 (2012). 1201.1394
Sedrakian, A.: Astron. Astrophys. **555**, L10 (2013). 1303.5380
Sedrakian, A.: Eur. Phys. J. A **52**, 44 (2016a). 1509.06986
Sedrakian, A.: Phys. Rev. D **93**, 065044 (2016b). 1512.07828
Sedrakian, A., Dieperink, A.: Phys. Lett. B **463**, 145 (1999). nucl-th/9905039
Sedrakian, A.D., Blaschke, D., Röpke, G., Schulz, H.: Phys. Lett. B **338**, 111 (1994)
Sedrakian, A., Müther, H., Schuck, P.: Phys. Rev. C **76**, 055805 (2007). astro-ph/0611676
Shahzamanian, M.A., Yavary, H.: Int. J. Mod. Phys. D **14**, 121 (2005)
Shalybkov, D.A., Urpin, V.A.: Mon. Not. R. Astron. Soc. **273**, 643 (1995)
Shapiro, S.L., Teukolsky, S.A.: Black holes, white dwarfs, and neutron stars: the physics of compact objects. Wiley, New York (1983)
Shen, G., Gandolfi, S., Reddy, S., Carlson, J.: Phys. Rev. C **87**, 025802 (2013). 1205.6499
Shovkovy, I.A., Wang, X.: New J. Phys. **13**, 045018 (2011). 1012.0354

Shternin, P.S.: J. Phys. A **41**, 205501 (2008a). 0803.3893

Shternin, P.S.: J. Exp. Theor. Phys. **107**, 212 (2008b)

Shternin, P.S.: ArXiv e-prints (2018). 1805.06000

Shternin, P.S., Yakovlev, D.G.: Phys. Rev. D **74**, 043004 (2006). astro-ph/0608371

Shternin, P.S., Yakovlev, D.G.: Phys. Rev. D **75**, 103004 (2007). 0705.1963

Shternin, P.S., Yakovlev, D.G.: Phys. Rev. D **78**, 063006 (2008a). 0808.2018

Shternin, P.S., Yakovlev, D.G.: Astron. Lett. **34**, 675 (2008b)

Shternin, P.S., Yakovlev, D.G.: Mon. Not. R. Astron. Soc. **446**, 3621 (2015). 1411.0150

Shternin, P.S., Yakovlev, D.G., Haensel, P., Potekhin, A.Y.: Mon. Not. R. Astron. Soc. **382**, L43 (2007). 0708.0086

Shternin, P.S., Yakovlev, D.G., Heinke, C.O., Ho, W.C.G., Patnaude, D.J.: Mon. Not. R. Astron. Soc. **412**, L108 (2011). 1012.0045

Shternin, P.S., Baldo, M., Haensel, P.: Phys. Rev. C **88**, 065803 (2013). 1311.4278

Shternin, P., Baldo, M., Schulze, H.: J. Phys. Conf. Ser. **932**, 012042 (2017)

Shternin, P.S., Baldo, M., Haensel, P.: ArXiv e-prints (2018). 1807.0656

Silin, V.P.: Sov. Phys. JETP **13**, 1244 (1961)

Sinha, M., Bandyopadhyay, D.: Phys. Rev. D **79**, 123001 (2009). 0809.3337

Shovkovy, I.A., Ellis, P.J.: Phys. Rev. C **66**, 015802 (2002). hep-ph/0204132

Son, D T.: ArXiv e-prints (2001). hep-ph/0108260

Son, D.T.: ArXiv e-prints (2002). hep-ph/0204199

Son, D.T., Wingate, M.: Ann. Phys. **321**, 197 (2006). cond-mat/0509786

Son, D.T., Surowka, P.: Phys. Rev. Lett. **103**, 191601 (2009). 0906.5044

Son, D.T., Yamamoto, N.: Phys. Rev. D **87**, 085016 (2013). 1210.8158

Stetina, S., Rrapaj, E., Reddy, S.: Phys. Rev. C **97**, 045801 (2018). 1712.05447

Steiner, A.W.: Phys. Rev. C **85**, 055804 (2012). 1202.3378

Steiner, A.W., Reddy, S.: Phys. Rev. C **79**, 015802 (2009). 0804.0593

Stephanov, M.A., Yin, Y.: Phys. Rev. Lett. **109**, 162001 (2012). 1207.0747

Stoica, S., Pastrav, B., Horvath, J. E., Allen, M.P.: Nucl. Phys. A **828**, 439 (2009). 0906.3134

Sykes, J., Brooker, G.A.: Ann. Phys. **56**, 1 (1970)

Szmaglinski, A., Wojcik, W., Kutschera, M.: Acta Phys. Pol. B **37**, 277 (2006). astro-ph/0602281

Tatsumi, T., Muto, T.: Phys. Rev. D **89**, 103005 (2014). 1403.1927

Tauber, G.E., Weinberg, J.W.: Phys. Rev. **122**, 1342 (1961)

Thorne, K.S., In: Gratton, L. (ed.) Proceedings of the International School of Physics "Enrico Fermi," Course XXXV, at Varenna, Italy, July 12–24, 1965. Academic Press, New York, (1966), pp. 166–280

Thorne, K.S.: Astrophys. J. **212**, 825 (1977)

Timmermans, R.G., Korchin, A.Y., van Dalen, E.N., Dieperink, A.E.: Phys. Rev. C **65**, 064007 (2002)

Tisza, L.: Nature **141**, 913 (1938)

Tolos, L., Manuel, C., Sarkar, S., Tarrus, J.: AIP Conf. Ser. **1701**, 080001 (2016)

Turolla, R., Zane, S., Watts, A.L.: Rep. Prog. Phys. **78**, 116901 (2015). 1507.02924

Umeda, H., Iwamoto, N., Tsuruta, S., Qin, L., Nomoto, K.: Workshop on Neutron Stars and Pulsars: Thirty Years After the Discovery, Tokyo, Japan, November 17–20, 1997 (1997). astro-ph/9806337

van Dalen, E.N., Dieperink, A.E., Tjon, J.A.: Phys. Rev. C **67**, 065807 (2003). nucl-th/0303037

van Dalen, E.N.E., Dieperink A.E.L.: Phys. Rev. C **69**, 025802 (2004). nucl-th/0311103

Vereshchagin, G.V., Aksenov, A.G.: Relativistic Kinetic Theory. With Applications in Astrophysics an Cosmology. Cambridge University Press, Cambridge (2017)

Vidaña, I.: Phys. Rev. C **85**, 045808 (2012). [Erratum: Phys. Rev. C **90**, 029901 (2014)], 1202.4731

Viganò, D., Rea, N., Pons, J.A., Perna, R., Aguilera, D.N., Miralles, J.A.: Mon. Not. R. Astron. Soc. **434**, 123 (2013). 1306.2156

Villain, L., Haensel, P.: Astron. Astrophys. **444**, 539 (2005). astro-ph/0504572

Vollhardt, D., Wölfle, P.: The Superfluid Phases of Helium 3. Taylor & Francis, Bristol (1990)

Voskresensky, D.N.: In: Blaschke, D., Glendenning, N.K., Sedrakian, A. (eds.) Physics of Neutron Star Interiors. Lecture Notes in Physics, vol. 578, p. 467. Springer, Berlin (2001). astro-ph/0101514

Voskresensky, D.N., Senatorov, A.V.: Sov. J. Nucl. Phys. **45**, 411 (1987)

Wambach, J., Ainsworth, T.L., Pines, D.: Nucl. Phys. A **555**, 128 (1993)

Wang, Q. D., Lu, T.: Phys. Lett. B **148**, 211 (1984)

Wang, X., Shovkovy, I.A.: Phys. Rev. D **82**, 085007 (2010). 1006.1293

Wang, Q., Wang, Z.-G., Wu, J.: Phys. Rev. D **74**, 014021 (2006). hep-ph/0605092

Wang, X., Malekzadeh, H., Shovkovy, I.A.: Phys. Rev. D **81**, 045021 (2010). 0912.3851

Weinberg, S.: Phys. Rev. Lett. **40**, 223 (1978)

Wiescher, M., Käppeler, F., Langanke, K.: Ann. Rev. Astron. Astrophys. **50**, 165 (2012)

Wilczek, F.: Phys. Rev. Lett. **40**, 279 (1978)

Yakovlev, D.G.: Astrophys. Space Sci. **98**, 37 (1984)

Yakovlev, D.G.: Mon. Not. R. Astron. Soc. **453**, 581 (2015). 1508.02603

Yakovlev, D.G., Pethick, C.J.: Ann. Rev. Astron. Astrophys. **42**, 169 (2004). astro-ph/0402143

Yakovlev, D.G., Shalybkov, D.A.: Astrophys. Space Sci. **176**, 171 (1991a)

Yakovlev, D.G., Shalybkov, D.A.: Astrophys. Space Sci. **176**, 191 (1991b)

Yakovlev, D.G., Urpin, V.A.: Sov. Astron. **24**, 303 (1980)

Yakovlev, D.G., Levenfish, K.P., Shibanov, Y.A.: Phys. Usp. **42**, 737 (1999). astro-ph/9906456

Yakovlev, D.G., Kaminker, A.D., Gnedin, O.Y., Haensel, P.. Phys. Rep. **354**, 1 (2001). astro-ph/0012122

Yakovlev, D.G., Gasques, L.R., Beard, M., Wiescher, M., Afanasjev, A.V.: Phys. Rev. C **74**, 035803 (2006). astro-ph/0608488

Yamamoto, N.: Phys. Rev. D **93**, 065017 (2016). 1511.00933

Young, D.A., Corey, E. M., Dewitt, H.E.: Phys. Rev. A **44**, 6508 (1991)

Zhang, H.F., Lombardo, U., Zuo, W.: Phys. Rev. C **82**, 015805 (2010). 1006.2656

Zhdanov, V.: Transport Processes in Multicomponent Plasma. Taylor & Francis, London (2002)

Zhitnitsky, A.R.: Sov. J. Nucl. Phys. **31**, 260 (1980). [Yad. Fiz.31,497(1980)]

Ziman, J.M.: Philos. Mag. **6**, 1013 (1961)

Ziman, J.M.: Electrons and Phonons. Oxford Classical Texts in the Physical Sciences. Clarendon Press/Oxford University Press, Oxford/New York (2001)

Zuo, W., Giansiracusa, G., Lombardo, U., Sandulescu, N., Schulze, H.-J.: Phys. Lett. B **421**, 1 (1998)

Chapter 10
Gravitational Waves from Merging Binary Neutron-Star Systems

Tanja Hinderer, Luciano Rezzolla, and Luca Baiotti

Abstract The merger of binary neutron-star systems is among the scientifically richest events in the universe: it involves extremes of matter and gravity, copious emission of gravitational waves, complex microphysics, and electromagnetic processes that can lead to astrophysical signatures observable at the largest redshifts. We review here the recent progress in understanding the gravitational-wave signal emitted in this process, focussing in particular on its properties during the inspiral and then after the merger.

10.1 Introduction

Neutron stars are believed to be born in supernova explosions triggered by the collapse of the iron core in massive stars. Many astronomical observations have revealed that binary neutron stars (BNSs) indeed exist (Kramer et al. 2004; Abbott et al. 2017). Despite this observational evidence of existence, the formation mechanisms of BNS systems are not known in detail. The general picture is that in a binary system made of two massive main-sequence stars of masses between approximately 8 and 25 $M_\odot$, the more massive one undergoes a supernova explosion and becomes a neutron star. This is followed by a very uncertain phase in which the neutron star and the main-sequence star evolve in a "common envelope", that is, with the neutron star orbiting in the extended outer layers of the secondary star

T. Hinderer
Department of Astrophysics/IMAPP, Radboud University, Nijmegen, The Netherlands

L. Rezzolla (✉)
Institute for Theoretical Physics, Frankfurt, Germany

Frankfurt Institute for Advanced Studies, Frankfurt, Germany
e-mail: rezzolla@itp.uni-frankfurt.de

L. Baiotti
Graduate School of Science, Osaka University, Toyonaka, Japan

© Springer Nature Switzerland AG 2018   575
L. Rezzolla et al. (eds.), *The Physics and Astrophysics of Neutron Stars*,
Astrophysics and Space Science Library 457,
https://doi.org/10.1007/978-3-319-97616-7_10

(Kiziltan et al. 2013; Ivanova et al. 2013; Özel and Freire 2016). At the end of this stage, also the second main-sequence star undergoes a supernova explosion and, if the stars are still bound after the explosions, a BNS system is formed. The common-envelope phase, though brief, is crucial because in that phase the distance between the stars becomes much smaller as a result of drag, and this allows the birth of BNS systems that are compact enough to merge within a Hubble time, following the dissipation of their angular momentum through the emission of gravitational radiation. It is also possible that during the common-envelope phase the neutron star collapses to a black hole, thus preventing the formation of a BNS. Another possible channel for the formation of BNS systems may be the interaction of two isolated neutron stars in dense stellar regions, such as globular clusters, in a process called "dynamical capture" (O'Leary et al. 2009; Lee et al. 2010; Thompson 2011). Dynamically formed binary systems are different from the others because they have higher ellipticities. It is presently not known what fraction of BNS systems would originate from dynamical capture, but it is expected that these binaries are only a small part of the whole population.

This is undoubtedly an exciting and dynamical time for research on BNS mergers, when many accomplishments have been achieved (especially since 2008), while many more need to be achieved in order to describe such fascinating objects and the related physical phenomena. The first direct detection through the advanced interferometric LIGO detectors (Harry et al. 2010) of the gravitational-wave (GW) signal from what has been interpreted as the inspiral, merger and ringdown of a binary system of black holes (The LIGO Scientific Collaboration and the Virgo Collaboration 2016) marks, in many respects, the beginning of GW astronomy and other detections of these systems have been made over the last few months (Abbott et al. 2016).

More importantly, however, a long-awaited event has taken place on August 17, 2017: the Advanced LIGO and Virgo (Accadia et al. 2011) network of GW detectors have recorded the signal from the inspiral and merger of a binary neutron-star (BNS) system: GW170817 (Abbott et al. 2017). The correlated electromagnetic signals that have been recorded by $\sim$70 astronomical observatories and satellites have provided the striking confirmation that such mergers can be associated directly with the observation of short gamma-ray bursts (SGRBs). Although the detection rate of these events is still very uncertain and spans three orders of magnitude, it is expected to be of several events per year (Abadie et al. 2010), so that the operation of additional advanced detectors, such as KAGRA (Aso et al. 2013) and LIGO India (see e.g., Fairhurst 2014), are likely to increase the number of detections in the near future.

This Chapter aims at providing a quick overview of the efforts made to date to model the GW signal produced by the (late) inspiral, merger and post-merger of BNS systems (see also Baiotti and Rezzolla (2017), Paschalidis (2017) for some recent reviews). Because the merger represents a natural divide—both in terms of the physics involved and of the methods employed to describe this signal—the report is organised in a first part dedicated to the inspiral and merger dynamics

(Sect. 10.3) and to a second part devoted instead to the post-merger dynamics (Sect. 10.4). Both of these parts are prefaced by a general broadbrush description of the whole process (Sect. 10.2).[1]

10.2 The Broadbrush Picture

General-relativistic hydrodynamical simulations of BNSs started being performed in Japan almost 20 years ago (Nakamura and Oohara 1998; Oohara and Nakamura 1999; Shibata 1999). Even if nowadays many state-of-the-art codes are able to solve more complex sets of equations (e.g., for the evolution of magnetic fields, neutrino emission, etc.), simulations involving only general-relativistic hydrodynamics are still the benchmark for any new code and the necessary testbed for more advanced codes. Furthermore, in many cases, results obtained with pure hydrodynamics, most notably, gravitational waveforms, provide already a wealth of information on BNS systems, especially during the inspiral. In many respects, the inspiral may be considered the *easiest* part of the problem, in which the stars spiral towards each other as a result of gravitational-radiation losses, being scarcely or not at all affected by magnetic fields or neutrinos. Its simplicity notwithstanding, this problem is still the object of continuous efforts and improvements, which are often carried out through the synergy of numerical simulations and analytical calculations based on post-Newtonian expansions or other approximation schemes. We describe progress on this topic in Sect. 10.3. The inspiral has also recently attracted renewed attention with the first simulations of arbitrarily spinning BNS systems (see Sect. 10.3).

In what follows, we give a general description of the BNS dynamics using the figures of Baiotti et al. (2008), which was one of the first to provide complete and accurate evolutions. Our description is here intentionally qualitative, as we focus on those aspects that are robust and independent of the EOS. As an aid to the discussion we show in Fig. 10.1 the various stages in the evolution of an equal-mass binary system of neutron stars as a function of the initial mass of the binary. More specifically, the diagram shows on the horizontal axis the progress of time during the evolution of the system (the intervals in square brackets indicate the expected duration range of each stage), while on the vertical axis it displays the ratio of the total (gravitational) mass of the binary (i.e., the sum of the gravitational masses of the stars composing the system), M, to the maximum mass of an isolated nonrotating star,[2] M_{TOV}. Because the EOS describing neutron stars is still unknown, the precise

---

[1] Much of the material presented in the second part relative to the post-merger dynamics has been presented elsewhere either in the form of original journal articles, as a Chapter in a textbook (Rezzolla and Zanotti 2013), or as a part of a Review of Progress in Physics (Baiotti and Rezzolla 2017).

[2] An isolated nonrotating neutron star is the solution of the Tolman-Oppenheimer-Volkoff (TOV) equation (Tolman 1939; Oppenheimer and Volkoff 1939) and so it is often called a "TOV" star.

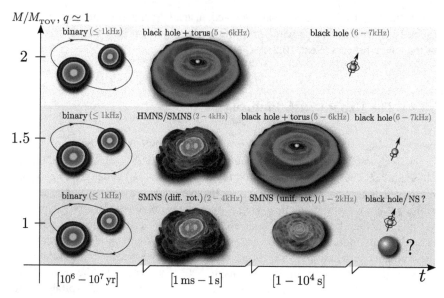

Fig. 10.1 Schematic diagram illustrating the various stages in the evolution of an equal-mass binary system of neutron stars as a function of the initial mass of the binary. Depending on the initial total mass of the binary M, and on how it relates to the maximum mass of a nonrotating neutron star M_{TOV}, the binary can either collapse promptly to a black hole surrounded by a torus (top row), or give rise to an hypermassive (HMNS) (or to a supramassive neutron star (SMNS) that ultimately collapses to a black hole and torus (middle row), or even lead to a SMNS (first differentially and subsequently uniformly rotating) neutron star that eventually yields a black hole or a nonrotating neutron star (bottom row). Also indicated in red are the typical frequencies at which gravitational waves are expected to be emitted [Adapted from Rezzolla and Zanotti (2013) by permission of Oxford University Press www.oup.com]

value of M_{TOV} cannot be determined, although a number of different studies have now converged on a possible upper limit of $M_{\mathrm{TOV}}/M_{\odot} \lesssim 2.16^{+0.17}_{-0.15}$ (Margalit and Metzger 2017; Rezzolla et al. 2018; Ruiz et al. 2018; Shibata et al. 2017). Furthermore, astronomical observations indicate that it should be larger than about two solar masses, since there are two different systems that have been measured to have masses in this range: PSR J0348+0432 with $M = 2.01 \pm 0.04 \, M_{\odot}$ (Antoniadis et al. 2013), and PSR J1614–2230 with $M = 1.97 \pm 0.04 \, M_{\odot}$ (Demorest et al. 2010).

For millions of years a comparatively slow inspiral progressively speeds up until the two neutron stars become so close that tidal waves produced by the (tidal) interaction start appearing on the stellar surface (these are clearly visible in the second and third panels of Fig. 10.2). Such waves are accompanied by emission of matter stripped from the surface and by shocks that represent the evolution of small sound waves that propagate from the central regions of the stars, steepening as they move outwards in regions of smaller rest-mass density (Stergioulas et al. 2004; Nagakura et al. 2014).

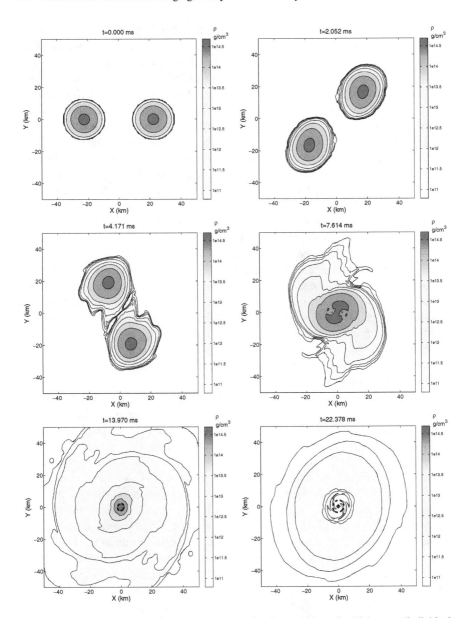

Fig. 10.2 Isodensity contours in the (x, y) plane for the evolution of a high-mass (individual stellar rest mass $1.625 M_\odot$) binary with an ideal-fluid EOS. The thick dashed lines in the lower panels show the location of the apparent horizon [Reprinted with permission from Baiotti et al. (2008). © (2008) by the American Physical Society]

At the merger, the two stars collide with a rather large impact parameter. A *vortex sheet* (or *shear interface*) develops, where the tangential component of the velocity exhibits a discontinuity. This condition is known to be unstable to very small perturbations and it can develop Kelvin-Helmholtz instability (KHI), which curls the interface forming a series of vortices at all wavelengths (Chandrasekhar 1981; Bodo et al. 1994). Even if this instability is purely hydrodynamical and it is likely to be important only for binaries with very similar masses, it can have strong consequences if the stars possess magnetic fields. It has in fact been shown that, in the presence of an initially poloidal magnetic field, this instability may lead to an exponential growth of the toroidal component (Price and Rosswog 2006; Giacomazzo et al. 2011; Rezzolla et al. 2011; Neilsen et al. 2014; Kiuchi et al. 2014, 2017). Such a growth is the result of the exponentially rapid formation of vortices that curl magnetic-field lines that were initially purely poloidal. The exponential growth caused by the KHI leads to an overall amplification of the magnetic field of about three orders of magnitude (Kiuchi et al. 2014). At the same time, high-resolution simulations in core-collapse supernovae find that parasitic instabilities quench the MRI, with a magnetic-field amplification factor of 100 at most, independently of the initial magnetic field strength (Rembiasz et al. 2016). Of course, KHI and MRI are two different instabilities, but the lesson these simulations provide is that parasitic instabilities may also appear during the development of the KHI and limit the overall magnetic-field amplification; such parasitic instabilities are at present not yet apparent because of the comparatively small resolutions employed when modelling BNS mergers.

The hypermassive neutron star (HMNS) produced from the merger may not collapse promptly to a black hole, but rather undergo large oscillations with variations such that the maximum of the rest-mass density may grow to be twice as large (or more) as the value in the original stars (see the right panel of Fig. 10.3). These oscillations have a dominant $m = 2$ non-axisymmetric character (Stergioulas et al. 2011) and will be discussed in detail in Sect. 10.4. As mentioned earlier, the formation and duration of the HMNS depends on the stellar masses, the EOS, the effects of radiative cooling, magnetic fields (Ravi and Lasky 2014; Rezzolla and Kumar 2015; Ciolfi and Siegel 2015), and the development of GW driven instabilities (Doneva et al. 2015). Furthermore, the equilibrium of the HMNS can also be modified by the losses of rest mass via winds that can be driven by shock heating (Sekiguchi et al. 2015; Bovard et al. 2017), by magnetic fields (Shibata et al. 2011; Kiuchi et al. 2012a; Siegel et al. 2014; Rezzolla and Kumar 2015; Ciolfi and Siegel 2015; Murguia-Berthier et al. 2017), or by viscosity and neutrino emission (Dessart et al. 2009; Perego et al. 2014; Just et al. 2015; Martin et al. 2015; Murguia-Berthier et al. 2014, 2017; Fujibayashi et al. 2017).

In essentially all cases when a black hole is formed, some amount of matter remains outside of it, having sufficient angular momentum to stay orbiting around the black hole on stable orbits. In turn, this leads to the formation of an accretion torus that may be rather dense ($\rho \sim 10^{12}$–10^{13} g cm^{-3}) and extended horizontally for tens of kilometres and vertically for a few tens of kilometres. Also this point will be discussed in more detail in Sect. 10.4.

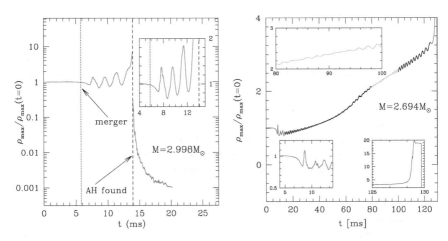

Fig. 10.3 *Left panel:* Evolution of the maximum rest-mass density normalized to its initial value for a high-mass (individual stellar rest mass $1.625 M_\odot$; gravitational mass of the system $2.998 M_\odot$) binary using an ideal-fluid EOS. Indicated with a dotted vertical line is the time at which the binary merges, while a vertical dashed line shows the time at which an apparent horizon is found. After this time, the maximum rest-mass density is computed in a region outside the apparent horizon [from Baiotti et al. (2008). © (2008) by the American Physical Society]. *Right panel:* The same as in the left panel but for a low-mass binary (individual stellar rest mass $1.456 M_\odot$; gravitational mass of the system $2.694 M_\odot$). Note that the evolution is much longer in this case and that different colours are used to denote the different parts of the evolution (see insets) [Adapted from Rezzolla et al. (2010). © IOP Publishing. Reproduced with permission. All rights reserved]

The dynamics of the inspiral and merger of a reference equal-mass binary system is summarised in Fig. 10.3, whose panels show the evolution of the maximum rest-mass density normalized to its initial value (after the formation of the apparent horizon, the curve shows the maximum rest-mass density in the region outside the apparent horizon). Note that together with the large oscillations, the rest-mass density also experiences a secular growth and the increased compactness eventually leads to the collapse to a rotating black hole. The differences in the two panels are essentially related to the initial mass of the system (i.e., $M = 2.998 \, M_\odot$ in the left panel and $M = 2.694 \, M_\odot$ in the right panel) and it can be seen that, for a given EOS (even a very simple one like the ideal-fluid EOS used in this case) smaller masses will yield systematically longer-lived HMNSs.

The matter dynamics described so far in the various stages of the evolution of a BNS system are imprinted in the GW signal, which then can be used to extract important information on the properties of the neutron stars. Different parts of the evolution will provide distinct pieces of information and with different overall signal-to-noise (SNR) ratios. For example, the post-merger signal would provide rather clear signatures but at such high frequencies that it may be difficult to measure them with present detectors. On the other hand, as we will discuss in detail in the following section, the inspiral signal does depend on the EOS much more weakly, but in a way that is still measurable because it comes at frequencies where the detectors are more sensitive.

10.3   Matter Effects During a Binary Inspiral

For inspiraling compact-object binary systems, GW measurements of the source parameters are based on matched filtering, where the datastream is cross-correlated with theoretically predicted template waveforms for different possible parameters within the wide physically plausible range. A detailed understanding and modelling of the effect of neutron-star matter on the binary dynamics and GWs is therefore essential to enhance the science gains from GW observations. For nonspinning compact objects the main imprint of their internal structure on the GW signal from the *inspiral* is due to tidal effects. The energy cost of the tidal deformation together with the contribution from the moving tidal bulges to gravitational radiation accelerate the inspiral inspiral and change the GWs compared to the signal from a black-hole binary. The main characteristic equation-of-state parameter imprinted in the GWs is the star's tidal deformability, as recently measured for GW170817 (Abbott et al. 2017) the ratio of the induced quadrupole moment to the perturbing tidal field.

Tidal effects in binary neutron-star systems are well-known in Newtonian gravity (Bildsten and Cutler 1992; Reisenegger and Goldreich 1994; Lai 1994; Kochanek 1992; Kokkotas and Schaefer 1995), and in post-Newtonian (PN) theory (Damour et al. 1992; Lombardi et al. 1997; Mora and Will 2004). A formal description of the coupling of the neutron-star's internal structure, described by full General Relativity (GR), to the orbital dynamics of a binary at large separation was developed in Flanagan (1998) and expounded upon in Racine and Flanagan (2005). This provided a rigorous proof that there are no new relativistic "star crushing" forces (Wilson and Mathews 1995) and that tidal interactions are the leading-order finite-size effects for nonspinning objects with arbitrarily strong self-gravity at large separation. The dominant tidal effects in the GWs and the associated relativistic tidal parameter were derived in Flanagan and Hinderer (2008), Hinderer (2008). Recent work computed the tidal parameters for higher multipole moments (Damour and Nagar 2009; Binnington and Poisson 2009), for a wide range of equations of state models (Hinderer et al. 2010; Postnikov et al. 2010), examined their physics content (Fattoyev et al. 2013, 2014; Steiner et al. 2015; Lattimer and Lim 2013; Van Oeveren and Friedman 2017) and the effect of stratification and elasticity (Penner et al. 2011), considered various other quantities characterising tidal deformations (Damour and Nagar 2009; Landry and Poisson 2014), the tidal parameters of black holes (Damour and Lecian 2009; Kol and Smolkin 2012; Porto 2016; Gurlebeck 2015) and of exotic objects (Cardoso et al. 2017; Sennett et al. 2017; Mendes and Yang 2017; Uchikata and Yoshida 2016; Pani 2015), and new tidal parameters that appear for slowly rotating neutron stars (Pani et al. 2015a; Landry and Poisson 2015a). Substantial recent interest has also focused on I-Love-Q relations (Yagi and Yunes 2013a,b) that link dimensionless parameters characterising various global properties of the neutron star in an approximately EOS-independent way (Lattimer and Lim 2013; Pappas and Apostolatos 2014; Yagi et al. 2014a; Pappas 2017, 2015; Haskell et al. 2014; Chakrabarti et al. 2014; Maselli et al. 2013a; AlGendy and Morsink 2014; Chirenti et al. 2015; Pannarale et al. 2015; Steiner et al. 2016; Breu and Rezzolla 2016; Silva et al. 2016; Yagi 2014; Reina et al. 2017; Chan et al. 2015, 2016); see also Chap. 13 of this book.

Significant progress has also been made on describing the tidal effects on the orbital dynamics and gravitational radiation, within post-Newtonian theory (Vines et al. 2011a,b; Bini et al. 2012; Steinhoff et al. 2016), a post-Newtonian affine approach (Ferrari et al. 2012; Maselli et al. 2012), the gravitational self-force formalism (Dolan et al. 2015; Bini and Damour 2014; Nolan et al. 2015; Shah and Pound 2015), effective field theory (Goldberger and Rothstein 2006), and the effective-one-body model (Damour and Nagar 2010; Vines et al. 2011a; Bini et al. 2012; Bini and Damour 2014; Damour et al. 2012; Bernuzzi et al. 2015a; Hinderer et al. 2016; Steinhoff et al. 2016). Comparisons and tests of these descriptions against numerical relativity simulations will be reported in Sect. 10.4 of this chapter. Based on the above models, several measurability studies of the tidal signature in the GW signal have been performed using Bayesian data analysis methods, both for double neutron-star binaries (Del Pozzo et al. 2013; Agathos et al. 2015; Lackey and Wade 2015; Wade et al. 2014; Chatziioannou et al. 2015; Markakis et al. 2009) and for neutron-star–black-hole systems (Lackey et al. 2012, 2014; Kumar et al. 2017), and a first result from LIGO and Virgo observations has recently been reported (Abbott et al. 2017).

This section will focus on theoretical models of matter effects during the inspiral epoch of a neutron-star binary system. We will focus on the main imprints from tidal effects, but in Sect. 10.3.5 will also point out references containing discussions of other effects including rotational deformations, the tidal excitation of a neutron star's various oscillation modes beyond the fundamental mode, nonlinear tidal effects, and other tidal interactions in general relativity that are not present in Newtonian gravity. To introduce the theoretical approaches for describing the inspiral, we will start by recalling tidal effects in Newtonian gravity in Sect. 10.3.1. The Newtonian discussion will be formulated in way that can be promoted to general relativity with appropriate modifications, as will be delineated in Sect. 10.3.2, where we will also review the information needed to compute the tidal parameters for a given equation of state model, and briefly outline approximate universal relations between these parameters and similar parameters characterising the rotational deformation and moment of inertia in Sect. 10.3.3. The effect on the GW signal from a binary inspiral will be considered in Sect. 10.3.4.

Conventions We will use units in which $G = c = 1$ unless otherwise indicated. Indices on tensors consisting of Greek letters $\alpha, \beta, \ldots$ denote four-dimensional spacetime quantities, while Latin indices $i, j, k, \ldots$ denote spatial, three-dimensional quantities. We will use overdots on quantities to denote derivatives with respect to coordinate time, e.g., $\dot{x} = dx/dt$.

10.3.1 Tidal Interactions in Newtonian Compact Binaries

In this section we will review Newtonian tidal interactions in a binary system, discuss the characteristic tidal parameters, and derive an effective action that compactly summarizes the dynamics. This formalism, with appropriate modifications, will carry over to the relativistic case discussed in Sect. 10.3.2. We will first review the multipole expansion of the self-field of a body and the gravitational potential of a binary system, then consider the equations of motion and a corresponding action principle. The discussion is based on Vines et al. (2011a) and Steinhoff et al. (2016); some of the introductory material can also be found in the book Gravity by Poisson and Will (2014).

10.3.1.1 Multipole Expansion of the Self-gravitational Potential

We consider two bodies labeled by A, B. In Newtonian gravity the gravitational potential generated by a mass distribution with density ρ_A at a field point x is a solution to Poisson's equation $\nabla^2 U_A = -4\pi\rho_A$ or

$$U_A(t, x) = \int d^3x' \rho_A(t, x') \frac{1}{|x - x'|} \tag{10.1}$$

Outside the body's mass distribution, for points $x > x'$, the potential can be written as a Taylor series expansion around a moving reference point $z_A(t)$ as

$$U_A = \int d^3x' \rho_A(t, x') \sum_{\ell=0}^{\infty} \frac{1}{\ell!} (x' - z_A)^L \left(\frac{\partial}{\partial x'^L} \frac{1}{|x - x'|} \right) |_{x'=z_A} \tag{10.2a}$$

$$= \int d^3x' \rho_A(t, x') \sum_{\ell=0}^{\infty} \frac{(-1)^\ell}{\ell!} (x' - z_A)^L \partial_L \frac{1}{|x - z_A|}. \tag{10.2b}$$

Here, the notation is that $L = a_1 a_2 \cdots a_\ell$ denotes a string of ℓ indices and

$$x^L = x^{a_1} x^{a_2} \cdots x^{a_\ell}, \qquad \partial_L = \frac{\partial}{\partial x^L} = \frac{\partial}{\partial x^{a_1}} \cdots \frac{\partial}{\partial x^{a_\ell}}, \tag{10.3}$$

Throughout this review, the summation over repeated indices is implied.

Similar to the definitions in electromagnetism, we define the body's Newtonian mass multipole moments by the following integrals

$$M_A = \int_A d^3x \, \rho_A(t, x), \qquad Q_A^L = \int_A d^3x \, \rho_A(t, x)(x - z_A)^{<L>}, \tag{10.4}$$

where M_A is the mass of the body and the integration is over a sphere surrounding the matter distribution. The angular brackets around the indices denote the symmetric and trace-free projection of the tensor, e.g., $x^{<ij>} = x^i x^j - \delta^{ij} |x|^2$; see Thorne (1980), Hartmann et al. (1994) for a pedagogical introduction to symmetric and trace-free tensors. The reason that the mass multipole moments defined in Eq. (10.4) are only the trace-free parts of the integrals that would be read off from (10.2b) is that the derivative $\partial_L |x - x'|^{-1}$ is a symmetric and trace-free tensor that projects out only trace-free piece Q^L in the contribution to the potential (10.2a). Finally, the mass dipole term ($\ell = 1$) has been omitted from the expansion (10.4) since it can always be made to vanish by choosing the reference point $z_A^i(t)$ for the multipole expansion to be the body's center of mass. With the definitions from Eq. (10.4), the potential outside the body becomes

$$U_A(t, x) = \frac{M_A}{|x - z_A|} + \sum_{\ell=2}^{\infty} \frac{(-1)^\ell}{\ell!} Q^L \partial_L \frac{1}{|x - z_A|}. \tag{10.5}$$

This form of the potential is convenient for computing the dynamics of a binary system. When considering the deformation of a single object, it is useful to work with spherical coordinates $(x - z_A)^i / |x - z_A| = (\sin\theta\cos\phi, \sin\theta\sin\phi, \cos\theta)$ and express (10.5) as a spherical-harmonic expansion

$$U_A(t, x) = \frac{M_A}{|x - z_A|} + \sum_{\ell=2}^{\infty} \sum_{m=-\ell}^{\ell} Q_{\ell m} \frac{Y_{\ell m}(\theta, \phi)}{|x - z_A|^{\ell+1}}. \tag{10.6}$$

The spherical-harmonic components of the multipole moments are related to the Cartesian multipole moments by (Thorne 1980)

$$Q_{\ell m}^A = \frac{4\pi}{2\ell+1} \mathcal{Y}_L^{*\ell m} Q_L \qquad Q_L = \frac{\ell!}{(2\ell-1)!!} \sum_{m=-\ell}^{\ell} Q_{\ell m} \mathcal{Y}_L^{\ell m}. \tag{10.7}$$

Here, the quantities $\mathcal{Y}_L^{\ell m}$ are symmetric-trace-free tensors consisting of complex coefficients that appear in the conversion between unit vectors n and spherical harmonics (Thorne 1980)

$$Y_{\ell m} = \mathcal{Y}_L^{\ell m} n^{<L>} \qquad n^{<L>} = \frac{4\pi \ell!}{(2\ell+1)!!} \sum_{m=-\ell}^{\ell} \mathcal{Y}_L^{\ell m} Y_{\ell m}^*. \tag{10.8}$$

10.3.1.2   Expansion of the Companion's Potential and Tidal Moments

Throughout the subsequent discussion, we will consider a binary system in the regime where the separation between the bodies is large compared to the characteristic size of the bodies. The potential due to external sources such as a companion in the binary that is felt by body A is denoted by U_A^{ext} and can be written as a Taylor expansion around A's center of mass in the form

$$U_A^{\text{ext}}(t, \boldsymbol{x}) = U_A^{\text{ext}}(t, z_A) + (x - z_A)^j \left[\partial_j U_A^{\text{ext}}(t, \boldsymbol{x}) \right]_{x=z_A} - \sum_{l=2}^{\infty} \frac{1}{\ell!} (x - z_A)^L \mathcal{E}_A^L,$$

(10.9a)

where "ext" denotes that the source of this part of the potential is external to the body. The coefficients $\mathcal{E}_A^L$ in the second line of (10.9) are the body A's tidal moments (Thorne and Hartle 1984). Assuming that the source of the external potential is the potential of body B in a binary system, denoted by U_B, the tidal moments are

$$\mathcal{E}_A^L = - \left(\frac{\partial}{\partial x^L} U_B(t, \boldsymbol{x}) \right)_{x=z_A(t)}.$$

(10.9b)

Similar to the self-field, the decomposition of the external potential in Eq. (10.9) can also be written as a spherical-harmonic expansion using the same conversion and thus not given explicitly here.

10.3.1.3 Equations of Motion and Action Principle

The total potential for the binary is $U = U_A + U_B$ and the equations of motion for the center-of-mass positions of either the bodies ($C = A$ or $C = B$) can be derived from Newton's second law:

$$M_C \ddot{z}_C^j = \int d^3x \, \rho_C \frac{\partial}{\partial x^j} U(t, \boldsymbol{x}) = M \frac{\partial}{\partial x^j} U_C^{\text{ext}}(t, \boldsymbol{x}) \mid_{x=z_C} - \sum_{l=2}^{\infty} \frac{1}{\ell!} Q_C^L \mathcal{E}_{jL},$$

(10.10)

where in the second equality we have used the multipole expansions described above and the fact that only the potential sourced by the companion contributes to the body's motion, as can be verified by direct calculation.

The dynamics can be conveniently summarised by an action principle constructed from the Lagrangian $\mathcal{L} = T - V$, where $T = T_A + T_B$ is the total kinetic energy and $V = V_A + V_B$ the potential energy for the binary system. As reviewed in detail in Vines et al. (2011a), each of these contributions can be split into a the center-of-mass motion of the body and an internal contribution:

$$T_A = \frac{1}{2} \int_A d^3x \rho_A \dot{z}_A^2 + T_A^{\text{int}}, \qquad V_A = \frac{1}{2} \int_A d^3x \rho_A U_B + V_{\text{int}}.$$

(10.11)

For simplicity, we will specialise the subsequent discussion to the case where only A is an extended body while B is a point mass. To linear order in the finite-size effects, the case of two extended objects can be recovered by adding the same contribution with A and B interchanged. Performing the expansions around A's center of mass, using the definitions of the multipole and tidal moments, adding the contributions T_B and V_B from the companion, and transforming to the barycentric frame of the binary system leads to

$$T = \frac{1}{2}\mu v^2 + T_{\text{int}}, \qquad V = -\frac{\mu M}{r} + \sum_{\ell \geq 2} \frac{1}{\ell!} Q_L \mathcal{E}_L + V_{\text{int}}. \qquad (10.12)$$

Here, we have defined the total mass $M = M_A + M_B$, the reduced mass $\mu = M_A M_B / M$, and the relative separation $\boldsymbol{r} = \boldsymbol{z}_A - \boldsymbol{z}_B$ whose magnitude we denote by $r = |\boldsymbol{r}|$, as well as the relative velocity $v^2 = \dot{\boldsymbol{r}} \cdot \dot{\boldsymbol{r}}$. The action is then given by (Flanagan and Hinderer 2008; Rathore et al. 2003; Lai 1994)

$$S = S_{\text{orbit}} + \int dt \sum_{\ell \geq 2} \left[-\frac{1}{\ell!} Q_L \mathcal{E}_L + \mathcal{L}^{\text{int}} \right], \qquad (10.13)$$

where $S_{\text{orbit}} = \int dt\, L_{\text{orbit}}$, with $L_{\text{orbit}} = (\mu/2)v^2 + \mu M / r$, describes the orbital motion of point-masses and $\mathcal{L}^{\text{int}}$ encapsulates the internal dynamics of the multipole moments mass multipole moment that still remain to be specified. We will discuss the form of $\mathcal{L}^{\text{int}}$ for the case where the multipole moments are tidally induced in the next subsection. The advantage of the formulation in Eq. (10.13) is that the information about the binary dynamics in summarised in a simple, single scalar function.

10.3.1.4   Tidally Induced Multipole Moments

We will now further specialise to the case of a body that would be spherically symmetric in isolation and whose multipole moments result only from the response to the companion's tidal field (Bildsten and Cutler 1992; Reisenegger and Goldreich 1994; Lai et al. 1993; Lai 1994; Zahn 1977, 1970; Kopal 1978; Kochanek 1992; Hansen 2006; Mora and Will 2004; Kokkotas and Schaefer 1995; Flanagan and Hinderer 2008; Ferrari et al. 2012; Damour et al. 1992; Shibata 1994). For a neutron star, the main dynamics of the tidally induced multipole moments can be described by its fundamental oscillation modes (f-modes) with an internal Lagrangian having the form of a harmonic oscillator (Flanagan and Hinderer 2008; Rathore et al. 2003; Lai 1994; Kokkotas and Schaefer 1995)

$$\mathcal{L}^{\text{int}} = \frac{1}{2\ell!\lambda_\ell \omega_{0\ell}^2} \left[\dot{Q}_L \dot{Q}^L - \omega_{0\ell}^2 Q_L Q^L \right]. \qquad (10.14)$$

An explicit derivation of this Lagrangian starting from the mode amplitudes of the fluid displacement and their relation to the multipole moments is exhibited in Chakrabarti et al. (2013). Here, the quantities $\omega_{0\ell}$ denote the f-mode frequencies, and only the contribution from the modes with no radial nodes have been included since higher modes contribute very little to the effect (McDermott et al. 1985). The parameters λ_ℓ are the tidal deformability coefficients that are defined by considering the *adiabatic* limit, where the body's internal time scales $\tau^{\text{int}} \sim \omega_{0\ell}^{-1} \sim \sqrt{R^3/M_A}$ are fast compared to the time scale of variations in the tidal field $\tau_{\text{orb}} \sim \sqrt{r^3/(M_A + M_B)}$. They characterize the equation-of-state-dependent ratio between the induced multipoles and the tidal field

$$Q_L^{\text{adiab}} = -\lambda_\ell \, \mathcal{E}_L, \tag{10.15}$$

and are sometimes also referred to as the tidal polarizability. The tidal parameters λ_ℓ are related to the body's tidal Love numbers k_ℓ (or apsidal constants) that were introduced by the British scientist A.E.H. Love in (1909) and its radius R by

$$\lambda_\ell = \frac{2}{(2\ell - 1)!!} k_\ell R^{2\ell+1}. \tag{10.16}$$

In many contexts it is useful to work with the dimensionless tidal deformabilities

$$\Lambda_\ell = \frac{\lambda_\ell}{M^{2\ell+1}} = \frac{2}{(2\ell - 1)!!} k_\ell C^{-(2\ell+1)}, \tag{10.17}$$

where $C = M/R$ is the star's compactness.

For adiabatically induced multipoles, $dQ_L/dt = 0$ and the internal Lagrangian is only the elastic potential energy associated with the deformation

$$\mathcal{L}_{\text{adiab}}^{\text{int}} = -\frac{1}{2\ell!\lambda_\ell} Q_L Q^L. \tag{10.18}$$

Using the relation (10.15), the finite size effects can be written entirely in terms of the orbital variables and λ_ℓ:

$$S_{\text{adiab}} = S_{\text{orbit}} + \int dt \left[\frac{\lambda_\ell}{2\ell!} \mathcal{E}_L \mathcal{E}^L \right], \tag{10.19}$$

where in this Newtonian context $\mathcal{E}_L = -M_B \partial_L r^{-1}$. The quantities λ_ℓ or k_ℓ depend on the details of the body's internal structure, and their computation therefore requires an explicit description of the perturbed interior.

For the Newtonian calculations of λ_ℓ we work in the rest frame of the body, in a region surrounding it that excludes the companion. The first step is to obtain an equilibrium configuration for the body in isolation, by solving to the Poisson equation for the gravitational potential together with the continuity and Euler's equations that express the conservation of mass and momentum:

$$\nabla^2 U = -4\pi\rho, \quad \frac{\partial\rho}{\partial t} + \nabla\cdot(\rho\boldsymbol{v}) = 0, \quad \frac{\partial v^i}{\partial t} + (\boldsymbol{v}\cdot\nabla)v^i = -\frac{\partial^i p}{\rho} + \partial_i U + a_{\text{ext}}^i, \tag{10.20}$$

where a^i_{ext} is the acceleration due to external forces. Next, we suppose that the star is disturbed by an external static tidal gravitational field, e.g., due to a distant companion. The tidal disturbance is characterized by the set of moments $\mathcal{E}_L$, that each cause the star to deform in response and settle down to a new static configuration which has a nonzero set of mass multipole moments Q_L. The gravitational potential outside the perturbed star is

$$U_{\text{total}} = U_{\text{self}} + U_{\text{tidal}} = \frac{M}{r} + \sum_{\ell=2}^{\infty} \sum_{m=-\ell}^{\ell} Y_{\ell m} \left[\frac{Q_{\ell m}}{r^{\ell+1}} - \frac{1}{(2\ell-1)!!} \mathcal{E}_{\ell m} r^\ell \right] \quad (10.21)$$

The multipole moments $Q_{\ell m}$ characterising the body's response are associated with the piece of the exterior potential that falls off as $1/r^{\ell+1}$, while the external tidal field is related to the terms that grow as r^ℓ. To linear order in the external tidal perturbation and in the adiabatic limit, the induced distortion will be linearly proportional to the tidal perturbation as in Eq. (10.15). Using the properties from (10.8) and an analogous decomposition as in (10.7) for $\mathcal{E}_L$ the relation (10.15) can be written as

$$Q_{\ell m} = -\lambda_\ell \mathcal{E}_{\ell m} \quad (10.22)$$

for each ℓth multipole. To compute λ_ℓ it is sufficient to consider a single value of m. The parameters are computed by matching the interior solution and the exterior description (10.21) at the surface of the star, as we briefly recall here. First, we compute the interior solution by noting that the perturbed neutron star is still described by (10.20) but with perturbed pressure $p = p_0 + \delta p$, density $\rho = \rho_0 + \delta \rho$, gravitational potential $U = U_0 + \delta U$, and an external acceleration $a_{\text{ext}} = \nabla U_{\text{tidal}}$. The fluid perturbation can be represented by a Lagrangian displacement $\xi(x, t)$ which is defined so that the fluid element at position x in the unperturbed star is at position $x + \xi(x, t)$ in the perturbed star. Expanding Euler's equation about the background star to linear order in the perturbations yields

$$\frac{d^2 \xi^i}{dt^2} = -\frac{\partial_i \delta p}{\rho_0} + \frac{\partial_i p_0}{\rho_0^2} \delta \rho + \partial_i \delta U + a^i_{\text{ext}}. \quad (10.23)$$

For a barotropic equation-of-state relation of the form $p = p(\rho)$ we can eliminate δp from (10.23) in terms of $\delta \rho$. After further specialising (10.23) to static perturbations ($\dot{\xi} = 0$), combining all terms that involve $\delta \rho$ into a total derivative, and integrating we obtain

$$\frac{1}{\rho_0} \frac{dp_0}{d\rho_0} \delta \rho - \delta U_{\text{tot}} = \text{const}, \quad (10.24)$$

where $\delta U_{\text{tot}} = \delta U + U_{\text{tidal}}$. Using the expansions

$$\delta\rho = f(r)Y_{\ell m}(\theta,\phi), \qquad \delta U_{\text{tot}} = H(r)Y_{\ell m}(\theta,\phi), \qquad (10.25)$$

and substituting this decomposition into the linearized Poisson's equation leads to

$$-4\pi f(r)Y_{\ell m} = Y_{\ell m}\frac{1}{r^2}\frac{d}{dr}r^2\frac{dH(r)}{dr} + H(r)\left[\frac{1}{r^2\sin\theta}\partial_\theta\sin\theta\partial_\theta + \frac{1}{r^2\sin^2\theta}\partial_\phi^2\right]Y_{\ell m} \quad (10.26)$$

$$= Y_{\ell m}\left[H'' + \frac{2}{r}H'\right] - H(r)\frac{\ell(\ell+1)}{r^2}Y_{\ell m}. \qquad (10.27)$$

Next, using (10.25) in Eq. (10.24) shows that the integration constant must vanish since the rest of the equation is purely $\ell \geq 2$, and we obtain the relation between the radial functions

$$\frac{1}{\rho_0}\frac{dp_0}{d\rho_0}f - H = 0. \qquad (10.28)$$

Combining Eqs. (10.27) and (10.28) leads to a single master equation for $H(r)$ in the region $r \leq R$:

$$H'' + \frac{2}{r}H' - \frac{\ell(\ell+1)}{r^2}H = -4\pi\left(\frac{1}{\rho_0}\frac{dp_0}{d\rho_0}\right)^{-1}H. \qquad (10.29)$$

Except for special choices of the equation of state (EOS), this ODE has to be integrated numerically in the interior of the object, with the boundary condition that ensures regularity at the center of the star, $H \propto r^\ell$ for $r \to 0$. For $r > R$ the exterior solution is

$$H^{\text{ext}} = \frac{Q_{\ell m}}{r^{\ell+1}} - \frac{1}{(2\ell-1)!!}\mathcal{E}_{\ell m}r^\ell = -\mathcal{E}_{\ell m}\left[\frac{\lambda_\ell}{r^{\ell+1}} + \frac{1}{(2\ell-1)!!}r^\ell\right]. \qquad (10.30)$$

To extract the Love numbers we eliminate $\mathcal{E}_{\ell m}$ by considering the logarithmic derivative

$$y(r) = \frac{rH'(r)}{H(r)}. \qquad (10.31)$$

We solve for k_ℓ by using (10.30) in (10.31) and matching the results for $y(R)$ obtained from the interior and exterior solutions at the stellar surface:

$$k_\ell = \frac{\ell - y(R)}{2\left[\ell+1+y(R)\right]|}. \qquad (10.32)$$

The strategy for practical computations of the Love number is thus the following: (1) obtain a solution for the background configuration, (2) compute the perturbed interior described by (10.29) and evaluate from it $y(R)$ at the surface, and (3) use this result in Eq. (10.32) to obtain the Love number. Exact solutions can be obtained for simple EOSs, for example for an incompressible $n = 0$ polytrope $k_\ell = 3/(4\ell - 4)$. The same general method applies in the relativistic case, where, however, one has to use Einstein's equations and a more general definition of the multipole moments and hence the solutions become more complex.

10.3.2 Tidal Effects in General Relativity

The Newtonian results for the binary dynamics of extended nonspinning objects (10.14) and computation of the characteristic tidal deformability coefficients can be turned into a relativistic result in the following way (Flanagan and Hinderer 2008; Hinderer 2008; Steinhoff et al. 2016), see also Bini et al. (2012) for the action in the adiabatic limit and Goldberger and Rothstein (2006) for an effective field theory approach. We still consider the regime where the separation between the bodies is large compared to their size, so that the description of the binary can be divided into several zones, each amenable to a different approximation method (Flanagan 1998; Racine and Flanagan 2005): (1) The "body-zone" extending over the neighbourhood of each body, where the presence of the companion produces small perturbations to its equilibrium structure but a fully relativistic description is required. (2) The "orbital zone" far from the bodies where the dynamics can be computed from post-Newtonian (PN) theory and is dominated by their point-mass contributions, with small corrections due to their finite size encoded in their multipole moments. (3) The "buffer zone", at distances large compared to the size of the body but small compared to the orbital separation, where both descriptions are connected by matching; this is also the region where the body's multipole and tidal moments are defined.

10.3.2.1 Definition of the Body's Tidal Moments

The generalisation of expanding the gravitational potential around the body's center of mass is the so-called worldline-skeleton description (Dixon 1970), where one considers a reference center-of-mass worldline $z^\mu(\sigma)$ with σ being a parameter along the worldline, together with a set of multipole moments of the body. The body's gravitoelectric tidal moments $\mathcal{E}_{\mu\nu}$ are given by projecting the "electric part" of the spacetime curvature due to the companion that is characterized by the Weyl tensor $C_{\mu\alpha\nu\beta}$ as (Thorne and Hartle 1984)

$$\mathcal{E}_{\mu\nu} = C_{\mu\alpha\nu\beta} \frac{u^\alpha u^\beta}{z^2} \tag{10.33}$$

where $u^\mu = dz^\mu/d\sigma$ is the tangent to the worldline and

$$z = \sqrt{-u^\gamma u_\gamma} \tag{10.34}$$

Since we are considering a region of spacetime that excludes the source of the curvature, the Weyl tensor is the same as the Riemann tensor and $R_{\mu\alpha\nu\beta}$ could equivalently be used in (10.33). The tensor $\mathcal{E}_{\mu\nu}$ has the properties that it is symmetric and trace free and is purely spatial in the body's rest frame, $\mathcal{E}_{\mu\nu}u^\nu = 0$. In the rest frame, $u^\mu = (-1, 0, 0, 0)$ and $\mathcal{E}_{ij} = C_{0i0j}$ replaces (10.33). Higher multipole tidal moments are defined in an analogous way from covariant derivatives of the Weyl tensor projection

$$\mathcal{E}_L = \frac{1}{(\ell - 2)!} C_{\langle 0a_1 0a_2; a_3 \cdots a_\ell \rangle}, \tag{10.35}$$

where a semicolon denotes a covariant derivative.

10.3.2.2 Definition of the Body's Multipole Moments

To define the body's multipole moments we consider a region of spacetime at distances outside the body that are large compared to the size of the object but small compared to the radius of curvature of the source of the tidal perturbations. In this zone, the body's multipole moments can be read off from the asymptotic metric expressed in a local asymptotic frame (Thorne 1980, 1998). For example, the time-time component of the metric can be written in terms of an effective potential U_{eff} that is analogous to the Newtonian gravitational potential:

$$g_{tt} = -(1 - 2U_{\text{eff}}), \tag{10.36}$$

where for a spherical body described by the Schwarzschild exterior spacetime $U_{\text{eff}} = M/r$ with r denoting the distance from the body. For a nonspherical body, the asymptotic form of this metric function is

$$\lim_{r \to \infty} U_{\text{eff}} = \frac{M}{r} + \frac{3n^{\langle ij \rangle} Q_{ij}}{2r^3} + \mathcal{O}(r^{-4}) - \frac{1}{2} n^{\langle ij \rangle} \mathcal{E}_{ij} r^2 + \mathcal{O}(r^3). \tag{10.37}$$

In this setting, the ℓth mass multipole moment is associated with the piece in the asymptotic expansion that falls off as $r^{-(\ell+1)}$. This method to define the multipole moments of a body is equivalent to the Geroch-Hansen multipole moments for stationary spacetimes (Guersel 1983). However, for tidally induced moments in a binary system a small ambiguity remains in the above definitions (Thorne and Hartle 1984).

Finally, the definition of λ_ℓ from Eq. (10.15) still applies for the relativistic definitions of $\mathcal{E}_L$ and Q_L.

Another new feature of General Relativity that is absent in Newtonian gravity is that there are also gravitomagnetic tidal fields that induce current multipoles. In the local frame of the star, the gravitomagnetic part of the curvature is given by

$$\mathcal{B}_L = \frac{3}{2(\ell + 1)(\ell - 2)!} \epsilon_{<a_1 jk} C^{jk}_{a_2 0; a_3 \cdots a_\ell >},$$
(10.38)

where ϵ_{ijk} is the completely antisymmetric permutation tensor. The induced current moments $\mathcal{S}_L$ appear in the time-space part of the asymptotic metric in a local asymptotic frame

$$g_{tj} = -\frac{8}{r^3} \epsilon_{jki} \mathcal{S}_{ki} n^{<ki>} + \mathcal{O}(r^{-4}) + \frac{2}{3} \epsilon_{jpq} \mathcal{B}^q_k r^2 n^{<pk>} + \mathcal{O}(r^3),$$
(10.39)

where $\mathcal{S}_{ij}$ is the body's current quadrupole moment and $\mathcal{B}_{ij}$ the quadrupolar gravitomagnetic tidal moment. Similar to the tidal deformability tidal deformability coefficients λ_ℓ, the relation between $\mathcal{S}_L$ and $\mathcal{B}_L$ is characterized by a set of gravitomagnetic Love numbers σ_ℓ

$$\mathcal{S}_L = -\sigma_\ell \mathcal{B}_L.$$
(10.40)

These have no Newtonian analogue but can also be written in terms of dimensionless Love numbers j_ℓ as $\sigma_\ell = (\ell - 1)/[4(\ell + 2)(2\ell - 1)!!] R^{2\ell+1} j_\ell$. See e.g. Damour and Nagar (2009) and Landry and Poisson (2015b) for further details.

10.3.2.3 Computation of Tidal Love Numbers in General Relativity

Before discussing tidally perturbed bodies we briefly review the construction of a spherically symmetric, isolated nonspinning neutron-star solution. The metric can be expressed as (Hartle 1967)

$$ds_0^2 = -e^{\nu(r)}dt^2 + e^{\gamma(r)}dr^2 + r^2(d\theta^2 + \sin^2\theta d\varphi^2),$$
(10.41)

and neutron-star matter is modeled by a perfect-fluid stress-energy tensor

$$T_{\mu\nu} = (\rho + p)u_\mu u_\nu + p g_{\mu\nu},$$
(10.42)

where p and ρ are the neutron-star's pressure and energy density and u^μ is the fluid's four-velocity. In the body's rest frame, the normalization condition $u_\mu u^\mu = -1$ implies that $u^\mu = (e^{-\nu/2}, 0, 0, 0)$. Substituting these expressions into the field equations $G_{\mu\nu} = 8\pi T_{\mu\nu}$ yields the Oppenheimer-Volkoff equations:

$$\frac{dm}{dr} = 4\pi r^2 \rho, \quad \frac{d\nu}{dr} = 2\frac{4\pi r^3 p + m}{r(r - 2m)}, \quad \frac{dp}{dr} = -\frac{(4\pi r^3 p + m)(\rho + p)}{r(r - 2m)},$$
(10.43)

where $m(r)$ is defined by

$$m(r) \equiv \frac{\left[1 - e^{-\gamma(r)}\right] r}{2}. \tag{10.44}$$

Outside the star, $m(r)$ becomes the body's constant gravitational mass M. To solve Eq. (10.43) requires specifying an EOS, $p = p(\rho)$. The interior solution is obtained by imposing regularity at the neutron-star center with a choice of central density ρ_c. The initial conditions close to the center $r \to 0$ are $\rho = \rho_c + \mathcal{O}(r^2)$, $p = p_c + \mathcal{O}(r^2)$, and $m = (4\pi/3)\rho_c r^3 + \mathcal{O}(r^5)$ where p_c is the central pressure. The neutron-star's surface $r = R$ corresponds to a vanishing pressure $p(R) = 0$.

We next consider linear, static perturbations to the equilibrium configuration described by the metric

$$ds^2 = ds_0^2 + h_{\mu\nu} dx^\mu dx^\nu. \tag{10.45}$$

We will work in the Regge-Wheeler gauge where $h_{\mu\nu}$ can be analysed into tensorial spherical harmonics (Regge and Wheeler 1957; Thorne and Campolattaro 1967; Ipser and Price 1991; Detweiler and Lindblom 1985). These are characterized by the mode integers (ℓ, m) and by a parity π which can be either $(-1)^\ell$ or $(-1)^{\ell+1}$. For small perturbations the (ℓ, m, π) modes are decoupled and the electric-type or even-parity $\pi = (-1)^\ell$ perturbations to the metric take the form

$$h_{\mu\nu}^e dx^\mu dx^\nu = \sum_{\ell,m} \left[-e^\nu H_0^{\ell m} dt^2 + 2H_1^{\ell m} dt dr + e^\gamma H_2^{\ell m} dr^2 + r^2 K^{\ell m} d\Omega^2 \right] Y^{\ell m} \tag{10.46}$$

Here, the functions H_0, H_2 and K generically depend on (t, r). However, for our purposes it is sufficient to consider static perturbations, where these functions depend only on r. The perturbations to the stress-energy tensor are given by (Thorne and Campolattaro 1967)

$$\delta T_0^0 = -\delta\rho_\ell Y_{\ell m}(\theta, \varphi) = -\frac{d\rho}{dp} \delta p_\ell Y_{\ell m}(\theta, \varphi), \qquad \delta T_i^i = \delta p_\ell Y_{\ell m}(\theta, \varphi). \tag{10.47}$$

We substitute the above decompositions into the Einstein field equations $G_\mu^\nu = 8\pi T_\mu^\nu$ and the stress-energy conservation $\nabla_\mu T^{\mu\nu} = 0$, and extract only the pieces that are linear in the perturbations in all the components. This leads to the relations

$$H_0 = H_2 \equiv H, \qquad H_1 = 0, \qquad \frac{\delta p}{\rho + p} = -\frac{1}{2} H. \tag{10.48}$$

Further, the (r, r) and (r, θ)-components can be used to algebraically eliminate K and K' in favour of H and its derivatives. Finally, the (t, t) component leads to the following second-order differential equation

$$
0 = \frac{d^2 H}{dr^2} + \left\{ \frac{2}{r} + e^{\gamma} \left[\frac{2M}{r^2} + 4\pi r (p - \rho) \right] \right\} \frac{dH}{dr}
$$

$$
+ \left\{ e^{\gamma} \left[-\frac{\ell(\ell+1)}{r^2} + 4\pi (\rho + p) \frac{d\rho}{dp} + 4\pi (5\rho + 9p) \right] - \left(\frac{dv}{dr} \right)^2 \right\} H . \quad (10.49)
$$

The initial condition at the center, for $r \to 0$, is $H \propto r^{\ell}$ to ensure regularity of the solution. The constant of proportionality is irrelevant in further calculations of the tidal deformabilities and can be chosen arbitrarily.

Outside the star, the metric perturbation reduces to the general form

$$
H_{\ell} = a_{\ell}^{Q} Q_{\ell 2}(x) + a_{\ell}^{P} P_{\ell 2}(x) , \quad (10.50)
$$

where $x \equiv r/M - 1$ and $P_{\ell 2}(x)$ and $Q_{\ell 2}(x)$ are the normalized associated Legendre functions of the first and second kinds respectively. The normalization is such that for $x \to \infty$ the asymptotic forms are $P_{\ell 2}(x) \sim x^{\ell}$ and $Q_{\ell 2}(x) \sim x^{-(\ell+1)}$. The constants a_{ℓ}^{P} and a_{ℓ}^{Q} are determined by matching the logarithmic derivative of the interior and exterior solutions,

$$
y_{\ell} \equiv \frac{r}{H_{\ell}} \frac{dH_{\ell}}{dr} , \quad (10.51)
$$

at the neutron-star surface. Comparing with the definition of Q_{ℓ} and $\mathcal{E}_{\ell}$ in the asymptotic metric (10.37) and the definition of the Love numbers enables writing the general expression for Λ_{ℓ} in the form given in Damour and Nagar (2009) (see Binnington and Poisson (2009) for an alternative expression):

$$
(2\ell - 1)!! \Lambda_{\ell} = - \left. \frac{P'_{\ell 2}(x) - Cy_{\ell} P_{\ell 2}(x)}{Q'_{\ell 2}(x) - Cy_{\ell} Q_{\ell 2}(x)} \right|_{x=1/C-1} , \quad (10.52)
$$

where $C = M/R$ is the neutron-star's compactness. The computation of Λ_{ℓ} thus proceeds by numerically solving for the background and perturbations in the interior, evaluating the results at the neutron-star surface, and using Eq. (10.52). For the dominant quadrupolar effect the explicit expression is

$$
\Lambda \mid_{\ell=2} = \frac{16}{15} (1 - 2C)^2 [2 + 2C(y - 1) - y]
$$

$$
\times \left\{ 2C[6 - 3y + 3C(5y - 8)] + 4C^3 [13 - 11y + C(3y - 2) + 2C^2(1 + y)] \right.
$$

$$
\left. + 3(1 - 2C)^2 [2 - y + 2C(y - 1)] \ln(1 - 2C) \right\}^{-1} , \quad (10.53)
$$

where y is evaluated at the surface $r = R$. Since H itself does not enter into Eq. (10.53) and only the combination of potentials (10.51) is needed it is more efficient to transform Eq. (10.49) into an equation for y. For $\ell = 2$ this becomes (Lindblom and Indik 2014; Landry and Poisson 2014):

$$
\frac{dy}{dr} = \frac{4(m + 4\pi r^3 p)^2}{r(r - 2m)^2} + \frac{6}{r - 2m} - \frac{y^2}{r} - \frac{r + 4\pi r^3(p - \rho)}{r(r - 2m)} y - \frac{4\pi r^2}{r - 2m}\left[5\rho + 9p + \frac{\rho + p}{(dp/d\rho)}\right].
\tag{10.54}
$$

Recasting the problem into the form (10.54) thus requires only integrating the first order differential equation with the boundary condition $y = 2$ at the center and evaluating the result at $r = R$. See Chakrabarti et al. (2013b) for a more general approach that simultaneously determines the tidal deformability and oscillation mode frequencies.

For an incompressible star with $\rho = \text{const}$ or $p = K\rho^{1 + 1/n}$ with $n = 0$, the density profile is a step function and the matching of the interior and exterior solutions must be modified in the following way (Damour and Nagar 2009). After obtaining a numerical solution to Eq. (10.54) in the interior the result is evaluated at the surface to determine $y^{\text{in}}(R)$. The step-function density discontinuity has a nonvanishing derivative at the neutron star surface, which must be taken into account and leads to a correction to the value of y just outside the star y^{out} that is computed from the relation, valid for any ℓ,

$$
y^{\text{out}}_{\text{incompressible}} = y^{\text{in}}_{\text{incompressible}} - 3.
\tag{10.55}
$$

10.3.2.4 Tidal Love Numbers for Current Multipoles

Gravitomagnetic tidal perturbations are described by the odd-parity sector of the metric. They can be decomposed as (Thorne and Campolattaro 1967)

$$
ds^2 = ds_0^2 - 2h_{0,\ell}(r)\frac{\partial_\varphi Y_{\ell m}(\theta, \varphi)}{\sin\theta} dt d\theta + 2h_{0,\ell}(r)\sin\theta\, \partial_\theta Y_{\ell m}(\theta, \varphi) dt d\varphi.
\tag{10.56}
$$

The perturbations to the stress-energy tensor depend on the assumptions on the fluid such as strict hydrostatic equilibrium as used in Damour and Nagar (2009) or that an irrotational configuration as a more realistic scenario studied in Landry and Poisson (2015b). The differential equations for the perturbed metric components can be derived similar to the procedure in the even-parity case. The master variable in this case is defined by

$$
h \equiv r^3 \frac{d}{dr}\left(\frac{h_{0,\ell}}{r^2}\right)
\tag{10.57}
$$

and it satisfies the differential equation (Landry and Poisson 2015b)

$$\frac{d^2h}{dr^2} + \frac{e^\gamma}{r^2}\left[2M + 4\pi(p-\rho)r^3\right]\frac{dh}{dr} - e^\gamma\left[\frac{\ell(\ell+1)}{r^2} - \frac{6M}{r^3} + (1-2\varepsilon)4\pi(\rho-p)\right]h = 0.$$
(10.58)

Here, the parameter ε characterises the assumptions on the fluid: $\varepsilon = 1$ for the irrotational case, and $\varepsilon = 0$ for strict hydrostatic equilibrium.

The gravitomagnetic Love numbers σ_ℓ are computed similar to the electric-type ones. The differential equation (10.58) is integrated numerically in the neutron-star interior with the initial condition at the center $h \propto r^{\ell+1}$ to ensure regularity, where the constant of proportionality is irrelevant for the final result. For the exterior solution, the asymptotic behavior of the two independent solutions at spatial infinity is $\hat{h}_\ell^P \sim (r/M)^{\ell+1}$ and $\hat{h}_\ell^Q \sim (r/M)^{-\ell}$. Specifically, for $\ell = 2$, the exterior solutions are given by $\hat{h}_2^P = (r/M)^3$ and $\hat{h}_2^Q = -(r/M)^3\partial_{(r/M)}[F(1,4;6;2M/r)M^4/r^4]/4$, where $F(a,b;c;z)$ is a hypergeometric function. As in the even-parity case, Love numbers are determined by matching $y_\ell^\sigma \equiv (r/h)(dh/dr)$ at the neutron-star surface, which leads to (Damour and Nagar 2009; Landry and Poisson 2015b)

$$\sigma_\ell = -\frac{(\ell-1)M^{2\ell+1}}{4(\ell+2)(2\ell-1)!!}\frac{\hat{h}_\ell^{P\prime}(C^{-1}) - Cy_\ell^\sigma(C^{-1})\hat{h}_\ell^P(C^{-1})}{\hat{h}_\ell^{Q\prime}(C^{-1}) - Cy_\ell^\sigma(C^{-1})\hat{h}_\ell^Q(C^{-1})}.$$
(10.59)

10.3.2.5   Love Numbers for Deformations of Isodensity Surfaces

The Love numbers discussed in the previous sections characterize the spacetime geometry at large distances from the deformed object. In the Newtonian limit, the multipole moments of the gravitational potential are related to the shape or surficial Love numbers h_ℓ characterising the deformation of the object's surface by $h_\ell = 1 + 2k_\ell$. In General Relativity, this relation becomes more complex and must be obtained by considering gauge-invariant quantities such as curvature scalars. Consider an initially spherical star of radius R whose surface deforms in response to a tidal disturbance. To linear order in the deformation, the Ricci scalar curvature of the surface is given by (Damour and Nagar 2009; Landry and Poisson 2014)

$$\mathcal{R} = \frac{1}{R^2}(2 + \delta\mathcal{R}), \qquad \delta\mathcal{R} = -2\sum_{\ell=2}^{\infty}\frac{\ell+2}{\ell}h_\ell\frac{R^{\ell+1}}{M}\mathcal{E}_L n^L.$$
(10.60)

The general-relativistic relation between shape and tidal Love numbers is (Landry and Poisson 2014)

$$h_\ell = \Gamma_1 + 2\Gamma_2 k_\ell,$$
(10.61)

where

$$\Gamma_1 = \frac{\ell+1}{\ell-1}(1 - M/R)F(-\ell, -\ell; -2\ell; 2M/R) - \frac{2}{\ell-1}F(-\ell, -\ell - 1; -2\ell; 2M/R),$$
(10.62a)

$$\Gamma_2 = \frac{\ell}{\ell+2}(1 - M/R)F(\ell + 1, \ell + 1; 2\ell + 2; 2M/R) + \frac{2}{\ell+2}F(\ell + 1, \ell; 2\ell + 2; 2M/R),$$
(10.62b)

where $F(a, b; c; z)$ is the hypergeometric function.

The relation of Eq. (10.61) can be applied directly to black holes, for which $M/R = 1/2$ and $k_\ell = 0$, and thus (Damour and Lecian 2009)

$$h_\ell^{\mathrm{BH}} = \frac{\ell+1}{2(\ell-1)} \frac{\ell!^2}{(2\ell)!},$$
(10.63)

10.3.2.6 Love Numbers for Tidally Perturbed Spinning Neutron Stars

For spinning neutron stars, new kinds of tidal couplings arise for which there are corresponding new Love numbers (Pani et al. 2015b; Landry 2017). The calculations become more complicated than for nonspinning objects because it leads to a coupling between spherical harmonic modes in the perturbation equations and because the identification of the Love numbers from asymptotic considerations becomes more subtle (Pani et al. 2015a,b; Landry and Poisson 2015a; Landry 2017). These issues have only recently received consideration and work is still ongoing to fully address the calculations of spin-tidal effects in a binary systems.

10.3.3 I-Love-Q Relations

The I-Love-Q relations are inter-relations between dimensionless quantities characterising the stellar moment of inertia I, the tidal parameter Λ, and spin-induced quadrupole moment Q that are insensitive to the equation of state to within a good approximation. These and similar relations in more general contexts beyond the inspiral and post-merger are discussed in Chap. 13. The computation of I and Q is reviewed in Yagi and Yunes (2017a) and will not be discussed in detail here.

The moment of inertia and spin-induced quadrupole moment are computed using a similar approach as for the computation of tidal deformabilities discussed above, by solving Einstein's equations and stress-energy conservation for small perturbations around an equilibrium stellar configuration. In this case the perturbations are due to the star's rotation instead of an external tidal field. At linear order in the spin the perturbation equations give the moment of inertia I relating the magnitude of the spin angular momentum S to the angular frequency Ω through $S = I\Omega$ and computed from (Hartle 1967; Kalogera and Psaltis 2000)

$$I = \frac{8\pi}{3}\frac{1}{\Omega}\int_0^R \frac{e^{-(\nu+\gamma)/2}r^5(\rho+p)\omega_1^{int}}{r-2m(r)}dr,$$ (10.64)

where ω_1^{int} is the solution to the following differential equation in the interior of the star:

$$\frac{d^2\omega_1}{dr^2} + \frac{4-4\pi r^2(\rho+p)e^\lambda}{r}\frac{d\omega_1}{dr} - 16\pi(\rho+p)e^\lambda\omega_1 = 0.$$ (10.65)

In the Newtonian limit, Eq. (10.64) reduces to (Hartle 1967) $I^N = (8\pi/3)\int_0^R r^4 \rho(r)dr$.

The spin-induced quadrupole moment Q determines the magnitude of the quadrupolar deformation of a star due to rotation and is obtained by carrying out the perturbative analysis to quadratic order in the spin. The details are described in Laarakkers and Poisson (1999), Mora and Will (2004), Berti et al. (2008) and will not be discussed here. Similar to the tidal Love numbers, the dimensionless measures of the spin-induced deformations are known as rotational Love numbers. In the Newtonian limit, the rotational and tidal Love numbers quadrupolar are exactly the same, however, relativistic effects break this degeneracy. Nevertheless for current models of neutron star EOSs, there exist mutual relations between the dimensionless parameters Λ and

$$\bar{I} = \frac{I}{M^3} \qquad \bar{Q} = -\frac{QM}{|\vec{S}|^2},$$ (10.66)

that are nearly independent of the EOS, to within percent-level accuracy. These take the empirical form that were tested for a large set of currently available EOSs (Yagi and Yunes 2017a)

$$\ln(\bar{Q}) = 0.1940 + 0.09163\ln\Lambda + 0.04812(\ln\Lambda)^2 - 4.283 \times 10^{-3}(\ln\Lambda)^3$$
$$+1.245 \times 10^{-4}(\ln\Lambda)^4$$ (10.67a)

$$\ln(\bar{I}) = 1.496 + 0.05951\ln\Lambda + 0.02238(\ln\Lambda)^2 - 6.953 \times 10^{-4}(\ln\Lambda)^3$$
$$+8.345 \times 10^{-6}(\ln\Lambda)^4$$ (10.67b)

$$\ln(\bar{I}) = 1.393 + 0.5471\ln\bar{Q} + 0.03028(\ln\bar{Q})^2 + 0.01926(\ln\bar{Q})^3$$
$$+4.434 \times 10^{-4}(\ln\bar{Q})^4.$$ (10.67c)

The above relations are only valid in the region where the fit was developed, roughly for the intervals $M \in [0.8, 2.4]M_\odot$, $\bar{I} \in [5, 30]$, $\Lambda \in [2, 4000]$, and $\bar{Q} \in [2, 10]$; see Yagi and Yunes (2013a,b) and Lattimer and Lim (2013).

The $I - Q$ universality encodes a relation between the neutron-star's spin or current dipole moment and its mass quadrupole moment that in some sense is reminiscent of a generalisation of the no-hair relations for black holes. A similar relation was also found for higher multipole moments (Pappas and Apostolatos 2014; Yagi et al. 2014a) that greatly simplify the structure of the exterior spacetime of a neutron star (Pappas 2017, 2015). The universal relations were also found to hold in the presence of weak magnetic fields (Haskell et al. 2014), and rapid rotation (Doneva et al. 2014; Chakrabarti et al. 2014). Universal relations between different neutron-star parameters had been found previously (Lattimer and Prakash 2001, 2004; Lattimer and Yahil 1989; Prakash et al. 1997; Andersson and Kokkotas 1998; Bejger and Haensel 2002; Carriere et al. 2003; Benhar et al. 2004; Tsui and Leung 2005; Lattimer and Schutz 2005; Morsink et al. 2007; Haensel et al. 2009; Lau et al. 2010; Urbanec et al. 2013; Bauböck et al. 2013). The work of Yagi and Yunes (2013a,b) prompted a number of further studies of universal relations (Maselli et al. 2013a; AlGendy and Morsink 2014; Chirenti et al. 2015; Pannarale et al. 2015; Steiner et al. 2016; Breu and Rezzolla 2016; Silva et al. 2016; Yagi 2014; Reina et al. 2017; Chan et al. 2015, 2016). From the studies in Majumder et al. (2015), Doneva et al. (2014), Pappas and Apostolatos (2014), Chakrabarti et al. (2014), Gagnon-Bischoff et al. (2018), from which it has become clear that universality holds only between dimensionless measures of the neutron-star properties, and that the choice of normalization for such appropriate quantities has an impact on the accuracy with which the interrelations hold. Approximate universal relations were recently also found for quantities characterising neutron-star binaries (Kiuchi et al. 2010; Kyutoku et al. 2010; Bauswein and Janka 2012; Takami et al. 2014, 2015; Bernuzzi et al. 2014a, 2015b; Bauswein and Stergioulas 2015; Yagi and Yunes 2016, 2017b; Rezzolla and Takami 2016; Maione et al. 2016); a more comprehensive review of universality in binaries will be discussed in Sect. 10.4. Finally, a list of work on (non-)universality in alternative theories of gravity and for exotic compact objects can be found in the review article (Yagi and Yunes 2017a) and also in Chap. 13 of this review.

Various possible reasons for the existence of universal relations for neutron stars have been considered, see e.g., the discussion in Yagi et al. (2014b). There is evidence that an approximate symmetry, specifically the self-similarity of radial profiles of iso-density surfaces in the interior of stars containing cold degenerate nuclear matter, is linked to the emergence of the universal relations in neutron stars and their absence in ordinary stars (Yagi et al. 2014b). The universal relations have several applications (Yagi and Yunes 2013a,b). For example, they be used to improve measurements by reducing the number of parameters, and for cross-comparisons between interpretations of measurements such as from GW and electromagnetic observations. The fact that the universal relations break down for exotic objects and in some alternative theories of gravity can also be used for tests of General Relativity and the nature of compact objects.

10.3.4   Binary Inspiral Dynamics and GWs

Having discussed the calculation of the relevant parameters we now return to the description of finite-size effects during a binary inspiral, specialising to nonspinning binaries where the models are currently well-developed. We will focus on the dominant mass-quadrupolar tidal effects, and only indicate extensions to other mass multipole moments and to gravitomagnetic interactions. As in the Newtonian context, unless explicitly indicated, we will take the system to be one extended object and a point-mass for simplicity and throughout work only to linear order in the tidal effects.

The action describing tidal interactions in a *relativistic* binary system can be obtained as follows. Recall that in the relativistic context we consider a worldline $z^\mu(\sigma)$ with tangent $u^\mu = dz^\mu/d\sigma$, where σ is an evolution parameter. We start by expressing the Newtonian result from Eq. (10.13) in a covariant form and inserting the appropriate redshift factors, defined in Eq. (10.34), to ensure invariance under re-parametrizations (Steinhoff et al. 2016):

$$S = S_{\text{orbit}} + \int d\sigma \left[-\frac{z}{2}\mathcal{E}_{\mu\nu}Q^{\mu\nu} + \mathcal{L}_{\text{rel}}^{\text{int}} \right], \qquad (10.68)$$

Here, $\mathcal{L}_{\text{rel}}^{\text{int}}$ denotes the relativistic internal Lagrangian. The action (10.68) can also be derived from an effective-field-theoretical approach (Goldberger and Rothstein 2006), by considering all possible terms that respect the symmetries (general covariance, parity, and time reversal) and re-defining variables to eliminate accelerations. For tidally induced quadrupoles due to the f-mode we can obtain $\mathcal{L}_{\text{rel}}^{\text{int}}$ from the Newtonian result given in Eq. (10.14) by replacing all time derivatives by covariant derivatives along the center-of-mass worldline and inserting appropriate factors of the redshift. This leads to (Steinhoff et al. 2016)

$$\mathcal{L}_{\text{rel}}^{\text{int}} = \frac{z}{4\lambda z^2 \omega_{02}^2} \left[\frac{DQ_{\mu\nu}}{d\sigma}\frac{DQ^{\mu\nu}}{d\sigma} - z^2\omega_{02}^2 Q_{\mu\nu}Q^{\mu\nu} \right], \qquad (10.69)$$

where

$$\frac{D}{d\sigma} = u^\beta \nabla_\beta \qquad (10.70)$$

and ∇_α is the covariant derivative. This neglects contributions from quadrupolar modes with higher radial nodes, as in the Newtonian case, and also omits other terms due to the incompleteness of the mode spectrum of relativistic compact objects. Contributions from higher multipoles can be described in a similar manner but are not given explicitly here. After decomposing Eq. (10.68) into the time and space components and imposing the constraints to isolate only the physical degrees of freedom, the action takes the form

$$S = S_{\text{orbit}} + \int d\sigma \left[-\frac{z}{2} \mathcal{E}_{ij} Q^{ij} + \frac{z}{4\lambda z^2 \omega_{02}^2} \left(\dot{Q}_{ij} \dot{Q}^{ij} - z^2 \omega_{02}^2 Q_{ij} Q^{ij} \right) + L_{\text{FD}} \right].$$

(10.71)

Here, the term L_{FD} describes relativistic frame-dragging effects that are contained in the kinematical term $(DQ^{\mu\nu}/d\sigma)^2$ when it is expressed with coordinate time as the evolution parameter. Specifically, L_{FD} describes the coupling of the orbital angular momentum to the angular momentum (or spin S_Q) associated with the quadrupole

$$S_Q^i = \frac{1}{\lambda \omega_{02}^2} \epsilon_{ijm} \left[Q^{kj} \dot{Q}^m{}_k - Q^{km} \dot{Q}^j{}_k \right].$$

(10.72)

To proceed with computing the dynamics and GWs requires explicit expressions for the various quantities such as $\mathcal{E}_{ij}$ and z and the frame-dragging terms appearing in the action. These have been computed in post-Newtonian (PN) theory (Vines et al. 2011a; Bini et al. 2012; Steinhoff et al. 2016), in the test-particle limit (Bini and Geralico 2015) or the gravitational self-force formalism (Dolan et al. 2015; Bini and Damour 2014; Nolan et al. 2015; Shah and Pound 2015). There also exists an alternative approach to the worldline-skeleton method termed the affine model. In the affine approach, the stars are described as triaxial ellipsoids and one solves a set of coupled ODEs for the evolution of the axes and the orbit. The most recent developments that take into account PN effects are derived in Ferrari et al. (2012), Maselli et al. (2012), where we refer the reader for more details about this model. Below, we will continue to work within the worldline-skeleton approach and outline the computation of tidal effects in the GW signal in PN theory before discussing their inclusion in more sophisticated GW models.

In PN theory, the GW signal can be computed by imposing that the power radiated by a binary system is balanced by a change in the energy of the binary. This enables computing the phase evolution of the orbital dynamics where at the leading order, the GW phase is twice the orbital phase. The radiated power is computed in a multipolar approximation that at leading order gives the quadrupole formula involving the total quadrupole moment of the system $Q_{ij}^{\text{T}} = \mu r^2 n^{<ij>} + Q_{ij}$. The energy of the binary is $E = E_{\text{pm}} + E_{\text{tidal}}$, where the subscript "pm" denotes point-masses. The computations starting from the action, are described in detail in Vines et al. (2011b) for the current state-of-the art complete knowledge at 1PN order in the tidal effects, and in Flanagan and Hinderer (2008), Hinderer et al. (2010) in Newtonian but more general contexts that include various other effects and estimate the size of the corrections. Partial information at higher PN orders is also available (Damour et al. 2012). The idea is to start from the action written out explicitly to 1PN order, derive from it the equations of motion, specialise to circular orbits $\ddot{r} = \dot{r} = 0$ and $\dot{\phi} = \Omega$, $\ddot{\phi} = 0$, and perturbatively solve from this for the orbital separation r in terms of the frequency variable $x = (M\Omega)^{2/3}$. The reason for wanting to eliminate r is that it is a gauge-dependent quantity whereas the frequency is an observable and hence less gauge-dependent. From the potential and kinetic energies contained in the action one can compute the energy of the system E and find the following tidal contribution in the limit of adiabatic tides (Vines et al. 2011a)

$$E_{\text{tidal}}(x) = -\frac{1}{2}\mu x \left[-9\frac{M_B}{M_A}\frac{\lambda_A x^5}{M^5} - \frac{11}{2}\left(\frac{3M}{M_A} - \frac{M_B}{M} - 3\frac{M_A^2}{M^2}\right)\frac{\lambda_A x^6}{M^5} \right] + (A \leftrightarrow B).$$

(10.73)

The factor outside the brackets is the result for Newtonian point masses. The results when including the finite f-mode frequency at Newtonian order can be found in Flanagan and Hinderer (2008), and the 1PN extension can be determined from the Hamiltonian and circular-orbit solutions given in Hinderer et al. (2016), Steinhoff et al. (2016). Next, the mass-quadrupole tidal corrections contribute to the power radiated in GWs as (Vines et al. 2011b)

$$P_{\text{GW}}^{\text{tidal}} = \frac{32\mu^2}{5M^2}x^{5/2}\left[\left(\frac{18M}{M_A} - 12\right)\frac{\lambda_A x^5}{M^5} + \left(\frac{643M_A}{4M} - \frac{176M}{7M_A} - \frac{1803}{28} - \frac{155M_A^2}{2M^2}\right)\frac{\lambda_A x^6}{M^5}\right].$$

(10.74)

By requiring that P_{GW} be balanced by a change in the energy E of the binary one can derive the evolution equations

$$\frac{d\phi}{dt} = \frac{x^{3/2}}{M} \qquad \frac{dx}{dt} = \frac{-P_{\text{GW}}}{dE/dx}$$

(10.75)

There are several ways to solve for ϕ in a PN approximation. For example, one can numerically solve Eq. (10.75) for $\phi(t)$ and $x(t)$ after first expanding the ratio $P_{\text{GW}}/(dE/dx)$ about $x = 0$ to the consistent PN order. These waveforms are known as TaylorT4 approximants as reviewed in Buonanno et al. (2009), where the point-mass terms are also given explicitly. The adiabatic quadrupolar tidal corrections that add linearly to the point-mass contributions are (Vines et al. 2011b):

$$\frac{dx}{dt}\bigg|_{\text{tidal}} = \frac{32}{5}\frac{M_B}{M^7}\lambda_A x^{10}\left[12\left(1 + 11\frac{M_B}{M}\right)\right.$$

$$\left. + x\left(\frac{4421}{28} - \frac{12263}{28}\frac{M_A}{M} + \frac{1893}{2}\frac{M_A^2}{M^2} - 661\frac{M_A^3}{M^3}\right)\right] + (A \leftrightarrow B). \quad (10.76)$$

Other possibilities to perturbatively solve (10.75) and obtain the tidal contributions to different approximants for the gravitational waveform are detailed in the Appendix of Wade et al. (2014). The reason why the phase evolution is the most important prediction for GW data analysis is that matched filtering is employed to identify and interpret signals, where the datastream is cross-correlated with theoretical predictions for the GWs (see e.g., Cutler and Flanagan 1994), thus making the process very sensitive to the phasing (Cutler et al. 1992). Besides the TaylorT4 approximants, another widely utilised class of template waveforms for data analysis are TaylorF2 waveforms. Their advantage is that they provide a fully analytic frequency-domain model and are thus very fast to generate. The derivation is explained e.g., in Cutler and Flanagan (1994) and in the stationary phase approximation leads to a Fourier transform of the signal, denoted by $\tilde{h}$, of the form

$$\tilde{h}(f) = \mathcal{A} f^{-7/6} \exp\left[i\left(\psi_{\text{pm}} + \psi_{\text{tidal}}\right)\right]. \tag{10.77}$$

Here f is the GW frequency, $\mathcal{A} \propto \mathcal{M}^{5/6}/D$, where $\mathcal{M}$ is the chirp mass $\mathcal{M} = \eta^{3/5} M$, and D is the distance between the GW detector and the binary. Extrinsic parameters of the source such as the location on the sky are also contained in $\mathcal{A}$, where higher PN order (and tidal) corrections to the amplitude also enter. The point-mass phase ψ_{pm} to the current best knowledge for nonspinning binaries is given e.g., in Eq. (3.18) of Buonanno et al. (2009). The mass-quadrupole adiabatic tidal contributions to ψ_{tidal} in the adiabatic limit can be expressed in the following form given explictly in Wade et al. (2014):

$$\delta\psi_{\text{tidal}} = \frac{3}{128\eta x^{5/2}}\left[-\frac{39}{2}\tilde{\Lambda}x^5 + \left(-\frac{3115}{64}\tilde{\Lambda} + \frac{6595}{364}\sqrt{1 - 4\frac{\mu}{M}}\,\delta\tilde{\Lambda}\right)x^6\right], \tag{10.78a}$$

where

$$\tilde{\Lambda} = \frac{16}{13}\left[\left(1 + \frac{12M_B}{M_A}\right)\frac{\Lambda_A M_A^5}{M^5} + \left(1 + \frac{12M_A}{M_B}\right)\frac{\Lambda_B M_B^5}{M^5}\right] \tag{10.78b}$$

$$\delta\tilde{\Lambda} = \left(1 - \frac{7996M_B}{1319M_A} - \frac{11005M_B^2}{1319M_A^2}\right)\frac{\Lambda_A M_A^6}{M^6} + \left(\frac{11005M_A^2}{1319M_B^2} + \frac{7996M_A}{1319M_B} - 1\right)\frac{\Lambda_B M_B^6}{M^6} \tag{10.78c}$$

In Eq. (10.78c) the assumption is that $M_A > M_B$. The parameter $\tilde{\Lambda}$ plays an analogous role as in GW measurements as the chirp mass $\mathcal{M}_{\text{chirp}} = \mu^{3/5} M^{2/5}$ as the most readily measurable combination of parameters. For equal-mass binary neutron stars $\tilde{\Lambda}$ reduces to Λ of the individual neutron stars, and the parameter $\delta\tilde{\Lambda}$ vanishes. Other combinations of the two parameters Λ_A and Λ_B are also in use and have advantages in different contexts, e.g., to characterize the dominant effect in the conservative dynamics (Damour and Nagar 2010), or to improve the measurability (Yagi and Yunes 2016, 2017b). The key point to note is that for a double neutron-star system GW measurements are most sensitive to a weighted average of the deformability parameters of the two objects.

From the discussion above, it is apparent that the fractional corrections to the Newtonian point-mass results due to tidal effects scale as a high power of the frequency, x^5 and higher, where $x = (\pi M f)^{2/3}$. This means that in a PN counting that is based on assigning to each power of x an additional PN order, tidal effects first enter effectively as 5PN corrections, although physically, they are Newtonian effects. Point-mass terms are currently only known to 4PN order in the dynamics

(Damour et al. 2016; Marchand et al. 2017) and only to 3.5PN order in the GW phasing (Blanchet 2006). This lack of complete information has raised concerns about systematic errors in GW measurements of tidal effects (Favata 2014; Yagi and Yunes 2014; Wade et al. 2014). However, there are two classes of effective or phenomenological models for black hole binaries that effectively include all PN orders in an approximate way. These are the effective one body (EOB) model (Buonanno and Damour 1999, 2000) and the so-called "Phenom" models (Ajith et al. 2007, 2008), both of which aim to combine the available information on the relativistic two-body problem from different regimes into a single framework to generate waveforms for data analysis.

The EOB approach is a framework to compute the dynamics and GWs from a binary by evolving a description of the coupled system of ODEs for the orbital motion, GW generation, and radiation-backreaction in the time-domain. The purpose of the Phenom models is to provide an efficient description of the dominant effects in the GW signal in the frequency-domain, through a generalisation of (10.77). Both approaches rely not only on analytical results but also include information from numerical relativity simulations for black hole binaries; the current state-of-the art refinements and calibrations of the models are described in Bohe et al. (2017), Babak et al. (2017) and Nagar et al. (2017, 2018) for an alternative version of the EOB model, and Khan et al. (2016), Schmidt et al. (2015) for the Phenom models. By design, these models therefore include high-PN order information, albeit only in an approximate and phenomenological manner. When tidal effects are included in such models one might expect the systematic uncertainties due to missing high-PN-order point-mass terms to be reduced. However, the level of remaining systematic errors must be assessed by testing the models in various ways, such as by comparing to numerical relativity simulations or comparing results from data analysis studies with the two different classes of models.

The Phenom models are frequency-domain models that prescribe an analytical expression for the amplitude and phase of $\tilde{h}$. In the Phenom models, the tidal terms from Eq. (10.78a) can directly be added to the black hole waveforms without further work. Alternative tidal phasing models based on a fit to numerical relativity results have recently also been developed (Dietrich et al. 2017, 2018; Kawaguchi et al. 2018) and can likewise be added on top of the black hole waveforms for the inspiral. There also exist surrogate models for EOB waveforms in the frequency domain (Purrer 2016; Lackey et al. 2017) where the tidal contributions can be directly added to the phasing as described for the Phenom models. The different kinds of frequency-domain models are the most efficient for data analysis and were used in the EOS inferences of GW170817 (Abbott et al. 2018a) as the analysis in Abbott et al. (2018b) indicated that statistical errors dominate over modeling uncertainties.

Time-domain models such as the EOB models are less computationally efficient than frequency-domain models but have other advantages, e.g. they describe both the binary dynamics and GWs and are based on additional theoretical considerations about the relativistic two-body problem. To include tidal effects in the

EOB approach at a fundamental level requires both the tidal corrections to the conservative dynamics and to the gravitational radiation. At present, tidal effects have only been fully included in the EOB model for nonspinning binaries. Spin effects for point masses as well as the spin-induced quadrupole effects are currently incorporated in the EOB models, however, including all spin-tidal interactions to the relevant order is the subject of ongoing work. Below, we will briefly summarise the different EOB tidal models available at present and refer the reader to the overviews of models in Dietrich and Hinderer (2017), Nagar et al. (2018) more specific details. For adiabatic tidal effects, tidal contributions to the conservative EOB dynamics were computed from a PN expansion of tidal effects in Damour and Nagar (2010), Vines et al. (2011a), Bini et al. (2012). These effects were also calculated within the gravitational self-force approximation in Bini and Damour (2014). The PN approximation assumes small corrections to Newtonian dynamics but is valid for any mass ratio, whereas the gravitational self-force formalism assumes linear-in-mass-ratio corrections to the strong-field test-particle limit. To a certain extent, the two approximations therefore provide different kinds of information. Tidal corrections to the GW amplitudes that are used in the EOB model to compute the emitted GWs and the radiation reaction forces on the orbital dynamics were given in Damour et al. (2012). These models, however, tend to underestimate finite-size effects when compared against numerical relativity simulations, see e.g. Bernuzzi et al. (2015a), Hinderer et al. (2016), Dietrich and Hinderer (2017) for recent studies. There are three main reasons to expect an enhancement of tidal effects relative to the information included in the models described above: relativistic corrections that lead to a stronger tidal field, an enhanced response of the neutron-star matter to tidal perturbations, and the fact that the tidal models are used to describe the binary including the nonlinear regimes at merger or tidal disruption. These considerations motivated two classes of improved EOB models. The first, discussed in Bernuzzi et al. (2015a), Nagar et al. (2018) is based on extrapolating the results of Bini and Damour (2014) to second-order in the mass ratio in a particular gauge. The second, discussed (Hinderer et al. 2016; Steinhoff et al. 2016) includes dynamical tidal effects due to the f-mode oscillations in the EOB model that can lead to a substantial tidal enhancement even if the mode resonance is not fully excited during the inspiral. Note that while the underlying EOB model for black holes are calibrated to numerical relativity simulations, the tidal part of the model of Hinderer et al. (2016), Steinhoff et al. (2016) is currently purely based on analytical results, without any calibrations.

10.3.5   Other Finite-Size Effects

The finite size of neutron stars in a binary system has a number of additional impacts on the dynamics and GWs, besides the tidal effects discussed above. For rotating neutron stars, the spin-induced quadrupole moment (and the higher moments) leads to a contribution to the GW signal that is quadratic in the neutron-star's spin (Poisson 1998; Laarakkers and Poisson 1999; Mora and Will 2004; Berti et al. 2008). As mentioned above, these effects from the rotational deformations are already included in the template models described in the previous subsection, both in the PN models (see Krishnendu et al. (2017) for the latest update) as well as in the EOB and Phenom models. The GW imprints from gravitomagnetic tidal effects (Banihashemi and Vines 2018) and spin-tidal interactions (Landry 2018; Jimenez-Forteza et al. 2018) have recently also been examined within PN theory.

Tidal interactions can also lead to the resonant excitation of various oscillation modes during the inspiral (Ho and Lai 1999; Flanagan and Racine 2007; Shibata 1994; Yu and Weinberg 2017; Lai 1994; Kokkotas and Schaefer 1995; Tsang 2013; Tsang et al. 2012), to nonlinear mode coupling effects (Xu and Lai 2017; Essick et al. 2016; Landry and Poisson 2015c), and to the full f-mode excitations for eccentric orbits (Gold et al. 2012; Chirenti et al. 2017). In neutron-star–black-hole binaries, depending on the parameters, the neutron star may get tidally disrupted, which leads to a sudden shutoff of the GW signal and contains additional equation-of-state information (Vallisneri 2000; Shibata and Taniguchi 2011; Pannarale et al. 2011; Ferrari et al. 2010; Maselli et al. 2013b; Foucart et al. 2014; Kawaguchi et al. 2017; Lackey et al. 2014). For a review article on GWs from neutron-star–black-hole binaries containing a comprehensive list of references on the topic see Shibata and Taniguchi (2011), and for the characteristics of possible associated electromagnetic counterparts see Fernández et al. (2017), Schnittman et al. (2018), Paschalidis et al. (2015a). An additional distinction between neutron stars and black holes is that the neutron star has a surface while a black hole possesses an event horizon that absorbs all incoming GWs; this effect is also imprinted in the GWs (Maselli et al. 2018).

As discussed above, there are several issues in modelling neutron-star binary inspirals that remain to be fully addressed and are an active area of research. Most of these concern currently unmodeled physics, which is also of great interest for enhancing the potential to extract more details about neutron-star interiors from GW observations. Examples of remaining work for the inspiral are to develop full waveform models that incorporate matter effects in spinning neutron-star binaries, to assess the importance of dynamical tides for various oscillation modes for a more realistic description of the neutron stars (e.g., including the effects of superfluidity (Gualtieri et al. 2014) and the effects of spins), and to analyze the effects of nonlinear couplings in a relativistic setting.

Another issue that could be improved concerns the fact that at present the models constructed for the inspiral of isolated, perturbed neutron stars are used up to merger, defined as the peak in the GW amplitude. The peak approximately coincides with the collision of the high-density neutron star cores, shortly after the neutron outer parts have already come into contact. The theoretical predictions of the very late stage in the evolution, could thus be improved by accounting for the main physical effects of the cores moving through the material of their former outer parts. Similarly, for neutron star – black hole binaries, developing an improved description that accounts for nonlinearities of the tidal disruption process remains an open issues.

Other ongoing efforts are focusing on developing a complete model that combines the information from the inspiral, merger and postmerger epochs, and likewise for the inspiral and possible tidal disruption in mixed binaries for more generic systems than considered to date. These studies must rely on interfacing the theoretical insights with numerical relativity and data analysis to test and improve the models and select the order of priorities for addressing the issues mentioned above. An important application of such models will not only be for the current network of GW detectors but also to inform the design and science case for future, third-generation detectors. More work is also required on optimising methods to combine information from the GWs with those from the electromagnetic counterparts (Baiotti and Rezzolla 2017; Paschalidis 2017), together with nuclear physics knowledge and other astrophysical measurements of neutron-star properties to maximise the overall scientific payoffs. Lastly, other areas of active research are to consider finite-size effects in alternative theories of gravity, for exotic objects, and for possible bound states of fundamental fields around black holes.

10.4   Post-merger Dynamics

Research on the post-merger phase has been undergoing intense development over the last few years because of its importance for linking numerical simulations and astrophysical observations. The (early) post-merger is also the phase in which most of the energy in GWs is emitted, as pointed out in Bernuzzi et al. (2016), even though the GWs emitted in this stage are not those that give the largest signal-to-noise ratio, because their frequency range is not in the best sensitivity zone of current interferometric detectors. The numerical description of this stage is far more challenging than the inspiral one because of the highly nonlinear dynamics and of the development of strong, large-scale shocks that inevitably reduce the convergence order, thus requiring far higher resolutions than the ones normally employed. As a result, the accuracy of some quantities computed after the merger is sometimes only marginal. The most notable example of these quantities is the lifetime of the remnant (be it an HMNS or an SMNS) before its collapse to black hole; since this object is only in metastable equilibrium, even small differences in resolution or even grid setup are sufficient to change its dynamical behaviour, accelerating or slowing down its collapse to a black hole. Fortunately, other quantities, such as the spectral properties of the GW post-merger emission appear far more robust and insensitive to the numerical details; we will discuss them later in this section.

Since the first general-relativistic simulations of BNS mergers, several works have studied the nature (neutron star or black hole) of the objects resulting from the mergers (Shibata and Uryū 2000, 2002; Shibata et al. 2005; Shibata and Taniguchi 2006; Yamamoto et al. 2008; Baiotti et al. 2008; Anderson et al. 2008; Giacomazzo et al. 2011). It is of course important to establish whether a black hole forms promptly after the merger or instead an HMNS forms and lives for long times (more than 0.1 s), because the post-merger GW signal in the two cases is clearly different. Anderson et al. (2008) and Giacomazzo et al. (2011) started investigating the dependence of the lifetime of the HMNS on the magnitude of the initial magnetic field in the case of magnetised binaries. However, as mentioned above, such investigations are extremely delicate since it is not straightforward to completely remove the influence of numerical artefacts on the lifetime of the remnant even in the absence of magnetic fields, at least with present resolutions.

In an alternative approach, Kaplan et al. (2014) have investigated the role of thermal pressure support in hypermassive merger remnants by computing sequences of axisymmetric uniformly and differentially rotating equilibrium solutions to the general-relativistic stellar structure equations and found that this too is a subtle issue: the role of thermal effects on the stability and lifetime of a given configuration depends sensitively and in a complicated way on its details, like central or mean rest-mass density, temperature distribution, degree of differential rotation and rotation rate (see discussion in Hanauske et al. 2016).

Clearly, the issue of the precise lifetime of the binary-merger product before it reaches its asymptotic state, especially when its equilibrium is mediated by the generation of magnetic fields or radiative losses is far from being solved and will require computational resources and/or methods not yet available.

Recently, Paschalidis, East and collaborators (Paschalidis et al. 2015b; East et al. 2016) pointed out that a one-arm spiral instability (Centrella et al. 2001; Watts et al. 2005; Baiotti et al. 2007; Corvino et al. 2010) can develop in HMNSs formed by dynamical-capture and that the $m = 1$ mode associated with this instability may become the dominant oscillation mode if the HMNS persists for long enough[3]; this instability has been subsequently studied also in quasi-circular BNSs (Radice et al. 2016; Lehner et al. 2016). The instability, is reminiscent of the shear instability that has been studied in detail for isolated stars (Baiotti et al. 2007; Corvino et al. 2010; Camarda et al. 2009; Franci et al. 2013; Muhlberger et al. 2014) and seems to be correlated with the generation of vortices near the surface of the HMNS that form due to shearing at the stellar surface. These vortices then spiral in toward the center of the star, creating an underdense region near the center. The growth of the $m = 1$ mode and so of the instability, could be related to the fact that the maximum density does not reside at the center of mass of the star (Saijo et al. 2003), or to the existence of a corotation band (Balbinski 1985; Luyten 1990; Watts et al. 2005). The instability has an imprint on the GW signal, but the prospects of detection are not encouraging, because of the small emitted power (Radice et al. 2016).

10.4.1 Gravitational-Wave Spectroscopy of the Post-merger Signal

Many researchers have taken up the challenge of studying the properties of the binary-merger product, because this may give indications on the ultra-high density EOS, the origin of SGRBs, and even the correct theory of gravity. In what follows

---

[3]The $m = 1$ mode had been studied previously together with the other modes, but it had never been found to become dominating (see, e.g., Dietrich et al. 2015a).

we will focus in particular on the determination of the EOS. While detectable differences between simulations that employed different EOSs already appear during the inspiral (see Sect. 10.3), the post-merger phase depends more markedly on the EOS (Bauswein et al. 2014; Takami et al. 2014, 2015; Bernuzzi et al. 2014a, 2015a,b; Rezzolla and Takami 2016; Maione et al. 2016). A note of caution is necessary here to say that post-merger waveforms are at rather high frequencies and thus probably only marginally measurable by detectors like Advanced LIGO. Third-generation detectors, such has ET (Punturo et al. 2010a), may provide the first realistic opportunity to use GWs to decipher the stellar structure and EOS (Andersson et al. 2011).

The first attempts to single out the influence of the EOS on the post-merger dynamics were done in Shibata et al. (2005), Shibata and Taniguchi (2006), Yamamoto et al. (2008), Baiotti et al. (2008). These works focused mostly on the dynamics of equal-mass binaries, as these are thought to be the most common (Osłowski et al. 2011) and are easier and faster to compute, since symmetries of the configuration can be exploited to save computational resources. The study of the effect of realistic EOSs in general-relativistic simulations has been subsequently brought forward by many groups. Kiuchi et al. (2009) made use of the Akmal-Pandharipande-Ravenhall (APR) EOS (Akmal et al. 1998).[4] This nuclear-physics EOS describes matter at zero temperature and so during the simulation it needs to be combined with a "thermal" part that accounts for the energy increase due to shock heating (this is mostly done through the addition of an ideal-fluid part to the EOS; see Rezzolla and Zanotti (2013) for a discussion). The resulting *"hybrid EOS"* appears to be appropriate for studying the inspiral and merger, but may not be satisfactory for studying the remnant formation and the evolution of the accretion disc around the formed black hole, because for such cases, effects associated with the thermal energy (finite temperature), neutrino cooling, and magnetic fields are likely to play an important role. In another work of the same group (Hotokezaka et al. 2011), the dependence of the dynamical behavior of BNS mergers on the EOS of the nuclear-density matter with piecewise-polytropic EOSs (Read et al. 2009) was studied.

One family of EOSs that has received special attention in the past years is that describing strange matter, namely matter containing hyperons, which are nucleons containing strange quarks. The strange-matter hypothesis (Witten 1984) considers the possibility that the absolute ground state of matter might not be formed by iron nuclei but by strange quark matter: a mixture of up, down, and strange quarks. This hypothesis introduced the possibility that compact stars could be stars made also of strange-quark matter, or strange stars (Haensel et al. 1986; Alcock et al. 1986). One of the astrophysical consequences of this is the possibility that collision events of

---

[4]The APR EOS, and many of the proposed EOSs, were later found to violate the light-speed constraint at very high densities and phenomenological constraints (Taranto et al. 2013), but no strong conclusions can be made to rule out such EOSs on this basis because the constraints themselves are affected by errors.

two strange stars lead to the ejection of strangelets, namely small lumps of strange quark matter.

Although the occurrence of hyperons at very large nuclear densities is rather natural, hyperonic EOSs are generally very soft and currently disfavoured by the observation of a $2M_\odot$ star (Antoniadis et al. 2013; Demorest et al. 2010), which they can hardly reproduce, except by fine tuning of the parameters (see, e.g., Alford et al. 2005; Rikovska-Stone et al. 2007; Weissenborn et al. 2011). This basic inconsistency between the expectations of many nuclear physicists and the observational evidence of very massive neutron stars is normally referred to as the "hyperon puzzle"; those supporting the use of hyperonic EOSs also state that the existence of exotic phases in strange stars remains unconstrained and could lead to higher masses (Bhowmick et al. 2014). Additional work is needed to settle this "hyperon puzzle" and we will present results on strange-star simulations setting these doubts aside.

The first investigations of binary strange stars were those of Bauswein et al. (2009, 2010), who employed the MIT bag model (Farhi and Jaffe 1984). Within this model, quarks are considered as a free or weakly interacting Fermi gas and the nonperturbative QCD interaction is simulated by a finite pressure of the vacuum, the *bag constant B*. Three-dimensional general-relativistic simulations with conformally flat gravity of the coalescence of strange stars were performed and the possibility to discriminate on the strange matter hypothesis by means of GW measurements was explored. The dynamics of mergers of strange stars, which are usually more compact, is different from those of neutron-star mergers, most notably in the tidal disruption during the merger. Furthermore, instead of forming dilute halo-structures around the binary-merger product, as in the case of neutron-star mergers, the coalescence of strange stars results in a differentially rotating hypermassive object with a sharp surface layer surrounded by a geometrically thin, clumpy high-density strange-quark-matter disc. It was found that in some cases (some types of EOS and stellar properties) the analysis of the GW signals emitted by strange-star mergers showed that it may be possible to discern whether strange-star or neutron-star mergers produced the emission. In particular, it was found that the maximal frequency during the inspiral and the frequency of the oscillations of the post-merger remnant are in general higher for strange-star mergers than for neutron-star mergers. In other cases, however, there remains a degeneracy among different models, and a conclusion about the strange-matter hypothesis could be reached only if other types of observations (e.g., of cosmic rays) were available.

Strange-matter EOSs were later studied with a fully general-relativistic code in a series of articles by Sekiguchi, Kiuchi and collaborators (Sekiguchi et al. 2011a, 2012; Kiuchi et al. 2012b,c), who showed results of simulations performed by incorporating both nucleonic and hyperonic finite-temperature EOSs (and neutrino cooling as well). It was found that also for the hyperonic EOS, an HMNS is first formed after the merger and subsequently collapses to a black hole. The radius of such an HMNS decreases in time because of the increase of the mass fraction of hyperons and the consequent decrease in pressure support. Such a shrinking is noticeably larger than the one simply due to angular-momentum loss through GW emission that is present also in nucleonic EOSs. These differences in the dynamics are clearly visible in the GW signal, whose characteristic peak frequency has an increase of 20–30% during the HMNS evolution. By contrast, for nucleonic EOSs, the peak GW frequency in the HMNS phase is approximately constant on the timescales considered. It was also stressed that these results raise a warning about using the peak frequency of the GW spectrum to extract information of the neutron-star matter (see below), because it may evolve and so make the relation of the peak frequency with the HMNS structure ambiguous. Finally, it was found that the torus mass for the hyperonic EOS is smaller than that for nucleonic EOSs, thus making hyperonic EOSs less favourable for the description of SGRBs.

More recently, Radice et al. (2017) performed numerical simulations of BNS mergers with one EOS, BHB$\Lambda\phi$ (Banik et al. 2014), that includes Λ-hyperons but that satisfies all presently known EOS constraints, including the astrophysical ones. They found that the rapid contraction of the merged object due to the appearance of hyperons does affect amplitude modulation and phase evolution of the gravitational waveform, but the peak frequencies in the PSD are very similar to those produced in case of a similar EOSs, DD2 (Hempel and Schaffner-Bielich 2010), that however has no phase transitions. It was also found that Advanced LIGO could distinguish the two EOSs considered in that work with the detection a single merger at a distance of up to about 20 Mpc (Radice et al. 2017).

10.4.2   Spectral Properties of the Signal

In addition to simulating BNS mergers with various EOSs, it is important to find ways to connect future GW observation with the EOS of the neutron stars. Recently there have been several suggestions on how to achieve this, based either on the signature represented by the tidal corrections to the orbital phase or on the power spectral density (PSD) of the post-merger gravitational waveforms or on the frequency evolution of the same. The first approach, described in Sect. 10.3, is reasonably well understood analytically (Flanagan and Hinderer 2008; Baiotti et al. 2010; Bernuzzi et al. 2012, 2015b; Read et al. 2013) and can be tracked accurately with advanced high-order numerical codes (Radice et al. 2014a,b). Here we describe works on the post-merger approach in some detail.

Hotokezaka et al. (2013) used their adaptive mesh-refinement (AMR) code[5] SACRA (Yamamoto et al. 2008) to perform a large number of simulations with a variety of mass ranges and EOSs (as done before, approximate finite-temperature effects were added to the cold EOSs through an additional ideal-fluid term), in order to find universal features of the frequency evolution of GWs emitted by the HMNS formed after the merger. In their analysis they found it convenient to decompose the merger and post-merger GW emission in four different parts: (1) a peak in frequency

---

[5]In previous works by this group, described above, a different code with a uniform grid had been used.

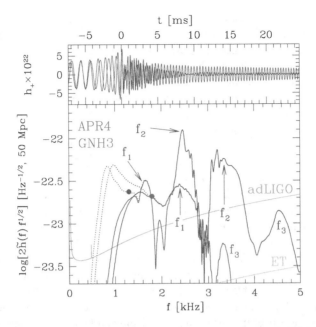

Fig. 10.4 Top sub-panel: evolution of h_+ for binaries with the APR4 and GNH3 EOSs (dark-red and blue lines, respectively) for optimally oriented sources at a distance of 50 Mpc. Bottom sub-panel: spectral density $2\tilde{h}(f)f^{1/2}$ windowed after the merger for the two EOSs and sensitivity curves of Advanced LIGO (green line) and ET (light-blue line); the dotted lines show the power in the inspiral, while the circles mark the contact frequency [Reprinted with permission from Takami et al. (2014). © (2014) by the American Physical Society]

and amplitude soon after the merger starts; (2) a decrease in amplitude during the merger and a new increase when the HMNS forms; (3) a damped oscillation of the frequency during the HMNS phase lasting for several oscillation periods and eventually settling to an approximately constant value (although a long-term secular change associated with the change of the state of the HMNSs is always present); (4) a final decrease in the amplitude during the HMNS phase, either monotonical or with modulations. Based on this, they find an optimal 13-parameters fitting function, using which it may be possible to constrain the neutron star radius with errors of about 1 km (Hotokezaka et al. 2013).

In contrast with this multi-stage, multi-parameter description of Hotokezaka et al. (2013), other groups have concentrated on the analysis of the full PSD of the post-merger signal, isolating those spectral features (i.e., peaks) that could be used to constrain the properties of the nuclear-physics EOSs. As a reference, we show in Fig. 10.4 the PSDs of some representative GWs when compared with the sensitivity curves of current and future GW detectors (Takami et al. 2014). More specifically, two examples are presented in Fig. 10.4, which refers to two equal-mass binaries with APR4 and GNH3 EOSs, and with individual gravitational masses at infinite separation of $\bar{M}/M_\odot = 1.325$, where $\bar{M}$ is the average of the initial gravitational mass of the two stars. The top sub-panel shows the evolution of the $\ell = m = 2$ plus polarization of the strain ($h_+ \sim h_+^{22}$), aligned at the merger for optimally oriented sources at a distance of 50 Mpc (dark-red and blue lines for the APR4 and GNH3 EOSs, respectively). The bottom panel, on the other hand, shows the spectral densities $2\tilde{h}(f)f^{1/2}$ windowed after the merger for the two EOSs, comparing them with the sensitivity curves of Advanced LIGO (2009) (green line) and of the Einstein Telescope (Punturo et al. 2010b; Sathyaprakash and Schutz 2009) (ET; light-blue line). The dotted lines refer to the whole time series and hence, where visible, indicate the power during the last phase of the inspiral, while the circles mark the "contact frequency" $f_{cont} = \mathcal{C}^{3/2}/(2\pi\bar{M})$ (Damour et al. 2012), where $\mathcal{C} := \bar{M}/\bar{R}$ is the average compactness, $\bar{R} := (R_1 + R_2)/2$, and $R_{1,2}$ are the radii of the nonrotating stars associated with each binary.

Note that besides the peak at low frequencies corresponding to the inspiral (cf., dashed lines), there is one prominent peak and several others of lower amplitudes. These are related to the oscillations of the HMNS and would be absent or much smaller if a black hole forms promptly, in which case the GW signal would terminate abruptly with a cutoff corresponding to the fundamental quasi-normal-mode frequency of the black hole (Kokkotas and Schmidt 1999). The behaviour summarised in Fig. 10.4 is indeed quite robust and has been investigated by a number of authors over the last decade (Oechslin and Janka 2007; Stergioulas et al. 2011; Bauswein and Janka 2012; Bauswein et al. 2012, 2014, 2016; Hotokezaka et al. 2013; Takami et al. 2014, 2015; Clark et al. 2014; Kastaun and Galeazzi 2015; Bernuzzi et al. 2015b; Bauswein and Stergioulas 2015; Dietrich et al. 2015b; Foucart et al. 2016; De Pietri et al. 2016; Rezzolla and Takami 2016; Maione et al. 2016; Bose et al. 2018).

Figure 10.5 provides a summarising view of some of the waveforms (i.e., of h_+ for sources at a distance of 50 Mpc) computed in this paper and that are combined with those of Takami et al. (2015) to offer a more comprehensive impression of the GW signal across different masses and EOSs. The figure is composed of 35 panels referring to the 35 equal-mass binaries with nuclear-physics EOSs that were simulated and that have a postmerger signal of at least 20 ms. Different rows refer to models with the same mass, while different columns select the five cold EOSs considered and colour-coded for convenience. It is then rather easy to see how small differences across the various EOSs during the inspiral become marked differences after the merger. In particular, it is straightforward to observe how the GW signal increases considerably in frequency after the merger and how low-mass binaries with stiff EOSs (e.g., top-left panel for the GNH3 EOS) show a qualitatively different behaviour from high-mass binaries with soft EOSs (e.g., bottom-right panel for the APR4 EOS). Also quite apparent is that, independently of the mass considered, the post-merger amplitude depends sensitively on the stiffness of the EOS, with stiff EOSs (e.g., GNH3) yielding systematically larger amplitudes than soft EOSs (e.g., APR4).

The first detailed description of a method for extracting information about the EOS of nuclear matter by carefully investigating the spectral properties of the post-merger signal was provided by Bauswein and Janka (2012), Bauswein et al. (2012).

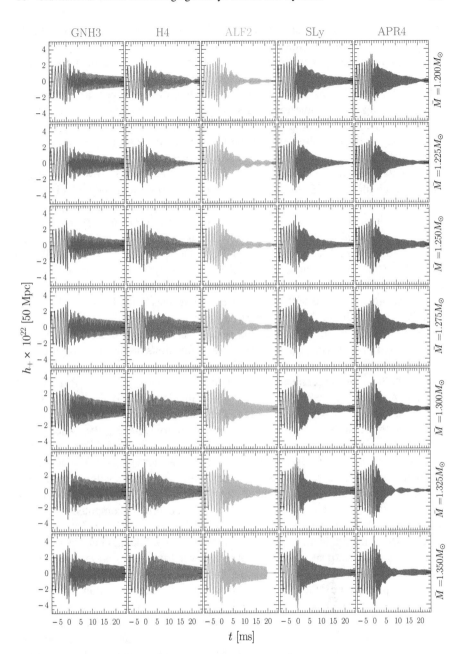

Fig. 10.5 Gravitational waveforms for some cold-EOS BNSs. The rows correspond to the gravitational masses, $\bar{M} = 1.200, 1.225, 1.250, 1.275, 1.300, 1.325, 1350\,M_\odot$, respectively, while each column refers to a given EOS. All models have formed long-lived HMNSs with $t > 20\,\text{ms}$ [Reprinted with permission from Rezzolla and Takami (2016). © (2016) by the American Physical Society]

After performing a large number of simulations using their conformally-flat SPH code, they pointed out that the largest peak in the PSD (whose frequency is dubbed there f_{peak}) correlates with the properties of the EOS, e.g., with the radius of the maximum-mass nonrotating star for the given EOS. The correlation found was rather tight, but this was partly due to the fact that their sample was restricted to binaries having all the same total mass (i.e., $2.7\,M_\odot$ in the specific case). It was shown that such a correlation can be used to gain information on the high-density EOS through GW measurements, if the masses of the neutron stars forming the binaries are known. Additionally, it was recognized that f_{peak} corresponds to a fundamental fluid mode of the HMNS with $\ell = 2 = m$ (Bauswein and Janka 2012; Stergioulas et al. 2011) and that the value of this frequency could also be used to set constraints on the maximum mass of the system and hence on the EOS (Bauswein et al. 2013, 2014). Subsequent analyses were performed by a number of groups with general-relativistic codes (Hotokezaka et al. 2013; Takami et al. 2014, 2015; Dietrich et al. 2015b; Foucart et al. 2016; De Pietri et al. 2016; Rezzolla and Takami 2016; Maione et al. 2016; Radice et al. 2017), which confirmed that the conformally flat approximation employed by Bauswein and collaborators provided a rather accurate estimate of the largest peak frequencies in the PSDs.

Takami et al. (2014, 2015) presented a more advanced method to use detected GWs for determining the EOS of matter in neutron stars. They used the results of a large number of accurate numerical-relativity simulations of binaries with different EOSs and different masses and identified two distinct and robust main spectral features in the post-merger phase. The first one is the largest peak in the PSD (whose frequency was called there f_2 and essentially coincides with the f_{peak} of Bauswein and Janka 2012; Bauswein et al. 2012). The functions describing the correlations of f_2 with the stellar properties (e.g., with the quantity $(\bar{M}/R_{\text{max}}^3)^{1/2}$, where R_{max} is the radius of the maximum-mass nonrotating star), which were first proposed by Bauswein and Janka (2012), are not universal, in the sense that different (linear) fits are necessary for describing the f_2-correlations for binaries with different total masses. This conclusion can be evinced by looking at Figs. 22–24 of Bauswein et al. (2012), but the different linear correlations were first explicitly computed by Takami et al. (2014, 2015) (see also Hotokezaka et al. 2013).

The second feature identified in all PSDs analysed by Takami et al. (2014, 2015) is the second-largest peak, which appears at lower frequencies and was called f_1 there. Clear indications were given about this low-frequency peak being related to the merger process (i.e., the first ≈ 3 ms after the merger). This was done by showing that the power in the peak is greatly diminished if the first few ms after the merger were removed from the waveform. Furthermore, a simple mechanical toy model was devised that can explain rather intuitively the main spectral features of the post-merger signal and therefore shed light on the physical interpretation of the origin of the various peaks. Despite its crudeness, the toy model was even able to reproduce the complex waveforms emitted right after the merger, hence possibly opening the way to an analytical modelling of a part of the signal (Takami et al. 2015).

More importantly, it was shown that the potential measurement of the f_1 frequency could reveal the EOS of the merging objects, since a correlation was found between the f_1-frequency and the average compactness of the two stars in the binary. Interestingly, this relation appears to be universal, that is, essentially valid for all EOSs and masses, and could therefore provide a powerful tool to set tight constraints on the EOS (Takami et al. 2014, 2015). Indeed, an analytic expression was suggested in Takami et al. (2015) to express the f_1 frequency via a third-order polynomial of the (average) stellar compactness, which reproduces reasonably well the numerical results. In addition to the correlations described above, Takami et al. (2014, 2015) also discussed additional correlations (24 in all), some of which had been already presented in the literature, e.g., in Read et al. (2013), Bernuzzi et al. (2014a), and some of which are presented there for the first time. Examples of these correlations are reported in Fig. 10.6, where different colours refer to different EOSs (see Fig. 1 of Takami et al. (2015) for a legend). The correlations refer to the f_{max}, f_1 and f_2 frequencies and the physical quantities of the binary system, e.g., the average compactness $\bar{M}/\bar{R}$, the average density $(\bar{M}/\bar{R}^3)^{1/2}$, the pseudo-average rest-mass density $(\bar{M}/R_{\mathrm{max}}^3)^{1/2}$, or the dimensionless tidal deformability tidal deformability $(\lambda/\bar{M}^5)^{1/5}$ (cf., also Fig. 15 of Takami et al. 2015). In confirmation of the accuracy of the computed frequencies, very similar values for the f_1 frequencies were also found by Dietrich et al. (2015b) in a distinct work aimed at determining the impact

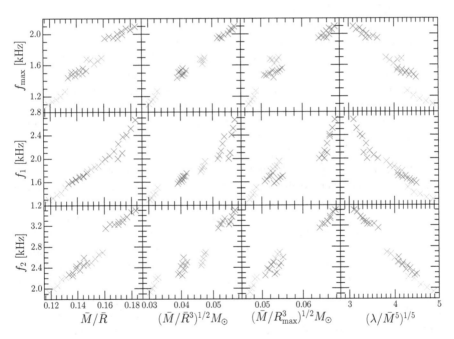

Fig. 10.6 Empirical correlations between the f_{max}, f_1 and f_2 frequencies and the physical quantities of the binary system, where different colours refer to different EOSs (see Fig. 1 of Takami et al. (2015) for a legend) [Reprinted with permission from Takami et al. (2015). © (2015) by the American Physical Society]

that conservative mesh-refinement techniques have on the accuracy of the post-merger dynamics.

Even though the toy model proposed by Takami et al. (2015) provides a simple and convincing explanation of the power associated to the f_1 frequency peak, alternative interpretations of the low-frequency part of the PSD have also been suggested. More specifically, Bauswein and Stergioulas (2015) claimed that the lower-frequency peak (i.e., the f_1 peak in Takami et al. 2014, 2015; Kastaun and Galeazzi 2015) is actually made of two separate peaks originating from different processes. One of these peaks is said to be produced by a nonlinear combination between the dominant quadrupolar oscillation (f_{peak} or f_2 in different notations) and the quasi-radial oscillation of the remnant and is named f_{2-0} (Stergioulas et al. 2011), while the other is said to be caused by a strong deformation initiated at the time of the merger, the pattern of which then rotates (in the inertial frame) more slowly than the inner cores of the remnant and lasts for a few rotational periods, while diminishing in amplitude. The GW emission associated with this motion then powers a peak that was named f_{spiral} in Bauswein and Stergioulas (2015). The connection between the f_{spiral} peak and the deformation was supported by showing that only PSDs computed from time intervals of the gravitational waveform that contain the deformation have the f_{spiral} peak. It was also claimed that the f_{spiral} peak can be roughly reproduced in a toy model, where two bulges orbit as point particles around the central double-core structure for a duration of few milliseconds, but no details were given in Bauswein and Stergioulas (2015).

In their analysis, Bauswein and Stergioulas (2015) also proposed an explanation for the low-frequency modulations seen in quantities like the lapse function at the stellar center, the maximum rest-mass density, and the separation between the two cores of the remnant. Such quantities are modulated according to the orientation of the antipodal bulges of the deformation with respect to the double central cores: the compactness is smaller, the central lapse function larger, and the GW amplitude maximal when the bulges and the cores are aligned, and viceversa.

Making use of a large set of simulations, Bauswein and Stergioulas (2015) were able to obtain empirical relations for both types of low-frequency peaks in terms of the compactness of nonrotating individual neutron stars. Different relations, however, were found for different sequences of constant total mass of the binary, in contrast with what found in Takami et al. (2014, 2015), where a different definition for the low-frequency peak was used. As discussed by Bauswein and Stergioulas (2015), the different behaviour could be due to the fact that the results of Takami et al. (2014, 2015) were based on a limited set of five EOSs of soft or moderate stiffness (with corresponding maximum masses of nonrotating neutron stars only up to $2.2\,M_\odot$), as well as on different chosen mass ranges for each EOS with a spread of only $0.2\,M_\odot$ in the total mass of the binary. In Bauswein and Stergioulas (2015), on the other hand, ten EOSs (including stiff EOSs with maximum masses reaching up to $2.8\,M_\odot$) and a larger mass range of 2.4–$3.0\,M_\odot$ were used. Overall, the differences between the results of the two groups are significant only for very low-mass neutron stars (i.e., $M = 1.2\,M_\odot$), which Takami et al. (2014, 2015) had not included in their sample because of the low statistical incidence they are thought to have (see also below).

One important consideration to bear in mind is that measuring the f_{spiral} frequencies through the motion of matter asymmetries via gauge-dependent quantities such as the rest-mass density is essentially impossible in genuine numerical-relativity calculations. This is because the spatial gauge conditions can easily distort the coordinate appearance of mass distributions and even the trajectories of the two stars during the inspiral (see Appendix A 2 of Baiotti et al. (2008) for some dramatic examples). In an attempt to clarify the different interpretations suggested in Takami et al. (2014, 2015), Bauswein and Stergioulas (2015) and to bring under a unified framework the spectral properties of the post-merger GW signal, Rezzolla and Takami (2016) have recently presented a comprehensive analysis of the GW signal emitted during the inspiral, merger and post-merger of 56 neutron-star binaries (Rezzolla and Takami 2016) (waveforms from this work are shown in Fig. 10.5). This sample of binaries, arguably the largest studied to date with realistic EOSs, spans across five different nuclear-physics EOSs and seven mass values, including the very low-mass binaries (e.g., with individual neutron-star masses of $1.2\,M_\odot$) that were suggested by Bauswein and Stergioulas (2015) to be lacking in the previous analysis of Takami et al. (2015). After a systematic analysis of the complete sample, it was possible to sharpen a number of arguments on the spectral properties of the post-merger GW signal. Overall it was found that: (1) for binaries with individual stellar masses differing no more than 20%, the frequency at the maximum of the GW amplitude is related quasi-universally with the tidal deformability of the two stars; (2) the spectral properties vary during the post-merger phase, with a transient phase lasting a few milliseconds after the merger and followed by a quasi-stationary phase; (3) when distinguishing the spectral peaks between these two phases, a number of ambiguities in the identification of the peaks disappear, leaving a simple and robust picture; (4) using properly identified frequencies, quasi-universal relations are found between the spectral features and the properties of the neutron stars; (5) for the most salient peaks analytic fitting functions can be obtained in terms of the stellar tidal deformability or compactness. Overall, the analysis of Rezzolla and Takami (2016) supports the idea that the EOS of nuclear matter can be constrained tightly when a signal in GWs from BNSs is detected.

An interesting extension of the work of Takami et al. (2014, 2015) was suggested by Bernuzzi et al. (2015b), who expressed the correlation between the peak frequencies f_2 with the tidal coupling constant κ_2^T instead of the tidal deformability parameter Λ, as done in Takami et al. (2014, 2015). As found in previous works by Bernuzzi et al. (2014a, 2015a) (see Sect. 10.3), the dimensionless GW frequency depends on the stellar EOS, binary mass, and mass ratio only through the tidal coupling constants κ_2^T and thus this is a better choice of parameter, also because it can be extended more straightforwardly to the case of unequal-mass binaries. The relation $f_2(\kappa_2^T)$ was found in Bernuzzi et al. (2015b) to be very weakly dependent on the binary total mass, mass ratio, EOS, and thermal effects (through the ideal-fluid index Γ_{th}). Relevant dependence on the stellar spins was instead found. This is shown in Fig. 10.7, which reports the dimensionless frequency $M f_2$ as a function

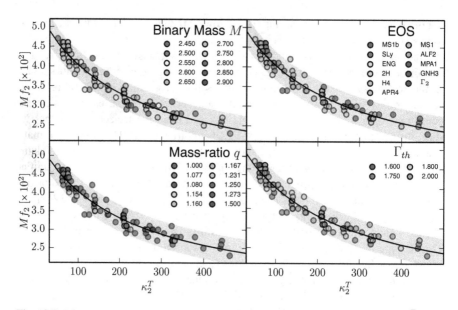

Fig. 10.7 Dimensionless frequency (Mf_2) as a function of the tidal coupling constant κ_2^T. Each panel shows the same data set; the colour code in each panel indicates the different values of binary mass (top left), EOS (top right), mass-ratio (bottom left), and Γ_{th} (bottom right). The black solid line is the fit, while the grey area marks the 95% confidence interval [Reprinted with permission from Bernuzzi et al. (2015b). © (2015) by the American Physical Society]

of the tidal coupling constant κ_2^T. Each panel shows the same data set; the colour code in each panel indicates the different values of binary mass (top left), EOS (top right), mass-ratio (bottom left), and Γ_{th} (bottom right). The black solid line is the fit obtained by Bernuzzi et al. (2015b), while the grey area marks the 95% confidence range.

Although not explicitly stated, all of the considerations made so far about the spectral properties of the post-merger signal refer to binaries that are initially irrotational. It is therefore natural to ask what changes, both qualitatively and quantitatively, when spinning binaries are considered. This was done in part by Bernuzzi et al. (2014b) and by Kastaun et al. (2013) and Kastaun and Galeazzi (2015). The first work considered in particular whether the main-peak frequency f_2 is influenced by the initial state of rotation and found that this is indeed the case at least for very rapidly rotating neutron stars, suggesting that spin effects may be more important than those found in Bauswein and Janka (2012). Kastaun and Galeazzi (2015), on the other hand, analysed the spectral changes induced by the initial spin on high-mass binaries and showed that the direct influence of the spin on the frequency f_2 is weak and comparable to the width of the corresponding peak. They also studied in detail the Fourier decomposition of the rest-mass density of the binary-merger product and its rotational profile, which is important for determining its lifetime, especially in view of the amplification of the magnetic field. A problem that needed to be tackled in their analysis is that of potential gauge artefacts. We recall, in fact, that rest-mass density distributions are gauge-dependent quantities and even when the system approaches an axisymmetric state after the merger, the spatial coordinates may not reflect this, because the gauge conditions employed in the evolution introduce local and global deformations. In order to exclude such systematic gauge effects, Kastaun and Galeazzi (2015) introduced a different coordinate system, used just for post-processing. In this new coordinate system, they found that, the Fourier decomposition is far more regular than in the coordinate system normally used in the evolution. Furthermore, they showed that, contrary to common assumptions, the law of differential rotation of the binary-merger product consists of a slowly rotating core with an extended and massive envelope rotating close to Keplerian velocity (see discussion in Hanauske et al. 2016). The latter result has been confirmed recently also for the binary-merger product produced by the merger of unequal-mass magnetised binaries (Endrizzi et al. 2016).

A rather different approach to analysing post-merger waveforms has been taken by Chatziioannou et al. (2017). In order to obtain results that rely less on numerical-relativity simulations, they use a morphology-independent Bayesian data analysis algorithm, BAYESWAVE (Cornish and Littenberg 2015; Littenberg and Cornish 2015), to reconstruct as a sum of wavelets injected post-merger GW signals that are assumed to have been measured (in the absence of observational data, data based on simulations were used). It was found that BAYESWAVE is capable of reconstructing the dominant features of the injected signal, in particular the dominant post-merger frequency, with an overlap between injected and reconstructed signals of above 90% for post-merger SNRs above 5. This allows f_2 to be measured at the 90% credible level with an error of about 36 (27) [45] Hz for a stiff (moderate) [soft] EOSs and so to set bounds on the NS radius obtained by the post-merger signal of order 100 m for a signal emitted at 20 Mpc. This accuracy is similar to that predicted by the other methods discussed above, which are completely based on numerical simulations. Actually also Chatziioannou et al. (2017) had to use empirical formulas from numerical simulations (Bauswein and Janka 2012; Bauswein et al. 2012, 2016) to relate f_2 to the radius and they indeed found that their error on the radius is dominated by the systematic uncertainty (scatter) in such a formula, rather than the statistical error of the reconstruction.

Before concluding this discussion on the post-merger GW signal we shall also make some additional important remarks.

- First, GW measurements at the expected frequencies and amplitudes are very difficult, namely limited to sources within ∼20 Mpc. This number can be easily estimated with back-of-the-envelope calculations, but it was confirmed through detailed analysis of the detectability of the dominant oscillation frequency in Clark et al. (2014), Bose et al. (2018), Yang et al. (2018) via large-scale Monte Carlo studies in which simulated post-merger GW signals are injected into realistic detector data that matches the design goals of Advanced LIGO and Advanced Virgo.

- Second, the post-merger frequencies evolve in time, albeit only slightly. Hence, the spectral properties of the GW signal can be asserted reliably only when the signal-to-noise ratio is sufficiently strong so that even these changes in time can be measured in the evolution of the PSDs (Kiuchi et al. 2012b; Hotokezaka et al. 2013; Takami et al. 2014; Shibata et al. 2014). In light of these considerations, the prospects for high-frequency searches for the post-merger signal are limited to rare nearby events. Yet, if such detections happen, the error in the estimate of the neutron star radius will be of the order of a few hundred metres (Clark et al. 2014, 2016; Bose et al. 2018; Yang et al. 2018).

- Third, viscous dissipation and energy transport in the post-merger phase may considerably affect gravitational waveforms. It has been recently pointed out that shear viscosity and thermal conductivity are not likely to play a major role in post-merger dynamics unless neutrino trapping occurs, which requires temperatures $T \gtrsim 10\,\mathrm{MeV}$, or flows that experience shear over short distances of the order of 0.01 km (Alford et al. 2018). By using the most likely values of the parameters describing shear viscosity other works had already estimated that the post-merger GW amplitude is affected, but its frequency peaks are not (Radice 2017; Shibata and Kiuchi 2017). On the other hand, bulk viscous dissipation could provide significant damping of the high-amplitude density oscillations observed right after merger, if modified-Urca processes (and not direct-Urca processes; see e.g. Lattimer et al. 1991) are those that establish flavor equilibrium (Alford et al. 2018). Hence, viscous dissipative processes deserve more careful investigation since they may well affect the spectral properties of the post-merger gravitational-wave signal, especially the f_1 and f_3 peaks that are produced right after the merger and that are dissipated rapidly (Takami et al. 2015). In addition, if viscous dissipation is active after the merger, it will also heat the merger product, possibly stabilising it on longer timescales via the extra thermal pressure (Baiotti et al. 2008; Sekiguchi et al. 2011b; Paschalidis et al. 2012; Kaplan et al. 2014). Finally, future gravitational-wave observations may also give indications about the fraction of merger material in which direct or modified Urca processes are dominant.

At the end of this long Section, we summarise the main finding on the spectral properties of the post-merger signal as follows:

- The most powerful methods to connect observations of the post-merger GW with the EOS of those neutron stars is based on analyses of the GW PSD.
- In general, there are three main peaks in the PSDs of the post-merger phase of binary mergers that do not result in a prompt collapse to a black hole. The frequencies of these peaks are named f_1, f_2, f_3 in Takami et al. (2014, 2015), Rezzolla and Takami (2016), while other works use different symbols, in particular the frequency of the highest peak, f_2, is referred to as f_{peak} in e.g., Bauswein and Janka (2012), Bauswein and Stergioulas (2015), Bauswein et al. (2016). The frequencies were found to roughly follow the relation $f_2 \simeq (f_1 + f_3)/2$.

- The f_2 frequencies correspond to the $\ell = 2 = m$ fundamental mode of the HMNS and hence are equal to twice the rotation frequency of the bar deformation of the HMNS. Their values change slightly in time (by $\sim$5%). The f_1 and f_3 frequencies are produced only in the first few milliseconds after the merger. A simple toy model was proposed in Takami et al. (2015) to explain their origin. See also Bauswein and Stergioulas (2015) for another model.
- The frequencies f_2 and f_1 were found to correlate well with properties of the stars in the BNS system. In particular, f_2 correlates well with the quantity $(\bar{M}/R_{\mathrm{max}}^3)^{1/2}$ for a given total mass of the BNS (Bauswein and Janka 2012), while f_1 correlates well with the average compactness of the two stars in the binary and such a relation seems valid for any total mass of the BNS, prompting Takami et al. (2014, 2015) to call the relation *universal*.
- Such correlations between post-merger frequencies and stellar properties can be used to estimate rather accurately the radius of the neutron stars and so to infer information on their EOS. There are, however, caveats against a simplistic use of post-merger frequencies, since they may vary in time, be affected by bulk viscosity (Alford et al. 2018) and by the spins of the stars in the binary (if very high) (Bernuzzi et al. 2014b), and since the post-merger GW signals have a small SNR in current detectors and therefore the chances of a detection are very small.

10.4.3 Spectral Properties and the Mass-Redshift Degeneracy

Besides providing information on the EOS, the spectral properties of the gravitational-wave post-merger signal can also be used in a completely different manner, namely, to remove the degeneracy in the determination of redshift and mass for cosmological investigations. Indeed, a well-known problem of the detection of gravitational waves from compact-object binaries at cosmological distances is the so-called "mass-redshift degeneracy". More precisely, given a source of (gravitational) mass M at a cosmological redshift z, a direct gravitational-wave observation provides information only on the combined quantity $M(1 + z)$, so that it is not possible to have an independent measurement of M and of z. The standard solution to this problem is to detect an electromagnetic counterpart to the gravitational-wave signal, so as to measure z and hence the mass M. However, this may be not easy in some cases.

In a recent investigation, Messenger et al. (2014) described how this degeneracy can be broken when exploiting information on the spectral properties of the post-merger gravitational-wave signal. More specifically, making use of numerically generated BNS waveforms, it was shown that it is possible to construct frequency-domain power-spectrum reference templates that capture the evolution of two of the primary spectral features in the post-merger stage of the waveforms as a function of the total gravitational mass.

This is summarised via a cartoon in Fig. 10.8, which shows how the information on the redshifted mass as a function of the redshift (blue stripe) can be correlated

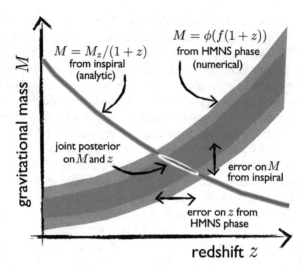

Fig. 10.8 A cartoon illustrating how the mass-redshift degeneracy is broken through the use of information from the inspiral and HMNS stages of a BNS merger event. Cross-correlating the information on the redshifted mass as a function of the redshift (blue stripe) with the information from the spectral properties of the HMNS phase (red stripe) will provide a localised range in mass and redshift, breaking the degeneracy [Reprinted with permission from Messenger et al. (2014). © (2014) by the American Physical Society]

with complementary information from the spectral properties of the HMNS phase. The overlap will provide a localised range in mass and redshift, breaking the degeneracy. A Bayesian inference method was then used to test the ability of the Einstein Telescope (Punturo et al. 2010b) to measure the characteristic frequencies in the post-merger stage of the signal, finding that redshift and gravitational mass can be determined separately, with uncertainties in the redshift of sources at $z = 0.01$–0.04 of 10–20% and in the gravitational mass of $<1\%$ in all cases.

10.5 Conclusions

As anticipated in the Introduction, there is little doubt that this is a particularly exciting and highly dynamical time for research on neutron stars, in general, and on BNS mergers, in particular. In about 10 years, i.e., starting approximately from 2008, a considerable effort by several groups across the world has obtained numerous important results about the dynamics of binary systems of neutron stars, employing a large variety of numerical (in most cases) and analytical (in a few cases) techniques and exploring this process with different degrees of approximation and realism.

Altogether, these works have revealed that the merger of a binary system of neutron stars is a marvellous physical laboratory. Indeed, BNS mergers are expected to be behind several fascinating physical processes, which we recall here: (1) they are significant sources of gravitational radiation; (2) they act as possible progenitors for short-gamma-ray bursts (SGRBs); (3) they have the potential to produce electromagnetic and neutrino emission that is visible from enormous distances; (4) they are likely responsible for the production of a good portion of the very heavy elements in the Universe. When viewed across this lens, it is quite natural to consider BNS mergers as Einstein's richest laboratory, binding in the same environment highly nonlinear gravitational dynamics with complex microphysical processes and astonishing astrophysical phenomena.

The huge progress accomplished over the last 10 years has helped trace a broadbrush picture of BNS mergers that has several sound aspects, among which the most robust in our opinion are the following ones[6]:

- Independently of the fine details of the EOS, of the mass ratio or of the presence of magnetic fields, the merger of a binary system of neutron stars eventually leads to a rapidly rotating black hole with dimensionless spin $J/M^2 \simeq$ 0.7–0.8 surrounded by a hot accretion torus with mass in the range $M_{torus} \sim$ 0.001–0.1 $M_\odot$. Only very low-mass progenitors whose total mass is below the maximum mass of a (nonrotating) neutron star would not produce a black hole. It is unclear whether such progenitors are statistically important.
- The complete GW signal from inspiralling and merging BNSs can be computed numerically with precision that is smaller but overall comparable with that available for black holes.
- When considering the inspiral-only part of the GW signal, semi-analytical approximations either in the post-Newtonian or EOB approximation, can reproduce the results of numerical-relativity calculations essentially up to the merger.
- The GW spectrum is marked by precise frequencies, either during the inspiral or after the merger that exhibit a "quasi-universal" behaviour. In other words, while the position of the peaks depends on the EOS, it can be easily factored out to obtain EOS-independent relations between the frequencies of the peaks and the properties of the progenitor stars.
- The result of the merger, i.e., the binary-merger product, is a highly massive and differentially rotating neutron star. The lifetime of the binary-merger product depends on a number of factors, including the mass of the progenitors, their mass ratio and EOS, as well as the role played by magnetic fields and neutrino losses. While sufficiently large initial masses can yield a prompt collapse at the merger, for smaller masses the object resulting from the merger can be a neutron star for at least some milliseconds or maybe forever.

Note that many of the aspects listed above are robust but have been addressed mostly at a rather qualitative level, with precisions that range from "a-factor-of-a-few" up to "order-of-magnitude" estimates. Furthermore, these results can be

---

[6]In this Section we will intentionally omit references to avoid cluttering the text; all the relevant references can be found in the various Sections covering the topics discussed here.

seen as the low-hanging fruits of a tree that still has a number of results to offer, although these will require an equal, if not larger, investment of effort, microphysical and numerical developments, and, of course, of computer time. In conclusion, if GW170817 and the first direct detection of the GW signals from binary systems of neutron stars has officially given birth to the era of multimessenger astronomy, the huge advances that are expected to come in the next few years on the physics and astrophysics of BNSs will help lift many of the veils that still cover Einstein's richest laboratory.

Acknowledgements It is a pleasure to thank Kentaro Takami and the relativistic-astrophysics group in Frankfurt for their help and input in preparing this Chapter.

Partial support has come from "NewCompStar", COST Action MP1304, from the LOEWE-Program in HIC for FAIR, the European Union's Horizon 2020 Research and Innovation Programme under grant agreement No. 671698 (call FETHPC-1–2014, project ExaHyPE), from the ERC Synergy Grant "BlackHoleCam - Imaging the Event Horizon of Black Holes" (Grant 610058), from JSPS Grant-in-Aid for Scientific Research(C) No. 18K036220, from the International Centre for Theoretical Sciences Code: ICTS/Prog-GWS/2017/07, and from the Excellence Initiative at Radboud University Nijmegen.

References

Abadie, J., et al.: Class. Quantum Grav. **27**, 173001 (2010). arXiv:1003.2480 [astro-ph.HE]

Abbott, B.P., Abbott, R., Abbott, T.D., Abernathy, M.R., Acernese, F., Ackley, K., Adams, C., Adams, T., Addesso, P., Adhikari, R.X., et al.: Phys. Rev. Lett. **116**, 241103 (2016). arXiv:1606.04855 [gr-qc]

Abbott, B.P., et al. (Virgo, LIGO Scientific): Phys. Rev. Lett. **119**, 161101 (2017). arXiv:1710.05832 [gr-qc]

Abbott, B.P., et al. (Virgo, LIGO Scientific): (2018a). arXiv:1805.11581

Abbott, B.P., et al. (Virgo, LIGO Scientific): (2018b). arXiv:1805.11579

Accadia, T., et al.: Class. Quantum Grav. **28**, 114002 (2011)

Advanced LIGO Anticipated Sensitivity Curves, LIGO Document No. T0900288-v3 (2009)

Agathos, M., Meidam, J., Del Pozzo, W., Li, T.G.F., Tompitak, M., Veitch, J., Vitale, S., Van Den Broeck, C.: Phys. Rev. D **92**, 023012 (2015). arXiv:1503.05405 [gr-qc]

Ajith, P., et al.: Class. Quantum Grav. **24**, S689 (2007). arXiv:0704.3764

Ajith, P., et al.: Phys. Re. D **77**, 104017 (2008). arXiv:0710.2335

Akmal, A., Pandharipande, V.R., Ravenhall, D.G.: Phys. Rev. C **58**, 1804 (1998). arXiv:hep-ph/9804388

Alcock, C., Farhi, E., Olinto, A.: Astrophys. J. **310**, 261 (1986)

Alford, M., Braby, M., Paris, M., Reddy, S.: Astrophys. J. **629**, 969 (2005). nucl-th/0411016

Alford, M.G., Bovard, L., Hanauske, M., Rezzolla, L., Schwenzer, K.: Phys. Rev. Lett. **120**, 041101 (2018). arXiv:1707.09475 [gr-qc]

AlGendy, M., Morsink, S.M.: Astrophys. J. **791**, 78 (2014). arXiv:1404.0609 [astro-ph.HE]

Anderson, M., Hirschmann, E.W., Lehner, L., Liebling, S.L., Motl, P.M., Neilsen, D., Palenzuela, C., Tohline, J.E.: Phys. Rev. Lett. **100**, 191101 (2008). arXiv:0801.4387 [gr-qc]

Andersson, N., Kokkotas, K.D.: Mon. Not. R. Astron. Soc. **299**, 1059 (1998). arXiv:gr-qc/9711088

Andersson, N., Ferrari, V., Jones, D.I., Kokkotas, K.D., Krishnan, B., Read, J.S., Rezzolla, L., Zink, B.: Gen. Relativ. Gravit. **43**, 409 (2011). arXiv:0912.0384 [astro-ph.SR]

Antoniadis, J., Freire, P.C.C., Wex, N., Tauris, T.M., Lynch, R.S., van Kerkwijk, M.H., Kramer, M., Bassa, C., Dhillon, V.S., Driebe, T., Hessels, J.W.T., Kaspi, V.M., Kondratiev, V.I., Langer,

N., Marsh, T.R., McLaughlin, M.A., Pennucci, T.T., Ransom, S.M., Stairs, I.H., van Leeuwen, J., Verbiest, J.P.W., Whelan, D.G.: Science **340**, 448 (2013). arXiv:1304.6875 [astro-ph.HE]

Aso, Y., Michimura, Y., Somiya, K., Ando, M., Miyakawa, O., Sekiguchi, T., Tatsumi, D., Yamamoto, H.: Phys. Rev. D **88**, 043007 (2013). arXiv:1306.6747 [gr-qc]

Babak, S., Taracchini, A., Buonanno, A.: Phys. Rev. **D95**, 024010 (2017). arXiv:1607.05661 [gr-qc]

Baiotti, L., Rezzolla, L.: Rep. Prog. Phys. **80**, 096901 (2017). arXiv:1607.03540 [gr-qc]

Baiotti, L., de Pietri, R., Manca, G.M., Rezzolla, L.: Phys. Rev. D **75**, 044023 (2007). astro-ph/0609473

Baiotti, L., Giacomazzo, B., Rezzolla, L.: Phys. Rev. D **78**, 084033 (2008). arXiv:0804.0594 [gr-qc]

Baiotti, L., Damour, T., Giacomazzo, B., Nagar, A., Rezzolla, L.: Phys. Rev. Lett. **105**, 261101 (2010). arXiv:1009.0521 [gr-qc]

Balbinski, E.: Mon. Not. R. Astron. Soc. **216**, 897 (1985)

Banihashemi, B., Vines, J. (2018). arXiv:1805.07266

Banik, S., Hempel, M., Bandyopadhyay, D.: Astrohys. J. Suppl. **214**, 22 (2014). arXiv:1404.6173 [astro-ph.HE]

Bauböck, M., Berti, E., Psaltis, D., Özel, F.: Astrophys. J. **777**, 68 (2013). arXiv:1306.0569 [astro-ph.HE]

Bauswein, A., Janka, H.-T.: Phys. Rev. Lett. **108**, 011101 (2012). arXiv:1106.1616 [astro-ph.SR]

Bauswein, A., Stergioulas, N.: Phys. Rev. D **91**, 124056 (2015). arXiv:1502.03176 [astro-ph.SR]

Bauswein, A., Janka, H.-T., Oechslin, R., Pagliara, G., Sagert, I., Schaffner-Bielich, J., Hohle, M.M., Neuhäuser, R.: Phys. Rev. Lett. **103**, 011101 (2009). arXiv:0812.4248

Bauswein, A., Oechslin, R., Janka, H.-T.: Phys. Rev. D **81**, 024012 (2010). arXiv:0910.5169 [astro-ph.SR]

Bauswein, A., Janka, H.-T., Hebeler, K., Schwenk, A.: Phys. Rev. D **86**, 063001 (2012). arXiv:1204.1888 [astro-ph.SR]

Bauswein, A., Baumgarte, T.W., Janka, H.-T.: Phys. Rev. Lett. **111**, 131101 (2013). arXiv:1307.5191 [astro-ph.SR]

Bauswein, A., Stergioulas, N., Janka, H.-T.: Phys. Rev. D **90**, 023002 (2014). arXiv:1403.5301 [astro-ph.SR]

Bauswein, A., Stergioulas, N., Janka, H.-T.: Eur. Phys. J. A **52**, 56 (2016). arXiv:1508.05493 [astro-ph.HE]

Bejger, M., Haensel, P.: Astron. Astrophys. **396**, 917 (2002). arXiv:astro-ph/0209151

Benhar, O., Ferrari, V., Gualtieri, L.: Phys. Rev. D **70**, 124015 (2004). arXiv:astro-ph/0407529

Bernuzzi, S., Nagar, A., Thierfelder, M., Brügmann, B.: Phys. Rev. D **86**, 044030 (2012). arXiv:1205.3403 [gr-qc]

Bernuzzi, S., Nagar, A., Balmelli, S., Dietrich, T., Ujevic, M.: Phys. Rev. Lett. **112**, 201101 (2014a). arXiv:1402.6244 [gr-qc]

Bernuzzi, S., Dietrich, T., Tichy, W., Brügmann, B.: Phys. Rev. D **89**, 104021 (2014b). arXiv:1311.4443 [gr-qc]

Bernuzzi, S., Nagar, A., Dietrich, T., Damour, T.: Phys. Rev. Lett. **114**, 161103 (2015a). arXiv:1412.4553 [gr-qc]

Bernuzzi, S., Dietrich, T., Nagar, A.: Phys. Rev. Lett. **115**, 091101 (2015b). arXiv:1504.01764 [gr-qc]

Bernuzzi, S., Radice, D., Ott, C.D., Roberts, L.F., Moesta, P., Galeazzi, F.: Phys. Rev. D **94**, 024023 (2016). arXiv:1512.06397 [gr-qc]

Berti, E., Iyer, S., Will, C.M.: Phys. Rev. D **77**, 024019 (2008). arXiv:0709.2589

Bhowmick, B., Bhattacharya, M., Bhattacharyya, A., Gangopadhyay, G.: Phys. Rev. **C89**, 065806 (2014). arXiv:1403.0341 [nucl-th]

Bildsten, L., Cutler, C.: Astrophys. J. **400**, 175 (1992)

Bini, D., Damour, T.: Phys. Rev. D **90**, 124037 (2014). arXiv:1409.6933 [gr-qc]

Bini, D., Geralico, A.: Phys. Rev. **D91**, 084012 (2015)

Bini, D., Damour, T., Faye, G.: Phys. Rev. **D85**, 124034 (2012). arXiv:1202.3565 [gr-qc]

Binnington, T., Poisson, E.: Phys. Rev. D **80**, 084018 (2009). arXiv:0906.1366 [gr-qc]

Blanchet, L.: Living Rev. Relativ. **9**, 4 (2006)

Bodo, G., Massaglia, S., Ferrari, A., Trussoni, E.: Astron. Astrophys. **283**, 655 (1994)

Bohe, A., et al.: Phys. Rev. **D95**, 044028 (2017). arXiv:1611.03703 [gr-qc]

Bose, S., Chakravarti, K., Rezzolla, L., Sathyaprakash, B.S., Takami, K.: Phys. Rev. Lett. **120**, 031102 (2018). arXiv:1705.10850 [gr-qc]

Bovard, L., Martin, D., Guercilena, F., Arcones, A., Rezzolla, L., Korobkin, O.: Phys. Rev. D **96**, 124005 (2017). arXiv:1709.09630 [gr-qc]

Breu, C., Rezzolla, L.: Mon. Not. R. Astron. Soc. **459**, 646 (2016). arXiv:1601.06083 [gr-qc]

Buonanno, A., Damour, T.: Phys. Rev. D **59**, 084006 (1999). gr-qc/9811091

Buonanno, A., Damour, T.: Phys. Rev. D **62**, 064015 (2000). gr-qc/0001013

Buonanno, A., Iyer, B., Ochsner, E., Pan, Y., Sathyaprakash, B.S.: Phys. Rev. **D80**, 084043 (2009). arXiv:0907.0700 [gr-qc]

Camarda, K.D., Anninos, P., Fragile, P.C., Font, J.A.: Astrophys. J. **707**, 1610 (2009). arXiv:0911.0670 [astro-ph.SR]

Cardoso, V., Franzin, E., Maselli, A., Pani, P., Raposo, G.: Phys. Rev. **D95**, 084014 (2017) [Addendum: Phys. Rev. **D95**(8), 089901 (2017)]. arXiv:1701.01116 [gr-qc]

Carriere, J., Horowitz, C.J., Piekarewicz, J.: Astrophys. J. **593**, 463 (2003). nucl-th/0211015

Centrella, J.M., New, K.C.B., Lowe, L.L., Brown, J.D.: Astrophys. J. **550**, L193 (2001). astro-ph/0010574

Chakrabarti, S., Delsate, T., Steinhoff, J.: Phys. Rev. **D88**, 084038 (2013a). arXiv:1306.5820 [gr-qc]

Chakrabarti, S., Delsate, T., Steinhoff, J. (2013b). arXiv:1304.2228

Chakrabarti, S., Delsate, T., Gürlebeck, N., Steinhoff, J.: Phys. Rev. Lett. **112**, 201102 (2014). arXiv:1311.6509 [gr-qc]

Chan, T.K., Chan, A.P.O., Leung, P.T.: Phys. Rev. **D91**, 044017 (2015). arXiv:1411.7141 [astro-ph.SR]

Chan, T.K., Chan, A.P.O., Leung, P.T.: Phys. Rev. **D93**, 024033 (2016). arXiv:1511.08566 [gr-qc]

Chandrasekhar, S.: Hydrodynamic and Hydromagnetic Stability. Dover Edition, New York (1981)

Chatziioannou, K., Yagi, K., Klein, A., Cornish, N., Yunes, N.: Phys. Rev. **D92**, 104008 (2015). arXiv:1508.02062 [gr-qc]

Chatziioannou, K., Clark, J.A., Bauswein, A., Millhouse, M., Littenberg, T.B., Cornish, N.: Phys. Rev. D **96**, 124035 (2017). arXiv:1711.00040 [gr-qc]

Chirenti, C., de Souza, G.H., Kastaun, W.: Phys. Rev. **D91**, 044034 (2015). arXiv:1501.02970 [gr-qc]

Chirenti, C., Gold, R., Miller, M.C.: Astrophys. J. **837**, 67 (2017). arXiv:1612.07097 [astro-ph.HE]

Ciolfi, R., Siegel, D.M.: Astrophys. J. **798**, L36 (2015). arXiv:1411.2015 [astro-ph.HE]

Clark, J., Bauswein, A., Cadonati, L., Janka, H.-T., Pankow, C., Stergioulas, N.: Phys. Rev. D **90**, 062004 (2014). arXiv:1406.5444 [astro-ph.HE]

Clark, J.A., Bauswein, A., Stergioulas, N., Shoemaker, D.: Class. Quantum Grav. **33**, 085003 (2016). arXiv:1509.08522 [astro-ph.HE]

Cornish, N.J., Littenberg, T.B.: Class. Quantum Grav. **32**, 135012 (2015). arXiv:1410.3835 [gr-qc]

Corvino, G., Rezzolla, L., Bernuzzi, S., De Pietri, R., Giacomazzo, B.: Class. Quantum Grav. **27**, 114104 (2010). arXiv:1001.5281 [gr-qc]

Cutler, C., Flanagan, É.E.: Phys. Rev. D **49**, 2658 (1994). gr-qc/9402014

Cutler, C., Apostolatos, T.A., Bildsten, L., Finn, L.S., Flanagan, E.E., Kennefick, D., Markovic, D.M., Ori, A., Poisson, E., Sussman, G.J., Thorne, K.S.: Phys. Rev. Lett. **70**, 2984 (1992)

Damour, T., Lecian, O.M.: Phys. Rev. D **80**, 044017 (2009). arXiv:0906.3003 [gr-qc]

Damour, T., Nagar, A.: Phys. Rev. D **80**, 084035 (2009). arXiv:0906.0096 [gr-qc]

Damour, T., Nagar, A.: Phys. Rev. D **81**, 084016 (2010). arXiv:0911.5041 [gr-qc]

Damour, T., Soffel, M., Xu, C.: Phys. Rev. D **45**, 1017 (1992)

Damour, T., Nagar, A., Villain, L.: Phys. Rev. D **85**, 123007 (2012)

Damour, T., Jaranowski, P., Schäfer, G.: Phys. Rev. **D93**, 084014 (2016). arXiv:1601.01283 [gr-qc]

De Pietri, R., Feo, A., Maione, F., Löffler, F.: Phys. Rev. D **93**, 064047 (2016). arXiv:1509.08804 [gr-qc]

Del Pozzo, W., Li, T.G.F., Agathos, M., Van Den Broeck, C., Vitale, S.: Phys. Rev. Lett. **111**, 071101 (2013). arXiv:1307.8338 [gr-qc]

Demorest, P.B., Pennucci, T., Ransom, S.M., Roberts, M.S.E., Hessels, J.W.T.: Nature **467**, 1081 (2010). arXiv:1010.5788 [astro-ph.HE]

Dessart, L., Ott, C.D., Burrows, A., Rosswog, S., Livne, E.: Astrophys. J. **690**, 1681 (2009). arXiv:0806.4380

Detweiler, S., Lindblom, L.: Astrophys. J. **292**, 12 (1985)

Dietrich, T., Hinderer, T.: Phys. Rev. **D95**, 124006 (2017). arXiv:1702.02053 [gr-qc]

Dietrich, T., Moldenhauer, N., Johnson-McDaniel, N.K., Bernuzzi, S., Markakis, C.M., Brügmann, B., Tichy, W.: Phys. Rev. D **92**, 124007 (2015a). arXiv:1507.07100 [gr-qc]

Dietrich, T., Bernuzzi, S., Ujevic, M., Brügmann, B.: Phys. Rev. D **91**, 124041 (2015b). arXiv:1504.01266 [gr-qc]

Dietrich, T., Bernuzzi, S., Tichy, W. (2017). arXiv:1706.02969 [gr-qc]

Dietrich, T., et al. (2018). arXiv:1804.02235

Dixon, W.G.: Proc. R. Soc. Lond. **A314**, 499 (1970)

Dolan, S.R., Nolan, P., Ottewill, A.C., Warburton, N., Wardell, B.: Phys. Rev. D **91**, 023009 (2015). arXiv:1406.4890 [gr-qc]

Doneva, D.D., Yazadjiev, S.S., Stergioulas, N., Kokkotas, K.D.: Astrophys. J. Lett. **781**, L6 (2014). arXiv:1310.7436 [gr-qc]

Doneva, D.D., Kokkotas, K.D., Pnigouras, P.: Phys. Rev. D **92**, 104040 (2015). arXiv:1510.00673 [gr-qc]

East, W.E., Paschalidis, V., Pretorius, F., Shapiro, S.L.: Phys. Rev. D **93**, 024011 (2016). arXiv:1511.01093 [astro-ph.HE]

Endrizzi, A., Ciolfi, R., Giacomazzo, B., Kastaun, W., Kawamura, T. (2016). arXiv:1604.03445 [astro-ph.HE]

Essick, R., Vitale, S., Weinberg, N.N.: Phys. Rev. **D94**, 103012 (2016). arXiv:1609.06362 [astro-ph.HE]

Fairhurst, S.: J. Phys. Conf. Ser. **484**, 012007 (2014). arXiv:1205.6611 [gr-qc]

Farhi, E., Jaffe, R.L.: Phys. Rev. D **30**, 2379 (1984)

Fattoyev, F.J., Carvajal, J., Newton, W.G., Li, B.-A.: Phys. Rev. **C87**, 015806 (2013). arXiv:1210.3402 [nucl-th]

Fattoyev, F.J., Newton, W.G., Li, B.-A.: Eur. Phys. J. **A50**, 45 (2014). arXiv:1309.5153 [nucl-th]

Favata, M.: Phys. Rev. Lett. **112**, 101101 (2014). arXiv:1310.8288 [gr-qc]

Fernández, R., Foucart, F., Kasen, D., Lippuner, J., Desai, D., Roberts, L.F.: Class. Quant. Grav. **34**, 154001 (2017). arXiv:1612.04829 [astro-ph.HE]

Ferrari, V., Gualtieri, L., Pannarale, F.: Phys. Rev. D **81**, 064026 (2010). arXiv:0912.3692 [gr-qc]

Ferrari, V., Gualtieri, L., Maselli, A.: Phys. Rev. **D85**, 044045 (2012). arXiv:1111.6607 [gr-qc]

Flanagan, E.E.: Phys. Rev. D **58**, 124030 (1998)

Flanagan, É.É., Hinderer, T.: Phys. Rev. D **77**, 021502 (2008). arXiv:0709.1915

Flanagan, E.E., Racine, E.: Phys. Rev. **D75**, 044001 (2007). arXiv:gr-qc/0601029 [gr-qc]

Foucart, F., Deaton, M.B., Duez, M.D., O'Connor, E., Ott, C.D., Haas, R., Kidder, L.E., Pfeiffer, H.P., Scheel, M.A., Szilagyi, B.: Phys. Rev. D **90**, 024026 (2014). arXiv:1405.1121 [astro-ph.HE]

Foucart, F., Haas, R., Duez, M.D., O'Connor, E., Ott, C.D., Roberts, L., Kidder, L.E., Lippuner, J., Pfeiffer, H.P., Scheel, M.A.: Phys. Rev. D **93**, 044019 (2016). arXiv:1510.06398 [astro-ph.HE]

Franci, L., De Pietri, R., Dionysopoulou, K., Rezzolla, L.: Phys. Rev. D **88**, 104028 (2013). arXiv:1308.3989 [gr-qc]

Fujibayashi, S., Sekiguchi, Y., Kiuchi, K., Shibata, M.: ArXiv e-prints (2017). arXiv:1703.10191 [astro-ph.HE]

Gagnon-Bischoff, J., Green, S.R., Landry, P., Ortiz, N.: Phys. Rev. **D97**(6), 064042 (2018). arXiv:1711.05694

Giacomazzo, B., Rezzolla, L., Baiotti, L.: Phys. Rev. D **83**, 044014 (2011). arXiv:1009.2468 [gr-qc]

Gold, R., Bernuzzi, S., Thierfelder, M., Brugmann, B., Pretorius, F.: Phys. Rev. **D86**, 121501 (2012). arXiv:1109.5128 [gr-qc]

Goldberger, W.D., Rothstein, I.Z.: Phys. Rev. **D73**, 104029 (2006). arXiv:hep-th/0409156 [hep-th]

Gualtieri, L., Kantor, E.M., Gusakov, M.E., Chugunov, A.I.: Phys. Rev. **D90**, 024010 (2014). arXiv:1404.7512 [gr-qc]

Guersel, Y.: Gen. Relativ. Gravit. **15**, 737 (1983)

Gurlebeck, N.: Phys. Rev. Lett. **114**, 151102 (2015). arXiv:1503.03240 [gr-qc]

Haensel, P., Zdunik, J.L., Schaefer, R.: Astron. Astrophys. **160**, 121 (1986)

Haensel, P., Zdunik, J.L., Bejger, M., Lattimer, J.M.: Astron. Astrophys. **502**, 605 (2009). arXiv:0901.1268 [astro-ph.SR]

Hanauske, M., Takami, K., Bovard, L., Rezzolla, L., Font, J.A., Galeazzi, F., Stöcker, H.: ArXiv e-prints (2016). arXiv:1611.07152 [gr-qc]

Hansen, D.: Gen. Relativ. Gravit. **38**, 1173 (2006). arXiv:gr-qc/0511033 [gr-qc]

Harry, G.M., et al.: Class. Quantum Grav. **27**, 084006 (2010)

Hartle, J.B.: Astrophys. J. **150**, 1005 (1967)

Hartmann, T., Soffel, M.H., Kioustelidis, T.: Celest. Mech. Dyn. Astron. **60**, 139 (1994)

Haskell, B., Ciolfi, R., Pannarale, F., Rezzolla, L.: Mon. Not. R. Astron. Soc. Lett. **438**, L71 (2014). arXiv:1309.3885 [astro-ph.SR]

Hempel, M., Schaffner-Bielich, J.: Nucl. Phys. A **837**, 210 (2010). arXiv:0911.4073 [nucl-th]

Hinderer, T.: Astrophys. J. **677**, 1216 (2008). arXiv:0711.2420

Hinderer, T., Lackey, B.D., Lang, R.N., Read, J.S.: Phys. Rev. D **81**, 123016 (2010). arXiv:0911.3535 [astro-ph.HE]

Hinderer, T., Taracchini, A., Foucart, F., Buonanno, A., Steinhoff, J., Duez, M., Kidder, L.E., Pfeiffer, H.P., Scheel, M.A., Szilagyi, B., Hotokezaka, K., Kyutoku, K., Shibata, M., Carpenter, C.W. (2016). arXiv:1602.00599 [gr-qc]

Ho, W.C.G., Lai, D.: Mon. Not. R. Astron. Soc. **308**, 153 (1999). arXiv:astro-ph/9812116 [astro-ph]

Hotokezaka, K., Kyutoku, K., Okawa, H., Shibata, M., Kiuchi, K.: Phys. Rev. D **83**, 124008 (2011). arXiv:1105.4370 [astro-ph.HE]

Hotokezaka, K., Kiuchi, K., Kyutoku, K., Muranushi, T., Sekiguchi, Y.-i., Shibata, M., Taniguchi, K.: Phys. Rev. D **88**, 044026 (2013). arXiv:1307.5888 [astro-ph.HE]

Ipser, J.R., Price, R.H.: Phys. Rev. D **43**, 1768 (1991)

Ivanova, N., Justham, S., Chen, X., De Marco, O., Fryer, C.L., Gaburov, E., Ge, H., Glebbeek, E., Han, Z., Li, X.-D., Lu, G., Marsh, T., Podsiadlowski, P., Potter, A., Soker, N., Taam, R., Tauris, T.M., van den Heuvel, E.P.J., Webbink, R.F.: Astron. Astrophys. Rev. **21**, 59 (2013). arXiv:1209.4302 [astro-ph.HE]

Jimenez-Forteza, X., Abdelsalhin, T., Pani, P., Gualtieri, L. (2018). arXiv:1807.08016

Just, O., Bauswein, A., Pulpillo, R.A., Goriely, S., Janka, H.-T.: Mon. Not. R. Astron. Soc. **448**, 541 (2015). arXiv:1406.2687 [astro-ph.SR]

Kalogera, V., Psaltis, D.: Phys. Rev. **D61**, 024009 (2000). arXiv:astro-ph/9903415 [astro-ph]

Kaplan, J.D., Ott, C.D., O'Connor, E.P., Kiuchi, K., Roberts, L., Duez, M.: Astrophys. J. **790**, 19 (2014). arXiv:1306.4034 [astro-ph.HE]

Kastaun, W., Galeazzi, F.: Phys. Rev. D **91**, 064027 (2015). arXiv:1411.7975 [gr-qc]

Kastaun, W., Galeazzi, F., Alic, D., Rezzolla, L., Font, J.A.: Phys. Rev. D **88**, 021501 (2013). arXiv:1301.7348 [gr-qc]

Kawaguchi, K., Kyutoku, K., Nakano, H., Shibata, M. (2017). arXiv:1709.02754 [astro-ph.HE]

Kawaguchi, K., Kiuchi, K., Kyutoku, K., Sekiguchi, Y., Shibata, M., Taniguchi, K.: Phys. Rev. **D97**(4), 044044 (2018). arXiv:1802.06518

Khan, S., Husa, S., Hannam, M., Ohme, F., Puerrer, M., Jimenez Forteza, X., Bohe, A.: Phys. Rev. **D93**, 044007 (2016). arXiv:1508.07253 [gr-qc]

Kiuchi, K., Sekiguchi, Y., Shibata, M., Taniguchi, K.: Phys. Rev. D **80**, 064037 (2009). arXiv:0904.4551 [gr-qc]

Kiuchi, K., Sekiguchi, Y., Shibata, M., Taniguchi, K.: Phys. Rev. Lett. **104**, 141101 (2010). arXiv:1002.2689 [astro-ph.HE]

Kiuchi, K., Kyutoku, K., Shibata, M.: Phys. Rev. D **86**, 064008 (2012a). arXiv:1207.6444 [astro-ph.HE]

Kiuchi, K., Sekiguchi, Y., Kyutoku, K., Shibata, M.: Class. Quantum Grav. **29**, 124003 (2012b). arXiv:1206.0509 [astro-ph.HE]

Kiuchi, K., Sekiguchi, Y., Kyutoku, K., Shibata, M.: In: Pogorelov, N.V., Font, J.A., Audit, E., Zank, G.P. (eds.) Numerical Modeling of Space Plasma Slows (ASTRONUM 2011), Astronomical Society of the Pacific Conference Series, vol. 459, p. 85 (2012c)

Kiuchi, K., Kyutoku, K., Sekiguchi, Y., Shibata, M., Wada, T.: Phys. Rev. D **90**, 041502 (2014). arXiv:1407.2660 [astro-ph.HE]

Kiuchi, K., Kyutoku, K., Sekiguchi, Y., Shibata, M. (2017). arXiv:1710.01311 [astro-ph.HE]

Kiziltan, B., Kottas, A., De Yoreo, M., Thorsett, S.E.: Astrophys. J. **778**, 66 (2013). arXiv:1011.4291 [astro-ph.GA]

Kochanek, C.S.: Astrophys. J. **398**, 234 (1992)

Kokkotas, K.D., Schaefer, G.: Mon. Not. R. Astron. Soc. **275**, 301 (1995). arXiv:gr-qc/9502034 [gr-qc]

Kokkotas, K., Schmidt, B.: Living Rev. Relativ. **2**, 2 (1999). gr-qc/9909058

Kol, B., Smolkin, M.: J. High Energy Phys. **02**, 010 (2012). arXiv:1110.3764 [hep-th]

Kopal, Z.: Dynamics of Close Binary Systems. D. Reidel Publishing Company, Dordrecht (1978)

Kramer, M., Lyne, A., Burgay, M., Possenti, A., Manchester, R., Camilo, F., McLaughlin, M., Lorimer, D., D'Amico, N., Joshi, B., Reynolds, J., Freire, P.: In: Rasio, F.A., Stairs, I.H. (eds.) Binary Pulsars. PSAP, Chicago (2004)

Krishnendu, N.V., Arun, K.G., Mishra, C.K.: Phys. Rev. Lett. **119**, 091101 (2017). arXiv:1701.06318 [gr-qc]

Kumar, P., Pürrer, M., Pfeiffer, H.P.: Phys. Rev. **D95**, 044039 (2017). arXiv:1610.06155 [gr-qc]

Kyutoku, K., Shibata, M., Taniguchi, K.: Phys. Rev. D **82**, 044049 (2010). arXiv:1008.1460 [astro-ph.HE]

Laarakkers, W.G., Poisson, E.: Astrophys. J. **512**, 282 (1999). arXiv:gr-qc/9709033 [gr-qc]

Lackey, B.D., Wade, L.: Phys. Rev. D **91**, 043002 (2015). arXiv:1410.8866 [gr-qc]

Lackey, B.D., Kyutoku, K., Shibata, M., Brady, P.R., Friedman, J.L.: Phys. Rev. D **85**, 044061 (2012). arXiv:1109.3402 [astro-ph.HE]

Lackey, B.D., Kyutoku, K., Shibata, M., Brady, P.R., Friedman, J.L.: Phys. Rev. D **89**, 043009 (2014). arXiv:1303.6298 [gr-qc]

Lackey, B.D., Bernuzzi, S., Galley, C.R., Meidam, J., Van Den Broeck, C.: Phys. Rev. **D95**, 104036 (2017). arXiv:1610.04742 [gr-qc]

Lai, D.: Mon. Not. R. Astron. Soc. **270**, 611 (1994). arXiv:astro-ph/9404062 [astro-ph]

Lai, D., Rasio, F.A., Shapiro, S.L.: Astrophys. J. Suppl. Ser. **88**, 205 (1993)

Landry, P.: Phys. Rev. **D95**, 124058 (2017). arXiv:1703.08168 [gr-qc]

Landry, P. (2018). arXiv:1805.01882

Landry, P., Poisson, E.: Phys. Rev. **D89**, 124011 (2014). arXiv:1404.6798 [gr-qc]

Landry, P., Poisson, E.: Phys. Rev. **D91**, 104018 (2015a). arXiv:1503.07366 [gr-qc]

Landry, P., Poisson, E.: Phys. Rev. **D91**, 104026 (2015b). arXiv:1504.06606 [gr-qc]

Landry, P., Poisson, E.: Phys. Rev. **D92**, 124041 (2015c). arXiv:1510.09170 [gr-qc]

Lattimer, J.M., Lim, Y.: Astrophys. J. **771**, 51 (2013). arXiv:1203.4286 [nucl-th]

Lattimer, J.M., Prakash, M.: Astrophys. J. **550**, 426 (2001)

Lattimer, J.M., Prakash, M.: Science **304**, 536 (2004). arXiv:astro-ph/0405262

Lattimer, J.M., Schutz, B.F.: Astrophys. J. **629**, 979 (2005). astro-ph/0411470

Lattimer, J.M., Yahil, A.: Astrophys. J. **340**, 426 (1989)

Lattimer, J.M., Pethick, C.J., Prakash, M., Haensel, P.: Phys. Rev. Lett. **66**, 2701 (1991)

Lau, H.K., Leung, P.T., Lin, L.M.: Astrophys. J. **714**, 1234 (2010). arXiv:0911.0131 [gr-qc]

Lee, W.H., Ramirez-Ruiz, E., van de Ven, G.: Astrophys. J. **720**, 953 (2010). arXiv:0909.2884 [astro-ph.HE]

Lehner, L., Liebling, S.L., Palenzuela, C., Motl, P.: ArXiv e-prints (2016). arXiv:1605.02369 [gr-qc]

Lindblom, L., Indik, N.M.: Phys. Rev. **D89**, 064003 (2014) [Erratum: Phys. Rev. **D93**(12), 129903 (2016)]. arXiv:1310.0803 [astro-ph.HE]

Littenberg, T.B., Cornish, N.J.: Phys. Rev. D **91**, 084034 (2015). arXiv:1410.3852 [gr-qc]

Lombardi, J.C., Rasio, F.A., Shapiro, S.L.: Phys. Rev. D **56**, 3416 (1997)

Love, A.E.H.: Proc. R. Soc. Lond. Ser. A **82**, 73 (1909)

Luyten, P.J.: Mon. Not. R. Astron. Soc. **242**, 447 (1990)

Maione, F., De Pietri, R., Feo, A., Löffler, F. (2016). arXiv:1605.03424 [gr-qc]

Majumder, B., Yagi, K., Yunes, N.: Phys. Rev. D **92**, 024020 (2015). arXiv:1504.02506 [gr-qc]

Marchand, T., Bernard, L., Blanchet, L., Faye, G. (2017). arXiv:1707.09289 [gr-qc]

Margalit, B., Metzger, B.D.: Astrophys. J. Lett. **850**, L19 (2017). arXiv:1710.05938 [astro-ph.HE]

Markakis, C., Read, J.S., Shibata, M., Uryu, K., Creighton, J.D.E., Friedman, J.L., Lackey, B.D.: J. Phys. Conf. Ser. **189**, 012024 (2009). arXiv:1110.3759 [gr-qc]

Martin, D., Perego, A., Arcones, A., Thielemann, F.-K., Korobkin, O., Rosswog, S.: Astrophys. J. **813**, 2 (2015). arXiv:1506.05048 [astro-ph.SR]

Maselli, A., Gualtieri, L., Pannarale, F., Ferrari, V.: Phys. Rev. D **86**, 044032 (2012)

Maselli, A., Cardoso, V., Ferrari, V., Gualtieri, L., Pani, P.: Phys. Rev. D **88**, 023007 (2013a). arXiv:1304.2052 [gr-qc]

Maselli, A., Gualtieri, L., Ferrari, V.: Phys. Rev. **D88**, 104040 (2013b). arXiv:1310.5381 [gr-qc]

Maselli, A., Pani, P., Cardoso, V., Abdelsalhin, T., Gualtieri, L., Ferrari, V.: Phys. Rev. Lett. **120**, 081101 (2018). arXiv:1703.10612 [gr-qc]

McDermott, P.N., van Horn, H.M., Hansen, C.J., Buland, R.: Astrophys. J. Lett. **297**, L37 (1985)

Mendes, R.F.P., Yang, H.: Class. Quant. Grav. **34**, 185001 (2017). arXiv:1606.03035 [astro-ph.CO]

Messenger, C., Takami, K., Gossan, S., Rezzolla, L., Sathyaprakash, B.S.: Phys. Rev. X **4**, 041004 (2014)

Mora, T., Will, C.M.: Phys. Rev. D **69**, 104021 (2004). arXiv:gr-qc/0312082

Morsink, S.M., Leahy, D.A., Cadeau, C., Braga, J.: Astrophys. J. **663**, 1244 (2007). astro-ph/0703123

Muhlberger, C.D., Nouri, F.H., Duez, M.D., Foucart, F., Kidder, L.E., Ott, C.D., Scheel, M.A., Szilágyi, B., Teukolsky, S.A.: Phys. Rev. D **90**, 104014 (2014). arXiv:1405.2144 [astro-ph.HE]

Murguia-Berthier, A., Montes, G., Ramirez-Ruiz, E., De Colle, F., Lee, W.H.: Astrophys. J. **788**, L8 (2014). arXiv:1404.0383 [astro-ph.HE]

Murguia-Berthier, A., Ramirez-Ruiz, E., Montes, G., De Colle, F., Rezzolla, L., Rosswog, S., Takami, K., Perego, A., Lee, W.H.: Astrophys. J. Lett. **835**, L34 (2017). arXiv:1609.04828 [astro-ph.HE]

Nagakura, H., Hotokezaka, K., Sekiguchi, Y., Shibata, M., Ioka, K.: Astrophys. J. **784**, L28 (2014). arXiv:1403.0956 [astro-ph.HE]

Nagar, A., Riemenschneider, G., Pratten, G.: Phys. Rev. **D96**, 084045 (2017). arXiv:1703.06814 [gr-qc]

Nagar, A., et al. (2018). arXiv:1806.01772

Nakamura, T., Oohara, K.: Numerical Astrophysics 1998 (NAP98) – Proceedings (1998). gr-qc/9812054

Neilsen, D.W., Liebling, S.L., Anderson, M., Lehner, L., O'Connor, E., Palenzuela, C.: Phys. Rev. D **89**, 104029 (2014). arXiv:1403.3680 [gr-qc]

Nolan, P., Kavanagh, C., Dolan, S.R., Ottewill, A.C., Warburton, N., Wardell, B.: Phys. Rev. **D92**, 123008 (2015). arXiv:1505.04447 [gr-qc]

Oechslin, R., Janka, H.-T.: Phys. Rev. Lett. **99**, 121102 (2007). astro-ph/0702228

O'Leary, R.M., Kocsis, B., Loeb, A.: Mon. Not. R. Astron. Soc. **395**, 2127 (2009). arXiv:0807.2638

Oohara, K.-i., Nakamura, T.: Black Holes and Gravitational Waves: New Eyes in the 21st Century. Proceedings, 9th Yukawa International Seminar, Kyoto, Japan, June 28-July 2, 1999. Prog. Theor. Phys. Suppl. **136**, 270 (1999). arXiv:astro-ph/9912085 [astro-ph]

Oppenheimer, J.R., Volkoff, G.: Phys. Rev. **55**, 374 (1939)

Osłowski, S., Bulik, T., Gondek-Rosińska, D., Belczyński, K.: Mon. Not. R. Astron. Soc. **413**, 461 (2011). arXiv:0903.3538 [astro-ph.GA]

Özel, F., Freire, P.: Annu. Rev. Astron. Astrophys. **54**, 401 (2016). arXiv:1603.02698 [astro-ph.HE]

Pani, P.: Phys. Rev. D **92**, 124030 (2015). arXiv:1506.06050 [gr-qc]

Pani, P., Gualtieri, L., Maselli, A., Ferrari, V.: Phys. Rev. **D92**, 024010 (2015a). arXiv:1503.07365 [gr-qc]

Pani, P., Gualtieri, L., Ferrari, V.: Phys. Rev. **D92**, 124003 (2015b). arXiv:1509.02171 [gr-qc]

Pannarale, F., Rezzolla, L., Ohme, F., Read, J.S.: Phys. Rev. D **84**, 104017 (2011). arXiv:1103.3526 [astro-ph.HE]

Pannarale, F., Berti, E., Kyutoku, K., Lackey, B.D., Shibata, M.: Phys. Rev. **D92**, 081504 (2015). arXiv:1509.06209 [gr-qc]

Pappas, G.: Mon. Not. R. Astron. Soc. **454**, 4066 (2015). arXiv:1506.07225 [astro-ph.HE]

Pappas, G.: Mon. Not. R. Astron. Soc. **466**, 4381 (2017). arXiv:1610.05370 [gr-qc]

Pappas, G., Apostolatos, T.A.: Phys. Rev. Lett. **112**, 121101 (2014). arXiv:1311.5508 [gr-qc]

Paschalidis, V.: Class. Quantum Grav. **34**, 084002 (2017). arXiv:1611.01519 [astro-ph.HE]

Paschalidis, V., Etienne, Z.B., Shapiro, S.L.: Phys. Rev. D **86**, 064032 (2012). arXiv:1208.5487 [astro-ph.HE]

Paschalidis, V., Ruiz, M., Shapiro, S.L.: Astrophys. J. Lett. **806**, L14 (2015a). arXiv:1410.7392 [astro-ph.HE]

Paschalidis, V., East, W.E., Pretorius, F., Shapiro, S.L.: Phys. Rev. D **92**, 121502 (2015b). arXiv:1510.03432 [astro-ph.HE]

Penner, A.J., Andersson, N., Samuelsson, L., Hawke, I., Jones, D.I.: Phys. Rev. **D84**, 103006 (2011). arXiv:1107.0669 [astro-ph.SR]

Perego, A., Rosswog, S., Cabezón, R.M., Korobkin, O., Käppeli, R., Arcones, A., Liebendörfer, M.: Mon. Not. R. Astron. Soc. **443**, 3134 (2014). arXiv:1405.6730 [astro-ph.HE]

Poisson, E.: Phys. Rev. D **57**, 5287 (1998). arXiv:gr-qc/9709032

Poisson, E., Will, C.M.: Gravity. Cambridge University Press, Cambridge (2014)

Porto, R.A.: Fortsch. Phys. **64**, 723 (2016). arXiv:1606.08895 [gr-qc]

Postnikov, S., Prakash, M., Lattimer, J.M.: Phys. Rev. **D82**, 024016 (2010). arXiv:1004.5098 [astro-ph.SR]

Prakash, M., Bombaci, I., Prakash, M., Ellis, P.J., Lattimer, J.M., Knorren, R.: Phys. Rep. **280**, 1 (1997). arXiv:nucl-th/9603042

Price, D.J., Rosswog, S.: Science **312**, 719 (2006). astro-ph/0603845

Punturo, M., et al.: Class. Quantum Grav. **27**, 084007 (2010a)

Punturo, M., et al.: Class. Quantum Grav. **27**, 194002 (2010b)

Purrer, M.: Phys. Rev. **D93**, 064041 (2016). arXiv:1512.02248 [gr-qc]

Racine, E., Flanagan, E.E.: Phys. Rev. **D71**, 044010 (2005). arXiv:gr-qc/0404101 [gr-qc]

Radice, D.: Astrophys. J. Lett. **838**, L2 (2017). arXiv:1703.02046 [astro-ph.HE]

Radice, D., Rezzolla, L., Galeazzi, F.: Mon. Not. R. Astron. Soc. L. **437**, L46 (2014a). arXiv:1306.6052 [gr-qc]

Radice, D., Rezzolla, L., Galeazzi, F.: Class. Quantum Grav. **31**, 075012 (2014b). arXiv:1312.5004 [gr-qc]

Radice, D., Bernuzzi, S., Ott, C.D. (2016). arXiv:1603.05726 [gr-qc]

Radice, D., Bernuzzi, S., Del Pozzo, W., Roberts, L.F., Ott, C.D.: Astrophys. J. Lett. **842**, L10 (2017). arXiv:1612.06429 [astro-ph.HE]

Rathore, Y., Broderick, A.E., Blandford, R.: Mon. Not. R. Astron. Soc. **339**, 25 (2003). arXiv:astro-ph/0209003 [astro-ph]

Ravi, V., Lasky, P.D.: Mon. Not. R. Astron. Soc. **441**, 2433 (2014)

Read, J.S., Lackey, B.D., Owen, B.J., Friedman, J.L.: Phys. Rev. D **79**, 124032 (2009). arXiv:0812.2163

Read, J.S., Baiotti, L., Creighton, J.D.E., Friedman, J.L., Giacomazzo, B., Kyutoku, K., Markakis, C., Rezzolla, L., Shibata, M., Taniguchi, K.: Phys. Rev. D **88**, 044042 (2013). arXiv:1306.4065 [gr-qc]

Regge, T., Wheeler, J.: Phys. Rev. **108**, 1063 (1957)

Reina, B., Sanchis-Gual, N., Vera, R., Font, J.A.: Mon. Not. R. Astron. Soc. **470**, L54 (2017). arXiv:1702.04568 [gr-qc]

Reisenegger, A., Goldreich, P.: Astrophys. J. **426**, 688 (1994)

Rembiasz, T., Guilet, J., Obergaulinger, M., Cerdá -Durán, P., Aloy, M.A., Müller, E.: Mon. Not. R. Astron. Soc. **460**, 3316 (2016). arXiv:1603.00466 [astro-ph.SR]

Rezzolla, L., Kumar, P.: Astrophys. J. **802**, 95 (2015). arXiv:1410.8560 [astro-ph.HE]

Rezzolla, L., Takami, K.: Phys. Rev. D **93**, 124051 (2016). arXiv:1604.00246 [gr-qc]

Rezzolla, L., Zanotti, O.: Relativistic Hydrodynamics. Oxford University Press, Oxford (2013)

Rezzolla, L., Baiotti, L., Giacomazzo, B., Link, D., Font, J.A.: Class. Quantum Grav. **27**, 114105 (2010). arXiv:1001.3074 [gr-qc]

Rezzolla, L., Giacomazzo, B., Baiotti, L., Granot, J., Kouveliotou, C., Aloy, M.A.: Astrophys. J. Lett. **732**, L6 (2011). arXiv:1101.4298 [astro-ph.HE]

Rezzolla, L., Most, E.R., Weih, L.R.: Astrophys. J. Lett. **852**, L25 (2018). arXiv:1711.00314 [astro-ph.HE]

Rikovska-Stone, J., Guichon, P.A., Matevosyan, H.H., Thomas, A.W.: Nucl. Phys. **A792**, 341 (2007). arXiv:nucl-th/0611030 [nucl-th]

Ruiz, M., Shapiro, S.L., Tsokaros, A.: Phys. Rev. D **97**, 021501 (2018). arXiv:1711.00473 [astro-ph.HE]

Saijo, M., Baumgarte, T.W., Shapiro, S.L.: Astrophys. J. **595**, 352 (2003). astro-ph/0302436

Sathyaprakash, B.S., Schutz, B.F.: Living Rev. Relativ. **12**, 2 (2009). arXiv:0903.0338 [gr-qc]

Schmidt, P., Ohme, F., Hannam, M.: Phys. Rev. **D91**, 024043 (2015). arXiv:1408.1810 [gr-qc]

Schnittman, J.D., Dal Canton, T., Camp, J., Tsang, D., Kelly, B.J.: Astrophys. J. **853**, 123 (2018). arXiv:1704.07886 [astro-ph.HE]

Sekiguchi, Y., Kiuchi, K., Kyutoku, K., Shibata, M.: Phys. Rev. Lett. **107**, 211101 (2011a). arXiv:1110.4442 [astro-ph.HE]

Sekiguchi, Y., Kiuchi, K., Kyutoku, K., Shibata, M.: Phys. Rev. Lett. **107**, 051102 (2011b). arXiv:1105.2125 [gr-qc]

Sekiguchi, Y., Kiuchi, K., Kyutoku, K., Shibata, M.: Prog. Theor. Exp. Phys. **2012**, 01A304 (2012). arXiv:1206.5927 [astro-ph.HE]

Sekiguchi, Y., Kiuchi, K., Kyutoku, K., Shibata, M.: Phys. Rev. D **91**, 064059 (2015). arXiv:1502.06660 [astro-ph.HE]

Sennett, N., Hinderer, T., Steinhoff, J., Buonanno, A., Ossokine, S.: Phys. Rev. **D96**, 024002 (2017). arXiv:1704.08651 [gr-qc]

Shah, A.G., Pound, A.: Phys. Rev. **D91**, 124022 (2015). arXiv:1503.02414 [gr-qc]

Shibata, M.: Prog. Theor. Phys. **91**, 871 (1994)

Shibata, M.: Phys. Rev. D **60**, 104052 (1999). gr-qc/9908027

Shibata, M., Kiuchi, K.: Phys. Rev. D **95**, 123003 (2017). arXiv:1705.06142 [astro-ph.HE]

Shibata, M., Taniguchi, K.: Phys. Rev. D **73**, 064027 (2006). astro-ph/0603145

Shibata, M., Taniguchi, K.: Living Rev. Rel. **14**, 6 (2011)

Shibata, M., Uryū, K.: Phys. Rev. D **61**, 064001 (2000). gr-qc/9911058

Shibata, M., Uryū, K.: Prog. Theor. Phys. **107**, 265 (2002). gr-qc/0203037

Shibata, M., Taniguchi, K., Uryū, K.: Phys. Rev. D **71**, 084021 (2005). gr-qc/0503119

Shibata, M., Suwa, Y., Kiuchi, K., Ioka, K.: Astrophys. J. Lett. **734**, L36 (2011). arXiv:1105.3302 [astro-ph.HE]

Shibata, M., Taniguchi, K., Okawa, H., Buonanno, A.: Phys. Rev. D **89**, 084005 (2014). arXiv:1310.0627 [gr-qc]

Shibata, M., Fujibayashi, S., Hotokezaka, K., Kiuchi, K., Kyutoku, K.: Sekiguchi, Y., Tanaka, M.: Phys. Rev. D **96**, 123012 (2017). arXiv:1710.07579 [astro-ph.HE]

Siegel, D.M., Ciolfi, R., Rezzolla, L.: Astrophys. J. **785**, L6 (2014). arXiv:1401.4544 [astro-ph.HE]

Silva, H.O., Sotani, H., Berti, E.: Mon. Not. R. Astron. Soc. **459**, 4378 (2016). arXiv:1601.03407 [astro-ph.HE]

Steiner, A.W., Gandolfi, S., Fattoyev, F.J., Newton, W.G.: Phys. Rev. **C91**, 015804 (2015). arXiv:1403.7546 [nucl-th]

Steiner, A.W., Lattimer, J.M., Brown, E.F.: Eur. Phys. J. A **52**, 18 (2016). arXiv:1510.07515 [astro-ph.HE]

Steinhoff, J., Hinderer, T., Buonanno, A., Taracchini, A.: Phys. Rev. **D94**, 104028 (2016). arXiv:1608.01907 [gr-qc]

Stergioulas, N., Apostolatos, T.A., Font, J.A.: Mon. Not. R. Astron. Soc. **352**, 1089 (2004). arXiv:astro-ph/0312648

Stergioulas, N., Bauswein, A., Zagkouris, K., Janka, H.-T.: Mon. Not. R. Astron. Soc. **418**, 427 (2011). arXiv:1105.0368 [gr-qc]

Takami, K., Rezzolla, L., Baiotti, L.: Phys. Rev. Lett. **113**, 091104 (2014). arXiv:1403.5672 [gr-qc]

Takami, K., Rezzolla, L., Baiotti, L.: Phys. Rev. D **91**, 064001 (2015). arXiv:1412.3240 [gr-qc]

Taranto, G., Baldo, M., Burgio, G.F.: Phys. Rev. C **87**, 045803 (2013). arXiv:1302.6882 [nucl-th]

The LIGO Scientific Collaboration and the Virgo Collaboration: Phys. Rev. Lett. **116**, 061102 (2016). arXiv:1602.03837 [gr-qc]

Thompson, T.A.: Astrophys. J. **741**, 82 (2011). arXiv:1011.4322 [astro-ph.HE]

Thorne, K.: Rev. Mod. Phys. **52**, 299 (1980)

Thorne, K.S.: Phys. Rev. D **58**, 124031 (1998). gr-qc/9706057

Thorne, K.S., Campolattaro, A.: Astrophys. J. **149**, 591 (1967)

Thorne, K.S., Hartle, J.B.: Phys. Rev. **D31**, 1815 (1984)

Tolman, R.C.: Phys. Rev. **55**, 364 (1939)

Tsang, D.: Astrophys. J. **777**, 103 (2013). arXiv:1307.3554 [astro-ph.HE]

Tsang, D., Read, J.S., Hinderer, T., Piro, A.L., Bondarescu, R.: Phys. Rev. Lett. **108**, 011102 (2012). arXiv:1110.0467 [astro-ph.HE]

Tsui, L.K., Leung, P.T.: Mon. Not. R. Astron. Soc. **357**, 1029 (2005). gr-qc/0412024

Uchikata, N., Yoshida, S.: Class. Quant. Grav. **33**, 025005 (2016). arXiv:1506.06485 [gr-qc]

Urbanec, M., Miller, J.C., Stuchlík, Z.: Mon. Not. R. Astron. Soc. **433**, 1903 (2013). arXiv:1301.5925 [astro-ph.SR]

Vallisneri, M.: Phys. Rev. Lett. **84**, 3519 (2000). arXiv:gr-qc/9912026

Van Oeveren, E.D., Friedman, J.L.: Phys. Rev. **D95**, 083014 (2017). arXiv:1701.03797 [gr-qc]

Vines, J., Flanagan, E.E., Hinderer, T.: Phys. Rev. D **83**, 084051 (2011a). arXiv:1101.1673 [gr-qc]

Vines, J., Hinderer, T., Flanagan, É.É. (2011b). arXiv:1101.1673 [gr-qc]

Wade, L., Creighton, J.D.E., Ochsner, E., Lackey, B.D., Farr, B.F., Littenberg, T.B., Raymond, V.: Phys. Rev. D **89**, 103012 (2014). arXiv:1402.5156 [gr-qc]

Watts, A.L., Andersson, N., Jones, D.I.: Astrophys. J. **618**, L37 (2005). astro-ph/0309554

Weissenborn, S., Sagert, I., Pagliara, G., Hempel, M., Schaffner-Bielich, J.: Astrophys. J. **740**, L14 (2011). arXiv:1102.2869 [astro-ph.HE]

Wilson, J.R., Mathews, G.J.: Phys. Rev. Lett. **75**, 4161 (1995)

Witten, E.: Phys. Rev. D **30**, 272 (1984)

Xu, W., Lai, D.: Phys. Rev. **D96**, 083005 (2017). arXiv:1708.01839 [astro-ph.HE]

Yagi, K.: Phys. Rev. **D89**, 043011 (2014). arXiv:1311.0872 [gr-qc]

Yagi, K., Yunes, N.: Science **341**, 365 (2013a). arXiv:1302.4499 [gr-qc]

Yagi, K., Yunes, N.: Phys. Rev. D **88**, 023009 (2013b). arXiv:1303.1528 [gr-qc]

Yagi, K., Yunes, N.: Phys. Rev. **D89**, 021303 (2014). arXiv:1310.8358 [gr-qc]

Yagi, K., Yunes, N.: Class. Quant. Grav. **33**, 13LT01 (2016). arXiv:1512.02639 [gr-qc]

Yagi, K., Yunes, N.: Phys. Rep. **681**, 1 (2017a). arXiv:1608.02582 [gr-qc]

Yagi, K., Yunes, N.: Class. Quant. Grav. **34**, 015006 (2017b). arXiv:1608.06187 [gr-qc]

Yagi, K., Kyutoku, K., Pappas, G., Yunes, N., Apostolatos, T.A.: Phys. Rev. D **89**, 124013 (2014a). arXiv:1403.6243 [gr-qc]

Yagi, K., Stein, L.C., Pappas, G., Yunes, N., Apostolatos, T.A.: Phys. Rev. D **90**, 063010 (2014b)

Yamamoto, T., Shibata, M., Taniguchi, K.: Phys. Rev. D **78**, 064054 (2008). arXiv:0806.4007 [gr-qc]

Yang, H., Paschalidis, V., Yagi, K., Lehner, L., Pretorius, F., Yunes, N.: Phys. Rev. D **97**, 024049 (2018). arXiv:1707.00207 [gr-qc]

Yu, H., Weinberg, N.N.: Mon. Not. R. Astron. Soc. **464**, 2622 (2017). arXiv:1610.00745 [astro-ph.HE]

Zahn, J.-P.: Astron. Astrophys. **4**, 452 (1970)

Zahn, J.-P.: Astron. Astrophys. **57**, 383 (1977)

Chapter 11
Electromagnetic Emission and Nucleosynthesis from Neutron Star Binary Mergers

Bruno Giacomazzo, Marius Eichler, and Almudena Arcones

Abstract In this chapter we provide a description of the current state of the art in the field of electromagnetic emission and nucleosynthesis from neutron star binaries. We will discuss binaries composed by two neutron stars or by a neutron star and a black hole. While we took the effort to represent what we believe are some of the most important results in this field, we strongly recommend the interested reader to check also the references cited in the mentioned publications for a more in detail view of each topic. As the reader will see from the following sections, this field has gained considerable attention in the latest years with an increasing number of publications on the topic. For each argument we will also discuss what we believe are the current limitations that will need to be addressed in the future.

11.1 Introduction

In the last years there has been a significant increase of theoretical and computational studies of electromagnetic (EM) emission from mergers of two neutron stars or of a neutron star and a black hole. While in the past one of the main motivations was the study of the connection between these systems and short gamma-ray bursts , nowadays there is a strong interest in understanding EM counterparts of gravitational wave (GW) signals. This increased interest, combined with new and

B. Giacomazzo (✉)
Physics Department, University of Trento, Trento, Italy

INFN-TIFPA, Trento Institute for Fundamental Physics and Applications, Trento, Italy
e-mail: bruno.giacomazzo@unitn.it

M. Eichler
Institut für Kernphysik, Technische Universität Darmstadt, Darmstadt, Germany

A. Arcones
Institut für Kernphysik, Technische Universität Darmstadt, Darmstadt, Germany

GSI Helmholtzzentrum für Schwerionenforschung, Darmstadt, Germany

© Springer Nature Switzerland AG 2018
L. Rezzolla et al. (eds.), *The Physics and Astrophysics of Neutron Stars*,
Astrophysics and Space Science Library 457,
https://doi.org/10.1007/978-3-319-97616-7_11

more accurate numerical codes, has led to a better understanding of all the possible
EM emission generated by these mergers, ranging from radio up to gamma rays.
These systems are also expected to emit neutrinos during and after merger, but due
to their large distances, a neutrino detection is not expected in the next years, even
with next-generation detectors such as Hyper-Kamiokande (Sekiguchi et al. 2011).

A major breakthrough in this field happened on August 17th 2017 with the first
detection of a GW signal from a binary neutron star (BNS) coalescence (Abbott
et al. 2017b). The GW detection was followed by a large number of electromagnetic
counterparts observed by ground- and space based telescopes (Abbott et al. 2017c;
Cowperthwaite et al. 2017; Drout et al. 2017; Smartt et al. 2017; Tanvir et al.
2017). Those confirm that (at least some type of) short gamma-ray bursts (SGRBs)
are indeed associated with binary neutron star mergers (Abbott et al. 2017a) and
provided evidence that such systems are also a source of the heaviest elements in
the universe produced via the rapid neutron capture process (r-process) in matter
ejected by the system (Abbott et al. 2017c; Metzger 2017; Rosswog et al. 2017).
Unfortunately the current sensitivity of the Virgo and LIGO detectors was not
sufficient to detect a post-merger GW signal (Abbott et al. 2017d) and therefore
it is not clear if a black hole (BH) or a neutron star (NS) were the result of the
merger. Nevertheless, preliminary constraints for the equation of state (EOS) of NS
matter were derived from tidal effects during the inspiral (Abbott et al. 2017b).

In this chapter we will provide an overview of our current understanding of EM
emission and nucleosynthesis from neutron star binary mergers. We will describe
the most important theoretical and computational results in this field and possible
future directions.

11.2   Review of Past Work

11.2.1   Overview of Electromagnetic Counterparts

In the merger of a binary neutron star (BNS) system or of a neutron star (NS) with
a black hole (BH) there are mainly two possible end results: the formation of a BH
surrounded by an accretion disk or of a long-lived NS (only in BNS mergers). Both
scenarios belong to the category of compact binary mergers (CBMs).

In particular, and as we will discuss later, the first scenario is the one that has been
strongly correlated with the central engine of short gamma-ray bursts (SGRBs). But
while SGRBs represent the strongest possible EM emission, there are other signals
that such mergers may emit, ranging from radio to X-rays. Some of these have also
the property of not being collimated, but more isotropic, and hence easier to detect.
This also depends on whether a BH or a long-lived NS is formed after the merger.
Observations of NSs with masses of up to $\sim 2M_\odot$ (Demorest et al. 2010; Antoniadis
et al. 2013) indicate that a significant fraction of BNS mergers may indeed lead
to the formation of a so-called supramassive NS (SMNS), i.e., of an NS with a

mass below the maximum mass for a uniformly rotating NS (Piro et al. 2017). This, combined with possible strong magnetic field amplifications, leads to the interesting possibility of forming magnetars after some BNS mergers (Giacomazzo and Perna 2013).

In the scenario where the end result is a spinning BH surrounded by an accretion disk (let us call it the "standard" scenario) and which is also the only possible outcome for the merger of an NS with a BH, all the possible EM emission investigated, at least theoretically, up to now are summarized in Fig. 11.1.

Figure 11.1 shows both the possible emission of a relativistic jet and the ejection of matter. The jet can be responsible for a short gamma-ray burst and its afterglow (from X-rays to radio).

The ejected matter is instead a promising site for the r-process (see Sect. 11.2.2) that can produce very neutron-rich heavy elements. Their decay to stability powers an EM transient, a signal that is known under the name "Kilonova" or "Macronova" (see Sect. 11.2.12) and that was detected after GW170817 (Abbott et al. 2017c; Pian et al. 2017) as well as in connection with other SGRBs (Tanvir et al. 2013; Yang

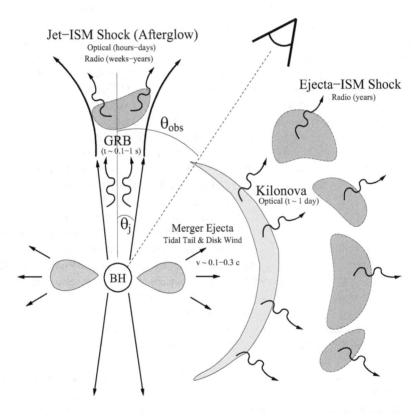

Fig. 11.1 Summary of potential EM emission from BNS and NS-BH mergers. Figure 1 in Metzger and Berger (2012)

et al. 2015; Jin et al. 2015, 2016). The EM signal is affected by the temperature and composition of the ejected matter which is strongly dependent also on whether the matter is ejected via tidal disruption, typical for NS-BH systems, or via shocks at merger, which is the dominant ejection process for NS-NS mergers. Therefore the detection of such a signal may be connected with the properties of the binary system.

On a longer timescale, the ejected matter may also shock with the interstellar medium giving rise to radio emission (Rosswog et al. 2014; Grossman et al. 2014).

We will now proceed on describing in more details some of the main processes taking place in NS-NS and NS-BH mergers.

11.2.2 Nucleosynthesis in Compact Binary Mergers

The electromagnetic afterglow of a compact binary merger (CBM) is thought to originate from the radioactive decay of neutron-rich heavy nuclei that are produced in the r-process . It is therefore essential to address the current status of nucleosynthesis research in these events, which is closely linked with the search for the cosmic r-process site(s). While regular core-collapse supernovae (CCSNe) were an early favourite and are still contenders to be a (maybe secondary) source of low- to intermediate-mass r-process isotopes up to Ag or higher, CBMs of two neutron stars (NS-NS) or a neutron star and a black hole (NS-BH) are now the most promising astrophysical scenario to host the r-process. It was always undisputed that the decompressing matter during a NS-NS or NS-BH merger would be extremely neutron-rich, but before the emergence of computer simulations it was all but impossible to estimate how much, if any, matter would become gravitationally unbound in such a scenario. Lattimer and Schramm (1974) were the first to propose that r-process material could be ejected from a NS-BH merger. Hydrodynamical simulations later confirmed this finding (Davies et al. 1994; Rosswog et al. 1999), followed by the first nucleosynthesis calculations based on trajectories obtained from these simulations (Freiburghaus et al. 1999). Since then, research in this field has mainly focused on two branches: (a) improving the physical treatment of e.g., neutrino interactions, general relativity, etc. in hydrodynamic models, and (b) understanding the nuclear properties (such as masses, excitation levels, decay rates) of the neutron-rich isotopes involved in the r-process path. The former is important for constraining the electron fraction and expansion timescales of the ejecta, while the latter is required in order to reduce uncertainties in the final abundances of r-process calculations and to compare the results with the r-process abundances in the solar system or metal-poor stars.

Recent simulations have shown that in addition to the *dynamic ejecta* which become unbound either by tidal forces or shocks during the merger phase, other mass loss channels are possible. A neutrino-driven baryonic wind component similar to the one in CCSNe (Perego et al. 2014; Just et al. 2015; Martin et al. 2015) has been found in the aftermath of CBM simulations, while material from the remnant torus that can form around the dense central object during the merger phase can later become unbound due to viscous dissipation and the release of recombination energy (Fernández and Metzger 2013; Just et al. 2015; Wu et al. 2016). The nucleosynthesis in these additional ejecta mechanisms has only been studied in the past few years, and it will be addressed further below in this chapter.

11.2.3 The r-Process

In order to explain the origin of stable nuclei heavier than iron, including some long-lived actinides (^{232}Th, ^{235,236,238}U, and ^{244}Pu), Burbidge et al. (1957) and Cameron (1957) proposed the *rapid neutron capture process* (r-process). As the name suggests, it operates on the basis of very fast neutron captures in comparison to β-decays, and runs close to the neutron-drip line. Wherever the path crosses isotopes with a closed neutron shell, both the neutron capture cross sections and β-decay rates become significantly smaller, leading to an accumulation of material in these nuclei, and the path moves closer to stability (see Fig. 11.2). After the supply of free neutrons has ceased, the extremely neutron-rich isotopes undergo a series of β-decays to stability. The mass numbers where the r-process path crossed the magic neutron numbers can be identified in the final abundance pattern by characteristic peaks, as is shown in Fig. 11.2.

The r-process can operate only if certain conditions are met. Neutron capture rates are dependent on the neutron density n_n, which means that material from the surface or the vicinity of neutron stars provides a good environment. Furthermore, the duration of the r-process and the maximum mass number of the final products is characterized by the ratio between the neutron abundance and the summed up abundances of all seed nuclei (usually iron group nuclei), Y_n/Y_{seed}. This ratio is a measure of the amount of neutrons every seed nucleus can capture before neutrons become depleted and the r-process stops. We can estimate the final average mass number $\langle A \rangle_f$ using

$$\langle A \rangle_f = \langle A \rangle_i + \left(\frac{Y_n}{Y_{seed}} \right)_i, \tag{11.1}$$

where i denotes the initial values. As mentioned above, iron group nuclei are typical seed nuclei, giving a typical value for $\langle A \rangle_i = 60$. Under these assumptions, a neutron-to-seed ratio of $Y_n/Y_{seed} = 184$ would be needed in order to reliably produce the heaviest long-lived actinide ^{244}Pu. Y_n/Y_{seed} is determined by several factors. A high Y_n can be achieved wherever the electron fraction Y_e is low (since charge neutrality is always assumed for astrophysical plasmas), making ejecta from the vicinity of neutron stars a natural environment favourable for the r-process. On the other hand, Y_n/Y_{seed} can reach large enough values if the abundances of seed nuclei are kept low. High entropies can counter the build-up of seed nuclei, with energetic photons dissociating them into nucleons and α-particles. Moreover,

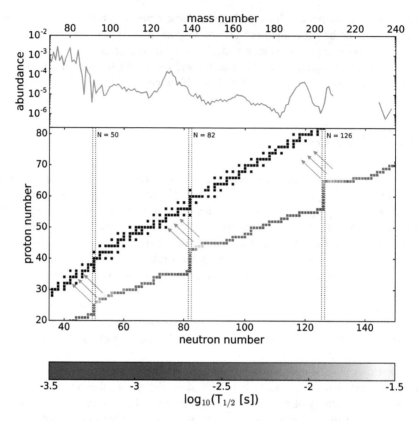

Fig. 11.2 Top panel: Solar r-process abundances as a function of nuclear mass number A. The values are taken from Sneden et al. (2008). Bottom panel: Typical r-process path in the nuclear chart and the corresponding β-decay half-lives according to Möller et al. (2003). Stable isotopes are marked in black and the magic neutron numbers are indicated by vertical dotted lines. The overlay of the two panels demonstrates how regions of large $T_{1/2}$ at the magic neutron numbers are responsible for the r-process abundance peaks after decay to stability

short dynamical timescales (i.e., fast expansions) also guarantee that the density decreases quickly past the point where seed nuclei can be efficiently produced from α-particles, leading to an α-rich freeze-out from charged-particle reactions. Qian and Woosley (1996), Otsuki et al. (2000), and Thompson et al. (2001) have explored the combinations of entropies, dynamical timescales, neutrino luminosities, and neutron star masses that result in the production of r-process nuclei of the second and the third peak, respectively. However, in hydrodynamic simulations of neutron star mergers, Y_e in the ejecta is so low that also the heaviest stable and long-lived nuclei are produced, independent of the other factors discussed here (Freiburghaus et al. 1999).

If the neutron-to-seed ratio is considerably larger than 180, very heavy and large nuclei are produced. While the attractive nuclear force has a limited range and acts mainly between two neighbouring nucleons, the repulsive Coulomb force of the protons has a larger range and its influence grows with increasing proton number, until it is strong enough to disrupt the nucleus and fission occurs. The freshly produced fragments continue capturing neutrons as long as the neutron density is high enough, and can eventually undergo fission themselves in a phenomenon that is called *fission cycling* (Beun et al. 2008). Assuming that each fission cycle divides $\langle A \rangle$ by two, we have to modify Eq. (11.1) as follows:

$$\langle A \rangle_f = \frac{\langle A \rangle_i + (Y_n/Y_{seed})_i}{2^{N_{cyc}}},$$ (11.2)

where N_{cyc} is the number of fission cycles. The products of each fission process depend on the nuclear properties of the parent nucleus as well as of the possible fragments. Furthermore, the fission rate is extremely sensitive to the height of the potential barrier that needs to be overcome (or tunnelled through). This makes the prediction of fission rates and the distribution of fission fragments a very challenging field of research (see e.g. Giuliani et al. 2017).

In a simplified approach, we neglect all reactions that are not (n, γ), (γ, n) reactions, or β^--decays, resulting in a set of differential equations involving all of the above reactions that either produce or destroy isotope i:

$$\dot{Y}_i = \sum_j N_j^i \lambda_j Y_j + \sum_j N_{j,n}^i \rho N_A \langle \sigma v \rangle_{j(n,\gamma)} Y_j Y_n,$$ (11.3)

where ρ is the density in units of g cm^{-3}, N_A is Avogadro's constant, $\langle \sigma v \rangle_{j(n,\gamma)}$ is the velocity integrated cross section of the reaction $j(n, \gamma)$, and the coefficients N_j^i and $N_{j,n}^i$ can take the values ± 1, determining if nucleus i is produced or destroyed. The first term on the right hand side contains β-decays and (γ, n) reactions, with the decay constant λ_j which relates to the half-live via

$$\lambda_j = \frac{\ln(2)}{\tau_{1/2}}.$$ (11.4)

The second term contains only two reactions in our example: the (n, γ) reaction producing nucleus i and the neutron capture on nucleus i itself. While the neutron capture rates mainly depend on the neutron density $n_n = \rho N_A Y_n$, the (γ, n) rates are affected by the temperature. Depending on the hydrodynamic trajectory, the (n, γ) reactions can therefore be in equilibrium with their reverse reactions (γ, n) for the majority of the duration of the r-process (this *hot r-process* scenario was first discussed by Seeger et al. 1965). The (n, γ)-(γ, n) equilibrium is established in every isotopic chain (isotopes with the same proton number Z) independently, and the abundance maximum in each chain is determined by the neutron separation energy S_n of the nuclei involved. The neutron separation energy of a nucleus (Z, A) is defined as the energy that is required to strip one neutron off of this nucleus, or, in other words, the negative Q-value of the (Z, A) (γ, n) reaction. As a trend, S_n

decreases with increasing neutron number $N = A - Z$, until it becomes 0 at the neutron-drip line, which means that no energy can be gained from further neutron captures.

Consider a neutron capture on a nucleus (Z, A) with a known reaction rate. We can find the rate of the reverse reaction $(Z, A + 1)$ (γ, n) from a concept called *detailed balance*. The equilibrium abundance ratio between the nuclei (Z, A) and $(Z, A + 1)$ is equal to the ratio of the (γ, n) and (n, γ) reaction rates:

$$\frac{Y(Z, A+1)}{Y(Z, A)} = n_n \frac{G(Z, A+1)}{2G(Z, A)} \left[\frac{A+1}{A}\right]^{3/2} \left[\frac{2\pi \hbar^2}{m_u kT}\right]^{3/2} \exp(S_n(A+1)/kT),$$

(11.5)

with the partition functions G that describe the thermally populated excited states, and the nuclear-mass unit m_u. This relation is valid for all nuclei in a given isotopic chain. For known n_n and T, the neutron separation energy at maximum abundance can be found by approximating $Y(A+1)/Y(Z, A) \approx 1$, identifying the most abundant isotopes in each isotopic chain. Note that this does not require the knowledge of the individual neutron capture cross sections, but only the nuclear masses which determine S_n. Adding the β-decay rates of these most abundant nuclei we can construct a simplified model for the r-process, the so-called *waiting point approximation*.

Reaction rates depend on the nuclear properties, such as masses or the energy levels of the excited states. Since the r-process path moves through very exotic neutron-rich region of the nuclear chart, these properties are not known experimentally for all nuclei involved. Thus, theoretical models are being developed which are fitted to the known nuclear data of less exotic isotopes and then extrapolated to all nuclei up to the neutron-drip line. Nowadays mass models belong to one of two categories: macroscopic-microscopic models which contain phenomenological terms to describe the overall behaviour of the nucleons and including microscopic corrections, such as shell effects or non-spherical ground-state configurations (e.g., Möller et al. 1995, 2012) and fully microscopic models. These are based on effective nucleon-nucleon interactions (e.g., Aboussir et al. 1995, Goriely et al. 2016, and references therein). Although there are many mass models of both categories which reproduce the experimental nuclear data very well, the mass predictions for very neutron-rich nuclei differ significantly, which has an impact on the reaction rates used in the nuclear network. Calculated r-process abundances are therefore heavily dependent on the mass model employed. For recent studies on the impact of nuclear masses on the calculated r-process abundances see, e.g., Arcones and Martínez-Pinedo (2011), Kratz et al. (2012), Martin et al. (2016), Mendoza-Temis et al. (2015), Mumpower et al. (2015), and Wanajo et al. (2005).

11.2.4 Observational Evidence for r-Process in CBMs

The r-process requires specific conditions to operate, such as large neutron densities, short dynamical timescales, etc. (see Sect. 11.2.3). These requirements point towards explosive environments containing decompressing deleptonized (e.g., neutron-star) material. Both CCSNe and CBMs can in principle meet the requirements. However, recent research suggests that regular CCSNe are ruled out as a source for the heaviest r-process nuclei. With the detection of GW170817, neutron star mergers have now been confirmed as an operating site of the r-process (Cowperthwaite et al. 2017; Drout et al. 2017; Kasen et al. 2017; Kasliwal et al. 2017; Kilpatrick et al. 2017; Rosswog et al. 2017; Smartt et al. 2017; Tanvir et al. 2017), as the light curve powered by the radioactive decays of heavy, neutron-rich nuclei has been observed (see Sect. 11.2.12). The ejected mass for this event is estimated to be at least $1.5 \times 10^{-2}\ M_\odot$, but may be larger if not all the energy released from radioactive decays ends up in the observed emission (Rosswog et al. 2017). In addition to the direct observation of a kilonova (Abbott et al. 2017c), a multitude of indirect observational evidence exists that points towards CBMs as a major r-process production site. The atmospheres of different extremely metalpoor (old) stars show the same abundance pattern for the heavy r-elements as the solar system abundances (e.g., Sneden et al. 2008), with very little scatter among them. This suggests that the r-process component producing the heaviest elements is robust and insensitive to small changes in the density or temperature evolution during the decompression of the ejecta. The robustness of the abundance pattern is a characteristic of the r-process in CBMs (Korobkin et al. 2012) because the environment is so neutron-rich that the reaction path runs along the neutron-drip line.

Regular CCSNe have a well-determined event rate in our galaxy, and in order to account for the accumulated r-process material in the present-day solar system a relatively low r-process ejecta mass per event is required (around $10^{-5}\ M_\odot$ of r-process material or even less). CBMs, on the other hand, are expected to occur at a much lower frequency, but ejecting around $10^{-3}\ M_\odot$ of r-process material. This degeneracy of high-rate/low-mass and low-rate/high-mass events can be broken by recent discoveries. For instance, the strong enhancement of r-elements in a group of stars belonging to the dwarf galaxy Reticulum II compared to other dwarf galaxies strongly supports a rare r-process event (Ji et al. 2016). From these data Beniamini et al. (2016) estimate an r-process event rate between 2.5×10^{-4} and 1.4×10^{-3} times lower than the event rate of regular CCSNe, and an ejected r-process mass between $6 \times 10^{-3}\ M_\odot$ and $4 \times 10^{-2}\ M_\odot$ per event, consistent with GW170817 and the values obtained in models of neutron star mergers. We can even find evidence in terrestrial deep sea sediments that exhibit the nuclear traces of the ejecta of nearby explosive events that rained down on earth millions of years ago. Studying longlived radioactive isotopes in these sediments does not only reveal the history of explosive events in our solar neighbourhood, but helps us disentangle the different nucleosynthesis processes at the same time. ^{60}Fe has a half-life of $T_{1/2} \approx 2.6$ Myr and it is produced in CCSNe. A sediment layer between 1.7 and 3.2 Myr old was found to contain increased amounts of ^{60}Fe, indicating that around that time the earth was polluted by the ejecta of a nearby CCSN (Wallner et al. 2016). Assuming that this CCSN also produced and ejected some r-process material, one should also find heavier long-lived r-process nuclei. However, the same ocean sediment samples contain significantly less ^{244}Pu ($T_{1/2} \approx 81$ Myr) than expected from a continuous enrichment from CCSN ejecta (Wallner et al. 2015). On the other hand, the measured abundances are compatible with the expected (lower) frequency and (higher) ejecta masses of CBMs (Hotokezaka et al. 2015).

Another hint comes from the observed frequency of SGRBs. The only observed kilonovae so far have been associated to GRB 170817A (Abbott et al. 2017a), SGRB 130603B (Tanvir et al. 2013), SGRB 060614 (Yang et al. 2015; Jin et al. 2015), and SGRB 050709 (Jin et al. 2016), and the observed frequency of SGRBs supports the required frequency of CBMs as the main r-process source.

It is noteworthy that all the pieces of observational evidence presented above are independent from each other, however the constraints presented by these observations to the r-process event rate all agree, pointing to a rate around 5×10^{-4} lower than the rate of CCSNe. The resulting ejected mass of r-process material needed to explain the origin of the r-process element abundances in the solar system (10^{-3}–$10^{-2} \, M_\odot$) agrees very well with the ejecta mass in CBM models. Upcoming gravitational wave detections from CBMs will provide further constraints to this current picture.

11.2.5 Nucleosynthesis in Dynamic Ejecta

While the first hydrodynamic simulations of CBMs have been performed with the intention to determine the mass of the ejecta (Davies et al. 1994), recent models aim to pin down the conditions for r-process nucleosynthesis. This includes the expansion timescales and the Y_e distribution of the ejecta, which are affected by the mass asymmetry of the merging objects, the equation of state (EOS), the treatment of gravity, the neutrino treatment, and the simulation method (grid, SPH). Both the EOS and the gravity prescription determine the radii of the neutron stars, setting the conditions for the actual merger phase. With smaller radii, the neutron stars are more compact and the collision is more energetic. This means that more matter from the center of the collision (the interaction region) is flung out and ejected, with very high initial temperatures. If the collision is less energetic, the component which becomes gravitationally unbound from tidal forces dominates. The matter from the tidal component is cooler than the interaction component. While other mass loss channels related to CBMs exist (see Sect. 11.2.6), the dynamic ejecta have the lowest Y_e and constitute the most promising source for heavy r-process elements. Typical velocities for the dynamic ejecta are around 0.1 c. It is therefore generally expected that the effect of neutrino irradiation is small and the Y_e stays low during the first second of the expansion.

Goriely et al. (2011) and Korobkin et al. (2012) both found narrow Y_e distributions of the dynamic ejecta centered around 0.015 (0.03) for a series of NS-NS and NS-BH mergers with different mass asymmetries. The latter further showed that the low Y_e values guarantee a robust r-process abundance pattern across the different models even for differing temperature and density evolutions, since the reaction flow proceeds along the neutron-drip line, in a (n, γ)-(γ, n) equilibrium (see also Eichler et al. 2015). In addition, the extremely neutron-rich conditions lead to several fission cycles, which further enhance the robustness in the abundance pattern beyond the second peak, while lighter elements are not produced in significant amounts. The insensitiveness of the abundances with respect to the temperature and density evolution means that the final abundance pattern is almost purely determined by the nuclear physics employed in the nuclear reaction network, providing an ideal testing ground for nuclear physics. Several studies have investigated the impact of nuclear mass models, β-decay rates, fission barriers, fission fragment distribution models, nuclear heating, and neutrino interactions (Panov et al. 2008; Surman et al. 2008; Panov et al. 2013; Bauswein et al. 2013; Goriely et al. 2013; Rosswog et al. 2014; Mumpower et al. 2015; Goriely et al. 2015; Mendoza-Temis et al. 2015; Eichler et al. 2015; Mumpower et al. 2017; Roberts et al. 2017).

As neutron capture rates and β-decay rates directly depend on the nuclear masses, so do the final abundances. This was shown in detail in Mendoza-Temis et al. (2015), where four independent mass models (FRDM (Möller et al. 1995), WS3 (Liu et al. 2011), DZ31 (Duflo and Zuker 1995), and HFB-21 (Goriely et al. 2010)) were employed to calculate the r-process abundances on the same hydrodynamic model of a neutron star merger. They found that mass models with small neutron separation energies around the $N = 130$ isotone exhibit a shift of the third abundance peak (compared to the solar abundances) due to a hold-up in the reaction flow in these nuclei. In addition, it was shown in Eichler et al. (2015) that the shift is amplified by late captures of neutrons which are emitted from fission reactions after the freeze-out from the (n, γ)-(γ, n) equilibrium (see also Caballero et al. 2014), while β-delayed neutrons also contribute. The strength of the effect depends on the timing of the last fission cycle with respect to the freeze-out from (n, γ)-(γ, n) equilibrium. The β-decay rates along the reaction path determine the waiting points where material is accumulated as well as the duration of the r-process. Therefore, they also have a say in the position of the third peak in the final abundances. Recent experimental (Domingo-Pardo et al. 2013; Kurtukian-Nieto et al. 2014) and theoretical (Suzuki et al. 2012; Zhi et al. 2013; Panov et al. 2015; Marketin et al. 2016) β-decay rate determinations find faster rates for the heaviest neutron-rich nuclei, decreasing the shift effect after freeze-out. Caballero-Folch et al. (2016) find that for nuclei beyond $N = 126$, however, the often used theoretical β-rates of Möller et al. (2003) give good predictions. Different fission fragment distribution models (FFDMs) were also tested within the framework of Eichler et al. (2015), revealing that the region around the second abundance peak is shaped by the fission fragments produced in the last fission cycle (see also Goriely 2015).

Late-time effects also affect the small peak around the rare-earth region (Surman and Engel 2001; Mumpower et al. 2012). An enlightening illustration of how nuclear masses shape the final rare-earth abundances is given in Mumpower et al. (2017). Their approach is fundamentally different, as they vary nuclear masses in Monte-Carlo calculations and from these determine new β-decay and neutron capture rates. The comparison of the obtained abundance pattern with the solar abundances in the region of the rare-earth peak allows to draw conclusions about the unknown masses of the nuclei far from stability in this mass range.

Depending on the setup of the hydrodynamical model, the Y_e distribution of the dynamic ejecta can be broader, which also allows for the production of first peak material in the dynamic ejecta (Wanajo et al. 2014; Sekiguchi et al. 2015, 2016; Radice et al. 2016), possibly at the price of the robustness discussed above. In order to settle this question, the neutrino treatment, in particular, has to be improved. As it is computationally very expensive, approximations are still used in simulations.

11.2.6 Disk Formation and Ejected Matter

In order to be able to asses the EM emission summarized in Fig. 11.1 and perform accurate nucleosynthesis calculations, it is crucial to make simulations as accurate as possible in order to compute the amount of mass that is ejected from these systems and its properties, such as temperature distribution and electron fraction. It is also important to compute the properties of the disks that may be formed after merger. Numerical simulations have in particular shown that, in the case of NS-NS mergers, larger disks with masses of up to $\sim 0.3 M_\odot$ are produced in mergers that form hypermassive neutron stars (i.e., NSs with masses larger than the maximum mass for an uniformly rotating NS and that collapse to a BH in less than ~ 1s after merger), while systems that produce a prompt collapse to BH are left with much smaller disks, with masses smaller than $\sim 10^{-2} M_\odot$ (Rezzolla et al. 2010; Giacomazzo et al. 2013). The mass of the disk can also be affected by the mass ratio of the system, with lower mass ratios producing larger disks. In the case of NS-BH merger, where mass ratios as low as $1/7$ can be expected, the BH spin plays an essential role in the disk formation, where disks as massive as $\sim 0.1 M_\odot$ may be formed if the BH was rapidly rotating with a spin of ~ 0.9 (Foucart 2012).

On timescales of ~ 100 ms a part of the disk becomes gravitationally unbound due to viscous dissipation and nuclear recombination. The fraction of the disk mass ejected this way depends sensitively on the survival time of the hypermassive NS (HMNS) before it collapses into a black hole, but according to simulations, around 20–25 % of the disk material can be ejected this way, making it an important factor for the overall ejecta. The timescales are much longer than for the dynamic ejecta, and the disk material, which eventually is ejected, can have a much larger Y_e due to neutrino interactions. Just et al. (2015) and Wu et al. (2016) show that in this component nuclei up to the third r-process peak (A=195) can be produced, and that it is a reliable source for second peak (A=130) material.

Similar to the neutrino-driven wind in core-collapse supernovae, a neutrino-driven wind component has been found in CBMs, where neutrinos can transfer enough energy to nuclei in the disk resulting in a baryonic wind. This mass loss channel mainly ejects matter in a direction usually perpendicular to the orbital plane, adding an angular dependence to light curve models. This mass loss channel relies on the absorption of neutrinos, which means that Y_e in the initially very neutron-rich material increases. Martin et al. (2015) find a Y_e distribution between 0.2 and 0.4, and a total ejecta mass of almost 10^{-2} $M_\odot$, comparable to the mass of the dynamic ejecta. The mass of the neutrino-driven wind depends on the total mass of the accretion disk. Due to the higher electron fractions, this channel is not thought to host a full r-process, but a weak r-process which produces first r-process peak nuclei.

In order to obtain the full picture, all the three mass loss channels need to be considered, adding complexity to the determination of the nuclear composition in the ejecta of a CBM. Figure 11.3 presents an overview on the different channels and their dynamical timescales.

The study of matter ejected in these systems has also provided new interesting results with general relativistic simulations showing that, in the case of NS-NS mergers, the shocks developed during merger have a mayor role in ejecting matter with respect to tidal disruption and that hence more compact NSs produce larger ejecta, in contradiction with previous Newtonian results that were underestimating

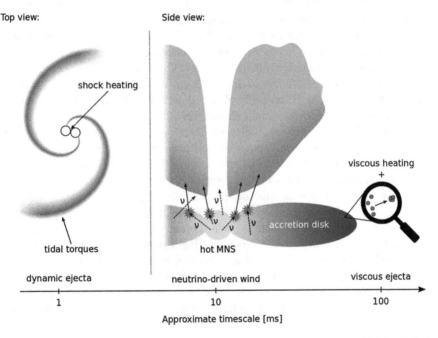

Fig. 11.3 Mass loss channels in a CBM, shown face-on (left) and edge-on (right). Typical timescales for the outflows are indicated at the bottom. Figure courtesy of D. Martin

the energy released in shocks. Tidal disruption is still instead the main mechanism for ejecting matter in NS-BH mergers, but also in this case the NS compactness, as well as the BH spin and mass ratio, play a fundamental role with less compact NS ejecting more matter, when keeping the mass ratio and BH spin fixed (Foucart 2012).

11.2.7 Short Gamma-Ray Bursts

One of the strongest EM signals expected from the merger of two NSs or of an NS and a BH are short gamma-ray bursts. Even before the simultaneous detection of GW170817 and GRB 170817A (Abbott et al. 2017a), there were already several indirect evidences that linked such systems to the central engine of SGRBs (Berger 2014). The main idea is that NS-BH and NS-NS mergers will produce a spinning BH surrounded by an accretion disk. Such a system could then launch a relativistic jet which, as in the case of long GRBs, will produce gamma-ray emission, e.g., via internal shocks.

The simultaneous detection of GW170817 and GRB 170817A proved that at least some type of SGRBs are indeed caused by BNS mergers. The detection was quite a surprise since it was thought to be quite unlikely. The most optimistic rate was estimated to be of up to ~ 2 simultaneous detections every year for the advanced LIGO and Virgo detectors at design sensitivities (Clark et al. 2015). This was due to the probably collimated nature of SGRBs, which are observed with median jet opening angles of 16 ± 10 degrees, but such measurements are very scarce being based on the few detected jet breaks (Fong et al. 2015). Moreover GRB 170817A was quite a peculiar SGRB with a delayed X-ray afterglow (Troja et al. 2017) that is thought to be due to the jet being observed off axis (Lazzati et al. 2017) or to a chocked jet (e.g., the cocoon model (Mooley et al. 2018)).

Numerical and theoretical studies are therefore still one of the best ways to try to assess the connection between SGRB properties and CBM. In the last few years there have been a number of efforts, as also described in more detail in the following sections, in order to try to explain current observations. As of today the main mechanism to produce the observed gamma-ray emission and at the same time explain short time-scale variability in the lightcurves is for the central engine to produce a strongly relativistic jet, with Lorentz factor of order ~ 100 or larger. Such jets need also to be collimated and have a sufficiently long (short) lifetime to fit the burst duration. Moreover, any theoretical model needs also to be able to explain the afterglow emission, including extended emission in the gamma rays and X-ray plateaus. All of this is clearly challenging and in the following sections we will describe our current understanding, thanks also to numerical simulations of CBMs.

11.2.8   Jet Formation

While it has been known for a while that accretion disks around spinning BHs can
produce relativistic jets via neutrino–antineutrino annihilation (Birkl et al. 2007) or
via magnetic fields (Blandford and Znajek 1977; Komissarov and Barkov 2009),
only very recently simulations of NS-NS and NS-BH mergers have started to show
the formation of such jets, even if with quite low Lorentz factors (lower than ~ 2).
In particular magnetic fields can give rise to relativistic jets via the Blandford-
Znajek (BZ) mechanism (Blandford and Znajek 1977). In the BZ mechanism, which
describes a spinning BH immersed in a magnetically dominated plasma, magnetic
fields can extract energy from the BH and produce a relativistic outflow.

The first simulation showing that NS-NS mergers may produce relativistic jets
was the so called "Missing Link" paper (Rezzolla et al. 2011) where the authors
showed for the first time that a collimated and mainly poloidal magnetic field
was produced at the end of a simulation of an equal-mass system. While the
formation of a collimated poloidal magnetic field structure is a necessary condition
for the production of a relativistic jet (De Villiers et al. 2005; McKinney 2006),
that simulation was not able to actually show the formation of a jet. The main
reason is that in order to activate the Blandford-Znajek mechanism (Blandford
and Znajek 1977), a magnetically dominated region needs to be formed in the BH
ergosphere (Komissarov and Barkov 2009). Very recently simulations of NS-NS
and NS-BH mergers performed with much higher magnetic fields (i.e., $\sim 10^{16}$ G
at the NS center) were instead able to see the formation of a mildly relativistic
jet (Paschalidis et al. 2015; Ruiz et al. 2016), even if only for one specific model
with a simple ideal-fluid EOS. Figure 11.4 shows the end result of those simulations
with the formation of an outflow along a magnetically dominated funnel. More
recently, following what was done in Rezzolla et al. (2011) and Kiuchi et al. (2014),
a different group performed simulations of different NS-NS systems with different
EOSs, mass ratios, and magnetic field orientations (Kawamura et al. 2016). Even if
in this case no jet was produced because of the much lower initial magnetic field
value (i.e., $\sim 10^{12}$ G), all the simulations showed the formation of a collimated and

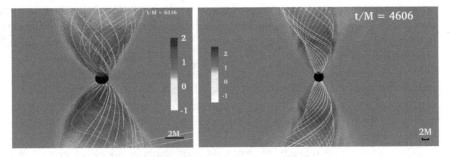

Fig. 11.4 Ratio of magnetic energy density over rest-mass energy density at the end of a NS-BH
merger (Paschalidis et al. 2015) and NS-NS merger (Ruiz et al. 2016) respectively

mainly poloidal magnetic field configuration at the end of the simulations. Therefore it seems that, at least in the case of NS-NS mergers, the results produced in Ruiz et al. (2016) should be general.

11.2.9 The Problem of Magnetic Field Amplification

Magnetic fields can play a crucial role in launching relativistic jets and in giving rise to SGRBs and other EM emission. It is therefore important to better understand their evolution in NS-NS and NS-BH mergers. While their evolution can be mainly neglected during the inspiral phase (Giacomazzo et al. 2009), hydrodynamic and magnetohydrodynamic instabilities and turbulence may strongly affect the magnetic field structure and strength during and after merger. There are two instabilities that play an important role in the magnetic field evolution: Kelvin-Helmholtz (KH) and the Magneto-Rotational Instability (MRI).

The KH instability develops when a shear layer is formed in the velocity field during the merger of two neutron stars. A series of vortices can be formed in this case and they can twist magnetic field lines. So, even when starting with a purely poloidal magnetic field configuration, the KH instability may lead to the production of a strong toroidal component and amplify the total magnetic field. How strong that amplification can be is still matter of debate. Preliminary Newtonian simulations (Price and Rosswog 2006) showed that the magnetic field can be amplified by several orders of magnitude up to at least $\sim 10^{15}$G. Such simulations seem to be supported by local simulations of turbulent fluid both in Newtonian (Obergaulinger et al. 2010) and Special Relativistic MHD (Zrake and MacFadyen 2013). In these local high-resolution simulations magnetic fields are shown to be amplified until the magnetic energy reaches equipartition with the kinetic energy of the turbulent flow. This means possible magnetic field amplifications up to values of $\sim 10^{16-17}$G.

Unfortunately in order to reproduce the results of such local simulations in a global simulation of NS-NS mergers one would need to run with much higher resolutions than those that can be currently afforded. A recent attempt was done by Kiuchi and collaborators who performed simulations with resolution of up to 17.5 meters (Kiuchi et al. 2015), ~ 1 order of magnitude higher that the typical resolution used in NS-NS mergers (~ 200 meters). Their simulations show an amplification in the magnetic energy of up to ~ 6 orders of magnitude and with magnetic fields growing from $\sim 10^{13}$ G to $\sim 10^{16}$ G. They also showed that the saturation of the magnetic energy should be at least $\sim 0.1\%$ of the bulk kinetic energy. Such simulations were possible thanks to the use of the K supercomputer in Japan, but their computational cost was so high that the group was not able to study what happens to the magnetic field after this initial amplification (the simulations were terminated soon after merger).

Other groups have recently proposed different subgrid models to mimic such large amplification without the need to use such expensive resolutions (Giacomazzo et al. 2015; Palenzuela et al. 2015).

While the KH instability and its related magnetic field amplification can be present only in NS-NS merges, MRI (Balbus and Hawley 1991) can affect the evolution of magnetic fields both in NS-NS and NS-BH mergers. In the former the magnetic field may be amplified both in the differentially rotating post-merger remnant and in the disk, while in the latter only in the disk.

11.2.10 Magnetar Formation

In case a long-lived NS is formed after a NS-NS merger, large magnetic fields may be able to give rise to a magnetar (Giacomazzo and Perna 2013). The production of a long-lived magnetar may be used to explain some EM observations, such as the X-ray plateaus observed in some SGRBs (Rowlinson et al. 2013). The main idea is that the angular momentum of the magnetar is extracted via magnetic fields and powers a long-duration EM emission. Unfortunately, while GRMHD simulations of NS-NS mergers have improved significantly the level of details with which such mergers are studied, they are still too expensive to be able to follow the magnetar for the long time scales ($\sim$ hours) required to match with current EM observations. Simple models have been proposed to overcome this issue and provide estimates for the EM emission that can be produced by post-merger NS remnants (Siegel and Ciolfi 2016a,b). In particular such models can provide an estimate of the X-ray emission and link it to the one observed in some SGRBs.

We stress though that the formation of a magnetar requires not only a large amplification of magnetic fields inside the NS remnant, but also the formation of a powerful poloidal magnetic field outside the NS remnant. It is not clear at the moment how the large magnetic field that is expected to be produced inside the remnant can emerge outside and form the typical poloidal configuration expected in magnetars.

11.2.11 The Time-Reversal Scenario

The "standard" scenario of SGRBs, depicted in Fig. 11.1, is based on the idea that the jet powering the SGRB is emitted from a BH surrounded by an accretion disk and that the afterglow is emitted by the interaction of the jet with the surrounding interstellar medium. An alternative model, the magnetar one, predicts instead that the SGRB is produced by the large magnetic energy contained in the magnetar formed soon after merger. The latter model excludes NS-BH as the source of SGRBs and it has the problem of explaining how highly relativistic jets may be emitted due to the baryon pollution problem. Numerical simulations show indeed that, when a NS is formed after the merger, it is surrounded by a dense baryonic region with densities of $\sim 10^{10}$ g cm^{-3} (Ciolfi et al. 2017). Such densities seem to be too high

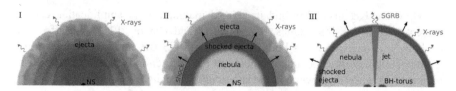

Fig. 11.5 The "time-reversal" scenario different evolution phases: (I) The differentially rotating supramassive NS ejects a baryon-loaded and highly isotropic wind; (II) The cooled-down and uniformly rotating NS emits spin-down radiation inflating a photon-pair nebula that drives a shock through the ejecta; (III) The NS collapses to a BH, a relativistic jet drills through the nebula and the ejecta shell and produces the prompt SGRB, while spin-down emission diffuses outwards on a much longer timescale. Figure and caption are taken from Ciolfi and Siegel (2015)

to allow the production of relativistic jets with Lorentz factors as high as those measured in SGRB jets (i.e., ~ 100–1000).

The former model instead has difficulties explaining the X-ray plateaus that seem to indicate that energy is injected into the system on time scales much larger than the typical survival time for an accretion disk (typically less than 1 s).

Therefore alternative models have been developed in order to try to solve both issues. In particular the two models presented in Rezzolla and Kumar (2015), Ciolfi and Siegel (2015) and called respectively "two-winds" and "time-reversal" scenario are both based on the idea that a long-lived magnetar is formed after merger. In Fig. 11.5 we present the main stages of the evolution of the system as described in the "time-reversal" scenario (Ciolfi and Siegel 2015).

Both models require a long-lived NS to be formed after merger. Therefore both models exclude NS-BH mergers as the source of SGRB (while they are not excluded by the "standard" scenario). The idea is that a baryon-rich wind is generated by the highly-magnetized remnant followed by the production of a faster but baryon-poor wind. The interaction between the two gives rise to an X-ray emission. Once the NS collapses to a BH and a BH-disk system is formed a relativistic jet may be finally launched and give rise to the SGRB. The wind will still produce the X-ray emission and therefore, what we observe as a plateau in the afterglow is actually an emission that originated from an earlier stage of the evolution with respect to the jet (from this the name "time-reversal" (Ciolfi and Siegel 2015)).

The problem with these models is that they require a disk to be formed after the collapse to BH of the long-lived NS. Simulations of uniformly rotating NS collapsing to BH do not show indeed the formation of disks (Baiotti et al. 2005, 2007). Other groups have expressed also doubts that such disks may be formed on the base also of simpler analytic models (Margalit et al. 2015). We note though that such models were based on the idea that the NS collapsing to a BH was not already surrounded by an accretion disk. If such disk was present at the moment of collapse it would not be destabilized by the collapse and hence form a BH-disk system (Giacomazzo and Perna 2012). Moreover magnetic fields may also be able to redistribute angular momentum in the rapidly rotating remnant and help the formation of a disk (Duez et al. 2006), even if this effect will be more relevant initially when the external regions of the star are still differentially rotating. Unfortunately such long timescales cannot be simulated and it is therefore difficult at the moment to state if a disk will be present or not and this is a strong constraint for both these scenarios.

11.2.12   Kilonovae/Macronovae

The r-process produces heavy and very neutron-rich nuclei which eventually decay radioactively to stability by means of β-, α-decays, and fission. Li and Paczyński (1998) were the first to suggest that these series of decays can lead to an electromagnetic transient, similar to the decay of ^{56}Ni in core-collapse SNe or Type Ia SNe. This was followed by light curve predictions based on real r-process compositions performed by Metzger et al. (2010), who assumed opacities in the ejecta similar to that in CCSN ejecta. Only later it was found that the ejecta in a CBM stay optically thick for a long time, even though they expand quickly and comprise only a small mass. This is due to the production of large amounts of lanthanides, which have a high line transition density and therefore a particularly high opacity (Kasen et al. 2013; Tanaka and Hotokezaka 2013). As a consequence, the emitted photons are absorbed and the ejecta are thermalized before radiation can break out eventually when the densities have sufficiently decreased. If only heavy r-process elements ($A > 130$) are synthesised, the high opacities cause the light curve to peak at a timescale of about 1 week instead of 1 day, at a lower luminosity and at a wavelength in the near-infrared region, as opposed to optical or ultra-violet. A good approximation for the time of peak luminosity is given e.g., in Metzger (2016):

$$t_{peak} \approx 1.6\,\mathrm{d} \left(\frac{M}{10^{-2}\,M_\odot} \right)^{1/2} \left(\frac{v}{0.1\,c} \right)^{1/2} \left(\frac{\kappa}{1\,\mathrm{cm}^2\,\mathrm{g}^{-1}} \right)^{1/2}, \qquad (11.6)$$

where M is the mass, v is the velocity, and κ is the opacity of the ejecta.

Lanthanide production is not guaranteed throughout the ejecta. In less neutron-rich conditions ($Y_e > 0.3$) the composition is almost lanthanide-free, with a direct consequence on the opacity. Electromagnetic emission from this lanthanide-free component would become visible already after about 1 day, and the spectrum would be at smaller wavelengths compared to the lanthanide-rich ejecta. Moreover, some simulations suggest that a small fraction of the ejecta is subject to extremely fast decompression, such that free neutrons are not captured but decay directly, producing a blue precursor to the kilonova that peaks on a timescale of a few hours (Metzger et al. 2015). A distinction can therefore be made into a red and a blue kilonova component, depending on the amount of lanthanides. Since the very neutron-rich dynamic ejecta provide large opacities along the equatorial plane (their emission is lanthanide-rich, therefore red), the blue component would only be visible if the event is viewed with a certain angle to the plane. The kilonova signal may therefore also depend on the viewing angle under which the event is observed.

Apart from the amount of lanthanides, the nuclear composition produced in the precedent r-process also sets the conditions for the nuclear heating rate in the ejecta. Since decay half-lives become longer as the composition approaches stability, the nuclear heating rate is another big factor in the temporal evolution of the kilonova light curve. The details of the heating rate depend on the nuclear composition and its decay history as well as the thermalization efficiency ϵ_{th} with which the decay products thermalize with the ejecta. However, a fitting formula can be found for a large sample decaying to stability. Korobkin et al. (2012) suggest the following fit for the specific heating rate:

$$\dot{\epsilon}(t) = \epsilon_0 \left(\frac{1}{2} - \frac{1}{\pi} \arctan \frac{t - t_0}{\sigma} \right)^\alpha \times \left(\frac{\epsilon_{th}}{0.5} \right) \tag{11.7}$$

with $\epsilon_0 = 2 \times 10^{18}$ erg g^{-1} s^{-1}, $t_0 = 1.3$ s, $\sigma = 0.11$ s, and $\alpha = 1.3$. The thermalization efficiency ϵ_{th} itself is time-dependent and varies significantly for different thermalization processes (e.g., photoionization, Compton scattering, Coulomb interactions, etc.), but again the collective behaviour can be approximated by (see Barnes et al. 2016)

$$\epsilon_{th}(t) = 0.36 \left[\exp(-at) + \frac{\ln(1 + 2b\,t^d)}{2b\,t^d} \right], \tag{11.8}$$

where $a, b,$ and d are parameters that depend on the mass and velocity of the matter in the layer of ejecta considered. Values for these parameters are given for a range of ejecta masses and velocities in Table 1 of Barnes et al. (2016).

The detailed dependence of the overall heating rate and the kilonova light curve on the initial electron fraction Y_e, the specific entropy e, and the expansion timescale τ is demonstrated in Lippuner and Roberts (2015). Furthermore, Barnes et al. (2016) and Hotokezaka et al. (2016) have investigated the nuclear heating rates from the individual decay types (α-decay, β-decay, and fission) and find a sensitivity to the initial nuclear composition, which in turn is strongly affected by the nuclear mass model employed for the r-process calculation. The discrepancies mainly originate from the different abundances of actinides ($220 \leq A \leq 250$) which usually undergo α-decay, as well as the large uncertainties in fission rates. The relationship between the nuclear composition before decay and the kilonova heating rates, if understood better, could ultimately lead to new independent constraints on the nuclear mass model for neutron-rich nuclei derived from kilonova light curves.

Such theoretical predictions is supported by recent detections in the afterglow emission of some SGRBs. The first claim for the possible detection of a kilonova was made for GRB 130603B (Tanvir et al. 2013), a SGRB at redshift $z = 0.356$. Unfortunately this claim was supported only by a single data point which showed a stronger emission in the near infrared with respect to what one would expect from an afterglow emission due to a decelerating jet. Nevertheless this was the first observation that could be explained as a kilonova powered by an ejected amount of matter between 10^{-2} and $10^{-1} M_\odot$. There were also other detections of kilonovae possibly associated with GRB 060614 (Yang et al. 2015; Jin et al. 2015) and with GRB 050709 (Jin et al. 2016). Now the observation of the kilonova GW170817 confirms the production of heavy r-process and lighter heavy elements in neutron star mergers. The electromagnetic followup of GW170817 shows a initial blue emission associated with lanthanide free ejecta and thus $Y_e > 0.3$ followed by a red/infrared late emission due to heavy r-process elements (e.g. Kilpatrick et al. 2017).

11.2.13 Electromagnetic Precursors

While in the previous part of this chapter we focused on EM emission powered after the merger, also because this is expected to be the most powerful one, there is also the possibility to have EM emission before merger.

Observations of SGRB precursors (Troja et al. 2010) have indeed triggered studies of possible EM emission generated before mergers. There are two main mechanisms that have been studied. The first one is related to the possibility that the NS crust may break because of tidal interaction between the NSs in a NS-NS merger (Tsang et al. 2012). The second one is instead due to the interaction between the NSs magnetosphere in a NS-NS binary (Palenzuela et al. 2013b,a) or between the magnetic field of the NS and a BH in the case of a NS-BH system (Paschalidis et al. 2013; McWilliams and Levin 2011). In both cases the expected luminosity may be quite low with maximum values between $\sim 10^{40}$ and $\sim 10^{43}$ erg/s in the hard X-ray part of the EM spectrum (Palenzuela et al. 2013b; Paschalidis et al. 2013).

No precursor was observed in the case of GW170817 (Abbott et al. 2017c), but such a possible detection would have required X-ray and gamma-ray observations of the site of the BNS before the detection of the GW peak. Therefore an early GW detection of the inspiral signal is required in order to be able to point EM telescopes before merger.

11.3 Present Challenges and Future Prospects

Numerical and theoretical modelling of EM counterparts and nucleosynthesis in CBM have made significant progresses in the last few years, but the number of questions that remain to be addressed is still large. In the following section we provide an overview of the main topics that need to be addressed in the next years.

11.3.1 Electromagnetic Emission

One of the main challenges is to provide accurate estimates of the lightcurves of the EM emission produced by NS-NS and NS-BH mergers. Most of the simulations (if not all of them) provide only preliminary estimates of the EM luminosity, often based simply on the integration of the Poynting flux. Such luminosity, while useful to get an idea of the possible maximum luminosity produced by such events, do not help in fitting current observation (e.g., SGRBs). It is therefore crucial to better address the long-term evolution of matter ejecta and magnetic field interaction.

In the case of matter ejected via tidal disruption during the late phase of inspiral or via shocks at merger, it is fundamental to take into account finite-temperature effects and neutrino emission and absorption. Very few simulations, as of today, have started to include both effects in full general relativity.

In the case of systems producing a BH surrounded by an accretion disk, there is preliminary evidence for the formation of jets, but how these jets depend on the properties of the binary has not been studied yet. Up to now there have been indeed only two papers (by the same group and with the same code) showing the production of a mildly relativistic outflow (Paschalidis et al. 2015; Ruiz et al. 2016). In both cases the initial magnetic field was several orders of magnitude larger than the typical magnetic field expected in merging neutron star binaries. While in the case of NS-NS binaries large magnetic fields may be produced during merger, but only in the region inside the HMNS/SMNS, such large amplification is not expected in NS-BH mergers, but it may only be produced later in the disk (if any). Therefore it is not clear yet if such results will apply also to simulations starting with lower and more realistic magnetic field values. Clearly these simulations will also need to properly resolve the turbulent scales on which the magnetic field is amplified and hence require large computational resources (Kiuchi et al. 2015).

In the case of systems producing a long-lived NS after merger, most of the studies have been focused on a detailed calculation of the GW emission and very few on the possible EM counterparts that such systems may emit (Siegel and Ciolfi 2016a,b). It is not currently possible to follow these systems numerically on the long time scales required to study their long-term EM emission. It will therefore be important to be able to better study the properties of post-merger remnants in order to develop better (analytic) models.

In all cases, it seems evident that a correct description of finite-temperature effects, magnetic field evolution, neutrino emission and reabsorption, and turbulence is crucial to provide more accurate models. Unfortunately all of this will require more robust and powerful computational algorithms and the use of faster supercomputers.

11.3.2 Nucleosynthesis

The young field of nucleosynthesis in CBMs has made considerable progress in the past few years, culminating in the observation of the kilonova related to GW170817. Nevertheless, many open questions remain, connected both to the hydrodynamical simulations and the nuclear physics. Since the r-process operates close to the neutron-drip line in the nuclear chart, no experimental data exist for the extremely neutron-rich isotopes that lie on the typical reaction path. The nuclear mass models that are used for nucleosynthesis calculations all reproduce the known nuclear data very well, but drastically diverge with increasing distance to the valley of stability, as is shown e.g., in Goriely et al. (2007).

Next-generation experimental facilities (e.g., RIKEN in Japan, while FAIR at GSI Darmstadt and FRIB at Michigan State University are currently being built at the time of writing) will push the boundaries of experimentally studied isotopes, in some mass regions almost reaching the typical r-process reaction path. The new data from mass measurements will significantly help to further constrain nuclear mass models.

On the theoretical side, more extensive models based on density functional theory are needed, e.g., fission barrier and fragment distribution calculations (Sadhukhan et al. 2016) or β-decay rates (Marketin et al. 2016). Efforts should be made to incorporate new experimental data and reach consistent treatments of (theoretically derived) nuclear properties. Recently, a lot of progress has been made in modelling the EM afterglow powered by radioactive decay of the r-process products. The link between the nuclear compositions achieved in the r-process calculations with different mass models and reaction rates and the radioactive heating rates should be studied further.

Equation (11.5) shows that for a given mass model the only factors that determine which nuclei lie on the r-process path in a hot r-process scenario are the temperature and the neutron density, with the latter directly dependent on the electron fraction Y_e. These are quantities that are evolved in the tracer particles used for nucleosynthesis post-processing, and in turn depend on the framework of the hydrodynamical setup, such as the EoS, the neutrino treatment, etc. The different Y_e and temperature distributions in the ejecta of different simulations therefore add another layer of uncertainties to any calculated r-process results. Further future detections of gravitational waves from binary neutron star mergers will greatly constrain the high-density EOS, since any gravitational wave signal carries the signature of the EOS (Radice et al. 2016; Kuroda et al. 2016; Clark et al. 2016; Chirenti et al. 2017; Richers et al. 2017; Bauswein et al. 2017). The three gravitational wave detectors LIGO-Hanford, LIGO-Livingston, and Virgo are currently being upgraded. Additional GW observatories (KAGRA, and LIGO India) are already being built or in the final planning stages. Future observing runs from these facilities will have a sensitivity high enough to detect several CBMs per year. The direction of any GW signal has to be determined by means of triangulation. Therefore the additional detectors are essential to constrain the coordinates of the GW event and enable follow-up observations of the ensuing kilonova, as it has been demonstrated by the LIGO-Virgo observation of GW170817 (Abbott et al. 2017b).

Parallel to gravitational wave signals, detections of kilonovae are needed to match theoretically derived light curves, which will provide new insights on nuclear aspects such as the dominant decay types and the end point of the r-process. These comparisons will allow for a quality assessment of the currently available predictions of nuclear properties far from stability (see Sect. 11.2.12). Furthermore, observed light curves will also let us draw conclusions about the amount of (dynamically) ejected matter. Observations including detailed study of features (i.e., absorption/emission lines) in kilonova spectra can also rise new insights about the composition of the ejecta in combination with the peak luminosity time and color (e.g., Pian et al. 2017).

11.4 Conclusions

Compact binary mergers are currently one of the most exciting astrophysical scenarios to be studied in the GW and EM spectrum, especially after the simultaneous observations of GW170817 and GRB170817A. That single event provided the first evidence that BNS mergers are among the main r-process site in our universe, but also established a link between them and at least some type of SGRBs. While the actual mechanism to produce the gamma-ray emission is not completely understood, the main idea is that a jet can be launched from the central BH formed after merger, emitting highly energetic gamma-rays, which would on earth be observed as a short GRB. Moreover, the radioactive decays of the fresh r-process products to stability power an electromagnetic transient known as "kilonova" (or "macronova") that peaks days after the actual merger in the optical or near-infrared wavelengths. In this chapter, the current status of research is presented with respect to nucleosynthesis in the ejecta of CBMs as well as the origins of the different kinds of electromagnetic counterparts that are expected in the aftermath of a merger. In particular, we have discussed the following aspects:

Possible Precursor EM Emission Before the merger takes place, the two neutron stars in a NS-NS merger affect each other by means of both their gravitational and magnetic fields, the former leading to a breaking up of the NS crust. Simulations have shown that these interactions can lead to EM emission before the merger phase. No precursor was observed in the case of GW170817, but such an observation would very probably require a very rapid EM search before the merger GW signal.

Dynamic Ejecta During the merger, matter can become gravitationally unbound mainly by two methods. First, a tidal tail forms behind each NS, ejecting cold, very neutron-rich matter at a high velocity. Second, at the collision interface matter gets heated and accelerated by the collision shock. Both components are expected to harbour a strong r-process, however the second component is considerably hotter than the tidal ejecta and subject to more neutrino interactions which alter the electron fraction. Therefore, the ratio of tidal ejecta to shocked ejecta can have an impact on the nucleosynthesis yields. Major uncertainties in the r-process nucleosynthesis yields are connected to the unknown properties of neutron-rich nuclei and the Y_e distribution of the ejecta. The radioactive decay of the heavy elements produced in the r-process later on powers the kilonova light curve (see below).

Short Gamma-Ray Bursts SGRBs are the result of highly relativistic jets that are launched from the central compact object shortly after the merger. Modern simulations with increased resolution or enhanced magnetic fields only recently have started to find favourable conditions for jet formation in NS-NS and NS-BH mergers, that is, a collimated and poloidal magnetic field configuration at the end of the simulation. The exact mechanism leading to the gamma-ray signal, however, is still uncertain. Different scenarios have been proposed and are summarized in this chapter. Apart from the "standard" scenario, where the relativistic jet is launched from a spinning BH surrounded by an accretion disk (which is the one possibly behind GRB170817A), other models rely on a long-lived NS emitting isotropic winds at different stages before its collapse to a BH and the resulting jet launch ("two-winds" and "time-reversal" scenarios).

The Fate of the Central Remnant In NS-NS mergers the dynamics of the post-merger remnant depends on the maximum stable NS mass ($M_{NS}^{max} \geq 2\ M_\odot$; still unknown today) and the initial masses of the two merging objects. If the resulting mass of the central object is larger than the maximum allowed NS mass, the system will collapse into a spinning BH. However, there is a possibility that the initial fast rotation can temporarily stabilize a central object that is only marginally more massive than M_{NS}^{max}, resulting in a hypermassive NS that can survive up to a few 100 ms. If the combined mass in a NS-NS merger is not larger than M_{NS}^{max}, a stable, supramassive NS is formed. In this case, simulations show that strong magnetic field amplifications can lead to the formation of a magnetar. Magnetars add complexity to the EM emission after a merger, as angular momentum can be extracted by the strong magnetic fields, powering a long-duration EM emission. This effect could explain some observed phenomena, e.g., X-ray plateaus in some SGRBs, and account for an additional heating source in the baryonic ejecta. However, more detailed and long-term simulations are needed in the future to pin down the role of this possible central engine.

Disk Formation and Additional Mass Loss The possible formation of a disk in the aftermath of a merger is linked to the fate of the central remnant. Simulations show that a prompt BH formation from a NS-NS merger leaves at most a small disk of around $10^{-2}\ M_\odot$, while HMNS or SMNS may be surrounded by much larger disks with up to 0.3 $M_\odot$. Fast spinning BHs in a NS-BH merger also allow for considerable disk formation. The existence of a disk opens up additional opportunities for the system to eject matter, which need to be considered for the successful modelling of CBMs. Namely, neutrinos from the central object can interact with the disk material and drive a baryonic wind (similar to the ν-driven wind scenario in CCSNe). According to simulations, material ejected this way amounts to about $10^{-2}\ M_\odot$ (although this number depends on the total mass of the accretion disk), comparable to the dynamic ejecta. However, the wind component is considerably less neutron-rich and therefore a good candidate for a weak r-process site which produces elements of the first r-process peak. On longer timescales, a fraction of the disk material becomes gravitationally unbound due to viscous dissipation and nuclear recombination. In this component, the whole range of r-process elements can be produced, as recent studies show.

EM Afterglow of the Ejecta (kilonova) The neutron-rich, heavy nuclei that have been freshly synthesized in the ejecta radioactively decay to stability, releasing energy that thermalizes the surroundings and leads to an EM transient. This phenomenon is known as a "kilonova" or "macronova", such as the one detected after GW170817. Depending on the opacity in the ejecta, the break-out of the EM radiation occurs on a timescale of days to weeks after the merger. The presence of lanthanides greatly increases the opacity and also leads to a red-shift in wavelength. The dependence of the EM signal on the nuclear r-process yields and the spatial configuration of the ejecta is the topic of ongoing research. Ultimately, future observations of kilonovae could provide more constraints on properties of nuclei along the r-process path. Since no significant amount of lanthanides is produced in the wind, the wind opacities are comparable to CCSNe, amounting to an EM emission component that peaks earlier and at smaller wavelengths.

Ejecta Shocks with the Interstellar Medium When the expanding baryonic and non-baryonic (jet) ejecta interact with the interstellar medium, the arising shock produces radio emission. This emission can potentially be observed even years after a CBM.

Synthesized r-Process Elements The ejecta will eventually mix with the interstellar medium and increase the content of heavy elements in new-born stars in the vicinity. Several independent observations of r-elements in metal-poor stars, the ocean crust, as well as the recent discovery of a dwarf galaxy that is strongly enhanced in r-elements point strongly toward a relatively rare r-process site ejecting about 10^{-2} $M_\odot$ of r-process material per event. These indications complement the observation of GW170817, suggesting that CBMs are a major production site for r-process nuclei, if not the only one.

Finally, we discuss future prospects for the research of CBMs in the context of nucleosynthesis and EM radiation signals. Most of the expected developments revolve around the new exciting opportunities arising with the emergence of multi-messenger astronomy. In particular, future follow-up EM observations of other CBM GW signals, including possibly a NS-BH merger, will provide more details about the properties of SGRBs central engines and the production of r-process material.

References

Abbott, B.P., Abbott, R., Abbott, T. D., Acernese, F., Ackley, K., Adams, C., Adams, T., Addesso, P., Adhikari, R.X., Adya, V.B., et al.: Gravitational waves and gamma-rays from a binary neutron star merger: GW170817 and GRB 170817A. Astrophys. J. Lett. **848**, L13 (2017)

Abbott, B.P., Abbott, R., Abbott, T. D., Acernese, F., Ackley, K., Adams, C., Adams, T., Addesso, P., Adhikari, R.X., Adya, V.B., et al.: GW170817: observation of gravitational waves from a binary neutron star inspiral. Phys. Rev. Lett. **119**, 161101 (2017)

Abbott, B.P., Abbott, R., Abbott, T. D., Acernese, F., Ackley, K., Adams, C., Adams, T., Addesso, P., Adhikari, R.X., Adya, V.B., et al.: Multi-messenger observations of a binary neutron star merger. Astrophys. J. Lett. **848**, L12 (2017)

Abbott, B.P., Abbott, R., Abbott, T. D., Acernese, F., Ackley, K., Adams, C., Adams, T., Addesso, P., Adhikari, R.X., Adya, V.B., et al.: Search for post-merger gravitational waves from the remnant of the binary neutron star merger GW170817. Astrophys. J. Lett. **851**, L16 (2017)

Aboussir, Y., Pearson, J.M., Dutta, A.K., Tondeur, F.: Nuclear mass formula via an approximation to the Hartree-Fock method. At. Data Nucl. Data Tables **61**, 127 (1995)

Antoniadis, J., Freire, P.C.C., Wex, N., Tauris, T.M., Lynch, R.S., van Kerkwijk, M.H., Kramer, M., Bassa, C., Dhillon, V.S., Driebe, T., Hessels, J.W.T., Kaspi, V.M., Kondratiev, V.I., Langer, N., Marsh, T.R., McLaughlin, M.A., Pennucci, T.T., Ransom, S.M., Stairs, I.H., van Leeuwen, J., Verbiest, J.P.W., Whelan, D.G.: A massive pulsar in a compact relativistic binary. Science **340**, 448 (2013)

Arcones, A., Martínez-Pinedo, G.: Dynamical r-process studies within the neutrino-driven wind scenario and its sensitivity to the nuclear physics input. Phys. Rev. C **83**, 045809 (2011)

Baiotti, L., Hawke, I., Montero, P. J., Löffler, F., Rezzolla, L., Stergioulas, N., Font, J.A., Seidel, E.: Three-dimensional relativistic simulations of rotating neutron star collapse to a Kerr black hole. Phys. Rev. D **71**, 024035 (2005)

Baiotti, L., Hawke, I., Rezzolla, L.: On the gravitational radiation from the collapse of neutron stars to rotating black holes. Class. Quantum Grav. **24**, S187–S206 (2007)

Balbus, S.A., Hawley, J.F.: A powerful local shear instability in weakly magnetized disks. I - Linear analysis. II - Nonlinear evolution. Astrophys. J. **376**, 214–233 (1991)

Barnes, J., Kasen, D., Wu, M.-R., Martínez-Pinedo, G.: Radioactivity and thermalization in the ejecta of compact object mergers and their impact on kilonova light curves. Astrophys. J. **829**, 110 (2016)

Bauswein, A., Goriely, S., Janka, H.-T.: Systematics of dynamical mass ejection, nucleosynthesis, and radioactively powered electromagnetic signals from neutron-star mergers. Astrophys. J. **773**, 78 (2013)

Bauswein, A., Just, O., Janka, H.-T., Stergioulas, N.: Neutron-star radius constraints from GW170817 and future detections. Astrophys. J. Lett. **850**, L34 (2017)

Beniamini, P., Hotokezaka, K., Piran, T.: r-process production sites as inferred from Eu abundances in dwarf galaxies. Astrophys. J. **832**, 149 (2016)

Berger, E.: Short-duration gamma-ray bursts. Annu. Rev. Astron. Astrophys. **52**, 43–105 (2014)

Beun, J., McLaughlin, G.C., Surman, R., Hix, W.R.: Fission cycling in a supernova r process. Phys. Rev. C **77**, 035804 (2008)

Birkl, R., Aloy, M.A., Janka, H.-T., Müller, E.: Neutrino pair annihilation near accreting, stellar-mass black holes. Astron. Astrophys. **463**, 51–67 (2007)

Blandford, R.D., Znajek, R.L.: Electromagnetic extraction of energy from Kerr black holes. Mon. Not. R. Astron. Soc. **179**, 433–456 (1977)

Burbidge, E.M., Burbidge, G.R., Fowler, W.A., Hoyle, F.: Synthesis of the elements in stars. Rev. Mod. Phys. **29**, 547–650 (1957)

Caballero, O.L., Arcones, A., Borzov, I.N., Langanke, K., Martinez-Pinedo, G.: Local and global effects of beta decays on r-process (2014). ArXiv e-prints

Caballero-Folch, R., Domingo-Pardo, C., Agramunt, J., Algora, A., Ameil, F., Arcones, A., Ayyad, Y., Benlliure, J., Borzov, I.N., Bowry, M., Calviño, F., Cano-Ott, D., Cortés, G., Davinson, T., Dillmann, I., Estrade, A., Evdokimov, A., Faestermann, T., Farinon, F., Galaviz, D., García, A.R., Geissel, H., Gelletly, W., Gernhäuser, R., Gómez-Hornillos, M.B., Guerrero, C., Heil, M., Hinke, C., Knöbel, R., Kojouharov, I., Kurcewicz, J., Kurz, N., Litvinov, Y.A., Maier, L., Marganiec, J., Marketin, T., Marta, M., Martínez, T., Martínez-Pinedo, G., Montes, F., Mukha, I., Napoli, D.R., Nociforo, C., Paradela, C., Pietri, S., Podolyák, Z., Prochazka, A., Rice, S., Riego, A., Rubio, B., Schaffner, H., Scheidenberger, C., Smith, K., Sokol, E., Steiger, K., Sun, B., Taín, J.L., Takechi, M., Testov, D., Weick, H., Wilson, E., Winfield, J.S., Wood, R., Woods, P., Yeremin, A.: First measurement of several β-delayed neutron emitting isotopes beyond N =126. Phys. Rev. Lett. **117**, 012501 (2016)

Cameron, A.G.W.: Nuclear reactions in stars and nucleogenesis. Publ. Astron. Soc. Pac. **69**, 201 (1957)

Chirenti, C., Gold, R., Miller, M.C.: Gravitational waves from F-modes excited by the inspiral of highly eccentric neutron star binaries. Astrophys. J. **837**, 67 (2017)

Ciolfi, R., Siegel, D.M.: Short gamma-ray bursts in the "Time-reversal" scenario. Astrophys. J. Lett. **798**, L36 (2015)

Ciolfi, R., Kastaun, W., Giacomazzo, B., Endrizzi, A., Siegel, D.M., Perna, R.: General relativistic magnetohydrodynamic simulations of binary neutron star mergers forming a long-lived neutron star. Phys. Rev. D **95**, 063016 (2017)

Clark, J., Evans, H., Fairhurst, S., Harry, I.W., Macdonald, E., Macleod, D., Sutton, P.J., Williamson, A.R.: Prospects for joint gravitational wave and short gamma-ray burst observations. Astrophys. J. **809**, 53 (2015)

Clark, J.A., Bauswein, A., Stergioulas, N., Shoemaker, D.: Observing gravitational waves from the post-merger phase of binary neutron star coalescence. Class. Quantum Grav. **33**, 085003 (2016)

Cowperthwaite, P.S., Berger, E., Villar, V.A., Metzger, B.D., Nicholl, M., Chornock, R., Blanchard, P.K., Fong, W., Margutti, R., Soares-Santos, M., Alexander, K.D., Allam, S., Annis, J., Brout, D., Brown, D. A., Butler, R.E., Chen, H.-Y., Diehl, H. T., Doctor, Z., Drout, M. R., Eftekhari, T., Farr, B., Finley, D.A., Foley, R.J., Frieman, J.A., Fryer, C.L., García-Bellido, J., Gill, M.S.S., Guillochon, J., Herner, K., Holz, D.E., Kasen, D., Kessler, R., Marriner, J., Matheson, T., Neilsen, E.H., Jr., Quataert, E., Palmese, A., Rest, A., Sako, M., Scolnic, D.M., Smith, N., Tucker, D.L., Williams, P.K.G., Balbinot, E., Carlin, J.L., Cook, E.R., Durret, F., Li, T.S., Lopes, P.A.A., Lourenço, A.C.C., Marshall, J.L., Medina, G.E., Muir, J., Muñoz, R.R., Sauseda, M., Schlegel, D.J., Secco, L.F., Vivas, A.K., Wester, W., Zenteno, A., Zhang, Y., Abbott, T.M.C., Banerji, M., Bechtol, K., Benoit-Lévy, A., Bertin, E., Buckley-Geer, E., Burke, D.L., Capozzi, D., Rosell, A.C., Kind, M.C., Castander, F.J., Crocce, M., Cunha, C.E., D'Andrea, C.B., da Costa, L.N., Davis, C., DePoy, D.L., Desai, S., Dietrich, J.P., Drlica-Wagner, A., Eifler, T.F., Evrard, A.E., Fernandez, E., Flaugher, B., Fosalba, P., Gaztanaga, E., Gerdes, D. W., Giannantonio, T., Goldstein, D.A., Gruen, D., Gruendl, R.A., Gutierrez, G., Honscheid, K., Jain, B., James, D.J., Jeltema, T., Johnson, M.W.G., Johnson, M.D., Kent, S., Krause, E., Kron, R., Kuehn, K., Nuropatkin, N., Lahav, O., Lima, M., Lin, H., Maia, M.A.G., March, M., Martini, P., McMahon, R.G., Menanteau, F., Miller, C.J., Miquel, R., Mohr, J.J., Neilsen, E., Nichol, R.C., Ogando, R.L.C., Plazas, A.A., Roe, N., Romer, A.K., Roodman, A., Rykoff, E.S., Sanchez, E., Scarpine, V., Schindler, R., Schubnell, M., Sevilla-Noarbe, I., Smith, M., Smith, R. C., Sobreira, F., Suchyta, E., Swanson, M.E.C., Tarle, G., Thomas, D., Thomas, R.C., Troxel, M.A., Vikram, V., Walker, A.R., Wechsler, R.H., Weller, J., Yanny, B., Zuntz, J.: The electromagnetic counterpart of the binary neutron star merger LIGO/VIRGO GW170817. II. UV, optical, and near-infrared light curves and comparison to kilonova models. Astrophys. J. Lett. **848**, L17 (2017)

Davies, M.B., Benz, W., Piran, T., Thielemann, F.K.: Merging neutron stars. 1. Initial results for coalescence of noncorotating systems. Astrophys. J. **431**, 742–753 (1994)

Demorest, P.B., Pennucci, T., Ransom, S.M., Roberts, M.S.E., Hessels, J.W.T.: A two-solar-mass neutron star measured using Shapiro delay. Nature **467**, 1081–1083 (2010)

De Villiers, J.-P., Hawley, J. F., Krolik, J. H., Hirose, S.: Magnetically driven accretion in the Kerr metric. III. Unbound outflows. Astrophys. J. **620**, 878–888 (2005)

Domingo-Pardo, C., Caballero-Folch, R., Agramunt, J., Algora, A., Arcones, A., Ameil, F., Ayyad, Y., Benlliure, J., Bowry, M., Calviño, F., Cano-Ott, D., Cortés, G., Davinson, T., Dillmann, I., Estrade, A., Evdokimov, A., Faestermann, T., Farinon, F., Galaviz, D., García-Rios, A., Geissel, H., Gelletly, W., Gernhäuser, R., Gómez-Hornillos, M.B., Guerrero, C., Heil, M., Hinke, C., Knöbel, R., Kojouharov, I., Kurcewicz, J., Kurz, N., Litvinov, Y., Maier, L., Marganiec, J., Marta, M., Martínez, T., Martínez-Pinedo, G., Meyer, B.S., Montes, F., Mukha, I., Napoli, D.R., Nociforo, C., Paradela, C., Pietri, S., Podolyák, Z., Prochazka, A., Rice, S., Riego, A., Rubio, B., Schaffner, H., Scheidenberger, C., Smith, K., Sokol, E., Steiger, K., Sun, B., Taín, J.L., Takechi, M., Testov, D., Weick, H., Wilson, E., Winfield, J.S., Wood, R., Woods, P., Yeremin, A., Approaching the precursor nuclei of the third r-process peak with RIBs (2013). ArXiv e-prints.

Drout, M.R., Piro, A.L., Shappee, B.J., Kilpatrick, C.D., Simon, J.D., Contreras, C., Coulter, D.A., Foley, R. J., Siebert, M.R., Morrell, N., Boutsia, K., Di Mille, F., Holoien, T.W.-S., Kasen, D., Kollmeier, J.A., Madore, B.F., Monson, A.J., Murguia-Berthier, A., Pan, Y.-C., Prochaska, J.X., Ramirez-Ruiz, E., Rest, A., Adams, C., Alatalo, K., Bañados, E., Baughman, J., Beers, T.C., Bernstein, R.A., Bitsakis, T., Campillay, A., Hansen, T.T., Higgs, C.R., Ji, A.P., Maravelias, G., Marshall, J.L., Bidin, C.M., Prieto, J.L., Rasmussen, K.C., Rojas-Bravo, C., Strom, A.L., Ulloa, N., Vargas-González, J., Wan, Z., Whitten, D.D.: Light curves of the neutron star merger gw170817/sss17a: implications for r-process nucleosynthesis. Science **358**(6370), 1570-1574 (2017)

Duez, M.D., Liu, Y.T., Shapiro, S.L., Shibata, M., Stephens, B.C.: Evolution of magnetized, differentially rotating neutron stars: Simulations in full general relativity. Phys. Rev. D **73**, 104015 (2006)

Duflo, J., Zuker, A.P.: Microscopic mass formulas. Phys. Rev. C **52**, R23–R27 (1995)

Eichler, M., Arcones, A., Kelic, A., Korobkin, O., Langanke, K., Marketin, T., Martinez-Pinedo, G., Panov, I., Rauscher, T., Rosswog, S., Winteler, C., Zinner, N.T., Thielemann, F.-K.: The role of fission in neutron star mergers and its impact on the r-process peaks. Astrophys. J. **808**, 30 (2015)

Fernández, R., Metzger, B.D.: Delayed outflows from black hole accretion tori following neutron star binary coalescence. Mon. Not. R. Astron. Soc. **435**, 502–517 (2013)

Fong, W., Berger, E., Margutti, R., Zauderer, B.A.: A decade of short-duration gamma-ray burst broadband afterglows: energetics, circumburst densities, and jet opening angles. Astrophys. J. **815**, 102 (2015)

Foucart, F.: Black-hole-neutron-star mergers: disk mass predictions. Phys. Rev. D **86**, 124007 (2012)

Freiburghaus, C., Rembges, J.-F., Rauscher, T., Kolbe, E., Thielemann, F.-K., Kratz, K.-L., Pfeiffer, B., Cowan, J.J.: The astrophysical r-process: a comparison of calculations following adiabatic expansion with classical calculations based on neutron densities and temperatures. Astrophys. J. **516**, 381–398 (1999)

Freiburghaus, C., Rosswog, S., Thielemann, F.-K.: R-process in neutron star mergers. Astrophys. J. Lett. **525**, L121–L124 (1999)

Giacomazzo, B., Perna, R.: General relativistic simulations of accretion induced collapse of neutron stars to black holes. Astrophys. J. Lett. **758**, L8 (2012)

Giacomazzo, B., Perna, R.: Formation of stable magnetars from binary neutron star mergers. Astrophys. J. Lett. **771**, L26 (2013)

Giacomazzo, B., Rezzolla, L., Baiotti, L.: Can magnetic fields be detected during the inspiral of binary neutron stars? Mon. Not. R. Astron. Soc. **399**, L164–L168 (2009)

Giacomazzo, B., Perna, R., Rezzolla, L., Troja, E., Lazzati, D.: Compact binary progenitors of short gamma-ray bursts. Astrophys. J. Lett. **762**, L18 (2013)

Giacomazzo, B., Zrake, J., Duffell, P., MacFadyen, A.I., Perna, R.: Producing magnetar magnetic fields in the merger of binary neutron stars. Astrophys. J. **809**, 39 (2015)

Giuliani, S.A., Martinez-Pinedo, G., Robledo, L.M.: Fission properties of superheavy nuclei for r-process calculations (2017). ArXiv e-prints.

Goriely, S.: The fundamental role of fission during r-process nucleosynthesis in neutron star mergers. Eur. Phys. J. A **51**, 22 (2015)

Goriely, S., Samyn, M., Pearson, J.M.: Further explorations of Skyrme-Hartree-Fock-Bogoliubov mass formulas. VII. Simultaneous fits to masses and fission barriers. Phys. Rev. C **75**, 064312 (2007)

Goriely, S., Chamel, N., Pearson, J.M.: Further explorations of Skyrme-Hartree-Fock-Bogoliubov mass formulas. XII. Stiffness and stability of neutron-star matter. Phys. Rev. C **82**, 035804 (2010)

Goriely, S., Bauswein, A., Janka, H.-T.: r-process nucleosynthesis in dynamically ejected matter of neutron star mergers. Astrophys. J. Lett. **738**, L32 (2011)

Goriely, S., Sida, J.-L., Lemaître, J.-F., Panebianco, S., Dubray, N., Hilaire, S., Bauswein, A., Janka, H.-T.: New fission fragment distributions and r-process origin of the rare-earth elements. Phys. Rev. Lett. **111**, 242502 (2013)

Goriely, S., Bauswein, A., Just, O., Pllumbi, E., Janka, H.-T.: Impact of weak interactions of free nucleons on the r-process in dynamical ejecta from neutron star mergers. Mon. Not. R. Astron. Soc. **452**, 3894–3904 (2015)

Goriely, S., Chamel, N., Pearson, J.M.: Further explorations of Skyrme-Hartree-Fock-Bogoliubov mass formulas. XVI. Inclusion of self-energy effects in pairing. Phys. Rev. C **93**, 034337 (2016)

Grossman, D., Korobkin, O., Rosswog, S., Piran, T.: The long-term evolution of neutron star merger remnants - II. Radioactively powered transients. Mon. Not. R. Astron. Soc. **439**, 757–770 (2014)

Hotokezaka, K., Piran, T., Paul, M.: Short-lived ^{244}Pu points to compact binary mergers as sites for heavy r-process nucleosynthesis. Nat. Phys. **11**, 1042 (2015)

Hotokezaka, K., Wanajo, S., Tanaka, M., Bamba, A., Terada, Y., Piran, T.: Radioactive decay products in neutron star merger ejecta: heating efficiency and γ-ray emission. Mon. Not. R. Astron. Soc. **459**, 35–43 (2016)

Ji, A.P., Frebel, A., Chiti, A., Simon, J.D.: R-process enrichment from a single event in an ancient dwarf galaxy. Nature **531**, 610–613 (2016)

Jin, Z.-P., Li, X., Cano, Z., Covino, S., Fan, Y.-Z., Wei, D.-M.: The light curve of the macronova associated with the long-short burst GRB 060614. Astrophys. J. **811**, L22 (2015)

Jin, Z.-P., Hotokezaka, K., Li, X., Tanaka, M., D'Avanzo, P., Fan, Y.-Z., Covino, S., Wei, D.-M., Piran, T.: The 050709 macronova and the GRB/macronova connection (2016). ArXiv e-prints

Just, O., Bauswein, A., Pulpillo, R.A., Goriely, S., Janka, H.-T.: Comprehensive nucleosynthesis analysis for ejecta of compact binary mergers. Mon. Not. R. Astron. Soc. **448**, 541–567 (2015)

Kasen, D., Badnell, N.R., Barnes, J.: Opacities and spectra of the r-process ejecta from neutron star mergers. Astrophys. J. **774**, 25 (2013)

Kasen, D., Metzger, B., Barnes, J., Quataert, E., Ramirez-Ruiz, E.: Origin of the heavy elements in binary neutron-star mergers from a gravitational-wave event. Nature **551**, 80–84 (2017)

Kasliwal, M.M., Nakar, E., Singer, L.P., Kaplan, D.L., Cook, D.O., Van Sistine, A., Lau, R.M., Fremling, C., Gottlieb, O., Jencson, J.E., Adams, S.M., Feindt, U., Hotokezaka, K., Ghosh, S., Perley, D.A., Yu, P.-C., Piran, T., Allison, J.R., Anupama, G.C., Balasubramanian, A., Bannister, K.W., Bally, J., Barnes, J., Barway, S., Bellm, E., Bhalerao, V., Bhattacharya, D., Blagorodnova, N., Bloom, J.S., Brady, P.R., Cannella, C., Chatterjee, D., Cenko, S.B., Cobb, B.E., Copperwheat, C., Corsi, A., De, K., Dobie, D., Emery, S.W.K., Evans, P.A., Fox, O.D., Frail, D.A., Frohmaier, C., Goobar, A., Hallinan, G., Harrison, F., Helou, G., Hinderer, T., Ho, A.Y.Q., Horesh, A., Ip, W.-H., Itoh, R., Kasen, D., Kim, H., Kuin, N.P.M., Kupfer, T., Lynch, C., Madsen, K., Mazzali, P.A., Miller, A.A., Mooley, K., Murphy, T., Ngeow, C.-C., Nichols, D., Nissanke, S., Nugent, P., Ofek, E.O., Qi, H., Quimby, R.M., Rosswog, S., Rusu, F., Sadler, E.M., Schmidt, P., Sollerman, J., Steele, I., Williamson, A.R., Xu, Y., Yan, L., Yatsu, Y., Zhang, C., Zhao, W.: Illuminating gravitational waves: a concordant picture of photons from a neutron star merger. Science **358**(6370), 1559–1565 (2017)

Kawamura, T., Giacomazzo, B., Kastaun, W., Ciolfi, R., Endrizzi, A., Baiotti, L., Perna, R.: Binary neutron star mergers and short gamma-ray bursts: effects of magnetic field orientation, equation of state, and mass ratio. Phys. Rev. D **94**, 064012 (2016)

Kilpatrick, C.D., Foley, R.J., Kasen, D., Murguia-Berthier, A., Ramirez-Ruiz, E., Coulter, D.A., Drout, M.R., Piro, A.L., Shappee, B.J., Boutsia, K., Contreras, C., Di Mille, F., Madore, B.F., Morrell, N., Pan, Y.-C., Prochaska, J.X., Rest, A., Rojas-Bravo, C., Siebert, M.R., Simon, J.D., Ulloa, N.: Electromagnetic evidence that SSS17a is the result of a binary neutron star merger. Science **358**(6370), 1583–1587 (2017)

Kiuchi, K., Kyutoku, K., Sekiguchi, Y., Shibata, M., Wada, T.: High resolution numerical relativity simulations for the merger of binary magnetized neutron stars. Phys. Rev. D **90**, 041502 (2014)

Kiuchi, K., Cerdá-Durán, P., Kyutoku, K., Sekiguchi, Y., Shibata, M.: Efficient magnetic-field amplification due to the Kelvin-Helmholtz instability in binary neutron star mergers. Phys. Rev. D **92**, 124034 (2015)

Komissarov, S.S., Barkov, M.V.: Activation of the Blandford-Znajek mechanism in collapsing stars. Mon. Not. R. Astron. Soc. **397**, 1153–1168 (2009)

Korobkin, O., Rosswog, S., Arcones, A., Winteler, C.: On the astrophysical robustness of the neutron star merger r-process. Mon. Not. R. Astron. Soc. **426**, 1940–1949 (2012)

Kratz, K.-L., Farouqi, K., Möller, P.: A high-entropy-wind r-process study based on nuclear-structure quantities from the new finite-range droplet model FRDM(2012). Astrophys. J. **792**, 6 (2014)

Kuroda, T., Kotake, K., Takiwaki, T.: A new gravitational-wave signature from standing accretion shock instability in supernovae. Astrophys. J. Lett. **829**, L14 (2016)

Kurtukian-Nieto, T., Benlliure, J., Schmidt, K.-H., Audouin, L., Becker, F., Blank, B., Borzov, I.N., Casarejos, E., Farget, F., Fernández-Ordóñez, M., Giovinazzo, J., Henzlova, D., Jurado, B., Langanke, K., Martínez-Pinedo, G., Pereira, J., Yordanov, O.: Beta-decay half-lives of new neutron-rich isotopes of Re, Os and Ir approaching the r-process path near N = 126. Eur. Phys. J. A **50**, 135 (2014)

Lattimer, J.M., Schramm, D.N.: Black-hole-neutron-star collisions. Astrophys. J. Lett. **192**, L145–L147 (1974)

Lazzati, D., Perna, R., Morsony, B.J., López-Cámara, D., Cantiello, M., Ciolfi, R., Giacomazzo, B., Workman, J.C.: Late time afterglow observations reveal a collimated relativistic jet in the ejecta of the binary neutron star merger GW170817. Phys. Rev. Lett. **120**, 241103 (2018)

Li, L.-X., Paczyński, B.: Transient events from neutron star mergers. Astrophys. J. Lett. **507**, L59–L62 (1998)

Lippuner, J., Roberts, L.F.: r-process lanthanide production and heating rates in kilonovae. Astrophys. J. **815**, 82 (2015)

Liu, M., Wang, N., Deng, Y., Wu, X.: Further improvements on a global nuclear mass model. Phys. Rev. C **84**, 014333 (2011)

Margalit, B., Metzger, B.D., Beloborodov, A.M.: Does the collapse of a supramassive neutron star leave a debris disk? Phys. Rev. Lett. **115**, 171101 (2015)

Marketin, T., Huther, L., Martínez-Pinedo, G.: Large-scale evaluation of β-decay rates of r-process nuclei with the inclusion of first-forbidden transitions. Phys. Rev. C **93**, 025805 (2016)

Martin, D., Perego, A., Arcones, A., Thielemann, F.-K., Korobkin, O., Rosswog, S.: Neutrino-driven winds in the aftermath of a neutron star merger: nucleosynthesis and electromagnetic transients. Astrophys. J. **813**, 2 (2015)

Martin, D., Arcones, A., Nazarewicz, W., Olsen, E.: Impact of nuclear mass uncertainties on the r process. Phys. Rev. Lett. **116**, 121101 (2016)

McKinney, J.C.: General relativistic magnetohydrodynamic simulations of the jet formation and large-scale propagation from black hole accretion systems. Mon. Not. R. Astron. Soc. **368**, 1561–1582 (2006)

McWilliams, S.T., Levin, J.: Electromagnetic extraction of energy from black-hole-neutron-star binaries. Astrophys. J. **742**, 90 (2011)

Mendoza-Temis, J.D.J., Wu, M.-R., Langanke, K., Martínez-Pinedo, G., Bauswein, A., Janka, H.-T.: Nuclear robustness of the r process in neutron-star mergers. Phys. Rev. C **92**, 055805 (2015)

Metzger, B.D.: The Kilonova Handbook (2016). ArXiv e-prints.

Metzger, B.D.: Welcome to the multi-messenger era! Lessons from a neutron star merger and the landscape ahead (2017). ArXiv e-prints: 1710.05931

Metzger, B.D., Berger, E.: What is the most promising electromagnetic counterpart of a neutron star binary merger?. Astrophys. J. **746**, 48 (2012)

Metzger, B.D., Martínez-Pinedo, G., Darbha, S., Quataert, E., Arcones, A., Kasen, D., Thomas, R., Nugent, P., Panov, I.V., Zinner, N.T.: Electromagnetic counterparts of compact object mergers powered by the radioactive decay of r-process nuclei. Mon. Not. R. Astron. Soc. **406**, 2650–2662 (2010)

Metzger, B.D., Bauswein, A., Goriely, S., Kasen, D.: Neutron-powered precursors of kilonovae. Mon. Not. R. Astron. Soc. **446**, 1115–1120 (2015)

Möller, P., Nix, J.R., Myers, W.D., Swiatecki, W.J.: Nuclear ground-state masses and deformations. At. Data Nucl. Data Tables **59**, 185 (1995)

Möller, P., Pfeiffer, B., Kratz, K.-L. : New calculations of gross β-decay properties for astrophysical applications: speeding-up the classical r process. Phys. Rev. C **67**, 055802 (2003)

Möller, P., Myers, W.D., Sagawa, H., Yoshida, S.: New finite-range droplet mass model and equation-of-state parameters. Phys. Rev. Lett. **108**, 052501 (2012)

Mooley, K. P., Nakar, E., Hotokezaka, K., Hallinan, G., Corsi, A., Frail, D.A., Horesh, A., Murphy, T., Lenc, E., Kaplan, D.L., de, K., Dobie, D., Chandra, P., Deller, A., Gottlieb, O., Kasliwal, M.M., Kulkarni, S.R., Myers, S.T., Nissanke, S., Piran, T., Lynch, C., Bhalerao, V., Bourke, S., Bannister, K.W., Singer, L.P.: A mildly relativistic wide-angle outflow in the neutron-star merger event GW170817. Nature **554**, 207–210 (2018)

Mumpower, M.R., McLaughlin, G.C., Surman, R.: Formation of the rare-earth peak: gaining insight into late-time r-process dynamics. Phys. Rev. C **85**, 045801 (2012)

Mumpower, M.R., Surman, R., Fang, D.-L., Beard, M., Möller, P., Kawano, T., Aprahamian, A.: Impact of individual nuclear masses on r -process abundances. Phys. Rev. C **92**, 035807 (2015)

Mumpower, M.R., McLaughlin, G.C., Surman, R., Steiner, A.W.: Reverse engineering nuclear properties from rare earth abundances in the r process. J. Phys. G Nucl. Phys. **44**, 034003 (2017)

Obergaulinger, M., Aloy, M.A., Müller, E.: Local simulations of the magnetized Kelvin-Helmholtz instability in neutron-star mergers. Astron. Astrophys. **515**, A30 (2010)

Otsuki, K., Tagoshi, H., Kajino, T., Wanajo, S.Y.: General relativistic effects on neutrino-driven winds from young, hot neutron stars and r-process nucleosynthesis. Astrophys. J. **533**, 424–439 (2000)

Palenzuela, C., Lehner, L., Liebling, S.L., Ponce, M., Anderson, M., Neilsen, D., Motl, P.: Linking electromagnetic and gravitational radiation in coalescing binary neutron stars. Phys. Rev. D **88**, 043011 (2013)

Palenzuela, C., Lehner, L., Ponce, M., Liebling, S. L., Anderson, M., Neilsen, D., Motl, P.: Electromagnetic and gravitational outputs from binary-neutron-star coalescence. Phys. Rev. Lett. **111**, 061105 (2013)

Palenzuela, C., Liebling, S.L., Neilsen, D., Lehner, L., Caballero, O.L., O'Connor, E., Anderson, M.: Effects of the microphysical equation of state in the mergers of magnetized neutron stars with neutrino cooling. Phys. Rev. D **92**, 044045 (2015)

Panov, I.V., Korneev, I.Y., Thielemann, F.-K.: The r-Process in the region of transuranium elements and the contribution of fission products to the nucleosynthesis of nuclei with A $\leq$ 130. Astron. Lett. **34**, 189–197 (2008)

Panov, I.V., Korneev, I.Y., Martinez-Pinedo, G., Thielemann, F.-K.: Influence of spontaneous fission rates on the yields of superheavy elements in the r-process. Astron. Lett. **39**, 150–160 (2013)

Panov, I.V., Lutostansky, Y.S., Thielemann, F.K.: Half-life of short lived neutron excess nuclei that participate in the r-process. Bull. Russ. Acad. Sci. Phys. **79**, 437–441 (2015)

Paschalidis, V., Etienne, Z.B., Shapiro, S.L.: General-relativistic simulations of binary black hole-neutron stars: precursor electromagnetic signals. Phys. Rev. D **88**, 021504 (2013)

Paschalidis, V., Ruiz, M., Shapiro, S.L.: Relativistic simulations of black hole-neutron star coalescence: the jet emerges. Astrophys. J. Lett. **806**, L14 (2015)

Perego, A., Rosswog, S., Cabezón, R.M., Korobkin, O., Käppeli, R., Arcones, A., Liebendörfer, M.: Neutrino-driven winds from neutron star merger remnants. Mon. Not. R. Astron. Soc. **443**, 3134–3156 (2014)

Pian, E., D'Avanzo, P., Benetti, S., Branchesi, M., Brocato, E., Campana, S., Cappellaro, E., Covino, S., D'Elia, V., Fynbo, J.P.U., Getman, F., Ghirlanda, G., Ghisellini, G., Grado, A., Greco, G., Hjorth, J., Kouveliotou, C., Levan, A., Limatola, L., Malesani, D., Mazzali, P.A., Melandri, A., Møller, P., Nicastro, L., Palazzi, E., Piranomonte, S., Rossi, A., Salafia, O.S., Selsing, J., Stratta, G., Tanaka, M., Tanvir, N.R., Tomasella, L., Watson, D., Yang, S., Amati, L., Antonelli, L.A., Ascenzi, S., Bernardini, M.G., Boër, M., Bufano, F., Bulgarelli, A., Capaccioli, M., Casella, P., Castro-Tirado, A. J., Chassande-Mottin, E., Ciolfi, R., Copperwheat, C. M., Dadina, M., De Cesare, G., di Paola, A., Fan, Y.Z., Gendre, B., Giuffrida, G., Giunta, A., Hunt, L.K., Israel, G.L., Jin, Z.-P., Kasliwal, M.M., Klose, S., Lisi, M., Longo, F., Maiorano, E.,

Mapelli, M., Masetti, N., Nava, L., Patricelli, B., Perley, D., Pescalli, A., Piran, T., Possenti, A., Pulone, L., Razzano, M., Salvaterra, R., Schipani, P., Spera, M., Stamerra, A., Stella, L., Tagliaferri, G., Testa, V., Troja, E., Turatto, M., Vergani, S.D., Vergani, D. : Spectroscopic identification of r-process nucleosynthesis in a double neutron-star merger. Nature **551**, 67–70 (2017)

Piro, A.L., Giacomazzo, B., Perna, R.: The fate of neutron star binary mergers. Astrophys. J. Lett. **844**, L19 (2017)

Price, D.J., Rosswog, S.: Producing ultrastrong magnetic fields in neutron star mergers. Science **312**, 719–722 (2006)

Qian, Y.-Z., Woosley, S.E.: Nucleosynthesis in neutrino-driven winds. I. The physical conditions. Astrophys. J. **471**, 331 (1996)

Radice, D., Bernuzzi, S., Ott, C.D.: One-armed spiral instability in neutron star mergers and its detectability in gravitational waves. Phys. Rev. D **94**, 064011 (2016)

Radice, D., Galeazzi, F., Lippuner, J., Roberts, L.F., Ott, C.D., Rezzolla, L.: Dynamical mass ejection from binary neutron star mergers. Mon. Not. R. Astron. Soc. **460**, 3255–3271 (2016)

Rezzolla, L., Kumar, P.: A novel paradigm for short gamma-ray bursts with extended X-ray emission. Astrophys. J. **802**, 95 (2015)

Rezzolla, L., Baiotti, L., Giacomazzo, B., Link, D., Font, J.A.: Accurate evolutions of unequal-mass neutron-star binaries: properties of the torus and short GRB engines. Class. Quantum Grav. **27**, 114105 (2010)

Rezzolla, L., Giacomazzo, B., Baiotti, L., Granot, J., Kouveliotou, C., Aloy, M.A. : The missing link: merging neutron stars naturally produce jet-like structures and can power short gamma-ray bursts. Astrophys. J. Lett. **732**, L6 (2011)

Richers, S., Ott, C.D., Abdikamalov, E., O'Connor, E., Sullivan, C.: Equation of state effects on gravitational waves from rotating core collapse (2017). ArXiv e-prints

Roberts, L.F., Lippuner, J., Duez, M.D., Faber, J.A., Foucart, F., Lombardi, J.C., Jr., Ning, S., Ott, C.D., Ponce, M.: The influence of neutrinos on r-process nucleosynthesis in the ejecta of black hole-neutron star mergers. Mon. Not. R. Astron. Soc. **464**, 3907–3919 (2017)

Rosswog, S., Liebendörfer, M., Thielemann, F.-K., Davies, M. B. , Benz, W., Piran, T.: Mass ejection in neutron star mergers. Astron. Astrophys. **341**, 499–526 (1999)

Rosswog, S., Korobkin, O., Arcones, A., Thielemann, F.-K., Piran, T.: The long-term evolution of neutron star merger remnants - I. The impact of r-process nucleosynthesis. Mon. Not. R. Astron. Soc. **439**, 744–756 (2014)

Rosswog, S., Sollerman, J., Feindt, U., Goobar, A., Korobkin, O., Fremling, C., Kasliwal, M.: The first direct double neutron star merger detection: implications for cosmic nucleosynthesis. A&A **615**, A132 (2018)

Rowlinson, A., O'Brien, P.T., Metzger, B.D., Tanvir, N.R., Levan, A.J.: Signatures of magnetar central engines in short GRB light curves. Mon. Not. R. Astron. Soc. **430**, 1061–1087 (2013)

Ruiz, M., Lang, R.N., Paschalidis, V., Shapiro, S.L.: Binary neutron star mergers: a jet engine for short gamma-ray bursts. Astrophys. J. **824**, L6 (2016)

Sadhukhan, J., Nazarewicz, W., Schunck, N.: Microscopic modeling of mass and charge distributions in the spontaneous fission of ^{240}Pu. Phys. Rev. C **93**, 011304 (2016)

Seeger, P.A., Fowler, W.A., Clayton, D.D.: Nucleosynthesis of heavy elements by neutron capture. Astrophys. J. Suppl. **11**, 121 (1965)

Sekiguchi, Y., Kiuchi, K., Kyutoku, K., Shibata, M.: Gravitational waves and neutrino emission from the merger of binary neutron stars. Phys. Rev. Lett. **107**, 051102 (2011)

Sekiguchi, Y., Kiuchi, K., Kyutoku, K., Shibata, M.: Dynamical mass ejection from binary neutron star mergers: Radiation-hydrodynamics study in general relativity. Phys. Rev. D **91**, 064059 (2015)

Sekiguchi, Y., Kiuchi, K., Kyutoku, K., Shibata, M., Taniguchi, K.: Dynamical mass ejection from the merger of asymmetric binary neutron stars: radiation-hydrodynamics study in general relativity. Phys. Rev. D **93**, 124046 (2016)

Siegel, D.M., Ciolfi, R.: Electromagnetic emission from long-lived binary neutron star merger remnants. I. Formulation of the problem. Astrophys. J. **819**, 14S (2016)

Siegel, D.M., Ciolfi, R.: Electromagnetic emission from long-lived binary neutron star merger remnants. II. Lightcurves and spectra. Astrophys. J. **819**, 15S (2016)

Smartt, S.J., Chen, T.-W., Jerkstrand, A., Coughlin, M., Kankare, E., Sim, S. A., Fraser, M., Inserra, C., Maguire, K., Chambers, K.C., Huber, M.E., Krühler, T., Leloudas, G., Magee, M., Shingles, L.J., Smith, K.W., Young, D. R., Tonry, J., Kotak, R., Gal-Yam, A., Lyman, J.D., Homan, D.S., Agliozzo, C., Anderson, J.P., Angus, C.R. , Ashall, C., Barbarino, C., Bauer, F.E., Berton, M., Botticella, M. T., Bulla, M., Bulger, J., Cannizzaro, G., Cano, Z., Cartier, R., Cikota, A., Clark, P., De Cia, A., Della Valle, M., Denneau, L., Dennefeld, M., Dessart, L., Dimitriadis, G., Elias-Rosa, N., Firth, R.E., Flewelling, H., Flörs, A., Franckowiak, A., Frohmaier, C., Galbany, L., González-Gaitán, S., Greiner, J., Gromadzki, M., Guelbenzu, A.N., Gutiérrez, C.P., Hamanowicz, A., Hanlon, L., Harmanen, J., Heintz, K.E., Heinze, A., Hernandez, M.-S., Hodgkin, S.T., Hook, I.M., Izzo, L., James, P.A., Jonker, P.A., Kerzendorf, W.E., Klose, S., Kostrzewa-Rutkowska, Z., Kowalski, M., Kromer, M., Kuncarayakti, H., Lawrence, A., Lowe, T.B., Magnier, E.A., Manulis, I., Martin-Carrillo, A., Mattila, S., McBrien, O., Müller, A., Nordin, J., O'Neill, D., Onori, F., Palmerio, J.T., Pastorello, A., Patat, F., Pignata, G., Podsiadlowski, P., Pumo, M.L., Prentice, S.J., Rau, A., Razza, A., Rest, A., Reynolds, T., Roy, R., Ruiter, A.J., Rybicki, K.A., Salmon, L., Schady, P., Schultz, A.S.B., Schweyer, T., Seitenzahl, I.R., Smith, M., Sollerman, J., Stalder, B., Stubbs, C.W., Sullivan, M., Szegedi, H., Taddia, F., Taubenberger, S., Terreran, G., van Soelen, B., Vos, J., Wainscoat, R.J., Walton, N.A., Waters, C., Weiland, H., Willman, M., Wiseman, P., Wright, D.E., Wyrzykowski, Ł., Yaron, O.: A kilonova as the electromagnetic counterpart to a gravitational-wave source. Nature **551**, 75–79 (2017)

Sneden, C., Cowan, J.J., Gallino, R.: Neutron-capture elements in the early galaxy. Ann. Rev. Astron. Astrophys. **46**, 241–288 (2008)

Surman, R., Engel, J.: Changes in r-process abundances at late times. Phys. Rev. C **64**, 035801 (2001)

Surman, R., McLaughlin, G.C., Ruffert, M., Janka, H.-T., Hix, W.R.: r-Process nucleosynthesis in hot accretion disk flows from black hole-neutron star mergers. Astrophys. J. Lett. **679**, L117 (2008)

Suzuki, T., Yoshida, T., Kajino, T., Otsuka, T.: β decays of isotones with neutron magic number of N=126 and r-process nucleosynthesis. Phys. Rev. C **85**, 015802 (2012)

Tanaka, M., Hotokezaka, K.: Radiative transfer simulations of neutron star merger ejecta. Astrophys. J. **775**, 113 (2013)

Tanvir, N.R., Levan, A.J., Fruchter, A.S., Hjorth, J., Hounsell, R.A., Wiersema, K., Tunnicliffe, R.L.: A 'kilonova' associated with the short-duration γ-ray burst GRB 130603B. Nature **500**, 547–549 (2013)

Tanvir, N.R., Levan, A.J., Gonzalez-Fernandez, C., Korobkin, O., Mandel, I., Rosswog, S., Hjorth, J., D'Avanzo, P., Fruchter, A.S., Fryer, C.L., Kangas, T., Milvang-Jensen, B., Rosetti, S., Steeghs, D., Wollaeger, R.T., Cano, Z., Copperwheat, C.M., Covino, S., D'Elia, V., de Ugarte Postigo, A., Evans, P.A., Even, W.P., Fairhurst, S., Jaimes, R.F., Fontes, C.J., Fujii, Y.I., Fynbo, J.P.U., Gompertz, B.P., Greiner, J., Hodosan, G., Irwin, M.J., Jakobsson, P., Jorgensen, U.G., Kann, D.A., Lyman, J.D., Malesani, D., McMahon, R.G., Melandri, A., O'Brien, P.T., Osborne, J.P., Palazzi, E., Perley, D.A., Pian, E., Piranomonte, S., Rabus, M., Rol, E., Rowlinson, A., Schulze, S., Sutton, P., Thoene, C.C., Ulaczyk, K., Watson, D., Wiersema, K., Wijers, R.A.M.J.: The emergence of a lanthanide-rich kilonova following the merger of two neutron stars. Astrophys. J. Lett. **848**, L27 (2017)

Thompson, T.A. , Burrows, A., Meyer, B.S.: The physics of proto-neutron star winds: implications for r-process nucleosynthesis. Astrophys. J. **562**, 887–908 (2001)

Troja, E., Rosswog, S., Gehrels, N.: Precursors of short gamma-ray bursts. Astrophys. J. **723**, 1711–1717 (2010)

Troja, E., et al.: The X-ray counterpart to the gravitational-wave event GW170817. Nature **551**, 71–74 (2017)

Tsang, D., Read, J.S., Hinderer, T., Piro, A.L., Bondarescu, R.: Resonant shattering of neutron star crusts. Phys. Rev. Lett. **108**, 011102 (2012)

Wallner, A., Faestermann, T., Feige, J., Feldstein, C., Knie, K., Korschinek, G., Kutschera, W., Ofan, A., Paul, M., Quinto, F., Rugel, G., Steier, P.: Abundance of live ^{244}Pu in deep-sea reservoirs on Earth points to rarity of actinide nucleosynthesis. Nat. Commun. **6**, 5956 (2015)

Wallner, A., Feige, J., Kinoshita, N., Paul, M., Fifield, L.K., Golser, R., Honda, M., Linnemann, U., Matsuzaki, H., Merchel, S., Rugel, G., Tims, S.G., Steier, P., Yamagata, T., Winkler, S.R.: Recent near-Earth supernovae probed by global deposition of interstellar radioactive ^{60}Fe. Nature **532**, 69–72 (2016)

Wanajo, S., Itoh, N., Goriely, S., Samyn, M., Ishimaru, Y.: The r-process in supernovae with new microscopic mass formulae. Nucl. Phys. A **758**, 671–674 (2005)

Wanajo, S., Sekiguchi, Y., Nishimura, N., Kiuchi, K., Kyutoku, K., Shibata, M.: Production of all the r-process nuclides in the dynamical ejecta of neutron star mergers. Astrophys. J. Lett. **789**, L39 (2014)

Wu, M.-R., Fernández, R., Martínez-Pinedo, G., Metzger, B.D.: Production of the entire range of r-process nuclides by black hole accretion disc outflows from neutron star mergers. Mon. Not. R. Astron. Soc. **463**, 2323–2334 (2016)

Yang, B., Jin, Z.-P., Li, X., Covino, S., Zheng, X.-Z., Hotokezaka, K., Fan, Y.-Z., Piran, T., Wei, D.-M.: A possible macronova in the late afterglow of the long-short burst GRB 060614. Nat. Commun. **6**, 7323 (2015)

Zhi, Q., Caurier, E., Cuenca-García, J. J., Langanke, K., Martínez-Pinedo, G., Sieja, K.: Shell-model half-lives including first-forbidden contributions for r-process waiting-point nuclei. Phys. Rev. C **87**, 025803 (2013)

Zrake, J., MacFadyen, A.I.: Magnetic energy production by turbulence in binary neutron star mergers. Astrophys. J. **769**, L29 (2013)

Chapter 12
Gravitational Waves from Single Neutron Stars: An Advanced Detector Era Survey

Kostas Glampedakis and Leonardo Gualtieri

Abstract With the doors beginning to swing open on the new gravitational wave astronomy, this review provides an up-to-date survey of the most important physical mechanisms that could lead to emission of potentially detectable gravitational radiation from isolated and accreting neutron stars. In particular we discuss the gravitational wave-driven instability and asteroseismology formalism of the f- and r-modes, the different ways that a neutron star could form and sustain a non-axisymmetric quadrupolar "mountain" deformation, the excitation of oscillations during magnetar flares and the possible gravitational wave signature of pulsar glitches. We focus on progress made in the recent years in each topic, make a fresh assessment of the gravitational wave detectability of each mechanism and, finally, highlight key problems and desiderata for future work.

12.1 Introduction

September 14, 2015 marked a milestone for gravitational physics with the first direct detection of gravitational waves (GWs) by the twin advanced LIGO observatories (Abbott et al. 2016a). Since then, and with the additional participation of the advanced Virgo detector, more detections have been announced (Abbott et al. 2016b, 2017a,b), with all observed signals having been identified as mergers of binary black hole systems. Finally, it was the turn of neutron stars to be "discovered" by the

K. Glampedakis (✉)
Departamento de Física, Universidad de Murcia, Murcia, Spain

Theoretical Astrophysics, University of Tübingen, Tübingen, Germany
e-mail: kostas@um.es

L. Gualtieri (✉)
Dipartimento di Fisica, "Sapienza" Università di Roma, Roma, Italy

Sezione INFN Roma1, Roma, Italy
e-mail: leonardo.gualtieri@roma1.infn.it

© Springer Nature Switzerland AG 2018   673
L. Rezzolla et al. (eds.), *The Physics and Astrophysics of Neutron Stars*,
Astrophysics and Space Science Library 457,
https://doi.org/10.1007/978-3-319-97616-7_12

interferometric telescopes of the new GW astronomy: on August 17, 2017, the first GW signal from neutron stars has been detected (Abbott et al. 2017c), together with transient counterparts across the entire electromagnetic spectrum (Abbott et al. 2017d). This signal came, as expected, from the same systems that indirectly established for the first time the very existence of gravitational radiation: binary neutron star systems, and in particular their final GW-driven merger (Abbott et al. 2017c). In fact, these catastrophic events had already been routinely observed by photon astronomy in the form of short gamma-ray bursts (GRBs).

Single neutron stars, the subject of this chapter, await their turn to be added to the catalogue of observed GW sources—what is still uncertain is if this goal will be reached by Advanced LIGO/Virgo or by the next generation instruments such as the Einstein Telescope (ET) (Punturo et al. 2010). The term "single" comprises (1) isolated neutron stars, that is, the bulk of the known neutron star population which includes normal radio pulsars, old recycled millisecond pulsars (MSPs), magnetars, etcetera, and (2) accreting systems, from which the neutron stars in low mass X-ray binaries (LMXBs) are the ones relevant to this review.

The purpose of this chapter is to provide a 2017 status report of our understanding of the small mosaic of known mechanisms that could turn single neutron stars into potentially detectable sources of GWs. More specifically, we discuss (1) the GW-driven instability of the fundamental f-mode and of the inertial r-mode (including a basic introduction to the CFS theory of unstable oscillations) (2) the GW asteroseismology formalism for stable and unstable f-modes (3) the different ways that non-axisymmetric quadrupolar deformations could be induced and sustained in neutron stars thereby making them continuous sources of GWs (4) the excitation of magneto-elastic oscillations and GW emission produced by magnetar flares, and finally (5) the GW detectability of pulsar glitches.

Although being mostly interested in providing a sober assessment of the prospects for detecting GWs from these various single neutron star sources, we also present in some detail key astrophysical "side effects" of GW emission such as the r-mode driven spin-temperature evolution in accreting neutron stars and the antagonism between the f-mode and magnetic dipole torque for controlling the dynamics of neutron star merger remnants.

Given the space limitations, our discussion is, of course, far from being comprehensive—the objective is to present and discuss key recent developments in each research topic and give pointers for future work rather than explaining things from scratch. References to textbooks, earlier review articles and selected research papers are plentifully given for the reader less familiar with the topics discussed here.

12.1.1 Summary, Conventions and Plan of This Review

Table 12.1 is a bird's eye view of the chapter: it provides a list of the various topics that are discussed in the following sections and of what we consider to be the most important recent progress and open problems in each area.

Table 12.1 Summary of topics discussed in this review

Topic	Recent progress	Key future goals	Section
Gravitational wave asteroseismology	Relativistic-Cowling formulae, alternative parametrisations based on moment of inertia/compactness instead of radius	Formalism with dynamical spacetime. Find "optimal" set of asteroseismology parameters	12.2
f-mode instability	Relativistic-Cowling instability windows with realistic equation of state. Saturation amplitude. Application to supramassive post-merger remnants and constraints from short gamma-ray bursts. Spin-temperature evolutions in Newtonian gravity	Models with dynamical spacetime. Obtain saturation amplitude using relativistic gravity and fast rotation	12.4
r-mode instability	Several r-mode unstable systems in minimum damping models. Upper limits on mode amplitude from low-mass X-ray binaries and millisecond pulsar data and implications. r-mode puzzle: small amplitude or enhanced damping? Multifluid r-mode modelling. Modified instability window due to mode resonances	Improve understanding of "conventional" damping: Ekman layer with elastic crust, magnetic field and paired matter. Vortex-fluxtube interactions. New saturation mechanisms	12.5
Magnetar oscillations	Models of magneto-elastic oscillations including superfluidity, crust phases, complex magnetic field structures. General relativistic magnetohydrodynamical simulations of magnetic field instability	Analytical expressions for magnetar asteroseismology with quasi-periodic oscillations. Models of giant flares	12.6
Mountains	Deformations of neutron stars with exotic matter and/or pinned superfluidity. Constraints from short gamma-ray bursts. Observational evidence suggesting the existence of thermal mountains in accreting neutron stars	Understanding stability of magnetic field configurations in neutron stars. Rigorous magnetic field modelling in the presence of exotic matter. Estimates of the actual size of thermal and core mountains	12.7

(continued)

Table 12.1 (continued)

Topic	Recent progress	Key future goals	Section
Pulsar glitches	Numerical simulations of pulsar glitches. Models of the post-glitch relaxation. Models of collective vortex motion using the Gross-Pitaevskii equation; extraction of the gravitational wave signal	Interface between Gross-Pitaevskii vortex simulations and hydrodynamical models, including magnetic field and crust-core coupling. Magnetohydrodynamical simulations of pulsar glitches. Modelling of post-glitch relaxation including magnetic fields and mutual friction	12.8

Table 12.2 Table of frequently used normalised parameters

Stellar mass	$M_{1.4}$	$M/1.4M_\odot$
Stellar radius	R_6	$R/10^6$ cm
Spin period	P_{-3}	$P/1$ ms
Spin period derivative	$\dot{P}_{-10}$	$\dot{P}/10^{-10}$
Proton fraction	x_{p1}	$x_p/0.01$
Magnetic field	B_n	$B/10^n$ G
Distance	D_1	$D/1$ kpc
Moment of inertia	I_{45}	$I/10^{45}$ g cm^2

Several equations presented below feature physical parameters normalised to some canonical value. The physical meaning of these parameters and their normalisations are summarised in Table 12.2.

Oscillation modes are assumed to be proportional to $e^{i\omega t + im\varphi}$. We also make frequent use of the term "canonical neutron star parameters". This refers to a non-rotating neutron star with a typical mass $M = 1.4M_\odot$ and typical radius $R = 12$ km. Although we discuss stellar models constructed with various equations of state (EOS) for matter, our frequently used canonical model will be that of a simple $n = 1$ Newtonian polytrope.

The remainder of the chapter is structured as follows: in Sect. 12.2 we discuss neutron star GW asteroseismology. Section 12.3 provides a basic "primer" introduction to the theory of unstable oscillation modes in rotating neutron stars. The two most important cases of unstable modes, the f-mode and the r-mode are discussed in detail in Sects. 12.4 and 12.5 respectively. Section 12.6 is devoted to magnetars and reviews the properties (including the emission of GWs) of the oscillations that are believed to be excited in these objects by their magnetic field activity. The physics and GW detectability of neutron stars with quadrupolar deformations (i.e. neutron star "mountains") is discussed in Sect. 12.7. GW emission associated with pulsar glitches is discussed in Sect. 12.8. Finally, our concluding remarks can be found in Sect. 12.9.

12.2   Neutron Star Asteroseismology

The premise behind the notion of neutron star GW "asteroseismology" is more or less the same as the one of the more traditional helioseismology: use GW observations of pulsating neutron stars as probes of their interiors. In fact, neutron star asteroseismology is a bit more specialised: the key idea is to parametrise an oscillation mode's angular frequency ω and GW damping timescale τ_{gw} in terms of the neutron star's three bulk parameters, that is, the mass M, the radius R and the angular frequency Ω. A clever combination of these parameters could then allow the construction of "universal" (i.e. EOS-insensitive) parametrised relations. Then, an actual observation of a pulsating neutron star, through the measurement of the ω, τ_{gw} pair for one or more modes and the inversion of the parametric relations, would in principle allow the inference of the three basic stellar parameters. Moreover, if the mass of the neutron star is known, a measurement of its oscillation modes would give direct information on the neutron star EOS: its stiffness, the presence of hyperons/quarks, etc. (Benhar et al. 2004).

The basic idea of neutron star asteroseismology was first put forward almost two decades ago (Andersson and Kokkotas 1998). This initial work was focused on the asteroseismology of the fundamental f-mode and that was done for a good reason: the f-mode is the oscillation most likely to be excited in violent processes such as neutron star mergers or neutron star formation by supernova core-collapse. The f-mode has the extra advantage of being a copious emitter of gravitational radiation—the same property is responsible for the mode's rapid damping.

The main drawback of the early astreroseismology models was the assumption of non-rotating systems—astrophysical neutron stars always have some degree of rotation. In fact, the most relevant scenarios from the point of view of GW detectability of pulsation modes are likely to involve rapidly rotating systems. It was thus imperative that the asteroseismology scheme should become "fast" so that it could be applied to rapidly spinning neutron stars. At the same time it should be able to adapt to the possibility of having modes growing in amplitude rather than decaying under the emission of GWs (this is the so-called CFS instability which is discussed in more detail in Sect. 12.3).

12.2.1   Fast Relativistic Asteroseismology

The problem of extending the asteroseismology scheme to rotating neutron stars was tackled just a few years ago (Gaertig and Kokkotas 2011; Doneva et al. 2013). This new effort was based on relativistic stellar models within the so-called Cowling approximation (where only the fluid is perturbed while the spacetime metric remains a fixed background) and was again focused on the f-mode. The resulting parametrised formulae for ω, τ_{gw} are constructed with a two-step procedure and are polynomial-type fits to the numerical data. The typical accuracy of these fits lies in the range between few percent and few tens of percent, the larger error typically associated with high spin rates.

In the first step we have an expression for the mode's frequency ω_0 of a non-rotating star of the same gravitational mass as the rotating one (Andersson and Kokkotas 1998; Gaertig and Kokkotas 2011; Doneva et al. 2013),

$$\frac{\omega_0(\text{kHz})}{2\pi} = \kappa_\ell + \mu_\ell \left(\frac{M_{1.4}}{R_6^3}\right)^{1/2}. \tag{12.1}$$

This fit is clearly inspired by the simple Newtonian relation between the f-mode's frequency and the mean stellar density, $\omega \sim \rho^{1/2}$. For the most interesting case of a quadrupolar ($\ell = 2$) mode the numerical coefficients are $(\kappa_2, \mu_2) = (0.498, 2.418)$.

In the second step, the mode frequency ω_r as measured in the stellar rotating frame is expressed in terms of Ω normalised by the mass shedding angular frequency Ω_K (i.e. the Kepler limit). The resulting parametric formula is the quadratic expression (Gaertig and Kokkotas 2011; Doneva et al. 2013),

$$\frac{\omega_r^{u,s}}{\omega_0} = 1 + a_\ell^{u,s} \frac{\Omega}{\Omega_K} + b_\ell^{u,s} \left(\frac{\Omega}{\Omega_K}\right)^2, \tag{12.2}$$

where the labels u, s stand for the $m > 0$ (retrograde and potentially unstable) and $m < 0$ (prograde and stable) f-mode branches, respectively. The stable branch turns out to be the one admitting the easiest parametrisation since *all* relevant $m = -\ell$ modes are well approximated by the single pair of coefficients $(\alpha_\ell^s, b_\ell^s) = (-0.235, -0.358)$. In contrast, the unstable branch requires different coefficients for each $m = \ell$ multipole. For example, for the lowest multipole we have $(a_2^u, b_2^u) = (0.402, -0.406)$. In all cases the mode frequency in the inertial frame can be recovered with the help of the general formula,

$$\omega_i = \omega_r - m\Omega. \tag{12.3}$$

The parametrisation of the f-mode's damping (or growth) timescale proceeds along the same lines. First, the damping timescale τ_0 of the non-rotating system's mode is parametrised as:

$$\frac{1}{\tau_0(s)} = \frac{M_{1.4}^{\ell+1}}{R_6^{\ell+2}} \left[\nu_\ell + \xi_\ell \left(\frac{M_{1.4}}{R_6}\right)\right]. \tag{12.4}$$

This expression makes direct contact with the f-mode timescale of a Newtonian uniform density star $\tau_{\text{gw}} \sim R(R/M)^{\ell+1}$ (Detweiler 1975). For the particular case of the $\ell = 2$ multipole the numerical coefficients are $(\nu_2, \xi_2) = (78.55, -46.71)$ (Doneva et al. 2013).

The timescale τ_{gw} for the modes of the rotating system are polynomial expansions with respect to the frequency ratio ω_r/ω_0 (or ω_i/ω_0). For the stable f-mode branch we have (Gaertig and Kokkotas 2011; Doneva et al. 2013)

$$\frac{\tau_{gw}}{\tau_0} = \left[c_\ell + d_\ell \left(\frac{\omega_r^s}{\omega_0} \right) + e_\ell \left(\frac{\omega_r^s}{\omega_0} \right)^2 + f_\ell \left(\frac{\omega_r^s}{\omega_0} \right)^3 \right]^{2\ell}, \tag{12.5}$$

where, unlike the previous frequency fit, different multipoles require different sets of coefficients. For example, for the $m = -2$ mode these take the values $(c_2, d_2, e_2, f_2) = (-0.127, 3.264, -5.486, 3.349)$.

This time it is the unstable branch that is well approximated by a single formula for all $\ell = m$ multipoles (Gaertig and Kokkotas 2011; Doneva et al. 2013)

$$\frac{\tau_{gw}}{\tau_0} = \mathrm{sgn}(\omega_i^u) \left(\frac{\omega_i^u}{\omega_0} \right)^{-2\ell} \left[0.9 - 0.057 \left(\frac{\omega_i^u}{\omega_0} \right) + 0.157 \left(\frac{\omega_i^u}{\omega_0} \right)^2 \right]^{-2\ell}, \tag{12.6}$$

where we can notice the appearance of the inertial frame frequency ω_i. This expression has the CFS instability of the f-mode (see Sect. 12.3) hardwired in it since the change of sign in τ_{gw} is synced with the change of sign in ω_i (or, equivalently, with the transition of the mode's pattern speed from retrograde to prograde with respect to the stellar rotation, see Eq. (12.14) below).

12.2.2 R or I Asteroseismology?

The previous f-mode formulae are constructed from the basic three parameters M, R and Ω. However, this may not be, after all, the best choice of parameters for building "universal" asteroseismology expressions. Seeking an optimal parametrisation with a higher degree of EOS-independence than the previous results, recent work has followed an alternative approach that relies on the use of the stellar moment of inertia I instead of the radius R. The original f-mode model for non-rotating stars (Lau et al. 2010) was subsequently generalised to rapidly rotating systems by Doneva and Kokkotas (2015). The final outcome of that work comprises expressions that directly relate the f-mode's inertial frame frequency and damping/growth time to M, Ω, I without the need to involve the mode properties of a non-rotating star. We thus have for the stable and unstable f-mode branches,

$$M_1 \omega_i^u (\mathrm{kHz}) = a_\ell^u + b_\ell^u M_1 \Omega_1 + c_\ell^u (M_1 \Omega_1)^2 + \left[d_\ell^u + e_\ell^u M_1 \Omega + f_\ell^u (M_1 \Omega)^2 \right] \eta, \tag{12.7}$$

$$M_1 \omega_i^s (\mathrm{kHz}) = a_\ell^s + b_\ell^s M_1 \Omega_1 + \left(d_\ell^s + e_\ell^s M_1 \Omega_1 \right) \eta, \tag{12.8}$$

where $M_1 = M/M_\odot$, $\Omega_1 = \Omega/\mathrm{kHz}$ and the effective compactness parameter

$$\eta^2 = M_1^3 I_{45}^{-1}, \tag{12.9}$$

serves as a proxy for the moment of inertia.[1] The numerical coefficients appearing in (12.7), (12.8) are tabulated in Ref. Doneva and Kokkotas (2015) and of course should not be confused with the ones of the preceding section. For the $m = \pm 2$ modes these coefficients take the values:

$$(a_2^u, b_2^u, c_2^u, d_2^u, e_2^u, f_2^u) = (-1.76, -0.143, -0.00665, 3.64, -0.0436, 0.002),$$

$$(12.10)$$

$$(a_2^s, b_2^s, d_2^s, e_2^s) = (-1.66, -0.249, 3.66, 0.0633).$$

$$(12.11)$$

Moving on to the τ_{gw} timescale, only the formula for the potentially unstable branch has been constructed (Doneva and Kokkotas 2015):

$$\frac{1}{\tau_{\mathrm{gw}}(s)} = \frac{\eta^{2(1-\ell)}}{M_1} \left[g_\ell^u M_1 \left(\frac{\omega_i^u}{\mathrm{kHz}} \right) + h_\ell^u M_1^2 \left(\frac{\omega_i^u}{\mathrm{kHz}} \right)^2 \right]^{2\ell}, \qquad (12.12)$$

where for the $\ell = m = 2$ mode the coefficients take the value $(g_2^u, h_2^u) = (0.644, 0.0207)$.

Clearly, these "I-asteroseismology" relations are algebraically simpler than their "R-asteroseismology" counterparts of the preceding section. This is combined with an enhanced universality with respect to the EOS of matter, although at the price of a lower overall accuracy caused by the adoption of the Cowling approximation (Doneva and Kokkotas 2015).

12.2.3 Future Directions

As described in the preceding sections, the last few years have seen an impressive advance in the modelling of f-mode asteroseismology in neutron stars. There is, however, much room for further improvement or extension. The main simplification affecting the accuracy of the entire asteroseismology structure is the adoption of the Cowling approximation in the numerical computation of the f-modes of relativistic stars. The error caused by this approximation can be easily gauged by calculating the f-mode frequency of a uniform non-rotating star in Newtonian gravity. The outcome of this straightforward exercise is $\omega^2 = 8\pi G\rho f_\ell/3$ with $f_\ell = \ell/2$ and $f_\ell = \ell(\ell-1)/(2\ell+1)$ when the Cowling approximation is used or not used, respectively. Although this simple calculation slightly overestimates the Cowling-induced error (which is typically $\sim 10 - 30\%$) it does tell us that the approximate result systematically lies above the exact one and that higher multipoles are less affected. These same trends are indeed found in more rigorous general relativistic

---

[1] As pointed out in Doneva and Kokkotas (2015), this asteroseismology formalism could equally well have been based on the compactness M/R rather than η.

(GR) f-mode calculations with or without the Cowling approximation (Zink et al. 2010). For a given error in the mode frequency the corresponding error in the GW timescale τ_{gw} is much more pronounced because of the steep dependence of this parameter with respect to ω. This is clearly exemplified by comparing the expressions for the $\ell = 2$ f-mode damping timescale in non-rotating stars. The coefficients in the Cowling-approximated formula (12.4) are about a factor three larger than the ones appearing in the non-Cowling expressions of Andersson and Kokkotas (1998), Gaertig and Kokkotas (2011). This large mismatch is caused by the $\sim \omega^{2\ell+2}$ dependence of the GW damping rate (see Eq. (12.15) below) and the aforementioned error in the mode frequency.

The upshot of this discussion is that a further improvement of the asteroseismology scheme should be based on the removal of the Cowling approximation and the use of full GR computations of f-modes in rapidly rotating stars. This effort could be combined with further experimentation with the variables used in the asteroseismology formulae along the lines discussed in this section. For example, is Ω the "best" parameter for representing rotation or is there a more suitable parametrisation (e.g. the Kerr parameter cJ/GM^2, where J is the stellar angular momentum)?

Neutron star asteroseismology should also be extended beyond the f-mode. Indeed, as the modelling of neutron stars improves including more and more physics, new classes of modes appear. Although the f-mode is likely to be the most relevant for GW detection at the birth and at the death of a neutron star, other modes could be relevant at different stages of its life. Magneto-elastic oscillations—a class of modes associated to the elastic strain of the crust and to the magnetic field in the crust and in the core—can be excited in giant flares of strongly magnetized neutron stars; they are discussed in detail in Sect. 12.6.1. Moreover, when a superfluid phase is present in the neutron star core, the multiplicity of the oscillation modes doubles (see the discussion at the end of Sect. 12.5.5). At characteristic values of the temperature (of the order of $\sim 10^8$ K), the damping times of these "superfluid" modes can become small enough to make them—at least in principle—efficient sources of GWs (Gualtieri et al. 2014). Although the superfluid phase is not present in the most violent stages of the neutron star life, superfluid g-modes (Gusakov and Kantor 2013; Kantor and Gusakov 2014; Passamonti et al. 2016; Dommes and Gusakov 2016) can be unstable by convection in the first year of neutron star's life, or—if hyperons are present—can be excited in neutron star's coalescences; superfluid r-modes (Gusakov et al. 2014a,b) will be discussed in Sect. 12.5.5.

Neutron star asteroseismology requires the knowledge of analytic expressions of the frequencies and damping times of the modes in terms of the fundamental parameters of the star, such as those in Eqs. (12.1), (12.2), (12.4)–(12.8), (12.12). In recent years, these quantities have been computed in specific models for magneto-elastic modes and superfluid modes, but we still need a better understanding of the underlying physics, in order to derive analytic expressions useful for asteroseismology.

12.3   Unstable Oscillation Modes in Rotating Stars: A CFS Theory Primer

The theory of instabilities in rotating self-gravitating fluid systems came of age in the 1970s with the seminal work of Chandrasekhar (1970), Friedman and Schutz (1978a,b), Friedman (1978) but its seeds had been sown much earlier with the development and completion of the Newtonian theory of classical ellipsoids (Chandrasekhar 1969). Nowadays this framework is widely known as the CFS theory or CFS instability mechanism and includes two types of instabilities, the secular and the dynamical. In reality they all are rotational "frame-dragging" instabilities that are based on the notion of perturbations with a conserved canonical energy that can become zero (dynamical case) or even negative (secular case) above a certain rotational frequency threshold. Apart from this similarity the two instability types are formally very different: the secular instability requires the fluid to be coupled to some dissipative mechanism such as GWs or viscosity while the dynamical instability can take place in an entirely dissipationless system.

Our own discussion of the CFS theory is focussed on the GW-driven secular instability of normal modes which is the most relevant mechanism for "normal" neutron stars. In contrast, the dynamical instability, which is briefly discussed at the end of this section, is known to require the presence of differential rotation in the system (unless we consider less realistic, uniform density stellar models) and therefore can take place for very brief periods of time and in very special environments such as neutron star mergers. Given the "primer" character of this section our presentation of the CFS theory is very far from being comprehensive; the reader can find more detailed reviews of the subject in Refs. Andersson and Kokkotas (2001), Andersson (2003), Friedman and Stergioulas (2013), Andersson and Comer (2007).

The rotational threshold for the onset of the secular CFS instability marks the transition of an initially counter-moving mode to a co-moving one as perceived by an inertial observer. In slightly more technical terms, a mode becomes CFS-unstable when its pattern speed (that is, the azimuthal propagation speed of a constant phase wavefront),

$$\sigma_p = \frac{d\varphi}{dt} = -\frac{\omega_i}{m},$$
(12.13)

changes sign, going from negative to positive. This is achieved when the stellar angular frequency Ω exceeds the mode's pattern speed in the rotating frame (see Eq. (12.3)):

$$\Omega > \frac{\omega_r}{m}.$$
(12.14)

This condition is the first of the two prerequisites for triggering the secular instability. The second one is the coupling of the oscillation to a dissipative mechanism which would allow the mode's negative canonical energy to grow (in absolute value) while more and more positive energy is removed from the system. Assuming coupling to GWs, the radiated power $\dot{E}_{gw}$ is equal and opposite to the change in the mode energy E_{mode} and can be calculated with the standard multipole moment formula (Thorne 1980):

$$\dot{E}_{gw} = -\dot{E}_{mode} = \omega_r \sum_{\ell \geq 2} \omega_i^{2\ell+1} N_\ell \left[|\mathcal{M}_{\ell m}|^2 + |\mathcal{S}_{\ell m}|^2 \right], \qquad (12.15)$$

where $\mathcal{M}_{\ell m}, \mathcal{S}_{\ell m}$ are the mass and current multipoles respectively and N_ℓ a positive constant. In fact, this same formula allows a quick "rediscovery" of the CFS instability since it predicts $\dot{E}_{mode} > 0$ for $\omega_i < 0$, i.e. the criterion (12.14). The GW timescale (for both stable and unstable modes) is defined as (see, e.g. Ipser and Lindblom 1991)

$$\tau_{gw} = \frac{2E_{mode}}{\dot{E}_{gw}}. \qquad (12.16)$$

It should be emphasised that the CFS instability is not strictly a GW-driven effect; electromagnetic radiation and/or normal shear viscosity can also drive unstable oscillations[2] but as it turns out gravitational radiation is almost always the dominant mechanism. In a realistic scenario, an oscillation mode can become CFS-unstable provided the mode-dragging condition (12.14) is satisfied and the growth rate due to GW emission is faster than the viscous damping rate. The damping timescale associated with a given viscosity mechanism is defined as

$$\tau_{visc} = \frac{2E_{mode}}{\dot{E}_{visc}}. \qquad (12.17)$$

where $\dot{E}_{visc}$ is the damping rate. In the most common case where several dissipative mechanisms are operating simultaneously the viscous timescales are combined according to the "parallel resistors" rule.

The resulting instability parameter space is commonly depicted as a "window" in the $\Omega - T$ plane and is typically a V-shaped region limited by bulk/shear viscosity at high/low temperatures and by the spin threshold (12.14) from below. Note that in this review we shall always consider (unless otherwise specified) the standard form for these viscosities, namely, shear viscosity due to electron-electron scattering and bulk viscosity due to the modified URCA reactions (the corresponding viscosity

---

[2]We note with some amusement that the literature contains a mechanical pendulum analogue of the viscosity-driven CFS instability in a 1908 paper by Lamb (1908). Another occurrence of a CFS-like instability in the context of wave dynamics can be found in the 1974 textbook by Pierce (1974, Chapter 11).

coefficients can be found in Sawyer (1989), Andersson et al. (2005a)). The boundary of the instability window (also known as the critical curve) is then defined by,

$$\tau_{gw}^{-1} = \tau_{sv}^{-1} + \tau_{bv}^{-1}, \tag{12.18}$$

where τ_{sv} and τ_{bv} are respectively the shear and bulk viscosity damping timescales and the growth timescale τ_{gw} is hereafter taken to be a positive parameter. The extension of (12.18) to cases with additional dissipation is straightforward.

Among the various neutron star oscillation modes, the ones most promising to become potential sources of GWs and play an important role in the dynamics of neutron stars via the CFS mechanism are the fundamental f-mode and the inertial r-mode. Appropriately, these have been the subject of most work in this area and we discuss them in some detail in Sects. 12.4 and 12.5.

The f-mode is also the one associated with the dynamical CFS instability[3] (for that reason it is nicknamed the "bar-mode" instability from the dominant $\ell = |m| = 2$ multipole). The instability erupts once the system reaches a critical value β_d of the rotational to gravitational binding energy ratio $\beta = T/|W|$. Mathematically this amounts to the merger of the $m = \pm \ell$ mode frequencies. The β_d threshold is significantly higher than the corresponding β_s for secular mode instabilities and for realistic models of rigidly rotating neutron stars it lies beyond the Kepler break-up limit. What this means in practice is that the dynamical f-mode instability is likely to appear in systems that are formed from the outset with fast and differential rotation so that $\beta > \beta_d$. This could happen in the immediate aftermath of binary neutron star merger—a scenario that is corroborated by numerical simulations, for some relatively recent work see e.g. Baiotti et al. (2007), Manca et al. (2007), Franci et al. (2013). The ensuing bar mode instability acts as a potent source of gravitational radiation but its duration is severely limited by the resulting rapid spin-down of the merger remnant and the quenching of differential rotation by the magnetic field (Shapiro 2000).

The neutron star arsenal of CFS instabilities can be more extensive than what has been described so far if we allow for a high degree of differential rotation or a superfluid component. In the former case a low-β (i.e. $\beta \ll \beta_d$) dynamical instability may arise via the system's quadrupole f-modes. This shear instability was first stumbled upon in numerical studies of the dynamics of differentially rotating neutron stars (Shibata et al. 2002) and is akin to shear instabilities in accretion disks. Since then low-β instability has been seen to be excited in core-collapse numerical simulations (see e.g. Kuroda and Umeda 2010). Follow up work (Watts et al. 2005; Passamonti and Andersson 2015) established the instability's

---

[3]Remarkably, a similar CFS dynamical instability is known to exist in an entirely different context, namely, that of a rotating liquid drop with surface tension playing the role of gravity. Laboratory experiments have probed the impact of the instability on the shape of the drop and have revealed a rich phenomenology, see e.g. Brown and Scriven (1980), Hill and Eaves (2008).

intimate connection with the presence of a corotation band (that is, a region where the mode's pattern speed matches the local Ω) in the stellar interior.

The presence of a superfluid opens the possibility for another type of r-mode CFS instability. This newly discovered "two-stream" instability (Glampedakis and Andersson 2009; Andersson et al. 2013) requires a spin lag between the neutron and proton fluids and is driven by vortex mutual friction instead of gravitational radiation. Although this instability could be a viable mechanism for triggering pulsar glitches (Glampedakis and Andersson 2009) it remains to be seen if it is of any relevance as a source of GWs.

12.4 The f-Mode Instability

12.4.1 Newtonian Origins and Recent Developments

The f-mode was chronologically the first to be studied in relation with the CFS instability, with initial work dating back to the 1980s (Friedman 1983; Ipser and Lindblom 1989, 1990, 1991). The main conclusion of those early papers was rather disappointing, predicting that in Newtonian stellar models and under the dissipative action of standard shear and bulk viscosity the instability sets in at a rotation rate very close to the Kepler limit, typically $\Omega \approx 0.95\Omega_K$. Subsequent work (Lindblom and Mendell 1995) showed that the instability is essentially wiped out in neutron stars with a superfluid core as a result of the dissipative coupling between the superfluid's vortex array and the electrons (this is what we call "standard mutual friction" Alpar et al. 1984; Andersson et al. 2006). This result has been considered as a showstopper for the f-mode instability since neutron stars are expected to become superfluid very soon after they are formed. This leaves as the only realistic candidates for harbouring unstable f-modes newly formed neutron stars with hot non-superfluid matter (and obviously with fast rotation).

The traditional astrophysical scenario where these requirements could be met is that of neutron star formation in the aftermath of a core-collapse supernova. The f-mode GW detectability of such an event was first analysed in detail in Lai and Shapiro (1994). This scenario is somewhat less favoured nowadays due to the uncertainty in forming near-Kepler limit spinning neutron stars (see Stergioulas (2003) and references therein). The alternative scenario that has attracted much attention in the recent years is that of binary neutron star mergers—a posterchild source of GWs. These violent events may provide the only suitable arena for the occurrence of the f-mode instability by playing the role of cosmic factories of metastable *supramassive* neutron stars.[4] As we are about to see in the following section it is this property of high mass, in combination with fast rotation and

---

[4]The term "supramassive" refers to a rapidly rotating neutron star with mass above the maximum allowed mass for spherical non-rotating neutron stars. The excess mass is supported by rotation which means that a supramassive system is dynamically stable only above a certain spin rate.

relativistic gravity, that makes these systems prime candidates for harbouring unstable f-modes.

But before opening the discussion on the f-mode instability in relativistic stars it is worth reviewing the recent progress achieved with the use of the "outdated" Newtonian framework. The first problem to be addressed was that of the coupled spin-temperature evolution of a neutron star undergoing an f-mode instability (Passamonti et al. 2013). Using a formalism similar to that previously employed for the r-mode instability (Owen et al. 1998), this work considered hot and rapidly rotating newly formed systems entering the instability window as the temperature drops below $\sim 10^{10}$ K—this value marks the regime where bulk viscosity becomes negligibly weak. The unstable f-mode is promptly saturated, at which point the star begins to spin down with emission of gravitational radiation at almost constant temperature until it exits the instability window.

A representative example of this evolution is shown in Fig. 12.1 for the most unstable $\ell = m = 4$ f-mode of a massive $M = 1.98\,M_\odot$ stellar model with a $n = 0.62$ polytropic EOS. A first noteworthy feature of these f-mode trajectories is that the system is unlikely to ever enter the region where neutrons pair to form a superfluid phase (this is expected to happen at a temperature $T \lesssim 10^9$ K) and the instability is suppressed by vortex mutual friction. Even more important is the interplay between the stellar magnetic field and the unstable f-mode. A neutron star with surface field $B \gg 10^{12}$ G will evolve along a shorter spin-temperature trajectory (see right panel of Fig. 12.1) with its spin evolution mostly driven by the magnetic dipole radiation rather than gravitational radiation, hence leading to a deteriorated GW detectability. A similar situation may arise if in parallel with the f-mode there is also an unstable r-mode present in the system—a not unlikely

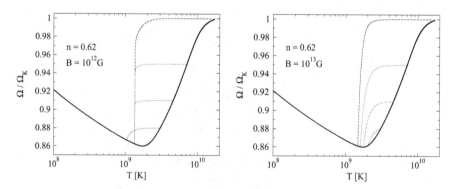

Fig. 12.1 *f-mode spin-temperature evolution*: adapted from Ref. Passamonti et al. (2013), this figure displays the coupled $\Omega - T$ evolution of a Newtonian $n = 0.62$ polytropic model with gravitational mass (in the non-rotating limit) $M = 1.98\,M_\odot$ under the combined influence of an unstable $\ell = m = 4$ f-mode and magnetic dipole torque. The thick solid curve represents the instability window (accounting for standard shear and bulk viscosity but not for superfluid mutual friction). Left panel: the various evolution trajectories correspond to different initial Ω and a surface dipole field $B = 10^{12}$ G. Right panel: the same initial Ω values as before for a surface dipole field $B = 10^{13}$ G

scenario given the much wider instability window of the latter, see Ref. Passamonti et al. (2013) for more details.

A second key recent development concerns saturation amplitude (or equivalently saturation energy) of the instability. This arduous calculation was undertaken in Pnigouras and Kokkotas (2015, 2016) by means of a nonlinear mode-coupling model similar to the one that had been employed before for the r-modes (see Sect. 12.5) but with the added property of non-uniform, stratified matter. According to the aforementioned work the mode's energy is primarily drained by the non-linear coupling to g-modes; this leads to a saturation energy that may fluctuate considerably across the instability's $\Omega - T$ parameter space. A representative maximum value for the saturation energy can be taken to be $E_{\text{sat}} \sim 10^{-6} Mc^2$ (Pnigouras and Kokkotas 2015, 2016). This result has obvious implications for the GW detectability of unstable f-modes and will be discussed in more detail below.

12.4.2 The f-Mode Instability in Relativistic Stars: A Story of Revival?

The last few years have seen a renewed interest in the f-mode instability with the objective of revising the earlier Newtonian results using relativistic and rapidly rotating neutron star models. This effort was spearheaded by Ref. Gaertig et al. (2011) which considered polytropic models and relativistic gravity in the Cowling approximation. These first relativistic results were promising, predicting a revised f-mode growth timescale about an order of magnitude shorter than the Newtonian value for the same canonical stellar parameters. Accordingly, the instability window was found to be larger than its Newtonian counterpart, with the $\ell = m = 4$ being the most unstable multipole, see Fig. 12.2 (left panel). Follow-up work (Doneva et al. 2013) established that the combination of a realistic EOS model with a higher stellar mass ($\approx 2M_\odot$) can support an even wider instability window—this is shown on the right panel of Fig. 12.2 and can be directly compared against the Newtonian window of Fig. 12.1 (which corresponds to a Newtonian polytrope of the same mass).

These calculations allow us to draw a clear conclusion: for a given rotation Ω, massive relativistic systems (which are also the most compact ones) have significantly enhanced f-mode instability properties as compared to their Newtonian and/or less massive counterparts.

The most dramatic manifestation of this conclusion may take place in the immediate aftermath of the merger of a binary neutron star system: for the expected range of initial masses the merger may not immediately produce a black hole but, instead, lead to a transient phase of a supramassive ($M \sim 2.5M_\odot$) and $\Omega \approx \Omega_K$ neutron star remnant.[5] Besides their central role in GW astronomy, these mergers have come to be seen as the leading theoretical model for the central engine

---

[5] A relatively low mass remnant may settle down to a normal neutron star existence without ever collapsing to a black hole.

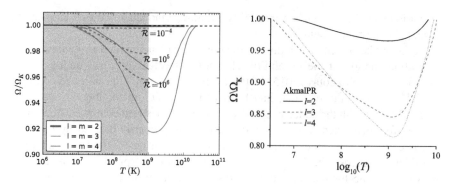

Fig. 12.2 *Relativistic f-mode instability window.* Left panel: this figure, taken from Ref. Gaertig et al. (2011), shows the f-mode instability window (solid curves) corresponding to the first three multipoles $\ell = m = 2, 3, 4$ and for a $n = 0.73$ relativistic polytrope with parameters $M = 1.48 M_\odot$ and $R = 10.47$ km (in the $\Omega = 0$ limit). The shaded area indicates the parameter space where neutron superfluidity is present (with a fiducial onset temperature $T_{cn} = 10^9$ K). The critical curves are discontinuous at T_{cn} as a result of the use of different shear viscosity coefficients in the superfluid and normal region. The dashed curves show the instability window of the $\ell = 4$ multipole with vortex mutual friction accounted for and the drag coefficient shifted from its canonical value $\mathcal{R} \approx 10^{-4}$ (which leads to a fully suppressed instability). Right panel: the instability window for the same f-mode multipoles (this time assuming non-superfluid matter) and for a neutron star model with a realistic EOS (in this particular example APR) and parameters $M = 2 M_\odot$, $R = 10.88$ km, see Ref. Doneva et al. (2013)

powering short GRBs (see e.g. Dai and Lu 1998; Zhang and Mészáros 2001; Rowlinson et al. 2013). The supramassive remnant, which now takes the form of a proto-magnetar as a result of strong magnetic field amplification (see e.g. Rezzolla et al. 2011; Kiuchi et al. 2014; Giacomazzo et al. 2015), is believed to power the burst's late-time emission and is associated with the X-ray plateau and power-law tail seen in the light curves of several of these events (Rowlinson et al. 2013; Metzger and Piro 2014). According to the GRB data, the remnant's lifetime spans a range $10^2 - 10^5$ s which is determined by the spin-down timescale due to magnetic dipole radiation (the same mechanism is responsible for powering the system's X-ray emission).

It is during this X-ray "afterglow"/supramassive phase where the f-mode instability is most likely to take place and become a potentially strong source of GWs. Viscosity is not likely to be an impeding factor in these circumstances since the system is expected to cool below 10^{10} K very shortly after the supramassive remnant has been formed, see e.g. Lasky and Glampedakis (2016). This scenario has been put forward in Ref. Doneva et al. (2015) and is backed up by f-mode calculations that suggest surprisingly short growth timescales, $\tau_{gw} \sim 10 - 100$ s, see Fig. 12.3 (left panel). However, a short τ_{gw} does not necessarily translate into a conspicuous f-mode GW signal. To what extent post-merger supramassive remnants could be realistic targets for present and next generation GW detectors is discussed in more detail in the following section.

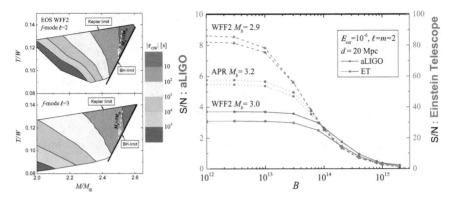

Fig. 12.3 *f-mode instability in supramassive neutron stars.* Left panel: this figure is adapted from Ref. Doneva et al. (2015) and shows the *f*-mode growth timescale (for the first two multipoles and assuming a WFF2 EOS stellar model) as a function of the supramassive neutron star's mass. Right panel: the GW signal-to-noise ratio as a function of the surface dipole field for a supramassive neutron star located at a fiducial distance 20 Mpc, spinning down under the combined influence of the *f*-mode instability (saturated at $E_{\text{sat}} = 10^{-6} M c^2$) and magnetic dipole radiation, for different choices of baryonic mass and EOS (figure adapted from Doneva et al. (2015))

12.4.3 The Observability of the *f*-Mode Instability

The overall amplitude of the *f*-mode signal is limited by the distance of the source and the mode's maximum saturation amplitude. As we have already seen, the latter parameter was recently obtained and expressed as a saturation energy, $E_{\text{sat}} \sim 10^{-6} M c^2$ (Pnigouras and Kokkotas 2015, 2016). A perhaps more intuitive way to quantify this result is via the *f*-mode-induced ellipticity in the stellar shape. A back-of-the-envelope calculation leads to Lasky and Glampedakis (2016),

$$\frac{\delta R}{R} \sim \left(\frac{E_{\text{sat}}}{M c^2}\right)^{1/2} \left(\frac{cR}{GM}\right)^{1/2}, \qquad (12.19)$$

which for a typical neutron star compactness returns $\delta R/R \sim 10^{-3}$. In other words, the mode gets saturated at a "linear" level. The GW observability of an unstable *f*-mode saturated at this maximum amplitude is shown in Fig. 12.3 (right panel) in the form of a signal-to-noise ratio (SNR) for the Advanced LIGO/Virgo and ET detectors as a function of the surface dipole field (Doneva et al. 2015). Detectability is strongly diminished in systems with magnetar-like fields for the simple reason that the spin-down timescale is controlled by magnetic dipole radiation and is much shorter than the duration of a GW-driven spin-down (see also Fig. 12.1). Less magnetised systems with $B \lesssim 10^{14}$ G are likely to be marginally detectable by Advanced LIGO but should be "in the bag" for ET.

A more empirical assessment of the *f*-mode's GW observability can be made with the help of the short GRB X-ray data (Lasky and Glampedakis 2016). The

$\sim t^{-2}$ late time decay profile seen in several light curves of these events can be taken as evidence of an electromagnetic radiation dominated spin-down, in accordance with the GRB proto-magnetar model. This information, in combination with the observed duration of the X-ray plateaus, can be used to set upper limits in the saturation amplitude of unstable f-modes. Interestingly, the resulting limit is similar to the theoretically predicted maximum amplitude—this result suggests that the f-mode instability could in principle play an important role in the dynamics of the post-merger remnant. Unfortunately, the predicted f-mode detectability is rather pessimistic even for ET, limited by the shortness of the spin-down timescale and the large distances ($\gtrsim 500$ Mpc) associated with short GRBs.

12.4.4   Future Directions

The f-mode calculations discussed in the preceding sections were carried out using the Cowling approximation. We have already pointed out in a previous section how this approximation affects the f-mode asteroseismology formalism. As far as the f-mode instability is concerned, it is known (Zink et al. 2010; Yoshida 2012) that the Cowling approximation makes relativistic stars less prone to the CFS instability by increasing the rotation threshold (12.14). It is therefore expected that the f-mode instability in neutron star models with fully relativistic dynamical spacetime will be enhanced but a detailed quantitative calculation of this modification is still lacking.

Another desideratum in this area should be the further improvement in the modelling of the f-mode's non-linear saturation physics. The very recent state-of-art calculation of Pnigouras and Kokkotas (2015, 2016) is "primitive" in the sense that it is based on a Newtonian framework and a slow rotation approximation. Taking this calculation to the next level (with relativistic gravity and/or fast rotation) is likely to prove a very challenging—but necessary—endeavour.

Finally, it should be borne in mind that the f-mode instability window could be significantly modified by viscosity due to the presence of exotic phases of matter in the interior of neutron stars, such as hyperons and quarks. Although this scenario has been exhaustively explored in the context of the r-mode instability (see Sect. 12.5) very little is known about its impact on the f-mode instability.

12.5   The r-Mode Instability

The discovery of the inertial r-mode CFS instability in 1998 came as something of a surprise to the neutron star community (Andersson 1998; Friedman and Morsink 1998). Since then this instability has received the lion's share of the published work on the subject of neutron star oscillations as a consequence of its potentially key role in the spin evolution of neutron stars and as a promising source of GWs (for early comprehensive reviews on the subject see Andersson and Kokkotas 2001; Friedman and Stergioulas 2013; a more specialised recent review can be found in Haskell (2015)). Accordingly, this r-mode section occupies a central place in this review.

Two key characteristics are associated with the r-mode instability:

(1) the mode frequency in the rotating frame is

$$\omega_r = \frac{2m\Omega}{\ell(\ell+1)},$$

(12.20)

so that with dissipation switched off, the mode becomes CFS-unstable as soon as the star acquires rotation (that is, the condition (12.14) is automatically satisfied for any Ω). With dissipation restored, the resulting $\Omega - T$ instability window is in general quite large.

(2) the mode's predominant axial geometry implies a nearly horizontal fluid flow pattern and leads to GW emission dominated by the current multipole S_{22} rather than the mass multipoles. The resulting growth timescale τ_{gw} exhibits a characteristic $\sim \Omega^{-6}$ dependence and is very short. For a canonical $n = 1$ Newtonian polytropic star this is Lindblom et al. (1998), Andersson and Kokkotas (2001),

$$\tau_{\text{gw}} \approx 50 \, M_{1.4}^{-1} R_6^{-4} P_{-3}^6 \, \text{s}.$$

(12.21)

The two above properties alone are sufficient to guarantee the astrophysical relevance of the r-mode instability. In addition, practitioners of neutron star dynamics enjoy the luxury of being able to do most of r-mode physics within a Newtonian/slow rotation framework rather than having to struggle with the complexity of rapidly rotating GR stars (as it was the case for the f-mode instability). Nonetheless, relativistic r-mode calculations have been performed (Lockitch et al. 2001; Ruoff and Kokkotas 2001; Yoshida and Lee 2002; Lockitch et al. 2003; Idrisy et al. 2015) and have demonstrated the corrections to the Newtonian mode eigenfunction GW growth timescale to be typically very small. At a qualitative level, GR changes the purely axial slow-rotation Newtonian r-mode into an axial-led inertial mode—this is similar to the modification caused by fast rotation in Newtonian theory. More pronounced is the relativistic correction to the mode frequency which is of the order of $\sim 20\%$. This is large enough to be accounted for in searches for GW signals from unstable r-modes (Owen 2010). The r-mode's insensitiveness extends to variations in the EOS of matter as well. This has been demonstrated by the very recent analysis of Idrisy et al. (2015) and is in agreement with earlier investigations that made use of polytropic models (Lockitch et al. 2003). The upshot is that r-mode calculations based on a canonical polytropic Newtonian model (or even a uniform density model) are sufficiently robust and accurate for most practical applications. This will be our benchmark model for the reminder of our discussion of the r-mode instability unless otherwise specified.

12.5.1 r-Mode Phenomenology

Most of the uncertainty related to the r-mode instability has to do with its damping. In a "minimum physics" (although not necessarily realistic!) model that assumes dissipation only due to standard shear and bulk viscosity the resulting instability window overlaps with the box-shaped region occupied by the known population of rapidly rotating neutron stars in LMXBs and MSPs. This r-mode window is shown in Fig. 12.4, where we have also included LMXB data tabulated in Mahmoodifar and Strohmayer (2013), Ho et al. (2011) (to be discussed below). The striking feature of Fig. 12.4 is that several known neutron stars reside well inside the minimum damping window and therefore should harbour unstable r-modes. This observation forms the basis of what we shall call the "r-mode puzzle" in the following section.

The spin distribution of these neutron stars has been something of a mystery since under the unhindered action of accretion, LMXBs (and their MSP descendants) should have been expected to straddle the Kepler frequency limit, $f_K \gtrsim 1$ kHz (see also e.g. the discussion in Haskell et al. (2012)). The apparent spin cut-off at a much lower frequency has been taken as evidence of the presence of a spin-down torque

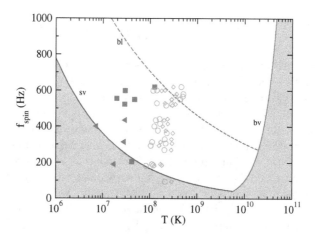

Fig. 12.4 *r-mode instability window.* This figure shows the r-mode spin-temperature instability window (in terms of the spin frequency $f_{spin} = \Omega/2\pi$) assuming a canonical neutron star model and accounting for standard shear (blue curve) and bulk (green curve) viscosity. The corresponding viscous damping timescales were taken from Lindblom et al. (1998) and Andersson and Kokkotas (2001) while Eq. (12.21) was used for the growth timescale. The shaded area represents the region of r-mode stability. Besides the minimum damping window, the figure also includes the critical curve due to Ekman boundary layer damping (red dashed curve) which is discussed in Sect. 12.5.6. The data points represent LMXBs with known spin frequencies. The core temperature is inferred by flux measurements during phases of quiescence Mahmoodifar and Strohmayer (2013) (filled squares and triangles for upper limits) or is theoretically predicted assuming a combined r-mode spin and thermal equilibrium during phases of accretion Ho et al. (2011) (open diamonds and circles for cooling dominated by mURCA reactions and Cooper pair processes respectively) see Sect. 12.5.3 for details

that counteracts accretion. A suggestion that has attracted much attention since its conception is that of a GW torque supplied either by a deformation in the stellar shape (i.e. a neutron star "mountain") or an unstable oscillation mode (Papaloizou and Pringle 1978; Bildsten 1998a; Andersson et al. 1999). This idea obviously combines well with the minimum dissipation r-mode model but we should not be too hasty in drawing conclusions. Spin equilibrium in LMXBs could be achieved by an alternative *non*-GW mechanism, namely, the coupling of the stellar magnetic field with the accretion disk (Ghosh and Lamb 1979; Wang 1995; Rappaport et al. 2004; Andersson et al. 2005b) and it is here fitting to open a parenthesis and discuss it.

In the disk coupling model the global stellar magnetic field threads the material of the disk and the field lines, being simultaneously anchored in the disk and on the stellar surface, provide a very efficient braking mechanism. With the input of canonical surface dipole fields, $B \sim 10^8 - 10^9$ G, and reasonable physical assumptions the predictions of the available phenomenological disk coupling models compare fairly well with LMXB spin data (Andersson et al. 2005b; Haskell and Patruno 2011; Patruno and Watts 2012) thus dispelling much of the mystery behind their spin distribution cut off. Although this is compelling evidence in favour of these models it is certainly premature to shelve the alternative GW-based mechanisms. In fact, recent work (Bhattacharyya and Chakrabarty 2017; Bhattacharyya 2017) suggests that the transient nature of accretion could seriously weaken the efficiency of the disk coupling model thus calling for an additional spindown torque. What can be said with certainty is that the existing data do not exclude the realistic possibility of having some r-mode activity (or indeed a neutron star mountain) in LMXB systems that could otherwise be dominated by magnetic disk coupling or by magnetic dipole spin-down when in quiescence. The disk coupling model does, however, remove the need to cling to solely GW-based spin equilibrium models. With this observation in mind we can resume our main r-mode discussion.

Once inside the instability window, the r-mode drives a coupled spin-temperature evolution the details of which are largely determined by the maximum (saturation) value of the mode amplitude α_r. This dimensionless parameter is commonly defined in the literature via the velocity field of the dominant $\ell = m$ mode (Owen et al. 1998)

$$\delta \mathbf{v} = \alpha_r \left(\frac{r}{R}\right)^\ell \Omega R \mathbf{Y}_{\ell\ell}^B e^{i\omega_r t}, \tag{12.22}$$

where $\mathbf{Y}_{\ell m}^B$ is a vector spherical harmonic of the magnetic type. More simply, we can think of the amplitude as the ratio $\alpha_r \approx \delta v / \Omega R$.

Considering first accreting neutrons stars, once the system moves across a $d\Omega/dT < 0$ segment of the instability curve during spin up it will undergo a cyclic thermal runaway (Levin 1999; Andersson et al. 2000; Bondarescu et al. 2007), see Fig. 12.5.

The cycle begins with the mode growing under the emission of gravitational radiation, rapidly reaching its maximum saturation value α_{sat}. When this happens, and assuming $\alpha_{sat} \ll 1$, the viscous timescale becomes comparable to the growth

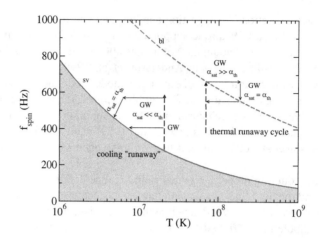

Fig. 12.5 *r-mode spin-temperature evolution.* The *r*-mode-driven $f_{\text{spin}} - T$ evolution of an accreting neutron star is determined by (1) the instability curve's slope at the point where the system, spun up by accretion, first enters the instability window (dashed vertical lines) and (2) the maximum saturation amplitude α_{sat}. In this figure we show the canonical scenario of a negative slopping instability curve due to shear viscosity (blue curve) or an Ekman layer (red curve). If the saturation amplitude is not too small ($\alpha_{\text{sat}} \gg \alpha_{\text{th}}$ initially), the system is forced into a cyclic thermal runaway loop in the vicinity of the instability curve where the GW emitting portions are indicated in the figure, see Sects. 12.5.1 and 12.5.3 for more details. The cycle's nearly horizontal path (*r*-mode heating) continues until thermal equilibrium is established, $\alpha_{\text{sat}} = \alpha_{\text{th}}$, followed by *r*-mode spin-down towards the instability curve. In the opposite scenario of a very small saturation amplitude (i.e. $\alpha_{\text{sat}} \ll \alpha_{\text{th}}$ initially) a weak *r*-mode instability is active during accretion and, once the latter has ended, cannot prevent the system from cooling (at nearly constant spin) towards a state of thermal equilibrium or until the instability curve is crossed again (this scenario is shown here with respect to the minimum damping curve)

timescale, $\tau_{\text{visc}} \approx \tau_{\text{gw}}$, (see the *r*-mode evolution equations in Owen et al. (1998), Levin (1999)). At the same time the heat deposited by shear viscosity lifts the stellar core temperature. This happens at almost constant spin frequency since the GW spin-down timescale, $\tau_{\text{sd}}^{\text{gw}} \equiv \Omega/|\dot{\Omega}|$, is always much longer than the heating timescale $\tau_{\text{heat}} \equiv T/|\dot{T}|$. These timescales are (see e.g. Levin 1999):

$$\tau_{\text{sd}}^{\text{gw}} \approx \frac{5.3}{\alpha_{\text{sat}}^2}\tau_{\text{gw}}, \qquad \tau_{\text{heat}} = \frac{C_v T}{\dot{E}_{\text{visc}}} \approx 7.8 \times 10^{-6} \frac{P_{-3}^2 T_8^2}{\alpha_{\text{sat}}^2 M_{1.4} R_6^2}\tau_{\text{gw}}. \qquad (12.23)$$

where in the last equation the damping rate $\dot{E}_{\text{visc}}$ is typically fixed by shear viscosity or a crust-core boundary layer (see below) and the heat capacity $C_v \approx 1.4 \times 10^{38}\, T_8$ erg/K is that of ordinary matter (Shapiro and Teukolsky 1986) (superfluidity would reduce τ_{heat} even further by decreasing C_v). Also note that we have explicitly used the timescale equality $\tau_{\text{visc}} \approx \tau_{\text{gw}}$ for a saturated mode.

The thermal runaway continues up to the point where stellar cooling can efficiently balance viscous heating. Assuming neutrino cooling due to the standard modified URCA process (with the corresponding emissivity $L_\nu \approx 7 \times 10^{31} T_8^8$ erg/s) Shapiro and Teukolsky (1986) and a minimum damping instability curve, we can convert thermal equilibrium into a "thermal" r-mode amplitude,

$$\dot{E}_{\mathrm{visc}} = \dot{E}_{\mathrm{sv}} = L_\nu \Rightarrow \alpha_{\mathrm{th}} \approx 1.4 \times 10^{-9} M_{1.4}^{-1} R_6^{-3} P_{-3}^4 T_8^4. \qquad (12.24)$$

Note that this result is indicative as it obviously depends on the assumed cooling physics. For the temperature range of LMXBs the cooling could be dominated by superfluid Cooper-pair processes (Ho et al. 2011) or even surface photon emission (Mahmoodifar and Strohmayer 2013).

At the stage of thermal equilibrium the temperature remains essentially constant and the r-mode-driven spin-down steers the system towards the instability curve; once the curve is crossed, the star becomes stable again and rapidly cools. Eventually, accretion will once again spin up the star preparing the ground for the next cycle. The duration of the GW-emitting portion of this cycle depends on α_{sat}. A large amplitude (the first papers to explore the implications of this r-mode evolution were somewhat optimistically assuming $\alpha_{\mathrm{sat}} \sim 0.1$–1, see e.g. Andersson et al. 2000) translates to fast GW-driven evolution but a tiny GW duty cycle, making LMXBs uninteresting sources of GWs. If α_{sat} is small, the system does not wander off much from the instability curve and the cycle's GW efficiency can improve dramatically. We can make this argument quantitative by approximating the cycle's GW efficiency as the ratio between the time the system spends emitting radiation and the typical LMXB lifetime Heyl (2002),

$$D_{\mathrm{cycle}} \approx \frac{t_{\mathrm{cycle}}}{10^7 \text{ year}} \approx \frac{10^{-11}}{\alpha_{\mathrm{sat}}^2}. \qquad (12.25)$$

Combining this with the estimated LMXB birth rate $\sim 10^{-5}$/year/galaxy, it is not too difficult to see that in order to have a system always switched on in our Galaxy,

$$D_{\mathrm{cycle}} \lesssim 10^{-2} \Rightarrow \alpha_{\mathrm{sat}} \lesssim 10^{-4}. \qquad (12.26)$$

Thus, somewhat counterintuitively, a relatively small amplitude is likely to improve the r-mode's GW detectability in LMXBs. Of course, the amplitude should not be too small for otherwise these sources would be too faint. As discussed below, the upper limit (12.26) is compatible with the theoretically predicted r-mode saturation amplitude.

In the thermal runaway scenario described above the cut off in the spin distribution of LMXBs is set by the spin frequency at which the systems enter the instability window. Taking into account that the expected temperature range for LMXBs is $T \sim 10^7 - 5 \times 10^8$ K it becomes immediately clear that the previously defined minimum damping window is in disagreement with the observed ~ 600 Hz cut off (see Fig. 12.4). The model does much better if we invoke a "canonical" r-mode instability window which, in addition to shear and bulk viscosity, accounts for dissipation due to a viscous Ekman boundary layer at the crust-core boundary. The implications of this Ekman layer could be crucial for the survivability of the r-mode instability and are discussed in detail below in Sects. 12.5.6 and 12.5.7.

It should also be emphasised that the cyclic evolution is *not* an unavoidable outcome of the r-mode evolution in LMXBs. For instance, the presence of exotic neutron star matter in the form of hyperons or quarks could lead to an instability curve with positive $d\Omega/dT$ slope in the temperature range relevant to LMXBs. As discussed by several authors, this configuration could effectively trap accreting systems near the critical curve and turn them into persistent sources of GWs (Andersson et al. 2002; Nayyar and Owen 2006; Haskell and Andersson 2010). Another way to prevent the cyclic evolution from happening is by invoking a saturation α_{sat} sufficiently small so that τ_{heat} (and consequently τ_{sd}^{gw}) exceeds the accretion timescale ($\sim 10^7 - 10^8$ year). In this scenario cooling dominates over r-mode heating, i.e. $\alpha_{sat} \ll \alpha_{th}$, and once accretion comes to an end the system undergoes a "cooling runaway", moving towards the low T part of the instability window until thermal equilibrium is established or the instability curve is crossed (Alford and Schwenzer 2015), see Fig. 12.5. We elaborate more on this small amplitude scenario below in Sect. 12.5.3.

The r-mode-driven evolution of rapidly rotating non-accreting neutron stars is somewhat simpler than the cyclic scenario of the preceding paragraphs. These systems can be either old MSPs or, more speculatively, very young neutron stars such as the central compact objects (CCOs) associated with supernova remnants. The latter objects have already been the target of broad band GW searches by LIGO and, in spite of their unknown spin frequencies (which are likely to be low), have led to direct upper limits on the r-mode amplitude (Aasi et al. 2015). The key parameter here is the spin-down timescale (12.23) which becomes

$$\tau_{sd}^{gw} \approx 800 \left(\frac{10^{-4}}{\alpha_{sat}}\right)^2 M_{1.4}^{-1} R_6^{-4} P_{-3}^6 \text{ year.} \tag{12.27}$$

According to this expression, a large amplitude r-mode drives a very rapid spin-down thus seriously diminishing the system's GW observability. In order to have a τ_{sd}^{gw} that is compatible with the estimated ages of the CCOs (i.e. $\sim 10^2 - 10^3$ year) we would need to invoke $\alpha_{sat} \sim 10^{-3}$. A much smaller amplitude is required if τ_{sd}^{gw} is associated with the observed spin-down age of MSPs. We will return to this point later, in Sect. 12.5.3.

12.5.2   An r-Mode Puzzle?

How do actual observations compare against the basic r-mode phenomenology described in the preceding section? As we have seen, the minimum damping model clearly predicts that the r-mode instability should be operating in a large portion of the LMXB and MSP populations (and perhaps in some young sources). The fact that these objects do *not* appear to show any evidence of strong r-mode activity should have important implications for the physics in their interior. For example, and as already pointed out, the data points shown in Fig. 12.4 are clearly incompatible with a thermal runaway cycle taking place near the standard shear viscosity instability curve. Similarly, the MSP timing data are incompatible with the r-mode instability unless α_{sat} is very small.

A clue is provided by the observed long-term spin-down of accreting millisecond X-ray pulsars (AMXPs) in quiescence (see Patruno and Watts 2012 for a review). For example, two of these objects, SAX J1808–3658 and IGRJ00291+5934, are likely to sit inside the instability window (given the uncertainties on the pulsar parameters), and therefore could experience r-mode-driven spin-down. However, the measured spin-down rate is consistent with that caused by a canonical LMXB magnetic field ($\sim 10^8$ G), hence suggesting that the r-mode instability is not the dominant effect here (Haskell and Patruno 2011; Pappito et al. 2011).

This apparent tension between the minimum damping r-mode model and the spin-temperature data of known rapidly rotating neutron stars may be dubbed the "r-mode puzzle".

There are two complementary ways to make theory and observations mutually compatible. The first one relies on the presence of additional damping mechanisms that could modify the instability window and render r-mode-stable the systems in question. The required extra damping could be provided, for example, by exotic matter in the neutron star core, strong superfluid vortex mutual friction or an Ekman-type viscous boundary layer at the crust-core interface. The second possibility is that of a small saturation amplitude r-mode. In this scenario the r-mode instability does operate in (at least) some rapidly rotating neutron stars but α_{sat} is sufficiently small so that the ensuing sluggish r-mode evolution is compatible with observations. We now can take a closer look at these two resolutions of the r-mode puzzle.

12.5.3   Small Amplitude r-Modes

Theoretical calculations already provide constraints on α_{sat} and obviously need to be incorporated in any realistic r-mode model. The most robust saturation mechanism is provided by non-linear couplings between the r-mode and other (primarily inertial) modes (Schenk et al. 2002; Arras et al. 2003; Bondarescu et al. 2007, 2009). A series of impressive *tour de force* calculations have revealed a complicated spin-temperature evolution pattern for α_{sat} but as a rule of the thumb estimate for the time-averaged amplitude we can take $\alpha_{sat} \sim 10^{-4} - 10^{-3}$ (Schenk et al. 2002; Arras et al. 2003; Bondarescu et al. 2007, 2009).

This level of saturation is obviously low but not low enough for the purposes of the small amplitude scenario! To illustrate this, we consider MSPs with measured spin-down rates that reside inside the minimum damping instability window. These data can be used to set upper limits on the r-mode amplitude (obviously, these limits make sense provided the systems in question are r-mode-unstable in the first place—this may not be the case in a more realistic enhanced damping scenario). Equating the total rate of change of the stellar rotational energy to the GW torque leads to a spin-down r-mode amplitude (see e.g. Owen 2010):

$$\alpha_{sd} \approx 5.7 \times 10^{-3} P_{-3}^{5/2} \dot{P}_{-10}^{1/2} M_{1.4}^{-1/2} R_6^{-2}. \tag{12.28}$$

The strongest constraints from the available MSP data imply a very small amplitude, $\alpha_{sd} \lesssim 10^{-7}$ (Alford and Schwenzer 2014a, 2015).

Given that MSPs are almost certainly spinning down via standard magnetic dipole radiation, it is meaningful to compare the relative $\dot{P}$ contribution of GW emission. It is an easy exercise to derive the following formula for the spin-down ratio $\dot{P}_{gw}/\dot{P}_{em}$ which is also equal to the spin-down age ratio $\tau_{sd}^{em}/\tau_{sd}^{gw}$:

$$\frac{\dot{P}_{gw}}{\dot{P}_{em}} = \frac{\tau_{sd}^{em}}{\tau_{sd}^{gw}} \approx 0.23 \left(\frac{\alpha_r}{10^{-7}}\right)^2 M_{1.4}^2 B_8^{-2} P_{-3}^{-4}. \tag{12.29}$$

This expression indeed verifies that for canonical MSP parameters the r-mode torque is much weaker than the electromagnetic one.

Two more key r-mode amplitudes can be calculated by making contact with LMXB observations. The first one comes from the assumption of spin equilibrium (discussed earlier). Using a simple fiducial spin up torque that ignores the effect of the magnetic field on the accretion dynamics leads to Brown and Ushomirsky (2000)

$$\alpha_{acc} \approx 1.2 \times 10^{-8} \left(\frac{L_{acc}}{10^{35} \text{ erg s}^{-1}}\right)^{1/2} P_{-3}^{7/2} \tag{12.30}$$

where L_{acc} is the accretion luminosity. This estimate is very close to the spin equilibrium amplitude obtained in Ref. Mahmoodifar and Strohmayer (2013) using a spin-up torque extracted observationally from the time-averaging over a succession of accretion episodes.

The second "handle" for estimating r-mode amplitudes is provided by the consideration of thermal equilibrium in LMXBs (Ho et al. 2011; Mahmoodifar and Strohmayer 2013). Measuring the luminosity of these objects in quiescence allows the inference of their surface and core temperature and in turn of their cooling rate. If heating is attributed to the dissipation of a steady-state unstable r-mode, the mode amplitude can be calculated by invoking thermal equilibrium. This is the approach taken in Mahmoodifar and Strohmayer (2013) and is of the same logic that led to the amplitude (12.24).

The quiescent LMXBs considered in Mahmoodifar and Strohmayer (2013) have $T \lesssim 10^8$ K (see data points in Fig. 12.4)—this implies a cooling dominated by surface photon emission rather that neutrinos. The resulting thermal equilibrium amplitudes lie in the range $\alpha_{th} \sim 10^{-8} - 10^{-7}$. A closer look at the tabulated data of Mahmoodifar and Strohmayer (2013) reveals that for sources with both spin and thermal equilibrium data $\alpha_{th} \approx 0.1\alpha_{acc}$. This means within the minimum damping model, and treating the inferred α_{th} as an empirical saturation amplitude, r-modes *cannot* balance the long-term accretion torque in LMXBs. Furthermore, using α_{th} in Eq. (12.29) we find $\dot{P}_{gw} \ll \dot{P}_{em}$. In other words, quiescent LMXBs are expected to predominantly spin-down via magnetic dipole radiation, in agreement with observations (Haskell and Patruno 2011).

A different angle of approach that makes more contact with the accreting phase of LMXBs rather that their quiescence assumes *both* r-mode thermal and spin equilibrium to take place at the same time. Then, the equality $\alpha_{acc} = \alpha_{th}$ relates T to the observables L_{acc} and P. As shown in Ho et al. (2011) the LMXB core temperatures calculated in this way lie in the range $T \sim 10^8 - 5 \times 10^8$ K and are consistently higher than those in Mahmoodifar and Strohmayer (2013) (see also Fig. 12.4). Of course this model would run into difficulties in explaining the quiescence data (at least for some systems).

What is noteworthy from the preceding discussion is that the three "observable" amplitudes $\alpha_{sd}, \alpha_{acc}, \alpha_{th}$ lie well below the predicted α_{sat} due to non-linear couplings. This is clearly problematic from a theoretical point of view and therefore other saturation mechanisms must be sought in order to fill the gap. A recently suggested alternative mechanism is based on the dissipative coupling between the superfluid vortex array and the quantised magnetic fluxtubes in regions of the star where a neutron superfluid co-exists with a proton superconductor (Haskell et al. 2014). The resulting saturation amplitude could be as small as $\alpha_{sat} \sim 10^{-5} - 10^{-6}$. Although this is much smaller than the mode-coupling α_{sat}, there is still some significant difference with the observable amplitudes.

The small r-mode amplitude scenario has been further explored in a recent series of papers by Alford and Schwenzer (2014b,a, 2015). Following an analysis similar to that of Bondarescu et al. (2007, 2009) but allowing for a very small $\alpha_{sat}(\Omega, T)$, they derive steady-state r-mode evolution trajectories for LMXBs/MSPs and young neutron stars. As mentioned earlier, in their model the cyclic thermal runaway in accreting systems does not take place since the r-mode is too feeble to heat up (or spin down) the star efficiently. After the end of accretion, the system simply cools until it reaches a steady state at a lower temperature (see Fig. 12.5). This scenario, however, cannot explain the cut off in the LMXB spin distribution since, in the absence of any other spin-down mechanism, these systems would be free to reach the Kepler limit.

At this point it is worth pausing to consider the actual GW detectability of r-mode-active neutron stars as a function of the amplitude and the spin frequency.

12.5.4 *r-Mode Detectability*

Given the broad scope of this review, the discussion of the r-mode's detectability will necessarily be brief—a more detailed recent analysis can be found in Kokkotas and Schwenzer (2016). The intrinsic GW strain associated with an r-mode can be computed with the help of Thorne's multipole moment formula (Thorne 1980). The contribution of the dominant $\ell = m = 2$ current multipole is (see e.g. Owen 2010):

$$h_0 \approx 4 \times 10^{-23}\, \alpha_r M_{1.4} R_6^3 \left(\frac{1\,\text{kpc}}{D}\right) \left(\frac{f_{\text{gw}}}{100\,\text{Hz}}\right)^3 \tag{12.31}$$

where we have normalised the distance D to a galactic source and used the equality between the GW frequency and the mode's inertial frame frequency,

$$f_{\text{gw}} = |f_{\text{mode}}| = \frac{4}{3} f_{\text{spin}}. \tag{12.32}$$

The detectability of the strain (12.31) is shown in Fig. 12.6 assuming a 1 year phase-coherent observation and a source at $D = 1\,\text{kpc}$. First of all, the smallness of the r-mode amplitude essentially eliminates extragalactic sources from being candidate targets for detection unless the source is located within our local group and the amplitude is much higher than the upper limits suggested by LMXB/MSP spin-down and thermal data. This could be a realistic possibility if the systems from which the data came from are not r-mode unstable—in that case it would make sense to use a fiducial amplitude $\alpha_r = 10^{-3}$, corresponding to a maximum saturation amplitude due to non-linear mode couplings. The GW strain for this optimistic

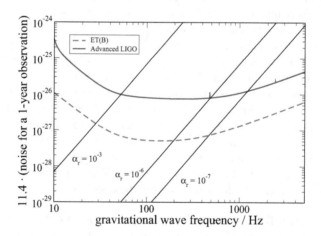

Fig. 12.6 *r-mode GW detectability.* We show the r-mode strain (12.31) assuming a 1-year Advanced LIGO/ET phase-coherent observation. For the source parameters we have set $M_{1.4} = 1$, $R_6 = 1.2$, $D_1 = 1$. The detector noise curves were taken from https://workarea.et-gw.eu/et/WG4-Astrophysics, sensitivity curve of ET; https://dcc.ligo.org/cgi-bin/DocDB/ShowDocument?docid=T0900288, sensitivity curve of Advanced LIGO. The three selected mode amplitudes represent approximate upper limits (discussed in the main text) for saturation due to (1) non-linear mode couplings ($\alpha_r = 10^{-3}$), (2) spin equilibrium in LMXBs ($\alpha_r = 10^{-6}$) and (3) MSP spin-down and thermal equilibrium in LMXBs ($\alpha_r = 10^{-7}$)

scenario could indeed be detectable by Advanced LIGO/Virgo (for sources located anywhere in the Galaxy) and of course by ET for sources located further way or rotating at a lower frequency, see Fig. 12.6. Such r-mode signals could be associated with very young neutron stars with still undepleted fast rotation (i.e. fast spinning CCOs), see discussion at the end of Sect. 12.5.1.

Taking at face value the aforementioned LMXB/MSP upper limits, the amplitude is constrained to be $\alpha_r \lesssim 10^{-7}$ and we can see that the prospects for detection of any r-mode signal look rather bleak. Only a next generation instrument like ET could score a detection provided the source is rapidly rotating and relatively close ($D \sim 1$ kpc).

Assuming a neutron star source with known spin frequency, the identification of an r-mode signal would be unmistakable due to the $f_{gw} - f_{spin}$ relation (12.32) which is unique among the various neutron star GW emission mechanisms. Making a source parameter estimation through the GW measured f_{mode} requires the use of the fully relativistic r-mode frequency. To leading post-Newtonian order, the correction to the Newtonian frequency comes from the combined effect of gravitational redshift and frame dragging and is of the order of the stellar compactness M/R; for a neutron star this translates to an appreciable $\sim 20\%$ frequency shift. With relativistic corrections accounted for, an r-mode GW detection would thus lead to a measurement of the compactness.

The importance of using the fully relativistic r-mode frequency was recently exemplified by the oscillation discovered in the light curve of a burst from the AMXP XTE 1751–305 (Strohmayer and Mahmoodifar 2014). The observed frequency is close to the expected Newtonian r-mode frequency for this neutron star's known spin frequency but an exact match for reasonable stellar mass and radius parameters is only possible if relativistic corrections are taken into account (Andersson et al. 2014). Unfortunately, the inferred r-mode amplitude is too large to be reconciled with the system's spin evolution, a result that hints at a different interpretation of the observed oscillation.

12.5.5 Beyond the Minimum Damping Model

The alternative scenario of enhanced damping can be seen as a pessimistic standpoint as it attempts to resolve the r-mode puzzle by invoking a much reduced instability parameter space that prevents much of the known rapidly rotating neutron stars from becoming r-mode active. With regard to the realism of this scenario, it can be safely stated that none of the additional damping mechanisms mentioned earlier (exotic matter, mutual friction, Ekman layer) can be ruled out given our present level of understanding (or ignorance!).

The situation is particularly murky when it comes to exotic matter. For example, the presence of hyperons in neutron star cores and their strong bulk viscosity was once thought to be lethal for the r-mode instability (Jones 2001; Lindblom and Owen 2002). The day was saved by the likely superfluidity of these particles that effectively shuts down their viscosity below a temperature $\sim 10^9$ K, see e.g. (Nayyar and Owen 2006; Haskell and Andersson 2010).

The degree of uncertainty is even higher when it comes to strange quark matter with the resulting instability windows being extremely sensitive to the details of quark pairing, see e.g. Madsen (1998, 2000). More recent work (Mannarelli et al. 2008; Andersson et al. 2010) focused on the multifluid aspect of strange stars with colour-flavor-locked paired quark matter and established that the r-mode instability suffers very little damping in this type of stars. Unfortunately, given that the very existence (let alone key properties such as viscosity and pairing) of such exotic phases of matter in neutron stars is still a matter of debate, we presently cannot make any reliable prediction.

In our view, it makes more sense to focus on damping mechanisms that rely on less exotic physics. For instance, taking the commonly accepted view that the outer core of mature neutron stars contains a mixture of superfluid neutrons and superconducting protons, the presence of vortex mutual friction is unavoidable. The standard (that is, best understood) form of this type of friction originates from the scattering of electrons by the superfluid's magnetised vortices and it has been shown to have a negligible effect on r-modes (Lindblom and Mendell 2000). However, additional mutual friction may originate from the direct interaction between the vortices and the magnetic fluxtubes threading the superconductor, see e.g. Ruderman et al., Link (2003). Our understanding of this mechanism is (at best) rudimentary, so far having been used in r-mode saturation amplitude calculations (Haskell et al. 2014). A more phenomenological approach to this problem is to consider the standard form for the mutual friction force and explore its impact on the r-mode by artificially increasing its strength (Haskell et al. 2009), see Fig. 12.7. It is then found that a factor ~ 100 increase in the mutual friction drag parameter is sufficient for suppressing the r-mode instability in a large portion of the parameter space. Vortex-fluxtube interactions could well lead to friction of this magnitude but more work is required in order to make a safe prediction.

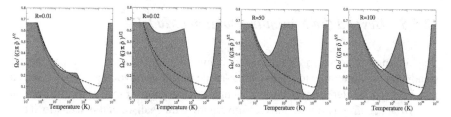

Fig. 12.7 *r-mode instability window with phenomenological mutual friction.* This figure represents the "strong superfluidity" model of Ref. Haskell et al. (2009) and shows the impact that an increasingly amplified vortex mutual friction force would have on the r-mode instability window. Note that standard mutual friction (electron scattering by vortices) has a drag coefficient $\mathcal{R} \sim 10^{-4}$ and leads to negligibly small r-mode damping. A much higher $\mathcal{R}$ could result from vortex-fluxtube interactions. The figure also shows instability curves (for a canonical stellar model) due to standard shear and bulk viscosity as well as due to a slippage-modified Ekman layer (dashed curve)

An entirely different kind of effect, taking place in superfluid neutron stars, could modify the r-mode instability's parameter space Gusakov et al. (2014a,b). In general superfluid matter with two fluid degrees of freedom supports twice as many oscillation modes as compared to ordinary matter. For r-modes in particular, this doubling leads to "ordinary" and "superfluid" modes, the real physical distinction between them being the relative amplitude of the two fluids' co-moving and counter-moving degrees of freedom. The ordinary (superfluid) r-mode is mostly co-moving (counter-moving). The mechanism proposed in Gusakov et al. (2014a,b) (see also Chugunov et al. 2017; Kantor and Gusakov 2017) is based on the observation that these two mode branches can experience resonant avoided crossings in a temperature range relevant for LMXBs. Close to these resonances the ordinary r-mode is mixed with its counter-moving counterpart and suffers strong damping from vortex mutual friction. The resulting r-mode instability window exhibits resonant "stability spikes" in the temperature range $T \sim 10^7 - 10^8$ K. The LMXB evolution model proposed in Gusakov et al. (2014a,b) represents an interesting modification of the earlier discussed runaway cycle, envisaging r-mode unstable systems climbing up these spikes while emitting GWs with the peak of a given spike setting the spin frequency upper limit. This mode-resonance mechanism is based on more or less conventional physics and, given its impact on the standard r-mode window, it deserves to be explored further.

The global interaction of a growing r-mode with the stellar magnetic field is another type of enhanced damping mechanism that needs to be discussed (there is also a local interaction with the field at the location of the crust-core Ekman layer, see Sect. 12.5.7 below). Early work put forward the scenario of a magnetic field "wind-up" by an unstable r-mode (Rezzolla et al. 2000, 2001a,b). This is a non-linear effect associated with the mode's differential rotation and the resulting Stokes drift experienced by the oscillating fluid elements. In the absence of any back-reaction from the field itself, this could potentially become a mechanism for generating a strong azimuthal component from an initial weaker poloidal field while sapping the mode's energy in the process (Rezzolla et al. 2000, 2001a,b; Cuofano et al. 2012). In practice, however, one would expect that at some stage the back-reaction of the perturbed field should kick in and self-regulate the process. This becomes obvious from the fact that once the magnetic energy becomes comparable to the mode energy one cannot even speak of an r-mode. Recent detailed work (Chugunov 2015; Friedman et al. 2016) suggests that, with back-reaction included, this mechanism is unlikely to suppress the r-mode instability and produce strongly magnetised neutron stars (in particular, Ref. Chugunov (2015) shows that r-mode activity in weakly magnetised systems such as LMXBs cannot amplify the field beyond $\sim 10^8 (\alpha_{\mathrm{sat}}/10^{-4})^2$ G).

12.5.6 The Role of the Crust

The remaining mechanism of enhanced r-mode dissipation, the viscous Ekman layer, may be considered as the most robust one since it relies on more or less conventional physics (Bildsten and Ushomirsky 2000). In its most basic form, the damping originates from the "rubbing" of the mode's flow against the solid crust and the thin viscous boundary layer formed at that region. The layer thickness is roughly given by $\delta_E \sim \sqrt{\nu/\Omega}$, where ν is the shear viscosity coefficient, and is of the order of a few centimetres. The Ekman damping timescale τ_E is related to that of shear viscosity by $\tau_E \sim \tau_{sv}/\sqrt{Re}$ where $Re \sim R^2\Omega/\nu$ is the characteristic Reynolds number. For neutron star matter $Re \gg 1$, suggesting that the Ekman layer can be strongly dissipative. Indeed, early calculations showed that the Ekman layer could dominate r-mode damping for any temperature $T \lesssim 10^{10}$ K (Bildsten and Ushomirsky 2000; Lindblom et al. 2000; Rieutord 2001).

This basic model was soon refined to account for the fluid's stratification and compressibility (Glampedakis and Andersson 2006a) and the crust's elasticity (Levin and Ushomirsky 2001; Glampedakis and Andersson 2006b). This latter property is crucial as it allows the crust to participate in the global r-mode oscillation albeit with a velocity jump at the crust-core interface. The resulting damping rate is weakened with respect to that of a solid crust by a factor $\mathcal{S}^2$, where $\mathcal{S}$ is the dimensionless crust-core "slippage" parameter (a typical value of which is $\mathcal{S} \approx 0.05$ (Levin and Ushomirsky 2001; Glampedakis and Andersson 2006b)).

The r-mode instability window produced by this "jelly" crust model with its slippage-modified Ekman layer can be considered as the canonical one. The Ekman layer curve shown in Figs. 12.4 and 12.5 assumes no slippage and was calculated using the formalism of Ref. Lindblom et al. (2000) for a canonical neutron star model with the crust-core boundary assumed at $R_c = 0.9R$ and the shear viscosity coefficient taken from Andersson et al. (2005b) (with the proton fraction set to $x_p = 0.03$). The resulting Ekman timescale is,

$$\tau_E \approx 154 \, P_{-3}^{1/2} T_8 \text{ s}. \tag{12.33}$$

The addition of slippage lowers the Ekman critical curve (i.e. $\tau_E \rightarrow \tau_E/\mathcal{S}^2$) and makes the instability window larger. Similar but much less pronounced would be the modification due to a more accurate shear viscosity coefficient (Shternin and Yakovlev 2008) (which is about a factor three lower than the one used here). The curve can also move up or down as a result of changing the location of the crust-core curve and the matter's symmetry energy (Wen et al. 2012).

Comparison of the Ekman layer-modified instability window against the LMXB quiescence data (Mahmoodifar and Strohmayer 2013) reveals that essentially all systems should be r-mode stable provided there is no crust-core slippage (i.e. $\mathcal{S} = 1$), see Fig. 12.4. In contrast, LMXBs with assumed r-mode spin and thermal equilibrium (Ho et al. 2011) are significantly hotter and some of them spill out of the instability window. If the slippage-modified Ekman layer damping is instead used, both sets of data would imply unstable r-modes in several sources. In practice, therefore, the slippage-modified τ_E leads to a "minimum damping" window and therefore belongs to the previous small amplitude scenario.

Slippage aside, the instability window may not be what shown in Fig. 12.4 because of the possibility of having *resonances* between the r-mode and the various crustal shear modes (Levin and Ushomirsky 2001; Glampedakis and Andersson 2006b). These resonances, by selectively amplifying damping near the resonant spin frequency, can lead to a "spiky" instability window with a large $\Omega - T$ instability swathe carved out in the region where LMXB and MSP may reside, see e.g. Ho et al. (2011).

12.5.7   Requiem for the r-Mode Instability?

The situation could change even more drastically by a more realistic modelling of the crust-core interface that takes into account the presence of a magnetic field threading the two regions. The discontinuity at the interface leads to a kink in the oscillating magnetic field lines and the launching of short wavelength Alfvén waves which are subsequently damped by viscosity. The physics of this *magnetised* Ekman layer was first explored in an early paper by Mendell (2001) albeit with the assumption of a solid crust. The main result of that work was that dissipation is significantly enhanced for field strengths $B \sim 10^{11} - 10^{13}$ G, hence leaving little or no room for the r-mode instability in systems like normal radio pulsars. On the other hand, weaker fields $B \lesssim 10^{10}$ G have a negligible impact on the Ekman layer suggesting that the magnetic field is not a factor for the r-mode instability of LMXBs or MSPs whose magnetic fields are typically concentrated around $\sim 10^{8}$ G.

This conclusion, however, could be premature. The outer core of these neutron stars is expected to be in a superconducting state which, among other things, means that the (squared) Alfvén speed is boosted by a factor $H_c/x_p B$ with respect to its ordinary value $v_A^2 = B^2/4\pi\rho$, where $H_c \approx 10^{15}$ G is the critical field for superconductivity, and at the same time the shear viscosity coefficient is rescaled as $\nu \to \nu/x_p$ (Glampedakis et al. 2011a). These modifications can be incorporated in Mendell's result (Mendell 2001) for the relation between the damping timescale τ_{mag} of the magnetised layer and the timescale τ_E of the ordinary Ekman layer:

$$\tau_{\text{mag}} \approx \tau_E \left(\frac{v_A^2}{\Omega\nu}\right)^{-1/2} \approx \tau_E \left(\frac{B_{\text{cr}}}{B}\right)^{1/2}. \tag{12.34}$$

The threshold $B_{\text{cr}} \approx 10^6 \rho_{14}^{3/2} x_{\text{p1}}^{3/2} T_8^{-2} P_{-3}^{-1}$ G marks the transition between a magnetic field-dominated Ekman layer ($B \gg B_{\text{cr}}$) and a non-magnetic one ($B \ll B_{\text{cr}}$). The above formula assumes the former limit and leads to a markedly shorter damping timescale for systems like LMXBs and MSPs. Crucially, this result remains accurate in a more realistic model which combines the magnetic field with an elastic crust (Glampedakis et al. in preparation). The dramatic increase in the local Alfvén speed stretches the boundary layer, increasing its thickness by a factor $\delta_{\text{mag}} \approx (B/B_{\text{crit}})^{3/2}\delta_E$.

The magnetic field has another local effect that has not been sufficiently stressed in the literature although it is commonly adopted in the modelling of magnetar oscillations (see e.g. Glampedakis et al. 2006; Gabler et al. 2011; Colaiuda and Kokkotas 2011; Gabler et al. 2012, 2016). As a consequence of the junction conditions satisfied by the Maxwell equations, the crust-core velocity slippage is suppressed[6] and the discontinuity now first appears in the radial velocity derivative.

The superconductivity-rescaled magnetic Ekman timescale (12.34), in combination with the suppression of the crust-core slippage, paints a very pessimistic picture for the viability of the r-mode instability in LMXBs and MSPs. The instability curve becomes a horizontal temperature-independent line located at the critical frequency (here we assume canonical stellar parameters),

$$f_{\mathrm{mag}} \approx 886\, B_8^{1/12}\,\mathrm{Hz}. \tag{12.35}$$

This amounts to an angular frequency $\Omega_{\mathrm{mag}} \approx 0.9 B_8^{1/12} \Omega_K$ and obviously implies the complete suppression of the r-mode instability in *all* known rapidly rotating neutron stars.

The significance of this result is obvious but we are still very far from reaching definitive conclusions. Much of the analysis presented here is preliminary and has been based on a simplified toy model calculation that does not fully take into account the rich superfluid/superconducting physics of the system (Kinney and Mendell 2003). Nor does it include the possibility of resonances between an r-mode and crustal modes or a finite thickness "pasta"-phase crust-core interface (Gearheart et al. 2011). All these effects could be important and ought to be included in a more realistic modelling of r-mode Ekman layer damping.

12.6 Magnetar Oscillations

Magnetars are strongly magnetized, isolated neutron stars, with dipole magnetic fields $\sim 10^{14} - 10^{15}$ G at the surface, and possibly even larger in the interior (Duncan and Thompson 1992). They are observed as soft γ-ray repeaters (SGRs) and anomalous X-ray pulsars (AXPs) (Woods and Thompson 2006). These extreme magnetic fields power an intense electromagnetic activity, with very energetic "giant flares", having peak luminosities as large as $10^{44} - 10^{47}$ erg/s. For a recent review on the subject, we refer the reader to Turolla et al. (2015).

---

[6]Starting from the perturbed Faraday's law, $\nabla \times \delta \mathbf{E} = -i\omega \delta \mathbf{B}/c$, and applying the usual "circuit" argument across the crust-core boundary leads to $\hat{\mathbf{n}} \times \langle \delta \mathbf{E} \rangle = 0$, where $\hat{\mathbf{n}}$ is the unit normal vector to the boundary and $\langle \ldots \rangle$ stands for the jump across the boundary. Using $c\delta \mathbf{E} = -\delta \mathbf{v} \times \mathbf{B}$ for the r-mode-induced electric field, we find $(\hat{\mathbf{n}} \cdot \mathbf{B})\langle \delta v \rangle = 0$ which for a general magnetic field implies a vanishing slippage.

12.6.1 Magneto-Elastic Oscillations

Quasi-periodic oscillations (QPOs) have been observed in the tails of giant flares from two magnetars, SGR 1804 − 20 (Israel et al. 2005; Strohmayer and Watts 2006) and SGR1900 + 14 (Strohmayer and Watts 2005). Similar QPOs have later been observed in less intense bursts from the same objects (Huppenkothen et al. 2014a) and from SGR J1550 − 5418 (Huppenkothen et al. 2014b). Most of these oscillations have frequencies between 18 Hz and 200 Hz, but two high-frequency QPOs (with frequencies of 625 Hz and 1840 Hz) have also been observed in SGR 1804 − 20.

QPOs have been interpreted as stellar oscillations. This means that we have already observed neutron star oscillations, and neutron star asteroseismology is in principle possible even before the first detection of GWs from neutron stars! In particular, once the QPO mechanism will be fully understood, it will be possible to extract information, from the frequencies of these oscillations, on the EOS of nuclear matter (see e.g. Samuelsson and Andersson 2007; Watts and Reddy 2007; Sotani et al. 2012, 2013a). An alternative explanation of QPOs is that they are oscillations of the neutron star magnetosphere, which is expected to be filled of plasma during a giant flare. However, the "standard model" based on neutron star oscillations is much more promising, since it allows to explain the details of QPO phenomenology.

After the first observations of magnetar QPOs a great effort has been devoted to the theoretical modelling of these oscillations (see e.g. Watts et al. (2016) and references therein). Magnetar QPOs have initially been identified with torsional (i.e. axial parity) elastic modes of the crust (Israel et al. 2005; Samuelsson and Andersson 2007), but it was soon understood that the crustal oscillations are strongly coupled with Alfvén modes, i.e. magnetic modes of the neutron star core (Levin 2006; Glampedakis et al. 2006). QPOs are now believed to be magneto-elastic oscillations, which involve both the core and the crust. The first models of torsional magneto-elastic oscillations of neutron stars have shown that Alfvén modes are not discrete, but instead have a continuous spectrum. However, the edges of this continuum can yield long-lived QPOs (Levin 2007; Sotani et al. 2008). Subsequently, several groups developed GR magnetohydrodynamics (MHD) numerical codes which allowed to model torsional magneto-elastic oscillations of magnetars, improving our understanding of this phenomenon. In van Hoven and Levin (2011), Colaiuda and Kokkotas (2011), van Hoven and Levin (2012) it was shown that torsional magneto-elastic oscillations of a poloidal magnetic field have continuous bands. The crustal modes falling in the bands are absorbed by the continuum (see also Gabler et al. 2012), while those falling in the gaps survive as discrete modes. Further discrete modes are given by the edges of the bands. Similar results were obtained in Asai and Lee (2014); Asai et al. (2016) using finite series expansions.

More recently, the models of magnetar oscillations—using GR MHD simulations—have been extended in two directions. Some works kept studying axial parity oscillations, including more and more physical contributions: super-conductivity in the neutron star interior (Passamonti and Lander 2013; Sotani et al.

2013b; Passamonti and Lander 2014; Glampedakis and Jones 2014; Gabler et al. 2013, 2016), the neutron star magnetosphere (Gabler et al. 2014; Glampedakis and Jones 2014),"pasta phases" in the neutron star crust (Passamonti and Pons 2016). The oscillation spectrum found in these works had the structure discussed above: continuous bands and discrete modes in the gaps. Other works, instead, studied axial-polar coupled oscillations. As shown in Colaiuda and Kokkotas (2012), if the background magnetic field is not purely poloidal (they considered the so-called "twisted torus" configuration, which is discussed in Sect. 12.7.1), oscillations with axial and polar parities are coupled. This coupling seems to destroy, or at least reduce, the Alfvén continuous spectrum, leaving a set of discrete modes. Similar results were found in Sotani (2015), Link and van Eysden (2016a,b), where a different choice of background magnetic field (the so-called "tilted torus") also removed the continuous spectrum.

We can now describe, at least qualitatively, low-frequency QPOs as magneto-elastic oscillations of magnetars. High-frequency QPOs are more difficult to model: if they are magneto-elastic oscillations they are expected to be damped in less than 1 s, but the observations show long-living oscillations. It has recently been suggested (Huppenkothen et al. 2014c) that the reason of this discrepancy could be the interpretation of the observations: high-frequency QPOs would die out in a short timescale, consistently with the theoretical model, but then they would be excited again, several times during the tail of the giant flare.

Our qualitative understanding of magnetar QPOs is not sufficient to carry on neutron star asteroseismology with these oscillations. Indeed, in order to extract information on the inner structure of the star from QPOs, we need (approximate) analytical expressions of the QPO frequencies in terms of the main features of the neutron star (mass, radius, magnetic field, etc.). We are very far away from deriving such expressions, for two reasons: the scarcity of observational data (we only detected few QPOs from three giant flares) and the limitation of the theoretical modelling, which still does not include all relevant physical processes and mechanisms. Hopefully, we will soon have more observational data [especially when new large-area X-ray detectors (Arzoumanian et al. 2014; Feroci et al. 2012; http://www.isdc.unige.ch/extp/) will be operating] and a deeper theoretical understanding of these phenomena.

12.6.2 Gravitational Waves from Giant Flares

Giant flares of magnetars are among the most luminous events in the universe. They are believed to be due to large-scale rearrangements of the magnetic fields, either in the stellar core (Thompson and Duncan 1995, 2001) or in the magnetosphere (Lyutikov 2006; Gill and Heyl 2010). Even a tiny part of their energy could excite non-radial oscillations of the neutron star—in particular, the f-mode—and thus be emitted through GWs. Preliminary estimates (Corsi and Owen 2011) were very optimistic on the detectability of the f-mode excited by giant flares, but they were based on the assumption that the energy emitted in GWs by f-mode excitation is comparable to the total magnetic energy of the star. More accurate estimates (Levin and van Hoven 2011) showed that only a small part of the magnetic energy is converted in GWs. Assuming that the released electromagnetic energy is of the order of the magnetic energy of the star, a giant flare at $D = 10$ kpc would excite an f-mode GW signal detectable by Advanced LIGO/Virgo with a SNR (Levin and van Hoven 2011)

$$\frac{S}{N} \lesssim 10^{-2} B_{15}^2 , \tag{12.36}$$

where B_{15} is the (normalized) magnetic field-strength at the poles. This result, based on a perturbative analysis, implies that the f-mode would definitely not be detectable by second-generation interferometers if $B \sim 10^{15}$ G, but it may be marginally detectable for $B \sim 10^{16}$ G.

Subsequently, giant flares and their coupling with the neutron star oscillations have been studied using general-relativistic MHD numerical simulations of the magnetic field instability in magnetars (Ciolfi and Rezzolla 2012; Zink et al. 2012). This is a toy-model which mimics the catastrophic magnetic field rearrangements during a giant flare, and allows to estimate the amplitude and the features of the GW emission. The numerical simulations show a power-law relation between the emitted gravitational signal and the surface magnetic field-strength (in these models the magnetic field-strength at the surface and in the interior are comparable) (Zink et al. 2012):

$$h \sim 7.6 \times 10^{-27} \frac{B_{15}^{3.3}}{D_1} , \tag{12.37}$$

where D_1 is the distance of the source normalized to 1 kpc. Most of the energy in the signal is in the f-mode, but also lower frequency modes (likely Alfvén and/or composition g-modes) are excited (see Fig. 12.8).

The results of numerical simulations (Ciolfi and Rezzolla 2012; Zink et al. 2012) are even more pessimistic than those of perturbation theory (Levin and van Hoven 2011) shown in Eq. (12.36). The SNR of the GW signal from an f-mode excited by a giant flare has been estimated to be (Ciolfi and Rezzolla 2012) $S/N \sim 10^{-4} B_{15}^2$ for Advanced LIGO/Virgo, and $S/N \sim 10^{-2} B_{15}^2$ for ET.

The numerical simulations of Zink et al. (2012) give similar results, see Fig. 12.8. The SNR can be read off the figure, as the ratio between the signal and the noise curve, for a given B_{pole} and for a given detector. As mentioned above, Fig. 12.8 also shows a low-frequency mode, whose SNR is comparable (but slightly larger) than that of the f-mode.[7]

---

[7]It is worth noting that the authors of Zink et al. (2012) fit the numerical data with a power-law relation, finding that the GW amplitude is proportional to $B^{3.3}$. The authors of Ciolfi and Rezzolla (2012), instead, have $h \sim B^2$, because they fit the data with a quadratic function; this choice is

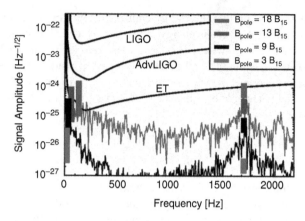

Fig. 12.8 *GW detectability of magnetar oscillations in giant flares.* We show the signal amplitude $\sqrt{T}|h|$, where h is the estimated dimensionless strain from magnetar oscillations in giant flares, and T is the estimated damping time of the oscillation. The giant flare is modeled as the instability of the neutron star magnetic field (see text). The colored boxes on the left represent the f-mode signal, assuming 50 ms$\leq T \leq$ 200 ms. The colored boxes on the right represent the signal of a lower-frequency mode (Alfvén and/or g-modes), assuming 10 ms$\leq T \leq$ 1 s. Different colors correspond to different values of the magnetic field at the poles. We also show the signal for the entire spectrum, assuming a damping time $T = 100$ ms (from Zink et al. 2012)

These studies confirm that the gravitational emission from mode excitation in magnetar flares can not be detected by advanced GW detectors. This signal is expected to be too weak to be detected even by third-generation GW experiments, such as ET, unless the surface magnetic field is $\gtrsim 10^{16}$ G (Zink et al. 2012; Ciolfi and Rezzolla 2012), i.e. an order of magnitude larger than the observed magnetic fields.

12.7 Neutron Star "Mountains"

Rotating neutron stars symmetric with respect to the rotation axis do not emit gravitational radiation. However, if a neutron star is not perfectly axisymmetric it emits GWs, mostly at frequencies f_{spin} and $2f_{\mathrm{spin}}$ (where f_{spin} is the rotation frequency). Most of the observed neutron stars have rotation frequencies in the range between ~ 10 Hz and 1 kHz, which is the range where ground-based interferometers such as Advanced LIGO/Virgo are most sensitive. Therefore, if the deviation from axisymmetry is large enough, rotating neutron stars can be promising sources of GWs.

---

correct as a first approximation, since in the perturbative results of Levin and van Hoven (2011) the GW amplitude is a quadratic function of the magnetic field strength.

The deviation from axisymmetry is described by the *quadrupole ellipticity*

$$\varepsilon = \frac{Q}{I},\tag{12.38}$$

where Q is the mass quadrupole moment associated to the distortion (i.e. excluding the contribution of rotation, which is necessarily axisymmetric and is not associated to GW emission), and I is the moment of inertia of the rotation axis. Such distortions are called *mountains*. When $\varepsilon > 0$ the star is oblate, while when $\varepsilon < 0$ it is prolate.

If the distortion has a symmetry axis forming an angle α (called *wobble angle*) with the rotation axis, the amplitude of the GW emission is Zimmermann and Szedenits (1979), Bonazzola and Gourgoulhon (1996)

$$h_0 \simeq \frac{4\pi^2 G}{Dc^4} f_{gw}^2 I \varepsilon \sin\alpha \simeq 4 \times 10^{-24} P_{-3}^{-2} D_1^{-1} I_{45} \left(\frac{\varepsilon}{10^{-6}}\right).\tag{12.39}$$

The deformation can be due to different mechanisms: from the magnetic field of the star (Bonazzola and Gourgoulhon 1996; Haskell et al. 2008; Ciolfi et al. 2010; "magnetic mountains"), to temperature gradients in the crust leading to local deformation sustained by the elastic strain (Bildsten 1998b; Ushomirsky et al. 2000) ("thermal mountains"). The former can arise both in isolated neutron stars (magnetars, but also in "ordinary" neutron stars), or in accreting neutron stars. The latter, instead, can only be formed in accreting neutron stars, because the temperature gradients are created by asymmetries in the matter accreting by a companion object.

Current searches for GWs from rotating pulsars using first generation LIGO/Virgo have shown no sign of such emission (see e.g. Aasi et al. (2014) and references therein). However, this negative result allows us to place upper limits on the quadrupole ellipticity of the pulsars under study; in the case of the Crab pulsar, $\varepsilon \lesssim 8.6 \times 10^{-4}$; in the case of other pulsars, the upper limits are significantly weaker. These limits are below the theoretical bounds (see below), but (in some cases) they exceed the spin-down limit on ε, which has been obtained with the unrealistic assumption that the observed pulsar spin-down is only due to GW emission.

In the following we shall first discuss current models of neutron stars deformed by a magnetic field, and then the different astrophysical scenarios for magnetic mountains. We shall then discuss thermal mountains, and other possible neutron star deformations in the context of neutron star models with exotic matter and/or pinned superfluidity.

12.7.1 Modelling Magnetic Mountains

Chandrasekhar and Fermi (1953) (see also Ferraro 1954) first pointed out that magnetic fields induce quadrupolar deformations on spherically symmetric stars, with ellipticities (which we hereafter denote by ε_B) of the order of the ratio between the magnetic energy E_B and the gravitational energy E_G, i.e.

$$\varepsilon_B \sim \frac{E_B}{E_G} \sim \frac{R^3 \langle B^2 \rangle}{GM^2/R} \sim (10^{-6} - 10^{-5})\langle B_{15}^2 \rangle , \tag{12.40}$$

where $\langle B^2 \rangle$ is the volume-averaged square field strength. It was later noted (Wentzel 1960; Ostriker and Gunn 1969) that the magnetic-induced quadrupole ellipticity ε_B can be positive or negative, i.e. the shape of the star can be oblate or prolate, depending on the magnetic field structure. Indeed, neutron star magnetic fields can have poloidal or toroidal structure,[8] and poloidal fields B_{pol} tend to deform the star to an oblate shape, while toroidal fields B_{tor} tend to deform it to a prolate shape. When poloidal and toroidal components are both present, the deformation has both positive and negative contributions. For instance, in a specific example of magnetic field geometry the deformation has been estimated to be (Ostriker and Gunn 1969)

$$\varepsilon_B \sim E_G^{-1} R^3 \left(3\langle B_{pol}^2 \rangle - \langle B_{tor}^2 \rangle \right) . \tag{12.41}$$

The relative contributions of poloidal and toroidal components can change if the magnetic field geometry is different.

Similar estimates have been derived for a neutron star with a superconducting phase (see the discussion below), in the case of a purely toroidal field (Cutler 2002):

$$\varepsilon_B \sim 10^{-6}\langle B_{15} \rangle \frac{H_{c1}}{10^{15}\,G} , \tag{12.42}$$

where H_{c1} (which is believed to be $\sim 10^{15}$ G in neutron star cores (Glampedakis et al. 2011a)) is the critical field strength characterizing superconductivity. These expressions have been computed for a constant density star, and assuming a dipole field configuration. However, as we discuss below, more sophisticated numerical analyses show that Eqs. (12.40), (12.41), (12.42) give good order-of-magnitude estimates for magnetic deformations of neutron stars.

When a neutron star is prolate (which requires a prevailing toroidal field), a spin-flip mechanism can take place (Jones 1975; Cutler 2002), in which the wobble angle grows until the symmetry axis and the rotation axis become orthogonal: an optimal geometry for GW emission (see Eq. (12.39)).

Presently, we do not have direct information on the internal structure of neutron star's magnetic fields. However, it is generally believed that they should have both a poloidal and a toroidal component. Indeed, analytical computations (Prendergast 1956; Tayler 1973) and numerical simulations (Braithwaite 2006, 2007) show that purely poloidal and purely toroidal configuration are unstable in an Alfvén

---

[8]In polar coordinates, the field-strength components B_r, B_ϑ are poloidal, while B_φ is toroidal.

timescale.[9] Conversely, poloidal-toroidal magnetic fields can develop from the evolution of arbitrary initial fields (Braithwaite and Spruit 2004). The fields found in Braithwaite and Spruit (2004), Braithwaite and Nordlund (2006) have a so-called "twisted-torus" structure: a poloidal magnetic field extending throughout the star and in the exterior, and a toroidal field confined to a torus-shaped region inside the star.

GR and Newtonian models of stationary neutron star configurations with a twisted-torus magnetic field have been carried out in Ciolfi et al. (2010), Lander and Jones (2009). They are based on the numerical integration of the GR MHD equations for an axisymmetric ideal fluid (which can be reduced to a single partial differential equation, the so-called Grad-Shafranov (GS) equation), and predict that the magnetic field induces a quadrupole ellipticity $\sim (10^{-6} - 10^{-5}) \, B_{15}^2$ (where B_{15} is the surface magnetic field),[10] consistent with the estimates (12.40), (12.41). Moreover, the results of Ciolfi et al. (2010), Lander and Jones (2009) show that in a twisted-torus configuration, the ratio of toroidal field energy over poloidal field energy is $\langle B_{tor}^2 \rangle / \langle B_{pol}^2 \rangle \lesssim 0.15$. Therefore, in these configurations the poloidal field prevails, the star is oblate, and the spin-flip mechanism does not occur. More recently, twisted-torus configurations (with $\langle B_{tor}^2 \rangle / \langle B_{pol}^2 \rangle \lesssim 0.1$) have also been described by solving the fully non-linear Einstein-Maxwell's equations, see e.g. Pili et al. (2014), Uryu et al. (2014).

A different class of twisted-torus configurations has later been found in Ciolfi and Rezzolla (2013), where—with an appropriate prescription of the azimuthal currents—the ratio $\langle B_{tor}^2 \rangle / \langle B_{pol}^2 \rangle$ can be as large as ~ 0.9, i.e. the toroidal field can prevail. In these models the deformation of the star is larger than the prediction of Eqs. (12.40), (12.41) by a factor ~ 100

$$\varepsilon_B \sim 10^{-4} B_{15}^2 . \tag{12.43}$$

The equilibrium properties of the magnetized star depend on whether the EOS is barotropic or non-barotropic. The models in Ciolfi et al. (2010), Lander and Jones (2009), Ciolfi and Rezzolla (2013) assume a barotropic EOS. As shown in Mastrano et al. (2011), Glampedakis and Lasky (2016), if the EOS is non-barotropic one is free to prescribe the magnetic field, and the ratio between toroidal and poloidal energy becomes a free parameter, even though it is likely to be constrained by the requirement of stability. In the model studied in Mastrano et al. (2011),

$$\varepsilon_B \sim 10^{-5} B_{15}^2 \left(1 - 0.64 \frac{\langle B_{tor}^2 \rangle}{\langle B_{pol}^2 \rangle} \right) . \tag{12.44}$$

---

[9]The instability of purely toroidal configurations is also shown in Akgun and Wasserman (2008) for a superconducting neutron star, by studying the variations of an energy functional.

[10]In these models the magnetic field-strength has as the same order of magnitude on the surface and in the interior.

If the ratio $\langle B_{tor}^2 \rangle / \langle B_{pol}^2 \rangle$ is sufficiently large, i.e. the toroidal field prevails, the deformation can be as large as that described by Eq. (12.43). In these configurations (such as in those studied in Ciolfi and Rezzolla (2013)) the neutron star is prolate, and the spin-flip mechanism can take place.

If higher-order multipoles of the magnetic field are present, the (prolate) deformation can be even larger (Mastrano et al. 2015). Alternative geometries with poloidal-toroidal fields have also been studied, such as the "tilted torus" configuration, in which the poloidal and the toroidal fields have misaligned symmetry axes (Lasky and Melatos 2013).

The models discussed above do not take into account the fact that outer cores of neutron star are expected to contain superconducting protons (and superfluid neutrons). In a (type II) superconducting phase, the magnetic field is quantized into fluxtubes surrounded by non-magnetised matter. The force exerted by the quantized field is not the Lorentz force, but a fluxtube tension force (Mendell 1991; Glampedakis et al. 2011a); therefore, the description based on the GS equation is not adequate to model the superconducting phase (even though they still obey a GS-type equation). In recent years, the deformation of magnetized superconducting neutron stars has been computed by numerical integrations of the equations for the fluxtube magnetic force for toroidal fields (Akgun and Wasserman 2008; Lander et al. 2012), for purely poloidal fields (Henriksson and Wasserman 2013), and for twisted-torus fields (Lander 2013). All these computations give deformations consistent with the analytical estimates (Cutler 2002) of Eq. (12.42).

We remark that $\varepsilon_B \sim B^2$ in non-superconducting neutron stars, while $\varepsilon_B \sim B$ in neutron stars with a superconducting phase. This is due to the different magnetic force acting in superconducting matter. Comparing Eqs. (12.40), (12.43), we can see that in magnetars (with $B \sim 10^{15}$ G) the magnetic deformations have the same order of magnitude regardless of whether a superconducting phase is present; conversely, in ordinary neutron stars—such as the observed pulsars—the magnetic field strength is $\sim 10^{12}$ G or smaller, and magnetic deformations are much larger if a (type II) superconducting phase is present.

12.7.2 Magnetic Mountains in Newly-Born Magnetars and in Known Pulsars

As we discussed in Sect. 12.6, observations of SGRs and AXPs, with their large spin-down rates and intense burst activity, suggest that these objects are magnetars (Duncan and Thompson 1992). Due to their very large magnetic fields, magnetars are expected to have the largest magnetic mountains. However, since the observed SGRs and AXPs have low spin frequencies, they cannot be GW sources for ground-based interferometers.

The standard magnetar model (Duncan and Thompson 1992; Goldreich and Reisenegger 1992) predicts that the strong magnetic fields form at the birth of the neutron star, through convection and dynamo effects, and can last for $\sim 10^4$–10^5

years (Goldreich and Reisenegger 1992; Pons and Geppert 2007). If this scenario is accurate, a non-negligible fraction ($\sim 10\%$) of newly-formed neutron stars can be born as magnetars (Kouveliotou et al. 1998; Woods and Thompson 2006), and, if their rotation rates are large enough, they can be potential sources for GW detectors (see e.g. Dall'Osso et al. 2009). Presently, it is not clear what the actual rotation rate of newly-born neutron stars is; current estimates suggest that it should not be larger than few hundreds of Hz (see e.g. Camelio et al. 2016 and references therein). A newly-born magnetar with $B \sim 10^{15}$ G could have a quadrupole deformation $\varepsilon_B \sim 10^{-6}$–$10^{-5}$ (12.40) (even larger if the toroidal field prevails, see Eq. (12.43)). If $D \sim 10$ kpc and $P \sim 10$ ms, it may yield a GW strain (12.39) $h_0 \sim 10^{-26}$, potentially detectable by ET. However, the event rate of galactic newly-born magnetars (comparable with that of galactic supernovae) is too low to make them a promising GW source.

We have observed, instead, several pulsars with magnetic field-strengths $\lesssim 10^{12}$ G, and rotation rates in the bandwidth of ground-based GW detectors. The GW emission from these objects is negligible, unless a type II superconducting phase is present. Indeed, when $B \sim 10^{12}$ the estimate for (type II) superconducting stars (12.42) is $\varepsilon_B \sim 10^{-9}$, while the estimate for non-superconducting stars (12.40) is a factor ~ 1000 smaller. As noted in Andersson et al. (2011), a (superconducting) MSP with $B \sim 10^{12}$ G at $D = 1$ kpc would then emit a GW strain $h_0 \sim 10^{-27}$, marginally detectable by ET. This estimate, however, is probably too optimistic: all known MSPs have magnetic fields much smaller than 10^{12} G, while pulsars with $B \sim 10^{12}$ have periods much larger than ~ 1 ms. In Fig. 12.9 we show the GW strain for the population of known pulsars, assuming that the quadrupole deformation is given by the estimate (12.42), corresponding to the presence of a type-II superconducting phase in the core. We can see that, under these assumptions, the expected signal from known pulsars would be well below the sensitivity curves of second- and third-generation interferometers.

12.7.3 Magnetic Mountains in Millisecond Magnetars from Compact Binary Mergers

Binary neutron star mergers can result in strongly magnetized neutron stars (Zhang 2013; Giacomazzo and Perna 2013). They can be meta-stable supramassive neutron stars living from few seconds to hours (Ravi and Lasky 2014), but can also be stable neutron stars (Giacomazzo and Perna 2013). These objects have been proposed to be the central engine of short GRB, in the so-called *millisecond magnetar model* (Dai and Lu 1998; Zhang and Mészáros 2001; Rowlinson et al. 2013). In this scenario, the energy injection in the GRB is due to dipole radiation from a spinning-down magnetar. Strong evidence supporting the millisecond magnetar model is its ability to explain the shape of the observed GRB X-ray spectra, which are characterized by a plateau of ~ 10–10^2 s followed, in most cases, by a power-law decay; in a subset of short GRBs, instead, the decay is much steeper. The X-ray spectrum from

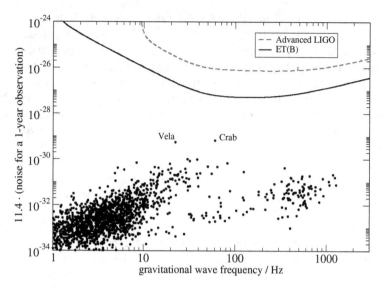

Fig. 12.9 *GW detectability of magnetic mountains in known pulsars.* GW strain from the population of known pulsars, assuming a 1-year phase coherent observation and quadrupole deformations given by Eq. (12.42). The detector noise curves were taken from https://workarea.et-gw.eu/et/WG4-Astrophysics, sensitivity curve of ET; https://dcc.ligo.org/cgi-bin/DocDB/ShowDocument?docid=T0900288, sensitivity curve of Advanced LIGO. All pulsar data were taken from the ATNF database (http://www.atnf.csiro.au/people/pulsar/psrcat/)

a spinning-down magnetar resulting from a compact binary coalescence would have the same structure: a plateau followed by a power-law decay if the neutron star is stable, by a steeper decay if it is a supramassive star eventually collapsing to a black hole.

The millisecond magnetar model is probably the most promising astrophysical scenario for GW detection from magnetic mountains. Indeed, numerical simulations show that the magnetars resulting from a binary neutron star coalescence would have periods of the order of milliseconds—significantly smaller than the expected periods of newly-born neutron stars. Moreover, in this scenario we have astrophysical data from these objects, since they would be the engine of short GRBs.

Preliminary estimates (Dall'Osso et al. 2015) show that the GW emission from these sources could be marginally detectable by second-generation interferometers, and detectable by third-generation interferometers such as ET (even more optimistic estimates have been derived in Corsi and Meszaros (2009)). However, these computations are based on the assumption that GW emission gives a significant contribution to the magnetar spin-down. Since the millisecond magnetars predicted in this model are progenitors of short GRBs, the assumptions of the model can be tested against GRB observations.

In Lasky and Glampedakis (2016), observations of the X-ray light curve of short GRBs have been used to constrain the ellipticity of the (stable) neutron stars produced in the merger, and then to set a bound on the GW emission from these objects. Indeed, the late time tail of the observed X-ray spectrum is $\sim t^{-2}$, suggesting that the spin-down is mainly due to dipole electromagnetic emission rather than to GW emission. Therefore, the comparison between the observed X-ray spectrum and the predictions of the model allows to set an observational upper limit on the neutron star ellipticities:

$$\varepsilon_B \lesssim 0.33\,\eta\,I_{45}^{1/2} \left(\frac{L_{\rm em}}{10^{49}\,{\rm erg/s}}\right)^{-1} \left(\frac{t_b}{100\,{\rm s}}\right)^{-3/2}, \qquad (12.45)$$

where $\eta \leq 1$ is the conversion efficiency of spin-down energy into X-ray luminosity, $L_{\rm em}$ is the short GRB luminosity in the initial plateau phase, and t_b is the plateau duration. This upper limit (through Eq. (12.39)) corresponds to an upper limit on the expected GW signal. The analysis of Lasky and Glampedakis (2016) shows that the agreement between the observed X-ray spectrum and the predictions of the model requires that the magnetar spin-down is mainly due to dipole emission, with a very small contribution from GW emission. Therefore, these signals are not expected to be detected by Advanced LIGO/Virgo, but are potentially detectable by third-generation interferometers such as ET. This can be seen in Fig. 12.10. In the left panel we show the upper limit (12.45) on the neutron star ellipticities, assuming $\eta = 0.1$ (black points with error bars) for eight observed short GRBs. In the right panel we show the corresponding upper limits on the GW strain amplitude evolution, for $\eta = 0.1$ (dotted lines) and for $\eta = 1$ (solid lines), compared with the sensitivity curves of Advanced LIGO and ET.

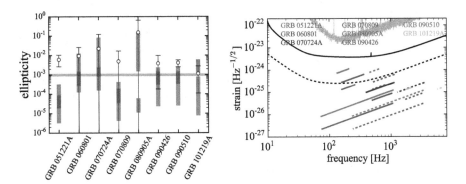

Fig. 12.10 *Observational upper limits on magnetar deformation and GW emission.* Left panel: Upper limits on the neutron star ellipticity from eight short GRBs, given by Eq. (12.45) (black point with error bars). The red bars show the theoretical estimate from Eq. (12.40). The blue bars show the range of maximal ellipticities for which the spin flip can take place. The green line represents the ellipticity which can be induced by f-mode excitation (see Sect. 12.4). Right panel: Upper limits on the GW strain amplitude evolution, compared with the sensitivity curves of Advanced LIGO and ET, for $\eta = 0.1$ (dotted lines) and for $\eta = 1$ (solid lines). The observation time is assumed to be 10^5 s for GRBs with power-law decay, while it coincides with the GRB emission time for those with steeper decay (figure from Lasky and Glampedakis (2016))

12.7.4 Magnetic Mountains in Accreting Systems

Magnetic mountains can also form in LMXBs. In this case, the mountain—a localized deformation—is formed by the accreting matter, but it is sustained by the magnetic field. Since the mountain is not sustained by crustal rigidity, its quadrupole ellipticity can be larger than the upper bound associated to the maximum crustal strain (Ushomirsky et al. 2000). In this scenario (Payne and Melatos 2004; Melatos and Payne 2005; Vigelius and Melatos 2008), accreted matter accumulates in a column over the polar cap, distorting the magnetic field and reducing the magnetic moment of the star. As the stars accretes, the exterior magnetic field evolves approximately as $B = B_*(1 + M_{\mathrm{acc}}/M_{\mathrm{c}})^{-1}$ where B_* the field when accretion starts, M_{acc} the accreted matter and M_{c} the critical accreted mass above which the magnetic moment starts to change (Shibazaki et al. 1989). The quadrupole ellipticity associated to magnetic mountains in LMXB is (Priymak et al. 2011; Haskell et al. 2015)

$$\varepsilon \sim \frac{M_{\mathrm{acc}}}{M_{\odot}} \left(1 + \frac{M_{\mathrm{acc}}}{M_{\mathrm{c}}}\right)^{-1}, \tag{12.46}$$

where $M_{\mathrm{c}} \sim 10^{-7}(B_*/10^{12}G)^{4/3}$. This formula is only accurate for $M_{\mathrm{acc}} \lesssim M_{\mathrm{c}}$: for larger values of the accreted mass, the quadrupole ellipticity saturates. Dynamical MHD simulations suggest that these deformations can persist for timescales of the order of $\sim 10^5 - 10^8$ years (Vigelius and Melatos 2009).

Although magnetic fields of LMXBs are much weaker than those of magnetars, the quadrupole ellipticities of local mountains can be much larger than those produced by a global magnetic deformation of the star. Moreover, LMXBs have rotation rates much larger than those of magnetars, well within the sensitivity band of ground-based interferometers. Fast rotation also enhances the GW signal from a deformed star. Therefore, as shown in Priymak et al. (2011), Haskell et al. (2015), a buried field $B_* \gtrsim 10^{12}$ G could generate a deformation, and then a GW emission, strong enough to be detected by Advanced LIGO/Virgo (Priymak et al. 2011; Haskell et al. 2015).

12.7.5 Thermal Mountains

Accreting neutron stars in LMXBs can have quadrupolar deformations due to temperature gradients in the accreted crust (Bildsten 1998b; Ushomirsky et al. 2000). Indeed, as the matter accretes, it is buried and compressed until nuclear reactions occur (electron capture, neutron emission, etc.) (Haensel and Zdunik 1990). These reactions heat the crust, and since the accreted matter is expected to be asymmetric, the temperature gradient produced by nuclear reactions is also asymmetric, giving rise to quadrupolar deformations in the neutron star. These deformations are called "thermal mountains"; they also belong to the broader class of "elastic mountains", i.e. deformations sustained by the elastic strain of the crust. Presently, thermal gradients are the only known viable mechanism to produce elastic mountains.

The deformation due to a thermal gradient with quadrupolar component δT_q (which we denote by ε_{th}) is (Ushomirsky et al. 2000; Haskell et al. 2015)

$$\varepsilon_{th} \sim 10^{-10} R_6^4 \left(\frac{\delta T_q}{10^5 \, K}\right) \left(\frac{Q}{30 \, MeV}\right)^3, \qquad (12.47)$$

where Q is the threshold energy of electron capture by nuclei. The quadrupolar thermal gradient is just a fraction of the total thermal gradient δT, which is expected to be, at most, of the order of 10^6 K (Ushomirsky and Rutledge 2001), if produced by an outburst. In the most optimistic scenario $\delta T_q \lesssim 0.1 \, \delta T \lesssim 10^5$ K, yielding a thermal mountain of the order of $\varepsilon_{th} \sim 10^{-10}$ (Haskell et al. 2015). The GW strain (12.39) from this deformation would be $h_0 \sim 10^{-27} - 10^{-28}$, too weak to be detected by Advanced LIGO/Virgo, but potentially detectable by third-generation GW intereferometers such as ET. This can be seen in Fig. 12.11, where we show the estimated GW strain from a set of observed LMXBs, computed from Eqs. (12.47) and (12.39), for different values of the electron capture threshold energy Q and assuming $\delta T_q \simeq 10^5$ K. In the figure, the Advanced LIGO and ET sensitivity curves are shown for different values of the integration time (1 month, which is a typical outburst duration, and 2 years) because the persistence timescale of the mountain,

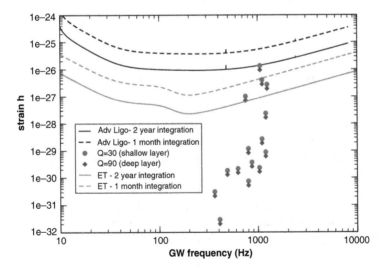

Fig. 12.11 *Detectability of thermal mountains.* We show the estimated GW strain from thermal mountains formed in a set of observed LMXB outbursts, for different values of the reaction threshold energy Q, assuming $\delta T_q \simeq 10^5$ K. The detector noise curves assume 1 month (dashed) or 2 years (solid) of phase-coherent observation (from Haskell et al. (2015))

and then the typical duration of the signal, is presently poorly constrained. In the most favourable case (the deformation persists after the outburst, $\delta T_q/\delta T \simeq 0.1$) some of the systems could emit a signal detectable by ET. We note, however, that this value of $\delta T_q/\delta T$ is an upper limit with no solid physical motivation; if the actual value of this ratio is smaller than ~ 0.1, even third-generation GW interferometers will not be able to detect thermal mountains.

The GW emission from thermal mountains, regardless of its direct detectability, may have a role in the spin evolution of LMXBs. Twenty years ago, thermal mountains have been proposed to explain the apparent spin cut-off in LMXBs (Bildsten 1998a). However (as discussed in detail in Sect. 12.5.1), it was later understood that this feature can probably be explained in terms of the coupling of the stellar magnetic field with the accretion disk. More recently, thermal mountains have been revived by the observations of the pulsar PSR J1023 + 0038, which shows a transition between a radio MSP state and a LMXB state (Archibald et al. 2009). Remarkably, the pulsar spins down faster (by $\sim 20\%$) during the LMXB state. It has been suggested (Haskell and Patruno 2017) that this enhanced spin-down is due to GW emission by a thermal mountain. To explain the additional spin-down, a deformation $\varepsilon_{th} \sim 5 \times 10^{-10}$ would be required, which—as we have discussed above—can indeed be a thermal mountain, and would emit gravitational radiation detectable by third-generation interferometers such as ET. If these results are confirmed, they will provide the first observational indirect evidence supporting the existence of GW emission from neutron star mountains.

Thermal mountains (and more generally, elastic mountains) are limited by the maximum stress that the crust can sustain before breaking (Ushomirsky et al. 2000; Haskell et al. 2006; Johnson-McDaniel and Owen 2013):

$$\varepsilon_{th} \lesssim \frac{\mu_{cr}\sigma_{br}V_{cr}}{GM^2/R} \sim 10^{-5}\left(\frac{\sigma_{br}}{0.1}\right), \tag{12.48}$$

where μ_{cr} is the shear modulus of the crust, V_{cr} is the volume of the crust, and σ_{br} is the crustal breaking strain, which could be as large as ~ 0.1 (Horowitz and Kadau 2009).[11]

We remark that this is just an upper bound: there is no reason to believe that neutron star have deformations close to this value. For instance, the thermal mountain which has been suggested in Haskell and Patruno (2017) to explain the observations of PSR J1023 + 0038 is much smaller than the deformation in Eq. (12.48). We also remark that this bound applies to thermal mountains (more generally, to elastic mountains), but it does not necessarily apply to magnetic mountains. Indeed, when the crust forms in a newly-born neutron star the magnetic deformation may already be present; in this case, the equilibrium shape of the crust

---

[11] It should be mentioned that the breaking strain found in Horowitz and Kadau (2009) is the result of numerical simulations with duration much shorter than the timescale associated with crust straining/relaxation.

would be non-spherical, and the crust would not have to sustain an elastic strain, at least while the magnetic field is present.

12.7.6 Exotic Mountains

The ellipticities associated with the neutron star deformations discussed so far are clearly pessimistic from the perspective of GW observability. A somewhat more promising situation may arise if we consider neutron stars with quark matter cores. If present, quarks are most likely to find themselves in a color-superconducting state, the exotic properties of which could accommodate significantly larger quadrupolar deformations (of both magnetic and elastic nature) as compared to conventional hadronic matter.

Among the various possible quark matter incarnations (for a review see Alford et al. 2008) the color-flavor-locked (CFL) phase, where all three quark species (u, d, s) are paired, appears to be the most favoured one. Another well-studied phase is the two-flavor superconducting (2SC) state where just the (u, d) quarks pair. Both of these phases come with remarkable magnetic permeability properties. The stellar magnetic field threading a CFL/2SC core is likely to do so by forming an array of quantised vortices in the same way that fluxtubes are formed in the more familiar type II protonic superconductor (Iida and Baym 2002; Alford and Sedrakian 2010). These vortices, however, are *color-magnetic* (rather than just magnetic) in the sense that they are carriers of an admixture of magnetic and gluonic field degrees of freedom. The latter component is the dominant one and, as a consequence, the energy per unit length $\mathcal{E}_X$ (where X labels the phase) of a color-magnetic vortex can be 2–3 orders of magnitude higher than that of a conventional protonic fluxtube (Iida and Baym 2002; Alford and Sedrakian 2010). This property entails an amplified vortex array tension and opens the possibility of creating a large deformation in a neutron star's CFL/2SC core (Glampedakis et al. 2012). The ellipticity of this *internal* color-magnetic mountain can be estimated by means of the ratio of the vortex array tension energy to the stellar gravitational energy:

$$\varepsilon_X \approx \frac{\mathcal{N}_X \mathcal{E}_X V_q}{GM^2/R}, \tag{12.49}$$

where $\mathcal{N}_X$ is the vortex surface density and V_q is the volume of the quark core. For a CFL core (and assuming canonical stellar parameters) this expression leads to (Glampedakis et al. 2012),

$$\varepsilon_{\text{cfl}} \approx 10^{-7} \langle B_{12} \rangle \left(\frac{V_q}{V_{\text{star}}} \right) \left(\frac{\mu_q}{400 \, \text{MeV}} \right)^2, \tag{12.50}$$

where μ_q is the strange quark chemical potential (normalised to a canonical value) and, as before, $\langle B \rangle$ represents the volume-averaged interior magnetic field.

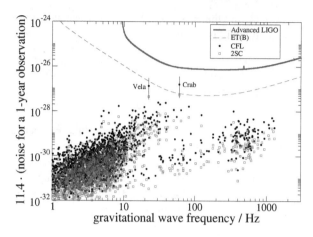

Fig. 12.12 *GW detectability of color-magnetic mountains.* In this figure, adapted from Glampedakis et al. (2012), we show the estimated GW strain (assuming a 1-year phase-coherent observation) from the population of known pulsars, assuming they have CFL (or 2SC) quark cores. We have used the fiducial parameters $\mu_q = 400$ MeV, $V_q = 0.5 V_{\text{star}}$, and $\langle B \rangle = 2 B_{\text{surf}}$

The GW strain associated with this deformation can be estimated with the help of Eq. (12.39), using reasonable values for the various parameters, i.e. $\mu_q = 400$ MeV, $V_q = 0.5 V_{\text{star}}$ and $\langle B \rangle = 2 B_{\text{surf}}$ (as suggested by models of MHD equilibria in neutron stars, see e.g. Ciolfi et al. 2009; Lander and Jones 2009). The resulting GW detectability by Advanced LIGO and ET of the pulsar population with known surface dipole fields B_{surf} is shown in Fig. 12.12. When compared to Fig. 12.9, it is evident that the presence of a color-magnetic deformation in neutron stars leads to a marked improvement in the strength of their GW signature. The results presented here suggest that young pulsars such as the Crab or the Vela could be detectable by ET. A much larger fraction of the pulsar population could become detectable (even by Advanced LIGO/Virgo) if we assume a higher magnetic field ratio $\langle B \rangle / B_{\text{surf}}$.

A drastically different phase of quark matter could be that of a *crystalline color-superconductor* (CCS) (Rajagopal and Sharma 2006; Mannarelli et al. 2007), effectively leading to a neutron star with a solid core. Given that the shear modulus μ_{ccs} of CCS matter is estimated to be much higher than that of the crustal bcc lattice (Mannarelli et al. 2007), the elastic deformation that could be sustained by a solid quark core could be significantly greater than that of normal hadronic neutron stars. For obtaining the ellipticity of the CCS mountain we can use a formula similar to Eq. (12.40). We find,

$$\varepsilon_{\text{ccs}} \approx \frac{\mu_{\text{ccs}} V_q \sigma_{\text{br}}}{GM^2/R} \approx 6 \times 10^{-4} \left(\frac{V_q}{V_{\text{star}}} \right) \left(\frac{\mu_q}{400 \, \text{MeV}} \right)^2 \left(\frac{\Delta_q}{10 \, \text{MeV}} \right)^2 \left(\frac{\sigma_{\text{br}}}{0.01} \right),$$

$$(12.51)$$

where Δ_q is a quark gap parameter (normalised to a canonical value) and the (highly uncertain) breaking strain of a crystalline quark core is normalized to the fiducial value $\sigma_{br} \sim 0.01$. This simple estimate is in good agreement with the results of more rigorous calculations (Owen 2005; B. Haskell, N. Andersson, D. I. Jones and L. Samuelsson 2007) and is about a factor $\mu_{ccs}/\mu_{cr} \sim 10^3$ higher than the maximum elastic deformation of the crust (Eq. (12.48)). As discussed in B. Haskell, N. Andersson, D. I. Jones and L. Samuelsson (2007), the presence of such a large internal deformation in neutron cores could in principle be tested by the upper limits set by GW searches. According to the recently published Advanced LIGO results (Abbott et al. 2017e) the most stringent upper limit is $\varepsilon \approx 10^{-7}$ which is well below the result (12.51). However, the non-detection of GWs can only be used as a constraint for the (unidentified) non-axisymmetric straining mechanism or the poorly known σ_{br} and not as evidence against the existence of CCS matter.

From our brief exposition of deformations related to the presence of quark matter in neutron star cores it should have become clear that observation of GWs from such sources could provide strong evidence in favour of the existence of these very exotic states of matter. But it should also be remembered that the properties of quark matter in neutron stars are largely uncertain and therefore any theoretical predictions based on them should be viewed in a similar way.

12.7.7 Mountains and Pinned Superfluidity

The presence of a pinned neutron superfluid component in the interior of neutron stars may lead to glitches (see next section) but could also have a number of interesting implications related to mountains. Vortex pinning could take place in the crust (with the pinning sites provided by the crustal lattice) and/or in the fluid core as a result of the interaction of the neutron vortices with the fluxtubes of the proton superconductor.

As discussed in Jones (2010), an elastically or magnetically deformed neutron star with a pinned superfluid component that has its angular momentum axis slightly misaligned with the stellar symmetry axes can modify the harmonic structure of the emitted GW signal by allowing emission at both frequencies 2Ω and Ω (whereas only the former should be present in the absence of pinning and assuming a non-precessing state).

Incidentally, it is worth mentioning that the same pinned superfluid acts as a fixed gyroscope and could easily stabilise the previously discussed spin-flip instability of a strong toroidal magnetic field component (Glampedakis and Jones 2010).

Finally, a pinned superfluid could itself act as a source of deformation and lead to GW mountain emission (Jones 2002). Since this "Magnus mountain" is the only deformation mechanism that has not been considered so far in detail it is worth discussing it a bit more. As the name suggests, the straining mechanism responsible for this mountain is the Magnus force acting on the pinned superfluid vortices due to the spin lag between the superfluid and the normal stellar component onto which the vortices are attached (see Ruderman 1991a,b for a more detailed discussion). Assuming both spin angular frequencies Ω_n, Ω to be aligned along the same axis (say the z-axis) the resulting Magnus force is purely radial in cylindrical coordinates and has a per unit volume magnitude,

$$f_m = 2\varpi \rho_n \Omega_{lag} \Omega, \tag{12.52}$$

where $\Omega_{lag} = \Omega_n - \Omega$ is the spin lag and ϖ is the standard cylindrical radius. In order to produce a non-axisymmetric quadrupolar deformation, the force (12.52) itself has to have a similar non-axisymmetry. In turn, this requires a non-uniform distribution of pinned vortices which would entail a non-rigid body rotational profile for the superfluid. Although detailed calculations for this non-axisymmetric Magnus force are still lacking it is not too difficult to see how this property could come about in a realistic neutron star model. For example, if pinning is provided by the magnetic fluxtubes then the generic non-axisymmetry of the stellar magnetic field would imply a similar property for the spatial distribution of the pinning sites. In our present back-of the-envelope-analysis the non-axisymmetric character of the Magnus force can be accounted for by multiplication of (12.52) with a phenomenological dimensionless factor $\lambda \lesssim 1$.

The ellipticity of the resulting Magnus mountain can be estimated by means of the ratio between the volume-integrated Magnus force $F_m = \int dV f_m$ and the gravitational binding force (Jones 2002):

$$\varepsilon_m \sim \frac{F_m}{F_{grav}} \sim \frac{2\lambda\Omega_{lag}\Omega R^3 \rho V_{pin}}{GM^2} \sim \frac{2\lambda\Omega_{lag}\Omega R^3 M_{pin}}{GM^2}, \tag{12.53}$$

where V_{pin} and M_{pin} represent the volume and mass of the pinned superfluid. Parametrising this result and assuming canonical stellar parameters we find,

$$\varepsilon_m \sim 5 \times 10^{-7} \lambda P_{-3}^{-1} \left(\frac{\Omega_{lag}}{0.01\,Hz}\right) \left(\frac{M_{pin}}{M_\odot}\right). \tag{12.54}$$

For the spin-lag we have used a representative value so that the Magnus force is at most comparable to the pinning force in the crust or in the core, see Link and Cutler (2002), Link (2003). Assuming $\lambda \sim 1$, we can see that a reasonable range for ellipticity could be $\varepsilon_m \sim 10^{-7} - 10^{-10}$, with the most favourable case coming from the scenario of vortex pinning in the core ($M_{pin} \sim M_\odot$). As also concluded in Jones (2002), the expected Magnus mountain could be comparable to that due to magnetic or elastic stresses and therefore could be of interest.

12.7.8 Future Directions

In recent years our understanding of the different processes which can lead to neutron star mountains greatly improved. However, we still do not know which mountains are actually present in the different kinds of neutron stars (magnetars, radio pulsars, LMXB, etc.). Indeed, past and present astrophysical observations mainly probe the surface and the exterior of the star, while we have very limited information from its interior.

For instance, spin-down measurements give us accurate estimates of the exterior magnetic field of magnetars (and of ordinary neutron stars), but if the field inside the star is much larger than in the exterior, we will need know the interior field to find the size and shape of magnetic mountains. Theoretical modelling can be improved, for instance to better understand the stability properties of magnetic field configurations (which can tell us how much of the field is toroidal and how much is poloidal). MHD calculations of neutron stars with exotic matter would also be very important, because these objects can accommodate larger deformations than stars with conventional hadronic matter; as discussed above, existing calculations of these systems are only back-of-the-envelope estimates. In any case, we need observational data to know the actual (volume averaged) strength of the interior magnetic fields.

The same holds for thermal mountains. Theoretical modelling can clarify the mechanism leading to temperature anisotropies and can give us estimates of their actual shape and magnitude. In particular, we need to know the size of quadrupole contribution to temperature anisotropies, δT_q. For instance, numerical simulations of the post-burst thermal evolution would help us understanding how large these mountains can be. However, even for thermal mountains, the power of theoretical modelling alone is limited: only observational data can tell us definitely how large actual mountains are.

The most promising observational probe to study neutron star mountains is the GW signal they emit as the star rotates. Once observed, this signal would provide a direct measurement of the mountain, shedding light on the formation mechanisms. Even an indirect GW observation, i.e., the observation in the electromagnetic spectrum of a process due to GW emission (such as the spin-down increase discussed in Haskell and Patruno (2017)) would give us valuable information to understand magnetic and thermal mountains and their formation.

12.8   Glitches

Radio pulsars have extremely stable rotation rates—they are the most precise natural clocks in the universe—but some of them exhibit sudden increases in the rotation rate, called *glitches*. After the first glitch observations in the Vela and Crab pulsars (Radhakrishnan and Manchester 1969; Reichley and Downs 1969; Boynton et al. 1969; Richards et al. 1969), more than three hundred glitches have been observed in $\sim$ 100 pulsars (Espinoza et al. 2011). More recently, glitches have also been observed in gamma-ray pulsars (Ray et al. 2011; Pletsch et al. 2012) and in magnetars (Dib et al. 2008) (MSPs can also have small glitches, but they are very rare (Cognard and Backer 2004)).

The glitch occurs in a very short timescale, still unresolved by observations (the best upper limit is ~ 40 s (Dodson et al. 2002)). The relative increase in the rotation rate $\Delta\Omega/\Omega$ ranges from $\sim 10^{-11}$ to $\sim 10^{-5}$ (Espinoza et al. 2011), and is generally followed by an increase in the spin-down rate. Most glitches have $\Delta\Omega/\Omega \lesssim 10^{-7}$, but few of them (such as those of the Vela pulsar) are significantly larger, with $\Delta\Omega/\Omega > 10^{-6}$.

The standard model of glitches is the superfluid model, in which glitches are due to superfluid neutrons in the neutron star interior, which store the angular momentum in a quantized array of vortices. The star spins down but its superfluid component does not, because the vortices are "pinned", i.e. their positions are fixed. Periodically, some of the vortices "unpin", move outwards and in this way the superfluid component removes the excess angular momentum, which is released in glitches.

An alternative model is the starquake model, in which glitches are due to the rigidity of the crust, which maintains its shape as the star spins down and reduces its quadrupole deformation; the strain on the crust increases and, periodically, there is a starquake, with a rearrangement of the moment of inertia and a glitch. This model can not explain large Vela-like glitches, but it is in principle possible that some small glitches are due to this mechanism. For a detailed discussion of the different glitch models, we refer the reader to Haskell and Melatos (2015) and references therein.

When a glitch occurs, there is a sudden rearrangement of the neutron star structure (in both the superfluid and the starquake models); therefore, glitches are expected to emit GWs. In 2006, during the fifth Science Run of LIGO, the Vela pulsar underwent a glitch. An analysis of the first-generation LIGO data looking for the excitation of stellar oscillations did not find a GW signal, thus setting an upper limit on the GW strain $h \leq 1.4 \times 10^{-20}$ (Abadie et al. 2011). However, second-generation detectors are expected to be about one order of magnitude more sensitive to this kind of signal.

Different mechanisms have been suggested for GW emission by pulsar glitches. They belong to two classes: the signals emitted during the glitch itself and those emitted during post-glitch relaxation.

- The energy released during the glitch can excite stellar oscillations (Sidery et al. 2010; Keer and Jones 2015; see also Sedrakian et al. 2003; Glampedakis and Andersson 2009; Santiago-Prieto et al. 2012). The GW strain can be expressed in terms of the total energy ΔE deposited into the mode as (Kokkotas et al. 2001)

$$h \sim 2 \times 10^{-23} D_1^{-1} \left(\frac{\Delta E}{10^{-12} M_\odot c^2} \right)^{1/2} \left(\frac{1\,\text{kHz}}{f_{\text{gw}}} \right) \left(\frac{0.1\,\text{s}}{\tau_{\text{gw}}} \right)^{1/2}, \qquad (12.55)$$

where f_{gw}, τ_{gw} are the frequency and damping time of the oscillation (and of the GW emission). A back-of-the-envelope estimate of the deposited energy, assuming that the rotational energy change associated to a glitch is entirely converted into non-radial oscillation, is $\Delta E \sim (2\pi)^2 I \Delta\Omega/\Omega \lesssim 10^{-12} M_\odot c^2$ (Andersson and Comer 2001). Then, assuming that most of the energy is deposited into the fundamental mode (as suggested by hydrodynamical simulations (Sidery et al. 2010; Keer and Jones 2015)), $f_{gw} \sim 1$ kHz, $\tau_{gw} \sim 0.1$ s and the GW strain (12.55) is too weak to be detected by Advanced LIGO/Virgo, but potentially detectable by an ET-class detector.

These estimates, however, are probably too optimistic, at least in the two-fluid scenario. Indeed, a detailed modelling (Sidery et al. 2010) shows that non-superfluid matter (ions in the crust and protons in the core) spin up, while superfluid neutrons spin down; therefore, the available energy is smaller by a factor $\sim 10^7$. This leads to a GW strain $\sim 10^{-26}$ or smaller, too weak to be detected even by third-generation interferometers. In the starquake scenario, instead, a careful analysis of the interplay between deformation energy and rotational kinetic energy shows that the GW strain can be as large as $\sim 10^{-23}$ (Keer and Jones 2015); however, as mentioned above, this scenario fails to describe Vela-like glitches.

- During a glitch, a large number of vortices move outwards, in a sort of "avalanche", transferring angular momentum to the crust. This process has been studied using a very sophisticated toy-model in which the superfluid neutron star is represented by a zero-temperature Bose condensate with dissipation described by a non-linear Schrödinger equation, the Gross-Pitaevskii (GP) equation (Warszawski and Melatos 2011, 2012; Warszawski et al. 2012; Warszawski and Melatos 2013). Numerical simulations of the GP equation model describe the collective vortex migration associated to a glitch. The simulations of the GP equations have then been extended using Monte-Carlo techniques, to include a larger number of vortices, compute the GW emission, and find the distribution of glitch parameters to be compared with observational data.

In the GP model the star is described as an infinite, rotating cylinder of fluid, with a rectangular grid of pinning sites. Applying an external spin-down torque, the vortices unpin and repin, with an average displacement of Δr (which is a key parameter of the model). The GP equation gives the evolution of the vortex distribution, as they unpin and repin. Each vortex generates a solenoidal velocity field, which is affected by the displacement process. Therefore, the velocity of the fluid acquires a non-axisymmetric component, which can be reconstructed by the vortex distribution. The non-axisymmetric velocity generates a time-varying current quadrupole moment, and then a GW emission with strain amplitude:

$$ h \sim 10^{-23} D_1^{-1} \left(\frac{\Delta r}{1 \, \text{cm}} \right)^{-1} \left(\frac{\Delta \Omega / \Omega}{10^{-5}} \right) \left(\frac{f_{\text{spin}}}{100 \, \text{Hz}} \right)^3 . \tag{12.56} $$

The gravitational signal decreases as the travel distance Δr increases. This is due to the fact that, for larger values of Δr (keeping fixed $\Delta \Omega / \Omega$), the number of vortices involved is smaller and the current quadrupole is also smaller. Recently, it has been suggested (Melatos et al. 2015) that part of the large scale non-axisymmetries in the velocity field produced by a vortex avalanche can persist in the inter-glitch recovery phase. If this is true, the GW signal can be as large as $h \sim 10^{-21}$, potentially detectable even by second-generation interferometers.

We remark that the GP model is based on simplifying assumptions, which may affect the GW detectability estimates: it is not an hydrodynamical model, it is assumed that the system is weakly interacting, the grid of pinning sites is defined *a priori*, and the effect of magnetic field, which can be relevant for vortex pinning in the core, is neglected. However, this model is very powerful, since it captures the collective behaviour of vortices, which is likely to have a fundamental role in pulsar glitches.

- After the glitch, the neutron star resumes its spin-down, entering in a recovery phase which can last from months to years. However, since—just after the glitch—the crust has a rotation rate larger than the superfluid interior, during the first part of the glitch recovery the superfluid interior spins-up due to viscous interactions, restoring corotation with the crust, and erasing the non-axisymmetries in the velocity field. In this "relaxation phase" viscous interactions act through the process of Ekman pumping (Benton and Clark Jr 1974), which operates on a timescale (the "Ekman time") which has been estimated to range from days to weeks. As the core spins up, the time-dependent mass and current quadrupole moments due to non-axisymmetric meridional circulation emit a continuous GW. Numerical simulations of the relaxation phase (assuming initial data in which the components with different azimuthal numbers have comparable amplitude) yield a GW strain (van Eysden and Melatos 2008; Bennett et al. 2010; Singh 2017)

$$h \sim 6 \times 10^{-27} D_1^{-1} \left(\frac{\Delta\Omega/\Omega}{10^{-5}} \right) \left(\frac{f_{\text{spin}}}{100\,\text{Hz}} \right)^3 . \qquad (12.57)$$

Although the GW signal emitted during the relaxation phase has a lower dimensionless strain amplitude than the burst emitted during the glitch, its duration is larger, and this increases the measured strain $\sim h\sqrt{T}$ (Prix et al. 2011). Therefore, the detectability of this signal crucially depends on its actual duration. Current models of the relaxation phase (van Eysden and Melatos 2008; Bennett et al. 2010; Singh 2017) find that the signal can last up to some weeks, and—if emitted by a large nearby neutron star after a large glitch—it may be marginally detectable even by second-generation GW interferometers. However, these models neglect the contribution of magnetic field and mutual friction, which couple the crust with the core faster than Ekman pumping, reducing the duration of the signal. These contributions could significantly reduce the actual detectability of the GW emission in the post-glitch recovery phase.

The estimates of the GW signal from a glitch have recently been extended, for most of the mechanisms discussed above, to the observed glitches in gamma-ray pulsars (most of which are radio-quiet), finding that the detectability of the GW signal from gamma-ray pulsar glitches is comparable with that of the signal from radio pulsar glitches (Stopnitzky and Profumo 2014).

Summarizing, GW emission from glitches is still not fully understood, and the predictions can be very different, from pessimistic (signal too weak even for third-generation interferometers, as in Sidery et al. (2010)), to moderately optimistic (signal potentially detectable by ET, as in Keer and Jones (2015), Warszawski and Melatos (2012)), to very optimistic (signal marginally detectable by Advanced LIGO/Virgo, as in van Eysden and Melatos (2008), Bennett et al. (2010), Prix et al. (2011), Melatos et al. (2015)). We remark that the models predicting detectability by second-generation interferometers require a persistent signal after the glitch—either due to crust-core viscous interaction, or to a persistent current quadrupole moment produced by a vortex avalanche—but the actual duration of the signal is still matter of debate.

12.8.1 Future Directions

Neutron star glitches are a very complex process, which we are just starting to understand. Our theoretical models probably are able to capture the main features of this process, but they are often qualitative, and based on simplifying assumptions. We still need to build a general model which captures the different aspects of pulsar glitches, and to perform MHD simulations based on this model.

In order to understand the glitch phase, we need a reliable description of the collective motion of vortices. To this aim, the GP model is a good starting point, but it needs to be extended and interfaced to an hydrodynamical model, including magnetic fields and crust-core couplings.

Concerning the post-glitch relaxation phase, our understanding of the mechanisms leading to density and current asymmetries (crust-core differential rotation, two-stream instabilities, cracks and tilts in the crust, etc. (van Eysden and Melatos 2008)) is only qualitative. We need an accurate model of these processes, including magnetic field and mutual friction, in order to understand the actual duration of the relaxation phase and then of the GW signal.

Only a detailed and quantitative description of the entire glitch and post-glitch relaxation phases, will allow us to determine the GW emission associated to this process, and to definitely assess its detectability by GW interferometers.

12.9 Concluding Remarks

In this chapter we have surveyed a number of physical processes in isolated and accreting neutron stars that could be interesting sources for the newborn field of GW astronomy. A general conclusion that can be drawn from the results summarised here is that the GW signals from these single neutron stars are not expected to be as loud as the ones produced by binary systems of black holes or neutron stars. This is mostly due to the basic fact that there is less gravitational mass "sloshing around" in single systems than in binary ones. This handicap, however, can be partially offset by the continuous character of the GW signal associated with some of the emission mechanisms (e.g. mountains) and/or the closeness to the source (e.g. known neutron stars in our Galaxy). All relevant factors accounted for, GWs from single neutron stars are more likely to be detected by future ET-class observatories.

It should be emphasized that in some cases our present level of understanding of a particular mechanism does not even allow us to make a safe prediction as to whether GW emission will operate in the first place. An example of this is provided by the r-mode instability where the mode's maximum amplitude and instability window are still largely uncertain factors. This situation is not surprising given the multi-faceted physics that has bearing on the problem.

There is one more key point worth considering here, namely, the possibility of GW emission mechanisms having an impact on photon astronomy observations of neutron stars. This has, of course, already happened when the orbital evolution of binary pulsar systems was found to be consistent with GW emission, thus providing evidence of the existence of GWs several decades before their first direct detection. A more recent remarkable example of this synergy may have been provided by the observed spin-down profile of PSR J1203+0038 which is a member of a LMXB system (Haskell and Patruno 2017). The enhanced spin-down rate of this pulsar immediately after an accretion phase has been attributed to the GW emission by a thermal mountain, but the required deformation is too small to be directly detected by Advanced LIGO. A similar situation of GWs "seen" in the electromagnetic channel may arise if, for example, a small amplitude r-mode drives the spin evolution of a MSP.

Acknowledgements We thank Nils Andersson, Daniela Doneva, Brynmor Haskell, Wynn Ho, Ian Jones, Kostas Kokkotas, Cristiano Palomba, George Pappas, Andrea Passamonti and Kai Schwenzer for useful discussions during the course of this work and for providing data and figures.

This work was supported by the H2020-MSCA-RISE-2015 Grant No. StronGrHEP-690904 and by the COST actions MP1304 and CA16104.

References

Asai, H., Lee, U.: Astrophys. J. **790**, 66 (2014)
Aasi, J., et al. (LIGO Scientific): Astrophys. J. **785**, 119 (2014)
Aasi, J., et al.: Astrophys. J. **813**, 16 (2015)
Abadie, J., et al. (LIGO): Phys. Rev. **D83**, 042001 (2011)
Abbott, B.P., et al.: Phys. Rev. Lett. **116**, 061102 (2016a)
Abbott, B.P., et al.: Phys. Rev. Lett. **116**, 241103 (2016b)
Abbott, B.P., et al.: Phys. Rev. Lett. **118**, 221101 (2017a)
Abbott, B.P., et al. (Virgo, LIGO Scientific): Astrophys. J. **851**, L35 (2017b)
Abbott, B., et al. (Virgo, LIGO Scientific): Phys. Rev. Lett. **119**, 161101 (2017c)
Abbott, B.P., et al. (GROND, SALT Group, OzGrav, DFN, INTEGRAL, Virgo, Insight-Hxmt, MAXI Team, Fermi-LAT, J-GEM, RATIR, IceCube, CAASTRO, LWA, ePESSTO, GRAWITA, RIMAS, SKA South Africa/MeerKAT, H.E.S.S., 1M2H Team, IKI-GW Follow-up, Fermi GBM, Pi of Sky, DWF (Deeper Wider Faster Program), Dark Energy Survey, MASTER, AstroSat Cadmium Zinc Telluride Imager Team, Swift, Pierre Auger, ASKAP, VINROUGE, JAGWAR, Chandra Team at McGill University, TTU-NRAO, GROWTH, AGILE

Team, MWA, ATCA, AST3, TOROS, Pan-STARRS, NuSTAR, ATLAS Telescopes, BOOTES, CaltechNRAO, LIGO Scientific, High Time Resolution Universe Survey, Nordic Optical Telescope, Las Cumbres Observatory Group, TZAC Consortium, LOFAR, IPN, DLT40, Texas Tech University, HAWC, ANTARES, KU, Dark Energy Camera GW-EM, CALET, Euro VLBI Team, ALMA): Astrophys. J. **848**, L12 (2017d)

Abbott, B.P., et al. (Virgo, LIGO Scientific): Astrophys. J. **839**, 12 (2017e)

Akgun, T., Wasserman, I.: Mon. Not. R. Astron. Soc. **383**, 1551 (2008)

Alford, M.G., Schwenzer, K.: Phys. Rev. Lett. **113**, 251102 (2014a)

Alford, M.G., Schwenzer, K.: Astrophys. J. **781**, 22 (2014b)

Alford, M.G., Schwenzer, K.: Mon. Not. R. Astron. Soc. **446**, 3631 (2015)

Alford, M.G., Sedrakian, A.: J. Phys. G **37**, 075202 (2010)

Alford, M.G., Schmitt, A., Rajagopal, K., Shäfer, T.: Rev. Mod. Phys. **80**, 1455 (2008)

Alpar, M.A., Langer, S.A., Sauls, J.A.: Astrophys. J. **282**, 533 (1984)

Andersson, N.: Astrophys. J. **502**, 708 (1998)

Andersson, N.: Classical Quantum Gravity **20**, R105 (2003)

Andersson, N., Comer, G.L.: Phys. Rev. Lett. **87**, 241101 (2001)

Andersson, N., Comer, G.L.: Living Rev. Relativ. **10**, 83 (2007)

Andersson, N., Kokkotas, K.D.: Mon. Not. R. Astron. Soc. **299**, 1059 (1998)

Andersson, N., Kokkotas, K.D.: Int. J. Mod. Phys. D **10**, 381 (2001)

Andersson, N., Kokkotas, K.D., Stergioulas, N.: Astrophys. J. **516**, 307 (1999)

Andersson, N., Jones, D.I., Kokkotas, K.D., Stergioulas, N.: Astrophys. J. **534**, L75 (2000)

Andersson, N., Jones, D.I., Kokkotas, K.D.: Mon. Not. R. Astron. Soc. **337**, 1224 (2002)

Andersson, N., Comer, G.L., Glampedakis, K.: Nucl. Phys. A **763**, 212 (2005a)

Andersson, N., Glampedakis, K., Haskell, B., Watts, A.L.: Mon. Not. R. Astron. Soc. **361**, 1153 (2005b)

Andersson, N., Sidery, T., Comer, G.L.: Mon. Not. R. Astron. Soc. **368**, 162 (2006)

Andersson, N., Haskell, B., Comer, G.L.: Phys. Rev. D **82**, 023007 (2010)

Andersson, N., Ferrari, V., Jones, D.I., Kokkotas, K.D., Krishnan, B., Read, J.S., Rezzolla, L., Zink, B.: Gen. Relativ. Gravit. **43**, 409 (2011)

Andersson, N., Glampedakis, K., Hogg, M.: Phys. Rev. D **87**, 063007 (2013)

Andersson, N., Jones, D.I., Ho, W.C.G.: Mon. Not. R. Astron. Soc. **442**, 1786 (2014)

Archibald, A.M., et al.: Science **324**, 1411 (2009)

Arras, P., Flanagan, E.E., Morsink, S.M., Schenk, A.K., Teukolsky, S.A., Wasserman, I.: Astrophys. J. **591**, 1129 (2003)

Arzoumanian, Z., Gendreau, K., Baker, C., Cazeau, T., Hestnes, Kellogg, J., Kenyon, S., Kozon, R., Liu, K.-C., Manthripragada, S., et al.: In: SPIE Astronomical Telescopes+ Instrumentation, pp. 914420–914420. International Society for Optics and Photonics, Bellingham (2014)

Asai, H., Lee, U., Yoshida, S.: Mon. Not. R. Astron. Soc. **455**, 2228 (2016)

Baiotti, L., De Pietri, R., Manca, G.M., Rezzolla, L.: Phys. Rev. D **75**, 044023 (2007)

Benhar, O., Ferrari, V., Gualtieri, L.: Phys. Rev. **D70**, 124015 (2004)

Bennett, M.F., van Eysden, C.A., Melatos, A.: Mon. Not. R. Astron. Soc. **409**, 1705 (2010)

Benton, E.R., Clark Jr, A.: Annu. Rev. Fluid Mech. **6**, 257 (1974)

Bhattacharyya, S.: Astrophys. J. **847**, 8 (2017)

Bhattacharyya, S., Chakrabarty, D.: Astrophys. J. **835**, 4 (2017)

Bildsten, L.: Astrophys. J. **501**, L89 (1998a)

Bildsten, L.: Astrophys. J. **501**, L89 (1998b)

Bildsten, L., Ushomirsky, G.: Astrophys. J. **529**, L33 (2000)

Bonazzola, S., Gourgoulhon, E.: Astron. Astrophys. **312**, 675 (1996)

Bondarescu, R., Teukolsky, S.A., Wasserman, I.: Phys. Rev. D **76**, 064019 (2007)

Bondarescu, R., Teukolsky, S.A., Wasserman, I.: Phys. Rev. D **79**, 104003 (2009)

Boynton, P., Groth III, E., Partridge, R., Wilkinson, D.T.: Astrophys. J. **157**, L197 (1969)

Braithwaite, J.: Astron. Astrophys. **453**, 687 (2006)

Braithwaite, J.: Astron. Astrophys. **469**, 275 (2007)

Braithwaite, J., Nordlund, A.: Astron. Astrophys. **450**, 1077 (2006)

Braithwaite, J., Spruit, H.C.: Nature **431**, 819 (2004)

Brown, R.A., Scriven, L.E.: Proc. R. Soc. Lond. A **371**, 331 (1980)

Brown, E.F., Ushomirsky, G.: Astrophys. J. **536**, 915 (2000)

Camelio, G., Gualtieri, L., Pons, J.A., Ferrari, V.: Phys. Rev. **D94**, 024008 (2016)

Chandrasekhar, S.: Ellipsoidal Figures of Equilibrium. Yale University Press, New Haven (1969)

Chandrasekhar, S.: Phys. Rev. Lett. **24**, 611 (1970)

Chandrasekhar, S., Fermi, E.: Astrophys. J. **118**, 116 (1953)

Chugunov, A.I.: Mon. Not. R. Astron. Soc. **451**, 2772 (2015)

Chugunov, A.I., Kantor, E.M., Gusakov, M.E.: Mon. Not. R. Astron. Soc. **468**, 291 (2017)

Ciolfi, R., Rezzolla, L.: Astrophys. J. **760**, 1 (2012)

Ciolfi, R., Rezzolla, L.: Mon. Not. R. Astron. Soc. **435**, L43 (2013)

Ciolfi, R., Ferrari, V., Gualtieri, L., Pons, J.A.: Mon. Not. R. Astron. Soc. **397**, 913 (2009)

Ciolfi, R., Ferrari, V., Gualtieri, L.: Mon. Not. R. Astron. Soc. **406**, 2540 (2010)

Cognard, I., Backer, D.C.: Astrophys. J. **612**, L125 (2004)

Colaiuda, A., Kokkotas, K.D.: Mon. Not. R. Astron. Soc. **414**, 3014 (2011)

Colaiuda, A., Kokkotas, K.D.: Mon. Not. R. Astron. Soc. **423**, 811 (2012)

Corsi, A., Meszaros, P.: Astrophys. J. **702**, 1171 (2009)

Corsi, A., Owen, B.J.: Phys. Rev. **D83**, 104014 (2011)

Cuofano, C., Dall'Osso, S., Drago, A., Stella, L.: Phys. Rev. D **86**, 044004 (2012)

Cutler, C.: Phys. Rev. **D66**, 084025 (2002)

Dai, Z.G., Lu, T.: Phys. Rev. Lett. **81**, 4301 (1998)

Dall'Osso, S., Shore, S.N., Stella, L.: Mon. Not. R. Astron. Soc. **398**, 1869 (2009)

Dall'Osso, S., Giacomazzo, B., Perna, R., Stella, L.: Astrophys. J. **798**, 25 (2015)

Detweiler, S.L.: Astrophys. J. **197**, 203 (1975)

Dib, R., Kaspi, V.M., Gavriil, F.P.: Astrophys. J. **673**, 1044-1061 (2008)

Dodson, R.G., McCulloch, P.M., Lewis, D.R.: Astrophys. J. **564**, L85 (2002)

Dommes, V.A., Gusakov, M.E.: Mon. Not. R. Astron. Soc. **455**, 2852 (2016)

Doneva, D.D., Kokkotas, K.D.: Phys. Rev. D **92**, 124004 (2015)

Doneva, D.D., Gaertig, E., Kokkotas, K.D., Krüger, C.: Phys. Rev. D **88**, 044052 (2013)

Doneva, D.D., Kokkotas, K.D., Pnigouras, P.: Phys. Rev. D **92**, 104040 (2015)

Duncan, R.C., Thompson, C.: Astrophys. J. Lett. **392**, L9 (1992)

Espinoza, C.M., Lyne, A.G., Stappers, B.W., Kramer, M.: Mon. Not. R. Astron. Soc. **414**, 1679 (2011)

Feroci, M., Stella, L., Van der Klis, M., Courvoisier, T.-L., Hernanz, M., Hudec, R., Santangelo, A., Walton, D., Zdziarski, A., Barret, D., et al.: Exp. Astron. **34**, 415 (2012)

Ferraro, V.C.A.: Astrophys. J. **119**, 407 (1954)

Franci, L., De Pietri, R., Dionysopoulou, K., Rezzolla, L.: Phys. Rev. **D88**, 104028 (2013)

Friedman, J.L.: Commun. Math. Phys. **62**, 247 (1978)

Friedman, J.L.: Phys. Rev. Lett. **51**, 11 (1983), [Erratum: Phys. Rev. Lett.51,718(1983)]

Friedman, J.L., Morsink, S.M.: Astrophys. J. **502**, 714 (1998)

Friedman, J.L., Schutz, B.F.: Astrophys. J. **221**, 937 (1978a)

Friedman, J.L., Schutz, B.F.: Astrophys. J. **222**, 281 (1978b)

Friedman, J.L., Stergioulas, N.: Rotating Relativistic Stars. Cambridge University Press, New York (2013)

Friedman, J.L., Lindblom, L., Lockitch, K.H.: Phys. Rev. D **93**, 024023 (2016)

Gabler, M., Cerda-Duran, P., Font, J.A., Muller, E., Stergioulas, N.: Mon. Not. R. Astron. Soc. **410**, 37 (2011)

Gabler, M., Duran, P.C., Stergioulas, N., Font, J.A., Muller, E.: Mon. Not. R. Astron. Soc. **421**, 2054 (2012)

Gabler, M., Cerdá-Durán, P., Stergioulas, N., Font, J.A., Müller, E.: Phys. Rev. Lett. **111**, 211102 (2013)

Gabler, M., Cerdá-Durán, P., Stergioulas, N., Font, J.A., Müller, E.: Mon. Not. R. Astron. Soc. **443**, 1416 (2014)

Gabler, M., Cerdá-Durán, P., Stergioulas, N., Font, J.A., Müller, E.: Mon. Not. R. Astron. Soc. **460**, 4242 (2016)

Gaertig, E., Kokkotas, K.D.: Phys. Rev. D **83**, 064031 (2011)

Gaertig, E., Glampedakis, K., Kokkotas, K.D., Zink, B.: Phys. Rev. Lett. **107**, 101102 (2011)

Gearheart, M., Newton, W.G., Hooker, J., Li, B.-A.: Mon. Not. R. Astron. Soc. **418**, 2343 (2011)

Ghosh, P., Lamb, F.K.: Astrophys. J. **234**, 296 (1979)

Giacomazzo, B., Perna, R.: Astrophys. J. **771**, L26 (2013)

Giacomazzo, B., Zrake, J., Duffell, P.C., MacFadyen, A.I., Perna, R.: Astrophys. J. **809**, 5 (2015)

Gill, R., Heyl, J.S.: Mon. Not. R. Astron. Soc. **407**, 1926 (2010)

Glampedakis, K., Andersson, N.: Mon. Not. R. Astron. Soc. **371**, 1311 (2006a)

Glampedakis, K., Andersson, N.: Phys. Rev. D **74**, 044040 (2006b)

Glampedakis, K., Andersson, N.: Phys. Rev. Lett. **102**, 141101 (2009)

Glampedakis, K., Jones, D.I.: Mon. Not. R. Astron. Soc. **405**, L6 (2010)

Glampedakis, K., Jones, D.I.: Mon. Not. R. Astron. Soc. **439**, 1522 (2014)

Glampedakis, K., Lasky, P.D.: Mon. Not. R. Astron. Soc. **463**, 2542 (2016)

Glampedakis, K., Samuelsson, L., Andersson, N.: Mon. Not. R. Astron. Soc. **371**, L74 (2006)

Glampedakis, K., Andersson, N., Samuelsson, L.: Mon. Not. R. Astron. Soc. **410**, 805 (2011)

Glampedakis, K., Jones, D.I., Samuelsson, L.: Phys. Rev. Lett. **109**, 081103 (2012)

Glampedakis, K., Andersson, N., Jones, D.I.: (in preparation)

Goldreich, P., Reisenegger, A.: Astrophys. J. **395**, 250 (1992)

Gualtieri, L., Kantor, E.M., Gusakov, M.E., Chugunov, A.I.: Phys. Rev. **D90**, 024010 (2014)

Gusakov, M.E., Kantor, E.M.: Phys. Rev. **D88**, 101302 (2013)

Gusakov, M.E., Chugunov, A.I., Kantor, E.M.: Phys. Rev. Lett. **112**, 151101 (2014a)

Gusakov, M.E., Chugunov, A.I., Kantor, E.M.: Phys. Rev. D **90**, 063001 (2014b)

Haensel, P., Zdunik, J.L.: Astron. Astrophys. **227**, 431 (1990)

Haskell, B.: Int. J. Mod. Phys. E **24**, 1541007 (2015)

Haskell, B., Andersson, N.: Mon. Not. R. Astron. Soc. **408**, 1897 (2010)

Haskell, B., Melatos, A.: Int. J. Mod. Phys. D **24**, 1530008 (2015)

Haskell, B., Patruno, A.: Astrophys. J. Lett. **738**, L14 (2011)

Haskell, B., Patruno, A.: (2017), arXiv:1703.08374 [astro-ph.HE]

Haskell, B., Jones, D.I., Andersson, N.: Mon. Not. R. Astron. Soc. **373**, 1423 (2006)

Haskell, B., Andersson, N., Jones, D.I., Samuelsson, L.: Phys. Rev. Lett. **99**, 231101 (2007)

Haskell, B., Samuelsson, L., Glampedakis, K., Andersson, N.: Mon. Not. R. Astron. Soc. **385**, 531 (2008)

Haskell, B., Andersson, N., Passamonti, A.: Mon. Not. R. Astron. Soc. **397**, 1464 (2009)

Haskell, B., Degenaar, N., Ho, W.C.G.: Mon. Not. R. Astron. Soc. **424**, 93 (2012)

Haskell, B., Glampedakis, K., Andersson, N.: Mon. Not. R. Astron. Soc. **441**, 1662 (2014)

Haskell, B., Priymak, M., Patruno, A., Oppenoorth, M., Melatos, A., Lasky, P.D.: Mon. Not. R. Astron. Soc. **450**, 2393 (2015)

Henriksson, K.T., Wasserman, I.: Mon. Not. R. Astron. Soc. **431**, 2986 (2013)

Heyl, J.S.: Astrophys. J. **574**, L57 (2002)

Hill, J.A., Eaves, L.: Phys. Rev. Lett. **101**, 234501 (2008)

Ho, W.C.G., Andersson, N., Haskell, B.: Phys. Rev. Lett. **107**, 101101 (2011)

Horowitz, C.J., Kadau, K.: Phys. Rev. Lett. **102**, 191102 (2009)

Huppenkothen, D., Heil, L.M., Watts, A.L., Göğüş, E.: Astrophys. J. **795**, 114 (2014a)

Huppenkothen, D., et al.: Astrophys. J. **787**, 128 (2014b)

Huppenkothen, D., Watts, A.L., Levin, Y.: Astrophys. J. **793**, 129 (2014c)

Idrisy, A., Owen, B.J., Jones, D.I.: Phys. Rev. D **91**, 024001 (2015)

Iida, K., Baym, G.: Phys. Rev. D **66**, 014015 (2002)

Ipser, J.R., Lindblom, L.: Phys. Rev. Lett. **62**, 2777 (1989)

Ipser, J.R., Lindblom, L.: Astrophys. J. **355**, 226 (1990)

Ipser, J.R., Lindblom, L.: Astrophys. J. **373**, 213 (1991)

Israel, G., Belloni, T., Stella, L., Rephaeli, Y., Gruber, D., Casella, P.G., Dall'Osso, S., Rea, N., Persic, M., Rothschild, R.: Astrophys. J. **628**, L53 (2005)

Johnson-McDaniel, N.K., Owen, B.J.: Phys. Rev. **D88**, 044004 (2013)

Jones, P.B.: Astrophys. Space Sci. **33**, 215 (1975)

Jones, P.B.: Phys. Rev. Lett. **86**, 1384 (2001)

Jones, D.I.: Classical Quantum Gravity **19**, 1255 (2002)

Jones, D.I.: Mon. Not. R. Astron. Soc. **402**, 2503 (2010)

Kantor, E.M., Gusakov, M.E.: Mon. Not. R. Astron. Soc. **442**, 90 (2014)

Kantor, E.M., Gusakov, M.E.: Mon. Not. R. Astron. Soc. **469**, 3928 (2017)

Keer, L., Jones, D.I.: Mon. Not. R. Astron. Soc. **446**, 865 (2015)

Kinney, J.B., Mendell, G.: Phys. Rev. D **67**, 024032 (2003)

Kiuchi, K., Kyutoku, K., Sekiguchi, Y., Shibata, M., Wada, T.: Phys. Rev. D **90**, 041502 (2014)

Kokkotas, K.D., Schwenzer, K.: Eur. Phys. J. A **52**, 15 (2016)

Kokkotas, K.D., Apostolatos, T.A., Andersson, N.: Mon. Not. R. Astron. Soc. **320**, 307 (2001)

Kouveliotou, C., et al.: Nature **393**, 235 (1998)

Kuroda, T., Umeda, H.: Astrophys. J. Suppl. **191**, 439 (2010)

Lai, D., Shapiro, S.L.: Astrophys. J. **442**, 259 (1994)

Lamb, H.: Proc. R. Soc. Lond. A **80**, 168 (1908)

Lander, S.K.: Phys. Rev. Lett. **110**, 071101 (2013)

Lander, S.K., Jones, D.I.: Mon. Not. R. Astron. Soc. **395**, 2162 (2009)

Lander, S.K., Andersson, N., Glampedakis, K.: Mon. Not. R. Astron. Soc. **419**, 732 (2012)

Lasky, P.D., Glampedakis, K.: Mon. Not. R. Astron. Soc. **458**, 1660 (2016)

Lasky, P.D., Melatos, A.: Phys. Rev. **D88**, 103005 (2013)

Lau, H.K., Leung, P.T., Lin, L.M.: Astrophys. J. **714**, 1234 (2010)

Levin, Y.: Astrophys. J. **517**, 328 (1999)

Levin, Y.: Mon. Not. R. Astron. Soc. **368**, L35 (2006)

Levin, Y.: Mon. Not. R. Astron. Soc. **377**, 159 (2007)

Levin, Y., Ushomirsky, G.: Mon. Not. R. Astron. Soc. **324**, 917 (2001)

Levin, Y., van Hoven, M.: Mon. Not. R. Astron. Soc. **418**, 659 (2011)

Lindblom, L., Mendell, G.: Astrophys. J. **444**, 804 (1995)

Lindblom, L., Mendell, G.: Phys. Rev. D **61**, 104003 (2000)

Lindblom, L., Owen, B.J.: Phys. Rev. D **65**, 063006 (2002)

Lindblom, L., Owen, B.J., Morsink, S.M.: Phys. Rev. Lett. **80**, 4843 (1998)

Lindblom, L., Owen, B.J., Ushomirsky, G.: Phys. Rev. D **62**, 084030 (2000)

Link, B.: Phys. Rev. Lett. **91**, 101101 (2003)

Link, B., Cutler, C.: Mon. Not. R. Astron. Soc. **336**, 211 (2002)

Link, B., van Eysden, C.A.: Astrophys. J. **823**(1), L1 (2016a)

Link, B., van Eysden, C.A.: Astrophys. J. **823**, L1 (2016b), [Erratum: Astrophys. J. Suppl.224,no.1,6(2016)]

Lockitch, K.H., Andersson, N., Friedman, J.L.: Phys. Rev. D **63**, 024019 (2001)

Lockitch, K.H., Andersson, N., Friedman, J.L.: Phys. Rev. D **68**, 124010 (2003)

Lyutikov, M.: Mon. Not. R. Astron. Soc. **367**, 1594 (2006)

Madsen, J.: Phys. Rev. Lett. **81**, 3311 (1998)

Madsen, J.: Phys. Rev. Lett. **85**, 10 (2000)

Mahmoodifar, S., Strohmayer, T.: Astrophys. J. **773**, 10 (2013)

Manca, G.M., Baiotti, L., De Pietri, R., Rezzolla, L.: Classical Quantum Gravity **24**, S171 (2007)

Mannarelli, M., Rajagopal, K., Sharma, R.: Phys. Rev. D **76**, 074026 (2007)

Mannarelli, M., Manuel, C., Sa'D, B.A.: Phys. Rev. Lett. **101**, 241101 (2008)

Mastrano, A., Melatos, A., Reisenegger, A., Akgun, T.: Mon. Not. R. Astron. Soc. **417**, 2288 (2011)

Mastrano, A., Suvorov, A.G., Melatos, A.: Mon. Not. R. Astron. Soc. **447**, 3475 (2015)

Melatos, A., Payne, D.J.B.: Astrophys. J. **623**, 1044 (2005)

Melatos, A., Douglass, J.A., Simula, T.P.: Astrophys. J. **807**, 132 (2015)

Mendell, G.: Astrophys. J. **380**, 515 (1991)

Mendell, G.: Phys. Rev. D **64**, 044009 (2001)

Metzger, B.D., Piro, A.L.: Mon. Not. R. Astron. Soc. **439**, 3916 (2014)

Nayyar, M., Owen, B.J.: Phys. Rev. D **73**, 084001 (2006)

Ostriker, J.P., Gunn, J.E.: Astrophys. J. **157**, 1395 (1969)

Owen, B.J.: Phys. Rev. Lett. **95**, 211101 (2005)

Owen, B.J.: Phys. Rev. D **82**, 104002 (2010)

Owen, B.J., Lindblom, L., Cutler, C., Schutz, B.F., Andersson, N.: Phys. Rev. D **58**, 084020 (1998)

Papaloizou, J., Pringle, J.E.: Mon. Not. R. Astron. Soc. **184**, 501 (1978)

Pappito, A., Riggio, A., Bunderi, L., di Salvo, T., D'Aí, A., Iaria, R.: Astron. Astrophys. **528**, 6 (2011)

Passamonti, A., Andersson, N.: Mon. Not. R. Astron. Soc. **446**, 555 (2015)

Passamonti, A., Lander, S.K.: Mon. Not. R. Astron. Soc. **429**, 767 (2013)

Passamonti, A., Lander, S.K.: Mon. Not. R. Astron. Soc. **438**, 156 (2014)

Passamonti, A., Pons, J.A.: Mon. Not. R. Astron. Soc. **463**, 1173 (2016)

Passamonti, A., Gaertig, E., Kokkotas, K.D., Doneva, D.: Phys. Rev. D **87**, 084010 (2013)

Passamonti, A., Andersson, N., Ho, W.C.G.: Mon. Not. R. Astron. Soc. **455**, 1489 (2016)

Patruno, A., Watts, A.L.: In: Belloni, T., Mendez, M., Zhang, C.M. (eds.) Timing Neutron Stars: Pulsations, Oscillations and Explosions. Springer, Berlin (2012)

Payne, D.J.B., Melatos, A.: Mon. Not. R. Astron. Soc. **351**, 569 (2004)

Pierce, J.R.: Almost All about Waves. MIT Press, Cambridge (1974)

Pili, A.G., Bucciantini, N., Del Zanna, L.: Mon. Not. R. Astron. Soc. **439**, 3541 (2014)

Pletsch, H.J., et al.: Astrophys. J. Lett. **755**, L20 (2012)

Pnigouras, P., Kokkotas, K.D.: Phys. Rev. D **92**, 084018 (2015)

Pnigouras, P., Kokkotas, K.D.: Phys. Rev. D **94**, 024053 (2016)

Pons, J.A., Geppert, U.: Astron. Astrophys. **470**, 303 (2007)

Prendergast, K.H.: Astrophys. J. **123**, 498 (1956)

Prix, R., Giampanis, S., Messenger, C.: Phys. Rev. **D84**, 023007 (2011)

Priymak, M., Melatos, A., Payne, D.J.B.: Mon. Not. R. Astron. Soc. **417**, 2696 (2011)

Punturo, M., et al.: Classical Quantum Gravity **27**, 084007 (2010)

Radhakrishnan, V., Manchester, R.N.: Nature **222**, 228 (1969)

Rajagopal, K., Sharma, R.: Phys. Rev. D **74**, 094019 (2006)

Rappaport, S.A., Fregeau, J.M., Spruit, H.: Astrophys. J. **606**, 436 (2004)

Ravi, V., Lasky, P.D.: Mon. Not. R. Astron. Soc. **441**, 2433 (2014)

Ray, P.S., et al.: Astrophys. J. Suppl. **194**, 17 (2011)

Reichley, P.E., Downs, G.S.: Nature **222**, 229 (1969)

Rezzolla, L., Lamb, F.K., Shapiro, S.L.: Astrophys. J. **531**, L139 (2000)

Rezzolla, L., Lamb, F.K., Markovic, D., Shapiro, S.L.: Phys. Rev. D **64**, 104013 (2001a)

Rezzolla, L., Lamb, F.K., Markovic, D., Shapiro, S.L.: Phys. Rev. D **64**, 104014 (2001b)

Rezzolla, L., Giacomazzo, B., Baiotti, L., Granot, J., Kouveliotou, C., Aloy, M.A.: Astrophys. J. Lett. **732**, L6 (2011)

Richards, D.W., Pettengill, G.H., Roberts, J.A., Counselman, C.C., Rankin, J.: IAU Circ. 2181 (1969)

Rieutord, M.: Astrophys. J. **550**, 443 (2001)

Rowlinson, A., O'Brien, P.T., Metzger, B.D., Tanvir, N.R., Levan, A.J.: Mon. Not. R. Astron. Soc. **430**, 1061 (2013)

Ruderman, M.: Astrophys. J. **382**, 576 (1991a)

Ruderman, M.: Astrophys. J. **382**, 587 (1991b)

Ruderman, M., Zhu, T., Chen, K.: Astrophys. J. **492**, 267

Ruoff, J., Kokkotas, K.D.: Mon. Not. R. Astron. Soc. **328**, 678 (2001)

Samuelsson, L., Andersson, N.: Mon. Not. R. Astron. Soc. **374**, 256 (2007)

Santiago-Prieto, I., Heng, I.S., Jones, D.I., Clark, J.: Proceedings of the 9th Amaldi Conference and NRDA Meeting, Cardiff, 2011, J. Phys. Conf. Ser. **363**, 012042 (2012)

Sawyer, R.F.: Phys. Rev. D **39**, 3804 (1989)

Schenk, A.K., Arras, P., Flanagan, E.E., Teukolsky, S.A., Wasserman, I.: Phys. Rev. D **65**, 024001 (2002)

Sedrakian, D.M., Benacquista, M., Shahabassian, K.M., Sadoyan, A.A., Hairapetyan, M.V.: Astrophysics **46**, 445 (2003)

Shapiro, S.L.: Astrophys. J. **544**, 397 (2000)
Shapiro, S.L., Teukolsky, S.A.: Black Holes, White Dwarfs and Neutron Stars: The Physics of Compact Objects. Wiley, Weinheim (1986)
Shibata, M., Karino, S., Eriguchi, Y.: Mon. Not. R. Astron. Soc. **334**, L27 (2002)
Shibazaki, N., Murakami, T., Shaham, J., Nomoto, K.: Nature **342**, 656 (1989)
Shternin, P.S., Yakovlev, D.G.: Phys. Rev. D **78**, 063006 (2008)
Sidery, T., Passamonti, A., Andersson, N.: Mon. Not. R. Astron. Soc. **405**, 1061 (2010)
Singh, A.: Phys. Rev. **D95**, 024022 (2017)
Sotani, H.: Phys. Rev. **D92**, 104024 (2015)
Sotani, H., Kokkotas, K.D., Stergioulas, N.: Mon. Not. R. Astron. Soc. **385**, L5 (2008)
Sotani, H., Nakazato, K., Iida, K., Oyamatsu, K.: Phys. Rev. Lett. **108**, 201101 (2012)
Sotani, H., Nakazato, K., Iida, K., Oyamatsu, K.: Mon. Not. R. Astron. Soc. **434**, 2060 (2013a)
Sotani, H., Nakazato, K., Iida, K., Oyamatsu, K.: Mon. Not. R. Astron. Soc. **428**, L21 (2013b)
Stergioulas, N.: Living Rev. Relativ. **6**, 109 (2003)
Stopnitzky, E., Profumo, S.: Astrophys. J. **787**, 114 (2014)
Strohmayer, T., Mahmoodifar, S.: Astrophys. J. Lett. **793**, L38 (2014)
Strohmayer, T.E., Watts, A.L.: Astrophys. J. **632**, L111 (2005)
Strohmayer, T.E., Watts, A.L.: Astrophys. J. **653**, 593 (2006)
Tayler, R.J.: Mon. Not. R. Astron. Soc. **161**, 365 (1973)
Thompson, C., Duncan, R.C.: Mon. Not. R. Astron. Soc. **275**, 255 (1995)
Thompson, C., Duncan, R.C.: Astrophys. J. **561**, 980 (2001)
Thorne, K.S.: Rev. Mod. Phys. **52**, 299 (1980)
Turolla, R., Zane, S., Watts, A.: Rept. Prog. Phys. **78**, 116901 (2015)
Uryu, K., Gourgoulhon, E., Markakis, C., Fujisawa, K., Tsokaros, A., Eriguchi, Y.: Phys. Rev. **D90**, 101501 (2014)
Ushomirsky, G., Rutledge, R.E.: Mon. Not. R. Astron. Soc. **325**, 1157 (2001)
Ushomirsky, G., Cutler, C., Bildsten, L.: Mon. Not. R. Astron. Soc. **319**, 902 (2000)
van Eysden, C.A., Melatos, A.: Classical Quantum Gravity **25**, 225020 (2008)
van Hoven, M., Levin, Y.: Mon. Not. R. Astron. Soc. **410**, 1036 (2011)
van Hoven, M., Levin, Y.: Mon. Not. R. Astron. Soc. **420**, 3035 (2012)
Vigelius, M., Melatos, A.: Mon. Not. R. Astron. Soc. **386**, 1294 (2008)
Vigelius, M., Melatos, A.: Mon. Not. R. Astron. Soc. **395**, 1985 (2009)
Wang, Y.-M.: Astrophys. J. Lett. **449**, L153 (1995)
Warszawski, L., Melatos, A.: Mon. Not. R. Astron. Soc. **415**, 1611 (2011)
Warszawski, L., Melatos, A.: Mon. Not. R. Astron. Soc. **423**, 2058 (2012)
Warszawski, L., Melatos, A.: Mon. Not. R. Astron. Soc. **428**, 1911 (2013)
Warszawski, L., Melatos, A., Berloff, N.G.: Phys. Rev. B **85**, 104503 (2012)
Watts, A.L., Reddy, S.: Mon. Not. R. Astron. Soc. **379**, L63 (2007)
Watts, A.L., Andersson, N., Jones, D.I.: Astrophys. J. **618**, L37 (2005)
Watts, A.L., et al.: Rev. Mod. Phys. **88**, 021001 (2016)
Wen, D.-H., Newton, W.G., Li, B.-A.: Phys. Rev. D **85**, 025801 (2012)
Wentzel, D.G.: Astrophys. J. Suppl. **5**, 187 (1960)
Woods, P.M., Thompson, C.: Compact Stellar X-Ray Sources, p. 547. Cambridge University Press, Cambridge (2006)
Yoshida, S.: Phys. Rev. D **86**, 104055 (2012)
Yoshida, S., Lee, U.: Astrophys. J. **567**, 1112 (2002)
Zhang, B.: Astrophys. J. **763**, L22 (2013)
Zhang, B., Mészáros, P.: Astrophys. J. **552**, L35 (2001)
Zimmermann, M., Szedenits, E.: Phys. Rev. **D20**, 351 (1979)
Zink, B., Korobkin, O., Schnetter, E., Stergioulas, N.: Phys. Rev. D **81**, 084055 (2010)
Zink, B., Lasky, P.D., Kokkotas, K.D.: Phys. Rev. **D85**, 024030 (2012)

Chapter 13
Universal Relations and Alternative Gravity Theories

Daniela D. Doneva and George Pappas

Abstract This is a review with the ambitious goal of covering the recent progress in: (1) universal relations (in general relativity and alternative theories of gravity), and (2) neutron star models in alternative theories. We also aim to be complementary to recent reviews in the literature.

13.1 Introduction

Neutron stars are complicated astrophysical objects with a lot of physics involved in their description, as one can see from the previous chapters in this volume. But when it comes to their overall structure and their stationary properties, gravity is the most important player. To determine the structure of a neutron star one specifies an equation of state and then solves the Einstein field equations together with the hydrostatic equilibrium equation (see Friedman and Stergioulas 2013; Paschalidis and Stergioulas 2016). The resulting models and their structure depend on the specific choice of the equation of state. At the moment there exists a large variety of realistic equations of state that come from different nuclear physics models (a result of our lack of knowledge with respect to the properties of matter at supra-nuclear densities), which in turn result to quite a large variety in neutron star models as Fig. 13.1 shows. Specifying the equation of state is therefore a very hot topic from both astrophysical perspective as well as nuclear physics perspective.

D. D. Doneva (✉)
Theoretical Astrophysics, Eberhard Karls University of Tübingen, Tübingen, Germany

INRNE - Bulgarian Academy of Sciences, Sofia, Bulgaria
e-mail: daniela.doneva@uni-tuebingen.de

G. Pappas (✉)
Departamento de Física, CENTRA, Instituto Superior Técnico, Universidade de Lisboa, Lisboa, Portugal

© Springer Nature Switzerland AG 2018   737
L. Rezzolla et al. (eds.), *The Physics and Astrophysics of Neutron Stars*,
Astrophysics and Space Science Library 457,
https://doi.org/10.1007/978-3-319-97616-7_13

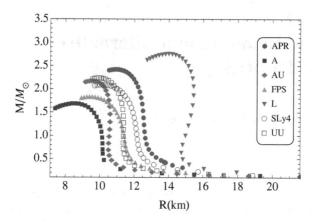

Fig. 13.1 Mass-Radius relation of non-rotating neutron stars for different equations of state. The mass is given in solar masses $M_\odot$

This large variety was always considered to be a natural outcome of having a lot of degrees of freedom available in the description of the equation of state and for this reason neutron stars were not considered to be useful objects for testing our theories of gravity. The conventional wisdom was that neutron stars were infested by matter that would hide any possible signal coming from modifications from general relativity. Nevertheless, there always existed an interest in finding ways to describe neutron stars in ways that were not very sensitive on the specific choice of the equation of state. The motives behind such attempts are twofold. On the one hand, having observables that are insensitive to the specifics of the equation of state can be very advantageous in our attempts to measure astrophysically the various properties of neutron stars, by reducing for example modelling uncertainties. This can be also important in solving the inverse problem of determining the equation of state from measurements of the bulk properties of neutron stars. On the other hand, such observables would open the window to observing deviations that are due to gravity and possible modifications to general relativity and not the equation of state uncertainty. All the more so, since neutron stars can be in some cases a much more suitable object for testing such modifications, as we shall discuss shortly.

Alternative theories of gravity attracted significant attention in the past decades. The reasons for this come both from theory and observations. Modifications of Einstein's theory of gravity for example are often employed as alternative explanation for the dark mater phenomenon or the accelerated expansion of the universe. The idea is that instead of attributing the accelerated expansion to unknown constituents of the Universe with rather unusual and strange properties such as dark energy, one can attribute it to our lack of understanding of gravity. Besides the observations, there are strong theoretical motivations for modifying general relativity. The standard Hilbert–Einstein action is by no means the only possible one but instead the simplest. That is why different generalizations are possible coming for example from the theories trying to unify all the interactions.

They predict the existence of a scalar field that can be considered as a new mediator of the gravitational interaction in addition to the spacetime metric. On the other hand the quantum corrections in the strong field regime which are introduced in order to make the theory renormalizable or to cure the singularities in the solutions, naturally lead to the fact that the Einstein–Hilbert action is supplemented with higher order terms. Last but not least, as the experience shows, studying in detail theories of gravity more general than Einstein's theory can often give as a better insight to general relativity itself.

Every viable theory of gravity should be consistent with the observations at all scales and regimes (Will 2014). The weak field experiments give strong constraints on the parameters of the theory, but the strong field regime is essentially unconstrained and leaves a lot of space for modifications. Moreover, there are theories of gravity that are completely indistinguishable from general relativity for weak fields but can lead to large deviations for strong fields. Therefore, constraining the strong field regime of gravity or even detecting possible deviations from pure Einstein's theory is a difficult but very important task.

Neutron stars and black holes in alternative theories of gravity have been explored for several decades because they offer the possibility to test the strong field regime of gravity at astrophysical scales. Black holes, though, have an important disadvantage because of the no-hair theorems. According to these theorems, the black hole solutions in some classes of alternative theories of gravity are the same as in general relativity.[1] Naturally, this poses obstacles to testing such theories of gravity via the astrophysical observations of black holes. Neutron stars on the other hand do not fall in the scope of the no-hair theorems because of the presence of matter. Thus, alternative theories of gravity can lead to large deviations from Einstein's theory. The neutron star matter, though, is a double edged sword—indeed it offers the possibility to have compact star solutions different from pure general relativity, but on the other hand the uncertainties in the high density nuclear matter equation of state are large. Moreover, there is a degeneracy between effects coming from modifying the equation of state and the theory of gravity. That is why testing the strong field regime of gravity via neutron star observations is also a subtle task. In order to address these problems one has to build a dense net of neutron star models and astrophysical predictions in various alternative theories that can help us either break the degeneracy or find effects that are stronger pronounced than the equation of state uncertainty. Again, as in general relativity, universal relations can play a significant part here by taking the subtleties of the equation of state out of the picture so that we can identify effects and deviations that are only due to modifications in gravity.

Neutron star solutions (both static and rotating) have been constructed in various alternative theories of gravity. The literature on the subject is vast and the size of the present chapter is clearly not enough to cover thoroughly the subject. Moreover,

---

[1]The perturbations, though, could be different and thus used for testing alternative theories of gravity.

we aim at a more pedagogical review. This is why instead of reviewing the whole subject in detail, we will concentrate on certain classes of alternative theories of gravity. Since the goal of the chapter is to cover both neutron stars in alternative theories and universal relations, a natural choice would be to focus mainly on those theories, in which universal relations have been derived. In the present chapter we will concentrate on scalar-tensor theories (STT), $f(R)$ theories, Einstein–dilaton–Gauss–Bonnet (EdGB) theories and the Chern–Simons (CS) theories of gravity. As a matter of fact they are amongst the most natural and widely used generalizations of Einstein's gravity and most of the neutron star studies in the literature have been done exactly in these alternative theories. In addition all of them fall into the same class of modifications of Einstein's theory, i.e. theories for which dynamical scalar fields are included as mediators of the gravitational interaction in addition to the metric tensor. A very nice recent review that covers a much larger spectrum of theories can be found in Berti et al. (2015).

In what follows we will start by discussing in Sect. 13.2.1 the history and present status of universal relations for neutron stars in general relativity. We will proceed in Sect. 13.2.2 to briefly review neutron stars in various modifications of general relativity. In Sect. 13.2.3 we will discuss the current status on the various extensions of the universal relations to alternative theories of gravity. In Sect. 13.3 we will talk about the work that needs to be done or could be done in the future as well as the various challenges that we will face in extending our current results. Finally we will close with some brief overview of the Chapter.

We will use geometric units ($G = c = 1$) throughout, unless it is specifically mentioned otherwise.

13.2 Review of Past Work

13.2.1 Universal Relations in GR

13.2.1.1 Prehistory

Even though the subject of neutron star universal relations, i.e., equation of state independent or insensitive relations, has received a lot of attention in recent years, as was mentioned above it has a longer history. Some first results have been presented in the literature already since the 90s when it was recognised in Lattimer and Yahil (1989) that the binding energy BE $\equiv (\mathcal{N} m_n - M)$ of a neutron star, where $\mathcal{N}$ is the nucleons number and m_n the corresponding nucleons mass, expressed in terms of the stellar compactness $\mathcal{C} \equiv M/R$ (the mass here is in geometric units) is insensitive to the equation of state. A later improved expression between the two quantities was given by Lattimer and Prakash (2001) and reads,

$$\mathrm{BE}/M = (0.60 \pm 0.05)\mathcal{C}(1 - \mathcal{C}/2)^{-1}. \tag{13.1}$$

Such an expression is motivated by the fact that analytic solutions like that of the incompressible fluid (ρ = const.), the Buchdahl solution (constructed using the equation of state $\rho = 12\sqrt{p_* P} - 5P$, where p_* is some constant with dimensions of pressure) or the Tolman VII model (which is constructed by assuming that inside the star the density goes like $\rho = \rho_c \left(1 - (r/R)^2\right)$, where ρ_c is the central density of the star), give similar expansions in terms of the compactness (see Lattimer and Prakash 2001).

On a different direction, on the topic of asteroseismology, in the late 90s it was recognised by Andersson and Kokkotas (1996); Andersson and Kokkotas (1998) that the f-mode and w-modes exhibit some equation of state independent behaviour. In particular it was shown that the f-mode frequency (in kHz) is related to the average density of a neutron star following the relation

$$\frac{\omega_f}{2\pi} = 0.78 + 1.635 \left(\frac{M_{1.4}}{R_{10}^3}\right)^{1/2}, \tag{13.2}$$

where different equations of state have a relative small spread around this common fit. In the above expression the mass is measured in units of $1.4 M_\odot$ and the radius of the star in units of 10 km. The f-mode is related to fluid motion inside the star that takes place in dynamic time scales. It is reasonable therefore to assume that the f-mode frequency will be related to the characteristic dynamic time which is proportional to the square root of the average density, $\tau_{dyn} \sim \sqrt{\bar{\rho}}$. It was also shown that the damping time (in s) of the f-mode, when scaled in the right way, is related to the compactness in a linear way,

$$\left(\tau_f \frac{M_{1.4}^3}{R_{10}^4}\right)^{-1} = 22.85 - 14.65\, \mathcal{C}_0, \tag{13.3}$$

where again the masses and radii are measured using the same units as above (this is the case also for $\mathcal{C}_0 \equiv M_{1.4}/R_{10}$). Finally for the first w-mode the corresponding relations for the frequency (in kHz) and the damping time (in ms) are,

$$R_{10}\left(\frac{\omega_w}{2\pi}\right) = 20.92 - 9.14\, \mathcal{C}_0, \tag{13.4}$$

$$M_{1.4}(\tau_w)^{-1} = 5.74 + 103\, \mathcal{C}_0 - 67.45\, \mathcal{C}_0^2, \tag{13.5}$$

where the mass and radius units are as before.

Returning to the topic of the structure of neutron stars, in the 2000s it was found by Lattimer and Prakash (2001); Bejger and Haensel (2002); Lattimer and Schutz (2005) that there is an equation of state insensitive relation between a neutron star's moment of inertia and its compactness,

$$\frac{I}{MR^2} = (0.237 \pm 0.008)(1 + 2.84\mathcal{C} + 18.9\mathcal{C}^4), \tag{13.6}$$

where the compactness is expressed in terms of the mass in geometric units. This expression is again motivated by the behaviour of the moment of inertia in the various analytic models, such as the Tolman VII model. Finally, another instance where relations of this sort were studied is in Urbanec et al. (2013) where for slowly rotating neutron stars, relations between the reduced moment of inertia $I/(MR^2)$ and the inverse compactness $x = 1/(2C)$ where derived in addition to a new relation between the reduced quadrupole $\tilde{q} \equiv QM/J^2$ and x. This last relation could be of additional interest since it distinguishes between neutron stars and quark stars. We give here a fitting formula for neutron stars' $\tilde{q}$ in terms of the compactness,

$$\tilde{q} = -0.2588C^{-1} + 0.2274C^{-2} + 0.0009528C^{-3} - 0.0007747C^{-4}, \quad (13.7)$$

as it is given in Yagi and Yunes (2017a).

One will notice a common theme in all of these relations, which is that the various quantities are expressed in terms of the compactness. Using the compactness was a sensible first choice for parameterising the various properties of neutron stars, since it is a dimensionless quantity that is representative of the overall structure of the star. It was a sensible choice also because, in results from analytic models, relations between various quantities would usually be expressed in terms of the compactness which would be a measure of how relativistic a particular model is.[2] One could think of these results as precursors of the later proliferation of universal relations. In what follows we will give an incomplete list of some of these relations. The interested reader can complement the material presented here with a recently published review on the subject (Yagi and Yunes 2017a).

13.2.1.2 I-Love-Q

Yagi and Yunes in 2013 (Yagi and Yunes 2013a; Yagi and Yunes 2013b), while studying slowly rotating and tidally deformed neutron and quark stars, found that there exist relations between the various pairs of the following three quantities, the moment of inertia I, the quadrupole moment Q and the quadrupolar tidal deformability or Love number λ, which are insensitive to the choice of the equation of state for both neutron as well as quark stars. We will present here a very brief outline of their calculation which comprises of two parts, first the calculation of the slowly rotating models and then the calculation of the tidally deformed models.

Initial I-Love-Q Formulation The slowly rotating models were calculated using the Hartle and Thorne slow rotation formalism (Hartle 1967; Hartle and Thorne 1968). In order to construct a slowly rotating model, one first calculates a non-rotating spherically symmetric model that has some mass M_* and radius R_*. The structure of the star is described by the Tolman–Oppenheimer–Volkoff (TOV)

---

[2]The newtonian limit is for $C \to 0$, while the relativistic limit is for $C \to 1/2$.

equation, while the metric has the form

$$ds^2 = -e^{\nu(r)}dt^2 + e^{\lambda(r)}dr^2 + r^2(d\theta^2 + \sin^2\theta d\phi^2), \tag{13.8}$$

where we can define the mass function $M(r) = r\left(1 - e^{-\lambda(r)}\right)/2$. The spacetime outside the star's surface is the Schwarzschild spacetime. From the mass and the radius of the star one calculates an angular frequency scale $\Omega_K = \sqrt{M_*/R_*^3}$ which is then used to introduce a small expansion parameter $\epsilon \equiv \Omega_*/\Omega_K$ that characterises rotation. Then one introduces corrections to the geometry due to rotation in the form of perturbations up to second order in the rotation,

$$ds^2 = -e^{\nu(r)}\left[1 + 2\epsilon^2\left(h_0 + h_2 P_2\right)\right]dt^2 + \frac{1 + 2\epsilon^2\left(m_0 + m_2 P_2\right)/\left(r - 2M(r)\right)}{1 - 2M(r)/r}dr^2$$
$$+r^2\left[1 + 2\epsilon^2 K_2 P_2\right]\left[d\theta^2 + \sin^2\theta\left(d\phi - \epsilon\omega dt\right)^2\right], \tag{13.9}$$

where $P_2 = (3\cos^2\theta - 1)/2$ is the second order Legendre polynomial and $\omega_1 \equiv \Omega_K - \omega$ is the angular velocity of the fluid relative to the local inertial frame. The equations for the perturbations are then solved on top of the background spherical non-rotating solution. The resulting perturbed configuration is deformed due to rotation and the deformation of the surfaces of constant density has the form, $\bar{r} = r + \epsilon^2\left(\xi_0(r) + \xi_2(r)P_2\right)$. Since all the perturbations scale with the angular velocity of the star, in practice one has only to calculate the model that rotates at Ω_K for $\epsilon = 1$ and then all the models with values of $\epsilon \in [0, 1]$ follow from that. Of course we should note that the results are accurate for values of ϵ that correspond to models that rotate as fast as a few milliseconds (Berti et al. 2005).

The various quantities of the rotating model will scale with rotation in term of the corresponding quantities of the $\epsilon = 1$ model as, angular velocity: $\Omega = \epsilon\Omega_K$, angular momentum: $J = \epsilon J_K$, stellar radius: $R = R_* + \epsilon^2\delta R_K$, and mass: $M = M_* + \epsilon^2\delta M_K$, and quadrupole moment: $Q = \epsilon^2 Q_K$. The moment of inertia for a rigidly rotating configuration is defined as $I \equiv J/\Omega = J_K/\Omega_K$ which means that for a calculation at the order of ϵ in J, the moment of inertia is independent of ϵ. The angular momentum and the quadrupole moment of the configuration can be evaluated from the form of the metric outside the star and in particular by it's asymptotic expansion in the appropriate coordinate system following Thorne's prescription (Thorne 1980; Quevedo 1990).[3] Therefore, the angular momentum can be calculated by the fact that outside the star the metric function ω_1 has the form

$$\omega_1(r) = \Omega_K - \frac{2J_K}{r^3}, \tag{13.10}$$

---

[3]There will be a further discussion on multipole moments when we talk about the 3-hair relations.

where the angular momentum J_K is a constant calculated by the matching of the interior metric to the exterior metric at the surface of the star. Similarly, in the exterior of the star the $h_2(r)$ perturbation functions has the form

$$h_2(r) = \frac{J_K^2}{M_* r^3}\left(1 + \frac{M_*}{r}\right) + A Q_2^2\left(\frac{r}{M_*} - 1\right), \qquad (13.11)$$

where $Q_2^2(x)$ is the associated Legendre polynomial of the second kind[4] and A is an integration constant that is determined by the matching of the interior solution to the exterior solution. From the asymptotic expansion of $-(1 + g_{tt})/2$, one can read the quadrupole as the coefficient in front of the P_2/r^3 term, which is $Q_K = -J_K^2/M_* - \frac{8}{5}A M_*^3$.

In a similar way to the slowly rotating models, the tidally deformed models are assumed to be slightly perturbed from sphericity due to the presence of an external deforming quadrupolar field. For an $l = 2$, static, even-parity perturbation, the perturbed metric will take the form,

$$ds^2 = -e^{\nu(r)}\left[1 + 2h_2 P_2\right]dt^2 + \frac{1 + 2m_2 P_2}{1 - 2M(r)/r}dr^2 + r^2\left[1 + 2K_2 P_2\right]\left(d\theta^2 + \sin^2\theta d\phi^2\right), \qquad (13.12)$$

where we have introduced again the perturbation functions and assumed zero rotation (Hinderer 2008). The tidal Love number is defined to be the deformability, i.e., the response, of a configuration for a given external deforming force. Specifically, for a given external tidal field $\mathcal{E}^{tid}$ that produces a quadrupolar deformation of the star Q^{tid}, the tidal Love number is defined to be,

$$\lambda \equiv -\frac{Q^{tid}}{\mathcal{E}^{tid}}. \qquad (13.13)$$

In addition to this definition, one can define the tidal apsidal constant and the dimensionless tidal Love number as

$$k_2 \equiv \frac{3}{2}\frac{\lambda}{R_*^5},$$

$$\bar{\lambda} \equiv \frac{\lambda}{M_*^5} = \frac{2}{3}k_2 C^{-5}, \qquad (13.14)$$

respectively. The quadrupolar response and the external quadrupole tidal field can both be extracted from the form of the metric perturbation $h_2(r)$ outside the star. One should be careful though in this case because the exterior to the star is not an asymptotically flat spacetime as it was in the rotating case. Instead, the

---

[4]The polynomial here is defined as $Q_2^2(x) = \frac{3}{2}(x^2 - 1)\ln\left(\frac{x+1}{x-1}\right) + \frac{5x - 3x^3}{x^2 - 1}$

asymptotic behaviour is determined by the external quadrupolar field that produces the deformation. In the exterior of the star the $h_2(r)$ perturbation function has in this case the form,

$$h_2(r) = 2c_1 Q_2^2 \left(\frac{r}{M_*} - 1 \right) + c_2 \left(\frac{r}{M_*} \right)^2 \left(1 - \frac{2M_*}{r} \right), \tag{13.15}$$

where again the constants c_1 and c_2 are integration constants that are determined from the matching conditions at the surface of the star. To identify the external tidal field and the quadrupolar response of the star, one again expands $-(1 + g_{tt})/2$ in powers of r. In this case, the expansion will have additional terms with positive powers of r due to the external tidal field.[5] The expansion will have the form,

$$-\frac{1 + g_{tt}}{2} = -\frac{M_*}{r} - \frac{Q}{r^3} P_2 + \cdots + \frac{1}{3} r^2 \mathcal{E}^{\text{tid}} P_2 + \cdots, \tag{13.16}$$

and by comparing this to the corresponding expansion of $-(1 + g_{tt})/2$, where g_{tt} is evaluated from Eqs. (13.12) and (13.15), one finds that $Q^{\text{tid}} = -\frac{16}{5} M_*^3 c_1$ and $\mathcal{E}^{\text{tid}} = 3M_*^{-2} c_2$. Therefore, from the definition of the tidal love number and the tidal apsidal constant, we have in terms of the integration constants c_1 and c_2 that,

$$\lambda = \frac{16}{15} M_*^5 \frac{c_1}{c_2} \Rightarrow \bar{\lambda} = \frac{16}{15} \frac{c_1}{c_2}, \quad \text{and} \quad k_2 = \frac{8}{5} \frac{c_1}{c_2} C^5, \tag{13.17}$$

where the ratio c_1/c_2 depends on the compactness and the quantity $y = R_* h_2'(R_*)/h_2(R_*)$ at the surface of the star.

Having all the relevant quantities at hand, Yagi and Yunes calculated sequences of neutron star models using various cold realistic equations of state[6] and found that the quantities, normalised moment of inertia: $\bar{I} \equiv I/M_*^3$, normalised Love number: $\bar{\lambda} \equiv \lambda/M_*^5$, and normalised quadrupole: $\bar{Q} \equiv -Q/(\chi^2 M_*^3)$,[7] where $\chi = J/M_*^2$, are related in an equation of state independent way following the relation

$$\ln y_i = a_i + b_i \ln x_i + c_i (\ln x_i)^2 + d_i (\ln x_i)^3 + e_i (\ln x_i)^4, \tag{13.18}$$

where the different coefficients depend on the pair of quantities to be related and are given in Table 13.1. The relations presented here, i.e., the I-Love, I-Q and Q-

---

[5] We should also note that the expansion in this case is not at infinity. It takes place at a buffer region outside the surface of the star and inside a radius given by the external field's characteristic curvature radius.

[6] See also Lattimer and Lim (2013) where the results were extended to a wider range of equations of state. We remind here that these results apply to quark stars as well.

[7] The minus sign here is due to the fact that rotating neutron stars tend to be oblate due to rotation which gives a negative value for the quadrupole. On the other hand, an object that is prolate has a positive quadrupole.

Table 13.1 Numerical coefficients for the fitting formula given in Eq. (13.18)

y_i	x_i	a_i	b_i	c_i	d_i	e_i
$\bar{I}$	λ	1.496	0.05951	0.02238	-6.953×10^{-4}	8.345×10^{-6}
$\bar{I}$	$\bar{Q}$	1.393	0.5471	0.03028	0.01926	4.434×10^{-4}
$\bar{Q}$	λ	0.1940	0.09163	0.04812	-4.283×10^{-3}	1.245×10^{-4}
$\delta\bar{M}$	λ	-0.703	0.255	-0.045	-5.707×10^{-4}	2.207×10^{-4}

This is an updated version of the table as it appears in Yagi and Yunes (2017a), where a very wide range of equations of state has been taken into account. In the table we also give the universal relation for the mass correction $\delta\bar{M}$ presented in Reina et al. (2017)

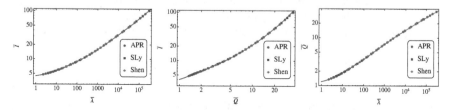

Fig. 13.2 I-Love, I-Q and Q-Love plots for three typical equations of state. The solid lines correspond to the fits given in Table 13.1

Love relations, are equation of state independent in the sense that for any given realistic equation of state, the calculated quantities follow the corresponding fits given by Eq. (13.18) and Table 13.1 to an accuracy better than $\mathcal{O}(1\%)$ in the range of applicability of the fit, which is for neutron stars with masses slightly less than $1 M_\odot$ up to the maximum mass of the given equation of state. The fits with the data from some typical equations of state are given in Fig. 13.2.

As it was previously mentioned, the moment of inertia by definition (within the slow rotation scheme) does not depend on the rotation parameter ϵ, while the normalised quadrupole $-Q/(\chi^2 M_*^3)$ is defined in such a way that the rotation parameter is scaled out. Similarly, the tidal love number is defined in such a way that it is independent of the strength of the external tidal field and characterises the given neutron star model. This means that essentially the quantities $\bar{I}$, λ, and $\bar{Q}$ form one parameter families characterised by the central density of the neutron star models. We should also note that the normalisation of the various quantities in the initial I-Love-Q formulation was performed with the masses M_* of the corresponding spherical configurations. From a practical perspective this could be considered to be problematic, since when one observes a neutron star and measures it's mass, that mass is not the mass M_* of the spherical configuration but instead it is the mass $M = M_* + \epsilon^2 \delta M_K$. Nevertheless, in the slowly rotating and small deformations case $\delta M_K/M_*$ is of the order of 10% and one can also assume values of ϵ up to 10%, which make any deviations from the given relations to be of the order of 10^{-3}. We will revisit this point later on when we will discuss rapidly rotating neutron stars. Nevertheless, recent work (Reina et al. 2017) has shown that the mass correction, normalised as $\delta\bar{M} \equiv M_*^3 \delta M_K/J_K^2$, also follows a universal relation, extending in this way the original family of I-Love-Q relations to include the mass correction as well.

Extensions Beyond the Initial I-Love-Q Formulation The discussion so far has been about unmagnetised isolated neutron stars, in the sense that we have taken into account deformations that are only due to rotation or due to some static external tidal field, the latter of which is not what one would expect if we assumed that the star was part of some binary for example and the source of the tidal field were a companion star. The effects of adding a magnetic field were studied in Haskell et al. (2014) while the effects of having the neutron star being part of a binary system were studied in Maselli et al. (2013).

In Haskell et al. (2014), Haskell et al. investigated three magnetic field configurations. The first was a purely poloidal magnetic field configuration, the second was a purely toroidal magnetic field configuration and finally the third was a twisted-torus configuration. It is known that magnetic fields can cause deformations to neutron stars which depend on the magnetic field configuration. It is also known that neutron stars can be strongly magnetised with magnetic fields at the surface as high as $10^{12}G$ for pulsars and $10^{15}G$ for magnetars, while the magnetic field in the interior can be even stronger than that on the surface. Therefore, for high enough magnetic fields and sufficiently slow rotation, the deformations due to magnetic fields can dominate those due to rotation.

When the magnetic field is purely poloidal, in the Newtonian limit and for an $n = 1$ rotating polytrope, the normalised quadrupole $\bar{Q}$ can be expressed in terms of the normalised moment of inertia $\bar{I}$ as,

$$\bar{Q} \approx 4.9\bar{I}^{1/2} + 10^{-3}\bar{I}\left(\frac{B_p}{10^{12}G}\right)^2\left(\frac{P}{s}\right)^2, \tag{13.19}$$

where the first term is the induced quadrupole due to rotation, while the second term is the induced quadrupole due to the magnetic field, B_p is the field at the pole and P is the rotation period of the star. Similarly for a purely toroidal magnetic field the reduced quadrupole is,

$$\bar{Q} \approx 4.9\bar{I}^{1/2} - 3 \times 10^{-5}\bar{I}\left(\frac{\langle B \rangle}{10^{12}G}\right)^2\left(\frac{P}{s}\right)^2, \tag{13.20}$$

where $\langle B \rangle$ is the field average over the volume of the star.[8] There are a few things that one notices from the above equations. The first is that for the purely toroidal case the magnetic field tends to make the star more prolate in contrast to rotation and the effect the magnetic field has in the purely poloidal case. The second is that the induced quadrupole is proportional to the square of the product $B \times P$ and therefore

---

[8]We should note that the numerical coefficients in these expressions depend on the specific configuration.

the effect is more prominent for larger periods, i.e., slower rotation rates. Finally, one notices that the effect of the magnetic field is suppressed in relation to the effect of rotation by factors of 10^{-3} and 10^{-5} respectively. These observations also hold in the relativistic case studied in Haskell et al. (2014) where it was found that the purely toroidal and purely poloidal magnetic field configurations give an approximately universal relation between $\bar{Q}$ and $\bar{I}$ which agrees with the unmagnetised case.

However, the purely toroidal or purely poloidal configurations are both known to be dynamically unstable (although, the crust could stabilise such configurations as long as it does not break). For this reason a more realistic configuration was also studied in Haskell et al. (2014), that of a twisted torus, where both toroidal and poloidal components of the magnetic field are present. In this case in addition to the strength of the magnetic field and the rotational period, the results also depend on the ratio of the toroidal-to-total magnetic field energy, i.e., on the particulars of the configuration of the magnetic field, while the $\bar{I} - \bar{Q}$ relation also acquires some equation of state dependence. Nevertheless, the I-Love-Q universality is preserved as long as the magnetic field is not too strong, i.e., $B \lesssim 10^{12}G$, and the neutron star is not rotating too slowly, i.e., $P \lesssim 10s$.

In Maselli et al. (2013) Maselli et al. investigate what are the effects on the $\bar{I} - \bar{\lambda}$ relation if one were to assume the more realistic situation of having dynamic tides caused by a companion star in a binary system. The method used to model tidal deformations in compact binaries was the Post-Newtonian-Affine approach. It was shown that the $\bar{I} - \bar{\lambda}$ relation is not the same as the one in the stationary case and that the new relation depends on the inspiral frequency. However, for any given inspiral frequency the $\bar{I} - \bar{\lambda}$ relation is insensitive to the equation of state with an accuracy of a few %. The fits for the $\bar{I} - \bar{\lambda}$ relations for the different gravitational wave frequencies f_{GW} that the system would emit, that are related to the binary inspiral frequency as $f_{GW} = 2f$, are given in Table 13.2. These fits are accurate to within 2% for every frequency, while there is also an overall fit given which is valid up to a gravitational wave frequency of $\sim 900\,\mathrm{Hz}$ and accurate to within 5% for any frequency in that range.

The tidal fields and the corresponding tidal deformations that we have discussed so far are of the so called "gravito-electric" type and they are the relativistic exten-

Table 13.2 Numerical coefficients for the binary $\bar{I} - \bar{\lambda}$ fitting formula, given in Eq. (13.18) with $y_i = \bar{I}$ and $x_i = \bar{\lambda}$

f_{GW}	a_i	b_i	c_i	d_i	e_i
170	1.54	-3.73×10^{-2}	5.49×10^{-2}	-4.78×10^{-3}	1.87×10^{-4}
300	1.58	-6.53×10^{-2}	6.26×10^{-2}	-5.68×10^{-3}	2.26×10^{-4}
500	1.60	-8.34×10^{-2}	6.83×10^{-2}	-6.39×10^{-3}	2.59×10^{-4}
700	1.64	-1.18×10^{-1}	7.89×10^{-2}	-7.69×10^{-3}	3.18×10^{-4}
800	1.68	-1.46×10^{-1}	8.76×10^{-2}	-8.77×10^{-3}	3.68×10^{-4}
Any	1.95	-3.73×10^{-1}	1.55×10^{-2}	-1.75×10^{-3}	7.75×10^{-4}

The different rows correspond to different inspiral frequencies, while the final row is an overall fit. The table is from (Maselli et al. 2013)

sions of their Newtonian counterparts. In general relativity though there can exist "gravito-magnetic" tidal fields and deformations which result to gravitomagnetic Love numbers. In order to allow for the fullest possible effect in the response of a compact object under a gravitomagnetic tidal field one needs to go beyond configurations that are in strict hydrostatic equilibrium (no internal motion of the fluid), i.e., allow for the fluid to be in an irrotational state. This is because one expects that the external gravitomagnetic tidal field in a binary system would also drive internal fluid motion in the star. Allowing for irrotational fluid flows inside the stars gives a dramatically different behaviour for the Love numbers with respect to the restricted hydrostatic case. The magnetic Love numbers for irrotational stars were studied by Landry and Poisson in (Landry and Poisson 2015) (where one can find further references to previous work). Furthermore Delsate explored in Delsate (2015) the existence of a universal relation between the $\ell = 2$ gravitomagnetic Love number and the moment of inertia. As in the case of gravitoelectric Love numbers, the gravitomagnetic Love numbers can be defined as the coefficients σ_ℓ that relate the tidally induced response for a given external tidal field. Delsate found that in the case of irrotational stars there exists a $|\bar{k}_2^{mag}| - \bar{I}$ universal relation, where $\bar{k}_2^{mag} = \bar{\sigma}_2(2C)^4$ is analogous to the tidal apsidal constant for the gravitoelectric Love number, which is equation of state independent with a variation less than 5%.

Another significant extension of the initial I-Love-Q analysis (performed in slow rotation) was the investigation of the $\bar{I} - \bar{Q}$ relation for rapidly rotating neutron stars with rotation rates as high as that of the mass shedding limit. Doneva et al. (2013a) explored the $\bar{I} - \bar{Q}$ relation for neutron and quark stars rotating at different rotation rates, from a few hundred Hz up to kHz frequencies close to the Kepler limit, using models numerically constructed with the RNS numerical code (Stergioulas and Friedman 1995). They found that the $\bar{I} - \bar{Q}$ relation changes with rotation frequency f and in addition for higher frequencies there is an increasing scattering of the different equations of state. Nevertheless Doneva et al. (2013a) produced a general fit that captures the behaviour of neutron star models constructed with modern realistic equations of state and for different rotation rates that has the form,

$$\ln \bar{I} = a_0 + a_1 \ln \bar{Q} + a_2 (\ln \bar{Q})^2, \quad \text{where } a_i = c_0 + c_1 \left(\frac{f}{\text{kHz}}\right) + c_2 \left(\frac{f}{\text{kHz}}\right)^2 + c_3 \left(\frac{f}{\text{kHz}}\right)^3.$$
(13.21)

The coefficients of the fit are given in Table 13.3, while in Fig. 13.3 one can see the fit for the rapidly rotating models for different frequencies compared to the

Table 13.3 Numerical coefficients for the fitting formula given in Eq. (13.21)

a_i	c_0	c_1	c_2	c_3
a_0	1.406	−0.051	0.154	−0.131
a_1	0.489	0.183	−0.562	0.471
a_2	0.098	−0.136	0.463	−0.273

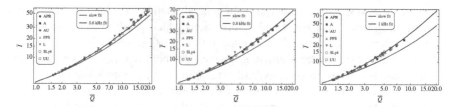

Fig. 13.3 I-Q plots for three rotation frequencies and various equations of state. The solid red lines correspond to the I-Q fit given in Doneva et al. (2013a), while the solid blue correspond to the slow rotation I-Q fit given in Table 13.1. One can see that there is some scatter between the different equations of state that was not present in the I-Q relation of Fig. 13.2

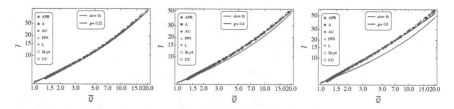

Fig. 13.4 I-Q plots for three spin parameters and various equations of state. The solid red lines correspond to the I-Q fit given in Pappas and Apostolatos (2014), while the solid blue correspond to the slow rotation I-Q fit given in Table 13.1. Unlike in Fig. 13.3 there is no scattering between the equations of state when the parameterisation is done with the spin parameter $\chi = J/M^2$

slow rotation fit. One can also notice the scattering around the fit for the different equations of state and the different rotation frequencies.

A different approach in parameterising rotation for rapidly rotating models was taken by Pappas and Apostolatos in Pappas and Apostolatos (2014), where the rotation was parameterised in terms of the spin parameter $\chi = J/M^2$, a dimensionless quantity, instead of the rotation frequency f. By comparing models of equal spin parameter, Pappas and Apostolatos found that the different equations of state exhibit no noticeable scattering as one can see in Fig. 13.4. They also produced a fit for the I-Q relation in terms of the spin parameter, which has the form,

$$\sqrt{\bar{I}} = 2.16 + (0.97 - 0.14\chi + 1.6\chi^2)\left(\sqrt{\bar{Q}} - 1.13\right) + (0.09 + 0.23\chi - 0.54\chi^2)\left(\sqrt{\bar{Q}} - 1.13\right)^2,$$

$$(13.22)$$

and is accurate to better than 1%. What becomes clear from the two different approaches is that the "universality" of the behaviour is sensitive to the choice that one makes in parameterising an effect. In this case, parameterising rotation with the spin parameter preserves the equation of state independence of the description, while the rotation frequency breaks it. Along these lines, Chakrabarti et al. explored in Chakrabarti et al. (2014) different ways of parameterising rotation so as to preserve the equation of state independence of the I-Q relation. Specifically, in addition to using the spin parameter, Chakrabarti et al. also explored relations that are parameterised with respect to $R \times f$ and $M \times f$, where R and M are the radii and masses of the corresponding models, and found that these dimensionless quantities are also good parameters for preserving equation of state independence, although the best choice remains the spin parameter.

At this point there is one more thing that we should note, with respect to an earlier discussion on the normalisation of the various quantities in the initial slow rotation approach and in view of the results from rapid rotation. As one can see in Fig. 13.4, the models with spin parameter $\chi = 0.22$ (left plot) are very close to the curve for the slow rotation I-Q fit.[9] But the normalisation for the moment of inertia and the quadrupole in this case is performed with the total mass of the numerically constructed model. Nevertheless, the agreement with the slow rotation results and the normalisation with respect to the non-rotating TOV mass is very good, providing thus more credibility to our earlier argument on the insensitivity of the result to whether one uses the TOV or the corrected mass in normalising the initial I-Love-Q relations.

Applications The existence of the I-Love-Q universal relations is of theoretical interest by itself, but the most interesting aspect is the potential applications of such relations. There are two ways in which these relations can be useful: (1) one could use them to make indirect measurements of quantities that are difficult to measure, i.e., assume that the relations hold and use them as tools to do physics and astrophysics, or (2) test their validity by measuring more than one of these quantities and in this way use them to test the assumptions on which the I-Love-Q relations are based. Both of these are very interesting prospects.

(1) Assuming the validity of the I-Love-Q relations, one could for example use a binary pulsar system to measure the moment of inertia (see for example Lattimer and Schutz 2005) and from that, using the I-Q relation infer the quadrupole of the neutron star. Such a quadrupole measurement together with the simultaneous measurement of the mass and the angular momentum of the neutron star could be used to constrain the equation of state, by taking advantage of the fact that different equations of state constitute a different surface in an $(M, \chi, \bar{Q})$ parameter space, as described in Pappas and Apostolatos (2014). Another application could be to use the Q-Love relation to break degeneracies between individual spins and quadrupoles[10] in the analysis of the waveforms of the gravitational waves emitted from the inspiral of neutron star binaries, along the lines given in Yagi and Yunes (2013a); Yagi and Yunes (2013b).

---

[9] In Chakrabarti et al. (2014) the authors arrive at the same result with χ values as low as 0.1, which get even closer to the slow rotation fit.

[10] The degeneracy is in the gravitational wave phase where a spin-spin coupling term has a contribution at the same order as the quadrupole term.

Returning to the equation of state measurement front, another application of I-Love-Q relations, as described by Silva et al. (2016), could be the estimation of parameters of the equation of state from electromagnetic observations of binary pulsars or gravitational wave observations of binary inspirals, from systems where the members are low mass neutron stars. Specifically, Silva et al. (using various equations of state) found that for low mass neutron stars, quantities like $\bar{I}$, $\bar{Q}$ and $\bar{\lambda}$ can be fitted by simple functions of the central density ρ_c and of the dimensional parameter $\eta = \left(K_0 L^2 \right)^{1/3}$, where K_0 is the incompressibility of symmetric nuclear matter and L is the slope of the symmetry energy at saturation density (all three parameters have units of energy). A measurement of any two of $\bar{I}$, $\bar{Q}$ or $\bar{\lambda}$ could be used to constrain η and ρ_c and in this way constrain the parameters of the equation of state. Alternatively, the measurement of one of the quantities and the use of the I-Love-Q relations could also provide similar constraints. Finally, one could use the simultaneous measurement of two or more of these quantities, the I-Love-Q relations and the relations in Silva et al. (2016) to perform consistency checks on the assumptions entering the modelling of the equation of state.

(2) Testing the validity of the I-Love-Q relations offers another possibility for testing the equation of state. As it has been demonstrated (Yagi et al. 2014a; Sham et al. 2015; Chan et al. 2015), the universality of the I-Love-Q relations comes from properties of the equation of state that are related to the fact that the equation of state for compact objects is close to being incompressible. Deviations from being incompressible would change the different relations and in principle could also introduce some spin dependence. Therefore, by measuring more than one of the $\bar{I}$, $\bar{Q}$ or $\bar{\lambda}$ one could test the validity of the I-Love-Q relations and in this way test whether the equation of state is close to our current models or not. One possibility for measuring more than one quantities is to combine astrophysical observations with gravitational wave observations. So for example, pulsar timing could provide a measurement of the moment of inertia while gravitational wave observations could provide a measurement of the tidal Love number (Yagi and Yunes 2013a). Another possibility has been recently proposed (Chirenti et al. 2017), where the analysis of gravitational waves from highly eccentric binary systems, where the f-mode is excited by close encounters between the members of the binary, could provide simultaneous measurements of the masses, moments of inertia, and tidal Love numbers of the members of the system.

Another prospect is to test general relativity by testing the I-Love-Q relations. In principle one would expect that for different theories of gravity, neutron stars could follow different I-Love-Q relations or no I-Love-Q relations. Therefore by measuring combinations of the $\bar{I}$, $\bar{Q}$ or $\bar{\lambda}$, one could test for deviations from general relativity or even identify an alternative theory of gravity. These prospects will be further discussed after the discussion of neutron stars in alternative theories of gravity.

13.2.1.3 Multipole Moments 3-Hair

The spacetime around a rotating neutron star is a stationary and axisymmetric spacetime, i.e., there exist two Killing vectors, one timelike ξ^a which characterises the spacetime's symmetry with respect to time translations and one spacelike η^a which characterises the spacetime's symmetry with respect to rotations around an axis which in this case is the stars' axis of rotation. In addition, for isolated stars, it is assumed that the spacetime is asymptotically flat. Under these general assumptions the line element for the spacetime of a rotating neutron star can take the form

$$ds^2 = -e^{2\nu}dt^2 + r^2(1 - \mu^2)B^2e^{-2\nu}(d\varphi - \omega dt)^2 + e^{2a}(dr^2 + \frac{r^2}{1 - \mu^2}d\mu^2)$$

$$(13.23)$$

where $\mu = \cos\theta$ and the metric functions ν, B, ω, and a are all functions of (r, μ). For such a spacetime one can define relativistic multipole moments, which can characterise the structure and the properties of the spacetime, as the multipole moments in Newtonian theory characterise a Newtonian potential. As one would expect, the moments in the two cases are not completely equivalent, with one difference between relativistic and Newtonian moments being that in the relativistic case there exist angular momentum or mass current moments in addition to the usual mass moments. Another one is that the relativistic moments are also sourced by geometry in addition to masses and currents, a consequence of the non-linear nature of the theory.

There exist a few different ways of defining the relativistic multipole moments of a spacetime, such as the Geroch and Hansen algorithm for stationary spacetimes (Geroch 1970a,b; Hansen 1974) which was later customised to axisymmetric spacetimes by Fodor et al. (1989), or alternatively the Thorne algorithm that was mentioned earlier (Thorne 1980). These formalisms give essentially the same moments related by a multiplicative factor as,[11]

$$M_\ell = (2\ell - 1)!!M_\ell^T \text{ and } J_\ell = \frac{2\ell(2\ell - 1)!!}{2\ell + 1}J_\ell^T \qquad (13.24)$$

(for a review see Quevedo 1990). While the Thorne formalism seems easier and more intuitive, because it is based on identifying the coefficients of the asymptotic expansion of the metric functions, which reminds the Newtonian case, in practice finding the appropriate coordinate system for calculating higher order moments can

---

[11] Multipole moments in general are tensorial quantities. The rank of the tensor is given by the order of the corresponding multipole moment. In the case that we also have axisymmetry, the moments are multiples of the symmetric trace-free tensor product of the axis vector n^a with itself. In this case therefore one can define scalar moments as $\mathcal{P}_\ell = 1/(\ell!)\mathcal{P}_{i1...i\ell}n^{i^1}...n^{i^\ell}$, where $\mathcal{P}_\ell$ is the ℓ-th order moment. Since we will discuss about stationary and axisymmetric spacetimes we will only talk about the scalar moments.

be very difficult. On the other hand the Geroch–Hansen formalism, even though it is more involved mathematically, in practice it is more algorithmic. Furthermore, it's reformulation in terms of the Ernst potential in the case of stationary and axisymmetric spacetimes by Fodor et al., makes the calculation of multipole moments even more straightforward. For the different applications in the literature so far, where the multipole moments of numerical neutron star spacetimes have been calculated, people have used the properties of circular equatorial geodesics and their relation to the moments (Laarakkers and Poisson 1999; Pappas and Apostolatos 2012a,b, 2013a; Yagi et al. 2014b). This has been based on Ryan's formalism (Ryan 1995) for relating the various relativistic precession frequencies between them and to other orbital properties, in terms of the multipole moments spectrum of the background spacetime. Although this approach has served us well in calculating the multipole moments of numerical spacetimes, it is somewhat limited and therefore here we will briefly present a more straightforward and rigorous way of calculating the moments of stationary and axisymmetric numerical spacetimes, that has the additional benefit of providing higher order moments in a less computationally expensive way.

As mentioned earlier, the spacetime around a neutron star is a stationary, axisymmetric, vacuum spacetime that admits the Killing vectors ξ^a (timelike) and η^a (spacelike). Using the timelike Killing vector one can define the two scalar quantities f and ψ through the equations,

$$f = -\xi^a \xi_a, \quad \psi_{,a} = \varepsilon_{abcd} \xi^b \xi^{c;d}, \tag{13.25}$$

where f is related to the norm of the Killing vector, while ψ is the scalar twist of the Killing vector. In the Ernst reformulation of the Einstein Field Equations these two scalar quantities define the complex Ernst potential $\mathcal{E} = f + i\psi$. If one has the Ernst potential for a given spacetime in terms of the Weyl–Papapetrou coordinates (ρ, z) then one can define along the axis of symmetry $\rho = 0$ the potential

$$\tilde{\xi}(\bar{z}) = (1/\bar{z}) \frac{1 - \mathcal{E}(\bar{z})}{1 + \mathcal{E}(\bar{z})} = \sum_{j=0}^{\infty} m_j \bar{z}^j, \tag{13.26}$$

in terms of the coordinate $\bar{z} = 1/z$ which is centred at infinity. The different coefficients in the expansion of $\tilde{\xi}(\bar{z})$ are the parameters m_i that give the moments of stationary and axisymmetric spacetimes as they were calculated by Fodor et al. (1989). The process of calculating the moments therefore involves initially two steps, first the calculation of the Ernst potential and then the transformation of the metric coordinates to Weyl–Papapetrou coordinates.

Returning to neutron star spacetimes, the metric functions in (13.23) have an asymptotic expansion outside the star of the form

$$\nu = \sum_{l=0}^{\infty} \left(\sum_{k=0}^{\infty} \frac{\nu_{2l,k}}{r^{2l+1+k}} \right) P_{2l}(\mu), \tag{13.27}$$

$$\omega = \sum_{l=1}^{\infty} \left(\sum_{k=0}^{\infty} \frac{\omega_{2l-1,k}}{r^{2l+1+k}} \right) \frac{d P_{2l-1}(\mu)}{d\mu}, \tag{13.28}$$

$$B = 1 + \left(\frac{\pi}{2} \right)^{1/2} \sum_{l=0}^{\infty} \frac{B_{2l}}{r^{2l+2}} T_{2l}^{1/2}(\mu), \tag{13.29}$$

where $P_l(\mu)$ are the Legendre polynomials, $T_{2l}^{1/2}(\mu)$ are the Gegenbauer polynomials,[12] and the various coefficients are not all independent, with the constraints coming from the field equations in vacuum, which in the frame of the zero angular momentum observers take the form,

$$\mathbf{D} \cdot (BD\nu) = \frac{1}{2} r^2 \sin^2 \theta B^3 e^{-4\nu} \mathbf{D}\omega \cdot \mathbf{D}\omega, \tag{13.30}$$

$$\mathbf{D} \cdot (r^2 \sin^2 \theta B^3 e^{-4\nu} \mathbf{D}\omega) = 0, \tag{13.31}$$

$$\mathbf{D} \cdot (r \sin\theta \mathbf{D}B) = 0, \tag{13.32}$$

where $\mathbf{D}$ is a flat space 3-dimensional derivative operator in spherical coordinates (see Butterworth and Ipser 1976). As an indicative example of how the coefficients are constrained, we have from the field equations that $\nu_{0,1} = 0$, which means essentially that there is no "mass dipole" contribution in the metric function ν, while the next coefficient is constrained to be $\nu_{0,2} = -\frac{1}{3} B_0 \nu_{0,0}$. We should note here that the coefficient $\nu_{0,0}$, as one can see from the asymptotic expansion of ν, gives the mass, i.e., $\nu_{0,0} = -M$, while similarly the angular momentum comes from the asymptotic expansion of ω, which gives $\omega_{1,0} = 2J$. Coefficients such as $\nu_{0,0}, \nu_{2,0}, \omega_{1,0}, \omega_{3,0}$, and so on, as well as all the B_{2l} coefficients, are not constrained by the field equations and are free parameters of the external spacetime that are determined by the characteristics of the fluid configuration.

Having the metric in terms of μ and as an expansion in $1/r$ one can proceed to calculate the scalar twist from Eq. (13.25). The definition of the scalar twist results in two equations, one for $\psi_{,r}$ and one for $\psi_{,\mu}$. Since the calculation is done using an expansion in $1/r$, the resulting scalar twist will be accurate up to some order in $1/r$ and can be evaluated as

$$\psi(r, \mu) = - \int_r^{\infty} \psi_{,r}|_{\mu=\text{const.}} dr, \tag{13.33}$$

where the asymptotic condition is that $\psi(r \to \infty, \mu) = 0$. With the scalar twist at hand, the Ernst potential will be given in terms of the angular coordinate μ and an expansion in inverse powers of r. By further setting $\mu = 1$ we have the Ernst

---

[12] The Gegenbauer polynomials are given by the definition $T_l^{1/2}(\mu) = \frac{(-1)^l \Gamma(l+2)}{2^{l+1/2} l! \Gamma(l+3/2)} (1 - \mu^2)^{-1/2} \frac{d^l}{d\mu^l} (1 - \mu^2)^{l+1/2}$.

potential along the axis. What remains is to express the r coordinate in terms of the Weyl–Papapetrou coordinate z along the axis of symmetry. This is done following the procedure given in Pappas and Apostolatos (2008). By integrating along curves of constant r from the equatorial plane up to the axis of symmetry and inverting the resulting expansion to solve for r we have,

$$z = \int_0^1 d\mu (r^2 B_{,r} + rB) \quad \Rightarrow \quad r = z + \frac{B_0}{z} + \frac{B_2 - B_0^2}{z^3} + \frac{2B_0^3 - 4B_0 B_2 + B_4}{z^5} + \dots$$

$$(13.34)$$

In this way we can calculate the Ernst potential along the axis of symmetry for a spacetime given in the form of a quasi-isotropic metric as in Eq. (13.23). From that Ernst potential one can calculate the moments from the coefficients of the expansion of $\tilde{\xi}$. The resulting first few multipole moments are,

$$M_0 = -v_{0,0}, \quad M_2 = \frac{1}{3}(4B_0 v_{0,0} + v_{0,0}^3 - 3v_{2,0}),$$

$$M_4 = -v_{4,0} - \frac{32}{21} B_0 v_{0,0}^3 - \frac{16}{5} B_0^2 v_{0,0} + \frac{64}{35} B_2 v_{0,0} + \frac{24}{7} B_0 v_{2,0} + \frac{3}{70} v_{0,0} \omega_{1,0}^2 - \frac{19}{105} v_{0,0}^5 + \frac{8}{7} v_{2,0} v_{0,0}^2,$$

$$J_1 = \frac{\omega_{1,0}}{2}, \quad J_3 = -\frac{3}{10}((4B_0 + v_{0,0}^2)\omega_{1,0} - 5\omega_{3,0}),$$

$$J_5 = \frac{5}{2}\omega_{5,0} + \frac{104}{63} B_0 v_{0,0}^2 \omega_{1,0} + \frac{24}{7} B_0^2 \omega_{1,0} - \frac{32}{21} B_2 \omega_{1,0} - \frac{20}{3} B_0 \omega_{3,0} + \frac{25}{126} v_{0,0}^4 \omega_{1,0} - \frac{5}{3} v_{0,0}^2 \omega_{3,0}$$

$$- \frac{5}{21} v_{2,0} v_{0,0} \omega_{1,0} - \frac{1}{28} \omega_{1,0}^3.$$

$$(13.35)$$

Multipole Moments of Neutron Stars and 3-Hair $(M, \chi, \bar{Q})$ Relations The calculation of the moments up to the mass hexadecapole M_4 has been numerically implemented so far for a wide variety of equations of state and for both slowly and rapidly rotating neutron (and quark) stars (Pappas and Apostolatos 2012a,b; Urbanec et al. 2013; Pappas and Apostolatos 2014; Yagi et al. 2014b). One very interesting property of the neutron star multipole moments that has been discovered is that they follow a simple scaling with the spin parameter χ and the mass M that is the same as the one that the moments of rotating black holes follow, i.e., the higher than the angular momentum moments behave as

$$M_2 = a\chi^2 M^3, \quad J_3 = \beta \chi^3 M^4, \quad M_4 = \gamma \chi^4 M^5, \quad (13.36)$$

where for Kerr black holes the mass moments behave as $M_{2n} = (i\chi)^{2n} M^{2n+1}$ and the mass current moments behave as $i J_{2n+1} = (i\chi)^{2n+1} M^{2n+2}$, where i is the imaginary unit. Table 13.4 gives the values of the coefficients a, β, and γ for some typical equations of state. As one can see, while for Kerr black holes these coefficients are equal to ± 1, in the case of neutron stars their magnitude can be quite larger than that. An immediate implication of this is that the spacetime around

Table 13.4 Multipole moments of some typical equations of state

$M/M_\odot$	APR			AU			SLy4			FPS			L			UU		
	a	β	γ	a	β	γ	a	β	γ	a	β	γ	a	β	γ	a	β	γ
0.9	−10.9	−23.8	295.2	−8.2	−17.2	160.9	−10.2	−21.9	252.7	−9.1	−19.7	205.6	−14.	−30.3	466.	−9.2	−19.5	203.1
1.	−9.3	−20.1	216.8	−6.9	−14.4	115.1	−8.6	−18.4	180.8	−7.8	−16.5	149.	−12.3	−26.5	359.	−7.8	−16.5	147.6
1.1	−7.7	−16.6	149.4	−6.	−12.2	85.1	−7.6	−16.1	141.5	−6.6	−13.9	108.3	−10.9	−23.1	280.3	−6.8	−14.1	110.6
1.2	−6.8	−14.2	112.9	−5.2	−10.5	63.9	−6.4	−13.4	99.8	−5.8	−12.2	85.1	−9.3	−19.8	206.4	−5.9	−12.1	83.6
1.3	−6.	−12.6	89.6	−4.4	−8.7	45.8	−5.6	−11.6	77.8	−5.	−10.3	61.8	−8.7	−18.2	177.8	−5.1	−10.3	61.6
1.4	−5.2	−10.5	64.9	−3.9	−7.8	37.2	−5.	−10.2	61.3	−4.3	−8.6	45.1	−7.6	−15.8	136.	−4.6	−9.2	50.4
1.5	−4.7	−9.5	54.	−3.5	−6.6	27.4	−4.3	−8.7	46.2	−3.7	−7.3	33.5	−6.6	−13.8	104.7	−3.9	−7.7	37.1
1.6	−4.1	−8.1	40.2	−3.1	−5.8	22.3	−3.8	−7.4	34.4	−3.2	−6.3	26.5	−6.2	−12.7	90.9	−3.4	−6.5	26.9
1.7	−3.8	−7.4	34.5	−2.7	−4.9	16.1	−3.4	−6.6	28.1	−2.8	−5.4	19.9	−5.5	−11.2	71.5	−3.1	−5.8	22.
1.8	−3.3	−6.2	25.1	−2.5	−4.4	13.7	−3.	−5.7	21.9	−2.4	−4.5	14.5	−5.	−10.1	59.2	−2.7	−4.9	16.3
1.9	−2.9	−5.4	19.7	−2.2	−3.7	10.2	−2.6	−4.8	16.2	−1.9	−3.3	9.5	−4.6	−9.1	49.3	−2.5	−4.5	14.3
2.	−2.7	−4.9	16.3	−2.	−3.4	8.6	−2.3	−4.1	12.3	−1.6	−2.4	5.1	−4.3	−8.5	43.5	−2.2	−3.8	10.5
2.1	−2.4	−4.1	12.3	−1.7	−2.8	6.2	−1.9	−3.4	9.	−1.5	−2.2	4.2	−3.9	−7.7	36.1	−1.9	−3.2	8.

The table has included equations of state that give neutron star models with masses above the maximum mass observed, i.e., $2M_\odot$, while the quantities shown are the coefficients $a = M_2/(\chi^2 M^3)$, $\beta = J_3/(\chi^3 M^4)$, and $\gamma = M_4/(\chi^4 M^5)$. The last three sets of coefficients for FPS (in italics) correspond to models with masses above the maximum mass of the non-rotating configurations. One can easily verify that the coefficients presented here fit the curves in Fig. 13.5

neutron stars can be quite different from that of Kerr black holes. One should notice from Table 13.4 that softer equations of state, like AU, FPS and UU, produce smaller values for the quadrupole and the higher order moments than the stiffer equations of state, like Sly4, APR and L. Also, as one increases the central density and approaches the models close to the maximum mass, then the deviations of the moments from their corresponding Kerr values become smaller, which means that in these cases the neutron star spacetime behaves more like a Kerr spacetime. We should note though that neutron star models never quite reach to the Kerr point where $-a = -\beta = \gamma = 1$.

If we further define the reduced moments, $\bar{Q} \equiv \bar{M}_2 \equiv -\frac{M_2}{\chi^2 M^3}$, $\bar{J}_3 \equiv -\frac{J_3}{\chi^3 M^4}$, and $\bar{M}_4 \equiv \frac{M_4}{\chi^4 M^5}$, then it was shown in Pappas and Apostolatos (2014); Yagi et al. (2014b) that both $\bar{J}_3$ and $\bar{M}_4$ are related to $\bar{Q}$ following universal relations, i.e., relations that are equation of state independent. These relations can be seen plotted in Fig. 13.5. As it can be seen from the two plots, the points seem to follow a power-law with some break appearing towards the lower values of the reduced quadrupole, which correspond to the most compact models close to the maximum mass limit. The two curves can be fitted with an expression of the form,

$$y_i = A + B_1 x^{n_1} + B_2 x^{n_2},\qquad(13.37)$$

where y_i can be either $(\bar{J}_3)^{1/3}$ or $(\bar{M}_4)^{1/4}$, while x is $(\bar{Q})^{1/2}$. The two fits presented in Table 13.5 have been performed for neutron star models only and deviate from the models by less than 4–5%. In Yagi et al. (2014b); Yagi and Yunes (2017a) one can find fits for the relations $\bar{J}_3 - \bar{Q}$ and $\bar{M}_4 - \bar{Q}$ that also include quark star models. Including quark stars results in a slightly wider spread of the data points, but nevertheless the relations between the moments remain remarkably equation of state independent. We should note here that the range of quadrupoles plotted in Fig. 13.5 is between the most compact neutron stars close to the maximum mass

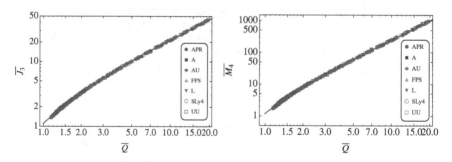

Fig. 13.5 Relations between the higher order moments and the mass quadrupole for different equations of state. The solid lines correspond to the best fits using a double power-law given in Table 13.5. The $\bar{J}_3$-$\bar{Q}$ fit comes from Pappas and Apostolatos (2014), while the $\bar{M}_4$-$\bar{Q}$ fit comes from Pappas (2017)

Table 13.5 Numerical coefficients for the fitting formula given in Eq. (13.37)

y_i	x	A	B_1	B_2	n_1	n_2
$(\bar{J}_3)^{1/3}$	$(\bar{Q})^{1/2}$	−4.82	5.83	0.024	0.205	1.93
$(\bar{M}_4)^{1/4}$	$(\bar{Q})^{1/2}$	−4.749	0.27613	5.5168	1.5146	0.22229

limit down to neutron stars with masses a little less than $1M_\odot$, i.e., within the entire observed mass range.

The 3-hair universal relations have been studied in both the slow rotation limit, using a Hartle and Thorne expansion in terms of rotation up to the fourth order so that M_4 can be included in the calculation, as well as for rapidly rotating neutron stars in full numerical relativity using both LORENE and RNS numerical codes. The results from all approaches have been in full agreement.

Newtonian Insights 3-hair relations where also studied in Newtonian theory by Stein et al. (2014), providing some insight on the possible origin of this universality, as well as a very elegant result. Stein et al. studied the Newtonian limit of multipole moments for a rotating neutron star by using the definitions,

$$M_\ell = 2\pi \int_0^\pi \int_0^{R(\theta)} \rho(r,\theta) P_\ell(\cos\theta) \sin\theta d\theta r^{\ell+2} dr, \qquad (13.38)$$

$$J_\ell = \frac{4\pi}{\ell+1} \int_0^\pi \int_0^{R(\theta)} \Omega\rho(r,\theta) \frac{dP_\ell(\cos\theta)}{d(\cos\theta)} \sin^3\theta d\theta r^{\ell+3} dr, \qquad (13.39)$$

where $R(\theta)$ is the surface of the star as a function of the polar angle θ and $\rho(r,\theta)$ is the density inside the star. Since the star is rotating it is assumed to have axial symmetry and in addition we also assume reflection symmetry with respect to the equatorial plane. These symmetries require that the odd mass moments and the even angular momentum moments are zero. Of course the angular momentum moments have no physical meaning in Newtonian theory but could be formally defined as the Newtonian limit of their relativistic counterparts.

Stein et al. in order to make progress with the analytic calculation of the moments introduced the following two assumptions, (1) the isodensity surfaces inside the star are self-similar ellipsoids with a constant eccentricity, and (2) the density as a function of the isodensity radius $\tilde{r}$ for a rotating configurations is the same as the corresponding radius of a non-rotating configuration with the same volume. For these two assumptions, the eccentricity is defined as $e = \sqrt{1 - (\text{semi-minor axis})^2/(\text{semi-major axis})^2}$, while the isodensity radius is defined as $\tilde{r} \equiv r/\Theta(\theta)$, where

$$\Theta(\theta) = \sqrt{\frac{1-e^2}{1-e^2\sin^2\theta}}. \qquad (13.40)$$

The first assumption is strictly true only for constant density stars, i.e., for $n = 0$ polytropes,[13] which give Maclaurin spheroides. In any other case the eccentricity of the iso-density surfaces varies with the radius inside the star. Nevertheless, for slowly rotating and compact objects, where the the deviations from sphericity are not large and the equation of state is close to the incompressible limit, these assumptions turn out to be good approximations.

Under the aforementioned assumptions, the integrals for the multipole moments separate to an angular and a radial part and can be given by the expressions,

$$M_\ell = 2\pi I_{\ell,3} R_\ell, \quad \text{and} \quad J_\ell = \frac{4\pi \ell}{2\ell + 1} \Omega (I_{\ell-1,5} - I_{\ell+1,3}) R_{\ell+1}, \tag{13.41}$$

where the radial and angular integrals are,

$$R_\ell \equiv \int_0^a \rho(\tilde{r}) \tilde{r}^{\ell+2} d\tilde{r}, \quad \text{and} \quad I_{\ell,k} \equiv \int_{-1}^1 \Theta(\mu)^{\ell+k} P_\ell(\mu) d\mu, \tag{13.42}$$

where $\mu \equiv \cos\theta$ and a is the equatorial radius of the surface. The angular integral depends only on the eccentricity e, while the radial integral depends on the eccentricity and the mass distribution. Using the second assumption, the density profile can be expressed in terms of a non-rotating model. Assuming a polytropic equation of state, the density will be given by solving the Lane-Emden equation

$$\frac{1}{\xi^2} \frac{d}{d\xi} \left(\xi^2 \frac{d\vartheta}{d\xi} \right) + \vartheta^n = 0, \tag{13.43}$$

where n is the polytropic index, the function ϑ is related to the density as $\rho = \rho_c \vartheta^n$, and the radial coordinate is $r = \alpha\xi$ with ξ being dimensionless and α being a length scale that depends on the equation of state.[14] For any choice of the polytropic index corresponds a solution of the Lane-Emden $\vartheta(\xi)$ that has a surface when $\vartheta(\xi_1) = 0$. Substituting such a solution in the integrals for the moments, we have that

$$2\ell + 2 = \frac{(-1)^{\ell+1}}{2\ell + 3} \frac{e^{2\ell+2}}{(1 - e^2)^{\frac{\ell+1}{3}}} \frac{R_{n,2+2\ell}}{\xi_1^{2\ell+4} |\vartheta'(\xi_1)|} \frac{M^{2\ell+3}}{C^{2\ell+2}},$$

$$J_{2\ell+1} = \frac{(-1)^\ell}{2\ell + 3} \frac{2\Omega e^{2\ell}}{(1 - e^2)^{\frac{\ell+1}{3}}} \frac{R_{n,2+2\ell}}{\xi_1^{2\ell+4} |\vartheta'(\xi_1)|} \frac{M^{2\ell+3}}{C^{2\ell+2}}, \tag{13.44}$$

---

[13] A polytropic equation of state has the form $P = k\rho^\Gamma$, where the exponent can be written as $\Gamma = 1 + 1/n$ in terms of the polytropic index n.

[14] For $n = 0$ Lane–Emden admits an exact solution which is $\vartheta = 1 - \frac{1}{6}\xi^2$ with a surface at $\xi_1 = \sqrt{6}$, while for $n = 1$ it admits the solution $\vartheta = \frac{\sin\xi}{\xi}$ with a surface at $\xi_1 = \pi$.

where we have defined the integral $\mathcal{R}_{n,\ell} \equiv \int_0^{\xi_1} \vartheta^n(\xi)\xi^{\ell+2}d\xi$, and the compactness $C = M/\bar{R}$ in terms of the mean radius $\bar{R} = a(1 - e^2)^{1/6}$. The expressions for the multipole moments can be combined so as to eliminate Ω and C giving the final result in terms of the normalised moments $\bar{M}_\ell = (-1)^{\ell/2}\frac{M_\ell}{\chi^\ell M^{\ell+1}}$ and $\bar{J}_\ell = (-1)^{(\ell-1)/2}\frac{J_\ell}{\chi^\ell M^{\ell+1}}$. The 3-hair relations for the Newtonian moments will be,

$$\bar{M}_{2\ell+2} + i\bar{J}_{2\ell+1} = B_{n,\ell}\bar{M}_2^\ell(\bar{M}_2 + i\bar{J}_1), \qquad (13.45)$$

where we note that $\bar{M}_0 = \bar{J}_1 = 1$ by definition. In this expression all the dependence on the equation of state is incorporated in the coefficient $B_{n,\ell}$ which has the form,

$$B_{n,\ell} \equiv \frac{3^{\ell+1}}{2\ell+3}\frac{\mathcal{R}_{n,0}^\ell \mathcal{R}_{n,2\ell+2}}{\mathcal{R}_{n,2}^{\ell+1}}, \qquad (13.46)$$

where we can see that everything depends on the polytropic index n and the corresponding $\vartheta(\xi)$ as well as the order ℓ of the moment. Therefore the universality of the 3-hair relations will depend on how sensitive the coefficients $B_{n,\ell}$ are to different choices of the equation of state. The numerical analysis in Stein et al. (2014) as well as analytic investigations performed by Chatziioannou et al. (2014) (where the coefficients $B_{n,\ell}$ are expanded around the $n = 0$ solution) show that the variation of the coefficients for $\ell \leq 2$ and for polytropic indices in the range of $0 \leq n \leq 1$ is less than 10% around a fiducial value obtained for $n = 0.6$. This means that the Newtonian moments up to M_6 and J_5 satisfy an approximate universal relation that expresses them in terms of the mass, the angular momentum and the quadrupole. Unfortunately, the variation in the coefficients $B_{n,\ell}$ increases with increasing ℓ, therefore for higher order moments the coefficients become more sensitive to the equation of state. These results have been also tested for piecewise polytropes by Chatziioannou et al. (2014) and have been found to be robust.

The question now is, what does the Newtonian analysis tells us about the relativistic 3-hair relations? The first thing we should note is that the Newtonian results are in good agreement with the relativistic results at the low compactness limit (Stein et al. 2014; Yagi et al. 2014a) for a polytropic index $n \approx 0.5$ and the agreement extends to the relativistic polytrope as well, which means that the Newtonian calculation captures the basic elements of the relativistic problem. Furthermore it is known (Lattimer and Prakash 2001) that realistic equations of state behave as polytropes with an effective polytropic index in the range $0 \leq n \leq 1$. As it was shown in Yagi et al. (2014a), the assumption in the Newtonian calculation to which the result is most sensitive is that of the isodensity surfaces being self-similar ellipses. If one were to violate this assumption by introducing a radial dependence to the eccentricity of the form $e(\tilde{r}) = e_0 f(\tilde{r})$, then the result would be to change the form that the $B_{n,\ell}$ coefficients would have, introducing a dependence on the function $f(\tilde{r})$. Specifically, it was shown in Yagi et al. (2014a) that the larger the variation of the eccentricity inside the star, the larger the variation in $B_{n,\ell}$ in the range $n \in (0, 1)$.

For Newtonian polytropes with $n \in (0, 1)$, one can calculate the radial profile of the eccentricity and see that the variation between the centre and the surface of the star ranges from 0% for incompressible $n = 0$ models up to 20% for $n = 1$ models, while for models with $n > 1.5$ it is larger than 40%. Therefore the small variation we see in the Newtonian case for $B_{n,\ell}$ is related to a less than 20% variation in the eccentricity profile. These results, i.e., regarding the eccentricity variation, generally also hold for relativistic polytropic stars, as well as realistic equations of state. The only difference with realistic equations of state is that lower compactness models can have a lower density atmosphere where the equation of state differs from the interior higher density equation of state and in this region the eccentricity varies more drastically. In these cases though, the outer parts of the star contribute a very small percentage of the mass of the star and have a small contribution to the various quantities. Therefore, having almost self-similar ellipses seems to be a key element in having universal 3-hair relations.

The above results can be seen from a slightly different point of view. As it has been presented in Yagi et al. (2014a); Chatziioannou et al. (2014); Sham et al. (2015); Chan et al. (2015), one could alternatively consider the 3-hair universal relations as well as the I-Love relation in terms of how close are realistic equations of state to being incompressible, as it was discussed earlier. Specifically, in Chatziioannou et al. (2014); Sham et al. (2015); Chan et al. (2015) various universal relations were expanded in terms of deviations from the incompressible equation of state ($n = 0$ limit) and it was shown that the results were insensitive to these deviations. We remind here that there is a relation between having an incompressible equation of state or one close to incompressible and the zero or small eccentricity variation inside the star, so in a sense statements about eccentricity variation and incompressibility are equivalent. These conclusions are further strengthened by the analysis of proto-neutron stars constructed with hot equations of state that was performed by Martinon et al. (2014). Martinon et al. found that in the first moments of the proto-neutron stars, where the equations of state have large entropy gradients, which correspond to models with high effective polytropic indices n and therefore deviate significantly from being incompressible, the ellipticity profiles inside the star vary significantly while the corresponding I-Love-Q relations differ from the relations that hold for cold starts.

Applications As with the I-Love-Q relations, the 3-hair relations could be used either as tools, by assuming their validity, to infer the higher order multipole moments from the mass, the angular momentum, and the quadrupole, or alternatively could be tested for their validity against observations.

In the context of general relativity, one could use the 3-hair relations to construct more accurate spacetimes for describing the exterior of rotating neutron stars (Manko et al. 1995; Berti and Stergioulas 2004; Pappas and Apostolatos 2013b; Teichmüller et al. 2011; Pappas 2015; Tsang and Pappas 2016; Pappas 2017). There exist general algorithms for constructing stationary and axisymmetric neutron star spacetimes that can be parameterised by the multipole moments. Having the moments of these spacetimes prescribed using the 3-hair relations one could have an accurate description of the neutron star spacetime that in addition depends on only three essential parameter, i.e., the first three multipole moments. This is of relevance to astrophysical observations in the electromagnetic spectrum. There is a plethora of astrophysical phenomena in the environment of neutron stars. One can observe phenomena related to matter accreting onto neutron stars that are members of X-ray binaries, such as quasi-periodic oscillations of the X-ray flux coming from the disc, the reflection spectra also coming from the disc, X-ray pulse profiles from matter accreting to the surface of the neutron star and so on. The accurate modelling of these phenomena depends on having an accurate description of the spacetime so that the motion of both matter and photons is described accurately. Furthermore, the analytic modelling in terms of only three parameters could improve our capability to solve the inverse problem of determining the equation of state from observations. As it has been discussed for the I-Love-Q case, one could in principle use combinations of observables to estimate the parameters (M, J, Q) for a given neutron star spacetime, which in turn can be used to constrain the equation of state for the matter inside neutron stars (Pappas 2012; Pappas and Apostolatos 2014; Pappas 2015).

The alternative use of the 3-hair relations will be to test their validity. As in the case of I-Love-Q relations this could provide a test for our models of realistic equations of state, but the most interesting possibility will be to test general relativity itself. The last prospect will be further discussed after we have discussed neutron stars in alternative theories of gravity.

13.2.1.4 Universality in Oscillation Frequencies

Asteroseismology is discussed more thoroughly in Chap. 12. Here we will only briefly present results that are related to the equation of state independent description of the various oscillation frequencies of neutron stars. We have already mentioned some early work (Andersson and Kokkotas 1996; Andersson and Kokkotas 1998) that indicated that there could be some equation of state independent description of f-modes and w-modes and their damping times in terms of some neutron star parameters. These results were later extended by including other modern realistic equations of state (Benhar et al. 2004). As it has become clear from our discussion so far, a key element to having a universal description is the appropriate choice of parameters.

In an attempt to extend earlier work, Tsui and Leung (2005a) and Lau et al. (2010) used a different type of normalisation, more similar to the I-Love-Q type of normalisations, that has demonstrated a higher degree of equation of state independence.[15] The approach by Lau et al. (2010), inspired by Lattimer and Schutz

---

[15]One should always keep in mind that at the end of the day what matters most is not exactly the errors of the fitting function but instead the accuracy of solving the inverse problem, i.e. determining the stellar parameters from the observed gravitational wave frequencies and damping times.

(2005) where a universal behaviour for the normalised moment of inertia was observed, turned out to be more useful since both neutron and quark stars could be fitted with a single relation. Lau et al. (2010) investigated the relation between the QNM frequency (both real ω_r and imaginary ω_i parts) of the (fundamental) f-mode oscillations to the mass and moment of inertia of compact stars. In contrast though to Tsui and Leung (2005a) and earlier work, Lau et al. convincingly argue that a better quantity instead of the compactness $\mathcal{C}$ to characterise models of different internal mass profiles would be the inverse square root of the normalised moment of inertia, $\eta \equiv \bar{I}^{-1/2} = \sqrt{M^3/I}$, which they call effective compactness. The reason is that this quantity is on the one hand a measure of the compactness[16] while on the other hand it also takes into account the distribution of mass inside the star and it could in this way counterbalance differences coming from having a stiffer or softer equation of state. The universal relations that Lau et al. found to describe the f-mode in terms of the effective compactness are,[17]

$$M\omega_r = -0.0047 + 0.133\eta + 0.575\eta^2, \quad \text{and} \quad \bar{I}^2\omega_i M = 0.00694 - 0.0256\eta^2,$$
(13.47)

and describe the real and imaginary parts of the $\ell = 2$ f-mode frequency with an accuracy that is better than $1 - 2\%$ for both neutron and quark stars. These two relations could be used to determine the mass and the moment of inertia of a compact object, if the f-mode was detected by gravitational waves. We should note here that this discussion is for non-rotating compact stars. Combining this work to the I-Love-Q results, Chan et al. (2014) extended the relation between the f-mode frequency and the moment of inertia to include also the Love number. They further proceeded to produce relations between higher order oscillation modes and higher order Love numbers (for a review see Yagi and Yunes 2017a).

The f-mode asteroseismology relations were further investigated and more equations of state were included by Blázquez-Salcedo et al. (2014) and Chirenti et al. (2015), where in addition an alternative relation for the f-mode damping time was proposed in terms of the compactness. The asteroseismology relations with w-modes were investigated further by Tsui and Leung (2005b,c) and Blázquez-Salcedo et al. (2013) with a special emphasis on the influence of the presence of hyperons and quarks in the neutron star core in the latter papers.

Up to this point the discussion has been about the quasi-normal modes of non-rotating stars. In the case of rotating stars things become a little more complicated. Due to rotation the frequencies of modes with the same spherical mode number ℓ but opposite azimuthal index $m = \pm|m|$, which correspond to co- and counterrotating modes, separate and become distinct. In this case, one needs to express the modes in the comoving frame in order to remove the complications that the rotation is

---

[16]If we approximate the moment of inertia by MR^2 then we can see that $\eta = M/R = \mathcal{C}$.

[17]See the relevant comment in Chan et al. (2014) for correcting some typos in the numerical coefficients given in Lau et al. (2010).

creating. Rapidly rotating models in the so-called Cowling approximation, where the spacetime is assumed to be frozen, were studied by Gaertig and Kokkotas in Gaertig and Kokkotas (2011) were the aforementioned behaviour was observed. For a star rotating at a frequency Ω, the relation between the mode frequencies in the inertial frame and the comoving frame is $\omega_{in} = \omega_{co} - m\Omega$. Gaertig and Kokkotas observed in Gaertig and Kokkotas (2011) that even though the mode frequencies in the inertial frame showed a large variation with the equation of state, the corresponding frequencies in the comoving frame were quite insensitive to the equation of state. The real part of the quasi-normal modes in the comoving frame for the $m = -2$ stable modes, ω_{co}^s, can be fitted with the expression

$$\frac{\omega_{co}^s}{\omega_0} = 1.0 - 0.27 \left(\frac{\Omega}{\Omega_K} \right) - 0.34 \left(\frac{\Omega}{\Omega_K} \right)^2 , \tag{13.48}$$

where the frequency is normalised with respect to the frequency of the non-rotating models ω_0, while the rotation frequency Ω is normalised with the Kepler limit frequency Ω_K. Similarly, the $m = 2$ potentially unstable modes, ω_{co}^u, frequency is fitted with the expression,

$$\frac{\omega_{co}^u}{\omega_0} = 1.0 + 0.47 \left(\frac{\Omega}{\Omega_K} \right) - 0.51 \left(\frac{\Omega}{\Omega_K} \right)^2 . \tag{13.49}$$

The non-rotating frequency ω_0 can be expressed in terms of the average density and has the form,

$$\frac{\omega_0}{2\pi} = 0.498 + 2.418 \left(\frac{M_{1.4}}{R_{10}^3} \right)^{1/2} , \tag{13.50}$$

where the mass is measured in units of $1.4 M_\odot$ and the radius of the star in units of 10 km. One will notice that Eq. (13.50) is slightly different than Eq. (13.2) from Andersson and Kokkotas (1998). This is because in Gaertig and Kokkotas (2011), the calculations are done in the Cowling approximation which tends to overestimate the values of the frequencies. Also, the reference to stable and potentially unstable modes here is with respect to the Chandrasekhar, Friedman and Schutz (CFS-)instability of non-axisymmetric pulsation modes, which is discussed in Chap. 12. The imaginary part of the quasi-normal modes will give the damping times for the different models. Things in this case are a little more complicated and damping times can be better expressed in terms of the frequencies in the inertial or the comoving frame. For the retrograde branch ($m > 0$) the fit for the damping times has the form,

$$\frac{\tau_0}{\tau} = sgn(\omega_{in}^u) \times 0.256 \left(\frac{\omega_{in}^u}{\omega_0} \right)^4 \times \left[1 + 0.048 \left(\frac{\omega_{in}^u}{\omega_0} \right) + 0.359 \left(\frac{\omega_{in}^u}{\omega_0} \right)^2 \right]^4 , \tag{13.51}$$

where τ_0 is the damping time of the non-rotating models and $sgn(x)$ is the sign function. Similarly for the prograde branch ($m < 0$) the fit for the damping times has the form,

$$\frac{\tau_0}{\tau} = -0.656 \times \left[1 - 7.33 \left(\frac{\omega_{co}^s}{\omega_0} \right) + 14.07 \left(\frac{\omega_{co}^s}{\omega_0} \right)^2 - 9.26 \left(\frac{\omega_{co}^s}{\omega_0} \right)^3 \right].$$

$$(13.52)$$

The damping time of the non-rotating models can be given by the equation

$$\left(\tau_0 \frac{M_{1.4}^3}{R_{10}^4} \right)^{-1} = 22.49 - 14.03 \left(\frac{M_{1.4}}{R_{10}} \right), \qquad (13.53)$$

where again the mass is measured in units of $1.4 M_\odot$ and the radius of the star in units of 10 km. One will notice that Eq. (13.53) is much closer to Eq. (13.3). These results were extended to higher order modes and realistic equations of state by Doneva et al. (2013b), where it was found that higher modes exhibit the same behaviour. The fitting formulas for these cases are given in Chap. 12.

A more recent approach that is closer to the I-Love-Q perspective using a parameterisation in terms of the effective compactness η was investigated by Doneva and Kokkotas (2015) and was found to be even more universal with respect to different equations of state. The detailed expressions for the frequencies and damping times are given in Chap. 12.

We should again stress at this point that these results are in the Cowling approximation and should be taken with a grain of salt. An obvious necessary extension of these results is to go beyond the Cowling approximation. Moreover, some improvement in the universality might also come, as we have seen in the case of I-Love-Q relations for rapidly rotating stars, if the rotation is parametrised in terms of the spin parameter $\chi = J/M^2$ instead of Ω.

13.2.1.5 Other Universal Properties

Having reviewed the three main classes of universal relations we will briefly mention here some additional results that either extend those that we have already discussed or go in a different direction.

One of the extensions of the I-Love-Q results was the multipole Love relations by Yagi (2014), where he produced universal relations among various ℓ-th (dimensionless) electric, magnetic and shape tidal deformabilities for neutron stars and quark stars, in the same sense as we have the various universal relations between the quadrupole and higher order multipole moments. His results are also reviewed in Yagi and Yunes (2017a). We should note that the tidal Love numbers were calculated for non-rotating compact objects.

Another extension of the I-Love-Q and 3-hair relations, with a more theoretical interest, is the one for anisotropic stars by Yagi and Yunes (2015a,b, 2016a). As we have noted, when we calculate the multipole moments for example of compact objects we see that as we increase the compactness up to the maximum possible value for these objects, we never quite get to the Black Hole limit of $\bar{Q} = \bar{J}_3 = \bar{M}_4 = \ldots = 1$. Instead there seems to be a gap between fluid configurations and Black Holes. Of course this is expected with respect to compactness, because a matter configuration cannot be more compact than the Buchdahl limit, i.e., $\mathcal{C} < 4/9 \approx 0.\bar{4}$ (Buchdahl 1959). The Buchdahl limit though is not a hard physical limit. One can go around it by using anisotropic pressure, which can produce objects with a compactness arbitrarily close to the Black Hole value of $\mathcal{C} = 1/2$ (Bowers and Liang 1974). For this reason Yagi and Yunes (2015a) used anisotropic stars to study how one approaches the Black Hole limit for the multipole moments and the other neutron star parameters. For anisotropic stars one has the radial pressure p and a tangential part of the pressure $p_\perp$ and the difference defines the anisotropy $\sigma = p - p_\perp$. In the model that they used, neutron stars with negative anisotropy reached continuously the Black Hole limit. The exact way that the Black Hole limit is reached though depends on the value of the anisotropy. The very interesting result was that there is a critical value of anisotropy above which the models reach the Black Hole limit directly, while below that value the models circle around the Black Hole limit until they almost spiral to it. Finally, in order to get to the Black Hole values one needs to have an "extreme" value of anisotropy. On a more practical side, there is also the question of how some anisotropy would affect the various universal relations of regular neutron stars. On that front Yagi and Yunes (2015b) find that anisotropy affects the universal relations only weakly, i.e., the relations become less universal by a factor of 1.5–3 relative to the isotropic case when anisotropy is maximal, but even then they remain approximately universal to 10%. They further find that this increase in variability is strongly correlated to an increase in the eccentricity variation of isodensity contours.

The results presented so far are mostly about isolated neutron stars, with the exception of the work by Maselli et al. (2013) where they studied I-Love-Q for a binary system. Extending this work Yagi and Yunes investigated the existence of universal relations between tidal parameters for binaries (Yagi and Yunes 2016b, 2017b) which they called binary Love relations. These relations are between the individual tidal Love numbers of the members of the binary $\bar{\lambda}_1, \bar{\lambda}_2$, as they are combined to form the symmetric $\bar{\lambda}_s$ and antisymmetric $\bar{\lambda}_a$ tidal parameters, and the so-called dimensionless chirp tidal deformability $\bar{\Lambda}$ and it's companion parameter $\delta\bar{\Lambda}$. The idea behind this is that the parameters $\bar{\Lambda}$ and $\delta\bar{\Lambda}$ enter the analysis of the waveforms from binary inspirals with the dimensionless chirp tidal deformability $\bar{\Lambda}$ being the dominant tidal parameter in analogy with the chirp mass being the dominant mass parameter in the waveform. Yagi and Yunes find universal relations that relate $\bar{\Lambda}$ to $\delta\bar{\Lambda}$ as well as $\bar{\lambda}_s$ to $\bar{\lambda}_a$, that depend on the symmetric mass ratio $X \equiv \frac{q}{(1+q)^2}$, where q is the mass ratio of the members of the binary. These results are also reviewed in Yagi and Yunes (2017a).

Continuing with binary inspirals, there is another recent interesting result. Initially Read et al. (2013) found, by modelling numerically the waveforms from binary inspirals and coalescences (spinning, equal mass, $q = 1$), that the instantaneous frequency at the moment of maximum amplitude in the waveform, which indicates the transition from the inspiral phase to the merger or coalescence phase, is related in an equation of state independent way with a parameter which is essentially the tidal Love number $\bar{\lambda}$ of one of the neutron stars defined in Eq. (13.14). Specifically they find the relation

$$\log_{10}(f_{GW}/\text{Hz}) = 8.51155 - 0.303350\,\bar{\lambda}^{1/5}. \tag{13.54}$$

Since in their numerical calculations they considered only equal mass inspirals, essentially one tidal Love number characterises both neutron stars. Expanding on this result, Bernuzzi et al. (2014) showed that for more general inspirals, where $q \neq 1$, a more appropriate quantity should be used, which is essentially an effective tidal Love number, called tidal coupling constant and is defined as,

$$\kappa_\ell \equiv 2\left[\frac{1}{q}\left(\frac{X_A}{C_A}\right)^{2\ell+1} k_\ell^A + q\left(\frac{X_B}{C_B}\right)^{2\ell+1} k_\ell^B\right], \tag{13.55}$$

where the definition is with respect to different ℓ-orders in tidal deformabilities and we define the mass ratio to be $q \equiv M_A/M_B \geq 1$, and for each member of the binary with respect to the total mass $X_A \equiv M_A/M$ and $X_B \equiv M_B/M$, where $M = M_A + M_B$ is the total mass, while $C_{A,B}$ are the respective compactnesses and $k_\ell^{A,B}$ the respective ℓth order tidal apsidal constants.[18] For $\ell = 2$ one can see that $\kappa \equiv \kappa_\ell \propto \bar{\lambda}$ and the results of Bernuzzi et al. reproduce the results of Read et al. quite accurately. Using the tidal coupling constant, Bernuzzi et al. produced an empirical relation for the mass scaled instantaneous frequency at peak amplitude in terms of κ, which is

$$M\omega_{GW} = 0.360\frac{1 + 2.59 \times 10^{-2}\kappa - 1.28 \times 10^{-5}\kappa^2}{1 + 7.49 \times 10^{-2}\kappa}, \tag{13.56}$$

and also produced a relation for the binding energy per reduced mass $\mu = M_A M_B/M$ in terms of κ, which is

$$E_b = -0.120\frac{1 + 2.62 \times 10^{-2}\kappa - 6.32 \times 10^{-6}\kappa^2}{1 + 6.18 \times 10^{-2}\kappa}. \tag{13.57}$$

These relations were found to be insensitive to variations in the mass ratio q and showed small sensitivity to the spin of the members of the binary, probably due to spin-orbit coupling (the calculations were performed for aligned spin binaries). These results were reproduced by Takami et al. (2014, 2015), where they also find

---

[18]Here we are using the naming convention employed in Sect. 13.2.1.2, which is different than the one used in Read et al. (2013); Bernuzzi et al. (2014).

by doing spectral analysis of the waveforms, that there is an equation of state independent relation between the lowest observed frequency f_1 of the spectrum and the average compactness $\bar{C} = \bar{M}/\bar{R}$, where $\bar{M} = (M_A + M_B)/2$ and $\bar{R} = (R_A + R_B)/2$, which is

$$\frac{f_1}{\text{kHz}} = (-22.0717 \pm 6.477) + (466.616 \pm 31.2)\bar{C} + (-3131.63 \pm 878.7)\bar{C}^2$$

$$+ (7210.01 \pm 1947)\bar{C}^3. \tag{13.58}$$

Returning to isolated neutron stars, there are some more results that we would like to conclude with. Bauböck et al. (2013) produced in the slow rotation approximation (using the Hartle and Thorne approach) analytic formulae that relate the ellipticity and eccentricity of the stellar surface to the compactness, the spin parameter, and the quadrupole moment of the neutron star. Defining the ellipticity ε_s and the eccentricity e_s as,

$$\varepsilon_s \equiv 1 - \frac{R_p}{R_{eq}}, \quad \text{and} \quad e_s \equiv \sqrt{\left(\frac{R_{eq}}{R_p}\right)^2 - 1}, \tag{13.59}$$

where R_p and R_{eq} are the polar and equatorial radius respectively, they found in terms of the compactness C, the spin parameter χ_K and the quadrupole $\bar{q}_K$, the relations

$$e_s^K(C, \chi_K, \bar{q}_K) = \left[1 - 4\chi_K C^{3/2} + \frac{15(\chi_K^2 - \bar{q}_K)(3 - 6C + 7C^2)}{8C^2} + C^2\chi_K^2(3 + 4C) \right.$$

$$\left. + \frac{45}{16C^2}(\bar{q}_K - \chi_K^2)(C - 1)(1 - 2C + 2C^2)\ln(1 - 2C) \right]^{1/2}, \tag{13.60}$$

$$\varepsilon_s^K(C, \chi_K, \bar{q}_K) = \frac{1}{32C^3} \left\{ 2C \left[8C^2 - 32\chi_K C^{7/2} + (\chi_K^2 - \bar{q}_K)(45 - 35C + 60C^2 + 30C^3) \right. \right.$$

$$\left. \left. + 24\chi_K^2 C^4 + 8\chi_K^2 C^5 - 48\chi_K^2 C^6 \right] + 45(\chi_K^2 - \bar{q}_K)(1 - 2C)^2 \ln(1 - 2C) \right\}. \tag{13.61}$$

In these relations, the compactness is defined in terms of the non-rotating models' masses and radii, while the spin parameter $\chi_K = J_K/M_*^2$, the quadrupole $\bar{q}_K = -Q_K/M_*^3$ and the ellipticity ε_s^K and eccentricity e_s^K correspond to the models that rotate at the frequency Ω_K, as it is discussed in Sect. 13.2.1.2. The ellipticities and eccentricities of models rotating with a rotation $\Omega < \Omega_K$ will be $\varepsilon_s = \epsilon^2 \varepsilon_s^K$ and $e_s = \epsilon e_s^K$ respectively, where $\epsilon = \Omega/\Omega_K$. The above exact expressions hold independently of the equation of state. All the equation of state information is encoded in the various parameters. AlGendy and Morsink (2014) also produce an equation of state independent fit for the ellipticity, where they use a slightly different parameterisation for the surface than the one used in the Hartle and Thorne approach. They parameterise the surface with a function of the form,[19]

$$R(\theta) = R_{eq}\left(1 + o_2(\mathcal{C}, \epsilon)\cos^2\theta\right),$$ (13.62)

where the function $o_2(\mathcal{C}, \epsilon)$, which is minus the ellipticity defined above, has the form

$$o_2(\mathcal{C}, \epsilon) = \epsilon^2(-0.788 + 1.030\mathcal{C}),$$ (13.63)

where $\epsilon = \Omega/\Omega_K$ as in the Hartle and Thorne case. Their results are in good agreement with previous results for rapidly rotating neutron stars. Additionally they find that the effective gravity at the surface of a rotating neutron star can be written as the simple function $g(\theta)/g_0 = c(\mathcal{C}, \epsilon^2, \theta)$, where g_0 is the acceleration due to gravity on the surface of a non-rotating relativistic star while the function c can be written in an equation of state independent form.

Finally, Breu and Rezzolla (2016) find that the mass of rotating configurations on the turning-point line shows a universal behaviour when expressed in terms of the spin parameter at the Kepler limit. In particular they find that the mass at the turning point, M_{crit} normalised by the maximum non rotating mass M_{TOV}, is expressed in terms of the spin parameter, χ normalised by the spin parameter at the Kepler limit χ_K, as

$$\frac{M_{crit}}{M_{TOV}} = 1 + 0.1316\left(\frac{\chi}{\chi_K}\right)^2 + 0.07111\left(\frac{\chi}{\chi_K}\right)^4.$$ (13.64)

This expression implies that the maximum mass for any given equation of state will be, for $\chi = \chi_K$, $M \simeq 1.203 M_{TOV}$.[20] Of course M_{TOV} will depend on the equation of state. In addition, they further explore the $\bar{I} = I/M^3$ relation to inverse powers of $\mathcal{C}$, where they find that there is a universal relation (that holds up to 10%) that also depends on the spin χ, as expected. In Breu and Rezzolla (2016) they give fitting coefficients for different values of χ, while here we will give a relations that is spin dependent, i.e.,

$$I/M^3 = (1.471 + 0.448\chi) - \frac{0.0802 + 0.27289\chi}{\mathcal{C}} + \frac{0.438 - 0.0346\chi}{\mathcal{C}^2}$$
$$- \frac{0.01694 + 0.0056\chi}{\mathcal{C}^3} + \frac{(3.316 + 1.57\chi) \times 10^{-4}}{\mathcal{C}^4}$$ (13.65)

---

[19]The radius here is expressed in terms of Schwarzschild-like coordinates which reduce to being circumferential at the non-rotating limit. The radial coordinates in the Hartle and Thorne approach also have this property.

[20]The interested reader might want to also have a look at the review by Hartle (1978).

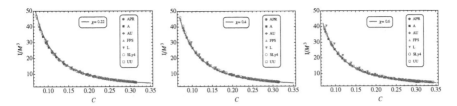

Fig. 13.6 I-C plots for three spin parameters and various equations of state. The solid lines correspond to the I-C fit given in Eq. (13.65)

which is also accurate to 10% and is in agreement with the relations given in Breu and Rezzolla (2016); Staykov et al. (2016) (Fig. 13.6).

13.2.2 Neutron Stars in Alternative Theories

Below we will review the basic theory behind the alternative theories of gravity we have chosen to concentrate on, the neutron star models in these theories and their astrophysical implications. We will consider the scalar-tensor theories, $f(R)$ theories, Einstein–dilaton–Gauss–Bonnet theories and Chern–Simons theories of gravity. These are all theories for which neutron stars universal relations have been considered. The strongest emphasis will be on the scalar-tensor theories that are ones of the simplest and most natural extensions of general relativity.

13.2.2.1 Alternative Theories of Gravity: Mathematical Setup and Overview

Scalar-Tensor Theories of Gravity The simplest representative of gravitational theories with an additional scalar field are the scalar-tensor theories of gravity. Historically, these are some of the first alternative theories to be constructed. The reason lies on one hand in their simplicity and the fact that such generalization of Einstein's theory seems quite natural. On the other hand there is a deep theoretical motivation coming from the theories trying to unify all the interactions such as Kaluza-Klein theories, string theories, etc. It can be easily shown for example that after a dimensional reduction of the five dimensional Kaluza-Klein theory action to four dimensions, an additional scalar field appears as a mediator of the gravitational interaction. The scalar-tensor theory of gravity can be formulated also independently and that was done first by Brans and Dicke on the basis of Mach's principle (Brans and Dicke 1961; Dicke 1962). According to it, the inertia of a particle is a consequence of the total mass distribution in the Universe. Therefore, the inertial mass is not a constant but also depends on the mass distribution, i.e. it depends on the interaction of the particle with some cosmological field Φ. According to the weak equivalence principle, which is verified extremely precisely, the interaction with this scalar field should be the same for all the particles up to some constant. Thus the mass of the particles should be

$$m = m_0 f(\Phi), \tag{13.66}$$

where $f(\Phi)$ describes the interaction of the particles with the scalar field. Using the above considerations and also the fact that the absolute scale of the masses of the particles can be measured only by their gravitational acceleration, we can conclude that the gravitational constant should depend on the total distribution of the matter in the universe, i.e. from the cosmological field Φ. Using all this, one can derive the scalar-tensor theory action and a very interesting fact is that its vacuum sector coincides with the action coming from the dimensional reduction of the Kaluza-Klein theory.

The most general form of the action in the scalar-tensor theories is Damour and Esposito-Farese (1992)

$$S = \frac{1}{16\pi G_*} \int d^4x \sqrt{-\tilde{g}} \left(F(\Phi)\tilde{R} - Z(\Phi)\tilde{g}^{\mu\nu}\partial_\mu\Phi\partial_\nu\Phi - 2U(\Phi) \right) + S_{\text{matter}}(\tilde{g}_{\mu\nu}, \chi), \tag{13.67}$$

where G_* is the bare gravitational constant and as a matter of fact the scalar field Φ can be interpreted as a "variable gravitational constant". The second and the third term in the action are the kinetic and potential terms for the scalar field respectively. The choice of functions $F(\Phi)$, $Z(\Phi)$, and $U(\Phi)$ determine the specific class of scalar-tensor theory. The requirement that the gravitons carry positive energy leads to the following condition $F(\Phi) > 0$, while the non-negativity of the scalar field energy requires that $2F(\Phi)Z(\Phi) + 3[dF(\Phi)/d\Phi]^2 \geq 0$. It is important to point out, that the action of the matter $S_{\text{matter}}(\tilde{g}_{\mu\nu}, \chi)$, where χ denotes collectively the matter fields, is the same as in general relativity. Therefore, there is no direct coupling between the matter and the scalar field, and the scalar field influences the matter only via the spacetime metric $\tilde{g}_{\mu\nu}$. Thus, the equation of motion is the same as in general relativity $\tilde{\nabla}_\alpha \tilde{T}^\alpha_\mu = 0$ and the weak equivalence principle is satisfied.

The action (13.67) is in the so-called Jordan frame which is the physical frame where distances, time, etc. are measured. The field equations, though, are quite complicated and difficult to work with in this frame. That is why a common practice is to introduce the so-called Einstein frame which simplifies the field equations considerably. The transition to the Einstein frame is made by performing a conformal transformation of the metric[21] and a re-definition of the scalar field:

$$g_{\mu\nu} = F(\Phi)\tilde{g}_{\mu\nu}, \tag{13.68}$$

$$\left(\frac{d\varphi}{d\Phi}\right)^2 = \frac{3}{4}\left(\frac{d\ln(F(\Phi))}{d\Phi}\right)^2 + \frac{Z(\Phi)}{2F(\Phi)}, \tag{13.69}$$

---

[21] For the interesting case of disformal coupling we refer the reader to (Minamitsuji and Silva 2016).

where φ and $g_{\mu\nu}$ are the Einstein frame scalar field and metric. After we introduce the functions $\mathcal{A}(\varphi)$ and $V(\varphi)$ defined as

$$\mathcal{A}(\varphi) = F^{-1/2}(\Phi), \tag{13.70}$$

$$V(\varphi) = \tfrac{1}{2}U(\Phi)F^{-2}(\Phi), \tag{13.71}$$

the Einstein frame action becomes

$$S = \frac{1}{16\pi G_*} \int d^4x \sqrt{-g} \left(R - 2g^{\mu\nu}\partial_\mu\varphi\partial_\nu\varphi - 4V(\varphi)\right) + S_{\text{matter}}(\mathcal{A}^2(\varphi)\tilde{g}_{\mu\nu}, \chi), \tag{13.72}$$

where R is the Ricci scalar curvature with respect to the spacetime metric $g_{\mu\nu}$. As we can see the action in the Einstein frame is much simpler and easier to work with, but everything comes with a price and the price we pay for this simplicity is that a direct coupling between the sources of gravity and the scalar field appears through the function $\mathcal{A}(\varphi)$. Nevertheless, the resulting field equations are easier to handle and most of the studies of compact objects in scalar-tensor theories adopt the Einstein frame. The specific choice of the scalar-tensor theory is completely determined by the function $\mathcal{A}(\varphi)$ and by the potential of the scalar field $V(\varphi)$.

The transformation between the two frames is regular practically for all of the physically relevant scalar-tensor theories. Thus, it is irrelevant in which frame we are going to perform our calculations as long as the final observable quantities are expressed in the physical Jordan frame.

The field equations in the Einstein frame are much simpler as well

$$R_{\mu\nu} = 8\pi G_*\left(T_{\mu\nu} - \tfrac{1}{2}g_{\mu\nu}T\right) + 2\partial_\mu\varphi\partial_\nu\varphi - \tfrac{1}{2}g_{\mu\nu}V(\varphi), \tag{13.73}$$

$$\nabla_\alpha\nabla^\alpha\varphi - \tfrac{1}{4}V'(\varphi) = -4\pi\alpha(\varphi)T, \tag{13.74}$$

where ∇_α is the covariant derivative with respect to the metric $g_{\mu\nu}$ and T is the trace of the Einstein frame energy momentum tensor $T_{\mu\nu}$. The function $\alpha(\varphi)$ is called coupling function and it is defined as

$$\alpha(\varphi) = \frac{d\ln A(\varphi)}{d\varphi}. \tag{13.75}$$

The connection between the Einstein frame and the Jordan frame energy momentum tensor is

$$T_{\mu\nu} = A^4(\varphi)\tilde{T}_{\mu\nu}. \tag{13.76}$$

Since we have a direct coupling between the matter and the scalar field in the Einstein frame, the equations of motion for the matter fields differ from that in pure general relativity:

$$\nabla_\alpha T^\alpha_\mu = \alpha(\varphi) T \partial_\mu \varphi. \tag{13.77}$$

Therefore, there will be an additional force acting on the particles in the Einstein frame which is proportional to the gradient of the scalar field and the particles will not move on the geodesic of the metric $g_{\mu\nu}$.

Let us briefly comment on the field equations and more precisely under what circumstances nontrivial scalar field can develop. For simplicity we will consider the case with zero scalar field potential. The right hand side of Eq. (13.74) is nonzero only for nonzero trace of the energy momentum tensor T. For isolated black holes $T = 0$ which roughly speaking leads to the fact that the solutions are the same as in pure general relativity, but for neutron stars T is nonzero and thus nontrivial scalar field can develop.

Let us now discuss briefly the parametrized post-Newtonian (PPN) expansion of the metric in scalar-tensor theories and the observational constraints one can impose, since in general different theories of gravity predict different values of the post-Newtonian parameters. In what follows we will again assume that the scalar field potential is zero for simplicity. The coupling function $\alpha(\varphi)$ on the other hand can be expanded in power series of the scalar field φ

$$\alpha(\varphi) = \alpha_0(\varphi - \varphi_0) + \frac{1}{2}\beta(\varphi - \varphi_0)^2 + O(\varphi - \varphi_0)^3, \tag{13.78}$$

where φ_0 is the background value of the scalar field and α_0 and β are constants. The case with $\beta = 0$ corresponds to the well known Brans–Dicke scalar-tensor theory. The PPN expansion of the Schwarzschild metric in isotropic coordinates takes the form (Damour and Esposito-Farese 1992; Will 2014):

$$-g_{00} = 1 - 2\frac{Gm}{rc^2} + 2\beta^{PPN}\left(\frac{Gm}{rc^2}\right)^2 + O\left(\frac{1}{c^6}\right), \tag{13.79}$$

$$g_{ij} = \delta_{ij}\left(1 + 2\gamma^{PPN}\frac{Gm}{rc^2}\right) + O\left(\frac{1}{c^4}\right). \tag{13.80}$$

As one can see, in scalar-tensor theories only two of the PPN parameters differ from GR—β^{PPN} and γ^{PPN} which were introduced for the first time by Eddington for the Schwarzschild metric in isotropic coordinates. They are connected to the coefficients in the expansion of the coupling function (13.70) in the following way:

$$\gamma^{PPN} - 1 = -\frac{2\alpha_0^2}{1 + \alpha_0^2}, \qquad \beta^{PPN} - 1 = \frac{1}{2}\frac{\alpha_0\beta\alpha_0}{(1 + \alpha_0^2)^2}. \tag{13.81}$$

Clearly, for $\alpha_0 = 0$ and $\beta = 0$ we have $\gamma^{PPN} = 0$ and $\beta^{PPN} = 0$ and the theory reduces to pure general relativity.

Different observations can impose constraints on the PPN parameters γ^{PPN} and β^{PPN} in the weak field regime, such as the rate of precession of the Mercury perihelia, deflection of light by the Sun, delay of the light travel time in the vicinity of the Sun and Lunar Laser Ranging experiment (for a review on this subject see Will 2014). One can easily show, though, that all these experiments set constraints only on the parameter α_0, while β remains essentially unconstrained. This is natural, since the case of $\alpha_0 = 0$ and $\beta \neq 0$ corresponds to a scalar-tensor theory that is perturbatively equivalent to general relativity, i.e. all the weak field experiments are automatically satisfied, but large deviations can be observed for strong fields. Currently, only the pulsars in close binary systems can impose constraints on β through observations of the shrinking of their orbits due to gravitational wave emission because the strong field effects in this case are non-negligible. More precisely, the shrinking would be different in general relativity and in scalar-tensor theories because in the latter case we have scalar gravitational radiation in addition to the standard gravitational waves. It turns out that the observations match very well to the predictions of pure general relativity and limit severely the emitted scalar radiation and thus the value of the scalar field. At the end, taking into account both the weak and the strong field experiments, we have $\alpha_0 < 10^{-4}$ and $\beta > -4.5$ (Freire et al. 2012; Antoniadis et al. 2013). The observational constraints on STT coming from both the weak field and the strong field experiments are plotted in Fig. 13.7. Such small values of α_0 make the neutron stars in Brans–Dicke scalar tensor theories practically indistinguishable from general relativity and leave little

Fig. 13.7 Solar-system and binary pulsar constraints on the constant α_0 and β in the expansion of the coupling function (13.78). LLR stands for lunar laser ranging, Cassini for the measurement of a Shapiro time-delay variation in the Solar System, and SEP for tests of the strong equivalence principle using a set of neutron star-white dwarf low-eccentricity binaries [Credit Ref. Freire et al. (2012)]

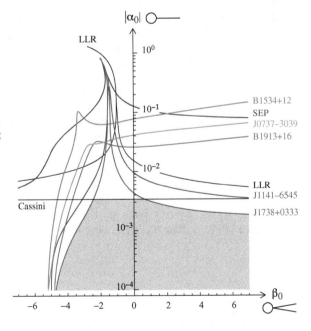

space for deviation even for the class of scalar-tensor theory with $\alpha_0 = 0$ and $\beta \neq 0$.[22]

The discussion above was concentrated on the case with zero scalar-field potential. The picture changes drastically if we consider a nontrivial potential. One of the simplest cases is to assume the following standard form of the Einstein frame potential:

$$V(\varphi) = 2m_\varphi^2\varphi^2 \tag{13.82}$$

that is equivalent to considering a scalar field with nonzero mass m_φ. Even though this might seems like a simple and straightforward extension, it has dramatic influence on the observational constraints of the theory because of the following reasons. Introducing a scalar-field mass means that we have introduced a new characteristic scale, namely the Compton wave-length of the scalar field $\lambda_\varphi = 2\pi/m_\varphi$, and the scalar field is loosely speaking confined inside λ_φ. More precisely, it drops exponentially at infinity and has non-negligible values only inside its Compton wave-length. This help us reconcile the theory with the observations for a much larger range of parameters compared to the massless case. For example let's consider the binary pulsar observations. If the Compton wavelength of the scalar field is smaller than the orbital separation between the two stars, no scalar gravitational radiation will be emitted and the predictions for the shrinking of the orbit would be the same as in pure general relativity. For such values of the scalar field mass the parameter β is essentially unconstrained (Ramazanoğlu and Pretorius 2016; Yazadjiev et al. 2016).

Below we will briefly discuss other alternative theories of gravity that are either equivalent to scalar-tensor theories, or fall into the same class of modifications of general relativity, i.e. we have inclusion of a dynamical scalar field. Exactly in these theories universal relations were built.

$f(R)$ *Theories of Gravity* We will continue with the $f(R)$ theories, since they are equivalent to a particular class of scalar-tensor theories with nonzero scalar field potential. The essence of these theories is that the Ricci scalar R in the Hilbert–Einstein actions is substituted with a function of R, thus $f(R)$ theories:

$$S = \frac{1}{16\pi G} \int d^4x\sqrt{-g}f(R) + S_{\text{matter}}(g_{\mu\nu}, \chi). \tag{13.83}$$

The viable $f(R)$ theories have to be free of tachyonic instabilities and the appearance of ghosts which is equivalent to Sotiriou and Faraoni (2010); De Felice

---

[22]As we will discuss below, this is true only in the static case. The rapid rotation magnifies the differences significantly and offers new possibilities for probing scalar-tensor theories of gravity.

and Tsujikawa (2010)

$$\frac{d^2 f}{dR^2} \geq 0, \quad \frac{df}{dR} > 0 \tag{13.84}$$

respectively. One can easily show that after certain transformations (see e.g. Yazadjiev et al. 2014) this action is equivalent to a scalar-tensor theory with zero kinetic term in the Jordan frame and nonzero potential. The gravitational scalar Φ and the potential $U(\Phi)$ are connected to the $f(R)$ function via the relations

$$\Phi = \frac{df(R)}{dR}, \quad U(\Phi) = R\frac{df}{dR} - f(R). \tag{13.85}$$

If one wants to express these quantities in terms of the Einstein frame function that appear in the action (13.72), then the coupling function becomes $\alpha(\varphi) = -1/\sqrt{3}$ and the potential $V(\varphi) = A^4(\varphi)U(\Phi(\varphi)) = [R(df/dR) - f(R)]/(df/dR)^2$. For example, for one of the simplest case—the R^2 gravity, where $f(R) = R + aR^2$, we have

$$V(\varphi) = \frac{1}{4a}\left(1 - e^{-\frac{2\varphi}{\sqrt{3}}}\right)^2. \tag{13.86}$$

$f(R)$ theories of gravity were mainly explored in cosmological context as an alternative explanation of the accelerated expansion of the Universe. Every viable theory of gravity should pass the observational test on astrophysical scales as well and that is why neutron stars in $f(R)$ theories attracted particular interest recently. The form of the $f(R)$ function, though, that is normally used to explain the dark energy phenomenon would not give significant influence on astrophysical scales. That is why most of the authors considered the problem the other way around— assume that indeed the gravitational theory is not the pure Einstein's theory, but instead we have an $f(R)$ type of modification of the action. Then one can ask the question what would be the additional terms in the $f(R)$ function that would give the dominant contribution on astrophysical scales. It is expected that this is exactly the R^2 term and that is why most of the compact star solutions were constructed in R^2 gravity.

Quadratic Gravity The idea of the quadratic gravity is to supplement the Hilbert–Einstein action with all the possible algebraic curvature invariants of second order. These invariants are: R^2, $R_{\mu\nu}^2 \equiv R_{\mu\nu}R^{\mu\nu}$, $R_{\mu\nu\rho\sigma}^2 \equiv R_{\mu\nu\rho\sigma}R^{\mu\nu\rho\sigma}$ and the Pontryagin scalar $*RR \equiv 1/2R_{\mu\nu\rho\sigma}\epsilon^{\nu\mu\lambda\kappa}R^{\rho\sigma}{}_{\lambda\kappa}$, where $\epsilon^{\nu\mu\lambda\kappa}$ is the Levi-Civita tensor. An extra dynamical scalar field is included as well that couples non-minimally to the second order curvature corrections.

The motivation behind such modifications lies in the fact that pure general relativity is not a renomalizable theory which naturally poses severe obstacles to the efforts of quantizing gravity. The modification of the action proposed by the quadratic gravity makes the theory renormalizable (Stelle 1977). The price we pay is the appearance of ghosts. A way to circumvent this problem is just by assuming that the quadratic gravity is a truncated effective theory of a more general one (such as the string theory where the effective action contains infinite series of higher curvature corrections and it is ghost free).

A general form of the action is Yunes and Stein (2011); Pani et al. (2011a):

$$S = \frac{1}{16\pi} \int \sqrt{-g} d^4 x \left[R - 2\nabla_\mu \varphi \nabla^\mu \varphi - V(\varphi) \right. \tag{13.87}$$

$$\left. + f_1(\varphi) R^2 + f_2(\varphi) R_{\mu\nu} R^{\mu\nu} + f_3(\varphi) R_{\mu\nu\rho\sigma} R^{\mu\nu\rho\sigma} + f_4(\varphi)^* RR \right] + S_{\text{matter}} \left[\chi, \gamma(\varphi) g_{\mu\nu} \right], \tag{13.88}$$

where $V(\varphi)$ is the scalar field potential, the coupling functions $f_1 .. f_4$ depend only on the scalar field, χ denotes collectively the matter fields in the matter action S_{matter} and we have a nonminimal coupling between the scalar field and the matter via the function $\gamma(\varphi)$.

The field equations derived from this action in their most general form are of order higher that two. This leads to problems in the theory such as the appearance of ghosts as we mentioned above. In some special cases, though, the field equations remain of second order as discussed below.

Two sectors of the quadratic gravity attracted particular interests—the Einstein–dilaton–Gauss–Bonnet gravity and the Chern–Simons gravity. As a matter of fact almost all of the studies of compact objects in quadratic gravity were made exactly in these sub-theories.

Einstein–dilaton–Gauss–Bonnet Gravity In the EdGB theory the function $f_4(\varphi) = 0$ and one choses a special combination between the other three, namely $f_1(\varphi) = f_3(\varphi) = -1/4 f_2(\varphi) \equiv f(\varphi)$ (Kanti et al. 1996). This combination is chosen in such a way that the resulting field equations are of second order. The action has the form

$$S = \frac{1}{16\pi} \int \sqrt{-g} d^4 x \left[R - 2\nabla_\mu \varphi \nabla^\mu \varphi - V(\varphi) + f_{GB}(\varphi)(R^2 - 4R_{\mu\nu}^2 + R_{\mu\nu\rho\sigma}^2) \right], \tag{13.89}$$

where $R_{GB}^2 \equiv R^2 - 4R_{\mu\nu}^2 + R_{\mu\nu\rho\sigma}^2$ is called Gauss–Bonnet scalar. Clearly the function $f(\varphi)$ has dimensions of inverse curvature and thus a characteristic scale can be introduced in the EdGB theory equal to $\sqrt{\alpha_{\text{GB}}}$ where we assumed that $f(\varphi)$ is proportional to α_{GB} times a dimensionless function of the scalar field. The strongest constraint on the parameters of the theory comes form the observations of black hole low-mass X-ray binaries (Yagi 2012), namely $\sqrt{|\alpha_{\text{GB}}|} \lesssim 5 \times 10^6$ cm. Another purely theoretical constraint comes from the requirement for the existence of black hole solutions, that is fulfilled when $\sqrt{\alpha_{BD}}$ is smaller than the black hole horizon size (Kanti et al. 1996). This leads to $\alpha_{\text{GB}}/M^2 \lesssim 0.691$ (Pani and Cardoso 2009). One should note that the scalar field potential is neglected in most of the studies of compact objects in EdGB gravity.

Chern–Simons Gravity The CS gravity considers a different sector of the quadratic gravity when $f_1 = f_2 = f_3 = 0$ and only the function f_4 is nontrivial (for a review on the CS theory see e.g. Alexander and Yunes 2009). This means that only the term proportional to the Pontryagin scalar $^*RR$ remains.

There are two versions of the theory. The first one is a non-dynamical version where the kinetic and the potential terms for the scalar field are omitted and the scalar field is externally prescribed, i.e. it does not evolve dynamically. This case is simpler and it was the first one to be considered. It turned out, though, that in this case the theory has certain problems and restrictions, such as the Pontryagin constraint $^*RR = 0$ (Grumiller and Yunes 2008; Alexander and Yunes 2009) That is why in the last several years another version of the theory, where the scalar field is dynamical, attracted much more attention. In order to distinguish the dynamical CS gravity from the non-dynamical version, the abbreviation dCS is used.

Thus the dCS action takes the form

$$S = \frac{1}{16\pi} \int \sqrt{-g} d^4x \left[R - 2\nabla_\mu \varphi \nabla^\mu \varphi - V(\varphi) + f_{CS}(\varphi)(^*RR) \right]. \qquad (13.90)$$

A very interesting property of the dCS gravity is that the static spherically symmetric solutions do not differ from Einstein's theory but rotation can introduce large deviations. As a matter of fact the dCS gravity is almost the only theory with such property which makes it very interesting to study.

In most of the studies, the function $f_{CS} = \alpha_{CS}\varphi$ and the potential of the scalar field $V(\varphi)$ is zero, which introduces a length scale of the theory $\sqrt{\alpha_{GB}}$. As we discussed above the general form of the quadratic gravity is prone to problems such as the appearance of ghost, because of the fact that the field equations contain higher order derivatives. This problem is circumvented in the EdGB theory because of the special combination of the curvature invariants, but this is not true for the dCS theory. Instead, in order to have a well posed theory, one can consider the decoupling limit, where the field equation are still of second order. Thus, in most paper dCS gravity is studied perturbatively in the coupling constant α_{GB}. In order to be able to apply the perturbative approach, the following condition should be fulfilled, $\alpha_{GB}/M^2 \ll 1$. Due to the fact that the static solutions are the same as in GR, the only weak field experiments that can impose constraints on α_{GB} are the ones measuring the frame dragging effects, such as the Gravity Probe B, which gives $\sqrt{\alpha_{GB}} \lesssim 10^{13}$ cm (Ali-Haimoud and Chen 2011) that is also in agreement with the results in Yagi et al. (2012).

13.2.2.2 Alternative Theories of Gravity: Neutron Star Models and Astrophysical Implications

Here, we will review the neutron star models constructed in the above discussed alternative theories of gravity and their astrophysical implications. We are not aiming towards an exhaustive review, but instead we will discuss the most important results with an emphasis on the recent achievements in the field.

Neutron Stars in (Massive) Scalar-Tensor Theories Historically, the compact star models in STT were some of the first to be considered (see e.g. Horbatsch and Burgess 2011 and the references therein) and they still attract significant attention, because of the fact that STT are ones of the most natural and widely used alternative theories of gravity. Most of the results fall into two categories based on the exact form of the Einstein frame coupling function $\alpha(\varphi)$ (see Eqs. (13.75) and (13.78))—neutron stars in Brans–Dicke scalar-tensor theories with constant coupling function $\alpha(\varphi) = \alpha_0$ and neutron stars in a theory with $\alpha(\varphi) = \beta\varphi$ that is perturbatively equivalent to general relativity in the weak field regime but can lead to large deviations for strong fields. As we discussed in the previous section, α_0 is severely limited by the weak field observations which leaves practically no space for any measurable deviations in the neutron star properties. This is not the case, though, with the second type of STT, where the only constraints come from the strong field experiments such as the observations of pulsars in binary systems. The current constraint on β, i.e. $\beta > -4.5$, still leaves space for non-negligible deviations from the pure Einstein's theory especially in the rapidly rotating case. That is why we will consider only the second class of scalar-tensor theory.

Neutron stars in scalar-tensor theories with the Einstein frame coupling function $\alpha(\varphi) = \beta\varphi$ were considered for the first time in Damour and Esposito-Farèse (1993) where an effect, called spontaneous scalarization, was found that consists of the following. There exists a range of stellar parameters where neutron stars can develop nontrivial scalar field even if the background cosmological value of the scalar field φ_0 is zero. As a matter of fact very similar phenomenon is observed also for black holes in the presence of nonlinear fields, such as charged black holes described by nonlinear electrodynamics (Stefanov et al. 2007a,b, 2009, 2008; Doneva et al. 2010), in the presence of a complex scalar field (Kleihaus et al. 2015) or black holes surrounded by matter (Cardoso et al. 2013a,b).

Sequences of scalarized solutions are plotted in Fig. 13.8 for several values of the parameter β and for both nonrotating star and stars rotating at the Kepler (mass-shedding) limit. Let us now discuss first the static case. This corresponds to the black, blue and green tick lines in Fig. 13.8. Note that the case with $\beta = -4.8$ is actually outside of the observational limit $\beta > -4.5$ but we plotted it so that one can have a better intuition about the qualitative behavior with the increase of β. One of the most important properties of this class of scalar-tensor theories is the nonuniqueness of the solutions. First, one should note that for $\alpha(\varphi) = \beta\varphi$ the field equations (13.73) and (13.74) always admit the solutions with zero scalar field (we will call them trivial solutions). This means that the pure general relativistic solutions are also solutions for this class of STT. As one can see from the left panel in Fig. 13.8, at some critical central energy density a new sequence of neutron star solutions with nontrivial scalar field (we will call them nontrivial solutions) branch out of the sequence of trivial solutions. This is the so-called spontaneous scalarization with the name coming from the analogy to the spontaneous magnetization in ferromagnets. Thus, for a certain range of parameters nonuniqueness of the solutions is present.

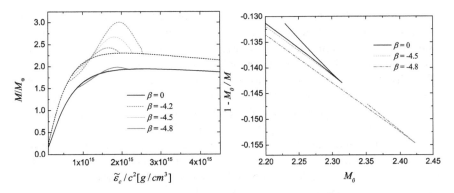

Fig. 13.8 Left panel: The neutron star mass as a function of the central energy density for static sequences of neutron stars (solid lines) and sequences of stars rotating at the mass-shedding limit (dotted lines) with different values of the coupling constant β. A polytropic equation of state with $N = 0.7463$ and $K = 1186$ is used. Right panel: The quantity $1 - M_0/M$, connected to the gravitational binding energy, as a function of the baryon rest mass M_0 for a fixed value of the angular momentum $cJ/(GM_\odot^2) = 1.38$ [Credit Ref. Doneva et al. (2013c)]

It is even more interesting that the scalarized neutron stars are energetically more favorable over the neutron stars with zero scalar field (Damour and Esposito-Farèse 1993; Doneva et al. 2013c) and they will be the ones that will be realised in practice. This is demonstrated also in the right panel of Fig. 13.8 where the quantity $1 - M_0/M$, that is connected to the gravitational binding energy, is plotted as a function of the baryon rest mass M_0. The sequences are calculated for the more general case of rotating stars with a fixed value of the angular momentum. As one can see, there is a cusp at the point where the mass reaches maximum which represents a turning point along the fixed-J sequence. This is the point where secular instability to collapse sets in. Moreover, the scalarized neutron stars have lower values of $1 - M_0/M$ and therefore, higher binding energy compared to the pure GR case, which makes them energetically more favorable. The stability of the scalarized neutron stars was examined in Harada (1997, 1998) and it was found that nontrivial scalar field develops for $\beta < -4.35$ in the non-rotating case. This result was derived in Harada (1998) for one particular polytropic EOS, but it turns out that the threshold value of β is quite similar for other realistic EOS (Novak 1998a). Since the current observational constraint is $\beta > -4.5$ there is not much space for deviations from pure general relativity.

The scalarization was further examined in Salgado et al. (1998) and the maximum mass limit was studied in Sotani and Kokkotas (2017). Slowly rotating neutron stars in STT where constructed in Damour and Esposito-Farèse (1996) where also the constrains on β coming from the binary pulsar observations where discussed for the first time. Rotational corrections up to first order in the rotational frequency were studied later in Sotani (2012) and up to second order in Pani and Berti (2014). What gives the largest difference, though, is the inclusion of rapid rotation (Doneva et al. 2013c) that enhances the differences between the pure general relativistic solutions and the scalarized ones, especially close to the Kepler limit, and also increases the range of parameters where nontrivial scalar field can developed. It was shown for example that scalarized solutions exist even for $\beta < -3.9$ close to the Kepler limit. As a matter of fact these were the first rapidly rotating models in alternative theories of gravity. The effect of rapid rotation is demonstrated in Fig. 13.8 where the dotted lines correspond to sequences of models rotating at the mass-shedding limit. As one can see for $\beta = -4.2$ no scalarized branch of solutions exists in the nonrotating limit, but such branch clearly appears close to the Kepler frequency with significant deviations from pure general relativity. Moreover, while the differences between the pure general relativistic solutions and the scalarized ones is quite small for the maximum allowed value $\beta = -4.5$ in the static case, the rapidly rotating models can reach large deviations for the same value of β.

The regime of rapid rotation opens a completely new window towards testing scalar tensor theories. Currently the fastest known pulsar rotates with a frequency of about 700 Hz (Hessels et al. 2006) where the rotational effects are important but they would not lead to significant enhancement of the differences with pure general relativity. Other objects, though, such as the binary neutron star merger remnants, will rotate with frequencies close to the Kepler limit shortly after their birth and they are supposed to be observed in the near future via their gravitational wave emission. This makes them perfect candidates for further testing of scalar-tensor theories and the strong field regime of gravity in general.

All the solutions discussed so far are in the negative β regime. It was argued recently that scalarization can occur for positive β as well for a limited set of equations of state that admit negative values of trace of the energy-momentum tensor inside the star (Mendes 2015; Palenzuela and Liebling 2016; Mendes and Ortiz 2016). This condition practically means that there are regions inside the star where the pressure surpasses one third of the energy density that can be translated to a threshold value of the compactness weakly dependent on the particular equation of state, i.e. $(M/R)_{min} \sim 0.265$ (Mendes 2015).

A lot of work has been done on astrophysical implications of the scalarized neutron stars. Possible ways to constrain scalar-tensor theories through measurements of surface atomic line redshifts was considered in DeDeo and Psaltis (2003). The orbital and epicyclic frequencies, and the innermost stable circular orbit for particles orbiting around scalarized neutron stars were calculated in DeDeo and Psaltis (2004) in the non-rotating case and in Doneva et al. (2014a) for rapidly rotating models, while general expressions in terms of the multipole moments have been produced in Pappas and Sotiriou (2015a). These quantities are related to the properties of accretion discs around compact stars and the observations of quasi-periodic oscillations in the emitted X-ray flux, and that is why they can be used to put constraints on the theory. The process of dynamical scalarization resulting for example after accretion of matter on the neutron star that brings it above the scalarization threshold, was considered in Novak (1998a). The collapse in scalar-tensor theories was examined in Harada et al. (1997); Novak and Ibanez (2000); Gerosa et al. (2016) and the collapse of a neutron star to black holes was considered in Novak (1998b). Almost all of these studies, with the exception of

Doneva et al. (2014a), are in the nonrotating limit. Even though some of them offered promising ways of constraining the coupling parameter β at that time, the very recent observations of pulsars in binary systems set very tight limit on β which makes the deviations from pure general relativity in these astrophysical scenario practically unmeasurable. Only the rapidly rotating case leads to larger deviations. If we restrict ourselves to objects rotating with frequencies up to 700 Hz, though, the deviations in the epicyclic frequencies and the position of the innermost stable circular orbit considered in Doneva et al. (2014a), would be still difficult to measure. That is why one needs to go to rotation close to the Kepler limit expected for example for the binary neutron star merger remnants.

As a matter of fact the neutron star merger in scalar-tensor theories was studied for the first time a few years ago in Barausse et al. (2013); Shibata et al. (2014). In these studies a phenomenon called "dynamical scalarization" was observed similar to the spontaneous scalarization. The essence of the phenomenon is that even though the two neutron stars would not be scalarized when they are isolated, they can develop nontrivial scalar field if they are close enough, i.e. as the orbital separation shrinks the stars undergo dynamical scalarization. This can have significant effect on the stellar dynamics and leads to some observable effects. Later a semi-analytical approach based on a modification of the post-Newtonian formalism was developed that takes into account the presence of a nontrivial scalar field and was proven to be in agreement with the previous fully nonlinear relativistic results (Palenzuela et al. 2014). This approach was recently improved in Sennett and Buonanno (2016). The advantage of the semi-analytical approach is that a much larger parameter space can be explored and template of waveforms can be generated. The problem of binary neutron star mergers in scalar-tensor theories was examined also via calculation of quasi-equilibrium sequences of equal-mass, irrotational binary neutron stars in Taniguchi et al. (2015). The question of detectability of possible (dynamically) scalarized stated by the Advanced LIGO, VIRGO and KAGRA was discussed in Sampson et al. (2014) and the results show that the gravitational wave signal is indeed detectable in certain case that are in agreement with the current observational constraints. In addition, the electromagnetic counterpart of the binary neutron star mergers was studied (Ponce et al. 2015) and it was concluded that the differences with pure general relativity are small but if they are combined with gravitational wave observations, constraints can be imposed on the deviations from Einstein's theory.

The oscillations of neutron stars in scalar-tensor theories were examined for the first time in Sotani and Kokkotas (2004) where the non-radial polar oscillations modes (f- and p-modes) were calculated in the Cowling approximation, i.e. when the spacetime and the scalar field perturbations are neglected. Even though this is a crude approximation it gives us good qualitative picture of the possible deviations from pure general relativity. Later, the full perturbation equations for both the axial and polar modes were derived and the frequencies of the axial modes were calculated (Sotani and Kokkotas 2005). In Sotani (2014) a different approach was undertaken—the radial oscillations of scalarized neutron stars were considered in the Cowling approximation, but without neglecting the scalar field perturbations.

The radial perturbations of course do not lead to gravitational wave emission but due to the presence of a scalar field, scalar waves are emitted (Morganstern and Chiu 1967). All these studies showed that the scalarization indeed changes significantly the oscillations spectrum leading to non-negligible deviations from general relativity for small enough values of β. The most recent constraints on β, though, limit the possible deviations from Einsteins' theory considerably similar to the other astrophysical scenarios. Torsional oscillations of scalarized neutron stars were examined in Silva et al. (2014). The results in Silva et al. (2014) showed that if one considers realistic values of the β the deviations due to scalarization are smaller than the uncertainties in microphysics and therefore there is no degeneracy between the two effects. Oscillations of rapidly rotating neutron stars were examined in Yazadjiev et al. (2017). The results show that the deviations from pure Einstein's theory can be significant especially in the case of nonzero scalar field mass.

Additional "ingredients" to the scalarized neutron stars were also explored, such as the presence of anisotropic pressure (Silva et al. 2015) both in the nonrotating and slowly rotating regimes using a quasi-local equation of state similar to the pure general relativistic case (Horvat et al. 2011; Doneva and Yazadjiev 2012). Depending on the "sign" of the anisotropy, i.e. whether the tangential pressure is larger than the radial one or the other way around, the deviations from pure general relativity due to the presence of nontrivial scalar field are either suppressed or magnified. This gives us hope that the binary pulsar observations can set constraints on the degree of anisotropy.

An interesting extension of the above work is to consider not only one scalar field coupled to the metric, but multiple scalar fields instead. This problems is of course much more involved and the first steps in this direction were undertaken in Horbatsch et al. (2015) (see also Damour and Esposito-Farese 1992). The simplest case is to consider two real scalar fields instead of one, that after a complexification can be cast to the problem of adding one complex scalar field. The criterion for neutron star scalarization were examined in this theory and the 3+1 formulation of the field equations was derived.

So far all the presented results are for the case when the potential of the scalar field $V(\varphi) = 0$. New and interesting effects arise, though, if we drop this assumption as commented in the previous subsection. For example the inclusion of scalar-field mass m_φ, via the specific form of the potential given by Eq. (13.82), leads to the fact that the scalar field is suppressed at length scales larger that its Compton wavelength λ_φ. Therefore, if properly chosen, the mass of the scalar field can help us reconcile the theory with the observations for a much larger range of parameters. One should note that the calculation of the neutron star models is numerically challenging because of the following reason. The presence of nonzero mass makes the field equation for the scalar field (13.74) stiff because this equation admits both exponentially decreasing and exponentially growing solutions at infinity. Clearly, only the exponentially decreasing solutions is physically relevant, but it is very difficult in certain cases to converge numerically to it and special techniques should be used.

Let us consider the same class of scalar-tenor theories with coupling function $\alpha(\varphi) = \beta_\varphi \varphi$ and impose the constraint that the Compton wavelength $\lambda_\varphi \ll 10^{10}$m (this is roughly the minimal observed orbital separation for close binary pulsars) or equivalently $m_\varphi \gg 10^{-16}$ eV. Then the emission of scalar gravitational waves will be suppressed which means that practically no constraints can be imposed on the parameter β. Thus, the deviations from pure general relativity can be very large. Neutron star models in such class of scalar-tensor theories were considered for the first time in Popchev (2015); Ramazanoğlu and Pretorius (2016) for the static case and in Yazadjiev et al. (2016) for the slowly rotating case. Rapidly rotating models in massive scalar-tensor theories were examined in Doneva and Yazadjiev (2016). Further studies on the astrophysical implications of neutron stars in massive scalar-tensor theories are needed because too large, negative β would clearly lead to dramatic changes in the neutron star structure that can be tested with the present astrophysical observations. Considering other forms of the scalar field potential would be interesting as well. The gravitational radiation of compact binaries in another class of massive scalar-tensor theories, the massive Brans–Dicke theory, was considered in Berti et al. (2012); Alsing et al. (2012).

Neutron Stars in $f(R)$ Theories of Gravity Neutron star models in $f(R)$ theories of gravity attracted considerable interest recently as a natural attempt to study the astrophysical applications of a class of alternative theories that gives promising results on cosmological scales, such as an alternative explanation of the accelerated expansion of the universe. Here we will not give a thorough review on the subject but instead focus mainly on the realistic astrophysically relevant models which were also used to construct the universal relations discussed in this chapter. We will concentrate on the particular case of R^2 gravity that is supposed to give the leading corrections for the neutron star structure. Moreover, the $f(R)$ theories of gravity are mathematically equivalent to a specific class of scalar-tensor theories with nonzero scalar field potential, as discussed in the previous subsection, which can simplify their treatment.

The initial work on the subject was concentrated mainly on discussing the existence of solutions and building such solutions (see for example Kobayashi and Maeda 2008; Upadhye and Hu 2009; Babichev and Langlois 2009; Jaime et al. 2011; Babichev and Langlois 2010). In the beginning there was some controversy on the question of whether compact star solutions exist in $f(R)$ theory but the overall studies showed that such stars can indeed be constructed. A drawback, though, is that solving the reduced field equations suffers from severe numerical instabilities. This can be easily demonstrated using the scalar-tensor formulation of $f(R)$ theory. Let us consider for example the case of R^2 gravity, i.e. when $f(R) = R + aR^2$. In this case the resulting scalar field potential will lead to a nonzero mass of the scalar field. Thus, similar to the massive scalar-tensor theories, the field equation for the scalar field (13.74) becomes stiff and thus leads to severe computational difficulties in certain cases. That is why, as a simplification, realistic neutron stars in R^2 gravity were first studied perturbatively (Cooney et al. 2010; Arapoglu et al. 2011) (see also Alavirad and Weller 2013 for the case of logarithmic $f(R)$ theory), i.e. instead of solving the full field equations a perturbative expansion in the parameter a was made. The studies in Yazadjiev et al. (2014) went beyond the perturbative approach calculating for the first time sequences of realistic neutron star models in R^2 gravity and comparing them with the observations. The calculations were performed using the mathematical equivalence to a particular class of scalar-tensor theories and they were later verified by deriving and calculating the full unperturbed field equations directly in $f(R)$ theories without going through scalar-tensor theories or different frames (Yazadjiev and Doneva 2015). The results showed that the nonperturbative results are not only quantitatively but also qualitatively very different from the results in the perturbative approach. Thus the perturbative approach is not applicable for $f(R)$ theories and in order to obtain correct results one has to calculate the full field equations.

The results in Yazadjiev et al. (2014) confirmed what was expected from the theory—in the limit when $a \to 0$ the solutions converge to the general relativistic ones and in the limit $a \to \infty$ the solutions tend to the case of massless Brans–Dicke theory with coupling function $\alpha(\varphi) = -1/\sqrt{3}$. Therefore, the neutron star solutions in R^2 gravity are bounded between two limiting cases which also puts an upper limit on the deviations from general relativity. This can be seen on Fig. 13.9 where different colors correspond to different dimensionless values of a and the case of $a = 10^4$ corresponds to nearly the maximum possible deviation from pure general relativity (we use the dimensionless parameter $a \to a/R_0^2$, where R_0 is one half of the solar gravitational radius $R_0 = 1.47664$ km, i.e. the solar mass in geometrical units). The Gravity Probe B experiment imposes the following constraints on the values of a, namely $a \lesssim 2.3 \times 10^5$ in the same dimension units (or $a \lesssim 5 \times 10^{11}$ m^2 in physical units) which means that all cases plotted on Fig. 13.9 fall into the allowed range of values of a. The studies for several equations of state in Yazadjiev et al. (2014) showed, though, that the deviations due to the change of the parameters a are of the same order of magnitude as the equation of state uncertainties which

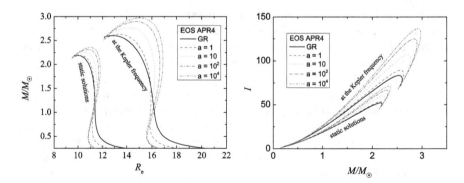

Fig. 13.9 The neutron star mass as a function of the radius (left panel) and the moment of inertia as a function of the mass (right panel) for static and rotating at the Kepler limit sequences of neutron stars in R^2 gravity. The results for several values of the parameter a are plotted for the APR4 EOS [Credit Ref. Yazadjiev et al. (2015)]

of course poses significant obstacle for testing the $f(R)$ theories. This is not the case, though, with the moment of inertia calculated in Staykov et al. (2014); Yazadjiev et al. (2015) where the differences with pure general relativity are better pronounced. Thus, the future observations of the neutron star moment of inertia can be used to discriminate between different gravitational theories in the strong field regime. Rapidly rotating models in R^2 gravity were calculated in Yazadjiev et al. (2015) and similar to the case of scalar-tensor theories, the rotation magnifies the deviations from pure general relativity considerably in comparison to the static case. The gravitational collapse in $f(R)$ theories was studied in Borisov et al. (2012); Cembranos et al. (2012). The astrophysical implication of the constructed neutron star solutions, such as the orbital and epicyclic frequencies, were considered in Staykov et al. (2015a) and the oscillations of neutron stars, including different asteroseismology relations, where examined in Staykov et al. (2015b). These studies showed that the deviations from Einstein's gravity are non-negligible, even though in some cases they are within the equation of state uncertainty. This gives us hope that the $f(R)$ theories would be better constrained in the future when the astrophysical observations limit further the allowed range of equations of state.

Neutron Stars in EdGB Gravity Neutron stars in the EdGB theories of gravity were constructed for the first time in Pani et al. (2011b) both in the static and the slowly rotating cases. The coupling function between the scalar field and the Gauss–Bonnet scalar in Eq. (13.89) is chosen to be

$$f_1(\varphi) = \alpha_{GB} e^{\beta \varphi}, \tag{13.91}$$

where α_{GB} and β are constants.

An interesting fact is that in the decoupling limit the monopole scalar charge, i.e. the coefficient in front of the $1/r$ term in the asymptotic of the scalar field at infinity, is zero unlike for example the massless scalar-tensor theories.[23] Thus there would be no scalar dipole radiation and it is not possible to impose constraints on the theory from the binary pulsar observations (Yagi et al. 2016).

In Figs. 13.10 the mass as a function of the central energy density and the moment of inertia as a function of the mass for neutron stars in EdGB gravity are plotted. As one can see, contrary to scalar-tensor and $f(R)$ theories, the maximum neutron star mass decreases with the increase of α_{GB} for fixed β. Even more, it was proven in Pani et al. (2011b) that for fixed $\alpha_{GB}\beta$ no neutron star solutions exist above some critical maximum central energy density (the exact limit is a quite lengthy expression and can be found in Pani et al. 2011b). This can serve as a way to impose constraint on the theory if the nuclear matter equation of state is known with a good accuracy—one should simply require that $\alpha_{GB}\beta$ is below some critical values chosen in such a way that the maximum mass is above the two solar mass barrier (Demorest et al. 2010; Antoniadis et al. 2013).

---

[23]The neutron stars, though, can still have nonzero higher scalar multipoles and thus nontrivial scalar field.

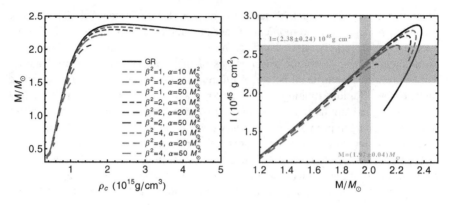

Fig. 13.10 The mass as a function of the central energy density (left panel) and the moment of inertia as a function of the mass (right panel) for neutron stars in EdGB gravity using the APR EOS. Models for different values of the Gauss–Bonnet coupling constants α (denoted by α_{GB} in the text) and β are shown. In the right panel the recent observation of a neutron star with $M \approx 2M_\odot$ and a possible future observation of the moment of inertia confirming general relativity within 10% Lattimer and Schutz (2005) are shown [Credit Ref. Pani et al. (2011b)]

As we mentioned above, one of the hopes to test the deviations from pure general relativity is via the future observations of the neutron star moment of inertia. Unfortunately, the studies showed that for EdGB gravity this would not be possible since the moment of inertia deviate from the one in pure Einstein's theory by a few percents at most, that is inside the expected observational error.

The axial quasinormal modes of neutron stars in EdGB were examined in Blázquez-Salcedo et al. (2016). The frequencies of the fundamental spacetime modes increase compare to pure general relativity. An interesting observation is that the universal (equation of state independent) relations for the oscillation modes that are available in Einstein's theory still hold for the EdGB gravity that can be used to put constraints on the parameters of the theory.

Rapidly rotating neutron star models in EdGB gravity were calculated in Kleihaus et al. (2014, 2016). The properties of the rotating compact stars are examined there in detail and the quadrupole moment is calculated as well. The mass-radius relation for rapidly rotating neutron stars is shown in Fig. 13.11.

Neutron Stars in dCS Gravity As we commented above, static neutron stars in dCS gravity are identical with the pure general relativistic ones and differences are present only in the rotating case. Up to now only models in the slow rotation approximation were calculated at first order in the rotation (Yunes et al. 2010; Ali-Haimoud and Chen 2011) and later at second order (Yagi et al. 2013). Up to leading order in rotation only the moment of inertia is affected by the dCS gravity, while the mass-radius relation remains the same as in Einstein's theory.

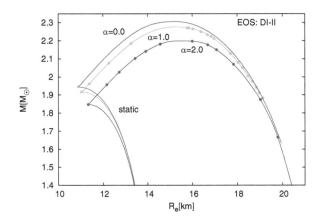

Fig. 13.11 The mass-radius relation for different values of the Gauss–Bonnet coupling constant $\alpha = 0$, 1 and 2 ($\alpha \equiv \alpha_{GB}/M_\odot^2$) for the EOS DI-II (the parameter β in Eq. (13.91) is fixed to $\beta = -1$). The chosen values of α are below the upper bound obtained from low mass x-ray binaries (Yagi 2012) which would correspond to $\alpha = 12$ when converted to the dimensionless α used in the graph. For a given α the left boundary curve represents the sequence of static solutions, while the right boundary curve represents the sequence of neutron stars rotating at the Kepler limit. Both are connected by the secular instability line [Credit Ref. Kleihaus et al. (2016)]

The change in the moment of inertia $\Delta I_{CS}/I_{GR}$ induced by the CS gravity as a function of the CS coupling strength is shown in Fig. 13.12. As one can see, depending on the Chern–Simons coupling constant, the moment of inertia can deviate significantly from general relativity and if we assume that the accuracy of the future observations of the neutron star moment of inertia are of the order of 10%, the Chern–Simons coupling constant can be constrained several orders of magnitude better than the current estimates.

The calculation of neutron stars in dCS gravity up to second order in rotation allowed to calculate the mass quadrupole moment and the rotational corrections to the mass and radius that is plotted in Fig. 13.13. Unfortunately, it turns out that these corrections are inside the EOS uncertainty and they can not be used to test the dCS gravity with the present observations. The corrections to the post-Keplerian parameters are also too small to be observable nowadays via double binary pulsars.

13.2.3 Universal Relations in Alternative Theories

Neutron star universal relations offer a very important tool for testing alternative theories of gravity, because the equation of state uncertainty, that causes a lot of problems as we have already mentioned, is taken out of the picture. There are three classes of universal relation in generalised theories of gravity that have been discussed in the literature so far. These are the gravitational waves asteroseismology relations, the I-Love-Q relations, and relations between the moment of inertia and

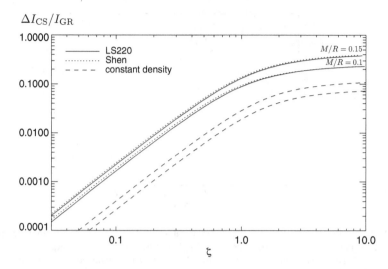

Fig. 13.12 Change in the moment of inertia $\Delta I_{CS}/I_{GR}$ induced by the CS modification as a function of the CS coupling strength ζ ($\zeta \equiv 4\sqrt{2}\alpha_{CS}M/R^3$), calculated using first order corrections in the rotational frequency for the two EOSs. The quantity $\Delta I_{CS}/I_{GR}$ is plotted for the values $M/R = 0.1$ and 0.15. The dashed lines show the analytic result for a constant density nonrelativistic star [Credit Ref. Ali-Haimoud and Chen (2011)]

the compactness of neutron stars. In general relativity we also discussed the 3-hair relations for the multipole moments but this topic has not been tackled yet in alternative theories of gravity. Below we will comment on the three former classes in further detail.

The oscillations of neutron stars are directly related to the emitted gravitational wave signal. That is why relations that connect the neutron star parameters to the oscillations frequencies and damping times, the so-called gravitational wave astereseismology relations, were extensively studied in pure Einstein's theory. They can be used in practice when gravitational waves from oscillating neutron stars are observed in the future (see discussion in Sect. 13.2.1.4). The extension to alternative theories of gravity was done in Sotani and Kokkotas (2004, 2005) for the case of massless scalar-tensor theories, in Staykov et al. (2015b) for $f(R)$ theories in the static case. The results there show, that if we restrict ourselves to values of the parameters that are in agreement with the current observational constraints, the relations are as equation of state independent as in pure general relativity but the deviations from pure Einstein's theory are very small in most of the cases. For example in R^2 gravity one can use an expression like the one in Eq. (13.47) with a small variation to the coefficients to fit the f-mode frequency in terms of η. Extensions to other alternative theories of gravity would be interesting, especially the massive scalar-tensor theory case that can produce very large differences. The first studies of oscillations modes of rapidly rotating neutron stars in alternative theories of gravity were performed very recently (Yazadjiev et al. 2017) in the case of (massive) scalar-tensor theories and the results show that the deviations from pure

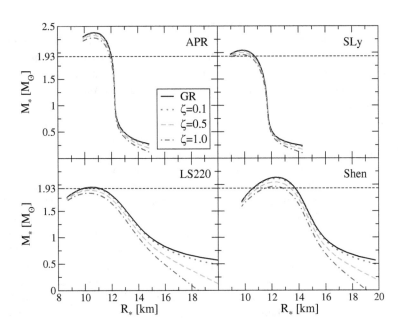

Fig. 13.13 The mass-radius relations for neutron stars in pure general relativity (solid curves) and dCS with different coupling strengths ζ ($\zeta \equiv 4\alpha_{CS}^2 M^2/R^6$) calculated using second order corrections in the rotational frequency. Results for several equations of state are shown. The mass-radius relations depend on the spin of the neutron star, which is set to match that for PSR J1614–2230 [Credit Ref. Yagi et al. (2013)]

general relativity can be large. Asteroseismology relations in the rapidly rotating case are still lacking.

Another set of relations that attracted a lot of attention recently are the I-Love-Q relations, that are discussed in a lot of detail in the rest of the chapter for the pure general relativistic case. A natural question is whether we can use them to constrain the strong field regime of gravity. If such universal relations hold in alternative theories as well and we are able to determine independently via observations two quantities from the I-Love-Q trio, then possible deviations from general relativity can be detected. With this motivation in mind, such relations were examined in Yagi and Yunes (2013a); Yagi and Yunes (2013b) for the dCS gravity, in Pani and Berti (2014); Doneva et al. (2014b) for massless and in Doneva and Yazadjiev (2016) for massive scalar-tensor theories, in Kleihaus et al. (2014) for the EdGB theory, in Sham et al. (2014) for the Eddington-inspired Born-Infeld (EiBI) theory and in Doneva et al. (2015) for $f(R)$ theories. From these studies only the papers (Doneva et al. 2014b; Kleihaus et al. 2014; Doneva et al. 2015; Doneva and Yazadjiev 2016) include the rapidly rotating case. It turned out that for all these theories the equation of state independence is preserved up to a large extend. The deviations from pure general relativity, though, are almost negligible for a big portion of the cases if one considers the allowed range of parameters of the corresponding theory. These are the massless STT, the EdGB and the EiBI theory. Larger differences on the other hand are observed for dCS gravity, massive STT and $f(R)$ theories. One should note that all these conclusions are true only for the normalized relations, if one considers the non-normalized relations then the deviations from pure general relativity can reach very large values. This means that the normalization that is used for the I-Love-Q relations is so good that not only the equation of state dependence is taken away, but in some cases also the dependence on the particular theory of gravity. Therefore, it is unclear so far up to what extend the I-Love-Q relations can be used to test the various alternative theories of gravity.

We should note here that, as with I-Love-Q, 3-hair relations could also be used to test gravity modifications, but as we have mentioned earlier these relations have not been studied yet in alternative theories of gravity. It is possible that the 3-hair relations in alternative theories can be quite different from their general relativity counterparts. The caveat here is that it might not be possible to define multipole moments in all classes of alternative theories of gravity, the way that they are defined in general relativity. This has been done so far only in scalar-tensor theory for a massless scalar field (Pappas and Sotiriou 2015b), where possible astrophysical applications in terms of these moments have been explored (Pappas and Sotiriou 2015a). We should note here that the quadrupole moment has been calculated in some additional cases such as in the case of EdGB theory by Kleihaus et al. (2016, 2014) as well as in R^2 gravity by Doneva et al. (2015), where the I-Q relations in these theories were compared against the corresponding relations in general relativity. Another case where a general definition of multipole moments has been given, is the case of $f(R)$ theories by Suvorov and Melatos (2016), although they investigate only Ricci flat cases where there are no extra degrees of freedom present and the corresponding multipole moments are equivalent to those of general relativity.

Finally, relations between different normalizations of the moment of inertia and the compactness of neutron stars, including also relations involving the maximum stellar mass, were examined for STT and $f(R)$ theories (Staykov et al. 2016) motivated by the work in pure general relativity (Breu and Rezzolla 2016; Lattimer and Schutz 2005).[24] The results showed that the equation of state universality is as good as in Einstein's theory of gravity but the deviations from general relativity can be large in certain cases,[25] especially for the relations involving the maximum stellar mass, that can be used as a probe of the gravitational theories. Specifically, Staykov et al. (2016) studied neutron star models using the following theories: (1) a scalar-tensor theory (denoted as STT) with a massless scalar field ($V(\varphi) = 0$) and a conformal factor, that relates the Einstein frame to the Jordan frame, of the

---

[24]I-C relations have been also studied in Horndenski and beyond-Horndeski gravity (Maselli et al. 2016; Babichev et al. 2016; Sakstein et al. 2017), as well as for theories with disformal couplings (Minamitsuji and Silva 2016), which are not covered here.

[25]For the I-C relations in EdGB, as we have noted earlier, this is not the case though. On the contrary we expect that the general relativistic relations to hold also in EdGB.

form $A(\varphi) = e^{\beta \varphi^2/2}$, with a chosen value for $\beta = -4.5$, which is within the range allowed by observational constrains and produces spontaneous scalarisation, (2) an $f(R)$ theory with the specific choice for the Lagrangian, $f(R) = R + aR^2$, where the coupling parameter a is in km^2 in geometric units and has the value $a = 10^4 M_\odot^2$, where the mass of the Sun is $M_\odot = 1.477$ km, and finally (3) general relativity. For these theories and the various models that they constructed, they fitted the moment of inertia in terms of the compactness $\mathcal{C}$ for two different choices of normalisation using two different polynomial forms following Lattimer and Schutz (2005) and Breu and Rezzolla (2016), i.e.,

$$\bar{I} = I/(MR^2) = \tilde{a}_0 + \tilde{a}_1 \mathcal{C} + \tilde{a}_2 \mathcal{C}^4, \quad \text{and} \quad \bar{I} = I/M^3 = \bar{a}_1 \mathcal{C}^{-1} + \bar{a}_2 \mathcal{C}^{-2} + \bar{a}_3 \mathcal{C}^{-3} + \bar{a}_4 \mathcal{C}^{-4},$$
(13.92)

respectively. The coefficients of the fits for the three theories and for three rotation rates are given in Table 13.6.

For the relations that concern the maximum stellar mass in the various theories, Staykov et al. (2016) used the following two fitting functions, one where the mass is normalised with respect to the maximum non-rotating mass, M_{TOV}, and another

Table 13.6 The fitting coefficients for the fit given by Eqs. (13.92)

$\bar{I} = I/(MR^2)$				$\bar{I} = I/M^3$			
	$\tilde{a}_0$	$\tilde{a}_1$	$\tilde{a}_2$	$\bar{a}_1$	$\bar{a}_2$	$\bar{a}_3$	$\bar{a}_4$
GR							
Slow. rot.	0.210	0.824	2.480	1.165	0.0538	0.0259	-0.00144
$\chi = 0.2$	0.211	0.788	3.135	1.077	0.100	0.0172	-9.830×10^{-4}
$\chi = 0.4$	0.200	0.823	2.469	1.024	0.123	0.0129	-7.985×10^{-4}
$\chi = 0.6$	0.176	0.839	2.393	0.943	0.143	0.00714	-5.539×10^{-4}
STT							
Slow. rot.	0.201	0.897	0.603	1.057	0.110	0.0173	-0.00103
$\chi = 0.2$	0.196	0.909	0.224	0.884	0.197	0.00270	-3.254×10^{-4}
$\chi = 0.4$	0.171	1.055	-2.985	0.664	0.313	-0.0163	5.647×10^{-4}
$\chi = 0.6$	0.127	1.256	-7.190	0.257	0.513	-0.0492	0.00202
$f(\mathcal{R})$							
Slow. rot.	0.221	0.981	-0.755	0.941	0.214	0.00521	-4.412×10^{-4}
$\chi = 0.2$	0.216	0.989	-1.255	0.853	0.243	0.00190	-3.841×10^{-4}
$\chi = 0.4$	0.208	1.011	-1.748	0.805	0.264	-0.00201	-1.978×10^{-4}
$\chi = 0.6$	0.201	0.998	-2.143	0.725	0.280	-0.00407	-1.706×10^{-4}

The results are given for the slowly rotating cases and rapidly rotating cases with $\chi = 0.2$, $\chi = 0.4$, $\chi = 0.6$, for the GR case, followed by the STT case and by the $f(R)$ case. The table is taken from Staykov et al. (2016)

Table 13.7 The fitting coefficients for the maximal mass for the fits (13.93)

	a_0	a_1	a_2
$M/M_{TOV}\ (J/J_{Kep})$			
GR	1	0.251	−0.0626
$f(\mathcal{R})$	1	0.382	−0.127
STT	–	–	–
$M/M_{Kep}\ (J/J_{Kep})$			
GR	0.844	0.203	−0.0484
$f(\mathcal{R})$	0.800	0.290	−0.0933
STT	0.724	0.403	−0.126

The table is taken from Staykov et al. (2016)

Fig. 13.14 The maximal mass normalised to the maximal Keplerian mass as a function of the angular momentum, normalised to the maximal Keplerian one for the three different theories. The curves are for the fit coefficients in Table 13.7

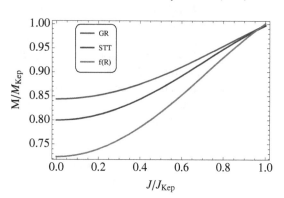

normalised with the maximum mass of the models rotating at the Kepler limit, M_{Kep},

$$\frac{M}{M_{TOV}} = 1 + a_1 \left(\frac{J}{J_{Kep}}\right)^2 + a_2 \left(\frac{J}{J_{Kep}}\right)^4, \quad \text{and} \quad \frac{M}{M_{Kep}} = a_0 + a_1 \left(\frac{J}{J_{Kep}}\right)^2 + a_2 \left(\frac{J}{J_{Kep}}\right)^4,$$

(13.93)

where the normalisation of the rotation is done with the angular momentum at the Kepler limit J_{Kep}. As one can see in Table 13.7, where the various coefficients of the fits are given, there is no fit for the STT case in the $\frac{M}{M_{TOV}}$ normalisation. This is because in this particular scalar-tensor theory the neutron star models get spontaneously scalarised, which means that the models of the theory are the same as the corresponding models in general relativity with a vanishing scalar field, until one gets to some critical value of central density and rotation and then the scalarised solutions appear. This causes a discontinuity in the behaviour of the M/M_{TOV} and the results are quite scattered, also because the point where spontaneous scalarisation starts depends on the equation of state. This is the reason why the second normalisation with respect to M_{Kep} was also considered in Staykov et al. (2016). This latter normalisation gives a nice behaviour without discontinuities and significantly less scattering in the STT case. The very interesting result is that the three different theories separate and are distinguishable as one can see in Fig. 13.14.

13.3   Present Challenges and Future Prospects

As we have seen, the topic of universal relations in general relativity is quite wide with a variety of relations having already been found. This does not mean that there are no more topics to be explored and progress to be made. One such case is the question of what happens in the presence of strong magnetic fields. We have seen already that there has been some work on I-Love-Q relations and magnetic fields (Haskell et al. 2014) but there has been no work on 3-hair relations and strong magnetic fields. In the I-Love-Q case we saw that strong magnetic fields and slow rotation can destroy universality, but it might be worth exploring the possibility that there is a more appropriate parameterisation that takes into account the magnetic field and in the end preserves universality. Universal relations that take into account magnetic fields could be useful in observationally constraining the strength of the magnetic field in neutron stars.

There is also a lot of work to be done on the quasi-normal modes universal relations. So far the calculations have been done mainly using the Cowling approximation. It is very important therefore that the quasi-normal modes are calculated beyond that approximation and it is verified that the universal relations that have been found for the various modes and decay times still hold. This will enable the more accurate determination of the various neutron star parameters that can be extracted by the quasi-normal modes observations from gravitational waves. Another question that could be of interest in the quasi-normal modes topic is to identify the theoretical/mathematical reason behind the existence of the universality in the modes.

Further expanding on this, it would be interesting to know if the various universal relations can be traced to a common mechanism or not. We should note here that relations like the I-Love-Q, the 3-hair or the quasi-normal modes universal relations are relations between "integral" or average quantities of the entire star. The question then is, are there any other quantities that are local and satisfy universal relations or is it that such relations can exist only between particularly weighted average quantities?

Finally, there is a lot of work to be done on the front of utilising universal relations for making measurements of astrophysical observables from neutron stars. Some work has already been done on using the I-Love-Q relations in order to break degeneracies in gravitational wave observations (see for example Yagi and Yunes 2013a; Yagi and Yunes 2013b) and there has been some preliminary work on using the 3-hair relations to model electromagnetic observables from X-ray binary systems in order to measure neutron star parameters (Pappas 2015; Tsang and Pappas 2016; Pappas 2017), but there is a lot more work to be done.

Turning to neutron stars in alternative theories, even though the topic has been studied for several decades, there is still a lot to be done in the field. One of the main reasons for this is the complexity of the equations especially if one wants to consider not only static but also rapidly rotating neutron stars. Thus the studies have to be extended in two main directions—constructing new neutron star models in alternative theories of gravity and studying their astrophysical implication in further detail. Below we will discuss both of them.

Constructing compact star models for a broader spectrum of alternative theories of gravity, such as more general or more realistic cases of tensor-multi scalar theories, Quadratic gravity, Horava gravity, Lorentz-violating theories, etc., is important since there are many extensions of Einstein's theory and the neutron stars offers a unique way of testing them. In addition, for some alternative theories of gravity a perturbative approach was used for constructing the solutions. This is a drawback since some strong field effects can be omitted or even worse, the perturbative approach can be misleading in certain cases such as the $f(R)$ gravity where it was explicitly shown that the non-perturbative results are not only quantitatively but also qualitatively different from the perturbative ones. As far as rotation is concerned, rapidly rotating neutron star solutions were constructed up to now only in three classes of alternative theories—the scalar-tensor theories, $f(R)$ theories and Einstein–dilaton–Gauss–Bonnet theory. All these studies show that the rapid rotation offer new phenomenology that should be further explored in other generalized theories of gravity. In addition, the full stability of neutron stars in alternative theories of gravity has not been studied yet, not even in the case of scalar-tensor theories.

The astrophysical implications of neutron stars, such as stellar collapse, neutron star mergers, quasiperiodic oscillations, absorption lines, etc. are scarcely studied in most of the alternative theories of gravity, with the exception of scalar-tensor theories and some sectors for the quadratic gravity. One of the problems there is that the observations themselves either suffer from large uncertainties or even there are multiple astrophysical models explaining the same observations that makes testing the strong field regime of gravity extremely difficult. Nevertheless, neutron stars are ones of the very few astrophysical objects where the strong field regime of gravity can be explored and that is why further studies in this directions are needed.

Last but not least, larger efforts should be put in the directions of breaking the degeneracy between the effects coming from the alternative theories of gravity and the uncertainties of the nuclear matter equation of state. Currently this degeneracy is plaguing a very large portion of the attempts to constrain the strong field regime of gravity. Some advance in this direction is actually expected to come in the near future since the astrophysical observations are narrowing the possible set of nuclear matter equations of state more and more. One thing we should always keep in mind, though, is that the interpretation of these observations is always done within general relativity, or even some type of Newtonian approximation. Therefore, an interesting study that deserved to be done is to explore whether the alternative theories of gravity could change the corresponding predictions.

On this note, the study of universal relations in alternative theories of gravity is very important, because in principle such relations could be used to test deviations from general relativity while evading the equation of state degeneracies. So far work has been done on the I-Love-Q relations in alternative theories of gravity and in most of the cases the I-Love-Q relations have been found to be identical to those that hold in general relativity. Exception to this are the cases of dCS gravity, massive scalar-tensor theories and $f(R)$ theories. At a first glance therefore one would say that the applicability of I-Love-Q relations in testing theories alternative to general relativity is limited. Still, more work needs to be done before we can definitively decide on the extent of the usefulness of I-Love-Q relations as tests of our theories of gravity.

Beyond the I-Love-Q relations, there are other relations, discussed in Sect. 13.2.3 that show more promise in distinguishing different theories. The exploration of these relations is only the beginning. There are many more relations discussed in Sect. 13.2.1 that should be also explored in other theories as well. Two important classes of relations that should be explored in alternative theories are the quasi-normal modes relations and the 3-hair relations. The quasi-normal modes relations are important to be studied in alternative theories especially in this moment in time since the opening of the gravitational waves observational window offers a unique opportunity to probe the structure of compact objects. So far, studies of gravitational waves emission in alternative theories are missing from the literature. These studies are very difficult but need to be made in order to better constrain possible deviations from general relativity.

3-hair relations in alternative theories will be easier to study and already some steps towards that direction are being made. For example multipole moments in some classes of alternative theories have been defined (Pappas and Sotiriou 2015b; Suvorov and Melatos 2016), while there exist numerical codes that can calculate neutron star models in these theories (Doneva et al. 2013c). Again a possible issue might be to properly define the normalised quantities that will enter the various universal relations in the different theories. Also, it will be necessary to have a good correspondence between the quantities in different theories. At this point there is an issue that will have to be addressed in order to make progress. Multipole moments are defined in general relativity as asymptotic quantities (these are the Geroch–Hansen multipole moments) that characterise fields that exist on asymptotically flat spacetimes. This definition will not be always possible in all classes of alternative theories of gravity. Therefore one should be careful when talking about moments in other theories, making sure that the asymptotic moments are meaningful quantities. Furthermore, in the cases that asymptotic moments (a la Geroch–Hansen) cannot be defined, alternative quantities should be considered. Possible alternatives could be source integrals like the ones given in the case of general relativity by Gürlebeck (2014) for example, or maybe other type of source integrals like those given by Hernández-Pastora et al. (2016). In that case one would have to re-evaluate all the relevant quantities in general relativity as well and find the new relations that they will satisfy. In addition one will also have to find a way to relate these new quantities to astrophysical observables as well as some of the usual neutron star properties, such as the moment of inertia for example.[26]

---

[26]It is also worth noting that one could take a theory independent approach in studying neutron stars and universal relations in alternative theories of gravity, such as the post-TOV approach (Glampedakis et al. 2015, 2016).

Finally there is an elephant in the room that we should address. As we place more constrains on the equation of state from astrophysics, the notion of universal relations will become meaningless. This is not expected to happen soon and probably there will be faster progress on universal relations than on the equation of state question, but eventually it will happen. Will universal relations become obsolete? The answer is, not necessarily. If the equation of state is determined, then it will be in all likelihood one that will preserve the universal relations. These relations can express quantities (neutron star parameters) in terms of other quantities and therefore will still be a tool for determining parameters that are difficult to measure directly. In addition, such relations can serve as consistency checks for our various astrophysical models that enter the "measurement" of different quantities from observations. Therefore they will continue to be a useful tool to do astrophysics. Furthermore, the class of these relations that can distinguish between different theories of gravity will be a particularly useful tool to do some fundamental physics by testing general relativity and even possibly selecting a likely alternative, or at least excluding unlikely ones.

13.4 Conclusions

This chapter has covered two quite wide topics, that of neutron star universal relations and that of neutron stars in alternative theories of gravity. We have tried to present the fullest possible spectrum of universal relations in general relativity, but inevitably some results were covered very briefly or not at all. Similarly, for neutron stars in alternative theories of gravity we have focused on the better studied classes of these theories, such as scalar-tensor gravity, $f(R)$, EdGB and CS. These are all theories for which neutron star universal relations have been considered. Our aim has been to present some of the basic ideas, giving some attention to the most important aspects of them, and then focus on the most important results.

As it was discussed, the notion of neutron star universal relations is not new, but the field gained a lot of momentum the last few years and has opened the way to circumvent the uncertainties of the equation of state. With respect to isolated neutron stars, one could say that the main classes of universal relations are three. The first class are the I-Love-Q relations which relate the moment of inertia, the quadrupolar love number and the mass quadrupole of a neutron star. The second class is the 3-hair relations that relate the higher order multipole moments of the spacetime around neutron stars to the first three non-zero multipole moments, i.e., the mass, the angular momentum and the mass quadrupole. Finally the third class are the universal relations of the oscillation frequencies, which relate the real parts and the imaginary parts of neutron star QNMs to its other parameters such as the moment of inertia or the compactness or the average density. Relations like the I-Love-Q and the 3-hair relations, characterise both neutron and quark stars and can have many useful applications. They can be used to measure neutron star properties that are hard to measure directly or they can be used to break degeneracies between parameters that enter the description of observables. They could also be used to test our assumptions or our models of astrophysical mechanisms, if one could measure one member of the I-Love-Q trio in one way and infer another member from another observation and an assumed model for example. Furthermore, by measuring the neutron star parameters from different channels, with the help of the universal relations, we could use our knowledge of these parameters to solve the inverse problem and constrain initially and maybe finally identify the equation of state for matter at supra-nuclear densities. In addition, with the opening of the gravitational wave observational window, we expect to learn many things for the structure of neutron stars from the mergers of binary systems and the universal relations of the oscillation frequencies will play an important part since the fundamental frequencies of the remnants will be some of our primary observables. In addition, apart from using the various universal relations, one could also test for their validity. In this way one could test the assumptions that enter the various relations, such as the assumptions on the equation of state, i.e., whether our current models are correct or not. Even more, one could test one of the most fundamental assumptions of all, that of the theory of gravity that we are using to construct neutron stars.

Modifications or extensions to general relativity are motivated for several reasons, such as cosmological and astrophysical considerations like the nature of dark matter and dark energy questions, as well as more theoretical considerations coming from attempts to create a theory of everything where additional degrees of freedom are introduced to the low energy description of such a theory. The resulting modifications can have the form of additional scalar fields or modifications of the Einstein–Hilbert action with the introduction of higher powers or arbitrary functions of the Ricci scalar or various contractions of the Riemann tensor. Apart from the cosmological implications of such modifications, the implications for the structure of compact objects in general and neutron stars in particular are of extreme interest and this is why neutron stars have been extensively studied in such theories. Here we focused on the most widely studied classes of these theories, i.e., scalar-tensor gravity, $f(R)$, EdGB and dCS, that are also the ones for which there have been studies of various universal relations beyond general relativity. When studying neutron stars in these theories one needs to also take into account the constrains that exist on various scales: from laboratory and solar system tests of gravity to constraints coming from astrophysical and cosmological observations, as well as viability restrictions to the theories. Although the aforementioned constrains can be very strong for some parameters of modified theories and the viability considerations limit the possible options, there is still the possibility to find theories that are viable and free of pathologies (like tachyonic instabilities and such), to which the various classes that we present here belong, and have neutron stars that are quite different from their general relativistic counterparts. One such example are neutron stars in massive scalar-tensor theory that exhibit spontaneous scalarisation and another example are neutron stars in R^2-gravity, a subclass of $f(R)$ theories. In both cases neutron stars can be quite different than their general relativistic counterparts, which makes them very interesting astrophysical targets. On the other hand, the deviations of neutron star masses and radii constructed in EdGB and dCS from their general relativistic counterparts are quite small, making them more difficult to distinguish with astrophysical observations (we should note though that in dCS only slow rotation models have been studied so far). The main source of uncertainty again is the unknown equation of state and any changes in the structure and the parameters of neutron stars in modified theories needs to be measured against the range of these parameters that is possible within general relativity and comes from choosing different equations of state.

This is where universal relations come into play. Comparing the relations in general relativity against their counterparts in modified theories of gravity has the potential to distinguish different theories even though the uncertainty due to the equation of state seems to give a mixed picture. Such success stories are the applications of the I-Love-Q relations in dCS and massive scalar-tensor or $f(R)$ theories, where the relevant relations have been found to differ from their general relativistic counterparts. In contrast, theories such as the massless scalar-tensor or EdGB seem to have the same relations as in general relativity, if we restrict ourselves to values of the parameters of the theory allowed by the observations. Apart from the I-Love-Q relations, some other classes of promising relations have emerged such as the relations that express the maximum neutron star mass in terms of the maximum non-rotating mass and the rotation rate, which seem to be more potent in distinguishing different theories of gravity. Finally, there is plenty of work to be done in other directions as well, such as the study of 3-hair relations in alternative theories, as well as extending work on the study of neutron star QNMs in alternative theories.

In conclusion, with respect to universal relations, although their usefulness now that the equation of state is still uncertain is quite obvious, it is important to stress that these relations will still be useful even after the equation of state has been determined. This is because they will still express useful relations between different properties of neutron stars, that can be utilised in the analysis of observations and in measuring the full range of the relevant neutron star parameters.

Acknowledgements We would like thank Kent Yagi for providing us data for some of the plots as well as useful comments on the manuscript. We would also like to thank for their many useful comments, Hector O. Silva, Kostas Glampedakis, Jutta Kunz and Stoytcho Yazadjiev.

G.P. is supported by FCT contract IF/00797/2014/CP1214/CT0012 under the IF2014 Programme.

D.D. would like to thank the European Social Fund, the "Ministry of Science, Research and the Arts" Baden-Wurttemberg and the Bulgarian NSF Grant DFNI T02/6 for the support. DD is also indebted to the Baden-Württemberg Stiftung for the financial support of by the Eliteprogramme for Postdocs.

References

Alavirad, H., Weller, J.M.: Phys. Rev. **D88**, 124034 (2013). arXiv:1307.7977 [gr-qc]

Alexander, S., Yunes, N.: Phys. Rep. **480**, 1 (2009). arXiv:0907.2562 [hep-th]

AlGendy, M., Morsink, S.M.: Astrophys. J. **791**, 78 (2014). arXiv:1404.0609 [astro-ph.HE]

Ali-Haimoud, Y. and Chen, Y.: Phys. Rev. **D84**, 124033 (2011). arXiv:1110.5329 [astro-ph.HE]

Alsing, J., Berti, E., Will, C.M., Zaglauer, H.: Phys. Rev. D **85**, 064041 (2012). arXiv:1112.4903 [gr-qc]

Andersson, N., Kokkotas, K.D.: Phys. Rev. Lett. **77**, 4134 (1996). gr-qc/9610035

Andersson, N., Kokkotas, K.D.: Mon. Not. R. Astron. Soc. **299**, 1059 (1998). gr-qc/9711088

Antoniadis, J., Freire, P.C., Wex, N., Tauris, T.M., Lynch, R.S., et al.: Science **340**, 6131 (2013). arXiv:1304.6875 [astro-ph.HE]

Arapoglu, A.S., Deliduman, C., Eksi, K.Y.: J. Cosmol. Astropart. Phys. **1107**, 020 (2011). arXiv:1003.3179 [gr-qc]

Babichev, E., Langlois, D.: Phys. Rev. **D80**, 121501 (2009) [Erratum: Phys. Rev. **D81**, 069901 (2010)]. arXiv:0904.1382 [gr-qc]

Babichev, E., Langlois, D.: Phys. Rev. **D81**, 124051 (2010). arXiv:0911.1297 [gr-qc]

Babichev, E., Koyama, K., Langlois, D., Saito, R., Sakstein, J.: Classical Quantum Gravity **33**, 235014 (2016). arXiv:1606.06627 [gr-qc]

Barausse, E., Palenzuela, C., Ponce, M., Lehner, L.: Phys. Rev. **D87**, 081506 (2013). arXiv:1212.5053 [gr-qc]

Bauböck, M., Berti, E., Psaltis, D., Özel, F.: Astrophys. J. **777**, 68 (2013). arXiv:1306.0569 [astro-ph.HE]

Bejger, M., Haensel, P.: Astron. Astrophys. **396**, 917 (2002). astro-ph/0209151

Benhar, O., Ferrari, V., Gualtieri, L.: Phys. Rev. D **70**, 124015 (2004). astro-ph/0407529

Bernuzzi, S., Nagar, A., Balmelli, S., Dietrich, T., Ujevic, M.: Phys. Rev. Lett. **112**, 201101 (2014). arXiv:1402.6244 [gr-qc]

Berti, E., Stergioulas, N.: Mon. Not. R. Astron. Soc. **350**, 1416 (2004). gr-qc/0310061

Berti, E., White, F., Maniopoulou, A., Bruni, M.: Mon. Not. R. Astron. Soc. **358**, 923 (2005). arXiv:gr-qc/0405146 [gr-qc]

Berti, E., Gualtieri, L., Horbatsch, M., Alsing, J.: Phys. Rev. D **85**, 122005 (2012). arXiv:1204.4340 [gr-qc]

Berti, E., Barausse, E., Cardoso, V., Gualtieri, L., Pani, P., Sperhake, U., Stein, L.C., Wex, N., Yagi, K., Baker, T., Burgess, C.P., Coelho, F.S., Doneva, D., De Felice, A., Ferreira, P.G., Freire, P.C.C., Healy, J., Herdeiro, C., Horbatsch, M., Kleihaus, B., Klein, A., Kokkotas, K., Kunz, J., Laguna, P., Lang, R.N., Li, T.G.F., Littenberg, T., Matas, A., Mirshekari, S., Okawa, H., Radu, E., O'Shaughnessy, R., Sathyaprakash, B.S., Van Den Broeck, C., Winther, H.A., Witek, H., Emad Aghili, M., Alsing, J., Bolen, B., Bombelli, L., Caudill, S., Chen, L., Degollado, J.C., Fujita, R., Gao, C., Gerosa, D., Kamali, S., Silva, H.O., Rosa, J.G., Sadeghian, L., Sampaio, M., Sotani, H., Zilhao, M.: Classical Quantum Gravity **32**, 243001 (2015). arXiv:1501.07274 [gr-qc]

Blázquez-Salcedo, J.L., González-Romero, L.M., Navarro-Lérida, F.: Phys. Rev. D **87**, 104042 (2013). arXiv:1207.4651 [gr-qc]

Blázquez-Salcedo, J.L., González-Romero, L.M., Navarro-Lérida, F.: Phys. Rev. D **89**, 044006 (2014). arXiv:1307.1063 [gr-qc]

Blázquez-Salcedo, J.L., González-Romero, L.M., Kunz, J., Mojica, S., Navarro-Lérida, F.: Phys. Rev. **D93**, 024052 (2016). arXiv:1511.03960 [gr-qc]

Borisov, A., Jain, B., Zhang, P.: Phys. Rev. **D85**, 063518 (2012). arXiv:1102.4839 [astro-ph.CO]

Bowers, R.L., Liang, E.P.T.: Astrophys. J. **188**, 657 (1974)

Brans, C., Dicke, R.H.: Phys. Rev. **124**, 925 (1961)

Breu, C., Rezzolla, L.: Mon. Not. R. Astron. Soc. **459**, 646 (2016). arXiv:1601.06083 [gr-qc]

Buchdahl, H.A.: Phys. Rev. **116**, 1027 (1959)

Butterworth, E.M., Ipser, J.R.: Astrophys. J. **204**, 200 (1976)

Cardoso, V., Carucci, I.P., Pani, P., Sotiriou, T.P.: Phys. Rev. Lett. **111**, 111101 (2013a), arXiv:1308.6587 [gr-qc]

Cardoso, V., Carucci, I.P., Pani, P., Sotiriou, T.P.: Phys. Rev. D **88**, 044056 (2013b). arXiv:1305.6936 [gr-qc]

Cembranos, J.A.R., de la Cruz-Dombriz, A., Montes Nunez, B.: J. Cosmol. Astropart. Phys. **1204**, 021 (2012). arXiv:1201.1289 [gr-qc]

Chakrabarti, S., Delsate, T., Gürlebeck, N., Steinhoff, J.: Phys. Rev. Lett. **112**, 201102 (2014). arXiv:1311.6509 [gr-qc]

Chan, T.K., Sham, Y.-H., Leung, P.T., Lin, L.-M.: Phys. Rev. D **90**, 124023 (2014). arXiv:1408.3789 [gr-qc]

Chan, T.K., Chan, A.P.O., Leung, P.T.: Phys. Rev. D **91**, 044017 (2015). arXiv:1411.7141 [astro-ph.SR]

Chatziioannou, K., Yagi, K., Yunes, N.: Phys. Rev. D **90**, 064030 (2014). arXiv:1406.7135 [gr-qc]

Chirenti, C., de Souza, G.H., Kastaun, W.: Phys. Rev. D **91**, 044034 (2015). arXiv:1501.02970 [gr-qc]

Chirenti, C., Gold, R., Miller, M.C.: Astrophys. J. **837**, 67 (2017). arXiv:1612.07097 [astro-ph.HE]

Cooney, A., DeDeo, S., Psaltis, D.: Phys. Rev. **D82**, 064033 (2010). arXiv:0910.5480 [astro-ph.HE]

Damour, T., Esposito-Farese, G.: Classical Quantum Gravity **9**, 2093 (1992)

Damour, T., Esposito-Farèse, G.: Phys. Rev. Lett. **70**, 2220 (1993)

Damour, T., Esposito-Farèse, G.: Phys. Rev. **D54**, 1474 (1996). arXiv:gr-qc/9602056 [gr-qc]

DeDeo, S., Psaltis, D.: Phys. Rev. Lett. **90**, 141101 (2003), arXiv:astro-ph/0302095 [astro-ph]

DeDeo, S., Psaltis, D.: ArXiv Astrophysics e-prints (2004). astro-ph/0405067

De Felice, A., Tsujikawa, S.: Living Rev. Relativ. **13**, 3 (2010). arXiv:1002.4928 [gr-qc]

Delsate, T.: Phys. Rev. **D92**, 124001 (2015). arXiv:1504.07335 [gr-qc]

Demorest, P., Pennucci, T., Ransom, S., Roberts, M., Hessels, J.: Nature **467**, 1081 (2010). arXiv:1010.5788 [astro-ph.HE]

Dicke, R.H.: Phys. Rev. **125**, 2163 (1962)

Doneva, D.D., Kokkotas, K.D.: Phys. Rev. D **92**, 124004 (2015). arXiv:1507.06606 [astro-ph.SR]

Doneva, D.D., Yazadjiev, S.S.: Phys. Rev. **D85**, 124023 (2012). arXiv:1203.3963 [gr-qc]

Doneva, D.D., Yazadjiev, S.S.: J. Cosmol. Astropart. Phys. **1611**, 019 (2016). arXiv:1607.03299 [gr-qc]

Doneva, D.D., Yazadjiev, S.S., Kokkotas, K.D., Stefanov, I.Z., Todorov, M.D.: Phys. Rev. D **81**, 104030 (2010). arXiv:1001.3569 [gr-qc]

Doneva, D.D., Yazadjiev, S.S., Stergioulas, N., Kokkotas, K.D.: Astrophys. J. **781**, L6 (2013a). arXiv:1310.7436 [gr-qc]

Doneva, D.D., Gaertig, E., Kokkotas, K.D., Krüger, C.: Phys. Rev. **D88**, 044052 (2013b). arXiv:1305.7197 [astro-ph.SR]

Doneva, D.D., Yazadjiev, S.S., Stergioulas, N., Kokkotas, K.D.: Phys. Rev. **D88**, 084060 (2013c), arXiv:1309.0605 [gr-qc]

Doneva, D.D., Yazadjiev, S.S., Stergioulas, N., Kokkotas, K.D., Athanasiadis, T.M.: Phys. Rev. **D90**, 044004 (2014a), arXiv:1405.6976 [astro-ph.HE]

Doneva, D.D., Yazadjiev, S.S., Staykov, K.V., Kokkotas, K.D.: Phys. Rev. **D90**, 104021 (2014b), arXiv:1408.1641 [gr-qc]

Doneva, D.D., Yazadjiev, S.S., Kokkotas, K.D.: Phys. Rev. **D92**, 064015 (2015). arXiv:1507.00378 [gr-qc]

Fodor, G., Hoenselaers, C., Perjés, Z.: J. Math. Phys. **30**, 2252 (1989)

Freire, P.C.C., Wex, N., Esposito-Farèse, G., Verbiest, J.P.W., Bailes, M., Jacoby, B.A., Kramer, M., Stairs, I.H., Antoniadis, J., Janssen, G.H.: Mon. Not. R. Astron. Soc. **423**, 3328 (2012). arXiv:1205.1450 [astro-ph.GA]

Friedman, J., Stergioulas, N.: Rotating Relativistic Stars. Cambridge Monographs on Mathematical Physics. Cambridge University Press, Cambridge (2013)

Gaertig, E., Kokkotas, K.D.: Phys. Rev. **D83**, 064031 (2011). arXiv:1005.5228 [astro-ph.SR]

Geroch, R.P.: J. Math. Phys. **11**, 1955 (1970a)

Geroch, R.P.: J. Math. Phys. **11**, 2580 (1970b)

Gerosa, D., Sperhake, U., Ott, C.D.: Classical Quantum Gravity **33**, 135002 (2016). arXiv:1602.06952 [gr-qc]

Glampedakis, K., Pappas, G., Silva, H.O., Berti, E.: Phys. Rev. D **92**, 024056 (2015). arXiv:1504.02455 [gr-qc]

Glampedakis, K., Pappas, G., Silva, H.O., Berti, E.: Phys. Rev. D **94**, 044030 (2016). arXiv:1606.05106 [gr-qc]

Grumiller, D., Yunes, N.: Phys. Rev. D **77**, 044015 (2008). arXiv:0711.1868 [gr-qc]

Gürlebeck, N.: Phys. Rev. D **90**, 024041 (2014)

Hansen, R.O.: J. Math. Phys. **15**, 46 (1974)

Harada, T.: Prog. Theor. Phys. **98**, 359 (1997). arXiv:gr-qc/9706014 [gr-qc]

Harada, T.: Phys. Rev. **D57**, 4802 (1998). arXiv:gr-qc/9801049 [gr-qc]

Harada, T.P. , Chiba, T., Nakao, K.-I., Nakamura, T.: Phys. Rev. **D55**, 2024 (1997). arXiv:gr-qc/9611031 [gr-qc]

Hartle, J.B.: Astrophys. J. **150**, 1005 (1967)

Hartle, J.B., Thorne, K.S.: Astrophys. J. **153**, 807 (1968)

Hartle, J.B.: Phys. Rep. **46**, 201 (1978)

Haskell, B., Ciolfi, R., Pannarale, F., Rezzolla, L.: Mon. Not. R. Astron. Soc. **438**, L71 (2014). arXiv:1309.3885 [astro-ph.SR]

Hernández-Pastora, J.L., Martín-Martín, J., Ruiz, E.: Classical Quantum Gravity **33**, 225009 (2016). arXiv:1604.07192 [gr-qc]

Hessels, J.W.T., Ransom, S.M., Stairs, I.H., Freire, P.C.C., Kaspi, V.M., Camilo, F.: Science **311**, 1901 (2006). astro-ph/0601337

Hinderer, T.: Astrophys. J. **677**, 1216-1220 (2008). arXiv:0711.2420

Horbatsch, M.W., Burgess, C.P.: J. Cosmol. Astropart. Phys. **8**, 027 (2011). arXiv:1006.4411 [gr-qc]

Horbatsch, M., Silva, H.O., Gerosa, D., Pani, P., Berti, E., Gualtieri, L., Sperhake, U.: Classical Quantum Gravity **32**, 204001 (2015). arXiv:1505.07462 [gr-qc]

Horvat, D., Ilijic, S., Marunovic, A.: Classical Quantum Gravity **28**, 025009 (2011). arXiv:1010.0878 [gr-qc]

Jaime, L.G., Patino, L., Salgado, M.: Phys. Rev. **D83**, 024039 (2011). arXiv:1006.5747 [gr-qc]

Kanti, P., Mavromatos, N.E., Rizos, J., Tamvakis, K., Winstanley, E.: Phys. Rev. D **54**, 5049 (1996). hep-th/9511071

Kleihaus, B., Kunz, J., Mojica, S.: Phys. Rev. **D90**, 061501 (2014). arXiv:1407.6884 [gr-qc]

Kleihaus, B., Kunz, J., Yazadjiev, S.: Phys. Lett. **B744**, 406 (2015). arXiv:1503.01672 [gr-qc]

Kleihaus, B., Kunz, J., Mojica, S., Zagermann, M.: Phys. Rev. **D93**, 064077 (2016). arXiv:1601.05583 [gr-qc]

Kobayashi, T., Maeda, K.-I.: Phys. Rev. **D78**, 064019 (2008). arXiv:0807.2503 [astro-ph]

Laarakkers, W.G., Poisson, E.: Astrophys. J. **512**, 282 (1999). gr-qc/9709033

Landry, P., Poisson, E.: Phys. Rev. D **91**, 104026 (2015). arXiv:1504.06606 [gr-qc]

Lattimer, J.M., Lim, Y.: Astrophys. J. **771**, 51 (2013). arXiv:1203.4286 [nucl-th]

Lattimer, J.M., PrakashM.: Astrophys. J. **550**, 426 (2001). astro-ph/0002232

Lattimer, J.M., Schutz, B.F.: Astrophys. J. **629**, 979 (2005). astro-ph/0411470

Lattimer, J.M., Yahil, A.: Astrophys. J. **340**, 426 (1989)

Lau, H.K., Leung, P.T., Lin, L.M.: Astrophys. J. **714**, 1234 (2010). arXiv:0911.0131 [gr-qc]

Manko, V.S., Martín, J., Ruiz, E.: J. Math. Phys. **36**, 3063 (1995)

Martinon, G., Maselli, A., Gualtieri, L., Ferrari, V.: Phys. Rev. D **90**, 064026 (2014). arXiv:1406.7661 [gr-qc]

Maselli, A., Cardoso, V., Ferrari, V., Gualtieri, L., Pani, P.: Phys. Rev. D **88**, 023007 (2013). arXiv:1304.2052 [gr-qc]

Maselli, A., Silva, H.O., Minamitsuji, M., Berti, E.: Phys. Rev. D **93**, 124056 (2016). arXiv:1603.04876 [gr-qc]

Mendes, R.F.P.: Phys. Rev. **D91**, 064024 (2015), arXiv:1412.6789 [gr-qc]

Mendes, R.F.P., Ortiz, N.: Phys. Rev. **D93**, 124035 (2016), arXiv:1604.04175 [gr-qc]

Minamitsuji, M., Silva, H.O.: Phys. Rev. D **93**, 124041 (2016). arXiv:1604.07742 [gr-qc]

Morganstern, R.E., Chiu, H.-Y.: Phys. Rev. **157**, 1228 (1967)

Novak, J.: Phys. Rev. **D58**, 064019 (1998a). arXiv:gr-qc/9806022 [gr-qc]

Novak, J.: Phys. Rev. **D57**, 4789 (1998b). arXiv:gr-qc/9707041 [gr-qc]

Novak, J., Ibanez, J.M.: Astrophys. J. **533**, 392 (2000). arXiv:astro-ph/9911298 [astro-ph]

Pani, P., Berti, E.: Phys. Rev. **D90**, 024025 (2014). arXiv:1405.4547 [gr-qc]

Pani, P., Cardoso, V.: Phys. Rev. D **79**, 084031 (2009). arXiv:0902.1569 [gr-qc]

Pani, P., Macedo, C.F.B., Crispino, L.C.B., Cardoso, V.: Phys. Rev. D **84**, 087501 (2011a). arXiv:1109.3996 [gr-qc]

Pani, P., Berti, E., Cardoso, V., Read, J.: Phys. Rev. **D84**, 104035 (2011b). arXiv:1109.0928 [gr-qc]

Pappas, G.: Mon. Not. R. Astron. Soc. **422**, 2581 (2012). arXiv:1201.6071 [astro-ph.HE]

Pappas, G.: Mon. Not. R. Astron. Soc. **454**, 4066 (2015). arXiv:1506.07225 [astro-ph.HE]

Pappas, G.: Mon. Not. R. Astron. Soc. **466**, 4381 (2017). arXiv:1610.05370 [gr-qc]

Pappas, G., Apostolatos, T.A.: Classical Quantum Gravity **25**, 228002 (2008). arXiv:0803.0602 [gr-qc]

Pappas, G., Apostolatos, T.A.: Phys. Rev. Lett. **108**, 231104 (2012a). arXiv:1201.6067 [gr-qc]

Pappas, G., Apostolatos, T.A.: ArXiv e-prints (2012b). arXiv:1211.6299 [gr-qc]

Pappas, G., Apostolatos, T.A.: Mon. Not. R. Astron. Soc. **429**, 3007 (2013a). arXiv:1209.6148 [gr-qc]

Pappas, G., Apostolatos, T.A.: Mon. Not. R. Astron. Soc. **429**, 3007 (2013b). arXiv:1209.6148 [gr-qc]

Pappas, G., Apostolatos, T.A.: Phys. Rev. Lett. **112**, 121101 (2014). arXiv:1311.5508 [gr-qc]

Pappas, G., Sotiriou, T.P.: Mon. Not. R. Astron. Soc. **453**, 2862 (2015a). arXiv:1505.02882 [gr-qc]

Pappas, G., Sotiriou, T.P.: Phys. Rev. D **91**, 044011 (2015b). arXiv:1412.3494 [gr-qc]

Palenzuela, C., Liebling, S.L.: Phys. Rev. **D93**, 044009 (2016). arXiv:1510.03471 [gr-qc]

Palenzuela, C., Barausse, E., Ponce, M., Lehner, L.: Phys. Rev. **D89**, 044024 (2014). arXiv:1310.4481 [gr-qc]

Paschalidis, V., Stergioulas, N.: ArXiv e-prints (2016). arXiv:1612.03050 [astro-ph.HE]

Ponce, M., Palenzuela, C., Barausse, E., Lehner, L.: Phys. Rev. **D91**, 084038 (2015). arXiv:1410.0638 [gr-qc]

Popchev, D.: Bifurcations of neutron stars solutions in massive scalar-tensor theories of gravity. Master's thesis, University of Sofia (2015)

Quevedo, H.: Fortschritte der Physik/Prog. Phys. **38**, 733 (1990)

Ramazanoğlu, F.M., Pretorius, F.: Phys. Rev. **D93**, 064005 (2016). arXiv:1601.07475 [gr-qc]

Read, J.S., Baiotti, L., Creighton, J.D.E., Friedman, J.L., Giacomazzo, B., Kyutoku, K., Markakis, C., Rezzolla, L., Shibata, M., Taniguchi, K.: Phys. Rev. D **88**, 044042 (2013). arXiv:1306.4065 [gr-qc]

Reina, B., Sanchis-Gual, N., Vera, R., Font, J.A.: ArXiv e-prints (2017). arXiv:1702.04568 [gr-qc]

Ryan, F.D.: Phys. Rev. D **52**, 5707 (1995)

Sakstein, J., Babichev, E., Koyama, K., Langlois, D., Saito, R.: Phys. Rev. D **95**, 064013 (2017). arXiv:1612.04263 [gr-qc]

Salgado, M., Sudarsky, D., Nucamendi, U.: Phys. Rev. **D58**, 124003 (1998). arXiv:gr-qc/9806070 [gr-qc]

Sampson, L., Yunes, N., Cornish, N., Ponce, M., Barausse, E., Klein, A., Palenzuela, C., Lehner, L.: Phys. Rev. D **90**, 124091 (2014). arXiv:1407.7038 [gr-qc]

Sennett, N., Buonanno, A.: Phys. Rev. **D93**, 124004 (2016). arXiv:1603.03300 [gr-qc]

Sham, Y.H., Lin, L.M., Leung, P.T.: Astrophys. J. **781**, 66 (2014). arXiv:1312.1011 [gr-qc]

Sham, Y.-H., Chan, T.K., Lin, L.-M., Leung, P.T.: Astrophys. J. **798**, 121 (2015). arXiv:1410.8271 [gr-qc]

Shibata, M., Taniguchi, K., Okawa, H., Buonanno, A.: Phys. Rev. **D89**, 084005 (2014). arXiv:1310.0627 [gr-qc]

Silva, H.O., Sotani, H., Berti, E., Horbatsch, M.: Phys. Rev. **D90**, 124044 (2014). arXiv:1410.2511 [gr-qc]

Silva, H.O., Macedo, C.F.B., Berti, E., Crispino, L.C.B.: Classical Quantum Gravity **32**, 145008 (2015). arXiv:1411.6286 [gr-qc]

Silva, H.O., Sotani, H., Berti, E.: Mon. Not. R. Astron. Soc. **459**, 4378 (2016). arXiv:1601.03407 [astro-ph.HE]

Sotani, H.: Phys. Rev. **D86**, 124036 (2012). arXiv:1211.6986 [astro-ph.HE]

Sotani, H.: Phys. Rev. **D89**, 064031 (2014). arXiv:1402.5699 [astro-ph.HE]

Sotani, H., Kokkotas, K.D.: Phys. Rev. **D70**, 084026 (2004). arXiv:gr-qc/0409066 [gr-qc]

Sotani, H., Kokkotas, K.D.: Phys. Rev. **D71**, 124038 (2005). arXiv:gr-qc/0506060 [gr-qc]

Sotani, H., Kokkotas, K.D.: Phys. Rev. **D95**, 044032 (2017). arXiv:1702.00874 [gr-qc]

Sotiriou, T.P., Faraoni, V.: Rev. Mod. Phys. **82**, 451 (2010). arXiv:0805.1726 [gr-qc]

Staykov, K.V., Doneva, D.D., Yazadjiev, S.S., Kokkotas, K.D.: J. Cosmol. Astropart. Phys. **1410**, 006 (2014). arXiv:1407.2180 [gr-qc]

Staykov, K.V., Doneva, D.D., Yazadjiev, S.S.: Eur. Phys. J. **C75**, 607 (2015a). arXiv:1508.07790 [gr-qc]

Staykov, K.V., Doneva, D.D., Yazadjiev, S.S., Kokkotas, K.D.: Phys. Rev. **D92**, 043009 (2015b). arXiv:1503.04711 [gr-qc]

Staykov, K.V., Doneva, D.D., Yazadjiev, S.S.: Phys. Rev. D **93**, 084010 (2016). arXiv:1602.00504 [gr-qc]

Stefanov, I.Z., Yazadjiev, S.S., Todorov, M.D.: Phys. Rev. D **75**, 084036 (2007a). arXiv:0704.3784 [gr-qc]

Stefanov, I.Z., Yazadjiev, S.S., Todorov, M.D.: Mod. Phys. Lett. A **22**, 1217 (2007b). arXiv:0708.3203 [gr-qc]

Stefanov, I.Z., Yazadjiev, S.S., Todorov, M.D.: Mod. Phys. Lett. A **23**, 2915 (2008). arXiv:0708.4141 [gr-qc]

Stefanov, I.Z., Yazadjiev, S.S., Todorov, M.D.: Classical Quantum Gravity **26**, 015006 (2009)

Stein, L.C., Yagi, K., Yunes, N.: Astrophys. J. **788**, 15 (2014). arXiv:1312.4532 [gr-qc]

Stelle, K.S.: Phys. Rev. **D16**, 953 (1977)

Stergioulas, N., Friedman, J.L.: Astrophys. J. **444**, 306 (1995). astro-ph/9411032

Suvorov, A.G., Melatos, A.: Phys. Rev. D **93**, 024004 (2016). arXiv:1512.02291 [gr-qc]

Takami, K., Rezzolla, L., Baiotti, L.: Phys. Rev. Lett. **113**, 091104 (2014). arXiv:1403.5672 [gr-qc]

Takami, K., Rezzolla, L., Baiotti, L.: Phys. Rev. D **91**, 064001 (2015). arXiv:1412.3240 [gr-qc]

Taniguchi, K., Shibata, M., Buonanno, A.: Phys. Rev. **D91**, 024033 (2015). arXiv:1410.0738 [gr-qc]

Teichmüller, C., Fröb, M.B., Maucher, F.: Classical Quantum Gravity **28**, 155015 (2011). arXiv:1102.5252 [gr-qc]

Thorne, K.S.: Rev. Mod. Phys. **52**, 299 (1980)

Tsang, D., Pappas, G.: Astrophys. J. **818**, L11 (2016). arXiv:1511.01948 [astro-ph.HE]

Tsui, L.K., Leung, P.T.: Mon. Not. R. Astron. Soc. **357**, 1029 (2005a). arXiv:gr-qc/0412024 [gr-qc]

Tsui, L.K., Leung, P.T.: Astrophys. J. **631**, 495 (2005b). gr-qc/0505113

Tsui, L.K., Leung, P.T.: Phys. Rev. Lett. **95**, 151101 (2005c). astro-ph/0506681

Upadhye, A., Hu, W.: Phys. Rev. **D80**, 064002 (2009). arXiv:0905.4055 [astro-ph.CO]

Urbanec, M., Miller, J.C., Stuchlik, Z.: Mon. Not. R. Astron. Soc. **433**, 1903 (2013). arXiv:1301.5925 [astro-ph.SR]

Will, C.M.: Living Rev. Relativ. **17**, 4 (2014). arXiv:1403.7377 [gr-qc]

Yagi, K.: Phys. Rev. **D86**, 081504 (2012). arXiv:1204.4524 [gr-qc]

Yagi, K.: Phys. Rev. D **89**, 043011 (2014). arXiv:1311.0872 [gr-qc]

Yagi, K., Yunes, N.: Science **341**, 365 (2013a). arXiv:1302.4499 [gr-qc]

Yagi, K., Yunes, N.: Phys. Rev. D **88**, 023009 (2013b). arXiv:1303.1528 [gr-qc]

Yagi, K., Yunes, N.: Phys. Rev. D **91**, 103003 (2015a). arXiv:1502.04131 [gr-qc]

Yagi, K., Yunes, N.: Phys. Rev. D **91**, 123008 (2015b). arXiv:1503.02726 [gr-qc]

Yagi, K., Yunes, N.: Classical Quantum Gravity **33**, 095005 (2016a). arXiv:1601.02171 [gr-qc]

Yagi, K., Yunes, N.: Classical Quantum Gravity **33**, 13LT01 (2016b). arXiv:1512.02639 [gr-qc]

Yagi, K., Yunes, N.: Phys. Rep. **681**, 1 (2017a). arXiv:1608.02582 [gr-qc]

Yagi, K., Yunes, N.: Classical Quantum Gravity **34**, 015006 (2017b). arXiv:1608.06187 [gr-qc]

Yagi, K., Yunes, N., Tanaka, T.: Phys. Rev. D **86**, 044037 (2012). arXiv:1206.6130 [gr-qc]
Yagi, K., Stein, L.C., Yunes, N., Tanaka, T.: Phys. Rev. **D87**, 084058 (2013) [Erratum: Phys. Rev. **D93**(8), 089909 (2016)]. arXiv:1302.1918 [gr-qc]
Yagi, K., Stein, L.C., Pappas, G., Yunes, N., Apostolatos, T.A.: Phys. Rev. D **90**, 063010 (2014a). arXiv:1406.7587 [gr-qc]
Yagi, K., Kyutoku, K., Pappas, G., Yunes, N., Apostolatos, T.A.: Phys. Rev. D **89**, 124013 (2014b). arXiv:1403.6243 [gr-qc]
Yagi, K., Stein, L.C., Yunes, N.: Phys. Rev. D **93**, 024010 (2016). arXiv:1510.02152 [gr-qc]
Yazadjiev, S.S., Doneva, D.D.: ArXiv e-prints (2015). arXiv:1512.05711 [gr-qc]
Yazadjiev, S.S., Doneva, D.D., Kokkotas, K.D., Staykov, K.V.: J. Cosmol. Astropart. Phys. **1406**, 003 (2014). arXiv:1402.4469 [gr-qc]
Yazadjiev, S.S., Doneva, D.D., Kokkotas, K.D.: Phys. Rev. **D91**, 084018 (2015). arXiv:1501.04591 [gr-qc]
Yazadjiev, S.S., Doneva, D.D., Popchev, D.: Phys. Rev. **D93**, 084038 (2016). arXiv:1602.04766 [gr-qc]
Yazadjiev, S.S., Doneva, D.D., Kokkotas, K.D.: ArXiv e-prints (2017). arXiv:1705.06984 [gr-qc]
Yunes, N., Stein, L.C.: Phys. Rev. D **83**, 104002 (2011). arXiv:1101.2921 [gr-qc]
Yunes, N., Psaltis, D., Ozel, F., Loeb, A.: Phys. Rev. **D81**, 064020 (2010). arXiv:0912.2736 [gr-qc]

Index

© Springer Nature Switzerland AG 2018

L. Rezzolla et al. (eds.), *The Physics and Astrophysics of Neutron Stars*,

Astrophysics and Space Science Library 457,

https://doi.org/10.1007/978-3-319-97616-7

CPSIA information can be obtained
at www.ICGtesting.com
Printed in the USA
LVHW080812081121
702733LV00002B/158